# Algebra Review

## Rules of Exponents

- **Multiplication of Exponents** To multiply $a^m$ by $a^n$, *add* the exponents: $a^m a^n = a^{m+n}$.

- **Division of Exponents** To divide $a^m$ by $a^n$, *subtract* the exponents: $\dfrac{a^m}{a^n} = a^{m-n}$.

- **Power of a Power** To find a power of a power, multiply the exponents: $(a^m)^n = a^{mn}$.

- **Product to a Power** To find $(ab)^n$, apply the exponent to *every* term inside the parentheses: $(ab)^n = a^n b^n$.

- **Quotient to a Power** To find $\left(\dfrac{a}{b}\right)^n$, apply the exponent to *both* the numerator and the denominator: $\left(\dfrac{a}{b}\right)^n = \dfrac{a^n}{b^n}$.

- **Zero Exponent** If $a$ is any nonzero real number, then $a^0 = 1$.

- **Negative Exponent** If $n$ is a natural number, and if $a \neq 0$, then $a^{-n} = \dfrac{1}{a^n}$.

- **Inversion Property** $\left(\dfrac{a}{b}\right)^{-n} = \left(\dfrac{b}{a}\right)^n$.

## Radicals

- **$n$th root of $a$** If $n$ is an even natural number and $a \geq 0$, or if $n$ is an odd natural number, $\sqrt[n]{a} = a^{1/n}$.

- For all rational numbers $m/n$ and all real numbers $a$ for which $\sqrt[n]{a}$ exists, $a^{m/n} = (\sqrt[n]{a})^m = \sqrt[n]{a^m}$.

- For any real number $a$ and any natural number $n$, $\sqrt[n]{a^n} = |a|$ if $n$ is even and $\sqrt[n]{a^n} = a$ if $n$ is odd.

- For all real numbers $a$ and $b$, and for positive integers $n$ for which all indicated roots exists, $\sqrt[n]{a} \cdot \sqrt[n]{b} = \sqrt[n]{ab}$ and $\dfrac{\sqrt[n]{a}}{\sqrt[n]{b}} = \sqrt[n]{\dfrac{a}{b}}$ $(b \neq 0)$.

## Factoring

- **Difference of Squares** $x^2 - y^2 = (x + y)(x - y)$.

- **Perfect Squares** $x^2 + 2xy + y^2 = (x + y)^2$ and $x^2 - 2xy + y^2 = (x - y)^2$.

- **Difference of Cubes** $x^3 - y^3 = (x - y)(x^2 + xy + y^2)$.

- **Sum of Cubes** $x^3 + y^3 = (x + y)(x^2 - xy + y^2)$.

## Quadratic Equation and Formulas

- **Square Root Property** If $b > 0$, then the solutions of $x^2 = b$ are $x = \sqrt{b}$ and $x = -\sqrt{b}$.

- **Quadratic Formula** The solutions of the quadratic equation $ax^2 + bx - c$, where $a \neq 0$, are given by $x = \dfrac{-b \pm \sqrt{b^2 - 4ac}}{2a}$.

- **Graph of a Quadratic Function** The graph of $f(x) = a(x - h)^2 + k$ is a parabola with vertex $(h, k)$. It opens upward when $a > 0$ and downward when $a < 0$. The graph of $f(x) = ax^2 + bx + c$ is a parabola with vertex $(h, k)$, where $h = \dfrac{-b}{2a}$ and $k = f(h)$.

## Equations of Lines

- **Slope** The slope of the line $m$ through the two points $(x_1, y_1)$ and $(x_2, y_2)$, where $x_1 \neq x_2$ is $m = \dfrac{y_2 - y_1}{x_2 - x_1}$.

- The slope of every **horizontal** line is 0, and the slope of every **vertical** line is undefined.

- **Slope–Intercept Form** If a line has slope $m$ and $y$-intercept $b$, then it is the graph of the equation $y = mx + b$.

- **Parallel Lines** Two nonvertical lines are parallel whenever they have the same slope.

- **Perpendicular Lines** Two nonvertical lines are perpendicular whenever the product of their slopes is $-1$ (in other words, they are negative reciprocals of each other).

- **Point–Slope Form** If a line has slope $m$ and passes through the point $(x_1, y_1)$, then $y - y_1 = m(x - x_1)$.

## Properties of Logarithms

- **Definition of Logarithms to the Base $a$** $y = \log_a x$ means $a^y = x$.

- Let $x$ and $a$ be any positive real numbers, with $a \neq 1$, and $r$ be any real number. Then $\log_a 1 = 0$; $\log_a a = 1$; $\log_a a^r = r$; $a^{\log_a x} = x$.

- **Product Property** $\log_a xy = \log_a x + \log_a y$

- **Quotient Property** $\log_a \dfrac{x}{y} = \log_a x - \log_a y$

- **Power Property** $\log_a x^r = r \log_a x$

# Mathematics with Applications

## IN THE MANAGEMENT, NATURAL, AND SOCIAL SCIENCES

**TWELFTH EDITION**

### Margaret L. Lial
*American River College*

### Thomas W. Hungerford
*Saint Louis University*

### John P. Holcomb, Jr.
*Cleveland State University*

### Bernadette Mullins
*Birmingham-Southern College*

 Pearson

**Director, Portfolio Management:** Deirdre Lynch
**Executive Editor:** Jeff Weidenaar
**Editorial Assistant:** Jennifer Snyder
**Content Producer:** Kathleen A. Manley
**Managing Producer:** Karen Wernholm
**Producer:** Jean Choe
**Manager, Courseware QA:** Mary Durnwald
**Manager, Content Development:** Kristina Evans

**Product Marketing Manager:** Emily Ockay
**Field Marketing Manager:** Evan St. Cyr
**Marketing Assistants:** Shannon McCormack, Erin Rush
**Senior Author Support/Technology Specialist:** Joe Vetere
**Manager, Rights and Permissions:** Gina Cheselka
**Text Design:** Studio Montage

**Production Coordination, Composition, and Illustrations:** iEnergizer Aptara®, Ltd.
**Cover Design:** Studio Montage
**Cover Image:** Starman963/Shutterstock
**About the Cover:** The Metropol Parasol in Seville, Spain, features a honeycomb of wooden panels supported by concrete bases. It is the world's largest wooden structure.

**Library of Congress Cataloging-in-Publication Data**

Names: Lial, Margaret L., author. | Hungerford, Thomas W., author. | Holcomb, John P., Jr., author. | Mullins, Bernadette, author.
Title: Mathematics with applications in the management, natural, and social sciences/Margaret L. Lial, American River College, Thomas W. Hungerford, Saint Louis University, John P. Holcomb, Jr., Cleveland State University, Bernadette Mullins, Birmingham-Southern College.
Description: Twelfth edition . | Boston : Pearson, [2018] | Includes indexes.
Identifiers: LCCN 2017036044 | ISBN 9780134767628 | ISBN 0134767624
Subjects: LCSH: Mathematics—Textbooks. | Social sciences—Mathematics—Textbooks.
Classification: LCC QA37.3 .L55 2018 | DDC 510—dc23 LC record available at https://lccn.loc.gov/2017036044

2    18

 **Pearson**

ISBN 13:    978-0-13-476762-8
ISBN 10:       0-13-476762-4

*We would like to dedicate this twelfth edition to Thomas W. Hungerford, who passed away from a sudden illness in 2014. Tom had a major impact on both our lives. For Bernadette, it was through his authorship of his classic graduate text,* Algebra*; for John it was through working closely with Tom on the ninth and tenth editions of this text. Tom was exact and thorough, and he was a deep thinker of how students and teachers would read and use the material in his books. He is greatly missed.*

*We would also like to thank Greg Webster and Stan Palla for their enthusiastic support of our work on this project. Without their understanding, we never would have been able to make enough time to complete the revision. We thank you from the bottoms of our hearts.*

# Contents

# Preface

*Mathematics with Applications* is an applications-centered text for students in business, management, and natural and social sciences. It covers the basics of college algebra, followed by topics in finite mathematics and concluding with a treatment of applied calculus. The text can be used for a variety of different courses, and the only prerequisite is a basic course in algebra. Chapter 1 provides a thorough review of basic algebra for those students who need it.

It has been our primary goal to present sound mathematics in an understandable manner, proceeding from the familiar to new material, and from concrete examples to general rules and formulas. There is an ongoing focus on real-world problem solving, and almost every section includes relevant, contemporary applications.

## New to This Edition

We have revised and added content, updated and added new applications, fine-tuned pedagogical devices, and evaluated and enhanced the exercise sets. In addition, both the functionality of MyLab™ Math and the resources within it have been greatly improved and expanded for this edition. These improvements were incorporated after careful consideration and much feedback from those who teach this course regularly. Following is a list of some of the more substantive revisions made to this edition.

- We updated or added real-world data for hundreds of examples and exercises. We have tried to make this text the most relevant and interesting of its kind, and the main way we do this is by immersing ourselves in the kinds of applications that we know from experience will motivate students. We realize that motivating reluctant learners is a major part of this course; the applications in this text are designed to give instructors a big advantage in facing this challenge.

- We analyzed aggregated student usage and performance data from MyLab Math for the previous edition of this text. The results of this analysis helped improve the quality and quantity of exercises that matter the most to instructors and students. Also, within exercise sets, we improved even-odd pairing of exercises and better gradated them by level.

- In Chapter 5, we changed the notation for financial formulas to match the notation used in the TVM Solver from the TI-84 calculator.

- We added weighted averages to Section 10.2 so that students can understand applications such as calculating their own final grade when different components of a course are weighted at different percentages.

- We moved the discussion about boxplots from Section 10.4 to Section 10.3 because boxplots are tools for visualizing variation, and variation is discussed in Section 10.3.

- Material in the previous edition on using the normal distribution to approximate the binomial distribution in Section 10.5 was consolidated and moved to Section 10.4. The normal approximation to the binomial is not as important a topic today as it once was because technology makes calculating exact binomial probabilities easy. However, the conceptual understanding of the ideas can be important for students to learn, so we condensed the material and put it in the section on the normal distribution.

- We added Section 13.3, an entirely new section on the topic of integration by parts. This was done in response to requests from users of the text. The section features six new examples, including antiderivatives that require one or two applications of integration by parts, the antiderivatives of the logarithmic function, and applications. It also features fifty new exercises, including applied problems involving energy consumption and pharmaceutical sales. The previous Section 13.6, Tables of Integrals, was deleted because users indicated they did not cover this section.

- Labels for the applications in the previous edition were very general (e.g., Business, Life Sciences). In this edition, we made them more specific (e.g., Google Profits) to pique student interest and to allow students to find applications that relate to their specific major and areas of interest.

The copy below is an example of the efforts put forth to update each chapter. **To see this type of detailed information for *all chapters* in the text, please see the "Features" portion of the Pearson online catalog for this text.**

## New to Chapter 4

- In Section 4.1, updated application Examples 5 and 6 and Checkpoint 7 with data on wine consumption and assets of AIG. Replaced or updated ten of the exercises with current data on the assets of Prudential Financial, Inc.; Netflix costs; GDP for China and the U.S.; asset management; imports from Vietnam; and subprime mortgages.

- Replaced Examples 1 and 2 in Section 4.2 with current data on debt in the U.S. and on sales of single-family homes. Updated Example 4 with more recent data on infant mortality rates. Replaced Example 6 and Checkpoint 4 with a current example on the price of scrap steel. Replaced or updated eleven of the application exercises with data on wind power, oil production, office rent, personal consumption, Medicare expenditures, Chinese assets in banks, Internet access in China, seat-belt use, death rates, food assistance, and labor force participation.

- In Section 4.3, added a graph of several logarithmic functions of different bases to help students visualize logarithmic functions better. Replaced an application example with a current logarithmic model function on wind energy generated in the U.S. Updated or replaced eight of the application exercises with data on health insurance costs, dairy expenditures, credit union assets, border patrol budgets, opioid deaths, iPhone sales, and vehicle miles traveled.

- In Section 4.4, utilized color to indicate nonpossible solutions to logarithmic equations more clearly. Replaced Example 7 and Checkpoint 7 with a new example on new jobs added to the U.S. economy. Also added a new example and checkpoint (using data on the digital grocery market) to illustrate solving for $x$ with logarithmic and exponential equations. Updated or replaced twelve of the application exercises with data on foreign earnings, nursing degrees, veterans' benefits, Snapchat users, wind energy, Japanese messaging, CVS Health earnings and revenue, the number of teachers in the U.S., Best Buy revenue, Twitter stock price, and outstanding loans in U.S. banks.

- In the Review Exercises, updated or replaced ten of the application exercises with current data on exports to Mexico, Royal Caribbean share price, the number of murders in Chicago, crude oil and coal futures, recent earthquakes, FedEx profits, Starbucks and Dunkin' Donuts App users, and bank capital.

- Updated Case Study examples and exercises with more recent data and graphs from gapminder.org.

## New to MyLab Math

Many improvements have been made to the overall functionality of MyLab Math since the previous edition. Beyond that, however, we have also increased and improved the content specific to this text.

- Instructors now have **more exercises** than ever to choose from when assigning homework. Most new questions are application-oriented. There are approximately 5900 assignable exercises in MyLab Math for this text. New exercise types include:
  - Additional Conceptual Questions provide support for assessing concepts and vocabulary. Many of these questions are application-oriented.

- Setup & Solve exercises require students to show how they set up a problem as well as the solution, which approximates more closely what is required of students on tests.

- The **videos are** *all new*, and they feature veteran instructors Thomas Hartfield (University of North Georgia), Mike Rosenthal (Florida International University), and Kate Haynes (Delaware Technical Community College).

  - Each section of the text now has an accompanying full lecture video. To make it easier for students to navigate to the content they need, each lecture video is segmented into shorter clips (labeled Introduction, Example, or Summary).

  - Both the video lectures and video segments are assignable within MyLab Math. We have included a Guide to Video-Based Assignments within the Instructor Resources section of MyLab Math that allows you to assign exercises for each video.

  - MathTalk and StatTalk videos highlight applications of the content of the course to business. The videos are supported by assignable exercises.

- A full suite of **Interactive Figures** has been added to support teaching and learning. The figures illustrate key concepts and allow manipulation. They have been designed to be used during lectures as well as by students working independently.

- An **Integrated Review** version of the MyLab Math course contains premade quizzes to assess the prerequisite skills needed for each chapter, plus personalized remediation for any gaps in skills that are identified.

- **Study Skills Modules** help students with the life skills that can make the difference between passing and failing.

- The **Graphing Calculator Manual** and **Excel Spreadsheet Manual**, both specific to this course, have been updated to support the TI-84 CE (color edition) and Excel 2016, respectively. Both manuals also contain additional topics to support the course.

- We heard from users that the Annotated Instructor Edition for the previous edition required too much flipping of pages to find answers, so MyLab Math now contains a downloadable **Instructor Answers document**—with *all answers in one place*. (This augments the downloadable Instructor Solutions Manual, which contains all solutions.)

# Continued Pedagogical Support

- **Real-Data Examples and Explanations:** Real-data exercises have long been a popular and integral aspect of this text. A significant number of new real-data examples and exercises have also been introduced into the text. Applications are noted with a **green header** to indicate the subject of the problem so instructors or students can focus on applications that are in line with students' majors.

- **Balanced Approach:** Multiple representations of a topic (symbolic, numerical, graphical, verbal) are given when appropriate. However, we do not believe that all representations are useful for all topics, so effective alternatives are discussed only when they are likely to increase student understanding.

- **Strong Algebra Foundation:** The text begins with four thorough chapters of college algebra that can be used in a variety of ways based on the needs of the students and the goals of the course. Take advantage of the content in these chapters as needed so students will be more successful with later topics and future courses.

- **Help for Skill Gaps:** The Prerequisite Skills Test (for Chapters 1–4) and Calculus Readiness Test (for Chapters 11–14) at the front of the text can help students determine where remediation is needed. The text contains solutions to the test exercises to help students remediate any gaps in basic skills.

- **Checkpoint** exercises are marked with icons such as ✓1 and provide an opportunity for students to stop, check their understanding of the specific concept at hand, and move forward with confidence. Answers to Checkpoint exercises are located at the end of the section to encourage students to work the problems before looking at the answers. (See pages 185 and 186.)

- **Caution** notes highlight common student difficulties or warn against frequently made mistakes. (See page 209.)
- **Exercises:** In addition to skill-based practice, conceptual, and application-based exercises, the text includes some specially marked exercises:
  - Writing Exercises ◣ (See page 188.)
  - Connection Exercises ☑ relate current topics to earlier sections (See page 213.)
  - Graphing Calculator Exercises ▦ (See page 205.)
  - Spreadsheet Exercises ☐ (See page 260.)
- **Example/Exercise Connection:** Selected exercises include a reference to related example(s) within the section (e.g., "See Examples 6 and 7") to facilitate what students do naturally when they use a book—that is, look for specific examples when they get stuck on a problem. Later exercises leave this information out and provide opportunities for mixed skill practice.
- **Graphing Calculators and Spreadsheets:** It is assumed that all students have a calculator that will handle exponential and logarithmic functions. Graphing calculator and spreadsheet references are highlighted in the text so that those who use the technology can easily incorporate it and those who do not can easily omit it. Examples and exercises that require a graphing calculator are marked with ▦ and those that require a spreadsheet are marked with ☐, making it obvious where technology is being included.
- **Technology Tips:** These tips are placed at appropriate points in the text to inform students of various features of their graphing calculator, spreadsheet, or other computer programs. Note that Technology Tips designed for TI-84 CE also apply to the TI-84 Plus, TI-83, and TI-Nspire.
- **End-of-chapter materials:** are designed to help students prepare for exams. These materials include a List of Key Terms and Symbols and Summary of Key Concepts, as well as a thorough set of Chapter Review Exercises.
- **Case Studies:** appear at the end of each chapter and offer contemporary, real-world applications of some of the mathematics presented in the chapter. Not only do these provide an opportunity for students to see the mathematics they are learning in action, but they also provide at least a partial answer to the question, "What is this stuff good for?" These have been expanded to include options for longer-term projects if the instructor should choose to use them.

## Course Flexibility

The content of the text is divided into three parts:

1. College Algebra (Chapters 1–4)
2. Finite Mathematics (Chapters 5–10)
3. Applied Calculus (Chapters 11–14)

This coverage of the material offers flexibility, making the text appropriate for a variety of courses, including:

- **Finite Mathematics and Applied Calculus** (one year or less). Use the entire book; cover topics from Chapters 1–4 as needed before proceeding to further topics.
- **Finite Mathematics** (one semester or two quarters). Use as much of Chapters 1–4 as needed, and then go into Chapters 5–10 as time permits and local needs require.
- **Applied Calculus** (one semester or quarter). Cover the precalculus topics in Chapters 1–4 as necessary, and then use Chapters 11–14.
- **College Algebra with Applications** (one semester or quarter). Use Chapters 1–8, with Chapters 7 and 8 being optional.

Pearson regularly produces custom versions of this text (and its accompanying MyLab Math course) to address the needs of specific course sequences. Custom versions can be produced for even smaller-enrollment courses due to advances in digital printing. Please contact your local Pearson representative for more details.

Chapter interdependence is as follows:

| Chapter | Prerequisite |
|---|---|
| 1 Algebra and Equations | None |
| 2 Graphs, Lines, and Inequalities | Chapter 1 |
| 3 Functions and Graphs | Chapters 1 and 2 |
| 4 Exponential and Logarithmic Functions | Chapter 3 |
| 5 Mathematics of Finance | Chapter 4 |
| 6 Systems of Linear Equations and Matrices | Chapters 1 and 2 |
| 7 Linear Programming | Chapters 3 and 6 |
| 8 Sets and Probability | None |
| 9 Counting, Probability Distributions, and Further Topics in Probability | Chapter 8 |
| 10 Introduction to Statistics | Chapter 8 |
| 11 Differential Calculus | Chapters 1–4 |
| 12 Applications of the Derivative | Chapter 11 |
| 13 Integral Calculus | Chapters 11 and 12 |
| 14 Multivariate Calculus | Chapters 11–13 |

# Acknowledgments

We wish to thank the following reviewers for their thoughtful feedback and helpful suggestions, without which we could not continue to improve this text.

John Altiere, Cleveland State University

Sviatoslav Archava, East Carolina University

Stephen Bast, Anne Arundel Community College

Phanuel Bediako, Delaware State University

Susan Bellini, Cleveland State University

Eric Erdmann, University of Minnesota, Duluth

Meghan Foster, American University

Lobna Mazzawi, Everett Community College

Thomas Milligan, University of Oklahoma

Margie Nowlin, Texas Christian University

David Stott, Sinclair Community College

Cong-Cong Xing, Nicholls State University

The following faculty members provided direction on the development of the MyLab Math course for this edition:

Vince Bander, Pierce College, Puyallup

Rachel Bates, Redlands Community College

Krista Blevins Cohlmia, Midland College

Larry Cook, Taft College

Stanislav Dubrovskiy, Sierra College

Jeffrey K. Dyess, Bishop State Community College

Shurron Farmer, University of the District of Columbia

Tonia Garrett, San Jacinto College

Abe Haje, Lone Star College, University Park

Ray Hendrickson, Bucks County Community College

MaryAlice Howe, Bucks County Community College

Edgar Jasso, North Seattle College

Dynechia Jones, Baton Rouge Community College

Tiffany Jones, Baylor University

Shelley Lenahan, University of Alabama, Huntsville

Steven Mark, Wilmington University

Maria Mathews, Baldwin Wallace University

Janet Noah, Brookhaven College

Margie Torres Nowlin, All Saints Episcopal School

Rose L. Pugh, Bellevue College

Peggy M. Slavik, Baldwin Wallace University

Kay Tekumalla, Prince George's Community College

R. Vafa, Temple University

Rick L. Wing, San Francisco State University

Cong-Cong Xing, Nicholls State University

Ju Zhou, Kutztown University

We gratefully acknowledge Birmingham-Southern College and Cleveland State University for their wholehearted support of this edition. Bernadette Mullins thanks two BSC students who checked new content: Amer Babi and Adam Alden Pratt.

We also thank our accuracy checkers, who did an excellent job of checking both text and exercise answers: Stephen Bast and Cong-Cong Xing. Thanks also to the supplements authors: Salvatore Sciandra, Chris True, and Stela Pudar-Hozo.

We want to thank the staff of Pearson Education for their assistance with and contributions to this book, particularly Jeff Weidenaar, Kathy Manley, Jenn Snyder, Emily Ockay, Kristina Evans, and Jean Choe. Finally, we wish to express our appreciation to Sherrill Redd of Aptara Corporation, who was a pleasure to work with.

*John P. Holcomb, Jr.*
*Bernadette Mullins*

# MyLab™ Math Online Course
## for *Mathematics with Applications in the Management, Natural, and Social Sciences*, 12e

### (access code required)

**Pearson** MyLab

MyLab™ Math is available to accompany Pearson's market-leading text offerings. To give students a consistent tone, voice, and teaching method, each text's flavor and approach are tightly integrated throughout the accompanying MyLab Math course, making learning the material as seamless as possible.

## PREPAREDNESS

One of the biggest challenges in applied math courses is making sure students are adequately prepared with the prerequisite skills needed to complete their course work successfully. MyLab Math supports students with just-in-time remediation and key-concept review.

**NEW! Integrated Review Course**   An Integrated Review version of the MyLab Math course contains premade, assignable quizzes to assess the prerequisite skills needed for each chapter, plus personalized remediation for any gaps in skills that are identified. Therefore, each student receives just the help that he or she needs—no more, no less.

**Study Skills Modules**   Study skills modules help students with the life skills that can make the difference between passing and failing.

## DEVELOPING DEEPER UNDERSTANDING

MyLab Math provides content and tools that help students build a deeper understanding of course content than would otherwise be possible.

### Exercises with Immediate Feedback
Homework and practice exercises for this text regenerate algorithmically to give students unlimited opportunity for practice and mastery. MyLab Math provides helpful feedback when students enter incorrect answers and includes the optional learning aids Help Me Solve This, View an Example, videos, and/or the eText.

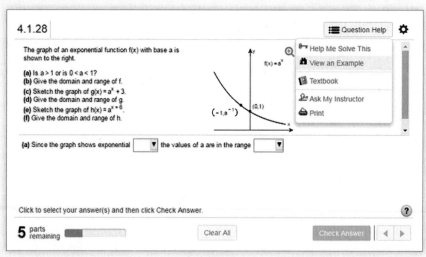

**pearson.com/mylab/math**

Pearson
MyLab

Solve the linear system by the Gauss-Jordan elimination method.

$$\begin{cases} x + y - z = -5 \\ -x + 4y + 16z = 7 \\ -4x + y + 4z = -7 \end{cases}$$

Without changing the order of any rows or columns, write a matrix that represents the system.

$$\begin{bmatrix} 1 & 1 & -1 & -5 \\ -1 & 4 & 16 & 7 \\ -4 & 1 & 4 & -7 \end{bmatrix}$$

(Do not simplify. Type an integer or simplified fraction for each matrix element.)

The solution of the system is $x = \dfrac{7}{3}$, $y = -\dfrac{27}{5}$, $z = \dfrac{29}{15}$.

(Simplify your answers. Type integers or fractions.)

## NEW! Setup & Solve Exercises
These exercises require students to show how they set up a problem as well as the solution, better mirroring what is required of students on tests.

## NEW! Additional Conceptual Questions
Additional Conceptual Questions provide support for assessing concepts and vocabulary. Many of these questions are application-oriented. They are clearly labeled "Conceptual" in the Assignment Manager.

## NEW! Interactive Figures
A full suite of Interactive Figures has been added to support teaching and learning. The figures illustrate key concepts and allow manipulation. They have been designed for use during lectures as well as by students working independently.

## NEW! Instructional Videos
The instructional videos for this course are *all new*, featuring veteran instructors Thomas Hartfield (University of North Georgia), Mike Rosenthal (Florida International University), and Kate Haynes (Delaware Technical Community College).

- Each section of the text now has an accompanying full lecture video. To make it easier for students to navigate to the content they need, each lecture video is segmented into shorter clips (labeled Introduction, Example, or Summary).

- Both the video lectures and video segments are assignable within MyLab Math. We have included a Guide to Video-Based Assignments within the Instructor Resources section of MyLab Math that allows you to assign exercises for each video.

In addition, the MathTalk and StatTalk videos connect the content of the course to business and management applications. The videos are supported by assignable exercises.

**pearson.com/mylab/math**

### Technology Manuals and Projects (downloadable)

- *Excel Spreadsheet Manual and Projects* by Stela Pudar-Hozo, Indiana University–Northwest
- *Graphing Calculator Manual and Projects* by Chris True, University of Nebraska

These manuals, both specific to this course, have been updated to support the TI-84 CE (color edition) and Excel 2016, respectively. Instructions are ordered by mathematical topic. The files can be downloaded from within MyLab Math.

### Student's Solutions Manual (softcover and downloadable)
ISBN: 0-13-477267-9 | 978-0-13-477267-7
Written by Salvatore Sciandra from Niagara County Community College, the Student's Solutions Manual contains worked-out solutions to all the odd-numbered exercises and all Chapter Review and Case Studies. This manual is available in print and can be download from within MyLab Math.

### A Complete eText    Students get unlimited access to the eText within any MyLab Math course using that edition of the textbook. The Pearson eText app allows existing subscribers to access their titles on an iPad or Android tablet for either online or offline viewing.

## SUPPORTING INSTRUCTION

MyLab Math comes from an experienced partner with educational expertise and an eye on the future. It provides resources to help you assess and improve students' results at every turn, and unparalleled flexibility to create a course tailored to you and your students.

### Learning Catalytics™    Now included in all MyLab Math courses, this student response tool uses students' smartphones, tablets, or laptops to engage them in more interactive tasks and thinking during lecture. Learning Catalytics fosters student engagement and peer-to-peer learning with real-time analytics. Access prebuilt exercises created specifically for this course.

### PowerPoint® Lecture Resources (downloadable)    Slides contain presentation resources such as figures, feature boxes, and key examples from the text. They can be downloaded from within MyLab Math or from Pearson's online catalog, **www.pearson.com**.

### Comprehensive Gradebook    The gradebook includes enhanced reporting functionality such as item analysis and a reporting dashboard to allow you to manage your course efficiently. Student performance data is presented at the class, section, and program levels in an accessible, visual manner so you will have the information you need to keep your students on track.

### TestGen    TestGen® (**www.pearson.com/testgen**) enables instructors to build, edit, print, and administer tests using a computerized bank of questions developed to cover all the objectives of the text. TestGen is algorithmically based, allowing instructors to create multiple but equivalent versions of the same question or test with the click of a button. Instructors can also modify test bank questions or add new questions. The software and test bank are available for download from Pearson's online catalog, **www.pearson.com**. The questions are also assignable in MyLab Math.

### Instructor's Solutions Manual (downloadable)    Written by Salvatore Sciandra from Niagara County Community College, the Instructor's Solutions Manual contains detailed solutions to all text exercises, suggested course outlines, and a chapter interdependence chart. It can be downloaded from within MyLab Math or from Pearson's online catalog, **www.pearson.com**.

### Instructor's Answers (downloadable)    These handy chapter-by-chapter documents provide answers to all Student Edition exercises in one place for easy reference by instructors. They are downloadable from within MyLab Math or from Pearson's online catalog, **www.pearson.com**.

### Accessibility    Pearson works continuously to ensure our products are as accessible as possible to all students. We are working toward achieving WCAG 2.0 Level AA and Section 508 standards, as expressed in the Pearson Guidelines for Accessible Educational Web Media, **www.pearson.com/mylab/math/accessibility**.

**pearson.com/mylab/math**

# To the Student

The key to succeeding in this course is to remember that *mathematics is not a spectator sport*. You can't expect to learn mathematics without *doing* mathematics any more than you could learn to swim without getting wet. You must take an active role and use all the resources at your disposal: your instructor, your fellow students, this book, and the supplements that accompany this book. Following are some general tips on how to be successful in the course and some specific tips on how to get the most out of this text and supplementary resources.

***Ask Questions!*** Remember the words of the great Hillel: "The bashful do not learn." There is no such thing as a "dumb question" (assuming, of course, that you have read the book and your class notes and attempted the homework). Your instructor will welcome questions that arise from a serious effort on your part. So get your money's worth: Ask questions!

***Read the Book Interactively!*** There is more to a math textbook than just the exercise sets. Each section introduces topics carefully with many examples—both mathematical and contextual. Take note of the "Caution" and "Note" comments, and bookmark pages with key definitions or formulas. After reading the example, try the Checkpoint exercise that appears next to it in the margin to check your understanding of the concept. This will help you solidify your understanding or diagnose if you do not fully understand the concept. The answers to the Checkpoint exercises are right after the homework exercises in each section. Resist the temptation to flip to the answer until you've worked the problem completely!

***Take Advantage of the Supplementary Material!*** Many resources are at your disposal within and outside the text. Take the time to interact with them and determine which resources suit your learning style best.

- If your instructor allows the use of graphing calculators and/or spreadsheets, work through the examples and exercises marked with ▦ or ▢. Some instructors may make this material part of the course, whereas others will not require you to use technology. If your instructor asks that you use technology, a Graphing Calculator Manual and an Excel Spreadsheet Manual are available in MyLab Math. In addition, there are ***Technology Tips*** throughout the text that describe the proper menu or keys to use for various procedures on a graphing calculator. Note that Technology Tips for the TI-84+ CE also apply to TI-83+, TI-Nspire, and usually TI-83.

- MyLab Math has a variety of types of resources to help you learn, including videos for every section of the text; Interactive Figures to help visualize difficult concepts; unlimited practice and assessment on newly learned or prerequisite skills; and access to the Student Solutions Manual, Graphing Calculator Manual, Excel Spreadsheet Manual, and a variety of helpful reference cards.

***Do Your Homework!*** Whether the homework is paper and pencil or assigned online, you must practice what you have learned. This is your opportunity to practice those essential skills needed for passing this course and those skills needed for application in future courses or your career.

We wish you the best in your efforts throughout this course, in future courses, and beyond school.

*John and Bernadette*

# Prerequisite Skills Test*

The following test is unlike your typical math test. Rather than testing your skills after you have worked on them, this test assesses skills that you should know from previous coursework and will use in this class. It is intended to diagnose any areas that you may need to remediate. Take advantage of the results of this test by checking your answers in Appendix B. The full solutions are included to remind you of the steps to answer the problem.

Find the most simplified answer for the given problems involving fractions:

1. $\dfrac{5}{2} - 6 =$

2. $\dfrac{1}{2} \div \dfrac{2}{5} =$

3. $\dfrac{1}{3} \div 3 =$

Simplify the given expression, keeping in mind the appropriate order of operations:

4. $7 + 2 - 3(2 \div 6) =$

5. $\dfrac{2 \times 3 + 12}{1 + 5} - 1 =$

Indicate whether each of the statements is true or false:

6. $\dfrac{4 + 3}{3} = 5$

7. $\dfrac{5}{7} + \dfrac{7}{5} = 1$

8. $\dfrac{3}{5} + 1 = \dfrac{6}{5}$

Translate each of the following written expressions into a corresponding mathematical statement. If possible, solve for the unknown value or values.

9. Alicia has $n$ pairs of shoes. Manuel has two more pairs of shoes than Alicia. How many pairs of shoes does Manuel have?

10. David's age and Selina's age, when added together, equals 42. Selena is 6 years older than David. What are David's and Selina's ages?

Solve the following problem.

11. The price of a sweater, originally sold for $72, is reduced by 20%. What is the new sale price of the sweater?

Given the following rectangular coordinate system, graph and label the following points:

12. A: $(3, -2)$    B: $(4, 0)$    C: $(-2, -3)$

13. D: $(3, 5)$    E: $(-1, 4)$    F: $(-4, -5)$

Round the following values as indicated:

14. (a) 4.27659 to the nearest tenth
    (b) 245.984 to the nearest unit (whole number)

15. (a) 16.38572 to the nearest hundredth
    (b) 1,763,304.42 to the nearest thousand

Write the number that corresponds with the given numerical statement:

16. (a) The Company's liabilities totaled 34 million dollars.
    (b) The total of investments was 2.2 thousand dollars.

17. (a) The population of a country is 17 hundred thousand.
    (b) The cost of the new airport could run as high as three and a quarter billion dollars.

Answer the following. If there is no solution, indicate that there is no solution and provide a reason.

18. $\dfrac{5}{0} =$

19. A car is traveling 60 miles per hour. Is the car traveling at 1 mile per minute?

20. Which number is greater, $-9$ or $-900$?

---

*Full Solutions to this test are provided in the Appendix B.

# Calculus Readiness Test*

1. Subtract the polynomials:
   $(7x^4 - 8x^3 + 2x^2 - 3) - (9x^4 - 2x^2 + 5x + 6)$.

2. Add the rational functions: $\dfrac{-6}{3x + 4} + \dfrac{2}{x}$.

3. Find the product: $(3x + 7)(5x - 8)$.

4. Find the product: $(7x + 3)(6x^2 + x - 8)$.

5. Simplify: $\dfrac{(x^5 y^{-5})^3}{(x^2 \sqrt{y})^6}$.

6. Multiply out and simplify: $(8x - 9)^2$.

7. Factor: $9x^2 - 49$.

8. Solve for $x$: $9x - 4 = 8 + 7(x - 5)$.

9. Solve for $x$: $3x^2 + 2x + 8 = 20x - 7$.

10. Solve the inequality for $x$: $x^2 - x - 6 < 0$.

11. Solve for $x$: $5e^{x-3} - 1 = 9$.

12. Solve for $x$: $\ln(5x + 1) = 2$.

13. Given that $f(x) = 5x^3 - 3x^2 + 9x + 60$, find $f(-2)$.

14. Given that $f(x) = 6x^2 - 7x - 20$, solve $f(x) = 0$.

15. Find an equation of a line through the point $(8, -3)$ with slope $-5$.

---

*Full Solutions to this test are provided in the Appendix B.

**xx**

# Algebra and Equations

CHAPTER 1

## CHAPTER OUTLINE

Mathematics is widely used in business, finance, and the biological, social, and physical sciences, from developing efficient production schedules for a factory to mapping the human genome. Mathematics also plays a role in determining interest on a loan from a bank, the growth of traffic on websites, and the study of falling objects.

Algebra and equations are the basic mathematical tools for handling many applications. Your success in this course will depend on your having the algebraic skills presented in this chapter.

## 1.1 The Real Numbers

Only real numbers will be used in this book.[*] The names of the most common types of real numbers are as follows.

### The Real Numbers

| | |
|---|---|
| **Natural (counting) numbers** | $1, 2, 3, 4, \ldots$ |
| **Whole numbers** | $0, 1, 2, 3, 4, \ldots$ |

---

[*] Not all numbers are real numbers. For example, $\sqrt{-1}$ is a number that is *not* a real number.

| Integers | $\ldots, -3, -2, -1, 0, 1, 2, 3, \ldots$ |
|---|---|
| **Rational numbers** | All numbers that can be written in the form $p/q$, where $p$ and $q$ are integers and $q \neq 0$ |
| **Irrational numbers** | Real numbers that are not rational |

As you can see, every natural number is a whole number, and every whole number is an integer. Furthermore, every integer is a rational number. For instance, the integer 7 can be written as the fraction $\frac{7}{1}$ and is therefore a rational number.

One example of an irrational number is $\pi$, the ratio of the circumference of a circle to its diameter. The number $\pi$ can be approximated as $\pi \approx 3.14159$ ($\approx$ means "is approximately equal to"), but there is no rational number that is exactly equal to $\pi$.

**EXAMPLE 1**   What kind of number is each of the following?

**(a)** 6

**Solution**  The number 6 is a natural number, a whole number, an integer, a rational number, and a real number.

**(b)** $\dfrac{3}{4}$

**Solution**  This number is rational and real.

**(c)** $3\pi$

**Solution**  Because $\pi$ is not a rational number, $3\pi$ is irrational and real.

All real numbers can be written in decimal form. A rational number, when written in decimal form, is either a terminating decimal, such as .5 or .128, or a repeating decimal, in which some block of digits eventually repeats forever, such as $1.3333\ldots$ or $4.7234234234\ldots$.[†] Irrational numbers are decimals that neither terminate nor repeat.

When a calculator is used for computations, the answers it produces are often decimal *approximations* of the actual answers; they are accurate enough for most applications. To ensure that your final answer is as accurate as possible,

*you should not round off any numbers during long calculator computations.*

It is usually OK to round off the final answer to a reasonable number of decimal places once the computation is finished.

The important basic properties of the real numbers are as follows.

## Properties of the Real Numbers

For all real numbers, $a$, $b$, and $c$, the following properties hold true:

| **Commutative properties** | $a + b = b + a$   and   $ab = ba.$ |
|---|---|
| **Associative properties** | $(a + b) + c = a + (b + c)$   and   $(ab)c = a(bc).$ |

---

[*]The use of Checkpoint exercises is explained in the "To the Student" section preceding this chapter.

[†]Some graphing calculators have a FRAC key that automatically converts some repeating decimals to fraction form.

| Identity properties | There exists a unique real number 0, called the **additive identity**, such that $$a + 0 = a \quad \text{and} \quad 0 + a = a.$$ There exists a unique real number 1, called the **multiplicative identity**, such that $$a \cdot 1 = a \quad \text{and} \quad 1 \cdot a = a.$$ |
|---|---|
| Inverse properties | For each real number $a$, there exists a unique real number $-a$, called the **additive inverse** of $a$, such that $$a + (-a) = 0 \quad \text{and} \quad (-a) + a = 0.$$ If $a \neq 0$, there exists a unique real number $1/a$, called the **multiplicative inverse** of $a$, such that $$a \cdot \frac{1}{a} = 1 \quad \text{and} \quad \frac{1}{a} \cdot a = 1.$$ |
| Distributive property | $$a(b + c) = ab + ac \quad \text{and} \quad (b + c)a = ba + ca.$$ |

The next five examples illustrate the properties listed in the preceding box.

**EXAMPLE 2**   The commutative property says that the order in which you add or multiply two quantities doesn't matter.

**(a)**  $(6 + x) + 9 = 9 + (6 + x) = 9 + (x + 6)$      **(b)**  $5 \cdot (9 \cdot 8) = (9 \cdot 8) \cdot 5$

**EXAMPLE 3**   When the associative property is used, the order of the numbers does not change, but the placement of parentheses does.

**(a)**  $4 + (9 + 8) = (4 + 9) + 8$      **(b)**  $3(9x) = (3 \cdot 9)x$ ✓2

**✓Checkpoint 2**

Name the property illustrated in each of the following examples.

**(a)**  $(2 + 3) + 9 = (3 + 2) + 9$

**(b)**  $(2 + 3) + 9 = 2 + (3 + 9)$

**(c)**  $(2 + 3) + 9 = 9 + (2 + 3)$

**(d)**  $(4 \cdot 6)p = (6 \cdot 4)p$

**(e)**  $4(6p) = (4 \cdot 6)p$

**EXAMPLE 4**   By the identity properties,

**(a)**  $-8 + 0 = -8$      **(b)**  $(-9) \cdot 1 = -9.$

⊞ **TECHNOLOGY TIP** To enter $-8$ on a calculator, use the negation key (labeled (−) or +/−), *not* the subtraction key. On most one-line scientific calculators, key in 8 +/−. On graphing calculators or two-line scientific calculators, key in either (−) 8 or +/− 8.

**✓Checkpoint 3**

Name the property illustrated in each of the following examples.

**(a)**  $2 + 0 = 2$

**(b)**  $-\frac{1}{4} \cdot (-4) = 1$

**(c)**  $-\frac{1}{4} + \frac{1}{4} = 0$

**(d)**  $1 \cdot \frac{2}{3} = \frac{2}{3}$

**EXAMPLE 5**   By the inverse properties, the statements in parts (a) through (d) are true.

**(a)**  $9 + (-9) = 0$      **(b)**  $-15 + 15 = 0$

**(c)**  $-8 \cdot \left(\frac{1}{-8}\right) = 1$      **(d)**  $\frac{1}{\sqrt{5}} \cdot \sqrt{5} = 1$

📄 **NOTE** There is no real number $x$ such that $0 \cdot x = 1$, so 0 has no multiplicative inverse. ✓3

**EXAMPLE 6**     By the distributive property,

**(a)** $9(6 + 4) = 9 \cdot 6 + 9 \cdot 4$

**(b)** $3(x + y) = 3x + 3y$

**(c)** $-8(m + 2) = (-8)(m) + (-8)(2) = -8m - 16$

**(d)** $(5 + x)y = 5y + xy.$ ✓ 4

✓ **Checkpoint 4**

Use the distributive property to rewrite each of the following.

**(a)** $4(-2 + 5)$

**(b)** $2(a + b)$

**(c)** $-3(p + 1)$

**(d)** $(8 - k)m$

**(e)** $5x + 3x$

## Order of Operations

Some complicated expressions may contain many sets of parentheses. To avoid ambiguity, the following procedure should be used.

### Parentheses

Work separately above and below any fraction bar. Within each set of parentheses or square brackets, start with the innermost set and work outward.

**EXAMPLE 7**     Simplify: $[(3 + 2) - 7]5 + 2([6 \cdot 3] - 13).$

**Solution**  On each segment, work from the inside out:

$$[(3 + 2) - 7]5 + 2([6 \cdot 3] - 13)$$
$$= [5 - 7]5 + 2(18 - 13)$$
$$= [-2]5 + 2(5)$$
$$= -10 + 10 = 0.$$

Does the expression $2 + 4 \times 3$ mean

$$(2 + 4) \times 3 = 6 \times 3 = 18?$$

Or does it mean

$$2 + (4 \times 3) = 2 + 12 = 14?$$

To avoid this ambiguity, mathematicians have adopted the following rules (which are also followed by almost all scientific and graphing calculators).

### Order of Operations

1. Find all powers and roots, working from left to right.

2. Do any multiplications or divisions in the order in which they occur, working from left to right.

3. Finally, do any additions or subtractions in the order in which they occur, working from left to right.

If sets of parentheses or square brackets are present, use the rules in the preceding box within each set, working from the innermost set outward.

According to these rules, multiplication is done *before* addition, so $2 + 4 \times 3 = 2 + 12 = 14$. Here are some additional examples.

EXAMPLE 8   Use the order of operations to evaluate each expression if $x = -2$, $y = 5$, and $z = -3$.

(a) $-4x^2 - 7y + 4z$

Solution   Use parentheses when replacing letters with numbers:

$$-4x^2 - 7y + 4z = -4(-2)^2 - 7(5) + 4(-3)$$
$$= -4(4) - 7(5) + 4(-3) = -16 - 35 - 12 = -63.$$

(b) $\dfrac{2(x-5)^2 + 4y}{z+4} = \dfrac{2(-2-5)^2 + 4(5)}{-3+4}$

$$= \frac{2(-7)^2 + 20}{1}$$

$$= 2(49) + 20 = 118. \quad \checkmark_5$$

✓ **Checkpoint 5**

Evaluate the following if $m = -5$ and $n = 8$.

(a) $-2mn - 2m^2$

(b) $\dfrac{4(n-5)^2 - m}{m+n}$

EXAMPLE 9   Use a calculator to evaluate

$$\frac{-9(-3) + (-5)}{3(-4) - 5(2)}.$$

Solution   Use extra parentheses (shown here in blue) around the numerator and denominator when you enter the number in your calculator, and be careful to distinguish the negation key from the subtraction key.

If you don't get $-1$ as the answer, then you are entering something incorrectly.   $\checkmark_6$

✓ **Checkpoint 6**

Use a calculator to evaluate the following.

(a) $4^2 \div 8 + 3^2 \div 3$

(b) $[-7 + (-9)] \cdot (-4) - 8(3)$

(c) $\dfrac{-11 - (-12) - 4 \cdot 5}{4(-2) - (-6)(-5)}$

(d) $\dfrac{36 \div 4 \cdot 3 \div 9 + 1}{9 \div (-6) \cdot 8 - 4}$

## Square Roots

There are two numbers whose square is 16, namely, 4 and $-4$. The positive one, 4, is called the **square root** of 16. Similarly, the square root of a nonnegative number $d$ is defined to be the *nonnegative* number whose square is $d$; this number is denoted $\sqrt{d}$. For instance,

$$\sqrt{36} = 6 \text{ because } 6^2 = 36, \quad \sqrt{0} = 0 \text{ because } 0^2 = 0, \text{ and}$$
$$\sqrt{1.44} = 1.2 \text{ because } (1.2)^2 = 1.44.$$

No negative number has a square root that is a real number. For instance, there is no real number whose square is $-4$, so $-4$ has no square root.

Every nonnegative real number has a square root. Unless an integer is a perfect square (such as $64 = 8^2$), its square root is an irrational number. A calculator can be used to obtain a rational approximation of these square roots.

▦ **TECHNOLOGY TIP**

On one-line scientific calculators, $\sqrt{40}$ is entered as 40 $\sqrt{\ }$. On graphing calculators and two-line scientific calculators, key in $\sqrt{\ }$ 40 ENTER (or EXE).

EXAMPLE 10   Estimate each of the given quantities. Verify your estimates with a calculator.

(a) $\sqrt{40}$

Solution   Since $6^2 = 36$ and $7^2 = 49$, $\sqrt{40}$ must be a number between 6 and 7. A typical calculator shows that $\sqrt{40} \approx 6.32455532$.

**✓ Checkpoint 7**

Estimate each of the following.

(a)  $\sqrt{73}$

(b)  $\sqrt{22} + 3$

(c)  Confirm your estimates in parts (a) and (b) with a calculator.

**(b)**  $5\sqrt{7}$

**Solution**  $\sqrt{7}$ is between 2 and 3 because $2^2 = 4$ and $3^2 = 9$, so $5\sqrt{7}$ must be a number between $5 \cdot 2 = 10$ and $5 \cdot 3 = 15$. A calculator shows that $5\sqrt{7} \approx 13.22875656$.  **✓₇**

**⚠ CAUTION**  If $c$ and $d$ are positive real numbers, then $\sqrt{c + d}$ is *not* equal to $\sqrt{c} + \sqrt{d}$. For example, $\sqrt{9 + 16} = \sqrt{25} = 5$, but $\sqrt{9} + \sqrt{16} = 3 + 4 = 7$.

## The Number Line

The real numbers can be illustrated geometrically with a diagram called a **number line.** Each real number corresponds to exactly one point on the line and vice versa. A number line with several sample numbers located (or **graphed**) on it is shown in Figure 1.1.  **✓₈**

**✓ Checkpoint 8**

Draw a number line, and graph the numbers $-4, -1, 0, 1, 2.5$, and $13/4$ on it.

**Figure 1.1**

When comparing the sizes of two real numbers, the following symbols are used.

| Symbol | Read | Meaning |
|---|---|---|
| $a < b$ | $a$ is less than $b$. | $a$ lies to the *left* of $b$ on the number line. |
| $b > a$ | $b$ is greater than $a$. | $b$ lies to the *right* of $a$ on the number line. |

Note that $a < b$ means the same thing as $b > a$. The inequality symbols are sometimes joined with the equals sign, as follows.

| Symbol | Read | Meaning |
|---|---|---|
| $a \leq b$ | $a$ is less than or equal to $b$. | either $a < b$ or $a = b$ |
| $b \geq a$ | $b$ is greater than or equal to $a$. | either $b > a$ or $b = a$ |

**▦ TECHNOLOGY TIP**

If your graphing calculator has inequality symbols (usually located on the TEST menu), you can key in statements such as "$5 < 12$" or "$-2 \geq 3$." When you press ENTER, the calculator will display 1 if the statement is true and 0 if it is false.

Only one part of an "either . . . or" statement needs to be true for the entire statement to be considered true. So the statement $3 \leq 7$ is true because $3 < 7$, and the statement $3 \leq 3$ is true because $3 = 3$.

**EXAMPLE 11**     Write *true* or *false* for each of the following.

**(a)**  $8 < 12$

**Solution**  This statement says that 8 is less than 12, which is true.

**(b)**  $-6 > -3$

**Solution**  The graph in Figure 1.2 shows that $-6$ is to the *left* of $-3$. Thus, $-6 < -3$, and the given statement is false.

**✓ Checkpoint 9**

Write *true* or *false* for the following.

(a)  $-9 \leq -2$

(b)  $8 > -3$

(c)  $-14 \leq -20$

**Figure 1.2**

**(c)**  $-2 \leq -2$

**Solution**  Because $-2 = -2$, this statement is true.

A number line can be used to draw the graph of a set of numbers, as shown in the next few examples.

---

**EXAMPLE 12**   Graph all real numbers $x$ such that $1 < x < 5$.

**Solution**   This graph includes all the real numbers between 1 and 5, not just the integers. Graph these numbers by drawing a heavy line from 1 to 5 on the number line, as in Figure 1.3. Parentheses at 1 and 5 show that neither of these points belongs to the graph.

**Figure 1.3**

---

✓**Checkpoint 10**

Graph all real numbers $x$ such that

**(a)** $-5 < x < 1$

**(b)** $4 < x < 7$.

A set that consists of all the real numbers between two points, such as $1 < x < 5$ in Example 12, is called an **interval.** A special notation called **interval notation** is used to indicate an interval on the number line. For example, the interval including all numbers $x$ such that $-2 < x < 3$ is written as $(-2, 3)$. The parentheses indicate that the numbers $-2$ and 3 are *not* included. If $-2$ and 3 are to be included in the interval, square brackets are used, as in $[-2, 3]$. The following chart shows several typical intervals, where $a < b$.

## Intervals

| Inequality | Interval Notation | Explanation |
|---|---|---|
| $a \leq x \leq b$ | $[a, b]$ | Both $a$ and $b$ are included. |
| $a \leq x < b$ | $[a, b)$ | $a$ is included; $b$ is not. |
| $a < x \leq b$ | $(a, b]$ | $b$ is included; $a$ is not. |
| $a < x < b$ | $(a, b)$ | Neither $a$ nor $b$ is included. |

Interval notation is also used to describe sets such as the set of all numbers $x$ such that $x \geq -2$. This interval is written $[-2, \infty)$. The set of all real numbers is written $(-\infty, \infty)$ in interval notation.

---

**EXAMPLE 13**   Graph the interval $[-2, \infty)$.

**Solution**   Start at $-2$ and draw a heavy line to the right, as in Figure 1.4. Use a square bracket at $-2$ to show that $-2$ itself is part of the graph. The symbol $\infty$, read "infinity," *does not* represent a number. This notation simply indicates that *all* numbers greater than $-2$ are in the interval. Similarly, the notation $(-\infty, 2)$ indicates the set of all numbers $x$ such that $x < 2$.

✓**Checkpoint 11**

Graph all real numbers $x$ in the given interval.

**(a)** $(-\infty, 4]$

**(b)** $[-2, 1]$

**Figure 1.4**

## Absolute Value

The **absolute value** of a real number $a$ is the distance from $a$ to 0 on the number line and is written $|a|$. For example, Figure 1.5 shows that the distance from 9 to 0 on the number line is 9, so we have $|9| = 9$. The figure also shows that $|-9| = 9$, because the distance from $-9$ to 0 is also 9.

**Figure 1.5**

The facts that $|9| = 9$ and $|-9| = 9 = -(-9)$ suggest the following algebraic definition of absolute value.

---

## Absolute Value

For any real number $a$,

$$|a| = a \qquad \text{if } a \geq 0$$
$$|a| = -a \qquad \text{if } a < 0.$$

---

The first part of the definition shows that $|0| = 0$ (because $0 \geq 0$). It also shows that the absolute value of any positive number $a$ is the number itself, so $|a|$ is positive in such cases. The second part of the definition says that the absolute value of a negative number $a$ is the *negative* of $a$. For instance, if $a = -5$, then $|-5| = -(-5) = 5$. So $|-5|$ is positive. The same thing works for any negative number—that is, its absolute value (the negative of a negative number) is positive. Thus, we can state the following:

*For every nonzero real number a, the number $|a|$ is positive.*

✓ **Checkpoint 12**

Find the following.

(a)  $|-6|$

(b)  $-|7|$

(c)  $-|-2|$

(d)  $|-3 - 4|$

(e)  $|2 - 7|$

**EXAMPLE 14**   Evaluate $|8 - 9|$.

**Solution**  First, simplify the expression within the absolute-value bars:

$$|8 - 9| = |-1| = 1. \quad \text{✓}_{12}$$

# 1.1 Exercises

*In Exercises 1 and 2, label the statement true or false. (See Example 1.)*

1.  Every integer is a rational number.

2.  Every real number is an irrational number.

3.  The decimal expansion of the irrational number $\pi$ begins 3.141592653589793 . . . . Use your calculator to determine which of the following rational numbers is the best approximation for the irrational number $\pi$:

$$\frac{22}{7}, \quad \frac{355}{113}, \quad \frac{103,993}{33,102}, \quad \frac{2,508,429,787}{798,458,000}.$$

Your calculator may tell you that some of these numbers are equal to $\pi$, but that just indicates that the number agrees with $\pi$ for as many decimal places as your calculator can handle (usually 10–14). No rational number is exactly equal to $\pi$.

*Identify the properties that are illustrated in each of the following. (See Examples 2–6.)*

4.  $0 + (-7) = -7 + 0$

5.  $6(t + 4) = 6t + 6 \cdot 4$

6.  $3 + (-3) = (-3) + 3$

7.  $-5 + 0 = -5$

8.  $(-4) \cdot \left( \dfrac{-1}{4} \right) = 1$

9.  $8 + (12 + 6) = (8 + 12) + 6$

10.  $1 \cdot (-20) = -20$

11.  How is the additive inverse property related to the additive identity property? the multiplicative inverse property to the multiplicative identity property?

12.  Explain the distinction between the commutative and associative properties.

*Evaluate each of the following if $p = -2, q = 3,$ and $r = -5$. (See Examples 7–9.)*

13.  $-3(p + 5q)$

14.  $2(q - r)$

15.  $\dfrac{q + r}{q + p}$

16.  $\dfrac{3q}{3p - 2r}$

**Interest Rates**  *The nominal annual percentage rate (APR) reported by lenders has the formula APR = 12r, where r is the monthly interest rate. Find the APR when*

**17.**  $r = 3.8$

**18.**  $r = 0.8$

*Find the monthly interest rate r when*

**19.**  $APR = 11$

**20.**  $APR = 13.2$

*Evaluate each expression, using the order of operations given in the text. (See Examples 7–9.)*

**21.**  $3 - 4 \cdot 5 + 5$

**22.**  $8 - (-4)^2 - (-12)$

**23.**  $(4 - 5) \cdot 6 + 6$

**24.**  $\dfrac{2(3 - 7) + 4(8)}{4(-3) + (-3)(-2)}$

**25.**  $8 - 4^2 - (-12)$

**26.**  $-(3 - 5) - [2 - (3^2 - 13)]$

**27.**  $\dfrac{2(-3) + 3/(-2) - 2/(-\sqrt{16})}{\sqrt{64} - 1}$

**28.**  $\dfrac{6^2 - 3\sqrt{25}}{\sqrt{6^2 + 13}}$

*Use a calculator to help you list the given numbers in order from smallest to largest. (See Example 10.)*

**29.**  $\dfrac{189}{37}, \quad \dfrac{4587}{691}, \quad \sqrt{47}, \quad 6.735, \quad \sqrt{27}, \quad \dfrac{2040}{523}$

**30.**  $\dfrac{385}{117}, \quad \sqrt{10}, \quad \dfrac{187}{63}, \quad \pi, \quad \sqrt{\sqrt{85}}, \quad 2.9884$

*Express each of the following statements in symbols, using $<$, $>$, $\leq$, or $\geq$.*

**31.**  12 is less than 18.5.

**32.**  $-2$ is greater than $-20$.

**33.**  $x$ is greater than or equal to 5.7.

**34.**  $y$ is less than or equal to $-5$.

**35.**  $z$ is at most 7.5.

**36.**  $w$ is negative.

*Fill in the blank with $<$, $=$, or $>$ so that the resulting statement is true.*

**37.**  $-6$ _____ $-2$

**38.**  $3.14$ _____ $\pi$

**39.**  $3/4$ _____ $.75$

**40.**  $1/3$ _____ $.33$

*Fill in the blank so as to produce two equivalent statements. For example, the arithmetic statement "a is negative" is equivalent to the geometric statement "the point a lies to the left of the point 0."*

| Arithmetic Statement | Geometric Statement |
|---|---|

**41.**  $a \geq b$  _____

**42.**  _____  $a$ lies $c$ units to the right of $b$

**43.**  _____  $a$ lies between $b$ and $c$, and to the right of $c$

**44.**  $a$ is positive  _____

*Graph the given intervals on a number line. (See Examples 12 and 13.)*

**45.**  $(-8, -1)$

**46.**  $[-1, 10]$

**47.**  $(-2, 3]$

**48.**  $[-2, 2)$

**49.**  $(-2, \infty)$

**50.**  $(-\infty, -2]$

*Evaluate each of the following expressions (see Example 14).*

**51.**  $|-9| - |-12|$

**52.**  $|8| - |-4|$

**53.**  $-|-4| - |-1 - 14|$

**54.**  $-|6| - |-12 - 4|$

*In each of the following problems, fill in the blank with either $=$, $<$, or $>$, so that the resulting statement is true.*

**55.**  $|5|$ _____ $|-5|$

**56.**  $-|-4|$ _____ $|4|$

**57.**  $|10 - 3|$ _____ $|3 - 10|$

**58.**  $|6 - (-4)|$ _____ $|-4 - 6|$

**59.**  $|-2 + 8|$ _____ $|2 - 8|$

**60.**  $|3| \cdot |-5|$ _____ $|3(-5)|$

**61.**  $|3 - 5|$ _____ $|3| - |5|$

**62.**  $|-5 + 1|$ _____ $|-5| + |1|$

*Write the expression without using absolute-value notation.*

**63.**  $|a - 7|$ if $a < 7$

**64.**  $|b - c|$ if $b \geq c$

**65.**  If $a$ and $b$ are any real numbers, is it always true that $|a + b| = |a| + |b|$? Explain your answer.

**66.**  If $a$ and $b$ are any two real numbers, is it always true that $|a - b| = |b - a|$? Explain your answer.

**67.**  For which real numbers $b$ does $|2 - b| = |2 + b|$? Explain your answer.

**68.**  **Health**  Data from the National Health and Nutrition Examination Study estimates that 95% of adult heights (inches) are in the following ranges for females and males. (Data from: www.cdc.gov/nchs/nhanes.htm.)

| Females | $63.5 \pm 8.4$ |
|---|---|
| Males | $68.9 \pm 9.3$ |

Express the ranges as an absolute-value inequality in which $x$ is the height of the person.

**Consumer Price Index**  *The Consumer Price Index (CPI) tracks the cost of a typical sample of a consumer goods. The following table shows the percentage increase in the CPI for each year in a 10-year period.*

| Year | 2006 | 2007 | 2008 | 2009 | 2010 |
|---|---|---|---|---|---|
| % Increase in CPI | 3.2 | 4.1 | .1 | 2.7 | 1.5 |

| Year | 2011 | 2012 | 2013 | 2014 | 2015 |
|---|---|---|---|---|---|
| % Increase in CPI | 3.0 | 1.7 | 1.5 | 1.6 | .1 |

*Let r denote the yearly percentage increase in the CPI. For each of the following inequalities, find the number of years during the given period that r satisfied the inequality. (Data from: U.S. Bureau of Labor Statistics.)*

**69.** $r > 3.2$

**70.** $r < 3.2$

**71.** $r \leq 3.2$

**72.** $r \leq 1.5$

**73.** $r \geq 2.1$

**74.** $r \geq 1.3$

**Stocks**  *The table presents the 2016 percentage change in the stock price for six well-known companies. (Data from: finance.yahoo.com.)*

| Company | Percentage Change |
|---|---|
| American Express | +10.6% |
| Coca-Cola | −1.0% |
| ExxonMobil | +14.9% |
| Hewlett Packard | +63.1% |
| Ford | −5.7% |
| Red Robin Gourmet Burgers | −8.0% |

*Suppose that we wish to determine the difference in percentage change between two of the companies in the table, and suppose that we are interested only in the magnitude, or absolute value, of this difference. Then we subtract the two entries and find the absolute value. For example, the difference in percentage change of stock price for American Express and Ford is* $|10.6\% - (-5.7\%)| = 16.3\%$.

*Find the absolute value of the difference between the two indicated changes in stock price.*

**75.** American Express and ExxonMobil

**76.** Hewlett Packard and Red Robin

**77.** Coca-Cola and Hewlett Packard

**78.** American Express and Ford

**79.** Ford and Red Robin

**80.** Coca-Cola and Ford

**Disposable Income**  *The following graph shows the real per capita amount of disposable income (in thousands of dollars) in the United States. (Data from: Federal Reserve Bank of St. Louis.)*

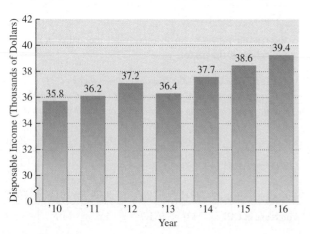

*For each of the following, determine the years for which the expression is true, where x is the per capita disposable income.*

**81.** $|x - 37{,}000| > 1000$

**82.** $|x - 32{,}000| > 3500$

**83.** $|x - 36{,}500| \leq 2200$

**84.** $|x - 38{,}500| \leq 1000$

## ✓ Checkpoint Answers

**1.** **(a)** Integer, rational, real
　　**(b)** Rational, real
　　**(c)** Irrational, real

**2.** **(a)** Commutative property
　　**(b)** Associative property
　　**(c)** Commutative property
　　**(d)** Commutative property
　　**(e)** Associative property

**3.** **(a)** Additive identity property
　　**(b)** Multiplicative inverse property
　　**(c)** Additive inverse property
　　**(d)** Multiplicative identity property

**4.** **(a)** $4(-2) + 4(5) = 12$
　　**(b)** $2a + 2b$
　　**(c)** $-3p - 3$
　　**(d)** $8m - km$
　　**(e)** $(5 + 3)x = 8x$

**5.** **(a)** 30　　　　　　　　　**(b)** $\dfrac{41}{3}$

**6.** **(a)** 5　　　　　　　　　**(b)** 40
　　**(c)** $\dfrac{19}{38} = \dfrac{1}{2} = .5$　　　**(d)** $-\dfrac{1}{4} = -.25$

**7.** **(a)** Between 8 and 9
　　**(b)** Between 7 and 8
　　**(c)** 8.5440; 7.6904

**8.**

**9.** **(a)** True　　**(b)** True　　**(c)** False

**10.** **(a)**　　　　　　　**(b)**

**11.** **(a)**　　　　　　　**(b)**

**12.** **(a)** 6　　　　　　　　**(b)** −7
　　**(c)** −2　　　　　　　　**(d)** 7
　　**(e)** 5

# 1.2 Polynomials

Polynomials are the fundamental tools of algebra and will play a central role in this course. In order to do polynomial arithmetic, you must first understand exponents. So we begin with them. You are familiar with the usual notation for squares and cubes, such as:

$$5^2 = 5\cdot 5 \quad \text{and} \quad 6^3 = 6\cdot 6\cdot 6.$$

We now extend this convenient notation to other cases.

> If $n$ is a natural number and $a$ is any real number, then
>
> $$a^n \quad \text{denotes the product} \quad a\cdot a\cdot a\cdots a\,(n\text{ factors}).$$
>
> The number $a$ is the **base**, and the number $n$ is the **exponent.**

**EXAMPLE 1**　$4^6$, which is read "four to the sixth," or "four to the sixth power," is the number

$$4\cdot 4\cdot 4\cdot 4\cdot 4\cdot 4 = 4096.$$

Similarly, $(-5)^3 = (-5)(-5)(-5) = -125$, and

$$\left(\frac{3}{2}\right)^4 = \frac{3}{2}\cdot\frac{3}{2}\cdot\frac{3}{2}\cdot\frac{3}{2} = \frac{81}{16}.$$

**EXAMPLE 2**　Use a calculator to approximate the given expressions.

**(a)** $(1.2)^8$

**Solution** Key in 1.2 and then use the $x^y$ key (labeled $\wedge$ on some calculators); finally, key in the exponent 8. The calculator displays the (exact) answer 4.29981696.

**(b)** $\left(\frac{12}{7}\right)^{23}$

**Solution** Don't compute 12/7 separately. Use parentheses and key in (12/7), followed by the $x^y$ key and the exponent 23 to obtain the approximate answer 242,054.822. ✓1

**✓ Checkpoint 1**

Evaluate the following.

**(a)** $6^3$

**(b)** $5^{12}$

**(c)** $1^9$

**(d)** $\left(\frac{7}{5}\right)^8$

⬣ **CAUTION** A common error in using exponents occurs with expressions such as $4\cdot 3^2$. The exponent of 2 applies only to the base 3, so that

$$4\cdot 3^2 = 4\cdot 3\cdot 3 = 36.$$

On the other hand,

$$(4\cdot 3)^2 = (4\cdot 3)(4\cdot 3) = 12\cdot 12 = 144,$$

so

$$4\cdot 3^2 \neq (4\cdot 3)^2.$$

Be careful to distinguish between expressions like $-2^4$ and $(-2)^4$:

$$-2^4 = -(2^4) = -(2\cdot 2\cdot 2\cdot 2) = -16$$
$$(-2)^4 = (-2)(-2)(-2)(-2) = 16,$$

so

$$-2^4 \neq (-2)^4. ✓2$$

**✓ Checkpoint 2**

Evaluate the following.

**(a)** $3\cdot 6^2$

**(b)** $5\cdot 4^3$

**(c)** $-3^6$

**(d)** $(-3)^6$

**(e)** $-2\cdot(-3)^5$

By the definition of an exponent,

$$3^4 \cdot 3^2 = (3 \cdot 3 \cdot 3 \cdot 3)(3 \cdot 3) = 3^6.$$

Since $6 = 4 + 2$, we can write the preceding equation as $3^4 \cdot 3^2 = 3^{4+2}$. This result suggests the following fact, which applies to any real number $a$ and natural numbers $m$ and $n$.

## Multiplication with Exponents

To multiply $a^m$ by $a^n$, *add* the exponents:

$$a^m \cdot a^n = a^{m+n}.$$

### EXAMPLE 3    Verify each of the following simplifications.

**(a)** $7^4 \cdot 7^6 = 7^{4+6} = 7^{10}$

**(b)** $(-2)^3 \cdot (-2)^5 = (-2)^{3+5} = (-2)^8$

**(c)** $(3k)^2 \cdot (3k)^3 = (3k)^5$

**(d)** $(m + n)^2 \cdot (m + n)^5 = (m + n)^7$  ✓₃

The multiplication property of exponents has a convenient consequence. By definition,

$$(5^2)^3 = 5^2 \cdot 5^2 \cdot 5^2 = 5^{2+2+2} = 5^6.$$

Note that $2 + 2 + 2$ is $3 \cdot 2 = 6$. This is an example of a more general fact about any real number $a$ and natural numbers $m$ and $n$.

## Power of a Power

To find a power of a power, $(a^m)^n$, *multiply* the exponents:

$$(a^m)^n = a^{mn}.$$

### EXAMPLE 4    Verify the following computations.

**(a)** $(x^3)^4 = x^{3 \cdot 4} = x^{12}.$

**(b)** $[(-3)^5]^3 = (-3)^{5 \cdot 3} = (-3)^{15}.$

**(c)** $[(6z)^4]^4 = (6z)^{4 \cdot 4} = (6z)^{16}.$  ✓₄

It will be convenient to give a zero exponent a meaning. If the multiplication property of exponents is to remain valid, we should have, for example, $3^5 \cdot 3^0 = 3^{5+0} = 3^5$. But this will be true only when $3^0 = 1$. So we make the following definition.

## Zero Exponent

If $a$ is any nonzero real number, then

$$a^0 = 1.$$

For example, $6^0 = 1$ and $(-9)^0 = 1$. Note that $0^0$ is *not* defined.

---

✓ **Checkpoint 3**

Simplify the following.

**(a)** $5^3 \cdot 5^6$

**(b)** $(-3)^4 \cdot (-3)^{10}$

**(c)** $(5p)^2 \cdot (5p)^8$

---

✓ **Checkpoint 4**

Compute the following.

**(a)** $(6^3)^7$

**(b)** $[(4k)^5]^6$

---

✓ **Checkpoint 5**

Evaluate the following.

**(a)** $17^0$

**(b)** $30^0$

**(c)** $(-10)^0$

**(d)** $-(12)^0$

## Polynomials

A **polynomial** is an algebraic expression such as

$$5x^4 + 2x^3 + 6x, \qquad 8m^3 + 9m^2 + \frac{3}{2}m + 3, \qquad -10p, \qquad \text{or} \qquad 8.$$

The letter used is called a **variable,** and a polynomial is a sum of **terms** of the form

$$(\text{constant}) \times (\text{nonnegative integer power of the variable}).$$

We assume that $x^0 = 1, m^0 = 1$, etc., so terms such as 3 or 8 may be thought of as $3x^0$ and $8x^0$, respectively. The constants that appear in each term of a polynomial are called the **coefficients** of the polynomial. The coefficient of $x^0$ is called the **constant term.**

**EXAMPLE 5** Identify the coefficients and the constant term of the given polynomials.

**(a)** $5x^2 - x + 12$

**Solution** The coefficients are 5, −1, and 12, and the constant term is 12.

**(b)** $7x^3 + 2x - 4$

**Solution** The coefficients are 7, 0, 2, and −4, because the polynomial can be written $7x^3 + 0x^2 + 2x - 4$. The constant term is −4.

A polynomial that consists only of a constant term, such as 15, is called a **constant polynomial.** The **zero polynomial** is the constant polynomial 0. The **degree** of a polynomial is the *exponent* of the highest power of $x$ that appears with a *nonzero* coefficient, and the nonzero coefficient of this highest power of $x$ is the **leading coefficient** of the polynomial. For example,

| Polynomial | Degree | Leading Coefficient | Constant Term |
|---|---|---|---|
| $6x^7 + 4x^3 + 5x^2 - 7x + 10$ | 7 | 6 | 10 |
| $-x^4 + 2x^3 + \frac{1}{2}$ | 4 | −1 | $\frac{1}{2}$ |
| $x^3$ | 3 | 1 | 0 |
| 12 | 0 | 12 | 12 |

The degree of the zero polynomial is *not defined*, since no exponent of $x$ occurs with a nonzero coefficient. First-degree polynomials are often called **linear polynomials.** Second- and third-degree polynomials are called **quadratics** and **cubics,** respectively. ✔6

### Addition and Subtraction

Two terms having the same variable with the same exponent are called **like terms;** other terms are called **unlike terms.** Polynomials can be added or subtracted by using the distributive property to combine like terms. Only like terms can be combined. For example,

$$12y^4 + 6y^4 = (12 + 6)y^4 = 18y^4$$

and

$$-2m^2 + 8m^2 = (-2 + 8)m^2 = 6m^2.$$

The polynomial $8y^4 + 2y^5$ has unlike terms, so it cannot be further simplified.

In more complicated cases of addition, you may have to eliminate parentheses, use the commutative and associative laws to regroup like terms, and then combine them.

✔ **Checkpoint 6**

Find the degree of each polynomial.

**(a)** $x^4 - x^2 + x + 5$
**(b)** $7x^5 + 6x^3 - 3x^8 + 2$
**(c)** 17
**(d)** 0

---

**EXAMPLE 6**　Add the following polynomials.

**(a)** $(8x^3 - 4x^2 + 6x) + (3x^3 + 5x^2 - 9x + 8)$

**Solution** $8x^3 - 4x^2 + 6x + 3x^3 + 5x^2 - 9x + 8$　　Eliminate parentheses.

$= \underline{(8x^3 + 3x^3)} + \underline{(-4x^2 + 5x^2)} + \underline{(6x - 9x)} + 8$　　Group like terms.

$= 11x^3 + x^2 - 3x + 8$　　Combine like terms.

**(b)** $(-4x^4 + 6x^3 - 9x^2 - 12) + (-3x^3 + 8x^2 - 11x + 7)$

**Solution**

$-4x^4 + 6x^3 - 9x^2 - 12 - 3x^3 + 8x^2 - 11x + 7$　　Eliminate parentheses.

$= -4x^4 + \underline{(6x^3 - 3x^3)} + \underline{(-9x^2 + 8x^2)} - 11x + \underline{(-12 + 7)}$　　Group like terms.

$= -4x^4 + 3x^3 - x^2 - 11x - 5$　　Combine like terms.

Care must be used when parentheses are preceded by a minus sign. For example, we know that

$$-(4 + 3) = -(7) = -7 = -4 - 3.$$

If you simply delete the parentheses in $-(4 + 3)$, you obtain $-4 + 3 = -1$, which is the wrong answer. This fact and the preceding examples illustrate the following rules.

## Rules for Eliminating Parentheses

Parentheses preceded by a plus sign (or no sign) may be deleted.

Parentheses preceded by a minus sign may be deleted *provided that* the sign of every term within the parentheses is changed.

---

**EXAMPLE 7**　Subtract: $(2x^2 - 11x + 8) - (7x^2 - 6x + 2)$.

**Solution** $2x^2 - 11x + 8 - 7x^2 + 6x - 2$　　Eliminate parentheses.

$= (2x^2 - 7x^2) + (-11x + 6x) + (8 - 2)$　　Group like terms.

$= -5x^2 - 5x + 6$　　Combine like terms. ☑7

✓**Checkpoint 7**

Add or subtract as indicated.

**(a)** $(-2x^2 + 7x + 9)$
$+ (3x^2 + 2x - 7)$

**(b)** $(4x + 6) - (13x - 9)$

**(c)** $(9x^3 - 8x^2 + 2x)$
$- (9x^3 - 2x^2 - 10)$

## Multiplication

The distributive property is also used to multiply polynomials. For example, the product of $8x$ and $6x - 4$ is found as follows:

$$8x(6x - 4) = 8x(6x) - 8x(4)$$　　Distributive property

$$= 48x^2 - 32x.$$　　$x \cdot x = x^2$

---

**EXAMPLE 8**　Use the distributive property to find each product.

**(a)** $2p^3(3p^2 - 2p + 5) = 2p^3(3p^2) + 2p^3(-2p) + 2p^3(5)$
$= 6p^5 - 4p^4 + 10p^3$

✓**Checkpoint 8**

Find the following products.

**(a)** $-6r(2r - 5)$

**(b)** $(8m + 3) \cdot (m^4 - 2m^2 + 6m)$

**(b)** $(3k - 2)(k^2 + 5k - 4) = 3k(k^2 + 5k - 4) - 2(k^2 + 5k - 4)$
$= 3k^3 + 15k^2 - 12k - 2k^2 - 10k + 8$
$= 3k^3 + 13k^2 - 22k + 8$ ☑8

**EXAMPLE 9**   The product $(2x - 5)(3x + 4)$ can be found by using the distributive property twice:

$$(2x - 5)(3x + 4) = 2x(3x + 4) - 5(3x + 4)$$
$$= 2x \cdot 3x + 2x \cdot 4 + (-5) \cdot 3x + (-5) \cdot 4$$
$$= 6x^2 + 8x - 15x - 20$$
$$= 6x^2 - 7x - 20.$$

Observe the pattern in the second line of Example 9 and its relationship to the terms being multiplied.

$$(2x - 5)(3x + 4) = 2x \cdot 3x + 2x \cdot 4 + (-5) \cdot 3x + (-5) \cdot 4$$

First terms

$(2x - 5)(3x + 4)$   Outside terms

$(2x - 5)(3x + 4)$   Inside terms

$(2x - 5)(3x + 4)$   Last terms

This pattern is easy to remember by using the acronym **FOIL** (**F**irst, **O**utside, **I**nside, **L**ast). The FOIL method makes it easy to find products such as this one mentally, without the necessity of writing out the intermediate steps.

**EXAMPLE 10**   Use FOIL to find the product of the given polynomials.

**(a)** $(3x + 2)(x + 5) = 3x^2 + 15x + 2x + 10 = 3x^2 + 17x + 10$

First   Outside   Inside   Last

**(b)** $(x + 3)^2 = (x + 3)(x + 3) = x^2 + 3x + 3x + 9 = x^2 + 6x + 9$

**(c)** $(2x + 1)(2x - 1) = 4x^2 - 2x + 2x - 1 = 4x^2 - 1$ ✓ 9

✓ **Checkpoint 9**

Use FOIL to find these products.

**(a)** $(5k - 1)(2k + 3)$

**(b)** $(7z - 3)(2z + 5)$

## Applications

In business, the *revenue* from the sales of an item is given by

**Revenue = (price per item) × (number of items sold).**

The *cost* to manufacture and sell these items is given by

**Cost = Fixed Costs + Variable Costs,**

where the fixed costs include such things as buildings and machinery (which do not depend on how many items are made) and variable costs include such things as labor and materials (which vary, depending on how many items are made). Then

**Profit = Revenue − Cost.**

**EXAMPLE 11**   **Profit** A manufacturer of scientific calculators sells calculators for \$12 each (wholesale) and can produce a maximum of 150,000. The variable cost of producing $x$ thousand calculators is $6995x - 7.2x^2$ dollars, and the fixed costs for the manufacturing operation are \$230,000. If $x$ thousand calculators are manufactured and sold, find expressions for the revenue, cost, and profit.

**Solution**   If $x$ thousand calculators are sold at \$12 each, then

$$\text{Revenue} = (\text{price per item}) \times (\text{number of items sold})$$
$$R = 12 \times 1000x = 12{,}000x,$$

where $x \le 150$ (because only 150,000 calculators can be made). The variable cost of making $x$ thousand calculators is $6995x - 7.2x^2$, so that

✓ **Checkpoint 10**

Suppose revenue is given by $7x^2 - 3x$, fixed costs are $500, and variable costs are given by $3x^2 + 5x - 25$. Write an expression for

(a) Cost

(b) Profit

$$\text{Cost} = \text{Fixed Costs} + \text{Variable Costs,}$$
$$C = 230{,}000 + (6995x - 7.2x^2) \quad (x \le 150).$$

Therefore, the profit is given by

$$P = R - C = 12{,}000x - (230{,}000 + 6995x - 7.2x^2)$$
$$= 12{,}000x - 230{,}000 - 6995x + 7.2x^2$$
$$P = 7.2x^2 + 5005x - 230{,}000 \quad (x \le 150). \; ✓_{10}$$

# 1.2 Exercises

*Use a calculator to approximate these numbers. (See Examples 1 and 2.)*

1. $11.2^6$

2. $(-6.54)^{11}$

3. $(-18/7)^6$

4. $(5/9)^7$

5. Explain how the value of $-3^2$ differs from $(-3)^2$. Do $-3^3$ and $(-3)^3$ differ in the same way? Why or why not?

6. Describe the steps used to multiply $4^3$ and $4^5$. Is the product of $4^3$ and $3^4$ found in the same way? Explain.

*Simplify each of the given expressions. Leave your answers in exponential notation. (See Examples 3 and 4.)*

7. $4^2 \cdot 4^3$

8. $(-4)^4 \cdot (-4)^6$

9. $(-6)^2 \cdot (-6)^5$

10. $(2z)^5 \cdot (2z)^6$

11. $[(5u)^4]^7$

12. $(6y)^3 \cdot [(6y)^5]^4$

*List the degree of the given polynomial, its coefficients, and its constant term. (See Example 5.)*

13. $6.2x^4 - 5x^3 + 4x^2 - 3x + 3.7$

14. $6x^7 + 4x^6 - x^3 + x$

*State the degree of the given polynomial.*

15. $1 + x + 2x^2 + 3x^3$

16. $5x^4 - 4x^5 - 6x^3 + 7x^4 - 2x + 8$

*Add or subtract as indicated. (See Examples 6 and 7.)*

17. $(3x^3 + 2x^2 - 5x) + (-4x^3 - x^2 - 8x)$

18. $(-2p^3 - 5p + 7) + (-4p^2 + 8p + 2)$

19. $(-4y^2 - 3y + 8) - (2y^2 - 6y + 2)$

20. $(7b^2 + 2b - 5) - (3b^2 + 2b - 6)$

21. $(2x^3 + 2x^2 + 4x - 3) - (2x^3 + 8x^2 + 1)$

22. $(3y^3 + 9y^2 - 11y + 8) - (-4y^2 + 10y - 6)$

*Find each of the given products. (See Examples 8–10.)*

23. $-9m(2m^2 + 6m - 1)$

24. $2a(4a^2 - 6a + 8)$

25. $(3z + 5)(4z^2 - 2z + 1)$

26. $(2k + 3)(4k^3 - 3k^2 + k)$

27. $(6k - 1)(2k + 3)$

28. $(8r + 3)(r - 1)$

29. $(3y + 5)(2y + 1)$

30. $(5r - 3s)(5r - 4s)$

31. $(9k + q)(2k - q)$

32. $(.012x - .17)(.3x + .54)$

33. $(6.2m - 3.4)(.7m + 1.3)$

34. $2p - 3[4p - (8p + 1)]$

35. $5k - [k + (-3 + 5k)]$

36. $(3x - 1)(x + 2) - (2x + 5)^2$

**Profit** *Find expressions for the revenue, cost, and profit from selling $x$ thousand items. (See Example 11.)*

| Item Price | Fixed Costs | Variable Costs |
|---|---|---|
| 37. $5.00 | $200,000 | $1800x$ |
| 38. $8.50 | $225,000 | $4200x$ |

39. **Profit** Beauty Works sells its cologne wholesale for $9.75 per bottle. The variable costs of producing $x$ thousand bottles is $-3x^2 + 3480x - 325$ dollars, and the fixed costs of manufacturing are $260,000. Find expressions for the revenue, cost, and profit from selling $x$ thousand items.

40. **Profit** A self-help guru sells her book *Be Happy in 45 Easy Steps* for $23.50 per copy. Her fixed costs are $145,000 and the estimate of the variable cost of printing, binding, and distributing $x$ thousand books is given by $-4.2x^2 + 3220x - 425$ dollars. Find expressions for the revenue, cost, and profit from selling $x$ thousand copies of the book.

*Work these problems.*

**Starbucks Earnings** *The accompanying bar graph shows the net earnings (in millions of dollars) for the Starbucks Corporation. The polynomial*

$$4.79x^3 - 122.5x^2 + 1104x - 2863$$

gives a good approximation of Starbuck's net earnings in year x, where x = 7 corresponds to 2007, and so on (7 ≤ x ≤ 16). For each of the given years,

(a) use the bar graph to determine the net earnings;

(b) use the polynomial to determine the net earnings.

(Data from: www.morningstar.com.)

**41.** 2007 **42.** 2015

**43.** 2012 **44.** 2013

Chart of Earnings

Assuming that the polynomial approximation in Exercises 41–44 remains accurate in later years, use it to estimate Starbucks's net earnings in each of the following years.

**45.** 2017 **46.** 2018 **47.** 2019

**48.** Do the estimates in Exercises 45–47 seem plausible? Explain.

**Starbucks Costs** *The costs (in millions of dollars) for the Starbucks Corporation can be approximated by the polynomial* $9.5x^3 - 401.6x^2 + 6122x - 25,598$, *where x = 7 corresponds to the year 2007. Determine whether each of the given statements is true or false. (Data from: www.morningstar.com.)*

**49.** The costs were higher than $5000 million in 2010.

**50.** The costs were higher than $7500 million in 2015.

**51.** The costs were higher in 2012 than in 2015.

**52.** The costs were lower in 2011 than in 2016.

**PepsiCo Profits** *The profits (in millions of dollars) for PepsiCo Inc. can be approximated by the polynomial* $-72.85x^3 + 2082x^2 -$ $16,532x + 59,357$, *where x = 7 corresponds to the year 2007. (Data from: www.morningstar.com.) Find the approximate profits for the following years.*

**53.** 2007 **54.** 2010 **55.** 2012 **56.** 2015

*Determine whether each of the given statements is true or false.*

**57.** Were profits higher in 2013 or 2009?

**58.** Were profits higher in 2012 or 2015?

**Business** *Use the table feature of a graphing calculator or use a spreadsheet to make a table of values for the profit expression in Example 11, with x = 0, 5, 10, . . . , 150. Use the table to answer the following questions.*

**59.** What is the profit or loss (negative profit) when 25,000 calculators are sold? when 60,000 are sold? Explain these answers.

**60.** Approximately how many calculators must be sold in order for the company to make a profit?

**61.** What is the profit from selling 100,000 calculators?

**62.** What is the profit from selling 150,000 calculators?

✓ **Checkpoint Answers**

1. (a) 216 (b) 244,140,625
   (c) 1 (d) 14.75789056

2. (a) 108 (b) 320 (c) −729
   (d) 729 (e) 486

3. (a) $5^9$ (b) $(-3)^{14}$ (c) $(5p)^{10}$

4. (a) $6^{21}$ (b) $(4k)^{30}$

5. (a) 1 (b) 1 (c) 1 (d) −1

6. (a) 4 (b) 8 (c) 0 (d) Not defined

7. (a) $x^2 + 9x + 2$
   (b) $-9x + 15$
   (c) $-6x^2 + 2x + 10$

8. (a) $-12r^2 + 30r$
   (b) $8m^5 + 3m^4 - 16m^3 + 42m^2 + 18m$

9. (a) $10k^2 + 13k - 3$ (b) $14z^2 + 29z - 15$

10. (a) $C = 3x^2 + 5x + 475$ (b) $P = 4x^2 - 8x - 475$

# 1.3 Factoring

The number 18 can be written as a product in several ways: $9 \cdot 2$, $(-3)(-6)$, $1 \cdot 18$, etc. The numbers in each product (9, 2, −3, etc.) are called **factors,** and the process of writing 18 as a product of factors is called **factoring.** Thus, factoring is the reverse of multiplication.

Factoring of polynomials is a means of simplifying many expressions and of solving certain types of equations. As is the usual custom, factoring of polynomials in this text will be restricted to finding factors with *integer* coefficients (otherwise there may be an infinite number of possible factors).

## Greatest Common Factor

The algebraic expression $15m + 45$ is made up of two terms: $15m$ and $45$. Each of these terms has 15 as a factor. In fact, $15m = 15 \cdot m$ and $45 = 15 \cdot 3$. By the distributive property,

$$15m + 45 = 15 \cdot m + 15 \cdot 3 = 15(m + 3).$$

Both 15 and $m + 3$ are factors of $15m + 45$. Since 15 is a factor of all terms of $15m + 45$ and is the largest such number, it is called the **greatest common factor** for the polynomial $15m + 45$. The process of writing $15m + 45$ as $15(m + 3)$ is called **factoring out** the greatest common factor.

---

**EXAMPLE 1**    Factor out the greatest common factor.

**(a)** $12p - 18q$

**Solution**  Both $12p$ and $18q$ are divisible by 6, and

$$12p - 18q = 6 \cdot 2p - 6 \cdot 3q$$
$$= 6(2p - 3q).$$

**(b)** $8x^3 - 9x^2 + 15x$

**Solution**  Each of these terms is divisible by $x$:

$$8x^3 - 9x^2 + 15x = (8x^2) \cdot x - (9x) \cdot x + 15 \cdot x$$
$$= x(8x^2 - 9x + 15).$$

**(c)** $5(4x - 3)^3 + 2(4x - 3)^2$

**Solution**  The quantity $(4x - 3)^2$ is a common factor. Factoring it out gives

$$5(4x - 3)^3 + 2(4x - 3)^2 = (4x - 3)^2[5(4x - 3) + 2]$$
$$= (4x - 3)^2(20x - 15 + 2)$$
$$= (4x - 3)^2(20x - 13). \ \boxed{✓}_1$$

---

✓ **Checkpoint 1**

Factor out the greatest common factor.

**(a)** $12r + 9k$

**(b)** $75m^2 + 100n^2$

**(c)** $6m^4 - 9m^3 + 12m^2$

**(d)** $3(2k + 1)^3 + 4(2k + 1)^4$

## Factoring Quadratics

If we multiply two first-degree polynomials, the result is a quadratic. For instance, using FOIL, we see that $(x + 1)(x - 2) = x^2 - x - 2$. Since factoring is the reverse of multiplication, factoring quadratics requires using FOIL backward.

---

**EXAMPLE 2**    Factor $x^2 + 9x + 18$.

**Solution**  We must find integers $b$ and $d$ such that

$$x^2 + 9x + 18 = (x + b)(x + d)$$
$$= x^2 + dx + bx + bd$$
$$x^2 + 9x + 18 = x^2 + (b + d)x + bd.$$

Since the constant coefficients on each side of the equation must be equal, we must have $bd = 18$; that is, $b$ and $d$ are factors of 18. Similarly, the coefficients of $x$ must be the same, so that $b + d = 9$. The possibilities are summarized in this table:

| Factors $b, d$ of 18 | Sum $b + d$ |
|---|---|
| $18 \cdot 1$ | $18 + 1 = 19$ |
| $9 \cdot 2$ | $9 + 2 = 11$ |
| $6 \cdot 3$ | $6 + 3 = 9$ |

✓ **Checkpoint 2**

Factor the following.

(a) $r^2 + 7r + 10$

(b) $x^2 + 4x + 3$

(c) $y^2 + 6y + 8$

There is no need to list negative factors, such as $(-3)(-6)$, because their sum is negative. The table suggests that 6 and 3 will work. Verify that

$$(x + 6)(x + 3) = x^2 + 9x + 18. \quad \checkmark_2$$

**EXAMPLE 3**    Factor $x^2 + 3x - 10$.

**Solution** As in Example 2, we must find factors $b$ and $d$ whose product is $-10$ (the constant term) and whose sum is 3 (the coefficient of $x$). The following table shows the possibilities.

| Factors $b, d$ of $-10$ | Sum $b + d$ |
|---|---|
| $1(-10)$ | $1 + (-10) = -9$ |
| $(-1)10$ | $-1 + 10 = 9$ |
| $2(-5)$ | $2 + (-5) = -3$ |
| $(-2)5$ | $-2 + 5 = 3$ |

The only factors with product $-10$ and sum 3 are $-2$ and 5. So the correct factorization is

$$x^2 + 3x - 10 = (x - 2)(x + 5),$$

as you can readily verify.

It is usually not necessary to construct tables as was done in Examples 2 and 3—you can just mentally check the various possibilities. The approach used in Examples 2 and 3 (with minor modifications) also works for factoring quadratic polynomials whose leading coefficient is not 1.

**EXAMPLE 4**    Factor $4y^2 - 11y + 6$.

**Solution** We must find integers $a$, $b$, $c$, and $d$ such that

$$4y^2 - 11y + 6 = (ay + b)(cy + d)$$
$$= acy^2 + ady + bcy + bd$$
$$4y^2 - 11y + 6 = acy^2 + (ad + bc)y + bd.$$

Since the coefficients of $y^2$ must be the same on both sides, we see that $ac = 4$. Similarly, the constant terms show that $bd = 6$. The positive factors of 4 are 4 and 1 or 2 and 2. Since the middle term is negative, we consider only negative factors of 6. The possibilities are $-2$ and $-3$ or $-1$ and $-6$. Now we try various arrangements of these factors until we find one that gives the correct coefficient of $y$:

✓ **Checkpoint 3**

Factor the following.

(a) $x^2 - 4x + 3$

(b) $2y^2 - 5y + 2$

(c) $6z^2 - 13z + 6$

$$(2y - 1)(2y - 6) = 4y^2 - 14y + 6 \quad \text{Incorrect}$$
$$(2y - 2)(2y - 3) = 4y^2 - 10y + 6 \quad \text{Incorrect}$$
$$(y - 2)(4y - 3) = 4y^2 - 11y + 6. \quad \text{Correct}$$

The last trial gives the correct factorization. $\checkmark_3$

**EXAMPLE 5**    Factor $6p^2 - 7pq - 5q^2$.

**Solution** Again, we try various possibilities. The positive factors of 6 could be 2 and 3 or 1 and 6. As factors of $-5$, we have only $-1$ and 5 or $-5$ and 1. Try different combinations of these factors until the correct one is found:

✓ **Checkpoint 4**

Factor the following.

(a) $r^2 - 5r - 14$

(b) $3m^2 + 5m - 2$

(c) $6p^2 + 13pq - 5q^2$

$$(2p - 5q)(3p + q) = 6p^2 - 13pq - 5q^2 \quad \text{Incorrect}$$
$$(3p - 5q)(2p + q) = 6p^2 - 7pq - 5q^2. \quad \text{Correct}$$

So $6p^2 - 7pq - 5q^2$ factors as $(3p - 5q)(2p + q)$. $\checkmark_4$

📄 **NOTE** In Examples 2–4, we chose positive factors of the positive first term. Of course, we could have used two negative factors, but the work is easier if positive factors are used.

---

**EXAMPLE 6**    Factor $x^2 + x + 3$.

**Solution** There are only two ways to factor 3, namely, $3 = 1 \cdot 3$ and $3 = (-1)(-3)$. They lead to these products:

$$(x + 1)(x + 3) = x^2 + 4x + 3 \quad \text{\small Incorrect}$$
$$(x - 1)(x - 3) = x^2 - 4x + 3. \quad \text{\small Incorrect}$$

Therefore, this polynomial cannot be factored.

## Factoring Patterns

In some cases, you can factor a polynomial with a minimum amount of guesswork by recognizing common patterns. The easiest pattern to recognize is the *difference of squares*.

---

### Difference of Squares

$$x^2 - y^2 = (x + y)(x - y).$$

---

To verify the accuracy of the preceding equation, multiply out the right side.

---

**EXAMPLE 7**    Factor each of the following.

**(a)** $4m^2 - 9$

**Solution** Notice that $4m^2 - 9$ is the difference of two squares, since $4m^2 = (2m)^2$ and $9 = 3^2$. Use the pattern for the difference of two squares, letting $2m$ replace $x$ and 3 replace $y$. Then the pattern $x^2 - y^2 = (x + y)(x - y)$ becomes

$$4m^2 - 9 = (2m)^2 - 3^2$$
$$= (2m + 3)(2m - 3).$$

**(b)** $128p^2 - 98q^2$

**Solution** First factor out the common factor of 2:

$$128p^2 - 98q^2 = 2(64p^2 - 49q^2)$$
$$= 2[(8p)^2 - (7q)^2]$$
$$= 2(8p + 7q)(8p - 7q).$$

**(c)** $x^2 + 36$

**Solution** The sum of two perfect squares cannot be factored with real numbers. To understand why this is true, let us say for the sake of argument that $x^2 + 36$ can be factored with real numbers. Then there would exist two real numbers $a$ and $b$ where we could write

$$x^2 + 36 = (x + a)(x + b).$$

The above expression is an equality. This means that for any real number that we substitute for $x$, the expressions on each side of the equals sign must evaluate to the same value. However, if we let $x = -a$, then we obtain:

$$(-a)^2 + 36 = (-a + a)(-a + b) \quad \text{\small Substituting } x = -a.$$
$$(-a)^2 + 36 = 0(-a + b) \quad \text{\small Additive inverse of } a$$
$$(-a)^2 + 36 = 0. \quad \text{\small Multiplication by 0}$$

The above is problematic. The left-hand side of the equation has to be greater than 0 because $(-a)^2$ has to be 0 or positive, and then, when added to 36, it will remain greater than 0. As we see above, the right-hand side has to be 0. This is a contradiction. Thus, our assumption that $x^2 + 36$ can be factored with real numbers must be incorrect.

When we see an expression of the form $x^2 + a^2$, we call this a sum of two squares. Such expressions cannot be factored with real numbers.

**(d)** $(x-2)^2 - 49$

**Solution** Since $49 = 7^2$, this is a difference of two squares. So it factors as follows:

$$(x-2)^2 - 49 = (x-2)^2 - 7^2$$
$$= [(x-2) + 7][(x-2) - 7]$$
$$= (x+5)(x-9). \checkmark_5$$

✓ **Checkpoint 5**

Factor the following.
**(a)** $9p^2 - 49$
**(b)** $y^2 + 100$
**(c)** $(x+3)^2 - 64$

Another common pattern is the *perfect square*. Verify each of the following factorizations by multiplying out the right side.

## Perfect Squares

$$x^2 + 2xy + y^2 = (x+y)^2$$
$$x^2 - 2xy + y^2 = (x-y)^2$$

Whenever you have a quadratic whose first and last terms are squares, it *may* factor as a perfect square. The key is to look at the middle term. To have a perfect square whose first and last terms are $x^2$ and $y^2$, the middle term must be $\pm 2xy$. To avoid errors, always check this.

**EXAMPLE 8** Factor each polynomial, if possible.

**(a)** $16p^2 - 40pq + 25q^2$

**Solution** The first and last terms are squares, namely, $16p^2 = (4p)^2$ and $25q^2 = (5q)^2$. So the second perfect-square pattern, with $x = 4p$ and $y = 5q$, might work. To have a perfect square, the middle term $-40pq$ must equal $-2(4p)(5q)$, which it does. So the polynomial factors as

$$16p^2 - 40pq + 25q^2 = (4p - 5q)(4p - 5q),$$

as you can easily verify.

**(b)** $9u^2 + 5u + 1$

**Solution** Again, the first and last terms are squares: $9u^2 = (3u)^2$ and $1 = 1^2$. The middle term is positive, so the first perfect-square pattern might work, with $x = 3u$ and $y = 1$. To have a perfect square, however, the middle term would have to be $2(3u) \cdot 1 = 6u$, which is *not* the middle term of the given polynomial. So it is not a perfect square—in fact, it cannot be factored.

**(c)** $169x^2 + 104xy^2 + 16y^4$

**Solution** This polynomial may be factored as $(13x + 4y^2)^2$, since $169x^2 = (13x)^2$, $16y^4 = (4y^2)^2$, and $2(13x)(4y^2) = 104xy^2$. $\checkmark_6$

✓ **Checkpoint 6**

Factor.
**(a)** $4m^2 + 4m + 1$
**(b)** $25z^2 - 80zt + 64t^2$
**(c)** $9x^2 + 15x + 25$

**EXAMPLE 9**    Factor each of the given polynomials.

**(a)** $12x^2 - 26x - 10$

**Solution**  Look first for a greatest common factor. Here, the greatest common factor is 2: $12x^2 - 26x - 10 = 2(6x^2 - 13x - 5)$. Now try to factor $6x^2 - 13x - 5$. Possible factors of 6 are 3 and 2 or 6 and 1. The only factors of $-5$ are $-5$ and 1 or 5 and $-1$. Try various combinations. You should find that the quadratic factors as $(3x + 1)(2x - 5)$. Thus,

$$12x^2 - 26x - 10 = 2(3x + 1)(2x - 5).$$

**(b)** $4z^2 + 12z + 9 - w^2$

**Solution**  There is no common factor here, but notice that the first three terms can be factored as a perfect square:

$$4z^2 + 12z + 9 - w^2 = (2z + 3)^2 - w^2.$$

Written in this form, the expression is the difference of squares, which can be factored as follows:

$$(2z + 3)^2 - w^2 = [(2z + 3) + w][(2z + 3) - w]$$
$$= (2z + 3 + w)(2z + 3 - w).$$

**(c)** $16a^2 - 100 - 48ac + 36c^2$

**Solution**  Factor out the greatest common factor of 4 first:

$$16a^2 - 100 - 48ac + 36c^2 = 4[4a^2 - 25 - 12ac + 9c^2]$$
$$= 4[(4a^2 - 12ac + 9c^2) - 25] \qquad \text{Rearrange terms and group.}$$
$$= 4[(2a - 3c)^2 - 25] \qquad \text{Factor.}$$
$$= 4(2a - 3c + 5)(2a - 3c - 5). \qquad \text{Factor the difference of squares.} \; ✓_7$$

✓ **Checkpoint 7**

Factor the following.

**(a)** $6x^2 - 27x - 15$

**(b)** $9r^2 + 12r + 4 - t^2$

**(c)** $18 - 8xy - 2y^2 - 8x^2$

🛑 **CAUTION**  Remember always to look first for a greatest common factor.

## Higher Degree Polynomials

Polynomials of degree greater than 2 are often difficult to factor. However, factoring is relatively easy in two cases: *the difference and the sum of cubes.* By multiplying out the right side, you can readily verify each of the following factorizations.

### Difference and Sum of Cubes

$$x^3 - y^3 = (x - y)(x^2 + xy + y^2)$$
$$x^3 + y^3 = (x + y)(x^2 - xy + y^2)$$

**EXAMPLE 10**    Factor each of the following polynomials.

**(a)** $k^3 - 8$

**Solution**  Since $8 = 2^3$, use the pattern for the difference of two cubes to obtain

$$k^3 - 8 = k^3 - 2^3 = (k - 2)(k^2 + 2k + 4).$$

**(b)** $m^3 + 125$

**Solution** $m^3 + 125 = m^3 + 5^3 = (m + 5)(m^2 - 5m + 25)$

**(c)** $8k^3 - 27z^3$

**Solution** $8k^3 - 27z^3 = (2k)^3 - (3z)^3 = (2k - 3z)(4k^2 + 6kz + 9z^2)$ ✓ 8

✓ **Checkpoint 8**

Factor the following.

**(a)** $a^3 + 1000$

**(b)** $z^3 - 64$

**(c)** $1000m^3 - 27z^3$

Substitution and appropriate factoring patterns can sometimes be used to factor higher degree expressions.

**EXAMPLE 11**  Factor the following polynomials.

**(a)** $x^8 + 4x^4 + 3$

**Solution** The idea is to make a substitution that reduces the polynomial to a quadratic or cubic that we can deal with. Note that $x^8 = (x^4)^2$. Let $u = x^4$. Then

$$
\begin{aligned}
x^8 + 4x^4 + 3 &= (x^4)^2 + 4x^4 + 3 && \text{Power of a power} \\
&= u^2 + 4u + 3 && \text{Substitute } x^4 = u. \\
&= (u + 3)(u + 1) && \text{Factor.} \\
&= (x^4 + 3)(x^4 + 1). && \text{Substitute } u = x^4.
\end{aligned}
$$

**(b)** $x^4 - y^4$

**Solution** Note that $x^4 = (x^2)^2$, and similarly for the $y$ term. Let $u = x^2$ and $v = y^2$. Then

$$
\begin{aligned}
x^4 - y^4 &= (x^2)^2 - (y^2)^2 && \text{Power of a power} \\
&= u^2 - v^2 && \text{Substitute } x^2 = u \text{ and } y^2 = v. \\
&= (u + v)(u - v) && \text{Difference of squares} \\
&= (x^2 + y^2)(x^2 - y^2) && \text{Substitute } u = x^2 \text{ and } v = y^2. \\
&= (x^2 + y^2)(x + y)(x - y). && \text{Difference of squares} \;✓\, 9
\end{aligned}
$$

✓ **Checkpoint 9**

Factor each of the following.

**(a)** $2x^4 + 5x^2 + 2$

**(b)** $3x^4 - x^2 - 2$

Once you understand Example 11, you can often factor without making explicit substitutions.

**EXAMPLE 12**  Factor $256k^4 - 625m^4$.

**Solution** Use the difference of squares twice, as follows:

$$
\begin{aligned}
256k^4 - 625m^4 &= (16k^2)^2 - (25m^2)^2 \\
&= (16k^2 + 25m^2)(16k^2 - 25m^2) \\
&= (16k^2 + 25m^2)(4k + 5m)(4k - 5m). \;✓\, 10
\end{aligned}
$$

✓ **Checkpoint 10**

Factor $81x^4 - 16y^4$.

# 1.3 Exercises

*Factor out the greatest common factor in each of the given polynomials. (See Example 1.)*

**1.** $12x^2 - 24x$

**2.** $5y - 65xy$

**3.** $r^3 - 5r^2 + r$

**4.** $t^3 + 3t^2 + 8t$

**5.** $6z^3 - 12z^2 + 18z$

**6.** $5x^3 + 55x^2 + 10x$

**7.** $3(2y - 1)^2 + 7(2y - 1)^3$

**8.** $(3x + 7)^5 - 4(3x + 7)^3$

**9.** $3(x + 5)^4 + (x + 5)^6$

**10.** $3(x + 6)^2 + 6(x + 6)^4$

*Factor the polynomial. (See Examples 2 and 3.)*

**11.** $x^2 + 5x + 4$

**12.** $u^2 + 7u + 6$

**13.** $x^2 + 7x + 12$

**14.** $y^2 + 8y + 12$

**15.** $x^2 + x - 6$

**16.** $x^2 + 4x - 5$

**17.** $x^2 + 2x - 3$

**18.** $y^2 + y - 12$

**19.** $x^2 - 3x - 4$

**20.** $u^2 - 2u - 8$

**21.** $z^2 - 9z + 14$

**22.** $w^2 - 6w - 16$

**23.** $z^2 + 10z + 24$

**24.** $r^2 + 16r + 60$

*Factor the polynomial. (See Examples 4–6.)*

**25.** $2x^2 - 9x + 4$

**26.** $3w^2 - 8w + 4$

**27.** $15p^2 - 23p + 4$

**28.** $8x^2 - 14x + 3$

**29.** $4z^2 - 16z + 15$

**30.** $12y^2 - 29y + 15$

**31.** $6x^2 - 5x - 4$

**32.** $12z^2 + z - 1$

**33.** $10y^2 + 21y - 10$

**34.** $15u^2 + 4u - 4$

**35.** $6x^2 + 5x - 4$

**36.** $12y^2 + 7y - 10$

*Factor each polynomial completely. Factor out the greatest common factor as necessary. (See Examples 2–9.)*

**37.** $3a^2 + 2a - 5$

**38.** $6a^2 - 48a - 120$

**39.** $x^2 - 81$

**40.** $x^2 + 17xy + 72y^2$

**41.** $9p^2 - 12p + 4$

**42.** $3r^2 - r - 2$

**43.** $r^2 + 3rt - 10t^2$

**44.** $2a^2 + ab - 6b^2$

**45.** $m^2 - 8mn + 16n^2$

**46.** $8k^2 - 16k - 10$

**47.** $4u^2 + 12u + 9$

**48.** $9p^2 - 16$

**49.** $25p^2 - 10p + 4$

**50.** $10x^2 - 17x + 3$

**51.** $4r^2 - 9v^2$

**52.** $x^2 + 3xy - 28y^2$

**53.** $x^2 + 4xy + 4y^2$

**54.** $16u^2 + 12u - 18$

**55.** $3a^2 - 13a - 30$

**56.** $3k^2 + 2k - 8$

**57.** $21m^2 + 13mn + 2n^2$

**58.** $81y^2 - 100$

**59.** $y^2 - 4yz - 21z^2$

**60.** $49a^2 + 9$

**61.** $121x^2 - 64$

**62.** $4z^2 + 56zy + 196y^2$

*Factor each of these polynomials. (See Example 10.)*

**63.** $a^3 - 64$

**64.** $b^3 + 216$

**65.** $8r^3 - 27s^3$

**66.** $1000p^3 + 27q^3$

**67.** $64m^3 + 125$

**68.** $216y^3 - 343$

**69.** $1000y^3 - z^3$

**70.** $125p^3 + 8q^3$

*Factor each of these polynomials. (See Examples 11 and 12.)*

**71.** $x^4 + 5x^2 + 6$

**72.** $y^4 + 7y^2 + 10$

**73.** $b^4 - b^2$

**74.** $z^4 - 3z^2 - 4$

**75.** $x^4 - x^2 - 12$

**76.** $4x^4 + 27x^2 - 81$

**77.** $16a^4 - 81b^4$

**78.** $x^6 - y^6$

**79.** $x^8 + 8x^2$

**80.** $x^9 - 64x^3$

**81.** When asked to factor $6x^4 - 3x^2 - 3$ completely, a student gave the following result:

$$6x^4 - 3x^2 - 3 = (2x^2 + 1)(3x^2 - 3).$$

Is this answer correct? Explain why.

**82.** When can the sum of two squares be factored? Give examples.

**83.** Explain why $(x + 2)^3$ is not the correct factorization of $x^3 + 8$, and give the correct factorization.

**84.** Describe how factoring and multiplication are related. Give examples.

✓**Checkpoint Answers**

1. **(a)** $3(4r + 3k)$  **(b)** $25(3m^2 + 4n^2)$
   **(c)** $3m^2(2m^2 - 3m + 4)$  **(d)** $(2k + 1)^3(7 + 8k)$

2. **(a)** $(r + 2)(r + 5)$  **(b)** $(x + 3)(x + 1)$
   **(c)** $(y + 2)(y + 4)$

3. **(a)** $(x - 3)(x - 1)$  **(b)** $(2y - 1)(y - 2)$
   **(c)** $(3z - 2)(2z - 3)$

4. **(a)** $(r - 7)(r + 2)$  **(b)** $(3m - 1)(m + 2)$
   **(c)** $(2p + 5q)(3p - q)$

5. **(a)** $(3p + 7)(3p - 7)$  **(b)** Cannot be factored
   **(c)** $(x + 11)(x - 5)$

6. **(a)** $(2m + 1)^2$  **(b)** $(5z - 8t)^2$
   **(c)** Does not factor

7. **(a)** $3(2x + 1)(x - 5)$
   **(b)** $(3r + 2 + t)(3r + 2 - t)$
   **(c)** $2(3 - 2x - y)(3 + 2x + y)$

8. **(a)** $(a + 10)(a^2 - 10a + 100)$
   **(b)** $(z - 4)(z^2 + 4z + 16)$
   **(c)** $(10m - 3z)(100m^2 + 30mz + 9z^2)$

9. **(a)** $(2x^2 + 1)(x^2 + 2)$
   **(b)** $(3x^2 + 2)(x + 1)(x - 1)$

10. $(9x^2 + 4y^2)(3x + 2y)(3x - 2y)$

# 1.4 Rational Expressions

A **rational expression** is an expression that can be written as the quotient of two polynomials, such as

$$\frac{8}{x - 1}, \qquad \frac{3x^2 + 4x}{5x - 6}, \qquad \text{and} \qquad \frac{2y + 1}{y^4 + 8}.$$

It is sometimes important to know the values of the variable that make the denominator 0 (in which case the quotient is not defined). For example, 1 cannot be used as a replacement for $x$ in the first expression above and $6/5$ cannot be used in the second one, since these values make the respective denominators equal 0. *Throughout this section, we assume that all denominators are nonzero,* which means that some replacement values for the variables may have to be excluded. ✓₁

✓**Checkpoint 1**

What value of the variable makes each denominator equal 0?

(a) $\dfrac{5}{x-3}$

(b) $\dfrac{2x-3}{4x-1}$

(c) $\dfrac{x+2}{x}$

(d) Why do we need to determine these values?

## Simplifying Rational Expressions

A key tool for simplification is the following fact.

### Cancellation Property

For all expressions $P$, $Q$, and $S$, with $Q \neq 0$ and $S \neq 0$,

$$\frac{PS}{QS} = \frac{P}{Q}.$$

**EXAMPLE 1** Write each of the following rational expressions in lowest terms (so that the numerator and denominator have no common factor with integer coefficients except 1 or $-1$).

(a) $\dfrac{12m}{-18}$

**Solution** Both $12m$ and $-18$ are divisible by 6. By the cancellation property,

$$\frac{12m}{-18} = \frac{2m \cdot 6}{-3 \cdot 6}$$
$$= \frac{2m}{-3}$$
$$= -\frac{2m}{3}.$$

(b) $\dfrac{8x+16}{4}$

**Solution** Factor the numerator and cancel:

$$\frac{8x+16}{4} = \frac{8(x+2)}{4} = \frac{4 \cdot 2(x+2)}{4} = \frac{2(x+2)}{1} = 2(x+2).$$

The answer could also be written as $2x+4$ if desired.

(c) $\dfrac{k^2+7k+12}{k^2+2k-3}$

**Solution** Factor the numerator and denominator and cancel:

$$\frac{k^2+7k+12}{k^2+2k-3} = \frac{(k+4)(k+3)}{(k-1)(k+3)} = \frac{k+4}{k-1}. ✓₂$$

✓**Checkpoint 2**

Write each of the following in lowest terms.

(a) $\dfrac{12k+36}{18}$

(b) $\dfrac{15m+30m^2}{5m}$

(c) $\dfrac{2p^2+3p+1}{p^2+3p+2}$

## Multiplication and Division

The rules for multiplying and dividing rational expressions are the same fraction rules you learned in arithmetic.

## Multiplication and Division Rules

For all expressions $P$, $Q$, $R$, and $S$, with $Q \neq 0$ and $S \neq 0$,

$$\frac{P}{Q} \cdot \frac{R}{S} = \frac{PR}{QS}$$

and

$$\frac{P}{Q} \div \frac{R}{S} = \frac{P}{Q} \cdot \frac{S}{R} \qquad (R \neq 0).$$

### EXAMPLE 2

**(a)** Multiply $\dfrac{2}{3} \cdot \dfrac{y}{5}$.

**Solution**  Use the multiplication rule. Multiply the numerators and then the denominators:

$$\frac{2}{3} \cdot \frac{y}{5} = \frac{2 \cdot y}{3 \cdot 5} = \frac{2y}{15}.$$

The result, $2y/15$, is in lowest terms.

**(b)** Multiply $\dfrac{3y + 9}{6} \cdot \dfrac{18}{5y + 15}$.

**Solution**  Factor where possible:

$$\frac{3y + 9}{6} \cdot \frac{18}{5y + 15} = \frac{3(y + 3)}{6} \cdot \frac{18}{5(y + 3)}$$

$$= \frac{3 \cdot 18(y + 3)}{6 \cdot 5(y + 3)} \qquad \text{Multiply numerators and denominators.}$$

$$= \frac{3 \cdot 6 \cdot 3(y + 3)}{6 \cdot 5(y + 3)} \qquad 18 = 6 \cdot 3$$

$$= \frac{3 \cdot 3}{5} \qquad \text{Write in lowest terms.}$$

$$= \frac{9}{5}.$$

**(c)** Multiply $\dfrac{m^2 + 5m + 6}{m + 3} \cdot \dfrac{m^2 + m - 6}{m^2 + 3m + 2}$.

**Solution**  Factor numerators and denominators:

$$\frac{(m + 2)(m + 3)}{m + 3} \cdot \frac{(m - 2)(m + 3)}{(m + 2)(m + 1)} \qquad \text{Factor.}$$

$$= \frac{(m + 2)(m + 3)(m - 2)(m + 3)}{(m + 3)(m + 2)(m + 1)} \qquad \text{Multiply.}$$

$$= \frac{(m - 2)(m + 3)}{m + 1} \qquad \text{Lowest terms}$$

$$= \frac{m^2 + m - 6}{m + 1}. \quad \text{✓}_3$$

### ✓ Checkpoint 3

Multiply.

**(a)** $\dfrac{3r^2}{5} \cdot \dfrac{20}{9r}$

**(b)** $\dfrac{y - 4}{y^2 - 2y - 8} \cdot \dfrac{y^2 - 4}{3y}$

EXAMPLE 3

**(a)** Divide $\dfrac{8x}{5} \div \dfrac{11x^2}{20}$.

**Solution**  Invert the second expression and multiply (division rule):

$$\frac{8x}{5} \div \frac{11x^2}{20} = \frac{8x}{5} \cdot \frac{20}{11x^2} \qquad \text{Invert and multiply.}$$

$$= \frac{8x \cdot 20}{5 \cdot 11x^2} \qquad \text{Multiply.}$$

$$= \frac{32}{11x}. \qquad \text{Lowest terms}$$

**(b)** Divide $\dfrac{9p - 36}{12} \div \dfrac{5(p-4)}{18}$.

**Solution**  We have

$$\frac{9p - 36}{12} \cdot \frac{18}{5(p-4)} \qquad \text{Invert and multiply.}$$

$$= \frac{9(p-4)}{12} \cdot \frac{18}{5(p-4)} \qquad \text{Factor.}$$

$$= \frac{27}{10}. \qquad \text{Cancel, multiply, and write in lowest terms.} \quad ✓_4$$

✓**Checkpoint 4**

Divide.

**(a)** $\dfrac{5m}{16} \div \dfrac{m^2}{10}$

**(b)** $\dfrac{2y-8}{6} \div \dfrac{5y-20}{3}$

**(c)** $\dfrac{m^2 - 2m - 3}{m(m+1)} \div \dfrac{m+4}{5m}$

## Addition and Subtraction

As you know, when two numerical fractions have the same denominator, they can be added or subtracted. The same rules apply to rational expressions.

---

### Addition and Subtraction Rules

For all expressions $P$, $Q$, $R$, with $Q \neq 0$,

$$\frac{P}{Q} + \frac{R}{Q} = \frac{P+R}{Q}$$

and

$$\frac{P}{Q} - \frac{R}{Q} = \frac{P-R}{Q}.$$

---

EXAMPLE 4   Add or subtract as indicated.

**(a)** $\dfrac{4}{5k} + \dfrac{11}{5k}$

**Solution**  Since the denominators are the same, we add the numerators:

$$\frac{4}{5k} + \frac{11}{5k} = \frac{4+11}{5k} = \frac{15}{5k} \qquad \text{Addition rule}$$

$$= \frac{3}{k}. \qquad \text{Lowest terms}$$

**(b)** $\dfrac{2x^2 + 3x + 1}{x^5 + 1} - \dfrac{x^2 - 7x}{x^5 + 1}$

**Solution** The denominators are the same, so we subtract numerators, paying careful attention to parentheses:

$$\dfrac{2x^2 + 3x + 1}{x^5 + 1} - \dfrac{x^2 - 7x}{x^5 + 1} = \dfrac{(2x^2 + 3x + 1) - (x^2 - 7x)}{x^5 + 1} \qquad \text{Subtraction rule}$$

$$= \dfrac{2x^2 + 3x + 1 - x^2 - (-7x)}{x^5 + 1} \qquad \text{Subtract numerators.}$$

$$= \dfrac{2x^2 + 3x + 1 - x^2 + 7x}{x^5 + 1}$$

$$= \dfrac{x^2 + 10x + 1}{x^5 + 1}. \qquad \text{Simplify the numerator.}$$

When fractions do not have the same denominator, you must first find a common denominator before you can add or subtract. A common denominator is a denominator that has each fraction's denominator as a factor.

---

**EXAMPLE 5**  Add or subtract as indicated.

**(a)** $\dfrac{7}{p^2} + \dfrac{9}{2p} + \dfrac{1}{3p^2}$

**Solution** These three denominators are different, so we must find a common denominator that has each of $p^2$, $2p$, and $3p^2$ as factors. Observe that $6p^2$ satisfies these requirements. Use the cancellation property to rewrite each fraction as one that has $6p^2$ as its denominator and then add them:

$$\dfrac{7}{p^2} + \dfrac{9}{2p} + \dfrac{1}{3p^2} = \dfrac{6 \cdot 7}{6 \cdot p^2} + \dfrac{3p \cdot 9}{3p \cdot 2p} + \dfrac{2 \cdot 1}{2 \cdot 3p^2} \qquad \text{Cancellation property}$$

$$= \dfrac{42}{6p^2} + \dfrac{27p}{6p^2} + \dfrac{2}{6p^2}$$

$$= \dfrac{42 + 27p + 2}{6p^2} \qquad \text{Addition rule}$$

$$= \dfrac{27p + 44}{6p^2}. \qquad \text{Simplify.}$$

**(b)** $\dfrac{k^2}{k^2 - 1} - \dfrac{2k^2 - k - 3}{k^2 + 3k + 2}$

**Solution** Factor the denominators to find a common denominator:

$$\dfrac{k^2}{k^2 - 1} - \dfrac{2k^2 - k - 3}{k^2 + 3k + 2} = \dfrac{k^2}{(k + 1)(k - 1)} - \dfrac{2k^2 - k - 3}{(k + 1)(k + 2)}.$$

A common denominator here is $(k + 1)(k - 1)(k + 2)$, because each of the preceding denominators is a factor of this common denominator. Write each fraction with the common denominator:

$$\frac{k^2}{(k + 1)(k - 1)} - \frac{2k^2 - k - 3}{(k + 1)(k + 2)}$$

$$= \frac{k^2(k + 2)}{(k + 1)(k - 1)(k + 2)} - \frac{(2k^2 - k - 3)(k - 1)}{(k + 1)(k - 1)(k + 2)}$$

$$= \frac{k^3 + 2k^2 - (2k^2 - k - 3)(k - 1)}{(k + 1)(k - 1)(k + 2)} \qquad \text{Subtract fractions.}$$

$$= \frac{k^3 + 2k^2 - (2k^3 - 3k^2 - 2k + 3)}{(k + 1)(k - 1)(k + 2)} \qquad \text{Multiply } (2k^2\text{-}k\text{-}3)(k\text{-}1).$$

$$= \frac{k^3 + 2k^2 - 2k^3 + 3k^2 + 2k - 3}{(k + 1)(k - 1)(k + 2)} \qquad \text{Polynomial subtraction}$$

$$= \frac{-k^3 + 5k^2 + 2k - 3}{(k + 1)(k - 1)(k + 2)}. \qquad \text{Combine terms.} \quad \checkmark_5$$

## ✓ Checkpoint 5

Add or subtract.

(a) $\dfrac{3}{4r} + \dfrac{8}{3r}$

(b) $\dfrac{1}{m - 2} - \dfrac{3}{2(m - 2)}$

(c) $\dfrac{p + 1}{p^2 - p} - \dfrac{p^2 - 1}{p^2 + p - 2}$

## Complex Fractions

Any quotient of rational expressions is called a **complex fraction.** Complex fractions are simplified as demonstrated in Example 6.

**EXAMPLE 6**   Simplify the complex fraction

$$\frac{6 - \dfrac{5}{k}}{1 + \dfrac{5}{k}}.$$

**Solution**  Multiply both numerator and denominator by the common denominator $k$:

$$\frac{6 - \dfrac{5}{k}}{1 + \dfrac{5}{k}} = \frac{k\left(6 - \dfrac{5}{k}\right)}{k\left(1 + \dfrac{5}{k}\right)} \qquad \text{Multiply by } \dfrac{k}{k}.$$

$$= \frac{6k - k\left(\dfrac{5}{k}\right)}{k + k\left(\dfrac{5}{k}\right)} \qquad \text{Distributive property}$$

$$= \frac{6k - 5}{k + 5}. \qquad \text{Simplify.}$$

# 1.4 Exercises

*Write each of the given expressions in lowest terms. Factor as necessary. (See Example 1.)*

**1.** $\dfrac{8x^2}{56x}$

**2.** $\dfrac{27m}{81m^3}$

**3.** $\dfrac{25p^2}{35p^3}$

**4.** $\dfrac{18y^4}{24y^2}$

**5.** $\dfrac{5m + 15}{4m + 12}$

**6.** $\dfrac{10z + 5}{20z + 10}$

**7.** $\dfrac{4(w - 3)}{(w - 3)(w + 6)}$

**8.** $\dfrac{-6(x + 2)}{(x + 4)(x + 2)}$

**9.** $\dfrac{3y^2 - 12y}{9y^3}$

**10.** $\dfrac{15k^2 + 45k}{9k^2}$

**11.** $\dfrac{m^2 - 4m + 4}{m^2 + m - 6}$

**12.** $\dfrac{r^2 - r - 6}{r^2 + r - 12}$

**13.** $\dfrac{x^2 + 2x - 3}{x^2 - 1}$

**14.** $\dfrac{z^2 + 4z + 4}{z^2 - 4}$

*Multiply or divide as indicated in each of the exercises. Write all answers in lowest terms. (See Examples 2 and 3.)*

**15.** $\dfrac{3a^2}{64} \cdot \dfrac{8}{2a^3}$

**16.** $\dfrac{2u^2}{8u^4} \cdot \dfrac{10u^3}{9u}$

**17.** $\dfrac{7x}{11} \div \dfrac{14x^3}{66y}$

**18.** $\dfrac{6x^2y}{2x} \div \dfrac{21xy}{y}$

**19.** $\dfrac{2a + b}{3c} \cdot \dfrac{15}{4(2a + b)}$

**20.** $\dfrac{4(x + 2)}{w} \cdot \dfrac{3w^2}{8(x + 2)}$

**21.** $\dfrac{15p - 3}{6} \div \dfrac{10p - 2}{3}$

**22.** $\dfrac{2k + 8}{6} \div \dfrac{3k + 12}{3}$

**23.** $\dfrac{9y - 18}{6y + 12} \cdot \dfrac{3y + 6}{15y - 30}$

**24.** $\dfrac{12r + 24}{36r - 36} \div \dfrac{6r + 12}{8r - 8}$

**25.** $\dfrac{4a + 12}{2a - 10} \div \dfrac{a^2 - 9}{a^2 - a - 20}$

**26.** $\dfrac{6r - 18}{9r^2 + 6r - 24} \cdot \dfrac{12r - 16}{4r - 12}$

**27.** $\dfrac{k^2 - k - 6}{k^2 + k - 12} \cdot \dfrac{k^2 + 3k - 4}{k^2 + 2k - 3}$

**28.** $\dfrac{n^2 - n - 6}{n^2 - 2n - 8} \div \dfrac{n^2 - 9}{n^2 + 7n + 12}$

**29.** In your own words, explain how to find the least common denominator of two fractions.

**30.** Describe the steps required to add three rational expressions. You may use an example to illustrate.

*Add or subtract as indicated in each of the following. Write all answers in lowest terms. (See Example 4.)*

**31.** $\dfrac{2}{7z} - \dfrac{1}{5z}$

**32.** $\dfrac{4}{3z} - \dfrac{5}{4z}$

**33.** $\dfrac{r + 2}{3} - \dfrac{r - 2}{3}$

**34.** $\dfrac{3y - 1}{8} - \dfrac{3y + 1}{8}$

**35.** $\dfrac{4}{x} + \dfrac{1}{5}$

**36.** $\dfrac{6}{r} - \dfrac{3}{4}$

**37.** $\dfrac{1}{m - 1} + \dfrac{2}{m}$

**38.** $\dfrac{8}{y + 2} - \dfrac{3}{y}$

**39.** $\dfrac{7}{b + 2} + \dfrac{2}{5(b + 2)}$

**40.** $\dfrac{4}{3(k + 1)} + \dfrac{3}{k + 1}$

**41.** $\dfrac{2}{5(k - 2)} + \dfrac{5}{4(k - 2)}$

**42.** $\dfrac{11}{3(p + 4)} - \dfrac{5}{6(p + 4)}$

**43.** $\dfrac{2}{x^2 - 4x + 3} + \dfrac{5}{x^2 - x - 6}$

**44.** $\dfrac{3}{m^2 - 3m - 10} + \dfrac{7}{m^2 - m - 20}$

**45.** $\dfrac{2y}{y^2 + 7y + 12} - \dfrac{y}{y^2 + 5y + 6}$

**46.** $\dfrac{-r}{r^2 - 10r + 16} - \dfrac{3r}{r^2 + 2r - 8}$

*In each of the exercises in the next set, simplify the complex fraction. (See Example 6.)*

**47.** $\dfrac{1 + \dfrac{1}{x}}{1 - \dfrac{1}{x}}$

**48.** $\dfrac{2 - \dfrac{2}{y}}{2 + \dfrac{2}{y}}$

**49.** $\dfrac{\dfrac{1}{x + h} - \dfrac{1}{x}}{h}$

**50.** $\dfrac{\dfrac{1}{(x + h)^2} - \dfrac{1}{x^2}}{h}$

*Work these problems.*

**Natural Science** *Each figure in the following exercises is a dartboard. The probability that a dart which hits the board lands in the shaded area is the fraction*

$$\frac{\text{area of the shaded region}}{\text{area of the dartboard}}.$$

*Note: The area of a circle is $\pi r^2$, the area of a square is bh, and the area of a triangle is $\frac{1}{2}bh$.*

**(a)** *Express the probability as a rational expression in x.*

**(b)** *Then reduce the expression to lowest terms.*

**51.**

**52.**

**53.**

**54.**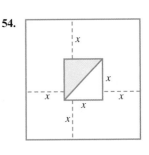

**Costs** *In Example 11 of Section 1.2, we saw that the cost C of producing x thousand calculators is given by*

$$C = -7.2x^2 + 6995x + 230{,}000 \quad (x \le 150).$$

**55.** The average cost per calculator is the total cost $C$ divided by the number of calculators produced. Write a rational expression that gives the average cost per calculator when $x$ thousand are produced.

**56.** Find the average cost per calculator for each of these production levels: 20,000, 50,000, and 125,000.

**Superbowl Advertising Costs** *The cost (in millions of dollars) for a 30-second ad during the TV broadcast of the Superbowl can be approximated by*

$$\frac{.505x^2 - 4.587x + 27.6}{x + 1}$$

*where $x = 6$ corresponds to the year 2006. (Data from: forbes.com and statistica.com.)*

**57.** How much did an ad cost in 2013?

**58.** How much did an ad cost in 2016?

**59.** If this trend continues, will the cost of an ad reach $7 million by 2020?

**60.** If this trend continues, will the cost of an ad reach $8 million by 2022?

**Health Care Costs** *The average company cost per hour of an employee's health insurance in year $x$ is approximated by*

$$\frac{.049x^2 + 2.40x + .83}{x + 2},$$

*where $x = 4$ corresponds to the year 2004. (Data from: U.S. Bureau of Labor Statistics.)*

**61.** What is the hourly health insurance cost in 2011?

**62.** What is the hourly health insurance cost in 2017?

**63.** Assuming that this model remains accurate and that an employee works 2100 hours per year, what is the annual company cost of her health care insurance in 2020?

**64.** Will annual costs reach $10,000 by 2023?

✓**Checkpoint Answers**

**1.** **(a)** 3      **(b)** 1/4      **(c)** 0
**(d)** Because division by 0 is undefined

**2.** **(a)** $\dfrac{2(k + 3)}{3}$   or   $\dfrac{2k + 6}{3}$

    **(b)** $3(1 + 2m)$   or   $3 + 6m$      **(c)** $\dfrac{2p + 1}{p + 2}$

**3.** **(a)** $\dfrac{4r}{3}$      **(b)** $\dfrac{y - 2}{3y}$

**4.** **(a)** $\dfrac{25}{8m}$      **(b)** $\dfrac{1}{5}$      **(c)** $\dfrac{5(m - 3)}{m + 4}$

**5.** **(a)** $\dfrac{41}{12r}$      **(b)** $\dfrac{-1}{2(m - 2)}$      **(c)** $\dfrac{-p^3 + p^2 + 4p + 2}{p(p - 1)(p + 2)}$

# 1.5 Exponents and Radicals

Exponents were introduced in Section 1.2. In this section, the definition of exponents will be extended to include negative exponents and rational-number exponents such as 1/2 and 7/3.

## Integer Exponents

Positive-integer and zero exponents were defined in Section 1.2, where we noted that

$$a^m \cdot a^n = a^{m+n}$$

for nonnegative integers $m$ and $n$. Now we develop an analogous property for quotients. By definition,

$$\frac{6^5}{6^2} = \frac{6 \cdot 6 \cdot 6 \cdot 6 \cdot 6}{6 \cdot 6} = 6 \cdot 6 \cdot 6 = 6^3.$$

Because there are 5 factors of 6 in the numerator and 2 factors of 6 in the denominator, the quotient has $5 - 2 = 3$ factors of 6. In general, we can make the following statement, which applies to any real number $a$ and nonnegative integers $m$ and $n$ with $m > n$.

> ## Division with Exponents
>
> To divide $a^m$ by $a^n$, *subtract* the exponents (assuming $a \neq 0$):
>
> $$\frac{a^m}{a^n} = a^{m-n}.$$

**EXAMPLE 1**    Compute each of the following.

**(a)** $\dfrac{5^7}{5^4} = 5^{7-4} = 5^3.$

**(b)** $\dfrac{(-8)^{10}}{(-8)^5} = (-8)^{10-5} = (-8)^5.$

**(c)** $\dfrac{(3c)^9}{(3c)^3} = (3c)^{9-3} = (3c)^6.$ ✔<sub></sub>1

✓**Checkpoint 1**

Evaluate each of the following.

**(a)** $\dfrac{2^{14}}{2^5}$

**(b)** $\dfrac{(-5)^9}{(-5)^5}$

**(c)** $\dfrac{(xy)^{17}}{(xy)^{12}}$

When an exponent applies to the product of two numbers, such as $(7 \cdot 19)^3$, use the definitions carefully. For instance,

$$(7 \cdot 19)^3 = (7 \cdot 19)(7 \cdot 19)(7 \cdot 19) = 7 \cdot 7 \cdot 7 \cdot 19 \cdot 19 \cdot 19 = 7^3 \cdot 19^3.$$

In other words, $(7 \cdot 19)^3 = 7^3 \cdot 19^3$. This is an example of the following fact, which applies to any real numbers $a$ and $b$ and any nonnegative-integer exponent $n$.

## Product to a Power

To find $(ab)^n$, apply the exponent to *every* factor inside the parentheses:

$$(ab)^n = a^n b^n.$$

**⚠ CAUTION**  A common mistake is to write an expression such as $(2x)^5$ as $2x^5$, rather than the correct answer $(2x)^5 = 2^5 x^5 = 32x^5$.

Analogous conclusions are valid for quotients (where $a$ and $b$ are any real numbers with $b \neq 0$ and $n$ is a nonnegative-integer exponent).

## Quotient to a Power

To find $\left(\dfrac{a}{b}\right)^n$, (assuming $b \neq 0$), apply the exponent to both numerator and denominator:

$$\left(\dfrac{a}{b}\right)^n = \dfrac{a^n}{b^n}.$$

**EXAMPLE 2**    Compute each of the following.

**(a)** $(5y)^3 = 5^3 y^3 = 125y^3$        Product to a power

**(b)** $(c^2 d^3)^4 = (c^2)^4 (d^3)^4$        Product to a power

  $\quad\quad\quad = c^8 d^{12}$        Power of a power

**(c)** $\left(\dfrac{x}{2}\right)^6 = \dfrac{x^6}{2^6} = \dfrac{x^6}{64}$        Quotient to a power

**(d)** $\left(\dfrac{a^4}{b^3}\right)^3 = \dfrac{(a^4)^3}{(b^3)^3}$        Quotient to a power

  $\quad\quad\quad = \dfrac{a^{12}}{b^9}$        Power of a power

**(e)** $\left(\dfrac{(rs)^3}{r^4}\right)^2$

**Solution**   Use several of the preceding properties in succession:

$$\left(\frac{(rs)^3}{r^4}\right)^2 = \left(\frac{r^3 s^3}{r^4}\right)^2 \qquad \text{Product to a power in numerator}$$

$$= \left(\frac{s^3}{r}\right)^2 \qquad \text{Cancel.}$$

$$= \frac{(s^3)^2}{r^2} \qquad \text{Quotient to a power}$$

$$= \frac{s^6}{r^2}. \qquad \text{Power of a power in numerator}$$

As is often the case, there is another way to reach the last expression. You should be able to supply the reasons for each of the following steps:

$$\left(\frac{(rs)^3}{r^4}\right)^2 = \frac{[(rs)^3]^2}{(r^4)^2} = \frac{(rs)^6}{r^8} = \frac{r^6 s^6}{r^8} = \frac{s^6}{r^2}. \quad \boxed{\checkmark}_2$$

✓ **Checkpoint 2**

Compute each of the following.

**(a)** $(3x)^4$

**(b)** $(r^2 s^5)^6$

**(c)** $\left(\dfrac{2}{z}\right)^5$

**(d)** $\left(\dfrac{3a^5}{(ab)^3}\right)^2$

## Negative Exponents

The next step is to define negative-integer exponents. If they are to be defined in such a way that the quotient rule for exponents remains valid, then we must have, for example,

$$\frac{3^2}{3^4} = 3^{2-4} = 3^{-2}.$$

However,

$$\frac{3^2}{3^4} = \frac{3 \cdot 3}{3 \cdot 3 \cdot 3 \cdot 3} = \frac{1}{3^2},$$

which suggests that $3^{-2}$ should be defined to be $1/3^2$. Thus, we have the following definition of a negative exponent.

---

### Negative Exponent

If $n$ is a natural number, and if $a \neq 0$, then

$$a^{-n} = \frac{1}{a^n}.$$

---

**EXAMPLE 3**   Evaluate the following.

✓ **Checkpoint 3**

Evaluate the following.

**(a)** $6^{-2}$

**(b)** $-6^{-3}$

**(c)** $-3^{-4}$

**(d)** $\left(\dfrac{5}{8}\right)^{-1}$

**(a)** $3^{-2} = \dfrac{1}{3^2} = \dfrac{1}{9}.$

**(b)** $5^{-4} = \dfrac{1}{5^4} = \dfrac{1}{625}.$

**(c)** $x^{-1} = \dfrac{1}{x^1} = \dfrac{1}{x}.$

**(d)** $-4^{-2} = -\dfrac{1}{4^2} = -\dfrac{1}{16}.$

**(e)** $\left(\dfrac{3}{4}\right)^{-1} = \dfrac{1}{\left(\dfrac{3}{4}\right)^1} = \dfrac{1}{\dfrac{3}{4}} = \dfrac{4}{3}. \quad \boxed{\checkmark}_3$

There is a useful property that makes it easy to raise a fraction to a negative exponent. Consider, for example,

$$\left(\frac{2}{3}\right)^{-4} = \frac{1}{\left(\frac{2}{3}\right)^4} = \frac{1}{\left(\frac{2^4}{3^4}\right)} = 1 \cdot \frac{3^4}{2^4} = \left(\frac{3}{2}\right)^4.$$

This example is easily generalized to the following property (in which $a/b$ is a nonzero fraction and $n$ a positive integer).

## Inversion Property

$$\left(\frac{a}{b}\right)^{-n} = \left(\frac{b}{a}\right)^{n}.$$

**EXAMPLE 4**   Use the inversion property to compute each of the following.

**(a)** $\left(\frac{2}{5}\right)^{-3} = \left(\frac{5}{2}\right)^3 = \frac{5^3}{2^3} = \frac{125}{8}.$

**(b)** $\left(\frac{3}{x}\right)^{-5} = \left(\frac{x}{3}\right)^5 = \frac{x^5}{3^5} = \frac{x^5}{243}.$ ✓4

✓**Checkpoint 4**

Compute each of the following.

**(a)** $\left(\frac{5}{8}\right)^{-1}$

**(b)** $\left(\frac{1}{2}\right)^{-5}$

**(c)** $\left(\frac{a^2}{b}\right)^{-3}$

When keying in negative exponents on a calculator, be sure to use the negation key (labeled $(-)$ or $+/-$), not the subtraction key. Calculators normally display answers as decimals, as shown in Figure 1.6. Some graphing calculators have a FRAC key that converts these decimals to fractions, as shown in Figure 1.7.

**Figure 1.6**          **Figure 1.7**

## Roots and Rational Exponents

There are two numbers whose square is 16: 4 and −4. As we saw in Section 1.1, the nonnegative one, 4, is called the *square root* (or second root) of 16. Similarly, there are two numbers whose fourth power is 16: 2 and −2. We call 2 the **fourth root** of 16. This suggests the following generalization.

> If $a$ is a nonnegative real number and $n$ is an even natural number, the ***n*th root of $a$** is the nonnegative real number whose $n$th power is $a$.

All nonnegative numbers have $n$th roots for every natural number $n$, but *no negative number has a real, even $n$th root.* For example, there is no real number whose square is −16, so −16 has no square root.

We say that the **cube root** (or third root) of 8 is 2 because $2^3 = 8$. Similarly, since $(-2)^3 = -8$, we say that −2 is the cube root of −8. Again, we can make the following generalization.

🖩 **TECHNOLOGY TIP**

The FRAC key is in the MATH menu of TI graphing calculators. Fractions can be displayed on some graphing calculators by changing the number display format (in the MODES menu) to "fraction" or "exact."

> If $a$ is a real number and $n$ is an odd natural number, the ***n*th root of *a*** is the real number whose $n$th power is $a$.

Every real number has an $n$th root for every *odd* natural number $n$.

We can now define rational exponents. If they are to have the same properties as integer exponents, we want $a^{1/2}$ to be a number such that

$$(a^{1/2})^2 = a^{1/2} \cdot a^{1/2} = a^{1/2+1/2} = a^1 = a.$$

Thus, $a^{1/2}$ should be a number whose square is $a$, and it is reasonable to *define* $a^{1/2}$ to be the square root of $a$ (if it exists). Similarly, $a^{1/3}$ is defined to be the cube root of $a$, and we have the following definition.

> If $a$ is a real number and $n$ is a natural number, then
>
> $$a^{1/n} \text{ is defined to be the } n\text{th root of } a \text{ (if it exists).}$$

**EXAMPLE 5**     Examine the reasoning used to evaluate the following roots.

**(a)** $36^{1/2} = 6$ because $6^2 = 36$.

**(b)** $100^{1/2} = 10$ because $10^2 = 100$.

**(c)** $-(225^{1/2}) = -15$ because $15^2 = 225$.

**(d)** $625^{1/4} = 5$ because $5^4 = 625$.

**(e)** $(-1296)^{1/4}$ is not a real number.

**(f)** $-1296^{1/4} = -6$ because $6^4 = 1296$.

**(g)** $(-27)^{1/3} = -3$ because $(-3)^3 = -27$.

**(h)** $-32^{1/5} = -2$ because $2^5 = 32$.  ☑ 5

✔ **Checkpoint 5**

Evaluate the following.

**(a)** $16^{1/2}$

**(b)** $16^{1/4}$

**(c)** $-256^{1/2}$

**(d)** $(-256)^{1/2}$

**(e)** $-8^{1/3}$

**(f)** $243^{1/5}$

A calculator can be used to evaluate expressions with fractional exponents. Whenever it is easy to do so, enter the fractional exponents in their equivalent decimal form. For instance, to find $625^{1/4}$, enter $625^{.25}$ into the calculator. When the decimal equivalent of a fraction is an infinitely repeating decimal, however, it is best to enter the fractional exponent directly. If you use a shortened decimal approximation (such as .333 for 1/3), you will not get the correct answers. Compare the incorrect answers in Figure 1.9 with the correct ones in Figure 1.8.

Figure 1.8          Figure 1.9

For other rational exponents, the symbol $a^{m/n}$ should be defined so that the properties for exponents still hold. For example, by the product property, we want

$$(a^{1/3})^2 = a^{1/3} \cdot a^{1/3} = a^{1/3+1/3} = a^{2/3}.$$

This result suggests the following definition.

> For all integers $m$ and all positive integers $n$, and for all real numbers $a$ for which $a^{1/n}$ is a real number,
>
> $$a^{m/n} = (a^{1/n})^m.$$

**EXAMPLE 6**     Verify each of the following calculations.

**(a)** $27^{2/3} = (27^{1/3})^2 = 3^2 = 9.$

**(b)** $32^{2/5} = (32^{1/5})^2 = 2^2 = 4.$

**(c)** $64^{4/3} = (64^{1/3})^4 = 4^4 = 256.$

**(d)** $25^{3/2} = (25^{1/2})^3 = 5^3 = 125.$  ✓ 6

✓ **Checkpoint 6**

Evaluate the following.

**(a)** $16^{3/4}$

**(b)** $25^{5/2}$

**(c)** $32^{7/5}$

**(d)** $100^{3/2}$

🚫 **CAUTION** When the base is negative, as in $(-8)^{2/3}$, some calculators produce an error message. On such calculators, you should first compute $(-8)^{1/3}$ and then square the result; that is, compute $[(-8)^{1/3}]^2$.

Since every terminating decimal is a rational number, decimal exponents now have a meaning. For instance, $5.24 = \frac{524}{100}$, so $3^{5.24} = 3^{524/100}$, which is easily approximated by a calculator (Figure 1.10).

Rational exponents were defined so that one of the familiar properties of exponents remains valid. In fact, it can be proved that *all* of the rules developed earlier for integer exponents are valid for rational exponents. The following box summarizes these rules, which are illustrated in Examples 7–9.

```
3^5.24
            316.3117863
 524
3 100
            316.3117863
```

**Figure 1.10**

## Properties of Exponents

For any rational numbers $m$ and $n$, and for any real numbers $a$ and $b$ for which the following exist,

**(a)** $a^m \cdot a^n = a^{m+n}$      Product property

**(b)** $\dfrac{a^m}{a^n} = a^{m-n}$      Quotient property

**(c)** $(a^m)^n = a^{mn}$      Power of a power

**(d)** $(ab)^m = a^m \cdot b^m$      Product to a power

**(e)** $\left(\dfrac{a}{b}\right)^m = \dfrac{a^m}{b^m}$      Quotient to a power

**(f)** $a^0 = 1$      Zero exponent

**(g)** $a^{-n} = \dfrac{1}{a^n}$      Negative exponent

**(h)** $\left(\dfrac{a}{b}\right)^{-n} = \left(\dfrac{b}{a}\right)^n.$      Inversion property

The power-of-a-power property provides another way to compute $a^{m/n}$ (when it exists):

$$a^{m/n} = a^{m(1/n)} = (a^m)^{1/n}. \tag{1}$$

For example, we can now find $4^{3/2}$ in two ways:

$$4^{3/2} = (4^{1/2})^3 = 2^3 = 8 \quad \text{or} \quad 4^{3/2} = (4^3)^{1/2} = 64^{1/2} = 8.$$

Definition of $a^{m/n}$           Statement (1)

**EXAMPLE 7**    Simplify each of the following expressions.

**(a)** $7^{-4} \cdot 7^6 = 7^{-4+6} = 7^2 = 49.$    Product property

**(b)** $5x^{2/3} \cdot 2x^{1/4} = 10x^{2/3}x^{1/4}$

$\qquad\qquad\quad = 10x^{2/3+1/4}$    Product property

$\qquad\qquad\quad = 10x^{11/12}.$    $\dfrac{2}{3} + \dfrac{1}{4} = \dfrac{8}{12} + \dfrac{3}{12} = \dfrac{11}{12}$

**(c)** $\dfrac{9^{14}}{9^{-6}} = 9^{14-(-6)} = 9^{20}.$    Quotient property

**(d)** $\dfrac{c^5}{2c^{4/3}} = \dfrac{1}{2} \cdot \dfrac{c^5}{c^{4/3}}$

$\qquad\quad = \dfrac{1}{2}c^{5-4/3}$    Quotient property

$\qquad\quad = \dfrac{1}{2}c^{11/3} = \dfrac{c^{11/3}}{2}.$    $5 - \dfrac{4}{3} = \dfrac{15}{3} - \dfrac{4}{3} = \dfrac{11}{3}$

**(e)** $\dfrac{27^{1/3} \cdot 27^{5/3}}{27^3} = \dfrac{27^{1/3+5/3}}{27^3}$    Product property

$\qquad\qquad\quad = \dfrac{27^2}{27^3} = 27^{2-3}$    Quotient property

$\qquad\qquad\quad = 27^{-1} = \dfrac{1}{27}.$    Definition of negative exponent ✓7

## Checkpoint 7

Simplify each of the following.

**(a)** $9^7 \cdot 9^{-5}$

**(b)** $3x^{1/4} \cdot 5x^{5/4}$

**(c)** $\dfrac{8^7}{8^{-3}}$

**(d)** $\dfrac{5^{2/3} \cdot 5^{-4/3}}{5^2}$

```
  1     5
27 3 *27 3
―――――――
   27 3
              .037037037
 1
――
27
              .037037037
```

**Figure 1.11**

You can use a calculator to check numerical computations, such as those in Example 7, by computing the left and right sides separately and confirming that the answers are the same in each case. Figure 1.11 shows this technique for part (e) of Example 7.

**EXAMPLE 8**    Perform the indicated operations.

**(a)** $(2^{-3})^{-4/7} = 2^{(-3)(-4/7)} = 2^{12/7}.$    Power of a power

**(b)** $\left(\dfrac{3m^{5/6}}{y^{3/4}}\right)^2 = \dfrac{(3m^{5/6})^2}{(y^{3/4})^2}$    Quotient to a power

$\qquad\qquad\quad = \dfrac{3^2(m^{5/6})^2}{(y^{3/4})^2}$    Product to a power

$\qquad\qquad\quad = \dfrac{9m^{(5/6)2}}{y^{(3/4)2}}$    Power of a power

$\qquad\qquad\quad = \dfrac{9m^{5/3}}{y^{3/2}}.$    $\dfrac{5}{6} \cdot 2 = \dfrac{10}{6} = \dfrac{5}{3}$   and   $\dfrac{3}{4} \cdot 2 = \dfrac{3}{2}$

**(c)** $m^{2/3}(m^{7/3} + 2m^{1/3}) = m^{2/3}m^{7/3} + m^{2/3}2m^{1/3}$    Distributive property

$\qquad\qquad\qquad\qquad = m^{2/3+7/3} + 2m^{2/3+1/3} = m^3 + 2m.$    Product rule ✓8

## Checkpoint 8

Simplify each of the following.

**(a)** $(7^{-4})^{-2} \cdot (7^4)^{-2}$

**(b)** $\dfrac{c^4c^{-1/2}}{c^{3/2}d^{1/2}}$

**(c)** $a^{5/8}(2a^{3/8} + a^{-1/8})$

**EXAMPLE 9** Simplify each expression in parts (a)–(c). Give answers with only positive exponents.

**(a)** $\dfrac{(m^3)^{-2}}{m^4} = \dfrac{m^{-6}}{m^4} = m^{-6-4} = m^{-10} = \dfrac{1}{m^{10}}$.

**(b)** $6y^{2/3} \cdot 2y^{-1/2} = 12y^{2/3-1/2} = 12y^{1/6}$.

**(c)** $\dfrac{x^{1/2}(x-2)^{-3}}{5(x-2)} = \dfrac{x^{1/2}}{5} \cdot \dfrac{(x-2)^{-3}}{x-2} = \dfrac{x^{1/2}}{5} \cdot (x-2)^{-3-1}$

$\qquad\qquad = \dfrac{x^{1/2}}{5} \cdot \dfrac{1}{(x-2)^4} = \dfrac{x^{1/2}}{5(x-2)^4}$.

**(d)** Write $a^{-1} + b^{-1}$ as a single quotient.

**Solution** Be careful here. $a^{-1} + b^{-1}$ does *not* equal $(a+b)^{-1}$; the exponent properties deal only with products and quotients, not with sums. However, using the definition of negative exponents and addition of fractions, we have

$$a^{-1} + b^{-1} = \frac{1}{a} + \frac{1}{b} = \frac{b+a}{ab}. \quad \checkmark_9$$

We can use some expressions of the form $ax^b$ where both $a$ and $b$ are constants and $b$ is also an exponent.

**EXAMPLE 10** **College Students** The total number of students (in millions) attending institutes of higher education can be approximated by the expression

$$6.1x^{.337} \quad (x \ge 10),$$

where $x = 10$ corresponds to the year 1990. Find the approximate number of students enrolled in higher education in 2017. (Data from: U.S. National Center for Education Statistics.)

**Solution** Since 2017 is 27 years after 1990 and $x = 10$ corresponds to 1990, we have that $x = 27 + 10 = 37$ corresponds to 2017. We then obtain

$$6.1(37)^{.337} \approx 20.6 \text{ million students.} \quad \checkmark_{10}$$

## Radicals

Earlier, we denoted the $n$th root of $a$ as $a^{1/n}$. An alternative notation for $n$th roots uses the radical symbol $\sqrt[n]{\ }$.

> If $n$ is an even natural number and $a \ge 0$, or if $n$ is an odd natural number,
>
> $$\sqrt[n]{a} = a^{1/n}.$$

In the radical expression $\sqrt[n]{a}$, $a$ is called the *radicand* and $n$ is called the *index*. When $n = 2$, the familiar square-root symbol $\sqrt{a}$ is used instead of $\sqrt[2]{a}$.

**EXAMPLE 11** Simplify the following radicals.

**(a)** $\sqrt[4]{16} = 16^{1/4} = 2$.

**(b)** $\sqrt[5]{-32} = -2$.

**(c)** $\sqrt[3]{1000} = 10$.

**(d)** $\sqrt[6]{\dfrac{64}{729}} = \left(\dfrac{64}{729}\right)^{1/6} = \dfrac{64^{1/6}}{729^{1/6}} = \dfrac{2}{3}$. $\quad \checkmark_{11}$

---

✓ **Checkpoint 9**

Simplify the given expressions. Give answers with only positive exponents.

**(a)** $(3x^{2/3})(2x^{-1})(y^{-1/3})^2$

**(b)** $\dfrac{(t^{-1})^2}{t^{-5}}$

**(c)** $\left(\dfrac{2k^{1/3}}{p^{5/4}}\right)^2 \cdot \left(\dfrac{4k^{-2}}{p^5}\right)^{3/2}$

**(d)** $x^{-1} - y^{-2}$

✓ **Checkpoint 10**

Assuming the model from Example 10 remains accurate, find the number of students for 2020.

✓ **Checkpoint 11**

Simplify.

**(a)** $\sqrt[3]{27}$

**(b)** $\sqrt[4]{625}$

**(c)** $\sqrt[6]{64}$

**(d)** $\sqrt[3]{\dfrac{64}{125}}$

Recall that $a^{m/n} = (a^{1/n})^m$ by definition and $a^{m/n} = (a^m)^{1/n}$ by statement (**1**) on page 36 (provided that all terms are defined). We translate these facts into radical notation as follows.

For all rational numbers $m/n$ and all real numbers $a$ for which $\sqrt[n]{a}$ exists,
$$a^{m/n} = (\sqrt[n]{a})^m \quad \text{or} \quad a^{m/n} = \sqrt[n]{a^m}.$$

Notice that $\sqrt[n]{x^n}$ cannot be written simply as $x$ when $n$ is even. For example, if $x = -5$, then
$$\sqrt{x^2} = \sqrt{(-5)^2} = \sqrt{25} = 5 \neq x.$$
However, $|-5| = 5$, so that $\sqrt{x^2} = |x|$ when $x$ is $-5$. This relationship is true in general.

For any real number $a$ and any natural number $n$,
$$\sqrt[n]{a^n} = |a| \quad \text{if } n \text{ is even}$$
and
$$\sqrt[n]{a^n} = a \quad \text{if } n \text{ is odd.}$$

To avoid the difficulty that $\sqrt[n]{a^n}$ is not necessarily equal to $a$, we shall assume that all variables in radicands represent only nonnegative numbers, as they usually do in applications. The properties of exponents can be written with radicals as follows.

For all real numbers $a$ and $b$, and for natural numbers $n$ for which all indicated roots exist,

(a) $\sqrt[n]{a} \cdot \sqrt[n]{b} = \sqrt[n]{ab}$ and

(b) $\dfrac{\sqrt[n]{a}}{\sqrt[n]{b}} = \sqrt[n]{\dfrac{a}{b}}$ $(b \neq 0).$

**EXAMPLE 12**   Simplify the following expressions.

(a) $\sqrt{6} \cdot \sqrt{54} = \sqrt{6 \cdot 54} = \sqrt{324} = 18.$

Alternatively, simplify $\sqrt{54}$ first:
$$\sqrt{6} \cdot \sqrt{54} = \sqrt{6} \cdot \sqrt{9 \cdot 6}$$
$$= \sqrt{6} \cdot 3\sqrt{6} = 3 \cdot 6 = 18.$$

(b) $\sqrt{\dfrac{7}{64}} = \dfrac{\sqrt{7}}{\sqrt{64}} = \dfrac{\sqrt{7}}{8}.$

(c) $\sqrt{75} - \sqrt{12}.$

**Solution**  Note that $12 = 4 \cdot 3$ and that 4 is a perfect square. Similarly, $75 = 25 \cdot 3$ and 25 is a perfect square. Consequently,
$$\sqrt{75} - \sqrt{12} = \sqrt{25 \cdot 3} - \sqrt{4 \cdot 3} \quad \text{Factor.}$$
$$= \sqrt{25}\sqrt{3} - \sqrt{4}\sqrt{3} \quad \text{Property (a)}$$
$$= 5\sqrt{3} - 2\sqrt{3} = 3\sqrt{3}. \quad \text{Simplify.}$$

✓**Checkpoint 12**

Simplify.

(a) $\sqrt{3} \cdot \sqrt{27}$

(b) $\sqrt{\dfrac{3}{49}}$

(c) $\sqrt{50} + \sqrt{72}$

!**CAUTION** When $a$ and $b$ are nonzero real numbers,

$$\sqrt[n]{a+b} \text{ is NOT equal to } \sqrt[n]{a} + \sqrt[n]{b}.$$

For example,

$$\sqrt{9+16} = \sqrt{25} = 5, \text{ but } \sqrt{9} + \sqrt{16} = 3 + 4 = 7,$$

so $\sqrt{9+16} \neq \sqrt{9} + \sqrt{16}$.

Multiplying radical expressions is much like multiplying polynomials.

**EXAMPLE 13**    Perform the following multiplications.

**(a)** $(\sqrt{2}+3)(\sqrt{8}-5) = \sqrt{2}(\sqrt{8}) - \sqrt{2}(5) + 3\sqrt{8} - 3(5)$    FOIL
$$= \sqrt{16} - 5\sqrt{2} + 3(2\sqrt{2}) - 15$$
$$= 4 - 5\sqrt{2} + 6\sqrt{2} - 15$$
$$= -11 + \sqrt{2}.$$

**(b)** $(\sqrt{7}-\sqrt{10})(\sqrt{7}+\sqrt{10}) = (\sqrt{7})^2 - (\sqrt{10})^2$
$$= 7 - 10 = -3.$$

✓**Checkpoint 13**

Multiply.
**(a)** $(\sqrt{5}-\sqrt{2})(3+\sqrt{2})$
**(b)** $(\sqrt{3}+\sqrt{7})(\sqrt{3}-\sqrt{7})$

## Rationalizing Denominators and Numerators

Before the invention of calculators, it was customary to **rationalize the denominators** of fractions (that is, write equivalent fractions with no radicals in the denominator), because this made many computations easier. Although there is no longer a computational reason to do so, rationalization of denominators (and sometimes numerators) is still used today to simplify expressions and to derive useful formulas.

**EXAMPLE 14**    Rationalize each denominator.

**(a)** $\dfrac{4}{\sqrt{3}}$

**Solution**  The key is to multiply by 1, with 1 written as a radical fraction:

$$\frac{4}{\sqrt{3}} = \frac{4}{\sqrt{3}} \cdot 1 = \frac{4}{\sqrt{3}} \cdot \frac{\sqrt{3}}{\sqrt{3}} = \frac{4\sqrt{3}}{3}.$$

**(b)** $\dfrac{1}{3-\sqrt{2}}$

**Solution**  The same technique works here, using $1 = \dfrac{3+\sqrt{2}}{3+\sqrt{2}}$:

$$\frac{1}{3-\sqrt{2}} = \frac{1}{3-\sqrt{2}} \cdot 1 = \frac{1}{3-\sqrt{2}} \cdot \frac{3+\sqrt{2}}{3+\sqrt{2}} = \frac{3+\sqrt{2}}{(3-\sqrt{2})(3+\sqrt{2})}$$
$$= \frac{3+\sqrt{2}}{9-2} = \frac{3+\sqrt{2}}{7}.$$

✓**Checkpoint 14**

Rationalize the denominator.
**(a)** $\dfrac{2}{\sqrt{5}}$
**(b)** $\dfrac{1}{2+\sqrt{3}}$

**EXAMPLE 15**    Rationalize the numerator of

$$\frac{2+\sqrt{5}}{1+\sqrt{3}}.$$

**Solution**  As in Example 14(b), we must write 1 as a suitable fraction. Since we want to rationalize the numerator here, we multiply by the fraction $1 = \dfrac{2-\sqrt{5}}{2-\sqrt{5}}$:

$$\frac{2 + \sqrt{5}}{1 + \sqrt{3}} = \frac{2 + \sqrt{5}}{1 + \sqrt{3}} \cdot \frac{2 - \sqrt{5}}{2 - \sqrt{5}} = \frac{4 - 5}{2 - \sqrt{5} + 2\sqrt{3} - \sqrt{3}\sqrt{5}}$$

$$= \frac{-1}{2 - \sqrt{5} + 2\sqrt{3} - \sqrt{15}}.$$

# 1.5 Exercises

*Perform the indicated operations and simplify your answer. (See Examples 1 and 2.)*

**1.** $\dfrac{7^5}{7^3}$

**2.** $\dfrac{(-6)^{14}}{(-6)^6}$

**3.** $(4c)^2$

**4.** $(-2x)^4$

**5.** $\left(\dfrac{2}{x}\right)^5$

**6.** $\left(\dfrac{5}{xy}\right)^3$

**7.** $(3u^2)^3(2u^3)^2$

**8.** $\dfrac{(5v^2)^3}{(2v)^4}$

*Perform the indicated operations and simplify your answer, which should not have any negative exponents. (See Examples 3 and 4.)*

**9.** $7^{-1}$

**10.** $10^{-3}$

**11.** $-6^{-5}$

**12.** $(-x)^{-4}$

**13.** $(-y)^{-3}$

**14.** $\left(\dfrac{1}{6}\right)^{-2}$

**15.** $\left(\dfrac{4}{3}\right)^{-2}$

**16.** $\left(\dfrac{x}{y^2}\right)^{-2}$

**17.** $\left(\dfrac{a}{b^3}\right)^{-1}$

**18.** Explain why $-2^{-4} = -1/16$, but $(-2)^{-4} = 1/16$.

*Evaluate each expression. Write all answers without exponents. Round decimal answers to two places. (See Examples 5 and 6.)*

**19.** $49^{1/2}$

**20.** $8^{1/3}$

**21.** $(5.71)^{1/4}$

**22.** $12^{5/2}$

**23.** $-64^{2/3}$

**24.** $-64^{3/2}$

**25.** $(8/27)^{-4/3}$

**26.** $(27/64)^{-1/3}$

*Simplify each expression. Write all answers using only positive exponents. (See Example 7.)*

**27.** $\dfrac{5^{-3}}{4^{-2}}$

**28.** $\dfrac{7^{-4}}{7^{-3}}$

**29.** $4^{-3} \cdot 4^6$

**30.** $9^{-9} \cdot 9^{10}$

**31.** $\dfrac{4^{10} \cdot 4^{-6}}{4^{-4}}$

**32.** $\dfrac{5^{-4} \cdot 5^6}{5^{-1}}$

*Simplify each expression. Assume all variables represent positive real numbers. Write answers with only positive exponents. (See Examples 8 and 9.)*

**33.** $\dfrac{z^6 \cdot z^2}{z^5}$

**34.** $\dfrac{k^6 \cdot k^9}{k^{12}}$

**35.** $\dfrac{3^{-1}(p^{-2})^3}{3p^{-7}}$

**36.** $\dfrac{(5x^3)^{-2}}{x^4}$

**37.** $(q^{-5}r^3)^{-1}$

**38.** $(2y^2z^{-2})^{-3}$

**39.** $(2p^{-1})^3 \cdot (5p^2)^{-2}$

**40.** $(4^{-1}x^3)^{-2} \cdot (3x^{-3})^4$

**41.** $(2p)^{1/2} \cdot (2p^3)^{1/3}$

**42.** $(5k^2)^{3/2} \cdot (5k^{1/3})^{3/4}$

**43.** $p^{2/3}(2p^{1/3} + 5p)$

**44.** $3x^{3/2}(2x^{-3/2} + x^{3/2})$

**45.** $\dfrac{(x^2)^{1/3}(y^2)^{2/3}}{3x^{2/3}y^2}$

**46.** $\dfrac{(c^{1/2})^3(d^3)^{1/2}}{(c^3)^{1/4}(d^{1/4})^3}$

**47.** $\dfrac{(7a)^2(5b)^{3/2}}{(5a)^{3/2}(7b)^4}$

**48.** $\dfrac{(4x)^{1/2}\sqrt{xy}}{x^{3/2}y^2}$

**49.** $x^{1/2}(x^{2/3} - x^{4/3})$

**50.** $x^{1/2}(3x^{3/2} + 2x^{-1/2})$

**51.** $(x^{1/2} + y^{1/2})(x^{1/2} - y^{1/2})$

**52.** $(x^{1/3} + y^{1/2})(2x^{1/3} - y^{3/2})$

*Match the rational-exponent expression in Column I with the equivalent radical expression in Column II. Assume that x is not zero.*

| I | II |
|---|---|
| **53.** $(-3x)^{1/3}$ | **(a)** $\dfrac{3}{\sqrt[3]{x}}$ |
| **54.** $-3x^{1/3}$ | **(b)** $-3\sqrt[3]{x}$ |
| **55.** $(-3x)^{-1/3}$ | **(c)** $\dfrac{1}{\sqrt[3]{3x}}$ |
| **56.** $-3x^{-1/3}$ | **(d)** $\dfrac{-3}{\sqrt[3]{x}}$ |
| **57.** $(3x)^{1/3}$ | **(e)** $3\sqrt[3]{x}$ |
| **58.** $3x^{-1/3}$ | **(f)** $\sqrt[3]{-3x}$ |
| **59.** $(3x)^{-1/3}$ | **(g)** $\sqrt[3]{3x}$ |
| **60.** $3x^{1/3}$ | **(h)** $\dfrac{1}{\sqrt[3]{-3x}}$ |

*Simplify each of the given radical expressions. (See Examples 11–13.)*

**61.** $\sqrt[3]{125}$

**62.** $\sqrt[6]{64}$

**63.** $\sqrt[4]{625}$

**64.** $\sqrt[7]{-128}$

**65.** $\sqrt{63}\sqrt{7}$

**66.** $\sqrt[3]{81} \cdot \sqrt[3]{9}$

**67.** $\sqrt{81 - 4}$

**68.** $\sqrt{49 - 16}$

**69.** $\sqrt{5}\sqrt{15}$

**70.** $\sqrt{8}\sqrt{96}$

**71.** $\sqrt{50} - \sqrt{72}$

**72.** $\sqrt{75} + \sqrt{192}$

**73.** $5\sqrt{20} - \sqrt{45} + 2\sqrt{80}$

**74.** $(\sqrt{3} + 2)(\sqrt{3} - 2)$

**75.** $(\sqrt{5} + \sqrt{2})(\sqrt{5} - \sqrt{2})$

**76.** What is wrong with the statement $\sqrt[3]{4} \cdot \sqrt[3]{4} = 4$?

*Rationalize the denominator of each of the given expressions. (See Example 14.)*

**77.** $\dfrac{3}{1 - \sqrt{2}}$

**78.** $\dfrac{2}{1 + \sqrt{5}}$

**79.** $\dfrac{9 - \sqrt{3}}{3 - \sqrt{3}}$

**80.** $\dfrac{\sqrt{3} - 1}{\sqrt{3} - 2}$

*Rationalize the numerator of each of the given expressions. (See Example 15.)*

**81.** $\dfrac{3 - \sqrt{2}}{3 + \sqrt{2}}$

**82.** $\dfrac{1 + \sqrt{7}}{2 - \sqrt{3}}$

*The following exercises are applications of exponentiation and radicals.*

**83. Lot Size** The theory of economic lot size shows that, under certain conditions, the number of units to order to minimize total cost is

$$x = \sqrt{\frac{kM}{f}},$$

where $k$ is the cost to store one unit for one year, $f$ is the (constant) setup cost to manufacture the product, and $M$ is the total number of units produced annually. Find $x$ for the following values of $f$, $k$, and $M$.

(a) $k = \$1, f = \$500, M = 100,000$

(b) $k = \$3, f = \$7, M = 16,700$

(c) $k = \$1, f = \$5, M = 16,800$

**84. Health** The threshold weight $T$ for a person is the weight above which the risk of death increases greatly. One researcher found that the threshold weight in pounds for men aged 40–49 is related to height in inches by the equation $h = 12.3T^{1/3}$. What height corresponds to a threshold of 216 pounds for a man in this age group?

**Box Office** *The annual domestic revenue (in billions of dollars) generated by the sale of movie tickets can be approximated by the expression*

$$6.67x^{.188} \quad (x \geq 5),$$

*where $x = 5$ corresponds to 2005. Assuming the model remains accurate, approximate the revenue in the following years. (Data from: www.the-numbers.com.)*

**85.** 2015

**86.** 2018

**87.** 2020

**88.** 2023

**Facebook Revenue** *The quarterly advertising revenue (in billions of dollars) earned by Facebook can be approximated by the expression*

$$1.08x^{.527} \quad (x \geq 1),$$

*where $x = 1$ corresponds to the first quarter of the year 2013. Assuming the model continues to be accurate, find the approximate quarterly revenue for the following quarters. (Data from: The Wall Street Journal.)*

**89.** Q1 2016 ($x = 13$)

**90.** Q2 2017 ($x = 18$)

**91.** Q4 2018

**92.** Q3 2019

**Part-Time College Students** *The number of students attending college or university (in millions) on a part-time basis can be approximated by the expression*

$$2.72x^{.238}, \quad (x \geq 10),$$

*where $x = 10$ corresponds to the year 1990. Assuming the model remains accurate, find the number of part-time students for the following years. (Data from: U.S. National Center for Education Statistics.)*

**93.** 2009

**94.** 2014

**95.** 2018

**96.** 2023

✓**Checkpoint Answers**

**1.** (a) $2^9$    (b) $(-5)^4$    (c) $(xy)^5$

**2.** (a) $81x^4$    (b) $r^{12}s^{30}$

    (c) $\dfrac{32}{z^5}$    (d) $\dfrac{9a^4}{b^6}$

**3.** (a) $1/36$    (b) $-1/216$

    (c) $-1/81$    (d) $8/5$

**4.** (a) $8/5$    (b) $32$    (c) $b^3/a^6$

**5.** (a) $4$    (b) $2$

    (c) $-16$    (d) Not a real number

    (e) $-2$    (f) $3$

**6.** (a) $8$    (b) $3125$

    (c) $128$    (d) $1000$

**7.** (a) $81$    (b) $15x^{3/2}$

    (c) $8^{10}$    (d) $5^{-8/3}$ or $1/5^{8/3}$

**8.** (a) $1$    (b) $c^2/d^{1/2}$    (c) $2a + a^{1/2}$

**9.** (a) $\dfrac{6}{x^{1/3} y^{2/3}}$    (b) $t^3$

    (c) $32/(p^{10}k^{7/3})$    (d) $\dfrac{y^2 - x}{xy^2}$

**10.** About 21.1 million

**11.** (a) $3$    (b) $5$

    (c) $2$    (d) $4/5$

**12.** (a) $9$    (b) $\dfrac{\sqrt{3}}{7}$    (c) $11\sqrt{2}$

**13.** (a) $3\sqrt{5} + \sqrt{10} - 3\sqrt{2} - 2$    (b) $-4$

**14.** (a) $\dfrac{2\sqrt{5}}{5}$    (b) $2 - \sqrt{3}$

# 1.6   First-Degree Equations

An **equation** is a statement that two mathematical expressions are equal; for example,

$$5x - 3 = 13, \qquad 8y = 4, \qquad \text{and} \qquad -3p + 5 = 4p - 8$$

are equations.

The letter in each equation is called the variable. This section concentrates on **first-degree equations,** which are equations that involve only constants and the first power of the variable. All of the equations displayed above are first-degree equations, but neither of the following equations is of first degree:

$$2x^2 = 5x + 6 \qquad \text{(the variable has an exponent greater than 1);}$$
$$\sqrt{x + 2} = 4 \qquad \text{(the variable is under the radical).}$$

A **solution** of an equation is a number that can be substituted for the variable in the equation to produce a true statement. For example, substituting the number 9 for $x$ in the equation $2x + 1 = 19$ gives

$$2x + 1 = 19$$
$$2(9) + 1 \stackrel{?}{=} 19 \qquad \text{Let } x = 9.$$
$$18 + 1 = 19. \qquad \text{True}$$

This true statement indicates that 9 is a solution of $2x + 1 = 19$. ✓₁

The following properties are used to solve equations.

## ✓ Checkpoint 1

Is $-4$ a solution of the equations in parts (a) and (b)?

(a)  $3x + 5 = -7$

(b)  $2x - 3 = 5$

(c)  Is there more than one solution of the equation in part (a)?

---

## Properties of Equality

**1.** The same number may be added to or subtracted from both sides of an equation:

$$\text{If } a = b, \text{ then } a + c = b + c \text{ and } a - c = b - c.$$

**2.** Both sides of an equation may be multiplied or divided by the same nonzero number:

$$\text{If } a = b \text{ and } c \neq 0, \text{ then } ac = bc \text{ and } \frac{a}{c} = \frac{b}{c}.$$

---

**EXAMPLE 1**   Solve the equation $5x - 3 = 12$.

**Solution**  Using the first property of equality, add 3 to both sides. This isolates the term containing the variable on one side of the equation:

$$5x - 3 = 12$$
$$5x - 3 + 3 = 12 + 3 \qquad \text{Add 3 to both sides.}$$
$$5x = 15.$$

Now arrange for the coefficient of $x$ to be 1 by using the second property of equality:

$$5x = 15$$
$$\frac{5x}{5} = \frac{15}{5} \qquad \text{Divide both sides by 5.}$$
$$x = 3.$$

The solution of the original equation, $5x - 3 = 12$, is 3. Check the solution by substituting 3 for $x$ in the original equation. ✓₂

## ✓ Checkpoint 2

Solve the following.

(a)  $3p - 5 = 19$

(b)  $4y + 3 = -5$

(c)  $-2k + 6 = 2$

**EXAMPLE 2**  Solve $2k + 3(k - 4) = 2(k - 3)$.

**Solution** First, simplify the equation by using the distributive property on the left-side term $3(k - 4)$ and right-side term $2(k - 3)$:

$$2k + 3(k - 4) = 2(k - 3)$$
$$2k + 3k - 12 = 2(k - 3) \quad \text{Distributive property on left side}$$
$$2k + 3k - 12 = 2k - 6 \quad \text{Distributive property on right side}$$
$$5k - 12 = 2k - 6. \quad \text{Collect like terms on left side.}$$

One way to proceed is to add $-2k$ to both sides:

$$5k - 12 + (-2k) = 2k - 6 + (-2k) \quad \text{Add } -2k \text{ to both sides.}$$
$$3k - 12 = -6$$
$$3k - 12 + 12 = -6 + 12 \quad \text{Add 12 to both sides.}$$
$$3k = 6$$
$$\frac{1}{3}(3k) = \frac{1}{3}(6) \quad \text{Multiply both sides by } \frac{1}{3}.$$
$$k = 2.$$

The solution is 2. Check this result by substituting 2 for $k$ in the original equation. ✓₃

✓ **Checkpoint 3**

Solve the following.

**(a)** $3(m - 6) + 2(m + 4)$
$= 4m - 2$

**(b)** $-2(y + 3) + 4y$
$= 3(y + 1) - 6$

**EXAMPLE 3**  **IRAs** The total amount of money $y$ (in trillions of dollars) invested in individual retirement accounts (IRAs) in the United States in year $x$ can be approximated by the equation

$$5(x - 2000) = 15y - 30.$$

Assuming this equation remains valid, use a calculator to determine the year when the total assets are \$9 trillion. (Data from: ProQuest Statistical Abstract of the United States: 2016.)

**Solution** Let $y = 9$ in the equation and solve for $x$. To avoid any rounding errors in the intermediate steps, it is often a good idea to do all the algebra first, before using the calculator:

$$5(x - 2000) = 15y - 30$$
$$5(x - 2000) = 15(9) - 30 \quad \text{Substitute } y = 9.$$
$$5x - 5(2000) = 15(9) - 30 \quad \text{Distributive property}$$
$$5x = 15(9) - 30 + 5(2000) \quad \text{Add 5(2000) to both sides.}$$
$$x = \frac{15(9) - 30 + 5(2000)}{5}. \quad \text{Divide both sides by 5.}$$

```
15(9)-30+5(2000)
       5
              2021
```

**Figure 1.12**

Now use a calculator to determine that $x = 2021$, as shown in Figure 1.12. So in 2021, the total invested in IRAs is \$9 trillion. ✓₄

✓ **Checkpoint 4**

In Example 3, what year were the total IRA assets \$6 trillion?

The next three examples show how to simplify the solution of first-degree equations involving fractions. We solve these equations by multiplying both sides of the equation by a *common denominator*. This step will eliminate the fractions.

**EXAMPLE 4**  Solve

$$\frac{r}{10} - \frac{2}{15} = \frac{3r}{20} - \frac{1}{5}.$$

**Solution** Here, the denominators are 10, 15, 20, and 5. Each of these numbers is a factor of 60; therefore, 60 is a common denominator. Multiply both sides of the equation by 60:

$$60\left(\frac{r}{10} - \frac{2}{15}\right) = 60\left(\frac{3r}{20} - \frac{1}{5}\right)$$

$$60\left(\frac{r}{10}\right) - 60\left(\frac{2}{15}\right) = 60\left(\frac{3r}{20}\right) - 60\left(\frac{1}{5}\right) \qquad \text{Distributive property}$$

$$6r - 8 = 9r - 12$$

$$6r - 8 + (-6r) + 12 = 9r - 12 + (-6r) + 12 \qquad \text{Add } -6r \text{ and } 12 \text{ to both sides.}$$

$$4 = 3r$$

$$r = \frac{4}{3}. \qquad \text{Multiply both sides by } 1/3.$$

 **Checkpoint 5**

Solve the following.

(a) $\dfrac{x}{2} - \dfrac{x}{4} = 6$

(b) $\dfrac{2x}{3} + \dfrac{1}{2} = \dfrac{x}{4} - \dfrac{9}{2}$

Check this solution in the original equation. ✔5

**⚠ CAUTION** Multiplying *both* sides of an *equation* by a number to eliminate fractions is valid. But multiplying a single fraction by a number to simplify it is not valid. For instance, multiplying $\dfrac{3x}{8}$ by 8 *changes* it to $3x$, which is *not equal* to $\dfrac{3x}{8}$.

The second property of equality (page 43) applies only to *nonzero* quantities. Multiplying or dividing both sides of an equation by a quantity involving the variable (which might be zero for some values) may lead to an **extraneous solution**—that is, a number that does not satisfy the original equation. To avoid errors in such situations, always *check your solutions in the original equation.*

**EXAMPLE 5**    Solve

$$\frac{4}{3(k + 2)} - \frac{k}{3(k + 2)} = \frac{5}{3}.$$

**Solution**  Multiply both sides of the equation by the common denominator $3(k + 2)$. Here, $k \neq -2$, since $k = -2$ would give a 0 denominator, making the fraction undefined. So, we have

$$3(k + 2) \cdot \frac{4}{3(k + 2)} - 3(k + 2) \cdot \frac{k}{3(k + 2)} = 3(k + 2) \cdot \frac{5}{3}.$$

Simplify each side and solve for $k$:

$$4 - k = 5(k + 2)$$

$$4 - k = 5k + 10 \qquad \text{Distributive property}$$

$$4 - k + k = 5k + 10 + k \qquad \text{Add } k \text{ to both sides.}$$

$$4 = 6k + 10$$

$$4 + (-10) = 6k + 10 + (-10) \qquad \text{Add } -10 \text{ to both sides.}$$

$$-6 = 6k$$

$$-1 = k. \qquad \text{Multiply both sides by } \frac{1}{6}.$$

The solution is $-1$. Substitute $-1$ for $k$ in the original equation as a check:

$$\frac{4}{3(-1 + 2)} - \frac{-1}{3(-1 + 2)} \overset{?}{=} \frac{5}{3}$$

$$\frac{4}{3} - \frac{-1}{3} \overset{?}{=} \frac{5}{3}$$

$$\frac{5}{3} = \frac{5}{3}.$$

 **Checkpoint 6**

Solve the equation

$$\frac{5p + 1}{3(p + 1)} = \frac{3p - 3}{3(p + 1)} + \frac{9p - 3}{3(p + 1)}.$$

The check shows that $-1$ is the solution. ✔6

**EXAMPLE 6** Solve

$$\frac{3x - 4}{x - 2} = \frac{x}{x - 2}.$$

**Solution** Multiplying both sides by $x - 2$ produces

$$3x - 4 = x$$
$$2x - 4 = 0 \qquad \text{Subtract } x \text{ from both sides.}$$
$$2x = 4 \qquad \text{Add 4 to both sides.}$$
$$x = 2. \qquad \text{Divide both sides by 2.}$$

Substituting 2 for $x$ in the original equation produces fractions with 0 denominators. Since division by 0 is not defined, $x = 2$ is an extraneous solution. So the original equation has no solution. ✓7

Sometimes an equation with several variables must be solved for one of the variables. This process is called **solving for a specified variable.**

**EXAMPLE 7** Solve for $x$: $3(ax - 5a) + 4b = 4x - 2$.

**Solution** Use the distributive property to get

$$3ax - 15a + 4b = 4x - 2.$$

Treat $x$ as the variable and the other letters as constants. Get all terms with $x$ on one side of the equation and all terms without $x$ on the other side:

$$3ax - 4x = 15a - 4b - 2 \qquad \text{Isolate terms with } x \text{ on the left.}$$
$$(3a - 4)x = 15a - 4b - 2 \qquad \text{Distributive property}$$
$$x = \frac{15a - 4b - 2}{3a - 4} \qquad \text{Multiply both sides by } \frac{1}{3a - 4}.$$

The final equation is solved for $x$, as required. ✓8

## Absolute-Value Equations

Ⓒ Recall from Section 1.1 that the absolute value of a number $a$ is either $a$ or $-a$, whichever one is nonnegative. For instance, $|4| = 4$ and $|-7| = -(-7) = 7$.

**EXAMPLE 8** Solve $|x| = 3$.

**Solution** Since $|x|$ is either $x$ or $-x$, the equation says that

$$x = 3 \qquad \text{or} \qquad -x = 3$$
$$x = -3.$$

The solutions of $|x| = 3$ are 3 and $-3$.

**EXAMPLE 9** Solve $|p - 4| = 2$.

**Solution** Since $|p - 4|$ is either $p - 4$ or $-(p - 4)$, we have

$$p - 4 = 2 \qquad \text{or} \qquad -(p - 4) = 2$$
$$p = 6 \qquad\qquad\qquad -p + 4 = 2$$
$$-p = -2$$
$$p = 2,$$

so that 6 and 2 are possible solutions. Checking them in the original equation shows that both are solutions. ✓9

✓ **Checkpoint 7**

Solve each equation.

(a) $\dfrac{3p}{p + 1} = 1 - \dfrac{3}{p + 1}$

(b) $\dfrac{8y}{y - 4} = \dfrac{32}{y - 4} - 3$

✓ **Checkpoint 8**

Solve for $x$.

(a) $2x - 7y = 3xk$

(b) $8(4 - x) + 6p = -5k - 11yx$

✓ **Checkpoint 9**

Solve each equation.

(a) $|y| = 9$

(b) $|r + 3| = 1$

(c) $|2k - 3| = 7$

---

**EXAMPLE 10**   Solve $|4m - 3| = |m + 6|$.

**Solution**  To satisfy the equation, the quantities in absolute-value bars must either be equal or be negatives of one another. That is,

$$4m - 3 = m + 6 \quad \text{or} \quad 4m - 3 = -(m + 6)$$
$$3m = 9 \qquad\qquad\qquad 4m - 3 = -m - 6$$
$$m = 3 \qquad\qquad\qquad\qquad 5m = -3$$
$$m = -\frac{3}{5}.$$

Check that the solutions for the original equation are 3 and $-3/5$. ✓₁₀

✓ **Checkpoint 10**

Solve each equation.

(a)  $|r + 6| = |2r + 1|$

(b)  $|5k - 7| = |10k - 2|$

## Applications

One of the main reasons for learning mathematics is to be able to use it to solve practical problems. There are no hard-and-fast rules for dealing with real-world applications, except perhaps to use common sense. However, you will find it much easier to deal with such problems if you do not try to do everything at once. After reading the problem carefully, attack it in stages, as suggested in the following guidelines.

### Solving Applied Problems

**Step 1**  Read the problem carefully, focusing on the facts you are given and the unknown values you are asked to find. Look up any words you do not understand. You may have to read the problem more than once, until you understand exactly what you are being asked to do.

**Step 2**  Identify the unknown. (If there is more than one, choose one of them, and see Step 3 for what to do with the others.) Name the unknown with some variable that you *write down*. Many students try to skip this step. They are eager to get on with the writing of the equation. But this is an important step. If you do not know what the variable represents, how can you write a meaningful equation or interpret a result?

**Step 3**  Decide on a variable expression to represent any other unknowns in the problem. For example, if $x$ represents the width of a rectangle, and you know that the length is one more than twice the width, then *write down* the fact that the length is $1 + 2x$.

**Step 4**  Draw a sketch or make a chart, if appropriate, showing the information given in the problem.

**Step 5**  Using the results of Steps 1–4, write an equation that expresses a condition that must be satisfied.

**Step 6**  Solve the equation.

**Step 7**  Check the solution in the words of the *original problem*, not just in the equation you have written.

---

**EXAMPLE 11**   **Finance**  A financial manager has $15,000 to invest for her company. She plans to invest part of the money in tax-free bonds at 5% interest and the remainder in taxable bonds at 8%. She wants to earn $1020 per year in interest from the investments. Find the amount she should invest at each rate.

**Solution**

**Step 1** We are asked to find how much of the $15,000 should be invested at 5% and how much at 8%, in order to earn the required interest.

**Step 2** Let $x$ represent the amount to be invested at 5%.

**Step 3** After $x$ dollars are invested, the remaining amount is $15,000 - x$ dollars, which is to be invested at 8%.

**Step 4** To find the amount of interest that is earned for one year, we multiply the rate (converted to a decimal) by the amount invested. For instance, when we invest $x$ dollars at 5%, we convert 5% to .05 and then multiply by $x$. This means that we calculate $.05x$. If we invest the remaining dollars at 8%, we have $.08(15,000 - x)$. The given information is summarized in the following chart.

| Investment | Amount Invested | Interest Rate | Interest Earned in One Year |
|---|---|---|---|
| Tax-free bonds | x | 5% = .05 | .05x |
| Taxable bonds | 15,000 − x | 8% = .08 | .08(15,000 − x) |
| Totals | 15,000 | | 1020 |

**Step 5** Because the total interest is to be $1020, the last column of the table shows that

$$.05x + .08(15,000 - x) = 1020.$$

**Step 6** Solve the preceding equation as follows:

$$.05x + .08(15,000 - x) = 1020$$
$$.05x + .08(15,000) - .08x = 1020$$
$$.05x + 1200 - .08x = 1020$$
$$-.03x = -180$$
$$x = 6000.$$

The manager should invest $6000 at 5% and $15,000 - \$6000 = \$9000$ at 8%.

**Step 7** Check these results in the original problem. If $6000 is invested at 5%, the interest is $.05(6000) = \$300$. If $9000 is invested at 8%, the interest is $.08(9000) = \$720$. So the total interest is $\$300 + \$720 = \$1020$ as required ✓₁₁

✓ **Checkpoint 11**

An investor owns two pieces of property. One, worth twice as much as the other, returns 6% in annual interest, while the other returns 4%. Find the value of each piece of property if the total annual interest earned is $8000.

**EXAMPLE 12** **Music Streaming** According to billboard.com, Apple Music acquired 10 million subscribers in the beginning of 2016. This number took only 6 months to reach, whereas Spotify took 6 years (72 months) to hit that same milestone. Apple Music's average rate of acquiring subscribers per month was 1.53 million more than Spotify's average rate per month. Find the average rate of subscriber acquisition per month for each company.

**Solution**

**Step 1** We must find the subscriber acquisition rate for Apple Music and Spotify.

**Step 2** Let $x$ represent the rate of subscriber acquisition rate for Spotify.

**Step 3** The acquisition rate for Apple Music is 1.53 million more than Spotify, so its rate is $x + 1.53$.

**Step 4** In general, the number of subscribers can be found by the formula:

Number of subscribers = number of months × rate of subscribers per month

For Spotify, the number of subscribers is $72x$ and for Apple Music, the number of subscribers is $6(x + 1.53)$.

We can collect these facts to make a chart that organizes the information given in the problem.

|  | Months | Rate | Number of Subscribers |
|---|---|---|---|
| **Spotify** | 72 | $x$ | $72x$ |
| **Apple Music** | 6 | $x + 1.53$ | $6(x + 1.53)$ |

**Step 5**  Because both companies obtained the *same number of subscribers*, the equation is

$$72x = 6(x + 1.53).$$

**Step 6**  If we distribute the 6 through the quantity on the right side of the equation, we obtain

$$72x = 6x + 9.18$$
$$66x = 9.18 \qquad \text{Add } -6x \text{ to each side.}$$
$$x \approx .139. \qquad \text{Divide each side by 66.}$$

**Step 7**  Since $x$ represents Spotify's rate, Spotify acquired, on average .139 million (139,000) users per month. Apple Music's rate is $x + 1.53$, or $.139 + 1.53 = 1.669$ million users (1,669,000) per month. ✓₁₂

✓ **Checkpoint 12**

In Example 12, suppose Apple Music had taken 9 months and Spotify had taken 75 months to reach the same number of subscribers. Assuming that Apple Music had acquired subscribers at a rate of 1.35 million per month more than Spotify, find the average rate at which Spotify and Apple Music had acquired subscribers.

**EXAMPLE 13**  **Business** An oil company needs to fill orders for 89-octane gas, but has only 87- and 93-octane gas on hand. The octane rating is the percentage of isooctane in the standard fuel. How much of each type should be mixed together to produce 100,000 gallons of 89-octane gas?

**Solution**

**Step 1**  We must find how much 87-octane gas and how much 93-octane gas are needed for the 100,000 gallon mixture.

**Step 2**  Let $x$ be the amount of 87-octane gas.

**Step 3**  Then $100,000 - x$ is the amount of 93-octane gas.

**Step 4**  We can summarize the relevant information in a chart.

| Type of Gas | Quantity | % Isooctane | Amount of Isooctane |
|---|---|---|---|
| 87-octane | $x$ | 87% | $.87x$ |
| 93-octane | $100,000 - x$ | 93% | $.93(100,000 - x)$ |
| Mixture | 100,000 | 89% | $.89(100,000)$ |

**Step 5**  The amount of isooctane satisfies this equation:

$$.87x + .93(100,000 - x) = .89(100,000).$$

**Step 6**  Solving this equation yields

$$.87x + 93,000 - .93x = .89(100,000) \quad \text{Distribute the left side.}$$
$$.87x + 93,000 - .93x = 89,000 \quad \text{Multiply on the right side.}$$
$$-.06x = -4000 \quad \text{Combine terms and add } -93,000 \text{ to each side.}$$
$$x = \frac{-4000}{-.06} \approx 66,667.$$

**Step 7**  So the distributor should mix 66,667 gallons of 87-octane gas with $100,000 - 66,667 = 33,333$ gallons of 93-octane gas. Then the amount of isooctane in the mixture is

$$.87(66,667) + .93(33,333)$$
$$\approx 58,000 + 31,000$$
$$= 89,000.$$

✓ **Checkpoint 13**

How much 89-octane gas and how much 94-octane gas are needed to produce 1500 gallons of 91-octane gas?

Hence, the octane rating of the mixture is $\frac{89,000}{100,000} = .89$ as required. ✓₁₃

# 1.6 Exercises

*Solve each equation. (See Examples 1–6.)*

1. $3x + 8 = 20$

2. $4 - 5y = 19$

3. $.6k - .3 = .5k + .4$

4. $2.5 + 5.04m = 8.5 - .06m$

5. $2a - 1 = 4(a + 1) + 7a + 5$

6. $3(k - 2) - 6 = 4k - (3k - 1)$

7. $2[x - (3 + 2x) + 9] = 3x - 8$

8. $-2[4(k + 2) - 3(k + 1)] = 14 + 2k$

9. $\dfrac{3x}{5} - \dfrac{4}{5}(x + 1) = 2 - \dfrac{3}{10}(3x - 4)$

10. $\dfrac{4}{3}(x - 2) - \dfrac{1}{2} = 2\left(\dfrac{3}{4}x - 1\right)$

11. $\dfrac{5y}{6} - 8 = 5 - \dfrac{2y}{3}$

12. $\dfrac{x}{2} - 3 = \dfrac{3x}{5} + 1$

13. $\dfrac{m}{2} - \dfrac{1}{m} = \dfrac{6m + 5}{12}$

14. $-\dfrac{3k}{2} + \dfrac{9k - 5}{6} = \dfrac{11k + 8}{k}$

15. $\dfrac{4}{x - 3} - \dfrac{8}{2x + 5} + \dfrac{3}{x - 3} = 0$

16. $\dfrac{5}{2p + 3} - \dfrac{3}{p - 2} = \dfrac{4}{2p + 3}$

17. $\dfrac{3}{2m + 4} = \dfrac{1}{m + 2} - 2$

18. $\dfrac{8}{3k - 9} - \dfrac{5}{k - 3} = 4$

*Use a calculator to solve each equation. Round your answer to the nearest hundredth. (See Example 3.)*

19. $9.06x + 3.59(8x - 5) = 12.07x + .5612$

20. $-5.74(3.1 - 2.7p) = 1.09p + 5.2588$

21. $\dfrac{2.63r - 8.99}{1.25} - \dfrac{3.90r - 1.77}{2.45} = r$

22. $\dfrac{8.19m + 2.55}{4.34} - \dfrac{8.17m - 9.94}{1.04} = 4m$

*Solve each equation for x. (See Example 7.)*

23. $4(a + x) = b - a + 2x$

24. $(3a - b) - bx = a(x - 2)$   $(a \neq -b)$

25. $5(b - x) = 2b + ax$   $(a \neq -5)$

26. $bx - 2b = 2a - ax$

*Solve each equation for the specified variable. Assume that all denominators are nonzero. (See Example 7.)*

27. $PV = k$ for $V$

28. $i = prt$ for $p$

29. $V = V_0 + gt$ for $g$

30. $S = S_0 + gt^2 + k$ for $g$

31. $A = \dfrac{1}{2}(B + b)h$ for $B$

32. $C = \dfrac{5}{9}(F - 32)$ for $F$

*Solve each equation. (See Examples 8–10.)*

33. $|2h - 1| = 5$

34. $|4m - 3| = 12$

35. $|6 + 2p| = 10$

36. $|-5x + 7| = 15$

37. $\left|\dfrac{5}{r - 3}\right| = 10$

38. $\left|\dfrac{3}{2h - 1}\right| = 4$

*Solve the following applied problems.*

**Natural Science**  *The equation that relates Fahrenheit temperature F to Celsius temperature C is*

$$C = \dfrac{5}{9}(F - 32).$$

*Find the Fahrenheit temperature corresponding to these Celsius temperatures.*

39. $-5$

40. $-15$

41. $22$

42. $36$

**Federal Debt**  *The gross federal debt y (in trillions of dollars) in year x can be approximated by*

$$y = 1.15x + 1.62,$$

*where x is the number of years after 2000. (Data from: www.treasurydirect.gov.) Find the federal debt in the following years.*

43. $2010$

44. $2015$

*Assuming the trend continues, in what year will the federal debt be the amount given in each instance below?*

45. $22.32$

46. $24.62$

**47.** 25.77          **48.** 30.37

**Health Economics** *The total health care expenditures E in the United States (in trillions of dollars) can be approximated by*

$$E = .108x + 1.517,$$

*where x is the number of years after 2000. Assuming the trend continues, determine the year in which health care expenditures are at the given level. (Data from: U.S. Centers for Medicare and Medicaid Services.)*

**49.** $2.705 trillion

**50.** $3.029 trillion

**51.** $3.461 trillion

**52.** $3.893 trillion

**Social Security** *The amount y (in billions of dollars) of income taken in by the U.S. Social Security Administration for old-age and survivor's insurance in year x can be approximated by the equation*

$$114.8(x - 2010) = 5y - 3390.5.$$

*Assuming the trend continues, find the year in which the given income amounts occurred. (Data from: www.ssa.gov.)*

**53.** 746.98

**54.** 815.86

**55.** 907.7

**56.** 1022.5

**Finance Charges** *When a loan is paid off early, a portion of the finance charge must be returned to the borrower. By one method of calculating the finance charge (called the rule of 78), the amount of unearned interest (finance charge to be returned) is given by*

$$u = f \cdot \frac{n(n + 1)}{q(q + 1)},$$

*where u represents unearned interest, f is the original finance charge, n is the number of payments remaining when the loan is paid off, and q is the original number of payments. Find the amount of the unearned interest in each of the given cases.*

**57.** Original finance charge = $800, loan scheduled to run 36 months, paid off with 18 payments remaining

**58.** Original finance charge = $1400, loan scheduled to run 48 months, paid off with 12 payments remaining

**Investing** *Solve the following investment problems. (See Example 11.)*

**59.** Joe Gonzalez received $52,000 profit from the sale of some land. He invested part at 5% interest and the rest at 4% interest. He earned a total of $2290 interest per year. How much did he invest at 5%?

**60.** Weijen Luan invests $20,000 received from an insurance settlement in two ways: some at 6% and some at 4%. Altogether, she makes $1040 per year in interest. How much is invested at 4%?

**61.** Maria Martinelli bought two plots of land for a total of $120,000. On the first plot, she made a profit of 15%. On the second, she lost 10%. Her total profit was $5500. How much did she pay for each piece of land?

**62.** Suppose $20,000 is invested at 5%. How much additional money must be invested at 4% to produce a yield of 4.8% on the entire amount invested?

*Solve the given applied problems. (See Example 12.)*

**Social Media** *According to data from businessinsider.com, two social media companies, Twitter and Instagram, had the same number of active users in the fourth quarter of 2014. Twitter took 35 quarters to acquire these users, while Instagram took 21 quarters. Instagram's rate of active user growth was 5.7 million more per quarter, on average, than Twitter's average growth per quarter.*

**63.** Find the average rate of growth of Twitter.

**64.** Find the average rate of growth of Instagram.

**65.** Approximately how many active users did Twitter acquire in 35 quarters?

**66.** Approximately how many active users did Instagram acquire in 21 quarters?

**Natural Science** *Using the same assumptions about octane ratings as in Example 13, solve the following problems.*

**67.** How many liters of 94-octane gasoline should be mixed with 200 liters of 99-octane gasoline to get a mixture that is 97-octane gasoline?

**68.** A service station has 92-octane and 98-octane gasoline. How many liters of each gasoline should be mixed to provide 12 liters of 96-octane gasoline for a chemistry experiment?

✓**Checkpoint Answers**

**1.** (a) Yes     (b) No     (c) No

**2.** (a) 8     (b) −2     (c) 2

**3.** (a) 8     (b) −3

**4.** 2012

**5.** (a) 24     (b) −12

**6.** 1

**7.** Neither equation has a solution.

**8.** (a) $x = \dfrac{7y}{2 - 3k}$     (b) $x = \dfrac{5k + 32 + 6p}{8 - 11y}$

**9.** (a) 9, −9     (b) −2, −4     (c) 5, −2

**10.** (a) 5, −7/3     (b) −1, 3/5

**11.** 6% return: $100,000; 4% return: $50,000

**12.** (a) Spotify: .184 million per month;
    (b) Apple Music: 1.534 million per month

**13.** 900 gallons of 89-octane gas; 600 gallons of 94-octane gas

# 1.7 Quadratic Equations

An equation that can be written in the form

$$ax^2 + bx + c = 0,$$

where $a$, $b$, and $c$ are real numbers with $a \neq 0$, is called a **quadratic equation.** For example, each of

$$2x^2 + 3x + 4 = 0, \qquad x^2 = 6x - 9, \qquad 3x^2 + x = 6, \qquad \text{and} \qquad x^2 = 5$$

is a quadratic equation. A solution of an equation that is a real number is said to be a **real solution** of the equation.

One method of solving quadratic equations is based on the following property of real numbers.

## Zero-Factor Property

If $a$ and $b$ are real numbers, with $ab = 0$, then $a = 0$ or $b = 0$ or both.

---

**EXAMPLE 1** Solve the equation $(x - 4)(3x + 7) = 0$.

**Solution** By the zero-factor property, the product $(x - 4)(3x + 7)$ can equal 0 only if at least one of the factors equals 0. That is, the product equals zero only if $x - 4 = 0$ or $3x + 7 = 0$. Solving each of these equations separately will give the solutions of the original equation:

$$x - 4 = 0 \qquad \text{or} \qquad 3x + 7 = 0$$
$$x = 4 \qquad \text{or} \qquad 3x = -7$$
$$x = -\frac{7}{3}.$$

The solutions of the equation $(x - 4)(3x + 7) = 0$ are 4 and $-7/3$. Check these solutions by substituting them into the original equation. ✓₁

✓ **Checkpoint 1**

Solve the following equations.

(a)  $(y - 6)(y + 2) = 0$

(b)  $(5k - 3)(k + 5) = 0$

(c)  $(2r - 9)(3r + 5) \cdot (r + 3) = 0$

---

**EXAMPLE 2** Solve $6r^2 + 7r = 3$.

**Solution** Rewrite the equation as

$$6r^2 + 7r - 3 = 0.$$

Now factor $6r^2 + 7r - 3$ to get

$$(3r - 1)(2r + 3) = 0.$$

By the zero-factor property, the product $(3r - 1)(2r + 3)$ can equal 0 only if

$$3r - 1 = 0 \qquad \text{or} \qquad 2r + 3 = 0.$$

Solving each of these equations separately gives the solutions of the original equation:

$$3r = 1 \quad \text{or} \quad 2r = -3$$
$$r = \frac{1}{3} \qquad\qquad r = -\frac{3}{2}.$$

Verify that both $1/3$ and $-3/2$ are solutions by substituting them into the original equation. ✔2

## ✓ Checkpoint 2

Solve each equation by factoring.

(a) $y^2 + 3y = 10$

(b) $2r^2 + 9r = 5$

(c) $4k^2 = 9k$

An equation such as $x^2 = 5$ has two solutions: $\sqrt{5}$ and $-\sqrt{5}$. This fact is true in general.

## Square-Root Property

If $b > 0$, then the solutions of $x^2 = b$ are $\sqrt{b}$ and $-\sqrt{b}$.

The two solutions are sometimes abbreviated $\pm\sqrt{b}$.

### EXAMPLE 3   Solve each equation.

(a) $m^2 = 17$

**Solution**  By the square-root property, the solutions are $\sqrt{17}$ and $-\sqrt{17}$, abbreviated $\pm\sqrt{17}$.

(b) $(y - 4)^2 = 11$

**Solution**  Using a generalization of the square-root property, we work as follows.

$$(y - 4)^2 = 11$$
$$y - 4 = \sqrt{11} \quad \text{or} \quad y - 4 = -\sqrt{11}$$
$$y = 4 + \sqrt{11} \qquad\qquad y = 4 - \sqrt{11}.$$

## ✓ Checkpoint 3

Solve each equation by using the square-root property.

(a) $p^2 = 21$

(b) $(m + 7)^2 = 15$

(c) $(2k - 3)^2 = 5$

Abbreviate the solutions as $4 \pm \sqrt{11}$. ✔3

When a quadratic equation cannot be easily factored, it can be solved by using the following formula, which you should memorize.*

## Quadratic Formula

The solutions of the quadratic equation $ax^2 + bx + c = 0$, where $a \neq 0$, are given by

$$x = \frac{-b \pm \sqrt{b^2 - 4ac}}{2a}.$$

**! CAUTION**  When using the quadratic formula, remember that the equation must be in the form $ax^2 + bx + c = 0$. Also, notice that the fraction bar in the quadratic formula extends under *both* terms in the numerator. Be sure to add $-b$ to $\pm\sqrt{b^2 - 4ac}$ *before* dividing by $2a$.

---

*A proof of the quadratic formula can be found in many College Algebra books.

**EXAMPLE 4**     Solve $x^2 + 1 = 4x$.

**Solution**  First add $-4x$ to both sides to get 0 alone on the right side:

$$x^2 - 4x + 1 = 0.$$

Now identify the values of $a$, $b$, and $c$. Here, $a = 1$, $b = -4$, and $c = 1$. Substitute these numbers into the quadratic formula to obtain

$$x = \frac{-(-4) \pm \sqrt{(-4)^2 - 4(1)(1)}}{2(1)}$$

$$= \frac{4 \pm \sqrt{16 - 4}}{2}$$

$$= \frac{4 \pm \sqrt{12}}{2}$$

$$= \frac{4 \pm 2\sqrt{3}}{2} \qquad \sqrt{12} = \sqrt{4 \cdot 3} = \sqrt{4} \cdot \sqrt{3} = 2\sqrt{3}$$

$$= \frac{2(2 \pm \sqrt{3})}{2} \qquad \text{Factor } 4 \pm 2\sqrt{3}.$$

$$x = 2 \pm \sqrt{3}. \qquad \text{Cancel 2.}$$

The $\pm$ sign represents the two solutions of the equation. First use $+$ and then use $-$ to find each of the solutions: $2 + \sqrt{3}$ and $2 - \sqrt{3}$. ☑4

✓**Checkpoint 4**

Use the quadratic formula to solve each equation.

(a)  $x^2 - 2x = 2$

(b)  $u^2 - 6u + 4 = 0$

```
2+√3
              3.732050808
2-√3
              .2679491924
```

**Figure 1.13**

Example 4 shows that the quadratic formula produces exact solutions. In many real-world applications, however, you must use a calculator to find decimal approximations of the solutions. The approximate solutions in Example 4 are

$$x = 2 + \sqrt{3} \approx 3.732050808 \qquad \text{and} \qquad x = 2 - \sqrt{3} \approx .2679491924,$$

as shown in Figure 1.13.

**EXAMPLE 5**     **Ford Revenue** Many companies in recent years saw their revenue follow a parabolic pattern because of the recession that began in 2008. One such example is Ford Motor Corp. The revenue $R$ for Ford (in millions of dollars) can be approximated by the equation

$$R = 1.451x^2 - 31.84x + 306, \quad (6 \leq x \leq 15)$$

where $x = 6$ corresponds to the year 2006. Use the quadratic formula and a calculator to find the first full year in which revenue recovered back to at least $145 million. (Data from: www.morningstar.com.)

**Solution**  To find the year $x$, solve the equation above when $R = 145$:

$$1.451x^2 - 31.84x + 306 = R$$
$$1.451x^2 - 31.84x + 306 = 145 \qquad \text{Substitute } R = 145.$$
$$1.451x^2 - 31.84x + 161 = 0. \qquad \text{Subtract 145 from both sides.}$$

If you are using a scientific calculator, first compute the radical part to apply the quadratic formula:

$$\sqrt{b^2 - 4ac} = \sqrt{(-31.84)^2 - 4(1.451)(161)} = \sqrt{79.3416}.$$

Then store $\sqrt{79.3416}$ (which we denote by $M$) in the calculator memory. (Check your instruction manual for how to store and recall numbers.) By the quadratic formula, the exact solutions to the equation are

$$x = \frac{-b \pm \sqrt{b^2 - 4ac}}{2a} = \frac{-b \pm M}{2a} = \frac{-(-31.84) \pm M}{2(1.451)} = \frac{31.84 \pm M}{2.902}.$$

**Figure 1.14**

✓ **Checkpoint 5**

Use the equation of Example 5 to find the year in which revenue was first $137 million.

📟 **TECHNOLOGY TIP**

You can approximate the solutions of quadratic equations on a graphing calculator by using a built-in quadratic equation solver if your calculator has one. Then you need only enter the values of the coefficients *a*, *b*, and *c* to obtain the approximate solutions.

✓ **Checkpoint 6**

Solve each equation.
(a) $9k^2 - 6k + 1 = 0$
(b) $4m^2 + 28m + 49 = 0$
(c) $2x^2 - 5x + 5 = 0$

✓ **Checkpoint 7**

Use the discriminant to determine the number of real solutions of each equation.
(a) $x^2 + 8x + 3 = 0$
(b) $2x^2 + x + 3 = 0$
(c) $x^2 - 194x + 9409 = 0$

The approximate solutions are

$$\frac{31.84 + M}{2.902} \approx 14.04 \quad \text{and} \quad \frac{31.84 - M}{2.902} \approx 7.90.$$

If you are using a graphing calculator, you can find the same results, as shown in Figure 1.14. We were told that revenue had recovered, so we use the solution later in time. The question also wants to know the "first full year" revenue recovered, and so we do not follow regular rounding rules. In this case round up to $x = 15$, which corresponds to 2015. ✓5

**EXAMPLE 6** Solve $9x^2 - 30x + 25 = 0$.

**Solution** Applying the quadratic formula with $a = 9$, $b = -30$, and $c = 25$, we have

$$x = \frac{-(-30) \pm \sqrt{(-30)^2 - 4(9)(25)}}{2(9)}$$

$$= \frac{30 \pm \sqrt{900 - 900}}{18} = \frac{30 \pm 0}{18} = \frac{30}{18} = \frac{5}{3}.$$

Therefore, the given equation has only one real solution. The fact that the solution is a rational number indicates that this equation could have been solved by factoring.

**EXAMPLE 7** Solve $x^2 - 6x + 10 = 0$.

**Solution** Apply the quadratic formula with $a = 1$, $b = -6$, and $c = 10$:

$$x = \frac{-(-6) \pm \sqrt{(-6)^2 - 4(1)(10)}}{2(1)}$$

$$= \frac{6 \pm \sqrt{36 - 40}}{2}$$

$$= \frac{6 \pm \sqrt{-4}}{2}.$$

Since no negative number has a square root in the real-number system, $\sqrt{-4}$ is not a real number. Hence, the equation has no real solutions. ✓6

Examples 4–7 show that the number of solutions of the quadratic equation $ax^2 + bx + c = 0$ is determined by $b^2 - 4ac$, the quantity under the radical, which is called the **discriminant** of the equation. ✓7

**The Discriminant**

The equation $ax^2 + bx + c = 0$ has either two, or one, or no real solutions.

If $b^2 - 4ac > 0$, there are two real solutions. (*Examples 4 and 5*)
If $b^2 - 4ac = 0$, there is one real solution. (*Example 6*)
If $b^2 - 4ac < 0$, there are no real solutions. (*Example 7*)

## Applications

Quadratic equations arise in a variety of settings, as illustrated in the next set of examples. Example 8 depends on the following useful fact from geometry.

## The Pythagorean Theorem

In a right triangle with legs of lengths $a$ and $b$ and hypotenuse of length $c$,

$$a^2 + b^2 = c^2.$$

**EXAMPLE 8** **Screen Size** The size of a flat screen television is the diagonal measurement of its screen. The height of the screen is approximately 63% of its width. Kathy claims that John's 39-inch flat screen has only half of the viewing area that her 46-inch flat screen television has. Is Kathy right?

**Figure 1.15**

**Solution** First, find the area of Kathy's screen. Let $x$ be its width. Then its height is 63% of $x$ (that is, $.63x$, as shown in Figure 1.15). By the Pythagorean theorem,

$$x^2 + (.63x)^2 = 46^2$$
$$x^2 + .3969x^2 = 2116 \qquad \text{Expand } (.63x)^2 \text{ and } 46^2.$$
$$(1 + .3969)x^2 = 2116 \qquad \text{Distributive property}$$
$$1.3969x^2 = 2116$$
$$x^2 = 1514.78 \qquad \text{Divide both sides by 1.3969.}$$
$$x = \pm\sqrt{1514.78} \qquad \text{Square-root property}$$
$$x \approx \pm 38.9.$$

We can ignore the negative solution, since $x$ is a width. Thus, the width is 38.9 inches and the height is $.63x = .63(38.9) \approx 24.5$ inches, so the area is

$$\text{Area} = \text{width} \times \text{height} = 38.9 \times 24.5 = 953.1 \text{ square inches.}$$

Next we find the area of John's television in a similar manner. Let $y$ be the width of John's television. Then its height is 63% of $y$. By the Pythagorean theorem,

$$y^2 + (.63y)^2 = 39^2$$
$$y^2 + .3969y^2 = 1521 \qquad \text{Expand } (.63y)^2 \text{ and } 39^2.$$
$$(1 + .3969)y^2 = 1521 \qquad \text{Distributive property}$$
$$1.3969y^2 = 1521$$
$$y^2 = 1088.84 \qquad \text{Divide both sides by 1.3969.}$$
$$y = \pm\sqrt{1088.84} \qquad \text{Square-root property}$$
$$y \approx \pm 33.0.$$

Again, we can ignore the negative solution, since $y$ is a width. Thus, the width is 33.0 inches and the height is $.63y = .63(33.0) \approx 20.8$ inches, so the area is

$$\text{Area} = \text{width} \times \text{height} = 33.0 \times 20.8 = 686.4 \text{ square inches.}$$

Since half of 953.1 square inches (the area of Kathy's television) is 476.55 square inches, Kathy is wrong because John's viewing area is 686.4 square inches. ✓ 8

**✓ Checkpoint 8**

If John's television was only a 26-inch flat screen television, would Kathy have been right?

**EXAMPLE 9** **Landscaping** A landscape architect wants to make an exposed gravel border of uniform width around a small shed behind a company plant. The shed is 10 feet by 6 feet. He has enough gravel to cover 36 square feet. How wide should the border be?

**Solution** A sketch of the shed with border is given in Figure 1.16. Let $x$ represent the width of the border. Then the width of the large rectangle is $6 + 2x$, and its length is $10 + 2x$.

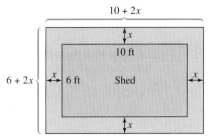

**Figure 1.16**

We must write an equation relating the given areas and dimensions. The area of the large rectangle is $(6 + 2x)(10 + 2x)$. The area occupied by the shed is $6 \cdot 10 = 60$. The area of the border is found by subtracting the area of the shed from the area of the large rectangle. This difference should be 36 square feet, giving the equation

$$(6 + 2x)(10 + 2x) - 60 = 36.$$

Solve this equation with the following sequence of steps:

$$60 + 32x + 4x^2 - 60 = 36 \qquad \text{Multiply out left side.}$$
$$4x^2 + 32x - 36 = 0 \qquad \text{Simplify.}$$
$$x^2 + 8x - 9 = 0 \qquad \text{Divide both sides by 4.}$$
$$(x + 9)(x - 1) = 0 \qquad \text{Factor.}$$
$$x + 9 = 0 \quad \text{or} \quad x - 1 = 0 \qquad \text{Zero-factor property.}$$
$$x = -9 \quad \text{or} \quad x = 1.$$

The number $-9$ cannot be the width of the border, so the solution is to make the border 1 foot wide. ✔9

**✓ Checkpoint 9**

The length of a picture is 2 inches more than the width. It is mounted on a mat that extends 2 inches beyond the picture on all sides. What are the dimensions of the picture if the area of the mat is 99 square inches?

**EXAMPLE 10**   **Physical Science**   If an object is thrown upward, dropped, or thrown downward and travels in a straight line subject only to gravity (with wind resistance ignored), the height $h$ of the object above the ground (in feet) after $t$ seconds is given by

$$h = -16t^2 + v_0 t + h_0,$$

where $h_0$ is the height of the object when $t = 0$ and $v_0$ is the initial velocity at time $t = 0$. The value of $v_0$ is taken to be positive if the object moves upward and negative if it moves downward. Suppose that a golf ball is thrown downward from the top of a 625-foot-high building with an initial velocity of 65 feet per second. How long does it take to reach the ground?

**Solution**   In this case, $h_0 = 625$ (the height of the building) and $v_0 = -65$ (negative because the ball is thrown downward). The object is on the ground when $h = 0$, so we must solve the equation

$$h = -16t^2 + v_0 t + h_0$$
$$0 = -16t^2 - 65t + 625. \qquad \text{Let } h = 0,\ v_0 = -65,\text{ and } h_0 = 625.$$

Using the quadratic formula and a calculator, we see that

$$t = \frac{-(-65) \pm \sqrt{(-65)^2 - 4(-16)(625)}}{2(-16)} = \frac{65 \pm \sqrt{44{,}225}}{-32} \approx \begin{cases} -8.60 \\ \quad \text{or} \quad . \\ 4.54 \end{cases}$$

Only the positive answer makes sense in this case. So it takes about 4.54 seconds for the ball to reach the ground.

In some applications, it may be necessary to solve an equation in several variables for a specific variable.

**EXAMPLE 11** Solve $v = mx^2 + x$ for $x$. (Assume that $m$ and $v$ are positive.)

**Solution** The equation is quadratic in $x$ because of the $x^2$ term. Before we use the quadratic formula, we write the equation in standard form:

$$v = mx^2 + x$$
$$0 = mx^2 + x - v.$$

Let $a = m$, $b = 1$, and $c = -v$. Then the quadratic formula gives

$$x = \frac{-1 \pm \sqrt{1^2 - 4(m)(-v)}}{2m}$$

$$x = \frac{-1 \pm \sqrt{1 + 4mv}}{2m}.$$ ✓10

✓ **Checkpoint 10**

Solve each of the given equations for the indicated variable. Assume that all variables are positive.

(a) $k = mp^2 - bp$ for $p$

(b) $r = \dfrac{APk^2}{3}$ for $k$

# 1.7 Exercises

*Use factoring to solve each equation. (See Examples 1 and 2.)*

1. $(x + 4)(x - 14) = 0$    2. $(p - 16)(p - 5) = 0$

3. $x(x + 6) = 0$    4. $x^2 - 2x = 0$

5. $2z^2 = 4z$    6. $x^2 - 64 = 0$

7. $y^2 + 15y + 56 = 0$    8. $k^2 - 4k - 5 = 0$

9. $2x^2 = 7x - 3$    10. $2 = 15z^2 + z$

11. $6r^2 + r = 1$    12. $3y^2 = 16y - 5$

13. $2m^2 + 20 = 13m$    14. $6a^2 + 17a + 12 = 0$

15. $m(m + 7) = -10$    16. $z(2z + 7) = 4$

17. $9x^2 - 16 = 0$    18. $36y^2 - 49 = 0$

19. $16x^2 - 16x = 0$    20. $12y^2 - 48y = 0$

*Solve each equation by using the square-root property. (See Example 3.)*

21. $(r - 2)^2 = 7$    22. $(b + 4)^2 = 27$

23. $(4x - 1)^2 = 20$    24. $(3t + 5)^2 = 11$

*Use the quadratic formula to solve each equation. If the solutions involve square roots, give both the exact and approximate solutions. (See Examples 4–7.)*

25. $2x^2 + 7x + 1 = 0$    26. $3x^2 - x - 7 = 0$

27. $4k^2 + 2k = 1$    28. $r^2 = 3r + 5$

29. $5y^2 + 5y = 2$    30. $2z^2 + 3 = 8z$

31. $6x^2 + 6x + 4 = 0$    32. $3a^2 - 2a + 2 = 0$

33. $2r^2 + 3r - 5 = 0$    34. $8x^2 = 8x - 3$

35. $2x^2 - 7x + 30 = 0$    36. $3k^2 + k = 6$

37. $1 + \dfrac{7}{2a} = \dfrac{15}{2a^2}$    38. $5 - \dfrac{4}{k} - \dfrac{1}{k^2} = 0$

*Use the discriminant to determine the number of real solutions of each equation. You need not solve the equations.*

39. $25t^2 + 49 = 70t$    40. $9z^2 - 12z = 1$

41. $13x^2 + 24x - 5 = 0$    42. $20x^2 + 19x + 5 = 0$

*Use a calculator and the quadratic formula to find approximate solutions of each equation. (See Example 5.)*

43. $4.42x^2 - 10.14x + 3.79 = 0$

44. $3x^2 - 82.74x + 570.4923 = 0$

45. $7.63x^2 + 2.79x = 5.32$

46. $8.06x^2 + 25.8726x = 25.047256$

*Solve the following problems. (See Examples 3, 4, and 5.)*

**Trucking Pay** *According to data from the National Transportation Institute, the average annual pay for truckers P (in thousands of dollars) can be approximated by*

$$P = .396x^2 - 7.09x + 74.9 \quad (5 \le x \le 15),$$

*where x is the number of years since 2000. Find the pay for the following years.*

47. 2010    48. 2015

*Assuming the trend continues, find the following.*

49. The first full year when annual pay was below $48 thousand.

50. The first full year when annual pay recovered to be above $48 thousand.

51. The first full year when annual pay recovered to be above $55 thousand.

52. The first full year when annual pay recovered to be above $60 thousand.

**Oil Prices** *According to data from the U.S. Energy Information Administration, the price O in real dollars for an imported barrel of crude oil can be approximated by*

$$O = .867x^2 - 9.31x + 63.32 \quad (1 \leq x \leq 8),$$

*where x = 1 corresponds to the first quarter of 2015, and so forth. Find the price of imported crude oil for the following quarters.*

53. Third quarter of 2015    54. Second quarter of 2016

55. Determine the most recent full quarter where the price was at least $45.

56. Determine the most recent full quarter where the price was at least $55.

*Solve the following problems. (See Examples 8 and 9).*

57. **Physical Science** A 13-foot-long ladder leans on a wall, as shown in the accompanying figure. The bottom of the ladder is 5 feet from the wall. If the bottom is pulled out 2 feet farther from the wall, how far does the top of the ladder move down the wall? [*Hint:* Draw pictures of the right triangle formed by the ladder, the ground, and the wall before and after the ladder is moved. In each case, use the Pythagorean theorem to find the distance from the top of the ladder to the ground.]

5 ft

58. **Physical Science** A 15-foot-long pole leans against a wall. The bottom is 9 feet from the wall. How much farther should the bottom be pulled away from the wall so that the top moves the same amount down the wall?

59. **Physical Science** Two trains leave the same city at the same time, one going north and the other east. The eastbound train travels 20 mph faster than the northbound one. After 5 hours, the trains are 300 miles apart. Determine the speed of each train, using the following steps.

    (a) Let $x$ denote the speed of the northbound train. Express the speed of the eastbound train in terms of $x$.

    (b) Write expressions that give the distance traveled by each train after 5 hours.

    (c) Use part (b) and the fact that the trains are 300 miles apart after 5 hours to write an equation. (A diagram of the situation may help.)

    (d) Solve the equation and determine the speeds of the trains.

60. **Physical Science** Tyrone and Miguel have received walkie-talkies for Christmas. If they leave from the same point at the same time, Tyrone walking north at 2.5 mph and Miguel walking east at 3 mph, how long will they be able to talk to each other if the range of the walkie-talkies is 4 miles? Round your answer to the nearest minute.

61. **Landscaping** An ecology center wants to set up an experimental garden. It has 300 meters of fencing to enclose a rectangular area of 5000 square meters. Find the length and width of the rectangle as follows.

    (a) Let $x =$ the length and write an expression for the width.

    (b) Write an equation relating the length, width, and area, using the result of part (a).

    (c) Solve the problem.

62. **Landscaping** A landscape architect has included a rectangular flower bed measuring 9 feet by 5 feet in her plans for a new building. She wants to use two colors of flowers in the bed, one in the center and the other for a border of the same width on all four sides. If she can get just enough plants to cover 24 square feet for the border, how wide can the border be?

63. **Physical Science** Joan wants to buy a rug for a room that is 12 feet by 15 feet. She wants to leave a uniform strip of floor around the rug. She can afford 108 square feet of carpeting. What dimensions should the rug have?

64. **Indy 500** In 2016, Alexander Rossi won the 500-mile Indianapolis 500 race. His speed (rate) was, on average, 92 miles per hour faster than that of the 1911 winner, Ray Harroun. Rossi completed the race in 3.72 hours less time than Harroun. Find Harroun's and Rossi's rates to the nearest tenth.

**Physical Science** *Use the height formula in Example 10 to work the given problems. Note that an object that is dropped (rather than thrown downward) has initial velocity $v_0 = 0$.*

65. How long does it take a baseball to reach the ground if it is dropped from the top of a 625-foot-high building? Compare the answer with that in Example 10.

66. After the baseball in Exercise 65 is dropped, how long does it take for the ball to fall 196 feet? (*Hint:* How high is the ball at that time?)

67. You are standing on a cliff that is 200 feet high. How long will it take a rock to reach the ground if

    (a) you drop it?

    (b) you throw it downward at an initial velocity of 40 feet per second?

    (c) How far does the rock fall in 2 seconds if you throw it downward with an initial velocity of 40 feet per second?

68. A rocket is fired straight up from ground level with an initial velocity of 800 feet per second.

    (a) How long does it take the rocket to rise 3200 feet?

    (b) When will the rocket hit the ground?

69. A ball is thrown upward from ground level with an initial velocity of 64 ft per second. In how many seconds will the ball be at the given height?

    (a) 64 ft              (b) 39 ft

    (c) Why are two answers possible in part (b)?

70. **Physical Science** A ball is thrown upward from ground level with an initial velocity of 100 feet per second. In how many seconds will the ball first reach the given height?

    (a) 50 ft              (b) 35 ft

*Solve each of the given equations for the indicated variable. Assume that all denominators are nonzero and that all variables represent positive real numbers. (See Example 11.)*

**71.** $S = \dfrac{1}{2}gt^2$ for $t$

**72.** $a = \pi r^2$ for $r$

**73.** $L = \dfrac{d^4 k}{h^2}$ for $h$

**74.** $F = \dfrac{kMv^2}{r}$ for $v$

**75.** $P = \dfrac{E^2 R}{(r + R)^2}$ for $R$

**76.** $S = 2\pi rh + 2\pi r^2$ for $r$

✓ **Checkpoint Answers**

**1. (a)** $6, -2$     **(b)** $3/5, -5$     **(c)** $9/2, -5/3, -3$

**2. (a)** $2, -5$     **(b)** $1/2, -5$     **(c)** $9/4, 0$

**3. (a)** $\pm\sqrt{21}$     **(b)** $-7 \pm \sqrt{15}$     **(c)** $(3 \pm \sqrt{5})/2$

**4. (a)** $x = 1 + \sqrt{3}$ or $1 - \sqrt{3}$
   **(b)** $u = 3 + \sqrt{5}$ or $3 - \sqrt{5}$

**5.** Just at the end of 2008

**6. (a)** $1/3$     **(b)** $-7/2$     **(c)** No real solutions

**7. (a)** $2$     **(b)** $0$     **(c)** $1$

**8.** Yes, area $= 304.92$ sq. in.

**9.** 5 inches by 7 inches

**10. (a)** $p = \dfrac{b \pm \sqrt{b^2 + 4mk}}{2m}$

   **(b)** $k = \pm\sqrt{\dfrac{3r}{AP}}$   or   $\dfrac{\pm\sqrt{3rAP}}{AP}$

# CHAPTER 1 Summary and Review

## Key Terms and Symbols

**1.1** $\approx$ is approximately equal to
$\pi$ pi
$|a|$ absolute value of $a$
real number
natural (counting) number
whole number
integer
rational number
irrational number
properties of real numbers
order of operations
square roots
number line
interval
interval notation
absolute value

**1.2** $a^n$ $a$ to the power $n$
exponent or power
multiplication with exponents
power of a power rule
zero exponent
base
polynomial
variable
coefficient
term
constant term
degree of a polynomial
zero polynomial
leading coefficient
quadratics
cubics
like terms
FOIL
revenue
fixed cost
variable cost
profit

**1.3** factor
factoring
greatest common factor
difference of squares
perfect squares
sum and difference of cubes

**1.4** rational expression
cancellation property
operations with rational expressions
complex fraction

**1.5** $a^{1/n}$ nth root of $a$
$\sqrt{a}$ square root of $a$
$\sqrt[n]{a}$ nth root of $a$
properties of exponents
radical
radicand
index
rationalizing the denominator
rationalizing the numerator

**1.6** first-degree equation
solution of an equation
properties of equality
extraneous solution
solving for a specified variable
absolute-value equations
solving applied problems

**1.7** quadratic equation
real solution
zero-factor property
square-root property
quadratic formula
discriminant
Pythagorean theorem

## Chapter 1 Key Concepts

**Factoring**

$x^2 + 2xy + y^2 = (x + y)^2$
$x^2 - 2xy + y^2 = (x - y)^2$
$x^2 - y^2 = (x + y)(x - y)$

$x^3 - y^3 = (x - y)(x^2 + xy + y^2)$
$x^3 + y^3 = (x + y)(x^2 - xy + y^2)$

**Properties of Radicals**

Let $a$ and $b$ be real numbers, $n$ be a positive integer, and $m$ be any integer for which the given relationships exist. Then

$a^{m/n} = \sqrt[n]{a^m} = (\sqrt[n]{a})^m$;     $\sqrt[n]{a^n} = |a|$ if $n$ is even;     $\sqrt[n]{a^n} = a$ if $n$ is odd;

$\sqrt[n]{a} \cdot \sqrt[n]{b} = \sqrt[n]{ab}$;     $\dfrac{\sqrt[n]{a}}{\sqrt[n]{b}} = \sqrt[n]{\dfrac{a}{b}}$   $(b \neq 0)$.

**Properties of Exponents**

Let $a$, $b$, $r$, and $s$ be any real numbers for which the following exist. Then

$$a^{-r} = \frac{1}{a^r} \qquad a^0 = 1 \qquad \left(\frac{a}{b}\right)^r = \frac{a^r}{b^r}$$

$$a^r \cdot a^s = a^{r+s} \qquad (a^r)^s = a^{rs} \qquad a^{1/r} = \sqrt[r]{a}$$

$$\frac{a^r}{a^s} = a^{r-s} \qquad (ab)^r = a^r b^r \qquad \left(\frac{a}{b}\right)^{-r} = \left(\frac{b}{a}\right)^r$$

**Absolute Value**

Assume that $a$ and $b$ are real numbers with $b > 0$.

The solutions of $|a| = b$ or $|a| = |b|$ are $a = b$ or $a = -b$.

**Quadratic Equations**

Facts needed to solve quadratic equations (in which $a$, $b$, and $c$ are real numbers):

**Factoring** If $ab = 0$, then $a = 0$ or $b = 0$ or both.

**Square-Root Property** If $b > 0$, then the solutions of $x^2 = b$ are $\sqrt{b}$ and $-\sqrt{b}$.

**Quadratic Formula** The solutions of $ax^2 + bx + c = 0$ (with $a \neq 0$) are

$$x = \frac{-b \pm \sqrt{b^2 - 4ac}}{2a}.$$

**Discriminant** There are two real solutions of $ax^2 + bx + c = 0$ if $b^2 - 4ac > 0$, one real solution if $b^2 - 4ac = 0$, and no real solutions if $b^2 - 4ac < 0$.

# Chapter 1 Review Exercises

*Name the numbers from the list $-12$, $-6$, $-9/10$, $-\sqrt{7}$, $-\sqrt{4}$, $0$, $1/8$, $\pi/4$, $6$, and $\sqrt{11}$ that are*

1. whole numbers
2. integers
3. rational numbers
4. irrational numbers

*Identify the properties of real numbers that are illustrated in each of the following expressions.*

5. $9[(-3)4] = 9[4(-3)]$

6. $7(4 + 5) = (4 + 5)7$

7. $6(x + y - 3) = 6x + 6y + 6(-3)$

8. $11 + (5 + 3) = (11 + 5) + 3$

*Express each statement in symbols.*

9. $x$ is at least 9.
10. $x$ is negative.

*Write the following numbers in numerical order from smallest to largest.*

11. $|6 - 4|, -|-2|, |8 + 1|, -|3 - (-2)|$

12. $\sqrt{7}, -\sqrt{8}, -|\sqrt{16}|, |-\sqrt{12}|$

*Write the following without absolute-value bars.*

13. $7 - |-8|$
14. $|-3| - |-9 + 6|$

*Graph each of the following on a number line.*

15. $x \geq -3$
16. $-4 < x \leq 6$

*Use the order of operations to simplify each of the following.*

17. $\dfrac{-9 + (-6)(-3) \div 9}{6 - (-3)}$
18. $\dfrac{20 \div 4 \cdot 2 \div 5 - 1}{-9 - (-3) - 12 \div 3}$

*Perform each of the indicated operations.*

19. $(3x^4 - x^2 + 5x) - (-x^4 + 3x^2 - 6x)$

20. $(-8y^3 + 8y^2 - 5y) - (2y^3 + 4y^2 - 10)$

21. $(5k - 2h)(5k + 2h)$
22. $(2r - 5y)(2r + 5y)$

23. $(3x + 4y)^2$
24. $(2a - 5b)^2$

*Factor each of the following as completely as possible.*

25. $2kh^2 - 4kh + 5k$
26. $2m^2n^2 + 6mn^2 + 16n^2$

27. $5a^4 + 12a^3 + 4a^2$
28. $24x^3 + 4x^2 - 4x$

29. $144p^2 - 169q^2$
30. $81z^2 - 25x^2$

31. $27y^3 - 1$
32. $125a^3 + 216$

*Perform each operation.*

33. $\dfrac{3x}{5} \cdot \dfrac{45x}{12}$
34. $\dfrac{5k^2}{24} - \dfrac{70k}{36}$

35. $\dfrac{c^2 - 3c + 2}{2c(c - 1)} \div \dfrac{c - 2}{8c}$

36. $\dfrac{p^3 - 2p^2 - 8p}{3p(p^2 - 16)} \div \dfrac{p^2 + 4p + 4}{9p^2}$

37. $\dfrac{2m^2 - 4m + 2}{m^2 - 1} \div \dfrac{6m + 18}{m^2 + 2m - 3}$

38. $\dfrac{x^2 + 6x + 5}{4(x^2 + 1)} \cdot \dfrac{2x(x + 1)}{x^2 - 25}$

*Simplify each of the given expressions. Write all answers without negative exponents. Assume that all variables represent positive real numbers.*

39. $5^{-3}$
40. $10^{-2}$

41. $-8^0$
42. $\left(-\dfrac{5}{6}\right)^{-2}$

43. $4^6 \cdot 4^{-3}$
44. $7^{-5} \cdot 7^{-2}$

45. $\dfrac{8^{-5}}{8^{-4}}$
46. $\dfrac{6^{-3}}{6^4}$

**47.** $5^{-1} + 2^{-1}$        **48.** $5^{-2} + 5^{-1}$

**49.** $\dfrac{5^{1/3} \cdot 5^{1/2}}{5^{3/2}}$        **50.** $\dfrac{2^{3/4} \cdot 2^{-1/2}}{2^{1/4}}$

**51.** $(3a^2)^{1/2} \cdot (3^2 a)^{3/2}$        **52.** $(4p)^{2/3} \cdot (2p^3)^{3/2}$

*Simplify each of the following expressions.*

**53.** $\sqrt[3]{27}$        **54.** $\sqrt[6]{-64}$

**55.** $\sqrt[3]{54p^3 q^5}$        **56.** $\sqrt[4]{64a^5 b^3}$

**57.** $3\sqrt{3} - 12\sqrt{12}$        **58.** $8\sqrt{7} + 2\sqrt{63}$

*Rationalize each denominator.*

**59.** $\dfrac{\sqrt{3}}{1 + \sqrt{2}}$        **60.** $\dfrac{4 + \sqrt{2}}{4 - \sqrt{5}}$

*Solve each equation.*

**61.** $3x - 4(x - 2) = 2x + 9$

**62.** $4y + 9 = -3(1 - 2y) + 5$

**63.** $\dfrac{2m}{m - 3} = \dfrac{6}{m - 3} + 4$

**64.** $\dfrac{15}{k + 5} = 4 - \dfrac{3k}{k + 5}$

*Solve for x.*

**65.** $8ax - 3 = 2x$        **66.** $b^2 x - 2x = 4b^2$

*Solve each equation.*

**67.** $\left| \dfrac{2 - y}{5} \right| = 8$        **68.** $|4k + 1| = |6k - 3|$

*Find all real solutions of each equation.*

**69.** $(b + 7)^2 = 5$        **70.** $(2p + 1)^2 = 7$

**71.** $2p^2 + 3p = 2$        **72.** $2y^2 = 15 + y$

**73.** $2q^2 - 11q = 21$        **74.** $3x^2 + 2x = 16$

**75.** $6k^4 + k^2 = 1$        **76.** $21p^4 = 2 + p^2$

*Solve each equation for the specified variable.*

**77.** $p = \dfrac{E^2 R}{(r + R)^2}$ for $r$        **78.** $p = \dfrac{E^2 R}{(r + R)^2}$ for $E$

**79.** $K = s(s - a)$ for $s$        **80.** $kz^2 - hz - t = 0$ for $z$

*Work these problems.*

**81.** **Stock Prices** The percent change in stock price over the course of 2016 for JCPenny was +27%. For Sears, the percent change in stock price was −55%. Find the difference in the percent change for these two companies in 2016. (Data from: www.morningstar.com.)

**82.** **Stock Prices** The two worst performers from the Dow Jones Index in 2016 were Nike and Coca-Cola. Nike's stock price decreased 22% and Coca-Cola's decreased 2.5%. Find the difference in the percent change for these two companies in 2016. (Data from: www.morningstar.com.)

**83.** **Computer Prices.** A new laptop computer is on sale for 20% off. The sale price is $1229. What was the original price?

**84.** **Stock Prices.** The stock price for eBay increased 27.3% in the last six months of 2016. If the price was $29.83 a share at the end of the year, what was the price six months earlier?

**Government Spending** *The amount of outlays O of the U.S. government per year (in trillions of dollars) can be approximated by the equation*

$$O = .11x + 2.2,$$

*where $x = 6$ corresponds to the year 2006. Find the year in which the outlays were the following. (Data from: U.S. Office of Management and Budget.)*

**85.** $3.63 trillion        **86.** $3.96 trillion

**Oil Output** *The crude oil output (in millions of barrels per day) from Russia can be approximated by the equation*

$$O = .2x + 8,$$

*where $x = 10$ corresponds to the year 2010. In what year was output at the following levels? (Data from: The Wall Street Journal.)*

**87.** 10.8 million barrels per day

**88.** 11.4 million barrels per day

**Vehicle Sales** *The number (in millions) of new vehicles V sold in the United States can be approximated by the equation*

$$V = .04x^2 - .47x + 18.38,$$

*where $x = 1$ corresponds to the month of January in 2016. (Data from tradingeconomics.com.)*

**89.** Find the number of vehicles sold in July 2016.

**90.** Find the number of vehicles sold in October 2016.

**91.** Find the first full month of the year in which sales were below 17.5 million.

**92.** Find the first full month where sales recovered to 17.5 million.

**Aircraft Departures** *The number (in millions) of flight departures F within the United States per year can be approximated by the equation*

$$F = \frac{.061x^2 + 6.09x + 38.8}{x + 1},$$

*where $x = 6$ corresponds to the year 2006. (Data from: U.S. Department of Transportation.)*

**93.** How many flights occurred in 2013?

**94.** If trends continue, how many flights will occur in 2020?

**95.** In what year did the number of flights first fall below 9.2 million?

**96.** If trends continue, in what year will the number of flights first fall below 8.9 million?

**Research Spending** *The total amount (in billions of dollars) spent by the U.S. government on basic research R can be approximated by*

$$R = 16.2x^{.26},$$

*where $x = 10$ corresponds to the year 2010. (Data from: nsf.gov/ statistics.) Find the spending in the following years.*

**97.** 2012        **98.** 2017

**99. Finance** Kinisha borrowed $2000 from a credit union at 12% annual interest and borrowed $500 from her aunt at 7% annual interest. What single rate of interest on $2500 results in the same total amount of interest?

**100. Business** A butcher makes a blend of stew meat from beef that sells for $2.80 a pound and pork that sells for $3.25 a pound. How many pounds of each meat should be used to get 30 pounds of a mix that can be sold for $3.10 a pound?

**Federal Debt** *The ratio of the U.S. federal debt D as a percentage of gross domestic product (GDP) can be approximated by the equation*

$$D = .047x^2 - 7.69x + 341.6,$$

*where x = 50 corresponds to the year 1950. (Data from: U.S. Office of Management and Budget.)*

**101.** What is the ratio of federal debt to GDP in 2012?

**102.** What is the most recent year in which the ratio of federal debt to GDP was approximately 55%?

**103. Landscaping** A concrete mixer needs to know how wide to make a walk around a rectangular fountain that is 10 by 15 feet. She has enough concrete to cover 200 square feet. To the nearest tenth of a foot, how wide should the walk be in order to use up all the concrete?

**104. Landscaping** A new homeowner needs to put up a fence to make his yard secure. The house itself forms the south boundary and the garage forms the west boundary of the property, so he needs to fence only 2 sides. The area of the yard is 4000 square feet. He has enough material for 160 feet of fence. Find the length and width of the yard (to the nearest foot).

**105. Physical Science** A projectile is launched from the ground vertically upward at a rate of 150 feet per second. How many seconds will it take for the projectile to be 200 feet from the ground on its downward trip?

**106. Physical Science** Suppose a tennis ball is thrown downward from the top of a 700-foot-high building with an initial velocity of 55 feet per second. How long does it take to reach the ground?

# Case Study 1   Energy Efficiency and Long-Term Cost Savings

In a series of pioneering studies, researchers conducted famous experiments in which children were tested regarding their ability for delayed gratification.[*] In these experiments, a child would be offered a marshmallow, cookie, or pretzel and told that if they did not eat the treat, they could have two of the treats when the researcher returned (after about 15 minutes). Of course, some children would eat the treat immediately, and others would wait and receive two of the treats.

Similar studies occur in studying consumer behavior among adults. Often when making purchases, consumers will purchase an appliance that initially costs less, but in the long run costs more money when one accounts for the operating expense. Imagine trying to decide whether to buy a new furnace for $4500 that is 80% efficient versus a new furnace that is 90% efficient that costs $5700. Each furnace is expected to last at least 20 years. The furnace with the 90% efficiency is estimated to cost $1800 a year to operate, while the 80% efficiency furnace is estimated to cost $2100 a year to operate—a difference of $300 a year. Thus, the original price differential of $1200 would be made up in four years. Over the course of 16 additional years, the 90% efficiency furnace would end up saving the buyer $4800.

For other appliances, the difference in total expenditures may not be as dramatic, but can still be substantial. The energy guide tags that often accompany new appliances help make it easier for consumers "to do the math" by communicating the yearly operating costs of an appliance. For example, a Kenmore new side-by-side 25.4 cubic foot refrigerator retails for $1128. The energy guide estimates the cost of operation for a year to be $72. A similarly sized new Frigidaire refrigerator retails for $1305, and the energy guide information estimates the annual cost to be $43. Over 10 years, which refrigerator costs the consumer more money?

We can write mathematical expressions to answer this question. With an initial cost of $1128, and an annual operating cost of $72, for $x$ years of operation the cost $C$ for the Kenmore refrigerator is

$$C = 1128 + 72x$$

For 10 years, the Kenmore would cost $C = 1128 + 72(10) = $1848$. For the Frigidaire model, the cost is

$$C = 1305 + 43x.$$

For 10 years, the Frigidaire would cost $C = 1305 + 43(10) = $1735$.

Thus, the Frigidaire refrigerator costs $113 less in total costs over 10 years.

Behavior when a consumer seeks the up-front savings sounds very similar to the delayed gratification studies among children. These kinds of behavior are of interest to both psychologists and economists. The implications are also very important in the marketing of energy-efficient appliances that often do have a higher initial cost, but save consumers money over the lifetime of the product.

One tool to help consumers make the best choice in the long run is simply to take the time to "just do the math," as the common expression goes. The techniques of this chapter can be applied to do just that!

[*]Mischel, Walter, Ebbe B. Ebbesen, and Antonette Raskoff Zeiss. (1972). "Cognitive and attentional mechanisms in delay of gratification." *Journal of Personality and Social Psychology*, pp. 204–218.

## Exercises

1. On the homedepot.com website, a General Electric 40-gallon electric hot water tank retails for $218 and the estimated annual cost of operation is $508. Write an expression for the cost to buy and run the hot water tank for $x$ years.

2. On the homedepot.com website, a General Electric 40-gallon gas hot water tank retails for $328 and the estimated annual cost of operation is $309. Write an expression for the cost to buy and run the hot water tank for $x$ years.

3. Over ten years, does the electric or gas hot water tank cost more? By how much?

4. In how many years will the total for the two hot water tanks be equal?

5. On the Lowes website, a Maytag 25-cubic-foot refrigerator was advertised for $1529.10 with an estimated cost per month in electricity of $50 a year. Write an expression for the cost to buy and run the refrigerator for $x$ years.

6. On the HomeDepot website, an LG Electronics 25-cubic-foot refrigerator was advertised for $1618.20 with an estimated cost per month in electricity of $44 a year. Write an expression for the cost to buy and run the refrigerator for $x$ years.

7. Over ten years, which refrigerator (the Maytag or the LG) costs the most? By how much?

8. In how many years will the total for the two refrigerators be equal.

# Extended Project

A high-end real estate developer needs to install 50 new washing machines in a building she leases to tenants (with utilities included in the monthly rent). She knows that energy-efficient front loaders cost less per year to run, but they can be much more expensive than traditional top loaders. Investigate by visiting an appliance store or using the Internet, the prices and energy efficiency of two comparable front-loading and top-loading washing machines.

1. Estimate how many years it will take for the total expenditure (including purchase and cost of use) for the 50 front-loading machines to equal the total expenditure for the 50 top-loading machines.

2. The energy guide ratings use data that is often several years old. If the owner believes that current energy costs are 20% higher than what is indicated with the energy guide, redo the calculations for (1) above and determine if it will take less time for the two total expenditures to be equal.

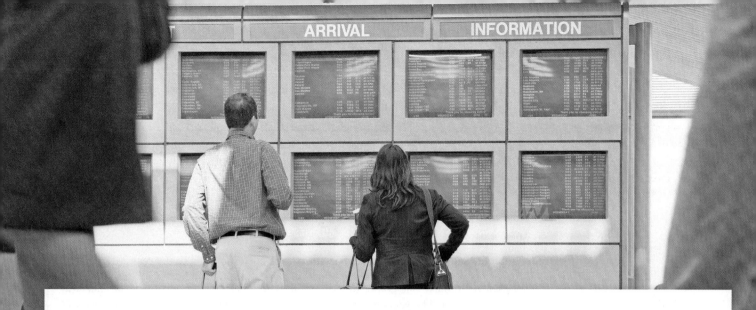

# Graphs, Lines, and Inequalities

**2**

## CHAPTER OUTLINE

Data from current and past events is often a useful tool in business and in the social and health sciences. Gathering data is the first step in developing mathematical models that can be used to analyze a situation and predict future performance. In this chapter, for example, we explore linear models in transportation, stock price, and higher education enrollment.

Graphical representations of data are commonly used in business and in the health and social sciences. Lines, equations, and inequalities play an important role in developing mathematical models from such data. This chapter presents both algebraic and graphical methods for dealing with these topics.

## 2.1 Graphs

Just as the number line associates the points on a line with real numbers, a similar construction in two dimensions associates points in a plane with *ordered pairs* of real numbers. A **Cartesian coordinate system,** as shown in Figure 2.1, consists of a horizontal number line (usually called the *x*-**axis**) and a vertical number line (usually called the *y*-**axis**). The point where the number lines meet is called the **origin.** Each point in a Cartesian coordinate system is labeled with an **ordered pair** of real numbers, such as $(-2, 4)$ or $(3, 2)$. Several points and their corresponding ordered pairs are shown in Figure 2.1.

For the point labeled $(-2, 4)$, for example, $-2$ is the *x*-**coordinate** and 4 is the *y*-**coordinate.** You can think of these coordinates as directions telling you how to move to this point from the origin: You go 2 horizontal units to the left (*x*-coordinate) and 4 vertical units upward

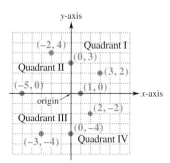

**Figure 2.1**

✓ **Checkpoint 1**

Locate $(-1, 6)$, $(-3, -5)$, $(4, -3)$, $(0, 2)$, and $(-5, 0)$ on a coordinate system.

*Answers to Checkpoint exercises are found at the end of the section.*

(*y*-coordinate). From now on, instead of referring to "the point labeled by the ordered pair $(-2, 4)$," we will say "the point $(-2, 4)$." ✓₁

**EXAMPLE 1**   **Unemployment Rate** The percentage of unemployed workers in the U.S. labor force in January of recent years is shown in the following table. (Data from: U.S. Department of Labor.)

| Year | 2012 | 2013 | 2014 | 2015 | 2016 | 2017 |
|------|------|------|------|------|------|------|
| Rate | 8.3  | 8.0  | 6.6  | 5.7  | 4.9  | 4.8  |

Graph the data points from the table with $x = 12$ corresponding to the year 2012.

**Solution** To make our graph, we need to determine the appropriate "window" that will display our data. The years go from 2012 to 2017, and using the correspondence, we will want our *x*-axis to go from 12 to 17. We see that our *y*-axis will need to go as high as 8.3. If we use a maximum value of 9, we can then plot the ordered pairs:

$$(12, 8.3), (13, 8.0), (14, 6.6), (15, 5.7), (16, 4.9), (17, 4.8),$$

as we see in Figure 2.2.

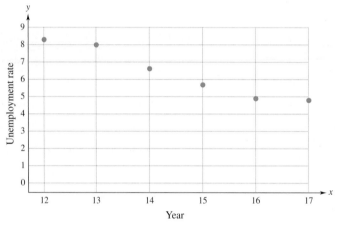

**Figure 2.2**

The *x*-axis and the *y*-axis divide the plane into four parts, or **quadrants,** which are numbered as shown in Figure 2.1. The points on the coordinate axes belong to no quadrant.

## Equations and Graphs

A **solution of an equation** in two variables, such as

$$y = -2x + 3$$

or

$$y = x^2 + 7x - 2,$$

is an ordered pair of numbers such that the substitution of the first number for *x* and the second number for *y* produces a true statement.

✓ **Checkpoint 2**

Which of the following are solutions of

$$y = x^2 + 5x - 3?$$

**(a)** $(1, 3)$

**(b)** $(-2, -3)$

**(c)** $(-1, -7)$

**EXAMPLE 2**   Which of the following are solutions of $y = -2x + 3$?

**(a)** $(2, -1)$

**Solution** This is a solution of $y = -2x + 3$ because "$-1 = -2 \cdot 2 + 3$" is a true statement.

**(b)** $(4, 7)$

**Solution** Since $-2 \cdot 4 + 3 = -5$, and not 7, the ordered pair $(4, 7)$ is not a solution of $y = -2x + 3$. ✓₂

Equations in two variables, such as $y = -2x + 3$, typically have an infinite number of solutions. To find one, choose a number for $x$ and then compute the value of $y$ that produces a solution. For instance, if $x = 5$, then $y = -2 \cdot 5 + 3 = -7$, so that the pair $(5, -7)$ is a solution of $y = -2x + 3$. Similarly, if $x = 0$, then $y = -2 \cdot 0 + 3 = 3$, so that $(0, 3)$ is also a solution.

The **graph** of an equation in two variables is the set of points in the plane whose coordinates (ordered pairs) are solutions of the equation. Thus, the graph of an equation is a picture of its solutions. Since a typical equation in two variables has infinitely many solutions, its graph has infinitely many points.

**✓ Checkpoint 3**

Graph $x = 5y$.

| EXAMPLE 3 | Sketch the graph of $y = -2x + 5$. |

**Solution** Since we cannot plot infinitely many points, we construct a table of $y$-values for a reasonable number of $x$-values, plot the corresponding points, and make an "educated guess" about the rest. The table of values and points in Figure 2.3a suggests that the graph is a straight line, as shown in Figure 2.3b. ✓₃

| $x$ | $-2x + 5$ |
|-----|-----------|
| $-1$ | 7 |
| 0 | 5 |
| 2 | 1 |
| 4 | $-3$ |
| 5 | $-5$ |

**Figure 2.3a**

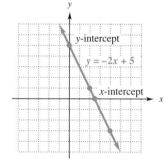

**Figure 2.3b**

An **x-intercept** of a graph is the $x$-coordinate of a point where the graph intersects the $x$-axis. The $y$-coordinate of this point is 0, since it is on the $x$-axis. Consequently, to find the $x$-intercepts of the graph of an equation, set $y = 0$ and solve for $x$. For instance, in Example 3, the $x$-intercept of the graph of $y = -2x + 5$ (see Figure 2.3b) is found by setting $y = 0$ and solving for $x$:

$$0 = -2x + 5$$
$$2x = 5$$
$$x = \frac{5}{2}.$$

Similarly, a **y-intercept** of a graph is the $y$-coordinate of a point where the graph intersects the $y$-axis. The $x$-coordinate of this point is 0. The $y$-intercepts are found by setting $x = 0$ and solving for $y$. For example, the graph of $y = -2x + 5$ in Figure 2.3b has $y$-intercept 5. ✓₄

**✓ Checkpoint 4**

Find the $x$- and $y$-intercepts of the graphs of these equations.

**(a)** $3x + 4y = 12$

**(b)** $5x - 2y = 8$

| EXAMPLE 4 | Find the $x$- and $y$-intercepts of the graph of $y = x^2 - 2x - 8$, and |

sketch the graph.

**Solution** To find the $y$-intercept, set $x = 0$ and solve for $y$:

$$y = x^2 - 2x - 8 = 0^2 - 2 \cdot 0 - 8 = -8.$$

The $y$-intercept is $-8$. To find the $x$-intercept, set $y = 0$ and solve for $x$.

$$x^2 - 2x - 8 = y$$
$$x^2 - 2x - 8 = 0 \qquad \text{Set } y = 0.$$
$$(x + 2)(x - 4) = 0 \qquad \text{Factor.}$$
$$x + 2 = 0 \quad \text{or} \quad x - 4 = 0 \qquad \text{Zero-factor property}$$
$$x = -2 \quad \text{or} \quad x = 4$$

The $x$-intercepts are $-2$ and $4$. Now make a table, using both positive and negative values for $x$, and plot the corresponding points, as in Figure 2.4. These points suggest that the entire graph looks like Figure 2.5.

| $x$ | $x^2 - 2x - 8$ |
|-----|----------------|
| $-3$ | $7$ |
| $-2$ | $0$ |
| $-1$ | $-5$ |
| $0$ | $-8$ |
| $1$ | $-9$ |
| $3$ | $-5$ |
| $4$ | $0$ |
| $5$ | $7$ |

**Figure 2.4**

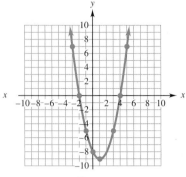

**Figure 2.5**

**EXAMPLE 5**    Sketch the graph of $y = \sqrt{x + 2}$.

**Solution**  Notice that $\sqrt{x + 2}$ is a real number only when $x + 2 \geq 0$—that is, when $x \geq -2$. Furthermore, $y = \sqrt{x + 2}$ is always nonnegative. Hence, all points on the graph lie on or above the $x$-axis and on or to the right of $x = -2$. Computing some typical values, we obtain the graph in Figure 2.6.

| $x$ | $\sqrt{x + 2}$ |
|-----|----------------|
| $-2$ | $0$ |
| $0$ | $\sqrt{2} \approx 1.414$ |
| $2$ | $2$ |
| $5$ | $\sqrt{7} \approx 2.646$ |
| $7$ | $3$ |
| $9$ | $\sqrt{11} \approx 3.317$ |

**Figure 2.6**

Example 3 shows that the solution of the equation $-2x + 5 = 0$ is the $x$-intercept of the graph of $y = -2x + 5$. Example 4 shows that the solutions of the equation $x^2 - 2x - 8 = 0$ are the $x$-intercepts of the graph $y = x^2 - 2x - 8$. Similar facts hold in the general case.

## Intercepts and Equations

The real solutions of a one-variable equation of the form

$$\text{expression in } x = 0$$

are the $x$-intercepts of the graph of

$$y = \text{same expression in } x.$$

## Graph Reading

Information is often given in graphical form, so you must be able to read and interpret graphs—that is, translate graphical information into statements in English.

EXAMPLE 6 **Finance** Newspapers and websites summarize activity of the S&P 500 Index in graphical form. The results for the 21 trading days for the month of December, 2016 are displayed in Figure 2.7. The first coordinate of each point on the graph is the trading day in December, and the second coordinate represents the closing price of the S&P 500 on that day. (Data from: www.morningstar.com.)

Figure 2.7

**(a)** What was the value of the S&P 500 Index on day 4 and day 15?

**Solution** If we examine the point when $x = 4$, we see that the $y$-value is a bit higher than the tick mark value of 2210. So we approximate the value for $y$ as 2212, and thus the value of the S&P 500 for trading day 4 of December 2016 was about 2212. Similarly, when $x = 15$, we see the $y$-value is roughly halfway between the tick mark values of 2260 and 2270. Thus, we approximate the value of the S&P 500 index to be 2265 on trading day 15.

**(b)** On which trading days was the value of the index above 2240?

**Solution** Look for points whose second coordinates are greater than 2240, that is, points that lie above the grid line for 2240 in Figure 2.7. The first coordinates of these points are the days when the index value was above 2240. We see these days occurred on days 5 through 20. ✓5

The next example deals with the basic business relationship that was introduced in Section 1.2:

$$\text{Profit} = \text{Revenue} - \text{Cost}.$$

✓**Checkpoint 5**

From Figure 2.7 determine when the S&P 500 had its highest point and its lowest point. What where the index values on those days?

Figure 2.8

EXAMPLE 7 **Revenue, Cost, and Profit** Monthly revenue and costs for the Webster Cell Phone Company are determined by the number $t$ of phones produced and sold, as shown in Figure 2.8.

**(a)** How many phones should be produced each month if the company is to make a profit (assuming that all phones produced are sold)?

**Solution** Profit is revenue minus cost, so the company makes a profit whenever revenue is greater than cost—that is, when the revenue graph is above the cost graph. Figure 2.8 shows that this occurs between $t = 12$ and $t = 48$—that is, when 12,000 to 48,000 phones are produced. If the company makes fewer than 12,000 phones, it will lose money (because costs will be greater than revenue.) It also loses money by making more than 48,000 phones. (One reason might be that high production levels require large amounts of overtime pay, which drives costs up too much.)

**(b)** Is it more profitable to make 40,000 or 44,000 phones?

**Solution** On the revenue graph, the point with first coordinate 40 has second coordinate of approximately 3.7, meaning that the revenue from 40,000 phones is about

3.7 million dollars. The point with first coordinate 40 on the cost graph is (40, 2), meaning that the cost of producing 40,000 phones is 2 million dollars. Therefore, the profit on 40,000 phones is about 3.7 − 2 = 1.7 million dollars. For 44,000 phones, we have the approximate points (44, 4) on the revenue graph and (44, 3) on the cost graph. So the profit on 44,000 phones is 4 − 3 = 1 million dollars. Consequently, it is more profitable to make 40,000 phones. ◢6

✓**Checkpoint 6**

In Example 7, find the profit from making

(a) 32,000 phones;

(b) 4000 phones.

## ⊞ Technology and Graphs

A graphing calculator or computer graphing program follows essentially the same procedure used when graphing by hand: The calculator selects a large number of $x$-values (95 or more), equally spaced along the $x$-axis, and plots the corresponding points, simultaneously connecting them with line segments. Calculator-generated graphs are generally quite accurate, although they may not appear as smooth as hand-drawn ones. The next example illustrates the basics of graphing on a graphing calculator. (Computer graphing software and cell phone apps operate similarly.)

**EXAMPLE 8** Use a graphing calculator to sketch the graph of the equation

$$2x^3 - 2y - 10x + 2 = 0.$$

**Solution** *First, set the viewing window*—the portion of the coordinate plane that will appear on the screen. Press the WINDOW key (labeled RANGE or PLOT-SETUP on some calculators) and enter the appropriate numbers, as in Figure 2.9 (which shows the screen from a TI-84+; other calculators are similar). Then the calculator will display the portion of the plane inside the dashed lines shown in Figure 2.10—that is, the points $(x, y)$ with $-9 \leq x \leq 9$ and $-6 \leq y \leq 6$.

In Figure 2.9, we have set Xscl = 2 and Yscl = 1, which means the **tick marks** on the $x$-axis are two units apart and the tick marks on the $y$-axis are one unit apart (as shown in Figure 2.10).

Figure 2.9          Figure 2.10

*Second, enter the equation to be graphed in the equation memory.* To do this, you must first solve the equation for $y$ (because a calculator accepts only equations of the form $y =$ expression in $x$):

$$2y = 2x^3 - 10x + 2$$
$$y = x^3 - 5x + 1.$$

Now press the Y = key (labeled SYMB on some calculators) and enter the equation, using the "variable key" for $x$. (This key is labeled X, T, $\theta$, $n$ or X, $\theta$, T or $x$-VAR, depending on the calculator.) Figure 2.11 shows the equation entered on a TI-84+; other calculators are similar. Now press GRAPH (or PLOT or DRW on some calculators), and obtain Figure 2.12.

**Figure 2.11**          **Figure 2.12**

*Finally, if necessary, change the viewing window to obtain a more readable graph.* It is difficult to see the *y*-intercept in Figure 2.12, so press WINDOW and change the viewing window (Figure 2.13); then press GRAPH to obtain Figure 2.14, in which the *y*-intercept at $y = 1$ is clearly shown. (It isn't necessary to reenter the equation.) ✓₇

          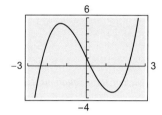

**Figure 2.13**          **Figure 2.14**

### ✓ Checkpoint 7

Use a graphing calculator to graph $y = 18x - 3x^3$ in the following viewing windows:

(a)  $-10 \le x \le 10$ and
     $-10 \le y \le 10$ with
     Xscl = 1, Yscl = 1;

(b)  $-5 \le x \le 5$ and
     $-20 \le y \le 20$ with
     Xscl = 1, Yscl = 5.

## ▦ Technology Tools

In addition to graphing equations, graphing calculators (and graphing software) provide convenient tools for solving equations and reading graphs. For example, when you have graphed an equation, you can readily determine the points the calculator plotted. Press **trace** (a cursor will appear on the graph), and use the left and right arrow keys to move the cursor along the graph. The coordinates of the point the cursor is on appear at the bottom of the screen.

Recall that the solutions of an equation, such as $x^3 - 5x + 1 = 0$, are the *x*-intercepts of the graph of $y = x^3 - 5x + 1$. (See the box on page 68.) A **graphical root finder** enables you to find these *x*-intercepts and thus to solve the equation.

### ✓ Checkpoint 8

Use a graphical root finder to approximate the third solution of the equation in Example 8.

| **EXAMPLE 9** | Use a graphical root finder to solve $x^3 - 5x + 1 = 0$. |

**Solution**  First, graph $y = x^3 - 5x + 1$. The *x*-intercepts of this graph are the solutions of the equation. To find these intercepts, look for "root" or "zero" in the appropriate menu.\* Check your instruction manual for the proper syntax. A typical root finder (see Figure 2.15) shows that two of the solutions (*x*-intercepts) are $x \approx .2016$ and $x \approx 2.1284$. For the third solution, see Checkpoint 8. ✓₈

          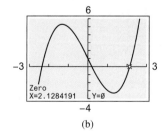

(a)          (b)

**Figure 2.15**

Many graphs have peaks and valleys (for instance, the graphs in Figure 2.15). A **maximum/minimum finder** provides accurate approximations of the locations of the "tops" of the peaks and the "bottoms" of the valleys.

---

\*CALC on TI-84+, GRAPH MATH on TI-86.

# 2.1 Exercises

*State the quadrant in which each point lies.*

**1.** $(1, -2), (-2, 1), (3, 4), (-5, -6)$

**2.** $(\pi, 2), (3, -\sqrt{2}), (4, 0), (-\sqrt{3}, \sqrt{3})$

*Determine whether the given ordered pair is a solution of the given equation. (See Example 2.)*

**3.** $(1, -3); 3x - y - 6 = 0$

**4.** $(2, -1); x^2 + y^2 - 6x + 8y = -15$

**5.** $(3, 4); (x - 2)^2 + (y + 2)^2 = 6$

**6.** $(1, -1); \dfrac{x^2}{2} + \dfrac{y^2}{3} = -4$

*Sketch the graph of each of these equations. (See Example 3.)*

**7.** $4y + 3x = 12$         **8.** $2x + 7y = 14$

**9.** $8x + 3y = 12$         **10.** $9y - 4x = 12$

**11.** $x = 2y + 3$          **12.** $x - 3y = 0$

*List the x-intercepts and y-intercepts of each graph.*

**13.**

**14.**

**15.**

**16.**
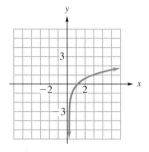

*Find the x-intercepts and y-intercepts of the graph of each equation. You need not sketch the graph. (See Example 4.)*

**17.** $3x + 4y = 12$        **18.** $x - 2y = 5$

**19.** $2x - 3y = 24$        **20.** $3x + y = 4$

**21.** $y = x^2 - 9$         **22.** $y = x^2 + 4$

**23.** $y = x^2 + x - 20$    **24.** $y = 5x^2 + 6x + 1$

**25.** $y = 2x^2 - 5x + 7$   **26.** $y = 3x^2 + 4x - 4$

*Sketch the graph of the equation. (See Examples 3–5.)*

**27.** $y = x^2$             **28.** $y = x^2 + 2$

**29.** $y = x^2 - 3$         **30.** $y = 2x^2$

**31.** $y = x^2 - 6x + 5$    **32.** $y = x^2 + 2x - 3$

**33.** $y = x^3$             **34.** $y = x^3 - 3$

**35.** $y = \sqrt{x + 4}$    **36.** $y = \sqrt{x - 2}$

*Sketch the graph of the information in the following tables. (See Example 1.)*

**37.** **Military Construction** The following table gives the amount of annual spending in the U.S. defense budget (in billions of dollars) for military construction. Let $x = 12$ correspond to the year 2012. (Data from: U.S. Office of Management and Budget.)

| Year | 2012 | 2013 | 2014 | 2015 | 2016 |
|------|------|------|------|------|------|
| Spending | 11.4 | 8.1 | 8.4 | 5.7 | 7.0 |

**38.** **Military Procurement** The following table gives the amount of annual spending in the U.S. defense budget (in billions of dollars) for military procurement. Let $x = 12$ correspond to the year 2012. (Data from: U.S. Office of Management and Budget.)

| Year | 2012 | 2013 | 2014 | 2015 | 2016 |
|------|------|------|------|------|------|
| Spending | 118 | 98 | 100 | 103 | 115 |

**39.** **Coal Mine Fatalities** The number of fatalities in coal mines in the United States for recent years is given in the following table. Let $x = 10$ correspond to the year 2010. (Data from: U.S. Mine Safety and Health Administration.)

| Year | 2010 | 2011 | 2012 | 2013 | 2014 | 2015 | 2016 |
|------|------|------|------|------|------|------|------|
| Number | 48 | 20 | 20 | 20 | 16 | 12 | 9 |

**40. All Mining Fatalities** The number of fatalities in all mining activities in the United States for recent years is given in the table. Let $x = 10$ correspond to the year 2010. (Data from: U.S. Mine Safety and Health Administration.)

| Year | 2010 | 2011 | 2012 | 2013 | 2014 | 2015 | 2016 |
|------|------|------|------|------|------|------|------|
| Number | 71 | 36 | 36 | 42 | 46 | 29 | 26 |

**ATM Fees** *An article in The Wall Street Journal on October 4, 2015, spoke of rising automated teller (ATM) fees. The graph below shows the average costs of noncustomer, out-of-network, and total fees from 1998 to 2016. (Data from: The Wall Street Journal and Bankrate.com.) (See Examples 6 and 7.)*

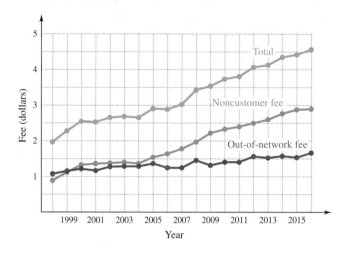

**41.** In what year were noncustomer and out-of-network fees almost the same amount?

**42.** In what year did the average out-of-network fee exceed $1.50 for the first time?

**43.** Which year was the last year that the average noncustomer fee was below $2.50?

**44.** Which year marked the first time that the average total fees rose above $3.00?

**Business** *Use the revenue and cost graphs for the Webster Cell Phone Company in Example 7 to do Exercises 45–48.*

**45.** Find the approximate cost of manufacturing the given number of phones.

  (a) 20,000   (b) 36,000   (c) 48,000

**46.** Find the approximate revenue from selling the given number of phones.

  (a) 12,000   (b) 24,000   (c) 36,000

**47.** Find the approximate profit from manufacturing the given number of phones.

  (a) 20,000   (b) 28,000   (c) 36,000

**48.** The company must replace its aging machinery with better, but much more expensive, machines. In addition, raw material prices increase, so that monthly costs go up by $250,000.

Owing to competitive pressure, phone prices cannot be increased, so revenue remains the same. Under these new circumstances, find the approximate profit from manufacturing the given number of phones.

  (a) 20,000   (b) 36,000   (c) 40,000

**Business** *The graph below gives the annual per-person retail availability of beef, chicken, and pork (in pounds) from 2007 to 2014. Use the graph to answer the following questions. (Data from: U.S. Department of Agriculture.)*

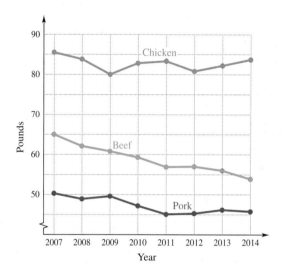

**49.** What is the approximate annual per-person retail availability of beef, chicken, and pork in 2012?

**50.** In what year was annual availability of beef the highest?

**51.** In what year was annual availability of chicken the lowest?

**52.** How much did the annual availability of beef decrease from 2007 to 2014?

**Netflix and Apple Stock Prices** *The graph below shows the opening share prices (in dollars) for Netflix, Inc., and Apple Inc. for the 21 trading days in December 2016. Use this graph to answer Exercises 53–62.*

**53.** What is the approximate share price of Apple stock on day 10?

**54.** What is the approximate share price of Netflix stock on day 8?

55. On which days is the share price of Netflix stock above $125?

56. On which days is the share price of Apple stock below $115?

57. On what day was the Apple share price at its highest, and what was the value?

58. On what day was the Netflix share price at its highest, and what was the value?

59. On what day was the Netflix share price at its lowest, and what was the value?

60. On what day was the Apple share price at its lowest, and what was the value?

61. Over the course of the month, what was the gain in share price for Netflix?

62. Over the course of the month, what was the gain in share price for Apple?

*Use a graphing calculator to find the graph of the equation. (See Example 8.)*

63. $y = x^2 + x + 1$

64. $y = 2 - x - x^2$

65. $y = (x - 3)^3$

66. $y = x^3 + 2x^2 + 2$

67. $y = x^3 - 3x^2 + x - 1$

68. $y = x^4 - 5x^2 - 2$

*Use a graphing calculator for Exercises 69–70.*

69. Graph $y = x^4 - 2x^3 + 2x$ in a window with $-3 \le x \le 3$. Is the "flat" part of the graph near $x = 1$ really a horizontal line segment? (*Hint*: Use the trace feature to move along the "flat" part and watch the $y$-coordinates. Do they remain the same [as they should on a horizontal segment]?)

70. (a) Graph $y = x^4 - 2x^3 + 2x$ in the **standard window** (the one with $-10 \le x \le 10$ and $-10 \le y \le 10$). Use the trace feature to approximate the coordinates of the lowest point on the graph.

    (b) Use a minimum finder to obtain an accurate approximation of the lowest point. How does this compare with your answer in part (a)?

*Use a graphing calculator to approximate all real solutions of the equation. (See Example 9.)*

71. $x^3 - 3x^2 + 5 = 0$

72. $x^3 + x - 1 = 0$

73. $2x^3 - 4x^2 + x - 3 = 0$

74. $6x^3 - 5x^2 + 3x - 2 = 0$

75. $x^5 - 6x + 6 = 0$

76. $x^3 - 3x^2 + x - 1 = 0$

✓**Checkpoint Answers**

1.

2. (a) and (c)

3.

4. (a) $x$-intercept 4, $y$-intercept 3

   (b) $x$-intercept 8/5, $y$-intercept $-4$

5. The highest was 2272 on Day 9

   The lowest was 2191 on Days 1 and 2

6. (a) About $1,200,000 (rounded)

   (b) About $-$500,000 (that is, a loss of $500,000)

7. (a)

   (b)

8. $x \approx -2.330059$

# 2.2   Equations of Lines

Straight lines, which are the simplest graphs, play an important role in a wide variety of applications. They are considered here from both a geometric and an algebraic point of view.

The key geometric feature of a nonvertical straight line is how steeply it rises or falls as you move from left to right. The "steepness" of a line can be represented numerically by a number called the *slope* of the line.

To see how the slope is defined, start with Figure 2.16, which shows a line passing through the two different points $(x_1, y_1) = (-5, 6)$ and $(x_2, y_2) = (4, -7)$. The difference in the two $x$-values,

$$x_2 - x_1 = 4 - (-5) = 9,$$

is called the **change in $x$.** Similarly, the **change in $y$** is the difference in the two $y$-values:

$$y_2 - y_1 = -7 - 6 = -13.$$

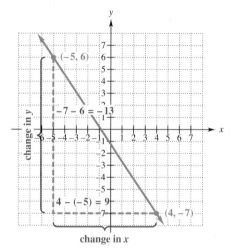

**Figure 2.16**

The **slope** of the line through the two points $(x_1, y_1)$ and $(x_2, y_2)$, where $x_1 \neq x_2$, is defined as the quotient of the change in $y$ and the change in $x$:

$$\text{slope} = \frac{\text{change in } y}{\text{change in } x} = \frac{y_2 - y_1}{x_2 - x_1}.$$

The slope of the line in Figure 2.16 is

$$\text{slope} = \frac{-7 - 6}{4 - (-5)} = -\frac{13}{9}.$$

Using similar triangles from geometry, we can show that the slope is independent of the choice of points on the line. That is, the same value of the slope will be obtained for *any* choice of two different points on the line.

**EXAMPLE 1**   Find the slope of the line through the points $(-6, 8)$ and $(5, 4)$.

**Solution**  Let $(x_1, y_1) = (-6, 8)$ and $(x_2, y_2) = (5, 4)$. Use the definition of slope as follows:

$$\text{slope} = \frac{y_2 - y_1}{x_2 - x_1} = \frac{4 - 8}{5 - (-6)} = \frac{-4}{11} = -\frac{4}{11}.$$

The slope can also by found by letting $(x_1, y_1) = (5, 4)$ and $(x_2, y_2) = (-6, 8)$. In that case,

$$\text{slope} = \frac{y_2 - y_1}{x_2 - x_1} = \frac{8 - 4}{-6 - 5} = \frac{4}{-11} = -\frac{4}{11},$$

which is the same answer.

✓**Checkpoint 1**

Find the slope of the line through the following pairs of points.

(a)  $(5, 9), (-5, -3)$

(b)  $(-4, 2), (-2, -7)$

⚠ **CAUTION** When finding the slope of a line, be careful to subtract the *x*-values and the *y*-values in the same order. For example, with the points (4, 3) and (2, 9), if you use $9 - 3$ for the numerator, you must use $2 - 4$ (*not* $4 - 2$) for the denominator.

**EXAMPLE 2**    Find the slope of the horizontal line in Figure 2.17.

**Solution** Every point on the line has the same *y*-coordinate, $-5$. Choose any two of them to compute the slope, say, $(x_1, y_1) = (-3, -5)$ and $(x_2, y_2) = (2, -5)$:

$$\text{slope} = \frac{-5 - (-5)}{2 - (-3)}$$

$$= \frac{0}{5}$$

$$= 0.$$

**Figure 2.17**

**EXAMPLE 3**    What is the slope of the vertical line in Figure 2.18?

**Solution** Every point on the line has the same *x*-coordinate, 4. If we attempt to compute the slope with two of these points, say, $(x_1, y_1) = (4, -2)$ and $(x_2, y_2) = (4, 1)$, we obtain

$$\text{slope} = \frac{1 - (-2)}{4 - 4}$$

$$= \frac{3}{0}.$$

**Figure 2.18**

Division by 0 is not defined, so the slope of this line is undefined.

The arguments used in Examples 2 and 3 work in the general case and lead to the following conclusion.

> The slope of every horizontal line is 0.
>
> The slope of every vertical line is undefined.

## Slope–Intercept Form

The slope can be used to develop an algebraic description of nonvertical straight lines. Assume that a line with slope $m$ has *y*-intercept $b$, so that it goes through the point $(0, b)$. (See Figure 2.19.) Let $(x, y)$ be any point on the line other than $(0, b)$. Using the definition of slope with the points $(0, b)$ and $(x, y)$ gives

$$m = \frac{y - b}{x - 0}$$

$$m = \frac{y - b}{x}$$

$$mx = y - b \qquad \text{Multiply both sides by } x.$$

$$y = mx + b. \qquad \text{Add } b \text{ to both sides. Reverse the equation.}$$

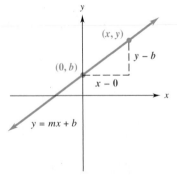

**Figure 2.19**

In other words, the coordinates of any point on the line satisfy the equation $y = mx + b$.

## Slope–Intercept Form

If a line has slope $m$ and $y$-intercept $b$, then it is the graph of the equation

$$y = mx + b.$$

This equation is called the **slope–intercept form** of the equation of the line.

**EXAMPLE 4**   Find an equation for the line with $y$-intercept $7/2$ and slope $-5/2$.

**Solution**   Use the slope–intercept form with $b = 7/2$ and $m = -5/2$:

$$y = mx + b$$
$$y = -\frac{5}{2}x + \frac{7}{2}.$$

✓ **Checkpoint 2**

Find an equation for the line with

(a)   $y$-intercept $-3$ and slope $2/3$;

(b)   $y$-intercept $1/4$ and slope $-3/2$.

**EXAMPLE 5**   Find the equation of the horizontal line with $y$-intercept 3.

**Solution**   The slope of the line is 0 (why?) and its $y$-intercept is 3, so its equation is

$$y = mx + b$$
$$y = 0x + 3$$
$$y = 3.$$

The argument in Example 5 also works in the general case.

If $k$ is a constant, then the graph of the equation $y = k$ is the horizontal line with $y$-intercept $k$.

**EXAMPLE 6**   Find the slope and $y$-intercept for each of the following lines.

**(a)**  $5x - 3y = 1$

**Solution**   Solve for $y$:

$$5x - 3y = 1$$
$$-3y = -5x + 1 \qquad \text{Subtract } 5x \text{ from both sides.}$$
$$y = \frac{5}{3}x - \frac{1}{3}. \qquad \text{Divide both sides by } -3.$$

This equation is in the form $y = mx + b$, with $m = 5/3$ and $b = -1/3$. So the slope is $5/3$ and the $y$-intercept is $-1/3$.

**(b)**  $-9x + 6y = 2$

**Solution**   Solve for $y$:

$$-9x + 6y = 2$$
$$6y = 9x + 2 \qquad \text{Add } 9x \text{ to both sides.}$$
$$y = \frac{3}{2}x + \frac{1}{3}. \qquad \text{Divide both sides by } 6.$$

✓ **Checkpoint 3**

Find the slope and $y$-intercept of

(a)   $x + 4y = 6$;

(b)   $3x - 2y = 1$.

The slope is $3/2$ (the coefficient of $x$), and the $y$-intercept is $1/3$.

The slope–intercept form can be used to show how the slope measures the steepness of a line. Consider the straight lines $A$, $B$, $C$, and $D$ given by the following equations, where each has $y$-intercept 0 and slope as indicated:

$$A: y = .5x; \qquad B: y = x; \qquad C: y = 3x; \qquad D: y = 7x.$$
$$\text{Slope } .5 \qquad \text{Slope } 1 \qquad \text{Slope } 3 \qquad \text{Slope } 7$$

For these lines, Figure 2.20a shows that the bigger the slope, the more steeply the line rises from left to right.

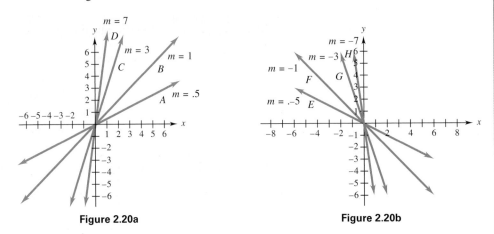

**Figure 2.20a**                    **Figure 2.20b**

If we consider straight lines, $E$, $F$, $G$, and $H$ given by the following equations

$$E: y = -.5x; \qquad F: y = -x; \qquad G: y = -3x; \qquad H: y = -7x.$$
$$\text{Slope } -.5 \qquad \text{Slope } -1 \qquad \text{Slope } -3 \qquad \text{Slope } -7$$

For these lines, Figure 2.20b shows the more negative the slope, the more steeply the line falls from left to right. ✔4

The preceding discussion and Checkpoint 4 may be summarized as follows.

✔ **Checkpoint 4**

(a) List the slopes of the following lines:

$E: y = -.3x; \quad F: y = -x;$

$G: y = -2x; \quad H: y = -5x.$

(b) Graph all four lines on the same set of axes.

(c) How are the slopes of the lines related to their steepness?

| Direction of Line (moving from left to right) | Slope |
|---|---|
| Upward | Positive (larger for steeper lines) |
| Horizontal | 0 |
| Downward | Negative (larger in absolute value for steeper lines) |
| Vertical | Undefined |

**EXAMPLE 7**     Sketch the graph of $x + 2y = 5$, and label the intercepts.

**Solution**  Find the $x$-intercept by setting $y = 0$ and solving for $x$:

$$x + 2 \cdot 0 = 5$$
$$x = 5.$$

The $x$-intercept is 5, and $(5, 0)$ is on the graph. The $y$-intercept is found similarly, by setting $x = 0$ and solving for $y$:

$$0 + 2y = 5$$
$$y = 5/2.$$

✔ **Checkpoint 5**

Graph the given lines and label the intercepts.

(a)  $3x + 4y = 12$

(b)  $5x - 2y = 8$

The $y$-intercept is $5/2$, and $(0, 5/2)$ is on the graph. The points $(5, 0)$ and $(0, 5/2)$ can be used to sketch the graph, as shown on the following page (Figure 2.21). ✔5

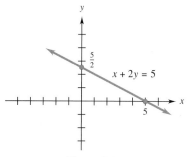

**Figure 2.21**

🖩 **TECHNOLOGY TIP** To graph a linear equation on a graphing calculator, you must first put the equation in slope–intercept form $y = mx + b$ so that it can be entered in the equation memory (called the Y = list on some calculators). Vertical lines cannot be graphed on most calculators.

## Slopes of Parallel and Perpendicular Lines

We shall assume the following facts without proof. The first one is a consequence of the fact that the slope measures steepness and that parallel lines have the same steepness.

> Two nonvertical lines are **parallel** whenever they have the same slope.
>
> Two nonvertical lines are **perpendicular** whenever the product of their slopes is $-1$.
>
> *Note:* An equivalent definition for **perpendicular lines** is nonvertical lines whose slopes are negative reciprocals of one another.

**EXAMPLE 8**    Determine whether each of the given pairs of lines are *parallel*, *perpendicular*, or *neither*.

**(a)** $2x + 3y = 5$ and $4x + 5 = -6y$.

**Solution** Put each equation in slope–intercept form by solving for $y$:

$$3y = -2x + 5 \qquad -6y = 4x + 5$$
$$y = -\frac{2}{3}x + \frac{5}{3} \qquad y = -\frac{2}{3}x - \frac{5}{6}.$$

In each case, the slope (the coefficient of $x$) is $-2/3$, so the lines are parallel. See Figure 2.22(a).

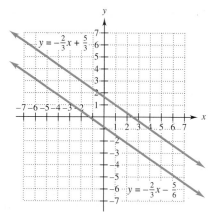

**Figure 2.22(a)**

**(b)** $3x = y + 7$ and $x + 3y = 4$.

**Solution** Put each equation in slope–intercept form to determine the slope of the associated line:

$$3x = y + 7 \qquad\qquad 3y = -x + 4$$

$$y = 3x - 7 \qquad\qquad y = -\frac{1}{3}x + \frac{4}{3}$$

$$\text{slope } 3 \qquad\qquad \text{slope } -1/3.$$

The slopes of 3 and $-1/3$ are negative reciprocals of one another. For this reason, and because $3(-1/3) = -1$, these lines are perpendicular. See Figure 2.22(b).

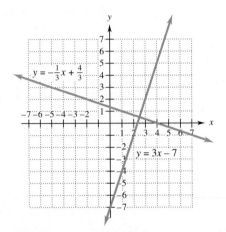

**Figure 2.22(b)**

**(c)** $x + y = 4$ and $x - 2y = 3$.

**Solution** Verify that the slope of the first line is $-1$ and the slope of the second is $1/2$. The slopes are not equal and their product is not $-1$, so the lines are neither parallel nor perpendicular. See Figure 2.22(c).

✓ **Checkpoint 6**

Tell whether the lines in each of the following pairs are *parallel, perpendicular*, or *neither*.

**(a)** $x - 2y = 6$ and $2x + y = 5$

**(b)** $3x + 4y = 8$ and $x + 3y = 2$

**(c)** $2x - y = 7$ and $2y = 4x - 5$

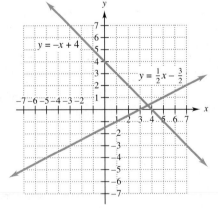

**Figure 2.22(c)**

⊞ **TECHNOLOGY TIP** Perpendicular lines may not appear perpendicular on a graphing calculator unless you use a *square window*—a window in which a one-unit segment on the *y*-axis is the same length as a one-unit segment on the *x*-axis. To obtain such a window on most calculators, use a viewing window in which the *y*-axis is about two-thirds as long as the *x*-axis. The SQUARE (or ZSQUARE) key in the ZOOM menu will change the current window to a square window by automatically adjusting the length of one of the axes.

## Point–Slope Form

The slope–intercept form of the equation of a line is usually the most convenient for graphing and for understanding how slopes and lines are related. However, it is not always the best way to *find* the equation of a line. In many situations (particularly in calculus), the slope and a point on the line are known and you must find the equation of the line. In such cases, the best method is to use the *point–slope form*, which we now explain.

Suppose that a line has slope $m$ and that $(x_1, y_1)$ is a point on the line. Let $(x, y)$ represent any other point on the line. Since $m$ is the slope, then, by the definition of slope,

$$\frac{y - y_1}{x - x_1} = m.$$

Multiplying both sides by $x - x_1$ yields

$$y - y_1 = m(x - x_1).$$

---

### Point–Slope Form

If a line has slope $m$ and passes through the point $(x_1, y_1)$, then

$$y - y_1 = m(x - x_1)$$

is the **point–slope form** of the equation of the line.

---

**Figure 2.23**

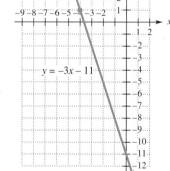

**Figure 2.24**

✓ **Checkpoint 7**

Find both the point–slope and the slope–intercept form of the equation of the line having the given slope and passing through the given point.

**(a)** $m = -3/5, (5, -2)$

**(b)** $m = 1/3, (6, 8)$

**EXAMPLE 9**    Find the equation of the line satisfying the given conditions.

**(a)**  Slope 2; the point $(1, 3)$ is on the line.

**Solution**  Use the point–slope form with $m = 2$ and $(x_1, y_1) = (1, 3)$. Substitute $x_1 = 1, y_1 = 3$, and $m = 2$ into the point–slope form of the equation.

$$y - y_1 = m(x - x_1)$$
$$y - 3 = 2(x - 1). \qquad \text{Point–slope form}$$

For some purposes, this form of the equation is fine; in other cases, you may want to rewrite it in the slope–intercept form.

Using algebra, we obtain the slope–intercept form of this equation:

$$y - 3 = 2(x - 1)$$
$$y - 3 = 2x - 2 \qquad \text{Distributive property}$$
$$y = 2x + 1. \qquad \text{Add 3 to each side to obtain slope-intercept form.}$$

See Figure 2.23 for the graph.

**(b)**  Slope $-3$; the point $(-4, 1)$ is on the line.

**Solution**  Use the point–slope form with $m = -3$ and $(x_1, y_1) = (-4, 1)$:

$$y - y_1 = m(x - x_1)$$
$$y - 1 = -3[x - (-4)]. \qquad \text{Point–slope form}$$

Using algebra, we obtain the slope–intercept form of this equation:

$$y - 1 = -3(x + 4)$$
$$y - 1 = -3x - 12 \qquad \text{Distributive property}$$
$$y = -3x - 11. \qquad \text{Slope–intercept form}$$

See Figure 2.24. ✓₇

The point–slope form can also be used to find an equation of a line, given two different points on the line. The procedure is shown in the next example.

**EXAMPLE 10**    Find an equation of the line through $(5, 4)$ and $(-10, -2)$.

**Solution**  Begin by using the definition of the slope to find the slope of the line that passes through the two points:

$$\text{slope} = m = \frac{-2 - 4}{-10 - 5} = \frac{-6}{-15} = \frac{2}{5}.$$

Use $m = 2/5$ and either of the given points in the point–slope form. If $(x_1, y_1) = (5, 4)$, then

$$y - y_1 = m(x - x_1)$$

$$y - 4 = \frac{2}{5}(x - 5) \qquad \text{Let } y_1 = 4, m = \frac{2}{5}, \text{ and } x_1 = 5.$$

$$y - 4 = \frac{2}{5}x - \frac{10}{5} \qquad \text{Distributive property}$$

$$y = \frac{2}{5}x + 2. \qquad \text{Add 4 to both sides and simplify.}$$

 **Checkpoint 8**

Find an equation of the line through

**(a)**   $(2, 3)$ and $(-4, 6)$;

**(b)**   $(-8, 2)$ and $(3, -6)$.

See Figure 2.25. Check that the results are the same when $(x_1, y_1) = (-10, -2)$. ✔8

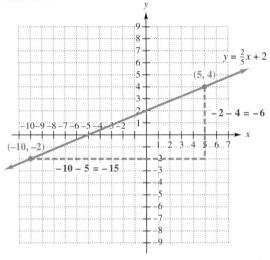

**Figure 2.25**

## Vertical Lines

The equation forms we just developed do not apply to vertical lines, because the slope is not defined for such lines. However, vertical lines can easily be described as graphs of equations.

**EXAMPLE 11**    Find the equation whose graph is the vertical line in Figure 2.26.

**Solution**  Every point on the line has $x$-coordinate $-1$ and hence has the form $(-1, y)$. Thus, every point is a solution of the equation $x + 0y = -1$, which is usually written simply as $x = -1$. Note that $-1$ is the $x$-intercept of the line.

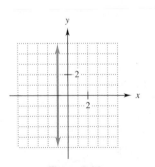

**Figure 2.26**

The argument in Example 11 also works in the general case.

If $k$ is a constant, then the graph of the equation $x = k$ is the vertical line with $x$-intercept $k$.

## Linear Equations

An equation in two variables whose graph is a straight line is called a **linear equation.** Linear equations have a variety of forms, as summarized in the following table.

| Equation | Description |
|---|---|
| $x = k$ | **Vertical line,** $x$-intercept $k$, no $y$-intercept (unless $x = 0$), undefined slope |
| $y = k$ | **Horizontal line,** $y$-intercept $k$, no $x$-intercept (unless $y = 0$), slope 0 |
| $y = mx + b$ | **Slope–intercept form,** slope $m$, $y$-intercept $b$ |
| $y - y_1 = m(x - x_1)$ | **Point–slope form,** slope $m$, the line passes through $(x_1, y_1)$ |
| $ax + by = c$ | **General form.** If $a \neq 0$ and $b \neq 0$, the line has $x$-intercept $c/a$, $y$-intercept $c/b$, and slope $-a/b$. |

Note that every linear equation can be written in general form. For example, $y = 4x - 5$ can be written in general form as $4x - y = 5$, and $x = 6$ can be written in general form as $x + 0y = 6$.

## Applications

Many relationships are linear or almost linear, so that they can be approximated by linear equations.

**EXAMPLE 12**    **Emerald Prices**  The price of emeralds has surged in recent years. The average price per carat for high-quality rough emeralds can be approximated by the linear equation

$$y = 10.2x - 84.4,$$

where $x = 9$ corresponds to the year 2009. The graph appears in Figure 2.27. (Data from: *The Wall Street Journal.*)

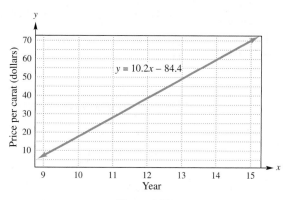

**Figure 2.27**

**(a)** What was the approximate price per carat for emeralds in 2014?

**Solution** Substitute $x = 14$ in the equation and compute $y$:

$$y = 10.2x - 84.4$$
$$y = 10.2(14) - 84.4 = 58.4.$$

The approximate price per carat was $58.40.

**(b)** What was the first full year that the price exceeded $50?

**Solution** Substitute $y = 50$ in the equation and solve for $x$:

$$50 = 10.2x - 84.4$$
$$50 + 84.4 = 10.2x \qquad \text{Add 84.4 to each side.}$$
$$x = \frac{50 + 84.4}{10.2} \qquad \text{Divide each side by 10.2 and reverse the equation.}$$
$$x \approx 13.2.$$

This implies that the first full year with a price $50 or higher was 2014. Note that we *round up* the year to 2014 to find the first time when the price was $50 for the *full year* (if we substitute $x = 13$ into the equation, we get $y = 10.2(13) - 84.4 = 48.2$, so the price was not $50 for the full year in 2013).

---

**EXAMPLE 13**   **Video Streaming** According to data from *The Wall Street Journal*, the online video audience in China was approximately 200 million people in the year 2008. In 2015, the audience had grown to 461.2 million.

**(a)** Find a linear equation for these data.

**Solution** Measure time along the $x$-axis and viewership along the $y$-axis. Then the $x$-coordinate of each point is a year and the $y$-coordinate is the viewership in that year. For convenience, let $x = 8$ correspond to 2008. The given data points are $(8, 200)$ and $(15, 461.2)$. The slope of the line joining these two points is

$$\frac{461.2 - 200}{15 - 8} = \frac{261.2}{7} \approx 37.3.$$

We can now use the point–slope form of a line with the point $(8, 200)$ to obtain the equation of the line:

$$y - 200 = 37.3(x - 8) \qquad \text{Point–slope form}$$
$$y - 200 = 37.3x - 298.4 \qquad \text{Distributive property}$$
$$y = 37.3x - 98.4. \qquad \text{Add 200 to each side.}$$

**Figure 2.28**

Figure 2.28 shows the derived equation.

**(b)** Use this equation to estimate the viewership in China in 2014.

**Solution** Because 2014 corresponds to $x = 14$, let $x = 14$ in the equation part of (a). Then

$$y = 37.3(14) - 98.4 = 423.8.$$

The viewership was approximately 423.8 million.

**(c)** Assuming the equation remains valid beyond 2015, estimate the viewership in 2019.

**Solution** The year 2019 corresponds to $x = 19$, so the viewership is

$$y = 37.3(19) - 98.4 = 610.3.$$

The viewership is estimated to be 610.3 million. ✓ 9

✓ **Checkpoint 9**

The online viewership in India was estimated to be 61.5 million in 2014. In 2016, the online viewership was estimated to be 81.7 million. (Data from: Statistica.com and cxotoday.com.)

**(a)** Let $x = 14$ correspond to the year 2014, and find a linear equation for the given data.

**(b)** Assuming that your equation remains accurate, estimate the online viewership in India in 2020.

# 2.2 Exercises

*Find the slope of the given line, if it is defined. (See Examples 1–3.)*

1. The line through (2, 5) and (0, 8)

2. The line through (9, 0) and (12, 12)

3. The line through (−4, 14) and (3, 0)

4. The line through (−5, −2) and (−4, 11)

5. The line through the origin and (−4, 10)

6. The line through the origin and (8, −2)

7. The line through (−1, 4) and (−1, 6)

8. The line through (−3, 5) and (2, 5)

*Find an equation of the line with the given y-intercept and slope m. (See Examples 4 and 5.)*

9. 5, m = 4

10. −3, m = −7

11. 1.5, m = −2.3

12. −4.5, m = 2.5

13. 4, m = −3/4

14. −3, m = 4/3

*Find the slope m and the y-intercept b of the line whose equation is given. (See Example 6.)*

15. $2x - y = 9$

16. $x + 2y = 7$

17. $6x = 2y + 4$

18. $4x + 3y = 24$

19. $6x - 9y = 16$

20. $4x + 2y = 0$

21. $2x - 3y = 0$

22. $y = 7$

23. $x = y - 5$

24. On one graph, sketch six straight lines that meet at a single point and satisfy this condition: one line has slope 0, two lines have positive slope, two lines have negative slope, and one line has undefined slope.

25. For which of the line segments in the figure is the slope

    (a) largest?

    (b) smallest?

    (c) largest in absolute value?

    (d) closest to 0?

26. Match each equation with the line that most closely resembles its graph. (*Hint*: Consider the signs of *m* and *b* in the slope–intercept form.)

    (a) $y = 3x + 2$

    (b) $y = -3x + 2$

    (c) $y = 3x - 2$

    (d) $y = -3x - 2$

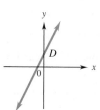

*Sketch the graph of the given equation and label its intercepts. (See Example 7.)*

27. $2x - y = -2$

28. $2y + x = 4$

29. $2x + 3y = 4$

30. $-5x + 4y = 3$

31. $4x - 5y = 2$

32. $3x + 2y = 8$

*Determine whether each pair of lines is parallel, perpendicular, or neither. (See Example 8.)*

33. $4x - 3y = 6$ and $3x + 4y = 8$

34. $2x - 5y = 7$ and $15y - 5 = 6x$

35. $3x + 2y = 8$ and $6y = 5 - 9x$

36. $x - 3y = 4$ and $y = 1 - 3x$

37. $4x = 2y + 3$ and $2y = 2x + 3$

38. $2x - y = 6$ and $x - 2y = 4$

39. (a) Find the slope of each side of the triangle with vertices (9, 6), (−1, 2), and (1, −3).

    (b) Is this triangle a right triangle? (*Hint*: Are two sides perpendicular?)

40. (a) Find the slope of each side of the quadrilateral with vertices (−5, −2), (−3, 1), (3, 0), and (1, −3).

    (b) Is this quadrilateral a parallelogram? (*Hint*: Are opposite sides parallel?)

*Find an equation of the line with slope m that passes through the given point. Put the answer in slope–intercept form. (See Example 9.)*

41. $(-3, 2), m = -2/3$

42. $(-5, -2), m = 4/5$

43. $(2, 3), m = 3$

44. $(3, -4), m = -1/4$

45. $(10, 1), m = 0$

46. $(-3, -9), m = 0$

47. $(-2, 12)$, undefined slope

48. $(1, 1)$, undefined slope

*Find an equation of the line that passes through the given points. (See Example 10.)*

49. (−1, 1) and (2, 7)

50. (2, 5) and (0, 6)

51. (1, 2) and (3, 9)

52. (−1, −2) and (2, −1)

*Find an equation of the line satisfying the given conditions.*

53. Through the origin with slope 5

54. Through the origin and horizontal

55. Through (6, 8) and vertical

56. Through (7, 9) and parallel to $y = 6$

57. Through (3, 4) and parallel to $4x - 2y = 5$

58. Through (6, 8) and perpendicular to $y = 2x - 3$

59. $x$-intercept 6; $y$-intercept $-6$

60. Through $(-5, 2)$ and parallel to the line through (1, 2) and (4, 3)

61. Through $(-1, 3)$ and perpendicular to the line through (0, 1) and (2, 3)

62. $y$-intercept 3 and perpendicular to $2x - y + 6 = 0$

**Depreciation** *The lost value of equipment over a period of time is called depreciation. The simplest method for calculating depreciation is straight-line depreciation. The annual straight-line depreciation D of an item that cost x dollars with a useful life of n years is $D = (1/n)x$. Find the depreciation for items with the given characteristics.*

63. Cost: $15,965; life 12 yr

64. Cost: $41,762; life 15 yr

65. Cost: $201,457; life 30 yr

66. Ral Corp. has an incentive compensation plan under which a branch manager receives 10% of the branch's income after deduction of the bonus, but before deduction of income tax. The income of a particular branch before the bonus and income tax was $165,000. The tax rate was 30%. What was the amount of the bonus?

**Light Beer Sales** *Data from The Wall Street Journal indicate that the percent change in beer shipments since the year 2007 for Miller Lite beer can be approximated by*

$$y = -3.9x + 27.3,$$

*where $x = 7$ corresponds to the year 2007. Find the percent change since 2007 that occurred by the following years.*

67. 2010

68. 2014

*Assume the model remains accurate, and find the first full year in which the percent change reaches the following values.*

69. $-32\%$

70. $-40\%$

**Global Malaria Cases** *Data from The Wall Street Journal indicate the number of global malaria cases has risen sharply since the year 2000. The equation*

$$y = 5.6x + 52$$

*approximates the number of global malaria cases y (in millions), where $x = 0$ corresponds to the year 2000. Find the number of global malaria cases in the following years.*

71. 2007

72. 2015

*Assume the model remains accurate, and find the first full year in which the number of global malaria cases per year achieves the following values.*

73. 150 million

74. 175 million

**Olympic 5000-Meters** *The accompanying graph shows the winning time (in minutes) at the Olympic Games from 1952 to 2016 for the men's 5000-meter run, together with a linear approximation of these data. (Data from: The World Almanac and Book of Facts: 2017.)*

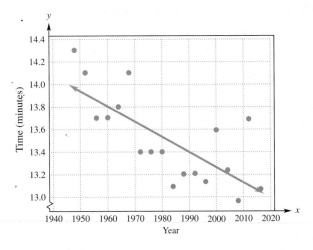

*The equation for the linear approximation is $y = -.0135x + 40.32$, where x is the actual year.*

75. What does the linear approximation predict for the winning time for the 2016 Olympics for this event? The actual time was 13.06 minutes. How close was the mathematical model's estimate?

76. What does the slope of the line represent? Why is it negative?

77. **Luxury Handbags** Michael Kors now dominates the handbag market in North America. In 2011, the company had 5% of the market; in 2015, it had 23%. (Data from: *The Wall Street Journal.*)

    (a) Write a linear equation expressing the percentage of the luxury handbag market $y$ that Michael Kors had $x$ years after 2000.

    (b) What was the percentage of the market that Michael Kors had in 2013?

    (c) If the trend continues, what will be the first full year in which Michael Kors had 30% of the market?

78. **Cocoa Production** In 2005, global cocoa bean production was approximately 3.3 million metric tons. In 2015, production was approximately 4.1 million metric tons. (Data from: *The Wall Street Journal.*)

    (a) Write a linear equation expressing cocoa bean production $y$ in terms of the number of years $x$ after 2000.

    (b) What was the cocoa bean production in 2012?

    (c) If the trend continues, what will be the first full year in which cocoa bean production exceeds 5 million metric tons?

**79. Employee Health Costs** In the year 2010, the average annual cost to employees for employer family health coverage was approximately $3500. In 2015, that amount was approximately $5000. (Data from: *The Wall Street Journal*.)

**(a)** Write a linear equation expressing the average annual cost to employees $y$ in terms of the number of years $x$ after 2000.

**(b)** What was the average annual cost in 2014?

**(c)** If the trend continues, what will be the first full year in which costs exceed $6000?

**80. Employer Health Costs** In the year 2010, the average annual cost to employers for employer family health coverage was approximately $10,000. In 2015, that amount was approximately $17,000. (Data from: *The Wall Street Journal*.)

**(a)** Write a linear equation expressing the average annual cost to employers $y$ in terms of the number of years $x$ after 2000.

**(b)** What was the average annual cost in 2012?

**(c)** If the trend continues, what will be the first full year in which costs exceed $25,000?

✓**Checkpoint Answers**

**1. (a)** $6/5$  **(b)** $-9/2$

**2. (a)** $y = \frac{2}{3}x - 3$  **(b)** $y = -\frac{3}{2}x + \frac{1}{4}$

**3. (a)** Slope $-1/4$; $y$-intercept $3/2$

  **(b)** Slope $3/2$; $y$-intercept $-1/2$

**4. (a)** Slope of $E = -.3$; slope of $F = -1$; slope of $G = -2$; slope of $H = -5$.

**(b)**

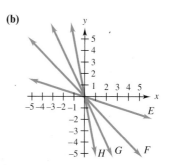

**(c)** The larger the slope in absolute value, the more steeply the line falls from left to right.

**5. (a)**  **(b)**

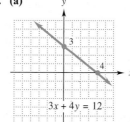

**6. (a)** Perpendicular **(b)** Neither  **(c)** Parallel

**7. (a)** $y + 2 = -\frac{3}{5}(x - 5)$; $y = -\frac{3}{5}x + 1$.

  **(b)** $y - 8 = \frac{1}{3}(x - 6)$; $y = \frac{1}{3}x + 6$.

**8. (a)** $y = -\frac{1}{2}x + 4$  **(b)** $y = -\frac{8}{11}x - \frac{42}{11}$

**9. (a)** $y = 10.1x - 79.9$  **(b)** $122.1$ million

# 2.3  Linear Models

In business and science, it is often necessary to make judgments on the basis of data from the past. For instance, a stock analyst might use a company's profits in previous years to estimate the next year's profits. Or a life insurance company might look at life expectancies of people born in various years to predict how much money it should expect to pay out in the next year.

In such situations, the available data is used to construct a mathematical model, such as an equation or a graph, which is used to approximate the likely outcome in cases where complete data is not available. In this section, we consider applications in which the data can be modeled by a linear equation.

The simplest way to construct a linear model is to use the line determined by two of the data points, as illustrated in the following example.

**EXAMPLE 1**  **Full-Time Faculty Members** The number of full-time faculty members at four-year colleges and universities (in thousands) in selected years is shown in the following table. (Data from: U.S. National Center for Education Statistics.)

| Year | 2005 | 2007 | 2009 | 2011 | 2013 |
|---|---|---|---|---|---|
| **Number of Faculty Members** | 917 | 990 | 1038 | 1115 | 1152 |

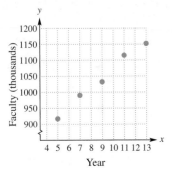

**Figure 2.29**

**(a)** Let $x = 5$ correspond to the year 2005. Plot the points $(x, y)$ where $x$ is the year and $y$ is the number of full-time faculty members.

**Solution** The data points are $(5, 917)$, $(7, 990)$, $(9, 1038)$, $(11, 1115)$, and $(13, 1152)$, as shown in Figure 2.29.

**(b)** Use the data points $(9, 1038)$ and $(11, 1115)$ to find a line that models these data.

**Solution** The slope of the line through $(9, 1038)$ and $(11, 1115)$ is $\dfrac{1115 - 1038}{11 - 9} = \dfrac{77}{2} = 38.5$. Using the point $(9, 1038)$ and the slope of $38.5$, we find the equation of this line is

$$y - 1038 = 38.5(x - 9) \qquad \text{Point–slope form for the equation of a line}$$
$$y - 1038 = 38.5x - 346.5 \qquad \text{Distributive property}$$
$$y = 38.5x + 691.5. \qquad \text{Slope–intercept form}$$

The line and the data points are shown in Figure 2.30. Although the line fits the two points that we used to calculate the slope perfectly, it seems to underestimate the first two points and overestimate the last point.

**(c)** Use the data points $(5, 917)$ and $(7, 990)$ to find a line that models these data.

**Solution** The slope of the line through $(5, 917)$ and $(7, 990)$ is $\dfrac{990 - 917}{7 - 5} = \dfrac{73}{2} = 36.5$. Using the point $(5, 917)$ and the slope of $36.5$, we find the equation of this line is

$$y - 917 = 36.5(x - 5) \qquad \text{Point–slope form for the equation of a line}$$
$$y - 917 = 36.5x - 182.5 \qquad \text{Distributive property}$$
$$y = 36.5x + 734.5. \qquad \text{Slope–intercept form}$$

The line and the data points are shown in Figure 2.31. Although the line fits the first two points perfectly, it seems to overestimate all three latter points. ✓ ₁

**Figure 2.30**

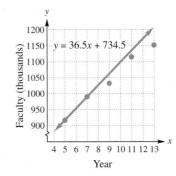

**Figure 2.31**

✓ **Checkpoint 1**

Use the points $(5, 917)$ and $(9, 1038)$ to find another model for the data in Example 1.

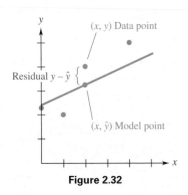

**Figure 2.32**

Opinions may vary as to which of the lines in Example 1 best fits the data. To make a decision, we might try to measure the amount of error in each model. One way to do this is to compute the difference between the number of faculty $y$ and the amount $\hat{y}$ given by the model. If the data point is $(x, y)$ and the corresponding point on the line is $(x, \hat{y})$, then the difference $y - \hat{y}$ measures the error in the model for that particular value of $x$. The number $y - \hat{y}$ is called a **residual**. As shown in Figure 2.32, the residual $y - \hat{y}$ is the vertical distance from the data point to the line (positive when the data point is above the line, negative when it is below the line, and 0 when it is on the line).

One way to determine how well a line fits the data points is to compute the sum of its residuals—that is, the sum of the individual errors. Unfortunately, however, the sum of the residuals of two different lines might be equal, thwarting our effort to decide which is the better fit. Furthermore, the residuals may sum to 0, which doesn't mean that there is no error, but only that the positive and negative errors (which might be quite large) add up to 0. (See Exercise 11 at the end of this section for an example.)

To avoid this difficulty, mathematicians use the sum of the *squares* of the residuals to measure how well a line fits the data points. When the sum of the squares is used, a smaller sum means a smaller overall error and hence a better fit. The error is 0 only when all the data points lie on the line (a perfect fit).

**EXAMPLE 2** **Full-Time Faculty Members** Two linear models for the number of full-time faculty members at four-year colleges and universities were constructed in Example 1:

$$y = 38.5x + 691.5 \quad \text{and} \quad y = 36.5x + 734.5.$$

For each model, determine the five residuals, square each residual, and sum the squares of the residuals.

**Solution** The information for each model is summarized in the following tables.

$$y = 38.5x + 691.5$$

| Data Point $(x, y)$ | Model Point $(x, \hat{y})$ | Residual $y - \hat{y}$ | Squared Residual $(y - \hat{y})^2$ |
|---|---|---|---|
| (5, 917) | (5, 884) | 33 | 1089 |
| (7, 990) | (7, 961) | 29 | 841 |
| (9, 1038) | (9, 1038) | 0 | 0 |
| (11, 1115) | (11, 1115) | 0 | 0 |
| (13, 1152) | (13, 1192) | −40 | 1600 |
| | | | Sum = 3530 |

$$y = 36.5x + 734.5$$

| Data Point $(x, y)$ | Model Point $(x, \hat{y})$ | Residual $y - \hat{y}$ | Squared Residual $(y - \hat{y})^2$ |
|---|---|---|---|
| (5, 917) | (5, 917) | 0 | 0 |
| (7, 990) | (7, 990) | 0 | 0 |
| (9, 1038) | (9, 1063) | −25 | 625 |
| (11, 1115) | (11, 1136) | −21 | 441 |
| (13, 1152) | (13, 1209) | −57 | 3249 |
| | | | Sum = 4315 |

According to this measure of the error, the line $y = 38.5x + 691.5$ is a better fit for these data because the sum of the squares of its residuals is smaller than the sum of the squares of the residuals for $y = 36.5x + 734.5$. ✓ 2

✓ **Checkpoint 2**

Another model for the data in Examples 1 and 2 is $y = 37x + 702$. Use this line to find

**(a)** the residuals and

**(b)** the sum of the squares of the residuals.

**(c)** Does this line fit the data better than the two lines in Example 2?

## ⊞ Linear Regression (Optional)*

Mathematical techniques from multivariable calculus can be used to prove the following result.

> For any set of data points, there is one, and only one, line for which the sum of the squares of the residuals is as small as possible.

---

*Examples 3–6 require either a graphing calculator or a spreadsheet program.

This *line of best fit* is called the **least-squares regression line,** and the computational process for finding its equation is called **linear regression.** Linear-regression formulas are quite complicated and require a large amount of computation. Fortunately, most graphing calculators and spreadsheet programs can do linear regression quickly and easily.

> 📝 **NOTE**  The process outlined here works for most TI graphing calculators. Other graphing calculators and spreadsheet programs operate similarly, but check your instruction manual.

**EXAMPLE 3**   **Full-Time Faculty Members**  Recall the number of full-time faculty members at four-year colleges and universities (in thousands) in selected years was as follows.

| Year | 2005 | 2007 | 2009 | 2011 | 2013 |
|---|---|---|---|---|---|
| **Number of Faculty Members** | 917 | 990 | 1038 | 1115 | 1152 |

Use a graphing calculator or spreadsheet to do the following:

**(a)**  Plot the data points with $x = 5$ corresponding to the year 2005.

**Solution**  The data points are (5, 917), (7, 990), (9, 1038), (11, 1115), and (13, 1152). Press STAT EDIT to bring up the statistics editor. Enter the $x$-coordinates as list $L_1$ and the corresponding $y$-coordinates as $L_2$, as shown in Figure 2.33. To plot the data points, go to the STAT PLOT menu, choose a plot (here it is Plot 1), choose ON, and enter the lists $L_1$ and $L_2$ as shown in Figure 2.34. Then set the viewing window as usual and press GRAPH to produce the plot in Figure 2.35.

**Figure 2.33**

**Figure 2.34**

**Figure 2.35**

**(b)**  Find the least squares regression line for these data.

**Solution**  Go to the STAT CALC menu and choose LINREG, which returns you to the home screen. As shown in Figure 2.36, enter the list names and the place where the equation of the regression line should be stored (here, $Y_1$ is chosen; it is on the VARS Y-VARS FUNCTION menu); then press ENTER. Figure 2.37 shows that the equation of the regression line is

$$y = 29.75x + 774.65$$

✓**Checkpoint 3**

Use the least-squares regression line $y = 29.75x + 774.65$ and the data points of Example 3 to find

**(a)**  the residuals and

**(b)**  the sum of the squares of the residuals.

**(c)**  How does this line compare with those in Example 1 and Checkpoint 2?

**Figure 2.36**

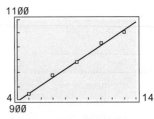

**Figure 2.37**

Figure 2.38

**(c)**  Graph the data points and the regression line on the same screen.

**Solution**  Press GRAPH to see the line plotted with the data points (Figure 2.38).

**EXAMPLE 4** **Full-Time Faculty Members** The following table gives the number (in thousands) of full-time faculty at two-year colleges. (Data from: U.S. National Center for Education Statistics.)

| Year | 2005 | 2007 | 2009 | 2011 | 2013 |
|---|---|---|---|---|---|
| **Number of Faculty** | 373 | 380 | 400 | 408 | 394 |

**(a)** Let $x = 5$ correspond to the year 2005. Use a graphing calculator or spreadsheet program to find the least-squares regression line that models the data in the table.

**Solution** The data points are (5, 373), (7, 380), (9, 400), (11, 408), and (13, 394). Enter the $x$-coordinates as list $L_1$ and the corresponding $y$-coordinates as list $L_2$ in a graphing calculator and then find the regression line as in Figure 2.39.

**(b)** Plot the data points and the regression line on the same screen.

**Solution** See Figure 2.40, which shows that the line is a reasonable model for the data.

**(c)** Assuming the trend continues, estimate the number of faculty in the year 2020.

**Solution** The year 2020 corresponds to $x = 20$. Substitute $x = 20$ into the regression-line equation:

$$y = 3.5(20) + 359.5 = 429.5.$$

This model estimates that the number of full-time faculty at two-year colleges will be approximately 429,500 in the year 2020.

**LinReg**
y=ax+b
a=3.5
b=359.5
$r^2$=.5946601942
r=.7711421362

**Figure 2.39**

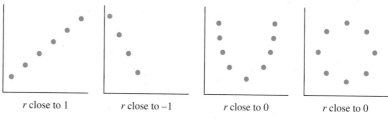

**Figure 2.40**

✓ **Checkpoint 4**

Using only the data from 2005, 2007, and 2009 in Example 4, find the equation of the least-squares regression line. Round the coefficients to two decimal places.

## Correlation

Although the "best fit" line can always be found by linear regression, it may not be a good model. For instance, if the data points are widely scattered, no straight line will model the data accurately. We calculate a value called the **correlation coefficient** to assess the fit of the regression line to the scatterplot. It measures how closely the data points fit the regression line and thus indicates how good the regression line is for predictive purposes.

The correlation coefficient $r$ is always between $-1$ and 1. When $r = \pm 1$, the data points all lie on the regression line (a perfect fit). When the absolute value of $r$ is close to 1, the line fits the data quite well, and when $r$ is close to 0, the line is a poor fit for the data (but some other curve might be a good fit). Figure 2.41 shows how the value of $r$ varies, depending on the pattern of the data points.

| $r$ close to 1 | $r$ close to $-1$ | $r$ close to 0 | $r$ close to 0 |
|---|---|---|---|

**Figure 2.41**

**EXAMPLE 5** **Unemployment Rate** The percentage of unemployed workers in the U.S. labor force in January of recent years is shown in the table. (Data from: U.S. Department of Labor.)

**Figure 2.42**

**Figure 2.43**

✓**Checkpoint 5**

Using only the data from 2010 and later in Example 5, find

**(a)** the equation of the least-squares regression line and

**(b)** the correlation coefficient.

**(c)** How well does this line fit the data?

| Year | Rate | Year | Rate |
|------|------|------|------|
| 2006 | 4.7 | 2012 | 8.3 |
| 2007 | 4.6 | 2013 | 8.0 |
| 2008 | 5.0 | 2014 | 6.6 |
| 2009 | 7.8 | 2015 | 5.7 |
| 2010 | 9.8 | 2016 | 4.9 |
| 2011 | 9.7 | | |

Determine whether or not a linear equation is a good model for these data.

**Solution** Let $x = 6$ correspond to the year 2006, and plot the data points (6, 4.7), and so on, either by hand or with a graphing calculator, as in Figure 2.42. They do not form a linear pattern (because unemployment rates tend to rise and fall). Alternatively, you can compute the regression equation for these data, as in Figure 2.43. Figure 2.43 shows the correlation coefficient is $r \approx .138$, which is a relatively low value. ✓5

**EXAMPLE 6**   **Higher Education Enrollment** The population (in thousands) and enrollment in all higher education institutions (in thousands) for 10 states in the year 2013 is given in the table below. (Data from: U.S. Center for Education Statistics and the U.S. Census Bureau.)

| State | Population | Enrollment | State | Population | Enrollment |
|-------|-----------|-----------|-------|-----------|-----------|
| Alabama | 4833 | 306 | Missouri | 6044 | 438 |
| Colorado | 5268 | 359 | New Jersey | 8899 | 437 |
| Georgia | 9992 | 533 | Ohio | 11,571 | 697 |
| Iowa | 3090 | 340 | Oklahoma | 8260 | 584 |
| Maryland | 5929 | 364 | South Carolina | 4775 | 258 |

**(a)** Use a graphing calculator or spreadsheet program to find a linear model that uses state population ($x$) to predict the enrollment in higher education ($y$).

**Solution** The least squares regression line (with coefficients rounded) is

$$y = .045x + 121.7,$$

as shown in Figure 2.44. The correlation coefficient is $r \approx .886$, which is quite high, and Figure 2.45 shows a linear trend. Thus, the line fits these data well.

**Figure 2.44**

**Figure 2.45**

**(b)** Michigan had a population of 9,896 thousand in 2013. Predict its enrollment in higher education.

**Solution** Let $x = 9896$ in the regression equation:

$$y = .045(9896) + 121.7 = 567.02.$$

Therefore, the estimated higher education enrollment for Michigan would be 567,020 students.

**(c)** According to this model, what is the state population that would predict a higher education enrollment of 800 thousand students?

**Solution** Let $y = 800$ and solve the regression equation for $x$:

$$800 = .045x + 121.7 \qquad \text{Let } y = 800.$$
$$678.3 = .045x \qquad \text{Subtract 121.7 from both sides.}$$
$$x \approx 15073.333. \qquad \text{Divide both sides by .045.}$$

Thus, the model predicts that a state with a population of 15,073,333 has an enrollment of 800,000 in higher education.

 **NOTE** The value $r^2$ seen on the TI calculator output is known as the coefficient of determination. It is simply the value of the correlation coefficient squared and $0 \le r^2 \le 1$. One interpretation of its value is that it describes the proportion of variation in the $y$-variable that is explained by variation in the $x$-variable.

# 2.3 Exercises

1. **Temperature** The following table shows equivalent Fahrenheit and Celsius temperatures:

| Degrees Fahrenheit | 32 | 68 | 104 | 140 | 176 | 212 |
|---|---|---|---|---|---|---|
| Degrees Celsius | 0 | 20 | 40 | 60 | 80 | 100 |

   **(a)** Choose any two data points and use them to construct a linear equation that models the data, with $x$ being Fahrenheit and $y$ being Celsius.

   **(b)** Use the model in part (a) to find the Celsius temperature corresponding to

   50° Fahrenheit and 75° Fahrenheit.

**Temperature** *Use the linear equation derived in Exercise 1 to work the following problems.*

2. Convert each temperature.

   **(a)** 58°F to Celsius  **(b)** 50°C to Fahrenheit

   **(c)** −10°C to Fahrenheit  **(d)** −20°F to Celsius

3. According to the *World Almanac and Book of Facts*, 2008, Venus is the hottest planet, with a surface temperature of 867° Fahrenheit. What is this temperature in degrees Celsius?

4. Find the temperature at which Celsius and Fahrenheit temperatures are numerically equal.

*In each of the next set of problems, assume that the data can be modeled by a straight line and that the trend continues indefinitely. Use two data points to find such a line and then estimate the requested quantities. (See Example 1.)*

5. **CPI** The Consumer Price Index (CPI), which measures the cost of a typical package of consumer goods, was 224.9 in the year 2011 and 240.0 in 2016. Let $x = 11$ correspond to the year 2011 and estimate the CPI in 2012 and 2015. (Data from: U.S. Bureau of Labor Statistics.)

6. **Tax Returns** The approximate number (in millions) of individual electronically filed tax returns filed with the Internal Revenue Service (IRS) in the year 2005 was 68.5, and it was 123.7 in 2016. Let $x = 5$ correspond to the year 2005, and estimate the number of electronically filed returns in 2010 and 2014. *Hint:* Round slope to two decimal places. (Data from: U.S. Internal Revenue Service.)

7. **Educational Services** The number (in millions) of employees working in educational services was 12.3 in 2005 and 13.3 in 2014. Let $x = 5$ correspond to the year 2005, and estimate the number of employees in 2012. *Hint:* Round slope to two decimal places. (Data from: U.S. Census Bureau.)

8. **Leisure and Hospitality** The number (in millions) of employees working in the leisure and hospitality industries was 12.1 in 2005 and 13.5 in 2014. Let $x = 5$ correspond to the year 2005, and estimate the number of employees in 2013. *Hint:* Round slope to two decimal places. (Data from: U.S. Census Bureau.)

9. **Baseball** Suppose a baseball is thrown at 85 miles per hour. The ball will travel 320 feet when hit by a bat swung at 50 miles per hour and will travel 440 feet when hit by a bat swung at 80 miles per hour. Let $y$ be the number of feet traveled by the ball when hit by a bat swung at $x$ miles per hour. (*Note*: The preceding data are valid for $50 \le x \le 90$, where the bat is 35 inches long, weighs 32 ounces, and strikes a waist-high pitch so that the place of the swing lies at 10° from the diagonal). [Data from: Robert K. Adair, *The Physics of Baseball* (HarperCollins, 1990)]. How much farther will a ball travel for each mile-per-hour increase in the speed of the bat?

10. **Skiing** Ski resorts require large amounts of water in order to make snow. Snowmass Ski Area in Colorado plans to pump at least 1120 gallons of water per minute for at least 12 hours a day from Snowmass Creek between mid-October and late December. Environmentalists are concerned about the effects on the ecosystem. Find the minimum amount of water pumped in 30 days.

[*Hint*: Let $y$ be the total number of gallons pumped $x$ days after pumping begins. Note that $(0, 0)$ is on the graph of the equation.] (Data from: York Snow, Inc.)

*In each of the next two problems, two linear models are given for the data. For each model,*

(a) *find the residuals and their sum;*

(b) *find the sum of the squares of the residuals;*

(c) *decide which model is the better fit. (See Example 2.)*

**11. Cisco Sales** The following table shows the estimated sales (in billions of dollars) in China for Cisco Systems, Inc. for several years.

| Year | 2006 | 2008 | 2010 | 2012 | 2014 |
|---|---|---|---|---|---|
| Sales | .7 | 1.22 | 1.53 | 2.02 | 1.81 |

Let $x = 6$ correspond to the year 2006. Two equations that model these data are $y = .07x + .83$ and $y = .2x - .38$ (Data from: *The Wall Street Journal*.)

**12. International Trade** The following table shows South Korea's petrochemical exports to China (in billions of dollars) for several years.

| Year | 2010 | 2011 | 2012 | 2013 | 2014 |
|---|---|---|---|---|---|
| Petrochemical exports | 16 | 21 | 21 | 23 | 21.4 |

Let $x = 10$ correspond to the year 2010. Two equations that model these data are $y = 1.35x + 2.5$ and $y = 2x - 3$ (Data from: *The Wall Street Journal*.)

*In each of the following problems, determine whether a straight line is a good model for the data. Do this visually by plotting the data points and by finding the correlation coefficient for the least-squares regression line. (See Examples 5 and 6.)*

**13. Credit Unions** The following table shows the number of operating federal credit unions in the United States for several years.

| Year | 2011 | 2012 | 2013 | 2014 | 2015 |
|---|---|---|---|---|---|
| Number of federal credit unions | 4444 | 4268 | 4100 | 3924 | 3762 |

Let $x = 11$ correspond to the year 2011. (Data from: Credit Union National Association.)

**14. Gas Prices** The following tables shows the average price (in dollars) for a gallon of regular unleaded gasoline on the first weekday of selected months in the year 2015. Let $x = 1$ correspond to the month of January, $x = 3$ correspond to the month of March, and so on. (Data from: U.S. Energy Information Administration.)

| Month | January | March | May | July | September | November |
|---|---|---|---|---|---|---|
| Price per gallon | $2.14 | $2.36 | $2.52 | $2.70 | $2.34 | $2.17 |

*In Exercises 15–18 find the required linear model using leastsquares regression. (See Examples 3–6.)*

**15. Credit Unions** Use the data on the number of federal credit unions in Exercise 13.

(a) Find a linear model for these data with $x = 11$ corresponding to the year 2011.

(b) Assuming the trend continues, estimate the number of federal credit unions in the year 2018.

**16. Student Loans** The accompanying table shows the approximate amount (in billions of dollars) of student loans in repayment in various quarters of 2015 and 2016. Let $x = 1$ correspond to the first quarter of 2015. (Data from: *The Wall Street Journal*.)

| Quarter | 1 | 2 | 3 | 4 | 5 | 6 | 7 |
|---|---|---|---|---|---|---|---|
| Loans in repayment | 510 | 525 | 570 | 580 | 610 | 630 | 650 |

(a) Find a linear model for these data.

(b) Assuming the trend continues, estimate the amount of loans in repayment in quarter 8.

**17. Google Ad Revenue** The accompanying table shows the approximate amount (in billions of dollars) of Google advertising revenue in various quarters of 2015 and 2016. Let $x = 1$ correspond to the first quarter of 2015. (Data from: *The Wall Street Journal*.)

| Quarter | 1 | 2 | 3 | 4 | 5 | 6 | 7 | 8 |
|---|---|---|---|---|---|---|---|---|
| Ad revenue | 15.1 | 16.2 | 16.9 | 18.0 | 17.5 | 18.0 | 19.2 | 22.4 |

(a) Find a linear model or these data.

(b) Assuming the trend continues, estimate the amount of ad revenue in the third quarter of 2017.

**18. Blood Pressure** Researchers wish to determine if a relationship exists between age and systolic blood pressure (SBP). The accompanying table gives the age and systolic blood pressure for 10 adults.

| Age ($x$) | 35 | 37 | 42 | 46 | 53 | 57 | 61 | 65 | 69 | 74 |
|---|---|---|---|---|---|---|---|---|---|---|
| SBP ($y$) | 102 | 127 | 120 | 131 | 126 | 137 | 140 | 130 | 148 | 147 |

(a) Find a linear model for these data using age ($x$) to predict systolic blood pressure ($y$).

(b) Use the results from part (a) to predict the systolic blood pressure for a person who is 42-years-old, 53-years-old, and 69-years-old. How well do the actual data agree with the predicted values?

(c) Use the results from part (a) to predict the systolic blood pressure for a person who is 45-years-old and 70-years-old.

*Work these problems.*

**19. Street and Highway Construction** The total amount of construction spending on highways and streets (in billions of dollars) is given in the accompanying table for the first six months of 2016.

| Month | January | February | March | April | May | June |
|---|---|---|---|---|---|---|
| Spending | 97 | 94 | 94 | 90 | 89 | 88 |

**(a)** Find the least squares regression line that models these data, with $x = 1$ corresponding to January 2016.

**(b)** Assuming the trend continues, estimate the amount of spending in December 2016.

**20. Cable Subscribers** The number of basic cable television subscribers (in millions) is shown in the following table for various years. (Data from: ProQuest Statistical Abstract of the United States 2016.)

| Year | 2006 | 2008 | 2010 | 2012 | 2014 |
|---|---|---|---|---|---|
| Subscribers | 65.8 | 64.8 | 61.5 | 57.8 | 54.4 |

**(a)** Find the least-squares regression line that models these data with $x = 6$ corresponding to the year 2006.

**(b)** Find the number of subscribers for the year 2013.

**(c)** If the trend continues indefinitely, determine the first full year when there will be at least 50 million subscribers.

**(d)** Find the correlation coefficient.

**21. Software Publishers** The revenue per quarter for the software publishing industry (in billions of dollars) for the 8 quarters of 2015 and 2016 is displayed in the accompanying table. Let $x = 1$ correspond to the first quarter of 2015. (Data from: U.S. Census Bureau.)

| Quarter | 1 | 2 | 3 | 4 | 5 | 6 | 7 | 8 |
|---|---|---|---|---|---|---|---|---|
| Revenue | 47.4 | 48.9 | 49.8 | 49.3 | 49.8 | 49.8 | 49.6 | 50.4 |

**(a)** Find the least-squares regression line that models these data.

**(b)** If the trend continues indefinitely, find the revenue for the second quarter of 2017.

**(c)** If the trend continues indefinitely, what is the first full quarter when revenue exceeds $52 billion?

**(d)** Find the correlation coefficient.

**22. Life Expectancy** The following table shows men's and women's life expectancy at birth (in years) for selected birth years in the United States: (Data from: U.S. Center for National Health Statistics.)

| Birth Year | Life Expectancy Men | Life Expectancy Women |
|---|---|---|
| 1970 | 67.1 | 74.7 |
| 1975 | 68.8 | 76.6 |
| 1980 | 70.0 | 77.4 |
| 1985 | 71.1 | 78.2 |
| 1990 | 71.8 | 78.8 |
| 1995 | 72.5 | 78.9 |
| 2000 | 74.1 | 79.3 |
| 2005 | 75.0 | 80.1 |
| 2010 | 76.2 | 81.0 |
| 2015 | 77.1 | 81.7 |

**(a)** Find the least-squares regression line for the men's data, with $x = 70$ corresponding to 1970.

**(b)** Find the least-squares regression line for the women's data, with $x = 70$ corresponding to 1970.

**(c)** Suppose life expectancy continues to increase as predicted by the equations in parts (a) and (b). Will men's life expectancy ever be higher than women's? If so, in what birth year will this occur?

## ✓ Checkpoint Answers

1. $y = 30.25x + 765.75$

2. **(a)** 30, 29, 3, 6, −31
   **(b)** 2747
   **(c)** Yes, because the sum of the squares of the residuals is lower.

3. **(a)** −6.4, 7.1, −4.4, 13.1, −9.4
   **(b)** 370.7
   **(c)** It fits the data best because the sum of the squares of its residuals is smallest.

4. $y = 6.75x + 337.08$

5. **(a)** $y = .87x + 18.9$
   **(b)** $r \approx -.988$
   **(c)** It fits reasonably well because $|r|$ is very close to 1 and the pattern is linear.

# 2.4  Linear Inequalities

An **inequality** is a statement that one mathematical expression is greater than (or less than) another. Inequalities are very important in applications. For example, a company wants revenue to be *greater than* costs and must use *no more than* the total amount of capital or labor available.

Inequalities may be solved by algebraic or geometric methods. In this section, we shall concentrate on algebraic methods for solving **linear inequalities**, such as

$$4 - 3x \le 7 + 2x \quad \text{and} \quad -2 < 5 + 3m < 20,$$

and absolute-value inequalities, such as $|x - 2| < 5$. The following properties are the basic algebraic tools for working with inequalities.

## Properties of Inequality

For real numbers $a$, $b$, and $c$,

**(a)** if $a < b$, then $a + c < b + c$;

**(b)** if $a < b$, and if $c > 0$, then $ac < bc$;

**(c)** if $a < b$, and if $c < 0$, then $ac > bc$.

Throughout this section, definitions are given only for $<$, but they are equally valid for $>$, $\le$, and $\ge$.

**⬡! CAUTION**   Pay careful attention to part (c) in the previous box; if both sides of an inequality are multiplied by a negative number, the direction of the inequality symbol must be reversed. For example, starting with the true statement $-3 < 5$ and multiplying both sides by the positive number 2 gives

$$-3 \cdot 2 < 5 \cdot 2,$$

or

$$-6 < 10,$$

still a true statement. However, starting with $-3 < 5$ and multiplying both sides by the negative number $-2$ gives a true result only if the direction of the inequality symbol is reversed:

$$-3(-2) > 5(-2)$$
$$6 > -10. \; ✔_1$$

### ✔ Checkpoint 1

**(a)** First multiply both sides of $-6 < -1$ by 4, and then multiply both sides of $-6 < -1$ by $-7$.

**(b)** First multiply both sides of $9 \ge -4$ by 2, and then multiply both sides of $9 \ge -4$ by $-5$.

**(c)** First add 4 to both sides of $-3 < -1$, and then add $-6$ to both sides of $-3 < -1$.

> **EXAMPLE 1**   Solve $3x + 5 > 11$. Graph the solution.
>
> **Solution**  First add $-5$ to both sides:
>
> $$3x + 5 + (-5) > 11 + (-5)$$
> $$3x > 6.$$
>
> Now multiply both sides by $1/3$:
>
> $$\frac{1}{3}(3x) > \frac{1}{3}(6)$$
> $$x > 2.$$

(Why was the direction of the inequality symbol not changed?) In interval notation (introduced in Section 1.1), the solution is the interval $(2, \infty)$, which is graphed on the number line in Figure 2.46. The parenthesis at 2 shows that 2 is not included in the solution.

**Figure 2.46**

As a partial check, note that 0, which is not part of the solution, makes the inequality false, while 3, which is part of the solution, makes it true:

### ✔ Checkpoint 2

Solve these inequalities. Graph each solution.

**(a)** $5z - 11 < 14$

**(b)** $-3k \le -12$

**(c)** $-8y \ge 32$

$$
\begin{array}{cc}
? & ? \\
3(0) + 5 > 11 & 3(3) + 5 > 11 \\
5 > 11 \quad \text{False} & 14 > 11. \quad \text{True} \; ✔_2
\end{array}
$$

**EXAMPLE 2**    Solve $4 - 3x \leq 7 + 2x$.

**Solution**  Add $-4$ to both sides:

$$4 - 3x + (-4) \leq 7 + 2x + (-4)$$
$$-3x \leq 3 + 2x.$$

Add $-2x$ to both sides (remember that *adding* to both sides never changes the direction of the inequality symbol):

$$-3x + (-2x) \leq 3 + 2x + (-2x)$$
$$-5x \leq 3.$$

Multiply both sides by $-1/5$. Since $-1/5$ is negative, change the direction of the inequality symbol:

$$-\frac{1}{5}(-5x) \geq -\frac{1}{5}(3)$$

$$x \geq -\frac{3}{5}.$$

✓ **Checkpoint 3**

Solve these inequalities. Graph each solution.

**(a)**  $8 - 6t \geq 2t + 24$

**(b)**  $-4r + 3(r + 1) < 2r$

Figure 2.47 shows a graph of the solution, $[-3/5, \infty)$. The bracket in Figure 2.47 shows that $-3/5$ is included in the solution. ✓₃

**Figure 2.47**

**EXAMPLE 3**    Solve $-2 < 5 + 3m < 20$. Graph the solution.

**Solution**  The inequality $-2 < 5 + 3m < 20$ says that $5 + 3m$ is between $-2$ and $20$. We can solve this inequality with an extension of the properties given at the beginning of this section. Work as follows, first adding $-5$ to each part:

$$-2 + (-5) < 5 + 3m + (-5) < 20 + (-5)$$
$$-7 < 3m < 15.$$

Now multiply each part by $1/3$:

$$-\frac{7}{3} < m < 5.$$

✓ **Checkpoint 4**

Solve each of the given inequalities. Graph each solution.

**(a)**  $9 < k + 5 < 13$

**(b)**  $-6 \leq 2z + 4 \leq 12$

A graph of the solution, $(-7/3, 5)$, is given in Figure 2.48. ✓₄

**Figure 2.48**

**EXAMPLE 4**    **Temperature Conversion** The formula for converting from Celsius to Fahrenheit temperature is

$$F = \frac{9}{5}C + 32.$$

What Celsius temperature range corresponds to the range from 32°F to 77°F?

**Solution**  The Fahrenheit temperature range is $32 < F < 77$. Since $F = (9/5)C + 32$, we have

$$32 < \frac{9}{5}C + 32 < 77.$$

Solve the inequality for C:

$$32 < \frac{9}{5}C + 32 < 77$$

$$0 < \frac{9}{5}C < 45 \qquad \text{Subtract 32 from each part.}$$

$$\frac{5}{9} \cdot 0 < \frac{5}{9} \cdot \frac{9}{5}C < \frac{5}{9} \cdot 45 \qquad \text{Multiply each part by } \frac{5}{9}.$$

$$0 < C < 25.$$

✓ **Checkpoint 5**

In Example 4, what Celsius temperatures correspond to the range from 5°F to 95°F?

The corresponding Celsius temperature range is 0°C to 25°C.  ✓₅

A product will break even or produce a profit only if the revenue $R$ from selling the product at least equals the cost $C$ of producing it—that is, if $R \geq C$.

**EXAMPLE 5**   **Revenue and Cost** A company analyst has determined that the cost to produce and sell $x$ units of a certain product is $C = 20x + 1000$. The revenue for that product is $R = 70x$. Find the values of $x$ for which the company will break even or make a profit on the product.

**Solution** Solve the inequality $R \geq C$:

$$R \geq C$$
$$70x \geq 20x + 1000 \qquad \text{Let } R = 70x \text{ and } C = 20x + 1000.$$
$$50x \geq 1000 \qquad \text{Subtract } 20x \text{ from both sides.}$$
$$x \geq 20. \qquad \text{Divide both sides by 50.}$$

The company must produce and sell 20 items to break even and more than 20 to make a profit.

**EXAMPLE 6**   **Break-Even Point** A pretzel manufacturer can sell a 6-ounce bag of pretzels to a wholesaler for $.35 a bag. The variable cost of producing each bag is $.25 per bag, and the fixed cost for the manufacturing operation is $110,000.* How many bags of pretzels need to be sold in order to break even or earn a profit?

**Solution** Let $x$ be the number of bags produced. Then the revenue equation is

$$R = .35x,$$

and the cost is given by

$$\text{Cost} = \text{Fixed Costs} + \text{Variable Costs}$$
$$C = 110{,}000 + .25x.$$

We now solve the inequality $R \geq C$:

$$R \geq C$$
$$.35x \geq 110{,}000 + .25x$$
$$.1x \geq 110{,}000$$
$$x \geq 1{,}110{,}000.$$

The manufacturer must produce and sell 1,110,000 bags of pretzels to break even and more than that to make a profit.

---

* Variable costs, fixed costs, and revenue were discussed on p. 15.

## Absolute-Value Inequalities

You may wish to review the definition of absolute value in Section 1.1 before reading the following examples, which show how to solve inequalities involving absolute values.

**EXAMPLE 7**  Solve each inequality.

**(a)** $|x| < 5$

**Solution**  Because absolute value gives the distance from a number to 0, the inequality $|x| < 5$ is true for all real numbers whose distance from 0 is less than 5. This includes all numbers between $-5$ and 5, or numbers in the interval $(-5, 5)$. A graph of the solution is shown in Figure 2.49.

**Figure 2.49**

**(b)** $|x| > 5$

**Solution**  The solution of $|x| > 5$ is given by all those numbers whose distance from 0 is *greater* than 5. This includes the numbers satisfying $x < -5$ or $x > 5$. A graph of the solution, all numbers in

$$(-\infty, -5) \quad \text{or} \quad (5, \infty),$$

is shown in Figure 2.50. ✔6

**Figure 2.50**

The preceding examples suggest the following generalizations.

> Assume that $a$ and $b$ are real numbers and that $b$ is positive.
>
> **1.** Solve $|a| < b$ by solving $-b < a < b$.
> **2.** Solve $|a| > b$ by solving $a < -b$ or $a > b$.

**EXAMPLE 8**  Solve $|x - 2| < 5$.

**Solution**  Replace $a$ with $x - 2$ and $b$ with 5 in property (1) in the box above. Now solve $|x - 2| < 5$ by solving the inequality

$$-5 < x - 2 < 5.$$

Add 2 to each part, getting the solution

$$-3 < x < 7,$$

which is graphed in Figure 2.51. ✔7

**EXAMPLE 9**  Solve $|2 - 7m| - 1 > 4$.

**Solution**  First add 1 to both sides:

$$|2 - 7m| > 5$$

Now use property (2) from the preceding box to solve $|2 - 7m| > 5$ by solving the inequality

$$2 - 7m < -5 \quad \text{or} \quad 2 - 7m > 5.$$

---

✔**Checkpoint 6**

Solve each inequality. Graph each solution.

**(a)** $|x| \le 1$

**(b)** $|y| \ge 3$

---

**Figure 2.51**

---

✔**Checkpoint 7**

Solve each inequality. Graph each solution.

**(a)** $|p + 3| < 4$

**(b)** $|2k - 1| \le 7$

Solve each part separately:

$$-7m < -7 \quad \text{or} \quad -7m > 3$$

$$m > 1 \quad \text{or} \quad m < -\frac{3}{7}.$$

## ✓ Checkpoint 8

Solve each inequality. Graph each solution.

**(a)** $|y - 2| > 5$

**(b)** $|3k - 1| \geq 2$

**(c)** $|2 + 5r| - 4 \geq 1$

The solution, all numbers in $\left(-\infty, -\frac{3}{7}\right)$ or $(1, \infty)$, is graphed in Figure 2.52. ✓ 8

**Figure 2.52**

## ✓ Checkpoint 9

Solve each inequality.

**(a)** $|5m - 3| > -10$

**(b)** $|6 + 5a| < -9$

**(c)** $|8 + 2r| > 0$

**EXAMPLE 10**   Solve $|3 - 7x| \geq -8$.

**Solution** The absolute value of a number is always nonnegative. Therefore, $|3 - 7x| \geq -8$ is always true, so the solution is the set of all real numbers. Note that the inequality $|3 - 7x| \leq -8$ has no solution, because the absolute value of a quantity can never be less than a negative number. ✓ 9

# 2.4 Exercises

**1.** Explain how to determine whether a parenthesis or a bracket is used when graphing the solution of a linear inequality.

**2.** The three-part inequality $p < x < q$ means "$p$ is less than $x$ and $x$ is less than $q$." Which one of the given inequalities is not satisfied by any real number $x$? Explain why.

**(a)** $-3 < x < 5$          **(b)** $0 < x < 4$

**(c)** $-7 < x < -10$      **(d)** $-3 < x < -2$

*Solve each inequality and graph each solution. (See Examples 1–3.)*

**3.** $-8k \leq 32$             **4.** $-4a \leq 36$

**5.** $-2b > 0$               **6.** $6 - 6z < 0$

**7.** $3x + 4 \leq 14$         **8.** $2y - 7 < 9$

**9.** $-5 - p \geq 3$         **10.** $5 - 3r \leq -4$

**11.** $7m - 5 < 2m + 10$     **12.** $6x - 2 > 4x - 10$

**13.** $m - (4 + 2m) + 3 < 2m + 2$

**14.** $2p - (3 - p) \leq -7p - 2$

**15.** $-2(3y - 8) \geq 5(4y - 2)$

**16.** $5r - (r + 2) \geq 3(r - 1) + 6$

**17.** $3p - 1 < 6p + 2(p - 1)$

**18.** $x + 5(x + 1) > 4(2 - x) + x$

**19.** $-7 < y - 2 < 5$      **20.** $-3 < m + 6 < 2$

**21.** $8 \leq 3r + 1 \leq 16$      **22.** $-6 < 2p - 3 \leq 5$

**23.** $-4 \leq \dfrac{2k - 1}{3} \leq 2$     **24.** $-1 \leq \dfrac{5y + 2}{3} \leq 4$

**25.** $\dfrac{3}{5}(2p + 3) \geq \dfrac{1}{10}(5p + 1)$

**26.** $\dfrac{8}{3}(z - 4) \leq \dfrac{2}{9}(3z + 2)$

*In the following exercises, write a linear inequality that describes the given graph.*

**27.**  ◄├┼┼┼┼┼┼┼[┼┼┼┼►
           −6 −4 −2  0  2  4  6

**28.**  ◄┼┼┼┼┤├┼┼┼┼┼┼┼►
           −6 −4 −2  0  2  4  6

**29.**  ◄├┼┼[┼┼┼┼┼┼┼]┼├►
           −6 −4 −2  0  2  4  6

**30.**  ◄├┼┼[┼┼┼┼┼┼]┼├┼►
           −6 −4 −2  0  2  4  6

**Break-Even Point** *In Exercises 31–36, find all values of x for which the given products will at least break even. (See Examples 5 and 6.)*

**31.** $C = 50x + 6000; R = 65x$.

**32.** $C = 100x + 6000; R = 500x$.

**33.** $C = 85x + 1000; R = 105x$

**34.** $C = 70x + 500; R = 60x$

**35.** $C = 1000x + 5000; R = 900x$

**36.** $C = 25{,}000x + 21{,}700{,}000; R = 102{,}500x$

*Solve each inequality. Graph each solution. (See Examples 7–10.)*

**37.** $|p| > 7$            **38.** $|m| < 2$

**39.** $|r| \leq 5$            **40.** $|a| < -2$

**41.** $|b| > -5$

**42.** $|2x + 5| < 1$

**43.** $\left|x - \dfrac{1}{2}\right| < 2$

**44.** $|3z + 1| \geq 4$

**45.** $|8b + 5| \geq 7$

**46.** $\left|5x + \dfrac{1}{2}\right| - 2 < 5$

*Work these problems.*

**Temperature** *The given inequality describes the monthly average high daily temperature T in degrees Fahrenheit in the given location. (Data from: Weatherbase.com.) What range of temperatures corresponds to the inequality?*

**47.** $|T - 83| \leq 7$; Miami, Florida

**48.** $|T - 63| \leq 27$; Boise, Idaho

**49.** $|T - 61| \leq 21$; Flagstaff, Arizona

**50.** $|T - 43| \leq 22$; Anchorage, Alaska

**51. Engagement at Work** According to the polling organization Gallup.com, 35.5% of American workers are meaningfully engaged at work. Taking into account the margin of error, we can write this result as $|P - 35.5| \leq 3$, where $P$ is the percentage of American workers engaged at work. Find the range of estimates for $P$.

**52. Social Science** When administering a standard intelligence quotient (IQ) test, we expect about one-third of the scores to be more than 12 units above 100 or more than 12 units below 100. Describe this situation by writing an absolute value inequality.

**53. Social Science** A Gallup poll found that among Americans aged 18–34 years who consume alcohol, from 35% to 43% prefer beer. Let $B$ represent the percentage of American alcohol consumers who prefer beer. Write the preceding information as an inequality. (Data from: Gallup.com.)

**54. Social Science** A Gallup poll in January 2017 found that from 13.1% to 15.1% of Americans considered themselves underemployed (working part-time or not working at all, but wanting to work full-time). Let $U$ represent the percentage of underemployed workers. Write the preceding information as an inequality. (Data from: Gallup.com.)

**55. Income Tax** The following table shows the 2017 federal income tax for a single person. (Data from: Internal Revenue Service.)

| If Taxable Income Is Over | But Not Over | Tax Rate Is |
|---|---|---|
| $0 | $9325 | 10% |
| $9325 | $37,950 | 15% |
| $37,950 | $91,900 | 25% |
| $91,900 | $191,650 | 28% |
| $191,650 | $416,700 | 33% |
| $416,700 | $418,400 | 35% |
| $418,400 | No Limit | 39.6% |

Let $x$ denote the taxable income. Write each of the seven income ranges in the table as an inequality.

**56. Income Tax** The following table shows the 2016 New York state income tax for a single person. (Data from: www.tax-brackets.org.)

| If Taxable Income Is Over | But Not Over | Tax Rate Is |
|---|---|---|
| $0 | $8450 | 4.00% |
| $8450 | $11,650 | 4.50% |
| $11,650 | $13,850 | 5.25% |
| $13,850 | $21,300 | 5.90% |
| $21,300 | $80,150 | 6.45% |
| $80,150 | $214,000 | 6.65% |
| $214,000 | $1,070,350 | 6.85% |
| $1,070,350 | No Limit | 8.82% |

Let $x$ denote the taxable income. Write each of the eight income ranges in the table as an inequality.

## ✓ Checkpoint Answers

**1.** (a) $-24 < -4; 42 > 7$    (b) $18 \geq -8; -45 \leq 20$
    (c) $1 < 3; -9 < -7$

**2.** (a) $z < 5$    (b) $k \geq 4$

  (c) $y \leq -4$

**3.** (a) $t \leq -2$    (b) $r > 1$

**4.** (a) $4 < k < 8$    (b) $-5 \leq z \leq 4$

**5.** $-15°C$ to $35°C$

**6.** (a) $[-1, 1]$

  (b) All numbers in $(-\infty, -3]$ or $[3, \infty)$

**7.** (a) $(-7, 1)$    (b) $[-3, 4]$

**8. (a)** All numbers in $(-\infty, -3)$ or $(7, \infty)$

**(b)** All numbers in $\left(-\infty, -\dfrac{1}{3}\right]$ or $[1, \infty)$

**(c)** All numbers in $\left(-\infty, -\dfrac{7}{5}\right]$ or $\left[\dfrac{3}{5}, \infty\right)$

**9. (a)** All real numbers

**(b)** No solution

**(c)** All real numbers except $-4$

# 2.5 Polynomial and Rational Inequalities

This section deals with the solution of polynomial and rational inequalities, such as

$$r^2 + 3r - 4 \geq 0, \quad x^3 - x \leq 0, \quad \text{and} \quad \frac{2x - 1}{3x + 4} < 5.$$

We shall concentrate on algebraic solution methods, but to understand why these methods work, we must first look at such inequalities from a graphical point of view.

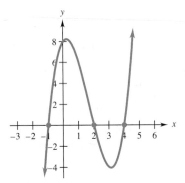

**Figure 2.53**

**EXAMPLE 1**   Use the graph of $y = x^3 - 5x^2 + 2x + 8$ in Figure 2.53 to solve each of the given inequalities.

**(a)** $x^3 - 5x^2 + 2x + 8 > 0$

**Solution** Each point on the graph has coordinates of the form $(x, x^3 - 5x^2 + 2x + 8)$. The number $x$ is a solution of the inequality exactly when the second coordinate of this point is positive—that is, when the point lies *above* the $x$-axis. So to solve the inequality, we need only find the first coordinates of points on the graph that are above the $x$-axis. This information can be read from Figure 2.53. The graph is above the $x$-axis when $-1 < x < 2$ and when $x > 4$. Therefore, the solutions of the inequality are all numbers $x$ in the interval $(-1, 2)$ or the interval $(4, \infty)$.

**(b)** $x^3 - 5x^2 + 2x + 8 < 0$

**Solution** The number $x$ is a solution of the inequality exactly when the second coordinate of the point $(x, x^3 - 5x^2 + 2x + 8)$ on the graph is negative—that is, when the point lies *below* the $x$-axis. Figure 2.53 shows that the graph is below the $x$-axis when $x < -1$ and when $2 < x < 4$. Hence, the solutions are all numbers $x$ in the interval $(-\infty, -1)$ or the interval $(2, 4)$.

The solution process in Example 1 depends only on knowing the graph and its $x$-intercepts (that is, the points where the graph intersects the $x$-axis). This information can often be obtained algebraically, without doing any graphing.

## Steps for Solving an Inequality Involving a Polynomial

1. Rewrite the inequality so that all the terms are on the left side and 0 is on the right side.
2. Find the $x$-intercepts by setting $y = 0$ and solving for $x$.
3. Divide the $x$-axis (number line) into regions using the solutions found in Step 2.
4. Test a point in each region by choosing a value for $x$ and substituting it into the equation for $y$.
5. Determine which regions satisfy the original inequality and graph the solution.

**EXAMPLE 2**   Solve each of the given quadratic inequalities.

**(a)** $x^2 - x < 12$

**Solution**

**Step 1**   Rewrite the inequality so that all the terms are on the left side and 0 is on the right side.

Hence we add $-12$ to each side and obtain $x^2 - x - 12 < 0$.

**Step 2**   Find the $x$-intercepts by setting $y = 0$ and solving for $x$.

We find the $x$-intercepts of $y = x^2 - x - 12$ by setting $y = 0$ and solving for $x$:

$$x^2 - x - 12 = 0$$
$$(x + 3)(x - 4) = 0$$
$$x + 3 = 0 \quad \text{or} \quad x - 4 = 0$$
$$x = -3 \qquad x = 4.$$

**Step 3**   Divide the $x$-axis (number line) into regions using the solutions found in Step 2. These numbers divide the $x$-axis into three regions, as indicated in Figure 2.54.

**Figure 2.54**

**Step 4**   Test a point in each region by choosing a value for $x$ and substituting it in to the equation for $y$.

In each region, the graph of $y = x^2 - x - 12$ is an unbroken curve, so it will be entirely above or entirely below the axis. It can pass from above to below the $x$-axis only at the $x$-intercepts. To see whether the graph is above or below the $x$-axis when $x$ is in region $A$, choose a value of $x$ in region $A$, say, $x = -5$, and substitute it into the equation:

$$y = x^2 - x - 12 = (-5)^2 - (-5) - 12 = 18.$$

Therefore, the point $(-5, 18)$ is on the graph. Since its $y$-coordinate 18 is positive, this point lies above the $x$-axis; hence, the entire graph lies above the $x$-axis in region $A$.

Similarly, we can choose a value of $x$ in region $B$, say, $x = 0$. Then

$$y = x^2 - x - 12 = 0^2 - 0 - 12 = -12,$$

so that $(0, -12)$ is on the graph. Since this point lies below the $x$-axis (why?), the entire graph in region $B$ must be below the $x$-axis. Finally, in region $C$, let $x = 5$. Then $y = 5^2 - 5 - 12 = 8$, so that $(5, 8)$ is on the graph, and the entire graph in region $C$ lies above the $x$-axis. We can summarize the results as follows:

| Interval | $x < -3$ | $-3 < x < 4$ | $x > 4$ |
|---|---|---|---|
| **Test value in interval** | $-5$ | $0$ | $5$ |
| **Value of $x^2 - x - 12$** | 18 | $-12$ | 8 |
| **Graph** | above $x$-axis | below $x$-axis | above $x$-axis |
| **Conclusion** | $x^2 - x - 12 > 0$ | $x^2 - x - 12 < 0$ | $x^2 - x - 12 > 0$ |

**Step 5**   Determine which regions satisfy the original inequality and graph the solution.

The last row shows that the only region where $x^2 - x - 12 < 0$ is region $B$, so the solutions of the inequality are all numbers $x$ with $-3 < x < 4$—that is, the interval $(-3, 4)$, as shown in the number line graph in Figure 2.55.

**Figure 2.55**

**(b)** $x^2 - x - 12 > 0$

**Solution** Use the chart in part (a). The last row shows that $x^2 - x - 12 > 0$ only when $x$ is in region $A$ or region $C$. Hence, the solutions of the inequality are all numbers $x$ with $x < -3$ or $x > 4$—that is, all numbers in the interval $(-\infty, -3)$ or the interval $(4, \infty)$.

✓ **Checkpoint 1**

Solve each inequality. Graph the solution on the number line.

**(a)** $x^2 - 2x < 15$

**(b)** $2x^2 - 3x - 20 < 0$

---

**EXAMPLE 3** Solve the quadratic inequality $r^2 + 3r \geq 4$.

**Solution**

**Step 1** First rewrite the inequality so that one side is 0:

$$r^2 + 3r \geq 4$$
$$r^2 + 3r - 4 \geq 0. \qquad \text{Add } -4 \text{ to both sides.}$$

**Step 2** Now solve the corresponding equation (which amounts to finding the $x$-intercepts of $y = r^3 + 3r - 4$):

$$r^2 + 3r - 4 = 0$$
$$(r - 1)(r + 4) = 0$$
$$r = 1 \qquad \text{or} \qquad r = -4.$$

**Step 3** These numbers separate the number line into three regions, as shown in Figure 2.56. Test a number from each region:

**Step 4** Let $r = -5$ from region $A$: $(-5)^2 + 3(-5) - 4 = 6 > 0$.
Let $r = 0$ from region $B$: $(0)^2 + 3(0) - 4 = -4 < 0$.
Let $r = 2$ from region $C$: $(2)^2 + 3(2) - 4 = 6 > 0$.

**Step 5** We want the inequality to be positive or 0. The solution includes numbers in region $A$ and in region $C$, as well as $-4$ and 1, the endpoints. The solution, which includes all numbers in the interval $(-\infty, -4]$ or the interval $[1, \infty)$, is graphed in Figure 2.56.

✓ **Checkpoint 2**

Solve each inequality. Graph each solution.

**(a)** $k^2 + 2k - 15 \geq 0$

**(b)** $3m^2 + 7m \geq 6$

Region $A$    Region $B$    Region $C$
$r < -4$    $-4 < r < 1$    $r > 1$

**Figure 2.56**

---

**EXAMPLE 4** Solve $q^3 - 4q > 0$.

**Solution**

**Step 1** Step 1 is already complete in the statement of the problem.

**Step 2** Solve the corresponding equation by factoring:

$$q^3 - 4q = 0$$
$$q(q^2 - 4) = 0$$
$$q(q + 2)(q - 2) = 0$$
$$q = 0 \quad \text{or} \quad q + 2 = 0 \quad \text{or} \quad q - 2 = 0$$
$$q = 0 \qquad\qquad q = -2 \qquad\qquad q = 2.$$

**Step 3** These three numbers separate the number line into the four regions shown in Figure 2.57.

**Step 4**    Test a number from each region:

$$A: \quad \text{If } q = -3, (-3)^3 - 4(-3) = -15 < 0.$$
$$B: \quad \text{If } q = -1, (-1)^3 - 4(-1) = 3 > 0.$$
$$C: \quad \text{If } q = 1, (1)^3 - 4(1) = -3 < 0.$$
$$D: \quad \text{If } q = 3, (3)^3 - 4(3) = 15 > 0.$$

**Step 5**    The numbers that make the polynomial positive are in the interval $(-2, 0)$ or the interval $(2, \infty)$, as graphed in Figure 2.57. ✓₃

✓ **Checkpoint 3**

Solve each inequality. Graph each solution.

**(a)**  $m^3 - 9m > 0$

**(b)**  $2k^3 - 50k \leq 0$

**Figure 2.57**

A graphing calculator can be used to solve inequalities without the need to evaluate at a test number in each interval. It is also useful for finding approximate solutions when the $x$-intercepts of the graph cannot be found algebraically.

**Figure 2.58**

✓ **Checkpoint 4** ▦

Use graphical methods to find approximate solutions of these inequalities.

**(a)**  $x^2 - 6x + 2 > 0$

**(b)**  $x^2 - 6x + 2 < 0$

▦ **EXAMPLE 5**    Use a graphing calculator to solve $x^3 - 5x^2 + x + 6 > 0$.

**Solution**  Begin by graphing $y = x^3 - 5x^2 + x + 6$ (Figure 2.58). Find the $x$-intercepts by solving $x^3 - 5x^2 + x + 6 = 0$. Since this cannot readily be done algebraically, use the graphical root finder to determine that the solutions ($x$-intercepts) are approximately $-.9254$, $1.4481$, and $4.4774$.

The graph is above the $x$-axis when $-.9254 < x < 1.4481$ and when $x > 4.4774$. Therefore, the approximate solutions of the inequality are all numbers in the interval $(-.9254, 1.4481)$ or the interval $(4.4774, \infty)$. ✓₄

▦ **EXAMPLE 6**    **Break-Even Point**  A company sells wholesale portable DVD players for \$39 each. The variable cost of producing $x$ thousand players is $5.5x - 4.9x^2$ (in thousands of dollars), and the fixed cost is \$550 (in thousands). Find the values of $x$ for which the company will break even or make a profit on the product.

**Solution**  If $x$ thousand DVD players are sold at \$39 each, then

$$R = 39 \times x = 39x.$$

The cost expression (in thousands of dollars) is

$$\text{Cost} = \text{Fixed Costs} + \text{Variable Costs}$$
$$C = 550 + (5.5x - 4.9x^2).$$

Therefore, to break even or earn a profit, we need Revenue ($R$) greater than or equal to Cost ($C$).

$$R \geq C$$
$$39x \geq 550 + 5.5x - 4.9x^2$$
$$4.9x^2 + 33.5x - 550 \geq 0.$$

Now graph $4.9x^2 + 33.5x - 550$. Since $x$ has to be positive in this situation (why?), we need only look at the graph in the right half of the plane. Here, and in other cases, you may have to try several viewing windows before you find one that shows what you need. Once you find a suitable window, such as in Figure 2.59, use the graphical root finder to determine the relevant $x$-intercept. Figure 2.59 shows the intercept is approximately $x \approx 7.71$. Hence, the company must manufacture at least $7.71 \times 1000 = 7710$ portable DVD players to make a profit.

**Figure 2.59**

## Rational Inequalities

Inequalities with quotients of algebraic expressions are called **rational inequalities**. These inequalities can be solved in much the same way as polynomial inequalities can.

> ## Steps for Solving Inequalities Involving Rational Expressions
>
> 1. Rewrite the inequality so that all the terms are on the left side and the 0 is on the right side.
> 2. Write the left side as a single fraction.
> 3. Set the numerator and the denominator equal to 0 and solve for $x$.
> 4. Divide the $x$-axis (number line) into regions using the solutions found in Step 3.
> 5. Test a point in each region by choosing a value for $x$ and substituting it into the equation for $y$.
> 6. Determine which regions satisfy the original inequality and graph the solution.

**EXAMPLE 7**    Solve the rational inequality

$$\frac{5}{x + 4} \geq 1.$$

**Solution**

**Step 1**    Write an equivalent inequality with one side equal to 0:

$$\frac{5}{x + 4} \geq 1$$

$$\frac{5}{x + 4} - 1 \geq 0.$$

**Step 2**    Write the left side as a single fraction:

$$\frac{5}{x + 4} - \frac{x + 4}{x + 4} \geq 0 \qquad \text{Obtain a common denominator.}$$

$$\frac{5 - (x + 4)}{x + 4} \geq 0 \qquad \text{Subtract fractions.}$$

$$\frac{5 - x - 4}{x + 4} \geq 0 \qquad \text{Distributive property}$$

$$\frac{1 - x}{x + 4} \geq 0.$$

**Step 3**    The quotient can change sign only at places where the denominator is 0 or the numerator is 0. (In graphical terms, these are the only places where the graph of $y = \dfrac{1 - x}{x + 4}$ can change from above the $x$-axis to below.) This happens when

$$1 - x = 0 \qquad \text{or} \qquad x + 4 = 0$$
$$x = 1 \qquad \text{or} \qquad x = -4.$$

**Step 4**    As in the earlier examples, the numbers $-4$ and 1 divide the $x$-axis into three regions:

$$x < -4, \quad -4 < x < 1, \quad x > 1.$$

**Step 5**  Test a number from each of these regions:

$$\text{Let } x = -5: \quad \frac{1 - (-5)}{-5 + 4} = -6 < 0.$$

$$\text{Let } x = 0: \quad \frac{1 - 0}{0 + 4} = \frac{1}{4} > 0.$$

$$\text{Let } x = 2: \quad \frac{1 - 2}{2 + 4} = -\frac{1}{6} < 0.$$

**Step 6**  The test shows that numbers in $(-4, 1)$ satisfy the inequality. With a quotient, the endpoints must be considered individually to make sure that no denominator is 0. In this inequality, $-4$ makes the denominator 0, while 1 satisfies the given inequality. Write the solution in interval notation as $(-4, 1]$ and graphically as in Figure 2.60.

**Figure 2.60**

> **⊘ CAUTION**  As suggested by Example 7, be very careful with the endpoints of the intervals in the solution of rational inequalities. ✓₅

✓ **Checkpoint 5**

Solve each inequality.

(a)  $\dfrac{3}{x - 2} \geq 4$

(b)  $\dfrac{p}{1 - p} < 3$

(c)  Why is 2 excluded from the solution in part (a)?

**EXAMPLE 8**    Solve

$$\frac{2x - 1}{3x + 4} < 5.$$

**Solution**

**Step 1**  Write an equivalent inequality with 0 on one side by adding $-5$ to each side.

$$\frac{2x - 1}{3x + 4} - 5 < 0 \qquad \text{Get 0 on the right side.}$$

**Step 2**  Write the left side as a single fraction by using $3x + 4$ as a common denominator.

$$\frac{2x - 1 - 5(3x + 4)}{3x + 4} < 0 \qquad \text{Obtain a common denominator.}$$

$$\frac{-13x - 21}{3x + 4} < 0 \qquad \text{Distribute in the numerator and combine terms.}$$

**Step 3**  Set the numerator and denominator each equal to 0 and solve the two equations:

$$-13x - 21 = 0 \qquad \text{or} \qquad 3x + 4 = 0$$

$$x = -\frac{21}{13} \qquad \text{or} \qquad x = -\frac{4}{3}$$

**Step 4**  The values $-21/13$ and $-4/3$ divide the $x$-axis into three regions:

$$x < -\frac{21}{13}, \quad -\frac{21}{13} < x < -\frac{4}{3}, \quad x > -\frac{4}{3}.$$

**Steps 5 and 6**  Testing points from each interval yields that the quotient is negative for numbers in the interval $(-\infty, -21/13)$ or $(-4/3, \infty)$. Neither endpoint satisfies the given inequality. ✓₆

**Figure 2.61**

✓ **Checkpoint 6**

Solve each rational inequality.

(a)  $\dfrac{3y - 2}{2y + 5} < 1$

(b)  $\dfrac{3c - 4}{2 - c} \geq -5$

✓ **Checkpoint 7**

(a)  Solve the inequality

$$\frac{10}{x + 2} \geq 3$$

by first multiplying both sides by $x + 2$.

(b)  Show that this method produces a wrong answer by testing $x = -3$.

> **⊘ CAUTION**  In problems like those in Examples 7 and 8, you should *not* begin by multiplying both sides by the denominator to simplify the inequality. Doing so will usually produce a wrong answer. For the reason, see Exercise 38. For an example, see Checkpoint 7. ✓₇

**▦ TECHNOLOGY TIP**  Rational inequalities can also be solved graphically. In Example 7, for instance, after rewriting the original inequality in the form $\dfrac{1 - x}{x + 4} \geq 0$, determine the

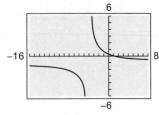

**Figure 2.62**

values of $x$ that make the numerator and denominator 0 (namely, $x = 1$ and $x = -4$). Then graph $\dfrac{1 - x}{x + 4}$. Figure 2.62 shows that the graph is above the $x$-axis when it is between the vertical asymptote at $x = -4$ and the $x$-intercept at $x = 1$. So the solution of the inequality is the interval $(-4, 1]$. (When the values that make the numerator and denominator 0 cannot be found algebraically, as they were here, you can use the root finder to approximate them.)

# 2.5 Exercises

*Solve each of these quadratic inequalities. Graph the solutions on the number line. (See Examples 2 and 3.)*

1. $(x + 4)(2x - 3) \leq 0$
2. $(5y - 1)(y + 3) > 0$
3. $r^2 + 4r > -3$
4. $z^2 + 6z > -8$
5. $4m^2 + 7m - 2 \leq 0$
6. $6p^2 - 11p + 3 \leq 0$
7. $4x^2 + 3x - 1 > 0$
8. $3x^2 - 5x > 2$
9. $x^2 \leq 36$
10. $y^2 \geq 9$
11. $p^2 - 16p > 0$
12. $r^2 - 9r < 0$

*Solve these inequalities. (See Example 4.)*

13. $x^3 - 9x \geq 0$
14. $p^3 - 25p \leq 0$
15. $(x + 7)(x + 2)(x - 2) \geq 0$
16. $(2x + 4)(x^2 - 9) \leq 0$
17. $(x + 5)(x^2 - 2x - 3) < 0$
18. $x^3 - 2x^2 - 3x \leq 0$
19. $6k^3 - 5k^2 < 4k$
20. $2m^3 + 7m^2 > 4m$

21. A student solved the inequality $p^2 < 16$ by taking the square root of both sides to get $p < 4$. She wrote the solution as $(-\infty, 4)$. Is her solution correct?

*Use a graphing calculator to solve these inequalities. (See Example 5.)*

22. $6x + 7 < 2x^2$
23. $.5x^2 - 1.2x < .2$
24. $3.1x^2 - 7.4x + 3.2 > 0$
25. $x^3 - 2x^2 - 5x + 7 \geq 2x + 1$
26. $x^4 - 6x^3 + 2x^2 < 5x - 2$
27. $2x^4 + 3x^3 < 2x^2 + 4x - 2$
28. $x^5 + 5x^4 > 4x^3 - 3x^2 - 2$

*Solve these rational inequalities. (See Examples 7 and 8.)*

29. $\dfrac{r - 4}{r - 1} \geq 0$
30. $\dfrac{z + 6}{z + 4} > 1$
31. $\dfrac{a - 2}{a - 5} < -1$
32. $\dfrac{1}{3k - 5} < \dfrac{1}{3}$
33. $\dfrac{1}{p - 2} < \dfrac{1}{3}$
34. $\dfrac{7}{k + 2} \geq \dfrac{1}{k + 2}$

35. $\dfrac{5}{p + 1} > \dfrac{12}{p + 1}$
36. $\dfrac{x^2 - 4}{x} > 0$
37. $\dfrac{x^2 - x - 6}{x} < 0$

38. Determine whether $x + 4$ is positive or negative when

   (a) $x > -4$;       (b) $x < -4$.

   (c) If you multiply both sides of the inequality $\dfrac{1 - x}{x + 4} \geq 0$ by $x + 4$, should you change the direction of the inequality sign? If so, when?

   (d) Explain how you can use parts (a)–(c) to solve $\dfrac{1 - x}{x + 4} \geq 0$ correctly.

*Use a graphing calculator to solve these inequalities. You may have to approximate the roots of the numerators or denominators.*

39. $\dfrac{2x^2 + x - 1}{x^2 - 4x + 4} \leq 0$
40. $\dfrac{x^3 - 3x^2 + 5x - 29}{x^2 - 7} > 3$

41. **Profit** An analyst has found that her company's profits, in hundreds of thousands of dollars, are given by $P = 2x^2 - 12x - 32$, where $x$ is the amount, in hundreds of dollars, spent on advertising. For what values of $x$ does the company make a profit?

42. **Profit** The commodities market is highly unstable; money can be made or lost quickly on investments in soybeans, wheat, and so on. Suppose that an investor kept track of his total profit $P$ at time $t$, in months, after he began investing, and he found that $P = 4t^2 - 30t + 14$. Find the time intervals during which he has been ahead.

43. **Profit** The manager of a 200-unit apartment complex has found that the profit is given by

   $$P = x^2 + 300x - 18,000,$$

   where $x$ is the number of apartments rented. For what values of $x$ does the complex produce a profit?

44. **Profit** A door-to-door knife salesman finds that his weekly profit can be modeled by the equation

   $$P = x^2 + 5x - 530$$

   where $x$ is the number of pitches he makes in a week. For what values of $x$ does the salesman need to make in order to earn a profit?

✓ **Checkpoint Answers**

1. **(a)** $(-3, 5)$

**(b)** $(-5/2, 4)$

2. **(a)** All numbers in $(-\infty, -5]$ or $[3, \infty)$

**(b)** All numbers in $(-\infty, -3]$ or $[2/3, \infty)$

3. **(a)** All numbers in $(-3, 0)$ or $(3, \infty)$

**(b)** All numbers in $(-\infty, -5]$ or $[0, 5]$

4. **(a)** All numbers in $(-\infty, .3542)$ or $(5.6458, \infty)$

**(b)** All numbers in $(.3542, 5.6458)$

5. **(a)** $(2, 11/4]$

**(b)** All numbers in $(-\infty, 3/4)$ or $(1, \infty)$

**(c)** When $x = 2$, the fraction is undefined.

6. **(a)** $(-5/2, 7)$

**(b)** All numbers in $(-\infty, 2)$ or $[3, \infty)$

7. **(a)** $x \leq \dfrac{4}{3}$

**(b)** $x = -3$ is a solution of $x \leq \dfrac{4}{3}$, but not of the original inequality $\dfrac{10}{x + 2} \geq 3$.

# CHAPTER 2 Summary and Review

## Key Terms and Symbols

**2.1** Cartesian coordinate system
x-axis
y-axis
origin
ordered pair
x-coordinate
y-coordinate
quadrant
solution of an equation
graph

x-intercept
y-intercept
[viewing window]
[trace]
[graphical root finder]
[maximum and minimum finder]
graph reading

**2.2** change in x
change in y

slope
slope–intercept form
parallel and perpendicular lines
point–slope form
linear equations
general form

**2.3** linear models
residual
[least-squares regression line]

[linear regression]
[correlation coefficient]

**2.4** linear inequality
properties of inequality
absolute-value inequality

**2.5** polynomial inequality
algebraic solution methods
[graphical solution methods]
rational inequality

## Chapter 2 Key Concepts

**Slope of a Line**

The **slope** of the line through the points $(x_1, y_1)$ and $(x_2, y_2)$, where $x_1 \neq x_2$, is $m = \dfrac{y_2 - y_1}{x_2 - x_1}$.

**Equation of a Line**

The line with equation $y = mx + b$ has slope $m$ and y-intercept $b$.
The line with equation $y - y_1 = m(x - x_1)$ has slope $m$ and goes through $(x_1, y_1)$.
The line with equation $ax + by = c$ (with $a \neq 0, b \neq 0$) has x-intercept $c/a$ and y-intercept $c/b$.
The line with equation $x = k$ is vertical, with x-intercept $k$, no y-intercept, and undefined slope.
The line with equation $y = k$ is horizontal, with y-intercept $k$, no x-intercept, and slope 0.

**Parellel and Perpendicular Lines**

Nonvertical **parallel lines** have the same slope, and **perpendicular lines,** if neither is vertical, have slopes with a product of $-1$.

---

*Terms in brackets deal with material in which a graphing calculator or other technology is used.

# Chapter 2 Review Exercises

*Which of the ordered pairs $(-2, 3), (0, -5), (2, -3), (3, -2), (4, 3),$*
*and $(7, 2)$ are solutions of the given equation?*

**1.** $y = x^2 - 2x - 5$        **2.** $x - y = 5$

*Sketch the graph of each equation.*

**3.** $5x - 3y = 15$        **4.** $2x + 7y - 21 = 0$

**5.** $y + 3 = 0$        **6.** $y - 2x = 0$

**7.** $y = .25x^2 + 1$        **8.** $y = \sqrt{x + 4}$

**9.** The following temperature graph was recorded in Bratenahl, Ohio:

Hours

**(a)** At what times during the day was the temperature over 55°?

**(b)** When was the temperature below 40°?

**10.** Greenville, South Carolina, is 500 miles south of Bratenahl, Ohio, and its temperature is 7° higher all day long. (See the graph in Exercise 9.) At what time was the temperature in Greenville the same as the temperature at noon in Bratenahl?

**11.** In your own words, define the slope of a line.

*In Exercises 12–21, find the slope of the line defined by the given conditions.*

**12.** Through $(-1, 3)$ and $(2, 6)$

**13.** Through $(4, -5)$ and $(1, 4)$

**14.** Through $(8, -3)$ and the origin

**15.** Through $(8, 2)$ and $(0, 4)$

**16.** $3x + 5y = 25$        **17.** $6x - 2y = 7$

**18.** $x - 2 = 0$        **19.** $y = -4$

**20.** Parallel to $3x + 8y = 0$

**21.** Perpendicular to $x = 3y$

**22.** Graph the line through $(0, 5)$ with slope $m = -2/3$.

**23.** Graph the line through $(-4, 1)$ with $m = 3$.

**24.** What information is needed to determine the equation of a line?

*Find an equation for each of the following lines.*

**25.** Through $(5, -1)$, slope $2/3$

**26.** Through $(8, 0)$, slope $-1/4$

**27.** Through $(5, -2)$ and $(1, 3)$

**28.** Through $(2, -3)$ and $(-3, 4)$

**29.** Undefined slope, through $(-1, 4)$

**30.** Slope 0, through $(-2, 5)$

**31.** $x$-intercept $-3$, $y$-intercept 5

**32.** Here is a sample SAT question: Which of the following is an equation of the line that has a $y$-intercept of 2 and an $x$-intercept of 3?

     **(a)** $-2x + 3y = 4$        **(b)** $-2x + 3y = 6$

     **(c)** $2x + 3y = 4$        **(d)** $2x + 3y = 6$

     **(e)** $3x + 2y = 6$

**33.** **Student Loans** In 2010, the total owed on federal student loans was approximately $.80 trillion. In 2015, that amount was approximately $1.2 trillion. (Data from: *The Wall Street Journal*.)

     **(a)** Assuming the increase in student loan debt is linear, write an equation that gives the total amount owed in year $x$, with $x = 10$ corresponding to the year 2010.

     **(b)** Is the slope of the line positive or negative? Why?

     **(c)** Assuming the linear trend continues, estimate the amount owed in 2017.

**34.** **Baseball** It appears that bunting is becoming more rare for nonpitchers. In 2011, the average number of plate appearances per bunt for nonpitchers was approximately 175. In 2016, that value was 337.

     **(a)** Assuming the trend is linear, write an equation that gives the average plate appearances per bunt in year $x$, with $x = 11$ corresponding to the year 2011.

     **(b)** Is the slope of the line positive or negative? Why?

     **(c)** Assuming the linear trend continues, estimate the average plate appearances per bunt in 2017.

**35.** **Median Income** The following table gives the median household income, meaning that half the households in the United States earned less than that amount and half the households earned more than that amount (see Chapter 10), for various years. (Data from: U.S. Census Bureau.)

| Year | 2005 | 2010 | 2014 | 2015 |
|---|---|---|---|---|
| **Household Income** | $46,326 | $49,276 | $53,657 | $56,516 |

     **(a)** Use the data from the years 2005 and 2010 to find a linear model for these data, with $x = 5$ corresponding to the year 2005.

     **(b)** Find the least-squares regression line for all these data.

     **(c)** Use the models in parts (a) and (b) to estimate the median household income for the year 2015 to the nearest dollar. Compare the estimates to the actual value for 2015.

     **(d)** Assuming the trend continues, use the model for part (b) to estimate the median household income for the year 2018 to the nearest dollar.

**36.** **Charitable Giving** The following table gives the total amount in charitable giving (in billions of dollars) for various years. (Data from: ProQuest Statistical Abstract of the United States: 2016.)

| Year | 2008 | 2011 | 2013 | 2014 |
|---|---|---|---|---|
| Charitable Giving | $299.6 | $298.5 | $334.5 | $358.4 |

(a) Use the data from the years 2008 and 2013 to find a linear model for these data, with $x = 8$ corresponding to the year 2008.

(b) Find the least-squares regression line for all these data.

(c) Use the models in parts (a) and (b) to estimate the total amount of charitable giving in 2014 to the nearest billion. Compare the estimates to the actual value for 2014.

(d) Assuming the trend continues, use the model for part (b) to estimate the total amount of charitable giving in 2019 to the nearest billion.

**37. Two-Year School Tuition** The following table shows the average annual tuition for public two-year institutions. (Data from: The College Board.)

| Year | 2012 | 2013 | 2014 | 2015 | 2016 |
|---|---|---|---|---|---|
| Tuition | $3154 | $3241 | $3336 | $3440 | $3520 |

(a) Let $x = 12$ correspond to the year 2012, and find the least-squares regression line for these data.

(b) Graph the data from the years 2012 to 2016 with the least-squares regression line.

(c) Does the least-squares regression line form a good fit?

(d) What is the correlation coefficient?

**38. Four-Year School Tuition** The following table shows the average annual tuition for public four-year institutions. (Data from: The College Board.)

| Year | 2012 | 2013 | 2014 | 2015 | 2016 |
|---|---|---|---|---|---|
| Tuition | $8646 | $8885 | $9145 | $9420 | $9650 |

(a) Let $x = 12$ correspond to the year 2012, and find the least-squares regression line for these data.

(b) Graph the data from the years 2012 to 2016 with the least-squares regression line.

(c) Does the least-squares regression line form a good fit?

(d) What is the correlation coefficient?

*Solve each inequality.*

**39.** $-6x + 3 < 2x$       **40.** $12z \geq 5z - 7$

**41.** $2(3 - 2m) \geq 8m + 3$    **42.** $6p - 5 > -(2p + 3)$

**43.** $-3 \leq 4x - 1 \leq 7$    **44.** $0 \leq 3 - 2a \leq 15$

**45.** $|b| \leq 8$    **46.** $|a| > 7$

**47.** $|2x - 7| \geq 3$    **48.** $|4m + 9| \leq 16$

**49.** $|5k + 2| - 3 \leq 4$    **50.** $|3z - 5| + 2 \geq 10$

**51. Natural Science** Here is a sample SAT question: For pumpkin carving, Mr. Sephera will not use pumpkins that weigh less than 2 pounds or more than 10 pounds. If $x$ represents the weight of a pumpkin (in pounds) that he will *not* use, which of the following inequalities represents all possible values of $x$?

(a) $|x - 2| > 10$     (b) $|x - 4| > 6$

(c) $|x - 5| > 5$     (d) $|x - 6| > 4$

(e) $|x - 10| > 4$

**52. Snow-Thrower Pricing** Prices at several retail outlets for a new 24-inch, 2-cycle snow thrower were all within $55 of $600. Write this information as an inequality, using absolute-value notation.

**53. Data Plans** One cellular company offers a data plan charging $45 a month plus $8 per gigabyte of data. Another cellular company offers a flat rate of $65 per month with unlimited data usage. For what range of data use (in gigabytes) is the first plan cheaper?

**54. Car Rental** One car rental firm charges $125 for a weekend rental (Friday afternoon through Monday morning) and gives unlimited mileage. A second firm charges $95 plus $.20 a mile. For what range of miles driven is the second firm cheaper?

*Solve each inequality.*

**55.** $r^2 + r - 6 < 0$    **56.** $y^2 + 4y - 5 \geq 0$

**57.** $2z^2 + 7z \geq 15$    **58.** $3k^2 \leq k + 14$

**59.** $(x - 3)(x^2 + 7x + 10) \leq 0$

**60.** $(x + 4)(x^2 - 1) \geq 0$    **61.** $\dfrac{m + 2}{m} \leq 0$

**62.** $\dfrac{q - 4}{q + 3} > 0$    **63.** $\dfrac{5}{p + 1} > 2$

**64.** $\dfrac{6}{a - 2} \leq -3$    **65.** $\dfrac{2}{r + 5} \leq \dfrac{3}{r - 2}$

**66.** $\dfrac{1}{z - 1} > \dfrac{2}{z + 1}$

# Case Study 2   Using Extrapolation and Interpolation for Prediction

One reason for developing a mathematical model is to make predictions. If your model is a least-squares line, you can predict the $y$-value corresponding to some new $x$-value by substituting that $x$-value into an equation of the form $\hat{y} = mx + b$. (We use $\hat{y}$ to remind us that we're getting a predicted value rather than an actual data value.) Data analysts distinguish between two very different kinds of prediction: *interpolation* and *extrapolation*. An interpolation uses a new $x$ inside the $x$-range of your original data. For example, if you have inflation data at five-year intervals from 1950 to 2010, estimating the rate of inflation in 1957 is an interpolation problem. But if you use the same data to estimate what the inflation rate was in 1920, or what it will be in 2020, you are extrapolating.

In general, interpolation is much safer than extrapolation, because data that are approximately linear over a short interval may be nonlinear over a larger interval. Let us examine a case of the dangers of extrapolation. Figure 1 shows the net income (in millions of dollars) for Chipotle Mexican Grill, Inc., from the years 2010–2015. Also on Figure 1, we see the least-squares regression line. (Data from: www.morningstar.com.)

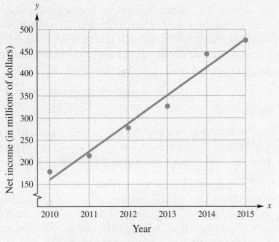

**Figure 1**

As we see from the graph, the fit is quite good and the linear trend seems quite clear. Additionally, the correlation coefficient is $r \approx .986$, which is very high. The least-squares regression equation is

$$y = 63.54x - 474.3,$$

where $x = 10$ corresponds to the year 2010. If we extrapolate to the next year (2016), we predict the net income for Chipotle to be

$$63.54(16) - 474.3 = \$542.34 \text{ million.}$$

The problem, however, is that Chipotle suffered from *E. coli* contamination in several states and a norovirus outbreak in the Boston, Massachusetts, area in late 2015. The net earnings in 2016 were a paltry $23 million—substantially lower than what our regression model would have predicted.

Using the past to predict the future is generally a dangerous business because we often cannot foresee events with major impact such as recessions, natural disasters, etc. Predictions into the future are often made, however, because planning has to occur and we use the best data we have available. One merely needs to realize that predictions into the future are highly likely to be inaccurate.

One way to determine if a model is likely to be inaccurate in the future is to examine the residuals. In Section 2.3, we defined a residual to be the difference between the actual value $y$ and its predicted value $\hat{y}$.

$$\text{Residual} = y - \hat{y}.$$

Graphing these residuals on the $y$-axis and the predictor variable on the $x$-axis can be illuminating. Let's look at another historical example. Figure 2 shows the revenue (in millions of dollars) for the social media company Twitter for the years 2010 through 2015. (Data from: www.morningstar.com.)

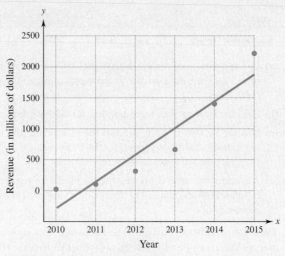

**Figure 2**

The regression equation for the least-squares line in Figure 2 is

$$y = 434x - 4635,$$

where $x = 10$ corresponds to the year 2010. Because the correlation coefficient is $r \approx .943$, our linear model appears to fit these data well. We notice, however, that the predictions underpredict, then overpredict, and then underpredict. We can get a better look at this pattern by plotting the residuals. To find them, we put each value of the independent variable into the regression equation, calculate the predicted value $\hat{y}$, and subtract it from the actual value of $y$, as in the following table.

| Year | 2010 | 2011 | 2012 | 2013 | 2014 | 2015 |
|---|---|---|---|---|---|---|
| $x$ | 10 | 11 | 12 | 13 | 14 | 15 |
| $y$ | 28 | 106 | 317 | 665 | 1403 | 2218 |
| $\hat{y}$ | −295.4 | 138.5 | 572.5 | 1006.5 | 1440.5 | 1874.4 |
| **Residual** $y - \hat{y}$ | 323.4 | −32.5 | −255.5 | −341.5 | −37.5 | 343.6 |

The residuals in Figure 3 indicate that our data have a nonlinear U-shaped component that is not captured by a linear model. When a residual plot shows a pattern such as this (or a similar pattern with an upside down U), extrapolating from the data is probably not a good idea.

**Figure 3**

## Exercises

**Stock Price** The following table gives the stock price for the large equipment manufacturer Caterpillar at the start of the given quarters. Let $x = 1$ correspond to the first quarter of the year 2015.

| Date | Quarter | Stock Price | Date | Quarter | Stock Price |
|---|---|---|---|---|---|
| January 2015 | 1 | $91.77 | January 2016 | 5 | $67.96 |
| April 2015 | 2 | $80.03 | April 2016 | 6 | $76.54 |
| July 2015 | 3 | $84.82 | July 2016 | 7 | $75.81 |
| October 2015 | 4 | $65.36 | October 2016 | 8 | $88.77 |

1. If you have access to the appropriate technology, verify that the least-squares regression line that models these data (with rounded coefficients) is $y = -.77x + 82.3$.

2. Use the model from Exercise 1 to calculate the predicted values $\hat{y}$ for each quarter in the table.

3. Use your answers from Exercise 2 to calculate the residuals for each quarter in the table. Make a graph of the residuals similar to Figure 3.

4. Do you think the linear model of Exercise 1 is a good fit for these data? Why or why not?

**Hourly Wages** The following table gives the average hourly earnings (in dollars) of U.S. production workers from 1970 to 2010. (Data from: U.S. Bureau of Labor Statistics.)

| Year | Average Hourly Wage (Dollars) |
|---|---|
| 1970 | 3.40 |
| 1975 | 4.73 |
| 1980 | 6.85 |
| 1985 | 8.74 |
| 1990 | 10.20 |
| 1995 | 11.65 |
| 2000 | 14.02 |
| 2005 | 16.13 |
| 2010 | 19.06 |
| 2015 | 21.03 |

5. If you have appropriate technology, verify that the least-squares regression line that models these data (with coefficients rounded) is $y = .3911x - 24.6$, where $y$ is the average hourly wage in year $x$ and $x = 0$ corresponds to 1900.

6. Use the model from Exercise 5 to interpolate a prediction of the hourly wage in 2012. The actual value was $14.97. How close was your prediction?

7. Use the model from Exercise 5 to extrapolate to 1960 and predict the hourly average wage. Why is this prediction nonsensical?

8. Using the model from Exercise 5, find the average hourly wage for each year in the table, and subtract it from the actual value in the second column. This gives you a table of the residuals. Plot your residuals as points on a graph.

9. What will happen if you try linear regression on the *residuals*? If you're not sure, use technology such as a graphing calculator to find the regression equation for the residuals. Why does this result make sense?

# Extended Project

Go the library or use the Internet to obtain the average amount of student loan debt for the most recent years available.

1. Create a scatterplot of the data with the amount on the $y$-axis and the year on the $x$-axis and assess the trend.

2. Fit a linear model to the data and predict the loan debt amount for each year.

3. Calculate the residuals from each year in the data set. Graph the residuals.

4. Do the residuals show a U-shape?

5. Find the correlation coefficient for your data.

# Functions and Graphs

# 3

## CHAPTER

The modern world is overwhelmed with data—from the cost of housing, utility bills, and student loans to trends in consumer confidence, corporate revenue, smoking rates, and alcohol consumption. Examples and exercises in this chapter investigate functions and graphs that describe these and other real-world phenomena. Functions enable us to construct mathematical models that can sometimes be used to estimate outcomes. Graphs of functions allow us to visualize a situation and to detect trends more easily.

Functions are an extremely useful way of describing many real-world situations in which the value of one quantity varies with, depends on, or determines the value of another. In this chapter, we will introduce functions, learn how to use functional notation, develop skills in constructing and interpreting the graphs of functions, and learn to apply this knowledge in a variety of situations.

## 3.1 Functions

To understand the origin of the concept of a function, we consider some real-life situations in which one numerical quantity depends on, corresponds to, or determines another. The way in which one quantity depends on another can be described verbally, graphically, numerically using a table of data, or algebraically by means of a formula.

> **EXAMPLE 1** **Cell Phone Data Plan** The amount of the bill you pay for Internet data usage on your cell phone *depends on* the number of gigabytes of data you use. The way in which the usage determines the bill is described in the data plan. In this example, the way in which one numerical quantity depends on another is described verbally.

**EXAMPLE 2**  **Stock Exchange** The graph and table of data in Figure 3.1 show the NASDAQ Composite Index from January 2011 through January 2017. The NASDAQ index is a statistical measure of how well a portion of the market is performing. The index tracks approximately 4000 stocks that are traded on the National Association of Securities Dealers Automated Quotations (NASDAQ) System electronic stock exchange. The graph and table display the index value *corresponding to* each year. In this example, the way in which one quantity corresponds to another is described graphically and numerically. (Data from: www.morningstar.com.)

| Year | NASDAQ |
|------|--------|
| 2011 | 2700.08 |
| 2012 | 2813.84 |
| 2013 | 3142.13 |
| 2014 | 4103.88 |
| 2015 | 4635.24 |
| 2016 | 4613.95 |
| 2017 | 5614.79 |

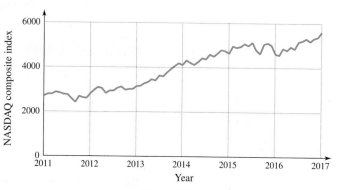

**Figure 3.1**

**EXAMPLE 3**  **Physical Science** Suppose a rock is dropped straight down from a high point. From physics, we know that the distance traveled by the rock in $t$ seconds is $16t^2$ feet. So the distance *depends on* the time, and this relationship is described by an algebraic formula.

These examples share a couple of features. First, each involves two sets of numbers, which we can think of as inputs and outputs. Second, in each case, there is a rule by which each input determines an output, as summarized here:

| | Set of Inputs | Set of Outputs | Rule |
|--|--|--|--|
| **Example 1** | All data usages | All bill amounts | The data plan |
| **Example 2** | Years | Index values | Graph or table |
| **Example 3** | Seconds elapsed after dropping the rock | Distances the rock travels | Distance $= 16t^2$ |

Each of these examples may be mentally represented by an idealized calculator that has a single operation key: A number is entered [*input*], the rule key is pushed [*rule*], and an answer is displayed [*output*]. The formal definition of a function incorporates these same common features (input–rule–output), with a slight change in terminology.

> A **function** consists of a set of inputs called the **domain,** a set of outputs called the **range,** and a rule by which each input determines *exactly one* output.

✓**Checkpoint 1**

Find the domain and range of the function in Example 3, assuming (unrealistically) that the rock can fall forever.

*Answers to Checkpoint exercises are found at the end of the section.*

**EXAMPLE 4**  Find the domains and ranges for the functions in Examples 1 and 2.

**Solution** In Example 1, the domain consists of all possible data usage amounts and the range consists of all possible bill amounts.

In Example 2, the domain is the set of years from 2011 to 2017, all real numbers in the interval [2011, 2017]. The range consists of the NASDAQ Composite Index values that actually occur during those years. ✓₁

Be sure that you understand the phrase "exactly one output" in the definition of the rule of a function. In Example 1, for instance, your data usage (input) determines exactly one bill amount (output)—you can't be charged two different amounts for the same data usage. However, it is quite possible for the amount of the bill (output) to be the same for using two different amounts of data (the input)—for example, using 2.9 gigabytes might cost the same as using 2.1 gigabytes of data. In other words, we can say the following:

> **In a function, each input produces a single output, but different inputs may produce the same output.**

**EXAMPLE 5** Which of the following rules describe functions?

**(a)** Use the optical reader at the checkout counter of the grocery store to convert codes to prices.

**Solution** For each code, the reader produces exactly one price, so this is a function.

**(b)** A loan officer at Bank of America asks each applicant to provide her or his Social Security number (customers must have a social security number in order to apply).

**Solution** Because each applicant (input) has exactly one Social Security number (output), this is a function.

**(c)** A loan officer at Bank of America asks each applicant to provide her or his credit card number.

**Solution** Because an applicant may have more than one credit card (or no credit cards), this is not a function.

**(d)** Enter a number in a calculator and press the $x^2$ key.

**Solution** This is a function, because the calculator produces just one number $x^2$ for each number $x$ that is entered.

**(e)** Assign to each number $x$ the number $y$ given by this table:

| $x$ | 1 | 1 | 2 | 2 | 3 | 3 |
|---|---|---|---|---|---|---|
| $y$ | 3 | −3 | −5 | −5 | 8 | −8 |

**Solution** Since $x = 1$ corresponds to more than one $y$-value (as does $x = 3$), this table does not define a function.

**(f)** Assign to each number $x$ the number $y$ given by the equation $y = 3x − 5$.

**Solution** Because the equation determines a unique value of $y$ for each value of $x$, it defines a function. ✔2

The equation $y = 3x − 5$ in part (f) of Example 5 defines a function, with $x$ as input and $y$ as output, because each value of $x$ determines a *unique* value of $y$. In such a case, the equation is said to **define y as a function of x**.

**EXAMPLE 6** Decide whether each of the following equations defines $y$ as a function of $x$.

**(a)** $y = −4x + 11$

**Solution** For a given value of $x$, calculating $−4x + 11$ produces exactly one value of $y$. (For example, if $x = −7$, then $y = −4(−7) + 11 = 39$). Because one value of the input variable leads to exactly one value of the output variable, $y = −4x + 11$ defines $y$ as a function of $x$.

✓ **Checkpoint 2**

Do the following define functions?

**(a)** The correspondence defined by the rule $y = x^2 + 5$, where $x$ is the input and $y$ is the output.

**(b)** The correspondence defined by entering a nonzero number in a calculator and pressing the $1/x$ key.

**(c)** The correspondence between a computer, $x$, and several users of the computer, $y$.

**(b)** $y^2 = x$

**Solution** Suppose $x = 36$. Then $y^2 = x$ becomes $y^2 = 36$, from which $y = 6$ or $y = -6$. Since one value of $x$ can lead to two values of $y$, $y^2 = x$ does *not* define $y$ as a function of $x$. ✓₃

Almost all the functions in this text are defined by formulas or equations, as in part (a) of Example 6. The domain of such a function is determined by the following **agreement on domains.**

> Unless otherwise stated, assume that the domain of any function defined by a formula or an equation is the largest set of real numbers (inputs) that each produces a real number as output.

**EXAMPLE 7**    Each of the given equations defines $y$ as a function of $x$. Find the domain of each function.

**(a)**  $y = x^4$

**Solution** Any number can be raised to the fourth power, so the domain is the set of all real numbers, which is sometimes written as $(-\infty, \infty)$.

**(b)**  $y = \sqrt{x - 6}$

**Solution** For $y$ to be a real number, $x - 6$ must be nonnegative. This happens only when $x - 6 \geq 0$—that is, when $x \geq 6$. So the domain is the interval $[6, \infty)$.

**(c)**  $y = \sqrt{4 - x}$

**Solution** For $y$ to be a real number here, we must have $4 - x \geq 0$, which is equivalent to $x \leq 4$. So the domain is the interval $(-\infty, 4]$.

**(d)**  $y = \sqrt[3]{4 - x}$

**Solution** We can take the cube root of any number, so the domain is the set of all real numbers, $(-\infty, \infty)$.

**(e)**  $y = \dfrac{1}{x + 3}$

**Solution** Because the denominator cannot be 0, $x \neq -3$, and the domain consists of all numbers in the intervals

$$(-\infty, -3) \quad \text{or} \quad (-3, \infty).$$

**(f)**  $y = \dfrac{\sqrt{x}}{x^2 - 3x + 2}$

**Solution** The numerator is defined only when $x \geq 0$. The domain cannot contain any numbers that make the denominator 0—that is, the numbers that are solutions of

$$x^2 - 3x + 2 = 0.$$
$$(x - 1)(x - 2) = 0 \qquad \text{Factor.}$$
$$x - 1 = 0 \quad \text{or} \quad x - 2 = 0 \qquad \text{Zero-factor property}$$
$$x = 1 \quad \text{or} \qquad x = 2.$$

Therefore, the domain consists of all nonnegative real numbers except 1 and 2. ✓₄

## Functional Notation

In actual practice, functions are seldom presented in the style of domain–rule–range, as they have been here. Functions are usually denoted by a letter such as $f$. If $x$ is an input, then $f(x)$ denotes the output number that the function $f$ produces from the input $x$. The symbol $f(x)$ is read "$f$ of $x$." The rule is usually given by a formula, such as $f(x) = \sqrt{x^2 + 1}$. This formula can be thought of as a set of directions:

Name of function    Input number

$$\underbrace{f(x)}_{\text{Output number}} = \sqrt{x^2 - 1}$$

Directions that tell you what to do with input $x$
in order to produce the corresponding output
$f(x)$—namely, "square it, subtract 1, and take
the square root of the result."

For example, to find $f(3)$ (the output number produced by the input 3), simply replace $x$ by 3 in the formula:

$$f(3) = \sqrt{3^2 - 1}$$
$$= \sqrt{8}.$$

Similarly, replacing $x$ by $-5$ shows that

$$f(-5) = \sqrt{(-5)^2 - 1} = \sqrt{25 - 1} = \sqrt{24}.$$

If we try to replace $x$ with 0, we get

$$f(0) = \sqrt{(0)^2 - 1} = \sqrt{-1}, \text{ which is not a real number.}$$

Therefore, $x = 0$ is not in the domain of the function, and we say that $f(0)$ is *not defined*.

These directions can be applied to any quantities, such as $a + b$ or $c^4$ (where $a$, $b$, and $c$ are real numbers). Thus, to compute $f(a + b)$, the output corresponding to input $a + b$, we square the input [obtaining $(a + b)^2$], subtract 1 [obtaining $(a + b)^2 - 1$], and take the square root of the result:

$$f(a + b) = \sqrt{(a + b)^2 - 1}$$
$$= \sqrt{a^2 + 2ab + b^2 - 1}.$$

Similarly, the output $f(c^4)$, corresponding to the input $c^4$, is computed by squaring the input $[(c^4)^2]$, subtracting 1 $[(c^4)^2 - 1]$, and taking the square root of the result:

$$f(c^4) = \sqrt{(c^4)^2 - 1}$$
$$= \sqrt{c^8 - 1}.$$

**EXAMPLE 8**    Let $g(x) = -x^2 + 4x - 5$. Find each of the given outputs.

**(a)** $g(-2)$

**Solution** Replace $x$ with $-2$:

$$g(-2) = -(-2)^2 + 4(-2) - 5$$
$$= -4 - 8 - 5$$
$$= -17.$$

**(b)** $g(x + h)$

**Solution** Replace $x$ by the quantity $x + h$ in the rule of $g$:

$$g(x + h) = -(x + h)^2 + 4(x + h) - 5$$
$$= -(x^2 + 2xh + h^2) + (4x + 4h) - 5$$
$$= -x^2 - 2xh - h^2 + 4x + 4h - 5.$$

**(c)** $g(x + h) - g(x)$

**Solution** Use the result from part (b) and the rule for $g(x)$:

$$g(x + h) - g(x) = \overbrace{(-x^2 - 2xh - h^2 + 4x + 4h - 5)}^{g(x+h)} - \overbrace{(-x^2 + 4x - 5)}^{g(x)}$$
$$= -2xh - h^2 + 4h.$$

**(d)** $\dfrac{g(x + h) - g(x)}{h}$ (assuming that $h \neq 0$)

**Solution** The numerator was found in part (c). Divide it by $h$ as follows:

$$\frac{g(x + h) - g(x)}{h} = \frac{-2xh - h^2 + 4h}{h}$$
$$= \frac{h(-2x - h + 4)}{h} \quad \text{Factor numerator.}$$
$$= -2x - h + 4. \quad \text{Simplify.}$$

The quotient found in Example 8(d),

$$\frac{g(x + h) - g(x)}{h},$$

is called the **difference quotient** of the function $g$. Difference quotients are important in calculus. ✔5

**Checkpoint 5**

Let $f(x) = 5x^2 - 2x + 1$. Find the following.

**(a)** $f(1)$
**(b)** $f(3)$
**(c)** $f(1 + 3)$
**(d)** $f(1) + f(3)$
**(e)** $f(m)$
**(f)** $f(x + h) - f(x)$
**(g)** $\dfrac{f(x + h) - f(x)}{h} \ (h \neq 0)$

**CAUTION** Functional notation is *not* the same as ordinary algebraic notation. You cannot simplify an expression such as $f(x + h)$ by writing $f(x) + f(h)$. To see why, consider the answers to Checkpoints 5(c) and (d), which show that

$$f(1 + 3) \neq f(1) + f(3).$$

## Applications

To describe some real-world situations, we need a function with a multipart rule, as the following example illustrates. Such a function is called a **piecewise-defined function**.

**EXAMPLE 9** **Income Tax** If you were a single person in Connecticut in 2017 with a taxable income of $x$ dollars and $x \leq \$50,000$, then your state income tax $T(x)$ was determined by the rule

$$T(x) = \begin{cases} .03x & \text{if } 0 \leq x \leq 10,000 \\ 300 + .05(x - 10,000) & \text{if } 10,000 < x \leq 50,000. \end{cases}$$

Find the income tax paid by a single person with the given taxable income. (Data from: www.tax-brackets.org.)

**(a)** $9200

**Solution** We must find $T(9200)$. Since 9200 is less than 10,000, the first piece of the rule applies (the first line of the piecewise-defined function):

$$T(x) = .03x$$
$$T(9200) = .03(9200) = \$276. \quad \text{Let } x = 9200.$$

**(b)** $30,000

**Solution** Now we must find $T(30,000)$. Since 30,000 is greater than $10,000, the second piece of the rule applies:

$$T(x) = 300 + .05(x - 10,000)$$
$$T(30,000) = 300 + .05(30,000 - 10,000) \quad \text{Let } x = 30,000.$$
$$= 300 + .05(20,000) \quad \text{Simplify.}$$
$$= 300 + 1000 = \$1300. \ ✔6$$

**Checkpoint 6**

Use Example 9 to find the tax on each of these incomes.

**(a)** $48,750
**(b)** $7345

**EXAMPLE 10**   **Sales** Suppose the projected sales (in thousands of dollars) of a small company over the next 10 years are approximated by the function

$$S(x) = .07x^4 - .05x^3 + 2x^2 + 7x + 62$$

where $x = 0$ corresponds to the current year.

**(a)** What are the projected sales for the current year?

**Solution** Since the current year corresponds to $x = 0$, the sales for this year are given by $S(0)$. Substituting 0 for $x$ in the rule for $S$, we see that $S(0) = 62$. So the current projected sales are $62,000.

**(b)** What will sales be in four years?

**Solution** The sales in four years from now are given by $S(4)$, which can be computed by hand or with a calculator:

$$S(x) = .07x^4 - .05x^3 + 2x^2 + 7x + 62$$
$$S(4) = .07(4)^4 - .05(4)^3 + 2(4)^2 + 7(4) + 62 \qquad \text{Let } x = 4.$$
$$= 136.72.$$

Thus, sales are projected to be $136,720. ✓ 7

✓**Checkpoint 7**

A developer estimates that the total cost of building $x$ large apartment complexes in a year is approximated by

$$A(x) = x^2 + 80x + 60,$$

where $A(x)$ represents the cost in hundred thousands of dollars. Find the cost of building

**(a)** 4 complexes;

**(b)** 10 complexes.

| X | Y1 |
|---|---|
| 5 | 184.5 |
| 6 | 255.92 |
| 7 | 359.92 |
| 8 | 507.12 |
| 9 | 709.82 |
| 10 | 982 |
| 11 | 1339.3 |

X=5

**Figure 3.2**

**EXAMPLE 11**   **Sales** Use the table feature of the graphing calculator to find the projected sales of the company in Example 10 for years 5 through 10.

**Solution** Enter the sales equation $y = .07x^4 - .05x^3 + 2x^2 + 7x + 62$ into the equation memory of the calculator (often called the Y = list). Check your instruction manual for how to set the table to start at $x = 5$ and go at least through $x = 10$. Then display the table, as in Figure 3.2. The figure shows that sales are projected to rise from $184,500 in year 5 to $982,000 in year 10.

# 3.1 Exercises

*For each of the following rules, state whether it defines y as a function of x or not. (See Examples 5 and 6.)*

**1.**

| x | 3 | 2 | 1 | 0 | −1 | −2 | −3 |
|---|---|---|---|---|---|---|---|
| y | 9 | 4 | 1 | 0 | 1 | 4 | 9 |

**2.**

| x | 10 | 20 | 30 | 40 | 50 | 60 |
|---|---|---|---|---|---|---|
| y | −1 | −1 | 0 | 0 | 1 | 1 |

**3.**

| x | 9 | 4 | 1 | 0 | 1 | 4 | 9 |
|---|---|---|---|---|---|---|---|
| y | 3 | 2 | 1 | 0 | −1 | −2 | −3 |

**4.**

| x | −1 | −1 | 0 | 0 | 1 | 1 |
|---|---|---|---|---|---|---|
| y | 10 | 20 | 30 | 40 | 50 | 60 |

**5.** $y = x^3$

**6.** $y = \sqrt{x - 1}$

**7.** $x = y^2 + 3$

**8.** $x = |y + 2|$

**9.** $y = \dfrac{-1}{x - 1}$

**10.** $y = \dfrac{4}{2x + 3}$

*State the domain of each function. (See Example 7.)*

**11.** $f(x) = 4x - 1$

**12.** $f(x) = 2x + 7$

**13.** $f(x) = x^4 - 1$

**14.** $f(x) = (2x + 5)^2$

**15.** $f(x) = \sqrt{5 - x}$

**16.** $f(x) = \sqrt{-x} + 3$

**17.** $f(x) = \dfrac{1}{x - 2}$

**18.** $g(x) = \dfrac{x}{x^2 + x - 2}$

**19.** $g(x) = \dfrac{x^2 + 4}{x^2 - 4}$

**20.** $g(x) = \dfrac{x^2 - 1}{x^2 + 1}$

**21.** $h(x) = \dfrac{\sqrt{x + 4}}{x^2 + x - 12}$

**22.** $\dfrac{\sqrt{x + 1}}{x^2 - x - 6}$

*For each of the following functions, find*

**(a)** $f(4)$;   **(b)** $f(-3)$;   **(c)** $f(2.7)$;   **(d)** $f(-4.9)$.

*(See Examples 8 and 9.)*

**23.** $f(x) = 8$

**24.** $f(x) = 0$

**25.** $f(x) = 2x^2 + 4x$

**26.** $f(x) = x^2 - 2x$

**27.** $f(x) = \sqrt{5 - x}$

**28.** $f(x) = \sqrt{x + 3}$

**29.** $f(x) = |x^2 - 6x - 4|$    **30.** $f(x) = |x^3 - x^2 + x - 1|$

**31.** $f(x) = \dfrac{\sqrt{x-1}}{x^2 - 1}$    **32.** $f(x) = \sqrt{-x} + \dfrac{2}{x+1}$

**33.** $f(x) = \begin{cases} -2x + 4 & \text{if } x \le 1 \\ 3 & \text{if } 1 < x < 4 \\ x + 1 & \text{if } x \ge 4 \end{cases}$

**34.** $f(x) = \begin{cases} x^2 & \text{if } x < 2 \\ 5x - 7 & \text{if } x \ge 2 \end{cases}$

*For each of the following functions, find*

**(a)** $f(p)$;    **(b)** $f(-r)$;    **(c)** $f(m+3)$.

*(See Example 8.)*

**35.** $f(x) = 6 - x$    **36.** $f(x) = 3x + 5$

**37.** $f(x) = \sqrt{4 - x}$    **38.** $f(x) = \sqrt{-2x}$

**39.** $x^2 - 1$    **40.** $1 - x^2$

**41.** $f(x) = \dfrac{3}{x-1}$    **42.** $f(x) = \dfrac{-1}{5+x}$

*For each of the following functions, find the difference quotient*

$$\frac{f(x+h) - f(x)}{h} \quad (h \ne 0).$$

*(See Example 8.)*

**43.** $f(x) = 2x - 4$    **44.** $f(x) = 2 + 4x$

**45.** $f(x) = x^2 + 1$    **46.** $f(x) = x^2 - x$

⊞ *Use a calculator to work these exercises. (See Examples 9 and 10.)*

**47. Gross Domestic Product** The gross domestic product (GDP) of the United States, which measures the overall size of the economy in billions of dollars, can be approximated by the function

$$f(x) = 589x + 8858,$$

where $x = 10$ corresponds to the year 2010. Estimate the GDP (to the nearest billion dollars) in the given years. (Data from: U.S. Bureau of Economic Analysis.)

**(a)** 2013    **(b)** 2015    **(c)** 2016

**48. Consumer Consumption** Personal consumption expenditures in billions of dollars in the United States can be approximated by the function

$$f(x) = 403x + 6061,$$

where $x = 10$ corresponds to the year 2010. Estimate personal consumption (to the nearest billion dollars) in the given years. (Data from: U.S. Bureau of Economic Analysis.)

**(a)** 2010    **(b)** 2015    **(c)** 2016

**49. Stock Prices** The closing stock price of Microsoft Corporation for the years 2010 to 2016 can be approximated by the function

$$f(x) = .8x^2 - 17x + 119,$$

where $x = 10$ corresponds to the start of the year 2010. Find the following function values and explain what your answers mean in practical terms. (Data from: www.morningstar.com.)

**(a)** $f(12)$    **(b)** $f(14)$    **(c)** $f(16)$

**50. Stock Prices** The closing stock price of American Express Company for the years 2013 to 2015 can be approximated by the function

$$f(x) = -15.27x^2 + 438.5x - 3061,$$

where $x = 13$ corresponds to the start of the year 2013. Find the following function values and explain what your answers mean in practical terms. (Data from: www.morningstar.com.)

**(a)** $f(13)$    **(b)** $f(14)$    **(c)** $f(15)$

**51. Taxes** Alabama state income tax for a single person in 2017 was determined by the rule

$$T(x) = \begin{cases} .02x & \text{if } 0 \le x \le 500 \\ 10 + .04(x - 500) & \text{if } 500 < x \le 3000 \\ 110 + .05(x - 3000) & \text{if } x > 3000, \end{cases}$$

where $x$ is the person's taxable income. Find the following function values and interpret your answers. (Data from: www.tax-brackets.org.)

**(a)** $T(100)$    **(b)** $T(2500)$    **(c)** $T(103,000)$

**52. Taxes** Mississippi state income tax for a single person in 2017 was determined by the rule

$$T(x) = \begin{cases} .03x & \text{if } 0 \le x \le 5000 \\ 150 + .04(x - 5000) & \text{if } 5000 < x \le 10{,}000 \\ 350 + .05(x - 10{,}000) & \text{if } x > 10{,}000, \end{cases}$$

where $x$ is the person's taxable income. Find the following function values and interpret your answers. (Data from: www.tax-brackets.org.)

**(a)** $T(100)$    **(b)** $T(7000)$    **(c)** $T(110,000)$

**53. Social Science** The mean age of a woman in the United States when her first child is born can be approximated by the function

$$f(x) = 21x^{.0565},$$

where $x = 10$ corresponds to the year 2010. Estimate the mean age of a woman at the birth of her first child in the following years. Round your answers to one decimal place. (Data from: The General Social Survey.)

**(a)** 2010    **(b)** 2012    **(c)** 2014

**54. Water Pollution** High concentrations of zinc ions in water are lethal to rainbow trout. The function

$$f(x) = \left(\frac{x}{1960}\right)^{-.833}$$

gives the approximate average survival time (in minutes) for trout exposed to $x$ milligrams per liter (mg/L) of zinc ions. Find the survival time (to the nearest minute) for the given concentrations of zinc ions.

**(a)** 110    **(b)** 525    **(c)** 1960    **(d)** 4500

**55. Physical Science** The distance from Chicago to Sacramento, California, is approximately 2050 miles. A plane flying directly to Sacramento passes over Chicago at noon. If the plane travels at 500 mph, find the rule of the function $f(t)$ that gives the distance of the plane from Sacramento at time $t$ hours (with $t = 0$ corresponding to noon).

56. **Physical Science** The distance from Toronto, Ontario to Dallas, Texas is approximately 1200 miles. A plane flying directly to Dallas passes over Toronto at 2 p.m. If the plane travels at 550 mph, find the rule of the function $f(t)$ that gives the distance of the plane from Dallas at time $t$ hours (with $t = 0$ corresponding to 2 p.m.).

57. **Cost, Revenue, and Profit** A pretzel factory has daily fixed costs of $1800. In addition, it costs 50 cents to produce each bag of pretzels. A bag of pretzels sells for $1.20.

   (a) Find the rule of the cost function $c(x)$ that gives the total daily cost of producing $x$ bags of pretzels.

   (b) Find the rule of the revenue function $r(x)$ that gives the daily revenue from selling $x$ bags of pretzels.

   (c) Find the rule of the profit function $p(x)$ that gives the daily profit from $x$ bags of pretzels. (*Hint:* Recall that profit = revenue − cost.)

58. **Cost, Revenue, and Profit** An aluminum can factory has daily fixed costs of $120,000 per day. In addition, it costs $.03 to produce each can. They can sell a can to a soda manufacturer for $.05 a can.

   (a) Find the rule of the cost function $c(x)$ that gives the total daily cost of producing $x$ aluminum cans.

   (b) Find the rule of the revenue function $r(x)$ that gives the daily revenue from selling $x$ aluminum cans.

   (c) Find the rule of the profit function $p(x)$ that gives the daily profit from $x$ aluminum cans.

🖩 *Use the table feature of a graphing calculator to do these exercises. (See Example 11.)*

59. **Imports** Total imports to the United States in billions of dollars for the period 2010 to 2014 can be approximated by the function

$$f(x) = 28.2x^3 - 208x^2 + 507x + 2348,$$

where $x = 0$ corresponds to the year 2010. Create a table that gives total imports for the years 2010 to 2014 (that is, $x = 0$ to $x = 4$). (Data from: U.S. Census Bureau.)

60. **Inflation** The inflation rate in the United States (expressed as a percentage) for the period 2010 to 2014 can be approximated by the function

$$f(x) = .28x^3 - 1.84x^2 + 2.88x + 1.68,$$

where $x = 0$ corresponds to the year 2010. Create a table that gives the inflation rate for the years 2010 to 2014 (that is, $x = 0$ to $x = 4$). (Data from: International Monetary Fund.)

✓**Checkpoint Answers**

1. The domain consists of all possible times—that is, all nonnegative real numbers. The range consists of all possible distances; thus, the range is also the set of all nonnegative real numbers.

2. (a) Yes      (b) Yes      (c) No

3. (a) Yes      (b) Yes      (c) No      (d) No

4. (a) $(-\infty, \infty)$   (b) $(-\infty, \infty)$   (c) $(-\infty, 0]$
   (d) All real numbers except 1 and −1

5. (a) 4       (b) 40       (c) 73       (d) 44
   (e) $5m^2 - 2m + 1$       (f) $10xh + 5h^2 - 2h$
   (g) $10x + 5h - 2$

6. (a) $2237.50       (b) $220.35

7. (a) $39,600,000       (b) $96,000,000

# 3.2  Graphs of Functions

The **graph** of a function $f$ is defined to be the graph of the *equation* $y = f(x)$. It consists of all points $(x, f(x))$—that is, every point whose first coordinate is an input number from the domain of $f$ and whose second coordinate is the corresponding output number.

Ⓒ **EXAMPLE 1**   The graph of the function $g(x) = .5x - 3$ is the graph of the equation $y = .5x - 3$. So the graph is a straight line with slope .5 and $y$-intercept −3. To graph the line, we plot just two points. Because we know the $y$-intercept, we already have the point $(0, -3)$. For a second point, we will find the $x$-intercept (although any other point would suffice). Set $y = 0$ and solve for $x$:

$$y = 0$$
$$.5x - 3 = 0$$
$$.5x = 3$$
$$x = 6.$$

✓**Checkpoint 1**

Graph $g(x) = 3 - .5x$.

Now we have the point $(6, 0)$, and we graph the line shown in Figure 3.3. For a refresher on graphing lines, see Section 2.2. ✓1

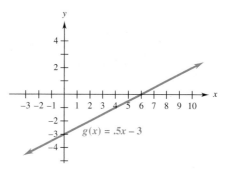

**Figure 3.3**

A function whose graph is a straight line, as in Example 1, is called a **linear function.** The rule of a linear function can always be put into the form

$$f(x) = ax + b$$

for some constants $a$ and $b$.

## Piecewise Linear Functions

We now consider functions whose graphs consist of straight-line segments. Such functions are called **piecewise linear functions** and are typically defined with different equations for different parts of the domain.

---

**EXAMPLE 2** Graph the following function:

$$f(x) = \begin{cases} x + 1 & \text{if } x \leq 2 \\ -2x + 7 & \text{if } x > 2. \end{cases}$$

**Solution** Consider the two parts of the rule of $f$. The graphs of $y = x + 1$ and $y = -2x + 7$ are straight lines. The graph of $f$ consists of

the part of the line $y = x + 1$ with $x \leq 2$ and

the part of the line $y = -2x + 7$ with $x > 2$.

Each of these line segments can be graphed by plotting two points in the appropriate interval, as shown in Figure 3.4.

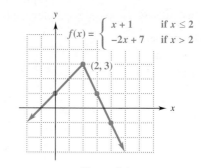

**Figure 3.4**

Note that the left and right parts of the graph each extend to the vertical line through $x = 2$, where the two halves of the graph meet at the point $(2, 3)$. ✓₂

---

✓ **Checkpoint 2**

Graph

$$f(x) = \begin{cases} x + 2 & \text{if } x < 0 \\ 2 - x & \text{if } x \geq 0. \end{cases}$$

**EXAMPLE 3** Graph the function

$$f(x) = \begin{cases} x - 2 & \text{if } x \leq 3 \\ -x + 8 & \text{if } x > 3. \end{cases}$$

**Solution** The graph consists of parts of two lines. To find the left side of the graph, choose two values of $x$ with $x \leq 3$, say, $x = 0$ and $x = 3$. Then find the corresponding points on $y = x - 2$, namely, $(0, -2)$ and $(3, 1)$. Use these points to draw the line segment to the left of $x = 3$, as in Figure 3.5. Next, choose two values of $x$ with $x > 3$, say, $x = 4$ and $x = 6$, and find the corresponding points on $y = -x + 8$, namely, $(4, 4)$ and $(6, 2)$. Use these points to draw the line segment to the right of $x = 3$, as in Figure 3.5.

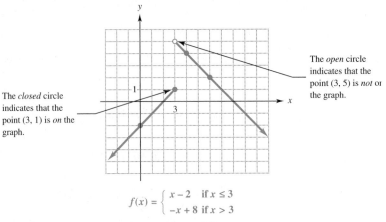

The *closed* circle indicates that the point $(3, 1)$ is *on* the graph.

The *open* circle indicates that the point $(3, 5)$ is *not* on the graph.

$$f(x) = \begin{cases} x - 2 & \text{if } x \leq 3 \\ -x + 8 & \text{if } x > 3 \end{cases}$$

**Figure 3.5**

Note that both line segments of the graph of $f$ extend to the vertical line through $x = 3$. Because the two graphs do not connect at the input $x = 3$, the graph has a break, or discontinuity, at $x = 3$. ✓ 3

✓ **Checkpoint 3**

Graph

$$f(x) = \begin{cases} -2x - 3 & \text{if } x < 1 \\ x - 2 & \text{if } x \geq 1. \end{cases}$$

**EXAMPLE 4**    Graph the **absolute-value function,** whose rule is $f(x) = |x|$.

**Solution** The definition of absolute value on page 8 shows that the rule of $f$ can be written as

$$f(x) = \begin{cases} x & \text{if } x \geq 0 \\ -x & \text{if } x < 0. \end{cases}$$

So the right half of the graph (that is, where $x \geq 0$) will consist of a portion of the line $y = x$. It can be graphed by plotting two points, say, $(0, 0)$ and $(1, 1)$. The left half of the graph (where $x < 0$) will consist of a portion of the line $y = -x$, which can be graphed by plotting $(-2, 2)$ and $(-1, 1)$, as shown in Figure 3.6. ✓ 4

✓ **Checkpoint 4**

Graph each function.

(a) $f(x) = |x - 4|$

(b) $f(x) = |2x - 4|$

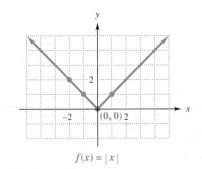

$$f(x) = |x|$$

**Figure 3.6**

Many real-world problems can be modeled using piecewise-defined functions. For example, every employee pays Social Security and Medicare taxes, and these formulas are piecewise-defined (see Exercises 37 and 38). The costs for many products and services are piecewise-defined due to either volume discounts or higher rates for excessive usage, as in the following example.

**EXAMPLE 5** **Utility Bills** The Birmingham Water Works charges its residential customers with a ⅝-inch water line a basic service charge of approximately $25 per month. The first 15 hundred cubic feet (CCF) of water cost about $2 per CCF. To encourage water conservation, excessive water usage is charged at a higher rate. It costs about $4 per CCF for each CCF above 15. (Data from: Birmingham Water Works.)

**(a)** What would be the amount of the water bill for a customer who used no water?

**Solution** The customer would just pay the $25 basic service charge.

**(b)** Find the water bill for a customer who used 15 CCF.

**Solution** The customer would pay the $25 service charge plus $2 for each CCF used, for a total of $25 + $2(15) = $25 + $30 = $55.

**(c)** How much would it cost to use 25 CCF of water?

**Solution** It would cost $55 for the first 15 CCF (as calculated above), plus $4 per CCF for each additional CCF above 15, for a total of $55 + $4(25 − 15) = $55 + $40 = $95.

**(d)** Find a rule for the function $f(x)$ that gives the amount of the water bill when $x$ CCF of water are used.

**Solution** For up to 15 CCF, the cost is the $25 fee plus $2 per CCF, for a total of $25 + 2x$. For usage above 15 CCF, the cost is $55 plus $4 per CCF for each additional CCF above 15, for a total of $55 + 4(x − 15)$. So the amount of the water bill is the piecewise-defined function

$$f(x) = \begin{cases} 25 + 2x & \text{if } 0 \leq x \leq 15 \\ 55 + 4(x - 15) & \text{if } x > 15. \end{cases}$$

**(e)** Graph $f(x)$.

**Solution** For $0 \leq x \leq 15$, the graph is part of the line $y = 25 + 2x$. We can easily graph this line segment by plotting and connecting the points $(0, 25)$ and $(15, 55)$. For $x > 15$, the graph is part of the line $y = 55 + 4(x - 15)$. Use the points $(15, 55)$ and $(25, 95)$ to sketch the portion of this line with $x > 15$. The graph is shown in Figure 3.7. ✓5

✓ **Checkpoint 5**

In Example 5, what would be the water bill for the following amounts of water?

**(a)** 5 CCF

**(b)** 30 CCF

**Figure 3.7**

⊞ **TECHNOLOGY TIP** To graph most piecewise linear functions on a graphing calculator, you must use a special syntax. For example, on TI calculators, the best way to obtain the graph in Example 3 is to graph two separate equations on the same screen:

$$y_1 = (x - 2)/(x \leq 3) \quad \text{and} \quad y_2 = (-x + 8)/(x > 3).$$

The inequality symbols are in the TEST (or CHAR) menu. However, most calculators will graph absolute-value functions directly. To graph $f(x) = |x + 2|$, for instance, graph the equation $y = abs(x + 2)$. "Abs" (for absolute value) is on the keyboard or in the MATH menu.

## Step Functions

The **greatest-integer function,** usually written $f(x) = [x]$, is defined by saying that $[x]$ denotes the largest integer that is less than or equal to $x$. For example, $[8] = 8$, $[7.45] = 7$, $[\pi] = 3$, $[-1] = -1$, $[-2.6] = -3$, and so on.

---

**EXAMPLE 6**    Graph the greatest-integer function $f(x) = [x]$.

**Solution** Consider the values of the function between each two consecutive integers—for instance,

| $x$ | $-2 \leq x < -1$ | $-1 \leq x < 0$ | $0 \leq x < 1$ | $1 \leq x < 2$ | $2 \leq x < 3$ |
|---|---|---|---|---|---|
| $[x]$ | $-2$ | $-1$ | $0$ | $1$ | $2$ |

Thus, between $x = -2$ and $x = -1$, the value of $f(x) = [x]$ is always $-2$, so the graph there is a horizontal line segment, all of whose points have second coordinate $-2$. The rest of the graph is obtained similarly (Figure 3.8). An open circle in that figure indicates that the endpoint of the segment is *not* on the graph, whereas a closed circle indicates that the endpoint *is* on the graph.

**Figure 3.8**

---

**Checkpoint 6**

Graph $y = [\tfrac{1}{2}x + 1]$.

---

Functions whose graphs resemble the graph of the greatest-integer function are sometimes called **step functions.**

Step functions occur in many real-life situations, such as the fare for an Uber ride, the cost of parking in a parking garage, and the bill for the data plan on your cell phone (for example, the cost might jump up by \$15 to buy an extra gigabyte of data). The following is another common example.

---

**EXAMPLE 7**    **Postal Rates** To ship a first-class package to Brazil in 2016, the U.S. Post Office charged a fee of \$13 for the first 8 ounces, \$9 for an additional 24 ounces or less, \$9 for the next 24 ounces or less, and then \$20 more for up to a maximum of 64 ounces. Let $D(x)$ represent the cost to send a package weighing $x$ ounces. Graph $y = D(x)$ for $x$ in the interval $(0,64]$. (Data from: www.usps.com.)

**Solution** For $x$ in the interval $(0, 8]$, $y = 13$. For $x$ in the interval $(8, 32]$, $y = 13 + 9 = 22$. For $x$ in $(32, 48]$, $y = 22 + 9 = 31$. Finally, for $x$ in $(48, 64]$, $y = 31 + 20 = 51$. The graph is a step function and is shown in Figure 3.9.

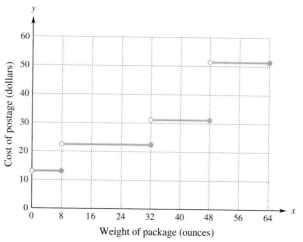

**Figure 3.9**

$$D(x) = \begin{cases} 13 & \text{if } 0 < x \le 8 \\ 22 & \text{if } 8 < x \le 32 \\ 31 & \text{if } 32 < x \le 48 \\ 51 & \text{if } 48 < x \le 64 \end{cases}$$

✓**Checkpoint 7**

To mail a letter to Canada in 2016, the U.S. Post Office charged $1.20 for up to and including the first two ounces, an additional $.47 for up to and including an additional ounce, and then an additional $.49 up to a maximum weight of 3.5 ounces. Let $L(x)$ represent the cost of sending a letter weighing $x$ ounces. Graph $L(x)$ for $x$ in the interval (0, 3.5].

⊞ **TECHNOLOGY TIP** On most graphing calculators, the greatest-integer function is denoted INT or FLOOR. (Look on the MATH menu or its NUM submenu.) When graphing these functions, put your calculator in "dot" graphing mode rather than the usual "connected" mode to avoid erroneous vertical line segments in the graph.

## Other Functions

The graphs of many functions do not consist only of straight-line segments. The most basic method for graphing functions by hand is to plot points. This method was introduced in Section 2.1 and is summarized here. In Sections 3.4–3.6, we will learn more techniques to aid with graphing. Calculus provides additional strategies.

### Graphing a Function by Plotting Points

1. Determine the domain of the function.
2. Select a few numbers in the domain of $f$ (include both negative and positive ones when possible), and compute the corresponding values of $f(x)$.
3. Plot the points $(x, f(x))$ computed in Step 2. Use these points and any other information you may have about the function to make an educated guess about the shape of the entire graph.
4. Unless you have information to the contrary, assume that the graph is continuous (unbroken) wherever it is defined.

This method was used to find the graphs of the functions $f(x) = x^2 - 2x - 8$ and $g(x) = \sqrt{x + 2}$ in Examples 4 and 5 of Section 2.1. Here are some more examples.

**EXAMPLE 8**    Graph $g(x) = \sqrt{x - 1}$.

**Solution**  Because the rule of the function is defined only when $x - 1 \ge 0$ (that is, when $x \ge 1$), the domain of $g$ is the interval [1, ∞). Use a calculator to make a table of values such as the one in Figure 3.10. Plot the corresponding points and connect them to get the graph in Figure 3.10. ✓8

✓**Checkpoint 8**

Graph $f(x) = \sqrt{5 - 2x}$.

| $x$ | $g(x) = \sqrt{x - 1}$ |
|---|---|
| 1 | 0 |
| 2 | 1 |
| 3 | $\sqrt{2} \approx 1.414$ |
| 5 | 2 |
| 7 | $\sqrt{6} \approx 2.449$ |
| 10 | 3 |

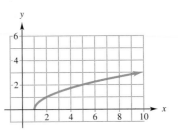

**Figure 3.10**

---

**EXAMPLE 9**    Graph the piecewise defined function

$$f(x) = \begin{cases} x^2 & \text{if } x \leq 2 \\ \sqrt{x - 1} & \text{if } x > 2. \end{cases}$$

**Solution**  When $x \leq 2$, the rule of the function is $f(x) = x^2$. Make a table of values such as the one in Figure 3.11. Plot the corresponding points and connect them to get the left half of the graph in Figure 3.11. When $x > 2$, the rule of the function is $f(x) = \sqrt{x - 1}$, whose graph is shown in Figure 3.10. In Example 8, the entire graph was given, beginning at $x = 1$. Here we use only the part of the graph to the right of $x = 2$, as shown in Figure 3.11. The open circle at (2, 1) indicates that this point is not part of the graph of $f$ (why?).

| $x$ | $x^2$ |
|---|---|
| $-2$ | 4 |
| $-1$ | 1 |
| 0 | 0 |
| 1 | 1 |
| 2 | 4 |

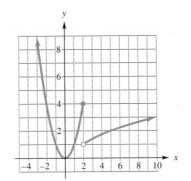

**Figure 3.11**

## Graph Reading

Graphs are often used in business and the social sciences to present data. It is just as important to know how to *read* such graphs as it is to construct them.

**EXAMPLE 10**    **Housing Prices**  Figure 3.12 shows the median sales prices (in thousands of dollars) of new privately owned, one-family houses by region of the United States (Northeast, Midwest, South, and West) for the years 2008 to 2015. (Data from: U.S. Department of Housing and Urban Development.)

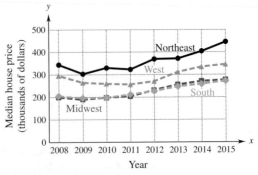

**Figure 3.12**

**(a)** How do prices in the West compare with those in the Midwest over the period shown in the graph?

**Solution** The prices of the houses in the West started out approximately $100,000 higher than those in the Midwest. In both regions, prices declined initially, but prices in both regions rebounded. By 2012, the gap between the two regions shrunk to less than $50,000 but widened again to about $75,000 by 2015.

**(b)** Which region typically had the highest prices?

**Solution** The Northeast had the highest median sales price for all years shown.

**EXAMPLE 11**   **Commodity Prices** The average price $f(x)$ of a barrel of crude oil (in dollars) in year $x$ is shown in Figure 3.13 for the years 2005 to 2015, with projected prices shown through the year 2025. (Data from: www.worldbank.org.)

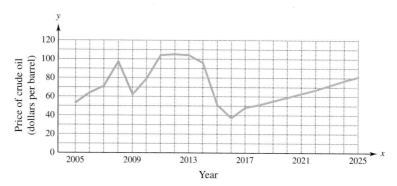

**Figure 3.13**

**(a)** Estimate the function values $f(2015)$ and $f(2025)$, and interpret your answers.

**Solution** The point (2015, 50) is on the graph, which means that $f(2015) = 50$. This tells us that, in the year 2015 (the input), the average price of a barrel of crude oil was $50 (the output). Similarly, we see that $f(2025) = 80$ because the point (2025, 80) is on the graph. This means that the price of crude oil is projected to be $80 per barrel in 2025.

**(b)** During what period did oil prices drop below $40 per barrel?

**Solution** Look for points on the graph whose second coordinates are 40 or below—that is, points on or below the horizontal line through 40. This occurs in the year 2016.

**(c)** During what period did oil prices soar above $100 per barrel?

**Solution** Look for points at or above the horizontal line through 100—that is, points with a second coordinate greater than or equal to 100. They occur from 2011 to 2013.

## The Vertical-Line Test

The following fact distinguishes the graphs of function from other graphs.

> ## Vertical-Line Test
>
> No vertical line intersects the graph of a function $y = f(x)$ at more than one point.

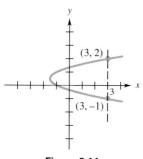

**Figure 3.14**

In other words, if a vertical line intersects a graph at more than one point, the graph is not the graph of a function. To see why this is true, consider the graph in Figure 3.14. The vertical line $x = 3$ intersects the graph at two points. If this were the graph of a function $f$, it would mean that $f(3) = 2$ (because (3, 2) is on the graph) and that $f(3) = -1$ (because (3, −1) is on the graph). This is impossible, because a *function* can have only one value when $x = 3$ (because each input determines exactly one output). Therefore, the graph in Figure 3.14 cannot be the graph of a function. A similar argument works in the general case.

**EXAMPLE 12**    Use the vertical-line test to determine which of the graphs in Figure 3.15 are the graphs of functions.

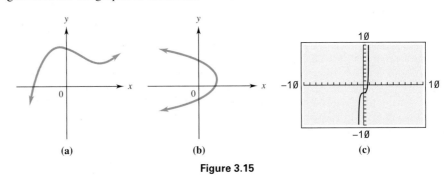

(a)                (b)                (c)

**Figure 3.15**

**Solution**  To use the vertical-line test, imagine dragging a ruler held vertically across the graph from left to right. If the graph is that of a function, the edge of the ruler would hit the graph only once for every $x$-value. If you do this for graph (a), every vertical line intersects the graph in at most one point, so this graph is the graph of a function. Many vertical lines (including the $y$-axis) intersect graph (b) twice, so it is not the graph of a function.

Graph (c) appears to fail the vertical-line test near $x = 1$ and $x = -1$, indicating that it is not the graph of a function. But this appearance is misleading because of the low resolution of the calculator screen. The table in Figure 3.16 and the very narrow segment of the graph in Figure 3.17 show that the graph actually rises as it moves to the right. The same thing happens near $x = -1$. So this graph *does* pass the vertical-line test and *is* the graph of a function. (Its rule is $f(x) = 15x^{11} - 2$.) The moral of this story is that you can't always trust images produced by a graphing calculator. When in doubt, try other viewing windows or a table to see what is really going on. ✓ 9

## ✓ Checkpoint 9

Find a viewing window that indicates the actual shape of the graph of the function $f(x) = 15x^{11} - 2$ of Example 12 near the point $(-1, -17)$.

**Figure 3.16**

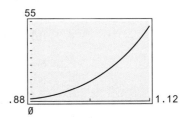

**Figure 3.17**

# 3.2 Exercises

*Graph each function. (See Examples 1–4.)*

**1.** $f(x) = -.5x + 2$

**2.** $g(x) = 3 - x$

**3.** $f(x) = \begin{cases} x + 3 & \text{if } x \le 1 \\ 4 & \text{if } x > 1 \end{cases}$

**4.** $g(x) = \begin{cases} 2x - 1 & \text{if } x < 0 \\ -1 & \text{if } x \ge 0 \end{cases}$

**5.** $y = \begin{cases} 4 - x & \text{if } x \le 0 \\ 3x + 4 & \text{if } x > 0 \end{cases}$

**6.** $y = \begin{cases} x + 5 & \text{if } x \le 1 \\ 2 - 3x & \text{if } x > 1 \end{cases}$

**7.** $f(x) = \begin{cases} |x| & \text{if } x < 2 \\ -2x & \text{if } x \ge 2 \end{cases}$

**8.** $g(x) = \begin{cases} -|x| & \text{if } x \le 1 \\ 2x & \text{if } x > 1 \end{cases}$

**9.** $f(x) = |x - 4|$       **10.** $g(x) = |4 - x|$

**11.** $f(x) = |3 - 3x|$       **12.** $g(x) = -|x|$

**13.** $y = -|x - 1|$       **14.** $f(x) = |x| - 2$

**15.** $y = |x - 2| + 3$

**16.** $|x| + |y| = 1$ (*Hint:* This is not the graph of a function, but is made up of four straight-line segments. Find them by using the definition of absolute value in these four cases: $x \geq 0$ and $y \geq 0$; $x \geq 0$ and $y < 0$; $x < 0$ and $y \geq 0$; $x < 0$ and $y < 0$).

*Graph each function. (See Examples 6 and 7.)*

**17.** $f(x) = [x - 3]$       **18.** $g(x) = [x + 3]$

**19.** $g(x) = [-x]$

**20.** $f(x) = [x] + [-x]$ (The graph contains horizontal segments, but is *not* a horizontal line.)

**21.** **Postal Rates** The accompanying table gives rates charged by the U.S. Postal Service for first-class letters in March, 2016. Graph the function $f(x)$ that gives the price of mailing a first-class letter, where $x$ represents the weight of the letter in ounces and $0 \leq x \leq 3.5$. (Data from: www.usps.com.)

| Weight Not Over | Price |
|---|---|
| 1 ounce | $ .49 |
| 2 ounces | $ .71 |
| 3 ounces | $ .93 |
| 3.5 ounces | $1.15 |

**22.** **Postal Rates** The accompanying table gives rates charged by the U.S. Postal Service for first-class large flat envelopes in March, 2016. Graph the function $f(x)$ that gives the price of mailing a first-class large flat envelope, where $x$ represents the weight of the envelope in ounces and $0 \leq x \leq 6$. (Data from: www.usps.com.)

| Weight Not Over | Price |
|---|---|
| 1 ounce | $ .98 |
| 2 ounces | $1.20 |
| 3 ounces | $1.42 |
| 4 ounces | $1.64 |
| 5 ounces | $1.86 |
| 6 ounces | $2.08 |

*Graph each function. (See Examples 8–9.)*

**23.** $g(x) = \sqrt{-x}$       **24.** $h(x) = \sqrt{x} - 1$

**25.** $f(x) = \begin{cases} x^2 & \text{if } x < 2 \\ -2x + 2 & \text{if } x \geq 2 \end{cases}$

**26.** $g(x) = \begin{cases} \sqrt{-x} & \text{if } x \leq -4 \\ \dfrac{x^2}{4} & \text{if } x > -4 \end{cases}$

*Determine whether each graph is a graph of a function or not. (See Example 12.)*

**27.**

**28.**

**29.**

**30.**

**31.**

**32.**

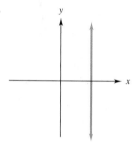

*Use a graphing calculator or other technology to graph each of the given functions. If the graph has any endpoints, indicate whether they are part of the graph or not.*

**33.** $f(x) = .2x^3 - .8x^2 - 4x + 9.6$

**34.** $g(x) = .1x^4 - .3x^3 - 1.3x^2 + 1.5x$

**35.** $g(x) = \begin{cases} 2x^2 + x & \text{if } x < 1 \\ x^3 - x - 1 & \text{if } x \geq 1 \end{cases}$

(*Hint:* See the Technology Tip on page 125.)

**36.** $f(x) = \begin{cases} x|x| & \text{if } x \leq 0 \\ -x^2|x| + 2 & \text{if } x > 0 \end{cases}$

*Work these exercises. (See Examples 2, 3, 10, and 11.)*

**37.** **Medicare** An employee must pay Medicare tax at a base rate plus an additional Medicare tax if wages exceed a certain threshold. For a single person in 2016, the employee's Medicare tax in dollars was determined by the piecewise linear function

$$f(x) = \begin{cases} 14.5x & \text{if } 0 \leq x \leq 200 \\ 2900 + 23.5(x - 200) & \text{if } x > 200 \end{cases},$$

where $x$ is the employee's Medicare wages in thousands of dollars. (Data from: Internal Revenue Service.)

(a) Find the function values $f(0)$, $f(200)$, $f(400)$.

(b) Graph this function for wages up to $400,000 (that is, $0 \leq x \leq 400$).

**38. Social Security** An employee must pay Social Security tax up to a maximum wage that is subject to the tax. In 2016, an employee's Social Security tax in dollars was determined by the piecewise linear function

$$f(x) = \begin{cases} 62x & \text{if } 0 \le x \le 118.5 \\ 7347 & \text{if } x > 118.5 \end{cases},$$

where $x$ is the employee's Social Security wages in thousands of dollars. (Data from: Internal Revenue Service.)

(a) Find the function values $f(0), f(118.5), f(400)$.

(b) Graph this function for wages up to \$400,000 (that is, $0 \le x \le 400$).

**39. Taxes** Mississippi state income tax for a single person in 2017 was determined by the rule

$$T(x) = \begin{cases} .03x & \text{if } 0 \le x \le 5000 \\ 150 + .04(x - 5000) & \text{if } 5000 < x \le 10,000, \\ 350 + .05(x - 10,000) & \text{if } x > 10,000 \end{cases}$$

where $x$ is the person's taxable income in dollars. Graph the function $T(x)$ for taxable incomes between 0 and \$15,000. (Data from: www.tax-brackets.org.)

**40. Taxes** The Alabama state income tax for a single person in 2017 was determined by the rule

$$T(x) = \begin{cases} .02x & \text{if } 0 \le x \le 500 \\ 10 + .04(x - 500) & \text{if } 500 < x \le 3000 \\ 110 + .05(x - 3000) & \text{if } x > 3000 \end{cases}$$

where $x$ is the person's taxable income in dollars. Graph the function $T(x)$ for taxable incomes between 0 and \$5000.

**41. Smoking Rates** The following graph shows the percentage of the adult population of the United States who smoke cigarettes. (Data from: www.cdc.gov.)

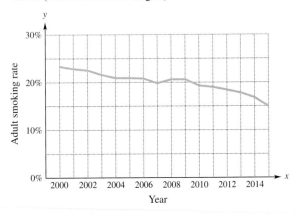

(a) According to the graph, approximately what was the adult smoking rate in 2015?

(b) Let $x = 0$ correspond to the year 2000. Approximate a linear function for the smoking rate (as a percentage) using the points (5, 21) and (15, 15).

(c) What does this model estimate for the smoking rate in 2000? In 2017?

(d) The Centers for Disease Control and Prevention (CDC) reports that a *Healthy People 2020* goal is that only 12% of the adult population will smoke cigarettes. If the trend continues, will that goal be met?

**42. Smoking Rates** Use the graph in the previous exercise to answer the following questions.

(a) What is the general trend for the adult smoking rate?

(b) Let $x = 0$ correspond to the year 2000. Approximate a linear function for the smoking rate (as a percentage) using the points (7, 19.8). and (15, 15).

(c) What does this model estimate for the smoking rate in 2000? In 2017?

**43. Consumer Price Index** The accompanying graph shows the annual percentage change in the consumer price indexes (CPIs) for various sectors of the economy. (Data from: Bureau of Labor Statistics.)

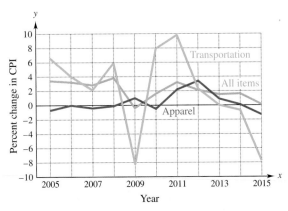

(a) During what year(s) was the change in CPI for transportation the largest?

(b) During what year(s) was the change in CPI for apparel the largest?

(c) During what year(s) was the change in CPI for all items approximately zero?

**44. Consumer Price Index** Use the graph in the previous exercise to answer the following questions.

(a) During what year(s) did the CPI for all items show a decrease?

(b) When was the CPI for transportation increasing at the greatest rate?

(c) During what year(s) did the CPI for apparel increase at a faster rate than the CPI for all items?

**45. Apple Shareholders' Equity** The following graph shows total assets and total liabilities for Apple Inc. in billions of dollars. (Data from: Apple Inc. Annual Report.)

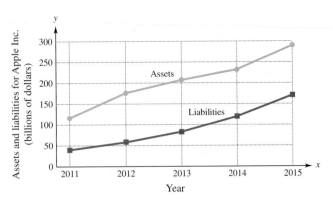

(a) Shareholders' equity is a corporation's total assets minus total liabilities. Estimate Apple's shareholders' equity in the year 2015.

(b) What is the domain of the assets function shown in the graph?

(c) Estimate the range of the liabilities function shown in the graph.

46. **iPhone Sales** The following graph shows iPhones sales in millions of units. (Data from: Apple Inc. Annual Report.)

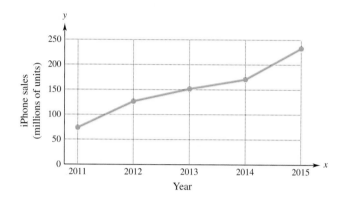

(a) Estimate and interpret the function value $f(2013)$.

(b) What is the domain shown in the graph?

(c) Estimate the range shown in the graph.

47. **Postal Rates** Whenever postage rates change, some newspaper publishes a graph like this one, which shows the price of a first-class stamp from 2006 to 2016:

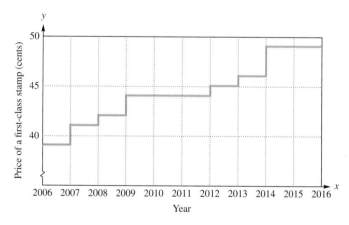

(a) Let $f$ be the function whose rule is $f(x) = $ cost of a first-class stamp in year $x$. Find $f(2010)$ and $f(2015)$.

(b) Explain why the graph in the figure is not the graph of the *function f*. What must be done to the figure to make it an accurate graph of the *function f*?

48. **Business** A chain-saw rental firm charges $20 per day or fraction of a day to rent a saw, plus a fixed fee of $7 for resharpening the blade. Let $S(x)$ represent the cost of renting a saw for $x$ days. Find each of the following.

(a) $S\left(\dfrac{1}{2}\right)$   (b) $S(1)$   (c) $S\left(1\dfrac{2}{3}\right)$   (d) $S\left(4\dfrac{3}{4}\right)$

(e) What does it cost to rent for $5\dfrac{7}{8}$ days?

(f) A portion of the graph of $y = S(x)$ is shown on the following page. Explain how the graph could be continued.

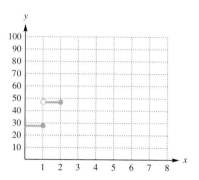

(g) What is the domain variable?

(h) What is the range variable?

(i) Write a sentence or two explaining what (c) and its answer represent.

(j) We have left $x = 0$ out of the graph. Discuss why it should or should not be included. If it were included, how would you define $S(0)$?

49. **Business** You need to rent a van to pick up a new couch you have purchased. The cost of the van is $19.99 for the first 75 minutes and then an additional $5 for each block of 15 minutes beyond 75. Find the cost to rent a van for

(a) 2 hours;   (b) 1.5 hours;

(c) 3.5 hours;   (d) 4 hours.

(e) Graph the ordered pairs (hours, cost).

50. **Business** A delivery company charges $25 plus 60¢ per mile or part of a mile. Find the cost for a trip of

(a) 3 miles;   (b) 4.2 miles;

(c) 5.9 miles;   (d) 8 miles.

(e) Graph the ordered pairs (miles, cost).

(f) Is this a function?

*Work these problems.*

51. **Bacteria Growth** A laboratory culture contains about 1 million bacteria at midnight. The culture grows very rapidly until noon, when a bactericide is introduced and the bacteria population plunges. By 4 p.m., the bacteria have adapted to the bactericide and the culture slowly increases in population until 9 p.m., when the culture is accidentally destroyed by the cleanup crew. Let $g(t)$ denote the bacteria population at time $t$ (with $t = 0$ corresponding to midnight). Draw a plausible graph of the function $g$. (Many correct answers are possible.)

52. **Flight Paths** A plane flies from Austin, Texas, to Cleveland, Ohio, a distance of 1200 miles. Let $f$ be the function whose rule is

$f(t) = $ distance (in miles) from Austin at time $t$ hours,

with $t = 0$ corresponding to the 4 p.m. takeoff. In each part of this exercise, draw a plausible graph of $f$ under the given circumstances. (There are many correct answers for each part.)

(a) The flight is nonstop and takes between 3.5 and 4 hours.

**(b)** Bad weather forces the plane to land in Dallas (about 200 miles from Austin) at 5 p.m., remain overnight, and leave at 8 a.m. the next day, flying nonstop to Cleveland.

**(c)** The plane flies nonstop, but due to heavy traffic it must fly in a holding pattern for an hour over Cincinnati (about 200 miles from Cleveland) and then go on to Cleveland.

**4. (a)**   **(b)**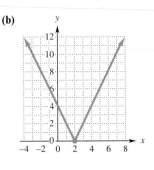

✓**Checkpoint Answers**

**5. (a)** $35    **(b)** $115

**1.**     **2.**

**6.**     **7.**

**3.**

**8.**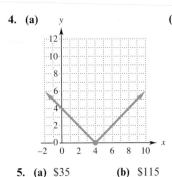

**9.** There are many correct answers, including, $-1.4 \leq x \leq -.6$ and $-30 \leq y \leq 0$.

# 3.3   Applications of Linear Functions

C Most of this section deals with the basic business relationships that were introduced in Section 1.2:

> **Revenue = (Price per item) × (Number of items);**
>
> **Cost = Fixed Costs + Variable Costs;**
>
> **Profit = Revenue − Cost.**

The examples will use only linear functions, but the methods presented here also apply to more complicated functions.

## Cost Analysis

Recall that fixed costs are for such things as buildings, machinery, real-estate taxes, and product design. Within broad limits, the fixed cost is constant for a particular product and does not change as more items are made. Variable costs are for labor, materials, shipping, and so on, and depend on the number of items made.

If $C(x)$ is the cost of making $x$ items, then the fixed cost (the cost that occurs even when no items are produced) can be found by letting $x = 0$. For example, for the cost function $C(x) = 45x + 250{,}000$, the fixed cost is

$$C(0) = 45(0) + 250{,}000 = \$250{,}000.$$

In this case, the variable cost of making $x$ items is $45x$—that is, $45 per item manufactured.

**EXAMPLE 1**   **Prescription Drugs** An anticlot drug can be made for $10 per unit. The total cost to produce 100 units is $1500.

**(a)** Assuming that the cost function is linear, find its rule.

**Solution** Since the cost function $C(x)$ is linear, its rule is of the form $C(x) = mx + b$. We are given that $m$ (the cost per item) is 10, so the rule is $C(x) = 10x + b$. To find $b$, use the fact that it costs $1500 to produce 100 units, which means that

$$C(100) = 1500$$
$$10(100) + b = 1500 \qquad C(x) = 10x + b.$$
$$1000 + b = 1500$$
$$b = 500.$$

So the rule of the cost function is $C(x) = 10x + 500$.

**(b)** What are the fixed costs?

✓ **Checkpoint 1**

The total cost of producing 10 calculators is $100. The variable costs per calculator are $4. Find the rule of the linear cost function.

**Solution** The fixed costs are $C(0) = 10(0) + 500 = \$500.$ ✓₁

If $C(x)$ is the total cost to produce $x$ items, then the **average cost** per item is given by

$$\overline{C}(x) = \frac{C(x)}{x}.$$

As more and more items are produced, the average cost per item typically decreases.

**EXAMPLE 2**   **Average Cost** Find the average cost of producing 100 and 1000 units of the anticlot drug in Example 1.

**Solution** The cost function is $C(x) = 10x + 500$, so the average cost of producing 100 units is

$$\overline{C}(100) = \frac{C(100)}{100} = \frac{10(100) + 500}{100} = \frac{1500}{100} = \$15.00 \text{ per unit.}$$

✓ **Checkpoint 2**

In Checkpoint 1, find the average cost per calculator when 100 are produced.

The average cost of producing 1000 units is

$$\overline{C}(1000) = \frac{C(1000)}{1000} = \frac{10(1000) + 500}{1000} = \frac{10{,}500}{1000} = \$10.50 \text{ per unit.}$$ ✓₂

| $x$ | $f(x) = 3x + 5$ |
|-----|------------------|
| 1 | 8 |
| 2 | 11 |
| 3 | 14 |
| 4 | 17 |
| 5 | 20 |

## Rates of Change

The rate at which a quantity (such as revenue or profit) is changing can be quite important. For instance, if a company determines that the rate of change of its revenue is decreasing, then sales growth is slowing down, a trend that may require a response.

The rate of change of a linear function is easily determined. For example, suppose $f(x) = 3x + 5$, and consider the table of values in the margin. The table shows that each time $x$ changes by 1, the corresponding value of $f(x)$ changes by 3. Thus, the rate of change of $f(x) = 3x + 5$ with respect to $x$ is 3, which is the slope of the line $y = 3x + 5$. The same thing happens for any linear function:

---

**The rate of change of a linear function $f(x) = mx + b$ is the slope $m$.**

---

In particular, the rate of change of a linear function is constant.

The value of a computer, or an automobile, or a machine *depreciates* (decreases) over time. **Linear depreciation** means that the value of the item at time $x$ is given by a linear function $f(x) = mx + b$. The slope $m$ of this line gives the rate of depreciation.

---

**EXAMPLE 3** **Automobile Depreciation** According to Kelly Blue Book, a Honda Civic four-door sedan that is worth $24,535 today will be worth $17,020 in three years (if it is in excellent condition with low mileage).

**(a)** Assuming linear depreciation, find the depreciation function for this car.

**Solution** We know the car is worth $24,535 now ($x = 0$) and will be worth $17,020 in three years ($x = 3$). So the points $(0, 24,535)$ and $(3, 17,020)$ are on the graph of the linear depreciation function and can be used to determine its slope:

$$m = \frac{17,020 - 24,535}{3 - 0} = \frac{-7515}{3} = -2505.$$

Using the point $(0, 24,535)$, we find that the equation of the line is

$$y - y_1 = m(x - x_1) \qquad \text{Point–slope form}$$
$$y - 24,535 = -2505(x - 0) \qquad \text{Substitute}$$
$$y = -2505x + 24,535. \qquad \text{Slope–intercept form}$$

Therefore, the rule of the depreciation function is $f(x) = -2505x + 24,535$.

**(b)** What will the car be worth five years from now?

**Solution** Evaluate $f$ when $x = 5$:

$$f(x) = -2505x + 24,535$$
$$f(5) = -2505(5) + 24,535 = -12,525 + 24,535 = \$12,010.$$

**(c)** At what rate is the car depreciating?

**Solution** The depreciation rate is given by the slope of $f(x) = -2505x + 24,535$, namely, $-2505$. This negative slope means that the car is decreasing in value an average of $2505 per year. ✓3

✓ **Checkpoint 3**

Using the information from Example 3, determine what the car will be worth in 6 years.

## Marginal Analysis

In economics, the rate of change of the cost function is called the **marginal cost.** Marginal cost is important to management in making decisions in such areas as cost control, pricing, and production planning. When the cost function is linear, say, $C(x) = mx + b$, the marginal cost is the number $m$ (the slope of the graph of $C$). Marginal cost can also be thought of as the cost of producing one more item, as the next example demonstrates.

---

**EXAMPLE 4** **Marginal Cost** An electronics company manufactures tablets. The cost function for one of its models is $C(x) = 160x + 750,000$.

**(a)** What are the fixed costs for this product?

**Solution** The fixed costs are $C(0) = 160(0) + 750,000 = \$750,000$.

**(b)**  What is the marginal cost?

**Solution**  The slope of $C(x) = 160x + 750,000$ is 160, so the marginal cost is \$160 per item.

**(c)**  After 50,000 units have been produced, what is the cost of producing one more?

**Solution**  The cost of producing 50,000 is

$$C(50,000) = 160(50,000) + 750,000 = \$8,750,000.$$

The cost of 50,001 units is

$$C(50,001) = 160(50,001) + 750,000 = \$8,750,160.$$

The cost of the additional unit is the difference

$$C(50,001) - C(50,000) = 8,750,160 - 8,750,000 = \$160.$$

Thus, the cost of one more item is the marginal cost. ✓4

Similarly, the rate of change of a revenue function is called the **marginal revenue**. When the revenue function is linear, the marginal revenue is the slope of the line, as well as the revenue from producing one more item.

**✓Checkpoint 4**

The cost in dollars to produce $x$ kilograms of chocolate candy is given by $C(x) = 3.5x + 800$. Find each of the following.

**(a)**  The fixed cost

**(b)**  The total cost for 12 kilograms

**(c)**  The marginal cost of the 40th kilogram

**(d)**  The marginal cost per kilogram

**EXAMPLE 5**  **Marginal Revenue**  The energy company New York State Electric and Gas charges each residential customer a basic fee for electricity of \$15.11, plus \$.0333 per kilowatt hour (kWh).

**(a)**  Assuming there are 700,000 residential customers, find the company's revenue function.

**Solution**  The monthly revenue from the basic fee is

$$15.11(700,000) = \$10,577,000.$$

If $x$ is the total number of kilowatt hours used by all customers, then the revenue from electricity use is .0333$x$. So the monthly revenue function is given by

$$R(x) = .0333x + 10,577,000.$$

**(b)**  What is the marginal revenue?

**Solution**  The marginal revenue (the rate at which revenue is changing) is given by the slope of the rate function: \$0.0333 per kWh. ✓5

Examples 4 and 5 are typical of the general case, as summarized here.

**✓Checkpoint 5**

Assume that the average customer in Example 5 uses 1600 kWh in a month.

**(a)**  What is the total number of kWh used by all customers?

**(b)**  What is the company's monthly revenue?

**Marginal cost** is the rate of change of the cost function. For a **linear cost function** $C(x) = mx + b$, the marginal cost is $m$ (the slope of the cost line) and the fixed cost is $b$ (the $y$-intercept of the cost line). The marginal cost is the cost of producing one more item.

**Marginal revenue** is the rate of change of the revenue function. For a **linear revenue function** $R(x) = kx + d$, the marginal revenue is $k$ (the slope of the revenue line), which is the revenue from selling one more item.

## Break-Even Analysis

A typical company must analyze its costs and the potential market for its product to determine when (or even whether) it will make a profit.

**EXAMPLE 6**   **Retail Profit** A company manufactures a 65-inch smart TV that sells to retailers for $550. The cost of making $x$ of these TVs for a month is given by the cost function $C(x) = 250x + 213,000$.

**(a)** Find the function $R$ that gives the revenue from selling $x$ TVs.

**Solution** Since revenue is the product of the price per item and the number of items, $R(x) = 550x$.

**(b)** What is the revenue from selling 600 TVs?

**Solution** Evaluate the revenue function $R$ at 600:

$$R(600) = 550(600) = \$330,000.$$

**(c)** Find the profit function $P$.

**Solution** Since Profit = Revenue − Cost,

$$P(x) = R(x) - C(x) = 550x - (250x + 213,000) = 300x - 213,000.$$

**(d)** What is the profit from selling 500 TVs?

**Solution** Evaluate the profit function at 500 to obtain

$$P(500) = 300(500) - 213,000 = -63,000,$$

that is, a loss of $63,000.

A company can make a profit only if the revenue on a product exceeds the cost of manufacturing it. The number of units at which revenue equals cost (that is, profit is 0) is the **break-even point.**

**EXAMPLE 7**   **Break-Even Point** Find the break-even point for the company in Example 6.

**Solution** The company will break even when revenue equals cost—that is, when

$$R(x) = C(x)$$
$$550x = 250x + 213,000$$
$$300x = 213,000 \qquad \text{Subtract 250x from both sides.}$$
$$x = 710.$$

The company breaks even by selling 710 TVs. The graphs of the revenue and cost functions and the break-even point (where $x = 710$) are shown in the Figure 3.18. The company must sell more than 710 TVs ($x > 710$) in order to make a profit. ☑ 6

✓ **Checkpoint 6**

For a certain newsletter, the cost equation is $C(x) = .90x + 1500$, where $x$ is the number of newsletters sold. The newsletter sells for $1.25 per copy. Find the break-even point.

**Figure 3.18**

700,000

0 | Intersection | 1000
X=710    Y=390500
0

**Figure 3.19**

🖩 **TECHNOLOGY TIP** The break-even point in Example 7 can be found on a graphing calculator by graphing the cost and revenue functions on the same screen and using the calculator's intersection finder, as shown in Figure 3.19. Depending on the calculator, the intersection finder is in the CALC or G-SOLVE menu or in the MATH or FCN submenu of the GRAPH menu.

## Supply and Demand

The supply of and demand for an item are usually related to its price. Producers will supply large numbers of the item at a high price, but consumer demand will be low. As the price of the item decreases, consumer demand increases, but producers are less willing to supply large numbers of the item. The curves showing the quantity that will be supplied at a given price and the quantity that will be demanded at a given price are called **supply and demand curves,** respectively. In supply-and-demand problems, we use $p$ for price and $q$ for quantity. We will discuss the economic concepts of supply and demand in more detail in later chapters.

---

**EXAMPLE 8**   **Economics** Suppose that an economist has studied the supply and demand for aluminum siding and has determined that the price per unit,[*] $p$, and the quantity demanded, $q$, are related by the linear equation

$$p = 60 - \frac{3}{4}q. \qquad \text{Demand Function}$$

**(a)** Find the demand at a price of $40 per unit.

**Solution** Let $p = 40$. Then we have

$$p = 60 - \frac{3}{4}q$$

$$40 = 60 - \frac{3}{4}q \qquad \textit{Let } p = 40.$$

$$-20 = -\frac{3}{4}q \qquad \textit{Add } -60 \textit{ to both sides.}$$

$$\frac{80}{3} = q. \qquad \textit{Multiply both sides by } -\frac{4}{3}.$$

At a price of $40 per unit, 80/3 (or $26\frac{2}{3}$) units will be demanded.

**(b)** Find the price if the demand is 32 units.

**Solution** Let $q = 32$. Then we have

$$p = 60 - \frac{3}{4}q$$

$$p = 60 - \frac{3}{4}(32) \qquad \textit{Let } q = 32.$$

$$p = 60 - 24$$

$$p = 36.$$

With a demand of 32 units, the price is $36.

**(c)** Graph $p = 60 - \frac{3}{4}q$.

**Solution** It is customary to use the horizontal axis for the quantity $q$ and the vertical axis for the price $p$. In part (a), we saw that 80/3 units would be demanded at a price of

---

[*]An appropriate unit here might be, for example, one thousand square feet of siding.

$40 per unit; this gives the ordered pair (80/3, 40). Part (b) shows that with a demand of 32 units, the price is $36, which gives the ordered pair (32, 36). Using the points (80/3, 40) and (32, 36) yields the demand graph depicted in Figure 3.20. Only the portion of the graph in Quadrant I is shown, because supply and demand are meaningful only for positive values of $p$ and $q$. ✔ 7

✓ **Checkpoint 7**

Suppose price and quantity demanded are related by
$p = 100 - 4q$.

(a) Find the price if the quantity demanded is 10 units.

(b) Find the quantity demanded if the price is $80.

(c) Write the corresponding ordered pairs.

**Figure 3.20**

(d) From Figure 3.20, at a price of $30, what quantity is demanded?

**Solution** Price is located on the vertical axis. Look for 30 on the $p$-axis, and read across to where the line $p = 30$ crosses the demand graph. As the graph shows, this occurs where the quantity demanded is 40.

(e) At what price will 60 units be demanded?

**Solution** Quantity is located on the horizontal axis. Find 60 on the $q$-axis, and read up to where the vertical line $q = 60$ crosses the demand graph. This occurs where the price is about $15 per unit.

(f) What quantity is demanded at a price of $60 per unit?

**Solution** The point $(0, 60)$ on the demand graph shows that the demand is 0 at a price of $60 (that is, there is no demand at such a high price).

**EXAMPLE 9**    **Economics** Suppose the economist in Example 8 concludes that the supply $q$ of siding is related to its price $p$ by the equation

$$p = .85q. \qquad \text{Supply Function}$$

(a) Find the supply if the price is $51 per unit.

**Solution**         $51 = .85q$    Let $p = 51$.

                    $60 = q$.    Divide both sides by .85.

If the price is $51 per unit, then 60 units will be supplied to the marketplace.

(b) Find the price per unit if the supply is 20 units.

**Solution**         $p = .85(20) = 17.$    Let $q = 20$.

If the supply is 20 units, then the price is $17 per unit.

(c) Graph the supply equation $p = .85q$.

**Solution** As with demand, each point on the graph has quantity $q$ as its first coordinate and the corresponding price $p$ as its second coordinate. Part (a) shows that the ordered pair $(60, 51)$ is on the graph of the supply equation, and part (b) shows that $(20, 17)$ is on the graph. Using these points, we obtain the supply graph in Figure 3.21.

**Figure 3.21**

**(d)** Use the graph in Figure 3.21 to find the approximate price at which 35 units will be supplied. Then use algebra to find the exact price.

**Solution** The point on the graph with first coordinate $q = 35$ is approximately $(35, 30)$. Therefore, 35 units will be supplied when the price is approximately \$30. To determine the exact price algebraically, substitute $q = 35$ into the supply equation:

$$p = .85q = .85(35) = \$29.75.$$

When the supply of a quantity exceeds the demand, we say there is a **surplus of supply**. When demand exceeds supply, we say there is a **shortage of supply**.

**EXAMPLE 10** **Economics** The supply and demand curves of Examples 8 and 9 are shown in Figure 3.22. Determine graphically whether there is a surplus or a shortage of supply at a price of \$40 per unit.

**Solution** Find 40 on the vertical axis in Figure 3.22 and read across to the point where the horizontal line $p = 40$ crosses the supply and demand graphs. At a price of \$40, the quantity supplied is about 47 units, and the quantity demanded is about 27 units. Since the quantity supplied is greater than the quantity demanded, there is a surplus of supply.

**Figure 3.22**

Supply and demand are equal at the point where the supply curve intersects the demand curve. This is the **equilibrium point.** Its second coordinate is the **equilibrium price,** the price at which the same quantity will be supplied as is demanded. Its first coordinate is the quantity that will be demanded and supplied at the equilibrium price; this number is called the **equilibrium quantity.**

**EXAMPLE 11** **Economics** In the situation described in Examples 8–10, what is the equilibrium quantity? What is the equilibrium price?

**Solution** The equilibrium point is where the supply and demand curves in Figure 3.22 intersect. To find the quantity $q$ at which the price given by the demand equation $p = 60 - .75q$ (Example 8) is the same as that given by the supply equation $p = .85q$ (Example 9), set these two expressions for $p$ equal to each other and solve the resulting equation:

$$\text{Demand} = \text{Supply}$$
$$60 - .75q = .85q$$
$$60 = 1.6q \qquad \text{Subtract .75q from both sides.}$$
$$37.5 = q. \qquad \text{Divide both sides by 1.6.}$$

Therefore, the equilibrium quantity is 37.5 units, the number of units for which supply will equal demand. Substituting $q = 37.5$ into either the demand or supply equation shows that

$$p = 60 - .75(37.5) = 31.875 \quad \text{or} \quad p = .85(37.5) = 31.875.$$

**Checkpoint 8**

The demand for a certain commodity is related to the price by $p = 80 - (2/3)q$. The supply is related to the price by $p = (4/3)q$. Find

**(a)** the equilibrium quantity;

**(b)** the equilibrium price.

So the equilibrium price is $31.875 (or $31.88, rounded). (To avoid error, it is a good idea to substitute into both equations, as we did here, to be sure that the same value of $p$ results; if it does not, a mistake has been made.) In this case, the equilibrium point—the point whose coordinates are the equilibrium quantity and price—is $(37.5, 31.875)$. ✔8

**⊞ TECHNOLOGY TIP** The equilibrium point $(37.5, 31.875)$ can be found on a graphing calculator by graphing the supply and demand curves on the same screen and using the calculator's intersection finder to locate their point of intersection.

# 3.3 Exercises

**Business** *Write a cost function for each of the given scenarios. Identify all variables used. (See Example 1.)*

1. A chain-saw rental firm charges $25, plus $5 per hour.

2. A trailer-hauling service charges $95, plus $8 per mile.

3. A parking garage charges $8.00, plus $2.50 per hour.

4. For a 1-day rental, a car rental firm charges $65, plus 45¢ per mile.

**Business** *Assume that each of the given situations can be expressed as a linear cost function. Find the appropriate cost function in each case. (See Examples 1 and 4.)*

5. Fixed cost, $200; 50 items cost $2000 to produce.

6. Fixed cost, $2000; 40 items cost $5000 to produce.

7. Marginal cost, $120; 100 items cost $15,800 to produce.

8. Marginal cost, $90; 150 items cost $16,000 to produce.

**Business** *In Exercises 9–12, a cost function is given. Find the average cost per item when the required numbers of items are produced. (See Example 2.)*

9. $C(x) = 12x + 1800$; 50 items, 500 items, 1000 items

10. $C(x) = 80x + 12{,}000$; 100 items, 1000 items, 10,000 items

11. $C(x) = 6.5x + 9800$; 200 items, 2000 items, 5000 items

12. $C(x) = 8.75x + 16{,}500$; 1000 items, 10,000 items, 75,000 items

**Business** *Work these exercises. (See Example 3.)*

13. **Automobile Depreciation** The manufacturer's suggested retail price (MSRP) for a base model 2016 Toyota Prius is $25,035, and it is expected to be worth $14,115 is 4 years.

    **(a)** Find a linear depreciation function for this car.

    **(b)** Estimate the value of the car 5 years from now.

    **(c)** At what rate is the car depreciating?

14. A computer that cost $1250 new is expected to depreciate linearly at a rate of $250 per year.

    **(a)** Find the depreciation function $f$.

    **(b)** Explain why the domain of $f$ is $[0, 5]$.

15. A machine is now worth $120,000 and will be depreciated linearly over an 8-year period, at which time it will be worth $25,000 as scrap.

    **(a)** Find the rule of the depreciation function $f$.

    **(b)** What is the domain of $f$?

    **(c)** What will the machine be worth in 6 years?

16. A house increases in value in an approximately linear fashion from $222,000 to $300,000 in 6 years.

    **(a)** Find the *appreciation function* that gives the value of the house in year $x$.

    **(b)** If the house continues to appreciate at this rate, what will it be worth 12 years from now?

**Business** *Work these problems. (See Examples 2 and 4.)*

17. The total cost (in dollars) of producing $x$ college algebra books is $C(x) = 42.5x + 80{,}000$.

    **(a)** What are the fixed costs?

    **(b)** What is the marginal cost per book?

    **(c)** What is the total cost of producing 1000 books? 32,000 books?

    **(d)** What is the average cost when 1000 are produced? when 32,000 are produced?

18. The total cost (in dollars) of producing $x$ DVDs is $C(x) = 6.80x + 450{,}000$.

    **(a)** What are the fixed costs?

    **(b)** What is the marginal cost per DVD?

    **(c)** What is the total cost of producing 50,000 DVDs? 600,000 DVDs?

    **(d)** What is the average cost per DVD when 50,000 are produced? when 500,000 are produced?

19. The owner of a food truck found that the cost of producing 50 tacos is $58, while the cost of producing 250 tacos is $142. Assume that the cost $C(x)$ is a linear function of $x$, the number of tacos produced.

    **(a)** Find a formula for $C(x)$.

    **(b)** What are the fixed costs?

    **(c)** Find the total cost of producing 1000 tacos.

    **(d)** Find the total cost of producing 1001 tacos.

    **(e)** Find the marginal cost of producing the 1001st taco.

    **(f)** Find the marginal cost of producing *any* taco.

20. In deciding whether to set up a new manufacturing plant, company analysts have determined that a linear function is a reasonable estimation for the total cost $C(x)$ in dollars of producing $x$ items. They estimate the cost of producing 10,000 items as $547,500 and the cost of producing 50,000 items as $737,500.

    (a) Find a formula for $C(x)$.

    (b) Find the total cost of producing 100,000 items.

    (c) Find the marginal cost of the items to be produced in this plant.

**Business** *Work these problems. (See Example 5.)*

21. **Water Utility Revenue** For the year 2016, a resident of the city of Dallas, Texas, with a ⅝-inch water meter pays $5.12 per month plus $1.87 per 1000 gallons of water used (up to 4000 gallons). If the city of Dallas has 550,000 customers that use less than 4000 gallons a month, find its monthly revenue function $R(x)$, where the total number of gallons $x$ is measured in thousands. (Data from: dallascityhall.com/departments/waterutilities.)

22. **Gas Utility Revenue** The Lacledge Gas Company in St. Louis, Missouri, charged customers $19.50 per month plus $.91686 per therm used (up to 30 terms) in the winter of 2017–2018. If the company has 150,000 customers that use less than 30 therms a month, find its monthly revenue function $R(x)$, where $x$ is the total number of terms sold. (Data from: http://www.lacledegas.com.)

**Business** *Assume that each row of the accompanying table has a linear cost function. Find (a) the cost function; (b) the revenue function; (c) the profit function; (d) the profit on 100 items. (See Example 6.)*

|  | Fixed Cost | Marginal Cost per Item | Item Sells For |
|---|---|---|---|
| 23. | $750 | $10 | $35 |
| 24. | $150 | $11 | $20 |
| 25. | $300 | $18 | $28 |
| 26. | $17,000 | $30 | $80 |

27. **Profit–Volume Graph** In the following profit–volume chart, $EF$ and $GH$ represent the profit–volume graphs of a single-product company for, 2016 and 2017, respectively. (Adapted from: Uniform CPA Examination, American Institute of Certified Public Accountants.)

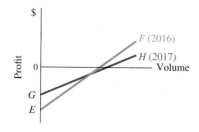

If the 2016 and 2017 unit sales prices are identical, how did the total fixed costs and unit variable costs of 2017 change compared with their values in 2016? Choose one:

|  | 2016 Total Fixed Costs | 2017 Unit Variable Costs |
|---|---|---|
| (a) | Decreased | Increased |
| (b) | Decreased | Decreased |
| (c) | Increased | Increased |
| (d) | Increased | Decreased |

28. Use the figure and assumptions from the previous exercise to answer the following questions.

    (a) Explain how you can tell from the graph which year had lower total fixed costs.

    (b) Explain how you can tell from the graph which year had lower unit variable costs.

*Use algebra to find the intersection points of the graphs of the given equations. (See Examples 7 and 11.)*

29. $2x - y = 7$ and $y = 8 - 3x$

30. $6x - y = 2$ and $y = 4x + 7$

31. $y = 3x - 7$ and $y = 7x + 4$

32. $y = 3x + 5$ and $y = 12 - 2x$

**Business** *Work the following problems. (See Example 7.)*

33. An insurance company claims that for $x$ thousand policies, its monthly revenue in dollars is given by $R(x) = 125x$ and its monthly cost in dollars is given by $C(x) = 100x + 5000$.

    (a) Find the break-even point.

    (b) Graph the revenue and cost equations on the same axes.

    (c) From the graph, estimate the revenue and cost when $x = 100$ (100,000 policies).

34. The owners of a parking lot have determined that their weekly revenue and cost in dollars are given by $R(x) = 80x$ and $C(x) = 50x + 2400$, where $x$ is the number of long-term parkers.

    (a) Find the break-even point.

    (b) Graph $R(x)$ and $C(x)$ on the same axes.

    (c) From the graph, estimate the revenue and cost when there are 60 long-term parkers.

35. The revenue (in millions of dollars) from the sale of $x$ units at a home supply outlet is given by $R(x) = .21x$. The profit (in millions of dollars) from the sale of $x$ units is given by $P(x) = .084x - 1.5$.

    (a) Find the cost equation.

    (b) What is the cost of producing 7 units?

    (c) What is the break-even point?

36. The profit (in millions of dollars) from the sale of $x$ million units of Blue Glue is given by $P(x) = .7x - 25.5$. The cost is given by $C(x) = .9x + 25.5$.

    (a) Find the revenue equation.

    (b) What is the revenue from selling 10 million units?

    (c) What is the break-even point?

**Business** *Suppose you are the manager of a firm. The accounting department has provided cost estimates, and the sales department sales estimates, on a new product. You must analyze the data they give you, determine what it will take to break even, and decide whether to go ahead with production of the new product. (See Example 7.)*

37. Cost is estimated by $C(x) = 80x + 7000$ and revenue is estimated by $R(x) = 95x$; no more than 400 units can be sold.

38. Cost is estimated by $C(x) = 37x + 9000$ and revenue is estimated by $R(x) = 53x$; no more than 500 items can be sold.

39. Cost is $C(x) = 125x + 42{,}000$ and revenue is $R(x) = 165.5x$; no more than 2000 units can be sold.

40. Cost is $C(x) = 1750x + 95{,}000$ and revenue is $R(x) = 1975x$; no more than 600 units can be sold.

41. **Female Workforce** The accompanying graph shows the percentage of the female population in the workforce for France and Spain for the years 2004 through 2013. Estimate the point at which the two countries had the same percentage of females in the workforce. (Data from: Organization for Economic Cooperation and Development.)

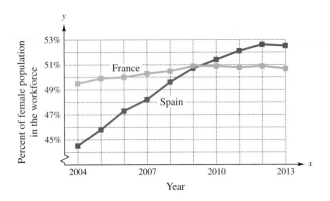

42. **Internet Users** The graph below shows the percentage of the population who are Internet users in Switzerland and the United Arab Emirates for the years 2006 through 2014. Find the point at which the two countries had the same percentage of Internet users. (Data from: www.worldbank.org.)

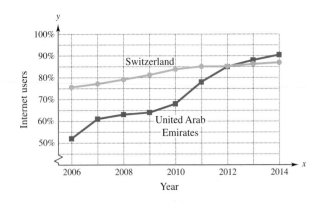

43. **Population Growth** The population (in thousands of people) of Arizona for the years 2010 through 2016 can be

approximated by the function $f(x) = 84x + 5540$. Similarly, the population (in thousands of people) of Massachusetts can be approximated by the function $g(x) = 48x + 6080$. For both of these models, let $x = 10$ correspond to the year 2010. (Data from: U.S. Census Bureau.)

(a) Graph both functions on the same coordinate axes for $x = 10$ to $x = 16$.

(b) Do the graphs intersect in this window?

(c) In what year were the populations of these two states approximately the same?

44. **Population Growth** The population (in thousands of people) of Florida for the years 2010 through 2016 can be approximated by the function $f(x) = 280x + 16{,}000$. Similarly, the population (in thousands of people) of New York can be approximated by the function $g(x) = 80x + 18{,}600$. For both of these models, let $x = 10$ correspond to the year 2010. (Data from: U.S. Census Bureau.)

(a) Graph both functions on the same coordinate axes for $x = 10$ to $x = 16$.

(b) Do the graphs intersect in this window?

(c) In what year were the populations of these two states approximately the same?

**Business** *Use the supply and demand curves in the accompanying graph to answer Exercises 45–48. (See Examples 8–11.)*

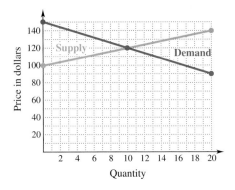

45. At what price are 20 items supplied?

46. At what price are 20 items demanded?

47. Find the equilibrium quantity.

48. Find the equilibrium price.

**Economics** *Work the following exercises. (See Examples 8–11.)*

49. Suppose that the demand and price for a certain brand of shampoo are related by

$$p = 16 - \frac{5}{4}q, \quad \text{(demand)}$$

where $p$ is price in dollars and $q$ is the quantity demanded. Find the price for a demand of

(a) 0 units;     (b) 4 units;     (c) 8 units.

Find the demand for the shampoo at a price of

**(d)** $6;        **(e)** $11;        **(f)** $16.

**(g)** Graph the demand curve $p = 16 - (5/4)q$.

Suppose the price and supply of the shampoo are related by

$$p = \frac{3}{4}q, \quad \text{(supply)}$$

where $q$ represents the quantity supplied and $p$ the price. Find the supply when the price is

**(h)** $0;        **(i)** $10;        **(j)** $20.

**(k)** Graph the supply curve $p = (3/4)q$ on the same axes used for part (g).

**(l)** Find the equilibrium quantity.

**(m)** Find the equilibrium price.

**50.** Let the supply and demand for radial tires in dollars be given by

$$\text{supply: } p = \frac{3}{2}q; \quad \text{demand: } p = 81 - \frac{3}{4}q.$$

**(a)** Graph these equations on the same axes.

**(b)** Find the equilibrium quantity.

**(c)** Find the equilibrium price.

**51.** Let the supply and demand for bananas in cents per pound be given by

$$\text{supply: } p = \frac{2}{5}q; \quad \text{demand: } p = 100 - \frac{2}{5}q.$$

**(a)** Graph these equations on the same axes.

**(b)** Find the equilibrium quantity.

**(c)** Find the equilibrium price.

**(d)** At a price of 40 cents per pound, is there a surplus or a shortage of supply?

**52.** Let the supply and demand for sugar be given by

$$\text{supply: } p = 1.4q - .6$$

and

$$\text{demand: } p = -2q + 3.2,$$

where $p$ is in dollars per pound.

**(a)** Graph these on the same axes.

**(b)** Find the equilibrium quantity.

**(c)** Find the equilibrium price.

**(d)** At a price of  $2 per pound, is there a surplus or a shortage of supply?

**53.** Explain why the graph of the (total) cost function is always above the $x$-axis and can never move downward as you go from left to right. Is the same thing true of the graph of the average cost function?

**54.** Explain why the graph of the profit function can rise or fall (as you go from left to right) and can be below the $x$-axis.

✓**Checkpoint Answers**

**1.** $C(x) = 4x + 60$.        **2.** $4.60        **3.** $9505

**4. (a)** $800        **(b)** $842

   **(c)** $3.50        **(d)** $3.50

**5. (a)** 1,120,000,000        **(b)** $47,873,000

**6.** 4286 newsletters

**7. (a)** $60        **(b)** 5 units        **(c)** (10, 60); (5, 80)

**8. (a)** 40        **(b)** $160/3 \approx 53.33$

# 3.4  Quadratic Functions and Applications

A **quadratic function** is a function whose rule is given by a quadratic polynomial, such as

$$f(x) = x^2, \quad g(x) = 3x^2 + 30x + 67, \quad \text{and} \quad h(x) = -x^2 + 4x.$$

Thus, a quadratic function is a function whose rule can be written in the form

$$f(x) = ax^2 + bx + c$$

for some constants $a$, $b$, and $c$, with $a \neq 0$.

**EXAMPLE 1**  Graph each of these quadratic functions:

$$f(x) = x^2; \quad g(x) = 4x^2; \quad h(x) = -.2x^2.$$

**Solution** In each case, choose several numbers (negative, positive, and 0) for $x$, find the values of the function at these numbers, and plot the corresponding points. Then connect the points with a smooth curve to obtain Figure 3.23, on the next page.

| $f(x) = x^2$ | | | | | |
|---|---|---|---|---|---|
| $x$ | $-2$ | $-1$ | 0 | 1 | 2 |
| $x^2$ | 4 | 1 | 0 | 1 | 4 |

| $g(x) = 4x^2$ | | | | | |
|---|---|---|---|---|---|
| $x$ | $-2$ | $-1$ | 0 | 1 | 2 |
| $4x^2$ | 16 | 4 | 0 | 4 | 16 |

| $h(x) = -.2x^2$ | | | | | |
|---|---|---|---|---|---|
| $x$ | $-5$ | $-3$ | $-1$ | 0 | 2 | 4 |
| $-.2x^2$ | $-5$ | $-1.8$ | $-.2$ | 0 | $-.8$ | $-3.2$ |

(a)

(b)

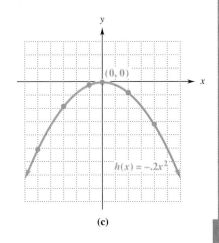

(c)

**Figure 3.23**

Each of the curves in Figure 3.23 is a **parabola.** It can be shown that the graph of every quadratic function is a parabola. Parabolas have many useful properties. Cross sections of radar dishes and spotlights form parabolas. Disks often visible on the sidelines of televised football games are microphones having reflectors with parabolic cross sections. These microphones are used by the television networks to pick up the signals shouted by the quarterbacks.

All parabolas have the same basic "cup" shape, although the cup may be broad or narrow and open upward or downward. The general shape of a parabola is determined by the coefficient of $x^2$ in its rule, as summarized here and illustrated in Example 1.

✓ **Checkpoint 1**

Graph each quadratic function.

**(a)**  $f(x) = x^2 - 4$

**(b)**  $g(x) = -x^2 + 4$

---

The graph of a quadratic function $f(x) = ax^2 + bx + c$ is a parabola.

If $a > 0$, the parabola opens upward.    [*Figure 3.23(a) and 3.23(b)*]

If $a < 0$, the parabola opens downward.    [*Figure 3.23(c)*]

If $|a| < 1$, the parabola appears wider than the graph of $y = x^2$. [*Figure 3.23(c)*]

If $|a| > 1$, the parabola appears narrower than the graph of $y = x^2$.
  [*Figure 3.23(b)*]

---

When a parabola opens upward [as in Figure 3.23(a), (b)], its lowest point is called the **vertex.** When a parabola opens downward [as in Figure 3.23(c)], its highest point is called the **vertex.** The vertical line through the vertex of a parabola is called the **axis of the parabola.** For example, (0, 0) is the vertex of each of the parabolas in Figure 3.23, and the axis of each parabola is the $y$-axis. If you were to fold the graph of a parabola along its axis, the two halves of the parabola would match exactly. This means that a parabola is *symmetric* about its axis. For this reason, the axis of the parabola is sometimes called the axis of symmetry.

In real-world problems, finding the vertex of a parabola is important because it tells us the maximum or minimum value, such as maximum profit or minimum cost. Although the vertex of a parabola can be approximated by a graphing calculator's maximum or minimum finder, its exact coordinates can be found algebraically, as in the following examples.

**EXAMPLE 2**    Consider the function $g(x) = 2(x - 3)^2 + 1$.

**(a)** Show that $g$ is a quadratic function.

**Solution**  Multiply out the rule of $g$ to show that it has the required form:

$$g(x) = 2(x - 3)^2 + 1$$
$$= 2[x^2 + 2(x)(-3) + (-3)^2] + 1 \qquad {\scriptstyle (a + b)^2 = a^2 + 2ab + b^2}$$
$$= 2(x^2 - 6x + 9) + 1$$
$$= 2x^2 - 12x + 18 + 1$$
$$g(x) = 2x^2 - 12x + 19.$$

According to the preceding box, the graph of $g$ is a somewhat narrow, upward-opening parabola.

**(b)** Show that the vertex of the graph of $g(x) = 2(x - 3)^2 + 1$ is $(3, 1)$.

**Solution**  Since $g(3) = 2(3 - 3)^2 + 1 = 0 + 1 = 1$, the point $(3, 1)$ is on the graph. The vertex of an upward-opening parabola is the lowest point on the graph, so we must show that $(3, 1)$ is the lowest point. Let $x$ be any number except 3 (so that $x - 3 \neq 0$). Then the quantity $2(x - 3)^2$ is positive, and hence

$$g(x) = 2(x - 3)^2 + 1 = (\text{a positive number}) + 1,$$

which means that $g(x) > 1$. Therefore, every point $(x, g(x))$ on the graph, where $x \neq 3$, has second coordinate $g(x)$ greater than 1. Hence $(x, g(x))$ lies *above* $(3, 1)$. In other words, $(3, 1)$ is the lowest point on the graph—the vertex of the parabola.

**(c)** Graph $g(x) = 2(x - 3)^2 + 1$.

**Solution**  Plot some points on both sides of the vertex $(3, 1)$ to obtain the graph in Figure 3.24. The vertical line $x = 3$ through the vertex is the axis of the parabola.

| $x$ | $y$ |
|---|---|
| 1 | 9 |
| 2 | 3 |
| 3 | 1 |
| 4 | 3 |
| 5 | 9 |

**Figure 3.24**

In Example 2, notice how the rule of the function $g$ is related to the coordinates of the vertex:

$$g(x) = 2(x - 3)^2 + 1. \qquad (3, 1).$$

Arguments similar to those in Example 2 lead to the following fact.

> The graph of the quadratic function $f(x) = a(x - h)^2 + k$ is a parabola with vertex $(h, k)$. It opens upward when $a > 0$ and downward when $a < 0$.

**EXAMPLE 3**    Determine algebraically whether the given parabola opens upward or downward, and find its vertex.

**(a)** $f(x) = -3(x - 4)^2 - 7$

**Solution** The rule of the function is in the form $f(x) = a(x - h)^2 + k$ (with $a = -3, h = 4$, and $k = -7$). The parabola opens downward ($a < 0$), and its vertex is $(h, k) = (4, -7)$.

**(b)**  $g(x) = 2(x + 3)^2 + 5$

**Solution**  Be careful here: The vertex is *not* (3, 5). To put the rule of $g(x)$ in the form $a(x - h)^2 + k$, we must rewrite it so that there is a minus sign inside the parentheses:

$$g(x) = 2(x + 3)^2 + 5$$
$$= 2(x - (-3))^2 + 5.$$

This is the required form, with $a = 2, h = -3$, and $k = 5$. The parabola opens upward, and its vertex is $(-3, 5)$.  ✓₂

✓ **Checkpoint 2**

Determine the vertex of each parabola, and graph the parabola.

**(a)**  $f(x) = (x + 4)^2 - 3$

**(b)**  $f(x) = -2(x - 3)^2 + 1$

**EXAMPLE 4**    Find the rule of a quadratic function whose graph has vertex (3, 4) and passes through the point (6, 22).

**Solution** The graph of $f(x) = a(x - h)^2 + k$ has vertex $(h, k)$. We want $h = 3$ and $k = 4$, so that $f(x) = a(x - 3)^2 + 4$. Since (6, 22) is on the graph, we must have $f(6) = 22$. Therefore,

$$f(x) = a(x - 3)^2 + 4$$
$$f(6) = a(6 - 3)^2 + 4$$
$$22 = a(3)^2 + 4$$
$$22 = 9a + 4$$
$$9a = 18$$
$$a = 2.$$

Thus, the graph of $f(x) = 2(x - 3)^2 + 4$ is a parabola with vertex (3, 4) that passes through (6, 22).

The vertex of each parabola in Examples 2 and 3 was easily determined because the rule of the function had the form

$$f(x) = a(x - h)^2 + k.$$

The rule of *any* quadratic function can be put in this form by using the technique of **completing the square**, which is illustrated in the next example.

**EXAMPLE 5**    Determine the vertex of the graph of $f(x) = x^2 - 4x + 2$. Then graph the parabola.

**Solution** In order to get $f(x)$ in the form $a(x - h)^2 + k$, we use the technique of completing the square. Take *half the coefficient of* $x$, namely $\frac{1}{2}(-4) = -2$, and *square it:* $(-2)^2 = 4$. Then proceed as follows.

$$f(x) = x^2 - 4x + 2$$

✓ **Checkpoint 3**

Rewrite the rule of each function by completing the square (add and subtract half the coefficient of $x$), and use this form to find the vertex of the graph.

**(a)**  $f(x) = x^2 + 6x + 5$

**(b)**  $g(x) = x^2 - 12x + 33$

$$= x^2 - 4x + \underline{\quad} + 2 \qquad \text{Leave space for the squared term and its negative.}$$
$$= x^2 - 4x + 4 - 4 + 2 \qquad \text{Add and subtract 4.}$$
$$= (x^2 - 4x + 4) - 4 + 2 \qquad \text{Insert parentheses.}$$
$$= (x - 2)^2 - 2 \qquad \text{Factor expression in parentheses and add.}$$

Adding and subtracting 4 did not change the rule of $f(x)$, but did make it possible to have a perfect square as part of its rule: $f(x) = (x - 2)^2 - 2$. Now we can see that the graph is an upward-opening parabola, as shown in Figure 3.25, on the following page.  ✓₃

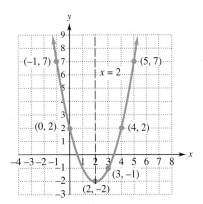

**Figure 3.25**

⚠ **CAUTION** The technique of completing the square only works when the coefficient of $x^2$ is 1. To find the vertex of a quadratic function such as

$$f(x) = 2x^2 + 12x - 19,$$

you must first factor out the coefficient of $x^2$ and write the rule as

$$f(x) = 2(x^2 + 6x - \tfrac{19}{2}).$$

Now complete the square on the expression in parentheses by adding and subtracting 9 (the square of half the coefficient of $x$), and proceed as in Example 5.

The technique of completing the square can be used to rewrite the general equation $f(x) = ax^2 + bx + c$ in the form $f(x) = a(x - h)^2 + k$. When this is done, we obtain a formula for the coordinates of the vertex.

---

The graph of the quadratic function $f(x) = ax^2 + bx + c$ is a parabola with vertex $(h, k)$, where

$$h = \frac{-b}{2a} \quad \text{and} \quad k = f(h).$$

---

Additionally, the fact that the vertex of a parabola is the highest or lowest point on the graph can be used in applications to find a maximum or minimum value.

**EXAMPLE 6**   **Microbrewery Profits** Shanese owns and operates her own microbrewery. She has hired a consultant to analyze her business operations. The consultant tells her that her daily profits from the sale of $x$ cases of beer are given by

$$P(x) = -x^2 + 120x.$$

**(a)** Find the vertex, determine if it is a maximum or minimum, write the equation of the axis of the parabola, compute the $x$- and $y$-intercepts, and sketch the graph of the profit function $P(x)$.

**Solution** Since $P(x) = -1x^2 + 120x + 0$ is a quadratic function with $a = -1$, $b = 120$, and $c = 0$, the preceding box tells us that the vertex is $(h, k)$ where the $x$-value of the vertex is

$$h = \frac{-b}{2a} = \frac{-120}{2(-1)} = 60.$$

The $y$-value of the vertex is

$$k = P(h) = P(60) = -(60)^2 + 120(60) = 3600.$$

The vertex is (60, 3600) and since $a$ is negative, it is a maximum because the parabola opens downward. The axis of the parabola is the vertical line through the vertex, namely, $x = 60$. The intercepts are found by setting $x$ and $y$ equal to 0.

| **x-intercepts** | **y-intercept** |
|---|---|
| Set $P(x) = y = 0$, so that | Set $x = 0$ to obtain |
| $0 = -x^2 + 120x$ | $P(0) = -(0)^2 + 120(0) = 0$ |
| $0 = x(-x + 120)$ | The $y$-intercept is 0. |
| $x = 0 \quad$ or $\quad -x + 120 = 0$ | |
| $\qquad\qquad\qquad -x = -120$ | |
| $\qquad\qquad\qquad\quad x = 120$ | |

The $x$–intercepts are 0 and 120.

Figure 3.26 shows the profit function $P(x)$.

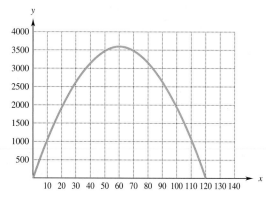

**Figure 3.26**

**(b)** Explain what the vertex tells us in this situation.

**Solution** Because the maximum occurs at the vertex, the *number of cases* of beer that should be sold to achieve the maximum profit is the $x$-value of the vertex, namely, 60 cases, and the resulting *maximum profit* is the $y$-value of the vertex, namely, \$3600.

 **Checkpoint 4**

When a company sells $x$ units of a product, its profit is $P(x) = -2x^2 + 40x + 280$. Find

**(a)** the number of units that should be sold so that maximum profit is received;

**(b)** the maximum profit.

**TECHNOLOGY TIP** The maximum or minimum finder on a graphing calculator can approximate the vertex of a parabola with a high degree of accuracy. The max–min finder is in the CALC menu or in the MATH or FCN submenu of the GRAPH menu. Similarly, the calculator's graphical root finder can approximate the $x$-intercepts of a parabola.

 Supply and demand curves were introduced in Section 3.3. Here is a quadratic example.

**EXAMPLE 7**   **Economics** Suppose that the price of and demand for an item are related by

$$p = 150 - 6q^2, \qquad \text{Demand function}$$

where $p$ is the price (in dollars) and $q$ is the number of items demanded (in hundreds). Suppose also that the price and supply are related by

$$p = 10q^2 + 2q, \qquad \text{Supply function}$$

where $p$ is the price (in dollars) and $q$ is the number of items supplied (in hundreds). Find the equilibrium quantity and the equilibrium price.

**Solution** The graphs of both of these equations are parabolas as seen in Figure 3.27, on the following page. Only those portions of the graphs which lie in the first quadrant are included, because none of supply, demand, or price can be negative.

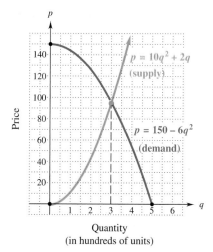

**Figure 3.27**

The point where the demand and supply curves intersect is the equilibrium point. Its first coordinate is the equilibrium quantity, and its second coordinate is the equilibrium price. These coordinates may be found in two ways.

**Algebraic Method**  At the equilibrium point, the second coordinate of the demand curve must be the same as the second coordinate of the supply curve, so that

$$\text{Demand} = \text{Supply}$$
$$150 - 6q^2 = 10q^2 + 2q.$$

Write this quadratic equation in standard form as follows:

$$0 = 16q^2 + 2q - 150 \qquad \text{Add } -150 \text{ and } 6q^2 \text{ to both sides.}$$
$$0 = 8q^2 + q - 75. \qquad \text{Multiply both sides by } \tfrac{1}{2}.$$

This equation can be solved by the quadratic formula, explained in Section 1.7. Here, $a = 8, b = 1,$ and $c = -75$:

$$q = \frac{-1 \pm \sqrt{1 - 4(8)(-75)}}{2(8)} \qquad \frac{-b \pm \sqrt{b^2 - 4ac}}{2a}$$

$$= \frac{-1 \pm \sqrt{1 + 2400}}{16} \qquad -4(8)(-75) = 2400$$

$$= \frac{-1 \pm 49}{16} \qquad \sqrt{1 + 2400} = \sqrt{2401} = 49$$

$$q = \frac{-1 + 49}{16} = \frac{48}{16} = 3 \quad \text{or} \quad q = \frac{-1 - 49}{16} = -\frac{50}{16} = -\frac{25}{8}.$$

It is not possible to make $-25/8$ units, so discard that answer and use only $q = 3$. Hence, the equilibrium quantity is 300. Find the equilibrium price by substituting 3 for $q$ in either the supply or the demand function (and check your answer by using the other one). Using the supply function gives

$$p = 10q^2 + 2q$$
$$p = 10(3)^2 + 2(3) \qquad \text{Let } q = 3.$$
$$= 10(9) + 6$$
$$p = \$96.$$

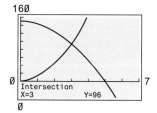

**Figure 3.28**

✔ **Checkpoint 5**

The price and demand for an item are related by $p = 32 - x^2$, while price and supply are related by $p = x^2$. Find

(a)  the equilibrium quantity;

(b)  the equilibrium price.

**Graphical Method**  Graph the two functions on a graphing calculator, and use the intersection finder to determine that the equilibrium point is (3, 96), as in Figure 3.28. ✔5

**EXAMPLE 8** **Rental Income** The rental manager of a small apartment complex with 16 units has found from experience that each $40 increase in the monthly rent results in an empty apartment. All 16 apartments will be rented at a monthly rent of $500. How many $40 increases will produce maximum monthly income for the complex?

**Solution** Let $x$ represent the number of $40 increases. Then the number of apartments rented will be $16 - x$. Also, the monthly rent per apartment will be $500 + 40x$. (There are $x$ increases of $40, for a total increase of $40x$.) The monthly income, $I(x)$, is given by the number of apartments rented times the rent per apartment, so

$$I(x) = (16 - x)(500 + 40x)$$

$$\underset{\text{FIRST OUTSIDE INSIDE LAST}}{= 8000 + 640x - 500x - 40x^2} \quad \text{FOIL method}$$

$$= 8000 + 140x - 40x^2.$$

Since $x$ represents the number of $40 increases and each $40 increase causes one empty apartment, $x$ must be a whole number. Because there are only 16 apartments, $0 \le x \le 16$.

The graph of $I(x) = 8000 + 140x - 40x^2$ is a downward-opening parabola (why?), and the value of $x$ that produces maximum income occurs at the vertex (see Figure 3.29). The box on page 149 shows that the vertex is $(1.75, 8122.50)$. Since $x$ must be a whole number, evaluate $I(x)$ at $x = 1$ and $x = 2$ to see which one gives the best result:

If $x = 1$, then $I(1) = -40(1)^2 + 140(1) + 8000 = 8100$.

If $x = 2$, then $I(2) = -40(2)^2 + 140(2) + 8000 = 8120$.

So maximum income occurs when $x = 2$. The manager should charge a rent of $500 + 2(40) = \$580$, leaving two apartments vacant.

**Figure 3.29**

## Quadratic Models

Real-world data can sometimes be used to construct a quadratic function that approximates the data. Such **quadratic models** can then be used to investigate the real-world scenarios.

**EXAMPLE 9** **Price of Crude Oil** The price of Dubai crude oil, in dollars per barrel, is given in the following table for the years 2010 through 2015. (Data from: Organization of Petroleum Exporting Countries.)

| Year | 2010 | 2011 | 2012 | 2013 | 2014 | 2015 |
|---|---|---|---|---|---|---|
| **Price per barrel** | $78 | $106 | $110 | $105 | $97 | $51 |

**(a)** Let $x = 0$ correspond to the year 2010. Display the information graphically.

**Solution**  Plot the points (0, 78), (1, 106), and so on, as shown in Figure 3.30.

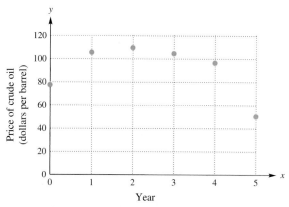

**Figure 3.30**

**(b)**  The shape of the data points in Figure 3.30 resembles a downward-opening parabola. Use the year 2012 as the maximum and the year 2010 as another data point to find a quadratic model $f(x) = a(x - h)^2 + k$ for these data.

**Solution**  Recall that when a quadratic function is written in the form $f(x) = a(x - h)^2 + k$, the vertex of the parabola is $(h, k)$. On the basis of Figure 3.30, let $(2, 110)$ be the vertex, so that

$$f(x) = a(x - h)^2 + k$$
$$f(x) = a(x - 2)^2 + 110.$$

To find $a$, use the fact that the price of oil was \$78 in 2010. Because $x = 0$ corresponds to the year 2010, the point (0, 78) lies on the parabola, therefore

$$f(x) = a(x - 2)^2 + 110$$
$$78 = a(0 - 2)^2 + 110 \qquad \text{Substitute 0 for } x \text{ and 78 for } f(x).$$
$$-32 = 4a \qquad \text{Subtract 110 from both sides.}$$
$$a = -8 \qquad \text{Divide both sides by 4 and reverse the equation.}$$

Therefore, $f(x) = -8(x - 2)^2 + 110$ is a quadratic model for these data. If we expand the form into $f(x) = ax^2 + bx + c$, we obtain

$$f(x) = -8(x - 2)^2 + 110$$
$$= -8(x^2 - 4x + 4) + 110 \qquad \text{Expand inside the parentheses.}$$
$$= -8x^2 + 32x - 32 + 110 \qquad \text{Distribute the } -8.$$
$$= -8x^2 + 32x + 78. \qquad \text{Add the last two terms.}$$

Figure 3.31 shows the original data with the graph of $f$. The graph appears to fit the data rather well except for the last two points. We will find a better fitting model in the next example.  ✔️₆

✔️ **Checkpoint 6**

Find another quadratic model in Example 9(b) by using (2, 110) as the vertex, but use the year 2013 for the other data point.

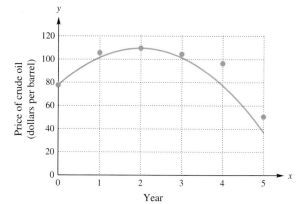

**Figure 3.31**

## ▦ Quadratic Regression

Linear regression was used in Section 2.3 to construct a linear function that modeled a set of data points. When the data points appear to lie on a parabola rather than on a straight line (as in Example 9), a similar least-squares regression procedure is available on most graphing calculators and spreadsheet programs to construct a quadratic model for the data. Simply follow the same steps as in linear regression, with one exception: Choose quadratic, rather than linear, regression. (Both options are on the same menu.)

▦ **EXAMPLE 10** Use a graphing calculator to find a quadratic-regression model for the data in Example 9.

**Solution** Enter the first coordinates of the data points as list $L_1$ and the second coordinates as list $L_2$ (go to the STAT EDIT menu). Performing a quadratic regression, as in Figure 3.32 (go to the STAT CALC menu and choose QuadReg), leads to the model

$$g(x) = -7.46x^2 + 32.55x + 78.21.$$

Figure 3.33 shows the data with the quadratic regression curve.

| Figure 3.32 | Figure 3.33 |

# 3.4 Exercises

*The graph of each of the functions in Exercises 1–4 is a parabola. Without graphing, determine whether the parabola opens upward or downward. (See Example 1.)*

**1.** $f(x) = x^2 - 3x - 12$

**2.** $g(x) = -x^2 + 5x + 15$

**3.** $h(x) = -3x^2 + 14x + 1$

**4.** $f(x) = 6.5x^2 - 7.2x + 4$

*Without graphing, determine the vertex of the parabola that is the graph of the given function. State whether the parabola opens upward or downward. (See Examples 2 and 3.)*

**5.** $f(x) = 2(x - 5)^2 + 7$

**6.** $g(x) = 7(x - 8)^2 - 3$

**7.** $h(x) = -4(x + 1)^2 - 9$

**8.** $f(x) = -8(x + 12)^2 + 9$

*Match each function with its graph, which is one of those shown. (See Examples 1–3.)*

**9.** $f(x) = x^2 + 2$

**10.** $g(x) = x^2 - 2$

**11.** $g(x) = (x - 2)^2$

**12.** $f(x) = -(x + 2)^2$

**13.** $f(x) = 2(x - 2)^2 + 2$

**14.** $g(x) = -2(x - 2)^2 - 2$

**15.** $g(x) = -2(x + 2)^2 + 2$

**16.** $f(x) = 2(x + 2)^2 - 2$

(a)

(b)

(c)

(d)

(e)

(f)

**(g)**

**(h)**

**(i)**

**(j)**

**(k)**

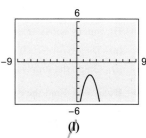
**(l)**

*Find the rule of a quadratic function whose graph has the given vertex and passes through the given point. (See Example 4.)*

**17.** Vertex $(1, 2)$; point $(5, 6)$

**18.** Vertex $(-3, 2)$; point $(2, 1)$

**19.** Vertex $(-1, -2)$; point $(1, 2)$

**20.** Vertex $(2, -4)$; point $(5, 2)$

*Without graphing, find the vertex of the parabola that is the graph of the given function. (See Examples 5 and 6.)*

**21.** $f(x) = x^2 + 6x - 3$    **22.** $g(x) = x^2 + 10x + 9$

**23.** $f(x) = 3x^2 - 12x + 5$    **24.** $g(x) = -4x^2 - 16x + 9$

*Without graphing, determine the x- and y-intercepts of each of the given parabolas. (See Example 6.)*

**25.** $f(x) = 3(x - 2)^2 - 3$    **26.** $g(x) = 2x^2 + 8x + 6$

**27.** $g(x) = x^2 - 10x + 20$    **28.** $f(x) = x^2 - 4x - 1$

*Graph each parabola and find its vertex and the axis of the parabola. (See Examples 1–6.)*

**29.** $f(x) = (x + 2)^2$    **30.** $f(x) = -(x + 5)^2$

**31.** $f(x) = x^2 - 4x + 6$    **32.** $f(x) = x^2 + 6x + 3$

*Work these problems. (See Example 6.)*

**33. McDonald's Revenue** The revenue for McDonald's Corporation (in billions of dollars) for the years 2010 to 2015 can be approximated by the quadratic function

$$R(x) = -\frac{1}{2}x^2 + 13x - 57,$$

where $x = 10$ corresponds to the year 2010. (Data from: www.morningstar.com.)

**(a)** Find the vertex.

**(b)** In what year during this period was revenue at a maximum?

**(c)** What was the maximum revenue, to the nearest billion dollars?

**34. Souvenir Sales** Janya owns a factory that manufactures souvenir key chains. Her weekly profit (in hundreds of dollars) is given by $P(x) = -2x^2 + 60x - 120$, where $x$ is the number of cases of key chains sold.

**(a)** What is the largest number of cases she can sell and still make a profit?

**(b)** Explain how it is possible for her to lose money if she sells more cases than your answer in part (a).

**(c)** How many cases should she make and sell in order to maximize her profits?

**35. Nerve Impulses** A researcher in physiology has decided that a good mathematical model for the number of impulses fired after a nerve has been stimulated is given by $y = -x^2 + 20x - 60$, where $y$ is the number of responses per millisecond and $x$ is the number of milliseconds since the nerve was stimulated.

**(a)** When will the maximum firing rate be reached?

**(b)** What is the maximum firing rate?

**36. Bullet Trajectory** A bullet is fired upward from ground level. Its height above the ground (in feet) at time $t$ seconds is given by

$$H(t) = -16t^2 + 960t.$$

**(a)** Find the vertex.

**(b)** Find the time at which the bullet reaches its maximum height.

**(c)** Find the maximum height.

**(d)** Find the time at which the bullet hits the ground (when the height is equal to zero).

**37. Automobile Fatalities** According to data from the National Safety Council, the fatal-accident rate per 100,000 licensed drivers can be approximated by the function $f(x) = .0328x^2 - 3.55x + 115$, where $x$ is the age of the driver ($16 \le x \le 88$). At what age is the rate the lowest?

**38. Population Dynamics** Using data from the U.S. Census Bureau, the population (in thousands) of Detroit, Michigan can be approximated by $g(x) = -.422x^2 + 48.84x + 248$, where $x = 0$ corresponds to the year 1900. In what year, according to the given model, did Detroit have its highest population?

*Work the following problems. (See Example 7.)*

**39. Business** Suppose the supply of and demand for a certain textbook are given by

$$\text{supply: } p = \frac{1}{5}q^2 \quad \text{and} \quad \text{demand: } p = -\frac{1}{5}q^2 + 40,$$

where $p$ is price and $q$ is quantity. How many books are demanded at a price of

(a) 10?     (b) 20?     (c) 30?     (d) 40?

How many books are supplied at a price of

(e) 5?     (f) 10?     (g) 20?     (h) 30?

(i) Graph the supply and demand functions on the same axes.

**40. Business** Find the equilibrium quantity and the equilibrium price in Exercise 39.

**41. Business** Suppose the price $p$ of widgets is related to the quantity $q$ that is demanded by

$$p = 640 - 5q^2,$$

where $q$ is measured in hundreds of widgets. Find the price when the number of widgets demanded is

(a) 0;     (b) 5;     (c) 10.

Suppose the supply function for widgets is given by $p = 5q^2$, where $q$ is the number of widgets (in hundreds) that are supplied at price $p$.

(d) Graph the demand function $p = 640 - 5q^2$ and the supply function $p = 5q^2$ on the same axes.

(e) Find the equilibrium quantity.

(f) Find the equilibrium price.

**42. Business** The supply function for a commodity is given by $p = q^2 + 200$, and the demand function is given by $p = -10q + 3200$

(a) Graph the supply and demand functions on the same axes.

(b) Find the equilibrium point.

(c) What is the equilibrium quantity? the equilibrium price?

**Business** *Find the equilibrium quantity and equilibrium price for the commodity whose supply and demand functions are given.*

**43.** Supply: $p = 45q$; demand: $p = -q^2 + 10,000$.

**44.** Supply: $p = q^2 + q + 10$; demand: $p = -10q + 3060$.

**45.** Supply: $p = q^2 + 20q$; demand: $p = -2q^2 + 10q + 3000$.

**46.** Supply: $p = .2q + 51$; demand: $p = \frac{3000}{q + 5}$.

**Business** *The revenue function $R(x)$ and the cost function $C(x)$ for a particular product are given. These functions are valid only for the specified domain of values. Find the number of units that must be produced to break even.*

**47.** $R(x) = 200x - x^2$; $C(x) = 70x + 2200$; $0 \le x \le 100$

**48.** $R(x) = 300x - x^2$; $C(x) = 65x + 7000$; $0 \le x \le 150$

**49.** $R(x) = 400x - 2x^2$; $C(x) = -x^2 + 200x + 1900$; $0 \le x \le 100$

**50.** $R(x) = 500x - 2x^2$; $C(x) = -x^2 + 270x + 5125$; $0 \le x \le 125$

**Business** *A store owner finds that at a price of $p(x)$ dollars, $x$ units of a certain item will be sold (the demand). Recall that revenue = demand × price. In Exercises 51 and 52, find the following:*

(a) *the revenue function;*

(b) *the number of items that leads to the maximum revenue;*

(c) *the maximum revenue.*

**51.** $p(x) = 84 - 2x$

**52.** $p(x) = 192 - 3x$

**Business** *Work each problem. (See Example 8.)*

**53.** A homeowner who listed her house for rent on Airbnb found that at a price of

$$p(x) = 300 - 10x$$

dollars per night, $x$ nights will be booked in a month.

(a) Find an expression for the total revenue from renting her house for $x$ nights. (*Hint:* Revenue = Demand × Price.)

(b) Find the number of rented nights that leads to the maximum revenue.

(c) Find the maximum revenue.

**54.** The manager of a bicycle shop has found that, at a price (in dollars) of $p(x) = 150 - \frac{x}{4}$ per bicycle, $x$ bicycles will be sold.

(a) Find an expression for the total revenue from the sale of $x$ bicycles. (*Hint:* Revenue = Demand × Price.)

(b) Find the number of bicycle sales that leads to maximum revenue.

(c) Find the maximum revenue.

**55.** A charter flight charges a fare of $200 per person, plus $4 per person for each unsold seat on the plane. If the plane holds 100 passengers and if $x$ represents the number of unsold seats, find the following:

(a) an expression for the total revenue received for the flight. (*Hint*: Multiply the number of people flying, $100 - x$, by the price per ticket, $200 + 4x$);

(b) the graph for the expression in part (a);

(c) the number of unsold seats that will produce the maximum revenue;

(d) the maximum revenue.

**56.** The revenue of a charter bus company depends on the number of unsold seats. If 100 seats are sold, the price is $50 per seat. Each unsold seat increases the price per seat by $1. Let $x$ represent the number of unsold seats.

(a) Write an expression for the number of seats that are sold.

(b) Write an expression for the price per seat.

(c) Write an expression for the revenue.

(d) Find the number of unsold seats that will produce the maximum revenue.

(e) Find the maximum revenue.

**57.** A farmer wants to find the best time to take her hogs to market. The current price is 88 cents per pound, and her hogs weigh an average of 90 pounds. The hogs gain 5 pounds per week, and the market price for hogs is falling each week by 2 cents per pound.

How many weeks should the farmer wait before taking her hogs to market in order to receive as much money as possible? At that time, how much money (per hog) will she get?

**58.** The manager of a peach orchard is trying to decide when to arrange for picking the peaches. If they are picked now, the average yield per tree will be 100 pounds, which can be sold for 40¢ per pound. Past experience shows that the yield per tree will increase about 5 pounds per week, while the price will decrease about 2¢ per pound per week.

(a) Let $x$ represent the number of weeks that the manager waits. Find the price per pound.

(b) Find the number of pounds per tree.

(c) Find the total revenue from a tree.

(d) When should the peaches be picked in order to produce the maximum revenue?

(e) What is the maximum revenue?

*Work these exercises. (See Examples 4 and 9.)*

**59. Federal Debt** The United States federal debt (in billions of dollars) is shown in the following table for the years 2000 through 2008. (Data from: U.S. Department of the Treasury.)

| Year | Debt (billions) |
| --- | --- |
| 2000 | 5674 |
| 2001 | 5807 |
| 2002 | 6228 |
| 2003 | 6783 |
| 2004 | 7379 |
| 2005 | 7933 |
| 2006 | 8507 |
| 2007 | 9008 |
| 2008 | 10,025 |

(a) Let $x = 0$ correspond to the year 2000. Use (0, 5674) as the vertex and the data for the year 2007 ($x = 7$) to find a quadratic function in the form $f(x) = a(x - h)^2 + k$ that models these data (round coefficients to the nearest integer).

(b) Use the model to estimate the federal debt in 2008. Is the value estimated by the model reasonably close to the actual value shown in the table?

**60. Federal Debt** The United States federal debt (in billions of dollars) is shown in the following table for the years 2009 through 2015. (Data from: U.S. Department of the Treasury.)

| Year | Debt (billions) |
| --- | --- |
| 2009 | 11,910 |
| 2010 | 13,562 |
| 2011 | 14,790 |
| 2012 | 16,066 |
| 2013 | 16,738 |
| 2014 | 17,824 |
| 2015 | 18,150 |

(a) Let $x = 0$ correspond to the year 2000. Use (15, 18150) as the vertex and the data for the year 2011 ($x = 11$) to find a quadratic function in the form $f(x) = a(x - h)^2 + k$ that models these data.

(b) Use the model to estimate the federal debt in 2014. Is the value estimated by the model reasonably close to the actual value shown in the table?

**61. Commodity Pricing** The price of soybean meal (in dollars per metric ton) is shown in the following table for the years 2011 to 2015. (Data from: U.S. Department of Agriculture.)

| Year | Price |
| --- | --- |
| 2011 | $398 |
| 2012 | $524 |
| 2013 | $545 |
| 2014 | $528 |
| 2015 | $395 |

(a) Let $x = 0$ correspond to the year 2000. Use (13, 545) as the vertex and the data for the year 2015 ($x = 15$) to find a quadratic function in the form $f(x) = a(x - h)^2 + k$ that models these data.

(b) Use the model to estimate the price of soybean meal in 2011. Is the value estimated by the model reasonably close to the actual value shown in the table?

**62. Forest Fires** The acreage (in millions) consumed by forest fires in the United States is given in the following table for selected years. (Data from: National Interagency Fire Center.)

| Year | Acres |
| --- | --- |
| 1985 | 2.9 |
| 1988 | 5.0 |
| 1991 | 3.0 |
| 1994 | 4.1 |
| 1997 | 2.9 |
| 2000 | 7.4 |
| 2003 | 4.0 |
| 2006 | 9.9 |
| 2009 | 5.9 |
| 2012 | 9.2 |
| 2015 | 10.1 |

(a) Let $x = 0$ correspond to the year 1985. Use (0, 2.9) as the vertex and the data for the year 2015 ($x = 30$) to find a quadratic function in the form $f(x) = a(x - h)^2 + k$ that models these data (round coefficients to three decimal places).

(b) Use the model to estimate the acreage destroyed by forest fires in 2010 ($x = 25$).

In Exercises 63–66, use quadratic regression to find a function that models these data. Plot the data and the quadratic regression curve. (See Example 10.)

63. Use the data in Exercise 59, with $x = 0$ corresponding to the year 2000. What does the graphing calculator model estimate for the year 2008? Compare your answer with that for Exercise 59.

64. Use the data in Exercise 60, with $x = 0$ corresponding to the year 2000. What does the graphing calculator model estimate for the year 2014? Compare your answer with that for Exercise 60.

65. Use the data in Exercise 61, with $x = 0$ corresponding to the year 2000. What does the graphing calculator model estimate for the year 2011? Compare your answer with that for Exercise 61.

66. Use the data in Exercise 62, with $x = 0$ corresponding to the year 1985. What does the graphing calculator model estimate for the year 2010 ($x = 25$)? Compare your answer with that for Exercise 62.

**Business** *Recall that profit equals revenue minus cost. In Exercises 67 and 68, find the following:*

(a) *The break-even point (to the nearest tenth)*

(b) *The x-value that makes profit a maximum*

(c) *The maximum profit*

(d) *For what x-values will a loss occur?*

(e) *For what x-values will a profit occur?*

67. $R(x) = 400x - 2x^2$ and $C(x) = 200x + 2000$, with $0 \le x \le 100$

68. $R(x) = 900x - 3x^2$ and $C(x) = 450x + 5000$, with $20 \le x \le 150$

 **Checkpoint Answers**

1. (a)

(b)

2. (a)

(b)

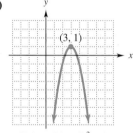

3. (a) $f(x) = (x + 3)^2 - 4; (-3, -4)$

   (b) $g(x) = (x - 6)^2 - 3; (6, -3)$

4. (a) 10 units      (b) $480

5. (a) 4      (b) 16

6. $f(x) = -5x^2 + 20x + 90$

# 3.5 Polynomial Functions

A **polynomial function of degree** $n$ is a function whose rule is given by a polynomial of degree $n$.[*] For example

$$f(x) = 3x - 2$$   polynomial function of degree 1;
$$g(x) = 3x^2 + 4x - 6$$   polynomial function of degree 2;
$$h(x) = x^4 + 5x^3 - 6x^2 + x - 3$$   polynomial function of degree 4.

---
[*] The degree of a polynomial was defined on page 13.

## Basic Graphs

The simplest polynomial functions are those whose rules are of the form $f(x) = ax^n$ (where $a$ is a constant).

> **EXAMPLE 1**    Graph $f(x) = x^3$.
>
> **Solution**  Make a table of data with ordered pairs belonging to the graph, as in Figure 3.34(a). Be sure to choose some negative $x$-values, $x = 0$, and some positive $x$-values in order to get representative ordered pairs. Find as many ordered pairs as you need in order to see the shape of the graph. Then plot the ordered pairs and draw a smooth curve through them to obtain the graph in Figure 3.34(b).  ✔₁

✓ **Checkpoint 1**

Graph $f(x) = -\dfrac{1}{2}x^3$

| $x$ | $f(x)$ |
|---|---|
| 2 | 8 |
| 1 | 1 |
| 0 | 0 |
| −1 | −1 |
| −2 | −8 |

**Figure 3.34(a)**          **Figure 3.34(b)**

> **EXAMPLE 2**    Graph
>
> $$f(x) = \frac{3}{2}x^4.$$
>
> **Solution**  The table in Figure 3.35(a) gives some typical ordered pairs and leads to the graph in Figure 3.35(b).  ✔₂

✓ **Checkpoint 2**

Graph $g(x) = -2x^4$.

| $x$ | $f(x)$ |
|---|---|
| −2 | 24 |
| −1 | 3/2 |
| 0 | 0 |
| 1 | 3/2 |
| 2 | 24 |

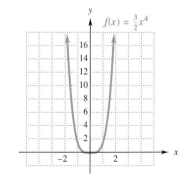

**Figure 3.35(a)**          **Figure 3.35(b)**

The graph of $f(x) = ax^n$ has one of the four basic shapes illustrated in Examples 1 and 2 and in Checkpoints 1 and 2. The basic shapes are summarized in the following box.

## Graph of $f(x) = ax^n$

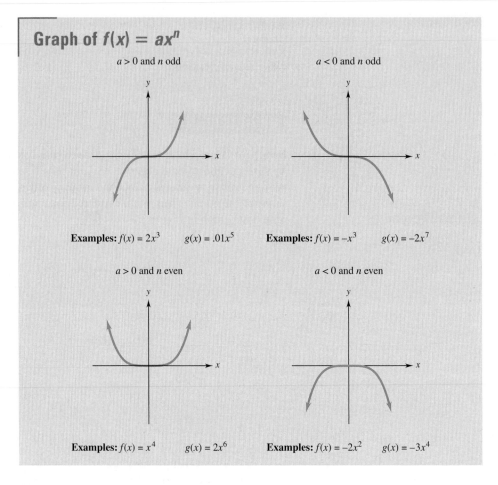

$a > 0$ and $n$ odd

$a < 0$ and $n$ odd

**Examples:** $f(x) = 2x^3$    $g(x) = .01x^5$

**Examples:** $f(x) = -x^3$    $g(x) = -2x^7$

$a > 0$ and $n$ even

$a < 0$ and $n$ even

**Examples:** $f(x) = x^4$    $g(x) = 2x^6$

**Examples:** $f(x) = -2x^2$    $g(x) = -3x^4$

## Properties of Polynomial Graphs

Unlike the graphs in the preceding figures, the graphs of more complicated polynomial functions may have several "peaks" and "valleys," as illustrated in Figure 3.36, on the following page. The locations of the peaks and valleys can be accurately approximated by a maximum or minimum finder on a graphing calculator. Calculus is needed to determine their exact location.

The total number of peaks and valleys in a polynomial graph, as well as the number of $x$-intercepts, depends on the degree of the polynomial, as shown in Figure 3.36 and summarized here.

| Polynomial | Degree | Number of peaks & valleys | Number of $x$-intercepts |
|---|---|---|---|
| $f(x) = x^3 - 4x + 2$ | 3 | 2 | 3 |
| $f(x) = x^5 - 5x^3 + 4x$ | 5 | 4 | 5 |
| $f(x) = 1.5x^4 + x^3 - 4x^2 - 3x + 4$ | 4 | 3 | 2 |
| $f(x) = -x^6 + x^5 + 2x^4 + 1$ | 6 | 3 | 2 |

In each case, the number of $x$-intercepts is *at most* the degree of the polynomial. The total number of peaks and valleys is at most *one less than* the degree of the polynomial. The same thing is true in every case.

1. The total number of peaks and valleys on the graph of a polynomial function of degree $n$ is at most $n - 1$.

2. The number of $x$-intercepts on the graph of a polynomial function of degree $n$ is at most $n$.

**(a)** $f(x) = x^3 - 4x + 2$
1 peak ⎫ total 2
1 valley ⎭
3 $x$-intercepts

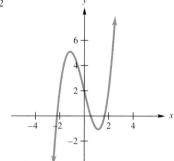

**(b)** $f(x) = x^5 - 5x^3 + 4x$
2 peaks ⎫ total 4
2 valleys ⎭
5 $x$-intercepts

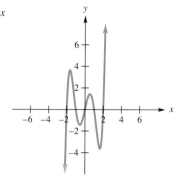

**(c)** $f(x) = 1.5x^4 + x^3 - 4x^2 - 3x + 4$
1 peak ⎫ total 3
2 valleys ⎭
2 $x$-intercepts

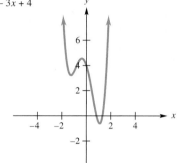

**(d)** $f(x) = -x^6 + x^5 + 2x^4 + 1$
2 peaks ⎫ total 3
1 (shallow) valley ⎭
2 $x$-intercepts

**Figure 3.36**

**Figure 3.37**

The domain of every polynomial function is the set of all real numbers, which means that its graph extends forever to the left and right. We can indicate this by the arrows on the ends of polynomial graphs, as shown in Figure 3.36.

Although there may be peaks, valleys, and bends in a polynomial graph, the far ends of the graph are easy to describe: *they look like the graph of the highest-degree term of the polynomial.* Consider, for example, $f(x) = 1.5x^4 + x^3 - 4x^2 - 3x + 4$, whose highest-degree term is $1.5x^4$ and whose graph is shown in Figure 3.36(c). The ends of the graph shoot upward, just as the graph of $y = 1.5x^4$ does in Figure 3.35. When $f(x)$ and $y = 1.5x^4$ are graphed in the same large viewing window of a graphing calculator (Figure 3.37), the graphs look almost identical, except near the origin. This is an illustration of the following facts.

> The graph of a polynomial function is a smooth, unbroken curve that extends forever to the left and right. When $|x|$ is large, the graph resembles the graph of its highest-degree term and moves sharply away from the $x$-axis.

**EXAMPLE 3**   Let $g(x) = x^3 - 11x^2 - 32x + 24$, and consider the graph in Figure 3.38.

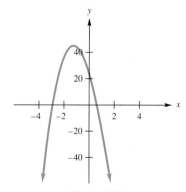

**Figure 3.38**

**(a)** Is Figure 3.38 a complete graph of $g(x)$; that is, does it show all the important features of the graph?

**Solution** The far ends of the graph of $g(x)$ should resemble the graph of its highest-degree term $x^3$. The graph of $f(x) = x^3$ in Figure 3.34 moves upward at the far right, but the graph in Figure 3.38 does not. So Figure 3.38 is *not* a complete graph.

 **(b)** Use a graphing calculator to find a complete graph of $g(x)$.

**Solution** Since the graph of $g(x)$ must eventually start rising on the right side (as does the graph of $x^3$), a viewing window that shows a complete graph must extend beyond $x = 4$. By experimenting with various windows, we obtain Figure 3.39. This graph shows a total of two peaks and valleys and three $x$-intercepts (the maximum possible for a polynomial of degree 3). At the far ends, the graph of $g(x)$ resembles the graph of $f(x) = x^3$. Therefore, Figure 3.39 is a complete graph of $g(x)$. ✔3

✔**Checkpoint 3**

Find a viewing window on a graphing calculator that shows a complete graph of
$f(x) = -.7x^4 + 119x^2 + 400$.
(*Hint*: The graph has two $x$-intercepts and the maximum possible number of peaks and valleys.)

**Figure 3.39**

## ⓒ Graphing Techniques

Accurate graphs of first- and second-degree polynomial functions (lines and parabolas) are easily found algebraically, as we saw in Sections 2.2 and 3.4. All polynomial functions of degree 3, and some of higher degree, can be accurately graphed by hand by using calculus and algebra to locate the peaks and valleys. When a polynomial can be completely factored, the general shape of its graph can be determined algebraically by using the basic properties of polynomial graphs, as illustrated in Example 4. Obtaining accurate graphs of other polynomial functions generally requires the use of technology.

**EXAMPLE 4** Graph $f(x) = (2x + 3)(x - 1)(x + 2)$.

**Solution** Note that $f(x)$ is a polynomial of degree 3. (If you don't see why, do Checkpoint 4.) Begin by finding any $x$-intercepts. Set $f(x) = 0$ and solve for $x$: ✔4

$$f(x) = 0$$
$$(2x + 3)(x - 1)(x + 2) = 0.$$

✔**Checkpoint 4**

Multiply out the expression for $f(x)$ in Example 4 and determine its degree.

Solve this equation by setting each of the three factors equal to 0:

$$2x + 3 = 0 \qquad \text{or} \qquad x - 1 = 0 \qquad \text{or} \qquad x + 2 = 0$$
$$x = -\frac{3}{2} \qquad\qquad x = 1 \qquad\qquad x = -2.$$

The three numbers $-3/2$, 1, and $-2$ divide the $x$-axis into four intervals:

$$x < -2, \quad -2 < x < -\frac{3}{2}, \quad -\frac{3}{2} < x < 1, \quad \text{and} \quad 1 < x.$$

These intervals are shown in Figure 3.40.

**Figure 3.40**

Since the graph is an unbroken curve, it can change from above the $x$-axis to below it only by passing through the $x$-axis. As we have seen, this occurs only at the $x$-intercepts: $x = -2, -3/2$, and 1. Consequently, in the interval between two intercepts (or to the left of $x = -2$ or to the right of $x = 1$), the graph of $f(x)$ must lie entirely above or entirely below the $x$-axis.

We can determine whether the graph lies above or below an interval by evaluating $f(x) = (2x + 3)(x - 1)(x + 2)$ at a number in that interval. For example, $x = -3$ is in the interval where $x < -2$, and

$$f(-3) = (2(-3) + 3)(-3 - 1)(-3 + 2)$$
$$= -12.$$

Therefore, $(-3, -12)$ is on the graph. Since this point lies below the $x$-axis, all points in this interval (that is, all points with $x < -2$) must lie below the $x$-axis. By testing numbers in the other intervals, we obtain the following table.

| Interval | $x < -2$ | $-2 < x < -3/2$ | $-3/2 < x < 1$ | $x > 1$ |
|---|---|---|---|---|
| Test Number | $-3$ | $-7/4$ | $0$ | $2$ |
| Value of $f(x)$ | $-12$ | $11/32$ | $-6$ | $28$ |
| Sign of $f(x)$ | Negative | Positive | Negative | Positive |
| Graph | Below $x$-axis | Above $x$-axis | Below $x$-axis | Above $x$-axis |

Since the graph intersects the $x$-axis at the intercepts $x = -2$ and $x = -3/2$, and is above the $x$-axis between these intercepts, there must be at least one peak there. Similarly, there must be at least one valley between $x = -3/2$ and $x = 1$, because the graph is below the $x$-axis there. A polynomial function of degree 3 can have a total of at most $3 - 1 = 2$ peaks and valleys. So there must be exactly one peak and exactly one valley on this graph.

Furthermore, when $|x|$ is large, the graph must resemble the graph of $y = 2x^3$ (the highest-degree term). The graph of $y = 2x^3$, like the graph of $y = x^3$ in Figure 3.34, moves upward to the right and downward to the left. Using these facts and plotting the $x$-intercepts shows that the graph must have the general shape shown in Figure 3.41. Plotting additional points leads to the reasonably accurate graph in Figure 3.42. We say "reasonably accurate" because we cannot be sure of the exact locations of the peaks and valleys on the graph without using calculus. ✓5

✓**Checkpoint 5**

Graph $f(x) =$
$.4(x - 2)(x + 3)(x - 4)$.

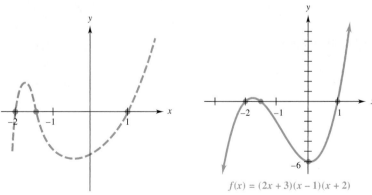

Figure 3.41                    Figure 3.42

$f(x) = (2x + 3)(x - 1)(x + 2)$

## ⓒ 🖩 Polynomial Models

Regression procedures similar to those presented for linear (degree 1) regression in Section 2.3 and quadratic (degree 2) regression in Section 3.5 can be used to find cubic (degree 3) and quartic (degree 4) polynomial models for appropriate data.

▦ **EXAMPLE 5** **Johnson & Johnson Profits** The following table shows the revenue and costs (in millions of dollars) for Johnson & Johnson. (Data from: www.morningstar.com.)

| Year | 2008 | 2009 | 2010 | 2011 | 2012 | 2013 | 2014 | 2015 |
|---|---|---|---|---|---|---|---|---|
| **Revenue** | 63,747 | 61,897 | 61,587 | 65,030 | 67,224 | 71,312 | 74,331 | 70,074 |
| **Costs** | 18,511 | 18,447 | 18,792 | 20,360 | 21,658 | 22,342 | 22,746 | 21,536 |

**(a)** Let $x = 8$ correspond to the year 2008, and use cubic regression to obtain models for the revenue data $R(x)$ and the costs $C(x)$.

**Solution** The procedure is the same as for linear regression, just choose cubic regression instead (go to the STAT CALC menu, and choose CubicReg). The functions obtained are:

$$R(x) = -234x^3 + 8139x^2 - 90{,}692x + 388{,}407 \text{ and}$$
$$C(x) = -73x^3 + 2461x^2 - 26{,}151x + 107{,}839.$$

**(b)** Plot both sets of data and graph $R(x)$ and $C(x)$ on the same screen.

**Solution** See Figure 3.43, which shows that the models fit these data well.

**Figure 3.43**

**(c)** Find the profit function $P(x)$ and show its graph. Were profits always increasing during this time period or did they fluctuate?

**Solution** The profit function is the difference between the revenue function and the cost function. We subtract the coefficients of the cost function from the respective coefficients of the revenue function:

$$\begin{aligned} P(x) &= R(x) - C(x) \\ &= -234x^3 + 8139x^2 - 90{,}692x + 388{,}407 \\ &\quad -(-73x^3 + 2461x^2 - 26{,}151x + 107{,}839) \\ &= -161x^3 + 5678x^2 - 64{,}541x + 280{,}568. \end{aligned}$$

The graph of $P(x)$ appears in Figure 3.44, which shows that profits for this company fluctuate over time.

**Figure 3.44**

▦ **EXAMPLE 6** **Personal Savings** The table below shows gross savings in the United States in trillions of dollars. (Data from: www.worldbank.org.)

| Year | Gross savings (trillions) |
|---|---|
| 2000 | 2126 |
| 2001 | 2076 |
| 2002 | 1998 |
| 2003 | 1998 |
| 2004 | 2155 |
| 2005 | 2349 |
| 2006 | 2659 |
| 2007 | 2512 |
| 2008 | 2281 |
| 2009 | 2082 |
| 2010 | 2272 |
| 2011 | 2450 |
| 2012 | 2868 |
| 2013 | 2949 |

**Figure 3.45**

**(a)** Plot these data on a graphing calculator or cell phone app with $x = 0$ corresponding to the year 2000.

**Solution** The points plotted in Figure 3.45 suggest the general shape of a fourth-degree polynomial. A fourth-degree polynomial is called a *quartic* function (not to be confused with a *quadratic* function, which is a second-degree polynomial whose graph is a parabola).

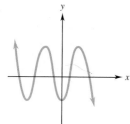

**Figure 3.46**

**(b)** Used quartic regression to obtain a model for these data and graph the function on the same screen as the data points.

**Solution** The function obtained from quartic regression (look for QuartReg in the STAT CALC menu) is

$$f(x) = .686x^4 - 16.4x^3 + 121.6x^2 - 245x + 2139.$$

The graph in Figure 3.46 shows this is a reasonable model for the data.

# 3.5 Exercises

*Graph each of the given polynomial functions. (See Examples 1 and 2.)*

**1.** $f(x) = x^4$

**2.** $g(x) = -.5x^6$

**3.** $h(x) = -.2x^5$

**4.** $f(x) = x^7$

*In Exercises 5–8, state whether the graph could possibly be the graph of (a) some polynomial function; (b) a polynomial function of degree 3; (c) a polynomial function of degree 4; (d) a polynomial function of degree 5. (See Example 3.)*

**5.**

**6.**

**7.**

**8.**

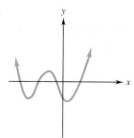

*In Exercises 9–14, match the given polynomial function to its graph [(a)–(f)], without using a graphing calculator. (See Example 3 and the two boxes preceding it.)*

**(a)**

**(b)**

**(c)**

**(d)**

**(e)**

**(f)**

**9.** $f(x) = x^3 - 7x - 9$

**10.** $f(x) = -x^3 + 4x^2 + 3x - 8$

**11.** $f(x) = x^4 - 5x^2 + 7$

**12.** $f(x) = x^4 + 4x^3 - 20$

**13.** $f(x) = -x^5 + 4x^4 + x^3 - 16x^2 + 12x + 5$

**14.** $f(x) = .7x^5 - 2.5x^4 - x^3 + 8x^2 + x + 2$

*Graph each of the given polynomial functions. (See Example 4.)*

**15.** $f(x) = (x + 3)(x - 4)(x + 1)$

**16.** $f(x) = (x - 5)(x - 1)(x + 1)$

**17.** $f(x) = x^2(x + 3)(x - 1)$

**18.** $f(x) = x^2(x + 2)(x - 2)$

**19.** $f(x) = x^3 - x^2 - 20x$

**20.** $f(x) = x^3 - 4x$

*Work these exercises.*

**21. Home Depot Revenue** The revenue for Home Depot Inc. (in billions of dollars) for the years 2004 to 2015 can be approximated by the function

$$R(x) = .19x^3 - 5.2x^2 + 44x - 40,$$

where $x = 4$ corresponds to the year 2004. (Data from: www.morningstar.com.)

(a) Find $R(4)$, round your answer to the nearest integer, and explain what your answer means in practical terms.

(b) Find the following (round your answers to the nearest integer): $R(6)$; $R(11)$; $R(14)$.

(c) Graph $R(x)$.

(d) Is revenue always increasing or does it fluctuate?

**22. Caterpillar Revenue** The revenue for Caterpillar Corporation (in billions of dollars) for the years 2008 to 2015 can be approximated by the function

$$R(x) = -.638x^3 + 21.1x^2 - 225x + 825,$$

where $x = 8$ corresponds to the year 2008. (Data from: www.morningstar.com.)

(a) Find the following (round your answers to the nearest integer): $R(8)$; $R(9)$; $R(13)$; $R(15)$.

(b) Graph $R(x)$.

(c) Is revenue always increasing or does it fluctuate?

**23. Home Depot Costs** The cost function for Home Depot Inc. (in billions of dollars) for the years 2004 to 2015 can be approximated by the function

$$C(x) = .13x^3 - 3.6x^2 + 31x - 34,$$

where $x = 4$ corresponds to the year 2004. (Data from: www.morningstar.com.)

(a) Find $C(4)$, round your answer to the nearest integer, and interpret your answer.

(b) Find the following (round your answers to the nearest integer): $C(6)$; $C(11)$; $C(14)$.

(c) Graph $C(x)$.

(d) Are costs always increasing or do they fluctuate?

**24. Caterpillar Costs** The cost function for Caterpillar Corporation (in billions of dollars) for the years 2008 to 2015 can be approximated by the function

$$C(x) = -.493x^3 + 16.4x^2 - 176x + 645,$$

where $x = 8$ corresponds to the year 2008. (Data from: www.morningstar.com.)

(a) Find the following (round your answers to the nearest integer): $C(8)$; $C(9)$; $C(13)$; $C(15)$.

(b) Graph $C(x)$.

(c) Are costs always increasing or do they fluctuate?

**25. Home Depot Profit** Find the profit function for Home Depot Inc. (in billions of dollars) by using the revenue function $R(x)$ from Exercise 21 and the cost function $C(x)$ from Exercise 23. Use your function to find the profit for the year 2014.

**26. Caterpillar Profit** Find the profit function for Caterpillar Corporation (in billions of dollars) by using the revenue function $R(x)$ from Exercise 22 and the cost function $C(x)$ from Exercise 24. Use your function to find the profit for the year 2015.

**27. Finance** An idealized version of the Laffer curve (originated by economist Arthur Laffer) is shown in the accompanying graph. According to this theory, decreasing the tax rate, say, from $x_2$ to $x_1$, may actually increase the total revenue to the government. The theory is that people will work harder and earn more if they are taxed at a lower rate, which means higher total tax revenues than would be the case at a higher rate. Suppose that the Laffer curve is given by the function

$$f(x) = \frac{x(x - 100)(x - 160)}{240} \quad (0 \le x \le 100),$$

where $f(x)$ is government revenue (in billions of dollars) from a tax rate of $x$ percent. Find the revenue from the given tax rates.

(a) 20%     (b) 40%     (c) 50%     (d) 70%

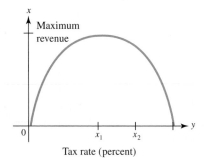

Tax rate (percent)

**28. Cardiac Output** A technique for measuring cardiac output depends on the concentration of a dye after a known amount is injected into a vein near the heart. In a normal heart, the concentration of the dye at time $x$ (in seconds) is given by the function defined by

$$g(x) = -.006x^4 + .140x^3 - .053x^2 + 1.79x.$$

Find the following: $g(0)$; $g(1)$; $g(2)$; $g(3)$.

**▦ Polynomial Models** *Use a graphing calculator to do the following cubic and quartic regression problems. (See Examples 5 and 6.)*

**29. Public Schools** Enrollment in public elementary and secondary schools (in thousands) in the United States is shown in the following table for selected years, including projected enrollment through the year 2024. (Data from: U.S. National Center for Education Statistics.)

| Year | Enrollment (thousands) |
|------|------------------------|
| 2000 | 47,204 |
| 2005 | 49,113 |
| 2010 | 49,484 |
| 2015 | 50,094 |
| 2016 | 50,229 |
| 2020 | 51,547 |
| 2024 | 52,920 |

**(a)** Plot the data with $x = 0$ corresponding to the year 2000.

**(b)** Use cubic regression to find a third-order polynomial function $f(x)$ that models these data (round coefficients to the nearest integer).

**(c)** Graph $f(x)$ on the same screen as the data points. Does the graph appear to fit the data well?

**(d)** According to the model, what is the project enrollment for the year 2023?

**30. Gross National Income** The following table shows the gross national income (GNI) in dollars per capita for middle- and upper-income countries for selected years. (Data from: Organization for Economic Cooperation and Development.)

| Year | 2000 | 2002 | 2004 | 2006 | 2008 | 2010 | 2012 | 2014 |
|------|------|------|------|------|------|------|------|------|
| GNI | 1768 | 1842 | 2335 | 3148 | 4403 | 5472 | 6949 | 7901 |

**(a)** Plot the data with $x = 0$ corresponding to the year 2000.

**(b)** Use cubic regression to find a third-order polynomial function $f(x)$ that models these data (round coefficients to one decimal place).

**(c)** Graph $y = f(x)$ on the same screen as the data points. Does the graph appear to fit the data well?

**(d)** What does the model predict for the GNI in the year 2017?

**31. Consumer Confidence** The Conference Board's consumer confidence index (CCI) is listed for the years 2008 to 2015 in the table below. (Data from: The Conference Board.)

| Year | 2008 | 2009 | 2010 | 2011 | 2012 | 2013 | 2014 | 2015 |
|------|------|------|------|------|------|------|------|------|
| CCI | 88 | 40 | 57 | 65 | 62 | 58 | 79 | 104 |

**(a)** Plot the data with $x = 8$ corresponding to the year 2008.

**(b)** Use quartic regression to find a fourth-order polynomial function $f(x)$ that models these data.

**(c)** Graph $y = f(x)$ on the same screen as the data points. Does the graph appear to be a reasonable model for these data?

**(d)** What does the model estimate for the CCI in 2015? Is this close to the actual index value shown in the table?

**32. American Express Revenue** The American Express Company reported its revenue (in billions) as shown in the following table. (Data from: www.morningstar.com.)

| Year | 2008 | 2009 | 2010 | 2011 | 2012 | 2013 | 2014 | 2015 |
|------|------|------|------|------|------|------|------|------|
| Revenue (billions) | 28.4 | 24.5 | 27.8 | 30.0 | 31.6 | 33.0 | 34.3 | 32.8 |

**(a)** Plot the data with $x = 8$ corresponding to the year 2008.

**(b)** Use cubic regression to find a third-order polynomial function $f(x)$ that models these data (round coefficients to two decimal places).

**(c)** Graph $f(x)$ on the same screen as the data points. Does the graph appear to fit the data well?

**(d)** What does the model estimate for the revenue in 2011? Is this close to the actual revenue shown in the table?

*Exercises 33–36 require a graphing calculator. Find a viewing window that shows a complete graph of the polynomial function (that is, a graph that includes all the peaks and valleys and that indicates how the curve moves away from the x-axis at the far left and far right). There are many correct answers. Consider your answer correct if it shows all the features that appear in the window given in the answers. (See Example 3.)*

**33.** $g(x) = x^3 - 3x^2 - 4x - 5$

**34.** $f(x) = x^4 - 10x^3 + 35x^2 - 50x + 24$

**35.** $f(x) = 2x^5 - 3.5x^4 - 10x^3 + 5x^2 + 12x + 6$

**36.** $g(x) = x^5 + 8x^4 + 20x^3 + 9x^2 - 27x - 7$

✓ **Checkpoint Answers**

**1.**

**2.**

**3.** Many correct answers, including $-15 \le x \le 15$ and $-2000 \le y \le 6000$.

**4.** $f(x) = 2x^3 + 5x^2 - x - 6$; degree 3.

**5.**

# 3.6 Rational Functions

A **rational function** is a function whose rule is the quotient of two polynomials, such as

$$f(x) = \frac{2}{1 + x}, \qquad g(x) = \frac{3x + 2}{2x + 4}, \qquad \text{and} \qquad h(x) = \frac{x^2 - 2x - 4}{x^3 - 2x^2 + x}.$$

Thus, a rational function is a function whose rule can be written in the form

$$f(x) = \frac{P(x)}{Q(x)},$$

where $P(x)$ and $Q(x)$ are polynomials, with $Q(x) \neq 0$. The function is undefined for any values of $x$ that make $Q(x) = 0$, so there are breaks in the graph at these numbers.

## Linear Rational Functions

We begin with rational functions in which both numerator and denominator are first-degree or constant polynomials. Such functions are sometimes called **linear rational functions.**

---

**EXAMPLE 1**    Graph the rational function defined by

$$f(x) = \frac{2}{1 + x}.$$

**Solution** This function is undefined for $x = -1$, since $-1$ leads to a 0 denominator. For that reason, the graph of this function will not intersect the vertical line $x = -1$. Since $x$ can take on any value except $-1$, the values of $x$ can approach $-1$ as closely as desired from either side of $-1$, as shown in the following table of values.

x approaches −1

| $x$ | −1.5 | −1.2 | −1.1 | −1.01 | −.99 | −.9 | −.8 | −.5 |
|---|---|---|---|---|---|---|---|---|
| $1 + x$ | −.5 | −.2 | −.1 | −.01 | .01 | .1 | .2 | .5 |
| $f(x) = \dfrac{2}{1 + x}$ | −4 | −10 | −20 | −200 | 200 | 20 | 10 | 4 |

$|f(x)|$ gets larger and larger

The preceding table suggests that as $x$ gets closer and closer to $-1$ from either side, the demoninator, $1 + x$, gets closer and closer to 0, so the absolute value of $f(x)$ gets larger and larger. The part of the graph near $x = -1$ in Figure 3.47, on the following page, shows this behavior. The vertical line $x = -1$ that is approached by the curve is called a *vertical asymptote*. For convenience, the vertical asymptote is indicated by a dashed line in Figure 3.47, but this line is *not* part of the graph of the function. However, the graph tends toward this vertical line as $x$ approaches $-1$.

As $|x|$ gets larger and larger, so does the absolute value of the denominator $1 + x$. Hence, $f(x) = 2/(1 + x)$ gets closer and closer to 0, as shown in the following table.

$|x|$ gets larger and larger

| $x$ | −101 | −11 | −2 | 0 | 9 | 99 |
|---|---|---|---|---|---|---|
| $1 + x$ | −100 | −10 | −1 | 1 | 10 | 100 |
| $f(x) = \dfrac{2}{1 + x}$ | −.02 | −.2 | −2 | 2 | .2 | .02 |

$f(x)$ approaches 0

✓**Checkpoint 1**

Graph the following.

(a)  $f(x) = \dfrac{3}{5 - x}$

(b)  $f(x) = \dfrac{-4}{x + 4}$

The horizontal line $y = 0$ is called a *horizontal asymptote* for this graph. Using the asymptotes and plotting the intercept and other points gives the graph of Figure 3.47. ✓₁

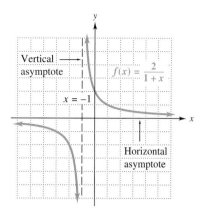

**Figure 3.47**

Example 1 suggests the following conclusion, which applies to all rational functions.

> If a number $c$ makes the denominator zero, while leaving the numerator nonzero, in the expression defining a rational function, then the line $x = c$ is a **vertical asymptote** for the graph of the function. The graph of the function approaches the vertical line as $x$ approaches $c$.

If the graph of a function approaches a horizontal line very closely when $x$ is very large or very small, we say that this line is a **horizontal asymptote** of the graph. In Example 1, the horizontal asymptote was the $x$-axis. This is not always the case, however, as the next example illustrates.

**EXAMPLE 2**    Graph

$$f(x) = \frac{3x + 2}{2x + 4}.$$

**Solution**  Find the vertical asymptote by setting the denominator equal to 0 and then solving for $x$:

$$2x + 4 = 0$$
$$x = -2.$$

In order to see what the graph looks like when $|x|$ is very large, we rewrite the rule of the function. When $x \neq 0$, dividing both numerator and denominator by $x$ does not change the value of the function:

$$f(x) = \frac{3x + 2}{2x + 4} = \frac{\dfrac{3x + 2}{x}}{\dfrac{2x + 4}{x}}$$

$$= \frac{\dfrac{3x}{x} + \dfrac{2}{x}}{\dfrac{2x}{x} + \dfrac{4}{x}} = \frac{3 + \dfrac{2}{x}}{2 + \dfrac{4}{x}}.$$

Now, when $|x|$ is very large, the fractions $2/x$ and $4/x$ are very close to 0. (For instance, when $x = 200, 4/x = 4/200 = .02$.) Therefore, the numerator of $f(x)$ is very close to $3 + 0 = 3$, and the denominator is very close to $2 + 0 = 2$. Hence, $f(x)$ is very close to $3/2$ when $|x|$ is large, so the line $y = 3/2$ is the horizontal asymptote of the graph, as shown in Figure 3.48, on the following page. ✔2

✓ **Checkpoint 2**

Graph the following.

(a)  $f(x) = \dfrac{2x - 5}{x - 2}$

(b)  $f(x) = \dfrac{3 - x}{x + 1}$

$$f(x) = \frac{3x+2}{2x+4}$$

Vertical asymptote

Horizontal asymptote

$$y = \frac{3}{2}$$

$$x = -2$$

**Figure 3.48**

**TECHNOLOGY TIP** Depending on the viewing window, a graphing calculator may not accurately represent the graph of a rational function. For example, the graph of $f(x) = \dfrac{3x+2}{2x+4}$ in Figure 3.49, which should look like Figure 3.48, has an erroneous vertical line at the place where the graph has a vertical asymptote. This problem can usually be avoided by using a window that has the vertical asymptote at the center of the $x$-axis, as in Figure 3.50.

**Figure 3.49**

**Figure 3.50**

The horizontal asymptotes of a linear rational function are closely related to the coefficients of the $x$-terms of the numerator and denominator, as illustrated in Examples 1 and 2:

| | **Function** | **Horizontal Asymptote** |
|---|---|---|
| **Example 1:** | $f(x) = \dfrac{2}{1+x} = \dfrac{0x+2}{1x+1}$ | $y = \dfrac{0}{1} = 0$   (the $x$-axis) |
| **Example 2:** | $f(x) = \dfrac{3x+2}{2x+4}$ | $y = \dfrac{3}{2}$ |

The same pattern holds in the general case.

The graph of $f(x) = \dfrac{ax+b}{cx+d}$ (where $c \neq 0$ and $ad \neq bc$) has a vertical asymptote at the root of the denominator (the value of $x$ that makes the denominator 0), and has horizontal asymptote $y = \dfrac{a}{c}$.

## Other Rational Functions

When the numerator or denominator of a rational function has degree greater than 1, the graph of the function can be more complicated than those in Examples 1 and 2. The graph might have no horizontal asymptote, it might have several vertical asymptotes or no vertical asymptote, and it might have peaks and valleys.

**EXAMPLE 3** Graph

$$f(x) = \frac{2x^2}{x^2 - 4}.$$

**Solution** Find the vertical asymptotes by setting the denominator equal to 0 and solving for $x$:

$$x^2 - 4 = 0 \qquad \text{Set denominator equal to 0.}$$
$$(x + 2)(x - 2) = 0 \qquad \text{Factor.}$$
$$x + 2 = 0 \quad \text{or} \quad x - 2 = 0 \qquad \text{Set each term equal to 0.}$$
$$x = -2 \quad \text{or} \quad x = 2. \qquad \text{Solve for } x.$$

Since neither of these numbers makes the numerator 0, the lines $x = -2$ and $x = 2$ are vertical asymptotes of the graph. The horizontal asymptote can be determined by dividing both the numerator and denominator of $f(x)$ by $x^2$ (the highest power of $x$ that appears in either one):

$$f(x) = \frac{2x^2}{x^2 - 4} = \frac{\dfrac{2x^2}{x^2}}{\dfrac{x^2 - 4}{x^2}} = \frac{\dfrac{2x^2}{x^2}}{\dfrac{x^2}{x^2} - \dfrac{4}{x^2}} = \frac{2}{1 - \dfrac{4}{x^2}}.$$

When $|x|$ is very large, the fraction $4/x^2$ is very close to 0, so the denominator is very close to $1 - 0 = 1$ and $f(x)$ is very close to 2. Hence, the line $y = 2$ is the horizontal asymptote of the graph. Using this information and plotting several points in each of the three regions determined by the vertical asymptotes, we obtain Figure 3.51. ✓3

✓ **Checkpoint 3**

List the vertical and horizontal asymptotes of the given function.

(a) $f(x) = \dfrac{3x + 5}{x + 5}$

(b) $g(x) = \dfrac{2 - x^2}{x^2 - 4}$

**Figure 3.51**

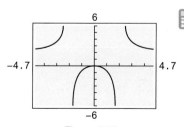

**Figure 3.52**

**TECHNOLOGY TIP** When a function whose graph has more than one vertical asymptote (as in Example 3) is graphed on a graphing calculator, erroneous vertical lines can sometimes be avoided by using a *decimal window* (with the $y$-range adjusted to show the graph). On TI, use (Z)DECIMAL in the ZOOM or VIEWS menu. Figure 3.52 shows the function of Example 3 graphed in a decimal window on a TI-84+. (The $x$-range may be different on other calculators.)

The arguments used to find the horizontal asymptotes in Examples 1–3 above work in the general case. Example 6 at the end of this section shows a graph with no horizontal asymptote. This leads to the following conclusion.

> If the numerator of a rational function $f(x)$ is of *smaller* degree than the denominator, then the $x$-axis (the line $y = 0$) is the horizontal asymptote of the graph. If the numerator and denominator are of the *same* degree, say, $f(x) = \dfrac{ax^n + \cdots}{cx^n + \cdots}$, then the line $y = \dfrac{a}{c}$ is the horizontal asymptote. If the numerator is of a *larger* degree than the denominator, the graph has no horizontal asymptote.[*]

## Applications

Rational functions have a variety of applications, some of which are explored next.

---

**EXAMPLE 4** **Pollution Control** In many situations involving environmental pollution, much of the pollutant can be removed from the air or water at a fairly reasonable cost, but the small amounts of the pollutant that remain can be very expensive to remove.

Cost as a function of the percentage of pollutant removed from the environment can be calculated for various percentages of removal, with a curve fitted through the resulting data points. This curve then leads to a function that approximates the situation. Rational functions often are a good choice for these **cost–benefit functions.**

For example, suppose a cost–benefit function is given by

$$f(x) = \frac{18x}{106 - x},$$

where $f(x)$ is the cost (in thousands of dollars) of removing $x$ percent of a certain pollutant. The domain of $x$ is the set of all numbers from 0 to 100, inclusive; any amount of pollutant from 0% to 100% can be removed. To remove 100% of the pollutant here would cost

$$f(100) = \frac{18(100)}{106 - 100} = 300,$$

or $300,000. Check that 95% of the pollutant can be removed for about $155,000, 90% for about $101,000, and 80% for about $55,000, as shown in Figure 3.53 (in which the displayed $y$-coordinates are rounded to the nearest integer). ✓4

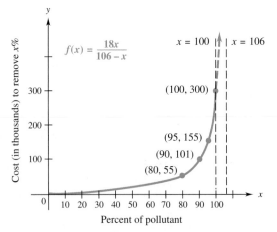

**Figure 3.53**

### Checkpoint 4

Using the function in Example 4, find the cost to remove the following percentages of pollutants.

(a) 70%

(b) 85%

(c) 98%

---

[*]In this case, the graph may have nonhorizontal lines or other curves as asymptotes; see Exercises 31 and 32 at the end of this section for examples.

In management, **product-exchange functions** give the relationship between quantities of two items that can be produced by the same machine or factory. For example, an oil refinery can produce gasoline, heating oil, or a combination of the two; a winery can produce red wine, white wine, or a combination of the two. The next example discusses a product-exchange function.

**EXAMPLE 5** **Agricultural Production** The product-exchange function for the Fruits of the Earth Winery for red wine $x$ and white wine $y$, in number of cases, is

$$y = \frac{150,000 - 75x}{1200 + x}.$$

Graph the function and find the maximum quantity of each kind of wine that can be produced.

**Solution** Only nonnegative values of $x$ and $y$ make sense in this situation, so we graph the function in the first quadrant (Figure 3.54). Note that the $y$-intercept of the graph (found by setting $x = 0$) is 125 and the $x$-intercept (found by setting $y = 0$ and solving for $x$) is 2000. Since we are interested only in the portion of the graph in Quadrant I, we can find a few more points in that quadrant and complete the graph, as shown in Figure 3.54.

**Figure 3.54**

The maximum value of $y$ occurs when $x = 0$, so the maximum amount of white wine that can be produced is 125 cases, as given by the $y$-intercept. The $x$-intercept gives the maximum amount of red wine that can be produced: 2000 cases. ✓5

✓**Checkpoint 5**

Rework Example 5 with the product-exchange function

$$y = \frac{70,000 - 10x}{70 + x}$$

to find the maximum amount of each wine that can be produced.

**EXAMPLE 6** **Business** A retailer buys 2500 specialty lightbulbs from a distributor each year. In addition to the cost of each bulb, there is a fee for each order, so she wants to order as few times as possible. However, storage costs are higher when there are fewer orders (and hence more bulbs per order to store). Past experience shows that the total annual cost (for the bulbs, ordering fees, and storage costs) is given by the rational function.

$$C(x) = \frac{.98x^2 + 1200x + 22,000}{x},$$

where $x$ is the number of bulbs ordered each time. How many bulbs should be ordered each time in order to have the smallest possible cost?

**Solution** Graph the cost function $C(x)$ in a window with $0 \le x \le 2500$ (because the retailer cannot order a negative number of bulbs and needs only 2500 for the year).

For each point on the graph in Figure 3.55,

the $x$-coordinate is the number of bulbs ordered each time;
the $y$-coordinate is the annual cost when $x$ bulbs are ordered each time.

**Figure 3.55**          **Figure 3.56**

Use the minimum finder on a graphing calculator (look in the CALC menu) to find the point with the smallest $y$-coordinate, which is approximately $(149.83, 1493.67)$, as shown in Figure 3.56. Since the retailer cannot order part of a lightbulb, she should order 150 bulbs each time, for an approximate annual cost of $1494.

# 3.6 Exercises

*Graph each function. Give the equations of the vertical and horizontal asymptotes. (See Examples 1–3.)*

**1.** $f(x) = \dfrac{1}{x + 5}$

**2.** $g(x) = \dfrac{-7}{x - 6}$

**3.** $f(x) = \dfrac{-3}{2x + 5}$

**4.** $h(x) = \dfrac{-4}{2 - x}$

**5.** $f(x) = \dfrac{3x}{x - 1}$

**6.** $g(x) = \dfrac{x - 2}{x}$

**7.** $f(x) = \dfrac{x - 3}{x + 5}$

**8.** $f(x) = \dfrac{x + 1}{x - 4}$

**9.** $f(x) = \dfrac{2 - x}{x - 3}$

**10.** $g(x) = \dfrac{3x - 2}{x + 3}$

**11.** $f(x) = \dfrac{4x - 8}{8x + 1}$

**12.** $f(x) = \dfrac{3x + 2}{2x + 4}$

**13.** $h(x) = \dfrac{x + 1}{x^2 + 2x - 8}$

**14.** $g(x) = \dfrac{1}{x(x + 2)^2}$

**15.** $f(x) = \dfrac{x^2 + 4}{x^2 - 4}$

**16.** $f(x) = \dfrac{x - 1}{x^2 - 2x - 3}$

*Find the equations of the vertical asymptotes of each of the given rational functions. (See Example 3.)*

**17.** $g(x) = \dfrac{x + 2}{x^2 - 1}$

**18.** $f(x) = \dfrac{x - 3}{x^2 + x - 2}$

**19.** $g(x) = \dfrac{x^2 + 2x}{x^2 - 4x - 5}$

**20.** $f(x) = \dfrac{x^2 - 2x - 4}{x^3 - 2x^2 + x}$

©**Average Cost**  *For Exercises 21 and 22, recall that if the cost of producing x units is C(x) then the average cost function (defined Section 3.3) is* $\overline{C}(x) = \dfrac{C(x)}{x}.$

**21.** An artist who makes handmade earrings has fixed costs of $100 for a table at the Portland Art Walk. The marginal cost (the cost of producing one additional pair of earrings) is $5 per pair.

  **(a)** Find the linear cost function $C(x)$.

  **(b)** Find the average cost function $\overline{C}(x)$.

  **(c)** Find $\overline{C}(5)$, and interpret your answer.

  **(d)** Find $\overline{C}(50)$, and interpret your answer.

  **(e)** Find the horizontal asymptote of $\overline{C}(x)$, and explain what it means in practical terms.

**22.** A taxi driver must pay a daily fee of $75 to the cab company and has marginal costs of $.50 per mile.

  **(a)** Find the daily cost function $C(x)$.

  **(d)** Find the average cost function $\overline{C}(x)$.

  **(c)** Find $\overline{C}(50)$, and interpret your answer.

  **(d)** Find $\overline{C}(250)$, and interpret your answer.

  **(e)** Find the horizontal asymptote of $\overline{C}(x)$, and explain what it means in practical terms.

*Work these problems. (See Example 2.)*

**23. Biological Modeling**  The function

$$f(x) = \frac{5x}{2 + x}$$

is used in biology to give the growth rate (as a percentage) of a certain population in the presence of a quantity of $x$ units of food (where $x \geq 0$). This is an example of the concept called Michaelis–Menten kinetics.[*]

  **(a)** Find $f(0), f(2), f(8),$ and $f(78)$, and explain what your answers mean.

---

[*]W. W. Chen, M. Neipel, and P. K. Sorger, "Classic and Contemporary Approaches to Modeling Biochemical Reactions," *Genes Dev* 24, no. 17 (2010): 1861–1875.

**(b)** Find the horizontal asymptote.

**(c)** Does the graph have any vertical asymptotes on the domain $x \geq 0$?

**(d)** Graph the function for $x \geq 0$.

**(e)** What do you think the horizontal asymptote represents?

**24. NASA** The failure of several O-rings in field joints was the cause of the fatal crash of the *Challenger* space shuttle. NASA data from 24 successful launches prior to *Challenger* suggested that O-ring failure was related to launch temperature by a function similar to

$$N(t) = \frac{600 - 7t}{4t - 100} \quad (50 \leq t \leq 85),$$

where $t$ is the temperature (in °F) at launch and $N$ is the approximate number of O-rings that fail. Assume that this function accurately models the number of O-ring failures that would occur at lower launch temperatures (an assumption NASA did not make).

**(a)** Does $N(t)$ have a vertical asymptote? At what value of $t$ does it occur?

**(b)** Without actually graphing the function, what would you conjecture that the graph would look like just to the right of the vertical asymptote? What does this suggest about the number of O-ring failures that might be expected near that temperature? (The temperature at the *Challenger* launching was 31°.)

**(c)** Confirm your conjecture by graphing $N(t)$ between the vertical asymptote and $t = 85$.

**25. Pollution Control** Suppose a cost–benefit model is given by

$$f(x) = \frac{4.3x}{100 - x},$$

where $f(x)$ is the cost, in thousands of dollars, of removing $x$ percent of a given pollutant. Find the cost of removing each of the given percentages of pollutants.

**(a)** 50%    **(b)** 70%    **(c)** 80%

**(d)** 90%    **(e)** 95%    **(f)** 98%

**(g)** 99%

**(h)** Is it possible, according to this model, to remove *all* the pollutant?

**(i)** Graph the function.

**26. Pollution Control** Suppose a cost–benefit model is given by

$$f(x) = \frac{6.2x}{112 - x},$$

where $f(x)$ is the cost, in thousands of dollars, of removing $x$ percent of a certain pollutant. Find the cost of removing the given percentages of pollutants.

**(a)** 0%    **(b)** 50%    **(c)** 80%

**(d)** 90%    **(e)** 95%    **(f)** 99%

**(g)** 100%    **(h)** Graph the function.

**Business** *Sketch the portion of the graph in Quadrant I of each of the functions defined in Exercises 27 and 28, and then find the x-intercept and y-intercept in order to estimate the maximum quantities of each product that can be produced. (See Example 5.)*

**27.** The product-exchange function for gasoline $x$ and heating oil $y$, in hundreds of gallons per day, is

$$y = \frac{125,000 - 25x}{125 + 2x}.$$

**28.** A drug factory found that the product-exchange function for a red tranquilizer $x$ and a blue tranquilizer $y$ is

$$y = \frac{900,000,000 - 30,000x}{x + 90,000}.$$

🖩 *Use a graphing calculator to do Exercises 29–32. (See Example 6.)*

**29. Finance** Another model of a Laffer curve (see Exercise 27 of Section 3.5) is given by

$$f(x) = \frac{300x - 3x^2}{10x + 200},$$

where $f(x)$ is government revenue (in billions of dollars) from a tax rate of $x$ percent. Find the revenue from the given tax rates.

**(a)** 16%    **(b)** 25%    **(c)** 40%

**(d)** 55%    **(e)** Graph $f(x)$.

**(f)** Use the maximum finder to find the tax rate that produces maximum revenue. What is the maximum revenue?

**30. Business** When no more than 110 units are produced, the cost of producing $x$ units is given by

$$C(x) = .2x^3 - 25x^2 + 1531x + 25,000.$$

Use the minimum finder to determine the number of units that should be produced in order to have the lowest possible average cost.

**31. (a)** Graph $f(x) = \dfrac{x^3 + 3x^2 + x + 1}{x^2 + 2x + 1}$.

**(b)** Does the graph appear to have a horizontal asymptote? Does the graph appear to have some nonhorizontal straight line as an asymptote?

**(c)** Graph $f(x)$ and the line $y = x + 1$ on the same screen. Does this line appear to be an asymptote of the graph of $f(x)$?

**32. (a)** Graph $g(x) = \dfrac{x^3 - 2}{x - 1}$ in the window with $-5 \leq x \leq 5$ and $-6 \leq y \leq 12$.

**(b)** Graph $g(x)$ and the parabola $y = x^2 + x + 1$ on the same screen. How do the two graphs compare when $|x| \geq 2$?

✓ **Checkpoint Answers**

**1. (a)**

**(b)**

$$f(x) = \frac{3}{5 - x} \qquad f(x) = \frac{-4}{x + 4}$$

**2.  (a)**

**(b)**

**3.  (a)**  Vertical, $x = -5$; horizontal, $y = 3$

  **(b)**  Vertical, $x = -2$ and $x = 2$; horizontal, $y = -1$

**4.  (a)**  \$35,000

  **(b)**  About \$73,000

  **(c)**  About \$221,000

**5.**  7000 cases of red, 1000 cases of white

# CHAPTER 3 Summary and Review

## Key Terms and Symbols

**3.1** function
    domain
    range
    functional notation
    piecewise-defined function

**3.2** graph
    linear function
    piecewise linear function
    absolute-value function
    greatest-integer function
    step function

graph reading
vertical-line test

**3.3** fixed costs
    variable cost
    average cost
    linear depreciation
    rate of change
    marginal cost
    marginal revenue
    linear cost function
    linear revenue function

break-even point
supply and demand curves
shortage of supply
surplus of supply
equilibrium point
equilibrium price
equilibrium quantity

**3.4** quadratic function
    parabola
    vertex
    axis of the parabola

quadratic model

**3.5** polynomial function
    graph of $f(x) = ax^n$
    properties of polynomial
      graphs
    polynomial models

**3.6** rational function
    linear rational function
    vertical asymptote
    horizontal asymptote

## Chapter 3 Key Concepts

**Functions**

A **function** consists of a set of inputs called the **domain,** a set of outputs called the **range,** and a rule by which each input determines exactly one output.

If a vertical line intersects a graph in more than one point, the graph is not that of a function.

**Linear Functions**

A **linear cost function** has equation $C(x) = mx + b$, where $m$ is the **marginal cost** (the cost of producing one more item) and $b$ is the **fixed cost.**

If $p = f(q)$ gives the price per unit when $q$ units can be supplied and $p = g(q)$ gives the price per unit when $q$ units are demanded, then the **equilibrium quantity** is the $q$-value such that $f(q) = g(q)$, and the **equilibrium price** is the corresponding $p$-value (on either the supply curve or demand curve because they are equal at that point). The **equilibrium point** is the point whose first coordinate is the equilibrium quantity and whose second coordinate is the equilibrium price.

**Quadratic Functions**

The **quadratic function** defined by $f(x) = a(x - h)^2 + k$ (with $a \neq 0$) has a graph that is a **parabola** with vertex $(h, k)$ and axis $x = h$. The parabola opens upward if $a > 0$ and downward if $a < 0$.

If the equation is in the form $f(x) = ax^2 + bx + c$, the vertex is $\left(-\dfrac{b}{2a},\ f\left(-\dfrac{b}{2a}\right)\right)$.

**Polynomial Functions**

When $|x|$ is large, the graph of a **polynomial function** resembles the graph of its highest-degree term $ax^n$. The graph of $f(x) = ax^n$ is described on page 160.

On the graph of a polynomial function of degree $n$,

  the total number of peaks and valleys is at most $n - 1$;

  the total number of $x$-intercepts is at most $n$.

**Rational Functions**

If a number $c$ makes the denominator of a **rational function** 0, but the numerator nonzero, then the line $x = c$ is a **vertical asymptote** of the graph.

Whenever the values of $f(x)$ approach, but do not equal, some number $k$ as $|x|$ gets larger and larger, the line $y = k$ is a **horizontal asymptote** of the graph.

If the numerator of a rational function is of *smaller* degree than the denominator, then the *x*-axis ($y = 0$) is the horizontal asymptote of the graph.

If the numerator and denominator of a rational function are of the *same* degree, say, $f(x) = \dfrac{ax^n + \cdots}{cx^n + \cdots}$, then the line $y = \dfrac{a}{c}$ is the horizontal asymptote of the graph.

If the numerator of a rational function is of a *larger* degree than the denominator, then the graph has no horizontal asymptote.

# Chapter 3 Review Exercises

*In Exercises 1–6, state whether or not the given rule defines y as a function x.*

**1.**

| x | 3 | 2 | 1 | 0 | 1 | 2 |
|---|---|---|---|---|---|---|
| y | 8 | 5 | 2 | 0 | −2 | −5 |

**2.**

| x | 2 | 1 | 0 | −1 | −2 |
|---|---|---|---|---|---|
| y | 5 | 3 | 1 | −1 | −3 |

**3.** $y = \sqrt{x}$  **4.** $x = |y|$

**5.** $x = y^2 + 1$  **6.** $y = 5x - 2$

*For the functions in Exercises 7–10, find*

(a) $f(6)$;  (b) $f(-2)$;  (c) $f(p)$;  (d) $f(r + 1)$.

**7.** $f(x) = 4x - 1$  **8.** $f(x) = 3 - 4x$

**9.** $f(x) = -x^2 + 2x - 4$  **10.** $f(x) = 8 - x - x^2$

**11.** Let $f(x) = 5x - 3$ and $g(x) = -x^2 + 4x$. Find each of the following:

(a) $f(-2)$  (b) $g(3)$  (c) $g(-k)$

(d) $g(3m)$  (e) $g(k - 5)$  (f) $f(3 - p)$

**12.** Let $f(x) = x^2 + x + 1$. Find each of the following:

(a) $f(3)$  (b) $f(1)$  (c) $f(4)$

(d) Based on your answers in parts (a)–(c), is it true that $f(a + b) = f(a) + f(b)$ for all real numbers $a$ and $b$?

*Graph the functions in Exercises 13–24.*

**13.** $f(x) = |x| - 3$  **14.** $f(x) = -|x| - 2$

**15.** $f(x) = -|x + 1| + 3$  **16.** $f(x) = 2|x - 3| - 4$

**17.** $f(x) = [x - 3]$  **18.** $f(x) = \left[\dfrac{1}{2}x - 2\right]$

**19.** $f(x) = \begin{cases} -4x + 2 & \text{if } x \le 1 \\ 3x - 5 & \text{If } x > 1 \end{cases}$

**20.** $f(x) = \begin{cases} 3x + 1 & \text{if } x < 2 \\ -x + 4 & \text{if } x \ge 2 \end{cases}$

**21.** $f(x) = \begin{cases} x^2 & \text{if } x \le 0 \\ x & \text{if } x > 0 \end{cases}$

**22.** $f(x) = \begin{cases} |x| & \text{if } x < 3 \\ 6 - x & \text{if } x \ge 3 \end{cases}$

**23.** $h(x) = \sqrt{x} + 2$  **24.** $f(x) = \sqrt{x^2}$

**25. Business** Let $f(x)$ be a function that gives the cost to rent a power washer for $x$ hours. The cost is a flat $45 for renting the washer, plus $20 per day or fraction of a day for using the washer.

(a) Graph $f(x)$.  (b) Give the domain and range.

(c) You want to rent the washer, but can spend no more than $90. What is the maximum number of days you can use it?

**26. Business** A tree removal service assesses a $400 fee and then charges $80 per hour or fraction of an hour for the time on an owner's property. Let $f(x)$ be the total cost for $x$ hours.

(a) Is $750 enough for 5 hours work?

(b) Graph $f(x)$.  (c) Give the domain and range.

**27. Alcohol Consumption** The following graph shows the percentage of twelfth graders in the United States by gender who had a drink of alcohol in the past 30 days. (Data from: Centers for Disease Control and Prevention.)

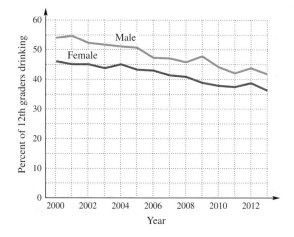

(a) What is the general trend for alcohol use by female twelfth graders during this time period?

(b) According to the graph, approximately what percentage of female twelfth graders drank alcohol in 2004?

(c) Let $x = 0$ correspond to the year 2000. Approximate a linear function for the percentage of female twelfth graders who drank alcohol using the points (4, 45) and (8, 41).

(d) What does this model estimate for this percentage in 2017?

(e) If the trend continues, when will this percentage drop to 33%?

**28. Alcohol Consumption** Answer the following questions by referring to the graph in Exercise 27, which shows the percentage of twelfth graders in the United States by gender who had a drink in the past 30 days.

**(a)** What is the general trend for alcohol use by male twelfth graders during this time period?

**(b)** According to the graph, approximately what percentage of male twelfth graders drank alcohol in 2001?

**(c)** Let $x = 0$ correspond to the year 2000. Approximate a linear function for the percentage of male twelfth graders who drank alcohol using the points $(1, 55)$ and $(12, 44)$.

**(d)** What does this model estimate for this percentage in 2017?

**(e)** If the trend continues, when will this percentage drop to 40%?

**Business** *In Exercises 29–32, find the following:*

**(a)** *the linear cost function;*

**(b)** *the marginal cost;*

**(c)** *the average cost per unit to produce 100 units.*

29. Eight units cost $300; fixed cost is $60.

30. Fixed cost is $2000; 36 units cost $8480.

31. Twelve units cost $445; 50 units cost $1585.

32. Thirty units cost $1500; 120 units cost $5640.

33. **Business** The cost of producing $x$ ink cartridges for a printer is given by $C(x) = 24x + 18{,}000$. Each cartridge can be sold for $28.

 **(a)** What are the fixed costs?

 **(b)** Find the revenue function.

 **(c)** Find the break-even point.

 **(d)** If the company sells exactly the number of cartridges needed to break even, what is its revenue?

34. **Business** The cost of producing $x$ laser printers is given by $C(x) = 325x + 2{,}300{,}000$. Each printer can be sold for $450.

 **(a)** What are the fixed costs?

 **(b)** Find the revenue function.

 **(c)** Find the break-even point.

 **(d)** If the company sells exactly the number of printers needed to break even, what is its revenue?

35. **Business** Suppose the demand and price for the HBO cable channel are related by $p = -.5q + 30.95$, where $p$ is the monthly price in dollars and $q$ is measured in millions of subscribers. If the price and supply are related by $p = .3q + 2.15$, what are the equilibrium quantity and price?

36. **Business** Suppose the supply and price for prescription-strength Tylenol are related by $p = .0015q + 1$, where $p$ is the price (in dollars) of a 30-day prescription. If the demand is related to price by $p = -.0025q + 64.36$, what are the equilibrium quantity and price?

*Without graphing, determine whether each of the following parabolas opens upward or downward, and find its vertex.*

37. $f(x) = 3(x - 2)^2 + 6$     38. $f(x) = 2(x + 3)^2 - 5$

39. $g(x) = -4(x + 1)^2 + 8$     40. $g(x) = -5(x - 4)^2 - 6$

*Graph each of the following quadratic functions, and label its vertex.*

41. $f(x) = x^2 - 9$     42. $f(x) = 5 - 2x^2$

43. $f(x) = x^2 + 2x - 6$     44. $f(x) = -x^2 + 8x - 1$

45. $f(x) = -x^2 - 6x + 5$     46. $f(x) = 5x^2 + 20x - 2$

47. $f(x) = 2x^2 - 12x + 10$     48. $f(x) = -3x^2 - 12x - 2$

*Determine whether the functions in Exercises 49–52 have a minimum or a maximum value, and find that value.*

49. $f(x) = x^2 + 6x - 2$     50. $f(x) = x^2 + 4x + 5$

51. $g(x) = -4x^2 + 8x + 3$     52. $g(x) = -3x^2 - 6x + 3$

*Solve each problem.*

53. **Youth Population** At the start of this millennium, the youth population in the United States was rising, but later it began to fall. The function

$$f(x) = -29x^2 + 522x + 71741$$

 models the number of people under 18 years old (in thousands) in year $x$, where $x = 0$ corresponds to the year 2000. (Data from: U.S. Census Bureau.)

 **(a)** According to this model, what was the youth population in 2000?

 **(b)** Find the vertex. Does a maximum or minimum occur at the vertex?

 **(c)** According to this model, what year did the youth population peak?

 **(d)** What was the maximum youth population with this model?

54. **Population of China** The declining birth rate in China has demographers believing that the population of China will soon peak. A model for the population of China (in millions) is

$$f(x) = -.2x^2 + 10.4x + 1263,$$

 where $x = 0$ corresponds to the year 2000. (Data from: U.S. Census Bureau.)

 **(a)** According to this model, what was the population of China in 2000?

 **(b)** Find the vertex. Does a maximum or minimum occur at the vertex?

 **(c)** According to this model, in what year does the population of China peak?

 **(d)** What is maximum projected population?

55. **Student Loans** Interest rates for subsidized Stafford loans can be approximated for the years 2007 to 2014 by the function

$$f(x) = .15x^2 - 3.6x + 25,$$

 where $x = 7$ corresponds to the year 2007. According to this model, during what year were interest rates lowest, and what was the lowest interest rate? (Data from: www.studentaid.ed.gov.)

56. **Natural Gas Pricing** The price of European natural gas (in dollars per million British thermal units [BTUs]) can be approximated for the years 2010 to 2015 by the function

$$f(x) = -.65x^2 + 16.9x - 98,$$

 where $x = 10$ corresponds to the year 2010. According to this model, during what year were European natural gas prices highest, and what was the highest price? (Data from: www.worldbank.org.)

57. **Netflix Revenue** Netflix Inc. reported revenue (in millions of dollars) as shown in the following table. (Data from: www.morningstar.com.)

| Year | Revenue |
|------|---------|
| 2006 | 997 |
| 2007 | 1205 |
| 2008 | 1365 |
| 2009 | 1670 |
| 2010 | 2163 |
| 2011 | 3205 |
| 2012 | 3609 |
| 2013 | 4375 |
| 2014 | 5505 |
| 2015 | 6780 |

(a) Let $x = 6$ correspond to the year 2006. Find a quadratic function in the form $f(x) = a(x - h)^2 + k$ that models these data using $(6, 997)$ as the vertex and the data for 2010.

(b) Estimate the revenue in 2016.

**58. Netflix Costs** Netflix Inc. reported costs (in millions of dollars) as shown in the following table. (Data from: www.morningstar.com.)

| Year | Costs |
|------|-------|
| 2006 | 627 |
| 2007 | 786 |
| 2008 | 910 |
| 2009 | 1079 |
| 2010 | 1357 |
| 2011 | 2040 |
| 2012 | 2626 |
| 2013 | 3083 |
| 2014 | 3753 |
| 2015 | 4591 |

(a) Let $x = 6$ correspond to the year 2006. Find a quadratic function in the form $f(x) = a(x - h)^2 + k$ that models these data using $(6, 627)$ as the vertex and the data for 2010.

(b) Estimate the costs in 2016.

**59.** Use quadratic regression and the data from Exercise 57 to find a function $R(x)$ that models these data.

(a) Give the function (round coefficients to the nearest integer).

(b) Graph the function with the data points. Does the model fit the data well?

(c) Find the estimated revenue with this model for the year 2016.

**60.** Use quadratic regression and the data from Exercise 58 to find a function $C(x)$ that models these data.

(a) Give the function (round coefficients to the nearest integer).

(b) Graph the function with the data points. Does the model fit the data well?

(c) Find the estimated costs with this model for the year 2016.

*Graph each of the following polynomial functions.*

**61.** $g(x) = x^3 - 4x$

**62.** $f(x) = x^4 - 5$

**63.** $f(x) = x(x - 4)(x + 1)$

**64.** $f(x) = (x - 1)(x + 2)(x - 3)$

**65.** $f(x) = x^4 - 5x^2 - 6$

**66.** $f(x) = x^4 - 7x^2 - 8$

*Use a graphing calculator to do Exercises 67–70.*

**67. Forward Industries Revenue** The following table shows the revenue (in millions of dollars) for Forward Industries Inc., a company that designs protective solutions for handheld electronic devices. (Data from: www.morningstar.com.)

| Year | Revenue |
|------|---------|
| 2006 | 31 |
| 2007 | 22 |
| 2008 | 20 |
| 2009 | 17 |
| 2010 | 19 |
| 2011 | 23 |
| 2012 | 29 |
| 2013 | 31 |
| 2014 | 33 |
| 2015 | 30 |

(a) Let $x = 6$ correspond to the year 2006. Use cubic regression to find a third-order polynomial function $R(x)$ that models these data (round coefficients to three decimal places).

(b) Graph $R(x)$ on the same screen as the data points. Does the graph appear to fit the data well?

(c) What does the model estimate for the revenue in 2012? Is it close to the actual revenue shown in the table?

**68. Forward Industries Costs** The following table shows the costs (in millions of dollars) for Forward Industries Inc. (Data from: www.morningstar.com.)

| Year | Costs |
|------|-------|
| 2006 | 23 |
| 2007 | 17 |
| 2008 | 16 |
| 2009 | 14 |
| 2010 | 15 |
| 2011 | 18 |
| 2012 | 25 |
| 2013 | 25 |
| 2014 | 27 |
| 2015 | 24 |

(a) Use cubic regression to find a third-order polynomial function $C(x)$ that models these data (round coefficients to three decimal places).

(b) Graph $C(x)$ on the same screen as the data points. Does the graph appear to fit the data well?

(c) What does the model estimate for the costs in 2014? Is it close to the actual costs shown in the table?

**69. Forward Industries Profit** Use the revenue function $R(x)$ found in Exercise 67 and the cost function $C(x)$ found in Exercise 68 to find the profit function $P(x)$ for Forward Industries, and estimate the profit in the year 2016.

**70. Netflix Profit** Use the revenue function $R(x)$ found in Exercise 57 and the cost function $C(x)$ found in Exercise 58 to find the profit function $P(x)$ for Netflix, and estimate the profit in the year 2016.

*List the vertical and horizontal asymptotes of each function, and sketch its graph.*

**71.** $f(x) = \dfrac{1}{x - 3}$

**72.** $f(x) = \dfrac{-2}{x + 4}$

**73.** $f(x) = \dfrac{-3}{2x - 4}$

**74.** $f(x) = \dfrac{5}{3x + 7}$

**75.** $g(x) = \dfrac{5x - 2}{4x^2 - 4x - 3}$

**76.** $g(x) = \dfrac{x^2}{x^2 - 1}$

**77. Business** The average cost per carton of producing $x$ cartons of cocoa is given by

$$A(x) = \frac{650}{2x + 40}.$$

Find the average cost per carton to make the given number of cartons.

(a) 10 cartons    (b) 50 cartons    (c) 70 cartons

(d) 100 cartons    (e) Graph $A(x)$.

**78. Business** The cost and revenue functions (in dollars) for a frozen-yogurt shop are given by

$$C(x) = \frac{400x + 400}{x + 4} \quad \text{and} \quad R(x) = 100x,$$

where $x$ is measured in hundreds of units.

(a) Graph $C(x)$ and $R(x)$ on the same set of axes.

(b) What is the break-even point for this shop?

(c) If the profit function is given by $P(x)$, does $P(1)$ represent a profit or a loss?

(d) Does $P(4)$ represent a profit or a loss?

**79. Business** The supply and demand functions for the yogurt shop in Exercise 78 are

$$\text{supply: } p = \frac{q^2}{4} + 25$$

and

$$\text{demand: } p = \frac{500}{q},$$

where $p$ is the price in dollars for $q$ hundred units of yogurt. Graph both functions on the same axes, and from the graph, estimate the equilibrium point.

**80. Business** A cost–benefit curve for pollution control is given by

$$y = \frac{9.2x}{106 - x},$$

where $y$ is the cost, in thousands of dollars, of removing $x$ percent of a specific industrial pollutant. Find $y$ for each of the given values of $x$.

(a) $x = 50$      (b) $x = 98$

(c) What percent of the pollutant can be removed for $22,000?

# Case Study 3 Maximizing Profit

For many people, visits to a coin-operated laundry are a regular part of life. In 2017, this industry generated nearly $5 billion in gross revenue from about 29,500 locations nationwide, the majority of which are individually owned and operated. (Data from: Coin Laundry Association.) Suppose that one business owner raised the price of a washing machine load from $2.50 to $2.75 and found that daily demand dropped from 700 loads to 650 loads. The business owner noted that revenue actually increased but wondered about bottom-line profits. The business owner is interested in learning about both the *maximum revenue* and the *maximum profit*. We will investigate both in this case study.

As usual, we express the price, $p$, as a function of the quantity demanded, $x$. We summarize the given data in the following table.

**Washing Machine Operations**

| Daily Demand, $x$ | Price per Load, $p$ |
| --- | --- |
| 650 | $2.75 |
| 700 | $2.50 |

Assuming linear demand, we can use these data to find the slope of the demand function:

$$m = \frac{\Delta p}{\Delta x} = \frac{2.50 - 2.75}{700 - 650} = \frac{-.25}{50} = -.005.$$

Using point–slope form, we find the equation of the linear demand function:

$$\begin{aligned}
y - y_1 &= m(x - x_1) && \text{Point–slope form using } x \text{ and } y \\
p - p_1 &= m(x - x_1) && \text{Point–slope form using } x \text{ and } p \\
p - 2.75 &= -.005(x - 650) && \text{Substitute 2.75 for } p_1, -.005 \\
& && \text{for } m, \text{ and 650 for } x_1. \\
p - 2.75 &= -.005x + 3.25 && \text{Distribute the } -.005. \\
p &= -.005x + 6 && \text{Add 2.75 to both sides.}
\end{aligned}$$

We find the revenue function by multiplying the quantity demanded by the price per item:

$$\begin{aligned}
R(x) &= xp \\
&= x(-.005x + 6) \\
&= -.005x^2 + 6x.
\end{aligned}$$

The graph of the revenue function is a downward-opening parabola, so the maximum occurs at the vertex $(h, k)$. The $x$-value of the vertex is

$$h = \frac{-b}{2a} = \frac{-6}{2(-.005)} = \frac{-6}{-.01} = 600.$$

The corresponding $y$-value of the vertex is

$$k = R(h) = -.005(600)^2 + 6(600) = 1800.$$

The vertex of the revenue function tells us that the maximum daily revenue is $1800, and it occurs when the demand for washing machines is 600 loads. The price per load corresponding to a demand of $x = 600$ is $p = -.005(600) + 6 = \$3$ per load.

To investigate the maximum profit, we ask the business owner about operating costs. The owner reports fixed costs of $1040 per day and variable costs of $.50 per washing machine load. Let's assume that there are about an equal number of washer and dryer loads, so we will attribute half of the fixed costs to washing machine operations. Therefore, the cost function for washing machines is

$$C(x) = 520 + .50x.$$

This leads to the profit function

$$
\begin{aligned}
P(x) &= R(x) - C(x) \\
&= (-.005x^2 + 6x) - (520 + .50x) \\
&= -.005x^2 + 5.50x - 520.
\end{aligned}
$$

## Exercises

1. Find the maximum profit and the number of washing machine loads that leads to the maximum profit.

2. Is the quantity of washing machine loads the same for the maximum revenue and the maximum profit? If not, which quantity should the owner prefer?

3. Based on this information, what price should the owner charge?

# Extended Project

Suppose the owner of the laundry has hired your consulting firm to determine an optimal price to charge per dryer load given the data below.

### Dryer Operations

| Daily Demand, $x$ | Price per Load, $p$ |
|---|---|
| 600 | $1.25 |
| 700 | $1.00 |

Assume half of the fixed costs are associated with dryer operations, and that variable costs are $.25 per dryer load. Write a report explaining your calculations and findings. Be sure to include the following:

(a) the number of loads corresponding to the maximum profit from dryers;

(b) the corresponding maximum profit from dryers; and

(c) the price per load to charge that corresponds to the maximum dryer profit.

# Exponential and Logarithmic Functions

## 4

### CHAPTER

Population growth (of humans, fish, bacteria, etc.), compound interest, radioactive decay, and a host of other phenomena can be described by exponential functions. The growth in the number of subscribers to a new tech service such as Netflix often follows an exponential pattern. Archeologists sometimes use carbon-14 dating to determine the approximate age of an artifact such as a mummy. This procedure involves using logarithms to solve an exponential equation. The Richter scale for measuring the magnitude of an earthquake is also a logarithmic function.

Exponential and logarithmic functions play a key role in management, economics, the social and physical sciences, and engineering. We begin with exponential growth and exponential decay functions.

## 4.1 Exponential Functions

In polynomial functions such as $f(x) = x^2 + 5x - 4$, the variable is raised to various constant exponents. In **exponential functions,** such as

$$f(x) = 10^x, \quad g(x) = 750(1.05^x), \quad h(x) = 3^{.6x}, \quad \text{and} \quad k(x) = 2^{-x^2},$$

the variable is in the exponent and the **base** is a positive constant. We begin with the simplest type of exponential function, whose rule is of the form $f(x) = a^x$, with $a > 0$.

**EXAMPLE 1** Graph $f(x) = 2^x$, and estimate the height of the graph when $x = 50$.

**Solution** Either use a graphing calculator or graph by hand: Make a table of values, plot the corresponding points, and join them by a smooth curve, as in Figure 4.1. The graph has $y$-intercept 1 and rises steeply to the right. Note that the graph gets very close to the $x$-axis on the left, but always lies *above* the axis (because *every* power of 2 is positive).

| $x$ | $y$ |
|---|---|
| $-3$ | $1/8$ |
| $-2$ | $1/4$ |
| $-1$ | $1/2$ |
| $0$ | $1$ |
| $1$ | $2$ |
| $2$ | $4$ |
| $3$ | $8$ |

**Figure 4.1**

The graph illustrates **exponential growth,** which is far more explosive than polynomial growth. At $x = 50$, the graph would be $2^{50}$ units high. Since there are approximately 6 units to the inch in Figure 4.1, and since there are 12 inches to the foot and 5280 feet to the mile, the height of the graph at $x = 50$ would be approximately

$$\frac{2^{50}}{6 \times 12 \times 5280} \approx 2,961,647,482 \text{ miles!}$$

When $a > 1$, the graph of the exponential function $h(x) = a^x$ has the same basic shape as the graph of $f(x) = 2^x$, as illustrated in Figure 4.2 and summarized in the next box.

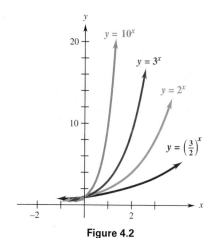

**Figure 4.2**

✓ **Checkpoint 1**

**(a)** Fill in this table:

| $x$ | $g(x) = 3^x$ |
|---|---|
| $-3$ | |
| $-2$ | |
| $-1$ | |
| $0$ | |
| $1$ | |
| $2$ | |
| $3$ | |

**(b)** Sketch the graph of $g(x) = 3^x$.

*Answers to Checkpoint exercises are found at the end of the section.*

When $a > 1$, the function $f(x) = a^x$ has the set of all real numbers as its domain. Its graph has the shape shown on the following page and all five of the properties listed below.

1. The graph is above the *x*-axis.
2. The *y*-intercept is 1.
3. The graph climbs steeply to the right.
4. The negative *x*-axis is a horizontal asymptote.
5. The larger the base *a*, the more steeply the graph rises to the right.

---

**EXAMPLE 2**    Consider the function $g(x) = 2^{-x}$.

**(a)** Rewrite the rule of *g* so that no minus signs appear in it.

**Solution** By the definition of negative exponents,

$$g(x) = 2^{-x} = \frac{1}{2^x} = \left(\frac{1}{2}\right)^x.$$

**(b)** Graph $g(x)$.

**Solution** Either use a graphing calculator or graph by hand in the usual way, as shown in Figure 4.3.

| $x$ | $y = 2^{-x}$ |
|-----|-----|
| $-3$ | $8$ |
| $-2$ | $4$ |
| $-1$ | $2$ |
| $0$ | $1$ |
| $1$ | $1/2$ |
| $2$ | $1/4$ |
| $3$ | $1/8$ |

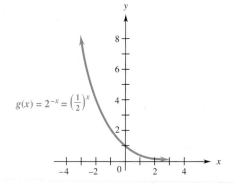

**Figure 4.3**

The graph falls sharply to the right, but never touches the *x*-axis, because every power of $\frac{1}{2}$ is positive. This is an example of **exponential decay.**

✓ **Checkpoint 2**

Graph $h(x) = (1/3)^x$.

When $0 < a < 1$, the graph of $k(x) = a^x$ has the same basic shape as the graph of $g(x) = (1/2)^x$, as illustrated in Figure 4.4 and summarized in the next box, on the following page.

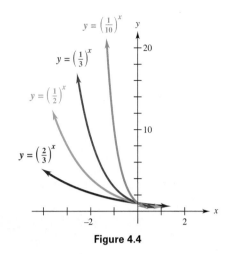

**Figure 4.4**

When $0 < a < 1$, the function $f(x) = a^x$ has the set of all real numbers as its domain. Its graph has the shape shown here and all five of the properties listed.

1. The graph is above the $x$-axis.
2. The $y$-intercept is 1.
3. The graph falls sharply to the right.
4. The positive $x$-axis is a horizontal asymptote.
5. The smaller the base $a$, the more steeply the graph falls to the right.

✓ **Checkpoint 3** ▦

Use a graphing calculator to graph $f(x) = 4^x$ and $g(x) = (\frac{1}{4})^x$ on the same screen.

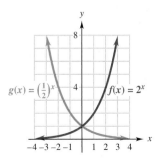

**Figure 4.5**

**EXAMPLE 3**    In each case, graph $f(x)$ and $g(x)$ on the same set of axes and explain how the graphs are related.

**(a)**  $f(x) = 2^x$ and $g(x) = (1/2)^x$

**Solution** The graphs of $f$ and $g$ are shown in Figures 4.1 and 4.3, above. Placing them on the same set of axes, we obtain Figure 4.5. It shows that the graph of $g(x) = (1/2)^x$ is the mirror image of the graph of $f(x) = 2^x$, with the $y$-axis as the mirror. ☑3

**(b)**  $f(x) = 3^{1-x}$ and $g(x) = 3^{-x}$

**Solution** Choose values of $x$ that make the exponent positive, zero, and negative, and plot the corresponding points. The graphs are shown in Figure 4.6, on the following page. The graph of $f(x) = 3^{1-x}$ has the same shape as the graph of $g(x) = 3^{-x}$, but is shifted 1 unit to the right, making the $y$-intercept $(0, 3)$ rather than $(0, 1)$.

**(c)**  $f(x) = 2^{.6x}$ and $g(x) = 2^{x}$

**Solution**  Comparing the graphs of $f(x) = 2^{.6x}$ and $g(x) = 2^{x}$ in Figure 4.7, we see that the graphs are both increasing, but the graph of $f(x)$ rises at a slower rate. This happens because of the .6 in the exponent. If the coefficient of $x$ were greater than 1, the graph would rise at a faster rate than the graph of $g(x) = 2^{x}$. ✔4

✔**Checkpoint 4**

Graph $f(x) = 2^{x+1}$.

Figure 4.6

Figure 4.7

When the exponent involves a nonlinear expression in $x$, the graph of an exponential function may have a much different shape than the preceding ones have.

**EXAMPLE 4**    Graph $f(x) = 2^{-x^2}$.

**Solution**  Either use a graphing calculator or plot points and connect them with a smooth curve, as in Figure 4.8. The graph is symmetric about the $y$-axis; that is, if the figure were folded on the $y$-axis, the two halves would match. This graph has the $x$-axis as a horizontal asymptote. The domain is still all real numbers, but here the range is $0 < y \leq 1$. Graphs such as this are important in probability, where the *normal* curve has an equation similar to $f(x)$ in this example. ✔5

✔**Checkpoint 5**

Graph $f(x) = \left(\frac{1}{2}\right)^{-x^2}$.

| $x$ | $y$ |
|-----|-----|
| $-2$ | $1/16$ |
| $-1.5$ | $.21$ |
| $-1$ | $1/2$ |
| $-.5$ | $.84$ |
| $0$ | $1$ |
| $.5$ | $.84$ |
| $1$ | $1/2$ |
| $1.5$ | $.21$ |
| $2$ | $1/16$ |

Figure 4.8

# The Number *e*

A certain irrational number, denoted $e$, plays an important role in many mathematical and scientific contexts. To 12 decimal places,

$$e \approx 2.718281828459.$$

Perhaps the single most useful exponential function is the function defined by $f(x) = e^{x}$.

**Figure 4.9**

✓**Checkpoint 6**

Evaluate the following powers of $e$.

**(a)**  $e^{.06}$

**(b)**  $e^{-.06}$

**(c)**  $e^{2.30}$

**(d)**  $e^{-2.30}$

In Figure 4.10, the functions defined by

$$g(x) = 2^x, \qquad f(x) = e^x, \qquad \text{and} \qquad h(x) = 3^x$$

are graphed for comparison.

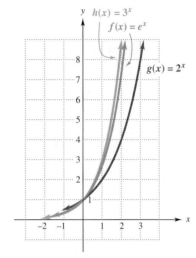

**Figure 4.10**

**EXAMPLE 5**   **Wine Consumption**  The annual amount of wine (in millions of gallons) consumed in the United States can be approximated by the function

$$f(x) = 34e^{.029x},$$

where $x = 50$ corresponds to the year 1950. (Data from: www.wineinstitute.org.)

**(a)**  How much wine was consumed in 1990?

**Solution**  Because 1990 corresponds to $x = 90$, we evaluate $f(90)$:

$$f(90) = 34e^{.029(90)} \approx 462 \text{ million gallons.}$$

So the consumption was approximately 462 million gallons.

**(b)**  How much wine was consumed in 2015?

**Solution**  Because 2015 corresponds to $x = 115$, we evaluate $f(115)$:

$$f(115) = 34e^{.029(115)} \approx 955 \text{ million gallons.}$$

**Figure 4.11**

✓**Checkpoint 7**

Per-person wine consumption (in gallons) can be approximated by $g(x) = .38e^{.018x}$, where $x = 50$ corresponds to the year 1950.

**(a)**  Estimate the per-person wine consumption in 2000.

**(b)**  Determine the first full year in which the per-person wine consumption exceeded 2.75 gallons.

**(c)**  Use a graphing calculator to determine the first full year in which consumption will exceed 1100 million (1.1 billion) gallons.

**Solution**  Because $f$ measures consumption in millions of gallons, we must solve the equation $f(x) = 1100$, that is,

$$34e^{.029x} = 1100.$$

One way to do this is to find the intersection point of the graphs of $y = 34e^{.029x}$ and $y = 1100$. (The calculator's intersection finder is in the same menu as its root [or zero] finder.) Figure 4.11 shows that this point is approximately (119.9, 1100). Hence, the first full year in which consumption will exceed 1.1 billion gallons is 2020. ✓7

**EXAMPLE 6**    **AIG Assets** The assets (in billions of dollars) for the multinational insurance company American International Group Inc. (AIG) can be approximated by the function $A(x) = 1756e^{-.091x}$, where $x = 7$ corresponds to the year 2007. (Data from: *The Wall Street Journal.*)

**(a)** What were the approximate assets of AIG in 2007 (prior to the great recession)? What about in 2015?

**Solution** Evaluate the function at $x = 7$ and $x = 15$:

$$A(7) = 1756e^{-.091(7)} \approx \$928.7 \text{ billion;}$$
$$A(15) = 1756e^{-.091(15)} \approx \$448.4 \text{ billion.}$$

**(b)** What was the first full year when AIG assets fell below $600 billion?

**Solution** We must solve the equation

$$1756e^{-.091x} = 600.$$

Using the method of Example 5(c), we must graph $y = 1756e^{-.091x}$ and $y = 600$, and find their intersection point. Figure 4.12 shows that this point is approximately (11.8, 600). So the first full year in which AIG assets fell below $600 billion is 2012.

**Figure 4.12**

# 4.1 Exercises

*Classify each function as linear, quadratic, or exponential.*

**1.** $f(x) = 6^x$

**2.** $g(x) = -5x$

**3.** $h(x) = 4x^2 - x + 5$

**4.** $k(x) = 4^{x+3}$

**5.** $f(x) = 675(1.055^x)$

**6.** $g(x) = 12e^{x^2+1}$

*Without graphing,*

**(a)** *describe the shape of the graph of each function;*

**(b)** *find the second coordinates of the points with first coordinates 0 and 1. (See Examples 1–3.)*

**7.** $f(x) = .6^x$

**8.** $g(x) = 4^{-x}$

**9.** $h(x) = 2^{.5x}$

**10.** $k(x) = 5(3^x)$

**11.** $f(x) = e^{-x}$

**12.** $g(x) = 3(16^{x/4})$

*Graph each function. (See Examples 1–3.)*

**13.** $f(x) = 3^x$

**14.** $g(x) = 3^{.5x}$

**15.** $f(x) = 2^{x/2}$

**16.** $g(x) = e^{x/4}$

**17.** $f(x) = (1/5)^x$

**18.** $g(x) = 2^{3x}$

**19.** Graph these functions on the same axes.

  **(a)** $f(x) = 2^x$    **(b)** $g(x) = 2^{x+3}$    **(c)** $h(x) = 2^{x-4}$

  **(d)** If $c$ is a positive constant, explain how the graphs of $y = 2^{x+c}$ and $y = 2^{x-c}$ are related to the graph of $f(x) = 2^x$.

**20.** Graph these functions on the same axes.

  **(a)** $f(x) = 3^x$    **(b)** $g(x) = 3^x + 2$    **(c)** $h(x) = 3^x - 4$

  **(d)** If $c$ is a positive constant, explain how the graphs of $y = 3^x + c$ and $y = 3^x - c$ are related to the graph of $f(x) = 3^x$.

*The accompanying figure shows the graphs of $y = a^x$ for $a = 1.8, 2.3, 3.2, .4, .75,$ and $.31$. They are identified by letter, but not necessarily in the same order as the values just given. Use your knowledge of how the exponential function behaves for various powers of a to match each lettered graph with the correct value of a.*

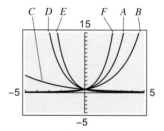

**21.** $A$

**22.** $B$

**23.** $C$

**24.** $D$

**25.** $E$

**26.** $F$

*In Exercises 27 and 28, the graph of an exponential function with base a is given. Follow the directions in parts (a)–(f) in each exercise.*

**27.**

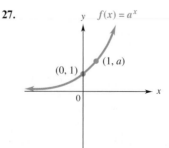

**(a)** Is $a > 1$ or is $0 < a < 1$?

**(b)** Give the domain and range of $f$.

(c) Sketch the graph of $g(x) = -a^x$.

(d) Give the domain and range of $g$.

(e) Sketch the graph of $h(x) = a^{-x}$.

(f) Give the domain and range of $h$.

28.

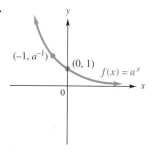

(a) Is $a > 1$ or is $0 < a < 1$?

(b) Give the domain and range of $f$.

(c) Sketch the graph of $g(x) = a^x + 2$.

(d) Give the domain and range of $g$.

(e) Sketch the graph of $h(x) = a^{x+2}$.

(f) Give the domain and range of $h$.

29. If $f(x) = a^x$ and $f(3) = 27$, find the following values of $f(x)$.

(a) $f(1)$   (b) $f(-1)$   (c) $f(2)$   (d) $f(0)$

30. Give a rule of the form $f(x) = a^x$ to define the exponential function whose graph contains the given point.

(a) $(3, 8)$   (b) $(-3, 64)$

*Graph each function. (See Example 4.)*

31. $f(x) = 2^{-x^2+2}$

32. $g(x) = 2^{x^2-2}$

33. $f(x) = x \cdot 2^x$

34. $f(x) = x^2 \cdot 2^x$

*Work the following exercises.*

35. **Finance** If \$1 is deposited into an account paying 6% per year compounded annually, then after $t$ years the account will contain

$$y = (1 + .06)^t = (1.06)^t$$

dollars.

(a) Use a calculator to complete the following table:

| t | 0 | 1 | 2 | 3 | 4 | 5 | 6 | 7 | 8 | 9 | 10 |
|---|---|---|---|---|---|---|---|---|---|---|----|
| y | 1 | | | | | 1.34 | | | | | 1.79 |

(b) Graph $y = (1.06)^t$.

36. **Finance** If money loses value at the rate of 3% per year, the value of \$1 in $t$ years is given by

$$y = (1 - .03)^t = (.97)^t.$$

(a) Use a calculator to complete the following table:

| t | 0 | 1 | 2 | 3 | 4 | 5 | 6 | 7 | 8 | 9 | 10 |
|---|---|---|---|---|---|---|---|---|---|---|----|
| y | 1 | | | | | .86 | | | | | .74 |

(b) Graph $y = (.97)^t$.

*Work these problems. (See Example 5.)*

37. **Finance** If money loses value, then as time passes, it takes more dollars to buy the same item. Use the results of Exercise 36(a) to answer the following questions.

(a) Suppose a house costs \$105,000 today. Estimate the cost of the same house in 10 years. (*Hint:* Solve the equation $.74t = \$105,000$.)

(b) Estimate the cost of a \$50 textbook in 8 years.

38. **Natural Science** Biologists have found that the oxygen consumption of yearling salmon is given by $g(x) = 100e^{.7x}$, where $x$ is the speed in feet per second. Find each of the following.

(a) the oxygen consumption when the fish are still;

(b) the oxygen consumption when the fish are swimming at a speed of 2 feet per second.

39. **Prudential Financial** The assets (in billions of dollars) for Prudential Financial, Inc. can be approximated by the function $A(x) = 315e^{.056x}$, where $x = 7$ corresponds to the year 2007. (Data from: *The Wall Street Journal.*) Find the assets in each of the following years.

(a) 2012   (b) 2015   (c) 2018

40. **Car Loan** The monthly payment on a car loan at 12% interest per year on the unpaid balance is given by

$$f(n) = \frac{P}{\dfrac{1 - 1.01^{-n}}{.01}},$$

where $P$ is the amount borrowed and $n$ is the number of months over which the loan is paid back. Find the monthly payment for each of the following loans.

(a) \$8000 for 48 months   (b) \$8000 for 24 months

(c) \$6500 for 36 months   (d) \$6500 for 60 months

41. **Natural Science** The amount of plutonium remaining from 1 kilogram after $x$ years is given by the function $W(x) = 2^{-x/24,360}$. How much will be left after

(a) 1000 years?   (b) 10,000 years?   (c) 15,000 years?

(d) Estimate how long it will take for the 1 kilogram to decay to half its original weight. Your answer may help to explain why nuclear waste disposal is a serious problem.

**Business** *The scrap value of a machine is the value of the machine at the end of its useful life. By one method of calculating scrap value, where it is assumed that a constant percentage of value is lost annually, the scrap value is given by*

$$S = C(1 - r)^n,$$

*where C is the original cost, n is the useful life of the machine in years, and r is the constant annual percentage of value lost. Find the scrap value for each of the following machines.*

42. Original cost, \$68,000; life, 10 years; annual rate of value loss, 8%

43. Original cost, \$244,000; life, 12 years; annual rate of value loss, 15%

44. Use the graphs of $f(x) = 2^x$ and $g(x) = 2^{-x}$ (not a calculator) to explain why $2^x + 2^{-x}$ is approximately equal to $2^x$ when $x$ is very large.

*Work the following problems. (See Examples 5 and 6.)*

**45. World Population** There were fewer than a billion people on Earth when Thomas Jefferson died in 1826, and there are now more than 6 billion. If the world population continues to grow as expected, the population (in billions) in year $t$ will be given by the function $P(t) = 4.834(1.01^{(t-1980)})$. (Data from: U.S. Census Bureau.) Estimate the world population in the following years.

    **(a)** 2015         **(b)** 2020         **(c)** 2040

    **(d)** What will the world population be when you reach 65 years old?

**46. Netflix Costs** The costs (in billions of dollars) related to streaming content for Netflix can be approximated by the function $C(x) = .35(1.26)^x$, where $x = 10$ corresponds to the year 2010. (Data from: *The Wall Street Journal.*)

    **(a)** Find the costs for streaming content in 2015.

    **(b)** Assuming the model stays accurate, what is the first full year when costs exceed $15 billion?

**GDP** *Use the following information to answer Exercises 47–50. The gross domestic product (GDP; in tens of trillions of U.S. dollars) for China and the United States can be approximated as follows:*

$$\text{China: } f(x) = .00011(1.102)^x,$$

$$\text{United States: } g(x) = .0132(1.067)^x,$$

*where $x = 60$ corresponds to the year 1960. (Data from: www.worldbank.org.)*

**47.** Find the projected GDP in 2015 and 2025 for China.

**48.** Find the projected GDP in 2015 and 2025 for the United States.

**49.** If the trend continues, what is the first full year in which China's GDP surpasses $25 trillion?

**50.** If the trend continues, what is the first full year in which the U.S. GDP surpasses $60 trillion?

**51. Asset Management** The amount of money (in trillions of dollars) that is invested in passively managed mutual funds can be approximated by the function

$$A(x) = .34(1.19)^x,$$

where $x = 1$ corresponds to the year 2001. (Data from: *The Wall Street Journal.*)

    **(a)** What was the amount of money in passively managed mutual funds in 2012 and 2016?

    **(b)** If the trend continues, what is the first full year when the amount passively managed exceeds $6.5 trillion?

**52. Imports from Vietnam** The value of U.S. imports from Vietnam (in billions of dollars) can be approximated by the function

$$f(x) = 2.48e^{.179x},$$

where $x = 9$ corresponds to the year 2009. (Data from: *The Wall Street Journal.*)

    **(a)** Find the value of U.S. imports in 2014.

    **(b)** If the trend continues, what is the first full year when imports exceed $60 billion in value?

**53. Subprime Mortgages** The amount of money (in billions of dollars) lent to customers with credit scores below 620 for subprime mortgages can be approximated by the function

$$g(x) = 243.5e^{-.25x},$$

where $x = 1$ corresponds to the year 2001. (Data from: *The Wall Street Journal.*)

    **(a)** Find the value of subprime mortgage lending in 2016 for the described customer base.

    **(b)** If the trend continues, what is the first full year in which subprime lending falls below $3 billion?

**54. Subprime Mortgages** The amount of money (in billions of dollars) lent to customers with a credit score below 660 for subprime mortgages can be approximated by the function

$$h(x) = 386e^{-.19x},$$

where $x = 1$ corresponds to the year 2001. (Data from: *The Wall Street Journal.*)

    **(a)** Find the value of subprime mortgage lending in 2017 for the described customer base.

    **(b)** If the trend continues, what is the first full year where subprime lending falls below $10 billion?

✓**Checkpoint Answers**

**1. (a)** The entries in the second column are 1/27, 1/9, 1/3, 1, 3, 9, 27, respectively.

    **(b)**

**2.**

**3.**

**4.**

**5.**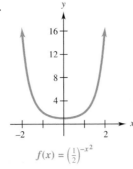

**6.** (a) 1.06184

(b) .94176

(c) 9.97418

(d) .10026

**7.** (a) About 2.3 gallons per person

(b) 2010 ($x \approx 109.95$)

# 4.2 Applications of Exponential Functions

In many situations in biology, economics, and the social sciences, a quantity changes at a rate proportional to the quantity present. For example, a country's population might be increasing at a rate of 1.3% a year. In such cases, the amount present at time $t$ is given by an **exponential growth function.**

It is understood that growth can involve either growing larger or growing smaller.

## Exponential Growth Function

Under normal conditions, growth can be described by a function of the form

$$f(t) = y_0 e^{kt} \quad \text{or} \quad f(t) = y_0 b^t,$$

where $f(t)$ is the amount present at time $t$, $y_0$ is the amount present at time $t = 0$, and $k$ and $b$ are constants that depend on the rate of growth.

When $f(t) = y_0 e^{kt}$, and $k > 0$, we describe $f(t)$ as modeling exponential growth, and when $k < 0$, we describe $f(t)$ as modeling exponential decay. When $f(t) = y_0 b^t$, and $b > 1$, we describe $f(t)$ as modeling exponential growth, and when $0 < b < 1$, we describe $f(t)$ as modeling exponential decay.

**NOTE** Writing an exponential function as $f(t) = y_0 e^{kt}$ or $f(t) = y_0 b^t$ is writing the same thing in two different ways. For example, if we let $k = 2.5$, and $b = e^{2.5}$, then notice that

$$e^{kt} = (e^k)^t = (e^{2.5})^t = b^t.$$

Sometimes it is helpful to write the function using $e$ in the formulation, and sometimes it is helpful to write the function using the formulation with $b = e^k$. Because $e$ is a constant itself, it becomes a different-valued constant that we simply represent with a different letter ($b$) when $e$ is raised to the $k$ power.

**EXAMPLE 1**    **Total Debt** The amount of debt across all sectors (in billions of dollars) can be approximated by the exponential function

$$f(t) = y_0 e^{.083t},$$

where $t$ is time in years, and $t = 0$ corresponds to the year 1970. (Data from: The Federal Reserve Bank of St. Louis.)

**(a)** If the amount of total debt was $1946 billion in the year 1970, find the total amount in the year 2017.

**Solution** Because $y_0$ represents the total debt when $t = 0$ (that is, in 1970), we have $y_0 = 1946$. So the growth function is $f(t) = 1946e^{.083t}$. To find the total debt in 2017, evaluate $f(t)$ at $t = 47$ (which corresponds to the year 2017):

$$f(t) = 1946e^{.083t}$$
$$f(47) = 1946e^{.083(47)} \approx 96{,}233.$$

Hence, the total debt in 2017 was approximately $96,233 billion (or $96.233 trillion).

**(b)** If the model remains accurate, what will the total debt be in the year 2021?

**Solution** Because 2021 corresponds to $t = 51$, evaluate $f(t)$ at $t = 51$:

$$f(51) = 1946e^{.083(51)} \approx \$134{,}126 \text{ billion.}$$

✓ **Checkpoint 1**

Suppose the number of bacteria in a culture at time $t$ is

$$y = 500e^{.4t},$$

where $t$ is measured in hours.

**(a)** How many bacteria are present initially?

**(b)** How many bacteria are present after 10 hours?

---

**EXAMPLE 2**  **Single-Family Houses** During the years from 2005 to 2011 (just before, during, and immediately after the recession years of 2007 to 2009), the number of single-family houses sold decreased at a rate that can be modeled well by an exponential function. The number of single-family houses sold (in thousands) can be approximated by the function

$$g(x) = 1483e^{-.28x},$$

where $x = 0$ corresponds to the year 2005. (Data from: *The Wall Street Journal.*)

**(a)** Find the number of houses sold in 2008.

**Solution** Evaluate $g(x)$ at $x = 3$ because 2008 corresponds to $x = 3$.

$$g(3) = 1483e^{-.28(3)} \approx 640 \text{ thousand.}$$

Thus, the number of houses sold in 2008 was approximately 640 thousand.

**(b)** What was the first full year in which the number of houses sold fell below 500 thousand?

**Solution** Graph $y = 1483e^{-.28x}$ and $y = 500$ on the same screen and find the $x$-coordinate of their intersection point. Figure 4.13 shows that $x \approx 3.88$, and we round up to the next full year of $x = 4$. Because $x = 0$ corresponds to 2005, $x = 4$ corresponds to 2009. Thus, 2009 is the first full year in which the number of houses sold fell below 500 thousand.

**Figure 4.13**

When a quantity is known to grow exponentially, it is sometimes possible to find a function that models its growth from a small amount of data.

---

**EXAMPLE 3**  **Finance** When money is placed in a bank account that pays compound interest, the amount in the account grows exponentially, as we shall see in Chapter 5. Suppose such an account grows from $1000 to $1316 in 7 years.

**(a)** Find a growth function of the form $f(t) = y_0b^t$ that gives the amount in the account at time $t$ years.

**Solution** The values of the account at time $t = 0$ and $t = 7$ are given; that is, $f(0) = 1000$ and $f(7) = 1316$. Solve the first of these equations for $y_0$:

$$f(0) = 1000$$
$$y_0b^0 = 1000 \qquad \text{Rule of } f$$
$$y_0 = 1000. \qquad b^0 = 1$$

So the rule of $f$ has the form $f(t) = 1000b^t$. Now solve the equation $f(7) = 1316$ for $b$:

$$f(7) = 1316$$

$$1000b^7 = 1316 \qquad \text{Rule of } f$$

$$b^7 = 1.316 \qquad \text{Divide both sides by 1000.}$$

$$b = (1.316)^{1/7} \approx 1.04. \qquad \text{Take the seventh root of each side.}$$

So the rule of the function is $f(t) = 1000(1.04)^t$.

**(b)** How much is in the account after 12 years?

**Solution** $f(12) = 1000(1.04)^{12} = \$1601.03.$ ✓2

## ✓ Checkpoint 2

Suppose an investment grows exponentially from \$500 to \$587.12 in three years.

**(a)** Find a function of the form $f(t) = y_0b^t$ that gives the value of the investment after $t$ years.

**(b)** How much is the investment worth after 10 years?

---

**EXAMPLE 4**    **Infant Mortality** Infant mortality rates in the United States are shown in the following table. (Data from: U.S. National Center for Health Statistics.)

| Year | Rate | Year | Rate |
|------|------|------|------|
| 1920 | 76.7 | 1980 | 12.6 |
| 1930 | 60.4 | 1990 | 9.2 |
| 1940 | 47.0 | 1995 | 7.6 |
| 1950 | 29.2 | 2000 | 6.9 |
| 1960 | 26.0 | 2010 | 6.1 |
| 1970 | 20.0 | 2014 | 5.8 |

**(a)** Let $t = 0$ correspond to 1920. Use the data for 1920 and 2014 to find a function of the form $f(t) = y_0b^t$ that models these data.

**Solution** Since the rate is 76.7 when $t = 0$, we have $y_0 = 76.7$. Hence, $f(t) = 76.7b^t$. Because 2014 corresponds to $t = 94$, we have $f(94) = 5.8$; that is,

$$76.7b^{94} = 5.8 \qquad \text{Rule of } f$$

$$b^{94} = \frac{5.8}{76.7} \qquad \text{Divide both sides by 76.7.}$$

$$b = \left(\frac{5.8}{76.7}\right)^{\frac{1}{94}} \approx .9729 \qquad \text{Take } 94^{th} \text{ roots on both sides.}$$

Therefore, the function is $f(t) = 76.7(.9729^t)$.

**(b)** Use exponential regression on a graphing calculator to find another model for these data.

**Solution** The procedure for entering the data and finding the function is the same as that for linear regression (just choose "exponential" instead of "linear"), as explained in Section 2.3. The calculator produces the equation

$$g(t) = 78.167(.9712^t).$$

This function fits the data very well, as shown in Figure 4.14.

**(c)** Use the preceding model, and assume it continues to remain accurate to estimate the infant mortality rate in the years 2018 and 2022.

**Solution** Evaluate the function $g(t)$ at $t = 98$ and $t = 102$. This yields

$$g(98) = 78.167(.9712^{98}) \approx 4.5$$

$$g(102) = 78.167(.9712^{102}) \approx 4.0.$$

**Figure 4.14**

## Other Exponential Models

When a quantity changes exponentially, but does not either grow very large or decrease practically to 0, as in Examples 1–3, different functions are needed.

> **EXAMPLE 5**    **Business** Sales of a new product often grow rapidly at first and then begin to level off with time. Suppose the annual sales of an inexpensive can opener are given by
>
> $$S(x) = 10,000(1 - e^{-.5x}),$$
>
> where $x = 0$ corresponds to the time the can opener went on the market.
>
> **(a)** What were the sales in each of the first three years?
>
> **Solution** At the end of one year ($x = 1$), sales were
>
> $$S(1) = 10,000(1 - e^{-.5(1)}) \approx 3935.$$
>
> Sales in the next two years were
>
> $$S(2) = 10,000(1 - e^{-.5(2)}) \approx 6321 \quad \text{and} \quad S(3) = 10,000(1 - e^{-.5(3)}) \approx 7769.$$
>
> **(b)** What were the sales at the end of the 10th year?
>
> **Solution** $S(10) = 10,000(1 - e^{-.5(10)}) \approx 9933.$
>
> **(c)** Graph the function $S$. What does it suggest?
>
> **Solution** The graph can be obtained by plotting points and connecting them with a smooth curve or by using a graphing calculator, as in Figure 4.15. The graph indicates that sales will level off after the 12th year, to around 10,000 can openers per year. ✓3

**Figure 4.15**

✓**Checkpoint 3**

Suppose the value of the assets (in thousands of dollars) of a certain company after $t$ years is given by

$$V(t) = 100 - 75e^{-.2t}.$$

**(a)** What is the initial value of the assets?

**(b)** What is the limiting value of the assets?

**(c)** Find the value after 10 years.

**(d)** Graph $V(t)$.

**Figure 4.16**

✓**Checkpoint 4**

In Example 6, find the price of scrap steel in 2013.

**TECHNOLOGY TIP**

Many graphing calculators can find a logistic model for appropriate data.

A variety of activities can be modeled by **logistic functions,** whose rules are of the form

$$f(x) = \frac{c}{1 + ae^{kx}}.$$

> **EXAMPLE 6**    **Price of Scrap Steel** The price (in dollars) for U.S. scrap steel per ton can be approximated by the function
>
> $$h(t) = \frac{496.7}{1 + .000048e^{.64t}},$$
>
> where $t = 11$ corresponds to the year 2011. (Data from: *The Wall Street Journal.*)
>
> **(a)** What was the price of U.S. scrap steel in 2011?
>
> **Solution** When $t = 11$,
>
> $$h(11) = \frac{496.7}{1 + .000048e^{.64(11)}} = \$470.90.$$
>
> So the price of scrap steel was \$470.90 a ton in 2011.
>
> **(b)** What was the price in 2016?
>
> **Solution** When $t = 16$,
>
> $$h(16) = \frac{496.7}{1 + .000048e^{.64(16)}} = \$211.90.$$
>
> So the price of scrap steel dropped to \$211.90 a ton by 2016.
>
> **(c)** Draw a graph of $h(t)$ from 2010 to 2017.
>
> **Solution** The TI-84 produces the graph shown in Figure 4.16. ✓4

# 4.2 Exercises

1. **Finance** Suppose you owe $800 on your credit card and you decide to make no new purchases and to make the minimum monthly payment on the account. Assuming that the interest rate on your card is 1% per month on the unpaid balance and that the minimum payment is 2% of the total (balance plus interest), your balance after $t$ months is given by

$$B(t) = 800(.9898^t).$$

   Find your balance at each of the given times.

   (a) six months

   (b) one year (remember that $t$ is in months)

   (c) five years

   (d) eight years

   (e) On the basis of your answers to parts (a)–(d), what advice would you give to your friends about minimum payments?

2. **Finance** Suppose you owe $1500 on your credit card and you decide to make no new purchases and to make the minimum monthly payment on the account. Assuming the interest rate on the card is 1.1% per month on the unpaid balance, and the minimum payment is 5% of the total (balance plus interest), your balance after $t$ months is given by

$$B(t) = 1500(.9604)^t.$$

   Find the balance at each of the given times.

   (a) five months

   (b) one year (remember that $t$ is in months)

   (c) Is the balance paid off in two years?

3. **Natural Gas Production** The annual amount of energy produced in the United States from dry natural gas (in trillion cubic feet) can be approximated by the function

$$g(t) = 14.66(1.042)^t,$$

   where $t = 10$ corresponds to the year 2010. (Data from: U.S. Energy Information Administration.)

   (a) Find the amount of dry natural gas energy produced in 2015.

   (b) If the model continues to be accurate, project the amount of dry natural gas energy produced in 2022.

4. **Oil Production** The annual amount of U.S. crude-oil production (in millions of barrels) can be approximated by the function

$$f(t) = 871(1.091)^t,$$

   where $t = 8$ corresponds to the year 2008. (Data from: U.S. Energy Information Administration.)

   (a) Find the amount of production in 2010.

   (b) If the trend continues, find the amount of production in 2020.

*In each of the following problems, find an exponential function of the form $f(t) = y_0 b^t$ to model the given data. (See Examples 3 and 4.)*

5. **Office Rent** The annual average price per square foot for office space in San Francisco was $30.48 in 2004, and it was $66.71 in 2015. (Data from: *San Francisco Business Times.*)

   (a) Find a model for these data in which $t = 0$ corresponds to the year 2004.

   (b) If the model remains accurate, what was the predicted rent per square foot in 2016?

   (c) By experimenting with different values (or using a graphing calculator to solve an appropriate equation), estimate the first full year in which the price per square foot exceeded $45.

6. **Population Growth** The U.S. Census Bureau predicts that the African-American population will increase from 35.3 million in 2000 to 61.8 million in 2060.

   (a) Find a model for these data, in which $t = 0$ corresponds to 2000.

   (b) What is the projected African-American population in 2020? in 2030?

   (c) By experimenting with different values of $t$ (or by using a graphing calculator to solve an appropriate equation), estimate the first full year in which the African-American population will exceed 55 million.

7. **Personal Consumption Expenditures** The total amount of expenditures in the United States on video and audio equipment, computers, and related services was approximately $9.6 billion in 1990 and $109.2 billion in 2015. (Data from: The Federal Reserve Bank of St. Louis.)

   (a) Let $t = 0$ correspond to the year 1990, and find a model for these data.

   (b) According to the model, what were the expenditures in 2014?

   (c) If the model remains accurate, estimate the expenditures for 2020.

8. **Medicare Expenditures** Medicare expenditures were $110 billion in 1990 and $646 billion in 2015. (Data from: Centers for Medicare and Medicaid Services.)

   (a) Let $t = 0$ correspond to the year 1990, and find a model for these data.

   (b) According to the model, what were Medicare expenditures in 2012?

   (c) If the model remains accurate, estimate Medicare expenditures in 2025.

*In the following exercises, find the exponential model as follows: If you do not have suitable technology, use the first and last data points to find a function. (See Examples 3 and 4.) If you have a graphing calculator or other suitable technology, use exponential regression to find a function. (See Example 4.)*

9. **Chinese Bank Assets** The table shows the total assets (in trillion yuan) held in Chinese banks for the given years.

| Year | 2000 | 2005 | 2007 | 2009 | 2011 | 2013 | 2015 |
|------|------|------|------|------|------|------|------|
| Assets | 17 | 41 | 53 | 79 | 118 | 151 | 199 |

   (a) Find the exponential model for these data, where $t = 0$ corresponds to the year 2000.

   (b) Assume that the model remains accurate, and predict the total assets in 2018 and 2021.

   (c) Use a graphing calculator (or trial and error) to determine the first full year in which the total assets exceed 450 trillion yuan.

**10. Atmospheric Pressure** The table shows the atmospheric pressure (in millibars) at various altitudes (in meters).

| Altitude | Pressure |
|---|---|
| 0 | 1,013 |
| 1000 | 899 |
| 2000 | 795 |
| 3000 | 701 |
| 4000 | 617 |
| 5000 | 541 |
| 6000 | 472 |
| 7000 | 411 |
| 8000 | 357 |
| 9000 | 308 |
| 10,000 | 265 |

(a) Find an exponential model for these data, in which $t$ is measured in thousands. (For instance, $t = 2$ is 2000 meters.)

(b) Use the function in part (a) to estimate the atmospheric pressure at 1500 meters and 11,000 meters. Compare your results with the actual values of 846 millibars and 227 millibars, respectively.

**11. Heart Disease** The table shows the age-adjusted death rates per 100,000 Americans for heart disease. (Data from: U.S. Center for Health Statistics.)

| Year | Death Rate |
|---|---|
| 2000 | 257.6 |
| 2002 | 240.8 |
| 2004 | 217.0 |
| 2006 | 210.2 |
| 2008 | 186.5 |
| 2010 | 178.5 |
| 2012 | 170.5 |
| 2014 | 167.0 |

(a) Find an exponential model for these data, where $t = 0$ corresponds to the year 2000.

(b) Assuming the model remains accurate, estimate the death rate in 2019 and 2023.

(c) Use a graphing calculator (or trial and error) to determine the first full year in which the death rate will be below 100.

**12. Business** The table shows outstanding consumer credit (in billions of dollars) at the beginning of various years. (Data from: U.S. Federal Reserve.)

| Year | Credit |
|---|---|
| 1980 | 350 |
| 1985 | 524 |
| 1990 | 798 |
| 1995 | 1010 |
| 2000 | 1539 |
| 2005 | 2197 |
| 2010 | 2544 |
| 2015 | 3330 |

(a) Find an exponential model for these data, where $t = 0$ corresponds to the year 1980.

(b) If this model remains accurate, what will the outstanding consumer credit be in 2020 and 2025?

(c) What is the first full year where consumer credit exceeds $6500 billion?

*Work the following problems. (See Example 5.)*

**13. Worker Productivity** Assembly-line operations tend to have a high turnover of employees, forcing the companies involved to spend much time and effort in training new workers. It has been found that a worker who is new to the operation of a certain task on the assembly line will produce $P(t)$ items on day $t$, where

$$P(t) = 25 - 25e^{-.3t}.$$

(a) How many items will be produced on the first day?

(b) How many items will be produced on the eighth day?

(c) According to the function, what is the maximum number of items the worker can produce?

**14. Words per Minute** The number of words per minute that an average person can type is given by

$$W(t) = 60 - 30e^{-.5t},$$

where $t$ is time in months after the beginning of a typing class. Find each of the following.

(a) $W(0)$  (b) $W(1)$
(c) $W(4)$  (d) $W(6)$

**Natural Science** *Newton's law of cooling says that the rate at which a body cools is proportional to the difference in temperature between the body and an environment into which it is introduced. Using calculus, we can find that the temperature $F(t)$ of the body at time $t$ after being introduced into an environment having constant temperature $T_0$ is*

$$F(t) = T_0 + Cb^t,$$

*where C and b are constants. Use this result in Exercises 15 and 16.*

**15.** Boiling water at 100° Celsius is placed in a freezer at −18° Celsius. The temperature of the water is 50° Celsius after 24 minutes. Find the temperature of the water after 76 minutes.

**16.** Paisley refuses to drink coffee cooler than 95°F. She makes coffee with a temperature of 170°F in a room with a temperature of 70°F. The coffee cools to 120°F in 10 minutes. What is the longest amount of time she can let the coffee sit before she drinks it?

**17. Internet Use in China** The percentage of Chinese residents who use the Internet can be approximated by the function

$$g(t) = \frac{49.8}{1 + 52.8e^{-.47t}},$$

where $t = 0$ corresponds to the year 2000. (Data from: *The Wall Street Journal.*)

(a) What percentage of Chinese residents used the Internet in 2006? In 2015?

(b) According to this model, will participation reach 50% by 2050?

**18. Seat-Belt Use** Data from the National Highway Traffic Safety Administration indicate that, in year $t$, the approximate percentage of people in the United States who wear seat belts when driving is given by

$$g(t) = \frac{88.2}{1 + .243e^{-.18t}},$$

where $t = 0$ corresponds to the year 2000.

(a) What percentage of drivers used seat belts in 2014?

(b) According to this model, does seat-belt use ever reach 100%?

**19. Food Assistance** The amount of money the U.S. government spent on food and nutrition assistance (in billions of dollars) can be approximated by the function

$$f(x) = \frac{120.9}{1 + 3.42e^{-.23x}},$$

where $x = 0$ corresponds to the year 2000. (Data from: U.S. Office of Management and Budget.)

(a) Find the amount spent in 2000 and 2015.

(b) Use the graph to determine the first full year in which expenditures for food and nutrition assistance was higher than $100 billion.

**20. Fish Population** The population of fish in a certain lake at time $t$ months is given by the function

$$p(t) = \frac{20,000}{1 + 24(2^{-.36t})}.$$

(a) Graph the population function from $t = 0$ to $t = 48$ (a four-year period).

(b) What was the population at the beginning of the period?

(c) Use the graph to estimate the one-year period in which the population grew most rapidly.

(d) When do you think the population will reach 25,000? What factors in nature might explain your answer?

**21. Labor Force Participation** The percentage of adults participating in the labor force can be approximated by the function

$$L(x) = \frac{67.9}{1 + .014e^{.132x}},$$

where $x = 0$ corresponds to the year 2000. (Data from: *The Wall Street Journal.*)

(a) Estimate the percentage of labor force participation in 2005 and 2015.

(b) Graph $L(x)$ from 2000 to 2015.

**22. Automobile Accidents** The probability $P(t)$ percent of having an automobile accident is related to the alcohol level $t$ of the driver's blood by the function $P(t) = e^{21.459t}$.

(a) Graph $P(t)$ in a viewing window with $0 \le t \le .2$ and $0 \le P(t) \le 100$.

(b) At what blood alcohol level is the probability of an accident at least 50%? What is the legal blood alcohol level in your state?

## ✓ Checkpoint Answers

1. (a) 500
   (b) About 27,300

2. (a) $f(t) = 500(1.055)^t$
   (b) $854.07

3. (a) $25,000
   (b) $100,000
   (c) $89,850
   (d)

4. $414.94

# 4.3 Logarithmic Functions

Until the development of computers and calculators, logarithms were the only effective tool for large-scale numerical computations. They are no longer needed for this purpose, but logarithmic functions still play a crucial role in calculus and in many applications.

## Common Logarithms

Logarithms are simply *a new language for old ideas*—essentially, a special case of exponents.

## Definition of Common (Base 10) Logarithms

$$y = \log x \qquad \text{means} \qquad 10^y = x.$$

**Log $x$**, which is read "the logarithm of $x$," is the answer to the question,

To what exponent must 10 be raised to produce $x$?

**EXAMPLE 1**    To find log 10,000, ask yourself, "To what exponent must 10 be raised to produce 10,000?" Since $10^4 = 10,000$, we see that log 10,000 = 4. Similarly,

$$\log 1 = 0 \qquad \text{because} \qquad 10^0 = 1;$$

$$\log .01 = -2 \qquad \text{because} \qquad 10^{-2} = \frac{1}{10^2} = \frac{1}{100} = .01;$$

$$\log \sqrt{10} = 1/2 \qquad \text{because} \qquad 10^{1/2} = \sqrt{10}.$$ ✓₁

**EXAMPLE 2**    Log $(-25)$ is the exponent to which 10 must be raised to produce $-25$. But every power of 10 is positive! So there is no exponent that will produce $-25$. *Logarithms of negative numbers and 0 are not defined.*

**EXAMPLE 3**    **(a)** We know that log 359 must be a number between 2 and 3 because $10^2 = 100$ and $10^3 = 1000$. By using the "log" key on a calculator, we find that log 359 (to four decimal places) is 2.5551. You can verify this statement by computing $10^{2.5551}$; the result (rounded) is 359. See the first two lines in Figure 4.17.

**(b)** When 10 is raised to a negative exponent, the result is a number less than 1. Consequently, the logarithms of numbers between 0 and 1 are negative. For instance, log .026 = $-1.5850$, as shown in the third line in Figure 4.17. ✓₂

📄 **NOTE** On most scientific calculators, enter the number followed by the log key. On graphing calculators, press the log key followed by the number, as in Figure 4.17.

## Natural Logarithms

The most widely used logarithm today is defined in terms of the number $e$ (whose decimal expansion begins 2.71828 . . .) rather than 10. It has a special name and notation.

## Definition of Natural (Base $e$) Logarithms

$$y = \ln x \qquad \text{means} \qquad e^y = x.$$

Thus, the number **ln $x$** (which is sometimes read "el-en $x$") is the exponent to which $e$ must be raised to produce the number $x$. For instance, ln 1 = 0 because $e^0 = 1$. Although logarithms to the base $e$ may not seem as "natural" as common logarithms, there are several reasons for using them, some of which are discussed in Section 4.4.

✓ **Checkpoint 1**

Find each common logarithm.

**(a)** log 100

**(b)** log 1000

**(c)** log .1

```
log(359)
           2.555094449
10^(2.5551)
             359.004589
log(.026)
           -1.585026652
```

**Figure 4.17**

✓ **Checkpoint 2**

Find each common logarithm.

**(a)** log 27

**(b)** log 1089

**(c)** log .00426

```
ln(85)
            4.442651256
e^4.4427
            85.0041433
ln(.38)
            -.9675840263
```

**Figure 4.18**

✓ **Checkpoint 3**

Find the following.

(a)  ln 6.1

(b)  ln 20

(c)  ln .8

(d)  ln .1

**EXAMPLE 4**    **(a)** To find ln 85, use the [LN] key of your calculator, as in Figure 4.18. The result is 4.4427. Thus, 4.4427 is the exponent (to four decimal places) to which $e$ must be raised to produce 85. You can verify this result by computing $e^{4.4427}$; the answer (rounded) is 85. See Figure 4.18.

**(b)** A calculator shows that ln .38 = −.9676 (rounded), which means that $e^{-.9676} \approx .38$. See Figure 4.18. ✓₃

**EXAMPLE 5**    You don't need a calculator to find ln $e^8$. Just ask yourself, "To what exponent must $e$ be raised to produce $e^8$?" The answer, obviously, is 8. So ln $e^8 = 8$.

## Other Logarithms

The procedure used to define common and natural logarithms can be carried out with any positive number $a \neq 1$ as the base in place of 10 or $e$.

### Definition of Logarithms to the Base $a$

$$y = \log_a x \qquad \text{means} \qquad a^y = x.$$

Read $y = \log_a x$ as "$y$ is the logarithm of $x$ to the base $a$." As was the case with common and natural logarithms, $\log_a x$ is an *exponent*; it is the answer to the question,

To what power must $a$ be raised to produce $x$?

For example, suppose $a = 2$ and $x = 16$. Then $\log_2 16$ is the answer to the question,

To what power must 2 be raised to produce 16?

It is easy to see that $2^4 = 16$, so $\log_2 16 = 4$. In other words, the exponential statement $2^4 = 16$ is equivalent to the logarithmic statement $4 = \log_2 16$.

In the definition of a logarithm to base $a$, note carefully the relationship between the base and the exponent:

Logarithmic form:    $y = \log_a x$

Exponential form:    $a^y = x$

Common and natural logarithms are the special cases when $a = 10$ and when $a = e$, respectively. Both log $u$ and $\log_{10} u$ mean the same thing. Similarly, ln $u$ and $\log_e u$ mean the same thing.

## Graphing Logarithmic Functions

The graph of the logarithmic function $h(x) = $ ln $x$ has the same basic shape as the graph of $f(x) = \log_a x$, as illustrated in Figure 4.19 and summarized in the next box.

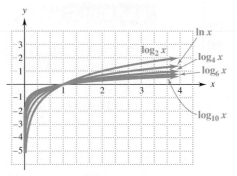

**Figure 4.19**

When $a > 0$ and $a \neq 1$, the function $f(x) = \log_a x$ has the set of nonnegative real numbers as its domain. Its graph has the shape shown below and the properties listed.

1. The graph is below the $x$-axis when $0 < x < 1$ and above the $x$-axis when $x > 1$.
2. The $x$-intercept is 1. In other words, $\log_a 1 = 0$ for all $a$ such that $a > 0$ and $a \neq 1$.
3. The graph climbs to the right and then grows more slowly.
4. The $y$-axis is a vertical asymptote.
5. The larger the base $a$, the less steeply the graph rises to the right.

**EXAMPLE 6**    This example shows several statements written in both exponential and logarithmic form.

| Exponential Form | Logarithmic Form |
|---|---|
| (a)  $3^2 = 9$ | $\log_3 9 = 2$ |
| (b)  $(1/5)^{-2} = 25$ | $\log_{1/5} 25 = -2$ |
| (c)  $10^5 = 100{,}000$ | $\log_{10} 100{,}000 \ (\text{or } \log 100{,}000) = 5$ |
| (d)  $4^{-3} = 1/64$ | $\log_4 (1/64) = -3$ |
| (e)  $2^{-4} = 1/16$ | $\log_2 (1/16) = -4$ |
| (f)  $e^0 = 1$ | $\log_e 1 \ (\text{or } \ln 1) = 0$ |

✓**4** ✓**5**

✓**Checkpoint 4**

Write the logarithmic form of

(a)  $5^3 = 125$;

(b)  $3^{-4} = 1/81$;

(c)  $8^{2/3} = 4$.

✓**Checkpoint 5**

Write the exponential form of

(a)  $\log_{16} 4 = 1/2$;

(b)  $\log_3 (1/9) = -2$;

(c)  $\log_{16} 8 = 3/4$.

## Properties of Logarithms

Some of the important properties of logarithms arise directly from their definition.

Let $x$ and $a$ be any positive real numbers, with $a \neq 1$, and $r$ be any real number. Then

(a)  $\log_a 1 = 0$;  (b)  $\log_a a = 1$;

(c)  $\log_a a^r = r$;  (d)  $a^{\log_a x} = x$.

**Figure 4.20**

Property (a) was discussed in Example 1 (with $a = 10$). Property (c) was illustrated in Example 5 (with $a = e$ and $r = 8$). Property (b) is property (c) with $r = 1$. To understand property (d), recall that $\log_a x$ is the exponent to which $a$ must be raised to produce $x$. Consequently, when you raise $a$ to this exponent, the result is $x$, as illustrated for common and natural logarithms in Figure 4.20.

The following properties are part of the reason that logarithms are so useful. They will be used in the next section to solve exponential and logarithmic equations.

## The Product, Quotient, and Power Properties

Let $x$, $y$, and $a$ be any positive real numbers, with $a \neq 1$. Let $r$ be any real number. Then

$$\log_a xy = \log_a x + \log_a y \qquad \text{Product property}$$

$$\log_a \frac{x}{y} = \log_a x - \log_a y \qquad \text{Quotient property}$$

$$\log_a x^r = r \log_a x \qquad \text{Power property}$$

Each of these three properties is illustrated on a calculator in Figure 4.21.

```
log(7*8)
          1.748188027
log(7)+log(8)
          1.748188027
```
```
ln(50/9)
          1.714798428
ln(50)-ln(9)
          1.714798428
```
```
log(12⁹)
          9.712631214
9*log(12)
          9.712631214
```

**Figure 4.21**

To prove the product property, let

$$m = \log_a x \qquad \text{and} \qquad n = \log_a y.$$

Then, by the definition of logarithm,

$$a^m = x \qquad \text{and} \qquad a^n = y.$$

Multiply to get

$$a^m \cdot a^n = x \cdot y,$$

or, by a property of exponents,

$$a^{m+n} = xy.$$

Use the definition of logarithm to rewrite this last statement as

$$\log_a xy = m + n.$$

Replace $m$ with $\log_a x$ and $n$ with $\log_a y$ to get

$$\log_a xy = \log_a x + \log_a y.$$

The quotient and power properties can be proved similarly.

✓ **Checkpoint 6**

Write each expression as a single logarithm, using the properties of logarithms.

(a)  $\log_a 5x + \log_a 3x^4$

(b)  $\log_a 3p - \log_a 5q$

(c)  $4 \log_a k - 3 \log_a m$

**EXAMPLE 7**   Using the properties of logarithms, we can write each of the following as a single logarithm:[*]

(a)  $\log_a x + \log_a (x - 1) = \log_a x(x - 1);$   Product property

(b)  $\log_a (x^2 + 4) - \log_a (x + 6) = \log_a \dfrac{x^2 + 4}{x + 6};$   Quotient property

(c)  $\log_a 9 + 5 \log_a x = \log_a 9 + \log_a x^5 = \log_a 9x^5.$   Product and power properties

---

[*]Here and elsewhere, we assume that variable expressions represent positive numbers and that the base $a$ is positive, with $a \neq 1$.

🛑 **CAUTION** There is no logarithm property that allows you to simplify the logarithm of a sum, such as $\log_a(x^2 + 4)$. In particular, $\log_a(x^2 + 4)$ is *not* equal to $\log_a x^2 + \log_a 4$. The product property of logarithms shows that $\log_a x^2 + \log_a 4 = \log_a 4x^2$.

---

**EXAMPLE 8** Assume that $\log_6 7 \approx 1.09$ and $\log_6 5 \approx .90$. Use the properties of logarithms to find each of the following:

**(a)** $\log_6 35 = \log_6 (7 \cdot 5) = \log_6 7 + \log_6 5 \approx 1.09 + .90 = 1.99$;

**(b)** $\log_6 5/7 = \log_6 5 - \log_6 7 \approx .90 - 1.09 = -.19$;

**(c)** $\log_6 5^3 = 3 \log_6 5 \approx 3(.90) = 2.70$;

**(d)** $\log_6 6 = 1$;

**(e)** $\log_6 1 = 0$. ✔₇

In Example 8, several logarithms to the base 6 were given. However, they could have been found by using a calculator and the following formula.

---

**✔ Checkpoint 7**

Use the properties of logarithms to rewrite and evaluate each expression, given that $\log_3 7 \approx 1.77$ and $\log_3 5 \approx 1.46$.

**(a)** $\log_3 35$

**(b)** $\log_3 7/5$

**(c)** $\log_3 25$

**(d)** $\log_3 3$

**(e)** $\log_3 1$

---

## Change-of-Base Theorem

For any positive numbers $a$ and $x$ (with $a \neq 1$),

$$\log_a x = \frac{\ln x}{\ln a}.$$

---

**EXAMPLE 9** To find $\log_7 3$, use the theorem with $a = 7$ and $x = 3$:

$$\log_7 3 = \frac{\ln 3}{\ln 7} \approx \frac{1.0986}{1.9459} \approx .5646.$$

You can check this result on your calculator by verifying that $7^{.5646} \approx 3$.

## Logarithmic Functions

For a given *positive* value of $x$, the definition of logarithm leads to exactly one value of $y$, so that $y = \log_a x$ defines a function.

---

If $a > 0$ and $a \neq 1$, the **logarithmic function** with base $a$ is defined as

$$f(x) = \log_a x.$$

---

The most important logarithmic function is the natural logarithmic function.

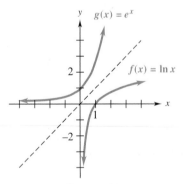

**Figure 4.22**

**EXAMPLE 10** Graph $f(x) = \ln x$ and $g(x) = e^x$ on the same axes.

**Solution** For each function, use a calculator to compute some ordered pairs. Then plot the corresponding points and connect them with a curve to obtain the graphs in Figure 4.22.

The dashed line in Figure 4.22 is the graph of $y = x$. Observe that the graph of $f(x) = \ln x$ is the mirror image of the graph of $g(x) = e^x$, with the line $y = x$ being the mirror. A pair of functions whose graphs are related in this way are said to be **inverses** of each other. A more complete discussion of inverse functions is given in many college algebra books. ✔₈

---

**✔ Checkpoint 8**

Verify that $f(x) = \log x$ and $g(x) = 10^x$ are inverses of each other by graphing $f(x)$ and $g(x)$ on the same axes.

When the base $a > 1$, the graph of $f(x) = \log_a x$ has the same basic shape as the graph of the natural logarithmic function in Figure 4.22.

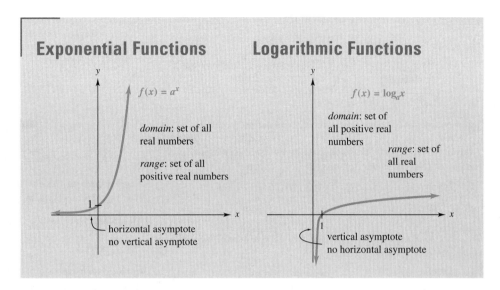

As the information in the box suggests, the functions $f(x) = \log_a x$ and $g(x) = a^x$ are inverses of each other. (Their graphs are mirror images of each other, with the line $y = x$ being the mirror.)

## Applications

Logarithmic functions are useful for, among other things, describing quantities that grow, but do so at a slower rate as time goes on.

**EXAMPLE 11**    **Wind Energy** Annual U.S. power generation from wind (in million megawatt-hours) can be approximated by the function

$$f(x) = -335.5 + 193.1 \ln x,$$

where $x = 6$ corresponds to the year 2006. (Data from: *The Wall Street Journal*.)

**(a)** Find the wind power generated in 2008, 2012, and 2015.

**Solution** These years correspond to $x = 8$, $x = 12$, and $x = 15$, respectively, so you can use a calculator to evaluate $f(x)$ at these numbers:

$$f(8) = -335.5 + 193.1 \ln(8) \approx 66.0;$$
$$f(12) = -335.5 + 193.1 \ln(12) \approx 144.3;$$
$$f(15) = -335.5 + 193.1 \ln(15) \approx 187.4.$$

So U.S. wind power generation was about 66.0 million megawatt-hours in 2008, 144.3 million megawatt-hours in 2012, and 187.4 million megawatt-hours in 2015.

**(b)** If this function remains accurate, what will be the first full year in which wind power production exceeds 225 million megawatt-hours?

**Solution** We must solve the equation $f(x) = 225$, that is,

$$-335.5 + 193.1 \ln x = 225.$$

In the next section, we shall see how to do this algebraically. For now, we solve the equation graphically by graphing $f(x)$ and $y = 225$ on the same screen and finding their intersection point. Figure 4.23 shows that the $x$-coordinate of this intersection point is $x \approx 18.22$. Thus, the first full year in which wind power will exceed 225 million megawatt-hours is 2019.

**Figure 4.23**

# 4.3 Exercises

*Complete each statement in Exercises 1–4.*

**1.** $y = \log_a x$ means $x = $ _____.

**2.** The statement $\log_5 125 = 3$ tells us that _____ is the power of _____ that equals _____.

**3.** What is wrong with the expression $y = \log_b$?

**4.** Logarithms of negative numbers are not defined because _____.

*Translate each logarithmic statement into an equivalent exponential statement. (See Examples 1, 5, and 6.)*

**5.** $\log 100{,}000 = 5$         **6.** $\log .001 = -3$

**7.** $\log_9 81 = 2$         **8.** $\log_2 (1/8) = -3$

*Translate each exponential statement into an equivalent logarithmic statement. (See Examples 5 and 6.)*

**9.** $10^{1.9823} = 96$         **10.** $e^{3.2189} = 25$

**11.** $3^{-2} = 1/9$         **12.** $16^{1/2} = 4$

*Without using a calculator, evaluate each of the given expressions. (See Examples 1, 5, and 6.)*

**13.** $\log 1000$         **14.** $\log .0001$

**15.** $\log_6 36$         **16.** $\log_3 81$

**17.** $\log_4 64$         **18.** $\log_5 125$

**19.** $\log_2 \dfrac{1}{4}$         **20.** $\log_3 \dfrac{1}{27}$

**21.** $\ln \sqrt{e}$         **22.** $\ln(1/e)$

**23.** $\ln e^{8.77}$         **24.** $\log 10^{74.3}$

*Use a calculator to evaluate each logarithm to three decimal places. (See Examples 3 and 4.)*

**25.** $\log 53$         **26.** $\log .005$

**27.** $\ln .0068$         **28.** $\ln 354$

**29.** Why does $\log_a 1$ always equal 0 for any valid base $a$?

*Write each expression as the logarithm of a single number or expression. Assume that all variables represent positive numbers. (See Example 7.)*

**30.** $\log 20 - \log 5$         **31.** $\log 6 + \log 8 - \log 2$

**32.** $3 \ln 2 + 2 \ln 3$         **33.** $2 \ln 5 - \frac{1}{2} \ln 25$

**34.** $5 \log x - 2 \log y$         **35.** $2 \log u + 3 \log w - 6 \log v$

**36.** $\ln(3x + 2) + \ln(x + 4)$         **37.** $2 \ln(x + 2) - \ln(x + 3)$

*Write each expression as a sum and/or a difference of logarithms, with all variables to the first degree.*

**38.** $\log 5x^2y^3$         **39.** $\ln \sqrt{6m^4n^2}$

**40.** $\ln \dfrac{3x}{5y}$         **41.** $\log \dfrac{\sqrt{xz}}{z^3}$

**42.** The calculator-generated table in the figure is for $y_1 = \log(4 - x)$. Why do the values in the $y_1$ column show ERROR for $x \geq 4$?

| X | Y1 |
|---|---|
| 0 | .60206 |
| 1 | .47712 |
| 2 | .30103 |
| 3 | 0 |
| ▮ | ERROR |
| 5 | ERROR |
| 6 | ERROR |
| X=4 | |

*Express each expression in terms of u and v, where $u = \ln x$ and $v = \ln y$. For example, $\ln x^3 = 3(\ln x) = 3u$.*

**43.** $\ln(x^2y^5)$         **44.** $\ln(\sqrt{x} \cdot y^2)$

**45.** $\ln(x^3/y^2)$         **46.** $\ln(\sqrt{x}/y)$

*Evaluate each expression. (See Example 9.)*

**47.** $\log_6 384$         **48.** $\log_{30} 78$

**49.** $\log_{35} 5646$         **50.** $\log_6 60 - \log_{60} 6$

*Find numerical values for b and c for which the given statement is FALSE.*

**51.** $\log(b + c) = \log b + \log c$         **52.** $\dfrac{\ln b}{\ln c} = \ln\left(\dfrac{b}{c}\right)$

*Graph each function. (See Example 11.)*

**53.** $y = \ln(x + 2)$         **54.** $y = \ln x + 2$

**55.** $y = \log(x - 3)$         **56.** $y = \log x - 3$

**57.** Graph $f(x) = \log x$ and $g(x) = \log(x/4)$ for $-2 \leq x \leq 8$. How are these graphs related? How does the quotient rule support your answer?

*In Exercises 58 and 59, the coordinates of a point on the graph of the indicated function are displayed at the bottom of the screen. Write the logarithmic and exponential equations associated with the display.*

**58.**

**59.**

**60.** Match each equation with its graph. Each tick mark represents one unit.

(a) $y = \log x$      (b) $y = 10^x$

(c) $y = \ln x$      (d) $y = e^x$

**(A)**

**(B)**

**(C)**

**(D)**

**61. Finance** The doubling function

$$D(r) = \frac{\ln 2}{\ln(1 + r)}$$

gives the number of years required to double your money when it is invested at interest rate $r$ (expressed as a decimal), compounded annually. How long does it take to double your money at each of the following rates?

(a) 4%    (b) 8%    (c) 18%    (d) 36%

(e) Round each of your answers in parts (a)–(d) to the nearest year, and compare them with these numbers: 72/4, 72/8, 72/18, and 72/36. Use this evidence to state a "rule of thumb" for determining approximate doubling time without employing the function $D$. This rule, which has long been used by bankers, is called the *rule of 72.*

**62. Flu Virus** Two people with the flu visited a college campus. The number of days, $T$, that it took for the flu virus to infect $n$ people is given by

$$T = -1.43 \ln\left(\frac{10{,}000 - n}{4998n}\right).$$

How many days will it take for the virus to infect

(a) 500 people?      (b) 5000 people?

**63. Health Insurance Costs** The average annual cost (in dollars) for health insurance in the United States can be approximated by the function

$$g(x) = -1736.6 + 1661.8 \ln x,$$

where $x = 6$ corresponds to the year 2006. (Data from: U.S. Bureau of Labor Statistics.)

(a) Estimate the average cost in 2010.

(b) Graph the function $g$ for the period 2006 to 2015.

(c) Assuming that the graph remains accurate, what does the shape of the graph suggest regarding the average cost of health insurance?

**64. Physical Science** The barometric pressure $p$ (in inches of mercury) is related to the height $h$ above sea level (in miles) by the equation

$$h = -5 \ln\left(\frac{p}{29.92}\right).$$

The pressure readings given in parts (a)–(c) were made by a weather balloon. At what heights were they made?

(a) 29.92 in.     (b) 20.05 in.     (c) 11.92 in.

(d) Use a graphing calculator to determine the pressure at a height of 3 miles.

**65. Dairy Expenditures** The average annual expenditures for the U.S. consumer on dairy products can be approximated by the function

$$f(x) = 310.4 + 40.7 \ln x,$$

where $x = 6$ corresponds to the year 2006. (Data from: U.S. Bureau of Labor Statistics.)

(a) Estimate the average expenditures for 2009 and 2014.

(b) Assuming that the model remains accurate, what is the first full year in which expenditures on dairy products exceed $430?

**66. Credit Union Assets** The total assets (in billions of dollars) held by U.S. credit unions can be approximated by the function

$$f(x) = -689.9 + 695.8 \ln x,$$

where $x = 10$ corresponds to the year 2010. (Data from: Credit Union National Association.)

(a) What were the total assets in 2011 and 2015?

(b) If the model remains accurate, what is the first full year in which the total assets in credit unions exceed $1450 billion?

**67. Border Patrol Budget** The amount (in billions) that the United States spent on border patrol can be approximated by the function

$$g(x) = .185 + 1.209 \ln x,$$

where $x = 2$ corresponds to the year 2002. (Data from: *USA Today*.)

**(a)** What was the amount spent in 2005 and 2016 on border patrol?

**(b)** If the model remains accurate, what is the first full year in which the amount spent on border patrol exceeds $4.0 billion?

**68. Opioid Deaths** The number of deaths from opioids in the United States can be approximated by the function

$$f(x) = 4760 + 4705 \ln x,$$

where $x = 1$ corresponds to the year 2001. (Data from: *The Wall Street Journal*.)

**(a)** What was the number of deaths from opioids in 2007 and 2013?

**(b)** If the model remains accurate, what is the first full year in which the number of opioid deaths exceeds 18,000?

**69. Apple iPhone Sales** The worldwide number (in millions) of iPhones sold can be approximated by the function

$$g(x) = -875 + 399 \ln x,$$

where $x = 11$ corresponds to the year 2011. (Data from: www.statistica.com.)

**(a)** What was the number of worldwide sales for iPhones in 2015?

**(b)** If the model continues to be accurate, what was the first full year in which iPhone sales exceed 250 million?

**70. Miles Traveled** The total number of miles traveled by vehicles on all roads and streets in the United States (in trillions of miles) can be approximated by the function

$$h(x) = 2.8 + .091 \ln x,$$

where $x = 1$ corresponds to the year 2001. (Data from: U.S. Department of Transportation.)

**(a)** What was the number of total vehicle miles traveled in 2015?

**(b)** According to this model, what was the first full year in which the total number of vehicle miles exceeded 2.95 trillion miles?

## ✓ Checkpoint Answers

1. **(a)** 2 **(b)** 3 **(c)** −1

2. **(a)** 1.4314 **(b)** 3.0370 **(c)** −2.3706

3. **(a)** 1.8083 **(b)** 2.9957 **(c)** −.2231
   **(d)** −2.3026

4. **(a)** $\log_5 125 = 3$ **(b)** $\log_3(1/81) = -4$ **(c)** $\log_8 4 = 2/3$

5. **(a)** $16^{1/2} = 4$ **(b)** $3^{-2} = 1/9$ **(c)** $16^{3/4} = 8$

6. **(a)** $\log_a 15x^5$ **(b)** $\log_a(3p/5q)$ **(c)** $\log_a(k^4/m^3)$

7. **(a)** 3.23 **(b)** .31 **(c)** 2.92
   **(d)** 1 **(e)** 0

8.

# 4.4 Logarithmic and Exponential Equations

Many applications involve solving logarithmic and exponential equations, so we begin with solution methods for such equations.

## Logarithmic Equations

When an equation involves only logarithmic terms, use the logarithm properties to write each side as a single logarithm. Then use the following fact.

> Let $u$, $v$, and $a$ be positive real numbers, with $a \neq 1$.
>
> If $\log_a u = \log_a v$, then $u = v$.

**EXAMPLE 1** Solve $\log x = \log(x + 3) - \log(x - 1)$.

**Solution** First, use the quotient property of logarithms to write the right side as a single logarithm:

$$\log x = \log(x + 3) - \log(x - 1)$$

$$\log x = \log\left(\frac{x + 3}{x - 1}\right). \qquad \text{Quotient property of logarithms}$$

The fact in the preceding box now shows that

$$x = \frac{x + 3}{x - 1}$$

$$x(x - 1) = x + 3 \qquad \text{Cross multiply.}$$

$$x^2 - x = x + 3 \qquad \text{Distributive property}$$

$$x^2 - 2x - 3 = 0 \qquad \text{Add } -x - 3 \text{ to each side and combine like terms.}$$

$$(x - 3)(x + 1) = 0 \qquad \text{Factor.}$$

$$x = 3 \quad \text{or} \quad x = -1. \qquad \text{Set each factor to 0 and solve for } x.$$

Since $\log x$ is not defined when $x = -1$, the only possible solution is $x = 3$. We indicate this by crossing the value that is not a solution

$$x = 3 \quad \text{or} \quad x \cancel{= -1}.$$

Use a calculator to verify that 3 is actually a solution.

When an equation involves constants and logarithmic terms, use algebra and the logarithm properties to write one side as a single logarithm and the other as a constant. Then use the following property of logarithms, which was discussed on pages 200–201.

> If $a$ and $u$ are positive real numbers, with $a \neq 1$, then
>
> $$a^{\log_a u} = u.$$

**EXAMPLE 2** Solve each equation.

**(a)** $\log_5 (2x - 3) = 2$

**Solution** Since the base of the logarithm is 5, raise 5 to the exponents given by the equation:

$$5^{\log_5 (2x - 3)} = 5^2$$

On the left side, use the fact in the preceding box (with $a = 5$ and $u = 2x - 3$) to conclude that

$$2x - 3 = 25$$

$$2x = 28$$

$$x = 14.$$

Verify that 14 is a solution of the original equation.

**(b)** $\log(x - 16) = 2 - \log(x - 1)$

**Solution** First rearrange the terms to obtain a single logarithm on the left side:

$$\log(x - 16) + \log(x - 1) = 2$$

$$\log[(x - 16)(x - 1)] = 2 \qquad \text{Product property of logarithms}$$

$$\log(x^2 - 17x + 16) = 2.$$

Since the base of the logarithm is 10, raise 10 to the given powers:

$$10^{\log(x^2 - 17x + 16)} = 10^2.$$

On the left side, apply the logarithm property in the preceding box (with $a = 10$ and $u = x^2 - 17x + 16$):

$$x^2 - 17x + 16 = 100$$

$$x^2 - 17x - 84 = 0 \qquad \text{Add } -100 \text{ to each side.}$$

$$(x + 4)(x - 21) = 0 \qquad \text{Factor.}$$

$$x = -4 \quad \text{or} \quad x = 21. \qquad \text{Set each factor} = 0 \text{ and solve for } x.$$

✓ **Checkpoint 1**

Solve each equation.

**(a)** $\log_2 (p + 9) - \log_2 p = \log_2 (p + 1)$

**(b)** $\log_3 (m + 1) - \log_3 (m - 1) = \log_3 m$

In the original equation, when $x = -4$, $\log(x - 16) = \log(-20)$, which is not defined. So $-4$ is not a solution. We indicate this by crossing the value that is not a solution

$$x \cancel{=} \cancel{-4} \quad \text{or} \quad x = 21.$$

You should verify that 21 is a solution of the original equation.

**(c)** $\log_2 x - \log_2 (x - 1) = 1$

**Solution** Proceed as before, with 2 as the base of the logarithm:

$$\log_2 \frac{x}{x - 1} = 1 \qquad \text{Quotient property of logarithms}$$

$$2^{\log_2 x/(x-1)} = 2^1 \qquad \text{Exponentiate to the base 2.}$$

$$\frac{x}{x - 1} = 2 \qquad \text{Use the fact in the preceding box.}$$

$$x = 2(x - 1) \qquad \text{Multiply both sides by } x - 1.$$

$$x = 2x - 2 \qquad \text{Distributive property}$$

$$-x = -2 \qquad \text{Multiply both sides by } -1.$$

$$x = 2.$$

Verify that 2 is a solution of the original equation. ✓₂

✓ **Checkpoint 2**

Solve each equation.

**(a)** $\log_5 x + 2 \log_5 x = 3$

**(b)** $\log_6 (a + 2) - \log_6 \dfrac{a - 7}{5} = 1$

## Exponential Equations

An equation in which all the variables are exponents is called an exponential equation. When such an equation can be written as two powers of the same base, it can be solved by using the following fact.

> Let $a$ be a positive real number, with $a \neq 1$.
>
> If $a^u = a^v$, then $u = v$.

**EXAMPLE 3** Solve $9^x = 27$.

**Solution** First, write both sides as powers of the same base. Since $9 = 3^2$ and $27 = 3^3$, we have

$$9^x = 27$$

$$(3^2)^x = 3^3 \qquad \text{Write 9 and 27 as powers of 3.}$$

$$3^{2x} = 3^3. \qquad \text{Power of a power}$$

Apply the fact in the preceding box (with $a = 3$, $u = 2x$, and $v = 3$):

$$2x = 3$$

$$x = \frac{3}{2}.$$

Verify that $x = 3/2$ is a solution of the original equation. ✓₃

✓ **Checkpoint 3**

Solve each equation.

**(a)** $8^{2x} = 4$

**(b)** $5^{3x} = 25^4$

**(c)** $36^{-2x} = 6$

Exponential equations involving different bases can often be solved by using the power property of logarithms, which is repeated here for natural logarithms.

If $u$ and $r$ are real numbers, with $u$ positive, then

$$\ln u^r = r \ln u.$$

Although natural logarithms are used in the following examples, logarithms to any base will produce the same solutions.

> **EXAMPLE 4**  Solve $3^x = 5$.
>
> **Solution**  Take natural logarithms on both sides:
>
> $$\ln 3^x = \ln 5.$$
>
> Apply the power property of logarithms in the preceding box (with $r = x$) to the left side:
>
> $$x \ln 3 = \ln 5$$
>
> $$x = \frac{\ln 5}{\ln 3} \approx 1.465. \qquad \text{Divide both sides by the constant ln 3.}$$
>
> To check, evaluate $3^{1.465}$; the answer should be approximately 5, which verifies that the solution of the given equation is 1.465 (to the nearest thousandth). ✓ 4

✓ **Checkpoint 4**

Solve each equation. Round solutions to the nearest thousandth.

**(a)**  $2^x = 7$

**(b)**  $5^m = 50$

**(c)**  $3^y = 17$

⚠ **CAUTION**  Be careful: $\dfrac{\ln 5}{\ln 3}$ is *not* equal to $\ln\left(\dfrac{5}{3}\right)$ or $\ln 5 - \ln 3$.

> **EXAMPLE 5**  Solve $3^{2x-1} = 4^{x+2}$.
>
> **Solution**  Taking natural logarithms on both sides gives
>
> $$\ln 3^{2x-1} = \ln 4^{x+2}.$$
>
> Now use the power property of logarithms and the fact that $\ln 3$ and $\ln 4$ are constants to rewrite the equation:
>
> $$(2x - 1)(\ln 3) = (x + 2)(\ln 4) \qquad \text{Power property}$$
>
> $$2x(\ln 3) - 1(\ln 3) = x(\ln 4) + 2(\ln 4) \qquad \text{Distributive property}$$
>
> $$2x(\ln 3) - x(\ln 4) = 2(\ln 4) + 1(\ln 3). \qquad \text{Collect terms with } x \text{ on one side.}$$
>
> Factor out $x$ on the left side to get
>
> $$[2(\ln 3) - \ln 4]x = 2(\ln 4) + \ln 3.$$
>
> Divide both sides by the coefficient of $x$:
>
> $$x = \frac{2(\ln 4) + \ln 3}{2(\ln 3) - \ln 4}.$$
>
> Using a calculator to evaluate this last expression, we find that
>
> $$x = \frac{2\ln 4 + \ln 3}{2\ln 3 - \ln 4} \approx 4.774. \quad ✓ 5$$

✓ **Checkpoint 5**

Solve each equation. Round solutions to the nearest thousandth.

**(a)**  $6^m = 3^{2m-1}$

**(b)**  $5^{6a-3} = 2^{4a+1}$

Recall that $\ln e = 1$ (because 1 is the exponent to which $e$ must be raised to produce $e$). This fact simplifies the solution of equations involving powers of $e$.

---

**EXAMPLE 6**   Solve $3e^{x^2} = 600$.

**Solution**   First divide each side by 3 to get

$$e^{x^2} = 200.$$

Now take natural logarithms on both sides; then use the power property of logarithms:

$$e^{x^2} = 200$$
$$\ln e^{x^2} = \ln 200 \qquad \text{Take natural logarithms on both sides.}$$
$$x^2 \ln e = \ln 200 \qquad \text{Power property}$$
$$x^2 = \ln 200 \qquad \ln e = 1.$$
$$x = \pm\sqrt{\ln 200} \qquad \text{Take the square root of each side.}$$
$$x \approx \pm 2.302.$$

Verify that the solutions are $\pm 2.302$, rounded to the nearest thousandth. (The symbol $\pm$ is used as a shortcut for writing the two solutions 2.302 and $-2.302$.) ✓6

### ✓ Checkpoint 6

Solve each equation. Round solutions to the nearest thousandth.

(a) $e^{.1x} = 11$

(b) $e^{3+x} = .893$

(c) $e^{2x^2-3} = 9$

---

📓 **TECHNOLOGY TIP**

Logarithmic and exponential equations can be solved on a graphing calculator. To solve $3^x = 5^{2x-1}$, for example, graph $y = 3^x$ and $y = 5^{2x-1}$ on the same screen. Then use the intersection finder to determine the $x$-coordinates of their intersection points. Alternatively, graph $y = 3^x - 5^{2x-1}$ and use the root finder to determine the $x$-intercepts of the graph.

## Applications

Some of the most important applications of exponential and logarithmic functions arise in banking and finance. They will be thoroughly discussed in Chapter 5. The applications here are from other fields.

---

**EXAMPLE 7**   **New Jobs**   The number of nonfarm jobs (in millions) added to the U.S. economy by quarter can be approximated by the function

$$f(x) = 12.1 + 2.03 \ln x,$$

where $x = 1$ corresponds to the first quarter of the year 2014. Find the first full quarter in which the number of jobs added to the economy exceeded 16 million. (Data from: The Bureau of Labor Statistics.)

**Solution**   We need to find the value for $x$ for which $f(x) = 16$. Thus, we need to solve the following equation for $x$:

$$12.1 + 2.03 \ln x = 16$$
$$2.03 \ln x = 3.9 \qquad \text{Subtract 12.1 from each side.}$$
$$\ln x = \frac{3.9}{2.03} \qquad \text{Divide each side by 2.03.}$$
$$e^{\ln x} = e^{3.9/2.03} \qquad \text{Exponentiate each side.}$$
$$x \approx 6.83.$$

Because $x$ is fractional and the question asks for the "first full quarter," we round up to $x = 7$, which would be the third quarter of 2015. ✓7

### ✓ Checkpoint 7

Use the function in Example 7 to determine the first full quarter in which the number of jobs added to the economy exceeded 17 million.

---

**EXAMPLE 8**   **Digital Groceries**   The amount of money U.S. consumers spend online for food (in billions) can be approximated by the function.

$$f(x) = .174(1.318)^x,$$

where $x = 11$ corresponds to the year 2011. Assuming that the model remains accurate, determine the first full year in which online grocery sales exceed $50 billion. (Data from: *The Wall Street Journal.*)

**Solution**  We need to find the value of $x$ for which $f(x) = 50$—that is, to solve the following equation:

$$.174(1.318)^x = 50$$

$$1.318^x = \frac{50}{.174}$$  Divide both sides by .174.

$$\ln(1.318^x) = \ln\left(\frac{50}{.174}\right)$$  Take logarithms on both sides.

$$x \ln(1.318) = \ln\left(\frac{50}{.174}\right)$$  Power property of logarithms

$$x = \frac{\ln(50/.174)}{\ln(1.318)}$$  Divide both sides by ln(1.318).

$$x \approx 20.5.$$

✓ **Checkpoint 8**

Use the function in Example 8 to determine the first full year in which online grocery sales exceeded $20 billion.

Thus, the first full year in which this model predicts that sales will exceed $50 billion is 2021. ✓ 8

The **half-life** of a radioactive substance is the time it takes for a given quantity of the substance to decay to one-half its original mass. The half-life depends only on the substance, not on the size of the sample. It can be shown that the amount of a radioactive substance at time $t$ is given by the function

$$f(t) = y_0\left(\frac{1}{2}\right)^{t/h} = y_0(.5^{t/h}),$$

where $y_0$ is the initial amount (at time $t = 0$) and $h$ is the half-life of the substance.

Radioactive carbon-14 is found in every living plant and animal. After the plant or animal dies, its carbon-14 decays exponentially, with a half-life of 5730 years. This fact is the basis for a technique called *carbon dating* for determining the age of fossils.

**EXAMPLE 9**    **Carbon Dating**  A round wooden table hanging in Winchester Castle (England) was alleged to have belonged to King Arthur, who lived in the fifth century. A recent chemical analysis showed that the wood had lost 9% of its carbon-14.[*] How old is the table?

**Solution**  The decay function for carbon-14 is

$$f(t) = y_0(.5^{t/5730}),$$

where $t = 0$ corresponds to the time the wood was cut to make the table. (That is when the tree died.) Since the wood has lost 9% of its carbon-14, the amount in the table now is 91% of the initial amount $y_0$ (that is, $.91y_0$). We must find the value of $t$ for which $f(t) = .91y_0$. So we must solve the equation

$$y_0(.5^{t/5730}) = .91y_0$$  Definition of $f(t)$

$$.5^{t/5730} = .91$$  Divide both sides by $y_0$.

$$\ln .5^{t/5730} = \ln .91$$  Take logarithms on both sides.

$$\left(\frac{t}{5730}\right)\ln .5 = \ln .91$$  Power property of logarithms

$$t \ln .5 = 5730 \ln .91$$  Multiply both sides by 5730.

$$t = \frac{5730 \ln .91}{\ln .5} \approx 779.63.$$  Divide both sides by ln .5.

✓ **Checkpoint 9**

How old is a skeleton that has lost 65% of its carbon-14?

The table is about 780 years old and therefore could not have belonged to King Arthur. ✓ 9

_____

[*]This is done by measuring the ratio of carbon-14 to nonradioactive carbon-12 in the table (a ratio that is approximately constant over long periods) and comparing it with the ratio in living wood.

Earthquakes are often in the news. The standard method of measuring their size, the **Richter scale,** is a logarithmic function (base 10).

---

**EXAMPLE 10** **Richter Scale** The intensity $R(i)$ of an earthquake, measured on the Richter scale, is given by

$$R(i) = \log\left(\frac{i}{i_0}\right),$$

where $i$ is the intensity of the ground motion of the earthquake and $i_0$ is the intensity of the ground motion of the so-called *zero earthquake* (the smallest detectable earthquake, against which others are measured). The underwater earthquake that caused the disastrous 2004 tsunami in Southeast Asia measured 9.1 on the Richter scale.

**(a)** How did the ground motion of this tsunami compare with that of the zero earthquake?

**Solution** In this case,

$$R(i) = 9.1$$

$$\log\left(\frac{i}{i_0}\right) = 9.1.$$

By the definition of logarithms, 9.1 is the exponent to which 10 must be raised to produce $i/i_0$, which means that

$$10^{9.1} = \frac{i}{i_0}, \quad \text{or equivalently,} \quad i = 10^{9.1} i_0.$$

So the earthquake that produced the tsunami had $10^{9.1}$ (about 1.26 *billion*) times more ground motion than the zero earthquake.

**(b)** What is the Richter-scale intensity of an earthquake with 10 times as much ground motion as the 2004 tsunami earthquake?

**Solution** From (a), the ground motion of the tsunami quake was $10^{9.1} i_0$. So a quake with 10 times that motion would satisfy

$$i = 10(10^{9.1} i_0) = 10^1 \cdot 10^{9.1} i_0 = 10^{10.1} i_0.$$

Therefore, its Richter scale measure would be

$$R(i) = \log\left(\frac{i}{i_0}\right) = \log\left(\frac{10^{10.1} i_0}{i_0}\right) = \log 10^{10.1} = 10.1.$$

Thus, a tenfold increase in ground motion increases the Richter scale measure by only 1. ✓ 10

✓ **Checkpoint 10**

Find the Richter-scale intensity of an earthquake whose ground motion is 100 times greater than the ground motion of the 2004 tsunami earthquake discussed in Example 10.

# 4.4 Exercises

*Solve each logarithmic equation. (See Example 1.)*

**1.** $\ln(x + 3) = \ln(2x - 5)$

**2.** $\ln(8k - 7) - \ln(3 + 4k) = \ln(9/11)$

**3.** $\ln(3x + 1) - \ln(5 + x) = \ln 2$

**4.** $\ln(5x + 2) = \ln 4 + \ln(x + 3)$

**5.** $2 \ln(x - 3) = \ln(x + 5) + \ln 4$

**6.** $\ln(k + 5) + \ln(k + 2) = \ln 18k$

*Solve each logarithmic equation. (See Example 2.)*

**7.** $\log_3 (6x - 2) = 2$

**8.** $\log_5 (3x - 4) = 1$

**9.** $\log x - \log(x + 4) = -1$

**10.** $\log m - \log(m + 4) = -2$

**11.** $\log_3 (y + 2) = \log_3 (y - 7) + \log_3 4$

**12.** $\log_8 (z - 6) = 2 - \log_8 (z + 15)$

**13.** $\ln(x + 9) - \ln x = 1$

**14.** $\ln(2x + 1) - 1 = \ln(x - 2)$

**15.** $\log x + \log(x - 9) = 1$

**16.** $\log(x - 1) + \log(x + 2) = 1$

*Solve each equation for c.*

**17.** $\log(3 + b) = \log(4c - 1)$    **18.** $\ln(b + 7) = \ln(6c + 8)$

**19.** $2 - b = \log(6c + 5)$      **20.** $8b + 6 = \ln(2c) + \ln c$

**21.** Suppose you overhear the following statement: "I must reject any negative answer when I solve an equation involving logarithms." Is this correct? Write an explanation of why it is or is not correct.

**22.** What values of $x$ cannot be solutions of the following equation?

$$\log_a (4x - 7) + \log_a (x^2 + 4) = 0.$$

*Solve these exponential equations without using logarithms. (See Example 3.)*

**23.** $2^{x-1} = 8$            **24.** $16^{-x+2} = 8$

**25.** $25^{-3x} = 3125$       **26.** $81^{-2x} = 3^{x-1}$

**27.** $6^{-x} = 36^{x+6}$         **28.** $16^x = 64$

**29.** $\left(\dfrac{3}{4}\right)^x = \dfrac{16}{9}$       **30.** $2^{x^2 - 4x} = \dfrac{1}{16}$

*Use logarithms to solve these exponential equations. (See Examples 4–6.)*

**31.** $2^x = 5$             **32.** $5^x = 8$

**33.** $2^x = 3^{x-1}$         **34.** $4^{x+2} = 5^{x-1}$

**35.** $3^{1-2x} = 5^{x+5}$       **36.** $4^{3x-1} = 3^{x-2}$

**37.** $e^{3x} = 6$            **38.** $e^{-3x} = 5$

**39.** $2e^{5a+2} = 8$         **40.** $10e^{3z-7} = 5$

*Solve each equation for c.*

**41.** $10^{4c-3} = d$        **42.** $3 \cdot 10^{2c+1} = 4d$

**43.** $e^{2c-1} = b$          **44.** $3e^{5c-7} = b$

*Solve these equations. (See Examples 1–6.)*

**45.** $\log_7 (r + 3) + \log_7 (r - 3) = 1$

**46.** $\log_4 (z + 3) + \log_4 (z - 3) = 2$

**47.** $\log_3 (a - 3) = 1 + \log_3 (a + 1)$

**48.** $\log w + \log(3w - 13) = 1$

**49.** $\log_2 \sqrt{2y^2} - 1 = 3/2$     **50.** $\log_2 (\log_2 x) = 1$

**51.** $\log_2 (\log_3 x) = 1$       **52.** $\dfrac{\ln(2x + 1)}{\ln(3x - 1)} = 2$

**53.** $5^{-2x} = \dfrac{1}{25}$         **54.** $5^{x^2+x} = 1$

**55.** $2^{|x|} = 16$            **56.** $5^{-|x|} = \dfrac{1}{25}$

**57.** $2^{x^2-1} = 10$         **58.** $3^{2-x^2} = 8$

**59.** $2(e^x + 1) = 10$          **60.** $5(e^{2x} - 2) = 15$

**61.** Explain why the equation $4^{x^2+1} = 2$ has no solutions.

**62.** Explain why the equation $\log(-x) = -4$ does have a solution, and find that solution.

*Work these problems. (See Examples 6, 7, and 8.)*

**63.** **Foreign Earnings** The amount of foreign earnings (in trillions of dollars) kept overseas by S&P 500 companies can be approximated by the function

$$f(x) = .227(1.17)^x,$$

where $x = 5$ corresponds to the year 2005. Find the first full year in which the foreign earnings kept overseas exceeds the following amounts. (Data from: *The Wall Street Journal.*)

**(a)** $1.5 trillion        **(b)** $2.5 trillion

**64.** **Nursing Degrees** The number of registered nurses (in thousands) enrolled in bachelor of nursing (BSN) degree completion programs in the United States can be approximated by the function

$$g(x) = 21.8(1.14)^x,$$

where $x = 4$ corresponds to the year 2004. Find the first full year in which the number of enrollees first exceeded the following values. (Data from: *The Wall Street Journal.*)

**(a)** 100,000        **(b)** 140,000

**65.** **Veterans Benefits** The total expenditures on benefits for U.S. veterans (in billions of dollars) can be approximated by the function

$$h(x) = 23.5(1.08)^x,$$

where $x = 5$ corresponds to the year 1995. (Data from: U.S. Department of Veterans Affairs.)

**(a)** What were the amount for total expenditures in 2013?

**(b)** What was the first full year in which expenditures exceeded $110 billion?

**66.** **Snapchat Users** The total number of global daily users (in millions) of Snapchat can be approximated by the function

$$g(x) = 44.2(1.12)^x,$$

where $x = 1$ corresponds to the first quarter of the year 2014. (Data from: *The Wall Street Journal.*)

**(a)** Give the number of global daily users for the third quarter of 2015.

**(b)** What was the first full quarter in which the number of daily global users exceeded 120 million?

**67.** **Wind Energy** As we saw in Example 12 of Section 4.3, annual U.S. power generation from wind (in million megawatt-hours) can be approximated by the function

$$f(x) = -335.5 + 193.1 \ln x,$$

where $x = 6$ corresponds to the year 2006. (Data from: *The Wall Street Journal.*) Assuming that the model remains accurate, find the first full year in which the wind power exceeds the following values.

**(a)** 250 million megawatt-hours

**(b)** 300 million megawatt-hours

**68. Japanese Messaging** The number of daily users for Japan's Line messaging service (in millions) can be approximated by the function

$$g(x) = 53.1 + 63.1 \ln(x),$$

where $x = 1$ corresponds to the first quarter of 2013. Numbering quarters from that date, and assuming the model remains accurate, find the first full quarter in which the number of users exceeded 220 million.

**69. CVS Health Earnings per Share** The earnings per share for CVS Health (in dollars) can be approximated by the function

$$g(x) = -5.3 + 3.54 \ln x,$$

where $x = 7$ corresponds to the year 2007. (Data from: www.morningstar.com.)

(a) Find the earnings per share for the year 2010.

(b) Find the first full year in which the earnings per share exceeded \$3.25.

**70. Number of Teachers** The total number of teachers in the United States (in millions) can be approximated by the function

$$h(x) = .5 + .84 \ln x,$$

where $x = 10$ corresponds to the year 1990. (Data from: U.S. National Center for Education Statistics.)

(a) Find the number of teachers in the year 2013.

(b) Find the first full year in which the number of teachers exceeded 3.0 million.

**71. CVS Health Revenue** The total revenue for CVS Health (in billions of dollars) can be approximated by the function

$$f(x) = 41.6e^{.088x},$$

where $x = 7$ corresponds to the year 2007. (Data from: www.morningstar.com.)

(a) What was the total revenue in 2011?

(b) What was the first full year in which revenue exceeded \$140 billion?

**72. Best Buy Revenue** The total revenue for Best Buy Co. Inc., (in billions of dollars) can be approximated by the function

$$h(x) = 94.6e^{-.054x},$$

where $x = 12$ corresponds to the year 2012. (Data from: www.morningstar.com.)

(a) What was the total revenue in 2017?

(b) What was the first full year in which revenue was below \$40 billion?

**73. Twitter Stock** The stock price to earnings ratio for Twitter can be approximated by the function

$$g(x) = 22e^{-.23x},$$

where $x = 1$ corresponds to the first quarter of the year 2015. What was the first full quarter where the ratio was lower than 5.0? (Data from: www.morningstar.com.)

**74. Outstanding Loans** The percentage of loans outstanding as a share of deposits in U.S. banks can be approximated by the function

$$h(x) = 120.7e^{-.04x},$$

where $x = 7$ corresponds to the year 2007. What was the first full year in which the percentage of loans outstanding was first below 75%? (Data from: Federal Deposit Insurance Corp.)

*Work these exercises. (See Example 9.)*

**75. Natural Science** The amount of cobalt-60 (in grams) in a storage facility at time $t$ is given by

$$C(t) = 25e^{-.14t},$$

where time is measured in years.

(a) How much cobalt-60 was present initially?

(b) What is the half-life of cobalt-60? (*Hint:* For what value of $t$ is $C(t) = 12.5$?)

**76. Carbon Dating** A Native American mummy was found recently. It had 73.6% of the amount of radiocarbon present in living beings. Approximately how long ago did this person die?

**77. Carbon Dating** How old is a piece of ivory that has lost 36% of its radiocarbon?

**78. Carbon Dating** A sample from a refuse deposit near the Strait of Magellan had 60% of the carbon-14 of a contemporary living sample. How old was the sample?

*Work these problems. (See Example 10.)*

**79. Richter Scale** In May 2008, Sichuan province, China suffered an earthquake that measured 7.9 on the Richter scale.

(a) Express the intensity of this earthquake in terms of $i_0$.

(b) In July of the same year, a quake measuring 5.4 on the Richter scale struck Los Angeles. Express the intensity of this earthquake in terms of $i_0$.

(c) How many times more intense was the China earthquake than the one in Los Angeles?

**80. Richter Scale** Find the Richter-scale intensity of earthquakes whose ground motion is

(a) $1000i_0$;               (b) $100,000i_0$;

(c) $10,000,000i_0$.

(d) Fill in the blank in this statement: Increasing the ground motion by a factor of $10^k$ increases the Richter intensity by _____ units.

**81. Sound** The loudness of sound is measured in units called decibels. The decibel rating of a sound is given by

$$D(i) = 10 \cdot \log\left(\frac{i}{i_0}\right),$$

where $i$ is the intensity of the sound and $i_0$ is the minimum intensity detectable by the human ear (the so-called *threshold sound*). Find the decibel rating of each of the sounds with the given intensities. Round answers to the nearest whole number.

**(a)** Whisper, $115i_0$

**(b)** Average sound level in the movie *Godzilla*, $10^{10}i_0$

**(c)** Jackhammer, $31{,}600{,}000{,}000i_0$

**(d)** Rock music, $895{,}000{,}000{,}000i_0$

**(e)** Jetliner at takeoff, $109{,}000{,}000{,}000{,}000i_0$

**82. Sound** Using information from Exercise 81, answer the following.

**(a)** How much more intense is a sound that measures 100 decibels than the threshold sound?

**(b)** How much more intense is a sound that measures 50 decibels than the threshold sound?

**(c)** How much more intense is a sound measuring 100 decibels than one measuring 50 decibels?

✓ **Checkpoint Answers**

**1.** **(a)** 3      **(b)** $1 + \sqrt{2} \approx 2.414$

**2.** **(a)** 5      **(b)** 52

**3.** **(a)** 1/3      **(b)** 8/3      **(c)** $-1/4$

**4.** **(a)** 2.807      **(b)** 2.431      **(c)** 2.579

**5.** **(a)** 2.710      **(b)** .802

**6.** **(a)** 23.979      **(b)** $-3.113$      **(c)** $\pm 1.612$

**7.** Fourth quarter of 2016 ($x \approx 11.18$)

**8.** 2018 ($x \approx 17.18$)

**9.** About 8679 years

**10.** 11.1

# CHAPTER 4 Summary and Review

## Key Terms and Symbols

**4.1** exponential function
exponential growth and decay
the number
$e \approx 2.71828 \ldots$

**4.2** exponential growth function
logistic function

**4.3** $\log x$   common (base-10) logarithm of $x$
$\ln x$   natural (base-$e$) logarithm of $x$
$\log_a x$   base-$a$ logarithm of $x$

product, quotient, and power properties of logarithms
change-of-base theorem
Inverse logarithmic functions

**4.4** logarithmic equations
exponential equations
half-life
Richter scale

## Chapter 4 Key Concepts

**Exponentialic Functions**

An important application of exponents is the **exponential growth function,** defined as $f(t) = y_0 e^{kt}$ or $f(t) = y_0 b^t$, where $y_0$ is the amount of a quantity present at time $t = 0$, $e \approx 2.71828$, and $k$ and $b$ are constants.

**Logarithmic Functions**

The **logarithm** of $x$ to the base $a$ is defined as follows: For $a > 0$ and $a \neq 1$, $y = \log_a x$ means $a^y = x$. Thus, $\log_a x$ is an *exponent*, the power to which $a$ must be raised to produce $x$.

**Properties of Logarithms**

Let $x$, $y$, and $a$ be positive real numbers, with $a \neq 1$, and let $r$ be any real number. Then

$$\log_a 1 = 0; \quad \log_a a = 1;$$
$$\log_a a^r = r; \quad a^{\log_a x} = x.$$

**Product property** $\log_a xy = \log_a x + \log_a y$

**Quotient property** $\log_a \dfrac{x}{y} = \log_a x - \log_a y$

**Power property** $\log_a x^r = r \log_a x$

**Solving Exponential and Logarithmic Equations**

Let $a > 0$, with $a \neq 1$.
If $\log_a u = \log_a v$, then $u = v$.
If $a^u = a^v$, then $u = v$.

# Chapter 4 Review Exercises

*Match each equation with the letter of the graph that most closely resembles the graph of the equation. Assume that $a > 1$.*

**1.** $y = a^{x+2}$

**2.** $y = a^x + 2$

**3.** $y = -a^x + 2$

**4.** $y = a^{-x} + 2$

**(a)**

**(b)**

**(c)**

**(d)**

*Consider the exponential function $y = f(x) = a^x$ graphed here. Answer each question on the basis of the graph.*

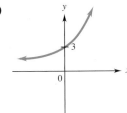

$y = f(x) = a^x$

**5.** What is true about the value of $a$ in comparison to 1?

**6.** What is the domain of $f$?

**7.** What is the range of $f$?

**8.** What is the value of $f(0)$?

*Graph each function.*

**9.** $f(x) = 4^x$

**10.** $g(x) = 4^{-x}$

**11.** $f(x) = \ln x + 5$

**12.** $g(x) = \log x - 3$

*Work these problems.*

**13. Exports to Mexico** The amount of U.S. exports to Mexico (in billions of dollars) can be approximated by the function

$$g(x) = 27.1(1.10)^x,$$

where $x = 0$ corresponds to the year 1990. (Data from: *The Wall Street Journal.*)

**(a)** Estimate the value of U.S. exports to Mexico in 2015.

**(b)** What was the first full year in which exports exceeded $150 billion in value according to this model?

**14. Royal Caribbean Share Price** The percent gain in share price for the cruise line Royal Caribbean can be approximated by the function

$$f(x) = .0007(2.29)^x,$$

where $x = 12$ corresponds to the year 2012. (Data from: *The Wall Street Journal.*)

**(a)** Find the percent gain in share price for the year 2013.

**(b)** According to this model, what was the first full year to have the share price percentage gain be greater than 100%?

*Translate each exponential statement into an equivalent logarithmic one.*

**15.** $10^{2.53148} = 340$

**16.** $5^4 = 625$

**17.** $e^{3.8067} = 45$

**18.** $7^{1/2} = \sqrt{7}$

*Translate each logarithmic statement into an equivalent exponential one.*

**19.** $\log 10{,}000 = 4$

**20.** $\log 26.3 = 1.4200$

**21.** $\ln 81.1 = 4.3957$

**22.** $\log_2 4096 = 12$

*Evaluate these expressions without using a calculator.*

**23.** $\ln e^5$

**24.** $\log \sqrt[3]{10}$

**25.** $10^{\log 8.9}$

**26.** $\ln e^{3t^2}$

**27.** $\log_8 2$

**28.** $\log_8 32$

*Write these expressions as a single logarithm. Assume all variables represent positive quantities.*

**29.** $\log 4x + \log 5x^5$

**30.** $4 \log u - 5 \log u^6$

**31.** $3 \log b - 2 \log c$

**32.** $7 \ln x - 3(\ln x^3 + 5 \ln x)$

*Solve each equation. Round to the nearest thousandth.*

**33.** $\ln(m + 8) - \ln m = \ln 3$

**34.** $2 \ln(y + 1) = \ln(y^2 - 1) + \ln 5$

**35.** $\log(m + 3) = 2$

**36.** $\log x^3 = 2$

**37.** $\log_2 (3k + 1) = 4$

**38.** $\log_5 \left( \dfrac{5z}{z - 2} \right) = 2$

**39.** $\log x + \log(x - 3) = 1$

**40.** $\log_2 r + \log_2 (r - 2) = 3$

**41.** $2^{3x} = \dfrac{1}{64}$

**42.** $\left(\dfrac{9}{16}\right)^x = \dfrac{3}{4}$

**43.** $9^{2y+1} = 27^y$

**44.** $\dfrac{1}{2} = \left(\dfrac{b}{4}\right)^{1/4}$

**45.** $8^p = 19$

**46.** $3^z = 11$

**47.** $5 \cdot 2^{-m} = 35$

**48.** $2 \cdot 15^{-k} = 18$

**49.** $e^{-5-2x} = 5$

**50.** $e^{3x-1} = 12$

**51.** $6^{2-m} = 2^{3m+1}$

**52.** $5^{3r-1} = 6^{2r+5}$

**53.** $(1 + .003)^k = 1.089$

**54.** $(1 + .094)^z = 2.387$

*Work these problems.*

**55. Population Growth** A population is increasing according to the growth law $y = 4e^{.03t}$, where $y$ is in thousands and $t$ is in months. Match each of the questions (a), (b), (c), and (d) with one of the solution (A), (B), (C), or (D).

**(a)** How long will it take for the population to double?

**(A)** Solve $4e^{.03t} = 12$ for $t$.

**(b)** When will the population reach 12 thousand?

**(B)** Evaluate $4e^{.03(72)}$.

**(c)** How large will the population be in 6 months?

**(C)** Solve $4e^{.03t} = 2 \cdot 4$ for $t$.

**(d)** How large will the population be in 6 years?

**(D)** Evaluate $4e^{.03(6)}$.

**56. Population Growth** A population is increasing according to the growth law $y = 2e^{.02t}$, where $y$ is in millions and $t$ is in years. Match each of the questions (a), (b), (c), and (d) with one of the solutions (A), (B), (C), or (D).

**(a)** How long will it take for the population to triple?

**(A)** Evaluate $2e^{.02(1/3)}$.

**(b)** When will the population reach 3 million?

**(B)** Solve $2e^{.02t} = 3 \cdot 2$ for $t$.

**(c)** How large will the population be in 3 years?

**(C)** Evaluate $2e^{.02(3)}$.

**(d)** How large will the population be in 4 months?

**(D)** Solve $2e^{.02t} = 3$ for $t$.

**57. Murders in Chicago** The number of murders in Chicago can be approximated by the function

$$g(x) = 846e^{-.034x},$$

where $x = 0$ corresponds to the year 1990. (Data from: Federal Bureau of Investigation.)

**(a)** What was the number of deaths according to this model in 2012?

**(b)** According to this model, what was the first full year in which the number of murders fell below 600?

**58. Natural Science** The amount of polonium (in grams) present after $t$ days is given by

$$A(t) = 10e^{-.00495t}.$$

**(a)** How much polonium was present initially?

**(b)** What is the half-life of polonium?

**(c)** How long will it take for the polonium to decay to 3 grams?

**59. Crude Oil Futures** The price per barrel (in dollars) for NYMEX-traded crude oil futures can be approximated by the function

$$g(x) = \dfrac{124}{1 + .118e^{.371x}},$$

where $x = 1$ corresponds to the first quarter of the year 2014. (Data from: *The Wall Street Journal.*)

**(a)** Find the futures price of crude oil for the first quarter of 2015.

**(b)** What quarter was the first full quarter in which the price of crude oil dropped below \$50?

**60. Coal Futures** The price of a metric ton of coal on the Newcastle futures market can be approximated by the function

$$h(x) = \dfrac{170}{1 + .009e^{.342x}},$$

where $x = 11$ corresponds to the year 2011. (Data from: *The Wall Street Journal.*)

**(a)** Find the futures price of coal for the year 2016.

**(b)** What was the first full year in which the price of coal fell below \$90 for a metric ton?

**61. Richter Scale** In April 2016, Ecuador suffered an earthquake that measured 7.8 on the Richter scale.

**(a)** Express the intensity of this earthquake in terms of $i_0$.

**(b)** That same month, a magnitude 6.0 earthquake struck offshore Mexico. Express the intensity of this earthquake in terms of $i_0$.

**(c)** How many times more intense was the earthquake in Ecuador than the one that struck offshore Mexico?

**62. Natural Science** One earthquake measures 4.6 on the Richter scale. A second earthquake has ground motion 1000 times greater than the first. What does the second one measure on the Richter scale?

**Natural Science** *Another form of Newton's law of cooling (see Section 4.2, Exercises 15 and 16) is $F(t) = T_0 + Ce^{-kt}$, where $C$ and $k$ are constants.*

**63.** A piece of metal is heated to 300° Celsius and then placed in a cooling liquid at 50° Celsius. After 4 minutes, the metal has cooled to 175° Celsius. Find its temperature after 12 minutes.

**64.** A frozen pizza has a temperature of 3.4° Celsius when it is taken from the freezer and left out in a room at 18° Celsius. After half an hour, its temperature is 7.2° Celsius. How long will it take for the pizza to thaw to 10° Celsius?

*In Exercises 65–68, do part (a) and skip part (b) if you do not have a graphing calculator. If you have a graphing calculator, then skip part (a) and do part (b).*

**65. FedEx Profit** The gross profit for FedEx (in billions of dollars) is given in the following table for the years 2011 to 2016. (Data from: *The Wall Street Journal.*)

| Year | Revenue (billions) |
|------|--------------------|
| 2011 | $25.0 |
| 2012 | $26.9 |
| 2013 | $27.8 |
| 2014 | $28.5 |
| 2015 | $30.5 |
| 2016 | $33.0 |

(a) Let $x = 0$ correspond to the year 2011. Use the data points from 2011 and 2016 to find a function of the form $f(x) = a(b^x)$ that models these data.

(b) Use exponential regression to find a function $g$ that models these data, with $x = 0$ corresponding to the year 2011.

(c) Assume the model remains accurate and estimate the gross profit for the year 2017.

(d) Assume the model remains accurate and find the first full year in which the gross profit will be higher than $40 billion.

66. **Starbucks App** The number of unique visitors (in millions) using the Starbucks app is given in the following table for 12 months starting in June 2014. (Data from: www.comscore.com.)

| Month | Average minutes |
|-------|-----------------|
| Jun 2014 | 10.0 |
| Jul 2014 | 10.4 |
| Aug 2014 | 10.2 |
| Sep 2014 | 11.2 |
| Oct 2014 | 10.1 |
| Nov 2014 | 11.6 |
| Dec 2014 | 11.8 |
| Jan 2015 | 13.1 |
| Feb 2015 | 14.4 |
| Mar 2015 | 15.7 |
| Apr 2015 | 14.9 |
| May 2015 | 15.9 |

(a) Let $x = 0$ correspond to June 2014. Use the data points from June 2014 and May 2015 to find a function of the form $f(x) = a(b^x)$ that models these data.

(b) Use exponential regression to find a function $h$ that models these data, with $x = 0$ corresponding to June 2014.

(c) Assume the model remains accurate and estimate the number of unique visitors using the app in June 2015.

(d) Assume the model remains accurate and find the first full month in which the number of unique visitors was higher than 18 million.

67. **Dunkin' Donuts App** The number of unique visitors (in millions) using the Dunkin' Donuts app is given in the following table for 12 months starting in June 2014. (Data from: www.comscore.com.)

| Month | Unique Visitors (millions) |
|-------|----------------------------|
| Jun 2014 | 4.9 |
| Jul 2014 | 4.9 |
| Aug 2014 | 5.6 |
| Sep 2014 | 7.1 |
| Oct 2014 | 6.9 |
| Nov 2014 | 6.0 |
| Dec 2014 | 6.8 |
| Jan 2015 | 6.7 |
| Feb 2015 | 6.9 |
| Mar 2015 | 6.2 |
| Apr 2015 | 6.0 |
| May 2015 | 6.6 |

(a) Let $x = 1$ correspond to June 2014. Use the data points from June 2014 and May 2015 to find a function of the form $f(x) = a + b \ln x$ that models these data. [*Hint:* Use (1, 4.9) to find $a$; then use (12, 6.6) to find $b$.]

(b) Use logarithmic regression to find a function $g$ that models these data, with $x = 1$ corresponding to June 2014.

(c) Assume the model remains accurate and estimate the number of unique visitors using the app in June 2015.

(d) Assume the model remains accurate and find the first full month in which the number of unique visitors was higher than 7.0 million.

68. **Bank Capital** The common equity capital ratio is given in the following table for the largest banks in the United States as a percentage of risk-weighted assets. (Data from: The Federal Reserve.)

| Year | Percent |
|------|---------|
| 2009 | 5.8 |
| 2010 | 8.1 |
| 2011 | 9.8 |
| 2012 | 10.2 |
| 2013 | 11.4 |
| 2014 | 12.1 |
| 2015 | 12.2 |

(a) Let $x = 1$ correspond to the year 2009. Use the data points from 2009 and 2015 to find a function of the form $f(x) = a + b \ln x$ that models these data. [*Hint:* See Exercise 67(a).]

(b) Use logarithmic regression to find a function $h$ that models these data, with $x = 9$ corresponding to the year 2009.

(c) Assume the model remains accurate and estimate the common equity capital ratio for the year 2016.

(d) Assume the model remains accurate and find the first full year in which the common equity capital ratio exceeds 13%.

# Case Study 4  Gapminder.org

The website www.gapminder.org is committed to displaying economic and health data for countries around the world in a visually compelling way. The goal is to make publicly available data easy for citizens of the world to understand so individuals and policy makers can comprehend trends and make data-driven conclusions. For example, the website uses the size of the bubble on a plot to indicate the population of the country, and the color indicates the region in which the country lies:

| | |
|---|---|
| Yellow | America |
| Orange | Europe and Central Asia |
| Green | Middle East and Northern Africa |
| Dark Blue | Sub-Saharan Africa |
| Red | East Asia and the Pacific |

Additionally, the website also uses animation to show how variables change over time.

Let's look at an example of two variables of interest as best we can on a printed page. We will use income per-person (in U.S. dollars and measured as gross domestic product per capita) as a measure of a country's wealth as the x-axis variable and life expectancy (in years) at birth as a measure of the health of a country on the y–axis. In Figure 4.24 we see a bubble for each country in the year 1800.

We see that all the countries are bunched together with quite low income per-person and average life expectancy ranging from 25–40 years. The two large bubbles you see are for the two most populous countries: China (red) and India (light blue). When you visit the website (please do so, it is very cool!) and press the "Play" button at the bottom left, you can see how income per-person and life expectancy change over time for each country. At its conclusion, you obtain Figure 4.25 on the following page, which shows the relationship between income per-person and average life expectancy in the year 2015.

We can see that a great deal has changed in 200 years! One great thing is that all countries now have average life expectancy

**Figure 4.24**   Relationship Between Income per-Person and Average Life Expectancy in 1800 from Gapminder Foundation. Reproduced by permission of Gapminder Foundation.

that is over 45 years. This is simply incredible. Also, we see there is a great range of income per-person values for a variety of countries. The country to the far right (the little orange dot) is the European country of Luxembourg. We see that many of the Sub-Sahara countries on the left (in dark blue) have the lowest per-person income and the lowest average life expectancy. The wealthier countries—

for example, countries whose per-person income is greater than $20,000—all have life expectancy over 65 years.

We can actually use the per-person income $x$ to predict the average life expectancy $f(x)$. Note that Gapminder.org has its default showing the income per-person on a logarithmic scale. When we see a linear pattern such as this where the x-axis is on a

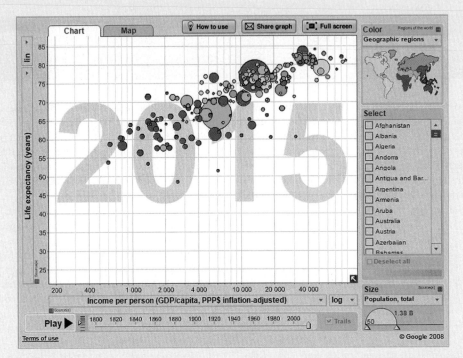

**Figure 4.25**    Relationship Between Income per-Person and Average Life Expectancy in 2015 from Gapminder Foundation. Reproduced by permission of Gapminder Foundation.

logarithmic scale and the $y$-axis is on the regular scale, a model of the form $f(x) = a + b \ln x$ often fits the data well. Since Gapminder.org makes all the data for their graphs available, we can fit a model to find values for $a$ and $b$ to use 2015 income per-person ($x$ = GDP per capita) to predict life expectancy ($y$ = years). Using regression, we obtain the following model:

$$f(x) = 23.97 + 5.221 \ln x. \tag{1}$$

**Figure 4.26**    Relationship Between Time and per Capita GDP in 2015 from Gapminder Foundation. Reproduced by permission of Gapminder Foundation.

**Figure 4.27**   Trajectory of United States from Gapminder Foundation. Reproduced by permission of Gapminder Foundation.

Gapminder.org will also show the economic development for an individual country. Figure 4.26 shows per capita GDP on the $y$-axis for the country of Brazil over time. We can see that economic gains began close to the year 1900. The graph shows per-person income on the $y$-axis (again on a logarithmic scale) and the trend seems approximately linear since 1900. When this is the case, a model of the form $a(b^x)$ usually fits the data well. Letting $x = 0$ correspond to the year 1900, we obtain the following:

$$g(x) = 1037(1.025)^x \qquad (2)$$

The trajectory of the United States (Figure 4.27) is a bit different, with rising incomes starting in the early 1800s and continuing to the present. Notice the dips that occur in the mid-1930s (the Great Depression) and the early 1940s (World War II). Fitting a model to the data for the United States (again with $x = 0$ corresponding to the year 1900) yields the model:

$$h(x) = 6260(1.019)^x \qquad (3)$$

## Exercises

For Exercises 1–6, use Equation **(1)** that provides a model using per-person income as the $x$-variable to predict life expectancy as the $y$-variable.

1. Create a graph of the function given in Equation **(1)** from $x = 1$ to 100,000.

2. Describe the shape of the graph.

3. The country furthest to the left in Figure 4.25 is the Central African Republic. In 2015, its income per-person was $599, and its life expectancy was 54. Suppose that economic development improved so that the Central African Republic has per-person

income level of $4000. What does the model of Equation **(1)** predict for life expectancy?

4. In 2015, Indonesia had per-person income of $10,500 and a life expectancy of 71 years. What does the model predict for life expectancy if the per-person income improved to $20,000?

5. According to the model, what per-person income level predicts a life expectancy of 60 years?

6. According to the model, what per-person income level predicts a life expectancy of 75 years?

For Exercises 7–10, use the model in Equation **(2)** that uses year (with $x = 0$ corresponding to the year 1900) to predict the per-person income for Brazil.

7. Graph the model of Equation **(2)** from $x = 0$ to 115.

8. Describe the shape of the model.

9. According to the model, what was the per-person income for Brazil in the year 1950? In the year 2000?

10. What is the first full year in which the model indicates that per-person income exceeded $4000?

For Exercises 11–14, use the model in Equation **(3)** that uses year (with $x = 0$ corresponding to the year 1900) to predict the per-person income for the United States.

11. Graph the model of Equation **(3)** from $x = 0$ to 115.

12. Describe the shape of the model.

13. According to the model, what was the per-person income for the United States in the year 1960? In the year 2010?

14. What is the first full year in which the model indicates that per-person income exceeded $30,000?

# ⊞ Extended Project

Go to the website www.gapminder.org and explore different relationships. Examine the relationship between a variety of variables that look at relationships among health, climate, disasters, economy, education, family, global trends, HIV, and poverty. (*Hint:* When you have the graphic display showing, click on the button for "Open Graph Menu" to see a list of options in these areas all ready to be viewed.) When observing the different graphs, notice when the *x*-axis, the *y*-axis, or both are on a logarithmic scale.

1. Find a relationship that appears linear when the *x*-axis is on the logarithmic scale and the *y*-axis is not. Now click on the button to the right of the *x*-axis and observe the graph on a linear scale (rather than the logarithmic scale). Describe the shape.

2. Find a relationship that appears linear when the *x*-axis is not on the logarithmic scale, but the *y*-axis is on the logarithmic scale. Now click on the button above the *y*-axis and observe the graph on a linear scale (rather than the logarithmic scale). Describe the shape.

3. Find a relationship that appears linear when the *x*-axis is on the logarithmic scale, and the *y*-axis is on the logarithmic scale. Now click on the button to the right of the *x*-axis and click on the button above the *y*-axis and observe the graph on a linear scale (rather than the logarithmic scale) for both axes. Describe the shape.

# Mathematics of Finance

## 5 CHAPTER

### CHAPTER OUTLINE

We should all plan for eventual retirement, and such savings can include savings accounts and annuities. Most people also take out a loan for a big purchase, such as a car, a major appliance, or a house. People who carry a balance on their credit cards are in effect borrowing money. Loan payments must be calculated accurately, and it may take some mathematical work to find the best deal. In this chapter we examine the mathematics of interest and compounding, with examples in the savings and bond markets. We explore annuities where payments are made into and out of different kinds of financial tools. Last, we cover loan amortization in the context of auto loans and home mortgages.

It is important for both businesspersons and consumers to understand the mathematics of finance in order to make sound financial decisions. Interest formulas for borrowing and investing money are introduced in this chapter.

📝 **NOTE** We try to present realistic, up-to-date applications in this text. Because interest rates change so frequently, however, it is very unlikely that the rates in effect when this chapter was written are the same as the rates today when you are reading it. Fortunately, the mathematics of finance is the same regardless of the level of interest rates. So we have used a variety of rates in the examples and exercises. Some will be realistic and some won't by the time you see them—but all of them have occurred in the past several decades.

# 5.1 Simple Interest and Discount

**Interest** is the fee paid to use someone else's money. Interest on loans of a year or less is frequently calculated as **simple interest,** which is paid only on the amount borrowed or invested and not on past interest. The amount borrowed or deposited is called the **principal.** We can also call the principal the present value of the amount borrowed or deposited. Throughout this chapter we will use $PV$ to represent present value. The **rate** of interest is given as a percent per year, expressed as a decimal. For example, $6\% = .06$ and $11\frac{1}{2}\% = .115$. The **time** during which the money is accruing interest is calculated in years. Simple interest is the product of the principal (present value), rate, and time.

> ## Simple Interest
>
> The simple interest $I$ on $PV$ dollars at a rate of interest $r$ per year for $t$ years is
>
> $$I = PVrt.$$

It is customary in financial problems to round interest to the nearest cent.

**EXAMPLE 1** **Personal Loan** To furnish her new apartment, Maggie Chan borrowed $4000 at 3% interest from her parents for 9 months. How much interest will she pay?

**Solution** Use the formula $I = PVrt$, with $PV = 4000, r = 0.03$, and $t = 9/12 = 3/4$ years:

$$I = PVrt$$

$$I = 4000 * .03 * \frac{3}{4} = 90.$$

Thus, Maggie pays a total of $90 in interest. ✓₁

✓ **Checkpoint 1**

Find the simple interest for each loan.

**(a)** $2000 at 8.5% for 10 months

**(b)** $3500 at 10.5% for $1\frac{1}{2}$ years

*Answers to Checkpoint exercises are found at the end of the section.*

Simple interest is normally used for loans with a term of one year or less. A significant exception is the case of **corporate bonds** and similar financial instruments. When an investor buys a corporate bond, she is essentially lending money to the corporation that issues the bond. The corporation promises to return the principal back to the investor at a specified maturity date *and* pay the investor a specified interest rate, usually semiannually (twice a year).

To determine the amount of money paid to the investor twice a year, we use the interest formula $I = PVrt$, with $PV$ as the amount invested in the bond (principal or present value), $r$ as the interest rate as a decimal, and $t = \frac{1}{2}$ because the bond pays twice a year.

**EXAMPLE 2** **Corporate Bond** Bank of America issued 10-year bonds with a maturity date in 2022 and an annual simple interest rate of 5.7%. The interest is paid semiannually. Santiago buys a $10,000 bond. Determine the following. (Data from: www.finra.org.)

**(a)** How much interest will Santiago earn every six months?

**Solution** Use the simple interest formula $I = PVrt$ with $PV = 10,000, r = .057$, and $t = \frac{1}{2}$.

$$I = PVrt = (10,000)(.057)\left(\frac{1}{2}\right) = \$285.$$

**(b)** How much interest will he earn over the 10-year life of the bond?

**Solution**  Either use the simple interest formula with $t = 10$, that is,

$$I = PVrt = (10,000)(.057)(10) = \$5700,$$

or take the answer in part (a), which will be paid out every six months for 10 years for a total of 20 times. Thus, Santiago would obtain $\$285(20) = \$5700$. ✔️₂

## Future Value

When you make a deposit at simple interest rate $r$ for $t$ years, the **future value** (or **maturity value**) is the sum of the deposit and the interest it has earned. In other words, the future value ($FV$) is the sum of the deposit (present value $PV$) and the interest $I$:

$$FV = PV + I$$
$$= PV + PVrt \qquad I = PVrt.$$
$$= PV(1 + rt). \qquad \text{Factor out } PV.$$

The following box summarizes this result.

> ## Future Value (or Maturity Value) for Simple Interest
>
> The future value (maturity value) $FV$ of $PV$ dollars for $t$ years at interest rate $r$ per year is
>
> $$FV = PV + I, \qquad \text{or} \qquad FV = PV(1 + rt).$$

**EXAMPLE 3**    **Personal Loan**  Find each maturity value and the amount of interest paid.

**(a)** Hai borrows $20,000 from his parents at 5.25% to add a room onto his house. He plans to repay the loan in 9 months with a bonus he expects to receive at that time.

**Solution**  The loan is for 9 months, or 9/12 of a year, so $t = .75$, $PV = 20,000$, and $r = .0525$. Use the formula to obtain

$$FV = PV(1 + rt)$$
$$= 20,000[1 + .0525(.75)]$$
$$= 20,787.50, \qquad \text{Use a calculator.}$$

or $20,787.50. The maturity value $FV$ is the sum of the principal $PV$ and the interest $I$, that is, $FV = PV + I$. To find the amount of interest paid, rearrange this equation:

$$I = FV - PV$$
$$I = \$20,787.50 - \$20,000 = \$787.50.$$

**(b)** A loan of $11,280 for 85 days at 9% interest.

**Solution**  Use the formula $FV = PV(1 + rt)$, with $PV = 11,280$ and $r = .09$. Unless stated otherwise, we assume a 365-day year, so the period in years is $t = 85/365$. The maturity value is

$$FV = PV(1 + rt)$$
$$FV = 11,280\left(1 + .09 * \frac{85}{365}\right)$$
$$\approx 11,280(1.020958904) \approx \$11,516.42.$$

As in part (a), the interest is

$$I = FV - PV = \$11,516.42 - \$11,280 = \$236.42. \quad ✔️₃$$

✔️**Checkpoint 2**

For the given bonds, find the semiannual interest payment and the total interest paid over the life of the bond.

**(a)** $7500 Time Warner Cable, Inc. 30-year bond at 7.3% annual interest.

**(b)** $15,000 Clear Channel Communications 10-year bond at 9.0% annual interest.

✔️**Checkpoint 3**

Find each future value.

**(a)** $1000 at 4.6% for 6 months

**(b)** $8970 at 11% for 9 months

**(c)** $95,106 at 9.8% for 76 days

**EXAMPLE 4**    **Personal Loan** Sally borrows $15,000 and is required to pay $15,315 in 4 months to pay off the loan and interest. What is the simple interest rate?

**Solution** One way to find the rate is to solve for $r$ in the future-value formula when $PV = 15,000$, $FV = 15,315$, and $t = 4/12 = 1/3$:

$$PV(1 + rt) = FV$$

$$15,000\left(1 + r * \frac{1}{3}\right) = 15,315$$

$$15,000 + \frac{15,000r}{3} = 15,315 \qquad \text{Multiply out left side.}$$

$$\frac{15,000r}{3} = 315 \qquad \text{Subtract 15,000 from both sides.}$$

$$15,000r = 945 \qquad \text{Multiply both sides by 3.}$$

$$r = \frac{945}{15,000} = .063. \qquad \text{Divide both sides by 15,000.}$$

Therefore, the interest rate is 6.3%. ✓4

✓ **Checkpoint 4**

You lend a friend $500. She agrees to pay you $520 in 6 months. What is the interest rate?

## Present Value

In interest problems, $PV$ always represents the amount at the beginning of the period, and $FV$ always represents the amount at the end of the period. To find a formula for $PV$, we begin with the future-value formula:

$$FV = PV(1 + rt).$$

Dividing each side by $1 + rt$ gives the following formula for the present value.

> ### Present Value for Simple Interest
>
> The **present value** $PV$ of a future amount of $FV$ dollars at a simple interest rate $r$ for $t$ years is
>
> $$PV = \frac{FV}{1 + rt}.$$

**EXAMPLE 5**    **Present Value** Find the present value of $32,000 in 4 months at 9% interest.

**Solution**

$$PV = \frac{FV}{1 + rt} = \frac{32,000}{1 + (.09)\left(\frac{4}{12}\right)} = \frac{32,000}{1.03} \approx 31,067.96.$$

A deposit of $31,067.96 today at 9% interest would produce $32,000 in 4 months. These two sums, $31,067.96 today and $32,000.00 in 4 months, are equivalent (at 9%) because the first amount becomes the second amount in 4 months. ✓5

✓ **Checkpoint 5**

Find the present value of the given future amounts. Assume 6% interest.

(a)  $7500 in 1 year

(b)  $89,000 in 5 months

(c)  $164,200 in 125 days

**EXAMPLE 6**    **Court Settlement** Because of a court settlement, Jalinda owes $5000 to Madison. The money must be paid in 10 months, with no interest. Suppose Jalinda wants to pay the money today and that Madison can invest it at an annual rate of 5%. What amount should Madison be willing to accept to settle the debt?

**Solution** The $5000 is the future value in 10 months. So Madison should be willing to accept an amount that will grow to $5000 in 10 months at 5% interest. In other words, she should accept an amount that is the present value of $5000 under these circumstances. Use the present-value formula with $FV = 5000$, $r = .05$, and $t = 10/12 = 5/6$:

$$PV = \frac{FV}{1 + rt} = \frac{5000}{1 + .05 * \dfrac{5}{6}} = 4800.$$

✓ **Checkpoint 6**

Jerrell Davis is owed $19,500 by Christine O'Brien. The money will be paid in 11 months, with no interest. If the current interest rate is 10%, how much should Davis be willing to accept today in settlement of the debt?

Madison should be willing to accept $4800 today in settlement of the debt.

---

**EXAMPLE 7**    **Selling a Note** Larry Parks owes $6500 to Virginia Donovan. The loan is payable in one year at 6% interest. Virginia needs cash to pay medical bills, so four months before the loan is due, she sells the note (loan) to the bank. If the bank wants a return of 9% on its investment, how much should it pay Virginia for the note?

**Solution** First find the maturity value of the loan—the amount (with interest) that Larry must pay Virginia:

$$
\begin{aligned}
FV &= PV(1 + rt) && \text{Maturity-value formula}\\
&= 6500(1 + .06 * 1) && \text{Let } PV = 6500, r = .06, \text{ and } t = 1.\\
&= 6500(1.06) = \$6890.
\end{aligned}
$$

In four months, the bank will receive $6890. Since the bank wants a 9% return, compute the present value of this amount at 9% for four months:

$$
PV = \frac{FV}{1 + rt} \qquad\qquad \text{Present-value formula}
$$

$$
= \frac{6890}{1 + .09\left(\dfrac{4}{12}\right)} = \$6689.32. \qquad \text{Let } FV = 6890, r = .09, \text{ and } t = 4/12.
$$

✓ **Checkpoint 7**

A firm accepts a $21,000 note due in 8 months, with interest of 10.5%. Two months before it is due, the firm sells the note to a broker. If the broker wants a 12.5% return on his investment, how much should he pay for the note?

The bank pays Virginia $6689.32 and four months later it is worth $6890.

## Discount

The preceding examples dealt with loans in which money is borrowed and simple interest is charged. For most simple interest loans, both the principal (amount borrowed) and the interest are paid at the end of the loan period. With a corporate bond (which is a loan to a company by the investor who buys the bond), interest is paid during the life of the bond and the principal is paid back at maturity. In both cases,

the borrower receives the principal,

but pays back the principal *plus* the interest.

In a **simple discount loan,** however, the interest is deducted in advance from the amount of the loan and the *balance* is given to the borrower. The *full value* of the loan must be paid back at maturity. Thus,

the borrower receives the principal *less* the interest,

but pays back the principal.

The most common examples of simple discount loans are U.S. Treasury bills (T-bills), which are essentially short-term loans to the U.S. government by investors. T-bills are sold at a **discount** from their face value and the Treasury pays back the face value of the T-bill at maturity. The discount amount is the interest deducted in advance from the face value. The Treasury receives the face value less the discount, but pays back the full face value. If you are interested in more information on purchasing T-bills, visit www.treasurydirect.gov.

### EXAMPLE 8    Treasury Bills An investor bought a 6-month $10,000 U.S. Treasury bill on July 14, 2016, that sold at a discount rate of .39%. What is the amount of the discount? What is the price of the T-bill? (Data from: www.treasurydirect.gov.)

**Solution** The discount rate on a T-bill is always a simple annual interest rate. Consequently, the discount (interest) is found with the simple interest formula, using $PV = 8000$ (face value), $r = .0039$ (discount rate), and $t = .5$ (because 6 months is half a year):

$$\text{Discount} = PVrt = 10,000 * .0039 * .5 = \$19.50.$$

So the price of the T-bill is:

$$\text{Face Value} - \text{Discount} = 10,000 - 19.50 = \$9980.50. \quad \checkmark_8$$

✓ **Checkpoint 8**

The maturity times and discount rates for $15,000 T-bills sold on June 23, 2016, are given. Find the discount amount and the price of each T-bill.

**(a)** one year; .56%

**(b)** six months; .34%

**(c)** three months; .26%

In a simple discount loan such as a T-bill, the discount rate is not the actual interest rate that the borrower pays. In Example 8, the discount rate of .39% was applied to the face value of $10,000 rather than the $9980.50 that the Treasury (the borrower) received.

### EXAMPLE 9    Treasury Bills Find the actual interest rate paid by the Treasury in Example 8.

**Solution** Use the formula for simple interest, $I = PVrt$, with $r$ as the unknown. Here, $PV = 9980.50$ (the amount the Treasury received) and $I = 19.50$ (the discount amount). Because this is a 6-month T-bill, $t = .5$, and we have

$$I = PVrt$$
$$19.50 = 9980.50(r)(.5)$$
$$19.50 = 4990.25r \qquad \text{Multiply the terms on the right side.}$$
$$r = \frac{19.5}{4990.25} \approx .0039076. \qquad \text{Divide both sides by 4990.25.}$$

✓ **Checkpoint 9**

Find the actual interest rate paid by the Treasury for each T-bill in Checkpoint 8.

The actual interest rate is then .39076%. $\checkmark_9$

# 5.1 Exercises

*Unless stated otherwise, "interest" means simple interest, and "interest rate" and "discount rate" refer to annual rates. Assume 365 days in a year.*

1. What factors determine the amount of interest earned on a fixed principal?

*Find the interest on each of these loans. (See Example 1.)*

2. $35,000 at 6% for 9 months

3. $2850 at 7% for 8 months

4. $1875 at 5.3% for 7 months

5. $3650 at 6.5% for 11 months

6. $5160 at 7.1% for 58 days

7. $2830 at 8.9% for 125 days

8. $8940 at 9%; loan made on May 7 and due September 19

9. $5328 at 8%; loan made on August 16 and due December 30

10. $7900 at 7%; loan made on July 7 and due October 25

**Corporate Bonds** *For each of the given corporate bonds, whose interest rates are provided, find the semiannual interest payment and the total interest earned over the life of the bond. (See Example 2, Data from: www.finra.org.)*

11. $5000 IBM, 30-year bond; 6.220%

12. $9000 Barrick Gold Corp., 10-year bond; 6.800%

13. $12,500 Morgan Stanley, 10-year bond; 4.100%

14. $4500 Goldman Sachs, 7-year bond; 2.134%

15. $6500 Amazon.com Corp, 5-year bond; 1.200%

16. $10,000 Wells Fargo, 12-year bond; 4.000%

*Find the future value of each of these loans. (See Example 3.)*

17. $12,000 loan at 3.5% for 3 months

18. $3475 loan at 7.5% for 6 months

19. $6500 loan at 5.25% for 8 months

20. $24,500 loan at 9.6% for 10 months

**21.** What is meant by the *present value* of money?

**22.** In your own words, describe the *maturity value* of a loan.

*Find the present value of each future amount. (See Examples 5 and 6.)*

**23.** $15,000 for 9 months; money earns 6%

**24.** $48,000 for 8 months; money earns 5%

**25.** $15,402 for 120 days; money earns 6.3%

**26.** $29,764 for 310 days; money earns 7.2%

**Treasury Bills** *The given U.S. Treasury bills were sold on July 1, 2016. Find (a) the price of the T-bill, and (b) the actual interest rate paid by the Treasury. (See Examples 8 and 9. Data from: www.treasury.gov.)*

**27.** Three-month $20,000 T-bill with discount rate of .27%

**28.** One-month $12,750 T-bill with discount rate of .23%

**29.** Six-month $15,500 T-bill with discount rate of .36%

**30.** One-year $7000 T-bill with discount rate of .44%

**Treasury Bills** *Historically, treasury bills offered higher rates. On January 2, 2007, the discount rates were substantially higher than in July 2016. For the following treasury bills bought in 2007, find (a) the price of the T-bill, and (b) the actual interest rate paid by the Treasury. (See Examples 8 and 9. Data from: www.treasury.gov.)*

**31.** Three-month $20,000 T-bill with discount rate of 5.07%

**32.** One-month $12,750 T-bill with discount rate of 4.74%

**33.** Six-month $15,500 T-bill with discount rate of 5.10%

**34.** Six-month $9000 T-bill with discount rate of 5.10%

*Work the following applied problems.*

**35. Tax Penalty** An accountant for a corporation forgot to pay the firm's income tax of $650,000 on time. The government charged a penalty of 5.2% simple interest for the 30 days that the money was late. Find the total amount (tax and penalty) that was paid.

**36. Family Loan** Wilma borrowed $12,000 from her uncle to buy a used car. Her uncle charged simple interest at 3.5% for 5 years. Find the total amount (principal and interest) that Wilma paid to her uncle.

**37. Savings** Gabriel Branson invested his summer earnings of $3000 in a savings account for college. The account pays 1.5% simple interest. What will be the value of this account after 2 years?

**38. Savings** Jada received an inheritance from her grandmother of $7500. She places the money in savings account paying simple interest at 1.8%. What will be the value of the account after 3.5 years?

**39. Money Market Account** An account invested in a money market fund grew from $67,081.20 to $67,359.39 in a month. What was the interest rate, to the nearest tenth?

**40. CD** A $100,000 certificate of deposit held for 60 days is worth $100,560.00. Assuming simple interest, what interest rate, to the nearest tenth of a percent, was earned?

**41. Student Loan** Danielle borrowed $7500 from her grandfather with simple interest of 7%. She eventually repaid $7675 (principal and interest). What was the time period of the loan?

**42. Auto Loan** Kiara received a $10,000 loan from her aunt to buy a car. Her aunt charged her 2.5% interest. How long was the term of the loan if the interest accrued was $125?

**43. Construction Payment** John Sun Yee will need $6000 to pay for remodeling work on his house. A contractor agrees to do the work in 10 months. What is the present value of $6000 at 3.6% simple interest to accumulate the $6000 in 10 months?

**44. Tuition Payment** Mariana will need to pay a tuition bill of $7500 for her daughter in 10 months. What dollar amount can she deposit today at 1.25% simple interest to have enough to pay the tuition bill?

**45. Computing Hardware Payment** An accounting firm has ordered 50 new computers at a cost of $1704 each. The machines will not be delivered for 7 months. What amount could the firm deposit today in an account paying 3.42% simple interest to have enough to pay for the machines in 7 months?

**46. Car Dealership** A car dealership has ordered 12 new vehicles from the auto manufacturer. The total amount of the purchase is $420,000. The vehicles will be delivered in three months. What amount could the dealership deposit today in an account paying 4.1% simple interest to have enough money to pay for the vehicles in three months?

**47. Treasury Bill** A six-month $4000 T-bill sold for $3930. What was the discount rate?

**48. Treasury Bill** A one-year $9600 T-bill sold for $9403. What was the discount rate?

**49. Treasury Bill** A three-month $7600 T-bill carries a discount of $80.75. What is the discount rate for this T-bill?

**50. Treasury Bill** A six-month $12,500 T-bill carries a discount of $120.25. What is the discount rate for this T-bill?

**Interest Rate** *Work the next set of problems, in which you are to find the annual simple interest rate. Consider any fees, dividends, or profits as part of the total interest.*

**51.** A stock that sold for $22 at the beginning of the year was selling for $24 at the end of the year. If the stock paid a dividend of $.50 per share, what is the simple interest rate on an investment in this stock? (*Hint:* Consider the interest to be the increase in value plus the dividend.)

**52. Loan** Jerry Ryan borrowed $8000 for nine months at an interest rate of 7%. The bank also charges a $100 processing fee. What is the actual interest rate for this loan?

**53. Refund Loan** You are due a tax refund of $760. Your tax preparer offers you a no-interest loan to be repaid by your refund check, which will arrive in four weeks. She charges a $60 fee for this service. What actual interest rate will you pay for this loan? (*Hint:* The time period of this loan is not 4/52, because a 365-day year is 52 weeks and 1 day. So use days in your computations.)

**54. Refund Loan** Your cousin is due a tax refund of $400 in six weeks. His tax preparer has an arrangement with a bank to get him the $400 now. The bank charges an administrative fee of $29 plus interest at 6.5%. What is the actual interest rate for this loan? (See the hint for Exercise 53.)

**Finance** *Work these problems. (See Example 7.)*

55. **Selling a Note** A building contractor gives a $13,500 promissory note to a plumber who has loaned him $13,500. The note is due in nine months with interest at 9%. Three months after the note is signed, the plumber sells it to a bank. If the bank gets a 10% return on its investment, how much will the plumber receive? Will it be enough to pay a bill for $13,650?

56. **Selling a Note** Shalia Johnson owes $7200 to the Eastside Music Shop. She has agreed to pay the amount in seven months at an interest rate of 10%. Two months before the loan is due, the store needs $7550 to pay a wholesaler's bill. The bank will buy the note, provided that its return on the investment is 11%. How much will the store receive? Is it enough to pay the bill?

57. **Future Value** Let $y_1$ be the future value after $t$ years of $100 invested at 8% annual simple interest. Let $y_2$ be the future value after $t$ years of $200 invested at 3% annual simple interest.

   (a) Think of $y_1$ and $y_2$ as functions of $t$ and write the rules of these functions.

   (b) Without graphing, describe the graphs of $y_1$ and $y_2$.

   (c) Verify your answer to part (b) by graphing $y_1$ and $y_2$ in the first quadrant.

   (d) What do the slopes and $y$-intercepts of the graphs represent (in terms of the investment situation that they describe)?

58. If $y = 16.25t + 250$ and $y$ is the future value after $t$ years of $P$ dollars at interest rate $r$, what are $P$ and $r$? (*Hint:* See Exercise 57.)

✓**Checkpoint Answers**

1. (a) $141.67          (b) $551.25

2. (a) $273.75; $16,425    (b) $675; $13,500

3. (a) $1023           (b) $9710.03        (c) $97,046.68

4. 8%

5. (a) $7075.47         (b) $86,829.27       (c) $160,893.96

6. $17,862.60

7. $22,011.43

8. (a) $84; $14,916      (b) $25.50; $14,974.50
   (c) $9.75; $14,990.25

9. (a) About .56314%    (b) About .34058%
   (c) About .26017%

# 5.2  Compound Interest

With annual simple interest, you earn interest each year on your original investment. With annual **compound interest,** however, you earn interest both on your original investment *and* on any previously earned interest. To see how this process works, suppose you deposit $1000 at 5% annual interest. The following chart shows how your account would grow with both simple and compound interest:

| End of Year | SIMPLE INTEREST | | COMPOUND INTEREST | |
| | Interest Earned | Balance | Interest Earned | Balance |
| --- | --- | --- | --- | --- |
| | *Original Investment:* $1000 | | *Original Investment:* $1000 | |
| 1 | 1000(.05) = $50 | $1050 | 1000(.05) = $50 | $1050 |
| 2 | 1000(.05) = $50 | $1100 | 1050(.05) = $52.50 | $1102.50 |
| 3 | 1000(.05) = $50 | $1150 | 1102.50(.05) = $55.13* | $1157.63 |

✓**Checkpoint 1**

Extend the chart in the text by finding the interest earned and the balance at the end of years 4 and 5 for **(a)** simple interest and **(b)** compound interest.

As the chart shows, simple interest is computed each year on the original investment, but compound interest is computed on the entire balance at the end of the preceding year. So simple interest always produces $50 per year in interest, whereas compound interest produces $50 interest in the first year and increasingly larger amounts in later years (because you earn interest on your interest). ✓₁

*Rounded to the nearest cent.

**EXAMPLE 1**   **Savings Account** If $7000 is deposited in a savings account that pays 4% interest compounded annually, how much money is in the account after nine years?

**Solution** After one year, the account balance is

$$7000 + 4\% \text{ of } 7000 = 7000 + (.04)7000$$
$$= 7000(1 + .04) \qquad \text{Distributive property}$$
$$= 7000(1.04) = \$7280.$$

The initial balance has grown by a factor of 1.04. At the end of the second year, the balance is

$$7280 + 4\% \text{ of } 7280 = 7280 + (.04)7280$$
$$= 7280(1 + .04) \qquad \text{Distributive property}$$
$$= 7280(1.04) = 7571.20.$$

Once again, the balance at the beginning of the year has grown by a factor of 1.04. This is true in general: If the balance at the beginning of a year is $PV$ dollars, then the balance at the end of the year is

$$PV + 4\% \text{ of } PV = PV + .04PV$$
$$= PV(1 + .04)$$
$$= PV(1.04).$$

So the account balance grows like this:

| Year 1 | Year 2 | Year 3 |
|--------|--------|--------|

$$7000 \rightarrow 7000(1.04) \rightarrow \underbrace{[7000(1.04)](1.04)}_{7000(1.04)^2} \rightarrow \underbrace{[7000(1.04)(1.040)](1.04)}_{7000(1.04)^3} \rightarrow \cdots.$$

At the end of nine years, the balance is

$$7000(1.04)^9 = \$9963.18 \qquad \text{(rounded to the nearest penny)}.$$

The argument used in Example 1 applies in the general case and leads to this conclusion.

## Compound Interest

If $PV$ dollars are invested at interest rate $i$ per period, then the **compound amount** (future value) $FV$ after $n$ compounding periods is

$$FV = PV(1 + i)^n.$$

In Example 1, for instance, we had $PV = 7000$, $n = 9$, and $i = .04$ (so that $1 + i = 1 + .04 = 1.04$).

**NOTE** Compare this future value formula for compound interest with the one for simple interest from the previous section, using $t$ as the number of years:

| Compound interest | $FV = PV(1 + r)^t$; |
|-------------------|---------------------|
| **Simple interest** | $FV = PV(1 + rt)$. |

The important distinction between the two formulas is that, in the compound interest formula, the number of years, $t$, is an *exponent*, so that money grows much more rapidly when interest is compounded.

**EXAMPLE 2**    **Savings Account** Suppose $1000 is deposited for six years in a savings account paying 8.31% per year compounded annually.

**(a)** Find the compound amount.

**Solution** In the formula above, $PV = 1000$, $i = .0831$, and $n = 6$. The compound amount is

$$FV = PV(1 + i)^n$$
$$FV = 1000(1.0831)^6$$
$$FV = \$1614.40.$$

**(b)** Find the amount of interest earned.

**Solution** Subtract the initial deposit from the compound amount:

$$\text{Amount of interest} = \$1614.40 - \$1000 = \$614.40. \quad \checkmark_2$$

✓ **Checkpoint 2**

Suppose $17,000 is deposited at 4% compounded annually for 11 years.

**(a)** Find the compound amount.

**(b)** Find the amount of interest earned.

**TECHNOLOGY TIP**

Spreadsheets are ideal for performing financial calculations. Figure 5.1 shows a Microsoft Excel spreadsheet with the results when $1000 is invested at an annual rate of 10% using compound and simple interest for 20 years in Columns B and C, respectively. The blue balloons show the Excel 2016 formulas for generating the results in Columns B and C that would actually be typed into Columns B and C. Notice how rapidly the compound amount increases compared to the maturity value with simple interest. For more details on the use of spreadsheets in the mathematics of finance, see the *Spreadsheet Manual* that is available with this text.

| | A | B | C |
|---|---|---|---|
| 1 | Period | Compound | Simple |
| 2 | 0 | $1000.00 | $1000.00 |
| 3 | 1 | $1100.00 | $1100.00 |
| 4 | 2 | $1210.00 | $1200.00 |
| 5 | 3 | $1331.00 | $1300.00 |
| 6 | 4 | $1464.10 | $1400.00 |
| 7 | 5 | $1610.51 | $1500.00 |
| 8 | 6 | $1771.56 | $1600.00 |
| 9 | 7 | $1948.72 | $1700.00 |
| 10 | 8 | $2143.59 | $1800.00 |
| 11 | 9 | $2357.95 | $1900.00 |
| 12 | 10 | $2593.74 | $2000.00 |
| 13 | 11 | $2853.12 | $2100.00 |
| 14 | 12 | $3138.43 | $2200.00 |
| 15 | 13 | $3452.27 | $2300.00 |
| 16 | 14 | $3797.50 | $2400.00 |
| 17 | 15 | $4177.25 | $2500.00 |
| 18 | 16 | $4594.97 | $2600.00 |
| 19 | 17 | $5054.47 | $2700.00 |
| 20 | 18 | $5559.92 | $2800.00 |
| 21 | 19 | $6115.91 | $2900.00 |
| 22 | 20 | $6727.50 | $3000.00 |

=B2+.10*B2

=C2+100

**Figure 5.1**

**EXAMPLE 3**    **Savings Account** If a $16,000 savings account grows to $50,000 in 18 years, what is the interest rate (assuming annual compounding)?

**Solution** Use the compound interest formula, with $PV = 16,000$, $FV = 50,000$, and $n = 18$, and solve for $i$:

$$PV(1 + i)^n = FV$$
$$16,000(1 + i)^{18} = 50,000$$
$$(1 + i)^{18} = \frac{50,000}{16,000} = 3.125 \qquad \text{Divide both sides by 16,000.}$$
$$\sqrt[18]{(1 + i)^{18}} = \sqrt[18]{3.125} \qquad \text{Take 18th roots on both sides.}$$
$$1 + i = \sqrt[18]{3.125}$$
$$i = \sqrt[18]{3.125} - 1 \approx .06535. \qquad \text{Subtract 1 from both sides.}$$

So the interest rate is about 6.535%.

Interest can be compounded more than once a year. Common **compounding periods** include

> *semiannually* (2 periods per year),
>
> *quarterly* (4 periods per year),
>
> *monthly* (12 periods per year), and
>
> *daily* (usually 365 periods per year).

When the annual interest $i$ is compounded $m$ times per year, the interest rate per period is understood to be $i/m$.

---

**EXAMPLE 4**  **Money Market Accounts** In July 2016, bankrate.com advertised a money market account at Ally Bank at an interest rate of .85%. Find the value of the account if $10,000 is invested for one year and interest is compounded according to the periods given below.

**(a)** Annually

**Solution** Apply the formula $FV = PV(1 + i)^n$ with $PV = 10{,}000$, $i = .0085$, and $n = 1$:

$$FV = PV(1 + i)^n = 10{,}000(1 + .0085)^1 = 10{,}000(1.0085) = \$10{,}085.$$

**(b)** Semiannually

**Solution** Use the same formula and value of $PV$. Here, interest is compounded twice a year, so the number of periods is $n = 2$, and the interest rate per period is $i = \dfrac{.0085}{2}$:

$$FV = PV(1 + i)^n = 10{,}000\left(1 + \frac{.0085}{2}\right)^2 = \$10{,}085.18.$$

**(c)** Quarterly

**Solution** Proceed as in part (b), but the new interest is compounded every 3 months for a total of 4 times a year. Thus, $n = 4$, and the interest rate per period is $i = \dfrac{.0085}{4}$:

$$FV = PV(1 + i)^n = 10{,}000\left(1 + \frac{.0085}{4}\right)^4 = \$10{,}085.27.$$

**(d)** Monthly

**Solution** Interest is compounded 12 times a year, so $n = 12$ and $i = \dfrac{.0085}{12}$:

$$FV = PV(1 + i)^n = 10{,}000\left(1 + \frac{.0085}{12}\right)^{12} = \$10{,}085.33.$$

**(e)** Daily

**Solution** Interest is compounded 365 times a year, so $n = 365$ and $i = \dfrac{.0085}{365}$:

$$FV = PV(1 + i)^n = 10{,}000\left(1 + \frac{.0085}{365}\right)^{365} = \$10{,}085.36.$$

✓ **Checkpoint 3**

Find the future value for these money market accounts.

**(a)** Sallie Mae: $1000 at 1.60% compounded monthly for 3 years.

**(b)** Discover Bank: $2500 at .8% compounded daily for 9 months (assume 30 days in each month).

Example 4 shows that the more often interest is compounded, the larger is the amount of interest earned. Since interest is rounded to the nearest penny, however, there is a limit on how much can be earned. In Example 4, part (e), for instance, that limit of $10,1085.36 has been reached. Nevertheless, the idea of compounding more and more frequently leads to a method of computing interest called **continuous compounding** that is used in certain financial situations. The formula for continuous compounding is given in the following box where $e = 2.7182818\ldots$, which was introduced in Chapter 4.

## Continuous Compound Interest

The compound amount $FV$ for a deposit of $PV$ dollars at an interest rate $r$ per year compounded continuously for $t$ years is given by

$$FV = PVe^{rt}.$$

### EXAMPLE 5   Savings Account

Suppose that $5000 is invested in a savings account at an annual interest rate of 3.1% compounded continuously for 4 years. Find the compound amount.

**Solution** In the formula for continuous compounding, let $PV = 5000, r = .031$, and $t = 4$. Then a calculator with an $e^x$ key shows that

$$FV = PVe^{rt} = 5000e^{.031(4)} = \$5660.08. \quad ✔_4$$

**TECHNOLOGY TIP** TI-84+ has a "TVM solver" for financial computations (in the TI APPS/financial menu); a similar one can be downloaded for TI-89. Figure 5.2 shows the solution of Example 4(e) on such a solver. The use of these solvers is explained in the next section. Most of the problems in this section can be solved just as quickly with an ordinary calculator.

Ordinary corporate or municipal bonds usually make semiannual simple interest payments. With a **zero-coupon bond,** however, there are no interest payments during the life of the bond. The investor receives a single payment when the bond matures, consisting of his original investment and the interest (compounded semiannually) that it has earned. Zero-coupon bonds are sold at a substantial discount from their face value, and the buyer receives the face value of the bond when it matures. The difference between the face value and the price of the bond is the interest earned.

### EXAMPLE 6   Bonds

Doug bought a 15-year zero-coupon bond paying 4.5% interest (compounded semiannually) for $12,824.50. What is the face value of the bond?

**Solution** Use the compound interest formula with $PV = 12,824.50$. Interest is paid twice a year, so the rate per period is $i = .045/2$, and the number of periods in 15 years is $n = 30$. The compound amount will be the face value:

$$FV = PV(1 + i)^n = 12,824.50(1 + .045/2)^{30} = 24,999.99618.$$

Rounding to the nearest cent, we see that the face value of the bond in $25,000.

### EXAMPLE 7   Inflation

Suppose that the inflation rate is 3.5% (which means that the overall level of prices is rising 3.5% a year). How many years will it take for the overall level of prices to double?

**Solution** We want to find the number of years it will take for $1 worth of goods or services to cost $2. Think of $1 as the present value and $2 as the future value, with an interest rate of 3.5%, compounded annually. Then the compound amount formula becomes

$$PV(1 + i)^n = FV$$

$$1(1 + .035)^n = 2,$$

---

### ✓ Checkpoint 4

Find the compound amount for $7500 invested at an annual interest rate of 2.07% compounded continuously for 3 years.

```
N=365
I%=.85
PV=-10000
PMT=0
•FV=10085.36128
P/Y=365
C/Y=365
PMT:ENC BEGIN
```

**Figure 5.2**

### ✓ Checkpoint 5

Find the face value of the zero coupon.

**(a)** 30-year bond at 6% sold for $2546

**(b)** 15-year bond at 5% sold for $16,686

which simplifies as

$$1.035^n = 2.$$

We must solve this equation for $n$. There are several ways to do this.

**Graphical** Use a graphing calculator (with $x$ in place of $n$) to find the intersection point of the graphs of $y_1 = 1.035^x$ and $y_2 = 2$. Figure 5.3 shows that the intersection point has (approximate) $x$-coordinate 20.14879. So it will take about 20.15 years for prices to double.

**Algebraic** The same answer can be obtained by using natural logarithms, as in Section 4.4:

$$1.035^n = 2$$
$$\ln 1.035^n = \ln 2 \qquad \text{Take the logarithm of each side.}$$
$$n \ln 1.035 = \ln 2 \qquad \text{Power property of logarithms}$$
$$n = \frac{\ln 2}{\ln 1.035} \qquad \text{Divide both sides by ln 1.035.}$$
$$n \approx 20.14879. \qquad \text{Use a calculator.}$$

Figure 5.3
Intersection
X=20.148792  Y=2

✓ **Checkpoint 6**

Using a calculator, find the number of years it will take for $500 to increase to $750 in an account paying 6% interest compounded semiannually.

## Effective Rate (Annual Percentage Yield)

If you invest $100 at 9%, compounded monthly, then your balance at the end of one year is

$$FV = PV(1 + i)^n = 100\left(1 + \frac{.09}{12}\right)^{12} = \$109.38.$$

You have earned $9.38 in interest, which is 9.38% of your original $100. In other words, $100 invested at 9.38% compounded *annually* will produce the same amount of interest (namely, $100 * .0938 = $9.38) as does 9% compounded monthly. In this situation, 9% is called the **nominal** or **stated rate,** while 9.38% is called the **effective rate** or **annual percentage yield (APY).**

In the discussion that follows, the nominal rate is denoted $r$ and the APY (effective rate) is denoted $r_E$.

### Effective Rate ($r_E$) or Annual Percentage Yield (APY)

The *APY* $r_E$ is the simple interest rate needed to produce the same amount of interest in one year as the nominal rate does with more frequent compounding over one year.

**EXAMPLE 8** **Money Market Accounts** In July 2016, American Express offered its customers a money market savings account at a rate of .90%. Assume the account compounded interest daily and the principal is $100,000. Find the APY. (Data from bankrate.com.)

**Solution** The definition given previously means that we must have the following:

$$\frac{\$100,000 \text{ at rate } r_E}{\text{simple interest}} = \frac{\$100,000 \text{ at .90,}}{\text{compounded daily}}$$
$$100,000(1 + r_E) = 100,000\left(1 + \frac{.0090}{365}\right)^{365} \qquad \text{Interest formulas}$$
$$(1 + r_E) = \left(1 + \frac{.0090}{365}\right)^{365} \qquad \text{Divide both sides by 100,000.}$$
$$r_E = \left(1 + \frac{.0090}{365}\right)^{365} - 1 \qquad \text{Subtract 1 from both sides.}$$
$$r_E \approx .00904.$$

So the APY is about .904%.

The argument in Example 8 can be carried out with 100,000 replaced by $PV$, .0090 by $r$, and 365 by $m$. The result is the effective rate formula.

## Effective Rate (APY)

The effective rate (APY) corresponding to a stated rate of annual interest $r$ compounded $m$ times per year is

$$r_E = \left(1 + \frac{r}{m}\right)^m - 1.$$

**EXAMPLE 9** **Money Market Accounts** When interest rates are low (as they were when this text went to press), the interest rate and the APY are insignificantly different. To see when the difference is more pronounced, we will find the APY for each of the given money market checking accounts, which were advertised in October 2008 when offered rates were higher.

**(a)** Imperial Capital Bank: 3.35% compounded monthly.

**Solution** Use the effective-rate formula with $r = .0335$ and $m = 12$:

$$r_E = \left(1 + \frac{r}{m}\right)^m - 1 = \left(1 + \frac{.0335}{12}\right)^{12} - 1 \approx .034019.$$

So the APY is about 3.40%, a slight increase over the nominal rate of 3.35%.

**(b)** U.S. Bank: 2.33% compounded daily.

**Solution** Use the formula with $r = .0233$ and $m = 365$:

$$r_E = \left(1 + \frac{r}{m}\right)^m - 1 = \left(1 + \frac{.0233}{365}\right)^{365} - 1 \approx .023572.$$

The APY is about 2.36%. ✔7

**✓ Checkpoint 7**

Find the APY corresponding to a nominal rate of

**(a)** 12% compounded monthly;

**(b)** 8% compounded quarterly.

**⊞ TECHNOLOGY TIP**
Effective rates (APYs) can be computed on TI-84+ by using "Eff" in the APPS financial menu, as shown in Figure 5.4 for Example 10.

```
▶Eff(1.8,1)
                    1.8
▶Eff(1.77,4)
              1.781783071
▶Eff(1.75,12)
              1.764104916
■
```

**Figure 5.4**

**✓ Checkpoint 8**

Find the APY corresponding to a nominal rate of

**(a)** 4% compounded quarterly;

**(b)** 7.9% compounded daily.

**EXAMPLE 10** **Money Market Accounts** Bank A offers a 3-year money market account with a rate of 1.80% compounded annually. Bank B offers a similar money market account with a rate of 1.77% compounded quarterly, and Bank C offers 1.75% compounded monthly. Which bank offers the highest APY?

**Solution** Compare the APYs:

$$\text{Bank A:} \quad \left(1 + \frac{.0180}{1}\right)^1 - 1 = .0180 = 1.80;$$

$$\text{Bank B:} \quad \left(1 + \frac{.0177}{4}\right)^4 - 1 \approx .0178 = 1.78;$$

$$\text{Bank C:} \quad \left(1 + \frac{.0175}{12}\right)^{12} - 1 \approx .0176 = 1.76.$$

Bank A offers the highest APY. ✔8

**📄 NOTE** Although you can find both the stated interest rate and the APY for most certificates of deposit and other interest-bearing accounts, most bank advertisements mention only the APY.

## Present Value for Compound Interest

The formula for compound interest, $FV = PV(1 + i)^n$, has four variables: $FV$, $PV$, $i$, and $n$. Given the values of any three of these variables, the value of the fourth can be found. In particular, if $FV$ (the future amount), $i$, and $n$ are known, then $PV$ can be found. Here, $PV$ is the amount that should be deposited today to produce $FV$ dollars in $n$ periods.

---

**EXAMPLE 11** **Lump Sum Payment** Keisha Jones must pay a lump sum of $6000 in 5 years. What amount deposited today at 6.2% compounded annually will amount to $6000 in 5 years?

**Solution** Here, $FV = 6000$, $i = .062$, $n = 5$, and $PV$ is unknown. Substituting these values into the formula for the compound amount gives

$$6000 = PV(1.062)^5$$

$$PV = \frac{6000}{(1.062)^5} = 4441.49,$$

**✓ Checkpoint 9**

Find $PV$ in Example 11 if the interest rate is

(a) 6%;

(b) 10%.

or $4441.49. If Keisha leaves $4441.49 for 5 years in an account paying 6.2% compounded annually, she will have $6000 when she needs it. To check your work, use the compound interest formula with $PV = \$4441.49$, $i = .062$, and $n = 5$. You should get $FV = \$6000.00$. ✓ 9

As Example 11 shows, $6000 in 5 years is (approximately) the same as $4441.49 today (if money can be deposited at 6.2% annual interest). An amount that can be deposited today to yield a given amount in the future is called the *present value* of the future amount. By solving $FV = PV(1 + i)^n$ for $PV$, we get the following general formula for present value.

---

## Present Value for Compound Interest

The **present value** of $FV$ dollars compounded at an interest rate $i$ per period for $n$ periods is

$$PV = \frac{FV}{(1 + i)^n}, \quad \text{or} \quad PV = FV(1 + i)^{-n}.$$

---

**EXAMPLE 12** **Bonds** A zero-coupon bond with face value $15,000 and a 6% interest rate (compounded semiannually) will mature in 9 years. What is a fair price to pay for the bond today?

**Solution** Think of the bond as a 9-year investment paying 6%, compounded semiannually, whose future value is $15,000. Its present value (what it is worth today) would be a fair price. So use the present-value formula with $FV = 15,000$. Since interest is compounded twice a year, the interest rate per period is $i = .06/2 = .03$ and the number of periods in 9 years is $n = 9(2) = 18$. Hence,

**✓ Checkpoint 10**

Find the fair price (present value) in Example 12 if the interest rate is 7.5%.

$$PV = \frac{FV}{(1 + i)^n} = \frac{15,000}{(1 + .03)^{18}} \approx 8810.919114.$$

So a fair price would be the present value of $8810.92. ✓ 10

**EXAMPLE 13**    **Inflation** The average annual inflation rate for the years 2012–2014 was 1.72%. How much did an item that sells for $1000 in early 2015 cost three years before? (Data from: inflationdata.com.)

**Solution** Think of the price three years prior as the present value $PV$ and $1000 as the future value $FV$. Then $i = .0172$, $n = 3$, and the present value is

$$PV = \frac{FV}{(1 + i)^n} = \frac{1000}{(1 + .0172)^3} = \$950.12.$$

The item cost $950.12 three years earlier. ✔11

✓ **Checkpoint 11**

What did a $1000 item sell for 5 years prior if the annual inflation rate has been 3.2%?

## Summary

At this point, it seems helpful to summarize the notation and the most important formulas for simple and compound interest. We use the following variables:

$PV =$ principal or present value;

$FV =$ future or maturity value;

$r =$ annual (stated or nominal) interest rate;

$t =$ number of years;

$m =$ number of compounding periods per year;

$i =$ interest rate per period;

$n =$ total number of compounding periods;

$r_E =$ effective rate (APY).

| Simple Interest | Compound Interest | Continuous Compounding |
|---|---|---|
| $FV = PV(1 + rt)$ | $FV = PV(1 + r)^n$ | $FV = PVe^{rt}$ |
| $PV = \dfrac{FV}{1 + rt}$ | $PV = \dfrac{FV}{(1 + i)^n} = FV(1 + i)^{-n}$ | $PV = \dfrac{FV}{e^{rt}}$ |
| | $r_E = \left(1 + \dfrac{r}{m}\right)^m - 1$ | |

# 5.2 Exercises

**Bonds** *Interest on the zero-coupon bonds here is compounded semiannually.*

1. In the preceding summary what is the difference between $r$ and $i$? between $t$ and $n$?

2. Explain the difference between simple interest and compound interest.

3. What factors determine the amount of interest earned on a fixed principal?

4. In your own words, describe the *maturity value* of a loan.

5. What is meant by the *present value* of money?

6. If interest is compounded more than once per year, which rate is higher, the stated rate or the effective rate?

**Savings Accounts** *Find the compound amount and the interest earned for each of the following deposits. (See Examples 1, 2, and 4.)*

7. $470 at 8% compounded monthly for 12 years

8. $15,000 at 4.6% compounded monthly for 11 years

9. $6500 at 4.5% compounded quarterly for 8 years

10. $9100 at 6.1% compounded quarterly for 4 years

**Money Market Accounts** *The following money market account rates were available at bankrate.com in July 2016. Find the compound amount and the interest earned for each of the following. (See Example 4.)*

11. Everbank: $10,000 at 1.11% compounded daily for one year

12. Alostar Bank of Commerce: $5000 at 1.05% compounded daily for one year

13. Goldman Sachs Bank: $15,000 at 1.00% compounded monthly for three years

14. Discover Bank: $60,000 at .95% compounded monthly for two years

15. Pacific National Bank: $75,000 at .90% compounded quarterly for four years

16. Colorado Federal: $80,000 at .85% compounded quarterly for 3 years

*Find the interest rate (with annual compounding) that makes the statement true. (See Example 3.)*

17. $3000 grows to $3606 in 5 years

18. $2550 grows to $3905 in 11 years

19. $8500 grows to $12,161 in 7 years

20. $9000 grows to $17,118 in 16 years

*Find the compound amount and the interest earned when the following investments have continuous compounding. (See Example 5.)*

21. $20,000 at 3.5% for 5 years

22. $15,000 at 2.9% for 10 years

23. $30,000 at 1.8% for 3 years

24. $100,000 at 5.1% for 20 years

**Bonds** *Find the face value (to the nearest dollar) of the zero-coupon bond. (See Example 6.)*

25. 15-year bond at 5.2%; price $4630

26. 10-year bond at 4.1%; price $13,328

27. 20-year bond at 3.5%; price $9992

28. How do the nominal, or stated, interest rate and the effective interest rate (APY) differ?

**APY** *Find the APY corresponding to the given nominal rates. (See Examples 8–10).*

29. 4% compounded semiannually

30. 6% compounded quarterly

31. 5% compounded quarterly

32. 4.7% compounded semiannually

**Present Value** *Find the present value of the given future amounts. (See Example 11.)*

33. $12,000 at 5% compounded annually for 6 years

34. $8500 at 6% compounded annually for 9 years

35. $17,230 at 4% compounded quarterly for 10 years

36. $5240 at 6% compounded quarterly for 8 years

**Bonds** *What price should you be willing to pay for each of these zero-coupon bonds? (See Example 12.)*

37. 5-year $5000 bond; interest at 3.5%

38. 10-year $10,000 bond; interest at 4%

39. 15-year $20,000 bond; interest at 4.7%

40. 20-year $15,000 bond; interest at 5.3%

**Inflation** *For Exercises 41 and 42, assume an annual inflation rate of 1.62% (the annual inflation rate of 2014 according to www.InflationData.com). Find the previous price of the following items. (See Example 13.)*

41. How much did an item that costs $5000 now cost 4 years prior?

42. How much did an item that costs $7500 now cost 5 years prior?

43. If the annual inflation rate is 3.6%, how much did an item that costs $500 now cost 2 years prior?

44. If the annual inflation rate is 1.18%, how much did an item that costs $1250 now cost 6 years prior?

45. If money can be invested at 8% compounded quarterly, which is larger, $1000 now or $1210 in 5 years? Use present value to decide.

46. If money can be invested at 6% compounded annually, which is larger, $10,000 now or $15,000 in 6 years? Use present value to decide.

*Work the following applied problems.*

47. **Investing** Lora Reilly has inherited $10,000 from her uncle's estate. She will invest the money for 2 years. She is considering two investments: a money market fund that pays a guaranteed 5.8% interest compounded daily and a 2-year Treasury note at 6% annual interest. Which investment pays the most interest over the 2-year period?

48. **Investing** As the prize in a contest, you are offered $1000 now or $1210 in 5 years. If money can be invested at 6% compounded annually, which is larger?

49. **Bonds** Which of these 20-year zero-coupon bonds will be worth more at maturity: one that sells for $4510, with a 6.1% interest rate, or one that sells for $5809, with a 4.8% interest rate?

50. **Bonds** Which of these 15-year zero-coupon bonds will be worth more at maturity: one that sells for $3890, with a 3.7% interest rate, or one that sells for $4510, with a 2.9% interest rate?

51. **Investing** A company has agreed to pay $2.9 million in 5 years to settle a lawsuit. How much must it invest now in an account paying 5% interest compounded monthly to have that amount when it is due?

52. **Investing** Bill Poole wants to have $20,000 available in 5 years for a down payment on a house. He has inherited $16,000. How much of the inheritance should he invest now to accumulate the $20,000 if he can get an interest rate of 5.5% compounded quarterly?

53. **Inflation** If inflation has been running at 3.75% per year and a new car costs $23,500 today, what would it have cost three years ago?

54. **Inflation** If inflation is 2.4% per year and a washing machine costs $345 today, what did a similar model cost five years ago?

**Economics** *Use the approach in Example 7 to find the time it would take for the general level of prices in the economy to double at the average annual inflation rates in Exercises 55–58.*

55. 3%      56. 4%      57. 5%      58. 5.5%

59. **Electricity Demand** The consumption of electricity has increased historically at 6% per year. If it continues to increase at this rate indefinitely, find the number of years before the electric utility companies will need to double their generating capacity.

**60. Electricity Demand** Suppose a conservation campaign coupled with higher rates causes the demand for electricity to increase at only 2% per year, as it has recently. Find the number of years before the utility companies will need to double their generating capacity.

**61. Investing** You decide to invest a $16,000 bonus in a money market fund that guarantees a 5.5% annual interest rate compounded monthly for 7 years. A one-time fee of $30 is charged to set up the account. In addition, there is an annual administrative charge of 1.25% of the balance in the account at the end of each year.

   **(a)** How much is in the account at the end of the first year?

   **(b)** How much is in the account at the end of the seventh year?

**62. Investing** Joe decides to invest $12,000 in a money market fund that guarantees a 4.6% annual interest rate compounded daily for 6 years. A one-time fee of $25 is charged to set up the account. In addition, there is an annual administration charge of .9% of the balance in the account at the end of each year.

   **(a)** How much is in the account at the end of the first year?

   **(b)** How much is in the account at the end of the sixth year?

*The following exercises are similar to professional examination questions.*

**63.** On January 1, 2015, Rhom Company exchanged equipment for a $200,000 non-interest-bearing note due on January 1, 2018. The prevailing rate of interest for a note of this type on January 1, 2015, was 10%. The present value of $1 at 10% for three periods is 0.75. What amount of interest revenue should be included in Rhom's 2016 income statement? (Adapted from the Uniform CPA Examination, American Institute of Certified Public Accountants.)

**64.** On January 1, 2015, Bartell Company exchanged property for a $450,000 non-interest-bearing note due on January 1, 2018. The prevailing rate of interest for a note of this type on January 1, 2015, was 7.7%. The present value of $1 at 7.7% for three periods is 0.80. What amount of interest revenue should be included in Bartell's 2016 income statement? (Adapted from the Uniform CPA Examination, American Institute of Certified Public Accountants.)

✓ **Checkpoint Answers**

**1. (a)**

| Year | Interest | Balance |
|---|---|---|
| 4 | $50 | $1200 |
| 5 | $50 | $1250 |

   **(b)**

| Year | Interest | Balance |
|---|---|---|
| 4 | $57.88 | $1215.51 |
| 5 | $60.78 | $1276.29 |

**2. (a)** $26,170.72     **(b)** $9170.72

**3. (a)** $1049.14     **(b)** $2514.84

**4.** $7980.52

**5. (a)** $15,000     **(b)** $35,000

**6.** About 7 years ($n = 6.86$)

**7. (a)** 12.68%     **(b)** 8.24%

**8. (a)** 4.06%     **(b)** 8.220%

**9. (a)** $4483.55     **(b)** $3725.53

**10.** $7732.24     **11.** $854.28

# 5.3 Annuities, Future Value, and Sinking Funds

So far in this chapter, only lump-sum deposits and payments have been discussed. Many financial situations, however, involve a sequence of payments at regular intervals, such as weekly deposits into a savings account or monthly payments on a mortgage or car loan. Such periodic payments are the subject of this section and the next.

The analysis of periodic payments will require an algebraic technique that we now develop. Suppose $x$ is a real number. For reasons that will become clear later, we want to find the product

$$(x - 1)(1 + x + x^2 + x^3 + \cdots + x^{11}).$$

Using the distributive property to multiply this expression out, we see that all but two of the terms cancel:

$$x(1 + x + x^2 + x^3 + \cdots + x^{11}) - 1(1 + x + x^2 + x^3 + \cdots + x^{11})$$
$$= (x + x^2 + x^3 + \cdots + x^{11} + x^{12}) - 1 - x - x^2 - x^3 - \cdots - x^{11}$$
$$= x^{12} - 1.$$

Hence, $(x - 1)(1 + x + x^2 + x^3 + \cdots + x^{11}) = x^{12} - 1$. Dividing both sides by $x - 1$, we have

$$1 + x + x^2 + x^3 + \cdots + x^{11} = \frac{x^{12} - 1}{x - 1}.$$

The same argument, with any positive integer $n$ in place of 12 and $n - 1$ in place of 11, produces the following result:

---

If $x$ is a real number and $n$ is a positive integer, then

$$1 + x + x^2 + x^3 + \cdots + x^{n-1} = \frac{x^n - 1}{x - 1}.$$

---

The above formula is a special case of a finite geometric series. For example, when $x = 5$ and $n = 7$, we see that

$$1 + 5 + 5^2 + 5^3 + 5^4 + 5^5 + 5^6 = \frac{5^7 - 1}{5 - 1} = \frac{78,124}{4} = 19,531.$$

A calculator can easily add up the terms on the left side, but it is faster to use the formula (Figure 5.5).

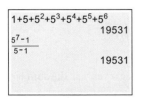

**Figure 5.5**

## Ordinary Annuities

A sequence of equal payments made at equal periods of time is called an **annuity.** The time between payments is the **payment period,** and the time from the beginning of the first payment period to the end of the last period is called the **term of the annuity.** Annuities can be used to accumulate funds—for example, when you make regular deposits into a savings account. Or they can be used to pay out funds—as when you receive regular payments from a pension plan after you retire.

Annuities that pay out funds are considered in the next section. This section deals with annuities in which funds are accumulated by regular payments into an account or investment that earns compound interest. The **future value** of such an annuity is the final sum on deposit—that is, the total amount of all deposits and all interest earned by them.

We begin with **ordinary annuities**—ones where the payments are made at the *end* of each period and the frequency of payments is the same as the frequency of compounding the interest.

**EXAMPLE 1**   **Saving Accounts** $1500 is deposited at the end of each year for the next 6 years in a savings account paying 8% interest compounded annually. Find the future value of this annuity.

**Solution**   Figure 5.6 shows the situation schematically.

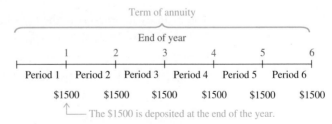

**Figure 5.6**

To find the future value of this annuity, look separately at each of the $1500 payments. The first $1500 is deposited at the end of period 1 and earns interest for the remaining 5 periods. From the formula in the box on page 231, the compound amount produced by this payment is

$$1500(1 + .08)^5 = 1500(1.08)^5.$$

The second $1500 payment is deposited at the end of period 2 and earns interest for the remaining 4 periods. So the compound amount produced by the second payment is

$$1500(1 + .08)^4 = 1500(1.08)^4.$$

Continue to compute the compound amount for each subsequent payment, as shown in Figure 5.7. Note that the last payment earns no interest.

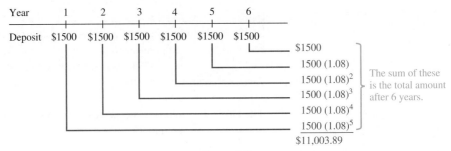

**Figure 5.7**

The last column of Figure 5.7 shows that the total amount after 6 years is the sum

$$1500 + 1500 \cdot 1.08 + 1500 \cdot 1.08^2 + 1500 \cdot 1.08^3 + 1500 \cdot 1.08^4 + 1500 \cdot 1.08^5$$
$$= 1500(1 + 1.08 + 1.08^2 + 1.08^3 + 1.08^4 + 1.08^5)$$
$$= \$11{,}003.89. \tag{1}$$

Now apply the algebraic fact in the box on page 241 to the expression in parentheses (with $x = 1.08$ and $n = 6$). It shows that the sum (the future value of the annuity) is

$$1500 \cdot \frac{1.08^6 - 1}{1.08 - 1} = \$11{,}003.89. ✔_1$$

Example 1 is the model for finding a formula for the future value of any annuity. Suppose that a payment of $PMT$ dollars is deposited at the end of each period for $n$ periods, at an interest rate of $i$ per period. Then the future value of this annuity can be found by using the procedure in Example 1, with these replacements:

| 1500 | .08 | 1.08 | 6 | 5 |
|:---:|:---:|:---:|:---:|:---:|
| ↓ | ↓ | ↓ | ↓ | ↓ |
| $PMT$ | $i$ | $1 + i$ | $n$ | $n - 1$ |

The future value $FV$ in Example 1 is the sum **(1)**, which now becomes

$$FV = PMT[1 + (1 + i) + (1 + i)^2 + \cdots + (1 + i)^{n-2} + (1 + i)^{n-1}].$$

Apply the algebraic fact in the box on page 241 to the expression in brackets (with $x = 1 + i$). Then we have

$$FV = PMT\left[\frac{(1 + i)^n - 1}{(1 + i) - 1}\right] = PMT\left[\frac{(1 + i)^n - 1}{i}\right].$$

The quantity in brackets in the right-hand part of the preceding equation is sometimes written $s_{\overline{n}|i}$ (read "*s*-angle-*n* at *i*"). So we can summarize as follows.

✔ **Checkpoint 1**

Complete these steps for an annuity of $2000 at the end of each year for 3 years. Assume interest of 6% compounded annually.

**(a)** The first deposit of $2000 produces a total of _____.

**(b)** The second deposit becomes _____.

**(c)** No interest is earned on the third deposit, so the total in the account is _____.

---

## Future Value of an Ordinary Annuity

The future value $FV$ of an ordinary annuity used to accumulate funds is given by

$$FV = PMT\left[\frac{(1 + i)^n - 1}{i}\right], \quad \text{or} \quad FV = PMT \cdot s_{\overline{n}|i},$$

where

$PMT$ is the payment at the end of each period;

$i = r/m$ is the interest rate per period, $r$ is the annual interest rate, $m$ is the number of periods per year; and

$n$ is the total number of periods.

📄 **NOTE**  In Example 1, 8% was the annual interest rate, and thus one payment was made per year. In cases like this, $m = 1$, and so $i = r/m = r/1 = r$.

```
N=6
I%=8
PV=0
PMT=-1500
•FV=11003.89356
P/Y=1
C/Y=1
PMT:END BEGIN
```
**Figure 5.8**

```
N=7
I%=4.1
PV=0
PMT=350000
•FV=-2772807.587
P/Y=1
C/Y=1
PMT:END BEGIN
```
**Figure 5.9**

✓**Checkpoint 2**

Johnson Building Materials deposits $2500 at the end of each year into an account paying 8% per year compounded annually. Find the total amount on deposit after

**(a)** 6 years;

**(b)** 10 years.

```
N=120
I%=1.11
PV=0
PMT=-950
•FV=120508.8653
P/Y=12
C/Y=12
PMT:END BEGIN
```
**Figure 5.10(a)**

⊞ **TECHNOLOGY TIP** Most computations with annuities can be done quickly with a spreadsheet program or a graphing calculator. On a calculator, use the TVM solver if there is one (see the Technology Tip on page 234).

Figure 5.8 shows how to do Example 1 on a TI-84+ TVM solver. First, enter the known quantities: N = number of payments, I% = interest rate, PV = present value, PMT = payment per period (entered as a negative amount), P/Y = number of payments per year, and C/Y = number of compoundings per year. At the bottom of the screen, set PMT: to "END" for ordinary annuities. Then put the cursor next to the unknown amount FV (future value), and press SOLVE.

*Note:* P/Y and C/Y should always be the same for problems in this text. If you use the solver for ordinary compound interest problems, set PMT = 0 and enter either PV or FV (whichever is known) as a negative amount.

**EXAMPLE 2** **Money Market Account** A rookie player in the National Football League just signed his first 7-year contract. To prepare for his future, he deposits $350,000 at the end of each year for 7 years in an account paying 4.1% compounded annually. How much will he have on deposit after 7 years?

**Solution** His payments form an ordinary annuity with $PMT = 350,000, n = 7$, and $i = .041$. The future value of this annuity (by the previous formula) is

$$FV = 350,000\left[\frac{(1.041)^7 - 1}{.041}\right] = \$2,772,807.59.$$

Figure 5.9 shows the result with the TVM solver.

**EXAMPLE 3** **Investing** Allyson opens a money market account at Silvergate Bank that in July 2016 offered a rate of 1.11%. Assume Allyson contributes $950 a month to the account and that the interest rate stays at 1.11% compounded monthly for the next 10 years. (Data from: www.bankrate.com.)

**(a)** What will be the value of the account in 10 years?

**Solution** Allyson's payments into the account form an annual annuity, with monthly payment $PMT = 950$. The interest per month is $i = \frac{r}{m} = \frac{.0111}{12}$, and the number of months in 10 years is $n = 10*12 = 120$. The future value of this annuity is

$$FV = PMT\left[\frac{(1+i)^n - 1}{i}\right] = 950\left[\frac{(1+.0111/12)^{120} - 1}{.0111/12}\right] = \$120,508.87.$$

Figure 5.10(a) shows the result with the TVM solver.

**(b)** Suppose that, after the 10 years, Allyson was able to invest the $120,508.87 in a money market account that paid 1.24% interest compounded monthly. If she continues to deposit $950 a month into this new account, how much would be the total after an additional five years?

**Solution** Deal separately with the two parts of her account (the $950 contributions in the future and the $120,508.87 already in the account). The contributions form an ordinary annuity, as they did in part (a). Now we have $PMT = 950, i = \frac{.0124}{12}$, and $n = 5*12 = 60$. The future value of this annuity is

$$FV = PMT\left[\frac{(1+i)^n - 1}{i}\right] = 950\left[\frac{(1+.0124/12)^{60} - 1}{.0124/12}\right] = \$58,772.78.$$

```
N=60
I%=1.24
PV=-120508.87
PMT=-950
•FV=186985.5751
P/Y=12
C/Y=12
PMT:END BEGIN
```
**Figure 5.10(b)**

✓**Checkpoint 3**

Find the total value of the account in part (b) of Example 3 if the fund's return for the last 5 years is 1.42%, compounded monthly.

Meanwhile, the $120,508.87 from the first 10 years is also earning interest at 1.24% compounded monthly. By the compound amount formula (Section 5.2), the future value of this money is

$$FV = (1 + i/12)^{n*12} = 120{,}508.87(1 + .0124/12)^{60} = \$128{,}212.80.$$

So the total amount in Allyson's account after 15 years is the sum

$$\$58{,}772.78 + \$128{,}212.80 = \$186{,}985.58.$$

This calculation can also be done in one step using the TVM solver. Figure 5.10(b) shows how the information is entered. Note that we enter the amount from part (a) as a negative value in the PV line. ✓₃

## Sinking Funds

A **sinking fund** is a fund set up to receive periodic payments. Corporations and municipalities use sinking funds to repay bond issues, to retire preferred stock, to provide for replacement of fixed assets, and for other purposes. If the payments are equal and are made at the end of regular periods, they form an ordinary annuity.

**EXAMPLE 4** **Investing** A business sets up a sinking fund so that it will be able to pay off bonds it has issued when they mature. If it deposits $12,000 at the end of each quarter in an account that earns 5.2% interest, compounded quarterly, how much will be in the sinking fund after 10 years?

**Solution** The sinking fund is an annuity, with $PMT = 12{,}000$, $i = r/m = .052/4$, and $n = 4(10) = 40$. The future value is

$$FV = PMT\left[\frac{(1 + i)^n - 1}{i}\right] = 12{,}000\left[\frac{(1 + .052/4)^{40} - 1}{.052/4}\right] = \$624{,}369.81.$$

So there will be about $624,370 in the sinking fund. Figure 5.11 shows the entries in the TVM solver.

```
N=40
I%=5.2
PV=0
PMT=-12000
•FV=624369.8089
P/Y=4
C/Y=4
PMT:END BEGIN
```
**Figure 5.11**

**EXAMPLE 5** **Investing** A firm borrows $6 million to build a small factory. The bank requires it to set up a $200,000 sinking fund to replace the roof after 15 years. If the firm's deposits earn 6% interest, compounded annually, find the payment it should make at the end of each year into the sinking fund.

**Solution** This situation is an annuity with future value $FV = 200{,}000$, interest rate $i = .06$, and $n = 15$. Solve the future-value formula for $PMT$:

$$FV = PMT\left[\frac{(1 + i)^n - 1}{i}\right]$$

$$200{,}000 = PMT\left[\frac{(1 + .06)^{15} - 1}{.06}\right] \quad \text{Let } FV = 200{,}000, i = .06, \text{ and } n = 15.$$

$$200{,}000 \approx PMT[23.27597] \quad \text{Compute the quantity in brackets.}$$

$$PMT = \frac{200{,}000}{23.27597} = \$8592.55. \quad \text{Divide both sides by 23.27597.}$$

So the annual payment is about $8593. Figure 5.12 shows the entries in the TVM solver. ✓₄

```
N=15
I%=6
PV=0
•PMT=-8592.5527...
FV=200000
P/Y=1
C/Y=1
PMT:END BEGIN
```
**Figure 5.12**

✓**Checkpoint 4**

Francisco Arce needs $8000 in 6 years so that he can go on an archaeological dig. He wants to deposit equal payments at the end of each quarter so that he will have enough to go on the dig. Find the amount of each payment if the bank pays

**(a)** 12% interest compounded quarterly;

**(b)** 8% interest compounded quarterly.

**EXAMPLE 6** **Incentive Bonus** As an incentive for a valued employee to remain on the job, a company plans to offer her a $100,000 bonus, payable when she retires in 20 years. If the company deposits $200 a month in a sinking fund, what interest rate must it earn, with monthly compounding, in order to guarantee that the fund will be worth $100,000 in 20 years?

**Solution** The sinking fund is an annuity with $PMT = 200$, $n = 12(20) = 240$, and future value $FV = 100,000$. We must find the interest rate. If $x$ is the annual interest rate in decimal form, then the interest rate per month is $i = x/12$. Inserting these values into the future-value formula, we have

$$PMT\left[\frac{(1 + i)^n - 1}{i}\right] = FV$$

$$200\left[\frac{(1 + x/12)^{240} - 1}{x/12}\right] = 100,000.$$

It is not possible to solve this equation algebraically. You can get a rough approximation by using a calculator and trying different values for $x$. With a graphing calculator, you can get an accurate solution by graphing

$$y_1 = 200\left[\frac{(1 + x/12)^{240} - 1}{x/12}\right] \quad \text{and} \quad y_2 = 100,000$$

and finding the $x$-coordinate of the point where the graphs intersect. Figure 5.13 shows that the company needs an interest rate of about 6.661%. The same answer can be obtained on a TVM solver (Figure 5.14). ✔ 5

**✓ Checkpoint 5**

Pete's Pizza deposits $5800 at the end of each quarter for 4 years.

**(a)** Find the final amount on deposit if the money earns 6.4% compounded quarterly.

**(b)** Pete wants to accumulate $110,000 in the 4-year period. What interest rate (to the nearest tenth) will be required?

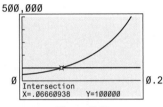

500,000

0 ........ 0.2

Intersection
X=.06660938      Y=100000

**Figure 5.13**

N=240
•I%=6.660938285
PV=0
PMT=-200
FV=100000
P/Y=12
C/Y=12
PMT:**END** BEGIN

**Figure 5.14**

## Annuities Due

The formula developed previously is for *ordinary annuities*—annuities with payments at the *end* of each period. The results can be modified slightly to apply to **annuities due**—annuities where payments are made at the *beginning* of each period.

An example will illustrate how this is done. Consider an annuity due in which payments of $100 are made for 3 years, and an ordinary annuity in which payments of $100 are made for 4 years, both with 5% interest, compounded annually. Figure 5.15 computes the growth of each payment separately (as was done in Example 1).

Annuity Due (payments at beginning of year for 3 years)

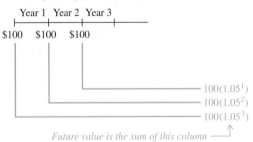

Year 1  Year 2  Year 3

$100    $100    $100

$100(1.05^1)$
$100(1.05^2)$
$100(1.05^3)$

*Future value is the sum of this column* →

Ordinary Annuity (payments at end of year for 4 years)

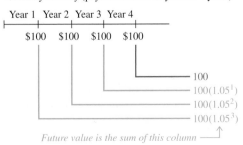

Year 1  Year 2  Year 3  Year 4

$100    $100    $100    $100

100
$100(1.05^1)$
$100(1.05^2)$
$100(1.05^3)$

*Future value is the sum of this column* →

**Figure 5.15**

Figure 5.15 shows that the future values are the same, *except* for one $100 payment on the ordinary annuity (shown in red). So we can use the formula on page 242 to find the future value of the 4-year ordinary annuity and then subtract one $100 payment to get the future value of the 3-year annuity due:

Future value of    =    Future value of    −    One payment
3-year annuity due        4-year ordinary annuity

$$FV = 100\left[\frac{1.05^4 - 1}{.05}\right] - 100 = \$331.01.$$

Essentially the same argument works in the general case.

## Future Value of an Annuity Due

The future value $FV$ of an annuity due used to accumulate funds is given by

$$FV = PMT\underbrace{\left[\frac{(1 + i)^{n+1} - 1}{i}\right]}_{\substack{\text{Future value of} \\ FV = \text{an ordinary annuity} - \text{One payment,} \\ \text{of } n + 1 \text{ payments}}} - PMT$$

where

    $PMT$ is the payment at the beginning of each period;

    $i = r/m$ is the interest rate per period, $r$ is the annual interest rate, $m$ is the number of periods per year; and

    $n$ is the total number of periods.

✓ **Checkpoint 6**

(a) Ms. Black deposits $800 at the beginning of each 6-month period for 5 years. Find the final amount if the account pays 6% compounded semiannually.

(b) Find the final amount if this account were an ordinary annuity.

```
N=28
I%=1.08
PV=0
PMT=-500
•FV=14561.65577
P/Y=4
C/Y=4
PMT:END BEGIN
```

**Figure 5.16**

**EXAMPLE 7** **Investing** Jillian will make payments of $500 at the beginning of each quarter for 7 years into a money market account at East Boston Savings Bank, which in July 2016 offered a rate of 1.08%. If interest is compounded quarterly, find the future value of this annuity due. (Data from: www.bankrate.com.)

**Solution** In 7 years, there are $n = 28$ quarterly periods. For an annuity due, add one period to get $n + 1 = 29$, and use the formula with $i = r/m = .0108/4 = .0027$:

$$FV = PMT\left[\frac{(1 + i)^{n+1} - 1}{i}\right] - PMT = 500\left[\frac{(1 + .0027)^{29} - 1}{.0027}\right] - 500 = \$14,561.66.$$

After 7 years, the account balance will be $14,561.66. ✓ 6

▦ **TECHNOLOGY TIP** When a TVM solver is used for annuities due, PMT: END at the bottom of the screen should be BEGIN. See Figure 5.16. Note that we use N = 28 periods with the calculator rather than 29 in the formula.

**EXAMPLE 8** **Investing** Shen Wong plans to have a fixed amount from his paycheck directly deposited into an account that pays 5.5% interest, compounded monthly. If he gets paid on the first day of the month and wants to accumulate $13,000 in the next three-and-a-half years, how much should he deposit each month?

**Solution** Shen's deposits form an annuity due whose future value is $FV = 13{,}000$. The interest rate is $i = r/m = .055/12$. There are 42 months in three-and-a-half years. Since this is an annuity due, add one period, so that $n + 1 = 43$. Then solve the future-value formula for the payment $PMT$:

$$PMT\left[\frac{(1 + i)^{n+1} - 1}{i}\right] - PMT = FV$$

$$PMT\left[\frac{(1 + .055/12)^{43} - 1}{.055/12}\right] - PMT = 13{,}000 \qquad \text{Let } i = .055/12, n = 43, \text{ and } FV = 13{,}000.$$

$$PMT\left(\left[\frac{(1 + .055/12)^{43} - 1}{.055/12}\right] - 1\right) = 13{,}000 \qquad \text{Factor out } PMT \text{ on left side.}$$

$$PMT(46.4103) = 13{,}000 \qquad \text{Compute left side.}$$

$$PMT = \frac{13{,}000}{46.4103} = 280.110. \qquad \text{Divide both sides by 46.4103.}$$

```
N=42
I%=5.5
PV=0
•PMT=-280.110136
FV=13000
P/Y=12
C/Y=12
PMT:END BEGIN
```
**Figure 5.17**

Shen should have \$280.11 deposited from each paycheck. The TVM solver entries are shown in Figure 5.17.

We have seen in this section that sinking funds and annuities due are simply special cases of annuities. Recall that an annuity is a sequence of equal payments made at equal periods of time. When the payment is made at the *end* of the payment period, we call it a sinking fund or ordinary annuity. If the payment is made at the *beginning* of the payment period, then it is called an annuity due.

# 5.3 Exercises

*Note: Unless stated otherwise, all payments are made at the end of the period.*

**Find each of these sums (to 4 decimal places).**

**1.** $1 + 1.05 + 1.05^2 + 1.05^3 + \cdots + 1.05^{14}$

**2.** $1 + 1.046 + 1.046^2 + 1.046^3 + \cdots + 1.046^{21}$

**Investing** *As of July 2016, bankrate.com advertised 3 money market rates of 1.10%, 1.05%, and 1.00% at 3 different banks. Find the future value for the ordinary annuities with the given payments and interest rates. (See Examples 1, 2, 3(a), and 4.)*

**3.** $PMT = \$2000$, 1.10% compounded monthly for 3 years.

**4.** $PMT = \$500$, 1.00% compounded quarterly for 3 years.

**5.** $PMT = \$750$, 1.05% compounded semiannually for 5 years.

**6.** $PMT = \$900$, 1.10% compounded monthly for 5 years.

**7.** $PMT = \$2500$, 1.00% compounded quarterly for 10 years.

**8.** $PMT = \$1100$, 1.05% compounded semiannually for 10 years.

**Investing** *Find the final amount (rounded to the nearest dollar) in each of these retirement accounts, in which the rate of return on the account and the regular contribution change over time. (See Example 3.)*

**9.** \$400 per month invested at 4%, compounded monthly, for 10 years; then \$600 per month invested at 6%, compounded monthly, for 10 years.

**10.** \$500 per month invested at 5%, compounded monthly, for 20 years; then \$1000 per month invested at 8%, compounded monthly, for 20 years.

**11.** \$1000 per quarter invested at 4.2%, compounded quarterly, for 10 years; then \$1500 per quarter invested at 7.4%, compounded quarterly, for 15 years.

**12.** \$1500 per quarter invested at 7.4%, compounded quarterly, for 15 years; then \$1000 per quarter invested at 4.2%, compounded quarterly, for 10 years. (Compare with Exercise 11.)

**Sinking Funds** *Find the amount of each payment to be made into a sinking fund to accumulate the given amounts. Payments are made at the end of each period. (See Example 5.)*

**13.** \$11,000; money earns 5% compounded semiannually for 6 years

**14.** \$65,000; money earns 6% compounded semiannually for $4\frac{1}{2}$ years

**15.** \$50,000; money earns 8% compounded quarterly for $2\frac{1}{2}$ years

**16.** \$25,000; money earns 9% compounded quarterly for $3\frac{1}{2}$ years

**17.** \$6000; money earns 6% compounded monthly for 3 years

**18.** \$9000; money earns 7% compounded monthly for $2\frac{1}{2}$ years

**Sinking Fund** *Find the interest rate needed for the sinking fund to reach the required amount. Assume that the compounding period is the same as the payment period. (See Example 6.)*

**19.** $50,000 to be accumulated in 10 years; annual payments of $3940.

**20.** $100,000 to be accumulated in 15 years; quarterly payments of $1200.

**21.** $38,000 to be accumulated in 5 years; quarterly payments of $1675.

**22.** $77,000 to be accumulated in 20 years; monthly payments of $195.

**23.** What is meant by a sinking fund? List some reasons for establishing a sinking fund.

**24.** Explain the difference between an ordinary annuity and an annuity due.

**Investing** *In 2009, the average savings account rate was .20%, which is a much higher rate than current savings rates when this book went to press. Find the future value of each annuity due with the given rate close to the 2009 average. (See Example 7. Data from: fred.stlouisfed.org.)*

**25.** Payments of $500 for 10 years at .25% compounded annually

**26.** Payments of $1050 for 8 years at .21% compounded annually

**27.** Payments of $16,000 for 11 years at .18% compounded monthly

**28.** Payments of $25,000 for 12 years at .16% compounded monthly

**29.** Payments of $1000 for 9 years at .15% compounded semiannually

**30.** Payments of $750 for 15 years at .27% compounded semiannually

**31.** Payments of $200 for 7 years at .30% compounded quarterly

**32.** Payments of $2800 for 20 years at .32% compounded quarterly

**Annuity Due** *Find the payment that should be used for the annuity due whose future value is given. Assume that the compounding period is the same as the payment period. (See Example 8.)*

**33.** $8000; quarterly payments for 3 years; interest rate 4.4%

**34.** $12,000; annual payments for 6 years; interest rate 5.1%

**35.** $55,000; monthly payments for 12 years; interest rate 5.7%

**36.** $125,000; monthly payments for 9 years; interest rate 6%

**37. Cigarette Costs** The website www.theawl.com estimates that a pack-a-day smoker in New York State spends $400 a month on cigarettes (the highest cost of the 50 states). Suppose the smoker invests that amount at the end of each month in a money market account that pays .85% interest compounded monthly. What would the account be worth after 30 years?

**38. Cigarette Costs** The website www.theawl.com estimates that a pack-a-day smoker in Virginia spends $150 a month on cigarettes (the lowest cost of the 50 states). Suppose the smoker invests that amount at the end of each month in a money market account that pays .85% interest compounded monthly. What would the account be worth after 30 years?

**39. Investing** Mary Dodge deposits $12,000 at the end of each year for 9 years in an investment account with a guaranteed interest rate of 6% compounded annually.

(a) Find the value of the account at the end of the 9 years.

(b) Mary's sister works for an investment firm that pays 5% compounded annually. If Mary deposits money with this firm instead of the one in part (a), how much will she have in her account at the end of 9 years?

(c) How much would Mary lose or gain over 9 years by investing in her sister's firm?

**40. Investing** Peter Healy deposits $800 at the end of each month for 5 years in an investment account with a guaranteed interest rate of 4.12% compounded monthly.

(a) Find the value of the account at the end of 5 years.

(b) A rival financial planner offers Peter an investment strategy of depositing $700 a month for 5 years with a guaranteed interest rate of 6.15% compounded monthly. What is the value of this investment strategy at the end of 5 years?

(c) How much more money is gained by investing in the better strategy described in part (a) or in part (b)?

**41. Investing** At the end of each quarter, a 50-year-old woman puts $1200 in a retirement account that pays 7.2% interest compounded quarterly.

(a) When she reaches age 60, what is the value of the account?

(b) If no further deposits or withdrawals are made to the account, what is the value of the account when she reaches age 65?

**42. Retirement Planning** Miguel Handley is 45 years old. At the end of each month, he deposits $300 in a retirement account that pays 5.15% interest compounded monthly.

(a) After ten years, what is the value of the account?

(b) If no further deposits or withdrawals are made to the account, what is the value of the account when Miguel reaches age 65?

**43. Investing** A mother opened an investment account for her son on the day he was born, investing $1000. Each year on his birthday, she deposits another $1000, making the last deposit on his 18th birthday. If the account paid a return rate of 5.6% compounded annually, how much is in the account at the end of the day on the son's 18th birthday?

**44. Investing** A grandmother opens an investment account for her only granddaughter on the day she was born, investing $500. Each year on her birthday, she deposits another $500, making the last deposit on her 25th birthday. If the account paid a return rate of 6.2% compounded annually, how much is in the account at the end of the day on the granddaughter's 25th birthday?

**45. Investing** Chuck Hickman deposits $10,000 at the beginning of each year for 12 years in an account paying 5% compounded annually. He then puts the total amount on deposit in another account paying 6% compounded semiannually for another 9 years. Find the final amount on deposit after the entire 21-year period.

**46. Investing** Suppose that the best rate that the company in Example 6 can find is 6.3%, compounded monthly (rather than the 6.661% it wants). Then the company must deposit more in the sinking fund each month. What monthly deposit will guarantee that the fund will be worth $100,000 in 20 years?

**47. Investing** Diego Sanchez needs $10,000 in 8 years.

(a) What amount should he deposit at the end of each quarter at 5% compounded quarterly so that he will have his $10,000?

(b) Find Diego's quarterly deposit if the money is deposited at 5.8% compounded quarterly.

**G48. Equipment Purchase** Harv's Meats knows that it must buy a new machine in 4 years. The machine costs $12,000. In order to accumulate enough money to pay for the machine, Harv decides to deposit a sum of money at the end of each 6 months in an account paying 6% compounded semiannually. How much should each payment be?

**49. Auto Purchase** Barbara Margolius wants to buy a $24,000 car in 6 years. How much money must she deposit at the end of each quarter in an account paying 5% compounded quarterly so that she will have enough to pay for her car?

**50. Museum Purchase** The Chinns agree to sell an antique vase to a local museum for $19,000. They want to defer the receipt of this money until they retire in 5 years (and are in a lower tax bracket). If the museum can earn 5.8%, compounded annually, find the amount of each annual payment it should make into a sinking fund so that it will have the necessary $19,000 in 5 years.

**51. Real Estate Purchase** Diane sells some land in Nevada. She will be paid a lump sum of $60,000 in 7 years. Until then, the buyer pays 8% simple interest quarterly.

(a) Find the amount of each quarterly interest payment.

(b) The buyer sets up a sinking fund so that enough money will be present to pay off the $60,000. The buyer wants to make semiannual payments into the sinking fund; the account pays 6% compounded semiannually. Find the amount of each payment into the fund.

**52. Stamp Purchase** Joe Seniw bought a rare stamp for his collection. He agreed to pay a lump sum of $4000 after 5 years. Until then, he pays 6% simple interest semiannually.

(a) Find the amount of each semiannual interest payment.

(b) Joe sets up a sinking fund so that enough money will be present to pay off the $4000. He wants to make annual payments into the fund. The account pays 8% compounded annually. Find the amount of each payment.

**53. Retirement Planning** To save for retirement, Karla Harby put $300 each month into an ordinary annuity for 20 years. Interest was compounded monthly. At the end of the 20 years, the annuity was worth $147,126. What annual interest rate did she receive?

**54. Real Estate** Destiny Green made payments of $250 per month at the end of each month to purchase a piece of property. After 30 years, she owned the property, which she sold for $330,000. What annual interest rate would she need to earn on an ordinary annuity for a comparable rate of return?

**55. Retirement Planning** When Joe and Sarah graduate from college, each expects to work a total of 45 years. Joe begins saving for retirement immediately. He plans to deposit $600 at the end of each quarter into an account paying 8.1% interest, compounded quarterly, for 10 years. He will then leave his balance in the account, earning the same interest rate, but make no further deposits for 35 years. Sarah plans to save nothing during the first 10 years and then begin depositing $600 at the end of each quarter in an account paying 8.1% interest, compounded quarterly, for 35 years.

(a) Without doing any calculations, predict which one will have the most in his or her retirement account after 45 years. Then test your prediction by answering the following questions (calculation required to the nearest dollar).

(b) How much will Joe contribute to his retirement account?

(c) How much will be in Joe's account after 45 years?

(d) How much will Sarah contribute to her retirement account?

(e) How much will be in Sarah's account after 45 years?

**56.** In a 1992 Virginia lottery, the jackpot was $27 million. An Australian investment firm tried to buy all possible combinations of numbers, which would have cost $7 million. In fact, the firm ran out of time and was unable to buy all combinations, but ended up with the only winning ticket anyway. The firm received the jackpot in 20 equal annual payments of $1.35 million. Assume these payments meet the conditions of an ordinary annuity. (Data from: *Washington Post*, March 10, 1992, p. A1.)

(a) Suppose the firm can invest money at 8% interest compounded annually. How many years would it take until the investors would be further ahead than if they had simply invested the $7 million at the same rate? (*Hint:* Experiment with different values of *n*, the number of years, or use a graphing calculator to plot the value of both investments as a function of the number of years.)

(b) How many years would it take in part (a) at an interest rate of 12%?

✓**Checkpoint Answers**

1. (a) $2247.20  (b) $2120.00  (c) $6367.20

2. (a) $18,339.82  (b) $36,216.41

3. $188,406.70

4. (a) $232.38  (b) $262.97

5. (a) $104,812.44  (b) 8.9%

6. (a) $9446.24  (b) $9171.10

# 5.4 Annuities, Present Value, and Amortization

In the annuities studied previously, regular deposits were made into an interest-bearing account and the value of the annuity increased from 0 at the beginning to some larger amount at the end (the future value). Now we expand the discussion to include annuities that begin with an amount of money and make regular payments out of the account each period until the value of the annuity decreases to 0. Examples of such annuities are lottery jackpots, structured settlements imposed by a court in which the party at fault (or his or her insurance company) makes regular payments to the injured party, and trust funds that pay the recipients a fixed amount at regular intervals.

In order to develop the essential formula for dealing with "payout annuities," we need another useful algebraic fact. If $x$ is a nonzero number and $n$ is a positive integer, verify the following equality by multiplying out the right-hand side:[*]

$$x^{-1} + x^{-2} + x^{-3} + \cdots + x^{-(n-1)} + x^{-n} = x^{-n}(x^{n-1} + x^{n-2} + x^{n-3} + \cdots + x^1 + 1).$$

Now use the sum formula in the box on page 241 to rewrite the expression in parentheses on the right-hand side:

$$x^{-1} + x^{-2} + x^{-3} + \cdots + x^{-(n-1)} + x^{-n} = x^{-n}\left(\frac{x^n - 1}{x - 1}\right)$$

$$= \frac{x^{-n}(x^n - 1)}{x - 1} = \frac{x^0 - x^{-n}}{x - 1} = \frac{1 - x^{-n}}{x - 1}.$$

We have proved the following result:

> If $x$ is a nonzero real number and $n$ is a positive integer, then
>
> $$x^{-1} + x^{-2} + x^{-3} + \cdots + x^{-n} = \frac{1 - x^{-n}}{x - 1}.$$

## Present Value

 In Section 5.2, we saw that the present value of $FV$ dollars at interest rate $i$ per period for $n$ periods is the amount that must be deposited today (at the same interest rate) in order to produce $FV$ dollars in $n$ periods. Similarly, the **present value of an annuity** is the amount that must be deposited today (at the same compound interest rate as the annuity) to provide all the payments for the term of the annuity. It does not matter whether the payments are invested to accumulate funds or are paid out to disperse funds; the amount needed to provide the payments is the same in either case. We begin with ordinary annuities.

A relative has funded an annuity for Timothy Carlo that pays him $1500 at the end of each year for six years. Let us assume that the interest rate is 8% compounded annually. We will now investigate the present value of this annuity. To do this, we look separately at each payment that Tim receives. We will then find the present value of each payment—the amount needed now in order to make the payment in the future. The sum of these present values will be the present value of the annuity because it will provide all the payments.

To find the first $1500 payment (due in one year), the present value of $1500 at 8% annual interest is needed now. According to the present-value formula for compound interest in the Summary of Section 5.2 (with $FV = 1500$, $i = .08$, and $n = 1$), this present value is

$$\frac{1500}{1 + .08} = \frac{1500}{1.08} = 1500(1.08^{-1}) \approx \$1388.89.$$

This amount will grow to $1500 in one year.

For the second $1500 payment (due in two years), we need the present value of $1500 at 8% interest, compounded annually for two years. The present-value formula for compound interest (with $FV = 1500$, $i = .08$, and $n = 2$) shows that this present value is

$$\frac{1500}{(1 + .08)^2} = \frac{1500}{1.08^2} = 1500(1.08^{-2}) \approx \$1286.01.$$

---

[*]Remember that powers of $x$ are multiplied by *adding* exponents and that $x^n x^{-n} = x^{n-n} = x^0 = 1$.

Less money is needed for the second payment because it will grow over two years instead of one.

A similar calculation shows that the third payment (due in three years) has present value $1500(1.08^{-3})$. Continue in this manner to find the present value of each of the remaining payments, as summarized in Figure 5.18.

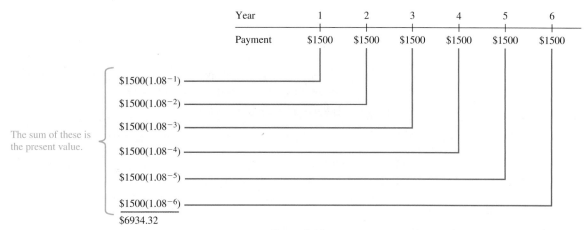

**Figure 5.18**

The left-hand column of Figure 5.18 shows that the present value is

$$1500 \cdot 1.08^{-1} + 1500 \cdot 1.08^{-2} + 1500 \cdot 1.08^{-3} + 1500 \cdot 1.08^{-4}$$
$$+ 1500 \cdot 1.08^{-5} + 1500 \cdot 1.08^{-6}$$
$$= 1500(1.08^{-1} + 1.08^{-2} + 1.08^{-3} + 1.08^{-4} + 1.08^{-5} + 1.08^{-6})$$
$$= \$6934.32. \tag{1}$$

Now apply the algebraic fact in the box on page 250 to the expression in parentheses (with $x = 1.08$ and $n = 6$). It shows that the sum (the present value of the annuity) is

$$1500 \left[ \frac{1 - 1.08^{-6}}{1.08 - 1} \right] = 1500 \left[ \frac{1 - 1.08^{-6}}{.08} \right] = \$6934.32.$$

This amount will provide for all six payments and leave a zero balance at the end of six years (give or take a few cents due to rounding to the nearest penny at each step).

We have just shown the model for finding a formula for the present value of any ordinary annuity. Suppose that a payment of *PMT* dollars is made at the end of each period for $n$ periods, at interest rate $i$ per period. Then the present value of this annuity can be found by using the procedure, with these replacements:

| 1500 | .08 | 1.08 | 6 |
|:---:|:---:|:---:|:---:|
| ↓ | ↓ | ↓ | ↓ |
| *PMT* | $i$ | $1 + i$ | $n$ |

The present value is the sum in equation **(1)**, which now becomes

$$PV = PMT[(1 + i)^{-1} + (1 + i)^{-2} + (1 + i)^{-3} + \cdots + (1 + i)^{-n}].$$

Apply the algebraic fact in the box on page 250 to the expression in brackets (with $x = 1 + i$). Then we have

$$PV = PMT \left[ \frac{1 - (1 + i)^{-n}}{(1 + i) - 1} \right] = PMT \left[ \frac{1 - (1 + i)^{-n}}{i} \right].$$

The quantity in brackets in the right-hand part of the preceding equation is sometimes written $a_{\overline{n}|i}$ (read "a-angle-$n$ at $i$"). So we can summarize as follows.

> ## Present Value of an Ordinary Annuity
>
> The present value $PV$ of an ordinary annuity is given by
>
> $$PV = PMT\left[\frac{1 - (1 + i)^{-n}}{i}\right], \quad \text{or} \quad PV = PMT \cdot a_{\overline{n}|i},$$
>
> where
>
> $PMT$ is the payment at the end of each period;
>
> $i = r/m$ is the interest rate per period, $r$ is the annual interest rate, $m$ is the number of periods per year; and
>
> $n$ is the total number of periods.

⚠ **CAUTION** Do not confuse the formula for the present value of an annuity with the one for the future value of an annuity. Notice the difference: The numerator of the fraction in the present-value formula is $1 - (1 + i)^{-n}$, but in the future-value formula, it is $(1 + i)^n - 1$.

```
N=84
I%=4.7
•PV=142924.7907
PMT=-2000
FV=0
P/Y=12
C/Y=12
PMT:END BEGIN
```

**Figure 5.19**

✓ **Checkpoint 1**

An insurance company offers to pay Jane Parks an ordinary annuity of $3000 per quarter for five years *or* the present value of the annuity now. If the interest rate is 6%, find the present value.

**EXAMPLE 1** **Insurance Settlement** Camila was in an auto accident. She sued the person at fault and was awarded a structured settlement in which an insurance company will pay her $2000 at the end of each month for the next seven years. How much money should the insurance company invest now at 4.7%, compounded monthly, to guarantee that all the payments can be made?

**Solution** The payments form an ordinary annuity. The amount needed to fund all the payments is the present value of the annuity. Apply the present-value formula with $PMT = 2000, n = 7 \cdot 12 = 84$, and $i = r/m = .047/12$ (the interest rate per month). The insurance company should invest

$$PV = PMT\left[\frac{1 - (1 + i)^{-n}}{i}\right] = 2000\left[\frac{1 - (1 + .047/12)^{-84}}{.047/12}\right] = \$142{,}924.79.$$

The entries in the TVM solver appear in Figure 5.19. ✓ 1

```
N=240
I%=5.2
PV=-400000
•PMT=2684.216207
FV=0
P/Y=12
C/Y=12
PMT:END BEGIN
```

**Figure 5.20**

✓ **Checkpoint 2**

Carl Dehne has $80,000 in an account paying 4.8% interest, compounded monthly. He plans to use up all the money by making equal monthly withdrawals for 15 years. If the interest rate is 4.8%, find the amount of each withdrawal.

**EXAMPLE 2** **Retirement Planning** To supplement his pension in the early years of his retirement, Hong plans to use $400,000 of his savings as an ordinary annuity that will make monthly payments to him for 20 years. If the interest rate is 5.2%, how much will each payment be?

**Solution** The present value of the annuity is $PV = \$400{,}000$, the monthly interest rate is $i = r/m = .052/12$, and $n = 12 \cdot 20 = 240$ (the number of months in 20 years). Solve the present-value formula for the monthly payment $PMT$:

$$PV = PMT\left[\frac{1 - (1 + i)^{-n}}{i}\right]$$

$$400{,}000 = PMT\left[\frac{1 - (1 + .052/12)^{-240}}{.052/12}\right]$$

$$PMT = \frac{400{,}000}{\left[\dfrac{1 - (1 + .052/12)^{-240}}{.052/12}\right]} = \$2684.22.$$

Hong will receive $2684.22 a month (about $51,382 per year) for 20 years. The entries in the TMV solver appear in Figure 5.20. ✓ 2

## Loans and Amortization

If you take out a car loan or a home mortgage, you repay it by making regular payments to the bank. From the bank's point of view, your payments are an annuity that is paying the bank a fixed amount each month. The present value of this annuity is the amount you borrowed.

We can use the present value of an ordinary annuity formula, as we did in Example 2, to determine the monthly payment needed for common loans such as auto loans, student loans, and mortgages. In these cases, $PV$ is the amount borrowed today, $i$ is the monthly interest rate (most loans consist of monthly payments), and $n$ is the number of payments.

**EXAMPLE 3**   **Auto Loan** In October 2016, Chase Bank advertised a used-car auto loan rate of 2.84% for a 5-year loan. Shelley wants to buy a used car for $18,700. Find the amount of each payment. (Data from: www.chase.com.)

**Solution** Use the present value formula of an ordinary annuity, with $PV = 18,700$, $n = 12*5 = 60$, $i = r/m = .0284/12$ (the monthly interest rate). Then solve for payment $PMT$.

$$PV = PMT\left[\frac{1 - (1 + i)^{-n}}{i}\right]$$

$$18,700 = PMT\left[\frac{1 - (1 + .0284/12)^{-60}}{.0284/12}\right]$$

$$PMT = \frac{18,700}{\left[\dfrac{1 - (1 + .0284/12)^{-60}}{.0284/12}\right]} \quad \text{Solve for } PMT.$$

$$PMT = \$334.69$$

The entries in the TVM solver appear in Figure 5.21.

```
N=60
I%=2.84
PV=18700
•PMT=-334.68657...
FV=0
P/Y=12
C/Y=12
PMT:END BEGIN
```

**Figure 5.21**

✓ **Checkpoint 3**

Yulanda uses a Chase Bank auto loan to purchase a new car priced at $35,650. Her 4-year loan has an interest rate of 2.54%. What is the monthly payment?

A loan is **amortized** if both the principal and interest are paid by a sequence of equal periodic payments. The periodic payment needed to amortize a loan may be found, as in Example 3, by solving the present-value formula for $PMT$.

### Amortization Payments

A loan of $PV$ dollars at interest rate $i$ per period may be amortized in $n$ equal periodic payments of $PMT$ dollars made at the end of each period, where

$$PMT = \frac{PV}{\left[\dfrac{1 - (1 + i)^{-n}}{i}\right]} = \frac{PVi}{1 - (1 + i)^{-n}}.$$

**EXAMPLE 4**   **Mortgage Loan** On October 10, 2016, the average rate for a 30-year fixed mortgage was 3.37%. Assume a down payment of 20% on a home purchase of $285,000. (Data from: Zillow.com.)

**(a)** Find the monthly payment needed to amortize this loan.

**Solution** The down payment is .20(285,000) = $57,000. Thus, the loan amount $PV$ is $285,000 - $57,000 = $228,000. We can now apply the formula in the preceding box, with $n = 12(30) = 360$ (the number of monthly payments in 30 years), and monthly interest rate $i = r/m = .0337/12$.[*]

$$PMT = \frac{PVi}{1 - (1 + i)^{-n}} = \frac{228,000(.0337/12)}{1 - (1 + .0337/12)^{-360}} = \$1007.35$$

---

[*]Mortgage rates are quoted in terms of annual interest, but it is always understood that the monthly rate is $\frac{1}{12}$ of the annual interest rate.

```
N=36Ø
 I%=3.37
 PV=228ØØØ
•PMT=-1ØØ7.34843
 FV=Ø
 P/Y=12
 C/Y=12
 PMT:END BEGIN
```

**Figure 5.22(a)**

```
N=24Ø
 I%= 3.37
 PV=17571Ø.2335
•PMT=-1ØØ7.35
 FV=Ø
 P/Y=12
 C/Y=12
 PMT:END BEGIN
```

**Figure 5.22(b)**

✓ **Checkpoint 4**

Find the remaining balance after 15 years.

Monthly payments of \$1007.34 are required to amortize the loan. The entries in the TVM solver appear in Figure 5.22(a).

**(b)** After 10, years approximately how much is owed on the mortgage?

**Solution** You may be tempted to say that, after 10 years of payments on a 30-year mortgage, the balance will be reduced by one-third. However, in the early months of repaying a loan, a significant portion of each payment goes to pay interest. Thus, much less than one-third of the mortgage is paid off in the first 10 years, as we now see.

After 10 years (120 payments), the 240 remaining payments can be thought of as an annuity. The present value for this annuity is the (approximate) remaining balance on the mortgage. Hence, we use the present value formula with $PMT = 1007.34$, $i = r/m = .0337/12$, and $n = 240$:

$$PV = 1007.35\left[\frac{1 - (1 + .0337/12)^{-240}}{(.0337/12)}\right] = \$175,710.23$$

So the remaining balance is about \$175,710.23. The actual balance probably differs slightly from this amount because payments and interest amounts are rounded to the nearest penny. The entries in the TVM solver appear in Figure 5.22(b). ✓ 4

Example 4(b) illustrates an important fact: Even though equal *payments* are made to amortize a loan, the loan *balance* does not decrease in equal steps. The method used to estimate the remaining balance in Example 4(b) works in the general case. If $n$ payments are needed to amortize a loan and $x$ payments have been made, then the remaining payments form an annuity of $n - x$ payments. So we apply the present-value formula with $n - x$ in place of $n$ to obtain this result.

## Remaining Balance

If a loan can be amortized by $n$ payments of $PMT$ dollars each at an interest rate $i$ per period, then the *approximate* remaining balance $B$ after $x$ payments is

$$B = PMT\left[\frac{1 - (1 + i)^{-(n-x)}}{i}\right].$$

## Amortization Schedules

The remaining-balance formula is a quick and convenient way to get a reasonable estimate of the remaining balance on a loan, but it is not accurate enough for a bank or business, which must keep its books exactly. To determine the exact remaining balance after each loan payment, financial institutions normally use an **amortization schedule,** which lists how much of each payment is interest, how much goes to reduce the balance, and how much is still owed after each payment.

**EXAMPLE 5**    **Personal Loan** Beth Hill borrows \$1000 for one year at 12% annual interest, compounded monthly.

**(a)** Find her monthly payment.

**Solution** Apply the amortization payment formula with $PV = 1000, n = 12$, and monthly interest rate $i = r/m = .12/12 = .01$. Her payment is

$$PMT = \frac{PVi}{1 - (1 + i)^{-n}} = \frac{1000(.01)}{1 - (1 + .01)^{-12}} = \$88.85.$$

**(b)** After making five payments, Beth decides to pay off the remaining balance. Approximately how much must she pay?

**Solution** Apply the remaining-balance formula just given, with $PMT = 88.85$, $i = .01$, and $n - x = 12 - 5 = 7$. Her approximate remaining balance is

$$B = PMT\left[\frac{1 - (1 + i)^{-(n-x)}}{i}\right] = 88.85\left[\frac{1 - (1 + .01)^{-7}}{.01}\right] = \$597.80.$$

**(c)** Construct an amortization schedule for Beth's loan.

**Solution** An amortization schedule for the loan is shown in the following table. It was obtained as follows: The annual interest rate is 12% compounded monthly, so the interest rate per month is $12\%/12 = 1\% = .01$. When the first payment is made, one month's interest, namely, $.01(1000) = \$10$, is owed. Subtracting this from the $88.85 payment leaves $78.85 to be applied to repayment. Hence, the principal at the end of the first payment period is $1000 - 78.85 = \$921.15$, as shown in the "payment 1" line of the table.

When payment 2 is made, one month's interest on the new balance of $921.15 is owed, namely, $.01(921.15) = \$9.21$. Continue as in the preceding paragraph to compute the entries in this line of the table. The remaining lines of the table are found in a similar fashion.

| Payment Number | Amount of Payment | Interest for Period | Portion to Principal | Principal at End of Period |
|---|---|---|---|---|
| 0 | — | — | — | $1000.00 |
| 1 | $88.85 | $10.00 | $78.85 | 921.15 |
| 2 | 88.85 | 9.21 | 79.64 | 841.51 |
| 3 | 88.85 | 8.42 | 80.43 | 761.08 |
| 4 | 88.85 | 7.61 | 81.24 | 679.84 |
| 5 | 88.85 | 6.80 | 82.05 | 597.79 |
| 6 | 88.85 | 5.98 | 82.87 | 514.92 |
| 7 | 88.85 | 5.15 | 83.70 | 431.22 |
| 8 | 88.85 | 4.31 | 84.54 | 346.68 |
| 9 | 88.85 | 3.47 | 85.38 | 261.30 |
| 10 | 88.85 | 2.61 | 86.24 | 175.06 |
| 11 | 88.85 | 1.75 | 87.10 | 87.96 |
| 12 | 88.84 | .88 | 87.96 | 0 |

Note that Beth's remaining balance after five payments differs slightly from the estimate made in part (b).

The final payment in the amortization schedule in Example 5(c) differs from the other payments. It often happens that the last payment needed to amortize a loan must be adjusted to account for rounding earlier and to ensure that the final balance will be exactly 0.

**TECHNOLOGY TIP** To create an amortization table with the TI-84 calculator, use the amortization table program. Spreadsheets are another useful tool for creating amortization tables. Microsoft Excel (Microsoft Corporation Excel © 2016) has a built-in feature for calculating monthly payments. Figure 5.23 shows an Excel amortization table for Example 5. For more details, see the *Spreadsheet Manual*, also available with this text.

| | A | B | C | D | E | F |
|---|---|---|---|---|---|---|
| 1 | Prnt# | Payment | Interest | Principal | End Principal | |
| 2 | 0 | | | | 1000 | |
| 3 | 1 | 88.85 | 10.00 | 78.85 | 921.15 | |
| 4 | 2 | 88.85 | 9.21 | 79.64 | 841.51 | |
| 5 | 3 | 88.85 | 8.42 | 80.43 | 761.08 | |
| 6 | 4 | 88.85 | 7.61 | 81.24 | 679.84 | |
| 7 | 5 | 88.85 | 6.80 | 82.05 | 597.79 | |
| 8 | 6 | 88.85 | 5.98 | 82.87 | 514.92 | |
| 9 | 7 | 88.85 | 5.15 | 83.70 | 431.22 | |
| 10 | 8 | 88.85 | 4.31 | 84.54 | 346.68 | |
| 11 | 9 | 88.85 | 3.47 | 85.38 | 261.30 | |
| 12 | 10 | 88.85 | 2.61 | 86.24 | 175.06 | |
| 13 | 11 | 88.85 | 1.75 | 87.10 | 87.96 | |
| 14 | 12 | 88.85 | 0.88 | 87.97 | -0.01 | |

**Figure 5.23**

When doing financial planning, it can be helpful to determine the amount of interest paid over the course of a loan. For example, in Example 4, borrowing $228,000 at 3.37% annual interest for a 30-year fixed mortgage yields a payment of $1007.35 per month. Over the lifetime of the loan, that yields

$$(30)(12)(1007.35) = \$362,646.00$$

in total payments to the bank. Thus, $362,646 - 228,000 = 134,646$ is paid in total interest to the bank.

Many financial planners recommend making extra payments toward the principal of the loan each month. One way to make an extra payment each year is by adding $\frac{1}{12}$ of the regular monthly payment to each monthly payment. This extra amount is then applied to the principal of the loan and thus reduces the amount of future interest payments. We can use the TVM solver to find out how much more quickly (and how much less interest we pay) with this approach. We simply enter the payment amount as a negative number and then solve for the number of payments. This method is described in Example 6.

**EXAMPLE 6**    Let us revisit the 30-year fixed mortgage loan described in Example 4. The loan amount was $228,000 at 3.37% annual interest. Each month, an extra amount of money equal to $\frac{1}{12}$ the monthly payment was added to the principal.

**(a)** How much quicker is the loan paid off?

**Solution** Paying $\frac{1}{12}$ of the required monthly payment extra each month is equivalent to making one extra mortgage payment in that year. In this case, $\frac{1}{12}$ of the monthly payment is:

$$\left(\frac{1}{12}\right)(1007.35) = \$83.95.$$

So now the monthly payment amount is $1007.35 + 83.95 = \$1091.30$. In the TVM solver, enter the information as in Figure 5.24(a). Put the curser in the N = section and solve for the number of payments. The solution appears in Figure 5.24(b).

The TVM solver tells us that it will take 316 (we round up) payments to amortize the loan. This represents 44 ($360 - 316$) fewer payments. Because 44 payments is equivalent to 3 years and 8 months of payments, that is how much more quickly the loan will be paid off.

**(b)** How much money in interest payments is saved?

**Solution** This problem is a little trickier. Notice that in part (a), the TVM solver states that 315.0971424 is the number of payments. This means that it takes .0971424 of a payment to pay off the remaining balance for the 316th payment. However, we can find this amount by multiplying the payment amount by this fraction of a payment, which is

$$(\$1091.30)(.0971424) = \$106.01 \text{ (rounded to the nearest cent)}.$$

```
N=
I%=3.37
PV=228000
PMT=-1091.3
FV=0
P/Y=12
C/Y=12
PMT:END BEGIN
```
**Figure 5.24(a)**

```
•N=315.0971424
 I%= 3.37
 PV= 228000
 PMT=-1091.3
 FV=0
 P/Y=12
 C/Y=12
 PMT:END BEGIN
```
**Figure 5.24(b)**

Now we can find the total amount of money paid over the loan by multiplying the full monthly payment amount ($1091.30) by 315 and then adding this extra amount for the last payment:

$$(\$1091.30)(315) + \$106.01 = \$343,865.51$$

We saw earlier that making 360 equal payments over 30 years yielded a total loan payment amount of $362,646.00. By adding an extra $83.95 per month to our payment, the borrower would save, in interest paid to the bank,

$$\$362,646.00 - 343,865.51 = \$18,780.49. \quad \checkmark_5$$

✔ **Checkpoint 5**

The same mortgage amount as Example 6 is paid off in 15 years at 2.75%.

**(a)** Find the monthly payment.

**(b)** If an extra $100 is paid each month toward the principal, how many payments are needed to pay off the loan?

**(c)** How much interest is saved by paying an extra $100 a month?

## Annuities Due

We want to find the present value of an annuity due in which 6 payments of *PMT* dollars are made at the *beginning* of each period, with interest rate *i* per period, as shown schematically in Figure 5.25.

**Figure 5.25**

The present value is the amount needed to fund all 6 payments. Since the first payment earns no interest, *PMT* dollars are needed to fund it. Now look at the last 5 payments by themselves in Figure 5.26.

**Figure 5.26**

If you think of these 5 payments as being made at the end of each period, you see that they form an ordinary annuity. The money needed to fund them is the present value of this ordinary annuity. So the present value of the annuity due is given by

$$PMT + \begin{array}{c} \text{Present value of the ordinary} \\ \text{annuity of 5 payments} \end{array}$$

$$PMT + PMT\left[\frac{1 - (1 + i)^{-5}}{i}\right].$$

Replacing 6 by *n* and 5 by *n* − 1, and using the argument just given, produces the general result that follows.

### Present Value of an Annuity Due

The present value *PV* of an annuity due is given by

$$PV = PMT + PMT\left[\frac{1 - (1 + i)^{-(n-1)}}{i}\right],$$

$$PV = \underset{\text{payment}}{\text{One}} \quad + \quad \begin{array}{c} \text{Present value of} \\ \text{an ordinary annuity} \\ \text{of } n - 1 \text{ payments} \end{array}$$

where

*PMT* is the payment at the beginning of each period;

$i = r/m$ is the interest rate per period, *r* is the annual interest rate, *m* is the number of periods per year;

*n* is the total number of periods.

```
N=20
I%=4.22
•PV=694593.4902
PMT=-50000
FV=0
P/Y=1
C/Y=1
PMT:END BEGIN
```
**Figure 5.27**

✓**Checkpoint 6**

If an Illinois player wins a Lotto jackpot of $60 million that would be paid out in 30 annual payments, and the state can earn 5.25% annual interest, what is the cash value of the jackpot?

**EXAMPLE 7**    **Lottery Winnings** The Illinois *Lottery Winners' Handbook* discusses the options of how to receive the winnings for a $1 million scratch ticket. One option is to take 20 annual payments of approximately $50,000, which is $1 million divided into 20 equal payments. The other option is to take a lump-sum payment (which is often called the cash value). If the Illinois Lottery Commission can earn 4.22% annual interest, how much is the cash value?

**Solution** The yearly payments form a 20-payment annuity due. An equivalent amount now is the present value of this annuity. Apply the present-value formula with $PMT = 50,000$, $i = .0422$, and $n = 20$:

$$PV = PMT + PMT\left[\frac{1 - (1 + i)^{-(n-1)}}{i}\right] = 50,000 + 50,000\left[\frac{1 - (1 + .0422)^{-19}}{.0422}\right]$$

$$= \$694,593.49.$$

The cash value is $694,593.49. The entries in the TVM solver appear in Figure 5.27. ✓6

# 5.4 Exercises

*Unless noted otherwise, all payments and withdrawals are made at the end of the period.*

1. Explain the difference between the present value of an annuity and the future value of an annuity.

**Ordinary Annuities** *Find the present value of each ordinary annuity. (See Examples 1 and 2.)*

2. Payments of $890 each year for 16 years at 6% compounded annually

3. Payments of $1400 each year for 8 years at 6% compounded annually

4. Payments of $10,000 semiannually for 15 years at 7.5% compounded semiannually

5. Payments of $50,000 quarterly for 10 years at 5% compounded quarterly

6. Payments of $15,806 quarterly for 3 years at 6.8% compounded quarterly

**Ordinary Annuities** *Find the amount necessary to fund the given withdrawals. (See Example 1.)*

7. Quarterly withdrawals of $650 for 5 years; interest rate is 4.9%, compounded quarterly.

8. Yearly withdrawals of $1200 for 14 years; interest rate is 5.6%, compounded annually.

9. Monthly withdrawals of $425 for 10 years; interest rate is 6.1%, compounded monthly.

10. Semiannual withdrawals of $3500 for 7 years; interest rate is 5.2%, compounded semiannually.

**Ordinary Annuities** *Find the payment made by the ordinary annuity with the given present value. (See Example 2.)*

11. $90,000; monthly payments for 22 years; interest rate is 4.9%, compounded monthly.

12. $45,000; monthly payments for 11 years; interest rate is 5.3%, compounded monthly.

13. $275,000; quarterly payments for 18 years; interest rate is 6%, compounded quarterly.

14. $330,000; quarterly payments for 30 years; interest rate is 6.1% compounded quarterly.

**Amortization** *Find the payment necessary to amortize each of the given loans. (See Example 4.)*

15. $2500; 8% interest; 6 quarterly payments

16. $41,000; 9% interest; 10 semiannual payments

17. $90,000; 7% interest; 12 annual payments

18. $140,000; 12% interest; 15 quarterly payments

19. $7400; 8.2% interest; 18 semiannual payments

20. $5500; 9.5% interest; 24 monthly payments

**Mortgages** *In October 2016, the mortgage interest rates listed in Exercises 21–24 for the given companies were listed at www.hsh.com. Find the monthly payment necessary to amortize the given loans. (See Example 4.)*

21. $225,000 at 3.68% for 30 years from loanDepot

22. $330,000 at 3.50% for 30 years from Quicken Loans

23. $140,000 at 2.63% for 15 years from CloseYourOwnLoan

24. $180,000 at 2.50% for 10 years from CloseYourOwnLoan

**Loans** *Find the monthly payment and estimate the remaining balance (to the nearest dollar). Assume interest is on the unpaid balance. The interest rates are from national averages from www.bankrate.com in October 2016. (See Examples 4 and 5.)*

25. Six-year new car loan for $26,799 at 3.74%; remaining balance after 2 years

**26.** Three-year used car loan for $15,875 at 2.89%; remaining balance after 1 year

**27.** Thirty-year mortgage for $210,000 at 3.63%; remaining balance after 12 years

**28.** Fifteen-year mortgage for $195,000 at 2.75%; remaining balance after 4.5 years

**Amortization** *Use the amortization table in Example 5(c) to answer the questions in Exercises 29–32.*

**29.** How much of the 5th payment is interest?

**30.** How much of the 10th payment is used to reduce the debt?

**31.** How much interest is paid in the first 5 months of the loan?

**32.** How much interest is paid in the last 5 months of the loan?

**Lottery** *Find the cash value of the lottery jackpot (to the nearest dollar). Yearly jackpot payments begin immediately (30 for Mega Millions and 30 for Powerball). Assume the lottery can invest at the given interest rate. (See Example 7.)*

**33.** Powerball: $57.6 million; 5.1% interest

**34.** Powerball: $207 million; 5.78% interest

**35.** Mega Millions: $41.6 million; 4.735% interest

**36.** Mega Millions: $23.4 million; 4.23% interest

**Work the following applied problems.**

**37. Financing** An auto stereo dealer sells a stereo system for $600 down and monthly payments of $30 for the next 3 years. If the interest rate is 1.25% *per month* on the unpaid balance, find

(a) the cost of the stereo system;

(b) the total amount of interest paid.

**38. Auto Loan** John Kushida buys a used car costing $16,000. He agrees to make payments at the end of each monthly period for 4 years. He pays 4.35% interest, in monthly payments.

(a) What is the amount of each payment?

(b) Find the total amount of interest Kushida will pay.

**39. Land Mortgage** A speculator agrees to pay $15,000 for a parcel of land; this amount, with interest, will be paid over 4 years with semiannual payments at an interest rate of 10%. Find the amount of each payment.

**40. Auto Loan** Alan Stasa buys a new car costing $26,750. What is the monthly payment if the interest rate is 4.2%, and the loan is for 60 months? Find the total amount of interest Alan will pay.

**Student Loan** *A student education loan has two repayment options. The standard plan repays the loan in 10 years with equal monthly payments. The extended plan allows from 12 to 30 years to repay the loan. A student borrows $35,000 at 7.43% interest.*

**41.** Find the monthly payment and total interest paid under the standard plan.

**42.** Find the monthly payment and total interest paid under the extended plan with 20 years to pay off the loan.

*Use the formula for the approximate remaining balance to work each problem. (See Examples 4(b) and 5(b).)*

**43. Business Loan** When Teresa Flores opened her law office, she bought $14,000 worth of law books and $7200 worth of office furniture. She paid $1200 down and agreed to amortize the balance with semiannual payments for 5 years at 12% interest.

(a) Find the amount of each payment.

(b) When her loan had been reduced below $5000, Flores received a large tax refund and decided to pay off the loan. How many payments were left at this time?

**44. Mortgage** Kareem Adams buys a house for $285,000. He pays $60,000 down and takes out a mortgage at 6.9% on the balance. Find his monthly payment and the total amount of interest he will pay if the length of the mortgage is

(a) 15 years;

(b) 20 years;

(c) 25 years.

(d) When will half the 20-year loan be paid off?

**45. Mortgage** Susan Carver will purchase a home for $257,000. She will use a down payment of 20% and finance the remaining portion at 3.9% interest for 30 years.

(a) What will be the monthly payment?

(b) How much will remain on the loan after making payments for 5 years?

(c) How much interest will be paid on the total amount of the loan over the course of 30 years?

**46. Mortgage** Mohsen Manouchehri will purchase a $230,000 home with a 20-year mortgage. If he makes a down payment of 20% and the interest rate is 3.3%,

(a) what will the monthly payment be?

(b) how much will he owe after making payments for 8 years?

(c) how much in total interest will he pay over the course of the 20-year loan?

*Work each problem.*

**47. Retirement Planning** Erin Pupillo and her employer contribute $400 at the end of each month to her retirement account, which earns 7% interest, compounded monthly. When she retires after 45 years, she plans to make monthly withdrawals for 30 years. If her account earns 5% interest, compounded monthly, then when she retires, what is her maximum possible monthly withdrawal (without running out of money)?

**48. Financial Planning** Jim Milliken won a $15,000 prize. On March 1, he deposited it in an account earning 5.2% interest, compounded monthly. On March 1 one year later, he begins to withdraw the same amount at the beginning of each month for a year. Assuming that he uses up all the money in the account, find the amount of each monthly withdrawal.

**49. Retirement Planning** Catherine Dohanyos plans to retire in 20 years. She will make 20 years of monthly contributions to her retirement account. One month after her last contribution, she will begin the first of 10 years of withdrawals. She wants to withdraw $2500 per month. How large must her monthly contributions be in order to accomplish her goal if the account earns interest of 7.1% compounded monthly for the duration of her contributions and the 120 months of withdrawals?

**50. Retirement Planning** Mateo plans to retire in 25 years. He will make 25 years of monthly contributions to his retirement account. One month after his last contribution, he will begin the first of 10 years of withdrawals. He wants to withdraw $3000 per month. How large must his monthly contributions be in order to accomplish his goal if the account earns interest of 6.8% compounded monthly for the duration of his contributions and the 120 months of withdrawals?

**51. Retirement Planning** Marcus plans to retire in 20 years. Marcus will make 10 years (120 months) of equal monthly payments into his account. Ten years after his last contribution, he will begin the first of 120 monthly withdrawals of $3400 per month. Assume that the retirement account earns interest of 8.2% compounded monthly for the duration of his contributions, the 10 years in between his contributions and the beginning of his withdrawals, and the 10 years of withdrawals. How large must Marcus's monthly contributions be in order to accomplish his goal?

**52. Retirement Planning** Joshua plans to retire in 25 years. He will make 15 years (180 months) of equal monthly payments into his account. Ten years after his last contribution, he will begin the first of 120 monthly withdrawals of $2900 per month. Assume that the retirement account earns interest of 5.4% compounded monthly for the duration of his contributions, the 10 years in between his contributions and the beginning of his withdrawals, and the 10 years of withdrawals. How large must Joshua's monthly contributions be in order to accomplish his goal?

**53. Auto Loan** Jessica Walters takes out a 6-year fixed-rate auto loan for $35,000 at 5.2% annual interest.

(a) Find the monthly payment for the loan.

(b) Find the total interest paid over the course of the loan.

(c) If Jessica pays an additional $100 per month toward the principal, how many payments will it take to pay off the loan?

(d) How much is the total interest paid on the loan when the additional $100 is paid each month?

**54. Mortgage** Ling Wu takes out a 30-year fixed rate mortgage for $325,000 at 3.9% annual interest.

(a) Find the monthly payment for the loan.

(b) Find the total interest paid over the course of the loan.

(c) If Ling pays an additional $150 per month toward the principal, how many payments will it take to pay off the loan?

(d) How much is the total interest of the loan when the additional $150 is paid each month?

*In Exercises 55–58, prepare an amortization schedule showing the first four payments for each loan. (See Example 5(c).)*

**55. Insurance Payment** An insurance firm pays $4000 for a new printer for its computer. It amortizes the loan for the printer in 4 annual payments at 8% interest.

**56. Commercial Auto Loan** Large semitrailer trucks cost $72,000 each. Ace Trucking buys such a truck and agrees to pay for it by a loan that will be amortized with 9 semiannual payments at 6% interest.

**57. Business Loan** One retailer charges $1048 for a certain computer. A firm of tax accountants buys 8 of these computers. It makes a down payment of $1200 and agrees to amortize the balance with monthly payments at 12% interest for 4 years.

**58. Business Loan** Joan Varozza plans to borrow $20,000 to stock her small boutique. She will repay the loan with semiannual payments for 5 years at 7% interest.

✓**Checkpoint Answers**

1. $51,505.92

2. $624.33

3. $781.86

4. $142,177.53

5. (a) $1547.26
   (b) 167
   (c) $3937.45

6. $31,456,910.67

# CHAPTER 5 Summary and Review

## Key Terms and Symbols

**5.1** simple interest
principal
rate
time
future value (maturity
　value)
present value
discount and
　T-bills

**5.2** compound interest
compound amount
compounding period
nominal rate (stated rate)
effective rate (APY)
present value

**5.3** annuity
payment period

term of an annuity
ordinary annuity
future value of an
　ordinary annuity
sinking fund
annuity due
future value of an
　annuity due

**5.4** present value of an
　ordinary annuity
amortization payments
remaining balance
amortization schedule
present value of an
　annuity due

## Chapter 5 Key Concepts

**A Strategy for Solving
Finance Problems**

We have presented a lot of new formulas in this chapter. By answering the following questions, you can decide which formula to use for a particular problem.

1. If simple interest is being used, what is being sought—interest amount, future value, present value, or discount?

2. If compound interest is being used, does it involve a lump sum (single payment) or an annuity (sequence of payments)?

　**(a)** For a lump sum,

　　**(i)** is ordinary compound interest involved?

　　**(ii)** what is being sought—present value, future value, number of periods, or effective rate (APY)?

　**(b)** For an annuity,

　　**(i)** is it an ordinary annuity (payment at the end of each period) or an annuity due (payment at the beginning of each period)?

　　**(ii)** what is being sought—present value, future value, or payment amount?

Once you have answered these questions, choose the appropriate formula and work the problem. As a final step, consider whether the answer you get makes sense. For instance, the amount of interest or the payments in an annuity should be fairly small compared with the total future value.

**Key Formulas**

**List of Variables**

$r$ is the annual interest rate.
$m$ is the number of periods per year.
$i$ is the interest rate per period.　　　　$i = \dfrac{r}{m}$
$t$ is the number of years.
$n$ is the number of periods.　　　　$n = tm$
$x$ is the number of remaining payments.
$PV$ is the principal or present value.
$FV$ is the future value of a lump sum.
$FV$ is the future value of an annuity.
$PMT$ is the periodic payment in an annuity.
$B$ is the remaining balance on a loan.

| Interest | Simple Interest | Compound Interest |
|---|---|---|
| Interest | $I = PVrt$ | $I = FV - PV$ |
| Future value | $FV = PV(1 + rt)$ | $FV = PV(1 + i)^n$ |
| Present value | $PV = \dfrac{FV}{1 + rt}$ | $PV = \dfrac{FV}{(1 + i)^n} = FV(1 + i)^{-n}$ |

$$\text{Effective rate (or APY) } r_E = \left(1 + \frac{r}{m}\right)^m - 1$$

**Discount**          If $D$ is the **discount** on a T-bill with face value $PV$ at simple interest rate $r$ for $t$ years, then $D = PVrt$.

**Continuous Interest**   If $PV$ dollars are deposited for $t$ years at interest rate $r$ per year, compounded continuously, the **compound amount (future value)** is $FV = PVe^{rt}$.

The **present value** $PV$ of $FV$ dollars at interest rate $r$ per year compounded continuously for $t$ years is

$$PV = \frac{FV}{e^{rt}}.$$

**Remaining Balance (Approximate)**   If a loan can be amortized by $n$ payments of $PMT$ dollars, each at an interest rate $i$ per period, then the approximate remaining balance $B$ after $x$ payments is

$$B = PMT\left[\frac{1 - (1 + i)^{-(n-x)}}{i}\right].$$

**Annuities**

Ordinary annuity   Future value   $FV = PMT\left[\dfrac{(1 + i)^n - 1}{i}\right] = PMT \cdot s_{\overline{n}|i}$

Present value   $PV = PMT\left[\dfrac{1 - (1 + i)^{-n}}{i}\right] = PMT \cdot a_{\overline{n}|i}$

Annuity due   Future value   $FV = PMT\left[\dfrac{(1 + i)^{n+1} - 1}{i}\right] - PMT$

Present value   $PV = PMT + PMT\left[\dfrac{1 - (1 + i)^{-(n-1)}}{i}\right]$

# Chapter 5 Review Exercises

**Loans** *Find the simple interest for the following loans.*

1. $4902 at 6.5% for 11 months

2. $42,368 at 9.22% for 5 months

3. $3478 at 7.4% for 88 days

4. $2390 at 8.7% from May 3 to July 28

**Bonds** *Find the semiannual (simple) interest payment and the total interest earned over the life of the bond.*

5. $12,000 Merck Company 6-year bond at 4.75% annual interest

6. $20,000 General Electric 9-year bond at 5.25% annual interest

**Loans** *Find the maturity value for each simple interest loan.*

7. $7750 at 6.8% for 4 months

8. $15,600 at 8.2% for 9 months

**Present Value** *Find the present value of the given future amounts; use simple interest.*

9. $3250 in 11 months; money earns 5.87%

10. $459.57 in 7 months; money earns 5.5%

11. $80,612 in 128 days; money earns 6.77%

12. A 9-month $7000 Treasury bill sells at a discount rate of 3.5%. Find the amount of the discount and the price of the T-bill.

13. A 6-month $10,000 T-bill sold at a 4% discount. Find the actual rate of interest paid by the Treasury.

14. For a given amount of money at a given interest rate for a given period greater than 1 year, does simple interest or compound interest produce more interest? Explain.

**Investing** *Find the compound amount and the amount of interest earned in each of the given scenarios.*

**15.** $2800 at 6% compounded annually for 12 years

**16.** $57,809.34 at 4% compounded quarterly for 6 years

**17.** $12,903.45 at 6.37% compounded quarterly for 29 quarters

**18.** $4677.23 at 4.57% compounded monthly for 32 months

**Investing** *Find the amount of compound interest earned by each deposit.*

**19.** $22,000 at 5.5%, compounded quarterly for 6 years

**20.** $2975 at 4.7%, compounded monthly for 4 years

**Bonds** *Find the face value (to the nearest dollar) of the zero-coupon bond.*

**21.** 5-year bond at 3.9%; price $12,366

**22.** 15-year bond at 5.2%; price $11,575

**APY** *Find the APY corresponding to the given nominal rate.*

**23.** 5% compounded semiannually

**24.** 6.5% compounded daily

**Present Value** *Find the present value of the given amounts at the given interest rate.*

**25.** $42,000 in 7 years; 12% compounded monthly

**26.** $17,650 in 4 years; 8% compounded quarterly

**27.** $1347.89 in 3.5 years; 6.2% compounded semiannually

**28.** $2388.90 in 44 months; 5.75% compounded monthly

**Bonds** *Find the price that a purchaser should be willing to pay for these zero-coupon bonds.*

**29.** 10-year $15,000 bond; interest at 4.4%

**30.** 25-year $30,000 bond; interest at 6.2%

**31.** What is meant by the future value of an annuity?

**Annuities** *Find the future value of each annuity.*

**32.** $1288 deposited at the end of each year for 14 years; money earns 7% compounded annually

**33.** $4000 deposited at the end of each quarter for 8 years; money earns 6% compounded quarterly

**34.** $233 deposited at the end of each month for 4 years; money earns 6% compounded monthly

**35.** $672 deposited at the beginning of each quarter for 7 years; money earns 5% compounded quarterly

**36.** $11,900 deposited at the beginning of each month for 13 months; money earns 7% compounded monthly

**37.** What is the purpose of a sinking fund?

**Sinking Funds** *Find the amount of each payment that must be made into a sinking fund to accumulate the given amounts. Assume payments are made at the end of each period.*

**38.** $6500; money earns 5% compounded annually; 6 annual payments

**39.** $57,000; money earns 6% compounded semiannually for $8\frac{1}{2}$ years

**40.** $233,188; money earns 5.7% compounded quarterly for $7\frac{3}{4}$ years

**41.** $56,788; money earns 6.12% compounded monthly for $4\frac{1}{2}$ years

**Annuities** *Find the present value of each ordinary annuity.*

**42.** Payments of $850 annually for 4 years at 5% compounded annually

**43.** Payments of $1500 quarterly for 7 years at 8% compounded quarterly

**44.** Payments of $4210 semiannually for 8 years at 5.6% compounded semiannually

**45.** Payments of $877.34 monthly for 17 months at 6.4% compounded monthly

**Annuities** *Find the amount necessary to fund the given withdrawals (which are made at the end of each period).*

**46.** Quarterly withdrawals of $800 for 4 years with interest rate 4.6%, compounded quarterly

**47.** Monthly withdrawals of $1500 for 10 years with interest rate 5.8%, compounded monthly

**48.** Yearly withdrawals of $3000 for 15 years with interest rate 6.2%, compounded annually

**Annuities** *Find the payment for the ordinary annuity with the given present value.*

**49.** $150,000; monthly payments for 15 years, with interest rate 5.1%, compounded monthly

**50.** $25,000; quarterly payments for 8 years, with interest rate 4.9%, compounded quarterly

**51.** Find the lump-sum deposit today that will produce the same total amount as payments of $4200 at the end of each year for 12 years. The interest rate in both cases is 4.5%, compounded annually.

**52.** If the current interest rate is 6.5%, find the price (to the nearest dollar) that a purchaser should be willing to pay for a $24,000 bond with coupon rate 5% that matures in 6 years.

**Amortization** *Find the amount of the payment necessary to amortize each of the given loans.*

**53.** $32,000 at 8.4% interest; 10 quarterly payments

**54.** $5607 at 7.6% interest; 32 monthly payments

**Mortgages** *Find the monthly house payments for the given mortgages.*

**55.** $95,000 at 3.67% for 30 years

**56.** $167,000 at 2.91% for 15 years

**57.** Find the approximate remaining balance after 5 years of payments on the loan in Exercise 55.

**58.** Find the approximate remaining balance after 7.5 years of payments on the loan in Exercise 56.

**Student Loans** *According to www.studentaid.ed.gov, in 2016 the interest rate for a direct unsubsidized student loan was 3.76%. A*

*portion of an amortization table is given here for a $15,000 direct unsubsidized student loan to be paid back in 10 years. Use the table to answer Exercises 59–62.*

| Payment Number | Amount of Payment | Portion to Principal | Interest for Period | Principal at End of Period |
|---|---|---|---|---|
| 0 | | | | 15000.00 |
| 1 | 150.16 | 103.16 | 47.00 | 14896.84 |
| 2 | 150.16 | 103.49 | 46.68 | 14793.35 |
| 3 | 150.16 | 103.81 | 46.35 | 14689.54 |
| 4 | 150.16 | 104.14 | 46.03 | 14585.41 |
| 5 | 150.16 | 104.46 | 45.70 | 14480.94 |
| 6 | 150.16 | 104.79 | 45.37 | 14376.16 |
| 7 | 150.16 | 105.12 | 45.04 | 14271.04 |
| 8 | 150.16 | 105.45 | 44.72 | 14165.59 |
| 9 | 150.16 | 105.78 | 44.39 | 14059.81 |
| 10 | 150.16 | 106.11 | 44.05 | 13953.71 |

**59.** How much of the seventh payment is interest?

**60.** How much of the fourth payment is used to reduce the debt?

**61.** How much interest is paid in the first 6 months of the loan?

**62.** How much has the debt been reduced at the end of the first 8 months?

*Work the following applied problems.*

**63. Lottery**  In May 2016, a New Jersey family won a Powerball lottery prize of $429,600,000. (Data from: www.powerball.com.)

  **(a)** If the family had chosen to receive the money in 30 yearly payments, beginning immediately, what would have been the amount of each payment?

  **(b)** The family chose the one-time lump-sum cash option. If the interest rate is 3.58%, approximately how much did the family receive?

**64. Business Loan**  A firm of attorneys deposits $15,000 of profit-sharing money in an account at 6% compounded semiannually for $7\frac{1}{2}$ years. Find the amount of interest earned.

**65. Retirement Planning**  Tom, a graduate student, is considering investing $500 now, when he is 23, or waiting until he is 40 to invest $500. How much more money will he have at age 65 if he invests now, given that he can earn 5% interest compounded quarterly?

**66. Historical CD Savings**  According to a financial Web site, on June 15, 2005, Frontenac Bank of Earth City, Missouri, paid 3.94% interest, compounded quarterly, on a 2-year CD, while E*TRADE Bank of Arlington, Virginia, paid 3.93% compounded daily. What was the effective rate for the two CDs, and which bank paid a higher effective rate? (Data from: www.bankrate.com.)

**67. Investing**  Chalon Bridges deposits semiannual payments of $3200, received in payment of a debt, in an ordinary annuity at 6.8% compounded semiannually. Find the final amount in the account and the interest earned at the end of 3.5 years.

**68. Retirement Benefits**  Each year, a firm must set aside enough funds to provide employee retirement benefits of $625,000,000 in 20 years. If the firm can invest money at 7.5% compounded monthly, what amount must be invested at the end of each month for this purpose?

**69. Charitable Giving**  A benefactor wants to be able to leave a bequest to the college she attended. If she wants to make a donation of $2,000,000 in 10 years, how much each month does she need to place in an investment account that pays an interest rate of 5.5%, compounded monthly?

**70. Retirement Planning**  Suppose you have built up a pension with $18,000 annual payments by working 10 years for a company when you leave to accept a better job. The company gives you the option of collecting half the full pension when you reach age 55 or the full pension at age 65. Assume an interest rate of 8% compounded annually. By age 75, how much will each plan produce? Which plan would produce the larger amount?

**71. Charitable Giving**  In 3 years, Ms. Nguyen must pay a pledge of $7500 to her favorite charity. What lump sum can she deposit today at 10% compounded semiannually so that she will have enough to pay the pledge?

**72. Home Improvement Loan**  To finance the $15,000 cost of their kitchen remodeling, the Chews will make equal payments at the end of each month for 36 months. They will pay interest at the rate of 7.2%. Find the amount of each payment.

**73. Business Loan**  To expand her business, the owner of a small restaurant borrows $40,000. She will repay the money in equal payments at the end of each semiannual period for 8 years at 9% interest. What payments must she make?

**74. Mortgage**  The Keller family bought a house for $210,000. They paid $42,000 down and took out a 30-year mortgage for the balance at 3.75%.

  **(a)** Find the monthly payment.

  **(b)** How much of the first payment is interest?

  After 15 years, the family sold their house for $255,000.

  **(c)** Estimate the current mortgage balance at the time of the sale.

  **(d)** Find the amount of money they received from the sale after paying off the mortgage.

**75. Investing**  Miranda Williams invests $325 at the end of each month in a savings account earning 6.8% annual interest. If interest is compounded monthly, what is the value of the account in 20 years?

**76. Retirement Planning**  Dan Hook deposits $400 a month to a retirement account that has an interest rate of 3.1%, compounded monthly. After making 60 deposits, Dan changes his job and stops making payments for 3 years during which time he earns 3.1% interest, compounded monthly. After 3 years, he starts making deposits again, but now he deposits $525 per month. What will the value of the retirement account be after Dan makes his $525 monthly deposits for 5 years?

**C 77. Inheritance** The proceeds of a $50,000 inheritance are left on deposit with an insurance company for 10 years at an interest rate of 4.2%, compounded annually. The balance at the end of 10 years is paid to the beneficiary in 120 equal monthly payments, with the first payment made immediately. During the payout period, interest is credited at an annual interest rate of 2.5%. What is the amount of the monthly payments?

**C 78. Retirement Planning** Eileen Gianiodis wants to retire on $75,000 per year for her life expectancy of 20 years after she retires. She estimates that she will be able to earn an interest rate of 10.1%, compounded annually, throughout her lifetime. To reach her retirement goal, Eileen will make annual contributions to her account for the next 30 years. One year after making her last deposit, she will receive her first retirement check. How large must her yearly contributions be?

# Case Study 5　Investing in Stocks and Using the Rule of 72

The future value formulas that we used throughout Chapter 5 are based on fixed interest rates that exist through the savings or investment period. Many people invest in the stock market, however, and wish to project the value of their money several years into the future. The challenge here is that individual stock returns and mutual fund returns are not fixed and are often unpredictable. Thus, projecting a rate of return into the future always requires assumptions and risks.

With these caveats in mind, financial planners and individual investors can still make projections on investment income in the future using formulas that we have already learned. For example, according to vanguard.com, the average return on equities (stocks) from 1930 through 2010 was 9.2%. Many index funds follow the return on the market quite closely. So let us assume that we have $5000 to invest in a market index fund. We will assume an annual return of 9.2%. What is the value of the fund in 20 years?

To answer this question, we can use the future value formula

$$FV = PV(1 + i)^n \qquad \text{(1)}$$

with $PV = \$5000$, $i = .092$, and $n = 20$. Thus,

$$FV = 5000(1 + .092)^{20} = 29,068.51.$$

Let us look more closely at the growth of the account value in this fund over time. The table below shows the value of the fund over the course of the first 10 years.

| Year | Account Value |
|------|---------------|
| 0 | $5000.00 |
| 1 | 5460.00 |
| 2 | 5962.32 |
| 3 | 6510.85 |
| 4 | 7109.85 |
| 5 | 7763.96 |
| 6 | 8478.24 |
| 7 | 9258.24 |
| 8 | 10,110.00 |
| 9 | 11,040.12 |
| 10 | 12,055.81 |

Notice that after 8 years, the fund's value has more than doubled from the original $5000 to $10,110.00. This is actually not surprising because investors often use an informal rule knows as the Rule of 72 to approximate how long it takes for a fund to double in value.

The Rule of 72 gives a rough estimate of how long it takes for an investment to double in value. The way it works is to take the number 72 and divide by the annual rate of return (let's round our return to 9%). So we take $72/9 = 8$, and 8 years is a rough approximation for the number of years it takes for an investment to double. The table below shows how close this rule actually works by comparing its estimated time to doubling with what is predicted by the future value formula in Equation **(1)**.

| Rate of Return | Rule of 72 | Actual Number of Years |
|----------------|-----------|------------------------|
| 4% | 18.0 | 17.7 |
| 6% | 12.0 | 11.9 |
| 8% | 9.0 | 9.0 |
| 10% | 7.2 | 7.3 |
| 12% | 6.0 | 6.1 |
| 14% | 5.1 | 5.3 |
| 16% | 4.5 | 4.7 |
| 18% | 4.0 | 4.2 |
| 20% | 3.6 | 3.8 |

As we can see, the Rule of 72 is most accurate when returns are between 6% and 12%. This is generally a good range of expectations from the stock market.

The power of doubling (or exponential growth) is the reason that many financial planners urge college graduates to begin saving for retirement as soon as they can. We can use the future value of an ordinary annuity formula from Section 5.3 for projecting the value of an account when regular deposits are made at the end of each time period. Even saving $100 a month can reap long-term benefits, as the following example shows.

**EXAMPLE 1** **Investing** Shanda graduates from college at age 22 and lands her first job. She manages to deposit $100 at the end of each month in an individual retirement account (IRA). Assume that the account earns an annual return of 8%. She makes these monthly deposits for 20 years. Determine the future value of her IRA.

**Solution** The future value of an ordinary annuity formula is

$$FV = PMT\left[\frac{(1 + i)^n - 1}{i}\right]$$

where $PMT = \$100$, $i = r/m = .08/12$, and $n = (20)(12) = 240$. Thus, we have

$$FV = PMT\left[\frac{(1 + i)^n - 1}{i}\right]$$

$$= 100\left[\frac{(1 + .08/12)^{240} - 1}{.08/12}\right] = \$58,902.04.$$

**EXAMPLE 2** **Investing** Miguel graduates at the same time as Shanda (see Example 1). However, he waits 10 years before he begins making deposits to an IRA account. When he does, he invests $200 a month at the end of each month. If we assume Miguel's IRA earns an average annual return of 8% as well, what is the value of his account in 10 years?

**Solution** Here we use the same future value of an ordinary annuity formula, but with $PMT = \$200$, $i = .08/12$, and $n = 10*12 = 120$. Thus, we have

$$FV = PMT\left[\frac{(1 + i)^n - 1}{i}\right]$$

$$= 200\left[\frac{(1 + .08/12)^{120} - 1}{.08/12}\right] = \$36,589.21.$$

So even though Miguel and Shanda invest the same amount of money in monthly payments (Sandra invests $100(20)(12) = \$24,000$ and Miguel invests $200(10)(12) = \$24,000$), Shanda has a much higher account value at the end of the 20 years than Miguel does. In fact, we can use the Rule of 72 to approximate the value of each account when they are ready to retire, as we see in the next example.

**EXAMPLE 3** Shanda and Miguel stop making monthly investments into their respective IRA accounts but continue to hold the money in an account earning 8%. Use the Rule of 72 to approximate the value of each account when both Shanda and Miguel turn 69 years old.

**Solution** In Examples 1 and 2, Shanda invested for 20 years starting age 22, and Miguel invested for 10 years starting at age 32. Thus, they are both 42 years old when they stop making monthly investments. The Rule of 72 tells us that, if we assume an 8% annual return,

then every 9 years, the value of the account will double. So the value of each of their accounts will double 3 times in the course of the 27 years when they reach age 69.

The table gives the amounts that Shanda and Miguel will earn after each 9-year interval.

| Time | Shanda | Miguel |
|---|---|---|
| First doubling | (2)($58,902.04) = $117,804.08 | (2)(36,589.21) = $73,178.42 |
| Second doubling | (2)(117,804.08) = $235,608.16. | (2)(73,178.42) = $146,356.84 |
| Third doubling | (2)(235,608.16) = $471,216.32 | (2)(146,356.84) = $292,713.68 |

We could have also multiplied each of the account values at age 42 by 8 [which is (2)(2)(2)] to obtain the value of three doublings, but the table shows in dramatic fashion how much more money Shanda will have in her account when she turns 69 years old than Miguel will.

## Exercises

1. **Investing** Shelia graduates from college at age 25 and obtains her first job. She manages to deposit $150 at the end of each month in an individual retirement account (IRA). Assume that the account earns an annual return of 6%. Shelia makes these monthly deposits for 30 years. Determine the future value of her IRA.

2. **Investing** Hector manages to deposit $300 at the end of each month in an individual retirement account (IRA) at age 30. Assume that the account earns an annual return of 12%. Hector makes these monthly deposits for 25 years. Determine the future value of his IRA.

3. **Investing** Suppose Shelia in Exercise 1 stops adding money to her account when she turns 55, but the account continues to earn a return of 6% per year.

   **(a)** If we use the Rule of 72, how old will Shelia be when the balance has doubled?

   **(b)** What will the balance be when it has doubled?

4. **Investing** Suppose Hector in Exercise 2 stops adding money to his account when he turns 55, but the account continues to earn a return of 12% per year.

   **(a)** If we use the Rule of 72, how old will Hector be when the balance has doubled?

   **(b)** What will the balance be when it has doubled?

5. **Investing** The rate of return can make a huge difference in how quickly money doubles. If $10,000 is invested for 36 years at 6%, how many times will the value double according to the Rule of 72? What will the value be at the end of 36 years?

6. **Investing** If $10,000 is invested for 36 years at 8%, how many times will the value double according to the Rule of 72? What will the value be at the end of 36 years?

# Extended Project

1. Investigate the interest rates for the subsidized and unsubsidized student loans. If you have taken out student loans or plan to take out student loans before graduating, calculate your own monthly payment and how much interest you will pay over the course of the repayment period. If you have not taken out, and you do not plan to take out a student loan, contact the financial aid office of your college or university to determine the median amount borrowed with student loans at your institution. Determine the monthly payment and how much interest is paid during repayment for the typical borrowing student.

2. Determine the best interest rate for a new car purchase for a 48-month loan at a bank near you. If you finance $25,999 with such a loan, determine the payment and the total interest paid over the course of the loan. Also, determine the best interest rate for a new car purchase for a 48-month loan at a credit union near you. Determine the monthly payment and total interest paid if the same auto loan is financed through the credit union. Is it true that the credit union would save you money?

3. Suppose an investor has $1,000,000 to invest for 20 years at an annual interest rate of 2.5%. Find the value of the account if the interest is compounded monthly and if it is compounded continuously. Is there an appreciable difference in the value? If the investor invested $100,000,000, would there be an appreciable difference in value then? What if the interest rate were only .85%?

# Systems of Linear Equations and Matrices

## 6

### CHAPTER

A variety of resource allocation problems involving many variables can be handled by solving an appropriate system of linear equations. Technology (such as graphing calculators, WolframAlpha.com, computer algebra systems, and smart phone apps) is very helpful for handling large systems. Smaller ones can easily be solved by hand.

This chapter deals with **linear** (or **first-degree**) **equations** such as

$$2x + 3y = 14 \quad \text{linear equation in two variables,}$$
$$4x - 2y + 5z = 8 \quad \text{linear equation in three variables,}$$

and so on. A **solution** of such an equation is an ordered list of numbers that, when substituted for the variables in the order they appear, produces a true statement. For instance, $(1, 4)$ is a solution of the equation $2x + 3y = 14$ because substituting $x = 1$ and $y = 4$ produces the true statement $2(1) + 3(4) = 14$.

Many applications involve **systems of linear equations,** such as these two:

| Two equations in two variables | Three equations in four variables |
|---|---|
| | $2x + y + z \qquad\;\; = 3$ |
| $5x - 3y = 7$ | $x + y + z + w = 5$ |
| $2x + 4y = 8$ | $-4x + \qquad z + w = 0.$ |

A **solution of a system** is a solution that satisfies *all* the equations in the system. For instance, in the right-hand system of equations on page 268, (1, 0, 1, 3) is a solution of all three equations (check it) and hence is a solution of the system. By contrast, (1, 1, 0, 3) is a solution of the first two equations, but not the third. Hence, (1, 1, 0, 3) is not a solution of the system.

This chapter presents methods for solving such systems, including matrix methods. Matrix algebra and other applications of matrices are also discussed.

# 6.1   Systems of Two Linear Equations in Two Variables

The graph of a linear equation in two variables is a straight line. The coordinates of each point on the graph represent a solution of the equation (Section 2.1). Thus, the solution of a system of two such equations is represented by the point or points where the two lines intersect. There are exactly three geometric possibilities for two lines: They intersect at a single point, or they coincide, or they are distinct and parallel. As illustrated in Figure 6.1, each of these geometric possibilities leads to a different number of solutions for the system.

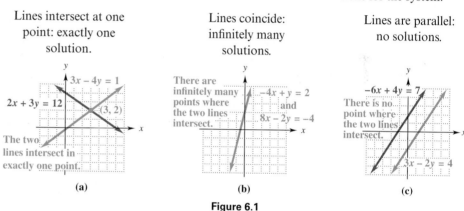

Lines intersect at one point: exactly one solution.

Lines coincide: infinitely many solutions.

Lines are parallel: no solutions.

$3x - 4y = 1$
$2x + 3y = 12$
$(3, 2)$
The two lines intersect in exactly one point.
**(a)**

There are infinitely many points where the two lines intersect.
$-4x + y = 2$
and
$8x - 2y = -4$
**(b)**

$-6x + 4y = 7$
There is no point where the two lines intersect.
$3x - 2y = 4$
**(c)**

**Figure 6.1**

## The Substitution Method

Example 1 illustrates the substitution method for solving a system of two equations in two variables.

---

**EXAMPLE 1**   Solve the system

$$2x - \ y = 1$$
$$3x + 2y = 4.$$

**Solution**   Begin by solving the first equation for $y$:

$$2x - y = 1$$
$$-y = -2x + 1 \qquad \text{Subtract } 2x \text{ from both sides.}$$
$$y = 2x - 1. \qquad \text{Multiply both sides by } -1.$$

Substitute this expression for $y$ in the second equation and solve for $x$:

$$3x + 2y = 4$$
$$3x + 2(2x - 1) = 4 \qquad \text{Substitute } 2x - 1 \text{ for } y.$$
$$3x + 4x - 2 = 4 \qquad \text{Multiply out the left side.}$$
$$7x - 2 = 4 \qquad \text{Combine like terms.}$$
$$7x = 6 \qquad \text{Add 2 to both sides.}$$
$$x = 6/7. \qquad \text{Divide both sides by 7.}$$

Therefore, every solution of the system must have $x = 6/7$. To find the corresponding solution for $y$, substitute $x = 6/7$ in one of the two original equations and solve for $y$. We will use the first equation.

$$2x - y = 1$$

$$2\left(\frac{6}{7}\right) - y = 1 \qquad \text{Substitute 6/7 for } x.$$

$$\frac{12}{7} - y = 1 \qquad \text{Multiply out the left side.}$$

$$-y = -\frac{12}{7} + 1 \qquad \text{Subtract 12/7 from both sides.}$$

$$y = \frac{12}{7} - 1 = \frac{5}{7}. \qquad \text{Multiply both sides by } -1.$$

Hence, the solution of the original system is $x = 6/7$ and $y = 5/7$. We would have obtained the same solution if we had substituted $x = 6/7$ in the second equation of the original system, as you can easily verify. ✔ 1

The substitution method is useful when at least one of the equations has a variable with coefficient 1. That is why we solved for $y$ in the first equation of Example 1. If we had solved for $x$ in the first equation or for $x$ or $y$ in the second, we would have had a lot more fractions to deal with.

▦ **TECHNOLOGY TIP** A graphing calculator can be used to solve systems of two equations in two variables. Solve each of the given equations for $y$ and graph them on the same screen. In Example 1, for instance, we would graph

$$y_1 = 2x - 1 \qquad \text{and} \qquad y_2 = \frac{-3x + 4}{2}.$$

Use the intersection finder (look in the CALC menu) to determine the point where the graphs intersect (the solution of the system), as shown in Figure 6.2. Note that this is an *approximate* solution, rather than the exact solution found algebraically in Example 1.

## The Elimination Method

The elimination method of solving systems of linear equations is often more convenient than substitution, as illustrated in the next four examples.

> **EXAMPLE 2** Solve the system
>
> $$5x + y = 4$$
> $$3x + 2y = 1.$$

**Solution** Multiply the first equation by $-2$, so that the coefficients of $y$ in the two equations are negatives of each other:

$$-10x - 2y = -8 \qquad \text{First equation multiplied by } -2.$$
$$3x + 2y = 1. \qquad \text{Second equation as given.}$$

Any solution of this system of equations must also be a solution of the sum of the two equations:

$$-10x - 2y = -8$$
$$\underline{\phantom{-}3x + 2y = \phantom{-}1}$$
$$-7x \phantom{- 2y} = -7 \qquad \text{Sum; variable } y \text{ is eliminated.}$$
$$x \phantom{- 2y} = 1. \qquad \text{Divide both sides by } -7.$$

✓ **Checkpoint 1**

Use the substitution method to solve this system:

$$x - 2y = 3$$
$$2x + 3y = 13.$$

*Answers to Checkpoint exercises are found at the end of the section.*

Intersection
X=.85714286  Y=.71428571

**Figure 6.2**

To find the corresponding value of $y$, substitute $x = 1$ in one of the original equations, say the first one:

$$5x + y = 4$$
$$5(1) + y = 4 \qquad \text{Substitute 1 for } x.$$
$$y = -1. \qquad \text{Subtract 5 from both sides.}$$

Therefore, the solution of the original system is $(1, -1)$. ✓₂

✓ **Checkpoint 2**

Use the elimination method to solve this system:
$$x + 2y = 4$$
$$3x - 4y = -8.$$

**EXAMPLE 3**   Solve the system
$$3x - 4y = 1$$
$$2x + 3y = 12.$$

**Solution**  Multiply the first equation by 2 and the second equation by $-3$ to get

$$6x - 8y = 2 \qquad \text{First equation multiplied by 2.}$$
$$-6x - 9y = -36. \qquad \text{Second equation multiplied by } -3.$$

The multipliers 2 and $-3$ were chosen so that the coefficients of $x$ in the two equations would be negatives of each other. Any solution of both these equations must also be a solution of their sum:

$$6x - 8y = \phantom{-}2$$
$$\underline{-6x - 9y = -36}$$
$$-17y = -34 \qquad \text{Sum; variable } x \text{ is eliminated.}$$
$$y = \phantom{-}2. \qquad \text{Divide both sides by } -17.$$

To find the corresponding value of $x$, substitute 2 for $y$ in either of the original equations. We choose the first equation:

$$3x - 4y \phantom{0} = 1$$
$$3x - 4(2) = 1 \qquad \text{Substitute 2 for } y.$$
$$3x - 8 \phantom{00} = 1 \qquad \text{Simplify.}$$
$$3x \phantom{0000} = 9 \qquad \text{Add 8 to both sides.}$$
$$x \phantom{00000} = 3. \qquad \text{Divide both sides by 3.}$$

Therefore, the solution of the system is $(3, 2)$. The graphs of both equations of the system are shown in Figure 6.1(a) at the beginning of this section. They intersect at the point $(3, 2)$, the solution of the system. ✓₃

✓ **Checkpoint 3**

Solve the system of equations
$$3x + 2y = -1$$
$$5x - 3y = 11.$$
Draw the graph of each equation on the same set of axes.

## Dependent and Inconsistent Systems

A system of equations that has a unique solution, such as those in Examples 1–3, is called an **independent system.** Now we consider systems that have infinitely many solutions or no solutions at all.

**EXAMPLE 4**   Solve the system
$$-4x + \phantom{0}y = 2$$
$$8x - 2y = -4.$$

**Solution**  If you solve each equation in the system for $y$, you see that the two equations are actually the same:

$$\begin{array}{ll} -4x + y = 2 & \qquad 8x - 2y = -4 \\ \phantom{-4x +} y = 4x + 2 & \qquad -2y = -8x - 4 \\ & \qquad \phantom{-2}y = \phantom{-}4x + 2. \end{array}$$

So the two equations have the same graph, as shown in Figure 6.1(b), and the system has infinitely many solutions (namely, all solutions of $y = 4x + 2$). A system such as this is said to be **dependent.**

You do not have to analyze a system as was just done in order to find out that the system is dependent. The elimination method will warn you. For instance, if you attempt to eliminate $x$ in the original system by multiplying both sides of the first equation by 2 and adding the results to the second equation, you obtain

$$\begin{array}{rl} -8x + 2y = & 4 \quad \text{First equation multiplied by 2.} \\ \underline{8x - 2y = -4} \quad \text{Second equation as given.} \\ 0 = & 0. \quad \text{Both variables are eliminated.} \end{array}$$

The equation "$0 = 0$" is an algebraic indication that the system is dependent and has infinitely many solutions. ✔4

## ✓ Checkpoint 4

Solve the following system:
$$3x - 4y = 13$$
$$12x - 16y = 52.$$

### EXAMPLE 5    Solve the system

$$3x - 2y = 4$$
$$-6x + 4y = 7.$$

**Solution** The graphs of these equations are parallel lines (each has slope $3/2$), as shown in Figure 6.1(c). Therefore, the system has no solution. However, you do not need the graphs to discover this fact. If you try to solve the system algebraically by multiplying both sides of the first equation by 2 and adding the results to the second equation, you obtain

$$\begin{array}{rl} 6x - 4y = & 8 \quad \text{First equation multiplied by 2.} \\ \underline{-6x + 4y = & 7} \quad \text{Second equation as given.} \\ 0 = & 15. \quad \text{False statement.} \end{array}$$

The false statement "$0 = 15$" is the algebraic signal that the system has no solution. A system with no solutions is said to be **inconsistent.** ✔5

## ✓ Checkpoint 5

Solve the system
$$x - y = 4$$
$$2x - 2y = 3.$$
Draw the graph of each equation on the same set of axes.

## Applications

In many applications, the answer is the solution of a system of equations. Solving the system is the easy part—*finding* the system, however, may take some thought.

### EXAMPLE 6    Basketball Tickets Eight hundred people attend a basketball game, and total ticket sales are $3102. If adult tickets are $6 and student tickets are $3, how many adults and how many students attended the game?

**Solution** Let $x$ be the number of adults and $y$ the number of students. Then

Number of adults + Number of students = Total attendance.

$$x + y = 800.$$

A second equation can be found by considering ticket sales:

Adult ticket sales   +   Student ticket sales   =   Total ticket sales

$$\left(\begin{array}{c}\text{Price}\\\text{per}\\\text{ticket}\end{array}\right) \times \left(\begin{array}{c}\text{Number}\\\text{of}\\\text{adults}\end{array}\right) + \left(\begin{array}{c}\text{Price}\\\text{per}\\\text{ticket}\end{array}\right) \times \left(\begin{array}{c}\text{Number}\\\text{of}\\\text{students}\end{array}\right) = 3102$$

$$6x + 3y = 3102.$$

**Checkpoint 6**

**(a)** Use substitution or elimination to solve the system of equations in Example 6.

**(b)** Use technology to solve the system. (*Hint:* Solve each equation for $y$, graph both equations on the same screen, and use the intersection finder.)

To find $x$ and $y$ we must solve this system of equations:

$$x + y = 800$$
$$6x + 3y = 3102.$$

The system can readily be solved by hand or by using technology. See Checkpoint 6, which shows that 234 adults and 566 students attended the game. ✓₆

# 6.1 Exercises

*Determine whether the given ordered list of numbers is a solution of the system of equations.*

**1.** $(-1, 3)$
$$2x + y = 1$$
$$-3x + 2y = 9$$

**2.** $(2, -.5)$
$$.5x + 8y = -3$$
$$x + 5y = -5$$

*Use substitution to solve each system. (See Example 1.)*

**3.** $3x - y = 1$
$$x + 2y = -9$$

**4.** $x + y = 7$
$$x - 2y = -5$$

**5.** $3x - 2y = 4$
$$2x + y = -1$$

**6.** $5x - 3y = -2$
$$-x - 2y = 3$$

*Use elimination to solve each system. (See Examples 2–5.)*

**7.** $x - 2y = 5$
$$2x + y = 3$$

**8.** $3x - y = 1$
$$-x + 2y = 4$$

**9.** $2x - 2y = 12$
$$-2x + 3y = 10$$

**10.** $3x + 2y = -4$
$$4x - 2y = -10$$

**11.** $x + 3y = -1$
$$2x - y = 5$$

**12.** $4x - 3y = -1$
$$x + 2y = 19$$

**13.** $2x + 3y = 15$
$$8x + 12y = 40$$

**14.** $2x + 5y = 8$
$$6x + 15y = 18$$

**15.** $2x - 8y = 2$
$$3x - 12y = 3$$

**16.** $3x - 2y = 4$
$$6x - 4y = 8$$

*In Exercises 17 and 18, multiply both sides of each equation by a common denominator to eliminate the fractions. Then solve the system.*

**17.** $\dfrac{x}{2} + \dfrac{y}{3} = 8$
$$\dfrac{2x}{3} + \dfrac{3y}{2} = 17$$

**18.** $\dfrac{x}{5} + 3y = 31$
$$2x - \dfrac{y}{5} = 8$$

**19. Millennials** The number of baby boomers has been decreasing, and the number of millennials has been increasing. Now millennials have surpassed boomers in number. The following equations approximate the population $y$ (in millions) of the

given generation in year $x$, where $x = 14$ corresponds to the year 2014. (Data from: Pew Research Center.)

$$\text{Boomers:} \quad 10x + 13y = 1125$$
$$\text{Millennials:} \quad -2x + 7y = 495$$

Solve this system of equations, and interpret your answer.

**20. Generation X** The size of generation X has been decreasing, but the size of the baby boomer generation has been decreasing even more rapidly, so the number of generation Xers is projected to overtake the number of baby boomers. The population $y$ (in millions) of the given generation in year $x$ is approximated by

$$\text{Boomers:} \quad 10x + 13y = 1125$$
$$\text{Gen X:} \quad x + 13y = 873,$$

where $x = 14$ corresponds to the year 2014. Solve this system of equations, and interpret your answer. (Data from: Pew Research Center.)

**21. Slow Midwestern Growth** According to U.S. Census Bureau projections through the year 2030, the population $y$ (in millions) of the given state in year $x$ is approximated by

$$\text{Ohio:} \quad -x + 150y = 1710$$
$$\text{North Carolina:} \quad -41x + 300y = 2430,$$

where $x = 0$ corresponds to the year 2000. In what year are the two states projected to have the same population? Round your answer to the nearest year.

**22. Booming Florida** At the start of the millennium, New York was the third most populous state in the country, followed by Florida. Since that time, Florida has experienced faster growth. The population $y$ (in millions) of the given state in year $x$ is approximated by

$$\text{Florida:} \quad 7y - 2x = 112$$
$$\text{New York:} \quad 14y - x = 266,$$

where $x = 0$ corresponds to the year 2000. In what year did Florida overtake New York in population? To the nearest million, what was the population of these states at that time? (Data from: U.S. Census Bureau.)

23. **Google Trends** According to Google Trends, popular interest in light-emitting diode (LED) lightbulbs has been soaring, while interest in compact florescent lightbulbs (CFLs) has been dropping. The following equations approximate the Google Trends rating (on a scale from 0 to 100) in year $x$, where $x = 10$ corresponds to the year 2010. (Data from: www.google.com/trends.)

$$\text{LED:} \quad -25x + 6y = 20$$
$$\text{CFL:} \quad 15x + 2y = 322$$

When did interest in LED lighting surpass CFL lighting? Round your answer to the nearest year.

24. **Heart Disease and Cancer Deaths** The number of deaths $y$ (in thousands) in year $x$ for diseases of the heart and malignant neoplasms (cancer) can be approximated by

$$\text{Heart disease:} \quad 100x + 13y = 9243$$
$$\text{Cancer:} \quad -64x + 26y = 14{,}379,$$

where $x = 0$ corresponds to the year 2000. According to these models, during what year did deaths from heart disease equal cancer deaths? (Data from: Centers for Disease Control and Prevention.)

25. **Workforce Participation for Women and Men** On the basis of projections for the year 2022, the number of women and men in the workforce (in millions) can be estimated by

$$\text{Women:} \quad -7x + 16y = 1070$$
$$\text{Men:} \quad -5x + 10y = 759,$$

where $x = 14$ corresponds to the year 2014. According to these models, will the number of women in the workforce equal the number of men during the time period from 2014 to 2022 (that is, $14 \le x \le 22$)? (Data from: U.S. Department of Labor.)

26. **Weekly Earnings for Women and Men** On the basis of data from the years 2000 through 2015, seasonally adjusted median weekly earnings $y$ in year $x$ can be approximated for women and men by

$$\text{Women:} \quad -791x + 50y = 24{,}850$$
$$\text{Men:} \quad -84x + 5y = 3210,$$

where $x = 0$ corresponds to the year 2000 and $y$ is in current dollars. If these equations remain valid in the future, will earnings for women catch up with men within the next decade? (Data from: Bureau of Labor Statistics.)

27. **Theater Tickets** A 200-seat theater charges $8 for adults and $5 for children. If all seats were filled and the total ticket income was $1435, how many adults and how many children were in the audience?

28. **Concession Sales** During intermission, the cash bar at a theater sold 72 drinks and took in a total of $663. If the theater sells beer for $7 each and wine for $10 a glass, how many servings of each beverage were sold?

29. **Physical Science** A plane flies 3000 miles from San Francisco to Boston in 5 hours, with a tailwind all the way. The return trip on the same route, now with a headwind, takes 6 hours. Assuming both remain constant, find the speed of the plane and the speed of the wind. [*Hint:* If $x$ is the plane's speed and $y$ the wind speed (in mph), then the

plane travels to Boston at $x + y$ mph because the plane and the wind go in the same direction; on the return trip, the plane travels at $x - y$ mph. (Why?) Then use the equation distance = speed · time.]

30. **Physical Science** A plane flying into a headwind travels 2200 miles in 5 hours. The return flight along the same route with a tailwind takes 4 hours. Find the wind speed and the plane's speed (assuming both are constant). (See the hint for Exercise 29.)

31. (a) Find the equation of the straight line through $(1, 2)$ and $(3, 4)$.

    (b) Find the equation of the line through $(-1, 1)$ with slope 3.

    (c) Find a point that lies on both of the lines in (a) and (b).

32. (a) Find an equation of the straight line through $(0, 9)$ and $(2, 1)$.

    (b) Find an equation of the straight line through $(1, 5)$ with slope 2.

    (c) Find a point that lies on both of the lines in (a) and (b).

## ✓ Checkpoint Answers

1. $(5, 1)$

2. $(0, 2)$

3. $(1, -2)$

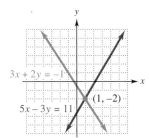

4. All ordered pairs that satisfy the equation $3x - 4y = 13$ (or $12x - 16y = 52$)

5. No solution

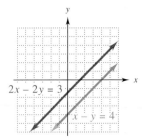

6. (a) $x = 234$ and $y = 566$

   (b)

# 6.2  Larger Systems of Linear Equations

Two systems of equations are said to be **equivalent** if they have the same solutions. The basic procedure for solving a large system of equations is to transform the system into a simpler, equivalent system and then solve this simpler system.

Three operations, called **elementary operations,** are used to transform a system into an equivalent one:

1. **Interchange any two equations in the system.**

   Changing the order of the equations obviously does not affect the solutions of the equations or the system.

2. **Multiply an equation in the system by a nonzero constant.**

   Multiplying an equation by a nonzero constant does not change its solutions. So it does not change the solutions of the system.

3. **Replace an equation in the system by the sum of itself and a nonzero constant multiple of another equation.**

   We are not saying here that the replacement equation has the same solutions as the one it replaces—it doesn't—only that the new *system* has the same solutions as the original system (that is, the systems are equivalent).[*]

Example 1 shows how to use elementary operations on a system to eliminate certain variables and produce an equivalent system that is easily solved. The italicized statements provide an outline of the procedure. Here and later on, we use $R_1$ to denote the first equation in a system, $R_2$ the second equation, and so on.

---

**EXAMPLE 1**    Solve the system

$$\begin{aligned} 2x + y - z &= 2 \\ x + 3y + 2z &= 1 \\ x + y + z &= 2. \end{aligned}$$

**Solution**  First, *use elementary operations to produce an equivalent system in which 1 is the coefficient of x in the first equation.* One way to do this is to interchange the first two equations (another would be to multiply both sides of the first equation by $\frac{1}{2}$):

$$\begin{aligned} x + 3y + 2z &= 1 \qquad \text{Interchange } R_1 \text{ and } R_2. \\ 2x + y - z &= 2 \\ x + y + z &= 2. \end{aligned}$$

Next, *use elementary operations to produce an equivalent system in which the x-term has been eliminated from the second and third equations.* To eliminate the $x$-term from the second equation, replace the second equation by the sum of itself and $-2$ times the first equation:

$$\begin{aligned} x + 3y + 2z &= 1 \\ -5y - 5z &= 0 \qquad {\scriptstyle -2R_1 + R_2} \\ x + y + z &= 2. \end{aligned}$$

$$\begin{array}{ll} -2R_1 & -2x - 6y - 4z = -2 \\ R_2 & 2x + y - z = 2 \\ \hline -2R_1 + R_2 & -5y - 5z = 0 \end{array}$$

To eliminate the $x$-term from the third equation of this last system, replace the third equation by the sum of itself and $-1$ times the first equation:

$$\begin{aligned} x + 3y + 2z &= 1 \\ -5y - 5z &= 0 \\ -2y - z &= 1. \qquad {\scriptstyle -1R_1 + R_3} \end{aligned}$$

$$\begin{array}{ll} -1R_1 & -x - 3y - 2z = -1 \\ R_3 & x + y + z = 2 \\ \hline -1R_1 + R_3 & -2y - z = 1 \end{array}$$

---

[*]This operation was used in a slightly different form in Section 6.1, when the sum of an equation and a constant multiple of another equation was found in order to eliminate one of the variables. See, for instance, Examples 2 and 3 of that section.

Now that $x$ has been eliminated from all but the first equation, we ignore the first equation and work on the remaining ones. *Use elementary operations to produce an equivalent system in which 1 is the coefficient of y in the second equation.* This can be done by multiplying the second equation in the system by $-\frac{1}{5}$:

$$
\begin{array}{rl}
x + 3y + 2z = 1 & \\
y + \ z = 0 & \quad -\frac{1}{5}R_2 \\
-2y - \ z = 1. &
\end{array}
$$

Then *use elementary operations to obtain an equivalent system in which y has been eliminated from the third equation.* Replace the third equation by the sum of itself and 2 times the second equation:

$$
\begin{array}{rl}
x + 3y + 2z = 1 & \\
y + \ z = 0 & \\
z = 1. & \quad 2R_2 + R_3
\end{array}
$$

The solution of the third equation is obvious: $z = 1$. Now work backward in the system. Substitute 1 for $z$ in the second equation and solve for $y$, obtaining $y = -1$. Finally, substitute 1 for $z$ and $-1$ for $y$ in the first equation and solve for $x$, obtaining $x = 2$. This process is known as **back substitution.** When it is finished, we have the solution of the original system, namely, $(2, -1, 1)$. It is always wise to check the solution by substituting the values for $x$, $y$, and $z$ in *all* equations of the original system.

The procedure used in Example 1 to eliminate variables and produce a system in which back substitution works can be carried out with any system, as summarized below. In this summary, the first variable that appears in an equation with a nonzero coefficient is called the **leading variable** of that equation, and its nonzero coefficient is called the **leading coefficient.**

## The Elimination Method for Solving Large Systems of Linear Equations

Use elementary operations to transform the given system into an equivalent one as follows:

1. Make the leading coefficient of the first equation 1 either by interchanging equations or by multiplying the first equation by a suitable constant.

2. Eliminate the leading variable of the first equation from each later equation by replacing the later equation by the sum of itself and a suitable multiple of the first equation.

3. Repeat Steps 1 and 2 for the second equation: Make its leading coefficient 1 and eliminate its leading variable from each later equation by replacing the later equation by the sum of itself and a suitable multiple of the second equation.

4. Repeat Steps 1 and 2 for the third equation, fourth equation, and so on, until it is not possible to go any further.

Then solve the resulting system by back substitution.

✓ **Checkpoint 1**

Use the elimination method to solve each system.

(a) $2x + \ y = -1$
$\ \ \ x + 3y = \ \ 2$

(b) $2x - \ y + 3z = \ \ 2$
$\ \ \ x + 2y - \ z = \ \ 6$
$\ -x - \ y + \ z = -5$

At various stages in the elimination process, you may have a choice of elementary operations that can be used. As long as the final result is a system in which back substitution can be used, the choice does not matter. To avoid unnecessary errors, choose elementary operations that minimize the amount of computation and, as far as possible, avoid complicated fractions. ✓₁

## Matrix Methods

You may have noticed that the variables in a system of equations remain unchanged during the solution process. So we need to keep track of only the coefficients and the constants. For instance, consider the system in Example 1:

$$
\begin{aligned}
2x + y - z &= 2 \\
x + 3y + 2z &= 1 \\
x + y + z &= 2.
\end{aligned}
$$

This system can be written in an abbreviated form without listing the variables as

$$
\begin{bmatrix}
2 & 1 & -1 & 2 \\
1 & 3 & 2 & 1 \\
1 & 1 & 1 & 2
\end{bmatrix}.
$$

Such a rectangular array of numbers, consisting of horizontal **rows** and vertical **columns,** is called a **matrix** (plural, **matrices**). Each number in the array is an **element,** or **entry.** To separate the constants in the last column of the matrix from the coefficients of the variables, we sometimes use a vertical line, producing the following **augmented matrix:**

$$
\left[
\begin{array}{ccc|c}
2 & 1 & -1 & 2 \\
1 & 3 & 2 & 1 \\
1 & 1 & 1 & 2
\end{array}
\right]. \quad \checkmark_2
$$

**Checkpoint 2**

**(a)** Write the augmented matrix of the following system:

$$
\begin{aligned}
4x - 2y + 3z &= 4 \\
3x + 5y + z &= -7 \\
5x - y + 4z &= 6.
\end{aligned}
$$

**(b)** Write the system of equations associated with the following augmented matrix:

$$
\left[
\begin{array}{cc|c}
2 & -2 & -2 \\
1 & 1 & 4 \\
3 & 5 & 8
\end{array}
\right].
$$

The rows of the augmented matrix can be transformed in the same way as the equations of the system, since the matrix is just a shortened form of the system. The following **row operations** on the augmented matrix correspond to the elementary operations used on systems of equations.

> Performing any one of the following **row operations** on the augmented matrix of a system of linear equations produces the augmented matrix of an equivalent system:
>
> **1.** Interchange any two rows.
>
> **2.** Multiply each element of a row by a nonzero constant.
>
> **3.** Replace a row by the sum of itself and a nonzero constant multiple of another row of the matrix.

**Checkpoint 3**

Perform the given row operations on the matrix

$$
\begin{bmatrix}
-1 & 5 \\
3 & -2
\end{bmatrix}.
$$

**(a)** Interchange $R_1$ and $R_2$.

**(b)** Replace $R_1$ by $2R_1$.

**(c)** Replace $R_2$ by $-3R_1 + R_2$.

**(d)** Replace $R_1$ by $2R_2 + R_1$.

Row operations on a matrix are indicated by the same notation we used for elementary operations on a system of equations. For example, $2R_3 + R_1$ indicates the sum of 2 times row 3 and row 1. $\checkmark_3$

**EXAMPLE 2**     Use matrices to solve the system

$$
\begin{aligned}
x - 2y &= 6 - 4z \\
x + 13z &= 6 - y \\
-2x + 6y - z &= -10.
\end{aligned}
$$

**Solution** First, put the system in the required form, with the constants on the right side of the equals sign and the terms with variables *in the same order* in each equation on the left side of the equals sign. Then write the augmented matrix of the system.

$$
\begin{aligned}
x - 2y + 4z &= 6 \\
x + y + 13z &= 6 \\
-2x + 6y - z &= -10
\end{aligned}
\qquad
\left[
\begin{array}{ccc|c}
1 & -2 & 4 & 6 \\
1 & 1 & 13 & 6 \\
-2 & 6 & -1 & -10
\end{array}
\right].
$$

The matrix method is the same as the elimination method, except that row operations are used on the augmented matrix instead of elementary operations on the corresponding system of equations, as shown in the following side-by-side comparison:

| Equation Method | Matrix Method |
|---|---|

**Equation Method**

Replace the second equation by the sum of itself and $-1$ times the first equation:

$$\begin{aligned} x - 2y + 4z &= 6 \\ 3y + 9z &= 0 \\ -2x + 6y - z &= -10. \end{aligned}$$

Replace the third equation by the sum of itself and 2 times the first equation:

$$\begin{aligned} x - 2y + 4z &= 6 \\ 3y + 9z &= 0 \\ 2y + 7z &= 2. \end{aligned}$$

Multiply both sides of the second equation by $\frac{1}{3}$:

$$\begin{aligned} x - 2y + 4z &= 6 \\ y + 3z &= 0 \\ 2y + 7z &= 2. \end{aligned}$$

Replace the third equation by the sum of itself and $-2$ times the second equation:

$$\begin{aligned} x - 2y + 4z &= 6 \\ y + 3z &= 0 \\ z &= 2. \end{aligned}$$

Now use back substitution:

$$z = 2 \qquad\qquad y + 3(2) = 0$$
$$y = -6$$

**Matrix Method**

Replace the second row by the sum of itself and $-1$ times the first row:

$$\begin{bmatrix} 1 & -2 & 4 & 6 \\ 0 & 3 & 9 & 0 \\ -2 & 6 & -1 & -10 \end{bmatrix}. \qquad \leftarrow -1R_1 + R_2 \rightarrow$$

Replace the third row by the sum of itself and 2 times the first row:

$$\begin{bmatrix} 1 & -2 & 4 & 6 \\ 0 & 3 & 9 & 0 \\ 0 & 2 & 7 & 2 \end{bmatrix}. \qquad \leftarrow 2R_1 + R_3 \rightarrow$$

Multiply each element of row 2 by $\frac{1}{3}$:

$$\begin{bmatrix} 1 & -2 & 4 & 6 \\ 0 & 1 & 3 & 0 \\ 0 & 2 & 7 & 2 \end{bmatrix}. \qquad \leftarrow \tfrac{1}{3}R_2 \rightarrow$$

Replace the third row by the sum of itself and $-2$ times the second row:

$$\begin{bmatrix} 1 & -2 & 4 & 6 \\ 0 & 1 & 3 & 0 \\ 0 & 0 & 1 & 2 \end{bmatrix}. \qquad \leftarrow -2R_2 + R_3 \rightarrow$$

$$x - 2(-6) + 4(2) = 6$$
$$x + 20 = 6$$
$$x = -14.$$

**✓ Checkpoint 4**

Complete the matrix solution of the system with this augmented matrix:

$$\begin{bmatrix} 1 & 1 & 1 & 2 \\ 1 & -2 & 1 & -1 \\ 0 & 3 & 1 & 5 \end{bmatrix}.$$

The solution of the system is $(-14, -6, 2)$. ✓ 4

A matrix, such as the last one in Example 2, is said to be in **row echelon form** when

all rows consisting entirely of zeros (if any) are at the bottom;

the first nonzero entry in each row is 1 (called a *leading* 1); and

each leading 1 appears to the right of the leading 1s in any preceding rows.

When a matrix in row echelon form is the augmented matrix of a system of equations, as in Example 2, the system can readily be solved by back substitution. So the matrix solution method amounts to transforming the augmented matrix of a system of equations into a matrix in row echelon form.

## The Gauss–Jordan Method

The **Gauss–Jordan method** is a variation on the matrix elimination method used in Example 2. It replaces the back substitution used there with additional elimination of variables, as illustrated in the next example.

**EXAMPLE 3**    Use the Gauss–Jordan method to solve the system in Example 2.

**Solution**  First, set up the augmented matrix. Then apply the same row operations used in Example 2 until you reach the final row echelon matrix, which is

$$\begin{bmatrix} 1 & -2 & 4 & | & 6 \\ 0 & 1 & 3 & | & 0 \\ 0 & 0 & 1 & | & 2 \end{bmatrix}.$$

Now use additional row operations to make the entries (other than the 1s) in columns two and three into zeros. Make the first two entries in column three 0 as follows:

$$\begin{bmatrix} 1 & -2 & 4 & | & 6 \\ 0 & 1 & \mathbf{0} & | & -6 \\ 0 & 0 & 1 & | & 2 \end{bmatrix} \qquad -3R_3 + R_2$$

$$\begin{bmatrix} 1 & -2 & \mathbf{0} & | & -2 \\ 0 & 1 & 0 & | & -6 \\ 0 & 0 & 1 & | & 2 \end{bmatrix} \qquad -4R_3 + R_1$$

Now make the first entry in column two 0.

$$\begin{bmatrix} 1 & \mathbf{0} & 0 & | & -14 \\ 0 & 1 & 0 & | & -6 \\ 0 & 0 & 1 & | & 2 \end{bmatrix}. \qquad 2R_2 + R_1$$

The last matrix corresponds to the system $x = -14, y = -6, z = 2$. So the solution of the original system is $(-14, -6, 2)$. This solution agrees with what we found in Example 2.

The final matrix in the Gauss–Jordan method is said to be in **reduced row echelon form,** meaning that it is in row echelon form *and* every column containing a leading 1 has zeros in all its other entries, as in Example 3. As we saw there, the solution of the system can be read directly from the reduced row echelon matrix. The *reduced* row echelon form of a matrix is unique. This is not true of a matrix that is simply in row echelon form. This is why a row echelon form produced by hand may differ from a row echelon form provided by a calculator, but each will lead to the same solution of the system of equations.

In the Gauss–Jordan method, row operations may be performed in any order, but it is best to transform the matrix systematically. Either follow the procedure in Example 3 (which first puts the system into a form in which back substitution can be used and then eliminates additional variables) or work column by column from left to right, as in Checkpoint 5. ✔5

## Technology and Systems of Equations

When working by hand, it is usually better to use the matrix elimination method (Example 2), because errors with back substitution are less likely to occur than errors in performing the additional row operations needed in the Gauss–Jordan method. When you are using technology, however, the Gauss–Jordan method (Example 3) is more efficient: The solution can be read directly from the final reduced row echelon matrix without any "hand work," as illustrated in the next example.

▦ **TECHNOLOGY TIP** To enter a matrix on a TI graphing calculator, go to the MATRIX menu and then to the EDIT submenu, and select the desired name of the matrix ([A], [B], and so on). Enter the number of rows and columns for the matrix. For example, the augmented matrix for the system in Example 2 is a 3 × 4 matrix because it has 3 rows and 4 columns. Next enter each element of the matrix. As usual, use QUIT to return to the home screen. To use a matrix in calculations, go to the MATRIX menu and then the NAMES submenu, and select the matrix of choice. For example, Figure 6.3, on the following page, shows the augmented matrix from Example 2 as matrix [A] on a TI-84+.

✓**Checkpoint 5**

Use the Gauss–Jordan method to solve the system

$$\begin{aligned} x + 2y &= 11 \\ -4x + y &= -8 \\ 5x + y &= 19 \end{aligned}$$

as instructed in parts (a)–(g). Give the shorthand notation for the required row operations in parts (b)–(f) and the new matrix in each step.

**(a)**  Set up the augmented matrix.

**(b)**  Get 0 in row 2, column 1.

**(c)**  Get 0 in row 3, column 1.

**(d)**  Get 1 in row 2, column 2.

**(e)**  Get 0 in row 1, column 2.

**(f)**  Get 0 in row 3, column 2.

**(g)**  List the solution of the system.

Figure 6.3

Figure 6.4

Figure 6.5

✓ **Checkpoint 6** ▦

On a graphing calculator, use the Gauss–Jordan method or the system solver to solve

$$x - y + 5z = -6$$
$$4x + 2y + 4z = 4$$
$$x - 5y + 8z = -17$$
$$x + 3y + 2z = 5.$$

▦ **EXAMPLE 4**    Use a graphing calculator to solve the system in Example 2.

**Solution**  There are three ways to proceed.

**Matrix Elimination Method**  Enter the augmented matrix of the system into the calculator (Figure 6.3). Use REF (in the MATH or OPS submenu of the MATRIX menu) to put this matrix in row echelon form (Figure 6.4). The system corresponding to this matrix is

$$x - 3y + \quad .5z = 5$$
$$y + 3.125z = .25$$
$$z = 2.$$

Because the calculator used a different sequence of row operations than was used in Example 2, it produced a different row echelon matrix (and corresponding system). However, back substitution shows that the solutions of this system are the same ones found in Example 2:

$$z = 2 \qquad y + 3.125(2) = .25 \qquad x - 3(-6) + .5(2) = 5$$
$$y + 6.25 = .25 \qquad x + 18 + 1 = 5$$
$$y = -6 \qquad x = -14.$$

As you can see, a significant amount of hand work is involved in this method, so it is not recommended.

**Gauss–Jordan Method**  Enter the augmented matrix (Figure 6.3). Use RREF (in same menu as REF) to produce the same reduced row echelon matrix obtained in Example 3 (Figure 6.5). The solution of the system can be read directly from the matrix: $x = -14$, $y = -6, z = 2$.

**System Solver Method**  This is essentially the same as the Gauss–Jordan method, with an extra step at the beginning. Call up the solver (see the Technology Tip below) and enter the number of variables and equations. The solver will display an appropriately sized matrix, with all entries 0. Change the entries so that the matrix becomes the augmented matrix of the system (Figure 6.6). Press SOLVE and the solution will appear (Figure 6.7). The calculator uses the variables $x_1$, $x_2$, and $x_3$ instead of $x, y,$ and $z$. ▦6

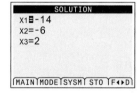

| Figure 6.6 | Figure 6.7 |

▦ **TECHNOLOGY TIP**  Many graphing calculators have system solvers. Select PLYSMLT2 in the APPS menu of TI-84+ or SIMULT on the TI-86 keyboard. *Note:* The TI-86 solver works only for systems that have the same number of variables as equations *and* have a unique solution (so they cannot be used in Checkpoint 6 or in Examples 5–8).

▦ **EXAMPLE 5**    Use a graphing calculator to solve the system

$$x + 2y - z = 0$$
$$3x - y + z = 6$$
$$7x + 28y - 8z = -5$$
$$5x + 3y - z = 6.$$

**Solution**  Enter the augmented matrix into the calculator (Figure 6.8). Then use RREF to put this matrix in reduced row echelon form (Figure 6.9). In Figure 6.9, you must use the arrow key to scroll over to see the full decimal expansions in the right-hand column. This inconvenience can be avoided for many matrices on TI calculators by using FRAC (in the MATH menu), as in Figure 6.10.

$$[A]$$
$$\begin{bmatrix} 1 & 2 & -1 & 0 \\ 3 & -1 & 1 & 6 \\ 7 & 28 & -8 & -5 \\ 5 & 3 & -1 & 6 \end{bmatrix}$$

**Figure 6.8**

$$\text{rref}([A])$$
$$\begin{bmatrix} 1 & 0 & 0 & 1.571428571 \\ 0 & 1 & 0 & -.2857142857 \\ 0 & 0 & 1 & 1 \\ 0 & 0 & 0 & 0 \end{bmatrix}$$

**Figure 6.9**

$$\text{rref}([A]) \blacktriangleright \text{Frac}$$
$$\begin{bmatrix} 1 & 0 & 0 & \frac{11}{7} \\ 0 & 1 & 0 & \frac{-2}{7} \\ 0 & 0 & 1 & 1 \\ 0 & 0 & 0 & 0 \end{bmatrix}$$

**Figure 6.10**

The answers can now be read from the last column of the matrix in Figure 6.10: $x = 11/7$, $y = -2/7, z = 1$. ✓7

## Dependent and Inconsistent Systems

Recall that a system of equations in two variables may have exactly one solution, an infinite number of solutions, or no solutions at all. This fact was illustrated geometrically in Figure 6.1. The same thing is true for systems with three or more variables (and the same terminology is used): A system has exactly one solution (an **independent system**), an infinite number of solutions (a **dependent system**), or no solutions at all (an **inconsistent system**).

**EXAMPLE 6**  Solve the system

$$2x + 4y = 4$$
$$3x + 6y = 8$$
$$2x + y = 7.$$

**Solution**  There are several possible ways to proceed.

**Manual Method**  Write the augmented matrix and perform row operations to obtain a first column whose entries (from top to bottom) are 1, 0, 0:

$$\begin{bmatrix} 2 & 4 & | & 4 \\ 3 & 6 & | & 8 \\ 2 & 1 & | & 7 \end{bmatrix}$$

$$\begin{bmatrix} 1 & 2 & | & 2 \\ 3 & 6 & | & 8 \\ 2 & 1 & | & 7 \end{bmatrix} \quad \tfrac{1}{2} R_1$$

$$\begin{bmatrix} 1 & 2 & | & 2 \\ 0 & 0 & | & 2 \\ 2 & 1 & | & 7 \end{bmatrix}. \quad -3R_1 + R_2$$

**Stop!** The second row of the matrix denotes the equation $0x + 0y = 2$. Since the left side of this equation is always 0 and the right side is 2, it has no solution. Therefore, the original system has no solution.

**Calculator Method**  Enter the augmented matrix into a graphing calculator and use RREF to put it into reduced row echelon form, as in Figure 6.11. The last row of that matrix corresponds to $0x + 0y = 1$, which has no solution. Hence, the original system has no solution. ✓8

Whenever the solution process produces a row whose elements are all 0 *except* the last one, as in Example 6, the system is inconsistent and has no solutions. However, if a row with a 0 for *every* entry is produced, it corresponds to an equation such as $0x + 0y + 0z = 0$, which has infinitely many solutions. So the system may have solutions, as in the next example.

**✓Checkpoint 7**

Use a graphing calculator to solve the following system:

$$x + 3y = 4$$
$$4x + 8y = 4$$
$$6x + 12y = 6.$$

$$\text{rref}([A])$$
$$\begin{bmatrix} 1 & 0 & 0 \\ 0 & 1 & 0 \\ 0 & 0 & 1 \end{bmatrix}$$

**Figure 6.11**

**✓Checkpoint 8**

Solve each system.

(a)  $x - y = 4$
     $-2x + 2y = 1$

(b)  $3x - 4y = 0$
     $2x + y = 0$

**EXAMPLE 7**    Solve the system

$$2x - 3y + 4z = 6$$
$$x - 2y + z = 9$$
$$y + 2z = -12.$$

**Solution**    Use matrix elimination as far as possible, beginning with the augmented matrix of the system:

$$\begin{bmatrix} 2 & -3 & 4 & \bigm| & 6 \\ 1 & -2 & 1 & \bigm| & 9 \\ 0 & 1 & 2 & \bigm| & -12 \end{bmatrix}$$

$$\begin{bmatrix} 1 & -2 & 1 & \bigm| & 9 \\ 2 & -3 & 4 & \bigm| & 6 \\ 0 & 1 & 2 & \bigm| & -12 \end{bmatrix} \quad \text{Interchange } R_1 \text{ and } R_2.$$

$$\begin{bmatrix} 1 & -2 & 1 & \bigm| & 9 \\ 0 & 1 & 2 & \bigm| & -12 \\ 0 & 1 & 2 & \bigm| & -12 \end{bmatrix} \quad -2R_1 + R_2$$

$$\begin{bmatrix} 1 & -2 & 1 & \bigm| & 9 \\ 0 & 1 & 2 & \bigm| & -12 \\ 0 & 0 & 0 & \bigm| & 0 \end{bmatrix}. \qquad \begin{aligned} x - 2y + z &= 9 \\ y + 2z &= -12. \end{aligned}$$
$$\qquad\qquad -R_2 + R_3$$

The last augmented matrix above represents the system shown to its right. Since there are only two nonzero rows in the matrix, it is not possible to continue the process. The fact that the corresponding system has one variable (namely, $z$) that is not the leading variable of an equation indicates a dependent system. To find its solutions, first solve the second equation for $y$:

$$y = -2z - 12.$$

Now substitute the result for $y$ in the first equation and solve for $x$:

$$x - 2y + z = 9$$
$$x - 2(-2z - 12) + z = 9$$
$$x + 4z + 24 + z = 9$$
$$x + 5z = -15$$
$$x = -5z - 15.$$

Each choice of a value for $z$ leads to values for $x$ and $y$. For example,

$$\text{if } z = 1, \quad \text{then } x = -20 \quad \text{and} \quad y = -14;$$
$$\text{if } z = -6, \quad \text{then } x = 15 \quad \text{and} \quad y = 0;$$
$$\text{if } z = 0, \quad \text{then } x = -15 \quad \text{and} \quad y = -12.$$

There are infinitely many solutions of the original system, since $z$ can take on infinitely many values. The solutions are all ordered triples of the form

$$(-5z - 15, -2z - 12, z),$$

where $z$ is any real number.    ✔ 9

✓ **Checkpoint 9**

Use the following values of $z$ to find additional solutions for the system of Example 7.

**(a)** $z = 7$

**(b)** $z = -14$

**(c)** $z = 5$

Since both $x$ and $y$ in Example 7 were expressed in terms of $z$, the variable $z$ is called a **parameter.** If we had solved the system in a different way, $x$ or $y$ could have been the parameter. The final system in Example 7 had one more variable than equations. If there are two more variables than equations, there usually will be two parameters, and so on.

**Figure 6.12**

**EXAMPLE 8** Row operations were used to reduce the augmented matrix of a system of three equations in four variables $(x, y, z,$ and $w)$ to the reduced row echelon matrix in Figure 6.12. Solve the system.

**Solution** First write out the system represented by the matrix:

$$x + \quad\quad 9w = -12$$
$$y - \quad w = \quad 4$$
$$z - 2w = \quad 1.$$

Let $w$ be the parameter. Solve the first equation for $x$, the second for $y$, and the third for $z$:

$$x = -9w - 12; \quad y = w + 4; \quad z = 2w + 1.$$

The solutions are given by $(-9w - 12, w + 4, 2w + 1, w)$, where $w$ is any real number.

**TECHNOLOGY TIP** When the system solver on a TI-86 produces an error message, the system might be inconsistent (no solutions) or dependent (infinitely many solutions). You must use RREF or manual methods to solve the system.

The TI-84+ solver usually solves dependent systems directly; Figure 6.13 shows its solutions for Example 7. When the message "no solution found" is displayed, as in Figure 6.14, select RREF at the bottom of the screen to display the reduced row echelon matrix of the system. From that, you can determine the solutions (if there are any) or that no solutions are possible.

**Figure 6.13**

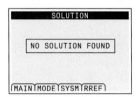

**Figure 6.14**

# 6.2 Exercises

*Obtain an equivalent system by performing the stated elementary operation on the system. (See Example 1.)*

**1.** Interchange equations 1 and 2.

$$2x - 4y + 5z = 1$$
$$x \quad\quad - 3z = 2$$
$$5x - 8y + 7z = 6$$
$$3x - 4y + 2z = 3$$

**2.** Interchange equations 1 and 3.

$$2x - 2y + \quad z = -6$$
$$3x + \quad y + 2z = \quad 2$$
$$x + \quad y - 2z = \quad 0$$

**3.** Multiply the third equation by $1/2$.

$$x + 2y + 4z = 3$$
$$x \quad\quad + 2z = 0$$
$$2x + 4y + \quad z = 3$$

**4.** Multiply the first equation by $-\dfrac{2}{3}$.

$$-\frac{3}{2}x + 6y - 5z = 12$$
$$9x \quad\quad + 7z = \quad 4$$

**5.** Replace the second equation by the sum of itself and $-2$ times the first equation.

$$x + y + 2z + 3w = 1$$
$$2x + y + 3z + 4w = 1$$
$$3x + y + 4z + 5w = 2$$

**6.** Replace the third equation by the sum of itself and $-1$ times the first equation.

$$x + 2y + 4z = \quad 6$$
$$y + \quad z = \quad 1$$
$$x + 3y + 5z = 10$$

**7.** Replace the third equation by the sum of itself and $-2$ times the second equation.

$$\begin{aligned} x + 12y - 3z + 4w &= 10 \\ 2y + 3z + w &= 4 \\ 4y + 5z + 2w &= 1 \\ 6y - 2z - 3w &= 0 \end{aligned}$$

**8.** Replace the third equation by the sum of itself and 3 times the second equation.

$$\begin{aligned} 2x + 2y - 4z + w &= -5 \\ 2y + 4z - w &= 2 \\ -6y - 4z + 2w &= 6 \\ 2y + 5z - 3w &= 7 \end{aligned}$$

*Solve the system by back substitution. (See Example 1.)*

**9.** $\begin{aligned} x + 3y - 4z + 2w &= 1 \\ y + z - w &= 4 \\ 2z + 2w &= -6 \\ 3w &= 9 \end{aligned}$

**10.** $\begin{aligned} x + 5z + 6w &= 10 \\ y + 3z - 2w &= 4 \\ z - 4w &= -6 \\ 2w &= 4 \end{aligned}$

**11.** $\begin{aligned} 2x + 2y - 4z + w &= -5 \\ 3y + 4z - w &= 0 \\ 2z - 7w &= -6 \\ 5w &= 15 \end{aligned}$

**12.** $\begin{aligned} 3x - 2y - 4z + 2w &= 6 \\ 2y + 5z - 3w &= 7 \\ 3z + 4w &= 0 \\ 3w &= 15 \end{aligned}$

*Write the augmented matrix of each of the given systems. Do not solve the systems. (See Example 2.)*

**13.** $\begin{aligned} 2x + y + z &= 3 \\ 3x - 4y + 2z &= -5 \\ x + y + z &= 2 \end{aligned}$

**14.** $\begin{aligned} 3x + 4y - 2z - 3w &= 0 \\ x - 3y + 7z + 4w &= 9 \\ 2x + 5z - 6w &= 0 \end{aligned}$

*Write the system of equations associated with each of the given augmented matrices. Do not solve the systems.*

**15.** $\begin{bmatrix} 2 & 3 & 8 & | & 20 \\ 1 & 4 & 6 & | & 12 \\ 0 & 3 & 5 & | & 10 \end{bmatrix}$

**16.** $\begin{bmatrix} 3 & 2 & 6 & | & 18 \\ 2 & -2 & 5 & | & 7 \\ 1 & 0 & 5 & | & 20 \end{bmatrix}$

*Use the indicated row operation to transform each matrix. (See Examples 2 and 3.)*

**17.** Interchange $R_1$ and $R_3$.

$$\begin{bmatrix} 5 & -1 & 4 & | & 8 \\ 2 & 2 & 9 & | & 3 \\ 1 & 7 & -3 & | & 5 \end{bmatrix}$$

**18.** Interchange $R_2$ and $R_3$.

$$\begin{bmatrix} 1 & 8 & 3 & | & -2 \\ 0 & 7 & -9 & | & 3 \\ 0 & 1 & 8 & | & -4 \end{bmatrix}$$

**19.** Replace $R_3$ by $\frac{1}{4}R_3$.

$$\begin{bmatrix} 2 & 5 & 1 & | & -1 \\ -4 & 0 & 4 & | & 6 \\ 6 & 0 & 8 & | & -4 \end{bmatrix}$$

**20.** Replace $R_1$ with $-\frac{1}{6}R_1$.

$$\begin{bmatrix} -6 & 3 & -8 & | & 12 \\ 7 & 4 & -5 & | & 25 \end{bmatrix}$$

**21.** Replace $R_2$ by $2R_1 + R_2$.

$$\begin{bmatrix} -4 & -3 & 1 & -1 & | & 2 \\ 8 & 2 & 5 & 0 & | & 6 \\ 0 & -2 & 9 & 4 & | & 5 \end{bmatrix}$$

**22.** Replace $R_3$ by $-3R_1 + R_3$.

$$\begin{bmatrix} 1 & 5 & 2 & 0 & | & -1 \\ 8 & 5 & 4 & 6 & | & 6 \\ 3 & 0 & 7 & 1 & | & -4 \end{bmatrix}$$

*In Exercises 23–28, the reduced row echelon form of the augmented matrix of a system of equations is given. Find the solutions of the system. (See Example 3.)*

**23.** $\begin{bmatrix} 1 & 0 & 0 & 0 & | & 3/2 \\ 0 & 1 & 0 & 0 & | & 17 \\ 0 & 0 & 1 & 0 & | & -5 \\ 0 & 0 & 0 & 1 & | & 0 \end{bmatrix}$

**24.** $\begin{bmatrix} 1 & 0 & 0 & 0 & | & 7 \\ 0 & 1 & 0 & 0 & | & 2 \\ 0 & 0 & 1 & 0 & | & -5 \\ 0 & 0 & 0 & 1 & | & 3 \\ 0 & 0 & 0 & 0 & | & 0 \\ 0 & 0 & 0 & 0 & | & 0 \end{bmatrix}$

**25.** $\begin{bmatrix} 1 & 0 & 0 & | & -5 \\ 0 & 1 & 0 & | & 9 \\ 0 & 0 & 0 & | & 1 \end{bmatrix}$

**26.** $\begin{bmatrix} 1 & 0 & 0 & 0 & 0 & | & 6 \\ 0 & 1 & 0 & 0 & 0 & | & 4 \\ 0 & 0 & 1 & 0 & 0 & | & 5 \\ 0 & 0 & 0 & 0 & 1 & | & 2 \\ 0 & 0 & 0 & 0 & 0 & | & 1 \end{bmatrix}$

**27.** $\begin{bmatrix} 1 & 0 & 0 & 1 & | & 12 \\ 0 & 1 & 0 & 2 & | & -3 \\ 0 & 0 & 1 & 0 & | & -5 \\ 0 & 0 & 0 & 0 & | & 0 \end{bmatrix}$

**28.** $\begin{bmatrix} 1 & 0 & -7 & | & 10 \\ 0 & 1 & 3 & | & 25 \end{bmatrix}$

*In Exercises 29–34, perform row operations on the augmented matrix as far as necessary to determine whether the system is independent, dependent, or inconsistent. (See Examples 6–8.)*

29.
$$\begin{aligned} x + 2y &= 0 \\ y - z &= 2 \\ x + y + z &= -2 \end{aligned}$$

30.
$$\begin{aligned} x + 2y + z &= 0 \\ y + 2z &= 0 \\ x + y - z &= 0 \end{aligned}$$

31.
$$\begin{aligned} x + 2y + 4z &= 6 \\ y + z &= 1 \\ x + 3y + 5z &= 10 \end{aligned}$$

32.
$$\begin{aligned} x + y + 2z + 3w &= 1 \\ 2x + y + 3z + 4w &= 1 \\ 3x + y + 4z + 5w &= 2 \end{aligned}$$

33.
$$\begin{aligned} a - 3b - 2c &= -3 \\ 3a + 2b - c &= 12 \\ -a - b + 4c &= 3 \end{aligned}$$

34.
$$\begin{aligned} 2x + 2y + 2z &= 6 \\ 3x - 3y - 4z &= -1 \\ x + y + 3z &= 11 \end{aligned}$$

*Write the augmented matrix of the system and use the matrix method to solve the system. (See Examples 2 and 4.)*

35.
$$\begin{aligned} -x + 3y + 2z &= 0 \\ 2x - y - z &= 3 \\ x + 2y + 3z &= 0 \end{aligned}$$

36.
$$\begin{aligned} 3x + 7y + 9z &= 0 \\ x + 2y + 3z &= 2 \\ x + 4y + z &= 2 \end{aligned}$$

37.
$$\begin{aligned} x - 2y + 4z &= 6 \\ x + 2y + 13z &= 6 \\ -2x + 6y - z &= -10 \end{aligned}$$

38.
$$\begin{aligned} x - 2y + 5z &= -6 \\ x + 2y + 3z &= 0 \\ x + 3y + 2z &= 5 \end{aligned}$$

39.
$$\begin{aligned} x + y + z &= 200 \\ x - 2y &= 0 \\ 2x + 3y + 5z &= 600 \\ 2x - y + z &= 200 \end{aligned}$$

40.
$$\begin{aligned} 2x - y + 2z &= 3 \\ -x + 2y - z &= 0 \\ 3y - 2z &= 1 \\ x + y - z &= 1 \end{aligned}$$

41.
$$\begin{aligned} x + y + z &= 5 \\ 2x + y - z &= 2 \\ x - y + z &= -2 \end{aligned}$$

42.
$$\begin{aligned} 2x + y + 3z &= 9 \\ -x - y + z &= 1 \\ 3x - y + z &= 9 \end{aligned}$$

*Use the Gauss–Jordan method to solve each of the given systems of equations. (See Examples 3–5.)*

43.
$$\begin{aligned} x + 2y + z &= 5 \\ 2x + y - 3z &= -2 \\ 3x + y + 4z &= -5 \end{aligned}$$

44.
$$\begin{aligned} 3x - 2y + z &= 6 \\ 3x + y - z &= -4 \\ -x + 2y - 2z &= -8 \end{aligned}$$

45.
$$\begin{aligned} x + 3y - 6z &= 7 \\ 2x - y + 2z &= 0 \\ x + y + 2z &= -1 \end{aligned}$$

46.
$$\begin{aligned} x &= 1 - y \\ 2x &= z \\ 2z &= -2 - y \end{aligned}$$

47.
$$\begin{aligned} x - 2y + 4z &= 9 \\ x + y + 13z &= 6 \\ -2x + 6y - z &= -10 \end{aligned}$$

48.
$$\begin{aligned} x - y + 5z &= -6 \\ 3x + 3y - z &= 10 \\ x + 2y + 3z &= 5 \end{aligned}$$

*Solve the system by any method.*

49.
$$\begin{aligned} x + 3y + 4z &= 14 \\ 2x - 3y + 2z &= 10 \\ 3x - y + z &= 9 \\ 4x + 2y + 5z &= 9 \end{aligned}$$

50.
$$\begin{aligned} x - y &= 2 \\ x + y &= 4 \\ 2x + 3y &= 9 \\ 3x - 2y &= 6 \end{aligned}$$

51.
$$\begin{aligned} 4x - 4y + 12z &= 28 \\ 3x - 2y - z &= 36 \end{aligned}$$

52.
$$\begin{aligned} 3x + y - z &= 0 \\ 2x - y + 3z &= -7 \end{aligned}$$

53.
$$\begin{aligned} -8x - 9y &= 11 \\ 24x + 34y &= 2 \\ 16x + 11y &= -57 \end{aligned}$$

54.
$$\begin{aligned} x + 2y &= 3 \\ 2x + 3y &= 4 \\ 3x + 4y &= 5 \\ 4x + 5y &= 6 \end{aligned}$$

55.
$$\begin{aligned} x - 2y + z &= 5 \\ 2x + y - z &= 2 \\ -2x + 4y - 2z &= 2 \end{aligned}$$

56.
$$\begin{aligned} 2x + y &= 7 \\ x - y &= 3 \\ x + 3y &= 4 \end{aligned}$$

57.
$$\begin{aligned} x + y - z &= -20 \\ 2x - y + z &= 11 \end{aligned}$$

58.
$$\begin{aligned} 4x + 3y + z &= 1 \\ -2x - y + 2z &= 0 \end{aligned}$$

59.
$$\begin{aligned} 2x + y + 3z - 2w &= -6 \\ 4x + 3y + z - w &= -2 \\ x + y + z + w &= -5 \\ -2x - 2y - 2z + 2w &= -10 \end{aligned}$$

60.
$$\begin{aligned} x + y + z + w &= -1 \\ -x + 4y + z - w &= 0 \\ x - 2y + z - 2w &= 11 \\ -x - 2y + z + 2w &= -9 \end{aligned}$$

61.
$$\begin{aligned} x + 2y - z &= 3 \\ 3x + y + w &= 4 \\ 2x - y + z + w &= 2 \end{aligned}$$

62.
$$\begin{aligned} 3x - 2y - 8z &= 1 \\ 9x - 6y - 24z &= -2 \\ x - y + z &= 1 \end{aligned}$$

63.
$$\begin{aligned} 2x + 3y + z &= 9 \\ 4x + y - 3z &= -7 \\ 6x + 2y - 4z &= -8 \end{aligned}$$

64.
$$\begin{aligned} x - 2y - z - 3w &= -3 \\ -x + y + z &= 2 \\ 4y + 3z - 6w &= -2 \end{aligned}$$

*Work these exercises.*

65. **Population Growth** The population $y$ in year $x$ of the county listed is approximated by:

$$\begin{aligned} \text{Calhoun County, IL:} \quad & x + 20y = 112 \\ \text{Martin County, TX:} \quad & -x + 10y = 38 \\ \text{Schley County, GA:} \quad & y = 5, \end{aligned}$$

where $x = 10$ corresponds to the year 2010 and $y$ is in thousands. Solve this system of equations, and interpret your answer. (Data from: U.S. Census Bureau.)

66. **Population Growth** The population $y$ in year $x$ of the greater metropolitan area listed is approximated by the given equation in which $x = 10$ corresponds to the year 2010 and $y$ is in thousands:

$$\begin{aligned} \text{Portland, OR:} \quad & -32x + y = 1906 \\ \text{Cambridge, MA:} \quad & -22x + y = 2026 \\ \text{Charlotte, NC:} \quad & -73x + 2y = 3704. \end{aligned}$$

Solve this system of equations, and interpret your answer. (Data from: U.S. Census Bureau.)

*Use a graphing calculator to solve these problems.*

67. **Concert Tickets** A band concert is attended by $x$ adults, $y$ teenagers, and $z$ preteen children. These numbers satisfy the following equations:

$$\begin{aligned} x + 1.25y + .25z &= 457.5 \\ x + .6y + .4z &= 390 \\ 3.16x + 3.48y + .4z &= 1297.2. \end{aligned}$$

How many adults, teenagers, and children were present?

**68. Credit Card Debt** The owner of a small business borrows money on three separate credit cards: $x$ dollars on Mastercard, $y$ dollars on Visa, and $z$ dollars on American Express. These amounts satisfy the following equations:

$$1.18x + 1.15y + 1.09z = 11{,}244.25$$
$$3.54x - .55y + .27z = 3{,}732.75$$
$$.06x + .05y + .03z = 414.75.$$

How much did the owner borrow on each card?

*Work these problems.*

**69.** Graph the equations in the given system. Then explain why the graphs show that the system is inconsistent:

$$2x + 3y = 8$$
$$x - y = 4$$
$$5x + y = 7.$$

**70.** Explain why the graphs of the equations in the given system suggest that the system is independent:

$$-x + y = 1$$
$$-2x + y = 0$$
$$x + y = 3.$$

**71.** Find constants $a$, $b$, and $c$ such that the points $(2, 3)$, $(-1, 0)$, and $(-2, 2)$ lie on the graph of the equation $y = ax^2 + bx + c$. (*Hint:* Since $(2, 3)$ is on the graph, we must have $3 = a(2^2) + b(2) + c$; that is, $4a + 2b + c = 3$. Similarly, the other two points lead to two more equations. Solve the resulting system for $a$, $b$, and $c$.)

**72.** Explain why a system with more variables than equations cannot have a unique solution (that is, be an independent system). (*Hint:* When you apply the elimination method to such a system, what must happen?)

✓ **Checkpoint Answers**

**1. (a)** $(-1, 1)$   **(b)** $(3, 1, -1)$

**2. (a)** $\begin{bmatrix} 4 & -2 & 3 & | & 4 \\ 3 & 5 & 1 & | & -7 \\ 5 & -1 & 4 & | & 6 \end{bmatrix}$

**(b)** 
$$2x - 2y = -2$$
$$x + y = 4$$
$$3x + 5y = 8$$

**3. (a)** $\begin{bmatrix} 3 & -2 \\ -1 & 5 \end{bmatrix}$

**(b)** $\begin{bmatrix} -2 & 10 \\ 3 & -2 \end{bmatrix}$

**(c)** $\begin{bmatrix} -1 & 5 \\ 6 & -17 \end{bmatrix}$

**(d)** $\begin{bmatrix} 5 & 1 \\ 3 & -2 \end{bmatrix}$

**4.** $(-1, 1, 2)$

**5. (a)** $\begin{bmatrix} 1 & 2 & | & 11 \\ -4 & 1 & | & -8 \\ 5 & 1 & | & 19 \end{bmatrix}$

**(b)** $4R_1 + R_2 \begin{bmatrix} 1 & 2 & | & 11 \\ 0 & 9 & | & 36 \\ 5 & 1 & | & 19 \end{bmatrix}$

**(c)** $-5R_1 + R_3 \begin{bmatrix} 1 & 2 & | & 11 \\ 0 & 9 & | & 36 \\ 0 & -9 & | & -36 \end{bmatrix}$

**(d)** $\frac{1}{9}R_2 \begin{bmatrix} 1 & 2 & | & 11 \\ 0 & 1 & | & 4 \\ 0 & -9 & | & -36 \end{bmatrix}$

**(e)** $-2R_2 + R_1 \begin{bmatrix} 1 & 0 & | & 3 \\ 0 & 1 & | & 4 \\ 0 & -9 & | & -36 \end{bmatrix}$

**(f)** $9R_2 + R_3 \begin{bmatrix} 1 & 0 & | & 3 \\ 0 & 1 & | & 4 \\ 0 & 0 & | & 0 \end{bmatrix}$

**(g)** $(3, 4)$

**6.** $(1, 2, -1)$

**7.** $(-5, 3)$

**8. (a)** No solution

**(b)** $(0, 0)$

**9. (a)** $(-50, -26, 7)$

**(b)** $(55, 16, -14)$

**(c)** $(-40, -22, 5)$

# 6.3 Applications of Systems of Linear Equations

There are no hard and fast rules for solving applied problems, but it is usually best to begin by identifying the unknown quantities and letting each be represented by a variable. Then look at the given data to find one or more relationships among the unknown quantities that lead to equations. If the equations are all linear, use the techniques of the preceding sections to solve the system.

**EXAMPLE 1**    **Truck Rentals** A rent-a-truck company plans to spend $5 million on 200 new vehicles. Each van will cost $20,000, each small truck $25,000, and each large truck $35,000. Past experience shows that the company needs twice as many vans as small trucks. How many of each kind of vehicle can the company buy?

**Solution** Let $x$ be the number of vans, $y$ the number of small trucks, and $z$ the number of large trucks. Then

$$\left(\begin{array}{c}\text{Number of}\\ \text{vans}\end{array}\right) + \left(\begin{array}{c}\text{Number of}\\ \text{small trucks}\end{array}\right) + \left(\begin{array}{c}\text{Number of}\\ \text{large trucks}\end{array}\right) = \text{Total number of vehicles}$$

$$x + y + z = 200. \tag{1}$$

Similarly,

$$\left(\begin{array}{c}\text{Cost of } x\\ \text{vans}\end{array}\right) + \left(\begin{array}{c}\text{Cost of } y\\ \text{small trucks}\end{array}\right) + \left(\begin{array}{c}\text{Cost of } z\\ \text{large trucks}\end{array}\right) = \text{Total cost}$$

$$20{,}000x + 25{,}000y + 35{,}000z = 5{,}000{,}000.$$

Dividing both sides by 5000 produces the equivalent equation

$$4x + 5y + 7z = 1000. \tag{2}$$

Finally, the number of vans is twice the number of small trucks; that is, $x = 2y$, or equivalently,

$$x - 2y = 0. \tag{3}$$

We must solve the system given by equations (1)–(3):

$$\begin{aligned} x + y + z &= 200 \\ 4x + 5y + 7z &= 1000 \\ x - 2y \phantom{{}+ 7z} &= 0. \end{aligned}$$

**Manual Method** Form the augmented matrix and transform it into row echelon form:

$$\begin{bmatrix} 1 & 1 & 1 & | & 200 \\ 4 & 5 & 7 & | & 1000 \\ 1 & -2 & 0 & | & 0 \end{bmatrix}$$

$$\begin{bmatrix} 1 & 1 & 1 & | & 200 \\ 0 & 1 & 3 & | & 200 \\ 0 & -3 & -1 & | & -200 \end{bmatrix} \quad \begin{array}{l} -4R_1 + R_2 \\ -R_1 + R_3 \end{array}$$

$$\begin{bmatrix} 1 & 1 & 1 & | & 200 \\ 0 & 1 & 3 & | & 200 \\ 0 & 0 & 8 & | & 400 \end{bmatrix} \quad 3R_2 + R_3$$

$$\begin{bmatrix} 1 & 1 & 1 & | & 200 \\ 0 & 1 & 3 & | & 200 \\ 0 & 0 & 1 & | & 50 \end{bmatrix}. \quad \tfrac{1}{8}R_3$$

This row echelon matrix corresponds to the system

$$\begin{aligned} x + y + z &= 200 \\ y + 3z &= 200 \\ z &= 50. \end{aligned}$$

**Figure 6.15**

✓ **Checkpoint 1**

In Example 1, suppose that the company can spend only $2 million on 150 new vehicles and the company needs three times as many vans as small trucks. Write a system of equations to express these conditions.

Use back substitution to solve this system:

$$z = 50 \qquad y + 3z = 200 \qquad x + y + z = 200$$
$$y + 3(50) = 200 \qquad x + 50 + 50 = 200$$
$$y + 150 = 200 \qquad x \qquad + 100 = 200$$
$$y = 50 \qquad x = 100.$$

Therefore, the company should buy 100 vans, 50 small trucks, and 50 large trucks.

**Calculator Method** Enter the augmented matrix of the system into the calculator. Use RREF to change it into reduced row echelon form, as in Figure 6.15. The answers may be read directly from the matrix: $x = 100$ vans, $y = 50$ small trucks, and $z = 50$ large trucks. ✓₁

---

**EXAMPLE 2**    **Finance** An investor plans to put a total of $100,000 in a money market account, a bond fund, an international stock fund, and a domestic stock fund. She wants 60% of her investment to be conservative (money market and bonds). She wants the amount in international stocks to be one-fourth of the amount in domestic stocks. Finally, she needs an annual return of $4000. Assuming she gets annual returns of 2.5% on the money market account, 3.5% on the bond fund, 5% on the international stock fund, and 6% on the domestic stock fund, how much should she put in each investment?

**Solution** Let $x$ be the amount invested in the money market account, $y$ the amount in the bond fund, $z$ the amount in the international stock fund, and $w$ the amount in the domestic stock fund. Then

$$x + y + z + w = \text{total amount invested} = 100,000. \qquad \textbf{(4)}$$

Use her annual return to get a second equation:

$$\left(\begin{array}{c}2.5\% \text{ return} \\ \text{on money} \\ \text{market}\end{array}\right) + \left(\begin{array}{c}3.5\% \text{ return} \\ \text{on bond} \\ \text{fund}\end{array}\right) + \left(\begin{array}{c}5\% \text{ return on} \\ \text{international} \\ \text{stock fund}\end{array}\right) + \left(\begin{array}{c}6\% \text{ return} \\ \text{on domestic} \\ \text{stock fund}\end{array}\right) = 4000$$

$$.025x \quad + \quad .035y \quad + \quad .05z \quad + \quad .06w \quad = 4000. \quad \textbf{(5)}$$

Since she wants the amount in international stocks to be one-fourth of the amount in domestic stocks, we have

$$z = \frac{1}{4}w, \qquad \text{or equivalently,} \qquad z - .25w = 0. \qquad \textbf{(6)}$$

Finally, the amount in conservative investments is $x + y$, and this quantity should be equal to 60% of $100,000—that is,

$$x + y = 60,000. \qquad \textbf{(7)}$$

✓ **Checkpoint 2**

**(a)** Write the augmented matrix of the system given by equations (4)–(7) of Example 2.

**(b)** List a sequence of row operations that transforms this matrix into row echelon form.

**(c)** Display the final row echelon form matrix.

Now solve the system given by equations (4)–(7):

$$x + y + z + w = 100,000$$
$$.025x + .035y + .05z + .06w = 4,000$$
$$z - .25w = 0$$
$$x + y = 60,000.$$

**Manual Method** Write the augmented matrix of the system and transform it into row echelon form, as in Checkpoint 2. ✓₂

The final matrix in Checkpoint 2 represented the system

$$x + y + z + w = 100{,}000$$
$$y + 2.5z + 3.5w = 150{,}000$$
$$z - .25w = 0$$
$$w = 32{,}000.$$

Back substitution shows that

$$w = 32{,}000$$
$$z = .25w = .25(32{,}000) = 8000$$
$$y = -2.5z - 3.5w + 150{,}000 = -2.5(8000) - 3.5(32{,}000) + 150{,}000 = 18{,}000$$
$$x = -y - z - w + 100{,}000 = -18{,}000 - 8000 - 32{,}000 + 100{,}000 = 42{,}000.$$

**Figure 6.16**

Therefore, she should put $42,000 in the money market account, $18,000 in the bond fund, $8000 in the international stock fund, and $32,000 in the domestic stock fund.

**Calculator Method**  Enter the augmented matrix of the system into the calculator. Use RREF to change it into reduced row echelon form, as in Figure 6.16. The answers can easily be read from this matrix: $x = 42{,}000$, $y = 18{,}000$, $z = 8000$, and $w = 32{,}000$.

---

**EXAMPLE 3**    **Livestock Dietary Requirements**  An animal feed is to be made from corn, soybeans, and cottonseed. Determine how many units of each ingredient are needed to make a feed that supplies 1800 units of fiber, 2800 units of fat, and 2200 units of protein, given that 1 unit of each ingredient provides the numbers of units shown in the table below. The table states, for example, that a unit of corn provides 10 units of fiber, 30 units of fat, and 20 units of protein.

| | Corn | Soybeans | Cottonseed | Totals |
|---|---|---|---|---|
| **Units of Fiber** | 10 | 20 | 30 | 1800 |
| **Units of Fat** | 30 | 20 | 40 | 2800 |
| **Units of Protein** | 20 | 40 | 25 | 2200 |

**Solution**  Let $x$ represent the required number of units of corn, $y$ the number of units of soybeans, and $z$ the number of units of cottonseed. Since the total amount of fiber is to be 1800, we have

$$10x + 20y + 30z = 1800.$$

The feed must supply 2800 units of fat, so

$$30x + 20y + 40z = 2800.$$

Finally, since 2200 units of protein are required, we have

$$20x + 40y + 25z = 2200.$$

Thus, we must solve this system of equations:

$$10x + 20y + 30z = 1800$$
$$30x + 20y + 40z = 2800 \qquad (8)$$
$$20x + 40y + 25z = 2200.$$

Now solve the system, either manually or with technology.

✓ **Checkpoint 3**

(a) Write the augmented matrix of system (8) of Example 3.

(b) List a sequence of row operations that transforms this matrix into row echelon form.

(c) Display the final row echelon form matrix.

$$
\begin{bmatrix}
10 & 20 & 30 & 1800 \\
30 & 20 & 40 & 2800 \\
20 & 40 & 25 & 2200
\end{bmatrix}
$$
rref([A])
$$
\begin{bmatrix}
1 & 0 & 0 & 30 \\
0 & 1 & 0 & 15 \\
0 & 0 & 1 & 40
\end{bmatrix}
$$

**Figure 6.17**

**Manual Method**  Write the augmented matrix and use row operations to transform it into row echelon form, as in Checkpoint 3. The resulting matrix represents the following system: ✓3

$$x + 2y + 3z = 180$$

$$y + \frac{5}{4}z = 65$$

$$z = 40. \qquad (9)$$

Back substitution now shows that

$$z = 40, \quad y = 65 - \frac{5}{4}(40) = 15, \quad \text{and} \quad x = 180 - 2(15) - 3(40) = 30.$$

Thus, the feed should contain 30 units of corn, 15 units of soybeans, and 40 units of cottonseed.

**Calculator Method**  Enter the augmented matrix of the system into the calculator (top of Figure 6.17). Use RREF to transform it into reduced row echelon form (bottom of Figure 6.17), which shows that $x = 30$, $y = 15$, and $z = 40$.

**EXAMPLE 4**    **Population Growth**  The table shows Census Bureau projections for the population of the United States (in millions).

| Year | 2020 | 2040 | 2050 |
|------|------|------|------|
| **U.S. Population** | 334 | 380 | 400 |

(a) Use the given data to construct a quadratic function that gives the U.S. population (in millions) in year $x$.

**Solution**  Let $x = 0$ correspond to the year 2000. Then the table represents the data points (20,334), (40,380), and (50,400). We must find a function of the form

$$f(x) = ax^2 + bx + c$$

whose graph contains these three points. If (20,334) is to be on the graph, we must have $f(20) = 334$; that is,

$$a(20)^2 + b(20) + c = 334$$
$$400a + 20b + c = 334.$$

The other two points lead to these equations:

$$f(40) = 380 \qquad\qquad f(50) = 400$$
$$a(40)^2 + b(40) + c = 380 \qquad a(50)^2 + b(50) + c = 400$$
$$1600a + 40b + c = 380 \qquad 2500a + 50b + c = 400.$$

Now work by hand or use technology to solve the following system:

$$400a + 20b + c = 334$$
$$1600a + 40b + c = 380$$
$$2500a + 50b + c = 400.$$

$$
\begin{bmatrix}
400 & 20 & 1 & 334 \\
1600 & 40 & 1 & 380 \\
2500 & 50 & 1 & 400
\end{bmatrix}
$$
rref([A])
$$
\begin{bmatrix}
1 & 0 & 0 & -.01 \\
0 & 1 & 0 & 2.9 \\
0 & 0 & 1 & 280
\end{bmatrix}
$$

**Figure 6.18**

The reduced row echelon form of the augmented matrix in Figure 6.18 shows that the solution is $a = -.01$, $b = 2.9$, and $c = 280$. So the function is

$$f(x) = -.01x^2 + 2.9x + 280.$$

**(b)** Use this model to estimate the U.S. population in the year 2030.

**Solution**  The year 2030 corresponds to $x = 30$, so the U.S. population is projected to be
$$f(30) = -.01(30)^2 + 2.9(30) + 280 = 358 \text{ million.}$$

**EXAMPLE 5**   **Delivery Logistics** A specialty wholesaler sells espresso machines. The EZ model weighs 10 pounds and comes in a 10-cubic-foot box. The compact model weighs 20 pounds and comes in an 8-cubic-foot box. The commercial model weighs 60 pounds and comes in a 28-cubic-foot box. Each delivery van has 248 cubic feet of space and can hold a maximum of 440 pounds. In order for a van to be fully loaded, how many of each model should it carry?

**Solution**  Let $x$ be the number of EZ, $y$ the number of compact, and $z$ the number of commercial models carried by a van. Then we can summarize the given information in this table.

| Model | Number | Weight | Volume |
|---|---|---|---|
| EZ | $x$ | 10 | 10 |
| Compact | $y$ | 20 | 8 |
| Commercial | $z$ | 60 | 28 |
| Total for a load | | 440 | 248 |

Since a fully loaded van can carry 440 pounds and 248 cubic feet, we must solve this system of equations:

$$10x + 20y + 60z = 440 \quad \text{Weight equation}$$
$$10x + 8y + 28z = 248. \quad \text{Volume equation}$$

The augmented matrix of the system is
$$\begin{bmatrix} 10 & 20 & 60 & | & 440 \\ 10 & 8 & 28 & | & 248 \end{bmatrix}.$$

If you are working by hand, transform the matrix into row echelon form and use back substitution. If you are using a graphing calculator, use RREF to transform the matrix into reduced row echelon form, as in Figure 6.19 (which also uses FRAC in the MATH menu to eliminate long decimals).

The system corresponding to Figure 6.19 is
$$x + \frac{2}{3}z = 12$$
$$y + \frac{8}{3}z = 16,$$

which is easily solved: $x = 12 - \frac{2}{3}z$ and $y = 16 - \frac{8}{3}z$.

**Figure 6.19**

Hence, all solutions of the system are given by $\left( 12 - \frac{2}{3}z, 16 - \frac{8}{3}z, z \right)$. The only solutions that apply in this situation, however, are those given by $z = 0, 3,$ and $6$, because all other values of $z$ lead to fractions or negative numbers. (You can't deliver part of a box or a negative number of boxes). Hence, there are three ways to have a fully loaded van:

| Solution | Van Load |
|---|---|
| $(12, 16, 0)$ | 12 EZ, 16 compact, 0 commercial |
| $(10, 8, 3)$ | 10 EZ, 8 compact, 3 commercial |
| $(8, 0, 6)$ | 8 EZ, 0 compact, 6 commercial. |

# 6.3 Exercises

*Use systems of equations to work these applied problems. (See Examples 1–5.)*

1. **Lyft Rates** Lyft charges customers a flat fee per ride (which is higher during rush hour traffic) plus charges for each minute and each mile. (Data from: www.lyft.com.) Suppose that, in a certain metropolitan area during rush hour, the flat fee is $4, the cost per minute is $.20, and the cost per mile is $1.40. Let $x$ be the number of minutes and $y$ the number of miles. At the end of a ride, the driver said that you owed $14 and remarked that the number of minutes was three times the number of miles. Find the number of minutes and the number of miles for this trip.

2. **Uber Rates** Uber charges customers a flat fee per ride (which is higher during rush hour traffic) plus charges for each minute and each mile. (Data from: www.uber.com.) For a certain trip in Birmingham, Alabama, in 2016, Uber charged a $3.05 flat fee plus $0.18 per minute and $1.20 per mile. Let $x$ be the number of minutes and $y$ the number of miles. At the end of the ride, the driver said that the bill was $10.85 and remarked that the number of minutes was twice the number of miles. Find the number of minutes and the number of miles for this trip.

3. **Euro and Pounds** Abigail Henderson has $1000 budgeted for spending money on an upcoming trip to London and Paris. The European currency is trading at $1.20 per euro, and British pounds sterling are trading at $1.40 per pound. She plans to spend more time in France, so she wants to have three times as many euros as pounds. Set up and solve a system of equations to model this problem, and explain what your answer means in practical terms.

4. **Pesos and Canadian Dollars** Emma Henderson is planning a North American odyssey and has $1500 budgeted for spending money. It costs $.60 to buy a peso and $.75 to buy a Canadian dollar. She expects her adventures in Mexico to take more time, so she wants to have five times as many pesos as Canadian dollars. Set up and solve a system of equations to model this problem, and explain what your answer means in practical terms.

5. **Stock Portfolio** You invested a total of $25,000 in Coca-Cola Company and Intel Corporation. You purchased Coca-Cola at $45 per share and Intel at $32 per share. If Coca-Cola doubles in value and Intel stock goes up by 50%, your portfolio will be worth $42,000. How many shares of each stock do you own?

6. **Bond Portfolio** You purchased a total of $10,000 in bond funds. The JP Morgan Emerging Markets bond fund costs $100 per share, and the Fidelity Conservative Income bond fund costs $10 per share. If the emerging markets fund goes up in value by 5% and the conservative bond fund increases by 10%, your portfolio will be worth $10,600. How many shares of each fund do you own?

7. **Basketball Statistics** In a championship basketball game, the center scored a total of 31 *points*. Let $x$ be the number of free-throw *baskets* made (worth 1 point each), $y$ the number

of regular *baskets* made (2 points each), and $z$ the number of *baskets* made from behind the three-point line (3 points each). The player made a total of 12 baskets from the floor (2-point and 3-point shots) and 4 baskets from the free-throw line.

(a) Set up a system of three equations to model this situation.

(b) Solve the system and interpret your answer.

8. **Baseball Statistics** In the final game of a baseball series that went to extra innings, the most valuable player (MVP) rounded a total of 12 *bases*. Let $x$ be the number of singles, $y$ the number of doubles, $z$ the number of triples, and $w$ the number of home runs hit. The MVP had 1 single and 4 hits for extra bases, including 1 home run.

(a) Set up a system of four equations to model this situation.

(b) Solve the system and interpret your answer.

9. **Truck Rentals** The rent-a-truck company of Example 1 learns that each van now costs $25,000, each small truck costs $30,000, and each large truck costs $40,000. The company still needs twice as many vans as small trucks and wants to spend $5 million, but it decides to buy only 175 vehicles. How many of each kind should it buy?

10. **Finance** Suppose that the investor in Example 2 finds that her annual return on the international stock fund will be only 4%. Now how much should she put in each investment?

11. **Livestock Dietary Requirements** To meet consumer demand, the animal feed in Example 3 now supplies only 2400 units of fat. How many units of corn, soybeans, and cottonseed are now needed?

12. **Delivery Logistics** Suppose that the wholesaler in Example 5 finds a way to pack the EZ model in an 8-cubic-foot box. Now how many of each model should a fully loaded van carry?

13. **Dry Cleaning** Mario took clothes to the cleaners three times last month. First, he brought 3 shirts and 1 pair of slacks and paid $10.96. Then he brought 7 shirts, 2 pairs of slacks, and a sport coat and paid $30.40. Finally, he brought 4 shirts and 1 sports coat and paid $14.45. How much was he charged for each shirt, each pair of slacks, and each sports coat?

14. **Online Retailing** Shipping charges at an online bookstore are $4 for one book, $6 for two books, and $7 for three to five books. Last week, there were 6400 orders of five or fewer books, and total shipping charges for these orders were $33,600. The number of shipments with $7 charges was 1000 less than the number with $6 charges. How many shipments were made in each category (one book, two books, three to five books)?

15. **Concert Tickets** Tickets to a band concert cost $5 for adults, $3 for teenagers, and $2 for preteens. There were 570 people at the concert, and total ticket receipts were $1950. Three-fourths as many teenagers as preteens attended. How many adults, teenagers, and preteens attended?

16. **Minor League Baseball** A minor league baseball team had 7000 fans in attendance, and total ticket receipts were $26,400.

Box seats cost $6, grandstand seats cost $4, and bleacher seats cost $2. If the number of tickets sold for box seats was one-third the number of bleacher seats sold, how many tickets of each type were sold?

17. **Finance** An investment firm recommends that a client invest in bonds rated AAA, A, and B. The average yield on AAA bonds is 6%, on A bonds 7%, and on B bonds 10%. The client wants to invest twice as much in AAA bonds as in B bonds. How much should be invested in each type of bond if the total investment is $25,000, and the investor wants an annual return of $1810 on the three investments?

18. **Finance** An investor wants to invest $30,000 in corporate bonds that are rated AAA, A, and B. The lower rated ones pay higher interest, but pose a higher risk as well. The average yield is 5% on AAA bonds, 6% on A bonds, and 10% on B bonds. Being conservative, the investor wants to have twice as much in AAA bonds as in B bonds. How much should she invest in each type of bond to have an interest income of $2000?

19. **Finance** An investor plans to invest $70,000 in a mutual fund, corporate bonds, and a fast-food franchise. She plans to put twice as much in bonds as in the mutual fund. On the basis of past performance, she expects the mutual fund to pay a 2% dividend, the bonds 6%, and the franchise 10%. She would like a dividend income of $4800. How much should she put in each of three investments?

20. **Finance** Makayla borrows $10,000. Some is from her friend at 8% annual interest, twice as much as that from her bank at 9%, and the remainder from her insurance company at 5%. She pays a total of $830 in interest for the first year. How much did she borrow from each source?

21. **Trail Mix Recipe** Pretzels cost $3 per pound, dried fruit $4 per pound, and nuts $8 per pound. How many pounds of each should be used to produce 140 pounds of trail mix costing $6 per pound in which there are twice as many pretzels (by weight) as dried fruit?

22. **Auto Manufacturing** An auto manufacturer sends cars from two plants, I and II, to dealerships A and B, located in a midwestern city. Plant I has a total of 28 cars to send, and plant II has 8. Dealer A needs 20 cars, and dealer B needs 16. Transportation costs based on the distance of each dealership from each plant are $220 from I to A, $300 from I to B, $400 from II to A, and $180 from II to B. The manufacturer wants to limit transportation costs to $10,640. How many cars should be sent from each plant to each of the two dealerships?

23. **Animal Feed** An animal breeder can buy four types of tiger food. Each case of Brand A contains 25 units of fiber, 30 units of protein, and 30 units of fat. Each case of Brand B contains 50 units of fiber, 30 units of protein, and 20 units of fat. Each case of Brand C contains 75 units of fiber, 30 units of protein, and 20 units of fat. Each case of Brand D contains 100 units of fiber, 60 units of protein, and 30 units of fat. How many cases of each brand should the breeder mix together to obtain a food that provides 1200 units of fiber, 600 units of protein, and 400 units of fat?

24. **Physical Science** The stopping distance for a car traveling 25 mph is 61.7 feet, and for a car traveling 35 mph it is 106 feet. The stopping distance in feet can be described by the equation

$y = ax^2 + bx$, where $x$ is the speed in mph. (Data from: National Traffic Safety Institute.)

(a) Find the values of $a$ and $b$.

(b) Use your answers from part (a) to find the stopping distance for a car traveling 55 mph.

*Work these exercises. (See Example 4.)*

25. **Female Population** The table shows Census Bureau projections for the female population of the United States (in millions).

| Year | 2015 | 2035 | 2040 |
|---|---|---|---|
| **Female Population** | 163 | 187 | 192 |

(a) Find a quadratic function $f(x) = ax^2 + bx + c$ that gives the female population (in millions) in year $x$, where $x = 0$ corresponds to the year 2000.

(b) Estimate the female population in the year 2020.

26. **Male Population** The table shows Census Bureau projections for the male population of the United States (in millions).

| Year | 2015 | 2035 | 2040 |
|---|---|---|---|
| **Male Population** | 158 | 183 | 188 |

(a) Find a quadratic function $f(x) = ax^2 + bx + c$ that gives the male population (in millions) in year $x$, where $x = 0$ corresponds to the year 2000.

(b) Estimate the male population in the year 2020.

27. **Fuel Consumption** At a pottery factory, fuel consumption for heating the kilns varies with the size of the order being fired. In the past, the company recorded the figures in the table.

| $x$ = Number of Platters | $y$ = Fuel Cost per Platter |
|---|---|
| 6 | $2.80 |
| 8 | 2.48 |
| 10 | 2.24 |

(a) Find an equation of the form $y = ax^2 + bx + c$ whose graph contains the three points corresponding to the data in the table.

(b) How many platters should be fired at one time in order to minimize the fuel cost per platter? What is the minimum fuel cost per platter?

28. **Alzheimer's Disease** The number of Alzheimer's cases in people 85 and older was 2 million in 2004 and is projected to be 3 million in 2025 and 8 million in 2050. (Data from: Alzheimer's Association.)

(a) Let $x = 0$ correspond to 2000. Find a quadratic function that models the given data.

(b) How many people 85 or older will have Alzheimer's disease in 2020 and in 2034?

(c) In the year you turn 85, how many people your age or older are expected to have Alzheimer's disease?

*A graphing calculator or other technology is recommended for the following exercises.*

**29. Health** The table shows the calories, sodium, and protein in one cup of various kinds of soup.

|  | Progresso™ Hearty Chicken Rotini | Healthy Choice™ Hearty Chicken | Campbell's™ Chunky Chicken Noodle |
|---|---|---|---|
| Calories | 100 | 130 | 110 |
| Sodium (mg) | 960 | 480 | 890 |
| Protein (g) | 7 | 9 | 8 |

How many cups of each kind of soup should be mixed together to produce 10 servings of soup, each of which provides 171 calories, 1158 milligrams of sodium, and 12.1 grams of protein? What is the serving size (in cups)? (*Hint:* In 10 servings, there must be 1710 calories, 11,580 milligrams of sodium, and 121 grams of protein.)

**30. Health** The table shows the calories, sodium, and fat in 1 ounce of various snack foods (all produced by Planters™):

|  | Sweet N' Crunchy Peanuts | Dry Roasted Honey Peanuts | Kettle Roasted Honey BBQ Peanuts |
|---|---|---|---|
| Calories | 140 | 160 | 180 |
| Sodium (mg) | 20 | 110 | 55 |
| Fat (g) | 8 | 13 | 15 |

How many ounces of each kind of snack should be combined to produce 10 servings, each of which provides 284 calories, 93 milligrams of sodium, and 20.6 grams of fat? What is the serving size?

✓ **Checkpoint Answers**

1. $x + y + z = 150$
   $4x + 5y + 7z = 400$
   $x - 3y = 0$

2. (a) $\begin{bmatrix} 1 & 1 & 1 & 1 & 100{,}000 \\ .025 & .035 & .05 & .06 & 4000 \\ 0 & 0 & 1 & -.25 & 0 \\ 1 & 1 & 0 & 0 & 60{,}000 \end{bmatrix}$

   (b) Many sequences are possible, including this one:
   Replace $R_2$ by $-.025R_1 + R_2$;
   replace $R_4$ by $-R_1 + R_4$;
   replace $R_2$ by $\dfrac{1}{.01}R_2$;
   replace $R_4$ by $R_3 + R_4$;
   replace $R_4$ by $\dfrac{-1}{1.25}R_4$.

   (c) $\begin{bmatrix} 1 & 1 & 1 & 1 & 100{,}000 \\ 0 & 1 & 2.5 & 3.5 & 150{,}000 \\ 0 & 0 & 1 & -.25 & 0 \\ 0 & 0 & 0 & 1 & 32{,}000 \end{bmatrix}$

3. (a) $\begin{bmatrix} 10 & 20 & 30 & 1800 \\ 30 & 20 & 40 & 2800 \\ 20 & 40 & 25 & 2200 \end{bmatrix}$

   (b) Many sequences are possible, including this one:
   Replace $R_1$ by $\dfrac{1}{10}R_1$;
   replace $R_2$ by $\dfrac{1}{10}R_2$;
   replace $R_3$ by $\dfrac{1}{5}R_3$;
   replace $R_2$ by $-3R_1 + R_2$;
   replace $R_3$ by $-4R_1 + R_3$;
   replace $R_2$ by $-\dfrac{1}{4}R_2$;
   replace $R_3$ by $-\dfrac{1}{7}R_3$.

   (c) $\begin{bmatrix} 1 & 2 & 3 & 180 \\ 0 & 1 & \dfrac{5}{4} & 65 \\ 0 & 0 & 1 & 40 \end{bmatrix}$

# 6.4 Basic Matrix Operations

Until now, we have used matrices only as a convenient shorthand to solve systems of equations. However, matrices are also important in the fields of management, natural science, engineering, and social science as a way to organize data, as Example 1 demonstrates.

**EXAMPLE 1** **Furniture Manufacturing** The EZ Life Company manufactures sofas and armchairs in three models: *A*, *B*, and *C*. The company has regional warehouses in New York, Chicago, and San Francisco. In its August shipment, the company sends 10 model *A* sofas, 12 model *B* sofas, 5 model *C* sofas, 15 model *A* chairs, 20 model *B* chairs, and 8 model *C* chairs to each warehouse.

These data might be organized by first listing them as follows:

| Sofas | 10 model $A$ | 12 model $B$ | 5 model $C$; |
|---|---|---|---|
| Chairs | 15 model $A$ | 20 model $B$ | 8 model $C$. |

Alternatively, we might tabulate the data:

|  |  | **MODEL** | | |
|---|---|---|---|---|
|  |  | $A$ | $B$ | $C$ |
| **FURNITURE** | Sofa | 10 | 12 | 5 |
|  | Chair | 15 | 20 | 8 |

With the understanding that the numbers in each row refer to the type of furniture (sofa or chair) and the numbers in each column refer to the model ($A$, $B$, or $C$), the same information can be given by a matrix as follows:

$$M = \begin{bmatrix} 10 & 12 & 5 \\ 15 & 20 & 8 \end{bmatrix}.$$

✓ **Checkpoint 1**

Rewrite matrix $M$ in Example 1 in a matrix with three rows and two columns.

A matrix with $m$ horizontal rows and $n$ vertical columns has dimension, or size, $m \times n$. The number of rows is always given first.

## EXAMPLE 2

(a) The matrix $\begin{bmatrix} 6 & 5 \\ 3 & 4 \\ 5 & -1 \end{bmatrix}$ is a $3 \times 2$ matrix.

(b) $\begin{bmatrix} 5 & 8 & 9 \\ 0 & 5 & -3 \\ -4 & 0 & 5 \end{bmatrix}$ is a $3 \times 3$ matrix.

(c) $\begin{bmatrix} 1 & 6 & 5 & -2 & 5 \end{bmatrix}$ is a $1 \times 5$ matrix.

(d) A graphing calculator displays a $4 \times 1$ matrix like this:

```
[A]
        [3 ]
        [-5]
        [0 ]
        [2 ]
```

✓ **Checkpoint 2**

Give the size of each of the following matrices.

(a) $\begin{bmatrix} 2 & 1 & -5 & 6 \\ 3 & 0 & 7 & -4 \end{bmatrix}$

(b) $\begin{bmatrix} 1 & 2 & 3 \\ 4 & 5 & 6 \\ 9 & 8 & 7 \end{bmatrix}$

A matrix with only one row, as in Example 2(c), is called a **row matrix,** or **row vector.** A matrix with only one column, as in Example 2(d), is called a **column matrix**, or **column vector.** A matrix with the same number of rows as columns is called a **square matrix.** The matrix in Example 2(b) is a square matrix, as are

✓ **Checkpoint 3**

Use the numbers 2, 5, −8, and 4 to write

(a) a row matrix;

(b) a column matrix;

(c) a square matrix.

$$A = \begin{bmatrix} -5 & 6 \\ 8 & 3 \end{bmatrix} \quad \text{and} \quad B = \begin{bmatrix} 0 & 0 & 0 & 0 \\ -2 & 4 & 1 & 3 \\ 0 & 0 & 0 & 0 \\ -5 & -4 & 1 & 8 \end{bmatrix}.$$

When a matrix is denoted by a single letter, such as the matrix $A$ above, the element in row $i$ and column $j$ is denoted $a_{ij}$. For example, $a_{21} = 8$ (the element in row 2, column 1). Similarly, in matrix $B$, $b_{42} = -4$ (the element in row 4, column 2).

## Addition

The matrix given in Example 1,

$$M = \begin{bmatrix} 10 & 12 & 5 \\ 15 & 20 & 8 \end{bmatrix},$$

shows the August shipment from the EZ Life plant to each of its warehouses. If matrix $N$ below gives the September shipment to the New York warehouse, what is the total shipment for each item of furniture to the New York warehouse for the two months?

$$N = \begin{bmatrix} 45 & 35 & 20 \\ 65 & 40 & 35 \end{bmatrix}$$

If 10 model $A$ sofas were shipped in August and 45 in September, then altogether 55 model $A$ sofas were shipped in the two months. Adding the other corresponding entries gives a new matrix, $Q$, that represents the total shipment to the New York warehouse for the two months:

$$Q = \begin{bmatrix} 55 & 47 & 25 \\ 80 & 60 & 43 \end{bmatrix}$$

It is convenient to refer to $Q$ as the *sum* of $M$ and $N$.

The way these two matrices were added illustrates the following definition of addition of matrices.

## Matrix Addition

The **sum** of two $m \times n$ matrices $X$ and $Y$ is the $m \times n$ matrix $X + Y$ in which each element is the sum of the corresponding elements of $X$ and $Y$.

It is important to remember that only matrices that are the same size can be added.

**EXAMPLE 3**    Find each sum if possible.

**(a)** $\begin{bmatrix} 5 & -6 \\ 8 & 9 \end{bmatrix} + \begin{bmatrix} -4 & 6 \\ 8 & -3 \end{bmatrix} = \begin{bmatrix} 5 + (-4) & -6 + 6 \\ 8 + 8 & 9 + (-3) \end{bmatrix} = \begin{bmatrix} 1 & 0 \\ 16 & 6 \end{bmatrix}.$

**(b)** The matrices

$$A = \begin{bmatrix} 5 & 8 \\ 6 & 2 \end{bmatrix} \quad \text{and} \quad B = \begin{bmatrix} 3 & 9 & 1 \\ 4 & 2 & 5 \end{bmatrix}$$

are of different sizes, so it is not possible to find the sum $A + B$. ✔ 4

✓ **Checkpoint 4**

Find each sum when possible.

**(a)** $\begin{bmatrix} 2 & 5 & 7 \\ 3 & -1 & 4 \end{bmatrix}$

$+ \begin{bmatrix} -1 & 2 & 0 \\ 10 & -4 & 5 \end{bmatrix}$

**(b)** $\begin{bmatrix} 1 \\ 2 \\ 3 \end{bmatrix} + \begin{bmatrix} 2 & -1 \\ 4 & 5 \\ 6 & 0 \end{bmatrix}$

**(c)** $[5 \quad 4 \quad -1] + [-5 \quad 2 \quad 3]$

▦ **TECHNOLOGY TIP** Graphing calculators can find matrix sums, as illustrated in Figure 6.20.

**Figure 6.20**

**EXAMPLE 4**　**Furniture Manufacturing** The September shipments of the three models of sofas and chairs from the EZ Life Company to the New York, San Francisco, and Chicago warehouses are given respectively in matrices $N$, $S$, and $C$ as follows:

$$N = \begin{bmatrix} 45 & 35 & 20 \\ 65 & 40 & 35 \end{bmatrix}; \quad S = \begin{bmatrix} 30 & 32 & 28 \\ 43 & 47 & 30 \end{bmatrix}; \quad C = \begin{bmatrix} 22 & 25 & 38 \\ 31 & 34 & 35 \end{bmatrix}.$$

What was the total amount shipped to the three warehouses in September?

**Solution** The total of the September shipments is represented by the sum of the three matrices $N$, $S$, and $C$:

$$N + S + C = \begin{bmatrix} 45 & 35 & 20 \\ 65 & 40 & 35 \end{bmatrix} + \begin{bmatrix} 30 & 32 & 28 \\ 43 & 47 & 30 \end{bmatrix} + \begin{bmatrix} 22 & 25 & 38 \\ 31 & 34 & 35 \end{bmatrix}$$

$$= \begin{bmatrix} 97 & 92 & 86 \\ 139 & 121 & 100 \end{bmatrix}.$$

For example, from this sum, the total number of model $C$ sofas shipped to the three warehouses in September was 86. ✔5

✓**Checkpoint 5**

From the result of Example 4, find the total number of the following shipped to the three warehouses.

(a) Model $A$ chairs

(b) Model $B$ sofas

(c) Model $C$ chairs

**EXAMPLE 5**　**Drug Testing** A drug company is testing 200 patients to see if Painoff (a new headache medicine) is effective. Half the patients receive Painoff and half receive a placebo. The data on the first 50 patients is summarized in this matrix:

Pain Relief Obtained

Yes　No

Patient took Painoff $\begin{bmatrix} 22 & 3 \\ 8 & 17 \end{bmatrix}$
Patient took placebo

For example, row 2 shows that, of the people who took the placebo, 8 got relief, but 17 did not. The test was repeated on three more groups of 50 patients each, with the results summarized by these matrices:

$$\begin{bmatrix} 21 & 4 \\ 6 & 19 \end{bmatrix}; \quad \begin{bmatrix} 19 & 6 \\ 10 & 15 \end{bmatrix}; \quad \begin{bmatrix} 23 & 2 \\ 3 & 22 \end{bmatrix}.$$

The total results of the test can be obtained by adding these four matrices:

$$\begin{bmatrix} 22 & 3 \\ 8 & 17 \end{bmatrix} + \begin{bmatrix} 21 & 4 \\ 6 & 19 \end{bmatrix} + \begin{bmatrix} 19 & 6 \\ 10 & 15 \end{bmatrix} + \begin{bmatrix} 23 & 2 \\ 3 & 22 \end{bmatrix} = \begin{bmatrix} 85 & 15 \\ 27 & 73 \end{bmatrix}.$$

The final matrix shows that 85 of 100 patients got relief with Painoff, but only 27 of 100 did so with the placebo. ✔6

✓**Checkpoint 6**

Later, it was discovered that the data in the last group of 50 patients in Example 5 was invalid. Use a matrix to represent the total test results after those data were eliminated.

## Subtraction

**Subtraction** of matrices can be defined in a manner similar to matrix addition.

### Matrix Subtraction

The **difference** of two $m \times n$ matrices $X$ and $Y$ is the $m \times n$ matrix $X - Y$ in which each element is the difference of the corresponding elements of $X$ and $Y$.

**EXAMPLE 6** Find the following.

(a) $\begin{bmatrix} 1 & 2 & 3 \\ 0 & -1 & 5 \end{bmatrix} - \begin{bmatrix} -2 & 3 & 0 \\ 1 & -7 & 2 \end{bmatrix}$

**Solution**

$$\begin{bmatrix} 1 & 2 & 3 \\ 0 & -1 & 5 \end{bmatrix} - \begin{bmatrix} -2 & 3 & 0 \\ 1 & -7 & 2 \end{bmatrix} = \begin{bmatrix} 1-(-2) & 2-3 & 3-0 \\ 0-1 & -1-(-7) & 5-2 \end{bmatrix}$$

$$= \begin{bmatrix} 3 & -1 & 3 \\ -1 & 6 & 3 \end{bmatrix}.$$

(b) $[8 \quad 6 \quad -4] - [3 \quad 5 \quad -8]$

**Solution**

$$[8 \quad 6 \quad -4] - [3 \quad 5 \quad -8] = [8-3 \quad 6-5 \quad -4-(-8)]$$

$$= [5 \quad 1 \quad 4].$$

(c) $\begin{bmatrix} -2 & 5 \\ 0 & 1 \end{bmatrix} - \begin{bmatrix} 3 \\ 5 \end{bmatrix}$

**Solution** The matrices are of different sizes and therefore cannot be subtracted. ✓7

✓**Checkpoint 7**

Find each of the following differences when possible.

(a) $\begin{bmatrix} 2 & 5 \\ -1 & 0 \end{bmatrix} - \begin{bmatrix} 6 & 4 \\ 3 & -2 \end{bmatrix}$

(b) $\begin{bmatrix} 1 & 5 & 6 \\ 2 & 4 & 8 \end{bmatrix} - \begin{bmatrix} 2 & 1 \\ 10 & 3 \end{bmatrix}$

(c) $[5 \quad -4 \quad 1] - [6 \quad 0 \quad -3]$

**EXAMPLE 7** **Furniture Manufacturing** During September, the Chicago warehouse of the EZ Life Company shipped out the following numbers of each model, where the entries in the matrix have the same meaning as given earlier:

$$K = \begin{bmatrix} 5 & 10 & 8 \\ 11 & 14 & 15 \end{bmatrix}.$$

What was the Chicago warehouse's inventory at the end of September, taking into account only the number of items received and sent out that month?

**Solution** The number of each kind of item received during September is given by matrix $C$ from Example 4; the number of each model sent out during September is given by matrix $K$. The inventory at the end of September is thus represented by the matrix $C - K$:

$$C - K = \begin{bmatrix} 22 & 25 & 38 \\ 31 & 34 & 35 \end{bmatrix} - \begin{bmatrix} 5 & 10 & 8 \\ 11 & 14 & 15 \end{bmatrix} = \begin{bmatrix} 17 & 15 & 30 \\ 20 & 20 & 20 \end{bmatrix}.$$

📱 **TECHNOLOGY TIP**

Graphing calculators can do matrix subtraction, as illustrated in Figure 6.21.

```
[B]
     [-4   5   0   2]
     [ 6   3  -7   1]
[C]
     [ 4  12   8  -5]
     [ 2   8   5   0]
```

```
[B]-[C]
     [-8  -7  -8   7]
     [ 4  -5 -12   1]
```

**Figure 6.21**

## Scalar Multiplication

Suppose one of the EZ Life Company warehouses receives the following order, written in matrix form, where the entries have the same meaning as given earlier:

$$\begin{bmatrix} 5 & 4 & 1 \\ 3 & 2 & 3 \end{bmatrix}.$$

Later, the store that sent the order asks the warehouse to send six more of the same order. The six new orders can be written as one matrix by multiplying each element in the matrix by 6, giving the product

$$6\begin{bmatrix} 5 & 4 & 1 \\ 3 & 2 & 3 \end{bmatrix} = \begin{bmatrix} 30 & 24 & 6 \\ 18 & 12 & 18 \end{bmatrix}.$$

In work with matrices, a real number, like the 6 in the preceding multiplication, is called a **scalar**.

[A]
$$\begin{bmatrix} 3 & -4 \\ 5 & 2 \end{bmatrix}$$
2[A]
$$\begin{bmatrix} 6 & -8 \\ 10 & 4 \end{bmatrix}$$

**Figure 6.22**

✓ **Checkpoint 8**

Find each product.

**(a)** $-3\begin{bmatrix} 4 & -2 \\ 1 & 5 \end{bmatrix}$

**(b)** $4\begin{bmatrix} 2 & 4 & 7 \\ 8 & 2 & 1 \\ 5 & 7 & 3 \end{bmatrix}$

**TECHNOLOGY TIP**

To compute the negative of matrix $A$ on a calculator, use the "negative" key (−), as shown in Figure 6.23.

[A]
$$\begin{bmatrix} 2 & 3 \\ 1 & 1.5 \end{bmatrix}$$
-[A]
$$\begin{bmatrix} -2 & -3 \\ -1 & -1.5 \end{bmatrix}$$

**Figure 6.23**

✓ **Checkpoint 9**

Let $A$ and $B$ be the matrices in Example 9. Find

**(a)** $A + (-B)$;

**(b)** $A - B$.

**(c)** What can you conclude from parts (a) and (b)?

## Scalar Multiplication

The **product** of a scalar $k$ and a matrix $X$ is the matrix $kX$ in which each element is $k$ times the corresponding element of $X$.

**EXAMPLE 8**

**(a)** $(-3)\begin{bmatrix} 2 & -5 \\ 1 & 7 \\ 4 & -6 \end{bmatrix} = \begin{bmatrix} -6 & 15 \\ -3 & -21 \\ -12 & 18 \end{bmatrix}$.

**(b)** Graphing calculators can also do scalar multiplication (Figure 6.22). ✓8

Recall that the *negative* of a real number $a$ is the number $-a = (-1)a$. The negative of a matrix is defined similarly.

The **negative** (or *additive inverse*) of a matrix $A$ is the matrix $(-1)A$ which is obtained by multiplying each element of $A$ by $-1$. It is denoted $-A$.

**EXAMPLE 9** Find $-A$ and $-B$ when

$$A = \begin{bmatrix} 1 & 2 & 3 \\ 0 & -1 & 5 \end{bmatrix} \quad \text{and} \quad B = \begin{bmatrix} -2 & 3 & 0 \\ 1 & -7 & 2 \end{bmatrix}.$$

**Solution** By the preceding definition,

$$-A = \begin{bmatrix} -1 & -2 & -3 \\ 0 & 1 & -5 \end{bmatrix} \quad \text{and} \quad -B = \begin{bmatrix} 2 & -3 & 0 \\ -1 & 7 & -2 \end{bmatrix}. \quad ✓9$$

A matrix consisting only of zeros is called a **zero matrix** (or *additive identity*) and is denoted $O$. There is an $m \times n$ zero matrix for each pair of values of $m$ and $n$—for instance,

$$\begin{bmatrix} 0 & 0 \\ 0 & 0 \end{bmatrix}; \qquad \begin{bmatrix} 0 & 0 & 0 & 0 \\ 0 & 0 & 0 & 0 \end{bmatrix}.$$

$2 \times 2$ zero matrix      $2 \times 4$ zero matrix

Matrices satisfy a number of properties that are similar to properties of real numbers, as summarized below and illustrated in Checkpoint 9 and Exercises 26–29.

If $A$, $B$, and $C$ are any $m \times n$ matrices and $O$ is the $m \times n$ zero matrix, then the following properties are satisfied.

| | |
|---|---|
| Commutative Property of Addition of Matrices | $A + B = B + A$ |
| Associative Property of Addition of Matrices | $A + (B + C) = (A + B) + C$ |
| Identity Property of Addition of Matrices | $A + O = A = O + A$ |
| Inverse Properties of Addition of Matrices | $A + (-A) = O = (-A) + A$ |
| | $A + (-B) = A - B$ |

# 6.4 Exercises

*Find the size of each of the given matrices. Identify any square, column, or row matrices. Give the negative (additive inverse) of each matrix. (See Examples 2 and 9.)*

1. $\begin{bmatrix} 7 & -8 & 4 \\ 0 & 13 & 9 \end{bmatrix}$

2. $\begin{bmatrix} -7 & 23 \\ 5 & -6 \end{bmatrix}$

3. $\begin{bmatrix} -3 & 0 & 11 \\ 1 & \frac{1}{4} & -7 \\ 5 & -3 & 9 \end{bmatrix}$

4. $[6 \quad -4 \quad \frac{2}{3} \quad 12 \quad 2]$

5. $\begin{bmatrix} 7 \\ 11 \end{bmatrix}$

6. $[-5]$

7. If $A$ is a $5 \times 3$ matrix and $A + B = A$, what do you know about $B$?

8. If $C$ is a $3 \times 3$ matrix and $D$ is a $3 \times 4$ matrix, then $C + D$ is _____.

*Perform the indicated operations where possible. (See Examples 3–7.)*

9. $\begin{bmatrix} 1 & 2 & 7 & -1 \\ 8 & 0 & 2 & -4 \end{bmatrix} + \begin{bmatrix} -8 & 12 & -5 & 5 \\ -2 & -3 & 0 & 0 \end{bmatrix}$

10. $\begin{bmatrix} 1 & 7 \\ 2 & -3 \\ 3 & 7 \end{bmatrix} + \begin{bmatrix} 2 & 8 \\ 6 & 8 \\ -1 & 9 \end{bmatrix}$

11. $\begin{bmatrix} -1 & -5 & 9 \\ 2 & 2 & 3 \end{bmatrix} + \begin{bmatrix} 4 & 4 & -7 \\ 1 & -1 & 2 \end{bmatrix}$

12. $\begin{bmatrix} 2 & 4 \\ -8 & 2 \end{bmatrix} + \begin{bmatrix} 9 & -5 \\ 8 & 5 \end{bmatrix}$

13. $\begin{bmatrix} 0 & -2 \\ 1 & 9 \\ -5 & -9 \end{bmatrix} - \begin{bmatrix} 8 & 6 \\ 9 & 17 \\ 3 & -1 \end{bmatrix}$

14. $\begin{bmatrix} -3 & -2 & 5 \\ 3 & 9 & 0 \end{bmatrix} - \begin{bmatrix} 1 & 5 & -2 \\ -3 & 6 & 8 \end{bmatrix}$

15. $\begin{bmatrix} 9 & 1 \\ 0 & -3 \\ 4 & 10 \end{bmatrix} - \begin{bmatrix} 1 & 9 & -4 \\ -1 & 1 & 0 \end{bmatrix}$

16. $\begin{bmatrix} 3 & -8 & 0 \\ 1 & 5 & -7 \end{bmatrix} - \begin{bmatrix} 2 & 6 \\ 9 & -4 \end{bmatrix}$

*Let $A = \begin{bmatrix} -2 & 0 \\ 5 & 3 \end{bmatrix}$ and $B = \begin{bmatrix} 0 & 2 \\ 4 & -6 \end{bmatrix}$. Find each of the following. (See Examples 8 and 9.)*

17. $2A$

18. $-3B$

19. $-4B$

20. $5A$

21. $-4A + 5B$

22. $3A - 10B$

*Let $A = \begin{bmatrix} 1 & -2 \\ 4 & 3 \end{bmatrix}$ and $B = \begin{bmatrix} 2 & -1 \\ 0 & 5 \end{bmatrix}$. Find a matrix $X$ satisfying the given equation.*

23. $2X = 2A + 3B$

24. $3X = A - 3B$

*Using matrices*

$$O = \begin{bmatrix} 0 & 0 \\ 0 & 0 \end{bmatrix}, P = \begin{bmatrix} m & n \\ p & q \end{bmatrix}, T = \begin{bmatrix} r & s \\ t & u \end{bmatrix}, \text{ and } X = \begin{bmatrix} x & y \\ z & w \end{bmatrix},$$

*verify that the statements in Exercises 25–29 are true.*

25. $X + T$ is a $2 \times 2$ matrix.

26. $X + T = T + X$ (commutative property of addition of matrices).

27. $X + (T + P) = (X + T) + P$ (associative property of addition of matrices).

28. $X + (-X) = O$ (inverse property of addition of matrices).

29. $P + O = P$ (identity property of addition of matrices).

30. Which of the preceding properties are valid for matrices that are not square?

*Work the following exercises. (See Example 1.)*

31. **Concessions and Merchandise** When ticket holders fail to attend, major league sports teams lose the money these fans would have spent on refreshments and merchandise. The percentage of fans who don't show up is 16% in basketball and hockey, 20% in football, and 18% in baseball. The lost revenue per fan is $18.20 in basketball, $18.25 in hockey, $19 in football, and $15.40 in baseball. The total annual lost revenue is $22.7 million in basketball, $35.8 million in hockey, $51.9 million in football, and $96.3 million in baseball. Express this information in matrix form; specify what the rows and columns represent. (Data from: American Demographics.)

32. **Earnings for College Graduates** In 2015, median weekly earnings were $493 for workers without a high school diploma, $678 for high school graduates, and $1137 for those who had earned a bachelor's degree. The unemployment rate was 2.8% for college graduates, 5.4% for high school graduates, and 8.0% for those with less than a high school diploma. Write this information as a $3 \times 2$ matrix, labeling rows and columns. (Data from: Bureau of Labor Statistics.)

33. **Organ Transplants** The shortage of organs for transplants is a continuing problem in the United States. In 2016, there were 4154 candidates waiting for a heart transplant, 1470 for a lung transplant, 14,794 for a liver transplant, and 100,467 for a kidney transplant. Corresponding figures for 2010 were 2874, 1759, 15,394, and 83,919, respectively. Express this information as a matrix, labeling rows and columns. (Data from: Organ Procurement and Transplant Network.)

34. **Household Debt** In 2015, only 38% of American households had credit card debt, but for households who *carried credit card*

*debt*, the average amount of debt per household was $15,762. Averages for households with other types of debt included $27,141 in auto loans, $48,172 in student loan, and $168,614 in mortgage debt. The corresponding total figures for all consumers in the United States are $733 billion, $1060 billion, $1230 billion, and $8250 billion, respectively. Write this information in a $4 \times 2$ matrix, labeling rows and columns. (Data from: Federal Reserve Bank of New York.)

**Work these exercises.**

**35. Convenience Stores** There are three convenience stores in Gambier. This week, Store I sold 88 loaves of bread, 48 quarts of milk, 16 jars of peanut butter, and 112 pounds of cold cuts. Store II sold 105 loaves of bread, 72 quarts of milk, 21 jars of peanut butter, and 147 pounds of cold cuts. Store III sold 60 loaves of bread, 40 quarts of milk, no peanut butter, and 50 pounds of cold cuts.

**(a)** Use a $3 \times 4$ matrix to express the sales information for the three stores.

**(b)** During the following week, sales on these products at Store I increased by 25%, sales at Store II increased by one-third, and sales at Store III increased by 10%. Write the sales matrix for that week.

**(c)** Write a matrix that represents total sales over the two-week period.

**36. Female and Male Graduates** The following table gives educational attainment as a percent of the U.S. population 25 years and older in various years. (Data from: U.S. Census Bureau.)

| | **MALE** | | **FEMALE** | |
|---|---|---|---|---|
| Year | Four Years of High School or More | Four Years of College or More | Four Years of High School or More | Four Years of College or More |
| 1940 | 22.7% | 5.5% | 26.3% | 3.8% |
| 1970 | 55.0% | 14.1% | 55.4% | 8.2% |
| 2000 | 84.2% | 27.8% | 84.0% | 23.6% |
| 2014 | 87.7% | 31.9% | 88.9% | 32.0% |

**(a)** Write a $2 \times 4$ matrix for the educational attainment of males.

**(b)** Write a $2 \times 4$ matrix for the educational attainment of females.

**(c)** Use the matrices from parts (a) and (b) to write a matrix showing how much more (or less) males have reached educational attainment than females.

✓**Checkpoint Answers**

**1.** $\begin{bmatrix} 10 & 15 \\ 12 & 20 \\ 5 & 8 \end{bmatrix}$

**2. (a)** $2 \times 4$      **(b)** $3 \times 3$

**3. (a)** $[2 \quad 5 \quad -8 \quad 4]$

**(b)** $\begin{bmatrix} 2 \\ 5 \\ -8 \\ 4 \end{bmatrix}$    **(c)** $\begin{bmatrix} 2 & 5 \\ -8 & 4 \end{bmatrix}$ or $\begin{bmatrix} 2 & -8 \\ 5 & 4 \end{bmatrix}$

(Other answers are possible.)

**4. (a)** $\begin{bmatrix} 1 & 7 & 7 \\ 13 & -5 & 9 \end{bmatrix}$    **(b)** Not possible

**(c)** $[0 \quad 6 \quad 2]$

**5. (a)** 139      **(b)** 92      **(c)** 100

**6.** $\begin{bmatrix} 62 & 13 \\ 24 & 51 \end{bmatrix}$

**7. (a)** $\begin{bmatrix} -4 & 1 \\ -4 & 2 \end{bmatrix}$    **(b)** Not possible

**(c)** $[-1 \quad -4 \quad 4]$

**8. (a)** $\begin{bmatrix} -12 & 6 \\ -3 & -15 \end{bmatrix}$    **(b)** $\begin{bmatrix} 8 & 16 & 28 \\ 32 & 8 & 4 \\ 20 & 28 & 12 \end{bmatrix}$

**9. (a)** $\begin{bmatrix} 3 & -1 & 3 \\ -1 & 6 & 3 \end{bmatrix}$    **(b)** $\begin{bmatrix} 3 & -1 & 3 \\ -1 & 6 & 3 \end{bmatrix}$

**(c)** $A + (-B) = A - B$.

# 6.5 Matrix Products and Inverses

To understand the reasoning behind the definition of matrix multiplication, look again at the EZ Life Company. Suppose sofas and chairs of the same model are often sold as sets, with matrix $W$ showing the number of each model set in each warehouse:

$$\begin{array}{c} \\ \text{New York} \\ \text{Chicago} \\ \text{San Francisco} \end{array} \begin{array}{ccc} \text{A} & \text{B} & \text{C} \\ \begin{bmatrix} 10 & 7 & 3 \\ 5 & 9 & 6 \\ 4 & 8 & 2 \end{bmatrix} \end{array} = W.$$

If the selling price of a model $A$ set is \$800, of a model $B$ set is \$1000, and of a model $C$ set is \$1200, find the total value of the sets in the New York warehouse as follows:

| Type | Number of Sets | | Price of Set | | Total |
|------|----------------|---|--------------|---|-------|
| $A$ | 10 | $\times$ | \$ 800 | $=$ | \$ 8,000 |
| $B$ | 7 | $\times$ | 1000 | $=$ | 7,000 |
| $C$ | 3 | $\times$ | 1200 | $=$ | 3,600 |
| | | | Total for New York | | \$18,600 |

✓**Checkpoint 1**

In this example of the EZ Life Company, find the total value of the New York sets if model $A$ sets sell for \$1200, model $B$ for \$1600, and model $C$ for \$1300.

The total value of the three kinds of sets in New York is \$18,600. ✓₁
The work done in the preceding table is summarized as

$$10(\$800) + 7(\$1000) + 3(\$1200) = \$18,600.$$

In the same way, the Chicago sets have a total value of

$$5(\$800) + 9(\$1000) + 6(\$1200) = \$20,200,$$

and in San Francisco, the total value of the sets is

$$4(\$800) + 8(\$1000) + 2(\$1200) = \$13,600.$$

The selling prices can be written as a column matrix $P$ and the total value in each location as a column matrix $V$:

$$P = \begin{bmatrix} 800 \\ 1000 \\ 1200 \end{bmatrix} \quad \text{and} \quad V = \begin{bmatrix} 18,600 \\ 20,200 \\ 13,600 \end{bmatrix}.$$

Consider how the first row of the matrix $W$ and the single column $P$ lead to the first entry of $V$:

$$\underset{W}{\begin{bmatrix} 10 & 7 & 3 \\ 5 & 9 & 6 \\ 4 & 8 & 2 \end{bmatrix}} \underset{P}{\begin{bmatrix} 800 \\ 1000 \\ 1200 \end{bmatrix}} \quad 10(800) + 7(1000) + 3(1200) = \quad \underset{V}{\begin{bmatrix} 18,600 \\ 20,200 \\ 13,600 \end{bmatrix}}.$$

Product of first entries + Product of second entries + Product of third entries

Similarly, adding the products of corresponding entries in the second row of $W$ and the column $P$ produces the second entry in $V$. The third entry in $V$ is obtained in the same way by using the third row of $W$ and column $P$. This suggests that it is reasonable to *define* the product $WP$ to be $V$:

$$WP = \begin{bmatrix} 10 & 7 & 3 \\ 5 & 9 & 6 \\ 4 & 8 & 2 \end{bmatrix} \begin{bmatrix} 800 \\ 1000 \\ 1200 \end{bmatrix} = \begin{bmatrix} 18,600 \\ 20,200 \\ 13,600 \end{bmatrix} = V$$

Note the sizes of the matrices here: The product of a $3 \times 3$ matrix and a $3 \times 1$ matrix is a $3 \times 1$ matrix.

## Multiplying Matrices

In order to define matrix multiplication in the general case, we first define the **product of a row of a matrix and a column of a matrix** (with the same number of entries in each) to be the *number* obtained by multiplying the corresponding entries (first by first, second by second, etc.) and adding the results. For instance,

$$[3 \quad -2 \quad 1] \cdot \begin{bmatrix} 4 \\ 5 \\ 0 \end{bmatrix} = 3 \cdot 4 + (-2) \cdot 5 + 1 \cdot 0 = 12 - 10 + 0 = 2.$$

Now **matrix multiplication** is defined as follows.

> ## Matrix Multiplication
>
> Let $A$ be an $m \times n$ matrix and let $B$ be an $n \times k$ matrix. The **product matrix** $AB$ is the $m \times k$ matrix whose entry in the $i$th row and $j$th column is
>
> the product of the $i$th row of $A$ and the $j$th column of $B$.

⚠ **CAUTION** Be careful when multiplying matrices. Remember that the number of *columns* of $A$ must equal the number of *rows* of $B$ in order to get the product matrix $AB$. The final product will have as many rows as $A$ and as many columns as $B$.

**EXAMPLE 1** Suppose matrix $A$ is $2 \times 2$ and matrix $B$ is $2 \times 4$. Can the product $AB$ be calculated? If so, what is the size of the product?

**Solution** The following diagram helps decide the answers to these questions:

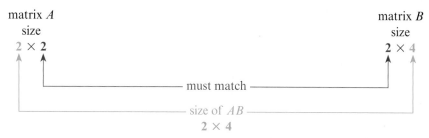

The product $AB$ can be calculated because $A$ has two columns and $B$ has two rows. The product will be a $2 \times 4$ matrix. ✔2

✓ **Checkpoint 2**

Matrix $A$ is $4 \times 6$ and matrix $B$ is $2 \times 4$.

**(a)** Can $AB$ be found? If so, give its size.

**(b)** Can $BA$ be found? If so, give its size.

**EXAMPLE 2** Find the product $CD$ when

$$C = \begin{bmatrix} -3 & 4 & 1 \\ 5 & 0 & 4 \end{bmatrix} \quad \text{and} \quad D = \begin{bmatrix} -6 & 4 \\ 2 & 3 \\ 3 & -2 \end{bmatrix}.$$

**Solution** Here, matrix $C$ is $2 \times 3$ and matrix $D$ is $3 \times 2$, so matrix $CD$ can be found and will be $2 \times 2$.

**Step 1** row 1, column 1

$$\begin{bmatrix} -3 & 4 & 1 \\ 5 & 0 & 4 \end{bmatrix} \begin{bmatrix} -6 & 4 \\ 2 & 3 \\ 3 & -2 \end{bmatrix} \quad (-3) \cdot (-6) + 4 \cdot 2 + 1 \cdot 3 = 29.$$

Hence, 29 is the entry in row 1, column 1, of $CD$, as shown in Step 5 on the following page.

**Step 2** row 1, column 2

$$\begin{bmatrix} -3 & 4 & 1 \\ 5 & 0 & 4 \end{bmatrix} \begin{bmatrix} -6 & 4 \\ 2 & 3 \\ 3 & -2 \end{bmatrix} \quad (-3) \cdot 4 + 4 \cdot 3 + 1 \cdot (-2) = -2.$$

So $-2$ is the entry in row 1, column 2, of $CD$, as shown in Step 5. Continue in this manner to find the remaining entries of $CD$.

**Step 3** row 2, column 1

$$\begin{bmatrix} -3 & 4 & 1 \\ 5 & 0 & 4 \end{bmatrix} \begin{bmatrix} -6 & 4 \\ 2 & 3 \\ 3 & -2 \end{bmatrix} \quad 5 \cdot (-6) + 0 \cdot 2 + 4 \cdot 3 = -18.$$

**Step 4**    row 2, column 2

$$\begin{bmatrix} -3 & 4 & 1 \\ \mathbf{5} & \mathbf{0} & \mathbf{4} \end{bmatrix} \begin{bmatrix} -6 & \mathbf{4} \\ 2 & \mathbf{3} \\ 3 & \mathbf{-2} \end{bmatrix} \quad \mathbf{5 \cdot 4 + 0 \cdot 3 + 4 \cdot (-2)} = 12.$$

**Step 5**    The product is

$$CD = \begin{bmatrix} -3 & 4 & 1 \\ 5 & 0 & 4 \end{bmatrix} \begin{bmatrix} -6 & 4 \\ 2 & 3 \\ 3 & -2 \end{bmatrix} = \begin{bmatrix} 29 & -2 \\ -18 & 12 \end{bmatrix}.$$

✓ **Checkpoint 3**

Find the product $CD$, given that

$$C = \begin{bmatrix} 1 & 3 & 5 \\ 2 & -4 & -1 \end{bmatrix}$$

and

$$D = \begin{bmatrix} 2 & -1 \\ 4 & 3 \\ 1 & -2 \end{bmatrix}.$$

✓ **Checkpoint 4**

Give the size of each of the following products, if the product can be found.

**(a)** $\begin{bmatrix} 2 & 4 \\ 6 & 8 \end{bmatrix} \begin{bmatrix} 1 & 2 & 3 \\ 0 & -1 & 2 \end{bmatrix}$

**(b)** $\begin{bmatrix} 1 & 2 \\ 5 & 10 \\ 12 & 7 \end{bmatrix} \begin{bmatrix} 2 & 4 \\ 3 & 6 \\ 9 & 1 \end{bmatrix}$

**(c)** $\begin{bmatrix} 5 \\ 2 \\ 4 \end{bmatrix} \begin{bmatrix} 1 & 0 & 6 \end{bmatrix}$

**EXAMPLE 3**    Find $BA$, given that

$$A = \begin{bmatrix} 1 & 7 \\ -3 & 2 \end{bmatrix} \quad \text{and} \quad B = \begin{bmatrix} 1 & 0 & -1 \\ 3 & 1 & 4 \end{bmatrix}.$$

Since $B$ is a $2 \times 3$ matrix and $A$ is a $2 \times 2$ matrix, the product $BA$ is not defined. ✓4

🖩 **TECHNOLOGY TIP** Graphing calculators can find matrix products (see Figure 6.24). However, if you use a graphing calculator to try to find the product in Example 3, the calculator will display an error message.

Matrix multiplication has some similarities to the multiplication of numbers, as summarized below and illustrated in Exercises 23 and 24.

> If $A$, $B$, and $C$ are any matrices such that all the indicated sums and products exist, then the following properties are satisfied.
>
> Associative Property of Multiplication of Matrices    $A(BC) = (AB)C$
>
> Distributive Properties    $A(B + C) = AB + AC$
>
> $(B + C)A = BA + CA$

However, there are important differences between matrix multiplication and the multiplication of numbers (see Exercises 19–22). In particular, matrix multiplication is *not* commutative.

> If $A$ and $B$ are matrices such that the products $AB$ and $BA$ exist,
>
> $$AB \text{ may not equal } BA.$$

Figure 6.24 shows an example of this situation.

**Figure 6.24**

**EXAMPLE 4** **Construction Costs** A contractor builds three kinds of houses, models $A$, $B$, and $C$, with a choice of two styles, Spanish or contemporary. Matrix $P$ shows the number of each kind of house planned for a new 100-home subdivision:

$$
\begin{array}{c}
\\
\text{Model } A \\
\text{Model } B \\
\text{Model } C
\end{array}
\begin{array}{cc}
\text{Spanish} & \text{Contemporary}
\end{array}
\left[
\begin{array}{cc}
0 & 30 \\
10 & 20 \\
20 & 20
\end{array}
\right] = P.
$$

The amounts for each of the exterior materials used depend primarily on the style of the house. These amounts are shown in matrix $Q$ (concrete is measured in cubic yards, lumber in units of 1000 board feet, brick in thousands, and shingles in units of 100 square feet):

$$
\begin{array}{c}
\\
\text{Spanish} \\
\text{Contemporary}
\end{array}
\begin{array}{cccc}
\text{Concrete} & \text{Lumber} & \text{Brick} & \text{Shingles}
\end{array}
\left[
\begin{array}{cccc}
10 & 2 & 0 & 2 \\
50 & 1 & 20 & 2
\end{array}
\right] = Q.
$$

Matrix $R$ gives the cost for each kind of material:

$$
\begin{array}{c}
\\
\text{Concrete} \\
\text{Lumber} \\
\text{Brick} \\
\text{Shingles}
\end{array}
\begin{array}{c}
\text{Cost per Unit}
\end{array}
\left[
\begin{array}{c}
20 \\
180 \\
60 \\
25
\end{array}
\right] = R.
$$

**(a)** What is the total cost for each model of house?

**Solution** First find the product $PQ$, which shows the amount of each material needed for each model of house:

$$
PQ = \left[
\begin{array}{cc}
0 & 30 \\
10 & 20 \\
20 & 20
\end{array}
\right]
\left[
\begin{array}{cccc}
10 & 2 & 0 & 2 \\
50 & 1 & 20 & 2
\end{array}
\right]
$$

$$
PQ = 
\begin{array}{cccc}
\text{Concrete} & \text{Lumber} & \text{Brick} & \text{Shingles}
\end{array}
\left[
\begin{array}{cccc}
1500 & 30 & 600 & 60 \\
1100 & 40 & 400 & 60 \\
1200 & 60 & 400 & 80
\end{array}
\right]
\begin{array}{l}
\text{Model } A \\
\text{Model } B. \\
\text{Model } C
\end{array}
$$

Now multiply $PQ$ and $R$, the cost matrix, to get the total cost for each model of house:

$$
\left[
\begin{array}{cccc}
1500 & 30 & 600 & 60 \\
1100 & 40 & 400 & 60 \\
1200 & 60 & 400 & 80
\end{array}
\right]
\left[
\begin{array}{c}
20 \\
180 \\
60 \\
25
\end{array}
\right]
=
\begin{array}{c}
\text{Cost} \\
\left[
\begin{array}{c}
72{,}900 \\
54{,}700 \\
60{,}800
\end{array}
\right]
\end{array}
\begin{array}{l}
\text{Model } A \\
\text{Model } B. \\
\text{Model } C
\end{array}
$$

So materials costs are \$72,900 for a Model A home, \$54,700 for a Model B home, and \$60,800 for a Model C home.

**(b)** How much of each of the four kinds of material must be ordered?

**Solution** The totals of the columns of matrix $PQ$ will give a matrix whose elements represent the total amounts of each material needed for the subdivision. Call this matrix $T$, and write it as a row matrix:

$$
T = [3800 \quad 130 \quad 1400 \quad 200].
$$

So the amount of material needed is 3800 units of concrete, 130 units of lumber, 1400 units of brick, and 200 units of shingles.

**(c)** What is the total cost for material?

**Solution** Find the total cost of all the materials by taking the product of matrix $T$, the matrix showing the total amounts of each material, and matrix $R$, the cost matrix. [To multiply these and get a $1 \times 1$ matrix representing total cost, we must multiply a $1 \times 4$ matrix by a $4 \times 1$ matrix. This is why $T$ was written as a row matrix in (b).] So, we have

$$TR = [3800 \quad 130 \quad 1400 \quad 200]\begin{bmatrix} 20 \\ 180 \\ 60 \\ 25 \end{bmatrix} = [188,400].$$

So the total cost of all materials is $188,400.

**(d)** Suppose the contractor builds the same number of homes in five subdivisions. What is the total amount of each material needed for each model in this case?

**Solution** Determine the total amount of each material for each model for all five subdivisions. Multiply $PQ$ by the scalar 5 as follows:

$$5\begin{bmatrix} 1500 & 30 & 600 & 60 \\ 1100 & 40 & 400 & 60 \\ 1200 & 60 & 400 & 80 \end{bmatrix} = \begin{bmatrix} 7500 & 150 & 3000 & 300 \\ 5500 & 200 & 2000 & 300 \\ 6000 & 300 & 2000 & 400 \end{bmatrix}.$$

So the amount of concrete needed for Model A homes in all five subdivisions is $7500 (see row 1, column 1), and so on.

We can introduce notation to help keep track of the quantities a matrix represents. For example, we can say that matrix $P$ from Example 4 represents models/styles, matrix $Q$ represents styles/materials, and matrix $R$ represents materials/cost. In each case, the meaning of the rows is written first and the columns second. When we found the product $PQ$ in Example 4, the rows of the matrix represented models and the columns represented materials. Therefore, we can say that the matrix product $PQ$ represents models/materials. The common quantity, styles, in both $P$ and $Q$ was eliminated in the product $PQ$. Do you see that the product $(PQ)R$ represents models/cost?

In practical problems, this notation helps decide in what order to multiply two matrices so that the results are meaningful. In Example 4(c), we could have found either product $RT$ or product $TR$. However, since $T$ represents subdivisions/materials and $R$ represents materials/cost, the product $TR$ gives subdivisions/cost. ✓5

## Identity Matrices

Recall from Section 1.1 that the real number 1 is the identity element for multiplication of real numbers: For any real number $a$, $a \cdot 1 = 1 \cdot a = a$. In this section, an **identity matrix $I$** is defined that has properties similar to those of the number 1.

If $I$ is to be the identity matrix for multiplication of matrices, the products $AI$ and $IA$ must both equal $A$. The $2 \times 2$ identity matrix that satisfies these conditions is

$$I = \begin{bmatrix} 1 & 0 \\ 0 & 1 \end{bmatrix}. \quad ✓6$$

To check that $I$ is really the $2 \times 2$ identity matrix, let

$$A = \begin{bmatrix} a & b \\ c & d \end{bmatrix}.$$

✓**Checkpoint 5**

Let matrix $A$ be

$$\begin{array}{c} \text{Vitamin} \\ \begin{array}{cc} & \text{C E K} \end{array} \\ \text{Brand} \begin{array}{c} X \\ Y \end{array} \begin{bmatrix} 2 & 7 & 5 \\ 4 & 6 & 9 \end{bmatrix} \end{array}$$

and matrix $B$ be

$$\begin{array}{c} \text{Cost} \\ \begin{array}{cc} X & Y \end{array} \\ \text{Vitamin} \begin{array}{c} C \\ E \\ K \end{array} \begin{bmatrix} 12 & 14 \\ 18 & 15 \\ 9 & 10 \end{bmatrix}. \end{array}$$

**(a)** What quantities do matrices $A$ and $B$ represent?

**(b)** What quantities does the product $AB$ represent?

**(c)** What quantities does the product $BA$ represent?

✓**Checkpoint 6**

Let $A = \begin{bmatrix} 3 & -2 \\ 4 & -1 \end{bmatrix}$ and

$I = \begin{bmatrix} 1 & 0 \\ 0 & 1 \end{bmatrix}.$

Find $IA$ and $AI$.

Then $AI$ and $IA$ should both equal $A$:

$$AI = \begin{bmatrix} a & b \\ c & d \end{bmatrix}\begin{bmatrix} 1 & 0 \\ 0 & 1 \end{bmatrix} = \begin{bmatrix} a(1) + b(0) & a(0) + b(1) \\ c(1) + d(0) & c(0) + d(1) \end{bmatrix} = \begin{bmatrix} a & b \\ c & d \end{bmatrix} = A;$$

$$IA = \begin{bmatrix} 1 & 0 \\ 0 & 1 \end{bmatrix}\begin{bmatrix} a & b \\ c & d \end{bmatrix} = \begin{bmatrix} 1(a) + 0(c) & 1(b) + 0(d) \\ 0(a) + 1(c) & 0(b) + 1(d) \end{bmatrix} = \begin{bmatrix} a & b \\ c & d \end{bmatrix} = A.$$

This verifies that $I$ has been defined correctly. (It can also be shown that $I$ is the only $2 \times 2$ identity matrix.)

The identity matrices for $3 \times 3$ matrices and $4 \times 4$ matrices are, respectively,

$$I = \begin{bmatrix} 1 & 0 & 0 \\ 0 & 1 & 0 \\ 0 & 0 & 1 \end{bmatrix} \quad \text{and} \quad I = \begin{bmatrix} 1 & 0 & 0 & 0 \\ 0 & 1 & 0 & 0 \\ 0 & 0 & 1 & 0 \\ 0 & 0 & 0 & 1 \end{bmatrix}.$$

By generalizing these findings, an identity matrix can be found for any $n$ by $n$ matrix. This identity matrix will have 1s on the main diagonal from upper left to lower right, with all other entries equal to 0.

### TECHNOLOGY TIP

An $n \times n$ identity matrix can be displayed on most graphing calculators by using IDENTITY $n$ or IDENT $n$ or IDENMAT($n$). Look in the MATH or OPS submenu of the MATRIX menu.

## Inverse Matrices

Recall that for every nonzero real number $a$, the equation $ax = 1$ has a solution, namely, $x = 1/a = a^{-1}$. Similarly, for a square matrix $A$, we consider the matrix equation $AX = I$. This equation does not always have a solution, but when it does, we use special terminology. If there is a matrix $A^{-1}$ satisfying

$$AA^{-1} = I,$$

(that is, $A^{-1}$ is a solution of $AX = I$), then $A^{-1}$ is called the **inverse matrix** of $A$. In this case, it can be proved that $A^{-1}A = I$ and that $A^{-1}$ is unique (that is, a square matrix has no more than one inverse). When a matrix has an inverse, it can be found by using the row operations given in Section 6.2, as we shall see later.

**CAUTION** Only square matrices have inverses, but not every square matrix has one. A matrix that does not have an inverse is called a **singular matrix**. Note that the symbol $A^{-1}$ (read "$A$-inverse") does *not* mean $1/A$; the symbol $A^{-1}$ is just the notation for the inverse of matrix $A$. There is no such thing as matrix division.

### EXAMPLE 5    Given matrices $A$ and $B$ as follows, determine whether $B$ is the inverse of $A$:

$$A = \begin{bmatrix} 1 & 2 \\ 4 & 6 \end{bmatrix}; \quad B = \begin{bmatrix} -3 & 1 \\ 2 & -\frac{1}{2} \end{bmatrix}.$$

**Solution** $B$ is the inverse of $A$ if $AB = I$, so we find that product:

$$AB = \begin{bmatrix} 1 & 2 \\ 4 & 6 \end{bmatrix}\begin{bmatrix} -3 & 1 \\ 2 & -\frac{1}{2} \end{bmatrix} = \begin{bmatrix} 1 & 0 \\ 0 & 1 \end{bmatrix} = I.$$

Therefore, $B$ is the inverse of $A$; that is, $A^{-1} = B$. (It is also true that $A$ is the inverse of $B$, or $B^{-1} = A$.) ✓ 7

✓ **Checkpoint 7**

Given $A = \begin{bmatrix} 2 & 1 \\ 3 & 8 \end{bmatrix}$ and

$B = \begin{bmatrix} -1 & 3 \\ 1 & -2 \end{bmatrix}$, determine whether they are inverses.

### EXAMPLE 6    Find the multiplicative inverse of

$$A = \begin{bmatrix} 2 & 4 \\ 1 & -1 \end{bmatrix}.$$

**Solution** Let the unknown inverse matrix be

$$A^{-1} = \begin{bmatrix} x & y \\ z & w \end{bmatrix}.$$

By the definition of matrix inverse, $AA^{-1} = I$, or

$$AA^{-1} = \begin{bmatrix} 2 & 4 \\ 1 & -1 \end{bmatrix} \begin{bmatrix} x & y \\ z & w \end{bmatrix} = \begin{bmatrix} 1 & 0 \\ 0 & 1 \end{bmatrix}.$$

Use matrix multiplication to get

$$\begin{bmatrix} 2x + 4z & 2y + 4w \\ x - z & y - w \end{bmatrix} = \begin{bmatrix} 1 & 0 \\ 0 & 1 \end{bmatrix}.$$

Setting corresponding elements equal to each other gives the system of equations

$$2x + 4z = 1 \qquad \qquad \textbf{(1)}$$
$$2y + 4w = 0 \qquad \qquad \textbf{(2)}$$
$$x - z = 0 \qquad \qquad \textbf{(3)}$$
$$y - w = 1. \qquad \qquad \textbf{(4)}$$

Since equations (1) and (3) involve only $x$ and $z$, while equations (2) and (4) involve only $y$ and $w$, these four equations lead to two systems of equations:

$$\begin{array}{ccc} 2x + 4z = 1 & & 2y + 4w = 0 \\ x - z = 0 & \text{and} & y - w = 1. \end{array}$$

Writing the two systems as augmented matrices gives

$$\begin{bmatrix} 2 & 4 & | & 1 \\ 1 & -1 & | & 0 \end{bmatrix} \quad \text{and} \quad \begin{bmatrix} 2 & 4 & | & 0 \\ 1 & -1 & | & 1 \end{bmatrix}.$$

Note that the row operations needed to transform both matrices into reduced row echelon form are the same because the first two columns of both matrices are identical. Consequently, we can save time by combining these matrices into the single matrix

$$\begin{bmatrix} 2 & 4 & | & 1 & 0 \\ 1 & -1 & | & 0 & 1 \end{bmatrix}. \qquad \qquad \textbf{(5)}$$

Columns 1–3 represent the first system and columns 1, 2, and 4 represent the second system. Now use row operations as follows:

$$\begin{bmatrix} 1 & -1 & | & 0 & 1 \\ 2 & 4 & | & 1 & 0 \end{bmatrix} \qquad \text{Interchange } R_1 \text{ and } R_2.$$

$$\begin{bmatrix} 1 & -1 & | & 0 & 1 \\ 0 & 6 & | & 1 & -2 \end{bmatrix} \qquad -2R_1 + R_2$$

$$\begin{bmatrix} 1 & -1 & | & 0 & 1 \\ 0 & 1 & | & \frac{1}{6} & -\frac{1}{3} \end{bmatrix} \qquad \frac{1}{6}R_2$$

$$\begin{bmatrix} 1 & 0 & | & \frac{1}{6} & \frac{2}{3} \\ 0 & 1 & | & \frac{1}{6} & -\frac{1}{3} \end{bmatrix}. \qquad R_2 + R_1 \qquad \qquad \textbf{(6)}$$

The left half of the augmented matrix in reduced row echelon form in equation (6) is the identity matrix, so the Gauss–Jordan process is finished and the solutions can be read from the right half of the augmented matrix. The numbers in the first column to the right of the vertical bar give the values of $x$ and $z$. The second column to the right of the bar gives the values of $y$ and $w$. That is,

$$\begin{bmatrix} 1 & 0 & | & x & y \\ 0 & 1 & | & z & w \end{bmatrix} = \begin{bmatrix} 1 & 0 & | & \frac{1}{6} & \frac{2}{3} \\ 0 & 1 & | & \frac{1}{6} & -\frac{1}{3} \end{bmatrix},$$

so that

$$A^{-1} = \begin{bmatrix} x & y \\ z & w \end{bmatrix} = \begin{bmatrix} \frac{1}{6} & \frac{2}{3} \\ \frac{1}{6} & -\frac{1}{3} \end{bmatrix}.$$

Check by multiplying $A$ and $A^{-1}$. The result should be $I$:

$$AA^{-1} = \begin{bmatrix} 2 & 4 \\ 1 & -1 \end{bmatrix}\begin{bmatrix} \frac{1}{6} & \frac{2}{3} \\ \frac{1}{6} & -\frac{1}{3} \end{bmatrix} = \begin{bmatrix} \frac{1}{3}+\frac{2}{3} & \frac{4}{3}-\frac{4}{3} \\ \frac{1}{6}-\frac{1}{6} & \frac{2}{3}+\frac{1}{3} \end{bmatrix} = \begin{bmatrix} 1 & 0 \\ 0 & 1 \end{bmatrix} = I.$$

Thus, the original augmented matrix in equation (5) has $A$ as its left half and the identity matrix as its right half, while the final augmented matrix in equation (6), at the end of the Gauss–Jordan process, has the identity matrix as its left half and the inverse matrix $A^{-1}$ as its right half:

$$[A \mid I] \rightarrow [I \mid A^{-1}]. \quad \text{✓}_8$$

The procedure in Example 6 can be generalized as follows.

✓ **Checkpoint 8** ▦

Carry out the process in Example 6 on a graphing calculator as follows: Enter matrix (5) in the example as matrix [B] on the calculator. Then find RREF [B]. What is the result?

## Inverse Matrix

To obtain an **inverse matrix** $A^{-1}$ for any $n \times n$ matrix $A$ for which $A^{-1}$ exists, follow these steps:

1. Form the augmented matrix $[A \mid I]$, where $I$ is the $n \times n$ identity matrix.
2. Perform row operations on $[A \mid I]$ to get a matrix of the form $[I \mid B]$.
3. Matrix $B$ is $A^{-1}$.

**EXAMPLE 7**    Find $A^{-1}$ if

$$A = \begin{bmatrix} 1 & 0 & 1 \\ 2 & -2 & -1 \\ 5 & 0 & 0 \end{bmatrix}.$$

**Solution**  First write the augmented matrix $[A \mid I]$:

$$[A \mid I] = \begin{bmatrix} 1 & 0 & 1 & 1 & 0 & 0 \\ 2 & -2 & -1 & 0 & 1 & 0 \\ 5 & 0 & 0 & 0 & 0 & 1 \end{bmatrix}.$$

Transform the left side of this matrix into the $3 \times 3$ identity matrix:

$$\begin{bmatrix} 1 & 0 & 1 & 1 & 0 & 0 \\ 0 & -2 & -3 & -2 & 1 & 0 \\ 0 & 0 & -5 & -5 & 0 & 1 \end{bmatrix} \quad \begin{matrix} \\ -2R_1 + R_2 \\ -5R_1 + R_3 \end{matrix}$$

$$\begin{bmatrix} 1 & 0 & 1 & 1 & 0 & 0 \\ 0 & 1 & \frac{3}{2} & 1 & -\frac{1}{2} & 0 \\ 0 & 0 & 1 & 1 & 0 & -\frac{1}{5} \end{bmatrix} \quad \begin{matrix} \\ -\frac{1}{2}R_2 \\ -\frac{1}{5}R_3 \end{matrix}$$

$$\begin{bmatrix} 1 & 0 & 0 & 0 & 0 & \frac{1}{5} \\ 0 & 1 & 0 & -\frac{1}{2} & -\frac{1}{2} & \frac{3}{10} \\ 0 & 0 & 1 & 1 & 0 & -\frac{1}{5} \end{bmatrix}. \quad \begin{matrix} -1R_3 + R_1 \\ -\frac{3}{2}R_3 + R_2 \\ \\ \end{matrix}$$

Looking at the right half of the preceding matrix, we see that

$$A^{-1} = \begin{bmatrix} 0 & 0 & \frac{1}{5} \\ -\frac{1}{2} & -\frac{1}{2} & \frac{3}{10} \\ 1 & 0 & -\frac{1}{5} \end{bmatrix}.$$

Verify that $AA^{-1}$ is $I$.

The best way to find matrix inverses on a graphing calculator is illustrated in the next example.

 **EXAMPLE 8** Use a graphing calculator to find the inverse of the following matrices (if they have inverses):

$$A = \begin{bmatrix} 1 & 0 & 1 \\ 2 & -2 & -1 \\ 5 & 0 & 0 \end{bmatrix} \quad \text{and} \quad B = \begin{bmatrix} 2 & 4 \\ 3 & 6 \end{bmatrix}.$$

**Solution** Enter matrix $A$ into the calculator [Figure 6.25(a)]. Then use the $x^{-1}$ key to find the inverse matrix, as in Figure 6.25(b). (Using $\wedge$ and $-1$ for the inverse results in an error message on most calculators.) Note that Figure 6.25(b) agrees with the answer we found working by hand in Example 7.

✓ **Checkpoint 9**

Use a graphing calculator to find the inverses of these matrices (if they exist).

(a) $A = \begin{bmatrix} 1 & 2 & 3 \\ 4 & -1 & 0 \\ 5 & 1 & 3 \end{bmatrix}$

(b) $B = \begin{bmatrix} 2 & 3 \\ -1 & 4 \end{bmatrix}$

**(a)** (b)

**Figure 6.25**

Now enter matrix $B$ into the calculator and use the $x^{-1}$ key. The result is an error message (Figure 6.26), which indicates that the matrix is singular; it does not have an inverse. (If you were working by hand, you would have found that the appropriate system of equations has no solution.) ✓ 9

**⊞ TECHNOLOGY TIP**

Because of round-off error, a graphing calculator may sometimes display an "inverse" for a matrix that doesn't actually have one. So always verify your results by multiplying $A$ and $A^{-1}$. If the product is not the identity matrix, then $A$ does not have an inverse.

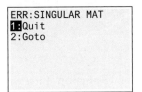

**Figure 6.26**

# 6.5 Exercises

*In Exercises 1–6, the sizes of two matrices A and B are given. Find the sizes of the product AB and the product BA whenever these products exist. (See Example 1.)*

1. $A$ is $2 \times 2$ and $B$ is $2 \times 2$.

2. $A$ is $3 \times 3$ and $B$ is $3 \times 2$.

3. $A$ is $3 \times 5$ and $B$ is $5 \times 3$.

4. $A$ is $4 \times 3$ and $B$ is $3 \times 6$.

5. $A$ is $4 \times 2$ and $B$ is $3 \times 4$.

6. $A$ is $7 \times 3$ and $B$ is $2 \times 7$.

7. To find the product matrix $AB$, the number of _____ of $A$ must be the same as the number of _____ of $B$.

8. The product matrix $AB$ has the same number of _____ as $A$ and the same number of _____ as $B$.

*Find each of the following matrix products, if they exist. (See Examples 2 and 3.)*

9. $\begin{bmatrix} 1 & 2 \\ 3 & 4 \end{bmatrix}\begin{bmatrix} -1 \\ 3 \end{bmatrix}$

10. $\begin{bmatrix} -2 & 5 \\ 7 & 0 \end{bmatrix}\begin{bmatrix} 2 \\ 6 \end{bmatrix}$

11. $\begin{bmatrix} 2 & 2 & -1 \\ 5 & 0 & 1 \end{bmatrix}\begin{bmatrix} 0 & -2 \\ -1 & 5 \\ 0 & 2 \end{bmatrix}$

12. $\begin{bmatrix} -9 & 3 & 1 \\ 3 & 0 & 0 \end{bmatrix}\begin{bmatrix} 2 \\ -1 \\ 4 \end{bmatrix}$

13. $\begin{bmatrix} -4 & 1 \\ 2 & -3 \end{bmatrix}\begin{bmatrix} 1 & 0 \\ 0 & 1 \end{bmatrix}$

14. $\begin{bmatrix} 1 & 0 \\ 0 & 1 \end{bmatrix}\begin{bmatrix} 3 & -2 \\ 1 & -5 \end{bmatrix}$

15. $\begin{bmatrix} 1 & 0 & 0 \\ 0 & 1 & 0 \\ 0 & 0 & 1 \end{bmatrix}\begin{bmatrix} 3 & -5 & 7 \\ -2 & 1 & 6 \\ 0 & -3 & 4 \end{bmatrix}$

16. $\begin{bmatrix} -8 & 9 \\ 3 & -4 \\ -1 & 6 \end{bmatrix}\begin{bmatrix} 1 & 0 & 0 \\ 0 & 1 & 0 \end{bmatrix}$

17. $\begin{bmatrix} 1 & 2 & 3 \\ 4 & 0 & 6 \\ 7 & 8 & 9 \end{bmatrix} \begin{bmatrix} -1 & 4 \\ 7 & 0 \\ 1 & 2 \end{bmatrix}$

18. $\begin{bmatrix} -2 & 0 & 3 \\ 4 & -3 & -1 \end{bmatrix} \begin{bmatrix} 2 & 0 & -1 & 5 \\ 0 & 1 & 0 & -1 \\ 4 & 2 & 5 & -4 \end{bmatrix}$

*In Exercises 19–21, use the matrices*

$$A = \begin{bmatrix} -3 & -9 \\ 2 & 6 \end{bmatrix} \quad and \quad B = \begin{bmatrix} 4 & 6 \\ 2 & 3 \end{bmatrix}.$$

19. Show that $AB \neq BA$. Hence, matrix multiplication is not commutative.

20. Show that $(A + B)^2 \neq A^2 + 2AB + B^2$.

21. Show that $(A + B)(A - B) \neq A^2 - B^2$.

22. Show that $D^2 = D$, where

$$D = \begin{bmatrix} 1 & 0 & 0 \\ \frac{1}{2} & 0 & \frac{1}{2} \\ 0 & 0 & 1 \end{bmatrix}.$$

*Given matrices*

$$P = \begin{bmatrix} m & n \\ p & q \end{bmatrix}, \quad X = \begin{bmatrix} x & y \\ z & w \end{bmatrix}, \quad and \quad T = \begin{bmatrix} r & s \\ t & u \end{bmatrix},$$

*verify that the statements in Exercises 23–26 are true.*

23. $(PX)T = P(XT)$ (associative property of multiplication of matrices)

24. $P(X + T) = PX + PT$ (distributive property)

25. $k(X + T) = kX + kT$ for any real number $k$

26. $(k + h)P = kP + hP$ for any real numbers $k$ and $h$

*Determine whether the given matrices are inverses of each other by computing their product. (See Example 5.)*

27. $\begin{bmatrix} 5 & 2 \\ 3 & -1 \end{bmatrix}$ and $\begin{bmatrix} -1 & 2 \\ 3 & -4 \end{bmatrix}$

28. $\begin{bmatrix} 3 & 2 \\ 7 & 5 \end{bmatrix}$ and $\begin{bmatrix} 5 & 0 \\ -7 & 3 \end{bmatrix}$

29. $\begin{bmatrix} 3 & -1 \\ -4 & 2 \end{bmatrix}$ and $\begin{bmatrix} 1 & \frac{1}{2} \\ 2 & \frac{3}{2} \end{bmatrix}$

30. $\begin{bmatrix} 3 & 5 \\ 7 & 9 \end{bmatrix}$ and $\begin{bmatrix} -\frac{9}{8} & \frac{5}{8} \\ \frac{7}{8} & -\frac{3}{8} \end{bmatrix}$

31. $\begin{bmatrix} 1 & 1 & 1 \\ 2 & 3 & 0 \\ 1 & 2 & 1 \end{bmatrix}$ and $\begin{bmatrix} 1.5 & .5 & -1.5 \\ -1 & 0 & 1 \\ .5 & -.5 & .5 \end{bmatrix}$

32. $\begin{bmatrix} 2 & 5 & 4 \\ 1 & 4 & 3 \\ 1 & 3 & 2 \end{bmatrix}$ and $\begin{bmatrix} 1 & 2 & 1 \\ -5 & 8 & 2 \\ 7 & -11 & -3 \end{bmatrix}$

*Find the inverse, if it exists, for each of the given matrices. (See Examples 6 and 7.)*

33. $\begin{bmatrix} 1 & 2 \\ 3 & 6 \end{bmatrix}$

34. $\begin{bmatrix} 2 & 1 \\ 1 & 1 \end{bmatrix}$

35. $\begin{bmatrix} -1 & 2 \\ 1 & -1 \end{bmatrix}$

36. $\begin{bmatrix} 1 & 2 \\ 3 & 4 \end{bmatrix}$

37. $\begin{bmatrix} 2 & -3 \\ -1 & 2 \end{bmatrix}$

38. $\begin{bmatrix} -3 & -5 \\ 6 & 10 \end{bmatrix}$

39. $\begin{bmatrix} 1 & -1 & 0 \\ -1 & 2 & 3 \\ 1 & 0 & 2 \end{bmatrix}$

40. $\begin{bmatrix} 0 & 1 & -1 \\ 2 & -2 & -1 \\ -1 & 1 & 1 \end{bmatrix}$

41. $\begin{bmatrix} 1 & -1 & 4 \\ 0 & 1 & 3 \\ 2 & -3 & 4 \end{bmatrix}$

42. $\begin{bmatrix} 1 & 4 & 3 \\ 1 & -3 & -2 \\ 2 & 5 & 4 \end{bmatrix}$

43. $\begin{bmatrix} 1 & 2 & 0 \\ 3 & -1 & 2 \\ -2 & 3 & -2 \end{bmatrix}$

44. $\begin{bmatrix} 1 & 0 & -5 \\ 4 & -7 & 3 \\ 3 & -7 & 8 \end{bmatrix}$

*Use a graphing calculator to find the inverse of each matrix. (See Example 8.)*

45. $\begin{bmatrix} 1 & 2 & 3 \\ 1 & 4 & 2 \\ 0 & 1 & -1 \end{bmatrix}$

46. $\begin{bmatrix} 2 & 2 & -4 \\ 2 & 6 & 0 \\ -3 & -3 & 5 \end{bmatrix}$

47. $\begin{bmatrix} 1 & 0 & -2 & 0 \\ -2 & 1 & 2 & 2 \\ 3 & -1 & -2 & -3 \\ 0 & 1 & 4 & 1 \end{bmatrix}$

48. $\begin{bmatrix} 1 & 1 & 0 & 2 \\ 2 & -1 & 1 & -1 \\ 3 & 3 & 2 & -2 \\ 1 & 2 & 1 & 0 \end{bmatrix}$

*Work these exercises. (See Example 4.)*

49. **Grocery Delivery** Several competing companies provide grocery home delivery services. You and your roommate agree to order together and are deciding between two local delivery companies, Timesaver and Wheels. Matrix $A$ shows the number of frozen pizza and quarts of ice cream that you each want to order, and matrix $B$ shows the prices for each company.

$$\begin{array}{cc} & \text{Pizza} \quad \text{Ice cream} \\ \begin{array}{c} \text{You} \\ \text{Roommate} \end{array} & \begin{bmatrix} 6 & 2 \\ 1 & 3 \end{bmatrix} = A \end{array}$$

$$\begin{array}{cc} & \text{Timesaver} \quad \text{Wheels} \\ \begin{array}{c} \text{Pizza} \\ \text{Ice cream} \end{array} & \begin{bmatrix} \$8 & \$7 \\ \$4 & \$5 \end{bmatrix} = B \end{array}$$

(a) Find the matrix product $AB$.

(b) What do the rows of $AB$ represent?

(c) What do the columns of $AB$ represent?

(d) What would be the total cost (for you and your roommate) if you order from Timesaver? From Wheels?

50. **Bulk Shipments** Boxed.com is a company that ships grocery items in bulk. As the purchasing manager for a start-up, you decide to compare Boxed.com prices with your current supplier for an order of Charmin toilet paper (30 count ultra soft) and Bounty paper towels (12 count double rolls). Matrix $A$ shows the quantities that the production and sales departments need

to order, and matrix $B$ shows the prices for each supplier. (Data from: www.boxed.com.)

$$\begin{array}{c} & \begin{array}{cc} \text{Charmin} & \text{Bounty} \end{array} \\ \begin{array}{c} \text{Production} \\ \text{Sales} \end{array} & \left[ \begin{array}{cc} 15 & 25 \\ 60 & 10 \end{array} \right] = A \end{array}$$

$$\begin{array}{c} & \begin{array}{cc} \text{Boxed.com} & \text{Current} \end{array} \\ \begin{array}{c} \text{Charmin} \\ \text{Bounty} \end{array} & \left[ \begin{array}{cc} \$23 & \$24 \\ \$20 & \$18 \end{array} \right] = B \end{array}$$

(a) Find the matrix product $AB$.

(b) What do the rows of $AB$ represent?

(c) What do the columns of $AB$ represent?

(d) What would the total cost be (for both departments) if you order from Boxed.com? From your current supplier?

**51. Food Truck** A restaurant has a main location and a traveling food truck. The first matrix $A$ shows the number of managers and associates employed. The second matrix $B$ shows the average annual cost of salary and benefits (in thousands of dollars).

$$\begin{array}{c} & \begin{array}{cc} \text{Managers} & \text{Associates} \end{array} \\ \begin{array}{c} \text{Restaurant} \\ \text{Food Truck} \end{array} & \left[ \begin{array}{cc} 5 & 25 \\ 1 & 4 \end{array} \right] = A \end{array}$$

$$\begin{array}{c} & \begin{array}{cc} \text{Salary} & \text{Benefits} \end{array} \\ \begin{array}{c} \text{Managers} \\ \text{Associates} \end{array} & \left[ \begin{array}{cc} 41 & 6 \\ 20 & 2 \end{array} \right] = B \end{array}$$

(a) Find the matrix product $AB$.

(b) Explain what $AB$ represents.

(c) According to matrix $AB$, what is the total cost of salaries for all employees (managers and associates) at the restaurant? The food truck? What is the total cost of benefits for all employees at the restaurant? The food truck?

**52. Teacher Salaries** The average annual wage (in thousands of dollars) for teachers in the United States at various grade levels is shown in matrix $W$. (Data from: Bureau of Labor Statistics.) A small school district has both city schools and county schools; matrix $N$ shows the number of teachers in each.

$$\begin{array}{c} & \text{Wage (thousands)} \\ \begin{array}{l} \text{Preschool and Kindergarten} \\ \text{Elementary and Middle School} \\ \text{Secondary School} \end{array} & \left[ \begin{array}{c} 34 \\ 61 \\ 51 \end{array} \right] = W \end{array}$$

$$\begin{array}{c} & \begin{array}{ccc} \text{Preschool and} & \text{Elementary and} & \text{Secondary} \\ \text{Kindergarten} & \text{Middle School} & \text{School} \end{array} \\ \begin{array}{c} \text{City} \\ \text{County} \end{array} & \left[ \begin{array}{ccc} 25 & 60 & 40 \\ 5 & 12 & 10 \end{array} \right] = N \end{array}$$

(a) Find the matrix product $NW$.

(b) Explain what $NW$ represents.

(c) According to matrix $NW$, what is the total cost of teacher wages in city schools? In county schools? In the entire school district?

**53. Gourmet Pops** Steel City Pops sells fruity, creamy, and cookie pops at locations in Texas, Kentucky, and Alabama (www.steelcitypops.com). Suppose that, for a special event, sales (in hundreds of pops) and prices (in dollars per hundred) were as shown in the following matrices.

$$\begin{array}{c} & \text{Price} \\ \begin{array}{l} \text{Fruity} \\ \text{Creamy} \\ \text{Cookie} \end{array} & \left[ \begin{array}{c} 250 \\ 300 \\ 350 \end{array} \right] = P \end{array}$$

$$\begin{array}{c} & \begin{array}{ccc} \text{Fruity} & \text{Creamy} & \text{Cookie} \end{array} \\ \begin{array}{l} \text{Texas} \\ \text{Kentucky} \\ \text{Alabama} \end{array} & \left[ \begin{array}{ccc} 8 & 7 & 2 \\ 4 & 5 & 2 \\ 6 & 9 & 4 \end{array} \right] = S \end{array}$$

(a) Is it possible to find the matrix product $PS$? If so, find it and explain its meaning.

(b) Is it possible to find the matrix product $SP$? If so, find it and explain its meaning.

(c) What was the total revenue from all locations for this special event?

**54. Franchise Costs** According to Franchise Business Review, the minimum investments (in thousands) to launch certain franchises are listed in matrix $M$ below. The numbers of franchises that your investment group would like to start this year in Pennsylvania and Indiana are shown in matrix $N$.

$$\begin{array}{c} & \text{Cost} \\ \begin{array}{l} \text{Deli Delicious} \\ \text{Taziki's Mediterranean Cafe} \\ \text{Your Pie (brick oven pizza)} \end{array} & \left[ \begin{array}{c} 150 \\ 313 \\ 400 \end{array} \right] = M \end{array}$$

$$\begin{array}{c} & \begin{array}{ccc} \text{Deli} & \text{Taziki's} & \text{Pie} \end{array} \\ \begin{array}{l} \text{Indiana} \\ \text{Pynnsylvania} \end{array} & \left[ \begin{array}{ccc} 3 & 0 & 5 \\ 2 & 7 & 1 \end{array} \right] = N \end{array}$$

(a) Is it possible to find the product $MN$? If so, find it and explain the meaning of each entry.

(b) Is it possible to find the product $NM$? If so, find it and explain the meaning of each entry.

(c) What is the total dollar amount that your group must invest for all locations?

📱 *Use a graphing calculator for Exercises 55 and 56.*

**55. Birth and Death Rates** Population estimates and projections (in millions) for the BRIC countries (Brazil, Russia, India, China) and birth and death rates per million people for those countries are shown in the tables. (Data from: United Nations Population Fund.)

|        | 2010 | 2020 |
|--------|------|------|
| **Brazil** | 195  | 210  |
| **Russia** | 143  | 141  |
| **India**  | 1225 | 1387 |
| **China**  | 1341 | 1388 |

|        | Births | Deaths |
|--------|--------|--------|
| **Brazil** | .016   | .006   |
| **Russia** | .011   | .014   |
| **India**  | .023   | .008   |
| **China**  | .013   | .007   |

(a) Write the data in the first table as a 2 × 4 matrix *A*.

(b) Write the data in the second table as a 4 × 2 matrix *B*.

(c) Find the matrix product *AB*.

(d) Explain what *AB* represents.

(e) According to matrix *AB*, what was the total number of deaths in the BRIC countries in 2010? What total number of births is projected for 2020?

**56. Birth and Death Rates** Population estimates and projections (in millions) for several regions of the world and the average birth and death rates per million people for those regions are shown in the tables below. (Data from: United Nations Population Fund.)

|       | Asia | Europe | Latin America | North America |
|-------|------|--------|---------------|---------------|
| **2010** | 4164 | 738 | 590 | 345 |
| **2020** | 4566 | 744 | 652 | 374 |

|               | Births | Deaths |
|---------------|--------|--------|
| **Asia**          | .019   | .007   |
| **Europe**        | .011   | .011   |
| **Latin America** | .019   | .006   |
| **North America** | .024   | .008   |

(a) Write the information in the first table as a 2 × 4 matrix *A*.

(b) Write the information in the second table as a 4 × 2 matrix *B*.

(c) Find the matrix product *AB*.

(d) Explain what *AB* represents.

(e) According to matrix *AB*, what was the total number of births in these four regions combined in 2010? What total number of deaths is projected for 2020?

✓**Checkpoint Answers**

1. $27,100

2. **(a)** No        **(b)** Yes; 2 × 6

3. $CD = \begin{bmatrix} 19 & -2 \\ -13 & -12 \end{bmatrix}$

4. **(a)** 2 × 3        **(b)** Not possible        **(c)** 3 × 3

5. **(a)** $A$ = brand/vitamin; $B$ = vitamin/cost.
   **(b)** $AB$ = brand/cost.
   **(c)** Not meaningful, although the product $BA$ can be found.

6. $IA = \begin{bmatrix} 3 & -2 \\ 4 & -1 \end{bmatrix} = A,$    and    $AI = \begin{bmatrix} 3 & -2 \\ 4 & -1 \end{bmatrix} = A.$

7. No, because $AB \neq I$.

8. RREF [B] is matrix (6) in the example.

9. **(a)** No inverse
   **(b)** Use the FRAC key (if you have one) to simplify the answer, which is
   $$B^{-1} = \begin{bmatrix} 4/11 & -3/11 \\ 1/11 & 2/11 \end{bmatrix}.$$

# 6.6  Applications of Matrices

This section gives a variety of applications of matrices.

## Solving Systems with Matrices

Consider this system of linear equations:

$$2x - 3y = 4$$
$$x + 5y = 2.$$

Let

$$A = \begin{bmatrix} 2 & -3 \\ 1 & 5 \end{bmatrix}, \qquad X = \begin{bmatrix} x \\ y \end{bmatrix}, \qquad \text{and} \qquad B = \begin{bmatrix} 4 \\ 2 \end{bmatrix}.$$

Since

$$AX = \begin{bmatrix} 2 & -3 \\ 1 & 5 \end{bmatrix} \begin{bmatrix} x \\ y \end{bmatrix} = \begin{bmatrix} 2x - 3y \\ x + 5y \end{bmatrix} \quad \text{and} \quad B = \begin{bmatrix} 4 \\ 2 \end{bmatrix},$$

the original system is equivalent to the single matrix equation $AX = B$. Similarly, any system of linear equations can be written as a matrix equation $AX = B$. The matrix $A$ is called the **coefficient matrix**, $X$ is called the **matrix of variables**, and $B$ is called the **matrix of constants**. ✓₁

A matrix equation $AX = B$ can be solved if $A^{-1}$ exists. Assuming that $A^{-1}$ exists and using the facts that $A^{-1}A = I$ and $IX = X$, along with the associative property of multiplication of matrices, gives

$$\begin{aligned} AX &= B \\ A^{-1}(AX) &= A^{-1}B \qquad \text{Multiply both sides by } A^{-1}. \\ (A^{-1}A)X &= A^{-1}B \qquad \text{Associative property} \\ IX &= A^{-1}B \qquad \text{Inverse property} \\ X &= A^{-1}B. \qquad \text{Identity property} \end{aligned}$$

When multiplying by matrices on both sides of a matrix equation, be careful to multiply in the same order on both sides, since multiplication of matrices is not commutative (unlike the multiplication of real numbers). This discussion is summarized below.

> Suppose that a system of equations with the same number of equations as variables is written in matrix form as $AX = B$, where $A$ is the square matrix of coefficients, $X$ is the column matrix of variables, and $B$ is the column matrix of constants. If $A$ has an inverse, then the unique solution of the system is $X = A^{-1}B$.*

✓ **Checkpoint 1**

Write the matrix of coefficients, the matrix of variables, and the matrix of constants for the system

$$\begin{aligned} 2x + 6y &= -14 \\ -x - 2y &= 3. \end{aligned}$$

---

**EXAMPLE 1** Consider this system of equations:

$$\begin{aligned} x + y + z &= 2 \\ 2x + 3y &= 5 \\ x + 2y + z &= -1. \end{aligned}$$

**(a)** Write the system as a matrix equation.

**Solution** We have these three matrices:

$$\underset{\text{Coefficient Matrix}}{A = \begin{bmatrix} 1 & 1 & 1 \\ 2 & 3 & 0 \\ 1 & 2 & 1 \end{bmatrix}}, \quad \underset{\text{Matrix of Variables}}{X = \begin{bmatrix} x \\ y \\ z \end{bmatrix}}, \quad \text{and} \quad \underset{\text{Matrix of Constants}}{B = \begin{bmatrix} 2 \\ 5 \\ -1 \end{bmatrix}}.$$

So the matrix equation is

$$AX = B$$

$$\begin{bmatrix} 1 & 1 & 1 \\ 2 & 3 & 0 \\ 1 & 2 & 1 \end{bmatrix} \begin{bmatrix} x \\ y \\ z \end{bmatrix} = \begin{bmatrix} 2 \\ 5 \\ -1 \end{bmatrix}.$$

---

*If $A$ does not have an inverse, then the system either has no solution or has an infinite number of solutions. Use the methods of Sections 6.1 or 6.2.

**(b)** Find $A^{-1}$ and solve the equation.

**Solution**  Use Exercise 31 of Section 6.5, technology, or the method of Section 6.5 to find that

$$A^{-1} = \begin{bmatrix} 1.5 & .5 & -1.5 \\ -1 & 0 & 1 \\ .5 & -.5 & .5 \end{bmatrix}.$$

Hence,

$$X = A^{-1}B = \begin{bmatrix} 1.5 & .5 & -1.5 \\ -1 & 0 & 1 \\ .5 & -.5 & .5 \end{bmatrix}\begin{bmatrix} 2 \\ 5 \\ -1 \end{bmatrix} = \begin{bmatrix} 7 \\ -3 \\ -2 \end{bmatrix}.$$

Thus, the solution of the original system is $(7, -3, -2)$. ✓₂

✓ **Checkpoint 2**

Use the inverse matrix to solve the system in Example 1 if the constants for the three equations are 12, 0, and 8, respectively.

**EXAMPLE 2**  Use the inverse of the coefficient matrix to solve the system

$$x + 1.5y = 8$$
$$2x + 3y = 10.$$

**Solution**  The coefficient matrix is $A = \begin{bmatrix} 1 & 1.5 \\ 2 & 3 \end{bmatrix}$. A graphing calculator will indicate that $A^{-1}$ does not exist. If we try to carry out the row operations, we see why:

$$\begin{bmatrix} 1 & 1.5 & | & 1 & 0 \\ 2 & 3 & | & 0 & 1 \end{bmatrix}$$

$$\begin{bmatrix} 1 & 1.5 & | & 1 & 0 \\ 0 & 0 & | & -2 & 1 \end{bmatrix}. \quad {\scriptstyle -2R_1 + R_2}$$

The next step cannot be performed because of the zero in the second row, second column. Verify that the original system has no solution. ✓₃

✓ **Checkpoint 3**

Solve the system in Example 2 if the constants are, respectively, 3 and 6.

## Input–Output Analysis

An interesting application of matrix theory to economics was developed by Nobel Prize winner Wassily Leontief. His application of matrices to the interdependencies in an economy is called **input–output analysis.** In practice, input–output analysis is very complicated, with many variables. We shall discuss only simple examples with just a few variables.

Input–output models are concerned with the production and flow of goods and services. A typical economy is composed of a number of different sectors (such as manufacturing, energy, transportation, agriculture, etc.). Each sector requires input from other sectors (and possibly from itself) to produce its output. For instance, manufacturing output requires energy, transportation, and manufactured items (such as tools and machinery). If an economy has $n$ sectors, then the inputs required by the various sectors from each other to produce their outputs can be described by an $n \times n$ matrix called the **input–output matrix** (or the **technological matrix**).

**EXAMPLE 3**  **Economics**  Suppose a simplified economy involves just three sectors—agriculture, manufacturing, and transportation—all in appropriate units. The production of 1 unit of agriculture requires $\frac{1}{2}$ unit of manufacturing and $\frac{1}{4}$ unit of transportation. The production of 1 unit of manufacturing requires $\frac{1}{4}$ unit of agriculture and $\frac{1}{4}$ unit of transportation. The production of 1 unit of transportation requires $\frac{1}{3}$ unit of agriculture and $\frac{1}{4}$ unit of manufacturing. Write the input–output matrix of this economy.

**Solution** Since there are three sectors in the economy, the input–output matrix $A$ is $3 \times 3$. Each row and each column is labeled by a sector of the economy, as shown below. The first column lists the units from each sector of the economy that are required to produce one unit of agriculture. The second column lists the units required from each sector to produce 1 unit of manufacturing, and the last column lists the units required from each sector to produce 1 unit of transportation.

$$
\begin{array}{c}
\textit{Output} \\
\begin{array}{ccc}
\text{Agriculture} & \text{Manufacturing} & \text{Transportation}
\end{array} \\
\textit{Input}\quad
\begin{array}{c}
\text{Agriculture} \\
\text{Manufacturing} \\
\text{Transportation}
\end{array}
\begin{bmatrix}
0 & \frac{1}{4} & \frac{1}{3} \\
\frac{1}{2} & 0 & \frac{1}{4} \\
\frac{1}{4} & \frac{1}{4} & 0
\end{bmatrix} = A. \quad \checkmark\ 4
\end{array}
$$

✓ **Checkpoint 4**

Write a $2 \times 2$ input–output matrix in which 1 unit of electricity requires $\frac{1}{2}$ unit of water and $\frac{1}{3}$ unit of electricity, while 1 unit of water requires no water but $\frac{1}{4}$ unit of electricity.

Example 3 is a bit unrealistic in that no sector of the economy requires any input from itself. In an actual economy, most sectors require input from themselves as well as from other sectors to produce their output. Nevertheless, it is easier to learn the basic concepts from simplified examples, so we shall continue to use them.

The input–output matrix gives only a partial picture of an economy. We also need to know the amount produced by each sector and the amount of the economy's output that is used up by the sectors themselves in the production process. The remainder of the total output is available to satisfy the needs of consumers and others outside the production system.

---

**EXAMPLE 4**    **Economics**  Consider the economy whose input–output matrix $A$ was found in Example 3.

**(a)** Suppose this economy produces 60 units of agriculture, 52 units of manufacturing, and 48 units of transportation. Write this information as a column matrix.

**Solution** Listing the sectors in the same order as the rows of input–output matrix, we have

$$
X = \begin{bmatrix} 60 \\ 52 \\ 48 \end{bmatrix}.
$$

The matrix $X$ is called the **production matrix.**

**(b)** How much from each sector is used up in the production process?

**Solution** Since $\frac{1}{4}$ unit of agriculture is used to produce each unit of manufacturing and there are 52 units of manufacturing output, the amount of agriculture used up by manufacturing is $\frac{1}{4} \times 52 = 13$ units. Similarly $\frac{1}{3}$ unit of agriculture is used to produce a unit of transportation, so $\frac{1}{3} \times 48 = 16$ units of agriculture are used up by transportation. Therefore $13 + 16 = 29$ units of agriculture are used up in the economy's production process.

A similar analysis shows that the economy's production process uses up

$$\underset{\substack{\text{agricul-}\\\text{ture}}}{\frac{1}{2} \times 60} + \underset{\substack{\text{transpor-}\\\text{tation}}}{\frac{1}{4} \times 48} = 30 + 12 = 42 \text{ units of manufacturing}$$

and

$$\underset{\substack{\text{agricul-}\\\text{ture}}}{\frac{1}{4} \times 60} + \underset{\substack{\text{manufac-}\\\text{turing}}}{\frac{1}{4} \times 52} = 15 + 13 = 28 \text{ units of transportation.}$$

**(c)** Describe the conclusions of part (b) in terms of the input–output matrix $A$ and the production matrix $X$.

**Solution** The matrix product $AX$ gives the amount from each sector that is used up in the production process, as shown here [with selected entries color coded as in part (b)]:

$$AX = \begin{bmatrix} 0 & \frac{1}{4} & \frac{1}{3} \\ \frac{1}{2} & 0 & \frac{1}{4} \\ \frac{1}{4} & \frac{1}{4} & 0 \end{bmatrix} \begin{bmatrix} 60 \\ 52 \\ 48 \end{bmatrix} = \begin{bmatrix} 0 \cdot 60 + \frac{1}{4} \cdot 52 + \frac{1}{3} \cdot 48 \\ \frac{1}{2} \cdot 60 + 0 \cdot 52 + \frac{1}{4} \cdot 48 \\ \frac{1}{4} \cdot 60 + \frac{1}{4} \cdot 52 + 0 \cdot 48 \end{bmatrix}$$

$$= \begin{bmatrix} \frac{1}{4} \cdot 52 + \frac{1}{3} \cdot 48 \\ \frac{1}{2} \cdot 60 + \frac{1}{4} \cdot 48 \\ \frac{1}{4} \cdot 60 + \frac{1}{4} \cdot 52 \end{bmatrix} = \begin{bmatrix} 29 \\ 42 \\ 28 \end{bmatrix}.$$

**(d)** Find the matrix $D = X - AX$ and explain what its entries represent.

**Solution** From parts (a) and (c), we have

$$D = X - AX = \begin{bmatrix} 60 \\ 52 \\ 48 \end{bmatrix} - \begin{bmatrix} 29 \\ 42 \\ 28 \end{bmatrix} = \begin{bmatrix} 31 \\ 10 \\ 20 \end{bmatrix}.$$

The matrix $D$ lists the amount of each sector that is *not* used up in the production process and hence is available to groups outside the production process (such as consumers). For example, 60 units of agriculture are produced and 29 units are used up in the process, so the difference $60 - 29 = 31$ is the amount of agriculture that is available to groups outside the production process. Similar remarks apply to manufacturing and transportation. The matrix $D$ is called the **demand matrix.** Matrices $A$ and $D$ show that the production of 60 units of agriculture, 52 units of manufacturing, and 48 units of transportation would satisfy an outside demand of 31 units of agriculture, 10 units of manufacturing, and 20 units of transportation. ✔5

✓ **Checkpoint 5**

**(a)** Write a $2 \times 1$ matrix $X$ to represent the gross production of 9000 units of electricity and 12,000 units of water.

**(b)** Find $AX$, using $A$ from Checkpoint 4.

**(c)** Find $D$, using $D = X - AX$.

Example 4 illustrates the general situation. In an economy with $n$ sectors, the input–output matrix $A$ is $n \times n$. The production matrix $X$ is a column matrix whose $n$ entries are the outputs of each sector of the economy. The demand matrix $D$ is also a column matrix with $n$ entries. This matrix is defined by

$$D = X - AX.$$

$D$ lists the amount from each sector that is available to meet the demands of consumers and other groups outside the production process.

In Example 4, we knew the input–output matrix $A$ and the production matrix $X$ and used them to find the demand matrix $D$. In practice, however, this process is reversed: The input–output matrix $A$ and the demand matrix $D$ are known, and we must find the production matrix $X$ needed to satisfy the required demands. Matrix algebra can be used to solve the equation $D = X - AX$ for $X$:

$$D = X - AX$$
$$D = IX - AX \qquad \text{Identity property}$$
$$D = (I - A)X. \qquad \text{Distributive property}$$

If the matrix $I - A$ has an inverse, then

$$X = (I - A)^{-1}D.$$

**EXAMPLE 5** **Economics** Suppose, in the three-sector economy of Examples 3 and 4, there is a demand for 516 units of agriculture, 258 units of manufacturing, and 129 units of transportation. What should production be for each sector?

**Solution** The demand matrix is

$$D = \begin{bmatrix} 516 \\ 258 \\ 129 \end{bmatrix}.$$

Find the production matrix by first calculating $I - A$:

$$I - A = \begin{bmatrix} 1 & 0 & 0 \\ 0 & 1 & 0 \\ 0 & 0 & 1 \end{bmatrix} - \begin{bmatrix} 0 & \frac{1}{4} & \frac{1}{3} \\ \frac{1}{2} & 0 & \frac{1}{4} \\ \frac{1}{4} & \frac{1}{4} & 0 \end{bmatrix} = \begin{bmatrix} 1 & -\frac{1}{4} & -\frac{1}{3} \\ -\frac{1}{2} & 1 & -\frac{1}{4} \\ -\frac{1}{4} & -\frac{1}{4} & 1 \end{bmatrix}.$$

Using a calculator with matrix capability or row operations, find the inverse of $I - A$:

$$(I - A)^{-1} = \begin{bmatrix} 1.3953 & .4961 & .5891 \\ .8372 & 1.3643 & .6202 \\ .5581 & .4651 & 1.3023 \end{bmatrix}.$$

(The entries are rounded to four decimal places.[*]) Since $X = (I - A)^{-1}D$.

$$X = \begin{bmatrix} 1.3953 & .4961 & .5891 \\ .8372 & 1.3643 & .6202 \\ .5581 & .4651 & 1.3023 \end{bmatrix}\begin{bmatrix} 516 \\ 258 \\ 129 \end{bmatrix} = \begin{bmatrix} 924 \\ 864 \\ 576 \end{bmatrix}$$

(rounded to the nearest whole numbers).

From the last result, we see that the production of 924 units of agriculture, 864 units of manufacturing, and 576 units of transportation is required to satisfy demands of 516, 258, and 129 units, respectively.

**EXAMPLE 6**    **Economics** An economy depends on two basic products: wheat and oil. To produce 1 metric ton of wheat requires .25 metric ton of wheat and .33 metric ton of oil. The production of 1 metric ton of oil consumes .08 metric ton of wheat and .11 metric ton of oil. Find the production that will satisfy a demand of 500 metric tons of wheat and 1000 metric tons of oil.

**Solution** The input–output matrix is

$$\begin{array}{cc} & \text{Wheat} \quad \text{Oil} \\ A = & \begin{bmatrix} .25 & .08 \\ .33 & .11 \end{bmatrix} \begin{array}{l} \text{Wheat} \\ \text{Oil} \end{array} \end{array},$$

and we also have

$$I - A = \begin{bmatrix} 1 & 0 \\ 0 & 1 \end{bmatrix} - \begin{bmatrix} .25 & .08 \\ .33 & .11 \end{bmatrix} = \begin{bmatrix} .75 & -.08 \\ -.33 & .89 \end{bmatrix}.$$

Next, use technology or the methods of Section 6.5 to calculate $(I - A)^{-1}$:

$$(I - A)^{-1} = \begin{bmatrix} 1.3882 & .1248 \\ .5147 & 1.1699 \end{bmatrix} \quad \text{(rounded)}.$$

The demand matrix is

$$D = \begin{bmatrix} 500 \\ 1000 \end{bmatrix}.$$

---

[*]Although we show the matrix $(I - A)^{-1}$ with entries rounded to four decimal places, we did not round off in calculating $(I - A)^{-1}D$. If the rounded figures are used, the numbers in the product may vary slightly in the last digit.

Consequently, the production matrix is

$$X = (I - A)^{-1}D = \begin{bmatrix} 1.3882 & .1248 \\ .5147 & 1.1699 \end{bmatrix} \begin{bmatrix} 500 \\ 1000 \end{bmatrix} = \begin{bmatrix} 819 \\ 1427 \end{bmatrix},$$

where the production numbers have been rounded to the nearest whole numbers. The production of 819 metric tons of wheat and 1427 metric tons of oil is required to satisfy the indicated demand. ✓6

**TECHNOLOGY TIP** If you are using a graphing calculator to determine $X$, you can calculate $(I - A)^{-1}D$ in one step without finding the intermediate matrices $I - A$ and $(I - A)^{-1}$.

## Code Theory

Governments need sophisticated methods of coding and decoding messages. One example of such an advanced code uses matrix theory. This code takes the letters in the words and divides them into groups. (Each space between words is treated as a letter; punctuation is disregarded.) Then, numbers are assigned to the letters of the alphabet. For our purposes, let the letter $a$ correspond to 1, $b$ to 2, and so on. Let the number 27 correspond to a space. Then we use matrix methods to make the code more secure, as the next example illustrates.

---

**EXAMPLE 7**   **Encoding** Encode the message "mathematics is hip" with no punctuation.

**Solution** The message

*mathematics is hip*

can be divided into groups of three letters each:

*mat   hem   ati   cs−   is−   hip*

(We used − to represent a space.) We now write a column matrix for each group of three symbols, using the corresponding numbers instead of letters. For example, the first four groups can be written as

$$\begin{array}{cccc} mat & hem & ati & cs- \\ \begin{bmatrix} 13 \\ 1 \\ 20 \end{bmatrix}, & \begin{bmatrix} 8 \\ 5 \\ 13 \end{bmatrix}, & \begin{bmatrix} 1 \\ 20 \\ 9 \end{bmatrix}, & \begin{bmatrix} 3 \\ 19 \\ 27 \end{bmatrix}. \end{array}$$

The entire message consists of six $3 \times 1$ column matrices:

$$\begin{bmatrix} 13 \\ 1 \\ 20 \end{bmatrix}, \begin{bmatrix} 8 \\ 5 \\ 13 \end{bmatrix}, \begin{bmatrix} 1 \\ 20 \\ 9 \end{bmatrix}, \begin{bmatrix} 3 \\ 19 \\ 27 \end{bmatrix}, \begin{bmatrix} 9 \\ 19 \\ 27 \end{bmatrix}, \begin{bmatrix} 8 \\ 9 \\ 16 \end{bmatrix}. \ ✓7$$

Although you could transmit these matrices, a simple substitution code such as this is very easy to break.

To get a more reliable code, we choose a $3 \times 3$ matrix $M$ that has an inverse. Suppose we choose

$$M = \begin{bmatrix} 1 & 3 & 3 \\ 1 & 4 & 3 \\ 1 & 3 & 4 \end{bmatrix}.$$

✓**Checkpoint 6**

A simple economy depends on just two products: beer and pretzels.

(a) Suppose $\frac{1}{2}$ unit of beer and $\frac{1}{2}$ unit of pretzels are needed to make 1 unit of beer, and $\frac{3}{4}$ unit of beer is needed to make 1 unit of pretzels. Write the input–output matrix $A$ for the economy.

(b) Find $I - A$.

(c) Find $(I - A)^{-1}$.

(d) Find the gross production $X$ that will be needed to get a net production of

$$D = \begin{bmatrix} 100 \\ 1000 \end{bmatrix}.$$

✓**Checkpoint 7**

Write the message "*when*" using $2 \times 1$ matrices.

Then encode each message group by multiplying by $M$—that is,

$$\begin{bmatrix} 1 & 3 & 3 \\ 1 & 4 & 3 \\ 1 & 3 & 4 \end{bmatrix}\begin{bmatrix} 13 \\ 1 \\ 20 \end{bmatrix} = \begin{bmatrix} 76 \\ 77 \\ 96 \end{bmatrix}, \quad \begin{bmatrix} 1 & 3 & 3 \\ 1 & 4 & 3 \\ 1 & 3 & 4 \end{bmatrix}\begin{bmatrix} 8 \\ 5 \\ 13 \end{bmatrix} = \begin{bmatrix} 62 \\ 67 \\ 75 \end{bmatrix},$$

$$\begin{bmatrix} 1 & 3 & 3 \\ 1 & 4 & 3 \\ 1 & 3 & 4 \end{bmatrix}\begin{bmatrix} 1 \\ 20 \\ 9 \end{bmatrix} = \begin{bmatrix} 88 \\ 108 \\ 97 \end{bmatrix},$$

and so on. The coded message consists of the six $3 \times 1$ column matrices

$$\begin{bmatrix} 76 \\ 77 \\ 96 \end{bmatrix}, \begin{bmatrix} 62 \\ 67 \\ 75 \end{bmatrix}, \begin{bmatrix} 88 \\ 108 \\ 97 \end{bmatrix}, \dots, \begin{bmatrix} 83 \\ 92 \\ 99 \end{bmatrix}.$$

The message would be sent as a string of numbers:

$$76, 77, 96, 62, 67, 75, 88, 108, 97, \dots, 83, 92, 99.$$

Note that the same letter may be encoded by different numbers. For instance, the first $a$ in "mathematics" is 77 and the second $a$ is 88. This makes the code harder to break. ✔ 8

## ✓ Checkpoint 8

Use the following matrix to find the $2 \times 1$ matrices to be transmitted for the message in Checkpoint 7:

$$\begin{bmatrix} 2 & 1 \\ 5 & 0 \end{bmatrix}.$$

**EXAMPLE 8**    **Decoding**  Decode the message sent in Example 7.

**Solution**  The receiving agent rewrites the message as the six $3 \times 1$ column matrices shown in blue in Example 7. The agent then decodes the message by multiplying each of these column matrices by the matrix $M^{-1}$. Verify that

$$M^{-1} = \begin{bmatrix} 7 & -3 & -3 \\ -1 & 1 & 0 \\ -1 & 0 & 1 \end{bmatrix}.$$

So the first two matrices of the coded message are decoded as

$$\begin{bmatrix} 7 & -3 & -3 \\ -1 & 1 & 0 \\ -1 & 0 & 1 \end{bmatrix}\begin{bmatrix} 76 \\ 77 \\ 96 \end{bmatrix} = \begin{bmatrix} 13 \\ 1 \\ 20 \end{bmatrix} \begin{matrix} m \\ a \\ t \end{matrix}$$

and

$$\begin{bmatrix} 7 & -3 & -3 \\ -1 & 1 & 0 \\ -1 & 0 & 1 \end{bmatrix}\begin{bmatrix} 62 \\ 67 \\ 75 \end{bmatrix} = \begin{bmatrix} 8 \\ 5 \\ 13 \end{bmatrix} \begin{matrix} h \\ e \\ m \end{matrix}.$$

The other blocks are decoded similarly.

## Routing

Many real-world applications involve networks such as connections on social media, airline routing maps, and networks of cell phone towers. The next example and the case study at the end of the chapter illustrate how matrices can be useful in these applications.

**EXAMPLE 9**    **Campus Pathways**  The diagram in Figure 6.27 shows the pathways connecting five buildings on a college campus.

**(a)** Represent the pathways using a matrix $A$, where the entries represent the number of pathways connecting two buildings without passing through another building.

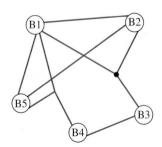

**Figure 6.27**

**Solution** From the diagram, we see that there are two pathways connecting building 1 (B1) to building 2 (B2) without passing through buildings 3, 4, or 5. This information is entered in row one, column two, and again in row two, column one of matrix $A$:

$$\begin{array}{c} \phantom{B1}\begin{array}{ccccc} \text{B1} & \text{B2} & \text{B3} & \text{B4} & \text{B5} \end{array} \\ \begin{array}{c} \text{B1} \\ \text{B2} \\ \text{B3} \\ \text{B4} \\ \text{B5} \end{array} \begin{bmatrix} 0 & 2 & 1 & 1 & 2 \\ 2 & 0 & 1 & 0 & 1 \\ 1 & 1 & 0 & 1 & 0 \\ 1 & 0 & 1 & 0 & 1 \\ 2 & 1 & 0 & 1 & 0 \end{bmatrix} = A. \end{array}$$

Note that there are no pathways connecting each building to itself, so the main diagonal of the matrix from upper left to lower right consists entirely of zeros.

**(b)** How many ways are there to go from building 1 to building 3 by going through exactly one other building?

**Solution** This can happen by going from building 1 to building 2 two ways, and then from building 2 to building 3. Another way is to go from building 1 to building 4 and then from building 4 to building 3. Thus, there are three total ways to go from building 1 to building 3 by going through exactly one other building.

**(c)** Given any pair of buildings, how many ways are there to go between them by going through exactly one other building?

**Solution** The matrix $A^2$ shows the number of ways to travel between any two buildings by going through exactly one other building.

$$A^2 = \begin{bmatrix} 10 & 3 & 3 & 3 & 3 \\ 3 & 6 & 2 & 4 & 4 \\ 3 & 2 & 3 & 1 & 4 \\ 3 & 4 & 1 & 3 & 2 \\ 3 & 4 & 4 & 2 & 6 \end{bmatrix}.$$

We ignore the values down the main diagonal for travel from one building into itself. We notice that the entry in row one, column three has a value of 3. This coincides with what we learned in part (b)—that there are three ways to go from building 1 to building 3 that go through exactly one other building.

In the previous example, we could have also found the number of ways to walk between any two buildings by passing through exactly two other buildings by calculating $A^3$. Additionally, $A + A^2$ gives the number of ways to walk between two buildings with at most one intermediate building.

The diagram can be given many other interpretations. For example, the lines could represent lines of mutual influence between people or nations, or they could represent communication lines, such as Ethernet cables.

✓ **Checkpoint 9** 🔢

Use a graphing calculator to find the following.

**(a)** $A^3$

**(b)** $A + A^2$

# 6.6 Exercises

*Solve the matrix equation $AX = B$ for $X$. (See Example 1.)*

**1.** $A = \begin{bmatrix} 1 & -1 \\ 5 & 6 \end{bmatrix}, B = \begin{bmatrix} 4 \\ -2 \end{bmatrix}$

**2.** $A = \begin{bmatrix} 1 & 2 \\ 1 & 5 \end{bmatrix}, B = \begin{bmatrix} -3 \\ 5 \end{bmatrix}$

**3.** $A = \begin{bmatrix} 3 & 1 \\ 4 & 2 \end{bmatrix}, B = \begin{bmatrix} 3 & 4 \\ 5 & 6 \end{bmatrix}$

**4.** $A = \begin{bmatrix} 1 & 4 \\ 2 & 7 \end{bmatrix}, B = \begin{bmatrix} 0 & 8 \\ 4 & 1 \end{bmatrix}$

**5.** $A = \begin{bmatrix} 2 & 1 & 0 \\ -4 & -1 & 3 \\ 3 & 1 & -2 \end{bmatrix}, B = \begin{bmatrix} 1 \\ 4 \\ 0 \end{bmatrix}$

**6.** $A = \begin{bmatrix} 3 & -1 & 0 \\ 0 & 1 & 2 \\ 9 & 0 & 5 \end{bmatrix}, B = \begin{bmatrix} -3 \\ 6 \\ 12 \end{bmatrix}$

*Use the inverse of the coefficient matrix to solve each system of equations. (See Example 1.)*

**7.**
$$x + 2y + 3z = 10$$
$$2x + 3y + 2z = 6$$
$$-x - 2y - 4z = -1$$

**8.**
$$-x + y - 3z = 3$$
$$2x + 4y - 4z = 6$$
$$-x + y + 4z = -1$$

**9.**
$$x + 4y + 3z = -12$$
$$x - 3y - 2z = 0$$
$$2x + 5y + 4z = 7$$

**10.**
$$5x + 2z = 3$$
$$2x + 2y + z = 4$$
$$-3x + y - z = 5$$

**11.**
$$x + 2y + 3z = 4$$
$$x + 4y + 2z = 8$$
$$y - z = -4$$

**12.**
$$2x + 2y - 4z = 12$$
$$2x + 6y = 16$$
$$-3x - 3y + 5z = -20$$

**13.**
$$x + y + 2w = 3$$
$$2x - y + z - w = 3$$
$$3x + 3y + 2z - 2w = 5$$
$$x + 2y + z = 3$$

**14.**
$$x - 2z = 4$$
$$-2x + y + 2z + 2w = -8$$
$$3x - y - 2z - 3w = 12$$
$$y + 4z + w = -4$$

*Write a system of equations, and use the inverse of the coefficient matrix to solve each system of equations. (See Example 1.)*

**15. E-Commerce** For online store sales, the payment processing service Square charges 2.9% of the total sales plus $.30 per transaction. For your online business, the cost of using Square last month was $291. A competitor to Square charges 3% of total sales plus $.25 per transaction, and your bill for the same month with that company would have been $295. Find the total sales amount and the number of transactions. (Data from: www.squareup.com.)

**16. Truck Rental** Budget Truck Rental charged $19.99 per day plus $.59 per mile to rent a 12-foot moving truck in Austin, Texas, in 2016. The total bill for your move was $86. The rates for a local moving company were $10 per day and $.95 per mile, and the bill for same trip (same number of miles in the same number of days) would have been $94.10. How many miles and how many days was this trip? (Data from: www.budgettruck.com.)

**17. Furniture Manufacturing** A Colorado-based company makes artisanal dining room furniture. A buffet requires 15 hours for cutting, 20 hours for assembly, and 5 hours for finishing. A chair requires 5 hours for cutting, 8 hours for assembly, and 5 hours for finishing. A table requires 10 hours for cutting, 6 hours for assembly, and 6 hours for finishing. The cutting department has 4900 hours of labor available each week, the assembly department has 6600 hours available, and the finishing department has 3900 hours available. How many pieces of each type of furniture should be produced each week if the factory is to run at full capacity?

**18. Nutrition** A hospital dietician is planning a special diet for a certain patient. The total amount per meal of food groups A, B,

and C must equal 400 grams. The diet should include one-third as much of group A as of group B, and the sum of the amounts of group A and group C should equal twice the amount of group B. How many grams of each food group should be included?

**19. Casual Clothing** A national chain of casual clothing stores recently sent shipments of jeans, jackets, sweaters, and shirts to its stores in various cities. The number of items shipped to each city and their total wholesale cost are shown in the table. Find the wholesale price of one pair of jeans, one jacket, one sweater, and one shirt.

| City | Jeans | Jackets | Sweaters | Shirts | Total Cost |
|------|-------|---------|----------|--------|-----------|
| Cleveland | 3000 | 3000 | 2200 | 4200 | $507,650 |
| St. Louis | 2700 | 2500 | 2100 | 4300 | 459,075 |
| Seattle | 5000 | 2000 | 1400 | 7500 | 541,225 |
| Phoenix | 7000 | 1800 | 600 | 8000 | 571,500 |

**20. Materials Mixing** A company produces three combinations of mixed vegetables, which sell in 1-kilogram packages. Italian style combines .3 kilogram of zucchini, .3 of broccoli, and .4 of carrots. French style combines .6 kilogram of broccoli and .4 of carrots. California style combines .2 kilogram of zucchini, .5 of broccoli, and .3 of carrots. The company has a stock of 16,200 kilograms of zucchini, 41,400 kilograms of broccoli, and 29,400 kilograms of carrots. How many packages of each style should the company prepare in order to use up its supplies?

**21. Bacteria Colonies** Three species of bacteria are fed three foods: I, II, and III. A bacterium of the first species consumes 1.3 units each of foods I and II and 2.3 units of food III each day. A bacterium of the second species consumes 1.1 units of food I, 2.4 units of food II, and 3.7 units of food III each day. A bacterium of the third species consumes 8.1 units of I, 2.9 units of II, and 5.1 units of III each day. If 16,000 units of I, 28,000 units of II, and 44,000 units of III are supplied each day, how many of each species can be maintained in this environment?

**22. Assisted Living** A 100-bed assisted living facility provides two levels of long-term care: regular and maximum. Patients at each level have a choice of a private room or a less expensive semiprivate room. The tables (on this page and the following page) show the number of patients in each category at various times last year. The total daily costs for all patients were $24,040 in January, $23,926 in April, $23,760 in July, and $24,042 in October. Find the daily cost of each of the following: a private room (regular care), a private room (maximum care), a semiprivate room (regular care), and a semiprivate room (maximum care).

| | REGULAR-CARE PATIENTS | |
|---|---|---|
| Month | Semiprivate | Private |
| January | 22 | 8 |
| April | 26 | 8 |
| July | 24 | 14 |
| October | 20 | 10 |

|  | MAXIMUM-CARE PATIENTS | |
|---|---|---|
| Month | Semiprivate | Private |
| January | 60 | 10 |
| April | 54 | 12 |
| July | 56 | 6 |
| October | 62 | 8 |

*Find the production matrix for the given input–output and demand matrices. (See Examples 3–6.).*

**23.** $A = \begin{bmatrix} .1 & .03 \\ .07 & .6 \end{bmatrix}, D = \begin{bmatrix} 5 \\ 10 \end{bmatrix}$

**24.** $A = \begin{bmatrix} \frac{1}{2} & \frac{2}{5} \\ \frac{1}{4} & \frac{1}{5} \end{bmatrix}, D = \begin{bmatrix} 2 \\ 4 \end{bmatrix}$

*Exercises 25 and 26 refer to Example 6.*

**25. Input–Output Analysis** If the demand is changed to 690 metric tons of wheat and 920 metric tons of oil, how many units of each commodity should be produced?

**26. Input–Output Analysis** Change the input–output matrix so that the production of 1 metric ton of wheat requires $\frac{1}{5}$ metric ton of oil (and no wheat) and the production of 1 metric ton of oil requires $\frac{1}{3}$ metric ton of wheat (and no oil). To satisfy the same demand matrix, how many units of each commodity should be produced?

*Work these problems. (See Examples 3–6.)*

**27. Input–Output Analysis** A primitive economy depends on two basic goods: yams and pork. The production of 1 bushel of yams requires $\frac{1}{4}$ bushel of yams and $\frac{1}{2}$ of a pig. To produce 1 pig requires $\frac{1}{6}$ bushel of yams. Find the amount of each commodity that should be produced to get

   **(a)** 1 bushel of yams and 1 pig;

   **(b)** 100 bushels of yams and 70 pigs.

**28. Input–Output Analysis** A two-segment economy consists of manufacturing and agriculture. To produce 1 unit of manufacturing output requires .40 unit of its own output and .20 unit of agricultural output. To produce 1 unit of agricultural output requires .30 unit of its own output and .40 unit of manufacturing output. If there is a demand of 240 units of manufacturing and 90 units of agriculture, what should be the output of each segment?

**29. Input–Output Analysis** A simplified economy has only two industries: the electric company and the gas company. Each dollar's worth of the electric company's output requires $.40 of its own output and $.50 of the gas company's output. Each dollar's worth of the gas company's output requires $.25 of its own output and $.60 of the electric company's output. What should the production of electricity and gas be (in dollars) if there is a $12 million demand for gas and a $15 million demand for electricity?

**30. Input–Output Analysis** A simplified economy is based on agriculture, manufacturing, and transportation. Each unit of agricultural output requires .4 unit of its own output, .3 unit of manufacturing output, and .2 unit of transportation output.

One unit of manufacturing output requires .4 unit of its own output, .2 unit of agricultural output, and .3 unit of transportation output. One unit of transportation output requires .4 unit of its own output, .1 unit of agricultural output, and .2 unit of manufacturing output. There is demand for 35 units of agricultural, 90 units of manufacturing, and 20 units of transportation output. How many units should each segment of the economy produce?

**31. Input–Output Analysis** In his work *Input–Output Economics*, Leontief provides an example of a simplified economy with just three sectors: agriculture, manufacturing, and households (that is, the sector of the economy that produces labor).[*] It has the following input–output matrix:

|  | Agriculture | Manufacturing | Households |
|---|---|---|---|
| Agriculture | .25 | .40 | .133 |
| Manufacturing | .14 | .12 | .100 |
| Households | .80 | 3.60 | .133 |

   **(a)** How many units from each sector does the manufacturing sector require to produce 1 unit?

   **(b)** What production levels are needed to meet a demand of 35 units of agriculture, 38 units of manufacturing, and 40 units of households?

   **(c)** How many units of agriculture are used up in the economy's production process?

**32. Input–Output Analysis** A much simplified version of Leontief's 42-sector analysis of the American economy has the following input–output matrix:[†]

|  | Agriculture | Manufacturing | Households |
|---|---|---|---|
| Agriculture | .245 | .102 | .051 |
| Manufacturing | .099 | .291 | .279 |
| Households | .433 | .372 | .011 |

   **(a)** What information about the needs of agricultural production is given by column 1 of the matrix?

   **(b)** Suppose the demand matrix (in billions of dollars) is

$$D = \begin{bmatrix} 2.88 \\ 31.45 \\ 30.91 \end{bmatrix}.$$

Find the amount of each commodity that should be produced.

**33. Input–Output Analysis** An analysis of the Israeli economy is simplified here by grouping the economy into three sectors, with the following input–output matrix:[‡]

|  | Agriculture | Manufacturing | Energy |
|---|---|---|---|
| Agriculture | .293 | 0 | 0 |
| Manufacturing | .014 | .207 | .017 |
| Energy | .044 | .010 | .216 |

   **(a)** How many units from each sector does the energy sector require to produce one unit?

---

[*]Wassily Leontief, *Input–Output Economics*, 2d ed. (Oxford University Press, 1986), pp. 19–27.
[†]Ibid., pp. 6–9.
[‡]Ibid., pp. 174–175.

**(b)** If the economy's production (in thousands of Israeli pounds) is 175,000 of agriculture, 22,000 of manufacturing, and 12,000 of energy, how much is available from each sector to satisfy the demand from consumers and others outside the production process?

**(c)** The actual demand matrix is

$$D = \begin{bmatrix} 138,213 \\ 17,597 \\ 1786 \end{bmatrix}.$$

How much must each sector produce to meet this demand?

**34. Input–Output Analysis** The Chinese economy can be simplified to three sectors: agriculture, industry and construction, and transportation and commerce.[*] The input–output matrix is

|  | Agri. | Industry/Constr. | Trans./Commerce |
|---|---|---|---|
| Agri. | .158 | .156 | .009 |
| Industry/Constr. | .136 | .432 | .071 |
| Trans./Commerce | .013 | .041 | .011 |

The demand [in 100,000 *renminbi* (RMB), the unit of money in China] is

$$D = \begin{bmatrix} 106,674 \\ 144,739 \\ 26,725 \end{bmatrix}.$$

**(a)** Find the amount each sector should produce.

**(b)** Interpret the economic value of an increase in demand of 1 RMB in agriculture exports.

**35. Input–Output Analysis** The economy of the state of Washington has been simplified to four sectors: natural resource, manufacturing, trade and services, and personal consumption. The input–output matrix is as follows:[†]

|  | Natural Resources | Manufacturing | Trade and Services | Personal Consumption |
|---|---|---|---|---|
| Natural Resources | .1045 | .0428 | .0029 | .0031 |
| Manufacturing | .0826 | .1087 | .0584 | .0321 |
| Trade and Services | .0867 | .1019 | .2032 | .3555 |
| Personal Consumption | .6253 | .3448 | .6106 | .0798 |

Suppose the demand (in millions of dollars) is

$$D = \begin{bmatrix} 450 \\ 300 \\ 125 \\ 100 \end{bmatrix}.$$

Find the amount each sector should produce.

**36. Input–Output Analysis** The economy of the state of Nebraska has been condensed to six sectors; livestock, crops, food products, mining and manufacturing, households, and other. The input–output matrix is as follows:[‡]

$$\begin{bmatrix} .178 & .018 & .411 & 0 & .005 & 0 \\ .143 & .018 & .088 & 0 & .001 & 0 \\ .089 & 0 & .035 & 0 & .060 & .003 \\ .001 & .010 & .012 & .063 & .007 & .014 \\ .141 & .252 & .088 & .089 & .402 & .124 \\ .188 & .156 & .103 & .255 & .008 & .474 \end{bmatrix}.$$

**(a)** Find the matrix $(I - A)^{-1}$ and interpret the value in row two, column one, of this matrix.

**(b)** Suppose the demand (in millions of dollars) is

$$D = \begin{bmatrix} 1980 \\ 650 \\ 1750 \\ 1000 \\ 2500 \\ 3750 \end{bmatrix}.$$

Find the dollar amount each sector should produce.

*Work these coding exercises. ( See Example 7 and 8.)*

**37. Encoding** Encode the message

*Anne is home.*

Break the message into groups of two letters and use the matrix

$$M = \begin{bmatrix} 1 & 3 \\ 2 & 7 \end{bmatrix}.$$

**38. Encoding** Use the matrix of Exercise 37 to encode the message

*Head for the hills!*

**39. Decoding** Decode the following message, which was encoded using the matrix $M$ of Exercise 37:

$$\begin{bmatrix} 57 \\ 129 \end{bmatrix}, \begin{bmatrix} 37 \\ 79 \end{bmatrix}.$$

**40. Decoding** Decode the following message, which was encoded by using the matrix $M$ of Exercise 37:

$$\begin{bmatrix} 90 \\ 207 \end{bmatrix}, \begin{bmatrix} 39 \\ 87 \end{bmatrix}, \begin{bmatrix} 26 \\ 57 \end{bmatrix}, \begin{bmatrix} 66 \\ 145 \end{bmatrix}, \begin{bmatrix} 61 \\ 142 \end{bmatrix}, \begin{bmatrix} 89 \\ 205 \end{bmatrix}.$$

*Work these routing problems. ( See Example 9.)*

**41. Highway Networks** The matrix

|  | City 1 | City 2 | City 3 | City 4 |
|---|---|---|---|---|
| City 1 | 0 | 1 | 2 | 2 |
| City 2 | 1 | 0 | 1 | 0 |
| City 3 | 2 | 1 | 0 | 1 |
| City 4 | 2 | 0 | 1 | 0 |

$= A$

[*]*Input–Output Tables of China: 1981*, China Statistical Information and Consultancy Service Centre, 1987, pp. 17–19.

[†]Robert Chase, Philip Bourque, and Richard Conway, Jr., *The 1987 Washington State Input–Output Study*, Report to the Graduate School of Business Administration, University of Washington, September 1993.

[‡]F. Charles Lamphear and Theodore Roesler, "1970 Nebraska Input–Output Tables," *Nebraska Economic and Business Report No. 10*, Bureau of Business Research, University of Nebraska, Lincoln.

shows the number of roads between various cities. Find $A^2$, then use it to answer the following questions: How many ways are there to travel from

(a) city 1 to city 3 by passing through exactly one city?

(b) city 2 to city 4 by passing through exactly one city?

(c) city 1 to city 3 by passing through at most one city?

(d) city 2 to city 4 by passing through at most one city?

**42. Highway Networks** For matrix $A$ in the previous exercise, find $A^3$, then use it to answer the following questions.

(a) How many ways are there to travel between cities 1 and 4 by passing through exactly two cities?

(b) How many ways are there to travel between cities 1 and 4 by passing through at most two cities?

**43. Food Webs** The figure shows a food web. The arrows indicate the food sources of each population. For example, cats feed on rats and on mice.

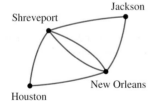

(a) Write a matrix $C$ in which each row and corresponding column represent a population in the food chain. Enter a 1 when the population in a given row feeds on the population in the given column.

(b) Calculate and interpret $C^2$.

**44. Airline Routing** The figure shows four southern cities served by Supersouth Airlines.

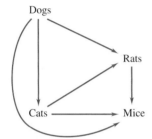

(a) Write a matrix to represent the number of nonstop routes between cities.

(b) Find the number of one-stop flights between Houston and Jackson.

(c) Find the number of flights between Houston and Shreveport that require at most one stop.

(d) Find the number of one-stop flights between New Orleans and Houston.

✓ **Checkpoint Answers**

1. $A = \begin{bmatrix} 2 & 6 \\ -1 & -2 \end{bmatrix}$, $X = \begin{bmatrix} x \\ y \end{bmatrix}$, and $B = \begin{bmatrix} -14 \\ 3 \end{bmatrix}$.

2. $(6, -4, 10)$

3. $(3 - 1.5y, y)$ for all real numbers $y$

4. 

|  | Elec. | Water |
|---|---|---|
| Elec. | $\frac{1}{3}$ | $\frac{1}{4}$ |
| Water | $\frac{1}{2}$ | $0$ |

5. (a) $\begin{bmatrix} 9000 \\ 12{,}000 \end{bmatrix}$   (b) $\begin{bmatrix} 6000 \\ 4500 \end{bmatrix}$   (c) $\begin{bmatrix} 3000 \\ 7500 \end{bmatrix}$

6. (a) $\begin{bmatrix} \frac{1}{2} & \frac{3}{4} \\ \frac{1}{2} & 0 \end{bmatrix}$   (b) $\begin{bmatrix} \frac{1}{2} & -\frac{3}{4} \\ -\frac{1}{2} & 1 \end{bmatrix}$

(c) $\begin{bmatrix} 8 & 6 \\ 4 & 4 \end{bmatrix}$   (d) $\begin{bmatrix} 6800 \\ 4400 \end{bmatrix}$

7. $\begin{bmatrix} 23 \\ 8 \end{bmatrix}, \begin{bmatrix} 5 \\ 14 \end{bmatrix}$

8. $\begin{bmatrix} 54 \\ 115 \end{bmatrix}, \begin{bmatrix} 24 \\ 25 \end{bmatrix}$

9. (a) 
```
[A]³
[18  26  16  16  26]
[26  12  13   9  16]
[16  13   6  10   9]
[16   9  10   6  13]
[26  16   9  13  12]
```

(b) 
```
[A]+[A]²
[10   5   4   4   5]
[ 5   6   3   4   5]
[ 4   3   3   2   4]
[ 4   4   2   3   3]
[ 5   5   4   3   6]
```

# CHAPTER 6 Summary and Review

## Key Terms and Symbols

**6.1** linear equation
system of linear equations
solution of a system
substitution method
elimination method
independent system

dependent system
inconsistent system
**6.2** equivalent systems
elementary operations
elimination method
row

column
matrix (matrices)
element (entry)
augmented matrix
row operations
row echelon form

Gauss–Jordan method
reduced row echelon form
independent system
inconsistent system
dependent system
parameter

| | | | |
|---|---|---|---|
| **6.3** applications of systems of linear equations | additive inverse of a matrix<br>zero matrix | identity matrix<br>inverse matrix | input–output analysis<br>input–output matrix |
| **6.4** row matrix (row vector)<br>column matrix (column vector)<br>square matrix | scalar<br>product of a scalar and a matrix<br>**6.5** product matrix | singular matrix<br>**6.6** coefficient matrix<br>matrix of variables<br>matrix of constants | production matrix<br>demand matrix<br>code theory<br>routing theory |

## Chapter 6 Key Concepts

**Solving Systems of Equations**

The following **elementary operations** are used to transform a system of equations into a simpler equivalent system:

1.  Interchange any two equations.
2.  Multiply both sides of an equation by a nonzero constant.
3.  Replace an equation by the sum of itself and a nonzero constant multiple of another equation in the system.

The **elimination method** is a systematic way of using elementary operations to transform a system into an equivalent one that can be solved by **back substitution.** See Section 6.2 for details.

The matrix version of the elimination method uses the following **matrix row operations,** which correspond to using elementary row operations with back substitution on a system of equations:

1.  Interchange any two rows.
2.  Multiply each element of a row by a nonzero constant.
3.  Replace a row by the sum of itself and a nonzero constant multiple of another row in the matrix.

The **Gauss–Jordan method** is an extension of the elimination method for solving a system of linear equations. It uses row operations on the augmented matrix of the system. See Section 6.2 for details.

**Operations on Matrices**

The **sum** of two $m \times n$ matrices $X$ and $Y$ is the $m \times n$ matrix $X + Y$ in which each element is the sum of the corresponding elements of $X$ and $Y$. The **difference** of two $m \times n$ matrices $X$ and $Y$ is the $m \times n$ matrix $X - Y$ in which each element is the difference of the corresponding elements of $X$ and $Y$.

The **product** of a scalar $k$ and a matrix $X$ is the matrix $kX$, with each element being $k$ times the corresponding element of $X$.

The **product matrix** $AB$ of an $m \times n$ matrix $A$ and an $n \times k$ matrix $B$ is the $m \times k$ matrix whose entry in the $i$th row and $j$th column is the product of the $i$th row of $A$ and the $j$th column of $B$.

The $n \times n$ **identity matrix** $I$ has 1s on the main diagonal from upper left to lower right, with all other entries equal to 0.

If there is a matrix $A^{-1}$ satisfying $AA^{-1} = I$, then $A^{-1}$ is called the **inverse matrix** of $A$. The inverse matrix $A^{-1}$ for any $n \times n$ matrix $A$ for which $A^{-1}$ exists is found as follows: Form the augmented matrix $[A|I]$, and perform row operations on $[A|I]$ to get the matrix $[I|A^{-1}]$.

## Chapter 6 Review Exercises

*Solve each of the following systems.*

1.  $-5x - 3y = -3$
    $2x + y = 4$

2.  $3x - y = 8$
    $2x + 3y = 6$

3.  $3x - 5y = 16$
    $2x + 3y = -2$

4.  $\dfrac{1}{4}x - \dfrac{1}{3}y = -\dfrac{1}{4}$
    $\dfrac{1}{10}x + \dfrac{2}{5}y = \dfrac{2}{5}$

5.  **Business** Juan Felipe plans to buy shares of two stocks. One costs $32 per share and pays dividends of $1.20 per share. The other costs $23 per share and pays dividends of $1.40 per share. He has $10,100 to spend and wants to earn dividends of $540. How many shares of each stock should he buy?

6.  **Business** Maria has money in two investment funds. Last year, the first fund paid a dividend of 8% and the second a dividend of 2%, and she received a total of $780. This year, the first fund paid a 10% dividend and the second only 1%, and she received $810. How much does she have invested in each fund?

*Solve each of the following systems.*

7.  $x - 2y = 1$
    $4x + 4y = 2$
    $10x + 8y = 4$

8.  $4x - y - 2z = 4$
    $x - y - \dfrac{1}{2}z = 1$
    $2x - y - z = 8$

**9.**  $3x + y - z = 3$
$\phantom{3}x \phantom{+ y} + 2z = 6$
$-3x - y + 2z = 9$

**10.**  $x + y - 4z = 0$
$2x + y - 3z = 2$

**25.**  $\begin{bmatrix} 6 & 8 & 10 \\ 5 & 3 & -2 \end{bmatrix}$

**26.**  $\begin{bmatrix} -9 \\ 15 \\ 4 \end{bmatrix}$

*Solve each of the following systems by using matrix methods.*

**11.**  $x + \phantom{y} z = -3$
$\phantom{x +} y - z = \phantom{-}6$
$2x + 3z = \phantom{-}5$

**12.**  $2x + 3y + 4z = \phantom{-}8$
$-x + \phantom{3}y - 2z = -9$
$2x + 2y + 6z = 16$

**13.**  $x - 2y + 5z = 3$
$4x + 3y - 4z = 1$
$3x + 5y - 9z = 7$

**14.**  $5x - \phantom{1}8y + \phantom{4}z = 1$
$3x - \phantom{1}2y + 4z = 3$
$10x - 16y + 2z = 3$

**15.**  $x - 2y + 3z = \phantom{-}4$
$2x + \phantom{2}y - 4z = \phantom{-}3$
$-3x + 4y - \phantom{4}z = -2$

**16.**  $3x + 2y - 6z = 3$
$x + \phantom{2}y + 2z = 2$
$2x + 2y + 5z = 0$

**17.** **Finance** You are given $144 in one-, five-, and ten-dollar bills. There are 35 bills. There are two more ten-dollar bills than five-dollar bills. How many bills of each type are there?

**18.** **Social Services** A social service agency provides counseling, meals, and shelter to clients referred by sources I, II, and III. Clients from source I require an average of $100 for food, $250 for shelter, and no counseling. Source II clients require an average of $100 for counseling, $200 for food, and nothing for shelter. Source III clients require an average of $100 for counseling, $150 for food, and $200 for shelter. The agency has funding of $25,000 for counseling, $50,000 for food, and $32,500 for shelter. How many clients from each source can be served?

**19.** **Manufacturing** A small business makes woven blankets, rugs, and skirts. Each blanket requires 24 hours for spinning the yarn, 4 hours for dying the yarn, and 15 hours for weaving. Rugs require 30, 5, and 18 hours, and skirts require 12, 3, and 9 hours, respectively. If there are 306, 59, and 201 hours available for spinning, dying, and weaving, respectively, how many of each item can be made? (*Hint:* Simplify the equations you write, if possible, before solving the system.)

**20.** **Manufacturing** Each week at a furniture factory, 2000 work hours are available in the construction department, 1400 work hours in the painting department, and 1300 work hours in the packing department. Producing a chair requires 2 hours of construction, 1 hour of painting, and 2 hours for packing. Producing a table requires 4 hours of construction, 3 hours of painting, and 3 hours for packing. Producing a chest requires 8 hours of construction, 6 hours of painting, and 4 hours for packing. If all available time is used in every department, how many of each item are produced each week?

*For each of the following, find the dimensions of the matrix and identify any square, row, or column matrices.*

**21.**  $\begin{bmatrix} 2 & 3 \\ 5 & 9 \end{bmatrix}$

**22.**  $\begin{bmatrix} 2 & -1 \\ 4 & 6 \\ 5 & 7 \end{bmatrix}$

**23.**  $[12 \quad 4 \quad -8 \quad -1]$

**24.**  $\begin{bmatrix} -7 & 5 & 6 & 4 \\ 3 & 2 & -1 & 2 \\ -1 & 12 & 8 & -1 \end{bmatrix}$

**27.** **Stock Prices** The opening stock price, high price, and closing price for the following companies were reported on April 12, 2016: Coca-Cola Bottling Company Consolidated, $162.45, $164.44, and $163.64; Intel Corporation: $31.88, $31.94, and $31.86; Verizon Communications Inc., $51.66, $52.19, and $51.95. Write these data as a $3 \times 3$ matrix. (Data from: www.morningstar.com.)

**28.** **Animal Behavior** The activities of a grazing animal can be classified roughly into three categories: grazing, moving, and resting. Suppose horses spend 8 hours grazing, 8 moving, and 8 resting; cattle spend 10 grazing, 5 moving, and 9 resting; sheep spend 7 grazing, 10 moving, and 7 resting; and goats spend 8 grazing, 9 moving, and 7 resting. Write this information as a $4 \times 3$ matrix.

*Given the matrices*

$$A = \begin{bmatrix} 4 & 6 \\ -2 & -2 \\ 5 & 9 \end{bmatrix}, \quad B = \begin{bmatrix} 1 & 2 & -3 \\ 2 & 3 & 0 \\ 0 & 1 & 4 \end{bmatrix}, \quad C = \begin{bmatrix} 5 & 0 \\ -1 & 3 \\ 4 & 7 \end{bmatrix},$$

$$D = \begin{bmatrix} 6 \\ 1 \\ 0 \end{bmatrix}, \quad E = [1 \quad 3 \quad -4], \quad F = \begin{bmatrix} -1 & 2 \\ 6 & 7 \end{bmatrix}, \quad and$$

$$G = \begin{bmatrix} 2 & 5 \\ 1 & 6 \end{bmatrix},$$

*find each of the following (if possible).*

**29.**  $-B$

**30.**  $-D$

**31.**  $A + 2C$

**32.**  $F + 3G$

**33.**  $2B - 5C$

**34.**  $D + E$

**35.**  $3A - 2C$

**36.**  $G - 2F$

**37.** **Trading Data** The change in price (in dollars) and volume of stocks traded (in millions of shares) for the following companies were reported on April 11, 2016: General Electric Company (GE), $-.08$ and 25.1M; Nike Inc. (NKE), $-1.50$ and 12.6M; United Health Group Incorporated (UNH), $.65 and 3.1M. On the following day, the changes in price and volume data for the same three companies were GE: $.10 and 27.2M, NKE: $.63 and 12.4M, UNH, $.47 and 2.9M. Write the daily data as $3 \times 2$ matrices, then use matrix addition to find a matrix representing the total change in price and volume of stock traded for these companies over the two days. (Data from: www.morningstar.com.)

**38.** **Oil Distribution** An oil refinery in Tulsa sent 110,000 gallons of oil to a Chicago distributor, 73,000 to a Dallas distributor, and 95,000 to an Atlanta distributor. Another refinery in New Orleans sent the following respective amounts to the same three distributors: 85,000, 108,000, and 69,000. The next month, the two refineries sent the same distributors new shipments of oil as follows: from Tulsa, 58,000 to Chicago, 33,000 to Dallas, and 80,000 to Atlanta; from New Orleans, 40,000, 52,000, and 30,000, respectively.

**(a)** Write the monthly shipments from the two distributors to the three refineries as $3 \times 2$ matrices.

**(b)** Use matrix addition to find the total amounts sent to the refineries from each distributor.

*Use the matrices shown before Exercise 29 to find each of the following (if possible).*

**39.** $AG$      **40.** $EB$      **41.** $GA$

**42.** $CA$      **43.** $AGF$      **44.** $EBD$

**45. Hockey Injuries** In a study, the numbers of head and neck injuries among hockey players wearing full face shields and half face shields were compared. Injury rates were recorded per 1000 athlete exposures for specific injuries that caused a player to miss one or more events.[*]

Players wearing a half shield had the following rates of injury:

> Head and face injuries, excluding concussions: 3.54;
> Concussions: 1.53;
> Neck injuries: .34;
> Other injuries: 7.53.

Players wearing a full shield had the following injury rates:

> Head and face injuries, excluding concussions: 1.41;
> Concussions: 1.57;
> Neck injuries: .29;
> Other injuries: 6.21.

If an equal number of players in a large league wear each type of shield and the total number of athlete exposures for the league in a season is 8000, use matrix operations to estimate the total number of injuries of each type.

**46. Manufacturing** An office supply manufacturer makes two kinds of paper clips: standard and extra large. To make a unit of standard paper clips requires $\frac{1}{4}$ hour on a cutting machine and $\frac{1}{2}$ hour on a machine that shapes the clips. A unit of extra-large paper clips requires $\frac{1}{3}$ hour on each machine.

**(a)** Write this information as a $2 \times 2$ matrix (size/machine).

**(b)** If 48 units of standard and 66 units of extra-large clips are to be produced, use matrix multiplication to find out how many hours each machine will operate. (*Hint:* Write the units as a $1 \times 2$ matrix.)

**47. Price and Dividends** The cost per share of stock was $60.94 for Starbucks Corporation, $113.22 for PepsiCo Inc., and $119.11 for Anheuser-Busch, as reported on May 8, 2017. The dividends per share were $1.00, $3.22, and $4.24, respectively. Your stock portfolio consists of 500 shares of Starbucks, 300 shares of PepsiCo, and 200 shares of Anheuser-Busch. (Data from: www.morningstar.com.)

**(a)** Write the cost per share and dividend per share data as a $3 \times 2$ matrix.

**(b)** Write the number of shares of each stock as a $1 \times 3$ matrix.

**(c)** Use matrix multiplication to find the total cost and total dividends for this portfolio.

**48. Price and Dividends** Stan Palla's stock portfolio consists of 1000 shares of Apple Inc., 500 shares of Texas Instruments Inc., and 700 shares of Microsoft Corporation. On May 8, 2017, the cost per share of stock was $153.01 for Apple, $79.46 for Texas Instruments, and $68.94 for Microsoft. The dividends per share

were $2.52, $2.00, and $1.56, respectively. (Data from: www.morningstar.com.)

**(a)** Write the cost per share and dividend per share data as a $3 \times 2$ matrix.

**(b)** Write the number of shares of each stock as a $1 \times 3$ matrix.

**(c)** Use matrix multiplication to find the total cost and total dividends for this portfolio.

**49.** If $A = \begin{bmatrix} 3 & 0 \\ 2 & 1 \end{bmatrix}$, find a matrix $B$ such that both $AB$ and $BA$ are defined and $AB \neq BA$.

**50.** Is it possible to do Exercise 49 if $A = \begin{bmatrix} 4 & 0 \\ 0 & 4 \end{bmatrix}$? Explain.

*Find the inverse of each of the following matrices, if an inverse exists for that matrix.*

**51.** $\begin{bmatrix} -2 & 2 \\ 0 & 5 \end{bmatrix}$      **52.** $\begin{bmatrix} 3 & -1 \\ -5 & 2 \end{bmatrix}$

**53.** $\begin{bmatrix} 6 & 4 \\ 3 & 2 \end{bmatrix}$      **54.** $\begin{bmatrix} 15 & -12 \\ 5 & -4 \end{bmatrix}$

**55.** $\begin{bmatrix} 2 & 0 & 6 \\ 1 & -1 & 0 \\ 0 & 1 & -3 \end{bmatrix}$      **56.** $\begin{bmatrix} 2 & -1 & 0 \\ 1 & 0 & 2 \\ 1 & -4 & 0 \end{bmatrix}$

**57.** $\begin{bmatrix} 2 & 3 & 5 \\ -2 & -3 & -5 \\ 1 & 4 & 2 \end{bmatrix}$      **58.** $\begin{bmatrix} 1 & 3 & 6 \\ 4 & 0 & 9 \\ 5 & 15 & 30 \end{bmatrix}$

**59.** $\begin{bmatrix} 1 & 3 & -2 & -1 \\ 0 & 1 & 1 & 2 \\ -1 & -1 & 1 & -1 \\ 1 & -1 & -3 & -2 \end{bmatrix}$      **60.** $\begin{bmatrix} 3 & 2 & 0 & -1 \\ 2 & 0 & 1 & 2 \\ 1 & 2 & -1 & 0 \\ 2 & -1 & 1 & 1 \end{bmatrix}$

*Refer again to the matrices shown before Exercise 29 to find each of the following (if possible).*

**61.** $F^{-1}$      **62.** $G^{-1}$      **63.** $(G - F)^{-1}$

**64.** $(F + G)^{-1}$      **65.** $B^{-1}$

**66.** Explain why the matrix $\begin{bmatrix} a & 0 \\ c & 0 \end{bmatrix}$, where $a$ and $c$ are nonzero constants, cannot have an inverse.

*Solve each of the following matrix equations $AX = B$ for X.*

**67.** $A = \begin{bmatrix} -3 & 4 \\ -1 & 2 \end{bmatrix}$, $B = \begin{bmatrix} 3 \\ -1 \end{bmatrix}$

**68.** $A = \begin{bmatrix} 1 & 3 \\ -2 & 4 \end{bmatrix}$, $B = \begin{bmatrix} 9 \\ 6 \end{bmatrix}$

**69.** $A = \begin{bmatrix} 1 & 0 & 2 \\ -1 & 1 & 0 \\ 3 & 0 & 4 \end{bmatrix}$, $B = \begin{bmatrix} 8 \\ 4 \\ -6 \end{bmatrix}$

**70.** $A = \begin{bmatrix} 2 & 4 & 0 \\ 1 & -2 & 0 \\ 0 & 0 & 3 \end{bmatrix}$, $B = \begin{bmatrix} 72 \\ -24 \\ 48 \end{bmatrix}$

---

[*]Brian Benson, Nicholas Nohtadi, Sarah Rose, and Willem Meeuwisse, "Head and Neck Injuries among Ice Hockey Players Wearing Full Face Shields vs. Half Face Shields," *JAMA*, 282, no. 24, (December 22/29, 1999): 2328–2332.

*Use the inverse of the coefficient matrix to solve each system of equations.*

**71.** $\begin{aligned} x + y &= -2 \\ 2x + 5y &= 2 \end{aligned}$

**72.** $\begin{aligned} 5x - 3y &= -2 \\ 2x + 7y &= -9 \end{aligned}$

**73.** $\begin{aligned} 2x + y &= 10 \\ 3x - 2y &= 8 \end{aligned}$

**74.** $\begin{aligned} x - 2y &= 7 \\ 3x + y &= 7 \end{aligned}$

**75.** $\begin{aligned} x + y + z &= 1 \\ 2x - y &= -2 \\ 3y + z &= 2 \end{aligned}$

**76.** $\begin{aligned} x &= -3 \\ y + z &= 6 \\ 2x - 3z &= -9 \end{aligned}$

**77.** $\begin{aligned} 3x - 2y + 4z &= 4 \\ 4x + y - 5z &= 2 \\ -6x + 4y - 8z &= -2 \end{aligned}$

**78.** $\begin{aligned} -2x + 3y - z &= 1 \\ 5x - 7y + 8z &= 4 \\ 6x - 9y + 3z &= 2 \end{aligned}$

*Solve each of the following problems by any method.*

**79. Wine Blending** A wine maker has two large casks of wine. One is 9% alcohol and the other is 14% alcohol. How many liters of each wine should be mixed to produce 40 liters of wine that is 12% alcohol?

**80. Metal Blending** A gold merchant has some 12-carat gold (12/24 pure gold) and some 22-carat gold (22/24 pure). How many grams of each could be mixed to get 25 grams of 15-carat gold?

**81. Materials Mixing** A chemist has a 40% acid solution and a 60% solution. How many liters of each should be used to get 40 liters of a 45% solution?

**82. Materials Mixing** How many pounds of tea worth $4.60 a pound should be mixed with tea worth $6.50 a pound to get 10 pounds of a mixture worth $5.74 a pound?

**83. Manufacturing** A machine in a pottery factory takes 3 minutes to form a bowl and 2 minutes to form a plate. The material for a bowl costs $.25, and the material for a plate costs $.20. If the machine runs for 8 hours and exactly $44 is spent for material, how many bowls and plates can be produced?

**84. Physical Science** A boat travels at a constant speed a distance of 57 kilometers downstream in 3 hours and then turns around and travels 55 kilometers upstream in 5 hours. What are the speeds of the boat and the current?

**85. Finance** Henry Miller invests $50,000 three ways—at 8%, $8\frac{1}{2}$%, and 11%. In total, he receives $4436.25 per year in interest. The interest on the 11% investment is $80 more than the interest on the 8% investment. Find the amount he has invested at each rate.

**86. Football Tickets** Tickets to the homecoming football game cost $4 for students, $10 for alumni, $12 for other adults, and $6 for children. The total attendance was 3750, and the ticket receipts were $29,100. Six times more students than children attended. The number of alumni was $\frac{4}{5}$ of the number of students. How many students, alumni, other adults, and children were at the game?

**87. Input–Output Analysis** Given the input–output matrix $A = \begin{bmatrix} 0 & \frac{1}{4} \\ \frac{1}{2} & 0 \end{bmatrix}$ and the demand matrix $D = \begin{bmatrix} 2100 \\ 1400 \end{bmatrix}$, find each of the following.

**(a)** $I - A$      **(b)** $(I - A)^{-1}$

**(c)** The production matrix $X = (I - A)^{-1}D$

**88. Input–Output Analysis** An economy depends on two commodities: goats and cheese. It takes $\frac{2}{3}$ unit of goats to produce 1 unit of cheese and $\frac{1}{2}$ unit of cheese to produce 1 unit of goats.

**(a)** Write the input–output matrix for this economy.

**(b)** Find the production required to satisfy a demand of 400 units of cheese and 800 units of goats.

*Use technology to do Exercises 89–92.*

**89. Input–Output Analysis** In a simple economic model, a country has two industries: agriculture and manufacturing. To produce $1 of agricultural output requires $.10 of agricultural output and $.40 of manufacturing output. To produce $1 of manufacturing output requires $.70 of agricultural output and $.20 of manufacturing output. If agricultural demand is $60,000 and manufacturing demand is $20,000, what must each industry produce? (Round answers to the nearest whole number.)

**90. Input–Output Analysis** Here is the input–output matrix for a small economy:

|  | Agriculture | Services | Mining | Manufacturing |
|---|---|---|---|---|
| Agriculture | .02 | .9 | 0 | .001 |
| Services | 0 | .4 | 0 | .06 |
| Mining | .01 | .02 | .06 | .07 |
| Manufacturing | .25 | .9 | .9 | .4 |

**(a)** How many units from each sector does the service sector require to produce 1 unit?

**(b)** What production levels are needed to meet a demand for 760 units of agriculture, 1600 units of services, 1000 units of mining, and 2000 units of manufacturing?

**(c)** How many units of manufacturing production are used up in the economy's production process?

**91. Input–Output Analysis** Use this input–output matrix to answer the questions below.

|  | Agriculture | Construction | Energy | Manufacturing | Transportation |
|---|---|---|---|---|---|
| Agriculture | .18 | .017 | .4 | .005 | 0 |
| Construction | .14 | .018 | .09 | .001 | 0 |
| Energy | .9 | 0 | .4 | .06 | .002 |
| Manufacturing | .19 | .16 | .1 | .008 | .5 |
| Transportation | .14 | .25 | .9 | .4 | .12 |

**(a)** How many units from each sector does the energy sector require to produce 1 unit?

**(b)** If the economy produces 28,067 units of agriculture, 9383 units of construction, 51,372 units of energy, 61,364 units of manufacturing, and 90,403 units of transportation, how much is available from each sector to satisfy the demand from consumers and others outside the production system?

**92. Input–Output Analysis** A new demand matrix for the economy in the previous exercise is

$$\begin{bmatrix} 2400 \\ 850 \\ 1400 \\ 3200 \\ 1800 \end{bmatrix}.$$

How much must each sector produce to meet this demand?

**93. Routing** Matrix $M$ represents the number of direct train lines between four stations in a metropolitan subway system.

$$\begin{array}{c} \\ A \\ B \\ C \\ D \end{array} \begin{array}{c} A\ B\ C\ D \\ \begin{bmatrix} 0 & 1 & 1 & 0 \\ 1 & 0 & 1 & 1 \\ 1 & 1 & 0 & 1 \\ 0 & 1 & 1 & 0 \end{bmatrix} \end{array} = M$$

**(a)** Find $M^2$.

**(b)** How many routes between station A and station D go through exactly one other station?

**(c)** Find $M^3$.

**(d)** How many routes between station A and station D go through exactly two other stations?

**94. Routing** The following matrix represents the number of direct flights between four cities:

$$\begin{array}{c} \\ A \\ B \\ C \\ D \end{array} \begin{array}{c} A\ B\ C\ D \\ \begin{bmatrix} 0 & 1 & 0 & 1 \\ 1 & 0 & 0 & 1 \\ 0 & 0 & 0 & 1 \\ 1 & 1 & 1 & 0 \end{bmatrix} \end{array}.$$

**(a)** Find the number of one-stop flights between cities $A$ and $C$.

**(b)** Find the total number of flights between cities $B$ and $C$ that are either direct or one stop.

**(c)** Find the matrix that gives the number of two-stop flights between these cities.

**95. Encoding** Use the matrix $M = \begin{bmatrix} 2 & 6 \\ 1 & 4 \end{bmatrix}$ to encode the message "leave now."

**96. Decoding** What matrix should be used to decode the answer to the previous exercise?

# Case Study 6  Airline Route Maps

Airline route maps are usually published on airline Web sites, as well as in in-flight magazines. The purpose of these maps is to show what cities are connected to each other by nonstop flights provided by the airline. We can think of these maps as another type of **graph,** and we can use matrix operations to answer questions of interest about the graph. To study these graphs, a bit of terminology will be helpful. In this context, a **graph** is a set of points called **vertices** or **nodes** and a set of lines called **edges** connecting some pairs of vertices. Two vertices connected by an edge are said to be **adjacent.** Suppose, for example, that Stampede Air is a short-haul airline serving the state of Texas whose route map is shown in Figure 1 below. The vertices are the cities to which Stampede Air flies, and two vertices are connected if there is a nonstop flight between them.

Some natural questions arise about graphs. It might be important to know if two vertices are connected by a sequence of two edges, even if they are not connected by a single edge. In the route map, Huston and Lubbock are connected by a two-edge sequence, meaning that a passenger would have to stop in Austin while flying between those cities on Stampede Air. It might be important to know if it is possible to get from a vertex to another vertex in a given number of flights. In the example, a passenger on Stampede Air can get from any city in the company's network to any other city, given enough flights. But how many flights are enough? This is another issue of interest: What is the minimum number of steps required to get from one vertex to another? What is the minimum number of steps required to get from any vertex

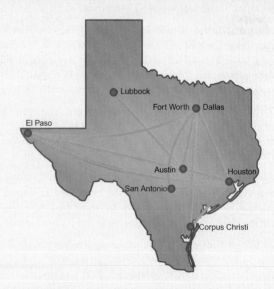

**Figure 1**  Stampede Air route map

on the graph to any other? While these questions are relatively easy to answer for a small graph, as the number of vertices and edges grows, it becomes harder to keep track of all the different ways the vertices are connected. Matrix notation and computation can help to answer these questions.

The **adjacency matrix** for a graph with $n$ vertices is an $n \times n$ matrix whose $(i, j)$ entry is 1 if the $i$th and $j$th vertices are connected and 0 if they are not. If the vertices in the Stampede Air graph respectively correspond to Austin (A), El Paso (E), Houston (H), Dallas-Fort Worth (D), San Antonio (S), Corpus Christi (C), and Lubbock (L), then the adjacency matrix for Stampede Air is as follows.

$$
A = \begin{array}{c} \\ \\ \\ \\ \\ \\ \\ \end{array}
\begin{array}{cccccccc}
\phantom{0} & A & E & H & D & S & C & L \\
\left[\begin{array}{ccccccc}
0 & 1 & 1 & 1 & 0 & 0 & 1 \\
1 & 0 & 1 & 1 & 0 & 0 & 0 \\
1 & 1 & 0 & 1 & 1 & 1 & 0 \\
1 & 1 & 1 & 0 & 1 & 1 & 0 \\
0 & 0 & 1 & 1 & 0 & 0 & 0 \\
0 & 0 & 1 & 1 & 0 & 0 & 0 \\
1 & 0 & 0 & 0 & 0 & 0 & 0
\end{array}\right] & \begin{array}{c} A \\ E \\ H \\ D \\ S \\ C \\ L \end{array}
\end{array}
$$

Adjacency matrices can be used to address the questions about graphs raised earlier. Which vertices are connected by a two-edge sequence? How many different two-edge sequences connect each pair of vertices? Consider the matrix $A^2$, which is $A$ multiplied by itself. For example, let the $(6, 2)$ entry in the matrix $A^2$ be named $b_{62}$. This entry in $A^2$ is the product of the 6th row of $A$ and the 2nd column of $A$, or

$$
\begin{aligned}
b_{62} &= a_{61}a_{12} + a_{62}a_{22} + a_{63}a_{32} + a_{64}a_{42} + a_{65}a_{52} + a_{66}a_{62} + a_{67}a_{72} \\
&= 0 \cdot 1 + 0 \cdot 0 + 1 \cdot 1 + 1 \cdot 1 + 0 \cdot 0 + 0 \cdot 0 + 0 \cdot 0 \\
&= 2,
\end{aligned}
$$

which happens to be the number of two-flight sequences that connect city 6 (Corpus Christi) and city 2 (El Paso). A careful look at Figure 1 confirms this fact. This calculation works because, in order for a two-flight sequence to occur between Corpus Christi and El Paso, Corpus Christi and El Paso must each connect to an intermediate city. Since Corpus Christi connects to Houston and Houston connects to El Paso, $a_{63}a_{32} = 1 \cdot 1 = 1$. Thus, there is one two-flight sequence between Corpus Christi and El Paso that passes through Houston. Since Corpus Christi does not connect to Austin (city 1), but Austin does connect with El Paso, $a_{61}a_{12} = 0 \cdot 1 = 0$. Hence, there is no two-flight sequence from Corpus Christi to El Paso that passes through Austin. To find the total number of two-flight sequences between Corpus Christi and El Paso, simply sum over all intermediate points. Notice that this sum, which is

$$
a_{61}a_{12} + a_{62}a_{22} + a_{63}a_{32} + a_{64}a_{42} + a_{65}a_{52} + a_{66}a_{62} + a_{67}a_{72},
$$

is just $b_{62}$, the $(6, 2)$ entry in the matrix $A^2$. So we see that the number of two-step sequences between vertex $i$ and vertex $j$ in a graph with adjacency matrix $A$ is the $(i, j)$ entry in $A^2$. A more general result is the following:

**The number of $k$-step sequences between vertex $i$ and vertex $j$ in a graph with adjacency matrix $A$ is the $(i, j)$ entry in $A^k$.**

If $A$ is the adjacency matrix for Figure 1, then

$$
A^2 = \begin{array}{c} \\ \\ \\ \\ \\ \\ \\ \end{array}
\begin{array}{cccccccc}
\phantom{0} & A & E & H & D & S & C & L \\
\left[\begin{array}{ccccccc}
4 & 2 & 2 & 2 & 2 & 2 & 0 \\
2 & 3 & 2 & 2 & 2 & 2 & 1 \\
2 & 2 & 5 & 4 & 1 & 1 & 1 \\
2 & 2 & 4 & 5 & 1 & 1 & 1 \\
2 & 2 & 1 & 1 & 2 & 2 & 0 \\
2 & 2 & 1 & 1 & 2 & 2 & 0 \\
0 & 1 & 1 & 1 & 0 & 0 & 1
\end{array}\right] & \begin{array}{c} A \\ E \\ H \\ D \\ S \\ C \\ L \end{array}
\end{array}
$$

and

$$
A^3 = \begin{array}{c} \\ \\ \\ \\ \\ \\ \\ \end{array}
\begin{array}{cccccccc}
\phantom{0} & A & E & H & D & S & C & L \\
\left[\begin{array}{ccccccc}
6 & 8 & 12 & 12 & 4 & 4 & 4 \\
8 & 6 & 11 & 11 & 4 & 4 & 2 \\
12 & 11 & 10 & 11 & 9 & 9 & 2 \\
12 & 11 & 11 & 10 & 9 & 9 & 2 \\
4 & 4 & 9 & 9 & 2 & 2 & 2 \\
4 & 4 & 9 & 9 & 2 & 2 & 2 \\
4 & 2 & 2 & 2 & 2 & 2 & 0
\end{array}\right] & \begin{array}{c} A \\ E \\ H \\ D \\ S \\ C \\ L \end{array}
\end{array}
$$

Since the $(6, 3)$ entry in $A^2$ is 1, there is one two-step sequence from Corpus Christi to Houston. Likewise, there are four three-step sequences between El Paso and San Antonio, since the $(2, 5)$ entry in $A^3$ is 4.

In observing the figure, note that some two-step or three-step sequences may not be meaningful. On the Stampede Air route map, Dallas-Fort Worth is reachable in two steps from Austin (via El Paso or Houston), but in reality this does not matter, since there is a nonstop flight between the two cities. A better question to ask of a graph might be, "What is the least number of edges that must be traversed to go from vertex $i$ to vertex $j$?"

To answer this question, consider the matrix $S_k = A + A^2 + \cdots + A^k$. The $(i, j)$ entry in this matrix tallies the number of ways to get from vertex $i$ to vertex $j$ in $k$ steps or less. If such a trip is impossible, this entry will be zero. Thus, to find the shortest number of steps between the vertices, continue to compute $S_k$ as $k$ increases; the first $k$ for which the $(i, j)$ entry in $S_k$ is nonzero is the shortest number of steps between $i$ and $j$. Note that although the shortest number of steps may be computed, the method does not determine what those steps are.

## Exercises

1. Which Stampede Air cities may be reached by a two-flight sequence from San Antonio? Which may be reached by a three-flight sequence?

2. It was shown previously that there are four three-step sequences between El Paso and San Antonio. Describe these three-step sequences.

3. Which trips in the Stampede Air network take the greatest number of flights?

4.  Suppose that Vacation Air is an airline serving Jacksonville, Orlando, West Palm Beach, Miami, and Tampa, Florida. The route map is given in Figure 2. Produce an adjacency matrix $B$ for this map, listing the cities in the order given.

5.  Find $B^2$, the square of the adjacency matrix found in Exercise 4. What does this matrix tell you?

6.  Find $B + B^2$. What can you conclude from this calculation?

**Figure 2**   Vacation Air route map

# Extended Project

Search the Internet for a small regional airline that serves your area and use their route map to produce an adjacency matrix. Determine which trips using this airline take the largest number of flights. How many flights did these trips take? Prepare a report that includes an overview of the airline, the route map, an explanation of the adjacency matrix, and your findings about which trips take the largest number of flights.

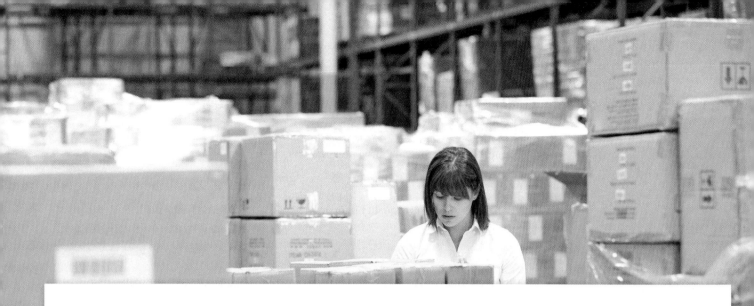

# Linear Programming

## 7 CHAPTER

### CHAPTER OUTLINE

Linear programming is one of the most remarkable (and useful) mathematical techniques developed in the last 65 years. It is used to deal with a variety of issues faced by businesses, financial planners, medical personnel, sports leagues, and others. Typical applications include) maximizing company profits by adjusting production schedules, minimizing shipping costs by locating warehouses efficiently, and maximizing pension income by choosing the best mix of financial products.

Many real-world problems involve inequalities. For example, a factory may have no more than 200 workers on a shift and must manufacture at least 3000 units at a cost of no more than $35 each. How many workers should it have per shift in order to produce the required number of units at minimal cost? *Linear programming* is a method for finding the optimal (best possible) solution for such problems—if there is one.

In this chapter, we shall study two methods of solving linear programming problems: the graphical method and the simplex method. The graphical method requires a knowledge of **linear inequalities,** which are inequalities involving only first-degree polynomials in $x$ and $y$. So we begin with a study of such inequalities.

# 7.1  Graphing Linear Inequalities in Two Variables

Examples of linear inequalities in two variables include

$$x + 2y < 4, \qquad 3x + 2y > 6, \qquad \text{and} \qquad 2x - 5y \geq 10.$$

A solution of a linear inequality is an ordered pair that satisfies the inequality. For example (4, 4) is a solution of

$$3x - 2y \leq 6.$$

(Check by substituting 4 for $x$ and 4 for $y$.) A linear inequality has an infinite number of solutions, one for every choice of a value for $x$. The best way to show these solutions is to sketch the **graph of the inequality,** which consists of all points in the plane whose coordinates satisfy the inequality.

---

**EXAMPLE 1**    Graph the inequality $3x - 2y \leq 6$.

**Solution**  First, solve the inequality for $y$:

$$3x - 2y \leq 6$$
$$-2y \leq -3x + 6$$
$$y \geq \frac{3}{2}x - 3 \qquad \text{Multiply by } -\tfrac{1}{2}; \text{ reverse the inequality.}$$
$$y \geq 1.5x - 3.$$

This inequality has the same solutions as the original one. To solve it, note that the points on the line $y = 1.5x - 3$ certainly satisfy $y \geq 1.5x - 3$. Plot some points, and graph this line, as in Figure 7.1.

The points on the line satisfy "$y$ equals $1.5x - 3$." The points satisfying "$y$ is *greater than* $1.5x - 3$" are the points *above* the line (because they have larger second coordinates than the points on the line; see Figure 7.2). Similarly, the points satisfying

$$y < 1.5x - 3$$

lie below the line (because they have smaller second coordinates), as shown in Figure 7.2. The line $y = 1.5x - 3$ is the **boundary line.**

Thus, the solutions of $y \geq 1.5x - 3$ are all points *on or above* the line $y = 1.5x - 3$. The line and the shaded region of Figure 7.3 make up the graph of the inequality $y \geq 1.5x - 3$.

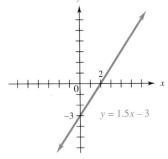

**Figure 7.1**

✓ **Checkpoint 1**

Graph the given inequalities.

**(a)**  $2x + 5y \leq 10$

**(b)**  $x - y \geq 4$

*Answers to Checkpoint exercises are found at the end of the section.*

Points satisfy $y > 1.5x - 3$.

Points satisfy $y < 1.5x - 3$.

**Figure 7.2**

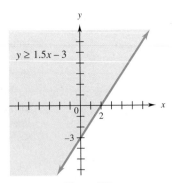

$y \geq 1.5x - 3$

**Figure 7.3**

**Figure 7.4**

 **Checkpoint 2**

Graph the given inequalities.

(a)  $2x + 3y > 12$

(b)  $3x - 2y < 6$

**EXAMPLE 2**    Graph $x + 4y < 4$.

**Solution**  First obtain an equivalent inequality by solving for $y$:

$$4y < -x + 4$$

$$y < -\frac{1}{4}x + 1$$

$$y < -.25x + 1.$$

The boundary line is $y = -.25x + 1$, but it is *not* part of the solution, since points *on* the line do not satisfy $y < -.25x + 1$. To indicate this, the line is drawn dashed in Figure 7.4. The points *below* the boundary line are the solutions of $y < -.25x + 1$, because they have smaller second coordinates than the points on the line $y = -.25x + 1$. The shaded region in Figure 7.4 (excluding the dashed line) is the graph of the inequality $y < -.25x + 1$. ✔2

Examples 1 and 2 show that the solutions of a linear inequality form a **half-plane** consisting of all points on one side of the boundary line (and possibly the line itself). When an inequality is solved for $y$, the inequality symbol immediately tells whether the points above ($>$), on ($=$), or below ($<$) the boundary line satisfy the inequality, as summarized here.

| Inequality | Solution Consists of All Points |
|---|---|
| $y \geq mx + b$ | *on or above* the line $y = mx + b$ |
| $y > mx + b$ | *above* the line $y = mx + b$ |
| $y \leq mx + b$ | *on or below* the line $y = mx + b$ |
| $y < mx + b$ | *below* the line $y = mx + b$ |

When graphing by hand, draw the boundary line $y = mx + b$ solid when it is included in the solution ($\geq$ or $\leq$ inequalities) and dashed when it is not part of the solution ($>$ or $<$ inequalities).

**⊞ TECHNOLOGY TIP** To shade the area above or below the graph of $Y_1$ on TI-84+, go to the Y = menu and move the cursor to the left of $Y_1$. Press ENTER until the correct shading pattern appears (◥ for above the line and ◣ for below the line). Then press GRAPH. On TI-86/89, use the STYLE key in the Y = menu instead of the ENTER key. For other calculators, consult your instruction manual.

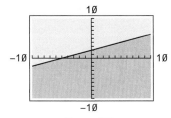

**Figure 7.5**

✔ **Checkpoint 3** ⊞

Use a graphing calculator to graph $2x < y$.

**EXAMPLE 3**    Graph $5y - 2x \leq 10$.

**Solution**  Solve the inequality for $y$:

$$5y \leq 2x + 10$$

$$y \leq \frac{2}{5}x + 2$$

$$y \leq .4x + 2.$$

The graph consists of all points on or below the boundary line $y \leq .4x + 2$, as shown in Figure 7.5. (See the Technology Tip.) ✔3

**❗ CAUTION** You cannot tell from a calculator-produced graph whether the boundary line is included. It is included in Figure 7.5, but not in the answer to Checkpoint 3.

**Figure 7.6**

**Figure 7.7**

✓ **Checkpoint 4**

Graph each of the following.

**(a)** $x \geq 3$

**(b)** $y - 3 \leq 0$

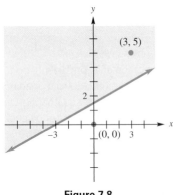

**Figure 7.8**

---

**EXAMPLE 4**    Graph each of the given inequalities.

**(a)** $y \geq 2$

**Solution**  The boundary line is the horizontal line $y = 2$. The graph consists of all points on or above this line (Figure 7.6).

**(b)** $x \leq -1$

**Solution**  This inequality does not fit the pattern discussed earlier, but it can be solved by a similar technique. Here, the boundary line is the vertical line $x = -1$, and it is included in the solution. The points satisfying $x < -1$ are all points to the left of this line (because they have $x$-coordinates smaller than $-1$). So the graph consists of the points that are *on or to the left of* the vertical line $x = -1$, as shown in Figure 7.7. ✓ 4

An alternative technique for solving inequalities that does not require solving for $y$ is illustrated in the next example. Feel free to use it if you find it easier than the technique presented in Examples 1–4.

**EXAMPLE 5**    Graph $4y - 2x \geq 6$.

**Solution**  The boundary line is $4y - 2x = 6$, which can be graphed by finding its intercepts:

$$x\text{-intercept: Let } y = 0. \qquad y\text{-intercept: Let } x = 0.$$
$$4(0) - 2x = 6 \qquad\qquad 4y - 2(0) = 6$$
$$x = -3. \qquad\qquad y = \frac{6}{4} = 1.5.$$

The graph contains the half-plane above or below this line. To determine which, choose a test point—any point not on the boundary line, say, $(0, 0)$. Letting $x = 0$ and $y = 0$ in the inequality produces

$$4(0) - 2(0) \geq 6, \qquad a \text{ } false \text{ statement.}$$

Therefore, $(0, 0)$ is not in the solution. So the solution is the half-plane that does *not* include $(0, 0)$, as shown in Figure 7.8. If a different test point is used, say, $(3, 5)$, then substituting $x = 3$ and $y = 5$ in the inequality produces

$$4(5) - 2(3) \geq 6, \qquad a \text{ } true \text{ statement.}$$

Therefore, the solution of the inequality is the half-plane containing $(3, 5)$, as shown in Figure 7.8.

📝 **NOTE**  When using the method of Example 5, $(0, 0)$ is the best choice for the test point because it makes the calculation very easy. The only time that $(0, 0)$ cannot be used is for inequalities of the form $ax + by \geq 0$ (or $>$ or $<$ or $\leq$); in such cases, $(0, 0)$ is on the line $ax + by = 0$.

## Systems of Inequalities

Real-world problems often involve many inequalities. For example, a manufacturing problem might produce inequalities resulting from production requirements, as well as inequalities about cost requirements. A set of at least two inequalities is called a **system of inequalities.** The **graph** of a system of inequalities is made up of all those points which satisfy *all* the inequalities of the system.

> **EXAMPLE 6** Graph the system
>
> $$3x + y \le 12$$
> $$x \le 2y.$$
>
> **Solution** First, solve each inequality for $y$:
>
> $$3x + y \le 12 \qquad\qquad x \le 2y$$
> $$y \le -3x + 12 \qquad y \ge \frac{x}{2}.$$
>
> Then the original system is equivalent to this one:
>
> $$y \le -3x + 12$$
> $$y \ge \frac{x}{2}.$$

✓ **Checkpoint 5**

Graph the system

$$x + y \le 6$$
$$2x + y \ge 4.$$

The solutions of the first inequality are the points *on or below* the line $y = -3x + 12$ (Figure 7.9). The solutions of the second inequality are the points *on or above* the line $y = x/2$ (Figure 7.10). So the solutions of the *system* are the points that satisfy both of these conditions, as shown in Figure 7.11. ✓ 5

**Figure 7.9**

**Figure 7.10**

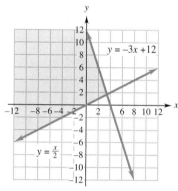

**Figure 7.11**

The shaded region in Figure 7.11 is sometimes called the **region of feasible solutions,** or just the **feasible region,** since it consists of all the points that satisfy (are feasible for) every inequality of the system.

> **EXAMPLE 7** Graph the feasible region for the system
>
> $$2x - 5y \le 10$$
> $$x + 2y \le 8$$
> $$x \ge 0, y \ge 0.$$
>
> **Solution** Begin by solving the first two inequalities for $y$:
>
> $$2x - 5y \le 10 \qquad\qquad x + 2y \le 8$$
> $$-5y \le -2x + 10 \qquad\qquad 2y \le -x + 8$$
> $$y \ge .4x - 2 \qquad\qquad y \le -.5x + 4.$$
>
> Then the original system is equivalent to this one:
>
> $$y \ge .4x - 2$$
> $$y \le -.5x + 4$$
> $$x \ge 0, y \ge 0.$$

The inequalities $x \geq 0$ and $y \geq 0$ restrict the graph to the first quadrant. So the feasible region consists of all points in the first quadrant that are on or above the line $y = .4x - 2$ *and* on or below the line $y = -.5x + 4$. In the calculator-generated graph of Figure 7.12, the feasible region is the darkest region. This is confirmed by the hand-drawn graph of the feasible region in Figure 7.13.

✓ **Checkpoint 6**

Graph the feasible region of the system

$$x + 4y \leq 8$$
$$x - y \geq 3$$
$$x \geq 0, y \geq 0.$$

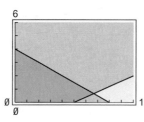

| Figure 7.12 | Figure 7.13 |

**EXAMPLE 8**    Graph the feasible region for the system

$$5y - 2x \leq 10$$
$$x \geq 3, y \geq 2.$$

**Solution** Solve the first inequality for $y$ (as in Example 3) to obtain an equivalent system:

$$y \leq .4x + 2$$
$$x \geq 3, y \geq 2.$$

As shown in Example 3, Checkpoint 4, and Example 4(a), the feasible region consists of all points that lie

> on or below the line $y \leq .4x + 2$ *and*
> on or to the right of the vertical line $x = 3$ *and*
> on or above the horizontal line $y = 2$,

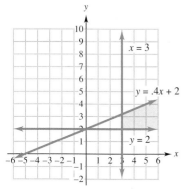

**Figure 7.14**

as shown in Figure 7.14.

**EXAMPLE 9**    **Bond Investing**  In 2017, the T. Rowe Price Tax-Free High-Yield Fund paid a 3-year return of 5%, and the BlackRock California Municipal Opportunities Fund paid a 3-year return of 4%. Sherri would like to invest up to $75,000 and obtain at least $2500 in interest over the 3-year period. Write a system of inequalities expressing these conditions, and graph the feasible region. (Data from: www.morningstar.com.)

**Solution** Let $x$ represent the amount invested in the T. Rowe Price fund and $y$ the amount invested in the BlackRock fund. Then make a chart that summarizes the given information.

| Funds | Value | Return |
|---|---|---|
| **T. Rowe Price** | $x$ | 5% = .05 |
| **BlackRock** | $y$ | 4% = .04 |
| **Maximum or minimum** | $75,000 | $2500 |

We must have $x \geq 0$ and $y \geq 0$ because Sherri cannot invest negative dollars into the two funds. In terms of investing up to $75,000, we have

$$x + y \leq 75,000$$
$$y \leq -x + 75,000. \quad \text{Add } -x \text{ to each side.}$$

Similarly, the interest requirement has that

$$.05x + .04y \geq 2500$$

$$5x + 4y \geq 250,000 \qquad \text{Multiply both sides by 100.}$$

$$4y \geq -5x + 250,000 \qquad \text{Add } -5x \text{ to each side.}$$

$$y \geq -\frac{5}{4}x + 62,500. \qquad \text{Divide both sides by 4.}$$

So we must solve the system

$$y \leq -x + 75,000,$$

$$y \geq -\frac{5}{4}x + 62,500,$$

$$x \geq 0, y \geq 0.$$

✓ **Checkpoint 7**

Mylene wishes to invest in two bond products. Bond A has an annual return of 4%, and Bond B has an annual return of 5.2%. She wants to earn at least $1800 annually. She can invest up to $40,000 in the two bonds. Write a system of inequalities expressing these conditions, and graph the feasible region.

The feasible region is shown in Figure 7.15. ✓₇

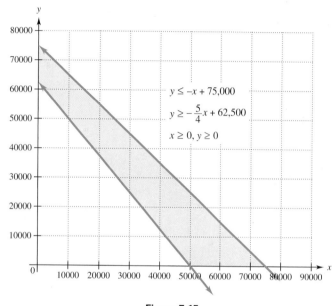

$$y \leq -x + 75,000$$
$$y \geq -\frac{5}{4}x + 62,500$$
$$x \geq 0, y \geq 0$$

**Figure 7.15**

# 7.1 Exercises

*Match the inequality with its graph, which is one of the ones shown.*

**1.** $y \geq -x - 2$

**2.** $y \leq 2x - 2$

**3.** $y \leq x + 2$

**4.** $y \geq x + 1$

**5.** $6x + 4y \geq -12$

**6.** $3x - 2y \geq -4$

**A.**

**B.**

**C.**

**D.**

**E.**

**F.**

*Graph each of the given linear inequalities. (See Examples 1–5.)*

**7.** $y < 5 - 2x$      **8.** $y < x + 3$

**9.** $x - 2y \geq 3$      **10.** $2x + 5y \geq 10$

**11.** $2x - y \leq 4$      **12.** $4x - 3y \leq 24$

**13.** $y \leq -4$      **14.** $x \geq -2$

**15.** $3x - 2y \geq 18$      **16.** $3x + 2y \geq -4$

**17.** $3x + 4y \geq 12$      **18.** $4x - 3y > 9$

**19.** $2x - 4y \leq 3$      **20.** $4x - 3y < 12$

**21.** $x \leq 5y$      **22.** $2x \geq y$

**23.** $-3x \leq y$      **24.** $-x \geq 6y$

**25.** $y \leq x$      **26.** $y > -2x$

**27.** In your own words, explain how to determine whether the boundary line of an inequality should be solid or dashed.

**28.** When graphing $y \leq 3x - 6$, would you shade above or below the line $y = 3x - 6$? Explain your answer.

*Graph the feasible region for the given systems of inequalities. (See Examples 6 and 7.)*

**29.** $y \geq 3x - 6$          **30.** $x + y \leq 4$
    $y \geq -x + 1$                $x - y \geq 2$

**31.** $2x + y \leq 5$          **32.** $x - y \geq 1$
    $x + 2y \leq 5$                $x \leq 3$

**33.** $2x + y \geq 8$          **34.** $4x + y \geq 9$
    $4x - y \leq 3$                $2x + 3y \leq 7$

**35.** $2x - y \leq 1$          **36.** $x + 3y \leq 6$
    $3x + y \leq 6$                $2x + 4y \geq 7$

**37.** $-x - y \leq 5$          **38.** $6x - 4y \geq 8$
    $2x - y \leq 4$                $3x + 2y \geq 4$

**39.** $3x + y \geq 6$          **40.** $2x + 3y \geq 12$
    $x + 2y \geq 7$                $x + y \geq 4$
    $x \geq 0$                        $x \geq 0$
    $y \geq 0$                        $y \geq 0$

**41.** $-2 \leq x \leq 3$       **42.** $-2 \leq x \leq 2$
    $-1 \leq y \leq 5$              $y \geq 1$
    $2x + y \leq 6$                $x - y \geq 0$

**43.** $2y - x \geq -5$         **44.** $2x + 3y \leq 12$
    $y \leq 3 + x$                  $2x + 3y \geq -6$
    $x \geq 0$                        $3x + y \leq 4$
    $y \geq 0$                        $x \geq 0$
                                           $y \geq 0$

**45.** $3x + 4y \geq 12$        **46.** $0 \leq x \leq 9$
    $2x - 3y \leq 6$                $x - 2y \geq 4$
    $0 \leq y \leq 2$                $3x + 5y \leq 30$
    $x \geq 0$                        $y \geq 0$

*Find a system of inequalities that has the given graph.*

**47.**

**48.**

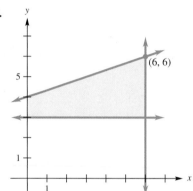

**C** *In Exercises 49 and 50, find a system of inequalities whose feasible region is the interior of the given polygon.*

**49.** Rectangle with vertices $(2, 3)$, $(2, -1)$, $(7, 3)$, and $(7, -1)$

**50.** Triangle with vertices $(2, 4)$, $(-4, 0)$, and $(2, -1)$

*In each of the following, write a system of inequalities that describes all the given conditions, and graph the feasible region of the system. (See Example 9.)*

**51. Investing** As of April 2017, the Nuveen Core Plus Bond C fund yielded an annual return of 3.6%, and the Nuveen Core Plus Bond R6 fund yielded an annual return of 5.0%. You would like to invest up to $50,000 and obtain at least $2100 in interest. Let $x$ be the amount invested in the C fund, and let $y$ be the amount invested in the R6 fund. (Data from: www.morningstar.com.)

**52. Investing** As of April 2017, the Guggenheim Total Return Bond A fund yielded an annual return of 6.0%, and the Quantified Managed Income Advisor fund yielded an annual return of 4.7%. You would like to invest up to $40,000 and earn at least $2250 in interest. Let $x$ be the amount invested in the Guggenheim fund, and let $y$ be the amount invested in the Quantified Managed fund. (Data from: www.morningstar.com.)

**53. Investing** As of April 2017, the Franklin Total Return Advisor fund yielded an annual return of 2.4%, and the Highland Fixed Income Y fund yielded an annual return of 4.2%. You would like to invest up to $60,000 and earn at least $2016 in interest. Let $x$ be the amount invested in the Franklin fund, and let $y$ be the amount invested in the Highland fund. (Data from: www.morningstar.com.)

**54. Investing** As of April 2017, the Brandes Core Plus Fixed Income fund yielded an annual return of 2.7%, and the Transamerica Institutional Asset Allocation Short Horizon fund yielded an annual return of 4.1%. You would like to invest up to $80,000 and earn at least $2214 in interest. Let $x$ be the amount invested in the Brandes fund, and let $y$ be the amount invested in the Transamerica fund. (Data from: www.morningstar.com.)

**55. Business** Joyce Blake is the marketing director for a new company selling a fashion collection for young women and she wishes to place ads in two magazines: *Vogue* and *Elle*. Joyce estimates that each one-page ad in *Vogue* will be read by 1.3 million people and each one-page ad in *Elle* will be read by 1.1 million people. Joyce wants to reach at least 8 million readers and to place at least 2 ads in each magazine. (Data from: www.nyjobsource.com.)

**56. Business** Corey Steinbruner is the marketing director for a new vitamin supplement and he wants to place ads in two magazines: *Sports Illustrated* and *Men's Health*. Corey estimates that each one-page ad in *Sports Illustrated* will be read by 3.2 million people and each one-page ad in *Men's Health* will be read by 1.9 million people. Corey wants to reach at least 20 million readers and to place at least 3 ads in each magazine. (Data from: www.nyjobsource.com.)

✓**Checkpoint Answers**

**1. (a)**

**(b)**

**2. (a)**

**(b)**

**3.**

**4. (a)**

**(b)**

**5.**

**6.**

**7.**

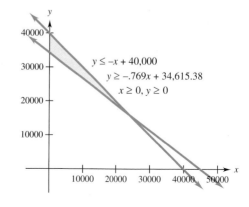

# 7.2 Linear Programming: The Graphical Method

Many problems in business, science, and economics involve finding the optimal value of a function (for instance, the maximum value of the profit function or the minimum value of the cost function), subject to various **constraints** (such as transportation costs, environmental protection laws, availability of parts, and interest rates). **Linear programming** deals with such situations. In linear programming, the function to be optimized, called the **objective function,** is linear and the constraints are given by linear inequalities. Linear programming problems that involve only two variables can be solved by the graphical method, explained in Example 1.

**EXAMPLE 1**   Find the maximum and minimum values of the objective function $z = 2x + 5y$, subject to the following constraints:

$$3x + 2y \leq 6$$
$$-2x + 4y \leq 8$$
$$x + y \geq 1$$
$$x \geq 0, y \geq 0.$$

**Solution**   First, graph the feasible region of the system of inequalities (Figure 7.16). The points in this region or on its boundaries are the only ones that satisfy all the constraints. However, each such point may produce a different value of the objective function. For instance, the points (.5, 1) and (1, 0) in the feasible region lead to the respective values

$$z = 2(.5) + 5(1) = 6 \quad \text{and} \quad z = 2(1) + 5(0) = 2.$$

We must find the points that produce the maximum and minimum values of $z$.

To find the maximum value, consider various possible values for $z$. For instance, when $z = 0$, the objective function is $0 = 2x + 5y$, whose graph is a straight line. Similarly, when $z$ is 5, 10, and 15, the objective function becomes (in turn)

$$5 = 2x + 5y, \quad 10 = 2x + 5y, \quad \text{and} \quad 15 = 2x + 5y.$$

These four lines are graphed in Figure 7.17. (All the lines are parallel because they have the same slope.) The figure shows that $z$ cannot take on the value 15, because the graph for $z = 15$ is entirely outside the feasible region. The maximum possible value of $z$ will be obtained from a line parallel to the others and between the lines representing the objective function when $z = 10$ and $z = 15$. The value of $z$ will be as large as possible, and all constraints will be satisfied, if this line just touches the feasible region. This occurs with the green line through point $A$.

**Figure 7.16**

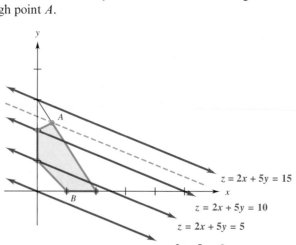

**Figure 7.17**

The point $A$ is the intersection of the graphs of $3x + 2y = 6$ and $-2x + 4y = 8$. (See Figure 7.16.) Its coordinates can be found either algebraically or using a graphing calculator.

**Algebraic Method**

Solve the system

$$3x + 2y = 6$$
$$-2x + 4y = 8,$$

as in Section 6.1, to get $x = \frac{1}{2}$ and $y = \frac{9}{4}$. Hence, $A$ has coordinates $(\frac{1}{2}, \frac{9}{4})$.

**Calculator Method**

Solve the two equations for $y$:

$$y = -1.5x + 3$$
$$y = .5x + 2.$$

Graph both equations on the same screen and use the intersection finder to find that the coordinates of the intersection point $A$ are (.5, 2.25).

The value of $z$ at point $A$ is

$$z = 2x + 5y$$

$$= 2\left(\frac{1}{2}\right) + 5\left(\frac{9}{4}\right) \qquad \text{Substitution}$$

$$= \frac{2}{2} + \frac{45}{4} \qquad \text{Multiplication}$$

$$= \frac{4}{4} + \frac{45}{4} \qquad \text{Addition of fractions}$$

$$= \frac{49}{4}.$$

Thus, the maximum possible value of $z$ is $49/4$. Similarly, the minimum value of $z$ occurs at point $B$, which has coordinates $(1, 0)$. The minimum value of $z$ is $2(1) + 5(0) = 2$. ✔1

Points such as $A$ and $B$ in Example 1 are called corner points. A **corner point** is a point in the feasible region where the boundary lines of two constraints cross. The feasible region in Figure 7.16 is **bounded** because the region is enclosed by boundary lines on all sides. Linear programming problems with bounded regions always have solutions. However, if Example 1 did not include the constraint $3x + 2y \le 6$, the feasible region would be **unbounded,** and there would be no way to *maximize* the value of the objective function.

Some general conclusions can be drawn from the method of solution used in Example 1. Figure 7.18 shows various feasible regions and the lines that result from graphing an objective function with various values of $z$. (Figure 7.18 shows the situation in which the lines are in order from left to right as $z$ increases.) In part (a) of Figure 7.18, the objective function takes on its minimum value at corner point $Q$ and its maximum value at corner point $P$. The

✔ **Checkpoint 1**

Suppose the objective function in Example 1 is changed to $z = 5x + 2y$.

**(a)** Sketch the graphs of the objective function when $z = 0, z = 5,$ and $z = 10$ on the region of feasible solutions given in Figure 7.16.

**(b)** From the graph, decide what values of $x$ and $y$ will maximize the objective function.

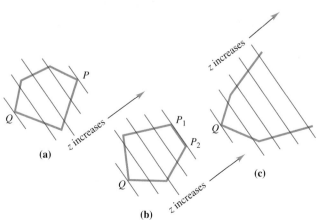

**Figure 7.18**

minimum is again at $Q$ in part (b), but the maximum occurs at $P_1$ or $P_2$, or any point on the line segment connecting them. Finally, in part (c), the minimum value occurs at $Q$, but the objective function has no maximum value because the feasible region is unbounded.

The preceding discussion suggests the **corner point theorem.**

## Corner Point Theorem

If the feasible region is bounded, then the objective function has both a maximum and a minimum value, and each occurs at one or more corner points.

If the feasible region is unbounded, the objective function may not have a maximum or minimum. But if a maximum or minimum value exists, it will occur at one or more corner points.

This theorem simplifies the job of finding an optimum value. First, graph the feasible region and find all corner points. Then test each point in the objective function. Finally, identify the corner point producing the optimum solution.

With the theorem, the problem in Example 1 could have been solved by identifying the five corner points of Figure 7.16: $(0, 1)$, $(0, 2)$, $(\frac{1}{2}, \frac{9}{4})$, $(2, 0)$, and $(1, 0)$. Then, substituting each of these points into the objective function $z = 2x + 5y$ would identify the corner points that produce the maximum and minimum values of $z$.

| Corner Point | Value of $z = 2x + 5y$ | |
|---|---|---|
| $(0, 1)$ | $2(0) + 5(1) = 5$ | |
| $(0, 2)$ | $2(0) + 5(2) = 10$ | |
| $(\frac{1}{2}, \frac{9}{4})$ | $2(\frac{1}{2}) + 5(\frac{9}{4}) = \frac{49}{4}$ | (maximum) |
| $(2, 0)$ | $2(2) + 5(0) = 4$ | |
| $(1, 0)$ | $2(1) + 5(0) = 2$ | (minimum) |

From these results, the corner point $(\frac{1}{2}, \frac{9}{4})$ yields the maximum value of $49/4$ and the corner point $(1, 0)$ gives the minimum value of 2. These are the same values found earlier. ✓2

A summary of the steps for solving a linear programming problem by the graphical method is given here.

### ✓Checkpoint 2

**(a)** Identify the corner points in the given graph.

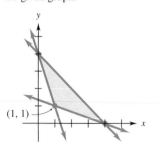

**(b)** Which corner point would minimize $z = 2x + 3y$?

## Solving a Linear Programming Problem Graphically

1. Write the objective function and all necessary constraints.
2. Graph the feasible region.
3. Determine the coordinates of each of the corner points.
4. Find the value of the objective function at each corner point.
5. If the feasible region is bounded, the solution is given by the corner point producing the optimum value of the objective function.
6. If the feasible region is an unbounded region in the first quadrant and both coefficients of the objective function are positive,* then the minimum value of the objective function occurs at a corner point and there is no maximum value.

---

*This is the only case of an unbounded region that occurs in the applications considered here.

**EXAMPLE 2** Sketch the feasible region for the following set of constraints:

$$3y - 2x \geq 0$$
$$y + 8x \leq 52$$
$$y - 2x \leq 2$$
$$x \geq 3.$$

Then find the maximum and minimum values of the objective function $z = 5x + 2y$.

**Solution** Graph the feasible region, as in Figure 7.19. To find the corner points, you must solve these four systems of equations:

| $A$ | $B$ | $C$ | $D$ |
|---|---|---|---|
| $y - 2x = 2$ | $3y - 2x = 0$ | $3y - 2x = 0$ | $y - 2x = 2$ |
| $x = 3$ | $x = 3$ | $y + 8x = 52$ | $y + 8x = 52$ |

The first two systems are easily solved by substitution, which shows that $A = (3, 8)$ and $B = (3, 2)$. The other two systems can be solved either with a calculator (as in Figure 7.20) or algebraically (see Checkpoint 3). Hence, $C = (6, 4)$ and $D = (5, 12)$. ✓₃

**✓ Checkpoint 3**

Use the elimination method (see Section 6.1) to solve the last system and find the coordinates of $D$ in Example 2.

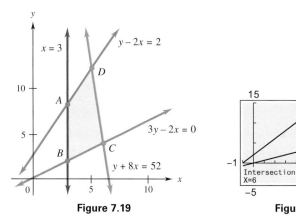

**Figure 7.19**          **Figure 7.20**

Use the corner points from the graph to find the maximum and minimum values of the objective function.

| Corner Point | Value of $z = 5x + 2y$ | |
|---|---|---|
| $(3, 8)$ | $5(3) + 2(8) = 31$ | |
| $(3, 2)$ | $5(3) + 2(2) = 19$ | (minimum) |
| $(6, 4)$ | $5(6) + 2(4) = 38$ | |
| $(5, 12)$ | $5(5) + 2(12) = 49$ | (maximum) |

The minimum value of $z = 5x + 2y$ is 19, at the corner point $(3, 2)$. The maximum value is 49, at $(5, 12)$. ✓₄

**✓ Checkpoint 4**

Use the region of feasible solutions in the accompanying sketch to find the given values.

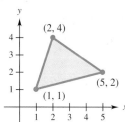

**(a)** The values of $x$ and $y$ that maximize $z = 2x - y$

**(b)** The maximum value of $z = 2x - y$

**(c)** The values of $x$ and $y$ that minimize $z = 4x + 3y$

**(d)** The minimum value of $z = 4x + 3y$

**EXAMPLE 3** Solve the following linear programming problem:

Minimize $\quad z = x + 3y$

subject to $\quad 2x + y \leq 10$
$$5x + 2y \geq 20$$
$$-x + 2y \geq 0$$
$$x \geq 0, y \geq 0.$$

**Solution** First, we graph the feasible region, as in Figure 7.21. To find the corner points, you must solve these three systems of equations:

|                  *A*          |        *B*        |       *C*        |
|:-----------------------------:|:-----------------:|:----------------:|
| $2x + y = 10$                 | $2x + y = 10$     | $5x + 2y = 20$   |
| $5x + 2y = 20$                | $-x + 2y = 0$     | $-x + 2y = 0$    |

Solving the two equations to find point *A* by substitution yields $A = (0, 10)$. Similarly, solving the two equations to find point *B* by substitution yields $B = (4, 2)$. Finding the last corner point *C* can also be found by substitution, and we obtain $C = \left(\frac{10}{3}, \frac{5}{3}\right)$.

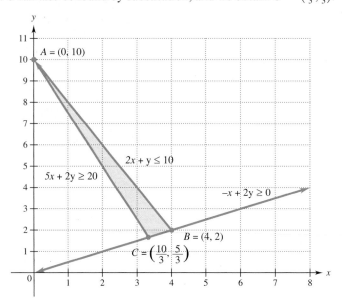

**Figure 7.21**

These corner points give the following values of *z*.

| Corner Point | Value of $z = x + 3y$ |
|:------------:|:---------------------:|
| $(0, 10)$ | $0 + 3(10) = 30$ |
| $(4, 2)$ | $4 + 3(2) = 10$ |
| $\left(\frac{10}{3}, \frac{5}{3}\right)$ | $\left(\frac{10}{3}\right) + 3\left(\frac{5}{3}\right) = \frac{25}{3} \approx 8.33$ (minimum) |

The minimum of *z* is $\frac{25}{3}$; it occurs at $\left(\frac{10}{3}, \frac{5}{3}\right)$. ✔5

✔ **Checkpoint 5**

The given sketch shows a feasible region. Let $z = x + 3y$. Use the sketch to find the values of *x* and *y* that

**(a)** minimize *z*;

**(b)** maximize *z*.

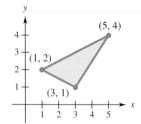

**EXAMPLE 4**    Solve the following linear programming problem:

$$\text{Minimize} \qquad z = 2x + 4y$$
$$\text{subject to} \qquad x + 2y \geq 10$$
$$3x + \ y \geq 10$$
$$x \geq 0, y \geq 0.$$

**Solution** Figure 7.22 shows the hand-drawn graph with corner points $(0, 10)$, $(2, 4)$, and $(10, 0)$, as well as the calculator graph with the corner point $(2, 4)$. Find the value of *z* for each point.

| Corner Point | Value of $z = 2x + 4y$ |     |
|:------------:|:----------------------:|:---:|
| $(0, 10)$ | $2(0) + 4(10) = 40$ | |
| $(2, 4)$ | $2(2) + 4(4) = 20$ | (minimum) |
| $(10, 0)$ | $2(10) + 4(0) = 20$ | (minimum) |

## ✓ Checkpoint 6

The sketch shows a region of feasible solutions. From the sketch, decide what ordered pair would minimize $z = 2x + 4y$.

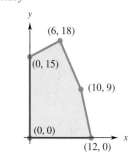

**Figure 7.22**

In this case, both (2, 4) and (10, 0), as well as all the points on the boundary line between them, give the same optimum value of $z$. So there is an infinite number of equally "good" values of $x$ and $y$ that give the same minimum value of the objective function $z = 2x + 4y$. The minimum value is 20. ✓ 6

# 7.2 Exercises

*Exercises 1–6 show regions of feasible solutions. Use these regions to find maximum and minimum values of each given objective function. (See Examples 1 and 2.)*

**1.** $z = 6x + y$

**2.** $z = 4x + y$

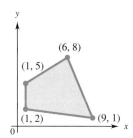

**3.** $z = .3x + .5y$

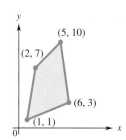

**4.** $z = .35x + 1.25y$

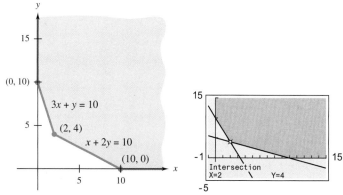

**5.** (a) $z = x + 5y$
   (b) $z = 2x + 3y$
   (c) $z = 2x + 4y$
   (d) $z = 4x + y$

**6.** (a) $z = 5x + 2y$
   (b) $z = 5x + 6y$
   (c) $z = x + 2y$
   (d) $z = x + y$

*Use graphical methods to solve Exercises 7–12. (See Examples 2–4.)*

**7.** Maximize $z = 4x + 3y$
   subject to $2x + 3y \le 6$
   $4x + y \le 6$
   $x \ge 0, y \ge 0.$

**8.** Minimize   $z = x + 2y$

subject to   $x + y \leq 10$

$3x + 2y \geq 6$

$x \geq 0, y \geq 0.$

**9.** Minimize   $z = 2x + y$

subject to   $3x - y \geq 12$

$x + y \leq 15$

$x \geq 2, y \geq 3.$

**10.** Maximize   $z = x + 3y$

subject to   $2x + 3y \leq 100$

$5x + 4y \leq 200$

$x \geq 10, y \geq 20.$

**11.** Maximize   $z = 5x + y$

subject to   $x - y \leq 10$

$5x + 3y \leq 75$

$x \geq 0, y \geq 0.$

**12.** Maximize   $z = 4x + 5y$

subject to   $10x - 5y \leq 100$

$20x + 10y \geq 150$

$x \geq 0, y \geq 0.$

*Find the minimum and maximum values of $z = 3x + 4y$ (if possible) for each of the given sets of constraints. (See Examples 2–4.)*

**13.**  $3x + 2y \geq 6$           **14.**   $2x + y \leq 20$

$x + 2y \geq 4$                $10x + y \geq 36$

$x \geq 0, y \geq 0$               $2x + 5y \geq 36$

**15.**   $x + y \leq 6$            **16.**  $-x + 2y \leq 6$

$-x + y \leq 2$               $3x + y \geq 3$

$2x - y \leq 8$               $x \geq 0, y \geq 0$

**17.** Find values of $x \geq 0$ and $y \geq 0$ that maximize $z = 10x + 12y$, subject to each of the following sets of constraints.

**(a)**  $x + y \leq 20$          **(b)**  $3x + y \leq 15$

$x + 3y \leq 24$                $x + 2y \leq 18$

**(c)**   $x + 2y \geq 10$

$2x + y \geq 12$

$x - y \leq 8$

**18.** Find values of $x \geq 0$ and $y \geq 0$ that minimize $z = 3x + 2y$, subject to each of the following sets of constraints.

**(a)**  $10x + 7y \leq 42$         **(b)**  $6x + 5y \geq 25$

$4x + 10y \geq 35$              $2x + 6y \geq 15$

**(c)**  $2x + 5y \geq 22$

$4x + 3y \leq 28$

$2x + 2y \leq 17$

**19.** Explain why it is impossible to maximize the function $z = 3x + 4y$ subject to the constraints

$x + y \geq 8$

$2x + y \leq 10$

$x + 2y \leq 8$

$x \geq 0, y \geq 0.$

**20.** Explain why it is impossible to maximize the function $z = 4x + 7y$ subject to the constraints

$8y + 5x \geq 40$

$4y + 9x \leq 36$

$11y + 2x \leq 22$

$x \geq 0, y \geq 0.$

✓**Checkpoint Answers**

**1. (a)**

**(b)**  $(2, 0)$

**2. (a)**  $(0, 4), (1, 1), (4, 0)$        **(b)**  $(1, 1)$

**3.**  $D = (5, 12)$

**4. (a)**  $(5, 2)$      **(b)**  8      **(c)**  $(1, 1)$      **(d)**  7

**5. (a)**  $(3, 1)$      **(b)**  $(5, 4)$

**6.**  $(1, 1)$

# 7.3   Applications of Linear Programming

In this section, we show several applications of linear programming with two variables.

**EXAMPLE 1**    **Office Purchases** An office manager needs to purchase new filing cabinets. He knows that Ace cabinets cost $40 each, require 6 square feet of floor space, and hold 8 cubic feet of files. On the other hand, each Excello cabinet costs $80, requires 8 square feet of floor space, and holds 12 cubic feet. His budget permits him to spend no more than $560, while the office has room for no more than 72 square feet of cabinets. The manager desires the greatest storage capacity within the limitations imposed by funds and space. How many of each type of cabinet should he buy?

**Solution** Let $x$ represent the number of Ace cabinets to be bought, and let $y$ represent the number of Excello cabinets. The information given in the problem can be summarized as follows.

|  | Number | Cost of Each | Space Required | Storage Capacity |
|---|---|---|---|---|
| **Ace** | $x$ | $40 | 6 sq ft | 8 cu ft |
| **Excello** | $y$ | $80 | 8 sq ft | 12 cu ft |
| **Maximum Available** |  | $560 | 72 sq ft |  |

The constraints imposed by cost and space are

$$40x + 80y \le 560 \quad \text{Cost}$$
$$6x + 8y \le 72. \quad \text{Floor space}$$

The number of cabinets cannot be negative, so $x \ge 0$ and $y \ge 0$. The objective function to be maximized gives the amount of storage capacity provided by some combination of Ace and Excello cabinets. From the information in the chart, the objective function is

$$z = \text{Storage capacity} = 8x + 12y.$$

In sum, the given problem has produced the following linear programming problem:

$$\text{Maximize} \quad z = 8x + 12y$$
$$\text{subject to} \quad 40x + 80y \le 560$$
$$6x + 8y \le 72$$
$$x \ge 0, y \ge 0.$$

## ✓ Checkpoint 1

Find the corner point labeled $P$ on the region of feasible solutions in the given graph.

A graph of the feasible region is shown in Figure 7.23. Three of the corner points can be identified from the graph as $(0, 0)$, $(0, 7)$, and $(12, 0)$. The fourth corner point, labeled $Q$ in the figure, can be found algebraically or with a graphing calculator to be $(8, 3)$. ✓₁

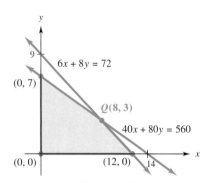

**Figure 7.23**

Use the corner point theorem to find the maximum value of $z$.

| Corner Point | Value of $z = 8x + 12y$ |  |
|---|---|---|
| $(0, 0)$ | 0 |  |
| $(0, 7)$ | 84 |  |
| $(8, 3)$ | 100 | (maximum) |
| $(12, 0)$ | 96 |  |

## ✓ Checkpoint 2

A popular cereal combines oats and corn. At least 27 tons of the cereal are to be made. For the best flavor, the amount of corn should be no more than twice the amount of oats. Oats cost $300 per ton, and corn costs $200 per ton. How much of each grain should be used to minimize the cost?

**(a)** Make a chart to organize the information given in the problem.

**(b)** Write an equation for the objective function.

**(c)** Write four inequalities for the constraints.

The objective function, which represents storage space, is maximized when $x = 8$ and $y = 3$. The manager should buy 8 Ace cabinets and 3 Excello cabinets. ✓₂

EXAMPLE 2    **Nutrition**  Certain laboratory animals must have at least 30 grams of protein and at least 20 grams of fat per feeding period. These nutrients come from food $A$, which costs 18¢ per unit and supplies 2 grams of protein and 4 of fat, and food $B$, with 6 grams of protein and 2 of fat, costing 12¢ per unit. Food $B$ is bought under a long-term contract requiring that at least 2 units of $B$ be used per serving. How much of each food must be bought to produce the minimum cost per serving?

**Solution**  Let $x$ represent the amount of food $A$ needed and $y$ the amount of food $B$. Use the given information to produce the following table.

| Food | Number of Units | Grams of Protein | Grams of Fat | Cost |
|------|-----------------|------------------|--------------|------|
| $A$ | $x$ | 2 | 4 | 18¢ |
| $B$ | $y$ | 6 | 2 | 12¢ |
| Minimum Required | | 30 | 20 | |

Use the table to develop the linear programming problem. Since the animals must have *at least* 30 grams of protein and 20 grams of fat, use $\geq$ in the constraint inequalities for protein and fat. The long-term contract provides a constraint not shown in the table, namely, $y \geq 2$. So we have the following problem:

$$\text{Minimize} \quad z = .18x + .12y \qquad \text{Cost}$$
$$\text{subject to} \quad 2x + 6y \geq 30 \qquad \text{Protein}$$
$$4x + 2y \geq 20 \qquad \text{Fat}$$
$$y \geq 2 \qquad \text{Contract}$$
$$x \geq 0, y \geq 0.$$

(The constraint $y \geq 0$ is redundant because of the constraint $y \geq 2$.) A graph of the feasible region with the corner points identified is shown in Figure 7.24.

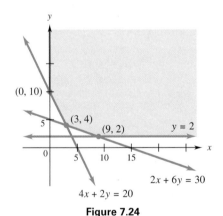

**Figure 7.24**

Use the corner point theorem to find the minimum value of $z$ as shown in the table.

✓**Checkpoint 3**

Use the information in Checkpoint 2 to do the following.

**(a)** Graph the feasible region and find the corner points.

**(b)** Determine the minimum value of the objective function and the point where it occurs.

**(c)** Is there a maximum cost?

| Corner Points | $z = .18x + .12y$ |
|---------------|-------------------|
| (0, 10) | $.18(0) + .12(10) = 1.20$ |
| (3, 4) | $.18(3) + .12(4) = 1.02$   (minimum) |
| (9, 2) | $.18(9) + .12(2) = 1.86$ |

The minimum value of 1.02 occurs at (3, 4). Thus, 3 units of food $A$ and 4 units of food $B$ will produce a minimum cost of $1.02 per serving. ✓₃

The feasible region in Figure 7.24 is an unbounded one: The region extends indefinitely to the upper right. With this region, it would not be possible to *maximize* the objective function, because the total cost of the food could always be increased by encouraging the animals to eat more.

One measure of the risk involved in investing in a stock or mutual fund is called the standard deviation. The standard deviation measures the volatility of investment returns relative to an historical average. If the return of an investment tool fluctuates a great deal from the historical average return, then there will be a higher standard deviation value for that stock. If an investment tool's value stays near the historical average, then it will have a small standard deviation value. Thus, a higher standard deviation for an investment tool can be one measure of higher risk. Investors often wish to obtain the highest profit while minimizing risk.

**EXAMPLE 3** **Investing** Mariana wants to invest up to $5000 in stocks. The share price for CVS Health Corp. is $81 and the share price for Target Corp. is $57. Based on the average of 10-year returns, CVS would produce in a year a profit of $11 per share, and Target would produce a profit of $4 per share. Mariana would like to obtain at least $520 in profit. A measure of risk for each of these stocks is the beta value. The lower the beta value, the less risk associated with that stock. If the beta value for CVS is .62 and the beta value for Target is .24, how many shares of each stock should Mariana purchase to minimize the risk as measured by the beta value? (Data from: www.morningstar.com.)

**Solution** Let $x$ represent the number of shares of CVS stock to be purchased, and let $y$ be the number of shares of Target stock to be purchased. The information in the problem can be summarized in the following table.

|  | Number of Shares | Cost of Each | Profit | Risk |
|---|---|---|---|---|
| CVS | $x$ | $81 | $11 | .62 |
| Target | $y$ | $57 | $4 | .24 |
| Constraints |  | $5000 | $520 |  |

The constraints imposed by the cost of the shares and the profits are

$$81x + 57y \leq 5000$$
$$11x + 4y \geq 520.$$

The number of stocks to be purchased cannot be negative, so $x \geq 0$ and $y \geq 0$. The objective function to be minimized gives the amount of risk provided by some combination of shares in CVS and Target stocks. From the information in the chart, the objective function is

$$z = \text{beta} = .62x + .24y.$$

In sum, the given problem has produced the following linear programming problem:

$$\text{Minimize} \quad z = .62x + .24y$$
$$\text{subject to} \quad 81x + 57y \leq 5000$$
$$11x + 4y \geq 520$$
$$x \geq 0, y \geq 0.$$

A graph of the feasible region with the corner points identified is shown in Figure 7.25a. Using the calculator method from the previous section to find the intersection point of the two lines shows the TI-84 screenshot in Figure 7.25b.

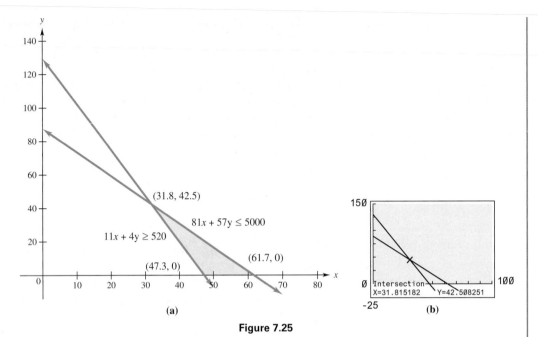

**Figure 7.25**

Use the corner point theorem to find the minimum $z$ as shown in the table.

| Corner Points | $z = .62x + .24y$ |
|---|---|
| (47.3, 0) | 29.326 |
| (61.7, 0) | 38.254 |
| (31.8, 42.5) | 29.916 |

The minimum value of 29.326 occurs at (47.3, 0). Mariana must buy a whole share of stock, so she would buy 47 shares of the CVS stock.

# 7.3 Exercises

*Write the constraints in Exercises 1–4 as linear inequalities and iden- tify all variables used. In some instances, not all of the information is needed to write the constraints. (See Examples 1–3.)*

1. A canoe requires 8 hours of fabrication and a rowboat 5 hours. The fabrication department has at most 110 hours of labor available each week.

2. Jayson Gilbert needs at least 2800 milligrams of vitamin C per day. Each Supervite pill provides 250 milligrams, and each Vita- health pill provides 350 milligrams.

3. A candidate can afford to spend no more than $9500 on radio and TV advertising. Each radio spot costs $250, and each TV ad costs $750.

4. A hospital dietician has two meal choices: one for patients on solid food that costs $2.75 and one for patients on liquids that costs $3.75. There is a maximum of 600 patients in the hospital.

*Solve these linear programming problems, which are somewhat simpler than the examples in the text.*

5. **Business** A chain saw requires 4 hours of assembly and a wood chipper 6 hours. A maximum of 48 hours of assembly time is available. The profit is $150 on a chain saw and $220

on a chipper. How many of each should be assembled for maximum profit?

6. **Nutrition** Samar likes to snack frequently during the day, but he wants his snacks to provide at least 24 grams of protein per day. Each Snack-Pack provides 4 grams of protein, and each Minibite provides 1 gram. Snack-Packs cost 50 cents each and Minibites 12 cents. How many of each snack should he use to minimize his daily cost?

7. **Coffee Production** Deluxe coffee is to be mixed with regular coffee to make at least 50 pounds of a blended coffee. The mix- ture must contain at least 10 pounds of deluxe coffee. Deluxe coffee costs $6 per pound and regular coffee $5 per pound. How many pounds of each kind of coffee should be used to minimize costs?

8. **Outdoor Recreation** The company in Exercise 1 cannot sell more than 10 canoes each week and always sells at least 6 row- boats. The profit on a canoe is $400, and the profit on a rowboat is $225. Assuming the same situation as in Exercise 1, how many of each should be made per week to maximize profits?

9. **Business** Alejandro is an auto mechanic. He spends 3 hours when he replaces the shocks on a car and 2 hours when he

replaces the brakes. He works no more than 48 hours a week. He routinely completes at least 2 shocks replacements and 6 brake replacements a week. If he charges $500 for labor replacing shocks and $300 in labor for replacing brakes, how many jobs of each type should he complete a week to maximize his income?

10. **Vitamin Supplements** Jayson Gilbert of Exercise 2 pays 3 cents for each Supervite pill and 4 cents for each Vitahealth pill. Because of its other ingredients, he cannot take more than 7 Supervite pills per day. Assuming the same conditions as in Example 2, how many of each pill should he take to provide the desired level of vitamin C at minimum cost?

11. **Advertising** The candidate in Exercise 3 wants to have at least 8 radio spots and at least 3 TV ads. A radio spot reaches 600 people, and a TV ad reaches 2000 people. Assuming the monetary facts given in Exercise 3, how many of each kind should be used to reach the largest number of people?

12. **Nutrition** The hospital in Exercise 4 always has at least 100 patients on solid foods and at least 100 on liquids. Assuming the facts in Exercise 4, what number of each type of patient would minimize food costs?

*Solve the following linear programming problems. (See Examples 1–3.)*

13. **Nutrition** Mike May has been told that each day he needs at least 16 units of vitamin A, at least 5 units of vitamin B-1, and at least 20 units of vitamin C. Each Brand X pill contain 8 units of vitamin A, 1 of vitamin B-1, and 2 of vitamin C, while each Brand Z pill contains 2 units of vitamin A, 1 of vitamin B-1, and 7 of vitamin C. A Brand X pill costs 15 cents, and a Brand Z pill costs 30 cents. How many pills of each brand should he buy to minimize his daily cost? What is the minimum cost?

14. **Oil and Gas** The manufacturing process requires that oil refineries manufacture at least 2 gallons of gasoline for every gallon of fuel oil. To meet the winter demand for fuel oil, at least 3 million gallons a day must be produced. The demand for gasoline is no more than 12 million gallons per day. It takes .25 hour to ship each million gallons of gasoline and 1 hour to ship each million gallons of fuel oil out of the warehouse. No more than 6.6 hours are available for shipping. If the refinery sells gasoline for $1.25 per gallon and fuel oil for $1 per gallon, how much of each should be produced to maximize revenue? Find the maximum revenue.

15. **Bolt Manufacturing** A machine shop manufactures two types of bolts. The bolts require time on each of three groups of machines, but the time required on each group differs, as shown in the table:

| | | MACHINE GROUP | | |
|---|---|---|---|---|
| | | I | II | III |
| **Bolts** | Type 1 | .1 min | .1 min | .1 min |
| | Type 2 | .1 min | .4 min | .02 min |

Production schedules are made up one day at a time. In a day, there are 240, 720, and 160 minutes available, respectively, on these machines. Type 1 bolts sell for 10¢ and type 2 bolts for 12¢. How many of each type of bolt should be manufactured per day to maximize revenue? What is the maximum revenue?

16. **Animal Nutrition** A zookeeper has a walrus with a nutritional deficiency. She wants to make sure the walrus consumes at least 2400 mg of iron, 2100 mg of vitamin B-1, and 1500 mg of vitamin B-2. One Maxivite pill contains 400 mg of iron, 100 mg of vitamin B-1, and 50 mg of vitamin B-2, and costs $.06. One Healthovite pill provides 100 mg of iron, 150 mg of vitamin B-1, and 150 mg of vitamin B-2, and costs $.08.

   (a) What combination of Maxivite and Healthovite pills will meet the walrus's nutrition requirements at lowest cost? What is the lowest cost?

   (b) In your solution to part (a), does the walrus receive more than the minimum amount needed of any vitamin? If so, which vitamin is it?

   (c) Is there a way for the zookeeper to avoid having the walrus receive more than the minimum needed, still meet the other constraints, and minimize the cost? Explain.

17. **Greeting Cards** A greeting card manufacturer has 500 boxes of a particular card in warehouse I and 290 boxes of the same card in warehouse II. A greeting card shop in San Jose orders 350 boxes of the card, and another shop in Memphis orders 250 boxes. The shipping costs per box to these shops from the two warehouses are shown in the following table:

| | | DESTINATION | |
|---|---|---|---|
| | | San Jose | Memphis |
| **Warehouse** | I | $.25 | $.22 |
| | II | $.23 | $.21 |

How many boxes should be shipped to each city from each warehouse to minimize shipping costs? What is the minimum cost? (*Hint:* Use $x$, $350 - x$, $y$, and $250 - y$ as the variables.)

18. **Magazine Distribution** *Hotnews Magazine* publishes a U.S. and a Canadian edition each week. There are 30,000 subscribers in the United States and 20,000 in Canada. Other copies are sold at newsstands. Postage and shipping costs average $80 per thousand copies for the United States and $60 per thousand copies for Canada. Surveys show that no more than 120,000 copies of each issue can be sold (including subscriptions) and that the number of copies of the Canadian edition should not exceed twice the number of copies of the U.S. edition. The publisher can spend at most $8400 a month on postage and shipping. If the profit is $200 for each thousand copies of the U.S. edition and $150 for each thousand copies of the Canadian edition, how many copies of each version should be printed to earn as large a profit as possible? What is that profit?

19. **Investing** A pension fund manager decides to invest at most $50 million in U.S. Treasury Bonds paying 4% annual interest and in mutual funds paying 6% annual interest. He plans to invest at least $20 million in bonds and at least $6 million in mutual funds. Bonds have an initial fee of $300 per million dollars, while the fee for mutual funds is $100 per million. The fund manager is allowed to spend no more than $8400 on fees. How much should be invested in each to maximize annual interest? What is the maximum annual interest?

20. **Ecology** A certain predator requires at least 10 units of protein and 8 units of fat per day. One prey of Species I provides

5 units of protein and 2 units of fat; one prey of Species II provides 3 units of protein and 4 units of fat. Capturing and digesting each Species II prey requires 3 units of energy, and capturing and digesting each Species I prey requires 2 units of energy. How many of each prey would meet the predator's daily food requirements with the least expenditure of energy?

*For Exercises 21–26, all prices are as of May 15, 2017.*

21. **Investing** General Electric Co. stock sells for $29 a share and has a 3-year average annual return of $2 a share. The beta value is 1.10. Procter & Gamble Co. sells for $86 a share and has a 3-year average annual return of $4 a share. The beta value is .53. Tori wants to spend no more than $12,000 investing in these two stocks, but she wants to earn at least $700 in annual revenue. Tori also wants to minimize the risk. Determine the number of shares of each stock that Tori should buy. (Data from: www.morningstar.com.)

22. **Investing** Deere and Co. stock sells for $113 a share and has a 3-year average annual return of $10 a share. The beta value is .68. Boeing Co. sells for $185 a share and has a 3-year average annual return of $26 a share. The beta value is 1.19. Jermaine wants to spend no more than $10,000 investing in these two stocks, but he wants to earn at least $1050 in annual revenue. Jermaine also wants to minimize the risk. Determine the number of shares of each stock that Jermaine should buy. (Data from: www.morningstar.com.)

23. **Investing** Visa Inc. stock sells for $92 a share and has a 3-year average annual return of $20 a share. The beta value is 1.06. JP Morgan Chase and Co. sells for $87 a share and has a 3-year average annual return of $17 a share. The beta value is 1.21. Derek wants to spend no more than $15,000 investing in these two stocks, but he wants to earn at least $2500 in annual revenue. Derek also wants to minimize the risk. Determine the number of shares of each stock that Derek should buy. (Data from: www.morningstar.com.)

24. **Investing** Microsoft stock sells for $69 a share and has a 3-year average annual return of $16 a share. The beta value is 1.38. McDonald's Corp. sells for $144 a share and has a 3-year average annual return of $16 a share. The beta value is .77. Roselyn wants to spend no more than $12,000 investing in these two stocks, but she wants to earn at least $2250 in annual revenue. Roselyn also wants to minimize the risk. Determine the number of shares of each stock that Roselyn should buy. (Data from: www.morningstar.com.)

25. **Investing** The Vanguard Total Bond Market Index Inv (VBMFX) fund sells for $11 a share and has a 3-year average

annual return of $.25 a share. The beta value is 1.07. The Fidelity Freedom® 2020 (FFFDX) fund sells for $16 a share and has a 3-year average annual return of $.92. The beta value is .77. Jonquin wants to spend no more than $3000 investing in these two funds, but he wants to obtain at least $100 in annual revenue. Jonquin also wants to minimize his risk. Determine the number of shares of each stock Jonquin should buy. (Data from: www.morningstar.com.)

26. **Investing** The American Funds Growth Fund of America A (AGTHX) fund sells for $47 a share and has a 3-year average annual return of $5.30 a share. The beta value is .94. The Fidelity® Puritan® Fund (FPURX) sells for $22 a share and has a 3-year average annual return of $1.70. The beta value is .66. Brittany wants to spend no more than $7500 investing in these two funds, but she wants to obtain at least $700 in annual revenue. Brittany also wants to minimize her risk. Determine the number of shares of each stock Brittany should buy. (Data from: www.morningstar.com.)

✓ **Checkpoint Answers**

1. $\left(\frac{8}{3}, \frac{4}{3}\right)$

2. **(a)**

| | Number of Tons | Cost/ Ton |
|---|---|---|
| Oats | $x$ | $300 |
| Corn | $y$ | $200 |
| | 27 | |

   **(b)** $z = 300x + 200y$

   **(c)** $x + y \geq 27$
   $y \leq 2x$
   $x \geq 0$
   $y \geq 0$

3. **(a)**

   Corner points: (27, 0), (9, 18)

   **(b)** $6300 at (9, 18)

   **(c)** No

# 7.4 The Simplex Method: Maximization

For linear programming problems with more than two variables or with two variables and many constraints, the graphical method is usually inefficient or impossible, so the **simplex method** is used. This method, which is introduced in this section, was developed for the U.S. Air Force by George B. Danzig in 1947. It is now used in industrial planning, factory design, product distribution networks, sports scheduling, truck routing, resource allocation, and a variety of other ways.

Because the simplex method is used for problems with many variables, it usually is not convenient to use letters such as $x$, $y$, $z$, or $w$ as variable names. Instead, the symbols $x_1$ (read "$x$-sub-one"), $x_2$, $x_3$, and so on, up to $x_n$, are used. In the simplex method, all constraints must be expressed in the linear form

$$a_1 x_1 + a_2 x_2 + a_3 x_3 + \cdots + a_n x_n \leq b,$$

where $x_1, x_2, x_3, \ldots, x_n$ are variables, $a_1, a_2, a_3, \ldots, a_n$ are coefficients, and $b$ is a constant.

We first discuss the simplex method for linear programming problems such as the following:

$$\begin{aligned}
\text{Maximize} \quad & z = 2x_1 + 3x_2 + x_3 \\
\text{subject to} \quad & x_1 + x_2 + 4x_3 \leq 100 \\
& x_1 + 2x_2 + x_3 \leq 150 \\
& 3x_1 + 2x_2 + x_3 \leq 320, \\
\text{with} \quad & x_1 \geq 0, x_2 \geq 0, x_3 \geq 0.
\end{aligned}$$

This example illustrates *standard maximum form*, which is defined as follows.

## Standard Maximum Form

A linear programming problem is in **standard maximum form** if

1. the objective function is to be maximized;
2. all variables are nonnegative ($x_i \geq 0$, $i = 1, 2, 3, \ldots$);
3. all constraints involve $\leq$;
4. the constants on the right side in the constraints are all nonnegative ($b \geq 0$).

Problems that do not meet all of these conditions are considered in Sections 7.6 and 7.7.

The "mechanics" of the simplex method are demonstrated in Examples 1–5. Although the procedures to be followed will be made clear, as will the fact that they result in an optimal solution, the reasons these procedures are used may not be immediately apparent. Examples 6 and 7 will supply these reasons and explain the connection between the simplex method and the graphical method used in Section 7.3.

## Setting up the Problem

The first step is to convert each constraint, a linear inequality, into a linear equation. This is done by adding a nonnegative variable, called a **slack variable,** to each constraint. For example, convert the inequality $x_1 + x_2 \leq 10$ into an equation by adding the slack variable $s_1$, to get

$$x_1 + x_2 + s_1 = 10, \qquad \text{where } s_1 \geq 0.$$

The inequality $x_1 + x_2 \leq 10$ says that the sum $x_1 + x_2$ is less than or equal to 10. The variable $s_1$ "takes up any slack" and represents the amount by which $x_1 + x_2$ fails to equal 10. For example, if $x_1 + x_2$ equals 8, then $s_1$ is 2. If $x_1 + x_2 = 10$, the value of $s_1$ is 0.

⚠ **CAUTION** A different slack variable must be used for each constraint.

**EXAMPLE 1** Restate the following linear programming problem by introducing slack variables:

$$\text{Maximize} \quad z = 2x_1 + 3x_2 + x_3$$
$$\text{subject to} \quad x_1 + x_2 + 4x_3 \leq 100$$
$$x_1 + 2x_2 + x_3 \leq 150$$
$$3x_1 + 2x_2 + x_3 \leq 320,$$
$$\text{with} \quad x_1 \geq 0, x_2 \geq 0, x_3 \geq 0.$$

**Solution** Rewrite the three constraints as equations by introducing nonnegative slack variables $s_1$, $s_2$, and $s_3$, one for each constraint. Then the problem can be restated as

$$\text{Maximize} \quad z = 2x_1 + 3x_2 + x_3$$
$$\text{subject to} \quad x_1 + x_2 + 4x_3 + s_1 \qquad\qquad = 100 \qquad \text{Constraint 1}$$
$$x_1 + 2x_2 + x_3 \qquad + s_2 \qquad = 150 \qquad \text{Constraint 2}$$
$$3x_1 + 2x_2 + x_3 \qquad\qquad + s_3 = 320, \qquad \text{Constraint 3}$$

with $x_1 \geq 0, x_2 \geq 0, x_3 \geq 0, s_1 \geq 0, s_2 \geq 0, s_3 \geq 0.$ ✓₁

Adding slack variables to the constraints converts a linear programming problem into a system of linear equations. These equations should have all variables on the left of the equals sign and all constants on the right. All the equations of Example 1 satisfy this condition except for the objective function, $z = 2x_1 + 3x_2 + x_3$, which may be written with all variables on the left as

$$-2x_1 - 3x_2 - x_3 + z = 0. \qquad \text{Objective Function}$$

Now the equations of Example 1 (with the constraints listed first and the objective function last) can be written as the following augmented matrix.

$$
\begin{array}{ccccccc}
x_1 & x_2 & x_3 & s_1 & s_2 & s_3 & z \\
\end{array}
$$

$$
\left[
\begin{array}{ccccccc|c}
1 & 1 & 4 & 1 & 0 & 0 & 0 & 100 \\
1 & 2 & 1 & 0 & 1 & 0 & 0 & 150 \\
3 & 2 & 1 & 0 & 0 & 1 & 0 & 320 \\
\hline
-2 & -3 & -1 & 0 & 0 & 0 & 1 & 0
\end{array}
\right]
$$

Constraint 1
Constraint 2
Constraint 3
Objective Function

Indicators

This matrix is the initial **simplex tableau.** Except for the last entries—the 1 and 0 on the right end—the numbers in the bottom row of a simplex tableau are called **indicators.** ✓₂

This simplex tableau represents a system of four linear equations in seven variables. Since there are more variables than equations, the system is dependent and has infinitely many solutions. Our goal is to find a solution in which all the variables are nonnegative and $z$ is as large as possible. This will be done by using row operations to replace the given system by an equivalent one in which certain variables are eliminated from some of the equations. The process will be repeated until the optimum solution can be read from the matrix, as explained next.

## Selecting the Pivot

Recall how row operations are used to eliminate variables in the Gauss–Jordan method: A particular nonzero entry in the matrix is chosen and changed to a 1; then all other entries in that column are changed to zeros. A similar process is used in the simplex method. The chosen entry is called the **pivot.** If we were interested only in solving the system, we could choose the various pivots in many different ways, as in Chapter 6. Here, however, it is not enough just to find a solution. We must find one that is nonnegative,

✓ **Checkpoint 1**

Rewrite the following set of constraints as equations by adding nonnegative slack variables:

$$x_1 + x_2 + x_3 \leq 12$$
$$2x_1 + 4x_2 \qquad \leq 15$$
$$x_2 + 3x_3 \leq 10.$$

✓ **Checkpoint 2**

Set up the initial simplex tableau for the following linear programming problem:

$$\text{Maximize} \quad z = 2x_1 + 3x_2$$
$$\text{subject to} \quad x_1 + 2x_2 \leq 85$$
$$2x_1 + x_2 \leq 92$$
$$x_1 + 4x_2 \leq 104,$$
with $x_1 \geq 0$ and $x_2 \geq 0.$
Locate and label the indicators.

satisfies all the constraints, *and* makes $z$ as a large as possible. Consequently, the pivot must be chosen carefully, as explained in the next example. The reasons this procedure is used and why it works are discussed in Example 7.

**EXAMPLE 2**    Determine the pivot in the simplex tableau for the problem in Example 1.

**Solution**  Look at the indicators (the last row of the tableau) and choose the most negative one:

$$
\begin{array}{ccccccc}
x_1 & x_2 & x_3 & s_1 & s_2 & s_3 & z \\
\end{array}
$$

$$
\left[
\begin{array}{ccccccc|c}
1 & 1 & 4 & 1 & 0 & 0 & 0 & 100 \\
1 & 2 & 1 & 0 & 1 & 0 & 0 & 150 \\
3 & 2 & 1 & 0 & 0 & 1 & 0 & 320 \\
\hline
-2 & \mathbf{-3} & -1 & 0 & 0 & 0 & 1 & 0 \\
\end{array}
\right].
$$

Most negative indicator

The most negative indicator identifies the variable that is to be eliminated from all but one of the equations (rows)—in this case, $x_2$. The column containing the most negative indicator is called the **pivot column.** Now, for each *positive* entry in the pivot column, divide the number in the far right column of the same row by the positive number in the pivot column:

$$
\begin{array}{ccccccc}
x_1 & x_2 & x_3 & s_1 & s_2 & s_3 & z \\
\end{array}
$$

$$
\left[
\begin{array}{ccccccc|c}
1 & \mathbf{1} & 4 & 1 & 0 & 0 & 0 & \mathbf{100} \\
1 & \mathbf{2} & 1 & 0 & 1 & 0 & 0 & \mathbf{150} \\
3 & \mathbf{2} & 1 & 0 & 0 & 1 & 0 & \mathbf{320} \\
\hline
-2 & -3 & -1 & 0 & 0 & 0 & 1 & 0 \\
\end{array}
\right].
$$

Quotients

$100/1 = 100$

$150/2 = 75 \leftarrow$ Smallest

$320/2 = 160$

The row with the smallest quotient (in this case, the second row) is called the **pivot row.** The entry in the pivot row and pivot column is the pivot:

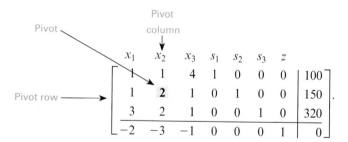

Pivot

Pivot column

Pivot row

$$
\begin{array}{ccccccc}
x_1 & x_2 & x_3 & s_1 & s_2 & s_3 & z \\
\end{array}
$$

$$
\left[
\begin{array}{ccccccc|c}
1 & 1 & 4 & 1 & 0 & 0 & 0 & 100 \\
1 & \mathbf{2} & 1 & 0 & 1 & 0 & 0 & 150 \\
3 & 2 & 1 & 0 & 0 & 1 & 0 & 320 \\
\hline
-2 & -3 & -1 & 0 & 0 & 0 & 1 & 0 \\
\end{array}
\right].
$$

⚠ **CAUTION**  In some simplex tableaus, the pivot column may contain zeros or negative entries. Only the positive entries in the pivot column should be used to form the quotients and determine the pivot row. If there are no positive entries in the pivot column (so that a pivot row cannot be chosen), then no maximum solution exists.  ✔3

✓ **Checkpoint 3**

Find the pivot for the following tableau:

$$
\begin{array}{cccccc}
x_1 & x_2 & s_1 & s_2 & s_3 & z \\
\end{array}
$$

$$
\left[
\begin{array}{cccccc|c}
0 & 1 & 1 & 0 & 0 & 0 & 50 \\
-2 & 3 & 0 & 1 & 0 & 0 & 78 \\
2 & 4 & 0 & 0 & 1 & 0 & 65 \\
\hline
-5 & -3 & 0 & 0 & 0 & 1 & 0 \\
\end{array}
\right].
$$

## Pivoting

Once the pivot has been selected, row operations are used to replace the initial simplex tableau by another simplex tableau in which the pivot column variable is eliminated from all but one of the equations. Since this new tableau is obtained by row operations, it represents an equivalent system of equations (that is, a system with the same solutions as the original system). This process, which is called **pivoting,** is explained in the next example.

**EXAMPLE 3** Use the indicated pivot, 2, to perform the pivoting on the simplex tableau of Example 2:

$$
\begin{bmatrix}
x_1 & x_2 & x_3 & s_1 & s_2 & s_3 & z \\
1 & 1 & 4 & 1 & 0 & 0 & 0 & 100 \\
1 & \mathbf{2} & 1 & 0 & 1 & 0 & 0 & 150 \\
3 & 2 & 1 & 0 & 0 & 1 & 0 & 320 \\
-2 & -3 & -1 & 0 & 0 & 0 & 1 & 0
\end{bmatrix}.
$$

**Solution** Start by multiplying each entry of row 2 by $\frac{1}{2}$ in order to change the pivot to 1:

$$
\begin{bmatrix}
x_1 & x_2 & x_3 & s_1 & s_2 & s_3 & z \\
1 & 1 & 4 & 1 & 0 & 0 & 0 & 100 \\
\frac{1}{2} & \mathbf{1} & \frac{1}{2} & 0 & \frac{1}{2} & 0 & 0 & 75 \\
3 & 2 & 1 & 0 & 0 & 1 & 0 & 320 \\
-2 & -3 & -1 & 0 & 0 & 0 & 1 & 0
\end{bmatrix}. \quad \frac{1}{2}R_2
$$

Now use row operations to make the entry in row 1, column 2, a 0:

$$
\begin{bmatrix}
x_1 & x_2 & x_3 & s_1 & s_2 & s_3 & z \\
\frac{1}{2} & 0 & \frac{7}{2} & 1 & -\frac{1}{2} & 0 & 0 & 25 \\
\frac{1}{2} & 1 & \frac{1}{2} & 0 & \frac{1}{2} & 0 & 0 & 75 \\
3 & 2 & 1 & 0 & 0 & 1 & 0 & 320 \\
-2 & -3 & -1 & 0 & 0 & 0 & 1 & 0
\end{bmatrix}. \quad -R_2 + R_1
$$

Change the 2 in row 3, column 2, to a 0 by a similar process:

$$
\begin{bmatrix}
x_1 & x_2 & x_3 & s_1 & s_2 & s_3 & z \\
\frac{1}{2} & 0 & \frac{7}{2} & 1 & -\frac{1}{2} & 0 & 0 & 25 \\
\frac{1}{2} & 1 & \frac{1}{2} & 0 & \frac{1}{2} & 0 & 0 & 75 \\
2 & 0 & 0 & 0 & -1 & 1 & 0 & 170 \\
-2 & -3 & -1 & 0 & 0 & 0 & 1 & 0
\end{bmatrix}. \quad -2R_2 + R_3
$$

Finally, add 3 times row 2 to the last row in order to change the indicator $-3$ to 0:

$$
\begin{bmatrix}
x_1 & x_2 & x_3 & s_1 & s_2 & s_3 & z \\
\frac{1}{2} & 0 & \frac{7}{2} & 1 & -\frac{1}{2} & 0 & 0 & 25 \\
\frac{1}{2} & 1 & \frac{1}{2} & 0 & \frac{1}{2} & 0 & 0 & 75 \\
2 & 0 & 0 & 0 & -1 & 1 & 0 & 170 \\
-\frac{1}{2} & 0 & \frac{1}{2} & 0 & \frac{3}{2} & 0 & 1 & 225
\end{bmatrix}. \quad 3R_2 + R_4
$$

The pivoting is now complete, because the pivot column variable $x_2$ has been eliminated from all equations except the one represented by the pivot row. The initial simplex tableau has been replaced by a new simplex tableau, which represents an equivalent system of equations.

✓**Checkpoint 4**

For the given simplex tableau,

(a) find the pivot;

(b) perform the pivoting and write the new tableau.

$$
\begin{bmatrix}
x_1 & x_2 & x_3 & s_1 & s_2 & z \\
1 & 2 & 6 & 1 & 0 & 0 & 16 \\
1 & 3 & 0 & 0 & 1 & 0 & 25 \\
-1 & -4 & -3 & 0 & 0 & 1 & 0
\end{bmatrix}
$$

🛑 **CAUTION** During pivoting, do not interchange rows of the matrix. Make the pivot entry 1 by multiplying the pivot row by an appropriate constant, as in Example 3. ✓4

When at least one of the indicators in the last row of a simplex tableau is negative (as is the case with the tableau obtained in Example 3), the simplex method requires that a new pivot be selected and the pivoting be performed again. This procedure is repeated until a simplex tableau with no negative indicators in the last row is obtained or a tableau is reached in which no pivot row can be chosen.

---

**EXAMPLE 4**    In the simplex tableau obtained in Example 3, select a new pivot and perform the pivoting.

**Solution** First, locate the pivot column by finding the most negative indicator in the last row. Then locate the pivot row by computing the necessary quotients and finding the smallest one, as shown here:

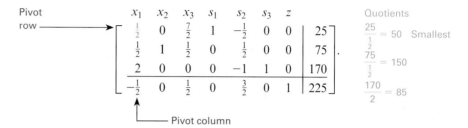

So the pivot is the number $\frac{1}{2}$ in row 1, column 1. Begin the pivoting by multiplying every entry in row 1 by 2. Then continue as indicated to obtain the following simplex tableau:

$$
\begin{array}{ccccccc}
x_1 & x_2 & x_3 & s_1 & s_2 & s_3 & z \\
\end{array}
$$

$$
\left[
\begin{array}{ccccccc|c}
1 & 0 & 7 & 2 & -1 & 0 & 0 & 50 \\
0 & 1 & -3 & -1 & 1 & 0 & 0 & 50 \\
0 & 0 & -14 & -4 & 1 & 1 & 0 & 70 \\
\hline
0 & 0 & 4 & 1 & 1 & 0 & 1 & 250 \\
\end{array}
\right]
\begin{array}{l}
2R_1 \\
-\frac{1}{2}R_1 + R_2 \\
-2R_1 + R_3 \\
\frac{1}{2}R_1 + R_4 \\
\end{array}
$$

Since there are no negative indicators in the last row, no further pivoting is necessary, and we call this the **final simplex tableau.**

## Reading the Solution

The next example shows how to read an optimal solution of the original linear programming problem from the final simplex tableau.

---

**EXAMPLE 5**    Solve the linear programming problem introduced in Example 1.

**Solution** Look at the final simplex tableau for this problem, which was obtained in Example 4:

$$
\begin{array}{ccccccc}
x_1 & x_2 & x_3 & s_1 & s_2 & s_3 & z \\
\end{array}
$$

$$
\left[
\begin{array}{ccccccc|c}
1 & 0 & 7 & 2 & -1 & 0 & 0 & 50 \\
0 & 1 & -3 & -1 & 1 & 0 & 0 & 50 \\
0 & 0 & -14 & -4 & 1 & 1 & 0 & 70 \\
\hline
0 & 0 & 4 & 1 & 1 & 0 & 1 & 250 \\
\end{array}
\right]
$$

The last row of this matrix represents the equation

$$4x_3 + s_1 + s_2 + z = 250, \qquad \text{or equivalently,} \qquad z = 250 - 4x_3 - s_1 - s_2.$$

If $x_3, s_1,$ and $s_2$ are all 0, then the value of $z$ is 250. If any one of $x_3, s_1,$ or $s_2$ is positive, then $z$ will have a smaller value than 250. (Why?) Consequently, since we want a solution for this system in which all the variables are nonnegative and $z$ is as large as possible, we must have

$$x_3 = 0, \qquad s_1 = 0, \qquad s_2 = 0.$$

When these values are substituted into the first equation (represented by the first row of the final simplex tableau), the result is

$$x_1 + 7 \cdot 0 + 2 \cdot 0 - 1 \cdot 0 = 50; \quad \text{that is,} \quad x_1 = 50.$$

Similarly, substituting 0 for $x_3$, $s_1$, and $s_2$ in the last three equations represented by the final simplex tableau shows that

$$x_2 = 50, \quad s_3 = 70, \quad \text{and} \quad z = 250.$$

Therefore, the maximum value of $z = 2x_1 + 3x_2 + x_3$ occurs when

$$x_1 = 50, \quad x_2 = 50, \quad \text{and} \quad x_3 = 0,$$

in which case $z = 2 \cdot 50 + 3 \cdot 50 + 0 = 250$. (The values of the slack variables are irrelevant in stating the solution of the original problem.)

In any simplex tableau, some columns look like columns of an identity matrix (one entry is 1 and the rest are 0). The variables corresponding to these columns are called **basic variables** and the variables corresponding to the other columns are referred to as **nonbasic variables**. In the tableau of Example 5, for instance, the basic variables are $x_1$, $x_2$, $s_3$, and $z$ (shown in blue), and the nonbasic variables are $x_3$, $s_1$, and $s_2$:

$$
\begin{array}{ccccccc}
x_1 & x_2 & x_3 & s_1 & s_2 & s_3 & z \\
\end{array}
$$

$$
\left[
\begin{array}{ccccccc|c}
1 & 0 & 7 & 2 & -1 & 0 & 0 & 50 \\
0 & 1 & -3 & -1 & 1 & 0 & 0 & 50 \\
0 & 0 & -14 & -4 & 1 & 1 & 0 & 70 \\
0 & 0 & 4 & 1 & 1 & 0 & 1 & 250 \\
\end{array}
\right].
$$

The optimal solution in Example 5 was obtained from the final simplex tableau by setting the nonbasic variables equal to 0 and solving for the basic variables. Furthermore, the values of the basic variables are easy to read from the matrix: Find the 1 in the column representing a basic variable; the last entry in that row is the value of that basic variable in the optimal solution. In particular, *the entry in the lower right-hand corner of the final simplex tableau is the maximum value of* $z$. ✔ 5

✔ **Checkpoint 5**

A linear programming problem with slack variables $s_1$ and $s_2$ has the following final simplex tableau:

$$
\begin{array}{cccccc}
x_1 & x_2 & x_3 & s_1 & s_2 & z \\
\end{array}
$$

$$
\left[
\begin{array}{cccccc|c}
0 & 3 & 1 & 5 & 2 & 0 & 9 \\
1 & -2 & 0 & 4 & 1 & 0 & 6 \\
0 & 5 & 0 & 1 & 0 & 1 & 21 \\
\end{array}
\right].
$$

What is the optimal solution?

⚠ **CAUTION** If there are two identical columns in a tableau, each of which is a column in an identity matrix, only one of the variables corresponding to these columns can be a basic variable. The other is treated as a nonbasic variable. You may choose either one to be the basic variable, unless one of them is $z$, in which case $z$ must be the basic variable.

The steps involved in solving a standard maximum linear programming problem by the simplex method have been illustrated in Examples 1–5 and are summarized here.

## Simplex Method

1. Determine the objective function.
2. Write down all necessary constraints.
3. Convert each constraint into an equation by adding a slack variable.
4. Set up the initial simplex tableau.

**5.** Locate the most negative indicator. If there are two such indicators, choose one. This indicator determines the pivot column.

**6.** Use the positive entries in the pivot column to form the quotients necessary for determining the pivot. If there are no positive entries in the pivot column, no maximum solution exists. If two quotients are equally the smallest, let either determine the pivot.[*]

**7.** Multiply every entry in the pivot row by the reciprocal of the pivot to change the pivot to 1. Then use row operations to change all other entries in the pivot column to 0 by adding suitable multiples of the pivot row to the other rows.

**8.** If the indicators are all positive or 0, you have found the final tableau. If not, go back to Step 5 and repeat the process until a tableau with no negative indicators is obtained.[†]

**9.** In the final tableau, the *basic* variables correspond to the columns that have one entry of 1 and the rest 0. The *nonbasic* variables correspond to the other columns. Set each nonbasic variable equal to 0 and solve the system for the basic variables. The maximum value of the objective function is the number in the lower right-hand corner of the final tableau.

The solution found by the simplex method may not be unique, especially when choices are possible in steps 5, 6, or 9. There may be other solutions that produce the same maximum value of the objective function. (See Exercises 37 and 38 at the end of this section.) ✔ 6

## ▦ The Simplex Method with Technology

▯ Unless indicated otherwise, the simplex method is carried out by hand in the examples and exercises of this chapter, so you can see how and why it works. Once you are familiar with the method, however,

> *we strongly recommend that you use technology to apply the simplex method.*

Doing so will eliminate errors that occur in manual computations. It will also give you a better idea of how the simplex method is used in the real world, where applications involve so many variables and constraints that the manual approach is impractical. Readily available technology includes the following:

**Graphing Calculators**  A program called "SIMPLEX" is available for the TI-84. It pauses after each round of pivoting, so you can examine the intermediate simplex tableau. The program and details on its use can be found at https://goo.gl/3stavd.

**Spreadsheets**  Most spreadsheets have a built-in simplex method program. Figures 7.26(a)–(c) show screenshots for using the Solver in Microsoft Excel (Microsoft Corporation Excel© 2016) to solve Example 1 in this section. Spreadsheets also provide a sensitivity analysis, which allows you to see how much the constraints can be varied without changing the maximal solution. For more information, see the *Spreadsheet Manual* available with this text.

**Other Computer Programs**  A variety of simplex method programs, many of which are free, can be downloaded on the Internet. Google "simplex method program" for some possibilities.

✓ **Checkpoint 6**

A linear programming problem has the following initial tableau:

$$
\begin{array}{c}
\begin{array}{ccccc} x_1 & x_2 & s_1 & s_2 & z \end{array} \\
\left[ \begin{array}{ccccc|c}
1 & 1 & 1 & 0 & 0 & 40 \\
2 & 1 & 0 & 1 & 0 & 24 \\
\hline
-300 & -200 & 0 & 0 & 1 & 0
\end{array} \right].
\end{array}
$$

Use the simplex method to solve the problem.

---

[*]It may be that the first choice of a pivot does not produce a solution. In that case, try the other choice.

[†]Some special circumstances are noted at the end of Section 7.7.

Figure 7.26(a)

Figure 7.26(b)

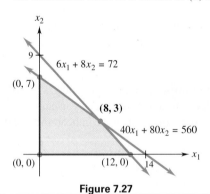

Figure 7.26(c)

## Geometric Interpretation of the Simplex Method

Although it may not be immediately apparent, the simplex method is based on the same geometrical considerations as the graphical method. This can be seen by looking at a problem that can be readily solved by both methods.

**EXAMPLE 6**    In Example 1 of Section 7.3, the following problem was solved graphically (using $x$ and $y$ instead of $x_1$ and $x_2$, respectively):

$$\text{Maximize} \quad z = 8x_1 + 12x_2$$
$$\text{subject to} \quad 40x_1 + 80x_2 \leq 560$$
$$6x_1 + 8x_2 \leq 72$$
$$x_1 \geq 0, x_2 \geq 0.$$

Graphing the feasible region (Figure 7.27) and evaluating $z$ at each corner point shows that the maximum value of $z$ occurs at (8, 3).

| Corner Point | Value of $z = 8x_1 + 12x_2$ | |
|---|---|---|
| (0, 0) | 0 | |
| (0, 7) | 84 | |
| (8, 3) | 100 | (maximum) |
| (12, 0) | 96 | |

Figure 7.27

To solve the same problem by the simplex method, add a slack variable to each constraint:

$$40x_1 + 80x_2 + s_1 \qquad = 560$$
$$6x_1 + 8x_2 \qquad + s_2 = 72.$$

Then write the initial simplex tableau:

$$
\begin{array}{ccccc}
x_1 & x_2 & s_1 & s_2 & z \\
\end{array}
$$
$$
\left[
\begin{array}{ccccc|c}
40 & 80 & 1 & 0 & 0 & 560 \\
6 & 8 & 0 & 1 & 0 & 72 \\
\hline
-8 & -12 & 0 & 0 & 1 & 0 \\
\end{array}
\right].
$$

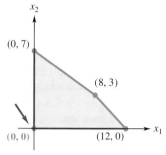

**Figure 7.28**

In this tableau, the basic variables are $s_1$, $s_2$, and $z$. (Why?) By setting the nonbasic variables (namely, $x_1$ and $x_2$) equal to 0 and solving for the basic variables, we obtain the following solution (which will be called a **basic feasible solution**):

$$x_1 = 0, \qquad x_2 = 0, \qquad s_1 = 560, \qquad s_2 = 72, \qquad \text{and} \qquad z = 0.$$

Since $x_1 = 0$ and $x_2 = 0$, this solution corresponds to the corner point at the origin in the graphical solution (Figure 7.28).

The basic feasible solution $(0, 0)$ given by the initial simplex tableau has $z = 0$, which is obviously not maximal. Each round of pivoting in the simplex method will produce another corner point, with a larger value of $z$, until we reach a corner point that provides the maximum solution.

The most negative indicator in the initial tableau is $-12$, and it determines the pivot column. Then we form the necessary quotients and determine the pivot row:

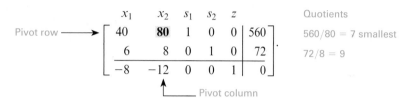

Thus, the pivot is 80 in row 1, column 2. Performing the pivoting leads to this tableau:

$$\begin{bmatrix} x_1 & x_2 & s_1 & s_2 & z & \\ \frac{1}{2} & 1 & \frac{1}{80} & 0 & 0 & 7 \\ 2 & 0 & -\frac{1}{10} & 1 & 0 & 16 \\ -2 & 0 & \frac{3}{20} & 0 & 1 & 84 \end{bmatrix} \qquad \begin{matrix} \frac{1}{80} R_1 \\ -8R_1 + R_2 \\ 12R_1 + R_3 \end{matrix}$$

The basic variables here are $x_2$, $s_2$, and $z$, and the basic feasible solution (found by setting the nonbasic variables equal to 0 and solving for the basic variables) is

$$x_1 = 0, \qquad x_2 = 7, \qquad s_1 = 0, \qquad s_2 = 16, \qquad \text{and} \qquad z = 84,$$

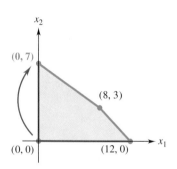

**Figure 7.29**

which corresponds to the corner point $(0, 7)$ in Figure 7.29. Note that the new value of the pivot variable $x_2$ is precisely the smallest quotient, 7, that was used to select the pivot row. Although this value of $z$ is better, further improvement is possible.

Now the most negative indicator is $-2$. We form the necessary quotients and determine the pivot as usual:

The pivot is 2 in row 2, column 1. Pivoting now produces the final tableau:

$$\begin{bmatrix} x_1 & x_2 & s_1 & s_2 & z & \\ 0 & 1 & \frac{3}{80} & -\frac{1}{4} & 0 & 3 \\ 1 & 0 & -\frac{1}{20} & \frac{1}{2} & 0 & 8 \\ 0 & 0 & \frac{1}{20} & 1 & 1 & 100 \end{bmatrix} \qquad \begin{matrix} -\frac{1}{2} R_2 + R_1 \\ \frac{1}{2} R_2 \\ 2R_2 + R_3 \end{matrix}$$

Here, the basic feasible solution is

$$x_1 = 8, \qquad x_2 = 3, \qquad s_1 = 0, \qquad s_2 = 0, \qquad \text{and} \qquad z = 100,$$

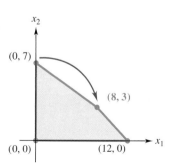

**Figure 7.30**

which corresponds to the corner point $(8, 3)$ in Figure 7.30. Once again, the new value of the pivot variable $x_1$ is the smallest quotient, 8, that was used to select the pivot. Because all the indicators in the last row of the final tableau are nonnegative, $(8, 3)$ is the maximum solution according to the simplex method. We know that this is the case, since this is the maximum solution found by the graphical method. An algebraic argument similar to the one in Example 5 could also be made.

As illustrated in Example 6, the basic feasible solution obtained from a simplex tableau corresponds to a corner point of the feasible region. Pivoting, which replaces one tableau with another, is a systematic way of moving from one corner point to another, each time improving the value of the objective function. The simplex method ends when a corner point that produces the maximum value of the objective function is reached (or when it becomes clear that the problem has no maximum solution).

When there are three or more variables in a linear programming problem, it may be difficult or impossible to draw a picture, but it can be proved that the optimal value of the objective function occurs at a basic feasible solution (corresponding to a corner point in the two-variable case). The simplex method provides a means of moving from one basic feasible solution to another until one that produces the optimal value of the objective function is reached.

## Explanation of Pivoting

The rules for selecting the pivot in the simplex method can be understood by examining how the first pivot was chosen in Example 6.

**EXAMPLE 7**   The initial simplex tableau of Example 6 provides a basic feasible solution with $x_1 = 0$ and $x_2 = 0$:

$$
\begin{array}{ccccc}
x_1 & x_2 & s_1 & s_2 & z \\
\end{array}
$$

$$
\left[
\begin{array}{ccccc|c}
40 & 80 & 1 & 0 & 0 & 560 \\
6 & 8 & 0 & 1 & 0 & 72 \\
\hline
-8 & -12 & 0 & 0 & 1 & 0 \\
\end{array}
\right].
$$

This solution certainly does not give a maximum value for the objective function $z = 8x_1 + 12x_2$. Since $x_2$ has the largest coefficient, $z$ will be increased most if $x_2$ is increased. In other words, the most negative indicator in the tableau (which corresponds to the largest coefficient in the objective function) identifies the variable that will provide the greatest change in the value of $z$.

To determine how much $x_2$ can be increased without leaving the feasible region, look at the first two equations,

$$
\begin{aligned}
40x_1 + 80x_2 + s_1 \quad\quad &= 560 \\
6x_1 + 8x_2 \quad\quad + s_2 &= 72,
\end{aligned}
$$

and solve for the basic variables $s_1$ and $s_2$:

$$
\begin{aligned}
s_1 &= 560 - 40x_1 - 80x_2 \\
s_2 &= 72 - 6x_1 - 8x_2.
\end{aligned}
$$

Now $x_2$ is to be increased while $x_1$ is to keep the value 0. Hence,

$$
\begin{aligned}
s_1 &= 560 - 80x_2 \\
s_2 &= 72 - 8x_2.
\end{aligned}
$$

Since $s_1 \geq 0$ and $s_2 \geq 0$, we must have

$$
\begin{array}{ll}
0 \leq s_1 & 0 \leq s_2 \\
0 \leq 560 - 80x_2 \quad \text{and} & 0 \leq 72 - 8x_2 \\
80x_2 \leq 560 & 8x_2 \leq 72 \\
x_2 \leq \dfrac{560}{80} = 7 & x_2 \leq \dfrac{72}{8} = 9.
\end{array}
$$

The right sides of these last inequalities are the quotients used to select the pivot row. Since $x_2$ must satisfy both inequalities, $x_2$ can be at most 7. In other words, the smallest quotient formed from positive entries in the pivot column identifies the value of $x_2$ that produces the largest change in $z$ while remaining in the feasible region. By pivoting with the pivot determined in this way, we obtain the second tableau and a basic feasible solution in which $x_2 = 7$, as was shown in Example 6.

An analysis similar to that in Example 7 applies to each occurrence of pivoting in the simplex method. The idea is to improve the value of the objective function by adjusting one variable at a time. The most negative indicator identifies the variable that will account for the largest increase in $z$. The smallest quotient determines the largest value of that variable which will produce a feasible solution. Pivoting leads to a solution in which the selected variable has this largest value.

# 7.4 Exercises

*In Exercises 1–4, (a) determine the number of slack variables needed; (b) name them; (c) use the slack variables to convert each constraint into a linear equation. (See Example 1.)*

1. Maximize $z = 32x_1 + 9x_2$
   subject to $4x_1 + 2x_2 \le 20$
   $5x_1 + x_2 \le 50$
   $2x_1 + 3x_2 \le 25$
   $x_1 \ge 0, x_2 \ge 0.$

2. Maximize $z = 3.7x_1 + 4.3x_2$
   subject to $2.4x_1 + 1.5x_2 \le 10$
   $1.7x_1 + 1.9x_2 \le 15$
   $x_1 \ge 0, x_2 \ge 0.$

3. Maximize $z = 8x_1 + 3x_2 + x_3$
   subject to $3x_1 - x_2 + 4x_3 \le 95$
   $7x_1 + 6x_2 + 8x_3 \le 118$
   $4x_1 + 5x_2 + 10x_3 \le 220$
   $x_1 \ge 0, x_2 \ge 0, x_3 \ge 0.$

4. Maximize $z = 12x_1 + 15x_2 + 10x_3$
   subject to $2x_1 + 2x_2 + x_3 \le 8$
   $x_1 + 4x_2 + 3x_3 \le 12$
   $x_1 \ge 0, x_2 \ge 0, x_3 \ge 0.$

*Introduce slack variables as necessary and then write the initial simplex tableau for each of these linear programming problems.*

5. Maximize $z = 5x_1 + x_2$
   subject to $2x_1 + 5x_2 \le 6$
   $4x_1 + x_2 \le 6$
   $5x_1 + 3x_2 \le 15$
   $x_1 \ge 0, x_2 \ge 0.$

6. Maximize $z = 5x_1 + 3x_2 + 7x_3$
   subject to $4x_1 + 3x_2 + 2x_3 \le 60$
   $3x_1 + 4x_2 + x_3 \le 24$
   $x_1 \ge 0, x_2 \ge 0, x_3 \ge 0.$

7. Maximize $z = x_1 + 5x_2 + 10x_3$
   subject to $x_1 + 2x_2 + 3x_3 \le 10$
   $2x_1 + x_2 + x_3 \le 8$
   $3x_1 + 4x_3 \le 6$
   $x_1 \ge 0, x_2 \ge 0, x_3 \ge 0.$

8. Maximize $z = 5x_1 - x_2 + 3x_3$
   subject to $3x_1 + 2x_2 + x_3 \le 36$
   $x_1 + 6x_2 + x_3 \le 24$
   $x_1 - x_2 - x_3 \le 32$
   $x_1 \ge 0, x_2 \ge 0, x_3 \ge 0.$

*Find the pivot in each of the given simplex tableaus. (See Example 2.)*

9.

| $x_1$ | $x_2$ | $x_3$ | $s_1$ | $s_2$ | $z$ | |
|---|---|---|---|---|---|---|
| 2 | 2 | 0 | 3 | 1 | 0 | 15 |
| 3 | 4 | 1 | 6 | 0 | 0 | 20 |
| −2 | −3 | 0 | 1 | 0 | 1 | 10 |

10.

| $x_1$ | $x_2$ | $x_3$ | $s_1$ | $s_2$ | $z$ | |
|---|---|---|---|---|---|---|
| 0 | 2 | 1 | 1 | 3 | 0 | 5 |
| 1 | −5 | 0 | 1 | 2 | 0 | 8 |
| 0 | −2 | 0 | −3 | 1 | 1 | 10 |

11.

| $x_1$ | $x_2$ | $x_3$ | $s_1$ | $s_2$ | $s_3$ | $z$ | |
|---|---|---|---|---|---|---|---|
| 6 | 2 | 1 | 3 | 0 | 0 | 0 | 8 |
| 0 | 2 | 0 | 1 | 0 | 1 | 0 | 7 |
| 6 | 1 | 0 | 3 | 1 | 0 | 0 | 6 |
| −3 | −2 | 0 | 2 | 0 | 0 | 1 | 12 |

12.

| $x_1$ | $x_2$ | $x_3$ | $s_1$ | $s_2$ | $s_3$ | $z$ | |
|---|---|---|---|---|---|---|---|
| 0 | 2 | 0 | 1 | 2 | 2 | 0 | 3 |
| 0 | 3 | 1 | 0 | 1 | 2 | 0 | 4 |
| 1 | 4 | 0 | 0 | 3 | 5 | 0 | 5 |
| 0 | −4 | 0 | 0 | 4 | 3 | 1 | 20 |

*In Exercises 13–16, use the indicated entry as the pivot and perform the pivoting. (See Examples 3 and 4.)*

**13.**

|       | $x_1$ | $x_2$ | $x_3$ | $s_1$ | $s_2$ | $z$ |     |
|-------|-------|-------|-------|-------|-------|-----|-----|
|       | 1     | 2     | 4     | 1     | 0     | 0   | 56  |
|       | 2     | **2** | 1     | 0     | 1     | 0   | 40  |
|       | −1    | −3    | −2    | 0     | 0     | 1   | 0   |

**14.**

|       | $x_1$ | $x_2$ | $x_3$ | $s_1$ | $s_2$ | $s_3$ | $z$ |     |
|-------|-------|-------|-------|-------|-------|-------|-----|-----|
|       | 2     | 2     | **1** | 1     | 0     | 0     | 0   | 12  |
|       | 1     | 2     | 3     | 0     | 1     | 0     | 0   | 45  |
|       | 3     | 1     | 1     | 0     | 0     | 1     | 0   | 20  |
|       | −2    | −1    | −3    | 0     | 0     | 0     | 1   | 0   |

**15.**

|       | $x_1$ | $x_2$ | $x_3$ | $s_1$ | $s_2$ | $s_3$ | $z$ |      |
|-------|-------|-------|-------|-------|-------|-------|-----|------|
|       | 1     | 1     | 1     | 1     | 0     | 0     | 0   | 60   |
|       | 3     | 1     | **2** | 0     | 1     | 0     | 0   | 100  |
|       | 1     | 2     | 3     | 0     | 0     | 1     | 0   | 200  |
|       | −1    | −1    | −2    | 0     | 0     | 0     | 1   | 0    |

**16.**

|       | $x_1$ | $x_2$ | $x_3$ | $s_1$ | $s_2$ | $s_3$ | $z$ |     |
|-------|-------|-------|-------|-------|-------|-------|-----|-----|
|       | 4     | 2     | 3     | 1     | 0     | 0     | 0   | 22  |
|       | 2     | 2     | **5** | 0     | 1     | 0     | 0   | 28  |
|       | 1     | 3     | 2     | 0     | 0     | 1     | 0   | 45  |
|       | −3    | −2    | −4    | 0     | 0     | 0     | 1   | 0   |

*For each simplex tableau in Exercises 17–20, (a) list the basic and the nonbasic variables, (b) find the basic feasible solution determined by setting the nonbasic variables equal to 0, and (c) decide whether this is a maximum solution. (See Examples 5 and 6.)*

**17.**

|       | $x_1$ | $x_2$ | $x_3$ | $s_1$ | $s_2$ | $z$ |     |
|-------|-------|-------|-------|-------|-------|-----|-----|
|       | 3     | 2     | 0     | −3    | 1     | 0   | 29  |
|       | 4     | 0     | 1     | −2    | 0     | 0   | 16  |
|       | −5    | 0     | 0     | −1    | 0     | 1   | 11  |

**18.**

|       | $x_1$ | $x_2$ | $x_3$         | $s_1$ | $s_2$ | $s_3$ | $z$ |     |
|-------|-------|-------|---------------|-------|-------|-------|-----|-----|
|       | −3    | 0     | $\frac{1}{2}$ | 1     | −2    | 0     | 0   | 22  |
|       | 2     | 0     | −3            | 0     | 1     | 1     | 0   | 10  |
|       | 4     | 1     | 4             | 0     | $\frac{3}{4}$ | 0     | 0   | 17  |
|       | −1    | 0     | 0             | 0     | 1     | 0     | 1   | 120 |

**19.**

|       | $x_1$ | $x_2$ | $x_3$ | $s_1$         | $s_2$ | $s_3$         | $z$ |     |
|-------|-------|-------|-------|---------------|-------|---------------|-----|-----|
|       | 1     | 0     | 2     | $\frac{1}{2}$ | 0     | $\frac{1}{3}$ | 0   | 6   |
|       | 0     | 1     | −1    | 5             | 0     | −1            | 0   | 13  |
|       | 0     | 0     | 1     | $\frac{3}{2}$ | 1     | $-\frac{1}{3}$| 0   | 21  |
|       | 0     | 0     | 2     | $\frac{1}{2}$ | 0     | 3             | 1   | 18  |

**20.**

|       | $x_1$ | $x_2$ | $x_3$ | $x_4$ | $s_1$ | $s_2$ | $s_3$          | $z$ |     |
|-------|-------|-------|-------|-------|-------|-------|----------------|-----|-----|
|       | −1    | 0     | 0     | 1     | 0     | 3     | −2             | 0   | 47  |
|       | 2     | 0     | 1     | 0     | 0     | 2     | $-\frac{1}{2}$ | 0   | 37  |
|       | 3     | 5     | 0     | 0     | 1     | −1    | 6              | 0   | 43  |
|       | 4     | 1     | 0     | 0     | 0     | 6     | 0              | 1   | 86  |

*Use the simplex method to solve Exercises 21–36.*

**21.** Maximize $z = x_1 + 3x_2$

subject to  $x_1 + x_2 \le 10$

$5x_1 + 2x_2 \le 20$

$x_1 + 2x_2 \le 36$

$x_1 \ge 0, x_2 \ge 0.$

**22.** Maximize $z = 5x_1 + x_2$

subject to  $2x_1 + 3x_2 \le 8$

$4x_1 + 8x_2 \le 12$

$5x_1 + 2x_2 \le 30$

$x_1 \ge 0, x_2 \ge 0.$

**23.** Maximize $z = 2x_1 + x_2$

subject to  $x_1 + 3x_2 \le 12$

$2x_1 + x_2 \le 10$

$x_1 + x_2 \le 4$

$x_1 \ge 0, x_2 \ge 0.$

**24.** Maximize $z = 4x_1 + 2x_2$

subject to  $-x_1 - x_2 \le 12$

$3x_1 - x_2 \le 15$

$x_1 \ge 0, x_2 \ge 0.$

**25.** Maximize $z = 5x_1 + 4x_2 + x_3$

subject to  $-2x_1 + x_2 + 2x_3 \le 3$

$x_1 - x_2 + x_3 \le 1$

$x_1 \ge 0, x_2 \ge 0, x_3 \ge 0.$

**26.** Maximize $z = 3x_1 + 2x_2 + x_3$

subject to  $2x_1 + 2x_2 + x_3 \le 10$

$x_1 + 2x_2 + 3x_3 \le 15$

$x_1 \ge 0, x_2 \ge 0, x_3 \ge 0.$

**27.** Maximize $z = 2x_1 + x_2 + x_3$

subject to  $x_1 - 3x_2 + x_3 \le 3$

$x_1 - 2x_2 + 2x_3 \le 12$

$x_1 \ge 0, x_2 \ge 0, x_3 \ge 0.$

**28.** Maximize $z = 4x_1 + 5x_2 + x_3$

subject to  $x_1 + 2x_2 + 4x_3 \le 10$

$2x_1 + 2x_2 + x_3 \le 10$

$x_1 \ge 0, x_2 \ge 0, x_3 \ge 0.$

**29.** Maximize $z = 2x_1 + 2x_2 - 4x_3$

subject to  $3x_1 + 3x_2 - 6x_3 \le 51$

$5x_1 + 5x_2 + 10x_3 \le 99$

$x_1 \ge 0, x_2 \ge 0, x_3 \ge 0.$

**30.** Maximize $z = 4x_1 + x_2 + 3x_3$

subject to  $x_1 + 3x_3 \le 6$

$6x_1 + 3x_2 + 12x_3 \le 40$

$x_1 \ge 0, x_2 \ge 0, x_3 \ge 0.$

**31.** Maximize $z = 300x_1 + 200x_2 + 100x_3$

subject to $x_1 + x_2 + x_3 \le 100$

$2x_1 + 3x_2 + 4x_3 \le 320$

$2x_1 + x_2 + x_3 \le 160$

$x_1 \ge 0, x_2 \ge 0, x_3 \ge 0.$

**32.** Maximize $z = x_1 + 5x_2 - 10x_3$

subject to $8x_1 + 4x_2 + 12x_3 \le 18$

$x_1 + 6x_2 + 2x_3 \le 45$

$5x_1 + 7x_2 + 3x_3 \le 60$

$x_1 \ge 0, x_2 \ge 0, x_3 \ge 0.$

**33.** Maximize $z = 4x_1 - 3x_2 + 2x_3$

subject to $2x_1 - x_2 + 8x_3 \le 40$

$4x_1 - 5x_2 + 6x_3 \le 60$

$2x_1 - 2x_2 + 6x_3 \le 24$

$x_1 \ge 0, x_2 \ge 0, x_3 \ge 0.$

**34.** Maximize $z = 3x_1 + 2x_2 - 4x_3$

subject to $x_1 - x_2 + x_3 \le 10$

$2x_1 - x_2 + 2x_3 \le 30$

$-3x_1 + x_2 + 3x_3 \le 40$

$x_1 \ge 0, x_2 \ge 0, x_3 \ge 0.$

**35.** Maximize $z = x_1 + 2x_2 + x_3 + 5x_4$

subject to $x_1 + 2x_2 + x_3 + x_4 \le 50$

$3x_1 + x_2 + 2x_3 + x_4 \le 100$

$x_1 \ge 0, x_2 \ge 0, x_3 \ge 0, x_4 \ge 0.$

**36.** Maximize $z = x_1 + x_2 + 4x_3 + 5x_4$

subject to $x_1 + 2x_2 + 3x_3 + x_4 \le 115$

$2x_1 + x_2 + 8x_3 + 5x_4 \le 200$

$x_1 + x_3 \le 50$

$x_1 \ge 0, x_2 \ge 0, x_3 \ge 0, x_4 \ge 0.$

**37.** The initial simplex tableau of a linear programming problem is

$$\begin{array}{ccccccc} x_1 & x_2 & x_3 & s_1 & s_2 & z & \\ \left[\begin{array}{cccccc|c} 1 & 1 & 1 & 1 & 0 & 0 & 12 \\ 2 & 1 & 2 & 0 & 1 & 0 & 30 \\ \hline -2 & -2 & -1 & 0 & 0 & 1 & 0 \end{array}\right]. \end{array}$$

**(a)** Use the simplex method to solve the problem with column 1 as the first pivot column.

**(b)** Now use the simplex method to solve the problem with column 2 as the first pivot column.

**(c)** Does this problem have a unique maximum solution? Why?

**38.** The final simplex tableau of a linear programming problem is

$$\begin{array}{cccccc} x_1 & x_2 & s_1 & s_2 & z & \\ \left[\begin{array}{ccccc|c} 1 & 1 & 2 & 0 & 0 & 24 \\ 2 & 0 & 2 & 1 & 0 & 8 \\ \hline 4 & 0 & 0 & 0 & 1 & 40 \end{array}\right]. \end{array}$$

**(a)** What is the solution given by this tableau?

**(b)** Even though all the indicators are nonnegative, perform one more round of pivoting on this tableau, using column 3 as the pivot column and choosing the pivot row by forming quotients in the usual way.

**(c)** Show that there is more than one solution to the linear programming problem by comparing your answer in part (a) with the basic feasible solution given by the tableau found in part (b). Does it give the same value of $z$ as the solution in part (a)?

✓**Checkpoint Answers**

**1.** $x_1 + x_2 + x_3 + s_1 = 12$

$2x_1 + 4x_2 + s_2 = 15$

$x_2 + 3x_3 + s_3 = 10$

**2.** $$\begin{array}{cccccc} x_1 & x_2 & s_1 & s_2 & s_3 & z & \\ \left[\begin{array}{cccccc|c} 1 & 2 & 1 & 0 & 0 & 0 & 85 \\ 2 & 1 & 0 & 1 & 0 & 0 & 92 \\ 1 & 4 & 0 & 0 & 1 & 0 & 104 \\ \hline -2 & -3 & 0 & 0 & 0 & 1 & 0 \end{array}\right] \end{array}$$
$\phantom{xxxxxxxxx}$ Indicators

**3.** 2 (in first column)

**4. (a)** 2

**(b)** $$\begin{array}{cccccc} x_1 & x_2 & x_3 & s_1 & s_2 & z & \\ \left[\begin{array}{cccccc|c} \frac{1}{2} & 1 & 3 & \frac{1}{2} & 0 & 0 & 8 \\ -\frac{1}{2} & 0 & -9 & -\frac{3}{2} & 1 & 0 & 1 \\ \hline 1 & 0 & 9 & 2 & 0 & 1 & 32 \end{array}\right] \end{array}$$

**5.** $z = 21$ when $x_1 = 6$, $x_2 = 0$, and $x_3 = 9$.

**6.** $x_1 = 0, x_2 = 24, s_1 = 16, s_2 = 0, z = 4800$

# 7.5 Maximization Applications

Applications of the simplex method are considered in this section. First, however, we make a slight change in notation. You have noticed that the column representing the variable $z$ in a simplex tableau never changes during pivoting. (Since all the entries except the last one in this column are 0, performing row operations has no effect on these entries—they remain 0.) Consequently, this column is unnecessary and can be omitted without causing any difficulty.

Hereafter in this text, the column corresponding to the variable $z$ (representing the objective function) will be omitted from all simplex tableaus.

---

**EXAMPLE 1**     **Farming** A farmer has 110 acres of available land he wishes to plant with a mixture of potatoes, corn, and cabbage. It costs him $400 to produce an acre of potatoes, $160 to produce an acre of corn, and $280 to produce an acre of cabbage. He has a maximum of $20,000 to spend. He makes a profit of $120 per acre of potatoes, $40 per acre of corn, and $60 per acre of cabbage.

**(a)** How many acres of each crop should he plant to maximize his profit?

**Solution** Let the number of acres alloted to each of potatoes, corn, and cabbage be $x_1$, $x_2$, and $x_3$, respectively. Then summarize the given information as follows:

| Crop | Number of Acres | Cost per Acre | Profit per Acre |
|---|---|---|---|
| Potatoes | $x_1$ | $400 | $120 |
| Corn | $x_2$ | $160 | $ 40 |
| Cabbage | $x_3$ | $280 | $ 60 |
| Maximum Available | 110 | $20,000 | |

The constraints can be expressed as

$$x_1 + x_2 + x_3 \leq 110 \qquad \text{Number of acres}$$
$$400x_1 + 160x_2 + 280x_3 \leq 20{,}000, \qquad \text{Production costs}$$

where $x_1$, $x_2$, and $x_3$ are all nonnegative. The first of these constraints says that $x_1 + x_2 + x_3$ is less than or perhaps equal to 110. Use $s_1$ as the slack variable, giving the equation

$$x_1 + x_2 + x_3 + s_1 = 110.$$

Here, $s_1$ represents the amount of the farmer's 110 acres that will not be used. ($s_1$ may be 0 or any value up to 110.)

In the same way, the constraint $400x_1 + 160x_2 + 280x_3 \leq 20{,}000$ can be converted into an equation by adding a slack variable $s_2$:

$$400x_1 + 160x_2 + 280x_3 + s_2 = 20{,}000.$$

The slack variable $s_2$ represents any unused portion of the farmer's $20,000 capital. (Again, $s_2$ may have any value from 0 to 20,000.)

The farmer's profit on potatoes is the product of the profit per acre ($120) and the number $x_1$ of acres, that is, $120x_1$. His profits on corn and cabbage are computed similarly. Hence, his total profit is given by

$$z = \text{profit on potatoes} + \text{profit on corn} + \text{profit on cabbage}$$
$$z = 120x_1 + 40x_2 + 60x_3.$$

The linear programming problem can now be stated as follows:

$$\text{Maximize} \quad z = 120x_1 + 40x_2 + 60x_3$$
$$\text{subject to} \quad x_1 + x_2 + x_3 + s_1 = 110$$
$$400x_1 + 160x_2 + 280x_3 + s_2 = 20{,}000,$$
$$\text{with} \quad x_1 \geq 0, x_2 \geq 0, x_3 \geq 0, s_1 \geq 0, s_2 \geq 0.$$

The initial simplex tableau (without the $z$ column) is

$$
\begin{array}{ccccc}
x_1 & x_2 & x_3 & s_1 & s_2 \\
\end{array}
$$

$$
\left[
\begin{array}{ccccc|c}
1 & 1 & 1 & 1 & 0 & 110 \\
400 & 160 & 280 & 0 & 1 & 20{,}000 \\
\hline
-120 & -40 & -60 & 0 & 0 & 0
\end{array}
\right].
$$

The most negative indicator is $-120$; column 1 is the pivot column. The quotients needed to determine the pivot row are $110/1 = 110$ and $20{,}000/400 = 50$. So the pivot is 400 in row 2, column 1. Multiplying row 2 by $1/400$ and completing the pivoting leads to the final simplex tableau:

$$
\begin{array}{ccccc}
x_1 & x_2 & x_3 & s_1 & s_2 \\
\end{array}
$$

$$
\left[
\begin{array}{ccccc|c}
0 & .6 & .3 & 1 & -.0025 & 60 \\
1 & .4 & .7 & 0 & .0025 & 50 \\
\hline
0 & 8 & 24 & 0 & .3 & 6000
\end{array}
\right].
\quad
\begin{array}{l}
-1R_2 + R_1 \\
\frac{1}{400}R_2 \\
120R_2 + R_3
\end{array}
$$

Setting the nonbasic variables $x_2$, $x_3$, and $s_2$ equal to 0, solving for the basic variables $x_1$ and $s_1$, and remembering that the value of $z$ is in the lower right-hand corner leads to this maximum solution:

$$x_1 = 50, \quad x_2 = 0, \quad x_3 = 0, \quad s_1 = 60, \quad s_2 = 0, \quad \text{and} \quad z = 6000.$$

Therefore, the farmer will make a maximum profit of \$6000 by planting 50 acres of potatoes and no corn or cabbage.

**(b)** If the farmer maximizes his profit, how much land will remain unplanted? What is the explanation for this?

**Solution** Since 50 of 110 acres are planted, 60 acres will remain unplanted. Alternatively, note that the unplanted acres of land are represented by $s_1$, the slack variable in the "number of acres" constraint. In the maximal solution found in part (a), $s_1 = 60$, which means that 60 acres are left unplanted.

The amount of unused cash is represented by $s_2$, the slack variable in the "production costs" constraint. Since $s_2 = 0$, all the available money has been used. By using the maximal solution in part (a), the farmer has used his \$20,000 most effectively. If he had more cash, he would plant more crops and make a larger profit.

**EXAMPLE 2** **Advertising** Ana Pott, who is a candidate for the state legislature, has \$96,000 to buy TV advertising time. Ads cost \$400 per minute on a local cable channel, \$4000 per minute on a regional independent channel, and \$12,000 per minute on a national network channel. Because of existing contracts, the TV stations can provide at most a total of 30 minutes of advertising time, with a maximum of 6 minutes on the national network channel. At any given time during the evening, approximately 100,000 people watch the cable channel, 200,000 the independent channel, and 600,000 the network channel. To get maximum exposure, how much time should Ana buy from each station?

**(a)** Set up the initial simplex tableau for this problem.

**Solution** Let $x_1$ be the number of minutes of ads on the cable channel, $x_2$ the number of minutes on the independent channel, and $x_3$ the number of minutes on the network channel. Exposure is measured in viewer-minutes. For instance, 100,000 people watching $x_1$ minutes of ads on the cable channel produces $100{,}000x_1$ viewer-minutes. The amount of exposure is given by the total number of viewer-minutes for all three channels, namely,

$$100{,}000x_1 + 200{,}000x_2 + 600{,}000x_3.$$

Since 30 minutes are available,

$$x_1 + x_2 + x_3 \leq 30.$$

The fact that only 6 minutes can be used on the network channel means that

$$x_3 \leq 6.$$

Expenditures are limited to $96,000, so

$$\text{Cable cost} + \text{independent cost} + \text{network cost} \leq 96,000$$
$$400x_1 + 4000x_2 + 12,000x_3 \leq 96,000.$$

Therefore, Ana must solve the following linear programming problem:

$$\text{Maximize} \quad z = 100,000x_1 + 200,000x_2 + 600,000x_3$$
$$\text{subject to} \quad x_1 + x_2 + x_3 \leq 30$$
$$x_3 \leq 6$$
$$400x_1 + 4000x_2 + 12,000x_3 \leq 96,000,$$
$$\text{with} \quad x_1 \geq 0, x_2 \geq 0, x_3 \geq 0.$$

Introducing slack variables $s_1$, $s_2$, and $s_3$ (one for each constraint), rewriting the constraints as equations, and expressing the objective function as

$$-100,000x_1 - 200,000x_2 - 600,000x_3 + z = 0$$

leads to the initial simplex tableau:

$$
\begin{array}{cccccc}
x_1 & x_2 & x_3 & s_1 & s_2 & s_3 \\
\end{array}
$$

$$
\left[
\begin{array}{cccccc|c}
1 & 1 & 1 & 1 & 0 & 0 & 30 \\
0 & 0 & 1 & 0 & 1 & 0 & 6 \\
400 & 4000 & 12,000 & 0 & 0 & 1 & 96,000 \\
\hline
-100,000 & -200,000 & -600,000 & 0 & 0 & 0 & 0
\end{array}
\right].
$$

**(b)** Use the simplex method to find the final simplex tableau.

**Solution** Work by hand, or use a graphing calculator's simplex program or a spreadsheet, to obtain this final tableau:

$$
\begin{array}{cccccc}
x_1 & x_2 & x_3 & s_1 & s_2 & s_3 \\
\end{array}
$$

$$
\left[
\begin{array}{cccccc|c}
1 & 0 & 0 & \frac{10}{9} & \frac{20}{9} & -\frac{25}{90,000} & 20 \\
0 & 0 & 1 & 0 & 1 & 0 & 6 \\
0 & 1 & 0 & -\frac{1}{9} & -\frac{29}{9} & \frac{25}{90,000} & 4 \\
\hline
0 & 0 & 0 & \frac{800,000}{9} & \frac{1,600,000}{9} & \frac{250}{9} & 6,400,000
\end{array}
\right].
$$

Therefore, the optimal solution is

$$x_1 = 20, \quad x_2 = 4, \quad x_3 = 6, \quad s_1 = 0, \quad s_2 = 0, \quad \text{and} \quad s_3 = 0.$$

Ana should buy 20 minutes of time on the cable channel, 4 minutes on the independent channel, and 6 minutes on the network channel.

**(c)** What do the values of the slack variables in the optimal solution tell you?

**Solution** All three slack variables are 0. This means that all the available minutes have been used ($s_1 = 0$ in the first constraint), the maximum possible 6 minutes on the national network have been used ($s_2 = 0$ in the second constraint), and all of the $96,000 has been spent ($s_3 = 0$ in the third constraint). ✓

✓ **Checkpoint 1**

In Example 2, what is the number of viewer-minutes in the optimal solution?

**EXAMPLE 3** **Manufacturing** A chemical plant makes three products—glaze, solvent, and clay—each of which brings in different revenue per truckload. Production is limited, first by the number of air pollution units the plant is allowed to produce each day and second by the time available in the evaporation tank. The plant manager wants to maximize the daily revenue. Using information not given here, he sets up an initial simplex tableau and uses the simplex method to produce the following final simplex tableau:

$$
\begin{array}{ccccc}
x_1 & x_2 & x_3 & s_1 & s_2 \\
\end{array}
$$

$$
\left[\begin{array}{ccccc|c}
-10 & -25 & 0 & 1 & -1 & 60 \\
3 & 4 & 1 & 0 & .1 & 24 \\
7 & 13 & 0 & 0 & .4 & 96
\end{array}\right].
$$

The three variables represent the number of truckloads of glaze, solvent, and clay, respectively. The first slack variable comes from the air pollution constraint and the second slack variable from the time constraint on the evaporation tank. The revenue function is given in hundreds of dollars.

**(a)** What is the optimal solution?

**Solution**

$$x_1 = 0, \quad x_2 = 0, \quad x_3 = 24, \quad s_1 = 60, \quad s_2 = 0, \quad \text{and} \quad z = 96.$$

**(b)** Interpret this solution. What do the variables represent, and what does the solution mean?

**Solution** The variable $x_1$ is the number of truckloads of glaze, $x_2$ the number of truckloads of solvent, $x_3$ the number of truckloads of clay to be produced, and $z$ the revenue produced (in hundreds of dollars). The plant should produce 24 truckloads of clay and no glaze or solvent, for a maximum revenue of $9600. The first slack variable, $s_1$, represents the number of air pollution units below the maximum number allowed. Since $s_1 = 60$, the number of air pollution units will be 60 less than the allowable maximum. The second slack variable, $s_2$, represents the unused time in the evaporation tank. Since $s_2 = 0$, the evaporation tank is fully used.

# 7.5 Exercises

*Set up the initial simplex tableau for each of the given problems. You will be asked to solve these problems in Exercises 19–22.*

1. **Juice Business** A juice blender has the following amounts of three kinds of juice: 90 gallons of pineapple juice, 80 gallons of orange juice, and 50 gallons of mango juice. To create its signature juice WildSplash, the juice blender uses 2 gallons of pineapple juice, 1 gallon of orange juice, and 1 gallon of mango juice. To create the new mixture TastyTreat, the juice blender uses 1, 2, and 1 gallon, respectively, of each type of juice. If the juice blender sells the mixtures wholesale at the price of $12 for a batch of WildSplash and $10 for a batch of TastyTreat, how many batches of each product should be created to obtain maximum gross income? What is the maximum gross income?

2. **Window Manufacturing** A window manufacturing company can make 3 windows in small, medium, and large. Each window is worked on by three technicians: Todd, Donna, and Jackson. Todd can work up to 25 hours per week, Donna can work up to 45 hours a week, and Jackson can work 40 hours a week. Todd works 2 hours on the small windows, 1 hour on the medium windows, and 1 hour on the large windows. Donna works 2 hours on the small windows, 3 hours on the medium windows, and 2 hours on the large windows. Jackson works 1 hour on the small windows, 2 hours on the medium windows, and 4 hours on the large windows. The manager is worried about getting overextended, so no more than 4 large windows can be done in a week. The small windows sell for $180, the medium windows for $200, and the large windows for $220. Find the number of each size of windows that should be made weekly in order to maximize profit. Find the maximum possible profit.

3. **Nutrition** A biologist has 500 kilograms of nutrient $A$, 600 kilograms of nutrient $B$, and 300 kilograms of nutrient $C$. These nutrients will be used to make 4 types of food—$P$, $Q$, $R$, and $S$—whose contents (in percent of nutrient per kilogram of food) and whose "growth values" are as shown in the following table:

|  | $P$ | $Q$ | $R$ | $S$ |
|---|---|---|---|---|
| $A$ | 0 | 0 | 37.5 | 62.5 |
| $B$ | 0 | 75 | 50 | 37.5 |
| $C$ | 100 | 25 | 12.5 | 0 |
| Growth Value | 90 | 70 | 60 | 50 |

How many kilograms of each food should be produced in order to maximize total growth value? Find the maximum growth value.

4. **Fish Nutrition** A lake is stocked each spring with three species of fish: $A$, $B$, and $C$. The average weights of the fish are 1.62, 2.12, and 3.01 kilograms for the three species, respectively. Three foods—I, II, and III—are available in the lake. Each fish of species $A$ requires 1.32 units of food I, 2.9 units of food II, and

1.75 units of food III, on the average, each day. Species *B* fish require 2.1 units of food I, .95 units of food II, and .6 units of food III daily. Species *C* fish require .86, 1.52, and 2.01 units of I, II, and III per day, respectively. If 490 units of food I, 897 units of food II, and 653 units of food III are available daily, how should the lake be stocked to maximize the weight of the fish it supports?

*In each of the given exercises, (a) use the simplex method to solve the problem and (b) explain what the values of the slack variables in the optimal solution mean in the context of the problem. (See Examples 1–3).*

5. **Business** A manufacturer of bicycles builds 1-, 3-, and 10-speed models. The bicycles are made of both aluminum and steel. The company has available 91,800 units of steel and 42,000 units of aluminum. The 1-, 3-, and 10-speed models need, respectively, 20, 30, and 40 units of steel and 12, 21, and 16 units of aluminum. How many of each type of bicycle should be made in order to maximize profit if the company makes $80 per 1-speed bike, $120 per 3-speed bike, and $240 per 10-speed bike? What is the maximum possible profit?

6. **Fund-Raising** Liz is working to raise money for breast cancer research by sending informational letters to local neighborhood organizations and church groups. She discovered that each church group requires 2 hours of letter writing and 1 hour of follow-up, while each neighborhood group needs 2 hours of letter writing and 3 hours of follow-up. Liz can raise $1000 from each church group and $2000 from each neighborhood organization, and she has a maximum of 16 hours of letter-writing time and a maximum of 12 hours of follow-up time available per month. Determine the most profitable mixture of groups she should contact and the most money she can raise in a month.

7. **Local News** A local news channel plans a 27-minute Saturday morning news show. The show will be divided into three segments involving sports, news, and weather. Market research has shown that the sports segment should be twice as long as the weather segment. The total time taken by the sports and weather segments should be twice the time taken by the news segment. On the basis of the market research, it is believed that 40, 60, and 50 (in thousands) viewers will watch the program for each minute the sports, news, and weather segments, respectively, are on the air. Find the time that should be allotted to each segment in order to get the maximum number of viewers. Find the number of viewers.

8. **Food Distribution** A food wholesaler has three kinds of individual bags of potato chips: regular, barbeque, and salt and vinegar. She wants to sells the bags of chips in bulk packages. The bronze package consists of 20 bags of regular and 10 bags of barbeque. The silver package contains 20 bags of regular, 10 bags

of barbeque, and 10 bags of salt and vinegar. The gold package consists of 30 bags of regular, 10 bags of barbeque, and 10 bags of salt and vinegar. The profit is $30 on each bronze package, $40 on each silver package, and $60 on each gold package. The food wholesaler has a total of 8000 bags of regular chips, 4000 bags of barbeque, and 2000 bags of salt and vinegar. Assuming all the packages will be sold, how many gold, silver, and bronze packages should be made up in order to maximize profit? What is the maximum profit?

9. **Tree Nursery** Mario Cekada owns a tree nursery business. Mario can sell a weeping Japanese maple of a certain size for $350 in profit. He can sell tricolor beech trees of a certain size for $500 profit. The travel time to acquire the Japanese maple trees is 5 hours while the travel time for the beech trees is 7 hours. The digging process takes 1 hour for the Japanese maple trees and 2 hours for the beech trees. Both kinds of trees require 4 hours of time to deliver to the client. In a particular season, Mario has available 3600 work hours for travel time to acquire the trees, 900 work-hours for digging, and 2600 work-hours for delivery to the clients. How many trees of each kind should he acquire to make a maximum profit? What is the maximum profit?

10. **Business** The Texas Poker Company assembles three different poker sets. Each Royal Flush poker set contains 1000 poker chips, 4 decks of cards, 10 dice, and 2 dealer buttons. Each Deluxe Diamond poker set contains 600 poker chips, 2 decks of cards, 5 dice, and 1 dealer button. The Full House poker set contains 300 poker chips, 2 decks of cards, 5 dice, and 1 dealer button. The Texas Poker Company has 2,800,000 poker chips, 10,000 decks of cards, 25,000 dice, and 6000 dealer buttons in stock. It earns a profit of $38 for each Royal Flush poker set, $22 for each Deluxe Diamond poker set, and $12 for each Full House poker set. How many of each type of poker set should it assemble to maximize profit? What is the maximum profit?

*Use the simplex method to solve the given problems. (See Examples 1–3.)*

11. **Advertising** The Fancy Fashions Store has $8000 available each month for advertising. Newspaper ads cost $400 each, and no more than 20 can be run per month. Radio ads cost $200 each, and no more than 30 can run per month. TV ads cost $1200 each, with a maximum of 6 available each month. Approximately 2000 women will see each newspaper ad, 1200 will hear each radio commercial, and 10,000 will see each TV ad. How much of each type of advertising should be used if the store wants to maximize its ad exposure?

12. **Candy Manufacturing** Caroline's Quality Candy Confectionery is famous for fudge, chocolate cremes, and pralines. Its candy-making equipment is set up to make 100-pound batches at a time. Currently there is a chocolate shortage, and the company can get only 120 pounds of chocolate in the next shipment. On a week's run, the confectionery's cooking and processing equipment is available for a total of 42 machine hours. During the same period, the employees have a total of 56 work hours available for packaging. A batch of fudge requires 20 pounds of chocolate, while a batch of cremes uses 25 pounds of chocolate. The cooking and processing take 120 minutes for fudge, 150 minutes for chocolate cremes, and 200 minutes for pralines. The packaging times, measured in minutes per 1-pound box, are 1, 2, and 3, respectively, for fudge, cremes, and pralines. Determine how many batches of each type of candy the confectionery

should make, assuming that the profit per 1-pound box is 50¢ on fudge, 40¢ on chocolate cremes, and 45¢ on pralines. Also, find the maximum profit for the week.

**13. Fund-Raising** A political party is planning its fund-raising activities for a coming election. It plans to raise money through large fund-raising parties, letters requesting funds, and dinner parties where people can meet the candidate personally. Each large fund-raising party costs $3000, each mailing costs $1000, and each dinner party costs $12,000. The party can spend up to $102,000 for these activities. From experience, it is known that each large party will raise $200,000, each letter campaign will raise $100,000, and each dinner party will raise $600,000. The party is able to carry out as many as 25 of these activities.

(a) How many of each should the party plan to raise the maximum amount of money? What is the maximum amount?

(b) Dinner parties are more expensive than letter campaigns, yet the optimum solution found in part (a) includes dinner parties, but no letter campaigns. Explain how this is possible.

**14. Bread Baking** A baker has 60 units of flour, 132 units of sugar, and 102 units of raisins. A loaf of raisin bread requires 1 unit of flour, 1 unit of sugar, and 2 units of raisins, while a raisin cake needs 2, 4, and 1 units, respectively. If raisin bread sells for $3 a loaf and a raisin cake for $4, how many of each should be baked so that the gross income is maximized? What is the maximum gross income?

**15. Exercise** Ramona, a fitness trainer, has an exercise regimen that includes running, biking, and walking. She has no more than 15 hours per week to devote to exercise, including at most 3 hours for running. She wants to walk at least twice as many hours as she bikes. A 130-pound person like Ramona will burn

on average 531 calories per hour running, 472 calories per hour biking, and 354 calories per hour walking. How many hours per week should Ramona spend on each exercise to maximize the number of calories she burns? What is the maximum number of calories she will burn? (*Hint:* Write the constraint involving walking and biking in the form $\leq 0$.)

**16. Exercise** Liam's exercise regimen includes light calisthenics, swimming, and playing the drums. He has at most 10 hours per week to devote to these activities. He wants the total time he does calisthenics and plays the drums to be at least twice as long as he swims. His neighbors, however, will tolerate no more than 4 hours per week on the drums. A 190-pound person like Liam will burn an average of 388 calories per hour doing calisthenics, 518 calories per hour swimming, and 345 calories per hour playing the drums. How many hours per week should Liam spend on each exercise to maximize the number of calories he burns? What is the maximum number of calories he will burn?

*Use a graphing calculator or a computer program for the simplex method to solve the given linear programming problems.*

**17.** Solve the problem in Exercise 1.

**18.** Exercise 2. Your final answer should consist of whole numbers (because you can't sell half a window).

**19.** Exercise 3

**20.** Exercise 4

✓ **Checkpoint Answer**

**1.** $z = 6{,}400{,}000$

# 7.6 The Simplex Method: Duality and Minimization

📄 **NOTE** Sections 7.6 and 7.7 are independent of each other and may be read in either order.

Here, we present a method of solving *minimization* problems in which all constraints involve $\geq$ and all coefficients of the objective function are positive. When it applies, this method may be more efficient than the method discussed in Section 7.7. However, the method in Section 7.7 applies to a wider variety of problems—both minimization and maximization problems, even those that involve *mixed constraints* ($\leq$, $=$, or $\geq$)—and it has no restrictions on the objective function.

We begin with a necessary tool from matrix algebra: if $A$ is a matrix, then the **transpose** of $A$ is the matrix obtained by interchanging the rows and columns of $A$.

---

**EXAMPLE 1**    Find the transpose of each matrix.

**(a)** $A = \begin{bmatrix} 2 & -1 & 5 \\ 6 & 8 & 0 \\ -3 & 7 & -1 \end{bmatrix}$.

**Solution** Write the rows of matrix $A$ as the columns of the transpose:

$$\text{Transpose of } A = \begin{bmatrix} 2 & 6 & -3 \\ -1 & 8 & 7 \\ 5 & 0 & -1 \end{bmatrix}.$$

**(b)** $A = \begin{bmatrix} 1 & 2 & 4 & 0 \\ 2 & 1 & 7 & 6 \end{bmatrix}.$

**Solution** The transpose of $\begin{bmatrix} 1 & 2 & 4 & 0 \\ 2 & 1 & 7 & 6 \end{bmatrix}$ is $\begin{bmatrix} 1 & 2 \\ 2 & 1 \\ 4 & 7 \\ 0 & 6 \end{bmatrix}.$ ✓1

### ✓ Checkpoint 1

Give the transpose of each matrix.

(a) $\begin{bmatrix} 2 & 4 \\ 6 & 3 \\ 1 & 5 \end{bmatrix}$

(b) $\begin{bmatrix} 4 & 7 & 10 \\ 3 & 2 & 6 \\ 5 & 8 & 12 \end{bmatrix}$

```
       [2    -1    5 ]
       [6     8    Ø ]
       [-3    7   -1 ]
[A]ᵀ
       [2     6   -3 ]
       [-1    8    7 ]
       [5     Ø   -1 ]
```

**Figure 7.31**

**TECHNOLOGY TIP** Most graphing calculators can find the transpose of a matrix. Look for this feature in the MATRIX MATH menu (TI). The transpose of matrix $A$ from Example 1(a) is shown in Figure 7.31.

We now consider linear programming problems satisfying the following conditions:

1. The objective function is to be minimized.
2. All the coefficients of the objective function are nonnegative.
3. All constraints involve $\geq$.
4. All variables are nonnegative.

The method of solving minimization problems presented here is based on an interesting connection between maximization and minimization problems: Any solution of a maximizing problem produces the solution of an associated minimizing problem, and vice versa. Each of the associated problems is called the **dual** of the other. Thus, duals enable us to solve minimization problems of the type just described by the simplex method introduced in Section 7.4.

When dealing with minimization problems, we use $y_1, y_2, y_3$, etc., as variables and denote the objective function by $w$. The next two examples show how to construct the dual problem. Later examples will show how to solve both the dual problem and the original one.

**EXAMPLE 2** Construct the dual of this problem:

$$\begin{aligned} \text{Minimize} \quad & w = 8y_1 + 16y_2 \\ \text{subject to} \quad & y_1 + 5y_2 \geq 9 \\ & 2y_1 + 2y_2 \geq 10 \\ & y_1 \geq 0, y_2 \geq 0. \end{aligned}$$

**Solution** Write the augmented matrix of the system of inequalities *and* include the coefficients of the objective function (not their negatives) as the last row of the matrix:

Constants

$$\begin{bmatrix} 1 & 5 & | & 9 \\ 2 & 2 & | & 10 \\ 8 & 16 & | & 0 \end{bmatrix}.$$

Objective function

Now form the transpose of the preceding matrix:

$$\begin{bmatrix} 1 & 2 & | & 8 \\ 5 & 2 & | & 16 \\ 9 & 10 & | & 0 \end{bmatrix}.$$

In this last matrix, think of the first two rows as constraints and the last row as the objective function. Then the dual maximization problem is as follows:

$$\text{Maximize} \quad z = 9x_1 + 10x_2$$
$$\text{subject to} \quad x_1 + 2x_2 \leq 8$$
$$5x_1 + 2x_2 \leq 16$$
$$x_1 \geq 0, x_2 \geq 0.$$

**EXAMPLE 3** Write the duals of the given minimization linear programming problems.

**(a)** Minimize $\quad w = 10y_1 + 8y_2$
$\quad$ subject to $\quad y_1 + 2y_2 \geq 2$
$$y_1 + y_2 \geq 5$$
$$y_1 \geq 0, y_2 \geq 0.$$

**Solution** Begin by writing the augmented matrix for the given problem:

$$\begin{bmatrix} 1 & 2 & | & 2 \\ 1 & 1 & | & 5 \\ \hline 10 & 8 & | & 0 \end{bmatrix}.$$

Form the transpose of this matrix to get

$$\begin{bmatrix} 1 & 1 & | & 10 \\ 2 & 1 & | & 8 \\ \hline 2 & 5 & | & 0 \end{bmatrix}.$$

The dual problem is stated from this second matrix as follows (using $x$ instead of $y$):

$$\text{Maximize} \quad z = 2x_1 + 5x_2$$
$$\text{subject to} \quad x_1 + x_2 \leq 10$$
$$2x_1 + x_2 \leq 8$$
$$x_1 \geq 0, x_2 \geq 0.$$

**(b)** Minimize $\quad w = 7y_1 + 5y_2 + 8y_3$
$\quad$ subject to $\quad 3y_1 + 2y_2 + y_3 \geq 10$
$$y_1 + y_2 + y_3 \geq 8$$
$$4y_1 + 5y_2 \geq 25$$
$$y_1 \geq 0, y_2 \geq 0, y_3 \geq 0.$$

**Solution** Find the augmented matrix for this problem, as in part (a). Form the dual matrix, which represents the following problem:

$$\text{Maximize} \quad z = 10x_1 + 8x_2 + 25x_3$$
$$\text{subject to} \quad 3x_1 + x_2 + 4x_3 \leq 7$$
$$2x_1 + x_2 + 5x_3 \leq 5$$
$$x_1 + x_2 \leq 8$$
$$x_1 \geq 0, x_2 \geq 0, x_3 \geq 0. \quad \boxed{2}$$

✓ **Checkpoint 2**

Write the dual of the following linear programming problem:

Minimize $\quad w = 2y_1 + 5y_2 + 6y_3$
subject to $\quad 2y_1 + 3y_2 + y_3 \geq 15$
$$y_1 + y_2 + 2y_3 \geq 12$$
$$5y_1 + 3y_2 \geq 10$$
$$y_1 \geq 0, y_2 \geq 0, y_3 \geq 0.$$

In Example 3, all the constraints of the minimization problems were $\geq$ inequalities, while all those in the dual maximization problems were $\leq$ inequalities. This is generally the case; inequalities are reversed when the dual problem is stated.

The following table shows the close connection between a problem and its dual.

| Given Problem | Dual Problem |
|---|---|
| *m* variables | *n* variables |
| *n* constraints | *m* constraints (*m* slack variables) |
| Coefficients from objective function | Constraint constants |
| Constraint constants | Coefficients from objective function |

Now that you know how to construct the dual problem, we examine how it is related to the original problem and how both may be solved.

**EXAMPLE 4**   Solve this problem and its dual:

$$\text{Minimize} \quad w = 8y_1 + 16y_2$$
$$\text{subject to} \quad y_1 + 5y_2 \geq 9$$
$$2y_1 + 2y_2 \geq 10$$
$$y_1 \geq 0, y_2 \geq 0.$$

**Solution**   In Example 2, we saw that the dual problem is

$$\text{Maximize} \quad z = 9x_1 + 10x_2$$
$$\text{subject to} \quad x_1 + 2x_2 \leq 8$$
$$5x_1 + 2x_2 \leq 16$$
$$x_1 \geq 0, x_2 \geq 0.$$

In this case, both the original problem and the dual may be solved geometrically, as in Section 7.2. Figure 7.32(a) shows the region of feasible solutions for the original minimization problem, and Checkpoint 3 shows that

the minimum value of *w* is 48 at the vertex (4, 1). ✔3

Figure 7.32(b) shows the region of feasible solutions for the dual maximization problem, and Checkpoint 4 shows that

the maximum value of *z* is 48 at the vertex (2, 3). ✔4

Even though the regions and the corner points are different, the minimization problem and its dual have the same solution, 48.

✔ **Checkpoint 3**

Use the corner points in Figure 7.32(a) to find the minimum value of $w = 8y_1 + 16y_2$ and where it occurs.

✔ **Checkpoint 4**

Use Figure 7.32(b) to find the maximum value of $z = 9x_1 + 10x_2$ and where it occurs.

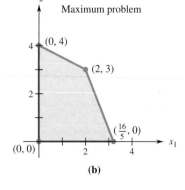

**Figure 7.32**

The next theorem, whose proof requires advanced methods, guarantees that what happened in Example 4 happens in the general case as well.

## Theorem of Duality

The objective function $w$ of a minimizing linear programming problem takes on a minimum value if, and only if, the objective function $z$ of the corresponding dual maximizing problem takes on a maximum value. The maximum value of $z$ equals the minimum value of $w$.

Geometric solution methods were used in Example 4, but the simplex method can also be used. In fact, the final simplex tableau shows the solutions for both the original minimization problem and the dual maximization problem, as illustrated in the next example.

**EXAMPLE 5**    Use the simplex method to solve the minimization problem in Example 4.

**Solution**  First, set up the dual problem, as in Example 4:

$$\text{Maximize}\quad z = 9x_1 + 10x_2$$
$$\text{subject to}\quad x_1 + 2x_2 \le 8$$
$$5x_1 + 2x_2 \le 16$$
$$x_1 \ge 0, x_2 \ge 0.$$

This is a maximization problem in standard form, so it can be solved by the simplex method. Use slack variables to write the inequalities and the objective function as equations:

$$x_1 + 2x_2 + s_1 \qquad\qquad = 8$$
$$5x_1 + 2x_2 \qquad + s_2 \qquad = 16$$
$$-9x_1 - 10x_2 \qquad\qquad + z = 0,$$

with $x_1 \ge 0$, $x_2 \ge 0$, $s_1 \ge 0$, and $s_2 \ge 0$. So the initial tableau is

$$
\begin{array}{cccc}
x_1 & x_2 & s_1 & s_2 \\
\end{array}
$$

$$
\left[
\begin{array}{cccc|c}
1 & \mathbf{2} & 1 & 0 & 8 \\
5 & 2 & 0 & 1 & 16 \\
\hline
-9 & -10 & 0 & 0 & 0
\end{array}
\right].
\qquad
\begin{array}{l}
\text{Quotients} \\
8/2 = 4 \\
16/2 = 8
\end{array}
$$

The quotients show that the first pivot is the 2 shaded in blue. Pivoting is done as follows:

$$
\begin{array}{cccc}
x_1 & x_2 & s_1 & s_2 \\
\end{array}
$$

$$
\left[
\begin{array}{cccc|c}
\frac{1}{2} & 1 & \frac{1}{2} & 0 & 4 \\
4 & 0 & -1 & 1 & 8 \\
\hline
-4 & 0 & 5 & 0 & 40
\end{array}
\right].
\qquad
\begin{array}{l}
\frac{1}{2}R_1 \\
-2R_1 + R_2 \\
10R_1 + R_3
\end{array}
$$

✓ **Checkpoint 5**

In the second tableau, find the next pivot.

Checkpoint 5 shows that the new pivot is the 4 in row 2, column 1. ✓₅ Pivoting leads to the final simplex tableau:

$$
\begin{array}{cccc}
x_1 & x_2 & s_1 & s_2 \\
\end{array}
$$

$$
\left[
\begin{array}{cccc|c}
0 & 1 & \frac{5}{8} & -\frac{1}{8} & 3 \\
1 & 0 & -\frac{1}{4} & \frac{1}{4} & 2 \\
\hline
0 & 0 & 4 & 1 & 48
\end{array}
\right].
\qquad
\begin{array}{l}
-\frac{1}{2}R_2 + R_1 \\
\frac{1}{4}R_2 \\
4R_2 + R_3
\end{array}
$$

The final simplex tableau shows that the maximum value of 48 occurs when $x_1 = 2$ and $x_2 = 3$. In Example 4, we saw that the minimum value of 48 occurs when $y_1 = 4$ and $y_2 = 1$. Note that this information appears in the last row (shown in blue). The minimum value of 48 is in the lower right-hand corner, and the values where this occurs ($y_1 = 4$ and $y_2 = 1$) are in the last row at the bottom of the slack-variable columns.

A minimization problem that meets the conditions listed after Example 1 can be solved by the method used in Example 5 and summarized here.

## Solving Minimization Problems with Duals

1. Find the dual standard maximization problem.[*]
2. Use the simplex method to solve the dual maximization problem.
3. Read the optimal solution of the original minimization problem from the final simplex tableau:

   $y_1$ is the last entry in the column corresponding to the first slack variable;

   $y_2$ is the last entry in the column corresponding to the second slack variable; and so on.

These values of $y_1, y_2, y_3$, etc., produce the minimum value of $w$, which is the entry in the lower right-hand corner of the tableau.

---

**EXAMPLE 6**    Minimize $w = 3y_1 + 2y_2$

subject to    $y_1 + 3y_2 \geq 6$

$2y_1 + y_2 \geq 3$

$y_1 \geq 0, y_2 \geq 0.$

**Solution**  Use the given information to write the matrix:

$$\left[\begin{array}{cc|c} 1 & 3 & 6 \\ 2 & 1 & 3 \\ \hline 3 & 2 & 0 \end{array}\right].$$

Transpose to get the following matrix for the dual problem:

$$\left[\begin{array}{cc|c} 1 & 2 & 3 \\ 3 & 1 & 2 \\ \hline 6 & 3 & 0 \end{array}\right].$$

Write the dual problem from this matrix as follows:

Maximize    $z = 6x_1 + 3x_2$

subject to    $x_1 + 2x_2 \leq 3$

$3x_1 + x_2 \leq 2$

$x_1 \geq 0, x_2 \geq 0.$

Solve this standard maximization problem by the simplex method. Start by introducing slack variables, giving the system

$$x_1 + 2x_2 + s_1 \qquad = 3$$
$$3x_1 + x_2 \qquad + s_2 \qquad = 2$$
$$-6x_1 - 3x_2 - 0s_1 - 0s_2 + z = 0,$$

with $x_1 \geq 0, x_2 \geq 0, s_1 \geq 0,$ and $s_2 \geq 0.$

---

[*]The coefficients of the objective function in the minimization problem are the constants on the right side of the constraints in the dual maximization problem. So when all these coefficients are nonnegative (condition 2), the dual problem is in standard maximum form.

The initial tableau for this system is

|  | $x_1$ | $x_2$ | $s_1$ | $s_2$ |  | Quotients |
|---|---|---|---|---|---|---|
|  | 1 | 2 | 1 | 0 | 3 | $3/1 = 3$ |
|  | **3** | 1 | 0 | 1 | 2 | $2/3$ |
|  | $-6$ | $-3$ | 0 | 0 | 0 |  |

with the pivot as indicated. Two rounds of pivoting produce the following final tableau:

$$\begin{bmatrix} x_1 & x_2 & s_1 & s_2 & \\ 0 & 1 & \frac{3}{5} & -\frac{1}{5} & \frac{7}{5} \\ 1 & 0 & -\frac{1}{5} & \frac{2}{5} & \frac{1}{5} \\ 0 & 0 & \frac{3}{5} & \frac{9}{5} & \frac{27}{5} \end{bmatrix}$$

$$\phantom{xxxx} y_1 \phantom{xx} y_2 \phantom{xx} w$$

As indicated in blue below the final tableau, the last entries in the columns corresponding to the slack variables ($s_1$ and $s_2$) give the values of the original variables $y_1$ and $y_2$ that produce the minimal value of $w$. This minimal value of $w$ appears in the lower right-hand corner (and is the same as the maximal value of $z$ in the dual problem). So the solution of the given minimization problem is as follows:

The minimum value of $w = 3y_1 + 2y_2$, subject to the given constraints, is $\frac{27}{5}$ and occurs when $y_1 = \frac{3}{5}$ and $y_2 = \frac{9}{5}$.  ✔ 6

✓**Checkpoint 6**

Minimize    $w = 10y_1 + 8y_2$

subject to    $y_1 + 2y_2 \geq 2$

$y_1 + \phantom{2}y_2 \geq 5$

$y_1 \geq 0, y_2 \geq 0.$

**EXAMPLE 7**    A minimization problem in three variables was solved by the use of duals. The final simplex tableau for the dual maximization problem is shown here:

$$\begin{bmatrix} x_1 & x_2 & x_3 & s_1 & s_2 & s_3 & \\ 3 & 1 & 1 & 0 & 9 & 0 & 1 \\ 13 & -1 & 0 & 1 & -2 & 0 & 10 \\ 9 & 10 & 0 & 0 & 7 & 1 & 7 \\ 5 & 1 & 0 & 4 & 1 & 7 & 28 \end{bmatrix}$$

**(a)**  What is the optimal solution of the dual minimization problem?

**Solution**  Looking at the bottom of the columns corresponding to the slack variables $s_1, s_2$, and $s_3$, we see that the solution of the minimization problem is

$y_1 = 4,$     $y_2 = 1,$     and     $y_3 = 7$, with a minimal value of $w = 28$.

**(b)**  What is the optimal solution of the dual maximization problem?

**Solution**  Since $s_1, s_2$, and $s_3$ are slack variables by part (a), the variables in the dual problem are $x_1, x_2$, and $x_3$. Read the solution from the final tableau, as in Sections 7.4 and 7.5:

$x_1 = 0,$     $x_2 = 0,$     and     $x_3 = 1,$     with a maximal value of $z = 28$.

## Further Uses of the Dual

The dual is useful not only in solving minimization problems, but also in seeing how small changes in one variable will affect the value of the objective function. For example, suppose an animal breeder needs at least 6 units per day of nutrient $A$ and at least 3 units of nutrient $B$ and that the breeder can choose between two different feeds: feed 1 and feed 2. Find the minimum cost for the breeder if each bag of feed 1 costs $3 and provides 1 unit of nutrient $A$ and 2 units of $B$, while each bag of feed 2 costs $2 and provides 3 units of nutrient $A$ and 1 of $B$.

If $y_1$ represents the number of bags of feed 1 and $y_2$ represents the number of bags of feed 2, the given information leads to the following minimization problem:

$$\text{Minimize} \quad w = 3y_1 + 2y_2$$
$$\text{subject to} \quad y_1 + 3y_2 \geq 6$$
$$2y_1 + y_2 \geq 3$$
$$y_1 \geq 0, y_2 \geq 0.$$

This minimization linear programming problem is the one we solved in Example 6 of this section. In that example, we formed the dual and reached the following final tableau:

$$
\begin{array}{cccc}
x_1 & x_2 & s_1 & s_2 \\
\end{array}
$$
$$
\left[
\begin{array}{cccc|c}
0 & 1 & \frac{3}{5} & -\frac{1}{5} & \frac{7}{5} \\
1 & 0 & -\frac{1}{5} & \frac{2}{5} & \frac{1}{5} \\
\hline
0 & 0 & \frac{3}{5} & \frac{9}{5} & \frac{27}{5}
\end{array}
\right].
$$

This final tableau shows that the breeder will obtain minimum feed costs by using $\frac{3}{5}$ bag of feed 1 and $\frac{9}{5}$ bags of feed 2 per day, for a daily cost of $\frac{27}{5} = 5.40$ dollars.

Now look at the data from the feed problem shown in the following table:

| | Units of Nutrient (per Bag) | | Cost per Bag |
|---|---|---|---|
| | **A** | **B** | |
| Feed 1 | 1 | 2 | $3 |
| Feed 2 | 3 | 1 | $2 |
| Minimum Nutrient Needed | 6 | 3 | |

If $x_1$ and $x_2$ are the cost *per unit* of nutrients $A$ and $B$, the constraints of the dual problem can be stated as follows (see pages 378–379):

$$\text{Cost of feed 1:} \quad x_1 + 2x_2 \leq 3$$
$$\text{Cost of feed 2:} \quad 3x_1 + x_2 \leq 2.$$

The solution of the dual problem, which maximizes nutrients, also can be read from the final tableau above:

$$x_1 = \frac{1}{5} = .20 \quad \text{and} \quad x_2 = \frac{7}{5} = 1.40.$$

This means that a unit of nutrient $A$ costs $\frac{1}{5}$ of a dollar $= \$.20$, while a unit of nutrient B costs $\frac{7}{5}$ dollars $= \$1.40$. The minimum daily cost, $5.40$, is found as follows.

$$
\begin{array}{ll}
(\$.20 \text{ per unit of } A) \times (6 \text{ units of } A) = & \$1.20 \\
+ (\$1.40 \text{ per unit of } B) \times (3 \text{ units of } B) = & \$4.20 \\
\hline
\text{Minimum daily cost} = & \$5.40.
\end{array}
$$

The numbers .20 and 1.40 are called the **shadow costs** of the nutrients. These two numbers from the dual, $.20 and $1.40, also allow the breeder to estimate feed costs for "small" changes in nutrient requirements. For example, an increase of 1 unit in the requirement for each nutrient would produce a total cost as follows:

| | |
|---|---|
| $5.40 | 6 units of A, 3 of B |
| .20 | 1 extra unit of A |
| 1.40 | 1 extra unit of B |
| $7.00 · | Total cost per day |

# 7.6 Exercises

*Find the transpose of each matrix. (See Example 1.)*

**1.** $\begin{bmatrix} 3 & -4 & 5 \\ 1 & 10 & 7 \\ 0 & 3 & 6 \end{bmatrix}$

**2.** $\begin{bmatrix} 3 & -5 & 9 & 4 \\ 1 & 6 & -7 & 0 \\ 4 & 18 & 11 & 9 \end{bmatrix}$

**3.** $\begin{bmatrix} 3 & 0 & 14 & -5 & 3 \\ 4 & 17 & 8 & -6 & 1 \end{bmatrix}$

**4.** $\begin{bmatrix} 15 & -6 & -2 \\ 13 & -1 & 11 \\ 10 & 12 & -3 \\ 24 & 1 & 0 \end{bmatrix}$

*State the dual problem for each of the given problems, but do not solve it. (See Examples 2 and 3.)*

**5.** Minimize $w = 3y_1 + 5y_2$
subject to $3y_1 + y_2 \geq 4$
$-y_1 + 2y_2 \geq 6$
$y_1 \geq 0, y_2 \geq 0.$

**6.** Minimize $w = 4y_1 + 7y_2$
subject to $y_1 + y_2 \geq 17$
$3y_1 + 6y_2 \geq 21$
$2y_1 + 4y_2 \geq 19$
$y_1 \geq 0, y_2 \geq 0.$

**7.** Minimize $w = 2y_1 + 8y_2$
subject to $y_1 + 7y_2 \geq 18$
$4y_1 + y_2 \geq 15$
$5y_1 + 3y_2 \geq 20$
$y_1 \geq 0, y_2 \geq 0.$

**8.** Minimize $w = y_1 + 2y_2 + 6y_3$
subject to $3y_1 + 4y_2 + 6y_3 \geq 8$
$y_1 + 5y_2 + 2y_3 \geq 12$
$y_1 \geq 0, y_2 \geq 0, y_3 \geq 0.$

**9.** Minimize $w = 5y_1 + y_2 + 3y_3$
subject to $7y_1 + 6y_2 + 8y_3 \geq 18$
$4y_1 + 5y_2 + 10y_3 \geq 20$
$y_1 \geq 0, y_2 \geq 0, y_3 \geq 0.$

**10.** Minimize $w = 4y_1 + 3y_2 + y_3$
subject to $y_1 + 2y_2 + 3y_3 \geq 115$
$2y_1 + y_2 + 8y_3 \geq 200$
$y_1 \quad - y_3 \geq 50$
$y_1 \geq 0, y_2 \geq 0, y_3 \geq 0.$

**11.** Minimize $w = 8y_1 + 9y_2 + 3y_3$
subject to $y_1 + y_2 + y_3 \geq 5$
$y_1 + y_2 \quad \geq 4$
$2y_1 + y_2 + 3y_3 \geq 15$
$y_1 \geq 0, y_2 \geq 0, y_3 \geq 0.$

**12.** Minimize $w = y_1 + 2y_2 + y_3 + 5y_4$
subject to $y_1 + y_2 + y_3 + y_4 \geq 50$
$3y_1 + y_2 + 2y_3 + y_4 \geq 100$
$y_1 \geq 0, y_2 \geq 0, y_3 \geq 0, y_4 \geq 0.$

*Use duality to solve the problem that was set up in the given exercise.*

**13.** Exercise 9    **14.** Exercise 8

**15.** Exercise 11    **16.** Exercise 12

*Use duality to solve the given problems. (See Examples 5 and 6.)*

**17.** Minimize $w = 2y_1 + y_2 + 3y_3$
subject to $y_1 + y_2 + y_3 \geq 100$
$2y_1 + y_2 \quad \geq 50$
$y_1 \geq 0, y_2 \geq 0, y_3 \geq 0.$

**18.** Minimize $w = 2y_1 + 4y_2$
subject to $4y_1 + 2y_2 \geq 10$
$4y_1 + y_2 \geq 8$
$2y_1 + y_2 \geq 12$
$y_1 \geq 0, y_2 \geq 0.$

**19.** Minimize $w = 3y_1 + y_2 + 4y_3$
subject to $2y_1 + y_2 + y_3 \geq 6$
$y_1 + 2y_2 + y_3 \geq 8$
$2y_1 + y_2 + 2y_3 \geq 12$
$y_1 \geq 0, y_2 \geq 0, y_3 \geq 0.$

**20.** Minimize $w = y_1 + y_2 + 3y_3$
subject to $2y_1 + 6y_2 + y_3 \geq 8$
$y_1 + 2y_2 + 4y_3 \geq 12$
$y_1 \geq 0, y_2 \geq 0, y_3 \geq 0.$

**21.** Minimize $w = 6y_1 + 4y_2 + 2y_3$
subject to $2y_1 + 2y_2 + y_3 \geq 2$
$y_1 + 3y_2 + 2y_3 \geq 3$
$y_1 + y_2 + 2y_3 \geq 4$
$y_1 \geq 0, y_2 \geq 0, y_3 \geq 0.$

**22.** Minimize $w = 12y_1 + 10y_2 + 7y_3$
subject to $2y_1 + y_2 + y_3 \geq 7$
$y_1 + 2y_2 + y_3 \geq 4$
$y_1 \geq 0, y_2 \geq 0, y_3 \geq 0.$

**23.** Minimize $w = 20y_1 + 12y_2 + 40y_3$
subject to $y_1 + y_2 + 5y_3 \geq 20$
$2y_1 + y_2 + y_3 \geq 30$
$y_1 \geq 0, y_2 \geq 0, y_3 \geq 0.$

**24.** Minimize $w = 4y_1 + 5y_2$
subject to $10y_1 + 5y_2 \geq 100$
$20y_1 + 10y_2 \geq 150$
$y_1 \geq 0, y_2 \geq 0.$

**25.** Minimize $w = 4y_1 + 2y_2 + y_3$
subject to $y_1 + y_2 + y_3 \geq 4$
$3y_1 + y_2 + 3y_3 \geq 6$
$y_1 + y_2 + 3y_3 \geq 5$
$y_1 \geq 0, y_2 \geq 0, y_3 \geq 0.$

**26.** Minimize   $w = 3y_1 + 2y_2$

subject to     $2y_1 + 3y_2 \geq 60$

$y_1 + 4y_2 \geq 40$

$y_1 \geq 0, y_2 \geq 0.$

**27.** **Nutrition** Glenn Russell, who is dieting, requires two food supplements: I and II. He can get these supplements from two different products—A and B—as shown in the following table:

|  | Supplement (Grams per Serving) | |
|---|---|---|
|  | **I** | **II** |
| Product **A** | 4 | 2 |
| Product **B** | 2 | 5 |

Glenn's physician has recommended that he include at least 20 grams of supplement I and 18 grams of supplement II in his diet. If product A costs 24¢ per serving and product B costs 40¢ per serving, how can he satisfy these requirements most economically?

**28.** **Animal Food** An animal food must provide at least 54 units of vitamins and 60 calories per serving. One gram of soybean meal provides at least 2.5 units of vitamins and 5 calories. One gram of meat by-products provides at least 4.5 units of vitamins and 3 calories. One gram of grain provides at least 5 units of vitamins and 10 calories. If a gram of soybean meal costs 8¢, a gram of meat by-products 9¢, and a gram of grain 10¢, what mixture of these three ingredients will provide the required vitamins and calories at minimum cost?

**29.** **Beer Production** A brewery produces regular beer and a lower-carbohydrate "light" beer. Steady customers of the brewery buy 12 units of regular beer and 10 units of light beer monthly. While setting up the brewery to produce the beers, the management decides to produce extra beer, beyond the need to satisfy the steady customers. The cost per unit of regular beer is $36,000, and the cost per unit of light beer is $48,000. Every unit of regular beer brings in $100,000 in revenue, while every unit of light beer brings in $300,000 in revenue. The brewery wants at least $7,000,000 in revenue. At least 20 additional units of beer can be sold. How much of each type of beer should be made so as to minimize total production costs?

**30.** **Business** Joan has a part-time job conducting public-opinion interviews. She has found that a political interview takes 45 minutes and a market interview takes 55 minutes. To allow more time for her full-time job, she needs to minimize the time she spends doing interviews. Unfortunately, to keep her part-time job, she must complete at least 8 interviews each week. Also, she must earn at least $60 per week at this job, at which she earns $8 for each political interview and $10 for each market interview. Finally, to stay in good standing with her supervisor, she must earn at least 40 bonus points per week; she receives 6 bonus points for each political interview and 5 points for each market interview. How many of each interview should she do each week to minimize the time spent?

**31.** **Feed Production** Refer to the end of this section in the text, on minimizing the daily cost of feeds.

(a) Find a combination of feeds that will cost $7.00 and give 7 units of A and 4 units of B.

(b) Use the dual variables to predict the daily cost of feed if the requirements change to 5 units of A and 4 units of B. Find a combination of feeds to meet these requirements at the predicted price.

**32.** Refer to Example 1 in Section 7.5.

(a) Give the dual problem.

(b) Use the shadow values to estimate the farmer's profit if land is cut to 90 acres, but capital increases to $21,000.

(c) Suppose the farmer has 110 acres, but only $19,000. Find the optimal profit and the planting strategy that will produce this profit.

**Toy Production** *For Exercises 33 and 34, use the following information. A small toy-manufacturing firm has 200 squares of felt, 600 ounces of stuffing, and 90 feet of trim available to make two types of toys: a small bear and a monkey. The bear requires 1 square of felt and 4 ounces of stuffing. The monkey requires 2 squares of felt, 3 ounces of stuffing, and 1 foot of trim. The firm makes $1 profit on each bear and $1.50 profit on each monkey. The linear programming problem to maximize profit is*

$$\text{Maximize} \quad z = x_1 + 1.5x_2$$

$$\text{subject to} \quad x_1 + 2x_2 \leq 200$$

$$4x_1 + 3x_2 \leq 600$$

$$x_2 \leq 90$$

$$x_1 \geq 0, x_2 \geq 0.$$

*The final simplex tableau is*

$$\begin{bmatrix} 1 & 0 & -.6 & .4 & 0 & 120 \\ 0 & 0 & -.8 & .2 & 1 & 50 \\ 0 & 1 & .8 & -.2 & 0 & 40 \\ 0 & 0 & .6 & .1 & 0 & 180 \end{bmatrix}.$$

**33.** (a) What is the corresponding dual problem?

(b) What is the optimal solution to the dual problem?

(c) Use the shadow values to estimate the profit the firm will make if its supply of felt increases to 210 squares.

**34.** How much profit will the firm make if its supply of stuffing is cut to 590 ounces and its supply of trim is cut to 80 feet?

✓**Checkpoint Answers**

**1.** (a) $\begin{bmatrix} 2 & 6 & 1 \\ 4 & 3 & 5 \end{bmatrix}$  (b) $\begin{bmatrix} 4 & 3 & 5 \\ 7 & 2 & 8 \\ 10 & 6 & 12 \end{bmatrix}$

**2.** Maximize   $z = 15x_1 + 12x_2 + 10x_3$

subject to     $2x_1 + x_2 + 5x_3 \leq 2$

$3x_1 + x_2 + 3x_3 \leq 5$

$x_1 + 2x_2 \leq 6$

$x_1 \geq 0, x_2 \geq 0, x_3 \geq 0.$

**3.** 48 when $y_1 = 4$ and $y_2 = 1$     **4.** 48 when $x_1 = 2$ and $x_2 = 3$

**5.** The 4 in row 2, column 1

**6.** $y_1 = 0$ and $y_2 = 5$, for a minimum of 40

# 7.7 The Simplex Method: Nonstandard Problems

📄 **NOTE** Section 7.7 is independent of Section 7.6 and may be read first, if desired.

So far, the simplex method has been used to solve problems in which the variables are non-negative and all the other constraints are of one type (either all $\leq$ or all $\geq$). Now we extend the simplex method to linear programming problems with nonnegative variables and mixed constraints ($\leq$, $=$, and $\geq$).

The solution method to be used here requires that all inequality constraints be written so that the constant on the right side is nonnegative. For instance, the inequality

$$4x_1 + 5x_2 - 12x_3 \leq -30$$

can be replaced by the equivalent one obtained by multiplying both sides by $-1$ and reversing the direction of the inequality sign:

$$-4x_1 - 5x_2 + 12x_3 \geq 30.$$

## Maximization with $\leq$ and $\geq$ Constraints

As is always the case when the simplex method is involved, each inequality constraint must be written as an equation. Constraints involving $\leq$ are converted to equations by adding a nonnegative slack variable, as in Section 7.4. Similarly, constraints involving $\geq$ are converted to equations by *subtracting* a nonnegative **surplus variable**. For example, the inequality $2x_1 - x_2 + 5x_3 \geq 12$ is written as

$$2x_1 - x_2 + 5x_3 - s_1 = 12,$$

where $s_1 \geq 0$. The surplus variable $s_1$ represents the amount by which $2x_1 - x_2 + 5x_3$ exceeds 12.

**EXAMPLE 1**    Restate the following problem in terms of equations, and write its initial simplex tableau:

$$\text{Maximize} \quad z = 4x_1 + 10x_2 + 6x_3$$
$$\text{subject to} \quad x_1 + 4x_2 + 4x_3 \geq 8$$
$$x_1 + 3x_2 + 2x_3 \leq 6$$
$$3x_1 + 4x_2 + 8x_3 \leq 22$$
$$x_1 \geq 0, x_2 \geq 0, x_3 \geq 0.$$

**Solution**  In order to write the constraints as equations, subtract a surplus variable from the $\geq$ constraint and add a slack variable to each $\leq$ constraint. So the problem becomes

$$\text{Maximize} \quad z = 4x_1 + 10x_2 + 6x_3$$
$$\text{subject to} \quad x_1 + 4x_2 + 4x_3 - s_1 \qquad\qquad = 8$$
$$x_1 + 3x_2 + 2x_3 \qquad + s_2 \qquad = 6$$
$$3x_1 + 4x_2 + 8x_3 \qquad\qquad + s_3 = 22$$
$$x_1 \geq 0, x_2 \geq 0, x_3 \geq 0, s_1 \geq 0, s_2 \geq 0, s_3 \geq 0.$$

✓ **Checkpoint 1**

(a) Restate this problem in terms of equations:

Maximize $\;z = 3x_1 - 2x_2$

subject to $\;2x_1 + 3x_2 \leq 8$

$6x_1 - 2x_2 \geq 3$

$x_1 + 4x_2 \geq 1$

$x_1 \geq 0, x_2 \geq 0.$

(b) Write the initial simplex tableau.

Write the objective function as $-4x_1 - 10x_2 - 6x_3 + z = 0$ and use the coefficients of the four equations to write the initial simplex tableau (omitting the $z$ column):

$$\begin{array}{cccccc}
x_1 & x_2 & x_3 & s_1 & s_2 & s_3 \\
\end{array}$$

$$\left[\begin{array}{cccccc|c}
1 & 4 & 4 & -1 & 0 & 0 & 8 \\
1 & 3 & 2 & 0 & 1 & 0 & 6 \\
3 & 4 & 8 & 0 & 0 & 1 & 22 \\
\hline
-4 & -10 & -6 & 0 & 0 & 0 & 0
\end{array}\right]. \quad ✓_1$$

The tableau in Example 1 resembles those which have appeared previously, and similar terminology is used. The variables whose columns have one entry that is $\pm 1$ and the rest that are 0 will be called **basic variables**; the other variables are nonbasic. A solution obtained by setting the nonbasic variables equal to 0 and solving for the basic variables (by looking at the constants in the right-hand column) will be called a **basic solution**. A basic solution that is feasible is called a **basic feasible solution**. In the tableau of Example 1, for instance, the basic variables are $s_1$, $s_2$, and $s_3$, and the basic solution is

$$x_1 = 0, \quad x_2 = 0, \quad x_3 = 0, \quad s_1 = -8, \quad s_2 = 6, \quad \text{and} \quad s_3 = 22.$$

However, because one variable is negative, this solution is not feasible. ✓2

The solution method for problems such as the one in Example 1 consists of two stages. **Stage I** consists of finding a basic *feasible* solution that can be used as the starting point for the simplex method. (This stage is unnecessary in a standard maximization problem, because the solution given by the initial tableau is always feasible.) There are many systematic ways of finding a feasible solution, all of which depend on the fact that row operations produce a tableau that represents a system with the same solutions as the original one. One such technique is explained in the next example. Since the immediate goal is to find a feasible solution, not necessarily an optimal one, the procedures for choosing pivots differ from those in the ordinary simplex method.

✓ **Checkpoint 2**

State the basic solution given by each tableau. Is it feasible?

(a)

| $x_1$ | $x_2$ | $s_1$ | $s_2$ | $s_3$ | |
|---|---|---|---|---|---|
| 3 | −5 | 1 | 0 | 0 | 12 |
| 4 | 7 | 0 | 1 | 0 | 6 |
| 1 | 3 | 0 | 0 | −1 | 5 |
| −7 | 4 | 0 | 0 | 0 | 0 |

(b)

| $x_1$ | $x_2$ | $x_3$ | $s_1$ | $s_2$ | |
|---|---|---|---|---|---|
| 9 | 8 | −1 | 1 | 0 | 12 |
| −5 | 3 | 0 | 0 | 1 | 7 |
| 4 | 2 | 3 | 0 | 0 | 0 |

**EXAMPLE 2**   Find a basic feasible solution for the problem in Example 1, whose initial tableau is

| $x_1$ | $x_2$ | $x_3$ | $s_1$ | $s_2$ | $s_3$ | |
|---|---|---|---|---|---|---|
| 1 | 4 | 4 | −1 | 0 | 0 | 8 |
| 1 | 3 | 2 | 0 | 1 | 0 | 6 |
| 3 | 4 | 8 | 0 | 0 | 1 | 22 |
| −4 | −10 | −6 | 0 | 0 | 0 | 0 |

**Solution**  In the basic solution given by this tableau, $s_1$ has a negative value. The only nonzero entry in its column is the −1 in row 1. Choose any *positive* entry in row 1 except the entry on the far right. The column that the chosen entry is in will be the pivot column. We choose the first positive entry in row 1: the 1 in column 1. The pivot row is determined in the usual way by considering quotients (the constant at the right end of the row, divided by the positive entry in the pivot column) in each row except the objective row:

$$\frac{8}{1} = 8, \qquad \frac{6}{1} = 6, \qquad \frac{22}{3} = 7\frac{1}{3}.$$

The smallest quotient is 6, so the pivot is the 1 in row 2, column 1. Pivoting in the usual way leads to the tableau

| $x_1$ | $x_2$ | $x_3$ | $s_1$ | $s_2$ | $s_3$ | | |
|---|---|---|---|---|---|---|---|
| 0 | 1 | 2 | −1 | −1 | 0 | 2 | $-R_2 + R_1$ |
| 1 | 3 | 2 | 0 | 1 | 0 | 6 | |
| 0 | −5 | 2 | 0 | −3 | 1 | 4 | $-3R_2 + R_3$ |
| 0 | 2 | 2 | 0 | 4 | 0 | 24 | $4R_2 + R_4$ |

and the basic solution

$$x_1 = 6, \qquad x_2 = 0, \qquad x_3 = 0, \qquad s_1 = -2, \qquad s_2 = 0, \qquad \text{and} \qquad s_3 = 4.$$

Since the basic variable $s_1$ is negative, this solution is not feasible. So we repeat the pivoting process. The $s_1$ column has a −1 in row 1, so we choose a positive entry in that row, namely, the 1 in row 1, column 2. This choice makes column 2 the pivot column. The pivot row is determined by the quotients $\frac{2}{1} = 2$ and $\frac{6}{3} = 2$. (Negative entries in the pivot column and the entry in the objective row are not used.) Since there is a tie, we can choose either row 1 or row 2. We choose row 1 and use the 1 in row 1, column 2, as the pivot. Pivoting produces the tableau

$$
\begin{array}{cccccc}
x_1 & x_2 & x_3 & s_1 & s_2 & s_3 \\
\end{array}
$$

$$
\left[
\begin{array}{cccccc|c}
0 & 1 & 2 & -1 & -1 & 0 & 2 \\
1 & 0 & -4 & 3 & 4 & 0 & 0 \\
0 & 0 & 12 & -5 & -8 & 1 & 14 \\
0 & 0 & -2 & 2 & 6 & 0 & 20 \\
\end{array}
\right]
\quad
\begin{array}{l}
\\
-3R_1 + R_2 \\
5R_1 + R_3 \\
-2R_1 + R_4 \\
\end{array}
$$

and the basic *feasible* solution

$$
x_1 = 0, \quad x_2 = 2, \quad x_3 = 0, \quad s_1 = 0, \quad s_2 = 0, \quad \text{and} \quad s_3 = 14.
$$

Once a basic feasible solution has been found, Stage I is ended. The procedures used in Stage I are summarized here.[*]

## Finding a Basic Feasible Solution

1. If any basic variable has a negative value, locate the $-1$ in that variable's column and note the row it is in.

2. In the row determined in Step 1, choose a positive entry (other than the one at the far right) and note the column it is in. This is the pivot column.

3. Use the positive entries in the pivot column (except in the objective row) to form quotients and select the pivot.

4. Pivot as usual, which results in the pivot column having one entry that is 1 and the rest that are 0's.

5. Repeat Steps 1–4 until every basic variable is nonnegative, so that the basic solution given by the tableau is feasible. If it ever becomes impossible to continue, then the problem has no feasible solution.

One way to make the required choices systematically is to choose the first possibility in each case (going from the top for rows and from the left for columns). However, any choice meeting the required conditions may be used. For maximum efficiency, it is usually best to choose the pivot column in Step 2, so that the pivot is in the same row chosen in Step 1, if this is possible. ✔3

In **Stage II**, the simplex method is applied as usual to the tableau that produced the basic feasible solution in Stage I. Just as in Section 7.4, each round of pivoting replaces the basic feasible solution of one tableau with the basic feasible solution of a new tableau in such a way that the value of the objective function is increased, until an optimal value is obtained (or it becomes clear that no optimal solution exists).

### ✓ Checkpoint 3

The initial tableau of a maximization problem is shown. Use column 1 as the pivot column for carrying out Stage I, and state the basic feasible solution that results.

$$
\begin{array}{cccc}
x_1 & x_2 & s_1 & s_2 \\
\end{array}
$$

$$
\left[
\begin{array}{cccc|c}
1 & 3 & 1 & 0 & 70 \\
2 & 4 & 0 & -1 & 50 \\
-8 & -10 & 0 & 0 & 0 \\
\end{array}
\right]
$$

**EXAMPLE 3**   Solve the linear programming problem in Example 1 of this section.

**Solution**  A basic feasible solution for this problem was found in Example 2 by using the tableau shown below. However, this solution is not maximal, because there is a negative indicator in the objective row. So we use the simplex method. The most negative indicator

---

[*]Except in rare cases that do not occur in this book, this method either eventually produces a basic feasible solution or shows that one does not exist. The *two-phase method* using artificial variables, which is discussed in more advanced texts, works in all cases and often is more efficient.

determines the pivot column, and the usual quotients determine that the number 2 in row 1, column 3, is the pivot:

$$
\begin{array}{c}
\begin{array}{cccccc} x_1 & x_2 & x_3 & s_1 & s_2 & s_3 \end{array} \\
\left[
\begin{array}{cccccc|c}
0 & 1 & \mathbf{2} & -1 & -1 & 0 & 2 \\
1 & 0 & -4 & 3 & 4 & 0 & 0 \\
0 & 0 & 12 & -5 & -8 & 1 & 14 \\
\hline
0 & 0 & -2 & 2 & 6 & 0 & 20
\end{array}
\right].
\end{array}
\qquad
\begin{array}{l}
\text{Quotients} \\
2/2 \leftarrow \text{Smallest} \\
\\
14/12
\end{array}
$$

Most negative indicator

Pivoting leads to the final tableau:

$$
\begin{array}{c}
\begin{array}{cccccc} x_1 & x_2 & x_3 & s_1 & s_2 & s_3 \end{array} \\
\left[
\begin{array}{cccccc|c}
0 & \frac{1}{2} & 1 & -\frac{1}{2} & -\frac{1}{2} & 0 & 1 \\
1 & 0 & -4 & 3 & 4 & 0 & 0 \\
0 & 0 & 12 & -5 & -8 & 1 & 14 \\
\hline
0 & 0 & -2 & 2 & 6 & 0 & 20
\end{array}
\right]
\end{array}
\qquad \tfrac{1}{2}R_1
$$

$$
\begin{array}{c}
\begin{array}{cccccc} x_1 & x_2 & x_3 & s_1 & s_2 & s_3 \end{array} \\
\left[
\begin{array}{cccccc|c}
0 & \frac{1}{2} & 1 & -\frac{1}{2} & -\frac{1}{2} & 0 & 1 \\
1 & 2 & 0 & 1 & 2 & 0 & 4 \\
0 & -6 & 0 & 1 & -2 & 1 & 2 \\
\hline
0 & 1 & 0 & 1 & 5 & 0 & 22
\end{array}
\right].
\end{array}
\qquad
\begin{array}{l}
\\
4R_1 + R_2 \\
-12R_1 + R_3 \\
2R_1 + R_4
\end{array}
$$

Therefore, the maximum value of $z$ occurs when $x_1 = 4$, $x_2 = 0$, and $x_3 = 1$, in which case $z = 22$. ✓ 4

✓ **Checkpoint 4**

Complete Stage II and find an optimal solution for Checkpoint 3 on page 385. What is the optimal value of the objective function $z$?

## Minimization Problems

When dealing with minimization problems, we use $y_1, y_2, y_3$, etc., as variables and denote the objective function by $w$. The two-stage method for maximization problems illustrated in Examples 1–3 also provides a means of solving minimization problems. To see why, consider this simple fact: When a number $t$ gets smaller, $-t$ gets larger, and vice versa. For instance, if $t$ goes from 6 down to $-8$, then $-t$ goes from $-6$ up to 8. Thus, if $w$ is the objective function of a linear programming problem, the feasible solution that produces the minimum value of $w$ also produces the maximum value of $-w$, and vice versa. Therefore, to solve a minimization problem with objective function $w$, we need only solve the maximization problem with the same constraints and objective function $z = -w$.

**EXAMPLE 4**    Minimize   $w = 2y_1 + y_2 - y_3$
subject to    $-y_1 - y_2 + y_3 \leq -4$
$y_1 + 3y_2 + 3y_3 \geq 6$
$y_1 \geq 0, y_2 \geq 0, y_3 \geq 0.$

**Solution** Make the constant in the first constraint positive by multiplying both sides by $-1$. Then solve this maximization problem:

Maximize   $z = -w = -2y_1 - y_2 + y_3$
subject to    $y_1 + y_2 - y_3 \geq 4$
$y_1 + 3y_2 + 3y_3 \geq 6$
$y_1 \geq 0, y_2 \geq 0, y_3 \geq 0.$

Convert the constraints to equations by subtracting surplus variables, and set up the first tableau:

$$
\begin{array}{ccccc}
y_1 & y_2 & y_3 & s_1 & s_2 \\
\end{array}
$$

$$
\left[\begin{array}{ccccc|c}
1 & 1 & -1 & -1 & 0 & 4 \\
1 & 3 & 3 & 0 & -1 & 6 \\
\hline
2 & 1 & -1 & 0 & 0 & 0
\end{array}\right].
$$

The basic solution given by this tableau, namely, $y_1 = 0$, $y_2 = 0$, $y_3 = 0$, $s_1 = -4$, and $s_2 = -6$, is not feasible, so the procedures of Stage I must be used to find a basic feasible solution. In the column of the negative basic variable $s_1$, there is a $-1$ in row 1; we choose the first positive entry in that row, so that column 1 will be the pivot column. The quotients $\frac{4}{1} = 4$ and $\frac{6}{1} = 6$ show that the pivot is the 1 in row 1, column 1. Pivoting produces this tableau:

$$
\begin{array}{ccccc}
y_1 & y_2 & y_3 & s_1 & s_2 \\
\end{array}
$$

$$
\left[\begin{array}{ccccc|c}
1 & 1 & -1 & -1 & 0 & 4 \\
0 & 2 & 4 & 1 & -1 & 2 \\
\hline
0 & -1 & 1 & 2 & 0 & -8
\end{array}\right]. \quad
\begin{array}{l}
-R_1 + R_2 \\[6pt]
-2R_1 + R_3
\end{array}
$$

The basic solution $y_1 = 4$, $y_2 = 0$, $y_3 = 0$, $s_1 = 0$, and $s_2 = -2$ is not feasible because $s_2$ is negative, so we repeat the process. We choose the first positive entry in row 2 (the row containing the $-1$ in the $s_2$ column), which is in column 2, so that column 2 is the pivot column. The relevant quotients are $\frac{4}{1} = 4$ and $\frac{2}{2} = 1$, so the pivot is the 2 in row 2, column 2. Pivoting produces a new tableau:

$$
\begin{array}{ccccc}
y_1 & y_2 & y_3 & s_1 & s_2 \\
\end{array}
$$

$$
\left[\begin{array}{ccccc|c}
1 & 1 & -1 & -1 & 0 & 4 \\
0 & 1 & 2 & \frac{1}{2} & -\frac{1}{2} & 1 \\
\hline
0 & -1 & 1 & 2 & 0 & -8
\end{array}\right] \quad \frac{1}{2}R_2
$$

$$
\begin{array}{ccccc}
y_1 & y_2 & y_3 & s_1 & s_2 \\
\end{array}
$$

$$
\left[\begin{array}{ccccc|c}
1 & 0 & -3 & -\frac{3}{2} & \frac{1}{2} & 3 \\
0 & 1 & 2 & \frac{1}{2} & -\frac{1}{2} & 1 \\
\hline
0 & 0 & -3 & \frac{5}{2} & -\frac{1}{2} & -7
\end{array}\right]. \quad
\begin{array}{l}
-R_2 + R_1 \\[6pt]
R_2 + R_3
\end{array}
$$

The basic solution $y_1 = 3$, $y_2 = 1$, $y_3 = 0$, $s_1 = 0$, and $s_2 = 0$ is feasible, so Stage I is complete. However, this solution is not optimal, because the objective row contains the negative indicator $-\frac{1}{2}$ in column 5. According to the simplex method, column 5 is the next pivot column. The only positive ratio, $3/\frac{1}{2} = 6$, is in row 1, so the pivot is $\frac{1}{2}$ in row 1, column 5. Pivoting produces the final tableau:

$$
\begin{array}{ccccc}
y_1 & y_2 & y_3 & s_1 & s_2 \\
\end{array}
$$

$$
\left[\begin{array}{ccccc|c}
2 & 0 & -6 & -3 & 1 & 6 \\
0 & 1 & 2 & \frac{1}{2} & -\frac{1}{2} & 1 \\
\hline
0 & 0 & 3 & \frac{5}{2} & -\frac{1}{2} & -7
\end{array}\right] \quad 2R_1
$$

$$
\begin{array}{ccccc}
y_1 & y_2 & y_3 & s_1 & s_2 \\
\end{array}
$$

$$
\left[\begin{array}{ccccc|c}
2 & 0 & -6 & -3 & 1 & 6 \\
1 & 1 & -1 & -1 & 0 & 4 \\
\hline
1 & 0 & 0 & 1 & 0 & -4
\end{array}\right]. \quad
\begin{array}{l}
\frac{1}{2}R_1 + R_2 \\[6pt]
\frac{1}{2}R_1 + R_3
\end{array}
$$

✓ **Checkpoint 5**

Minimize $w = 2y_1 + 3y_2$
subject to $y_1 + y_2 \geq 10$
$2y_1 + y_2 \geq 16$
$y_1 \geq 0$, $y_2 \geq 0$.

Since there are no negative indicators, the solution given by this tableau ($y_1 = 0$, $y_2 = 4$, $y_3 = 0$, $s_1 = 0$, and $s_2 = 6$) is optimal. The maximum value of $z = -w$ is $-4$. Therefore, the minimum value of the original objective function $w$ is $-(-4) = 4$, which occurs when $y_1 = 0$, $y_2 = 4$, and $y_3 = 0$. ✓ 5

## Equation Constraints

Recall that, for any real numbers $a$ and $b$,

$$a = b \qquad \text{exactly when } a \geq b \text{ and simultaneously } a \leq b.$$

Thus, an equation such as $y_1 + 3y_2 + 3y_3 = 6$ is equivalent to this pair of inequalities:

$$y_1 + 3y_2 + 3y_3 \geq 6$$
$$y_1 + 3y_2 + 3y_3 \leq 6.$$

In a linear programming problem, each equation constraint should be replaced in this way by a pair of inequality constraints. Then the problem can be solved by the two-stage method.

**EXAMPLE 5**

$$\text{Minimize} \quad w = 2y_1 + y_2 - y_3$$
$$\text{subject to} \quad -y_1 - y_2 + y_3 \leq -4$$
$$y_1 + 3y_2 + 3y_3 = 6$$
$$y_1 \geq 0, y_2 \geq 0, y_3 \geq 0.$$

**Solution** Multiply the first inequality by $-1$ and replace the equation by an equivalent pair of inequalities, as just explained, to obtain this problem:

$$\text{Maximize} \quad z = -w = -2y_1 - y_2 + y_3$$
$$\text{subject to} \quad y_1 + y_2 - y_3 \geq 4$$
$$y_1 + 3y_2 + 3y_3 \geq 6$$
$$y_1 + 3y_2 + 3y_3 \leq 6$$
$$y_1 \geq 0, y_2 \geq 0, y_3 \geq 0.$$

Convert the constraints to equations by subtracting surplus variables $s_1$ and $s_2$ from the first two inequalities and adding a slack variable $s_3$ to the third. Then the first tableau is

$$
\begin{array}{cccccc}
y_1 & y_2 & y_3 & s_1 & s_2 & s_3 \\
\end{array}
$$
$$
\left[
\begin{array}{cccccc|c}
1 & 1 & -1 & -1 & 0 & 0 & 4 \\
1 & 3 & 3 & 0 & -1 & 0 & 6 \\
1 & 3 & 3 & 0 & 0 & 1 & 6 \\
\hline
2 & 1 & -1 & 0 & 0 & 0 & 0 \\
\end{array}
\right].
$$

The basic solution given by this tableau is $y_1 = 0, y_2 = 0, y_3 = 0, s_1 = -4, s_2 = -6$, and $s_3 = 6$, which is not feasible. So we begin Stage I. The basic variable $s_1$ has a negative value because of the $-1$ in row 1. We choose the first positive entry in that row. So column 1 will be the pivot column. The quotients are $\frac{4}{1} = 4$ in row 1 and $\frac{6}{1} = 6$ in rows 2 and 3, which means that the pivot is the 1 in row 1, column 1. Pivoting produces this tableau:

$$
\begin{array}{cccccc}
y_1 & y_2 & y_3 & s_1 & s_2 & s_3 \\
\end{array}
$$
$$
\left[
\begin{array}{cccccc|c}
1 & 1 & -1 & -1 & 0 & 0 & 4 \\
0 & 2 & 4 & 1 & -1 & 0 & 2 \\
0 & 2 & 4 & 1 & 0 & 1 & 2 \\
\hline
0 & -1 & 1 & 2 & 0 & 0 & -8 \\
\end{array}
\right].
$$

Now the basic variable $s_2$ has a negative value because of the $-1$ in row 2. We choose the first nonzero entry in this row and form the quotients $\frac{4}{1} = 4$ in row 1 and $\frac{2}{2} = 1$ in rows 2 and 3. Thus, there are two choices for the pivot, and we take the 2 in row 2, column 2. Pivoting produces the tableau

$$
\begin{array}{cccccc}
y_1 & y_2 & y_3 & s_1 & s_2 & s_3 \\
\end{array}
$$
$$
\left[
\begin{array}{cccccc|c}
1 & 0 & -3 & -1.5 & .5 & 0 & 3 \\
0 & 1 & 2 & .5 & -.5 & 0 & 1 \\
0 & 0 & 0 & 0 & 1 & 1 & 0 \\
\hline
0 & 0 & 3 & 2.5 & -.5 & 0 & -7
\end{array}
\right].
$$

This tableau gives the basic feasible solution $y_1 = 3, y_2 = 1, y_3 = 0, s_1 = 0, s_2 = 0$, and $s_3 = 0$, so Stage I is complete. Now apply the simplex method. One round of pivoting produces the final tableau:

$$
\begin{array}{cccccc}
y_1 & y_2 & y_3 & s_1 & s_2 & s_3 \\
\end{array}
$$
$$
\left[
\begin{array}{cccccc|c}
1 & 0 & -3 & -1.5 & 0 & -.5 & 3 \\
0 & 1 & 2 & .5 & 0 & .5 & 1 \\
0 & 0 & 0 & 0 & 1 & 1 & 0 \\
\hline
0 & 0 & 3 & 2.5 & 0 & .5 & -7
\end{array}
\right].
$$

Therefore, the minimum value of $w = -z$ is $w = -(-7) = 7$, which occurs when $y_1 = 3, y_2 = 1$, and $y_3 = 0$.

You may have noticed that Example 5 is just Example 4 with the last inequality constraint replaced by an equation constraint. Note, however, that the optimal solutions are different in the two examples. The minimal value of $w$ found in Example 4 is smaller than the one found in Example 5, but does not satisfy the equation constraint in Example 5.

The two-stage method used in Examples 1–5 is summarized here.

## Solving Nonstandard Problems

1. Replace each equation constraint by an equivalent pair of inequality constraints.
2. If necessary, write each constraint with a positive constant.
3. Convert a minimization problem to a maximization problem by letting $z = -w$.
4. Add slack variables and subtract surplus variables as needed to convert the constraints into equations.
5. Write the initial simplex tableau.
6. Find a basic feasible solution for the problem if such a solution exists (Stage I).
7. When a basic feasible solution is found, use the simplex method to solve the problem (Stage II).

**NOTE** It may happen that the tableau which gives the basic feasible solution in Stage I has no negative indicators in its last row. In this case, the solution found is already optimal and Stage II is not necessary.

## Applications

Many real-world applications of linear programming involve mixed constraints. Since they typically include a large number of variables and constraints, technology is normally required to solve such problems.

**EXAMPLE 6** **Supply Management** A college textbook publisher has received orders from two colleges: $C_1$ and $C_2$. $C_1$ needs 500 books, and $C_2$ needs 1000. The publisher can supply the books from either of two warehouses. Warehouse $W_1$ has 900 books available, and warehouse $W_2$ has 700. The costs to ship a book from each warehouse to each college are as follows:

| From | | To | |
|---|---|---|---|
| | | $C_1$ | $C_2$ |
| | $W_1$ | $1.20 | $1.80 |
| | $W_2$ | $2.10 | $1.50 |

How many books should be sent from each warehouse to each college to minimize the shipping costs?

**Solution** To begin, let

$$y_1 = \text{the number of books shipped from } W_1 \text{ to } C_1;$$
$$y_2 = \text{the number of books shipped from } W_2 \text{ to } C_1;$$
$$y_3 = \text{the number of books shipped from } W_1 \text{ to } C_2;$$
$$y_4 = \text{the number of books shipped from } W_2 \text{ to } C_2.$$

$C_1$ needs 500 books, so $y_1 + y_2 = 500$, which is equivalent to this pair of inequalities:

$$y_1 + y_2 \geq 500$$
$$y_1 + y_2 \leq 500.$$

Similarly, $y_3 + y_4 = 1000$, which is equivalent to

$$y_3 + y_4 \geq 1000$$
$$y_3 + y_4 \leq 1000.$$

Since $W_1$ has 900 books available and $W_2$ has 700 available,

$$y_1 + y_3 \leq 900 \quad \text{and} \quad y_2 + y_4 \leq 700.$$

The company wants to minimize shipping costs, so the objective function is

$$w = 1.20y_1 + 2.10y_2 + 1.80y_3 + 1.50y_4.$$

Now write the problem as a system of linear equations, adding slack or surplus variables as needed, and let $z = -w$:

$$
\begin{aligned}
y_1 + y_2 \quad\quad\quad\quad\quad - s_1 \quad\quad\quad\quad\quad\quad\quad &= 500 \\
y_1 + y_2 + \quad\quad\quad\quad\quad s_2 \quad\quad\quad\quad\quad\quad &= 500 \\
y_3 + y_4 \quad\quad - s_3 \quad\quad\quad\quad &= 1000 \\
y_3 + y_4 + \quad\quad\quad s_4 \quad\quad\quad &= 1000 \\
y_1 + \quad\quad y_3 + \quad\quad\quad\quad\quad s_5 \quad &= 900 \\
y_2 + \quad\quad\quad y_4 + \quad\quad\quad\quad\quad\quad s_6 &= 700 \\
1.20y_1 + 2.10y_2 + 1.80y_3 + 1.50y_4 \quad\quad\quad\quad\quad\quad\quad + z &= 0.
\end{aligned}
$$

Set up the initial simplex tableau:

| $y_1$ | $y_2$ | $y_3$ | $y_4$ | $s_1$ | $s_2$ | $s_3$ | $s_4$ | $s_5$ | $s_6$ | |
|---|---|---|---|---|---|---|---|---|---|---|
| 1 | 1 | 0 | 0 | −1 | 0 | 0 | 0 | 0 | 0 | 500 |
| 1 | 1 | 0 | 0 | 0 | 1 | 0 | 0 | 0 | 0 | 500 |
| 0 | 0 | 1 | 1 | 0 | 0 | −1 | 0 | 0 | 0 | 1000 |
| 0 | 0 | 1 | 1 | 0 | 0 | 0 | 1 | 0 | 0 | 1000 |
| 1 | 0 | 1 | 0 | 0 | 0 | 0 | 0 | 1 | 0 | 900 |
| 0 | 1 | 0 | 1 | 0 | 0 | 0 | 0 | 0 | 1 | 700 |
| 1.20 | 2.10 | 1.80 | 1.50 | 0 | 0 | 0 | 0 | 0 | 0 | 0 |

The basic solution here is not feasible, because $s_1 = -500$ and $s_3 = -1000$. Stages I and II could be done by hand here, but because of the large size of the matrix, it is more efficient to use technology, such as the program in the Graphing Calculator Appendix. Stage I takes four rounds of pivoting and produces the feasible solution in Figure 7.33.

$$
\begin{array}{cccccc}
y_1 & y_2 & y_3 & y_4 & s_1 & s_2 \\
\end{array}
$$
$$
\begin{bmatrix}
1 & 1 & 0 & 0 & -1 & 0 \\
0 & 0 & 0 & 0 & 1 & 1 \\
0 & 1 & 0 & 1 & 0 & 0 \\
0 & 0 & 0 & 0 & 0 & 0 \\
0 & -1 & 1 & 0 & 0 & 0 \\
0 & 0 & 0 & 0 & 1 & 0 \\
0 & 1.2 & 0 & 0 & 1.2 & 0
\end{bmatrix}
$$

$$
\begin{array}{cccccc}
y_1 & y_2 & y_3 & y_4 & s_1 & s_2 \\
\end{array}
$$
$$
\begin{bmatrix}
1 & 1 & 0 & 0 & -1 & 0 \\
0 & 0 & 0 & 0 & 1 & 1 \\
-1 & 0 & 0 & 1 & 0 & 0 \\
0 & 0 & 0 & 0 & 0 & 0 \\
1 & 0 & 1 & 0 & 0 & 0 \\
0 & 0 & 0 & 0 & 1 & 0 \\
-1.2 & 0 & 0 & 0 & 2.1 & 0
\end{bmatrix}
$$

$$
\begin{array}{cccc}
s_3 & s_4 & s_5 & s_6 \\
\end{array}
$$
$$
\begin{bmatrix}
0 & 0 & 0 & 0 & 500 \\
0 & 0 & 0 & 1 & 0 \\
-1 & 0 & -1 & 0 & 100 \\
1 & 1 & 0 & 0 & 0 \\
0 & 0 & 1 & 0 & 900 \\
1 & 0 & 1 & 1 & 100 \\
1.5 & 0 & -.3 & 0 & -2820
\end{bmatrix}
$$

**Figure 7.33**

$$
\begin{array}{cccc}
s_3 & s_4 & s_5 & s_6 \\
\end{array}
$$
$$
\begin{bmatrix}
0 & 0 & 0 & 0 & 500 \\
0 & 0 & 0 & 0 & 0 \\
0 & 0 & 0 & 1 & 700 \\
1 & 1 & 0 & 0 & 0 \\
-1 & 0 & 0 & -1 & 300 \\
1 & 0 & 1 & 1 & 100 \\
1.8 & 0 & 0 & .3 & -2190
\end{bmatrix}
$$

**Figure 7.34**

Because of the small size of a calculator screen, you must scroll to the right to see the entire matrix. Now Stage II begins. Two rounds of pivoting produce the final tableau (Figure 7.34).

The optimal solution is $y_1 = 500, y_2 = 0, y_3 = 300$, and $y_4 = 700$, which results in a minimum shipping cost of \$2190. (Remember that the optimal value for the original minimization problem is the negative of the optimal value of the associated maximization problem.)

# 7.7 Exercises

*In Exercises 1–4, (a) restate the problem in terms of equations by introducing slack and surplus variables and (b) write the initial simplex tableau. (See Example 1.)*

**1.** Maximize $z = -5x_1 + 4x_2 - 2x_3$

subject to
$$-2x_2 + 5x_3 \geq 8$$
$$4x_1 - x_2 + 3x_3 \leq 12$$
$$x_1 \geq 0, x_2 \geq 0, x_3 \geq 0.$$

**2.** Maximize $z = -x_1 + 4x_2 - 2x_3$

subject to
$$2x_1 + 2x_2 + 6x_3 \leq 10$$
$$-x_1 + 2x_2 + 4x_3 \geq 7$$
$$x_1 \geq 0, x_2 \geq 0, x_3 \geq 0.$$

**3.** Maximize $z = 2x_1 - 3x_2 + 4x_3$

subject to
$$x_1 + x_2 + x_3 \leq 100$$
$$x_1 + x_2 + x_3 \geq 75$$
$$x_1 + x_2 \qquad \geq 27$$
$$x_1 \geq 0, x_2 \geq 0, x_3 \geq 0.$$

**4.** Maximize $z = -x_1 + 5x_2 + x_3$

subject to
$$2x_1 \qquad + x_3 \leq 40$$
$$x_1 + x_2 \qquad \geq 18$$
$$x_1 \qquad + x_3 = 20$$
$$x_1 \geq 0, x_2 \geq 0, x_3 \geq 0.$$

*Convert Exercises 5–8 into maximization problems with positive constants on the right side of each constraint, and write the initial simplex tableau. (See Examples 4 and 5.)*

**5.** Minimize $w = 2y_1 + 5y_2 - 3y_3$

subject to
$$y_1 + 2y_2 + 3y_3 \geq 115$$
$$2y_1 + y_2 + y_3 \leq 200$$
$$y_1 \qquad + y_3 \geq 50$$
$$y_1 \geq 0, y_2 \geq 0, y_3 \geq 0.$$

**6.** Minimize $w = 7y_1 + 6y_2 + y_3$

subject to
$$y_1 + y_2 + y_3 \geq 5$$
$$-y_1 + y_2 \qquad \leq -4$$
$$2y_1 + y_2 + 3y_3 \geq 15$$
$$y_1 \geq 0, y_2 \geq 0, y_3 \geq 0.$$

**7.** Minimize $w = 10y_1 + 8y_2 + 15y_3$

subject to
$$y_1 + y_2 + y_3 \geq 12$$
$$5y_1 + 4y_2 + 9y_3 \geq 48$$
$$y_1 \geq 0, y_2 \geq 0, y_3 \geq 0.$$

**8.** Minimize $w = y_1 + 2y_2 + y_3 + 5y_4$

subject to
$$-y_1 - y_2 + y_3 - y_4 \leq -50$$
$$3y_1 + y_2 + 2y_3 + y_4 = 100$$
$$y_1 \geq 0, y_2 \geq 0, y_3 \geq 0, y_4 \geq 0.$$

*Use the two-stage method to solve Exercises 9–20. (See Examples 1–5.)*

9. Maximize $z = 12x_1 + 10x_2$
   subject to $x_1 + 2x_2 \geq 24$
   $$x_1 + x_2 \leq 40$$
   $$x_1 \geq 0, x_2 \geq 0.$$

10. Find $x_1 \geq 0, x_2 \geq 0,$ and $x_3 \geq 0$ such that
    $$x_1 + x_2 + x_3 \leq 150$$
    $$x_1 + x_2 + x_3 \geq 100$$
    and $z = 2x_1 + 5x_2 + 3x_3$ is maximized.

11. Find $x_1 \geq 0, x_2 \geq 0,$ and $x_3 \geq 0$ such that
    $$x_1 + x_2 + 2x_3 \leq 38$$
    $$2x_1 + x_2 + x_3 \geq 24$$
    and $z = 3x_1 + 2x_2 + 2x_3$ is maximized.

12. Maximize $z = 6x_1 + 8x_2$
    subject to $3x_1 + 12x_2 \geq 48$
    $$2x_1 + 4x_2 \leq 60$$
    $$x_1 \geq 0, x_2 \geq 0.$$

13. Find $x_1 \geq 0$ and $x_2 \geq 0$ such that
    $$x_1 + 2x_2 \leq 18$$
    $$x_1 + 3x_2 \geq 12$$
    $$2x_1 + 2x_2 \leq 30$$
    and $z = 5x_1 + 10x_2$ is maximized.

14. Find $y_1 \geq 0, y_2 \geq 0$ such that
    $$10y_1 + 5y_2 \geq 100$$
    $$20y_1 + 10y_2 \geq 160$$
    and $w = 4y_1 + 5y_2$ is minimized.

15. Minimize $w = 3y_1 + 2y_2$
    subject to $2y_1 + 3y_2 \geq 60$
    $$y_1 + 4y_2 \geq 40$$
    $$y_1 \geq 0, y_2 \geq 0.$$

16. Minimize $w = 3y_1 + 4y_2$
    subject to $y_1 + 2y_2 \geq 10$
    $$y_1 + y_2 \geq 8$$
    $$2y_1 + y_2 \leq 22$$
    $$y_1 \geq 0, y_2 \geq 0.$$

17. Maximize $z = 3x_1 + 2x_2$
    subject to $x_1 + x_2 = 50$
    $$4x_1 + 2x_2 \geq 120$$
    $$5x_1 + 2x_2 \leq 200$$
    with $x_1 \geq 0, x_2 \geq 0.$

18. Maximize $z = 10x_1 + 9x_2$
    subject to $x_1 + x_2 = 30$
    $$x_1 + x_2 \geq 25$$
    $$2x_1 + x_2 \leq 40$$
    with $x_1 \geq 0, x_2 \geq 0.$

19. Minimize $w = 32y_1 + 40y_2$
    subject to $20y_1 + 10y_2 = 200$
    $$25y_1 + 40y_2 \leq 500$$
    $$18y_1 + 24y_2 \geq 300$$
    with $y_1 \geq 0, y_2 \geq 0.$

20. Minimize $w = 15y_1 + 12y_2$
    subject to $y_1 + 2y_2 \leq 12$
    $$3y_1 + y_2 \geq 18$$
    $$y_1 + y_2 = 10$$
    with $y_1 \geq 0, y_2 \geq 0.$

*Use the two-stage method to solve the problem that was set up in the given exercise.*

21. Exercise 1
22. Exercise 2
23. Exercise 3
24. Exercise 4
25. Exercise 5
26. Exercise 6
27. Exercise 7
28. Exercise 8

*In Exercises 29–32, set up the initial simplex tableau, but do not solve the problem. (See Example 6.)*

29. **Gasoline Additive** A company is developing a new additive for gasoline. The additive is a mixture of three liquid ingredients: I, II, and III. For proper performance, the total amount of additive must be at least 10 ounces per barrel of gasoline. However, for safety reasons, the amount of additive should not exceed 15 ounces per barrel of gasoline. At least $\frac{1}{4}$ ounce of ingredient I must be used for every ounce of ingredient II, and at least 1 ounce of ingredient III must be used for every ounce of ingredient I. If the costs of I, II, and III are $.30, $.09, and $.27 per ounce, respectively, find the mixture of the three ingredients that produces the minimum cost of the additive. What is the minimum cost?

30. **Soft Drink Production** A popular soft drink called Sugarlo, which is advertised as having a sugar content of no more than 10%, is blended from five ingredients, each of which has some sugar content. Water may also be added to dilute the mixture. The sugar content of the ingredients and their costs per gallon are given in the table:

| | Ingredient | | | | | |
|---|---|---|---|---|---|---|
| | **1** | **2** | **3** | **4** | **5** | **Water** |
| **Sugar content (%)** | .28 | .19 | .43 | .57 | .22 | 0 |
| **Cost ($/gal.)** | .48 | .32 | .53 | .28 | .43 | .04 |

At least .01 of the content of Sugarlo must come from ingredient 3 or 4, .01 must come from ingredient 2 or 5, and .01 from ingredient 1 or 4. How much of each ingredient should be used in preparing at least 15,000 gallons of Sugarlo to minimize the cost? What is the minimum cost?

31. **Supply Management** The manufacturer of a popular personal computer has orders from two dealers. Dealer $D_1$ wants 32 computers, and dealer $D_2$ wants 20 computers. The manufacturer can fill the orders from either of two warehouses, $W_1$ or $W_2$. $W_1$ has 25 of the computers on hand, and $W_2$ has 30. The

costs (in dollars) to ship one computer to each dealer from each warehouse are as follows:

| From | To | |
|---|---|---|
| | $D_1$ | $D_2$ |
| $W_1$ | $14 | $22 |
| $W_2$ | $12 | $10 |

How should the orders be filled to minimize shipping costs? What is the minimum cost?

32. **Nutrition** Mark, who is ill, takes vitamin pills. Each day, he must have at least 16 units of vitamin A, 5 units of vitamin $B_1$, and 20 units of vitamin C. He can choose between pill 1, which costs 10¢ and contains 8 units of vitamin A, 1 unit of vitamin $B_1$, and 2 units of vitamin C, and pill 2, which costs 20¢ and contains 2 units of vitamin A, 1 unit of vitamin $B_1$, and 7 units of vitamin C. How many of each pill should he buy in order to minimize his cost?

*Use the two-stage method to solve Exercises 33–40. (See Examples 5 and 6.)*

33. **Supply Management** Southwestern Oil supplies two distributors in the Northwest from two outlets: $S_1$ and $S_2$. Distributor $D_1$ needs at least 3000 barrels of oil, and distributor $D_2$ needs at least 5000 barrels. The two outlets can each furnish up to 5000 barrels of oil. The costs per barrel to ship the oil are given in the table:

| | Distributors | |
|---|---|---|
| | $D_1$ | $D_2$ |
| Outlets $S_1$ | $30 | $20 |
| $S_2$ | $25 | $22 |

There is also a shipping tax per barrel as given in the table below.

| | $D_1$ | $D_2$ |
|---|---|---|
| $S_1$ | $2 | $6 |
| $S_2$ | $5 | $4 |

Southwestern Oil is determined to spend no more than $40,000 on shipping tax. How should the oil be supplied to minimize shipping costs?

34. **Supply Management** Change Exercise 33 so that the two outlets each furnish exactly 5000 barrels of oil, with everything else the same. Solve the problem as in Example 5.

35. **Seed Production** Topgrade Turf lawn seed mixture contains three types of seeds: bluegrass, rye, and Bermuda. The costs per pound of the three types of seed are 12¢, 15¢, and 5¢, respectively. In each batch, there must be at least 20% bluegrass seed, and the amount of Bermuda seed must be no more than two-thirds the amount of rye seed. To fill current orders, the company must make at least 5000 pounds of the mixture. How much of each kind of seed should be used to minimize costs?

36. **Seed Production** Change Exercise 35 so that the company must make exactly 5000 pounds of the mixture. Solve the problem as in Example 5.

37. **Loan Allocation** A bank has set aside a maximum of $25 million for commercial and home loans. Every million dollars in commercial loans requires 2 lengthy application forms, while every million dollars in home loans requires 3 lengthy application forms. The bank cannot process more than 72 application forms at this time. The bank's policy is to loan at least four times as much for home loans as for commercial loans. Because of prior commitments, at least $10 million will be used for these two types of loans. The bank earns 12% on home loans and 10% on commercial loans. What amount of money should be allotted for each type of loan to maximize the interest income?

38. **Investing** Virginia has decided to invest a $100,000 inheritance in government securities that earn 7% per year, municipal bonds that earn 6% per year, and mutual funds that earn an average of 10% per year. She will spend at least $40,000 on government securities, and she wants at least half the inheritance to go to bonds and mutual funds. Government securities have an initial fee of 2%, municipal bonds an initial fee of 1%, and mutual funds an initial fee of 3%. Virginia has $2400 available to pay initial fees. How much money should go into each type of investment to maximize the interest while meeting the constraints? What is the maximum interest she can earn?

39. **Beer Production** A brewery produces regular beer and a lower-carbohydrate "light" beer. Steady customers of the brewery buy 12 units of regular beer and 10 units of light beer. While setting up the brewery to produce the beers, the management decides to produce extra beer, beyond that needed to satisfy the steady customers. The cost per unit of regular beer is $36,000, and the cost per unit of light beer is $48,000. The number of units of light beer should not exceed twice the number of units of regular beer. At least 20 additional units of beer can be sold. How much of each type of beer should be made so as to minimize total production costs?

40. **Supply Management** The chemistry department at a local college decides to stock at least 800 small test tubes and 500 large test tubes. It wants to buy at least 1500 test tubes to take advantage of a special price. Since the small tubes are broken twice as often as the large, the department will order at least twice as many small tubes as large. If the small test tubes cost 15¢ each and the large ones, made of a cheaper glass, cost 12¢ each, how many of each size should the department order to minimize cost?

*Use technology to solve the following exercises, whose initial tableaus were set up in Exercises 29–32.*

| 41. Exercise 29 | 42. Exercise 30 |
|---|---|
| 43. Exercise 31 | 44. Exercise 32 |

## ✓ Checkpoint Answers

1. (a) Maximize $z = 3x_1 - 2x_2$
subject to
$$2x_1 + 3x_2 + s_1 = 8$$
$$6x_1 - 2x_2 - s_2 = 3$$
$$x_1 + 4x_2 - s_3 = 1$$
$$x_1 \geq 0, x_2 \geq 0, s_1 \geq 0, s_2 \geq 0, s_3 \geq 0.$$

$$
\text{(b)} \quad
\begin{array}{ccccc}
x_1 & x_2 & s_1 & s_2 & s_3 \\
\end{array}
\left[
\begin{array}{ccccc|c}
2 & 3 & 1 & 0 & 0 & 8 \\
6 & -2 & 0 & -1 & 0 & 3 \\
1 & 4 & 0 & 0 & -1 & 1 \\
\hline
-3 & 2 & 0 & 0 & 0 & 0 \\
\end{array}
\right]
$$

**2. (a)** $x_1 = 0$, $x_2 = 0$, $s_1 = 12$, $s_2 = 6$, $s_3 = -5$; no

   **(b)** $x_1 = 0$, $x_2 = 0$, $x_3 = 0$, $s_1 = 12$, $s_2 = 7$; yes

**3.**
$$
\begin{array}{cccc}
x_1 & x_2 & s_1 & s_2 \\
\end{array}
\left[
\begin{array}{cccc|c}
0 & 1 & 1 & \frac{1}{2} & 45 \\
1 & 2 & 0 & -\frac{1}{2} & 25 \\
\hline
0 & 6 & 0 & -4 & 200 \\
\end{array}
\right]
$$
$x_1 = 25$, $x_2 = 0$, $s_1 = 45$, and $s_2 = 0$.

**4.** The optimal value $z = 560$ occurs when $x_1 = 70$, $x_2 = 0$, $s_1 = 0$, and $s_2 = 90$

**5.** $y_1 = 10$ and $y_2 = 0$; $w = 20$

# CHAPTER 7 Summary and Review

## Key Terms and Symbols

**7.1** linear inequality
  graphs of linear
    inequalities
  boundary
  half-plane
  system of inequalities
  region of feasible solutions
    (feasible region)

**7.2** linear programming
  objective function

constraints
corner point
bounded feasible region
unbounded feasible region
corner point theorem

**7.3** applications of linear
  programming

**7.4** standard maximum form
  slack variable
  simplex tableau

indicator
pivot and pivoting
final simplex tableau
basic variables
nonbasic variables
basic feasible solution

**7.6** transpose of a matrix
  dual
  theorem of duality
  shadow costs

**7.7** surplus variable
  basic variables
  basic solution
  basic feasible solution
  Stage I
  Stage II

## Chapter 7 Key Concepts

**Graphing a Linear Inequality**

Graph the boundary line as a solid line if the inequality includes "or equal to," and as a dashed line otherwise. Shade the half-plane for which the inequality is true. The graph of a system of inequalities, called the **region of feasible solutions**, includes all points that satisfy all the inequalities of the system at the same time.

**Solving Linear Programming Problems**

**Graphical Method:** Determine the objective function and all necessary constraints. Graph the region of feasible solutions. The maximum or minimum value will occur at one or more of the corner points of this region.

**Simplex Method:** Determine the objective function and all necessary constraints. Convert each constraint into an equation by adding slack variables. Set up the initial simplex tableau. Locate the most negative indicator. Form the quotients to determine the pivot. Use row operations to change the pivot to 1 and all other numbers in that column to 0. If the indicators are all positive or 0, this is the final tableau. If not, choose a new pivot and repeat the process until no indicators are negative. Read the solution from the final tableau. The optimum value of the objective function is the number in the lower right corner of the final tableau. For problems with **mixed constraints**, replace each equation constraint by a pair of inequality constraints. Then add slack variables and subtract surplus variables as needed to convert each constraint into an equation. In Stage I, use row operations to transform the matrix until the solution is feasible. In Stage II, use the simplex method as just described. For **minimization** problems, let the objective function be $w$ and set $-w = z$. Then proceed as with mixed constraints.

**Solving Minimization Problems with Duals**

Find the dual maximization problem. Solve the dual problem with the simplex method. The minimum value of the objective function $w$ is the maximum value of the dual objective function $z$. The optimal solution is found in the entries in the bottom row of the columns corresponding to the slack variables.

# Chapter 7 Review Exercises

*Graph each of the given linear inequalities.*

**1.** $y \le 3x + 2$

**2.** $2x - y \ge 6$

**3.** $3x + 4y \ge 12$

**4.** $y \le 4$

*Graph the solution of each of the given systems of inequalities.*

**5.**  $x + y \le 6$
    $2x - y \ge 3$

**6.** $4x + \;\; y \ge 8$
    $2x - 3y \le 6$

**7.** $2 \le x \le 5$
    $1 \le y \le 6$
    $x - y \le 3$

**8.**  $x + 2y \le 4$
    $2x - 3y \le 6$
    $x \ge 0, y \ge 0$

*Set up a system of inequalities for each of the given problems, and then graph the region of feasible solutions.*

**9. Consulting** Linda Quinn is a statistical consultant who creates two kind of reports: summary reports and inference reports. Each summary report requires 2 hours of analysis time and 1.5 hours to write the report. Each inference report needs 4 hours of analysis time and 2 hours of writing time. Suppose Linda has only 15 hours available for analysis and 9 hours for writing.

**10. Pizza Production** A company makes two kinds of pizza: basic (cheese and tomatoes) and margherita (cheese, tomatoes, and basil). The company sells at least 60 units a day of the pizza margherita and 40 units a day of the basic. The cost of the tomatoes and basil is $2 per unit for the pizza margherita and the cost of tomatoes is $1 per unit for the basic. No more than $320 a day can be spent on tomatoes and basil together. The cheese used for the pizza margherita is $5 per unit, and the cheese for the basic is $4 per unit. The company spends no more than $1000 a day on cheese.

*Use the given regions to find the maximum and minimum values of the objective function $z = 3x + 4y$.*

**11.**

**12.**

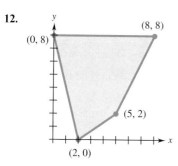

*Use the graphical method to solve Exercises 13–16.*

**13.** Maximize   $z = 6x + 2y$
    subject to   $2x + 7y \le 14$
        $2x + 3y \le 10$
        $x \ge 0, y \ge 0.$

**14.** Find $x \ge 0$ and $y \ge 0$ such that
$$8x + 9y \ge 72$$
$$6x + 8y \ge 72$$
    and $w = 2x + 10y$ is minimized.

**15.** Find $x \ge 0$ and $y \ge 0$ such that
$$x + \;\; y \le 50$$
$$2x + \;\; y \ge 20$$
$$x + 2y \ge 30$$
    and $w = 5x + 2y$ is minimized.

**16.** Maximize   $z = 5x - 2y$
    subject to   $3x + 2y \le 12$
        $5x + \;\; y \ge 5$
        $x \ge 0, y \ge 0.$

**17. Consulting** How many of each kind of statistical report (summary or inference) should Linda Quinn create in Exercise 9 in order to maximize profit if the summary reports earn a profit of $500 and the inference reports produce a profit of $750? (Assume it is possible to complete a fractional part of a project.)

**18. Pizza Production** How many units of each pizza in Exercise 10 should the company make in order to maximize revenue if the basic pizza sells for $15 per unit and the pizza margherita sells for $20 per unit.

**19. Finance** The BlackRock Equity Dividend Fund sells at $24 a share and has a 3-year average annual return of $3 per share. The risk measure of beta is .95. The Columbia Dividend Income Fund sells for $20 a share and has a 3-year average annual return of $2 a share. The risk measure of beta is .89. Donald wants to spend no more than $10,000 investing in these two funds, but he wants to obtain at least $1100 in annual revenue. Donald also wants to minimize his risk. Determine how many shares of each fund Donald should buy. (Data from: www.morningstar.com.)

**20. Finance** The Alger Spectra Z Fund sells at $15 a share and has a 3-year average annual return of $2 per share. The risk measure of beta is .96. The JPMorgan U.S. Dynamic Plus Fund sells for $22 a share and has a 3-year average annual return of $3.50 a share. The risk measure of beta is 1.14. Ming-Na wants to spend no more than $5016 inventing in these two funds, but she wants to obtain at least $700 in revenue. Ming-Na also wants to minimize risk. Determine how many shares of each fund Ming-Na should buy. (Data from: www.morningstar.com.)

*For each of the following problems, (a) add slack variables and (b) set up the initial simplex tableau.*

**21.** Maximize $z = 5x_1 + 6x_2 + 3x_3$

subject to $x_1 + x_2 + x_3 \leq 100$

$2x_1 + 3x_2 \qquad \leq 500$

$x_1 \qquad + 2x_3 \leq 350$

$x_1 \geq 0, x_2 \geq 0, x_3 \geq 0.$

**22.** Maximize $z = 2x_1 + 9x_2$

subject to $3x_1 + 5x_2 \leq 47$

$x_1 + x_2 \leq 25$

$5x_1 + 2x_2 \leq 35$

$2x_1 + x_2 \leq 30$

$x_1 \geq 0, x_2 \geq 0.$

**23.** Maximize $z = x_1 + 8x_2 + 2x_3$

subject to $x_1 + x_2 + x_3 \leq 90$

$2x_1 + 5x_2 + x_3 \leq 120$

$x_1 + 3x_2 \qquad \leq 80$

$x_1 \geq 0, x_2 \geq 0, x_3 \geq 0.$

**24.** Maximize $z = 15x_1 + 12x_2$

subject to $2x_1 + 5x_2 \leq 50$

$x_1 + 3x_2 \leq 25$

$4x_1 + x_2 \leq 18$

$x_1 + x_2 \leq 12$

$x_1 \geq 0, x_2 \geq 0.$

*For each of the following, use the simplex method to solve the maximization linear programming problems with initial tableaus as given.*

**25.**
$$\begin{array}{ccccc} x_1 & x_2 & x_3 & s_1 & s_2 \\ \left[\begin{array}{ccccc|c} 1 & 2 & 3 & 1 & 0 & 28 \\ 2 & 4 & 8 & 0 & 1 & 32 \\ \hline -5 & -2 & -3 & 0 & 0 & 0 \end{array}\right] \end{array}$$

**26.**
$$\begin{array}{cccc} x_1 & x_2 & s_1 & s_2 \\ \left[\begin{array}{cccc|c} 2 & 1 & 1 & 0 & 10 \\ 9 & 3 & 0 & 1 & 15 \\ \hline -2 & -3 & 0 & 0 & 0 \end{array}\right] \end{array}$$

**27.**
$$\begin{array}{cccccc} x_1 & x_2 & x_3 & s_1 & s_2 & s_3 \\ \left[\begin{array}{cccccc|c} 1 & 2 & 2 & 1 & 0 & 0 & 50 \\ 4 & 24 & 0 & 0 & 1 & 0 & 20 \\ 1 & 0 & 2 & 0 & 0 & 1 & 15 \\ \hline -5 & -3 & -2 & 0 & 0 & 0 & 0 \end{array}\right] \end{array}$$

**28.**
$$\begin{array}{ccccc} x_1 & x_2 & s_1 & s_2 & s_3 \\ \left[\begin{array}{ccccc|c} 1 & -2 & 1 & 0 & 0 & 38 \\ 1 & -1 & 0 & 1 & 0 & 12 \\ 2 & 1 & 0 & 0 & 1 & 30 \\ \hline -1 & -2 & 0 & 0 & 0 & 0 \end{array}\right] \end{array}$$

*Use the simplex method to solve the problem that was set up in the given exercise.*

**29.** Exercise 21 **30.** Exercise 22

**31.** Exercise 23 **32.** Exercise 24

*For Exercises 33–36, (a) select appropriate variables, (b) write the objective function, and (c) write the constraints as inequalities.*

**33. Sales** Roberta Hernandez sells three items—*A*, *B*, and *C*—in her gift shop. Each unit of *A* costs her $2 to buy, $1 to sell, and $2 to deliver. For each unit of *B*, the costs are $3, $2, and $2, respectively, and for each unit of *C*, the costs are $6, $2, and $4, respectively. The profit on *A* is $4, on *B* it is $3, and on *C* it is $3. How many of each item should she order to maximize her profit if she can spend $1200 to buy, $800 to sell, and $500 to deliver?

**34. Investing** An investor is considering three types of investment: a high-risk venture into oil leases with a potential return of 15%, a medium-risk investment in bonds with a 9% return, and a relatively safe stock investment with a 5% return. He has $50,000 to invest. Because of the risk, he will limit his investment in oil leases and bonds to 30% and his investment in oil leases and stock to 50%. How much should he invest in each to maximize his return, assuming investment returns are as expected?

**35. Wine Production** The Aged Wood Winery makes two white wines—Fruity and Crystal—from two kinds of grapes and sugar. The wines require the amounts of each ingredient per gallon and produce a profit per gallon as shown in the following table:

| | Grape *A* (bushels) | Grape *B* (bushels) | Sugar (pounds) | Profit (dollars) |
|---|---|---|---|---|
| **Fruity** | 2 | 2 | 2 | 12 |
| **Crystal** | 1 | 3 | 1 | 15 |

The winery has available 110 bushels of grape *A*, 125 bushels of grape *B*, and 90 pounds of sugar. How much of each wine should be made to maximize profit?

**36. Packaging** A company makes three sizes of plastic bags: 5 gallon, 10 gallon, and 20 gallon. The production time in hours for cutting, sealing, and packaging a unit of each size is as follows:

| Size | Cutting | Sealing | Packaging |
|---|---|---|---|
| 5 gallon | 1 | 1 | 2 |
| 10 gallon | 1.1 | 1.2 | 3 |
| 20 gallon | 1.5 | 1.3 | 4 |

There are at most 8 hours available each day for each of the three operations. If the profit per unit is $1 for 5-gallon bags, $.90 for 10-gallon bags, and $.95 for 20-gallon bags, how many of each size should be made per day to maximize the profit?

**37.** When is it necessary to use the simplex method rather than the graphical method?

**38.** What types of problems can be solved with the use of slack variables and surplus variables?

**39.** What kind of problem can be solved with the method of duals?

**40.** In solving a linear programming problem, you are given the following initial tableau:

$$\begin{bmatrix} 4 & 2 & 3 & 1 & 0 & | & 9 \\ 5 & 4 & 1 & 0 & 1 & | & 10 \\ \hline -6 & -7 & -5 & 0 & 0 & | & 0 \end{bmatrix}.$$

(a) What is the problem being solved?

(b) If the 1 in row 1, column 4, were a $-1$ rather than a 1, how would it change your answer to part (a)?

(c) After several steps of the simplex algorithm, the following tableau results:

$$\begin{bmatrix} .6 & 0 & 1 & .4 & -.2 & | & 1.6 \\ 1.1 & 1 & 0 & -.1 & .3 & | & 2.1 \\ \hline 4.7 & 0 & 0 & 1.3 & 1.1 & | & 22.7 \end{bmatrix}.$$

What is the solution? (List only the values of the original variables and the objective function. Do not include slack or surplus variables.)

(d) What is the dual of the problem you found in part (a)?

(e) What is the solution of the dual you found in part (d)? [Do not perform any steps of the simplex algorithm; just examine the tableau given in part (c).]

*The tableaus in Exercises 41–43 are the final tableaus of minimization problems solved by the method of duals. State the solution and the minimum value of the objective function for each problem.*

**41.**
$$\begin{bmatrix} 1 & 0 & 0 & 3 & 1 & 2 & | & 12 \\ 0 & 0 & 1 & 4 & 5 & 3 & | & 5 \\ 0 & 1 & 0 & -2 & 7 & -6 & | & 8 \\ \hline 0 & 0 & 0 & 5 & 7 & 3 & | & 172 \end{bmatrix}$$

**42.**
$$\begin{bmatrix} 0 & 0 & 1 & 6 & 3 & 1 & | & 2 \\ 1 & 0 & 0 & 4 & -2 & 2 & | & 8 \\ 0 & 1 & 0 & 10 & 7 & 0 & | & 12 \\ \hline 0 & 0 & 0 & 9 & 5 & 8 & | & 62 \end{bmatrix}$$

**43.**
$$\begin{bmatrix} 1 & 0 & 7 & -1 & | & 100 \\ 0 & 1 & 1 & 3 & | & 27 \\ \hline 0 & 0 & 7 & 2 & | & 640 \end{bmatrix}$$

*Use the method of duals to solve these minimization problems.*

**44.** Minimize $w = 5y_1 + 2y_2$

subject to $2y_1 + 3y_2 \geq 6$

$2y_1 + y_2 \geq 7$

$y_1 \geq 0, y_2 \geq 0.$

**45.** Minimize $w = 18y_1 + 10y_2$

subject to $y_1 + y_2 \geq 17$

$5y_1 + 8y_2 \geq 42$

$y_1 \geq 0, y_2 \geq 0.$

**46.** Minimize $w = 4y_1 + 5y_2$

subject to $10y_1 + 5y_2 \geq 100$

$20y_1 + 10y_2 \geq 150$

$y_1 \geq 0, y_2 \geq 0.$

*Write the initial simplex tableau for each of these mixed-constraint problems.*

**47.** Maximize $z = 20x_1 + 30x_2$

subject to $5x_1 + 10x_2 \leq 120$

$10x_1 + 15x_2 \geq 200$

$x_1 \geq 0, x_2 \geq 0.$

**48.** Minimize $w = 4y_1 + 2y_2$

subject to $y_1 + 3y_2 \geq 6$

$2y_1 + 8y_2 \leq 21$

$y_1 \geq 0, y_2 \geq 0.$

**49.** Minimize $w = 12y_1 + 20y_2 - 8y_3$

subject to $y_1 + y_2 + 2y_3 \geq 48$

$y_1 + y_2 \leq 12$

$y_3 \geq 10$

$3y_1 + y_3 \geq 30$

$y_1 \geq 0, y_2 \geq 0, y_3 \geq 0.$

**50.** Maximize $w = 6x_1 - 3x_2 + 4x_3$

subject to $2x_1 + x_2 + x_3 \leq 112$

$x_1 + x_2 + x_3 \geq 80$

$x_1 + x_2 \leq 45$

$x_1 \geq 0, x_2 \geq 0, x_3 \geq 0.$

*The given tableaus are the final tableaus of minimization problems solved by letting $w = -z$. Give the solution and the minimum value of the objective function for each problem.*

**51.**
$$\begin{bmatrix} 0 & 1 & 0 & 2 & 5 & 0 & | & 17 \\ 0 & 0 & 1 & 3 & 1 & 1 & | & 25 \\ 1 & 0 & 0 & 4 & 2 & \frac{1}{2} & | & 8 \\ \hline 0 & 0 & 0 & 2 & 5 & 0 & | & -427 \end{bmatrix}$$

**52.**
$$\begin{bmatrix} 0 & 0 & 2 & 1 & 0 & 6 & 6 & | & 92 \\ 1 & 0 & 3 & 0 & 0 & 0 & 2 & | & 47 \\ 0 & 1 & 0 & 0 & 0 & 1 & 0 & | & 68 \\ 0 & 0 & 4 & 0 & 1 & 0 & 3 & | & 35 \\ \hline 0 & 0 & 5 & 0 & 0 & 2 & 9 & | & -1957 \end{bmatrix}$$

*Use the two-stage method to solve these mixed-constraint problems.*

**53.** Exercise 47          **54.** Exercise 48

**55.** Minimize $w = 4y_1 - 8y_2$

subject to $y_1 + y_2 \leq 50$

$2y_1 - 4y_2 \geq 20$

$y_1 - y_2 \leq 22$

$y_1 \geq 0, y_2 \geq 0.$

**56.** Maximize   $z = 2x_1 + 4x_2$

   subject to   $3x_1 + 2x_2 \le 12$

   $5x_1 + x_2 \ge 5$

   $x_1 \ge 0, x_2 \ge 0.$

**Business** *Solve the following maximization problems, which were begun in Exercises 33–36.*

**57.** Exercise 33          **58.** Exercise 34

**59.** Exercise 35          **60.** Exercise 36

*Solve the following minimization problems.*

**61. Vegetable Production** Cauchy Canners produces canned corn, beans, and carrots. Demand for vegetables requires the company to produce at least 1000 cases per month. Based on past sales, it should produce at least twice as many cases of corn as of beans and at least 340 cases of carrots. It costs $10 to produce a case of corn, $15 to produce a case of beans, and $25 to produce a case of carrots. How many cases of each vegetable should be produced to minimize costs? What is the minimum cost?

**62. Construction** A contractor builds boathouses in two basic models: the Atlantic and the Pacific. Each Atlantic model requires 1000 feet of framing lumber, 3000 cubic feet of concrete, and $2000 for advertising. Each Pacific model requires 2000 feet of framing lumber, 3000 cubic feet of concrete, and $3000 for advertising. Contracts call for using at least 8000 feet of framing lumber, 18,000 cubic feet of concrete, and $15,000 worth of advertising. If the total spent on each Atlantic model is $3000 and the total spent on each Pacific

model is $4000, how many of each model should be built to minimize costs?

*Solve these mixed-constraint problems.*

**63. Food Production** Brand X Cannery produces canned whole tomatoes and tomato sauce. This season, the company has available 3,000,000 kilograms of tomatoes for these two products. To meet the demands of regular customers, it must produce at least 80,000 kilograms of sauce and 800,000 kilograms of whole tomatoes. The cost per kilogram is $4 to produce canned whole tomatoes and $3.25 to produce tomato sauce. Labor agreements require that at least 110,000 person-hours be used. Each 1-kilogram can of sauce requires 3 minutes for one worker to produce, and each 1-kilogram can of whole tomatoes requires 6 minutes for one worker. How many kilograms of tomatoes should Brand X use for each product to minimize cost? (For simplicity, assume that the production of $y_1$ kilograms of canned whole tomatoes and $y_2$ kilograms of tomato sauce requires $y_1 + y_2$ kilograms of tomatoes.)

**64. Steel Production** A steel company produces two types of alloys. A run of type I requires 3000 pounds of molybdenum and 2000 tons of iron ore pellets, as well as $2000 in advertising. A run of type II requires 3000 pounds of molybdenum and 1000 tons of iron ore pellets, as well as $3000 in advertising. Total costs are $15,000 on a run of type I and $6000 on a run of type II. The company has on hand 18,000 pounds of molybdenum and 7000 tons of iron ore pellets and wants to use all of it. It plans to spend at least $14,000 on advertising. How much of each type should be produced to minimize cost? What is the minimum cost?

# Case Study 7  Cooking with Linear Programming

Constructing a nutritious recipe can be a difficult task. The recipe must produce food that tastes good, and it must also balance the nutrients that each ingredient brings to the dish. This balancing of nutrients is very important in several diet plans that are currently popular. Many of these plans restrict the intake of certain nutrients (usually fat) while allowing for large amounts of other nutrients (protein and carbohydrates are popular choices). The number of calories in the dish is also often minimized. Linear programming can be used to help create recipes that balance nutrients.

In order to develop solutions to this type of problem, we will need to have nutritional data for the ingredients in our recipes. This data can be found in the U.S. Department of Agriculture's USDA Nutrient Database for Standard Reference, available at www.ndb.nal.usda.gov. This database contains the nutrient levels for hundreds of basic foods. The nutrient levels are given per 100 grams of food. Unfortunately, grams are not often used in recipes; instead, kitchen measures like cups, tablespoons, and fractions of vegetables are used. Table 1 shows the conversion factors from grams to more familiar kitchen units and gives serving sizes for various food.

Table 1    Serving Sizes of Various Food

| Food | Serving Size |
|---|---|
| Beef | 6 oz = 170 g |
| Egg | 1 egg = 61 g |
| Feta Cheese | $\frac{1}{4}$ cup = 38 g |
| Lettuce | $\frac{1}{2}$ cup = 28 g |
| Milk | 1 cup = 244 g |
| Oil | 1 Tbsp = 13.5 g |
| Onion | 1 onion = 110 g |
| Salad Dressing | 1 cup = 250 g |
| Soy Sauce | 1 Tbsp = 18 g |
| Spinach | 1 cup = 180 g |
| Tomato | 1 tomato = 123 g |

Consider creating a recipe for a spinach omelet from eggs, milk, vegetable oil, and spinach. The nutrients of interest will be protein, fat, and carbohydrates. Calories will also be monitored. The amounts of the nutrients and calories for these ingredients are given in Table 2.

Table 2   Nutritional Values per 100 g of Food

| Nutrient (units) | Eggs | Milk | Oil | Spinach |
|---|---|---|---|---|
| Calories (kcal) | 152 | 61.44 | 884 | 23 |
| Protein (g) | 10.33 | 3.29 | 0 | 2.9 |
| Fat (g) | 11.44 | 3.34 | 100 | .26 |
| Carbohydrates (g) | 1.04 | 4.66 | 0 | 3.75 |

Let $x_1$ be the number of 100-gram units of eggs to use in the recipe, $x_2$ be the number of 100-gram units of milk, $x_3$ be the number of 100-gram units of oil, and $x_4$ be the number of 100-gram units of spinach. We will want to minimize the number of calories in the dish while providing at least 15 grams of protein, 4 grams of carbohydrates, and 20 grams of fat. The cooking technique specifies that at least $\frac{1}{8}$ of a cup of milk (30.5 grams) must be used in the recipe. We should thus minimize the objective function (using 100-gram units of food)

$$z = 152x_1 + 61.44x_2 + 884x_3 + 23x_4$$

subject to

$$10.33x_1 + 3.29x_2 + \phantom{0}0x_3 + 2.90x_4 \geq 15$$
$$11.44x_1 + 3.34x_2 + 100x_3 + \phantom{0}.26x_4 \geq 20$$
$$1.04x_1 + 4.66x_2 + \phantom{00}0x_3 + 3.75x_4 \geq 4$$
$$0x_1 + \phantom{0}1x_2 + \phantom{00}0x_3 + \phantom{00}0x_4 \geq .305$$

Of course, all variables are subject to nonnegativity constraints:

$$x_1 \geq 0, \quad x_2 \geq 0, \quad x_3 \geq 0, \quad \text{and} \quad x_4 \geq 0.$$

Using a graphing calculator or a computer with linear programming software, we get the following solution:

$$x_1 = 1.2600, \quad x_2 = .3050, \quad x_3 = .0448, \quad \text{and} \quad x_4 = .338.$$

This recipe produces an omelet with 257.63 calories. The amounts of each ingredient are 126 grams of eggs, 30.5 grams of milk, 4.48 grams of oil, and 33.8 grams of spinach. Converting to kitchen units using Table 1, we find the recipe to be approximately 2 eggs, $\frac{1}{8}$ cup milk, 1 teaspoon oil, and $\frac{1}{4}$ cup spinach.

## Exercises

1. Consider preparing a high-carbohydrate Greek salad using feta cheese, lettuce, salad dressing, and tomato. The amount of

carbohydrates in the salad should be maximized. In addition, the salad should have less than 260 calories, over 210 milligrams of calcium, and over 6 grams of protein. The salad should also weigh less than 400 grams and be dressed with at least 2 tablespoons ($\frac{1}{8}$ cup) of salad dressing. The amounts of the nutrients and calories for these ingredients are given in Table 3.

Table 3   Nutritional Values per 100 g of Food

| Nutrient (units) | Feta Cheese | Lettuce | Salad Dressing | Tomato |
|---|---|---|---|---|
| Calories (kcal) | 263 | 14 | 448.8 | 21 |
| Calcium (mg) | 492.5 | 36 | 0 | 5 |
| Protein (g) | 10.33 | 1.62 | 0 | .85 |
| Carbohydrates (g) | 4.09 | 2.37 | 2.5 | 4.64 |

Use linear programming to find the number of 100-gram units of each ingredient in such a Greek salad, and convert to kitchen units by using Table 1. (*Hint:* Since the ingredients are measured in 100-gram units, the constant in the weight contraint is 4, not 400.)

2. Consider preparing a stir-fry using beef, oil, onion, and soy sauce. A low-calorie stir-fry is desired, which contains less than 10 grams of carbohydrates, more than 50 grams of protein, and more than 3.5 grams of vitamin C. In order for the wok to function correctly, at least one teaspoon (or 4.5 grams) of oil must be used in the recipe. The amounts of the nutrients and calories for these ingredients are given in Table 4.

Table 4   Nutritional Values per 100 g of Food

| Nutrient (units) | Beef | Oil | Onion | Soy Sauce |
|---|---|---|---|---|
| Calories (kcal) | 215 | 884 | 38 | 60 |
| Protein (g) | 26 | 0 | 1.16 | 10.51 |
| Carbohydrates (g) | 0 | 1 | 8.63 | 5.57 |
| Vitamin C (g) | 0 | 0 | 6.4 | 0 |

Use linear programming to find the number of 100-gram units of each ingredient to be used in the stir-fry, and convert to kitchen units by using Table 1.

# Extended Project

Use the USDA Nutrient Database for Standard Reference to determine the calorie content and the amount of protein, fat, and carbohydrates for a recipe of your choice. You can set your own constraints on the total amounts of calories, protein, fat, and car-

bohydrates desired. Determine the amount of each of the ingredients that make up the recipe that satisfies your constraints. Do you think the meal would taste good if you actually made the recipe?

# Sets and Probability

8

CHAPTER

## CHAPTER OUTLINE

We often use the relative frequency of an event from a study or survey to estimate unknown probabilities. For example, we can estimate the probability that a U.S. worker is highly satisfied with his or her job. Other applications of probability occur in business, marketing, health, and the social sciences. Examples include estimating the percentage of subscribers to different cable companies, whether a stock pays a dividend yield, the effectiveness of a medication, and marital happiness, just to name a few.

Federal officials cannot predict exactly how the number of traffic deaths is affected by the trend toward fewer drunken drivers and the increased use of seat belts. Economists cannot tell exactly how stricter federal regulations on bank loans affect the U.S. economy. The number of traffic deaths and the growth of the economy are subject to many factors that cannot be predicted precisely.

Probability theory enables us to deal with uncertainty. The basic concepts of probability are discussed in this chapter, and applications of probability are discussed in the next chapter. Sets and set operations are the basic tools for the study of probability, so we begin with them.

# 8.1  Sets

Think of a **set** as a well-defined collection of objects. A set of coins might include one of each type of coin now put out by the U.S. government. Another set might be made up of all the students in your English class. By contrast, a collection of young adults does not constitute a set unless the designation "young adult" is clearly defined. For example, this set might be defined as those aged 18–29.

In mathematics, sets are often made up of numbers. The set consisting of the numbers 3, 4, and 5 is written as

$$\{3, 4, 5\},$$

where **set braces,** { }, are used to enclose the numbers belonging to the set. The numbers 3, 4, and 5 are called the **elements,** or **members,** of this set. To show that 4 is an element of the set {3, 4, 5}, we use the symbol $\in$ and write

$$4 \in \{3, 4, 5\},$$

read, "4 is an element of the set containing 3, 4, and 5."

Also, $5 \in \{3, 4, 5\}$. Place a slash through the symbol $\in$ to show that 8 is *not* an element of this set:

$$8 \notin \{3, 4, 5\}.$$

This statement is read, "8 is not an element of the set {3, 4, 5}."

Sets are often named with capital letters, so that if

$$B = \{5, 6, 7\},$$

then, for example, $6 \in B$ and $10 \notin B$.

Sometimes a set has no elements. Some examples are the set of female presidents of the United States in the period 1789–2016, the set of natural numbers less than 1, and the set of men more than 10 feet tall. A set with no elements is called the **empty set.** The symbol $\varnothing$ is used to represent the empty set.

> **⚠ CAUTION** Be careful to distinguish between the symbols 0, $\varnothing$, and {0}. The symbol 0 represents a *number*; $\varnothing$ represents a *set* with no elements; and {0} represents a *set* with one element, the number 0. Do not confuse the empty set symbol $\varnothing$ with the zero symbol $\mathbf{0}$ on a computer screen or printout.

Two sets are **equal** if they contain exactly the same elements. The sets {5, 6, 7}, {7, 6, 5}, and {6, 5, 7} all contain exactly the same elements and are equal. In symbols,

$$\{5, 6, 7\} = \{7, 6, 5\} = \{6, 5, 7\}.$$

This means that the ordering of the elements in a set is unimportant. Sets that do not contain exactly the same elements are *not equal*. For example, the sets {5, 6, 7} and {5, 6, 7, 8} do not contain exactly the same elements and are not equal. We show this by writing

$$\{5, 6, 7\} \neq \{5, 6, 7, 8\}.$$

Sometimes we describe a set by a common property of its elements rather than by a list of its elements. This common property can be expressed with **set-builder notation;** for example,

$$\{x | x \text{ has property } P\}$$

(read, "the set of all elements $x$ such that $x$ has property $P$") represents the set of all elements $x$ having some property $P$.

---

**✓ Checkpoint 1**

Indicate whether each statement is *true* or *false*.

**(a)** $9 \in \{8, 4, -3, -9, 6\}$.

**(b)** $4 \notin \{3, 9, 7\}$.

**(c)** If $M = \{0, 1, 2, 3, 4\}$, then $0 \in M$.

*Answers to Checkpoint exercises are found at the end of the section.*

---

**✓ Checkpoint 2**

List the elements in the given sets.

**(a)** $\{x | x$ is a natural number more than 5 and less than 8$\}$

**(b)** $\{x | x$ is an integer and $-3 < x \leq 1\}$

---

| **EXAMPLE 1** | List the elements belonging to each of the given sets. |

**(a)** $\{x | x$ is a natural number less than 5$\}$

**Solution**  The natural numbers less than 5 make up the set {1, 2, 3, 4}.

**(b)** $\{x | x$ is a state that borders Florida$\}$

**Solution**  The states that border Florida make up the set {Alabama, Georgia}.

The **universal set** in a particular discussion is a set that contains all of the objects being discussed. In primary school arithmetic, for example, the set of whole numbers might be the universal set, whereas in a college calculus class the universal set might be the set of all real numbers. When it is necessary to consider the universal set being used, it will be clearly specified or easily understood from the context of the problem.

Sometimes, every element of one set also belongs to another set. For example, if

$$A = \{3, 4, 5, 6\}$$

and

$$B = \{2, 3, 4, 5, 6, 7, 8\},$$

then every element of $A$ is also an element of $B$. This is an example of the following definition.

---

A set $A$ is a **subset** of a set $B$ (written $A \subseteq B$) provided that every element of $A$ is also an element of B. A set $A$ is said to be a **proper subset** of a set B (written $A \subset B$) if every element of $A$ is an element of $B$, but $B$ contains at least one element that is not a member of $A$. When $A$ contains an element that is not in $B$, then $A$ is not a subset of $B$, and we write $A \nsubseteq B$

---

Thus, for the sets $A$ and $B$ above, we note that there are elements of $B$ not in A (namely $\{2, 7, 8\}$), so $A$ is a proper subset of $B$, which is written $A \subset B$.

---

**EXAMPLE 2**    **Stock Performance** For each case, decide if $M \subseteq N$, $M \subset N$, or $M \nsubseteq N$.

**(a)** $M$ is the set of all stocks that at least doubled in value over the course of a year. $N$ is the set of all stocks that at least tripled in value over the course of a year.

**Solution** Some stocks that at least doubled in value did not achieve the threshold of also tripling in value, so there are elements in $M$ that are not in $N$. Thus, $M$ is a proper subset of $N$, written $M \subset N$.

**(b)** $M$ is the set of all stocks that increased in value over the course of the year. $N$ is the set of all stocks that increased in value by a penny or more as listed on a stock exchange over the course of the year.

**Solution** Because stock prices are listed to the nearest penny on stock exchanges, for a stock to increase in value it must increase by a penny or more. Thus, $M$ is a subset of $N$. Because all the stocks that increased in value by a penny or more are stocks that increased in value, there are no stocks in $N$ that are not also in $M$. For this case, we can write $M \subseteq N$.

**(c)** $M$ is the set of all stocks priced $50.00 or higher on January 3, 2017. $N$ is the set of stocks priced $50.00 or higher on January 2, 2018.

**Solution** Invariably, some stocks that were at $50.00 or higher will fall below $50.00. So there are elements in $M$ that are not in $N$. Thus, $M$ is not a subset of $N$, and we write $M \nsubseteq N$.

---

Every set $A$ is a subset of itself, because the statement "every element of $A$ is also an element of $A$" is always true. It is also true that the empty set is a subset of every set.[*]

---

[*]This fact is not intuitively obvious to most people. If you wish, you can think of it as a convention that we agree to adopt in order to simplify the statements of several results later.

For any set $A$,

$$\varnothing \subset A \quad \text{and} \quad A \subseteq A.$$

---

**EXAMPLE 3**    Decide whether $E \subset F$.

**(a)**  $E = \{2, 4, 6, 8\}$ and $F = \{1, 2, 3, 4, 5, 6, 7, 8, 9, 10\}$.

**Solution**  Since each element of $E$ is an element of $F$ and $F$ contains several elements not in $E$, $E \subset F$.

**(b)**  $E$ is the set of registered voters in Texas. $F$ is the set of adults aged 18 years or older.

**Solution**  To register to vote, one must be at least 18 years old. Not all adults at least 18 years old, however, are registered. Thus, every element of $E$ is contained in $F$ and $F$ contains elements not in $E$. Therefore, $E \subset F$.

**(c)**  $E$ is the set of diet soda drinks. $F$ is the set of diet soda drinks sweetened with Nutrasweet®.

**Solution**  Some diet soda drinks are sweetened with the sugar substitute Splenda®. In this case, $E$ is not a subset of $F$ ($E \not\subseteq F$). ✓ 3

---

**✓ Checkpoint 3**

Indicate whether each statement is *true* or *false*.

**(a)**  $\{10, 20, 30\} \subseteq \{20, 30, 40, 50\}$.

**(b)**  $\{x | x \text{ is a minivan}\} \subseteq \{x | x \text{ is a motor vehicle}\}$.

**(c)**  $\{a, e, i, o, u\} \subset \{a, e, i, o, u, y\}$.

**(d)**  $\{x | x \text{ is a U.S. state that begins with the letter "A"}\} \subset$ {Alabama, Alaska, Arizona, Arkansas}.

**✓ Checkpoint 4**

List all subsets of $\{w, x, y, z\}$.

---

**EXAMPLE 4**    List all possible subsets for each of the given sets.

**(a)**  $\{7, 8\}$

**Solution**  A good way to find the subsets of $\{7, 8\}$ is to use a **tree diagram**—a systematic way of listing all the subsets of a given set. The tree diagram in Figure 8.1(a) shows there are four subsets of $\{7, 8\}$:

$$\varnothing, \quad \{7\}, \quad \{8\}, \quad \text{and} \quad \{7, 8\}.$$

**(b)**  $\{a, b, c\}$

**Solution**  The tree diagram in Figure 8.1(b) shows that there are 8 subsets of $\{a, b, c\}$:

$$\varnothing, \quad \{a\}, \quad \{b\}, \quad \{c\}, \quad \{a, b\}, \quad \{a, c\}, \quad \{b, c\}, \quad \text{and} \quad \{a, b, c\}. \text{ ✓ 4}$$

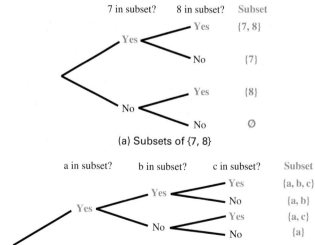

(a) Subsets of {7, 8}

(b) Subsets of {a, b, c}

**Figure 8.1**

By using the fact that there are two possibilities for each element (either it is in the subset or it is not), we have found that a set with 2 elements has 4 ($=2^2$) subsets and a set with 3 elements has 8 ($=2^3$) subsets. Similar arguments work for any finite set and lead to the following conclusion.

---

A set of $n$ distinct elements has $2^n$ subsets.

---

✓**Checkpoint 5**

Find the number of subsets for each of the given sets.

**(a)**  $\{x|x$ is a season of the year$\}$

**(b)**  $\{-6, -5, -4, -3, -2, -1, 0\}$

**(c)**  $\{6\}$

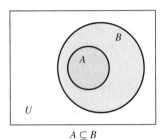

$A \subseteq B$

**Figure 8.2**

> **EXAMPLE 5**   Find the number of subsets for each of the given sets.
>
> **(a)**  $\{$blue, brown, hazel, green$\}$
>
> **Solution**  Since this set has 4 elements, it has $2^4 = 16$ subsets.
>
> **(b)**  $\{x|x$ is a month of the year$\}$
>
> **Solution**  This set has 12 elements and therefore has $2^{12} = 4096$ subsets.
>
> **(c)**  $\varnothing$
>
> **Solution**  Since the empty set has 0 elements, it has $2^0 = 1$ subset, $\varnothing$ itself. ✓₅

**Venn diagrams** are sometimes used to illustrate relationships among sets. The Venn diagram in Figure 8.2 shows a set $A$ that is a proper subset of a set $B$, because $A$ is entirely in $B$. (The areas of the regions are not meant to be proportional to the sizes of the corresponding sets.) The rectangle represents the universal set $U$. ✓₆

Some sets have infinitely many elements. We often use the notation "..." to indicate such sets. One example of an infinite set is the set of natural numbers, $\{1, 2, 3, 4, \ldots\}$. Another infinite set is the set of integers, $\{\ldots, -3, -2, -1, 0, 1, 2, 3, \ldots\}$.

✓**Checkpoint 6**

Refer to sets $A$, $B$, $C$, and $U$ in the diagram.

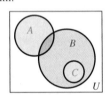

**(a)**  Is $A \subseteq B$?

**(b)**  Is $C \subset B$?

**(c)**  Is $C \subset U$?

**(d)**  Is $\varnothing \subset A$?

## Operations on Sets

Given a set $A$ and a universal set $U$, the set of all elements of $U$ that do *not* belong to $A$ is called the **complement** of set $A$. For example, if $A$ is the set of all the female students in your class and $U$ is the set of all students in the class, then the complement of $A$ would be the set of all male students in the class. The complement of set $A$ is written $A'$ (read "A-prime"). The Venn diagram in Figure 8.3 shows a set $B$. Its complement, $B'$, is shown in color.

Some textbooks use $\overline{A}$ to denote the complement of $A$. This notation conveys the same meaning as $A'$.

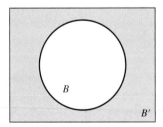

**Figure 8.3**

✓**Checkpoint 7**

Let $U = \{a, b, c, d, e, f, g\}$, with
$K = \{c, d, f, g\}$ and
$R = \{a, c, d, e, g\}$. Find

**(a)**  $K'$;

**(b)**  $R'$.

> **EXAMPLE 6**   Let $U = \{1, 2, 3, 4, 5, 6, 7\}$, $A = \{1, 3, 5, 7\}$, and $B = \{3, 4, 6\}$. Find the given sets.
>
> **(a)**  $A'$
>
> **Solution**  Set $A'$ contains the elements of $U$ that are not in $A$:
>
> $$A' = \{2, 4, 6\}.$$
>
> **(b)**  $B'$
>
> **Solution**  $B' = \{1, 2, 5, 7\}$.
>
> **(c)**  $\varnothing'$ and $U'$
>
> **Solution**  $\varnothing' = U$ and $U' = \varnothing$. ✓₇

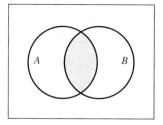

$A \cap B$

**Figure 8.4**

Given two sets $A$ and $B$, the set of all elements belonging to *both* set $A$ and set $B$ is called the **intersection** of the two sets, written $A \cap B$. For example, the elements that belong to both $A = \{1, 2, 4, 5, 7\}$ and $B = \{2, 4, 5, 7, 9, 11\}$ are 2, 4, 5, and 7, so

$$A \text{ and } B = A \cap B$$
$$= \{1, \mathbf{2}, \mathbf{4}, \mathbf{5}, 7\} \cap \{\mathbf{2}, \mathbf{4}, \mathbf{5}, 7, 9, 11\}$$
$$= \{2, 4, 5, 7\}.$$

The Venn diagram in Figure 8.4 shows two sets $A$ and $B$, with their intersection, $A \cap B$, shown in color.

**EXAMPLE 7** Find the given sets.

**(a)** $\{9, 15, 25, 36\} \cap \{15, 20, 25, 30, 35\}$

**Solution** $\{15, 25\}$. The elements 15 and 25 are the only ones belonging to both sets.

**(b)** $\{x | x \text{ is a teenager}\} \cap \{x | x \text{ is a senior citizen}\}$

**Solution** $\varnothing$ since no teenager is a senior citizen.

✓ **Checkpoint 8**

Find $\{1, 2, 3, 4\} \cap \{3, 5, 7, 9\}$.

Two sets that have no elements in common are called **disjoint sets.** For example, there are no elements common to both $\{50, 51, 54\}$ and $\{52, 53, 55, 56\}$, so these two sets are disjoint, and

$$\{50, 51, 54\} \cap \{52, 53, 55, 56\} = \varnothing.$$

The result of this example can be generalized as follows.

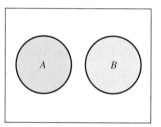

$A$ and $B$ are disjoint sets.

**Figure 8.5**

For any sets $A$ and $B$,

if $A$ and $B$ are disjoint sets, then $A \cap B = \varnothing$.

Figure 8.5 is a Venn diagram of disjoint sets.

The set of all elements belonging to set $A$ or to set $B$, or to both sets, is called the **union** of the two sets, written $A \cup B$. For example, for sets $A = \{1, 3, 5\}$ and $B = \{3, 5, 7, 9\}$,

$$A \text{ or } B = A \cup B$$
$$= \{1, 3, 5\} \cup \{3, 5, 7, 9\}$$
$$= \{1, 3, 5, 7, 9\}.$$

The Venn diagram in Figure 8.6 shows two sets $A$ and $B$, with their union, $A \cup B$, shown in color.

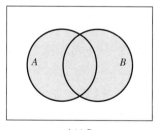

$A \cup B$

**Figure 8.6**

**EXAMPLE 8** Find the given sets.

**(a)** $\{1, 2, 5, 9, 14\} \cup \{1, 3, 4, 8\}$.

**Solution** Begin by listing the elements of the first set, $\{1, 2, 5, 9, 14\}$. Then include any elements from the second set *that are not already listed.* Doing this gives

$$\{1, 2, 5, 9, 14\} \cup \{1, 3, 4, 8\} = \{1, 2, \mathbf{3}, \mathbf{4}, 5, \mathbf{8}, 9, 14\}.$$

**(b)** {terriers, spaniels, chows, dalmatians} $\cup$ {spaniels, collies, bulldogs}

✓ **Checkpoint 9**

Find $\{a, b, c\} \cup \{a, c, e\}$.

**Solution** {terriers, spaniels, chows, dalmatians, collies, bulldogs}.

Finding the complement of a set, the intersection of two sets, or the union of two sets is an example of a *set operation.*

## Operations on Sets

Let $A$ and $B$ be any sets, with $U$ signifying the universal set. Then

the **complement** of $A$, written $A'$, is

$$A' = \{x \mid x \notin A \text{ and } x \in U\};$$

the **intersection** of $A$ and $B$ is

$$A \cap B = \{x \mid x \in A \textbf{ and } x \in B\};$$

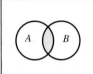

the **union** of $A$ and $B$ is

$$A \cup B = \{x \mid x \in A \textbf{ or } x \in B \textbf{ or both}\}.$$

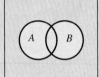

**CAUTION** As shown in the preceding definitions, an element is in the intersection of sets $A$ and $B$ if it is in *both* $A$ and $B$ at the same time, but an element is in the union of sets $A$ and $B$ if it is in *either* set $A$ or set $B$, *or* in both sets $A$ and $B$.

**EXAMPLE 9** **Stock Prices** The following table gives the 52-week high and low prices, the last price, and the change in price from the day before for six stocks on a recent day.

| Stock | High | Low | Last | Change |
|---|---|---|---|---|
| First Solar | 51.33 | 25.56 | 36.38 | −.24 |
| Ford | 14.04 | 10.90 | 10.92 | −.09 |
| IBM | 182.79 | 142.50 | 150.37 | −.28 |
| Netflix | 161.10 | 84.50 | 160.81 | +2.27 |
| Procter & Gamble | 92.00 | 79.41 | 86.19 | +.02 |
| Verizon | 56.95 | 45.76 | 45.84 | −.18 |

Let the universal set $U$ consist of the six stocks listed in the table. Let $A$ contain all stocks with a high price greater than $50, $B$ all stocks with a last price between $25 and $80, and $C$ all stocks with a negative value for the change. Find the results of the given set operations. (Data from: www.morningstar.com.)

**(a)** $B'$

**Solution** Set $B$ consists of First Solar and Verizon. Set $B'$ contains all the listed stocks that are not in set $B$, so

$$B' = \{\text{Ford, IBM, Netflix, Procter \& Gamble}\}.$$

**(b)** $A \cap C$

**Solution** Set $A$ consists of First Solar, IBM, Netflix, Procter & Gamble, and Verizon, and set $C$ consists of First Solar, Ford, IBM, and Verizon. Hence the stocks in both $A$ and $C$ are

$$A \cap C = \{\text{First Solar, IBM, Verizon}\}.$$

**(c)** $A \cup B$

**Solution** $A \cup B = \{\text{First Solar, IBM, Netflix, Procter \& Gamble, and Verizon}\}$

✓ **Checkpoint 10**

In Example 9, find the given set of stocks.

(a) $B \cap C$

(b) $B \cup C$

# 8.1 Exercises

*Write true or false for each statement. (See Example 1.)*

1. $3 \in \{2, 5, 7, 9, 10\}$

2. $6 \in \{-2, 6, 9, 5\}$

3. $9 \notin \{2, 1, 5, 8\}$

4. $3 \notin \{7, 6, 5, 4\}$

5. $\{2, 5, 8, 9\} = \{2, 5, 9, 8\}$

6. $\{3, 7, 12, 14\} = \{3, 7, 12, 14, 0\}$

7. {all whole numbers greater than 7 and less than 10} $= \{8, 9\}$

8. {all natural numbers not greater than 3} $= \{0, 1, 2\}$

9. $\{x \mid x \text{ is an odd integer}, 6 \le x \le 18\} = \{7, 9, 11, 15, 17\}$

10. $\{x \mid x \text{ is a vowel}\} = \{a, e, i, o, u\}$

11. The elements of a set may be sets themselves, as in $\{1, \{1, 3\}, \{2\}, 4\}$. Explain why the set $\{\varnothing\}$ is not the same set as $\{0\}$.

12. What is set-builder notation? Give an example.

*Let $A = \{-3, 0, 3\}$, $B = \{-2, -1, 0, 1, 2\}$, $C = \{-3, -1\}$, $D = \{0\}$, $E = \{-2\}$, and $U = \{-3, -2, -1, 0, 1, 2, 3\}$. Insert $\subset$ or $\not\subset$ to make the given statements true. (See Example 2.)*

13. $A$_____$U$

14. $E$_____$A$

15. $A$_____$E$

16. $B$_____$C$

17. $\varnothing$_____$A$

18. $\{0, 2\}$_____$D$

19. $D$_____$B$

20. $A$_____$C$

*Find the number of subsets of the given set. (See Example 5.)*

21. $\{A, B, C\}$

22. {red, yellow, blue, black, white}

23. $\{x \mid x \text{ is an integer strictly between 0 and 8}\}$

24. $\{x \mid x \text{ is a whole number less than 4}\}$

*Find the complement of each set. (See Example 6.)*

25. The set in Exercise 23 if $U$ is the set of all integers.

26. The set in Exercise 24 if $U$ is the set of all whole numbers.

27. Describe the intersection and union of sets. How do they differ?

*Insert $\cap$ or $\cup$ to make each statement true. (See Examples 7 and 8.)*

28. $\{5, 7, 9, 19\}$_____$\{7, 9, 11, 15\} = \{7, 9\}$

29. $\{8, 11, 15\}$_____$\{8, 11, 19, 20\} = \{8, 11\}$

30. $\{2, 1, 7\}$_____$\{1, 5, 9\} = \{1\}$

31. $\{6, 12, 14, 16\}$_____$\{6, 14, 19\} = \{6, 14\}$

32. $\{3, 5, 9, 10\}$_____$\varnothing = \varnothing$

33. $\{3, 5, 9, 10\}$_____$\varnothing = \{3, 5, 9, 10\}$

34. $\{1, 2, 4\}$_____$\{1, 2, 4\} = \{1, 2, 4\}$

35. $\{1, 2, 4\}$_____$\{1, 2\} = \{1, 2, 4\}$

36. Is it possible for two nonempty sets to have the same intersection and union? If so, give an example.

*Let $U = \{a, b, c, d, e, f, 1, 2, 3, 4, 5, 6\}$, $X = \{a, b, c, 1, 2, 3\}$, $Y = \{b, d, f, 1, 3, 5\}$, and $Z = \{b, d, 2, 3, 5\}$.*

*List the members of each of the given sets, using set braces. (See Examples 6–8.)*

37. $X \cap Y$

38. $X \cup Y$

39. $X'$

40. $Y'$

41. $X' \cap Y'$

42. $X' \cap Z$

43. $X \cup (Y \cap Z)$

44. $Y \cap (X \cup Z)$

*Let $U = \{\text{all students in this school}\}$,*
*$M = \{\text{all students taking this course}\}$,*
*$N = \{\text{all students taking accounting}\}$, and*
*$P = \{\text{all students taking philosophy}\}$.*

*Describe each of the following sets in words.*

45. $M'$

46. $M \cup N$

47. $N \cap P$

48. $N' \cap P'$

49. Refer to the sets listed in the directions for Exercises 13–20. Which pairs of sets are disjoint?

50. Refer to the sets listed in the directions for Exercises 37–44. Which pairs of sets are disjoint?

*Refer to Example 9 in the text. Describe each of the sets in Exercises 51–54 in words; then list the elements of each set.*

**51.** $A'$                    **52.** $B \cup C'$

**53.** $A' \cap B'$            **54.** $B' \cup C$

**Business** *A stock broker classifies her clients by sex, marital status, and employment status. Let the universal set be the set of all clients, M be the set of male clients, S be the set of single clients, and E be the set of employed clients. Describe the following sets in words.*

**55.** $M \cap E$              **56.** $M' \cap S$

**57.** $M' \cup S'$            **58.** $E' \cup S'$

**Food and Drink Sales** *Sales (in millions of dollars) in the United States at eating and drinking establishments in the years 2014 and 2015 are shown in the following table. (Data from: National Restaurant Association.)*

| Type of Place or Service | 2014 | 2015 |
|---|---|---|
| Table-service restaurants ($T$) | 213,498 | 219,690 |
| Quick-service and fast-casual restaurants ($Q$) | 192,836 | 201,128 |
| Cafeterias, grill-buffets, and buffets ($C$) | 8,429 | 8,336 |
| Snack and nonalcoholic beverage bars ($S$) | 31,221 | 32,845 |
| Bars and taverns ($B$) | 19,943 | 20,562 |
| Other ($O$) | 8,637 | 9,076 |

*List the elements of each set.*

**59.** The set of eating and drinking establishments that generated more than $200,000 million in revenue in both 2014 and 2015.

**60.** The set of eating and drinking establishments that generated less than $20,000 million in revenue in 2014 or 2015.

**61.** The set of eating and drinking establishments that had revenues fall from 2014 to 2015.

**62.** The set of eating and drinking establishments that had revenues rise from 2014 to 2015 and had less than $35,000 million in revenue.

**Cable Subscribers** *The top six basic cable television providers in the year 2016 are listed in the following table. Use this information for Exercises 63–72.*

| Rank | Cable Provider | Subscribers (millions) |
|---|---|---|
| 1 | Direct TV | 25.3 |
| 2 | Comcast | 22.5 |
| 3 | Charter | 17.5 |
| 4 | Dish | 13.9 |
| 5 | Verizon FIOS | 4.7 |
| 6 | Altice | 4.6 |

*List the elements of the following sets. (Data from: www.ncta.com.)*

**63.** $F$, the set of cable providers with more than 5 million subscribers.

**64.** $G$, the set of cable providers with between 10 million and 20 million subscribers.

**65.** $H$, the set of cable providers with less than 20 million subscribers.

**66.** $I$, the set of cable providers with more than 15 million subscribers.

**67.** $F \cup G$      **68.** $H \cap F$      **69.** $I'$      **70.** $I' \cap H$

**71.** $J$, the set of cable providers with more than 30 million subscribers

**72.** $(F \cup H)'$

**Farm Products** *The following table gives the amount (in millions of metric tons) of several farm products exported from the United States in 2011 and 2016. (Data from: U.S. Department of Agriculture.)*

| Product | 2011 | 2016 |
|---|---|---|
| Wheat | 28.6 | 28.2 |
| Corn | 39.4 | 56.5 |
| Rice | 3.2 | 3.6 |
| Barley | .2 | .1 |

*List the elements of the following sets.*

**73.** The set of farm products where exports increased from 2011 to 2016

**74.** The set of farm products that had more than 40 million metric tons of exports in both 2011 and 2016

**75.** The set of farm products that had more than 50 million metric tons of exports in both 2011 and 2016

**76.** The set of farm products that had less than 40 million metric tons of exports in either 2011 or 2016

**77.** The set of farm products that had more than 3.5 million metric tons of exports in either 2011 or 2016

**78.** The set of farm products that decreased in exports from 2011 to 2016 and had more than 30 million metric tons of exports in 2016

✓**Checkpoint Answers**

**1.** (a) False        (b) True        (c) True

**2.** (a) $\{6, 7\}$        (b) $\{-2, -1, 0, 1\}$

**3.** (a) False    (b) True    (c) True    (d) False

**4.** $\varnothing$, $\{w\}$, $\{x\}$, $\{y\}$, $\{z\}$, $\{w, x\}$, $\{w, y\}$, $\{w, z\}$, $\{x, y\}$, $\{x, z\}$, $\{y, z\}$, $\{w, x, y\}$, $\{w, x, z\}$, $\{w, y, z\}$, $\{x, y, z\}$, $\{w, x, y, z\}$

**5.** (a) 16        (b) 128        (c) 2

**6.** (a) No        (b) Yes        (c) Yes        (d) Yes

**7.** (a) $\{a, b, e\}$        (b) $\{b, f\}$

**8.** $\{3\}$

**9.** $\{a, b, c, e\}$

**10.** (a) $\{$First Solar, Verizon$\}$

(b) $\{$First Solar, Ford, IBM, Verizon$\}$

# 8.2 Applications of Venn Diagrams and Contingency Tables

We used Venn diagrams in the previous section to illustrate set union and intersection. The rectangular region in a Venn diagram represents the universal set $U$. Including only a single set $A$ inside the universal set, as in Figure 8.7, divides $U$ into two nonoverlapping regions. Region 1 represents $A'$, those elements outside set $A$, while region 2 represents those elements belonging to set $A$. (The numbering of these regions is arbitrary.)

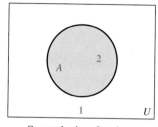

One set leads to 2 regions.
(Numbering is arbitrary.)

**Figure 8.7**

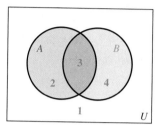

Two sets lead to 4 regions.
(Numbering is arbitrary.)

**Figure 8.8**

The Venn diagram of Figure 8.8 shows two sets inside $U$. These two sets divide the universal set into four nonoverlapping regions. As labeled in Figure 8.8, region 1 includes those elements outside both set $A$ and set $B$. Region 2 includes those elements belonging to $A$ and not to $B$. Region 3 includes those elements belonging to both $A$ and $B$. Which elements belong to region 4? (Again, the numbering is arbitrary.)

---

**EXAMPLE 1** Draw a Venn diagram similar to Figure 8.8, and shade the regions representing the given sets.

**(a)** $A' \cap B$

**Solution** Set $A'$ contains all the elements outside set $A$. As labeled in Figure 8.8, $A'$ is represented by regions 1 and 4. Set $B$ is represented by the elements in regions 3 and 4. The intersection of sets $A'$ and $B$, the set $A' \cap B$, is given by the region common to regions 1 and 4 and regions 3 and 4. The result, region 4, is shaded in Figure 8.9.

**(b)** $A' \cup B'$

**Solution** Again, set $A'$ is represented by regions 1 and 4 and set $B'$ by regions 1 and 2. To find $A' \cup B'$, identify the region that represents the set of all elements in $A'$, $B'$, or both. The result, which is shaded in Figure 8.10, includes regions 1, 2, and 4. ✔1

✓ **Checkpoint 1**

Draw Venn diagrams for the given set operations.

**(a)** $A \cup B'$

**(b)** $A' \cap B'$

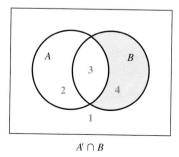

$A' \cap B$

**Figure 8.9**

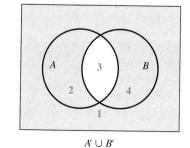

$A' \cup B'$

**Figure 8.10**

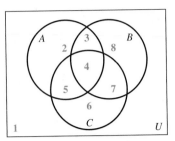

Three sets lead to 8 regions.

**Figure 8.11**

Venn diagrams also can be drawn with three sets inside $U$. These three sets divide the universal set into eight nonoverlapping regions that can be numbered (arbitrarily) as in Figure 8.11.

**EXAMPLE 2**  Shade $A' \cup (B \cap C')$ in a Venn diagram.

**Solution**  First find $B \cap C'$. Set $B$ is represented by regions 3, 4, 7, and 8, and set $C'$ by regions 1, 2, 3, and 8. The overlap of these regions, regions 3 and 8, represents the set $B \cap C'$. Set $A'$ is represented by regions 1, 6, 7, and 8. The union of regions 3 and 8 and regions 1, 6, 7, and 8 contains regions 1, 3, 6, 7, and 8, which are shaded in Figure 8.12.  ✓2

✓ **Checkpoint 2**

Draw Venn diagrams for the given set operations.

(a)  $(B' \cap A) \cup C$

(b)  $(A \cup B)' \cap C$

Venn diagrams can be used to solve problems that result from surveying groups of people.

**EXAMPLE 3**  **Consumer Profiling** A market researcher collecting data on 100 household finds that

81 have cable television (CT);

65 have high-speed Internet (HSI);

56 have both.

The researcher wants to answer the following questions:

(a)  How many households do not have high-speed Internet?

(b)  How many households have neither cable television nor high-speed Internet?

(c)  How many have cable television, but not high-speed Internet?

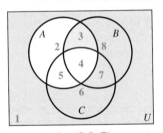

$A' \cup (B \cap C')$

**Figure 8.12**

**Solution**  A Venn diagram like the one in Figure 8.13 will help sort out the information. In Figure 8.13(a), we place the number 56 in the region common to both cable television and high-speed Internet, because 56 households have both. Of the 81 with cable television, $81 - 56 = 25$ do not have high-speed Internet, so in Figure 8.13(b) we place 25 in the region for cable television, but not high-speed Internet. Similarly, $65 - 56 = 9$ households have high-speed Internet, but not cable television, so we place 9 in that region. Finally, the diagram shows that $100 - 9 - 56 - 25 = 10$ have neither high-speed Internet nor cable television. Now we can answer the questions.

(a)

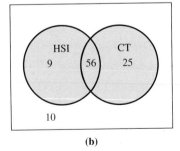

(b)

**Figure 8.13**

✓ **Checkpoint 3**

(a)  Place numbers in the regions on a Venn diagram if the data on the 100 households in Example 3 showed

78 with cable television;

52 with high-speed Internet;

48 with both.

(b)  How many have high-speed Internet, but not cable television?

(a)  $10 + 25 = 35$ do not have high-speed Internet.

(b)  10 have neither.

(c)  25 have cable television, but not high-speed Internet.  3

**EXAMPLE 4**   **Media**   A researcher surveyed 67 business majors at a university to determine which business-focused media websites the students regularly visited. The results are summarized below.

21 of the students visited *Business Week*;

20 visited the *Wall Street Journal*;

46 visited *Fortune*;

12 visited *Business Week* and the *Journal*;

11 visited the *Journal* and *Fortune*;

13 visited *Business Week* and *Fortune*;

9 visited all three websites.

Use the preceding data to answer the following questions:

(a) How many students visited none of the websites?

(b) How many visited only *Fortune*?

(c) How many visited *Business Week* and the *Journal*, but not *Fortune*?

**Solution**   Once again, use a Venn diagram to represent the data. Since 9 students visited all three websites, begin by placing 9 in the area in Figure 8.14(a) that belongs to all three regions.

Of the 13 students who visited *Business Week* and *Fortune*, 9 also visited the *Journal*. Therefore, only $13 - 9 = 4$ students visited just *Business Week* and *Fortune*. So place a 4 in the region common only to *Business Week* and *Fortune* readers, as in Figure 8.14(b).

**Figure 8.14(a)**

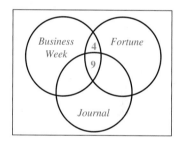

**Figure 8.14(b)**

In the same way, place a 3 in the region of Figure 8.14(c) common only to *Business Week* and the *Journal* visitors, and 2 in the region common only to *Fortune* and the *Journal* visitors.

The data shows that 21 students visited *Business Week*. However, $4 + 9 + 3 = 16$ visitors have already been placed in the *Business Week* region. The balance of this region in Figure 8.14(d) will contain only $21 - 16 = 5$ students. These 5 students visited *Business Week* only—not *Fortune* and not the *Journal*.

**Figure 8.14(c)**

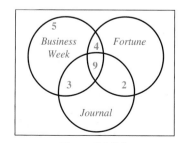

**Figure 8.14(d)**

In the same way, 6 students visited only the *Journal* and 31 visited only *Fortune*, as shown in Figure 8.14(e).

A total of $5 + 4 + 31 + 3 + 9 + 2 + 6 = 60$ students are placed in the various regions of Figure 8.14(e). Since 67 students were surveyed, $67 - 60 = 7$ students visited none of the three websites, and so 7 is placed outside the other regions in Figure 8.14(f).

**Figure 8.14(e)**

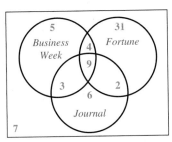

**Figure 8.14(f)**

Figure 8.14(f) can now be used to answer the questions asked at the beginning of this example:

**(a)** There are 7 students who did not visit any of the 3 websites.

**(b)** There are 31 students who visited only *Fortune*.

**(c)** The overlap of the regions representing *Business Week* and the *Journal* shows that 3 students regularly visited the sites for *Business Week* and the *Journal*, but not *Fortune*. ✔4

---

✔**Checkpoint 4**

In Example 4, how many students visited exactly,

**(a)** 1 of the websites?

**(b)** 2 of the websites?

---

**EXAMPLE 5** **Health** Mark McCloney, M.D., saw 100 patients exhibiting flu symptoms such as fever, chills, and headache. Dr. McCloney reported the following information on patients exhibiting symptoms:

Of the 100 patients,

74 reported a fever;

72 reported chills;

67 reported a headache;

55 reported both a fever and chills;

47 reported both a fever and a headache;

49 reported both chills and a headache;

35 reported all three;

3 thought they had the flu, but did not report fever, chills, or headache.

Create a Venn diagram to represent this data. It should show the number of people in each region.

**Solution** Begin with the 35 patients who reported all three symptoms. This leaves $55 - 35 = 20$ who reported fever and chills, but not headache; $47 - 35 = 12$ who reported fever and headache, but not chills; and $49 - 35 = 14$ who reported chills and headache, but not fever. With this information, we have $74 - (35 + 20 + 12) = 7$ who reported fever alone; $72 - (35 + 20 + 14) = 3$ with chills alone; and $67 - (35 + 12 + 14) = 6$ with headache alone. The remaining 3 patients who thought they had the flu, but did not report fever, chills, or headache are denoted outside the 3 circles. See Figure 8.15. ✔5

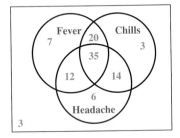

**Figure 8.15**

---

✔**Checkpoint 5**

In Example 5, suppose 75 patients reported a fever and only 2 thought they had the flu, but did not report fever, chills, or headache. Then how many

**(a)** reported only a fever?

**(b)** reported a fever or chills?

**(c)** reported a fever, chills, or headache?

---

📄 **NOTE** In all the preceding examples, we started in the innermost region with the intersection of the categories. This is usually the best way to begin solving problems of this type.

We use the symbol **n(A)** to denote the *number* of elements in *A*. For instance, if $A = \{w, x, y, z\}$, then $n(A) = 4$. Next, we prove the following useful fact.

## Addition Rule for Counting

$$n(A \cup B) = n(A) + n(B) - n(A \cap B).$$

For example, if $A = \{r, s, t, u, v\}$ and $B = \{r, t, w\}$, then $A \cap B = \{r, t\}$, so that $n(A) = 5$, $n(B) = 3$, and $n(A \cap B) = 2$. By the formula in the box, $n(A \cup B) = 5 + 3 - 2 = 6$, which is certainly true, since $A \cup B = \{r, s, t, u, v, w\}$.

Here is a proof of the statement in the box: Let *x* be the number of elements in *A* that are not in *B*, *y* be the number of elements in $A \cap B$, and *z* be the number of elements in *B* that are not in *A*, as indicated in Figure 8.16. That diagram shows that $n(A \cup B) = x + y + z$. It also shows that $n(A) = x + y$ and $n(B) = y + z$, so that

$$n(A) + n(B) - n(A \cap B) = (x + y) + (z + y) - y$$
$$= x + y + z$$
$$= n(A \cup B).$$

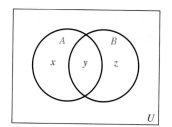

**Figure 8.16**

### EXAMPLE 6   College Majors

A group of 10 students meets to plan a school function. All are majoring in accounting or economics or both. Five of the students are economics majors, and 7 are majors in accounting. How many major in both subjects?

**Solution** Let *A* represent the set of economics majors and *B* represent the set of accounting majors. Use the union rule, with $n(A) = 5$, $n(B) = 7$, and $n(A \cup B) = 10$. We must find $n(A \cap B)$:

$$n(A \cup B) = n(A) + n(B) - n(A \cap B)$$
$$10 = 5 + 7 - n(A \cap B).$$

So,

$$n(A \cap B) = 5 + 7 - 10 = 2. \quad \boxed{✓6}$$

✓ **Checkpoint 6**

If $n(A) = 10$, $n(B) = 7$, and $n(A \cap B) = 3$, find $n(A \cup B)$.

In addition to Venn diagrams, we can use a *contingency table*, which is sometimes called a *cross tabulation*, to summarize counts from several different groups. A contingency table is a table in a matrix format that gives the frequency distribution of several variables. With such tables, we can compute the number of elements in sets of interest, the intersection of two sets, and the union of two sets.

### EXAMPLE 7   Employment Status

The following contingency table gives the number (in thousands) of males (denoted with *M*) and females (denoted with *F*) who were working full-time (denoted *A*), working part-time (denoted *B*), or looking for work (denoted *C*) in the year 2016. (Data from: U.S. Census Bureau.)

| | A Worked Full-Time | B Worked Part-Time | C Looked for Work |
|---|---|---|---|
| *M* Males | 70,567 | 10,002 | 4187 |
| *F* Females | 53,194 | 17,674 | 3565 |

Find the number of people in the given sets.

**(a)** *A*

**Solution** The set *A* consists of males and females who worked full-time. From the table, we see that there were 70,567 thousand males and 53,194 thousand females, which yields 123,761 thousand people who worked full-time in the year 2016.

**(b)** $F \cap A$

**Solution** The set $F \cap A$ consists of all the people who are female *and* who worked full-time. We see that that there were 53,194 thousand females who worked full-time.

**(c)** $M \cup C$

**Solution** The set $M \cup C$ consists of people who are male or people who were looking for work. Using the addition rule for counting, we have

$$n(M \cup C) = n(M) + n(C) - n(M \cap C).$$

From the table, we have $n(M) = 70{,}567 + 10{,}002 + 4187 = 84{,}756$ thousand males. Also from the table we have $n(C) = 4187 + 3565 = 7752$ thousand people who are looking for work. We also have $n(M \cap C)$, the number of people who are male and are looking for work, is 4187 thousand. Thus,

$$n(M \cup C) = n(M) + n(C) - n(M \cap C)$$
$$= 84{,}756 + 7752 - 4187$$
$$= 88{,}321 \text{ thousand people}.$$

✓ **Checkpoint 7**

Refer to Example 7 and find the number of people in the following sets.

**(a)** $A \cup B$

**(b)** $F \cup B$

**(d)** $(B \cup C) \cap F$

**Solution** Begin with the set $B \cup C$, which contains all those who worked part-time or were looking for work. Of this set, take only those who were females, for a total of $17{,}674 + 3565 = 21{,}239$ thousand. ✓₇

# 8.2 Exercises

*Sketch a Venn diagram like the one shown, and use shading to show each of the given sets. (See Example 1.)*

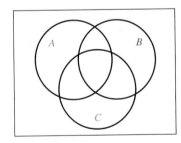

1. $A \cap B'$

2. $A \cup B'$

3. $B' \cup A'$

4. $A' \cap B'$

5. $B' \cup (A \cap B')$

6. $(A \cap B) \cup A'$

7. $U'$

8. $\varnothing'$

9. Three sets divide the universal set into at most _____ regions.

⬛10. What does the notation $n(A)$ represent?

*Sketch a Venn diagram like the one shown, and use shading to show each of the given sets. (See Example 2.)*

11. $(A \cap C') \cup B$

12. $A \cap (B \cup C')$

13. $A' \cap (B \cap C)$

14. $(A' \cap B') \cap C$

15. $(A \cap B') \cup C$

16. $(A \cap B') \cap C$

*Use Venn diagrams to answer the given questions. (See Examples 3, 4, and 5.)*

17. **Movies** In 2016, among the top 100 grossing movies in the United States, 15 were rated PG-13 and earned over $100 million. The number of movies that were rated PG-13 that earned less than $100 million was 38. The number of movies that were

not rated PG-13 that earned less than $100 million was 34. What was the number of movies that earned over $100 million and that were not rated PG-13? (Data from: www.the-numbers.com.)

18. **Movies** In 2016, among the top 100 grossing movies in the United States, 20 were classified as adventure and earned more than $75 million. The number of movies that earned over $75 million was 40. The number of movies that were not adventure movies and earned less than $75 million was 55. What was the number of movies that were adventure movies that earned less than $75 million? (Data from: www.the-numbers.com.)

19. **Job Applicants** The human resources director for a commercial real estate company received the following numbers of applications from people with the given information:

    66 with sales experience;

    40 with a college degree;

    23 with a real estate license;

    26 with sales experience and a college degree;

    16 with sales experience and a real estate license;

    15 with a college degree and a real estate license;

    11 with sales experience, a college degree, and a real estate license;

    22 with neither sales experience, a college degree, nor a real estate license.

    **(a)** How many applicants were there?

    **(b)** How many applicants did not have sales experience?

    **(c)** How many had sales experience and a college degree, but not a real estate license?

    **(d)** How many had only a real estate license?

20. **Product Sales** A pet store keeps track of the purchases of customers over a four-hour period. The store manager classifies purchases as containing a dog product, a cat product, a fish product, or a product for a different kind of pet. She found that:

    83 customers purchased a dog product;

    101 customers purchased a cat product;

    22 customers purchased a fish product;

    31 customers purchased a dog and a cat product;

    8 customers purchased a dog and a fish product;

    10 customers purchased a cat and a fish product;

    6 customers purchased a dog, a cat, and a fish product;

    34 customers purchased a product for a pet other than a dog, cat, or fish.

    **(a)** How many purchases were for a dog product only?

    **(b)** How many purchases were for a cat product only?

    **(c)** How many purchases were for a dog or fish product?

    **(d)** How many purchases were there in total?

21. **Recreational Fishing** A marine biologist surveys people who fish on Lake Erie and caught at least one fish to determine whether they had caught a walleye, a smallmouth bass, or a yellow perch in the last year. He finds:

    124 caught at least one walleye;

    133 caught at least one smallmouth bass;

    146 caught at least one yellow perch;

    75 caught at least one walleye and at least one smallmouth bass;

    67 caught at least one walleye and at least one yellow perch;

    79 caught at least one smallmouth bass and at least one yellow perch;

    45 caught all three.

    **(a)** Find the total number of people surveyed.

    **(b)** How many caught at least one walleye or at least one smallmouth bass?

    **(c)** How many caught only walleye?

22. **Blood Type** Human blood can contain either no antigens, the A antigen, the B antigen, or both the A and B antigens. A third antigen, called the Rh antigen, is important in human reproduction and, like the A and B antigens, may or may not be present in an individual. Blood is called type A positive if the individual has the A and Rh antigens, but not the B antigen. A person having only the A and B antigens is said to have type AB-negative blood. A person having only the Rh antigen has type O-positive blood. Other blood types are defined in a similar manner. Identify the blood type of the individuals in regions (a)–(g) of the Venn diagram.

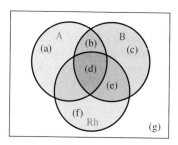

23. **Blood Type** Use the diagram from Exercise 22. In a certain hospital, the following data was recorded:

    25 patients had the A antigen;

    17 had the A and B antigens;

    27 had the B antigen;

    22 had the B and Rh antigens;

    30 had the Rh antigen;

    12 had none of the antigens;

    16 had the A and Rh antigens;

    15 had all three antigens.

    How many patients

    **(a)** were represented?

    **(b)** had exactly one antigen?

    **(c)** had exactly two antigens?

    **(d)** had O-positive blood?

    **(e)** had AB-positive blood?

    **(f)** had B-negative blood?

    **(g)** had O-negative blood?

    **(h)** had A-positive blood?

24. **Portfolio Management** In reviewing the portfolios of 365 of its clients, a mutual funds company categorized whether the clients were invested in international stock funds, domestic stock funds, or bond funds. It found that,

   125 were invested in domestic stocks, international stocks, and bond funds;

   145 were invested in domestic stocks and bond funds;

   300 were invested in domestic stocks;

   200 were invested in international and domestic stocks;

   18 were invested in international stocks and bond funds, but not domestic stocks;

   35 were invested in bonds, but not in international or domestic stocks;

   87 were invested in international stocks, but not in bond funds.

   (a) How many were invested in international stocks?

   (b) How many were invested in bonds, but not international stocks?

   (c) How many were not invested in bonds?

   (d) How many were invested in international or domestic stocks?

*For Exercises 25–30, use the given contingency tables. (See Example 7.)*

**Partnerships** *In Exercises 25 and 26, the given contingency table lists the cross tabulation of the number of business partnerships by industry and whether the partnership reported net income or net loss for the year 2014. (Data from: U.S. Internal Revenue Service.)*

| Industry | Number of Partnerships | |
| --- | --- | --- |
| | With Net Income ($A$) | With Net Loss ($B$) |
| Construction ($C$) | 75,071 | 67,561 |
| Manufacturing ($M$) | 24,389 | 42,386 |
| Wholesale Trade ($W$) | 37,023 | 45,369 |
| Retail Trade ($R$) | 86,749 | 81,878 |

*Using the letter given in the table, find the number of partnerships in each set.*

25. (a) $A \cup C$   (b) $M \cup W$
    (c) $R \cap B$   (d) $C'$

26. (a) $B \cup R$   (b) $W \cup R$
    (c) $M \cap A$   (d) $W'$

**Life Classification** *In Exercises 27 and 28, the given contingency table lists the cross tabulation of whether adults (for whom answering the question was appropriate) feel their life is exciting, routine, or dull for males and females among the 1873 appropriate adults in the 2016 General Social Survey. (Data from: gss.norc.org.)*

| Life Classification | Males ($M$) | Females ($F$) |
| --- | --- | --- |
| Exciting ($E$) | 435 | 510 |
| Routine ($R$) | 357 | 478 |
| Dull ($D$) | 38 | 55 |

*Using the letters in the table, find the number of respondents in each set.*

27. (a) $E$   (b) $D \cap F$
    (c) $E \cup M$   (d) $E'$

28. (a) $D$   (b) $R \cap M$
    (c) $D \cup F$   (d) $R'$

**Job Satisfaction** *In Exercises 29–32, the given contingency table lists the cross tabulation of how satisfied or dissatisfied adults (for whom answering the question was appropriate) are with their job for males and females among the 933 appropriate adults in the 2016 General Social Survey. (Data from: gss.norc.org.)*

| Job Satisfaction | Males ($M$) | Females ($F$) |
| --- | --- | --- |
| Completely Satisfied ($CS$) | 86 | 78 |
| Very Satisfied ($VS$) | 176 | 167 |
| Fairly Satisfied ($FS$) | 149 | 144 |
| Neither Satisfied nor Dissatisfied ($N$) | 26 | 28 |
| Fairly Dissatisfied ($FD$) | 20 | 32 |
| Very Dissatisfied ($VD$) | 9 | 11 |
| Completely Dissatisfied ($CD$) | 3 | 4 |

*Using the letters given in the table, find the number of respondents in each set.*

29. (a) $CS \cup VS$   (b) $(CS \cup VS) \cap F$
    (c) Describe your answer to part (b) in words.

30. (a) $VD \cup CD$   (b) $(VD \cup CD) \cap M$
    (c) Describe your answer to part (b) in words.

31. (a) $CS \cup VS \cup FS$   (b) $(CS \cup VS \cup FS) \cap M$
    (c) Describe your answer to part (b) in words.

32. (a) $FD \cup VD \cup CD$   (b) $(FD \cup VD \cup CD) \cap F$
    (c) Describe your answer to part (b) in words.

33. Restate the union rule in words.

*Use Venn diagrams to answer the given questions. (See Example 6.)*

34. If $n(A) = 5, n(B) = 8$, and $n(A \cap B) = 4$, what is $n(A \cup B)$?

35. If $n(A) = 12, n(B) = 27$, and $n(A \cup B) = 30$, what is $n(A \cap B)$?

36. Suppose $n(B) = 7, n(A \cap B) = 3$, and $n(A \cup B) = 20$. What is $n(A)$?

37. Suppose $n(A \cap B) = 5, n(A \cup B) = 35$, and $n(A) = 13$. What is $n(B)$?

*Draw a Venn diagram and use the given information to fill in the number of elements for each region.*

**38.** $n(U) = 48, n(A) = 26, n(A \cap B) = 12, n(B') = 30$

**39.** $n(A) = 28, n(B) = 12, n(A \cup B) = 30, n(A') = 19$

**40.** $n(A \cup B) = 17, n(A \cap B) = 3, n(A) = 8, n(A' \cup B') = 21$

**41.** $n(A') = 28, n(B) = 25, n(A' \cup B') = 45, n(A \cap B) = 12$

**42.** $n(A) = 28, n(B) = 34, n(C) = 25, n(A \cap B) = 14, n(B \cap C) = 15, n(A \cap C) = 11, n(A \cap B \cap C) = 9, n(U) = 59$

**43.** $n(A) = 54, n(A \cap B) = 22, n(A \cup B) = 85, n(A \cap B \cap C) = 4, n(A \cap C) = 15, n(B \cap C) = 16, n(C) = 44, n(B') = 63$

*In Exercises 44–47, show that the statements are true by drawing Venn diagrams and shading the regions representing the sets on each side of the equals signs.*[*]

**44.** $(A \cup B)' = A' \cap B'$

**45.** $(A \cap B)' = A' \cup B'$

**46.** $A \cap (B \cup C) = (A \cap B) \cup (A \cap C)$

**47.** $A \cup (B \cap C) = (A \cup B) \cap (A \cup C)$

**48.** Explain in words the statement about sets in question 44.

**49.** Explain in words the statement about sets in question 45.

**50.** Explain in words the statement about sets in question 46.

**51.** Explain in words the statement about sets in question 47.

---

[*]The statements in Exercises 44 and 45 are known as De Morgan's laws. They are named for the English mathematician Augustus De Morgan (1806–1871).

✓**Checkpoint Answers**

**1.** (a)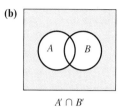

(b)

$A \cup B'$

$A' \cap B'$

**2.** (a)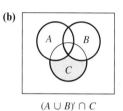

(b)

$(B' \cap A) \cup C$

$(A \cup B)' \cap C$

**3.** (a)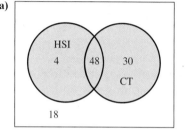

HSI 4 | 48 | 30 CT

18

(b) 4

**4.** (a) 42     (b) 9

**5.** (a) 8     (b) 92     (c) 98

**6.** 14

**7.** (a) 151,437 thousand

(b) 84,435 thousand

# 8.3 Introduction to Probability

If you go to a pizzeria and order two large pizzas at $14.99 each, you can easily find the *exact* price of your purchase: $29.98. For the manager at the pizzeria, however, it is impossible to predict the *exact* number of pizzas to be purchased daily. The number of pizzas purchased during a day is *random*: The quantity cannot be predicted exactly. A great many problems that come up in applications of mathematics involve random phenomena—phenomena for which exact prediction is impossible. The best that we can do is determine the *probability* of the possible outcomes.

## Random Experiments and Sample Spaces

A **random experiment** (sometimes called a random phenomenon) has outcomes that we cannot predict, but that nonetheless have a regular distribution in a large number of repetitions. We call a repetition from a random experiment a **trial**. The possible results of each trial are called **outcomes**. For instance, when we flip a coin, the outcomes are heads and tails. We do not know whether a particular flip will yield heads or tails, but we do know that if we flip the coin a large a number of times, about half the flips will be heads and half will be tails. Each flip of the coin is a trial. The **sample space** (denoted by $S$) for a random experiment is the set of all possible outcomes. For the coin flipping, the sample space is

$$S = \{\text{heads, tails}\}.$$

**EXAMPLE 1** Give the sample space for each random experiment.

**(a)** Use the spinner in Figure 8.17.

**Figure 8.17**

**Solution** The 7 outcomes are 1, 2, 3, . . . 7, so the sample space is

$$\{1, 2, 3, 4, 5, 6, 7\}.$$

**(b)** For the purposes of a public opinion poll, respondents are classified as young, middle aged, or senior and as male or female.

**Solution** A sample space for this poll could be written as a set of ordered pairs:

{(young, male), (young, female), (middle aged, male),

(middle aged, female), (senior, male), (senior, female)}.

**(c)** An experiment consists of studying the numbers of boys and girls in families with exactly 3 children. Let *b* represent *boy* and *g* represent *girl*.

**Solution** For this experiment, drawing a tree diagram can be helpful. First, we draw two branches starting on the left to indicate that the first child can be either a boy or a girl. From each of those outcomes, we draw two branches to indicate that the second child can be either a boy or girl. Last, we draw two branches from each of those outcomes to indicate that after the second child, the third child can be either a boy or a girl. The result is the tree in Figure 8.18.

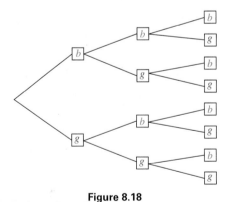

**Figure 8.18**

We can now easily list the members of the sample space *S*. We follow the eight paths of the branches to yield

$$S = \{bbb, bbg, bgb, bgg, gbb, gbg, ggb, ggg\}.$$ ✓₁

✓ **Checkpoint 1**

Draw a tree diagram for the random experiment of flipping a coin two times, and determine the sample space.

## Events

An **event** is an outcome, or a set of outcomes, of a random experiment. Thus, an event is a subset of the sample space. For example, if the sample space for tossing a coin is $S = \{h, t\}$, then one event is $E = \{h\}$, which represents the outcome "heads."

An ordinary die is a cube whose six different faces show the following numbers of dots: 1, 2, 3, 4, 5, and 6. If the die is fair (not "loaded" to favor certain faces over others), then any one of the faces is equally likely to come up when the die is rolled. The sample space for the experiment of rolling a single fair die is $S = \{1, 2, 3, 4, 5, 6\}$. Some possible events are as follows:

The die shows an even number: $E_1 = \{2, 4, 6\}$.

The die shows a 1: $E_2 = \{1\}$.

The die shows a number less than 5: $E_3 = \{1, 2, 3, 4\}$.

The die shows a multiple of 3: $E_4 = \{3, 6\}$.

**EXAMPLE 2**     For the sample space $S$ in Example 1(c) on the previous page, write the given events in set notation.

**(a)** Event $H$: The family has exactly two girls.

**Solution**  Families with three children can have exactly two girls with either *bgg*, *gbg*, or *ggb*, so that event $H$ is

$$H = \{bgg, gbg, ggb\}.$$

**(b)** Event $K$: The three children are the same sex.

**Solution**  Two outcomes satisfy this condition: all boys and all girls, or

$$K = \{bbb, ggg\}.$$

**(c)** Event $J$: The family has three girls.

**Solution**  Only *ggg* satisfies this condition, so

$$J = \{ggg\}.$$

**✓ Checkpoint 2**

Suppose a die is tossed. Write the given events in set notation.

**(a)** The number showing is less than 3.

**(b)** The number showing is 5.

**(c)** The number showing is 8.

If an event $E$ equals the sample space $S$, then $E$ is a **certain event.** If event $E = \varnothing$, then $E$ is an **impossible event.**

**EXAMPLE 3**     Suppose a fair die is rolled. Then the sample space is $\{1, 2, 3, 4, 5, 6\}$. Find the requested events.

**(a)** The event "the die shows a 4."

**Solution**  $\{4\}$.

**(b)** The event "the number showing is less than 10."

**Solution**  The event is the entire sample space $\{1, 2, 3, 4, 5, 6\}$. This event is a certain event; if a die is rolled, the number showing (either 1, 2, 3, 4, 5, or 6) must be less than 10.

**(c)** The event "the die shows a 7."

**Solution**  The empty set, $\varnothing$; this is an impossible event.

**✓ Checkpoint 3**

Which of the events listed in Checkpoint 2 is

**(a)** certain?

**(b)** impossible?

Since events are sets, we can use set operations to find unions, intersections, and complements of events. Here is a summary of the set operations for events.

## Set Operations for Events

Let $E$ and $F$ be events for a sample space $S$. Then

$E \cap F$ occurs when both $E$ **and** $F$ occur;

$E \cup F$ occurs when $E$ **or** $F$ **or both** occur;

$E'$ occurs when $E$ does **not** occur.

**EXAMPLE 4**     A study of college students grouped the students into various categories that can be interpreted as events when a student is selected at random. Consider the following events:

$E$: The student is under 20 years old;

$F$: The student is male;

$G$: The student is a business major.

Describe each of the following events in words.

**(a)** $E'$

**Solution** $E'$ is the event that the student is 20 years old or older.

**(b)** $F' \cap G$

**Solution** $F' \cap G$ is the event that the student is not male and the student is a business major—that is, the student is a female business major.

**(c)** $E' \cup G$

**Solution** $E' \cup G$ is the event that the student is 20 or over or is a business major. Note that this event includes all students 20 or over, regardless of major. ✔ 4

Two events that cannot both occur at the same time, such as getting both a head and a tail on the same toss of a coin, are called **disjoint events.** (Disjoint events are sometimes referred to as *mutually exclusive events.*)

✔ **Checkpoint 4**

Write the set notation for the given events for the experiment of rolling a fair die if $E = \{1, 3\}$ and $F = \{2, 3, 4, 5\}$.

**(a)** $E \cap F$

**(b)** $E \cup F$

**(c)** $E'$

✔ **Checkpoint 5**

In Example 5, let $F = \{2, 4, 6\}$, $K = \{1, 3, 5\}$, and $G$ remain the same. Are the given events disjoint?

**(a)** $F$ and $K$

**(b)** $F$ and $G$

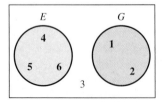

**Figure 8.19**

## Disjoint Events

Events $E$ and $F$ are disjoint events if $E \cap F = \varnothing$.

For any event $E$, $E$ and $E'$ are disjoint events.

**EXAMPLE 5**     Let $S = \{1, 2, 3, 4, 5, 6\}$, the sample space for tossing a die. Let $E = \{4, 5, 6\}$, and let $G = \{1, 2\}$. Are $E$ and $G$ disjoint events?

**Solution** Yes, because they have no outcomes in common; $E \cap G = \varnothing$. See Figure 8.19. ✔ 5

## Probability

For sample spaces with *equally likely* outcomes, the probability of an event is defined as follows.

---

### Basic Probability Principle

Let $S$ be a sample space of equally likely outcomes, and let event $E$ be a subset of $S$. Then the **probability that event $E$ occurs** is

$$P(E) = \frac{n(E)}{n(S)}.$$

---

By this definition, the **probability of an event** is a number that indicates the relative likelihood of the event.

**⚠ CAUTION** The basic probability principle applies only when the outcomes are equally likely.

---

**EXAMPLE 6**   Suppose a single fair die is rolled, with the sample space $S = \{1, 2, 3, 4, 5, 6\}$. Give the probability of each of the following events.

**(a)** $E$: The die shows an even number.

**Solution** Here, $E = \{2, 4, 6\}$, a set with three elements. Because $S$ contains six elements,

$$P(E) = \frac{3}{6} = \frac{1}{2}.$$

**(b)** $F$: The die shows a number less than 10.

**Solution** Event $F$ is a certain event, with

$$F = \{1, 2, 3, 4, 5, 6\},$$

so that

$$P(F) = \frac{6}{6} = 1.$$

**(c)** $G$: The die shows an 8.

**Solution** This event is impossible, so

$$P(G) = \frac{0}{6} = 0. \quad ✔_6$$

✓ **Checkpoint 6**

A fair die is rolled. Find the probability of rolling

**(a)** an odd number;

**(b)** 2, 4, 5, or 6;

**(c)** a number greater than 5;

**(d)** the number 7.

A standard deck of 52 cards has four suits—hearts (♥), clubs (♣), diamonds (♦), and spades (♠)—with 13 cards in each suit. The hearts and diamonds are red, and the spades and clubs are black. Each suit has an ace (A), a king (K), a queen (Q), a jack (J), and cards numbered from 2 to 10. The jack, queen, and king are called face cards and for many purposes can be thought of as having values 11, 12, and 13, respectively. The ace can be thought of as the low card (value 1) or the high card (value 14). See Figure 8.20. We will refer to this standard deck of cards often in our discussion of probability.

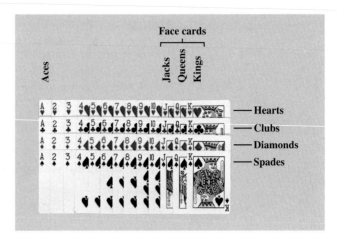

**Figure 8.20**

**EXAMPLE 7**    If a single card is drawn at random from a standard, well-shuffled, 52-card deck, find the probability of each of the given events.

**(a)** Drawing an ace

**Solution**   There are 4 aces in the deck. The event "drawing an ace" is

{heart ace, diamond ace, club ace, spade ace}.

Therefore,

$$P(\text{ace}) = \frac{4}{52} = \frac{1}{13}.$$

**(b)** Drawing a face card

**Solution**   Since there are 12 face cards,

$$P(\text{face card}) = \frac{12}{52} = \frac{3}{13}.$$

**(c)** Drawing a spade

**Solution**   The deck contains 13 spades, so

$$P(\text{spade}) = \frac{13}{52} = \frac{1}{4}.$$

**(d)** Drawing a spade or a heart

**Solution**   Besides the 13 spades, the deck contains 13 hearts, so

$$P(\text{spade or heart}) = \frac{26}{52} = \frac{1}{2}. \quad ✓ _7$$

✓ **Checkpoint 7**

A single playing card is drawn at random from an ordinary 52-card deck. Find the probability of drawing

**(a)**   a queen;

**(b)**   a diamond;

**(c)**   a red card.

In the preceding examples, the probability of each event was a number between 0 and 1, inclusive. The same thing is true in general. Any event $E$ is a subset of the sample space $S$, so $0 \leq n(E) \leq n(S)$. Since $P(E) = n(E)/n(S)$, it follows that $0 \leq P(E) \leq 1$.

For any event $E$,

$$0 \leq P(E) \leq 1.$$

In many real-life problems, the events in the sample space are not all equally likely. When this is the case, we can estimate probabilities by determining the long-run proportion that an outcome of interest will occur given many repetitions under identical, and independent, circumstances. We do this often when we perform a statistical study. The long-run proportion is called the **relative frequency probability.** Estimates based on relative frequency probability are sometimes called *empirical probabilities.* Independence in this context refers to the idea that what occurs in one run has no effect on the outcome of a subsequent run.

For example, imagine that we want to determine the probability that a newly manufactured porcelain sink contains a defect. We examine the first sink before it is shipped and see that it has no defects. So we estimate the probability of a defect as 0 because the relative frequency of defects is 0 out of 1 trial. The second sink, however, has a defect. Now our number of defects is 1 out of two trials, so our relative frequency is 1 out of 2, which is .5. The table shows the results for these two sinks and eight additional sinks.

| Sink Number | 1 | 2 | 3 | 4 | 5 | 6 | 7 | 8 | 9 | 10 |
|---|---|---|---|---|---|---|---|---|---|---|
| Defect Y/N | N | Y | N | N | N | Y | Y | N | N | N |
| Relative Frequency | $0/1 =$ 0 | $1/2 =$ .5 | $1/3 \approx$ .333 | $1/4 =$ .25 | $1/5 =$ .2 | $2/6 \approx$ .333 | $3/7 \approx$ .429 | $3/8 =$ .375 | $3/9 \approx$ .333 | $3/10 =$ .3 |

Notice that the relative frequency fluctuates a great deal. To get a better estimate of the probability of a defect, we need to examine the *long-run* proportion of defects. When the same conditions are repeated a large number of times, the relative frequency will stabilize to the long-run frequency. Figure 8.21 shows how the relative frequency stabilizes to some degree after 50 trials and Figure 8.22 shows the relative frequency after 1000 trials. We can see in Figure 8.22 that the long-run proportion of defects is close to .25.

**Figure 8.21**

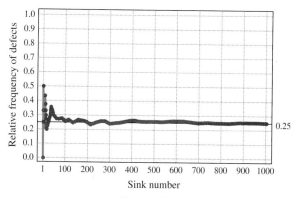

**Figure 8.22**

**NOTE** Independence here pertains to the assumption that one sink having, or not having, a defect has no influence on whether another sink has a defect or not. We will investigate independence in greater depth in Section 8.5.

Thus, when we examine many trials of a given phenomenon of interest we can obtain a long-run estimate of the probability of interest. The next example shows one approach to finding such relative frequency probabilities.

**EXAMPLE 8** **Standard of Living** The General Social Survey picks U.S. residents at random and asks them a great many questions. One of the questions they asked respondents in 2016 was "Do you feel your standard of living is much better, somewhat better, about the same, somewhat worse or much worse than your parents?" The table categorizes the responses.

| Response | Frequency |
|---|---|
| Much Better | 573 |
| Somewhat Better | 555 |
| About the Same | 466 |
| Somewhat Worse | 238 |
| Much Worse | 100 |

(a) Estimate the probability a U.S. resident feels his or her standard of living is much better than that of his or her parents.

**Solution** Let us define the set $A$ to be the event that a resident feels his or her standard of living is much better. To find the relative frequency probability, we first need to find the total number of respondents. This is $573 + 555 + 466 + 238 + 100 = 1932$. We then divide the frequency in the much better category (in this case 573) by 1932 to obtain

$$P(A) = \frac{573}{1932} \approx .2966.$$

We use this relative frequency as our estimate of the long-run frequency and say that the estimated probability that a U.S. resident feels his or her standard of living is much better is approximately .2966.

(b) Estimate the probability of the event $B$ that a U.S. resident feels that his or her standard of living is somewhat worse or much worse.

**Solution** Here, we add together the number of respondents in the categories of somewhat worse and much worse ($238 + 100 = 338$) to obtain

$$P(B) = \frac{238 + 100}{1932} = \frac{338}{1932} \approx .1749. \; ✓_8$$

✓ **Checkpoint 8**

From the data given in Example 8, estimate the probability that a U.S. resident believes his or her standard of living is much better, somewhat better, or about the same as that of his or her parents.

After conducting a study such as the General Social Survey where respondents are chosen at random, we can assume independence because one person's response should have no relation to another person's response. We also use the term "estimated probability" or just "probability" rather than "relative frequency probability."

A table of frequencies, as in Example 8, sets up a probability distribution; that is, for each possible outcome of an experiment, a number, called the probability of that outcome, is assigned. This assignment may be done in any reasonable way (on a relative frequency basis, as in Example 8, or by theoretical reasoning, as in Example 6), provided that it satisfies the following conditions.

## Properties of Probability

Let $S$ be a sample space consisting of $n$ distinct outcomes $s_1, s_2, \ldots, s_n$. An acceptable probability assignment consists of assigning to each outcome $s_i$ a number $p_i$ (the probability of $s_i$) according to the following rules:

1. The probability of each outcome is a number between 0 and 1:

$$0 \le p_1 \le 1, \quad 0 \le p_2 \le 1, \ldots, \quad 0 \le p_n \le 1.$$

2. The sum of the probabilities of all possible outcomes is 1:

$$p_1 + p_2 + p_3 + \cdots + p_n = 1.$$

# 8.3 Exercises

1. What is meant by a "fair" coin or die?

2. What is the sample space for a random experiment?

*Write sample spaces for the random experiments in Exercises 3–9. (See Example 1.)*

3. A month of the year is chosen for a wedding.

4. A day in April is selected for a bicycle race.

5. A student is asked how many points she earned on a recent 80-point test.

6. A person is asked the number of hours (to the nearest hour) he watched television yesterday.

7. The management of an oil company must decide whether to go ahead with a new oil shale plant or to cancel it.

8. A coin is tossed and a die is rolled.

9. The quarter of the year in which a company's profits were highest.

10. Define an event.

11. Define disjoint events in your own words.

*Decide whether the events are disjoint. (See Example 5.)*

12. Owning an SUV and owning a Jeep

13. Wearing a hat and wearing glasses

14. Being married and being under 30 years old

15. Being a doctor and being under 5 years old

16. Being male and being a nurse

17. Being female and being a pilot

*For the random experiments in Exercises 18–20, write out an equally likely sample space, and then write the indicated events in set notation. (See Examples 2 and 3.)*

18. A marble is drawn at random from a bowl containing 3 yellow, 4 white, and 8 blue marbles.

    (a) A yellow marble is drawn.

    (b) A blue marble is drawn.

    (c) A white marble is drawn.

    (d) A black marble is drawn.

19. Six people live in a dorm suite. Two are to be selected to go to the campus café to pick up a pizza. Of course, no one wants to go, so the six names (Connie, Kate, Lindsey, Jackie, Taisa, and Nicole) are placed in a hat. After the hat is shaken, two names are selected.

    (a) Taisa is selected.

    (b) The two names selected have the same number of letters.

20. An unprepared student takes a three-question true-or-false quiz in which he flips a coin to guess the answers. If the coin is heads, he guesses true, and if the coin is tails, he guesses false.

    (a) The student guesses true twice and guesses false once.

    (b) The student guesses all false.

    (c) The student guesses true once and guesses false twice.

*In Exercises 21–24, write out the sample space and assume each outcome is equally likely. Then give the probability of the requested outcomes. (See Examples 7 and 8.)*

21. In deciding what color and style to paint a room, Greg has narrowed his choices to three colors—forest sage, evergreen whisper, and opaque emerald—and two styles—rag painting and colorwash.

    (a) Greg picks a combination with colorwash.

    (b) Greg picks a combination with opaque emerald or rag painting.

22. Tami goes shopping and sees three kinds of shoes: flats, 2″ heels, and 3″ heels. They come in two shades of beige (light and dark) and black.

    (a) The shoe selected has a heel and is black.

    (b) The shoe selected has no heel and is beige.

    (c) The shoe selected has a heel and is beige.

23. Doug is shopping for a new patio umbrella. There is a 10-foot and a 12-foot model, and each is available in beige, forest green, and rust.

    (a) Doug buys a 12-foot forest green umbrella.

    (b) Doug buys a 10-foot umbrella.

    (c) Doug buys a rust-colored umbrella.

24. Sharonda is deciding between four brands of cell phones: Samsung, Motorola, Nokia, and Sony. Each of these phones comes with the option of insurance or no insurance.

    (a) Sharonda picks a phone with insurance.

    (b) Sharonda picks a phone that begins with the letter "S" and no insurance.

*A single fair die is rolled. Find the probabilities of the given events. (See Example 6.)*

25. Getting a number less than 4

26. Getting a number greater than 4

27. Getting a 2 or a 5

28. Getting a multiple of 3

*Alejandro wants to adopt a puppy from an animal shelter. At the shelter, he finds eight puppies that he likes: a male and female puppy from each of the four breeds of beagle, boxer, collie, and Labrador. The puppies are each so cute that Alejandro cannot make up his mind, so he decides to pick the dog randomly.*

29. Write the sample space for the outcomes, assuming each outcome is equally likely.

*Find the probability that Alejandro chooses the given puppy.*

30. A male boxer         31. A male puppy

**32.** A collie

**33.** A female Labrador

**34.** A beagle or a boxer

**35.** Anything except a Labrador

**36.** Anything except a male beagle or male boxer.

**Fatal Work Injuries** *The following table gives the number of fatal work injuries categorized by cause from 2015. (Data from: U.S. Bureau of Labor Statistics.)*

| Cause | Number of Fatalities |
|---|---|
| Transportation accidents | 2054 |
| Assaults and violent acts | 703 |
| Contacts with objects and equipment | 722 |
| Falls | 800 |
| Exposure to harmful substances or environments | 424 |
| Fires and explosions | 121 |
| Other | 12 |

*Find the probability that a randomly chosen work fatality had the given cause. (See Example 8.)*

**37.** A fall

**38.** Fires and explosions

**39. Religious Services** Respondents to the 2016 General Social Survey (GSS) indicated the following categorizations pertaining to attendance at religious services: (Data from: gss.norc.org.)

| Attendance | Number of Respondents |
|---|---|
| Never | 723 |
| Less than once a year | 173 |
| Once a year | 385 |
| Several times a year | 312 |
| Once a month | 190 |
| 2–3 times a month | 250 |
| Nearly every week | 126 |
| Every week | 481 |
| More than once a week | 210 |
| Don't know/no answer | 17 |
| **Total** | **2867** |

Find the probability that a randomly chosen person in the United States attends religious services

**(a)** several times a year;

**(b)** 2–3 times a month;

**(c)** nearly every week or more frequently.

**40. Gender Roles** Respondents to the 2016 General Social Survey (GSS) indicated whether they strongly agreed, agreed, disagreed, or strongly disagreed with the question, "It is better for

men to work and for women to tend to the home." The responses are categorized in the following table. (Data from: gss.norc.org.)

| Response | Number of Respondents |
|---|---|
| Strongly Agree | 110 |
| Agree | 405 |
| Disagree | 853 |
| Strongly Disagree | 504 |
| Don't Know/No Answer | 995 |

Find the probability that a person in the United States

**(a)** strongly agreed or agreed with the statement;

**(b)** strongly disagreed with the statement;

**(c)** did not strongly agree with the statement.

**S&P 500** *As of March 7, 2017, the 505 stocks in the S&P 500 were classified in 11 different sectors. The following table gives the numbers of stocks in each sector. (Data from: www.morningstar.com.)*

| Sector | Number of Stocks in Sector |
|---|---|
| Consumer Discretionary | 86 |
| Consumer Staples | 37 |
| Energy | 35 |
| Financials | 65 |
| Health Care | 60 |
| Industrials | 66 |
| Information Technology | 68 |
| Materials | 25 |
| Real Estate | 30 |
| Telecommunications Services | 5 |
| Utilities | 28 |
| **Total** | **505** |

*Find the probability that a stock in the S&P 500 is in the requested sector.*

**41.** Energy

**42.** Financials

**43.** Real Estate or Utilities

**44.** Health Care or Information Technology

**S&P 500** *The following table gives the counts for three price categories of the 505 stocks in the S&P 500. (Data from: www.morningstar.com.)*

| Price | Count |
|---|---|
| $0–$49.99 | 159 |
| $50–$99.99 | 207 |
| $100 or higher | 139 |

*Find the probability that a stock in the S&P 500 is in the requested price range.*

**45.** Below $50

**46.** Below $100

*An experiment is conducted for which the sample space is $S = \{s_1, s_2, s_3, s_4, s_5\}$. Which of the probability assignments in Exercises 47–52 is possible for this experiment? If an assignment is not possible, tell why.*

**47.**

| Outcomes | $s_1$ | $s_2$ | $s_3$ | $s_4$ | $s_5$ |
|---|---|---|---|---|---|
| Probabilities | .09 | .32 | .21 | .25 | .13 |

**48.**

| Outcomes | $s_1$ | $s_2$ | $s_3$ | $s_4$ | $s_5$ |
|---|---|---|---|---|---|
| Probabilities | .92 | .03 | 0 | .02 | .03 |

**49.**

| Outcomes | $s_1$ | $s_2$ | $s_3$ | $s_4$ | $s_5$ |
|---|---|---|---|---|---|
| Probabilities | 1/3 | 1/4 | 1/6 | 1/8 | 1/10 |

**50.**

| Outcomes | $s_1$ | $s_2$ | $s_3$ | $s_4$ | $s_5$ |
|---|---|---|---|---|---|
| Probabilities | 1/5 | 1/3 | 1/4 | 1/5 | 1/10 |

**51.**

| Outcomes | $s_1$ | $s_2$ | $s_3$ | $s_4$ | $s_5$ |
|---|---|---|---|---|---|
| Probabilities | .64 | −.08 | .30 | .12 | .02 |

**52.**

| Outcomes | $s_1$ | $s_2$ | $s_3$ | $s_4$ | $s_5$ |
|---|---|---|---|---|---|
| Probabilities | .05 | .35 | .5 | .2 | −.3 |

✓ **Checkpoint Answers**

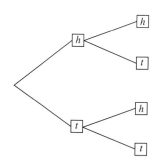

**1.** $\{hh, ht, th, tt\}$

**2.** (a) $\{1, 2\}$    (b) $\{5\}$    (c) $\varnothing$

**3.** (a) None    (b) Part (c)

**4.** (a) $\{3\}$    (b) $\{1, 2, 3, 4, 5\}$    (c) $\{2, 4, 5, 6\}$

**5.** (a) Yes    (b) No

**6.** (a) $1/2$    (b) $2/3$    (c) $1/6$    (d) 0

**7.** (a) $1/13$    (b) $1/4$    (c) $1/2$

**8.** About .8251

# 8.4  Basic Concepts of Probability

We determine the probability of more complex events in this section.

To find the probability of the union of two sets $E$ and $F$ in a sample space $S$, we use the union rule for counting given in Section 8.2:

$$n(E \cup F) = n(E) + n(F) - n(E \cap F).$$

Dividing both sides by $n(S)$ yields

$$\frac{n(E \cup F)}{n(S)} = \frac{n(E)}{n(S)} + \frac{n(F)}{n(S)} - \frac{n(E \cap F)}{n(S)}$$

$$P(E \cup F) = P(E) + P(F) - P(E \cap F).$$

This discussion is summarized in the next rule.

## Addition Rule for Probability

For any events $E$ and $F$ from a sample space $S$,

$$P(E \cup F) = P(E) + P(F) - P(E \cap F).$$

In words, we have

$$P(E \text{ or } F) = P(E) + P(F) - P(E \text{ and } F).$$

(Although the addition rule applies to any events $E$ and $F$ from any sample space, the derivation we have given is valid only for sample spaces with equally likely simple events.)

> **EXAMPLE 1** When playing American roulette, the croupier (attendant) spins a marble that lands in one of the 38 slots in a revolving turntable. The slots are numbered 1 to 36, with two additional slots labeled 0 and 00 that are painted green. Half the remaining slots are colored red, and half are black. (See Figure 8.23.)

**Figure 8.23**

In roulette, the slots labeled 0 or 00 are not considered as even or odd. Find the probability that the marble will land in a red or even number.

**Solution** Let $R$ represent the event of the marble landing in a red slot and $E$ the event of the marble landing in an even-numbered slot. There are 18 slots that are colored red, so $P(R) = 18/38$. There are also 18 even numbers between 1 and 36, so $P(E) = 18/38$. In order to use the addition rule, we also need to know the number of slots that are red and even numbered. Looking at Figure 8.23, we can see there are 8 such slots, which implies that $P(R \cap E) = 8/38$. Using the addition rule, we find the probability that the marble will land in a slot that is red or even numbered is

$$P(R \cup E) = P(R) + P(E) - P(R \cap E)$$
$$= \frac{18}{38} + \frac{18}{38} - \frac{8}{38} = \frac{28}{38} = \frac{14}{19}.$$

**✓ Checkpoint 1**

If an American roulette wheel is spun, find the probability of the marble landing in a black slot or a slot (excluding 0 and 00) whose number is divisible by 3.

> **EXAMPLE 2** Suppose two fair dice (plural of *die*) are rolled. Find each of the given probabilities.

**(a)** The first die shows a 2 or the sum of the results is 6 or 7.

**Solution** The sample space for the throw of two dice is shown in Figure 8.24, on the following page, where 1-1 represents the event "the first die shows a 1 and the second die shows a 1," 1-2 represents the event "the first die shows a 1 and the second die shows a 2," and so on. Let $A$ represent the event "the first die shows a 2" and $B$ represent the event "the sum of the results is 6 or 7." These events are indicated in color in Figure 8.24.

From the diagram, event $A$ has 6 elements, $B$ has 11 elements, and the sample space has 36 elements. Thus,

$$P(A) = \frac{6}{36}, \quad P(B) = \frac{11}{36}, \quad \text{and} \quad P(A \cap B) = \frac{2}{36}.$$

By the addition rule,

$$P(A \cup B) = P(A) + P(B) - P(A \cap B),$$

$$P(A \cup B) = \frac{6}{36} + \frac{11}{36} - \frac{2}{36} = \frac{15}{36} = \frac{5}{12}.$$

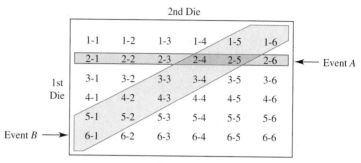

**Figure 8.24**

**(b)** The sum is 11 or the second die shows a 5.

**Solution** $P(\text{sum is } 11) = 2/36$, $P(\text{second die shows } 5) = 6/36$, and $P(\text{sum is } 11 \text{ and second die shows } 5) = 1/36$, so

$$P(\text{sum is } 11 \text{ or second die shows } 5) = \frac{2}{36} + \frac{6}{36} - \frac{1}{36} = \frac{7}{36}. \quad \boxed{2}$$

If events $E$ and $F$ are disjoint, then $E \cap F = \varnothing$ by definition; hence, $P(E \cap F) = 0$. Applying the addition rule yields the useful fact that follows.

## Addition Rule for Disjoint Events

For disjoint events $E$ and $F$,

$$P(E \cup F) = P(E) + P(F).$$

**EXAMPLE 3**    Assume that the probability of a couple having a baby boy is the same as the probability of the couple having a baby girl. If the couple has 3 children, find the probability that at least 2 of them are girls.

**Solution** The event of having at least 2 girls is the union of the disjoint events $E$ = "the family has exactly 2 girls" and $F$ = "the family has exactly 3 girls." Using the equally likely sample space

$$\{ggg, ggb, gbg, bgg, gbb, bgb, bbg, bbb\},$$

where $b$ represents a boy and $g$ represents a girl, we see that $P(2 \text{ girls}) = 3/8$ and $P(3 \text{ girls}) = 1/8$. Therefore,

$$P(\text{at least 2 girls}) = P(2 \text{ girls}) + P(3 \text{ girls})$$

$$= \frac{3}{8} + \frac{1}{8} = \frac{1}{2}. \quad \boxed{3}$$

---

✓**Checkpoint 2**

In the random experiment of Example 2, find the given probabilities.

**(a)** The sum is 5 or the second die shows a 3.

**(b)** Both dice show the same number, or the sum is at least 11.

---

✓**Checkpoint 3**

In Example 3, find the probability of having no more than 2 girls.

By definition of $E'$, for any event $E$ from a sample space $S$,

$$E \cup E' = S \quad \text{and} \quad E \cap E' = \varnothing.$$

Because $E \cap E' = \varnothing$, events $E$ and $E'$ are disjoint, so that

$$P(E \cup E') = P(E) + P(E').$$

However, $E \cup E' = S$, the sample space, and $P(S) = 1$. Thus,

$$P(E \cup E') = P(E) + P(E') = 1.$$

Rearranging these terms gives the following useful rule.

## Complement Rule

For any event $E$,

$$P(E') = 1 - P(E) \quad \text{and} \quad P(E) = 1 - P(E').$$

**EXAMPLE 4**   If a fair die is rolled, what is the probability that any number but 5 will come up?

**Solution**  If $E$ is the event that 5 comes up, then $E'$ is the event that any number but 5 comes up. $P(E) = 1/6$, so we have $P(E') = 1 - 1/6 = 5/6$. ✓ 4

✓ **Checkpoint 4**

(a) Let $P(K) = 2/3$. Find $P(K')$.

(b) If $P(X') = 3/4$, find $P(X)$.

**EXAMPLE 5**   If two fair dice are rolled, find the probability that the sum of the numbers showing is greater than 3.

**Solution**  To calculate this probability directly, we must find each of the probabilities that the sum is 4, 5, 6, 7, 8, 9, 10, 11, and 12 and then add them. It is much simpler to first find the probability of the complement, the event that the sum is less than or equal to 3:

$$P(\text{sum} \le 3) = P(\text{sum is } 2) + P(\text{sum is } 3)$$

$$= \frac{1}{36} + \frac{2}{36} = \frac{3}{36} = \frac{1}{12}.$$

Now use the fact that $P(E) = 1 - P(E')$ to get

$$P(\text{sum} > 3) = 1 - P(\text{sum} \le 3) = 1 - \frac{1}{12} = \frac{11}{12}.$$ ✓ 5

✓ **Checkpoint 5**

In Example 5, find the probability that the sum of the numbers rolled is at least 5.

## Odds

Sometimes probability statements are given in terms of **odds**: a comparison of $P(E)$ with $P(E')$. For example, suppose $P(E) = \frac{4}{5}$. Then $P(E') = 1 - \frac{4}{5} = \frac{1}{5}$. These probabilities predict that $E$ will occur 4 out of 5 times and $E'$ will occur 1 out of 5 times. Then we say that the **odds in favor** of $E$ are 4 to 1, or 4:1.

## Odds

The **odds in favor** of an event $E$ are defined as the ratio of $P(E)$ to $P(E')$, or

$$\frac{P(E)}{P(E')}, \quad P(E') \neq 0.$$

**EXAMPLE 6**   Suppose the weather forecaster says that the probability of rain tomorrow is 1/3. Find the odds in favor of rain tomorrow.

**Solution**   Let $E$ be the event "rain tomorrow." Then $E'$ is the event "no rain tomorrow." Since $P(E) = 1/3$, $P(E') = 2/3$. By the definition of odds, the odds in favor of rain are

$$\frac{1/3}{2/3} = \frac{1}{2}, \quad \text{written 1 to 2 or 1:2.}$$

On the other hand, the odds that it will *not* rain, or the odds *against* rain, are

$$\frac{2/3}{1/3} = \frac{2}{1}, \quad \text{written 2 to 1.}$$

If the odds in favor of an event are, say, 3 to 5, then the probability of the event is 3/8, while the probability of the complement of the event is 5/8. (Odds of 3 to 5 indicate 3 outcomes in favor of the event out of a total of 8 outcomes.) The above example suggests the following generalization:

If the odds favoring event $E$ are $m$ to $n$, then

$$P(E) = \frac{m}{m + n} \quad \text{and} \quad P(E') = \frac{n}{m + n}.$$

**EXAMPLE 7**   Often, weather forecasters give probability in terms of percentage. Suppose the weather forecaster says that there is a 40% chance that it will snow tomorrow. Find the odds of snow tomorrow.

**Solution**   In this case, we can let $E$ be the event "snow tomorrow." Then $E'$ is the event "no snow tomorrow." Now, we have $P(E) = .4 = 4/10$ and $P(E') = .6 = 6/10$. By the definition of odds in favor, the odds in favor of snow are

$$\frac{4/10}{6/10} = \frac{4}{6} = \frac{2}{3}, \quad \text{written 2 to 3 or 2:3.}$$

It is important to put the final fraction into lowest terms in order to communicate the odds. ✔6

**✓ Checkpoint 6**

In Example 7, suppose $P(E) = 9/10$. Find the odds

(a)   in favor of $E$;

(b)   against $E$.

Suppose the chance of snow is 80%. Find the odds

(c)   in favor of snow;

(d)   against snow.

**EXAMPLE 8**   The odds that a particular bid will be the low bid are 4 to 5.

(a)   Find the probability that the bid will be the low bid.

**Solution**   Odds of 4 to 5 show 4 favorable chances out of $4 + 5 = 9$ chances altogether, so

$$P(\text{bid will be low bid}) = \frac{4}{4 + 5} = \frac{4}{9}.$$

(b)   Find the odds against that bid being the low bid.

**Solution**   There is a 5/9 chance that the bid will not be the low bid, so the odds against a low bid are

$$\frac{P(\text{bid will not be low})}{P(\text{bid will be low})} = \frac{5/9}{4/9} = \frac{5}{4},$$

**✓ Checkpoint 7**

If the odds in favor of event $E$ are 1 to 5, find

(a)   $P(E)$;

(b)   $P(E')$.

or 5:4. ✔7

## Applications

**EXAMPLE 9**    **Fatal Vehicle Accidents** Let $A$ represent that the driver in a fatal vehicle accident was age 24 or younger and let $B$ represent that the driver had a blood alcohol content (BAC) of .08% or higher. From data on fatal crashes in the year 2015 from the U.S. National Highway Traffic Safety Administration, we have the following probabilities.

$$P(A) = .1882, \quad P(B) = .1648, \quad P(A \cap B) = .0330.$$

**(a)** Find the probability that the driver was 25 or older and had a BAC level below .08%.

**Solution** Place the given information on a Venn diagram, starting with .0330 in the intersection of the regions $A$ and $B$. (See Figure 8.25.) As stated earlier, event $A$ has probability .1882. Since .0330 has already been placed inside the intersection of $A$ and $B$,

$$.1882 - .0330 = .1552$$

goes inside region $A$, but outside the intersection of $A$ and $B$. In the same way,

$$.1648 - .0330 = .1318$$

goes inside region $B$, but outside the overlap.

Since being age 25 or older is the complement of event $A$ and having a BAC below .08% is the complement of event $B$, we need to find $P(A' \cap B')$. From the Venn diagram in Figure 8.25, the labeled regions have a total probability of

$$.1552 + .0330 + .1318 = .3200.$$

Since the entire region of the Venn diagram must have probability 1, the region outside $A$ and $B$, namely $A' \cap B'$, has the probability

$$1 - .3200 = .6800.$$

The probability that a driver of a fatal crash is 25 or older and has a BAC below .08% is .6800.

**(b)** Find the probability that a driver in a fatal crash is 25 or older or has a BAC below .08%.

**Solution** The corresponding region for $A' \cup B'$ from Figure 8.25 has probability

$$.6800 + .1318 + .1552 = .9670. \quad \text{✓}_8$$

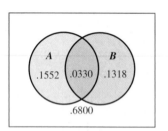

**Figure 8.25**

✓ **Checkpoint 8**

Using the data from Example 9, find the probability that a driver in a fatal crash

**(a)** has a BAC below .08%;

**(b)** is age 25 or older and has a BAC of .08% or higher.

**EXAMPLE 10**    **S&P 500 Stocks** Data from March 7, 2017, for all 505 stocks in the S&P 500 regarding the price of the stock and the dividend yield are cross classified in the table below. The price of the stock is categorized into three groups: $0–$49.99, $50.00–$99.99, and $100.00 or higher. The dividend yield is categorized as to whether there was no dividend, a dividend yield between 0.01% and 1.99%, and a dividend yield of 2.00% or higher. (Data from: www.okfn.org.)

| Price of Stock | Dividend Yield | | | Total |
|---|---|---|---|---|
| | No Dividend | .01%–1.99% | 2.00% or higher | |
| $0–$49.99 | 21 | 69 | 68 | 158 |
| $50.00–$99.99 | 18 | 87 | 102 | 207 |
| $100.00 or higher | 26 | 69 | 45 | 140 |
| Total | 65 | 225 | 215 | 505 |

**35.** The odds that tax revenue comes from property taxes are 31:69.

**36.** The odds that tax revenue comes from sales and gross receipts are 7:13.

**37.** The odds that tax revenue comes from individual income tax are 23:77.

**38.** The odds that tax revenue comes from corporate income tax are 37:963.

**S&P 500 Stocks** *For Exercises 39–46, use the following table, which cross classifies the price of the 505 stocks in the S&P 500 on March 7, 2017, with whether the earnings per share ratio was positive or not.*

| Price of Stock | Earnings per Share | |
|---|---|---|
| | Negative or 0 | Positive |
| $0–$49.99 | 30 | 128 |
| $50.00–$99.99 | 18 | 189 |
| $100.00 or higher | 5 | 135 |

*Find the probability of the described events. (Data from: www.okfn.org.)*

**39.** The earnings per share ratio is positive.

**40.** The earnings per share ratio is not positive.

**41.** The price of the stock is less than $50 and the earnings per share ratio is negative or 0.

**42.** The price of the stock is $100 or higher and the earning per share ratio is positive.

**43.** The price of the stock is $50.00–$99.99 or the earning per share ratio is positive.

**44.** The price of the stock is $100 or higher or the earnings per share ratio is positive.

**45.** The price of the stock is not below $50.00 and the earnings per share ratio is positive.

**46.** The price of the stock is not $100 or higher and the earnings per share ratio is negative or 0.

**Hours Worked** *Data from the 2016 General Social Survey can allow us to estimate how many hours people who classified their work status as part-time or full-time actually worked in the last week. Use the following table to find the probabilities of the events in Exercises 47–54. (Data from: gss.norc.org.)*

| Labor Force Status | Hours Worked in the Last Week | | | | | |
|---|---|---|---|---|---|---|
| | 0–19 | 20–29 | 30–39 | 40–49 | 50 or More | Total |
| Full-Time Worker | 26 | 43 | 137 | 691 | 377 | 1274 |
| Part-Time Worker | 74 | 82 | 87 | 41 | 24 | 308 |
| Total | 100 | 125 | 224 | 732 | 401 | 1582 |

**47.** Full-time worker

**48.** Part-time worker

**49.** Worked 50 hours or more

**50.** Worked 0–19 hours

**51.** Part-time worker and someone who worked 40 hours or more

**52.** Full-time worker and someone who worked less than 40 hours

**53.** Full-time worker or someone who worked 40 hours or more

**54.** Part-time worker or someone who worked 29 hours or less

*One way to solve a probability problem is to repeat the experiment many times, keeping track of the results. Then the probability can be approximated by using the basic definition of the probability of an event E, which is $P(E) = n(E)/n(S)$, where E occurs $n(E)$ times out of $n(S)$ trials of an experiment. This is called the Monte Carlo method of finding probabilities. If physically repeating the experiment is too tedious, it may be simulated with the use of a random-number generator, available on most computers and scientific or graphing calculators. To simulate a coin toss or the roll of a die on a graphing calculator, change the setting to fixed decimal mode with 0 digits displayed. To simulate multiple tosses of a coin, press RAND (or RANDOM or RND#) in the PROB submenu of the MATH (or OPTN) menu, and then press ENTER repeatedly. Interpret 0 as a head and 1 as a tail. To simulate multiple rolls of a die, press RAND × 6 + .5, and then press ENTER repeatedly.*

**55.** Suppose two dice are rolled. Use the Monte Carlo method with at least 50 repetitions to approximate the given probabilities. Compare them with the results of Exercise 10.

(a) $P$(the sum is 9 or more)

(b) $P$(the sum is less than 7)

**56.** Suppose two dice are rolled. Use the Monte Carlo method with at least 50 repetitions to approximate the given probabilities. Compare them with the results of Exercise 11.

(a) $P$(the sum is not more than 5)

(b) $P$(the sum is not less than 8)

**57.** Suppose three dice are rolled. Use the Monte Carlo method with at least 100 repetitions to approximate the given probabilities.

(a) $P$(the sum is 5 or less)

(b) $P$(neither a 1 nor a 6 is rolled)

**58.** Suppose a coin is tossed 5 times. Use the Monte Carlo method with at least 50 repetitions to approximate the given probabilities.

(a) $P$(exactly 4 heads)

(b) $P$(2 heads and 3 tails)

**Genetics** *Gregor Mendel, an Austrian monk, was the first to use probability in the study of genetics. In an effort to understand the mechanism of characteristic transmittal from one generation to the next in plants, he counted the number of occurrences of various characteristics. Mendel found that the flower color in certain pea plants obeyed this scheme:*

*Pure red crossed with pure white produces red.*

*From its parents, the red offspring received genes for both red (R) and white (W), but in this case red is dominant and white recessive, so the offspring exhibits the color red. However, the offspring still carries both genes, and when two such offspring are crossed, several things can happen in the third generation. The following table, called a Punnett square, shows the equally likely outcomes:*

|  |  | Second Parent | |
|---|---|---|---|
|  |  | **R** | **W** |
| **First Parent** | **R** | RR | RW |
|  | **W** | WR | WW |

*Use the fact that red is dominant over white to find each of the given probabilities. Assume that there are an equal number of red and white genes in the population.*

**59.** *P*(a flower is red)

**60.** *P*(a flower is white)

# 8.5 Conditional Probability and Independent Events

In Section 8.4, we examined the data from March 7, 2017, for all 505 stocks in the S&P 500 regarding the price of the stock and the dividend yield. (Data from: www.okfn.org.)

| | Dividend Yield | | | |
|---|---|---|---|---|
| **Price of Stock** | **No Dividend** | **.01%–1.99%** | **2.00% or Higher** | **Total** |
| **$0–$49.99** | 21 | 69 | 68 | 158 |
| **$50.00–$99.99** | 18 | 87 | 102 | 207 |
| **$100.00 or Higher** | 26 | 69 | 45 | 140 |
| **Total** | 65 | 225 | 215 | 505 |

Suppose we want to examine that data further to learn if higher-priced stocks are more likely to pay a dividend yield of 2.00% or higher. To find this out, we will need to calculate what are called *conditional probabilities*.

First, let us review. Let *A* be the event that the dividend yield is 2.00% or higher, and let *B* be the event that the stock price is $100 or higher. Because 215 of the stocks out of the total 505 have a dividend yield of 2.00% or higher, and 140 of the 505 stocks are priced $100.00 or higher, we have

$$P(A) = \frac{215}{505} \approx .4257 \text{ and } P(B) = \frac{140}{505} \approx 2772. \quad \boxed{✓_1}$$

✓ **Checkpoint 1**

Use the data in the table to find

**(a)** $P(A \cap B)$;

**(b)** $P(A')$;

**(c)** $P(B')$.

Now let's calculate the probability that a stock priced $100.00 or more has a dividend yield of 2.00% or higher. From the highlighted row in the table below, we see there are a total of 140 stocks priced $100.00 or higher, of which 45 have a dividend yield of 2.0% or higher (notice that the number 45 is circled).

| | Dividend Yield | | | |
|---|---|---|---|---|
| **Price of Stock** | **No Dividend** | **.01%–1.99%** | **2.00% or higher** | **Total** |
| **$0–$49.99** | 21 | 69 | 68 | 158 |
| **$50.00–$99.99** | 18 | 87 | 102 | 207 |
| **$100.00 or Higher** | 26 | 69 | (45) | 140 |
| **Total** | 65 | 225 | 215 | 505 |

Thus,

$$P(\text{a stock priced \$100.00 or more has a dividend yield 2.00\% or higher}) = \frac{45}{140} \approx 3214.$$

This probability (.3214) is a different value from the probability, in general, of a stock with a dividend yield of 2.00% or higher (.4257) because *we have additional information* (that the stock is priced $100 or higher) *that has reduced the sample size.* In other words, we found the probability the stock has a dividend yield of 2.00% or higher, *A*, given the additional information that the stock sells for $100.00 or higher, *B*. This is called the *conditional probability* of event *A*, given that event *B* has occurred, which is written $P(A|B)$ and read as "the probability of *A* given *B*."

In the preceding discussion,

$$P(A|B) = \frac{45}{140} \approx .3214.$$

If we divide the numerator and denominator by 505 (the number of stocks in the S&P 500), this can be written as

$$P(A|B) = \frac{\dfrac{45}{505}}{\dfrac{140}{505}} = \frac{P(A \cap B)}{P(B)},$$

where $P(A \cap B)$ represents, as usual, the probability that both *A* and *B* occur.

To generalize this result, assume that *E* and *F* are two events for a particular experiment. Assume also that the sample space *S* for the experiment has *n* possible equally likely outcomes. Suppose event *F* has *m* elements and $E \cap F$ has *k* elements ($k \leq m$). Then, using the fundamental principle of probability yields

$$P(F) = \frac{m}{n} \quad \text{and} \quad P(E \cap F) = \frac{k}{n}.$$

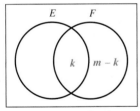

Event *F* has a total of *m* elements.

**Figure 8.26**

We now want to find $P(E|F)$: the probability that *E* occurs, given that *F* has occurred. Since we assume that *F* has occurred, we reduce the sample space to *F*; that is, we look only at the *m* elements inside *F*. (See Figure 8.26.) Of these *m* elements, there are *k* elements for which *E* also occurs, because $E \cap F$ has *k* elements. This yields

$$P(E|F) = \frac{k}{m}.$$

Divide numerator and denominator by *n* to get

$$P(E|F) = \frac{k/n}{m/n} = \frac{P(E \cap F)}{P(F)}.$$

The last result motivates the following definition of conditional probability. The **conditional probability** of an event *E*, given event *F*, written $P(E|F)$, is

$$P(E|F) = \frac{P(E \cap F)}{P(F)} = \frac{P(E \text{ and } F)}{P(F)}, P(F) \neq 0.$$

This definition tells us that, for equally likely outcomes, conditional probability is found by *reducing the sample space to event F* and then finding the number of outcomes in *F* that are also in event *E*. Thus,

$$P(E|F) = \frac{n(E \cap F)}{n(F)}.$$

Although the definition of conditional probability was motivated by an example with equally likely outcomes, it is valid in all cases. For an intuitive explanation, think of the formula as giving the probability that both *E* and *F* occur, compared with the entire probability of *F* occuring.

EXAMPLE 1   **S&P 500 Stocks**  Use the information at the beginning of this section from the table on the price of S&P 500 stocks and dividend yields to find the following probabilities.

**(a)** Let event $C$ be that the price of the stock was $0 to $49.99. Find $P(A|C)$.

**Solution**  In words, $P(A|C)$ is the probability that a stock has a dividend yield of 2.00% or higher given that the price of the stock is between $0 and $49.99.

$$P(A|C) = \frac{n(A \cap C)}{n(C)} = \frac{68}{158} \approx .4304$$

**(b)** Let event $D$ be that the price of the stock was $50.00 to $99.99. Find $P(A|D)$.

**Solution**  In words, $P(A|D)$ is the probability that a stock has a dividend yield of 2.00% or higher given that the price of the stock is between $50.00 and $99.99.

$$P(A|D) = \frac{n(A \cap D)}{n(D)} = \frac{102}{207} \approx .4928$$

Notice that each stock price has a very different probability of having a dividend yield of 2.00% or higher. If the price of the stock is low ($0–$49.99), we have $P(A|C) \approx .4304$. If the price of the stock is in the middle category ($50.00–$99.99), we have $P(A|D) \approx .4928$. Finally, if the price of the stock is high ($100.00 or higher), $P(A|B) \approx .3214$. The additional information of knowing the price of the stock changes our estimate of the probability of having a dividend yield of 2.00% or higher.

**(c)** Find $P(B|A)$.

**Solution**  It is important to note that $P(B|A)$ does not usually equal $P(A|B)$. In this case, $P(B|A)$ in words is the probability that a stock has a price of $100.00 or more given that the dividend yield is 2.00% or higher. The count for the numerator of the number of stocks with a price of $100 or more and a dividend yield of 2.00% or higher is still 45, but now our denominator is the number of stocks with a dividend yield of 2.00% or higher (215). Thus, we have

$$P(B|A) = \frac{n(B \cap A)}{n(A)} = \frac{n(A \cap B)}{n(A)} = \frac{45}{215} \approx .2093.$$

**(d)** Find $P(A'|B')$

**Solution**  In words, we are looking for the probability that the dividend yield is less than 2.00% given that the price of the stock is less than $100. To find $n(A' \cap B')$ and $n(B')$, we examine the table below. The numbers colored in pink indicate those where the stock price is less than $100. The circled numbers indicate, of those, which do not have dividend yield of 2.00% or higher.

| Price of Stock | Dividend Yield | | | |
|---|---|---|---|---|
| | No Dividend | .01%–1.99% | 2.00% or Higher | Total |
| $0–$49.99 | 21 | 69 | 68 | 158 |
| $50.00–$99.99 | 18 | 87 | 102 | 207 |
| $100.00 or Higher | 26 | 69 | 45 | 140 |
| Total | 65 | 225 | 215 | 505 |

Thus, we have

$$P(A'|B') = \frac{n(A' \cap B')}{n(B')} = \frac{21 + 69 + 18 + 87}{158 + 207} = \frac{195}{365} \approx .5342.$$

---

**✓Checkpoint 2**

The table shows the results of the 2016 General Social Survey regarding happiness for married and never married respondents.

| | Very Happy | Pretty Happy or Not Too Happy | Total |
|---|---|---|---|
| **Married** | 465 | 743 | 1208 |
| **Never Married** | 156 | 647 | 803 |
| **Total** | 621 | 1390 | 2011 |

Let $M$ represent married respondents and $V$ represent very happy respondents. Find each of the given probabilities.

**(a)** $P(V|M)$

**(b)** $P(V|M')$

**(c)** $P(M|V)$

**(d)** $P(M'|V')$

**(e)** State the probability of part (d) in words.

**Figure 8.27**

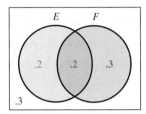

**Figure 8.28**

✓ **Checkpoint 3**

Find $P(F|E)$ if $P(E) = .3$, $P(F)=.4$, and $P(E \cup F) = .6$.

✓ **Checkpoint 4**

In Example 3, find the probability that exactly one coin showed a head, given that at least one was a head.

Venn diagrams can be used to illustrate problems in conditional probability. A Venn diagram for Example 1, in which the probabilities are used to indicate the number in the set defined by each region, is shown in Figure 8.27. In the diagram, $P(B|A)$ is found by *reducing the sample space to just set A*. Then $P(B|A)$ is the ratio of the number in that part of set $B$ which is also in $A$ to the number in set $A$, or $.0891/(.0891 + .3366) = .0891/.4257 \approx .2093$. This agrees with part (c) of Example 1.

**EXAMPLE 2**   Given $P(E) = .4$, $P(F) = .5$, and $P(E \cup F) = .7$, find $P(E|F)$.

**Solution** Find $P(E \cap F)$ first. Then use a Venn diagram to find $P(E|F)$. By the addition rule,

$$P(E \cup F) = P(E) + P(F) - P(E \cap F)$$
$$.7 = .4 + .5 - P(E \cap F)$$
$$P(E \cap F) = .2.$$

Now use the probabilities to indicate the number in each region of the Venn diagram in Figure 8.28. $P(E|F)$ is the ratio of the probability of that part of $E$ which is in $F$ to the probability of $F$, or

$$P(E|F) = \frac{P(E \cap F)}{P(F)} = \frac{.2}{.5} = \frac{2}{5} = .4. \; \checkmark_3$$

**EXAMPLE 3**   Two fair coins were tossed, and it is known that at least one was a head. Find the probability that both were heads.

**Solution** The sample space has four equally likely outcomes: $S = \{hh, ht, th, tt.\}$ Define two events

$$E_1 = \text{at least 1 head} = \{hh, ht, th\}$$

and

$$E_2 = 2 \text{ heads} = \{hh\}.$$

Because there are four equally likely outcomes, $P(E_1) = 3/4$. Also, $P(E_1 \cap E_2) = 1/4$. We want the probability that both were heads, given that at least one was a head; that is, we want to find $P(E_2|E_1)$. Because of the condition that at least one coin was a head, the reduced sample space is

$$\{hh, ht, th\}.$$

Since only one outcome in this reduced sample space is two heads,

$$P(E_2|E_1) = \frac{1}{3}.$$

Alternatively, use the definition given earlier:

$$P(E_2|E_1) = \frac{P(E_2 \cap E_1)}{P(E_1)} = \frac{1/4}{3/4} = \frac{1}{3}. \; \checkmark_4$$

It is important not to confuse $P(A|B)$ with $P(B|A)$. For example, in a criminal trial, a prosecutor may point out to the jury that the probability of the defendant's DNA profile matching that of a sample taken at the scene of the crime, given that the defendant is innocent, is very small. What the jury must decide, however, is the probability that the defendant is innocent, given that the defendant's DNA profile matches the sample. Confusing the two is an error sometimes called "the prosecutor's fallacy," and the 1990 conviction of a rape suspect in England was overturned by a panel of judges who ordered a retrial, because the fallacy made the original trial unfair.* This mistake is often called "confusion of the inverse."

In the next section, we will see how to compute $P(A|B)$ when we know $P(B|A)$.

## Product Rule

If $P(E) \neq 0$ and $P(F) \neq 0$, then the definition of conditional probability shows that

$$P(E|F) = \frac{P(E \cap F)}{P(F)} \quad \text{and} \quad P(F|E) = \frac{P(F \cap E)}{P(E)}.$$

Using the fact that $P(E \cap F) = P(F \cap E)$, and solving each of these equations for $P(E \cap F)$, we obtain the following rule.

> ### Product Rule of Probability
>
> If $E$ and $F$ are events, then $P(E \cap F)$ may be found by either of these formulas:
>
> $$P(E \cap F) = P(F) \cdot P(E|F) \quad \text{or} \quad P(E \cap F) = P(E) \cdot P(F|E).$$

The **product rule** gives a method for finding the probability that events $E$ and $F$ both occur. Here is a simple way to remember the ordering of $E$ and $F$ in the probability rule:

$$P(E \cap F) = P(F) \cdot P(E|F) \quad \text{or} \quad P(E \cap F) = P(E) \cdot P(F|E).$$

**EXAMPLE 4**  **Business** According to data from the U.S. Census Bureau, we can estimate the probability that a business is female owned as .3595. The probability that the business has one to four employees given that it is female owned is .5709. What is the probability that a business is female owned *and* has one to four employees?

**Solution** Let $F$ represent the event of "having a female-owned business" and $E$ represent the event of "having one to four employees." We want to find $P(F \cap E)$. By the product rule,

$$P(F \cap E) = P(F)P(E|F).$$

From the given information, $P(F) = .3595$ and the probability that a business has one to four employees given that it is a female-owned business is $P(E|F) = .5709$. Thus,

$$P(F \cap E) = (.3595)(.5709) \approx .2052 \quad \boxed{5}$$

✓ **Checkpoint 5**

In a litter of puppies, 3 were female and 4 were male. Half the males were black. Find the probability that a puppy chosen at random from the litter would be a black male.

In Section 8.1, we used a tree diagram to find the number of subsets of a given set. By including the probabilities for each branch of a tree diagram, we convert it to a **probability tree**. The following examples show how a probability tree is used with the product rule to find the probability of a sequences of events.

**EXAMPLE 5**  **Hiring Decision** A company needs to hire a new director of advertising. It has decided to try to hire either person $A$ or person $B$, both of whom are assistant advertising directors for its major competitor. To decide between $A$ and $B$, the company does research on the campaigns managed by $A$ or $B$ (none are managed by both) and finds that $A$ is in charge of twice as many advertising campaigns as $B$. Also, $A$'s campaigns have yielded satisfactory results three out of four times, while $B$'s campaigns have

*David Pringle, "Who's the DNA Fingerprinting Pointing At?," *New Scientist*, January 29, 1994, pp. 51–52.

yielded satisfactory results only two out of five times. Suppose one of the competitor's advertising campaigns (managed by $A$ or $B$) is selected randomly.

We can represent this situation schematically as follows: Let $A$ denote the event "Person $A$ does the job" and $B$ the event "Person $B$ does the job." Let $S$ be the event "satisfactory results" and $U$ the event "unsatisfactory results." Then the given information can be summarized in the probability tree in Figure 8.29. Since $A$ does twice as many jobs as $B$, $P(A) = 2/3$ and $P(B) = 1/3$, as noted on the first-stage branches of the tree. When $A$ does a job, the probability of satisfactory results is $3/4$ and of unsatisfactory results $1/4$, as noted on the second-stage branches. Similarly, the probabilities when $B$ does the job are noted on the remaining second-stage branches. The composite branches labeled 1–4 represent the four disjoint possibilities for the running and outcome of the campaign.

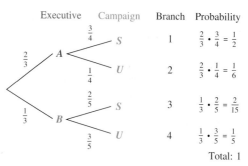

**Figure 8.29**

**(a)** Find the probability that $A$ is in charge of a campaign and it produces satisfactory results.

**Solution**  We are asked to find $P(A \cap S)$. We know that when $A$ does the job, the probability of success is $3/4$; that is, $P(S|A) = 3/4$. Hence, by the product rule,

$$P(A \cap S) = P(A) \cdot P(S|A) = \frac{2}{3} \cdot \frac{3}{4} = \frac{1}{2}.$$

The event $(A \cap S)$ is represented by branch 1 of the tree, and as we have just seen, its probability is the product of the probabilities that make up that branch.

**(b)** Find the probability that $B$ runs a campaign and it produces satisfactory results.

**Solution**  We must find $P(B \cap S)$. This event is represented by branch 3 of the tree, and as before, its probability is the product of the probabilities of the pieces of that branch:

$$P(B \cap S) = P(B) \cdot P(S|B) = \frac{1}{3} \cdot \frac{2}{5} = \frac{2}{15}.$$

**(c)** What is the probability that the selected campaign is satisfactory?

**Solution**  The event $S$ is the union of the disjoint events $A \cap S$ and $B \cap S$, which are represented by branches 1 and 3 of the tree diagram. By the addition rule,

$$P(S) = P(A \cap S) + P(B \cap S) = \frac{1}{2} + \frac{2}{15} = \frac{19}{30}.$$

Thus, the probability of an event that appears at the end of several branches is the sum of the probabilities of each of these branches.

**(d)** What is the probability that the selected campaign is unsatisfactory?

**Solution**  $P(U)$ can be read from branches 2 and 4 of the tree:

$$P(U) = \frac{1}{6} + \frac{1}{5} = \frac{11}{30}.$$

Alternatively, because $U$ is the complement of $S$,

$$P(U) = 1 - P(S) = 1 - \frac{19}{30} = \frac{11}{30}.$$

**(e)** Find the probability that either $A$ runs the campaign or the results are satisfactory (or possibly both).

**Solution**  Event $A$ combines branches 1 and 2, while event $S$ combines branches 1 and 3, so use branches 1, 2, and 3:

$$P(A \cup S) = \frac{1}{2} + \frac{1}{6} + \frac{2}{15} = \frac{4}{5}. \quad \checkmark_6$$

✓ **Checkpoint 6**

Find each of the given probabilities for the scenario in Example 5.

**(a)** $P(U|A)$

**(b)** $P(U|B)$

---

**EXAMPLE 6**    **Jury Selection**  Suppose 6 potential jurors remain in a jury pool and 2 are to be selected to sit on the jury for the trial. The races of the 6 potential jurors are 1 Hispanic, 3 Caucasian, and 2 African-American. If we select one juror at a time, find the probability that one Caucasian and one African-American are drawn.

**Solution**  A probability tree showing the various possible outcomes is given in Figure 8.30 on the next page. In this diagram, $C$ represents the event "selecting a Caucasian juror" and $A$ represents "selecting an African-American juror." On the first draw, $P(C$ on the 1st$) = 3/6 = 1/2$ because three of the six jurors are Caucasian. On the second draw, $P(A$ on the 2nd$|C$ on the 1st$) = 2/5$. One Caucasian juror has been removed, leaving 5, of which 2 are African-American.

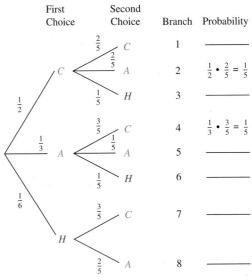

**Figure 8.30**

We want to find the probability of selecting exactly one Caucasian and exactly one African-American. Two events satisfy this condition: selecting a Caucasian first and then selecting an African-American (branch 2 of the tree) and drawing an African-American juror first and then selecting a Caucasian juror (branch 4). For branch 2,

✓ **Checkpoint 7**

In Example 6, find the probability of selecting an African-American juror and then a Caucasian juror.

$$P(C \text{ on 1st}) \cdot P(A \text{ on 2nd}|C \text{ on 1st}) = \frac{1}{2} \cdot \frac{2}{5} = \frac{1}{5}. \quad \checkmark_7$$

For branch 4, on which the African-American juror is selected first,

$$P(A \text{ first}) \cdot P(C \text{ second}|A \text{ first}) = \frac{1}{3} \cdot \frac{3}{5} = \frac{1}{5}.$$

Since these two events are disjoint, the final probability is the sum of the two probabilities.

$$P(\text{one } C \text{ and one } A) = P(C \text{ on 1st}) \cdot P(A \text{ on 2nd}|C \text{ on 1st})$$

$$+ P(A \text{ on 1st}) \cdot P(C \text{ on 2nd}|A \text{ on 1st}) = \frac{2}{5}.$$

The product rule is often used in dealing with *stochastic processes*, which are mathematical models that evolve over time in a probabilistic manner. For example, selecting different jurors is such a process, because the probabilities change with each successive selection. (Particular stochastic processes are studied further in Section 9.5.)

**EXAMPLE 7**    Two cards are drawn without replacement from an ordinary deck (52 cards). Find the probability that the first card is a heart and the second card is red.

**Solution**    Start with the probability tree of Figure 8.31. (You may wish to refer to the deck of cards shown on page 422.) On the first draw, since there are 13 hearts in the 52 cards, the probability of drawing a heart first is $13/52 = 1/4$. On the second draw, since a (red) heart has been drawn already, there are 25 red cards in the remaining 51 cards. Thus, the probability of drawing a red card on the second draw, given that the first is a heart, is $25/51$. By the product rule of probability,

$$P(\text{heart on 1st and red on 2nd})$$
$$= P(\text{heart on 1st}) \cdot P(\text{red on 2nd}|\text{heart on 1st})$$
$$= \frac{1}{4} \cdot \frac{25}{51} = \frac{25}{204} \approx .1225.$$

**EXAMPLE 8**    Three cards are drawn, without replacement, from an ordinary deck. Find the probability that exactly 2 of the cards are red.

**Solution**    Here, we need a probability tree with three stages, as shown in Figure 8.32. The three branches indicated with arrows produce exactly 2 red cards from the draws. Multiply the probabilities along each of these branches and then add:

$$P(\text{exactly 2 red cards}) = \frac{26}{52} \cdot \frac{25}{51} \cdot \frac{26}{50} + \frac{26}{52} \cdot \frac{26}{51} \cdot \frac{25}{50} + \frac{26}{52} \cdot \frac{26}{51} \cdot \frac{25}{50}$$

$$= \frac{50,700}{132,600} = \frac{13}{34} \approx .3824.$$

**Figure 8.31**

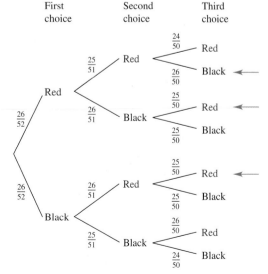

**Figure 8.32**

## Independent Events

Suppose in Example 7 that we draw the two cards *with* replacement rather than without replacement. (That is, we put the first card back, shuffle the deck, and then draw the second card.) If the first card is a heart, then the probability of drawing a red card on the second draw is 26/52, rather than 25/51, because there are still 52 cards in the deck, 26 of them red. In this case, $P$(red second|heart first) is the same as $P$(red second). The value of the second card is not affected by the value of the first card. We say that the event that the second card is red is *independent* of the event that the first card is a heart, since knowledge of the first card does not influence what happens to the second card. On the other hand, when we draw *without* replacement, the events that the first card is a heart and the second is red are *dependent* events. The fact that the first card is a heart means that there is one less red card in the deck, influencing the probability that the second card is red.

As another example, consider tossing a fair coin twice. If the first toss shows heads, the probability that the next toss is heads is still $1/2$. Coin tosses are independent events, since the outcome of one toss does not influence the outcome of the next toss. Similarly, rolls of a fair die are independent events. On the other hand, the events "the milk is old" and "the milk is sour" are dependent events: If the milk is old, there is an increased chance that it is sour. Also, in the example at the beginning of this section, the events $A$ (the stock's dividend yield is 2.00% or higher) and $B$ (the stock price is $100 or higher) are dependent events, because $P(A|B)$ is different from $P(A)$.

If events $E$ and $F$ are independent, then the knowledge that $E$ has occurred gives no (probability) information about the occurrence or nonoccurrence of event $F$. That is, $P(F)$ is exactly the same as $P(F|E)$, or

$$P(F|E) = P(F).$$

This, in fact, is the formal definition of independent events.

---

$E$ and $F$ are **independent events** if

$$P(F|E) = P(F) \quad \text{or} \quad P(E|F) = P(E).$$

---

If the events are not independent, they are **dependent events.**

When $E$ and $F$ are independent events, $P(F|E) = P(F)$, and the product rule becomes

$$P(E \cap F) = P(E) \cdot P(F|E) = P(E) \cdot P(F).$$

Conversely, if this equation holds, it follows that $P(F) = P(F|E)$. Consequently, we have the useful rule that follows.

---

## Product Rule for Independent Events

$E$ and $F$ are independent events if and only if

$$P(E \cap F) = P(E) \cdot P(F).$$

---

**EXAMPLE 9** **Manufacturing** A USB flash drive requires a two-step manufacturing process. The first step involves the circuit board, and let us assume that 98% of the circuit board assemblies are satisfactory. The second step involves the plastic casing. Assume that 96.5% of the casings are satisfactory. Assume the manufacturing process of the circuit boards has no relationship with the manufacturing of the plastic casings, and find the probability that both components are satisfactory.

**Solution** Since the manufacturing processes of the two components have no relationship, we can assume that event $A$ (the circuit board is satisfactory) and event $B$ (the plastic

✓ **Checkpoint 11**

Find the probability of getting 4 successive heads on 4 tosses of a fair coin.

casing is satisfactory) are two independent events. Thus, the probability that both components are satisfactory is

$$P(A \cap B) = P(A) \cdot P(B) = (.98)(.965) = .9457.$$ ✓11

⚠ **CAUTION** It is common to confuse the ideas of *disjoint* events and *independent* events. Events $E$ and $F$ are disjoint if $E \cap F = \emptyset$. For example, if a family has exactly one child, the only possible outcomes are $B = \{boy\}$ and $G = \{girl\}$. The events $B$ and $G$ are disjoint and $P(B) = P(G) = .5$. However, the events are *not* independent, since $P(G|B) = 0$ (if a family with only one child has a boy, the probability that it has a girl is then 0). Since $P(G|B) \neq P(G)$, the events are not independent. Of all the families with exactly *two* children, the events $G_1 = \{first child is a girl\}$ and $G_2 = \{second child is a girl\}$ are independent, because $P(G_2|G_1)$ equals $P(G_2)$. However, $G_1$ and $G_2$ are not disjoint, since $G_1 \cap G_2 = \{both children are girls\} \neq \emptyset$.

To show that two events $E$ and $F$ are independent, we can show that $P(F|E) = P(F)$, that $P(E|F) = P(E)$, or that $P(E \cap F) = P(E) \cdot P(F)$. Another way is to observe that knowledge of one outcome does not influence the probability of the other outcome, as we did for coin tosses.

📝 **NOTE** In some cases, it may not be apparent from the physical description of the problem whether two events are independent or not. For example, it is not obvious whether the event that a baseball player gets a hit tomorrow is independent of the event that he got a hit today. In such cases, it is necessary to use the definition and calculate whether $P(F|E) = P(F)$, or, equivalently, whether $P(E \cap F) = P(E) \cdot P(F)$.

**EXAMPLE 10**    **Types of Businesses** Among partnerships and corporations that file tax returns in the United States, the probability that the enterprise consists of a firm dedicated to real estate, rentals, or leasing is .2498. The probability that the firm is a partnership is .3672. The probability that a firm is dedicated to real estate, rentals, or leasing or is a partnership is .4367. Are the events of a firm being dedicated to real estate, rentals, or leasing and a firm being a partnership independent?

**Solution** Let $E$ represent the event the firm is dedicated to real estate, rentals, or leasing and let $F$ represent the event the firm is a partnership. We must determine whether

$$P(E|F) = P(E) \quad \text{or} \quad P(F|E) = P(F).$$

We know that $P(E) = .2498$, $P(F) = .3672$, and $P(E \cup F) = .4367$. By the addition rule, we know that

$$P(E) + P(F) - P(E \cap F) = P(E \cup F)$$
$$.2498 + .3672 - P(E \cap F) = .4367$$
$$.6170 - P(E \cap F) = .4367$$
$$P(E \cap F) = .1803.$$

Therefore,

$$P(E|F) = \frac{P(E \cap F)}{P(F)} = \frac{.1803}{.3672} \approx .4910 \neq .2498 = P(E).$$

Also,

✓ **Checkpoint 12**

The probability of living in Texas is .08, the probability of speaking English at home is .83, and the probability of living in Texas or speaking English at home is .854. Are the events "speaking English at home" and "living in Texas" independent?

$$P(F|E) = \frac{P(E \cap F)}{P(E)} = \frac{.1803}{.2498} \approx .7218 \neq .3672 = P(F).$$

Since $P(E|F) \neq P(E)$ and $P(F|E) \neq P(F)$ these two events are not independent. ✓12

📝 **NOTE** Although we showed that $P(E|F) \neq P(E)$ and $P(F|E) \neq P(F)$ in Example 10, only one of these results is needed to establish independence.

# 8.5 Exercises

*If a single fair die is rolled, find the probability of rolling the given events. (See Examples 1 and 2.)*

1. 3, given that the number rolled was odd

2. 5, given that the number rolled was even

3. An odd number, given that the number rolled was 3

*If two fair dice are rolled (recall the 36-outcome sample space), find the probability of rolling the given events.*

4. A sum of 8, given that the sum was greater than 7

5. A sum of 6, given that the roll was a "double" (two identical numbers)

6. A double, given that the sum was 9

*If two cards are drawn without replacement from an ordinary deck, find the probabilities of the given event. (See Example 7.)*

7. The second is a heart, given that the first is a heart

8. The second is black, given that the first is a spade

9. A jack and a 10 are drawn.

10. An ace and a 4 are drawn.

11. In your own words, explain how to find the conditional probability $P(E|F)$.

12. Your friend asks you to explain how the product rule for independent events differs from the product rule for dependent events. How would you respond?

13. Another friend asks you to explain how to tell whether two events are dependent or independent. How would you reply? (Use your own words.)

14. A student reasons that the probability in Example 3 of both coins being heads is just the probability that the other coin is a head—that is, 1/2. Explain why this reasoning is wrong.

*Decide whether the two events listed are independent.*

15. $S$ is the event that it is snowing and $L$ is the event that the instructor is late for class.

16. $R$ is the event that four semesters of theology are required to graduate from a certain college and $A$ is the event that the college is a religiously affiliated school.

17. $R$ is the event that it rains in the Amazon jungle and $H$ is the event that an instructor in New York City writes a difficult exam.

18. $T$ is the event that Tom Cruise's next movie grosses over $200 million and $R$ is the event that the Republicans have a majority in Congress in the year 2024.

19. In a two-child family, if we assume that the probabilities of a child being male and a child being female are each .5, are the events "the children are the same sex" and "at most one child is male" independent? Are they independent for a three-child family?

20. Let $A$ and $B$ be independent events with $P(A) = \dfrac{1}{4}$ and $P(B) = \dfrac{1}{5}$. Find $P(A \cap B)$ and $P(A \cup B)$.

**Firm Size** *According to data from the U.S. Small Business Administration, the probability that a firm has one to four employees given that it is a construction firm is .6930. The probability that a firm has one to four employees given that it is a manufacturing firm is .4196. The probability the firm is a construction firm is .1135, and the probability that a firm is a manufacturing firm is .0449. The probability that a firm has one to four employees is .6214. Use this information to find the probabilities for Exercises 21–24. (See Example 4.)*

21. A firm has one to four employees and is a construction firm.

22. A firm has one to four employees and is a manufacturing firm.

23. A firm is a construction firm given that it has one to four employees.

24. A firm is a manufacturing firm given that it has one to four employees.

**Hours Worked** *Data from the 2016 General Social Survey can allow us to estimate how many hours people who classified their work status as part-time or full-time actually worked in the last week. Use the following table to find the probabilities of the events in Exercises 25–30. (Data from: gss.norc.org.)*

| Labor Force Status | Hours Worked in the Last Week | | | | | |
|---|---|---|---|---|---|---|
| | 0–19 | 20–29 | 30–39 | 40–49 | 50 or More | Total |
| **Full-Time Worker** | 26 | 43 | 137 | 691 | 377 | 1274 |
| **Part-Time Worker** | 74 | 82 | 87 | 41 | 24 | 308 |
| **Total** | 100 | 125 | 224 | 732 | 401 | 1582 |

25. The worker was full-time given that he or she worked 50 or more hours.

26. The worker was part-time given that she or he worked 50 or more hours.

27. The hours worked were 0–19 given that the worker was full-time.

28. The hours worked were 20–29 given that the worker was part-time.

29. The hours worked were 20 or less given that the worker was part-time.

30. The hours worked were 40 or more given that the worker was full-time.

**Consumer Confidence** *The 2016 General Social Survey asked respondents about their confidence in banks and financial institutions as well as their confidence in major corporations. Each question allowed three responses: "a great deal, only some, or hardly any." The table below summarizes the responses.*

| Confidence in Banks and Financial Institutions | Confidence in Major Corporations | | | |
|---|---|---|---|---|
| | **A Great Deal** | **Only Some** | **Hardly Any** | **Total** |
| **A Great Deal** | 125 | 119 | 25 | 269 |
| **Only Some** | 172 | 759 | 113 | 1044 |
| **Hardly Any** | 43 | 362 | 201 | 606 |
| **Total** | 340 | 1240 | 339 | 1919 |

*Find the probability of the described events. ( Data from: gss.norc.org. )*

**31.** A respondent has a great deal of confidence in banks and financial institutions given that he or she has hardly any confidence in major corporations.

**32.** A respondent has hardly any confidence in banks and financial institutions given that he or she has hardly any confidence in major corporations.

**33.** A respondent has a great deal of confidence in major corporations given that he or she has only some confidence in banks and financial institutions.

**34.** A respondent has hardly any confidence in major corporations given that he or she has hardly any confidence in banks and financial institutions.

**Color Blindness** *The following table shows frequencies for red–green color blindness, where M represents that a person is male and C represents that a person is color blind.*

| | **M** | **M'** | **Totals** |
|---|---|---|---|
| **C** | .042 | .007 | .049 |
| **C'** | .485 | .466 | .951 |
| **Totals** | .527 | .473 | 1.000 |

*Use the table to find the given probabilities.*

**35.** $P(M)$

**36.** $P(C)$

**37.** $P(M \cap C)$

**38.** $P(M \cup C)$

**39.** $P(M|C)$

**40.** $P(M'|C)$

**41.** Are the events $C$ and $M$ dependent? Recall that two events $E$ and $F$ are dependent if $P(E|F) \neq P(E)$. (See Example 10.)

**42.** Are the events $M'$ and $C$ dependent?

**Color Blindness and Deafness** *A scientist wishes to determine whether there is any dependence between color blindness (C) and deafness (D). Use the probabilities in the table to answer Exercises 43 and 44.*

| | **D** | **D'** | **Total** |
|---|---|---|---|
| **C** | .0004 | .0796 | .0800 |
| **C'** | .0046 | .9154 | .9200 |
| **Total** | .0050 | .9950 | 1.0000 |

**43.** Find $P(C)$, $P(D)$, $P(C \cap D)$, and $P(C|D)$. Does $P(C) = P(C|D)$? What does that result imply regarding independence of $C$ and $D$?

**44.** Find $P(D')$, $P(C)$, $P(D' \cap C)$, and $P(D'|C)$. Does $P(D') = P(D'|C)$? What does that result imply regarding independence of $D'$ and $C$?

**Driver's Test** *The Motor Vehicle Department in a certain state has found that the probability of a person passing the test for a driver's license on the first try is .75. The probability that an individual who fails on the first test will pass on the second try is .80, and the probability that an individual who fails the first and second tests will pass the third time is .70. Find the probability of the given event.*

**45.** A person fails both the first and second tests

**46.** A person will fail three times in a row

**Airlines** *The number of on-time flights and the number of delayed or cancelled flights for four major airlines for March 2017 is given in the table below.*

| Airline | On Time | Delayed or Cancelled |
|---|---|---|
| American | 62,642 | 15,472 |
| Delta | 68,721 | 10,349 |
| Southwest | 91,664 | 23,480 |
| United | 39,262 | 9235 |

*Find the probability of the described event. ( Data from: www.transportation.gov. )*

**47.** An on-time flight given the airline was American.

**48.** An on-time flight given the airline was Delta.

**49.** An on-time flight given the airline was Southwest.

**50.** An on-time flight given the airline was United.

**51.** The airline was United given that the flight was on time.

**52.** The airline was American given that the flight was on time.

**53.** The airline was not Delta given that the flight was delayed or cancelled.

**54.** The airline was not American given that the flight was delayed or cancelled.

**55.** Which airline had the highest probability of an on-time flight?

**56.** Which airline had the lowest probability of an on-time flight?

*Suppose the probability that the first record by a singing group will be a hit is .32. If the first record is a hit, so are all the group's subsequent records. If the first record is not a hit, the probability of the group's second record and all subsequent ones being hits is .16. If the first two records are not hits, the probability that the third is a hit is .08. The probability that a record is a hit continues to decrease by half with each successive nonhit record. Find the probability of the given event.*

**57.** The group will have at least one hit in its first four records.

**58.** The group will have a hit in its first six records if the first three are not hits.

*Work the given problems on independent events. (See Examples 9 and 10.)*

**59. Computer Failure** Corporations such as banks, where a computer is essential to day-to-day operations, often have a second, backup computer in case of failure by the main computer. Suppose that there is a .003 chance that the main computer will fail in a given period and a .005 chance that the backup computer will fail while the main computer is being repaired. Suppose these failures represent independent events, and find the fraction of the time the corporation can assume that it will have computer service.

**60. Multiple Flights** According to data from the U.S. Department of Transportation, Delta Airlines was on time approximately 87% of the time in March 2017. Use this information, and assume that the event that a given flight takes place on time is independent of the event that another flight is on time to answer the following questions

(a) Elisabeta plans to visit her company's branch offices; her journey requires 3 separate flights on Delta Airlines. What is the probability that all of these flights will be on time?

(b) How reasonable do you believe it is to suppose the independence of being on time from flight to flight?

**61. Rocket Failure** The probability that a key component of a space rocket will fail is .03.

(a) How many such components must be used as backups to ensure that the probability that at least one of the components will work is .999999?

(b) Is it reasonable to assume independence here?

**62. Medication Effectiveness** A medical experiment showed that the probability that a new medicine is effective is .75, the probability that a patient will have a certain side effect is .4, and the probability that both events will occur is .3. Decide whether these events are dependent or independent.

**Consumer Confidence** *Refer to the table included for Exercises 31–34.*

**63.** Are the events of having hardly any confidence in banks and financial institutions and having hardly any confidence in major corporations independent?

**64.** Are the events of having a great deal of confidence in banks and financial institutions and having a great deal of confidence in major corporations independent?

✓**Checkpoint Answers**

1. (a) About .0891     (b) About .5743     (c) About .7228

2. (a) About .3849     (b) About .1943     (c) About .7488
   (d) About .4655
   (e) The probability of never being married, given that the person is partially or not happy

3. 1/3          4. 2/3          5. 2/7

6. (a) 1/4          (b) 3/5

7. 1/5          8. 1/10          9. $13/102 \approx .1275$

10. $13/34 \approx .3824$     11. 1/16          12. No

# 8.6   Bayes' Formula

Suppose the probability that a person gets lung cancer, given that the person smokes a pack or more of cigarettes daily, is known. For a research project, it might be necessary to know the probability that a person smokes a pack or more of cigarettes daily, given that the person has lung cancer. More generally, if $P(E|F)$ is known for two events $E$ and $F$, can $P(F|E)$ be found? The answer is yes, we can find $P(F|E)$ by using the formula to be developed in this section. To develop this formula, we can use a probability tree to find $P(F|E)$. Since $P(E|F)$ is known, the first outcome is either $F$ or $F'$. Then, for each of these outcomes, either $E$ or $E'$ occurs, as shown in Figure 8.33.

**Figure 8.33**

The four cases have the probabilities shown on the right. By the definition of conditional probability and the product rule,

$$P(E) = P(F \cap E) + P(F' \cap E),$$

$$P(F \cap E) = P(F) \cdot P(E|F), \quad \text{and} \quad P(F' \cap E) = P(F') \cdot P(E|F').$$

By substitution,

$$P(E) = P(F) \cdot P(E|F) + P(F') \cdot P(E|F') \tag{1}$$

and

$$P(F|E) = \frac{P(F \cap E)}{P(E)} = \frac{P(F) \cdot P(E|F)}{P(F) \cdot P(E|F) + P(F') \cdot P(E|F')}.$$

We have proved a special case of Bayes' formula, which is generalized later in this section.  ✔ 1

✓ **Checkpoint 1**

Use the special case of Bayes' formula to find $P(F|E)$ if $P(F) = .2$, $P(E|F) = .1$, and $P(E|F') = .3$. [*Hint:* $P(F') = 1 - P(F)$.]

## Bayes' Formula (Special Case)

$$P(F|E) = \frac{P(F) \cdot P(E|F)}{P(F) \cdot P(E|F) + P(F') \cdot P(E|F')}.$$

**EXAMPLE 1**    **Start-Up Success** Data reported by the *Harvard Business Review* states that start-ups with at least one female founder are more likely to succeed and that 18% of start-ups now have at least one female founder. Additionally, the probability that a start-up succeeds given that there is at least one female founder is .15. The probability that a start-up succeeds given there are only male founders is .09.

**(a)** Find the probability that a start-up succeeds.

**Solution** Let $E$ represent the event that the start-up succeeds, and let $F$ represent the event of having at least one female founder. From the given information, we have

$$P(F) = .18, \ P(E|F) = .15, P(E|F') = .09.$$

We can use the complement rule so that we have

$$P(F') = 1 - .18 = .82, P(E'|F) = 1 - .15 = .85, P(E'|F') = 1 - .09 = .91.$$

These probabilities are shown on the probability tree in Figure 8.34.

**Figure 8.34**

From Equation **(1)** earlier, we have

$$P(E) = P(F) \cdot P(E|F) + P(F') \cdot P(E|F').$$

Substituting values from our tree diagram in Figure 8.34, we have

$$P(E) = (.18)(.15) + (.82)(.09) = .1008.$$

Thus, the probability that a start-up is successful is about .1.

**(b)** Find the probability that there is at least one female founder given that the start-up succeeds.

**Solution** Using our notation, we want to find $P(F|E)$. Applying Bayes' formula, we find that

$$P(F|E) = \frac{P(F) \cdot P(E|F)}{P(F) \cdot P(E|F) + P(F') \cdot P(E|F')}$$

$$= \frac{(.18)(.15)}{(.18)(.15) + (.82)(.09)} \approx .2679.$$

The probability that a start-up has at least one female founder given that the start-up succeeds is about .2679. ✓ 2

✓ **Checkpoint 2**

In Example 1, find $P(F'|E)$.

If we rewrite the special case of Bayes' formula, replacing $F$ by $F_1$ and $F'$ by $F_2$, then it says:

$$P(F_1|E) = \frac{P(F_1) \cdot P(E|F_1)}{P(F_1) \cdot P(E|F_1) + P(F_2) \cdot P(E|F_2)}.$$

Since $F_1 = F$ and $F_2 = F'$ we see that $F_1$ and $F_2$ are disjoint and that their union is the entire sample space. The generalization of Bayes' formula to more than two possibilities follows this same pattern.

## Bayes' Formula

Suppose $F_1, F_2, \ldots, F_n$ are pairwise disjoint events (meaning that any two of them are disjoint) whose union is the sample space. Then for an event $E$ and for each $i$ with $1 \leq i \leq n$,

$$P(F_i|E) = \frac{P(F_i) \cdot P(E|F_i)}{P(F_1) \cdot P(E|F_1) + \cdots + P(F_n) \cdot P(E|F_n)}.$$

This result is known as Bayes' formula, after the Reverend Thomas Bayes (1702–61), whose paper on probability was published in 1764 after his death.

The statement of Bayes' formula can be daunting. It may be easier to remember the formula by thinking of the probability tree that produced it. Go through the following steps.

## Using Bayes' Formula

**Step 1**  Start a probability tree with branches representing events $F_1, F_2, \ldots, F_n$. Label each branch with its corresponding probability.

**Step 2**  From the end of each of these branches, draw a branch for event $E$. Label this branch with the probability of getting to it, or $P(E|F_i)$.

**Step 3**  There are now $n$ different paths that result in event $E$. Next to each path, put its probability: the product of the probabilities that the first branch occurs, $P(F_i)$, and that the second branch occurs, $P(E|F_i)$: that is, $P(F_i) \cdot P(E|F_i)$.

**Step 4**  $P(F_i|E)$ is found by dividing the probability of the branch for $F_i$ by the sum of the probabilities of all the branches producing event $E$.

Example 2 illustrates this process.

**EXAMPLE 2**  **Marital Status and Number of Children** The 2016 General Social Survey of women who are age 18 or older indicated that 86% of married women have one or more children, 42% of never married women have one or more children, and 90% of women who are divorced, separated, or widowed have one or more children. The survey also indicated that 41% of women age 18 or older were currently married, 25% had never been married, and 34% were divorced, separated, or widowed (labeled "other"). Find the probability that a woman who has one or more children is married.

**Solution** Let $E$ represent the event "having one or more children," with $F_1$ representing "married women," $F_2$ representing "never married women," and $F_3$ "other." Then

$$P(F_1) = .41; \quad P(E|F_1) = .86;$$
$$P(F_2) = .25; \quad P(E|F_2) = .42;$$
$$P(F_3) = .34; \quad P(E|F_3) = .90.$$

We need to find $P(F_1|E)$, the probability that a woman is married, given that she has one or more children. First, draw a probability tree using the given information, as in Figure 8.35. The steps leading to event $E$ are shown.

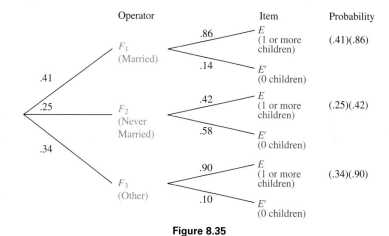

**Figure 8.35**

Find $P(F_1|E)$, using the top branch of the tree shown in Figure 8.35, by dividing the probability of this branch by the sum of the probabilities of all the branches leading to $E$:

$$P(F_1|E) = \frac{(.41)(.86)}{(.41)(.86) + (.25)(.42) + (.34)(.90)} \approx .4618.$$

✓ **Checkpoint 3**

In Example 2, find
(a)  $P(F_2|E)$;
(b)  $P(F_3|E)$.

**EXAMPLE 3**   **Business** A manufacturer buys items from six different suppliers. The fraction of the total number of items obtained from each supplier, along with the probability that an item purchased from that supplier is defective, is shown in the following table:

| Supplier | Fraction of Total Supplied | Probability of Being Defective |
|---|---|---|
| 1 | .05 | .04 |
| 2 | .12 | .02 |
| 3 | .16 | .07 |
| 4 | .23 | .01 |
| 5 | .35 | .03 |
| 6 | .09 | .05 |

Find the probability that a defective item came from supplier 5.

**Solution** Let $F_1$ be the event that an item came from supplier 1, with $F_2$, $F_3$, $F_4$, $F_5$, and $F_6$ defined in a similar manner. Let $E$ be the event that an item is defective. We want to find $P(F_5|E)$. By Bayes' formula,

$$P(F_5|E) = \frac{(.35)(.03)}{(.05)(.04) + (.12)(.02) + (.16)(.07) + (.23)(.01) + (.35)(.03) + (.09)(.05)}$$

$$= \frac{.0105}{.0329} \approx .3191.$$

✓ **Checkpoint 4**

In Example 3, find the probability that the defective item came from

**(a)** supplier 3;

**(b)** supplier 6.

There is about a 32% chance that a defective item came from supplier 5. Even though supplier 5 has only 3% defectives, his probability of being "guilty" is relatively high, about 32%, because of the large fraction of items he supplies. ✓ 4

# 8.6 Exercises

*For two events M and N, $P(M) = .4$, $P(N|M) = .3$, and $P(N|M') = .4$. Find each of the given probabilities. (See Example 1.)*

**1.** $P(M|N)$

**2.** $P(M'|N)$

*For disjoint events $R_1$, $R_2$, and $R_3$, $P(R_1) = .05$, $P(R_2) = .6$, and $P(R_3) = .35$. In addition, $P(Q|R_1) = .40$, $P(Q|R_2) = .30$, and $P(Q|R_3) = .60$. Find each of the given probabilities. (See Examples 2 and 3.)*

**3.** $P(R_1|Q)$

**4.** $P(R_2|Q)$

**5.** $P(R_3|Q)$

**6.** $P(R_1'|Q)$

*Suppose three jars have the following contents: 2 black balls and 1 white ball in the first; 1 black ball and 2 white balls in the second; 1 black ball and 1 white ball in the third. If the probability of selecting one of the three jars is 1/2, 1/3, and 1/6, respectively, find the probability that if a white ball is drawn, it came from the given jar.*

**7.** The second jar

**8.** The third jar

**S&P 500 Stocks** *On March 7, 2017, the probability that a stock was in the Information Technology sector was .1347. The probability that the stock had a dividend yield of 2.00% or higher given that it was in the Information Technology sector was .2941. The probability that a stock had a dividend yield of 2.00% or higher given that it was not in the Information Technology sector was .4462. Find the following probabilities. (Data from: www.okfn.org.)*

**9.** The stock was in the Information Technology sector given that it had a dividend yield of 2.00% or higher.

**10.** The stock was not in the Information Technology sector given that it did not have a dividend yield of 2.00% or higher.

**Educational Attainment and Earnings** *Data from the U.S. Census Bureau indicates that for the year 2015, 33.3% of the labor force age 25 and over had a high school diploma or fewer years of education, 27.9% had some college or an associate's degree, and 38.9% had a bachelor's degree or more education. Of those with a high school diploma or fewer years of education, 7.4% earned $75,000 or more. Of those with some college or an associate's degree, 13.5% earned $75,000 or more, and of those with a bachelor's degree or more education, 37.0% earned $75,000 or more. Find the following probabilities that a randomly chosen*

*labor force participant has the given characteristics. (Data from: www.census.gov.)*

**11.** Some college or an associate's degree, given that he or she earns $75,000 or more annually

**12.** A bachelor's degree or more education, given that he or she earns $75,000 or more annually

**13.** A high school diploma or less education, given that he or she earns *less* than $75,000 annually

**14.** A bachelor's degree or more education, given that he or she earns *less* than $75,000 annually

**Airlines** *In March 2017, the probability that a flight was an American Airlines flight was .1599, the probability that it was a Delta flight was .1618, and the probability that it was another airline was .6783. The probability that the flight was on time given that it was an American Airlines flight was .8019, given that it was a Delta flight was .8691, and given that it was another airline was .7820. Find the probability of the following events. (Data from: www.transportation.gov.)*

**15.** American Airlines given that it was on time.

**16.** Delta given that it was on time.

**17.** An airline not American or Delta given that it was not on time.

**18.** Delta given that it was not on time.

**Health** *In a test for toxemia, a disease that affects pregnant women, a woman lies on her left side and then rolls over on her back. The test is considered positive if there is a 20-mm rise in her blood pressure within 1 minute. The results have produced the following probabilities, where T represents having toxemia at some time during the pregnancy and N represents a negative test:*

$$P(T'|N) = .90 \quad and \quad P(T|N') = .75.$$

*Assume that $P(N') = .11$, and find each of the given probabilities.*

**19.** $P(N|T)$

**20.** $P(N'|T)$

**Business** *In April 2017, the total sales from General Motors, Ford, or Chrysler was 630,596 cars or light trucks. The probability that the vehicle sold was made by General Motors was .3873, by Ford .3385, and by Chrysler .2742. Additionally, the probability that a General Motors* *vehicle sold was a car was .2765, a Ford vehicle sold was a car was .2336, and a Chrysler vehicle sold was a car was .1253.*

**21.** Given the vehicle sold was a car, find the probability it was made by General Motors.

**22.** Given the vehicle sold was a car, find the probability it was made by Chrysler.

**23.** Given the vehicle sold was not a car, find the probability it was made by Ford.

**24.** Given the vehicle sold was not a car, find the probability it was made by General Motors.

**Business** *At the close of the markets on May 19, 2017, there were 6100 companies listed with the New York Stock Exchange (NYSE) and NASDAQ. For the day May 19, 2017, based on data from the Wall Street Journal, the probabilities of advancing, declining, and unchanged company stocks appears below for the two exchanges.*

| Exchange | Percentage of Companies Listed | Probability of Stock Advancing | Probability of Stock Declining | Probability of Stock Remaining Unchanged |
|---|---|---|---|---|
| NYSE | .5085 | .7240 | .2453 | .0306 |
| NASDAQ | .4915 | .6024 | .3436 | .0540 |

*Find the probability of each of the following events.*

**25.** The stock was listed on the NYSE given that it advanced.

**26.** The stock was listed on NASDAQ given that it declined.

**27.** The stock was listed on the NYSE given that it remained unchanged.

**28.** The stock was listed on NASDAQ given that it remained unchanged.

**29.** The stock was listed on NASDAQ given that it advanced.

**30.** The stock was listed on the NYSE given that it declined.

**Happiness** *The proportion of adult males is .492 and the proportion of adult females is .508. Data from the 2016 General Social Survey has the following probabilities of happiness for each sex. (Data from: U.S. Census Bureau and gss.norc.org.)*

| Sex | Proportion | Very Happy | Pretty Happy | Not Too Happy |
|---|---|---|---|---|
| Male | .492 | .287 | .564 | .149 |
| Female | .508 | .277 | .558 | .165 |

*Find the probability of the described events.*

**31.** Given that the person is very happy, the respondent is male.

**32.** Given that the person is pretty happy, the respondent is female.

**33.** Given that the person is not too happy, the respondent is female.

**34.** Given that the person is very happy, the respondent is female.

**35.** Given that the person is pretty happy, the respondent is male.

**36.** Given that the person is not too happy, the respondent is male.

✓**Checkpoint Answers**

**1.** $1/13 \approx .077$

**2.** About .7321

**3.** (a) About .1375    (b) About .4007

**4.** (a) About .3404    (b) About .1368

# CHAPTER 8 Summary and Review

## Key Terms and Symbols

| | |
|---|---|
| { } | set braces |
| $\in$ | is an element of |
| $\notin$ | is not an element of |
| $\varnothing$ | empty set |
| $\subseteq$ | is a subset of |
| $\nsubseteq$ | is not a subset of |
| $\subset$ | is a proper subset of |
| $A'$ | complement of set $A$ |
| $\cap$ | set intersection |
| $\cup$ | set union |
| $P(E)$ | probability of event $E$ |
| $P(F\mid E)$ | probability of event $F$, given that event $E$ has occurred |

**8.1** set
element (member)
empty set
set-builder notation
universal set
subset
set operations
tree diagram
Venn diagram
complement
intersection
disjoint sets
union

**8.2** addition rule for counting

**8.3** experiment
trial
outcome
sample space
event
certain event
impossible event
disjoint events
basic probability principle
relative frequency
    probability
probability distribution

**8.4** addition rule for
    probability
complement rule
odds

**8.5** conditional probability
product rule of probability
probability tree
independent events
dependent events

**8.6** Bayes' formula

## Chapter 8 Key Concepts

**Sets**

Set $A$ is a **subset** of set $B$ if every element of $A$ is also an element of $B$. A set of $n$ elements has $2^n$ subsets.

Let $A$ and $B$ be any sets with universal set $U$.

The **complement** of $A$ is $A' = \{x \mid x \notin A \text{ and } x \in U\}$.
The **intersection** of $A$ and $B$ is $A \cap B = \{x \mid x \in A \text{ and } x \in B\}$.
The **union** of $A$ and $B$ is $A \cup B = \{x \mid x \in A \text{ or } x \in B \text{ or both}\}$.
$n(A \cup B) = n(A) + n(B) - n(A \cap B)$, where $n(X)$ is the number of elements in set $X$.

**Basic Probability Principle**

Let $S$ be a sample space of equally likely outcomes, and let event $E$ be a subset of $S$. Then the probability that event $E$ occurs is

$$P(E) = \frac{n(E)}{n(S)}.$$

**Addition Rule**

For any events $E$ and $F$ from a sample space $S$,

$$P(E \cup F) = P(E) + P(F) - P(E \cap F).$$

For disjoint events $E$ and $F$,

$$P(E \cup F) = P(E) + P(F).$$

**Complement Rule**

$P(E) = 1 - P(E')$ and $P(E') = 1 - P(E)$.

**Odds**

The odds in favor of event $E$ are $\dfrac{P(E)}{P(E')}$, $P(E') \neq 0$.

**Properties of Probability**

1.  For any event $E$ in sample space $S$, $0 \leq P(E) \leq 1$.
2.  The sum of the probabilities of all possible distinct outcomes is 1.

**Conditional Probability**

The conditional probability of event $E$, given that event $F$ has occurred, is

$$P(E \mid F) = \frac{P(E \cap F)}{P(F)}, \quad \text{where} \quad P(F) \neq 0.$$

For equally likely outcomes, conditional probability is found by reducing the sample space to event $F$; then

$$P(E \mid F) = \frac{n(E \cap F)}{n(F)}.$$

**Product Rule of Probability**

If $E$ and $F$ are events, then $P(E \cap F)$ may be found by either of these formulas:

$$P(E \cap F) = P(F) \cdot P(E \mid F) \quad \text{or} \quad P(E \cap F) = P(E) \cdot P(F \mid E).$$

**Product Rule for Independent Events**

$E$ and $F$ are independent events if and only if

$$P(E \cap F) = P(E)P(F).$$

**Bayes' Formula**

$$P(F_i \mid E) = \frac{P(F_i) \cdot P(E \mid F_i)}{P(F_1) \cdot P(E \mid F_1) + P(F_2) \cdot P(E \mid F_2) + \cdots + P(F_n) \cdot P(E \mid F_n)}.$$

## Chapter 8 Review Exercises

*Write true or false for each of the given statements.*

**1.** $9 \in \{8, 4, -3, -9, 6\}$

**2.** $4 \in \{3, 9, 7\}$

**3.** $2 \notin \{0, 1, 2, 3, 4\}$

**4.** $0 \notin \{0, 1, 2, 3, 4\}$

**5.** $\{3, 4, 5\} \subset \{2, 3, 4, 5, 6\}$

**6.** $\{1, 2, 5, 8\} \subseteq \{1, 2, 5, 10, 11\}$

**7.** $\{1, 5, 9\} \subset \{1, 5, 6, 9, 10\}$

**8.** $0 \subseteq \varnothing$

*List the elements in the given sets.*

**9.** $\{x \mid x \text{ is a national holiday}\}$

**10.** $\{x \mid x \text{ is an integer, } -3 \leq x < 1\}$

**11.** $\{\text{all natural numbers less than 5}\}$

**12.** $\{x \mid x \text{ is a leap year between 1999 and 2018}\}$

Let $U$ = {*Vitamins A, B₁, B₂, B₃, B₆, B₁₂, C, D, E*}, $M$ = {*Vitamins A, C, D, E*}, and $N$ = {*Vitamins A, B₁, B₂, C, E*}. *Find the given sets.*

13. $M'$

14. $N'$

15. $M \cap N$

16. $M \cup N$

17. $M \cup N'$

18. $M' \cap N$

*Consider these sets:*

$U$ = {*Students taking Intermediate Accounting*};
$A$ = {*Females*};
$B$ = {*Finance majors*};
$C$ = {*Students older than 22*};
$D$ = {*Students with a GPA* $> 3.5$}.

*Describe each of the following in words.*

19. $A \cap C$

20. $B \cap D$

21. $A \cup D$

22. $A' \cap D$

23. $B' \cap C'$

*Draw a Venn diagram and shade the given set in it.*

24. $B \cup A'$

25. $A' \cap B$

26. $A' \cap (B' \cap C)$

**Box Office Receipts** *As of May 2017, the 50 highest grossing movies as measured by domestic box office receipts could be categorized as follows. (Data from: www.the-movie-times.com.)*

*28 were action movies;*
*30 were rated PG-13;*
*36 were made in 2005 or more recently;*
*24 were action and PG-13;*
*23 were action and made in 2005 or more recently;*
*23 were PG-13 and made in 2005 or more recently;*
*20 were action, PG-13, and made in 2005 or more recently.*

27. How many movies were action movies and not rated PG-13?

28. How many movies were action or rated PG-13?

29. How many movies were made in the year 2005 or more recently or rated PG-13?

30. How many movies were not action, not rated PG-13, and made prior to the year 2005?

*Write sample spaces for the given scenarios.*

31. A die is rolled and the number of dots showing is noted.

32. A color is selected from the set {red, blue, green}, and then a number is chosen from the set {10, 20, 30}.

**Music Selection** *A student purchases a digital music player and installs 10 songs on the device to see how it works. The genres of the songs are rock (3 songs), pop (4 songs), and alternative (3 songs). She listens to the first two songs on shuffle mode.*

33. Write the sample space for the genre if shuffle mode picks songs at random. (*Note:* Shuffle mode is allowed to play the same song twice in a row.)

34. Are the outcomes in the sample space for Exercise 33 equally likely?

**Business** *A customer wants to purchase a computer and printer. She has narrowed her selection among 2 Dell models, 1 Apple model, and 2 HP models for the computer and 2 Epson models and 3 HP models for the printer.*

35. Write the sample space for the brands among which she can choose for the computer and printer.

36. Are the outcomes in the sample space for Exercise 35 equally likely?

**Business** *A company sells computers and copiers. Let E be the event "a customer buys a computer," and let F be the event "a customer buys a copier." In Exercises 37 and 38, write each of the given scenarios, using ∩, ∪, or ' as necessary.*

37. A customer buys neither a computer nor a copier.

38. A customer buys at least one computer or copier.

39. A student gives the answer to a probability problem as 6/5. Explain why this answer must be incorrect.

40. Describe what is meant by disjoint sets, and give an example.

41. Describe what is meant by mutually exclusive events, and give an example.

42. How are disjoint sets and mutually exclusive events related?

**Stock Fund Allocation** *The American Century Global Small Cap Fund had the following sector allocations as of May 2017. (Data from: personal.vanguard.com.)*

| Sector | Percentage |
|---|---|
| Consumer Discretionary | 17.46 |
| Consumer Staples | 4.11 |
| Energy | 4.54 |
| Financials | 16.16 |
| Health Care | 9.89 |
| Industrials | 19.06 |
| Information Technology | 15.94 |
| Materials | 9.89 |
| Real Estate | 1.76 |
| Telecommunication Services | 0 |
| Utilities | 0 |

*Find the probability that an investment selected at random from this fund is in the given sector.*

43. Consumer Discretionary or Consumer Staples

44. Information Technology or Industrials

45. Not Financials

46. Not Information Technology

47. Neither Consumer Discretionary nor Consumer Staples

48. Neither Materials nor Industrials

**Confidence in Major Corporations** *The following table cross classifies years of education and confidence in major companies from adult participants in the 2016 General Social Survey. (Data from: gss.norc.org.)*

| Years of Education | Confidence in Major Corporations | | | Total |
|---|---|---|---|---|
| | A Great Deal | Only Some | Hardly Any | |
| Less than 12 | 42 | 149 | 57 | 248 |
| 12 Years | 95 | 389 | 82 | 566 |
| 13–16 Years | 153 | 502 | 153 | 808 |
| More Than 16 Years | 49 | 197 | 45 | 291 |
| Total | 339 | 1237 | 337 | 1913 |

*If a participant is chosen at random, find the probability that he or she has the following characteristics.*

49. 12 years or less of school

50. A great deal of confidence or only some confidence in major corporations

51. More than 16 years of education and only some confidence in major corporations

52. Less than 12 years of education and hardly any confidence in major corporations

53. 13–16 years of education or only some confidence in major corporations

54. More than 16 years of education or hardly any confidence in major corporations

**Confidence in Major Corporations** *Use the table above to determine if the probability changes substantially that someone has a great deal of confidence in major corporations depending on her or his years of education. Do this by finding the following.*

55. The probability a person has a great deal of confidence in major corporations given that the person has less than 12 years of school

56. The probability a person has a great deal of confidence in major corporations given that the person has 12 years of school

57. The probability a person has a great deal of confidence in major corporations given that the person has 13–16 years of school

58. The probability a person has a great deal of confidence in major corporations given that the person has more than 16 years of school

59. Which of your answers in 55–58 is the lowest? Which is the highest?

60. Do you think years of education seems to influence whether someone has a great deal of confidence in major corporations? Explain.

**Health** *The partial table shows the four possible (equally likely) combinations when both parents are carriers of the sickle-cell anemia trait. Each carrier parent has normal cells (N) and trait cells (T).*

| | | 2nd Parent | |
|---|---|---|---|
| | | $N_2$ | $T_2$ |
| 1st Parent | $N_1$ | | $N_1 T_2$ |
| | $T_1$ | | |

61. Complete the table.

62. If the disease occurs only when two trait cells combine, find the probability that a child born to these parents will have sickle-cell anemia.

63. Find the probability that the child will carry the trait, but not have the disease, if a normal cell combines with a trait cell.

64. Find the probability that the child neither is a carrier nor has the disease.

*Find the probabilities for the given sums when two fair dice are rolled.*

65. 8

66. No more than 4

67. At least 9

68. Odd and greater than 8

69. 2, given that the sum is less than 4

70. 7, given that at least one die shows a 4

*Suppose $P(E) = .62$, $P(F) = .45$, and $P(E \cap F) = .28$. Find each of the given probabilities.*

71. $P(E \cup F)$

72. $P(E \cap F')$

73. $P(E' \cup F)$

74. $P(E' \cap F')$

*For the events E and F, $P(E) = .2$, $P(E|F) = .3$, and $P(F) = .4$. Find each of the given probabilities.*

75. $P(E'|F)$

76. $P(E|F')$

77. Define independent events, and give an example of one.

78. Are independent events always disjoint? Are they ever disjoint? Give examples.

79. **Stock Market Performance** On May 12, 2017, the odds of a stock on the NYSE declining in value were 419:353. What is the probability that a stock declined in value?

80. **Stock Market Performance** On May 12, 2017, the odds of a stock on the NASDAQ declining in value were 1635:1348. What is the probability that a stock declined in value?

81. **Stock Market Performance** On May 18, 2017, the probability that a stock on the NASDAQ advanced in value was .5452. What are the odds that a stock increased in value?

82. **Stock Market Performance** On May 18, 2017, the probability that a stock on the NYSE advanced in value was .4879. What are the odds that a stock advanced in value?

**Business** *Of the appliance repair shops listed in the phone book, 80% are competent and 20% are not. A competent shop can repair an appliance correctly 95% of the time; an incompetent shop can repair an appliance correctly 60% of the time. Suppose an appliance was repaired correctly. Find the probability that it was repaired by the given type of shop.*

83. A competent shop

84. An incompetent shop

*Suppose an appliance was repaired incorrectly. Find the probability that it was repaired by the given type of shop.*

85. A competent shop

86. An incompetent shop

**Manufacturing** *A manufacturer buys items from four different suppliers. The fraction of the total number of items that is obtained* *from each supplier, along with the probability that an item purchased from that supplier is defective, is shown in the following table:*

| Supplier | Fraction of Total Supplied | Probability of Defective |
|----------|---------------------------|--------------------------|
| 1 | .17 | .04 |
| 2 | .39 | .02 |
| 3 | .35 | .07 |
| 4 | .09 | .03 |

**87.** Find the probability that a defective item came from supplier 4.

**88.** Find the probability that a defective item came from supplier 2.

**Titanic Survival** *The following tables list the number of passengers who were on the Titanic and the number of passengers who survived, according to class of ticket:*[*]

| | CHILDREN | | WOMEN | |
|---|---|---|---|---|
| | On | Survived | On | Survived |
| First Class | 6 | 6 | 144 | 140 |
| Second Class | 24 | 24 | 165 | 76 |
| Third Class | 79 | 27 | 93 | 80 |
| Total | 109 | 57 | 402 | 296 |

---

[*] Sandra L. Takis, "Titanic: A Statistical Exploration," *Mathematics Teacher* 92, no. 8, (November 1999): pp. 660–664. Reprinted with permission. ©1999 by the National Council of Teachers of Mathematics. All rights reserved.

| | MEN | | TOTALS | |
|---|---|---|---|---|
| | On | Survived | On | Survived |
| First Class | 175 | 57 | 325 | 203 |
| Second Class | 168 | 14 | 357 | 114 |
| Third Class | 462 | 75 | 634 | 182 |
| Total | 805 | 146 | 1316 | 499 |

*Use this information to determine the given probabilities. (Round answers to three decimal places.)*

**89.** What is the probability that a randomly selected passenger was in second class?

**90.** What is the overall probability of surviving?

**91.** What is the probability of a first-class passenger surviving?

**92.** What is the probability of a child who was in third class surviving?

**93.** Given that a survivor is from first class, what is the probability that she was a woman?

**94.** Given that a male has survived, what is the probability that he was in third class?

**95.** Are the events "third-class survival" and "male survival" independent events? What does this imply?

**96.** Are the events "first-class survival" and "child survival" independent events? What does this imply?

# Case Study 8  Medical Diagnosis

When patients undergo medical testing, a positive test result for a disease or condition can be emotionally devastating. In many cases, however, testing positive does not necessarily imply that the patient actually has the disease. Bayes' formula can be very helpful in determining the probability of actually having the disease when a patient tests positive.

Let us label the event of having the disease as $D$ and not having the disease as $D'$. We will denote testing positive for the disease as $T$ and testing negative as $T'$. Suppose a medical test is calibrated on patients so that we know that among patients with the disease, the test is positive 99.95% of the time. (This quantity is often called the **sensitivity** of the test.) Among patients known not to have the disease, 99.90% of the time the test gave a negative result. (This quantity is often called the **specificity** of the test.) In summary, we have

Sensitivity = $P(T|D)$ = .9995 and Specificity = $P(T'|D')$ = .9990.

Using the complement rule, we find that the probability the test will give a negative result when a patient has the disease is

$$P(T'|D) = 1 - P(T|D) = 1 - .9995 = .0005.$$

Similarly, for those patients without the disease, the probability of testing positive is .0010, calculated by

$$P(T|D') = 1 - P(T'|D') = 1 - .9990 = .0010.$$

These results do not yet answer the question of interest: If a patient tests positive for the disease, what is the probability the patient actually has the disease? Using our notation, we want to know $P(D|T)$. There are two steps to finding this probability. The first is that we need an estimate of the prevalence of the disease in the general population. Let us assume that one person in a thousand has the disease. We can then calculate that

$$P(D) = \frac{1}{1000} = .001 \quad \text{and} \quad P(D') = 1 - .001 = .999.$$

With this information, and the previous results from testing, we can now use Bayes' formula to find $P(D|T)$:

$$P(D|T) = \frac{P(D)P(T|D)}{P(D)P(T|D) + P(D')P(T|D')}.$$

Using $P(D) = .001$, $P(D') = .999$, the sensitivity $P(T|D) = .9995$ and the complement to the specificity $P(T|D') = .0010$, we have

$$P(D|T) = \frac{(.001)(.9995)}{(.001)(.9995) + (.999)(.0010)} = \frac{.0009995}{.0019985} \approx .5001.$$

Hence, the probability the patient actually has the disease after testing positive for the disease is only about .5. This is approximately the same probability as guessing "heads" when flipping a coin. It seems paradoxical that a test which has such high sensitivity (in this case, .9995) and specificity (in this case, .999) could lead to a probability of merely .5 that a person who tests positive for the disease actually has the disease. This is why it is imperative to have confirmatory tests run after testing positive.

The other factor in the calculation is the prevalence of the disease among the general population. In our example, we used $P(D) = .001$. Often, it is very difficult to know how prevalent a disease is among the general population. If the disease is more prevalent, such as 1 in 100, or $P(D) = .01$, we find the probability of a patient's having the disease, given that the patient tests positive, as

$$P(D|T) = \frac{(.01)(.9995)}{(.01)(.9995) + (.99)(.0010)} = \frac{.009995}{.010985} \approx .9099.$$

So when the disease has higher prevalence, then the probability of having the disease after testing positive is also higher. If the disease has a lower prevalence (as in the case of our first example), then the probability of having the disease after testing positive could be much lower than one might otherwise think.

## Exercises

1. Suppose the specificity of a test is 0.999. Find $P(T|D')$.

2. If the sensitivity of a test for a disease is .99 and the prevalence of the disease is .005, use your answer to Exercise 1 to find the probability of a patient's having the disease, given that the patient tested positive, or $P(D|T)$.

3. Recalculate your answer to Exercise 2 using a prevalence of disease of .0005.

4. According to some experts, the sensitivity of a mammogram in detecting breast cancer can range from .75 to .85, while the specificity might range from .85 to .95. If the prevalence of actually having breast cancer is .0145, find the probability of having the disease, given that the sensitivity is .8, the specificity is .9, and the patient tests positive. (Data from: www.cancer.gov and breast-cancer.ca/est-incidn.)

5. Use the data from Exercise 4, but assume that the sensitivity is .90 and the specificity is .95.

# Extended Project

For a disease or ailment of interest, find in the medical literature the prevalence of the disease or ailment among the general population, the specificity of a particular test to detect the disease or ailment, and the sensitivity of the same test. Then find the probability that a patient actually has the disease, given that the patient tested positive with the particular test. Did your results surprise you? Why or why not?

# Counting, Probability Distributions, and Further Topics in Probability

**9**

CHAPTER

Probability has applications to quality control in manufacturing and to decision making in business. It plays a role in testing new medications, in evaluating DNA evidence in criminal trials, and in a host of other situations. Sophisticated counting techniques are often necessary for determining the probabilities used in these applications.

Probability distributions enable us to compute the "average value" or "expected outcome" when an experiment or process is repeated a number of times. These distributions are introduced in Section 9.1 and used in Sections 9.4 and 9.6. The other focus of this chapter is the development of effective ways to count the possible outcomes of an experiment without actually listing them all (which can be *very* tedious when large numbers are involved). These counting techniques are introduced in Section 9.2 and are used to find probabilities throughout the rest of the chapter.

## 9.1  Probability Distributions and Expected Value

Probability distributions were introduced briefly in Section 8.3. Now we take a more complete look at them. In this section, we shall see that the *expected value* of a probability distribution is a type of average. A probability distribution depends on the idea of a *random variable*, so we begin with that.

## Random Variables

One of the questions asked in the 2016 General Social Survey (GSS) had to do with respondents' daily hours of TV viewing. For those who watched 6 hours or less, respondent's were given the choice of choosing the values 0 through 6. We will label their response as $x$. Since the value of $x$ is random, $x$ is called a random variable.

---

### Random Variable

A **random variable** is a function that assigns a real number to each outcome of an experiment.

---

The following table gives each possible outcome of the study question on TV and video use together with the probability $P(x)$ of each outcome $x$. (Data from: gss.norc.org.)

| $x$ | 0 | 1 | 2 | 3 | 4 | 5 | 6 |
|---|---|---|---|---|---|---|---|
| $P(x)$ | .10 | .21 | .28 | .17 | .13 | .06 | .05 |

A table that lists all the outcomes with the corresponding probabilities is called a **probability distribution.** The sum of the probabilities in a probability distribution must always equal 1. (The sum in some distributions may vary slightly from 1 because of rounding.)

Instead of writing the probability distribution as a table, we could write the same information as a set of ordered pairs:

$$\{(0, .10), (1, .21), (2, .28), (3, .17), (4, .13), (5, .06), (6, .05)\}.$$

There is just one probability for each value of the random variable.

The information in a probability distribution is often displayed graphically as a special kind of bar graph called a **histogram.** The bars of a histogram all have the same width, usually 1 unit. The heights of the bars are determined by the probabilities. A histogram for the data in the probability distribution presented above is given in Figure 9.1. A histogram shows important characteristics of a distribution that may not be readily apparent in tabular form, such as the relative sizes of the probabilities and any symmetry in the distribution.

**Figure 9.1**

**Figure 9.2**

The area of the bar above $x = 0$ in Figure 9.1 is the product of 1 and .10, or $1 \cdot .10 = .10$. Since each bar has a width of 1, its area is equal to the probability that corresponds to its $x$-value. The probability that a particular value will occur is thus given by the area of the appropriate bar of the graph. For example, the probability that one or more hours are spent watching TV is the sum of the areas for $x = 1, x = 2, x = 3, x = 4, x = 5,$ and $x = 6$. This area, shown in red in Figure 9.2, corresponds to .90 of the total area, since

$$P(x \geq 1) = P(x = 1) + P(x = 2) + P(x = 3) + P(x = 4)$$
$$+ P(x = 5) + P(x = 6)$$
$$= .10 + .21 + .28 + .17 + .13 + .06 + .05$$
$$= .90$$

## EXAMPLE 1

**(a)** Give the probability distribution for the number of heads showing when two coins are tossed.

**Solution** Let $x$ represent the random variable "number of heads." Then $x$ can take on the value 0, 1, or 2. Now find the probability of each outcome. When two coins are tossed, the sample space is {TT, TH, HT, HH}. So the probability of getting one head is $2/4 = 1/2$. Similar analysis of the other cases produces this table.

| $x$ | 0 | 1 | 2 |
|---|---|---|---|
| $P(x)$ | 1/4 | 1/2 | 1/4 |

**(b)** Draw a histogram for the distribution in the table. Find the probability that at least one coin comes up heads.

**Solution** The histogram is shown in Figure 9.3. The portion in red represents

$$P(x \geq 1) = P(x = 1) + P(x = 2)$$
$$= \frac{3}{4}.$$

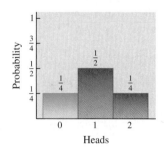

**Figure 9.3**

**Checkpoint 1**

**(a)** Give the probability distribution for the number of heads showing when three coins are tossed.

**(b)** Draw a histogram for the distribution in part (a). Find the probability that no more than one coin comes up heads.

*Answers to Checkpoint exercises are found at the end of the section.*

**TECHNOLOGY TIP** Virtually all graphing calculators can produce histograms. The procedures differ on various calculators, but you usually are required to enter the outcomes in one list and the corresponding frequencies in a second list. For specific details, check your instruction manual under "statistics graphs" or "statistical plotting." To get the histogram in Figure 9.3 with a TI-84+ calculator, we entered the outcomes 0, 1, and 2 in the first list and entered the probabilities .25, .5, and .25 in a second list. Two versions of the histogram are shown in Figure 9.4. They differ slightly because different viewing windows were used. With some calculators, the probabilities must be entered as integers, so make the entries in the second list 1, 2, and 1 (corresponding to 1/4, 2/4, and 1/4, respectively), and use a window with $0 \leq y \leq 4$.

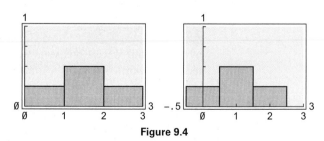

**Figure 9.4**

## Expected Value

In working with probability distributions, it is useful to have a concept of the typical or average value that the random variable takes on. In Example 1, for instance, it seems reasonable that, on the average, one head shows when two coins are tossed. This does not tell what will happen the next time we toss two coins; we may get two heads, or we may get none. If we tossed two coins many times, however, we would expect that, in the long run, we would average about one head for each toss of two coins.

A way to solve such problems in general is to imagine flipping two coins 4 times. Based on the probability distribution in Example 1, we would expect that 1 of the 4 times we would get 0 heads, 2 of the 4 times we would get 1 head, and 1 of the 4 times we would get 2 heads. The total number of heads we would get, then, is

$$0 \cdot 1 + 1 \cdot 2 + 2 \cdot 1 = 4.$$

The expected number of heads per toss is found by dividing the total number of heads by the total number of tosses:

$$\frac{0 \cdot 1 + 1 \cdot 2 + 2 \cdot 1}{4} = 0 \cdot \frac{1}{4} + 1 \cdot \frac{1}{2} + 2 \cdot \frac{1}{4} = 1.$$

Notice that the expected number of heads turns out to be the sum of the three values of the random variable $x$, multiplied by their corresponding probabilities. We can use this idea to define the *expected value* of a random variable as follows.

## Expected Value

Suppose that the random variable $x$ can take on the $n$ values $x_1, x_2, x_3, \ldots, x_n$. Suppose also that the probabilities that these values occur are, respectively, $p_1, p_2, p_3, \ldots, p_n$. Then the **expected value** of the random variable is

$$E(x) = x_1 p_1 + x_2 p_2 + x_3 p_3 + \cdots + x_n p_n.$$

**EXAMPLE 2** **TV Viewing** In the case of TV viewing at the beginning of this section, find the expected number of hours per day of viewing.

**Solution** Multiply each outcome in the table on page 460 by its probability, and sum the products:

$$E(x) = (0)(.10) + (1)(.21) + (2)(.28) + (3)(.17) + (4)(.13) + (5)(.06) + (6)(.05)$$
$$= 2.4.$$

✓ **Checkpoint 2**

Find the expected value of the number of heads showing when four coins are tossed.

On the average, a respondent of the survey will indicate 2.4 hours of TV viewing.

Physically, the expected value of a probability distribution represents a balance point. If we think of the histogram in Figure 9.1 as a series of weights with magnitudes represented by the heights of the bars, then the system would balance if supported at the point corresponding to the expected value.

**EXAMPLE 3** **Fund-Raising** Suppose a local symphony decides to raise money by raffling off concert box seats worth $400, a dinner for two worth $100, and two CD recordings worth $20 each. A total of 2000 tickets are sold at $1 each. Find the expected value of winning for a person who buys one ticket in the raffle.

**Solution** Here, the random variable represents the possible amounts of net winnings, where net winnings = amount won − cost of ticket. For example, the net winnings of the person winning the box seats are $400 (amount won) − $1 (cost of ticket) = $399, and the net winnings for each losing ticket are $0 − $1 = −$1.

The net winnings of the various prizes, as well as their respective probabilities, are shown in the table on the next page. The probability of winning $19 is 2/2000, because there are 2 prizes worth $20. (We have not reduced the fractions in order to keep all the denominators equal.) Because there are 4 winning tickets, there are 1996 losing tickets, so the probability of winning $-\$1$ is $1996/2000$.

| $x$ | $399 | $99 | $19 | $-\$1$ |
|------|------|-----|-----|--------|
| $P(x)$ | 1/2000 | 1/2000 | 2/2000 | 1996/2000 |

The expected winnings for a person buying one ticket are

$$399\left(\frac{1}{2000}\right) + 99\left(\frac{1}{2000}\right) + 19\left(\frac{2}{2000}\right) + (-1)\left(\frac{1996}{2000}\right) = -\frac{1460}{2000}$$
$$= -.73.$$

On average, a person buying one ticket in the raffle will lose $.73, or 73¢.

It is not possible to lose 73¢ in this raffle: Either you lose $1, or you win a prize worth $400, $100, or $20, minus the $1 you paid to play. But if you bought tickets in many such raffles over a long time, you would lose 73¢ per ticket, on average. It is important to note that the expected value of a random variable may be a number that can never occur in any one trial of the experiment. ✔ 3

📝 **NOTE** An alternative way to compute expected value in this and other such examples is to calculate the expected amount won and then subtract the cost of the ticket afterward. The amount won is either $400 (with probability 1/2000), $100 (with probability 1/2000), $20 (with probability 2/2000), or $0 (with probability 1996/2000). The expected winnings for a person buying one ticket are then

$$400\left(\frac{1}{2000}\right) + 100\left(\frac{1}{2000}\right) + 20\left(\frac{2}{2000}\right) + 0\left(\frac{1996}{2000}\right) - 1 = -\frac{1460}{2000}$$
$$= -.73.$$

**EXAMPLE 4**    Each day, Lynette and Tanisha toss a coin to see who buys coffee (at $1.75 a cup). One tosses, while the other calls the outcome. If the person who calls the outcome is correct, the other buys the coffee; otherwise the caller pays. Find Lynette's expected winnings.

**Solution** Assume that an honest coin is used, that Tanisha tosses the coin, and that Lynette calls the outcome. The possible results and corresponding probabilities are shown in the following table:

| | Possible Results | | | |
|---|---|---|---|---|
| **Result of Toss** | Heads | Heads | Tails | Tails |
| **Call** | Heads | Tails | Heads | Tails |
| **Caller Wins?** | Yes | No | No | Yes |
| **Probability** | 1/4 | 1/4 | 1/4 | 1/4 |

Lynette wins a $1.75 cup of coffee whenever the results and calls match, and she loses $1.75 when there is no match. Her expected winnings are

$$1.75\left(\frac{1}{4}\right) + (-1.75)\left(\frac{1}{4}\right) + (-1.75)\left(\frac{1}{4}\right) + 1.75\left(\frac{1}{4}\right) = 0.$$

On the average, over the long run, Lynette breaks even. ✔ 4

A game with an expected value of 0 (such as the one in Example 4) is called a **fair game**. Casinos do not offer fair games. If they did, they would win (on the average) $0 and have a hard time paying their employees! Casino games have expected winnings for the house that vary from 1.5 cents per dollar to 60 cents per dollar. The next example examines the popular game of roulette.

**EXAMPLE 5** **Roulette** As we saw in Chapter 8, an American roulette wheel has 38 slots. Two of the slots are marked 0 and 00 and are colored green. The remaining slots are numbered 1–36 and are colored red and black (18 slots are red and 18 slots are black). One simple wager is to bet $1 on the color red. If the marble lands in a red slot, the player gets his or her dollar back, plus $1 of winnings. Find the expected winnings for a $1 bet on red.

**Solution** For this bet, there are only two possible outcomes: winning or losing. The random variable has outcomes $+1$ if the marble lands in a red slot and $-1$ if it does not. We need to find the probability for these two outcomes. Since there are 38 total slots, 18 of which are colored red, the probability of winning a dollar is $18/38$. The player will lose if the marble lands in any of the remaining 20 slots, so the probability of losing the dollar is $20/38$. Thus, the probability distribution is

| $x$ | $-1$ | $+1$ |
|------|------|------|
| $P(x)$ | $\dfrac{20}{38}$ | $\dfrac{18}{38}$ |

The expected winnings are

$$E(x) = -1\left(\frac{20}{38}\right) + 1\left(\frac{18}{38}\right) = -\frac{2}{38} \approx -.053.$$

The winnings on a dollar bet for red average out to losing about a nickel on every spin of the roulette wheel. In other words, a casino earns, on average, 5.3 cents on every dollar bet on red. ✔5

Exercises 17–20 at the end of the section ask you to find the expected winnings for other bets on games of chance. The idea of expected value can be very useful in decision making, as shown by the next example.

✓ **Checkpoint 5**

A gambling game requires a $5 bet. If the player wins, she gets $1000, but if she loses, she loses her $5. The probability of winning is .001. What are the expected winnings of this game?

**EXAMPLE 6** **Real Estate Development** A real estate developer has two potential projects. With Project 1, there is a 30% chance of earning $1.25 million in profits and a 70% chance of earning $1.10 million in profits. With Project 2, there is a 25% chance of earning $1.45 million in profits and a 75% chance of earning $1.05 million in profits. Based strictly on expected value, which job should be developed?

**Solution** The probability distribution for Project 1 is shown below.

| Profits $x$ | 1.10 | 1.25 |
|-------------|------|------|
| $P(x)$ | .7 | .3 |

The expected value for Project 1 is then

$$E(x) = 1.10(.7) + 1.25(.3)$$
$$= 1.145.$$

The probability distribution for Project 2 is shown below.

| Profits $x$ | 1.05 | 1.45 |
|-------------|------|------|
| $P(x)$ | .75 | .25 |

The expected value for Project 2 is

$$E(x) = 1.05(.75) + 1.45(.25)$$
$$= 1.15.$$

The expected value for Project 2 of $1.15 million is slightly higher than Project 1. The developer should develop Project 2 based only on expected value. ✔6

✓ **Checkpoint 6**

After college, a person is offered two jobs. With job $A$, after five years, there is a 50% chance of making $60,000 per year and a 50% chance of making $45,000. With job $B$, after five years, there is a 30% chance of making $80,000 per year and a 70% chance of making $35,000. Based strictly on expected value, which job should be taken?

**EXAMPLE 7**    **Social Science** The table gives the probability distribution for the number of children of respondents to the 2016 General Social Survey, for those with 7 or fewer children. (Data from: gss.norc.org.)

| $x$ | 0 | 1 | 2 | 3 | 4 | 5 | 6 | 7 |
|------|------|------|------|------|------|------|------|------|
| $P(x)$ | .281 | .162 | .258 | .165 | .075 | .032 | .018 | .009 |

Find the expected value for the number of children.

**Solution** Using the formula for the expected value, we have

$$E(x) = 0(.281) + 1(.162) + 2(.258) + 3(.165) + 4(.075) + 5(.032) + 6(.018) + 7(.009)$$
$$= 1.804.$$

For those respondents with 7 or fewer children, the number of children, on average, is 1.804.

# 9.1 Exercises

*For each of the experiments described, let x determine a random variable and use your knowledge of probability to prepare a probability distribution. (Hint: Use a tree diagram.)*

**1.** Four children are born, and the number of boys is noted. (Assume an equal chance of a boy or a girl for each birth.)

**2.** Two dice are rolled, and the total number of dots is recorded.

**3.** Three cards are drawn from a deck. The number of Queens are counted.

**4.** Two names are drawn from a hat, signifying who should go pick up pizza. Three of the names are on the swim team and two are not. The number of swimmers selected is counted.

*Draw a histogram for each of the given exercises, and shade the region that gives the indicated probability. (See Example 1.)*

**5.** Exercise 1; $P(x \le 2)$

**6.** Exercise 2; $P(x \ge 11)$

**7.** Exercise 3; $P(\text{at least one queen})$

**8.** Exercise 4; $P(\text{fewer than two swimmers})$

*Find the expected value for each random variable. (See Example 2.)*

**9.**

| $x$ | 1 | 3 | 5 | 7 |
|------|------|------|------|------|
| $P(x)$ | .1 | .5 | .2 | .2 |

**10.**

| $y$ | 0 | 15 | 30 | 40 |
|------|------|------|------|------|
| $P(y)$ | .15 | .20 | .40 | .25 |

**11.**

| $z$ | 0 | 2 | 4 | 8 | 16 |
|------|------|------|------|------|------|
| $P(z)$ | .21 | .24 | .21 | .17 | .17 |

**12.**

| $x$ | 5 | 10 | 15 | 20 | 25 |
|------|------|------|------|------|------|
| $P(x)$ | .40 | .30 | .15 | .10 | .05 |

*Find the expected values for the random variables x whose probability functions are graphed.*

**13.**

**14.**

**15.**

**16.**

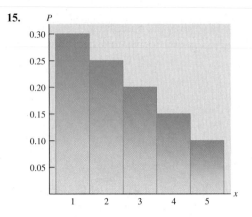

**Games of Chance** *Find the expected winnings for the games of chance described in Exercises 17–20. (See Example 5.)*

**17.** In one form of roulette, you bet $1 on "even." If one of the 18 positive even numbers comes up, you get your dollar back, plus another one. If one of the 20 other numbers (18 odd, 0, and 00) comes up, you lose your dollar.

**18.** Repeat Exercise 17 if there are only 19 noneven numbers (no 00).

**19.** *Numbers* is a game in which you bet $1 on any three-digit number from 000 to 999. If your number comes up, you get $500.

**20.** In one form of the game Keno, the house has a pot containing 80 balls, each marked with a different number from 1 to 80. You buy a ticket for $1 and mark one of the 80 numbers on it. The house then selects 20 numbers at random. If your number is among the 20, you get $3.20 (for a net winning of $2.20).

**21. Online Gambling** An online gambling site offers a first prize of $50,000 and two second prizes of $10,000 each for registered users when they place a bet. A random bet will be selected over a 24-hour period. Two million bets are received in the contest. Find the expected winnings if you can place one registered bet of $1 in the given period.

**22. Online Gambling** An online gambling site offers a lottery with a first prize of $25,000, a second prize of $10,000, and a third prize of $5000. It takes $5 to enter and a winner is selected at random. What are the expected winnings if 20,000 people enter the lottery?

**Prize Promotion** *A contest at a fast-food restaurant offered the following cash prizes and probabilities of winning when one buys a large order of French fries for $1.89:*

| Prize | Probability |
|-------|-------------|
| $100,000 | 1/8,504,860 |
| $50,000 | 1/302,500 |
| $10,000 | 1/282,735 |
| $1000 | 1/153,560 |
| $100 | 1/104,560 |
| $25 | 1/9,540 |

**23.** Find the expected winnings if the player buys one large order of French fries.

**24.** Find the expected winnings if the player buys 25 large orders of French fries in multiple visits.

**25. Unbanked Households** According to the Federal Deposit Insurance Corporation (FDIC), 28.3% of all U.S. households conduct some or all of their financial transactions outside the mainstream banking system. Such people are sometimes called the "unbanked." If we select four households at random, and let *x* denote the number of unbanked households, the probability distribution of *x* is:

| x | 0 | 1 | 2 | 3 | 4 |
|---|---|---|---|---|---|
| P(x) | .2643 | .4173 | .2470 | .0650 | .0064 |

Find the expected value for the number of unbanked households.

**26. Retirement Funds** Approximately 54% of U.S. residents own individual stocks, stock mutual funds, or stocks in a retirement fund. If we select five U.S. residents at random and let *x* denote the number who own stocks, the probability distribution of *x* is:

| x | 0 | 1 | 2 | 3 | 4 | 5 |
|---|---|---|---|---|---|---|
| P(x) | .0206 | .1209 | .2838 | .3332 | .1956 | .0459 |

Find the expected value for the number of U.S. residents who own stocks. (Data from: www.gallup.com.)

*For Exercises 27–30, determine whether the probability distributions are valid or not. If not, explain why.*

**27.**

| x | 5 | 10 | 15 | 20 | 25 | 30 | 35 |
|---|---|----|----|----|----|----|----|
| P(x) | .01 | .09 | .25 | .45 | .05 | .20 | −.05 |

**28.**

| x | −2 | −1 | 0 | 1 | 2 | 3 | 4 |
|---|----|----|---|---|---|---|---|
| P(x) | .05 | .10 | .75 | .02 | .03 | .04 | .01 |

**29.**

| x | 1 | 3 | 5 | 7 | 9 | 11 |
|---|---|---|---|---|---|----|
| P(x) | .01 | .02 | .03 | .04 | .05 | .85 |

**30.**

| x | −10 | −5 | 0 | 5 | 10 |
|---|---|---|---|---|---|
| P(x) | .50 | .10 | −.20 | .30 | .30 |

*For Exercises 31–35, fill in the missing value(s) to make a valid probability distribution.*

**31.**

| x | 5 | 10 | 15 | 20 | 25 | 30 |
|---|---|---|---|---|---|---|
| P(x) | .01 | .09 | .25 | .45 | .05 | |

**32.**

| x | −3 | −2 | −1 | 0 | 1 | 2 | 3 |
|---|---|---|---|---|---|---|---|
| P(x) | | .15 | .15 | .15 | .15 | .15 | .15 |

**33.**

| x | 10 | 20 | 30 | 40 |
|---|---|---|---|---|
| P(x) | .20 | | .25 | .30 |

**34.**

| x | −50 | −40 | −30 | −20 | −10 | 0 | 10 |
|---|---|---|---|---|---|---|---|
| P(x) | .05 | .25 | .10 | .10 | .05 | | |

**35.**

| x | 1 | 2 | 3 | 6 | 12 | 24 | 48 |
|---|---|---|---|---|---|---|---|
| P(x) | .10 | .10 | .20 | .25 | .05 | | |

**36. Retail Returns** During the month of July, a home improvement store sold a great many air-conditioning units, but some were returned. The following table shows the probability distribution for the daily number of returns of air-conditioning units sold in July:

| x | 0 | 1 | 2 | 3 | 4 | 5 |
|---|---|---|---|---|---|---|
| P(x) | .55 | .31 | .08 | .04 | .01 | .01 |

Find the expected number of returns per day.

**37. Life Insurance** An insurance company has written 100 policies of $15,000, 250 of $10,000, and 500 of $5000 for people age 20. If experience shows that the probability that a person will die in the next year at age 20 is .0007, how much can the company expect to pay out during the year after the policies were written?

**38. Business** A market researcher came upon a recent survey by the Pew Research Center that found that 73% of U.S. teenagers owned a smartphone. If six teens are selected at random, the probability distribution for x, the number of teens with smartphones, is as follows.

| x | 0 | 1 | 2 | 3 | 4 | 5 | 6 |
|---|---|---|---|---|---|---|---|
| P(x) | .0004 | .0063 | .0425 | .1531 | .3105 | .3358 | .1513 |

Find the expected number of teens with smartphones.

**39. Auto Sales** In April 2017, General Motors captured 17.1% of the light vehicles (cars and light trucks) sales market. If three new

vehicle sales are selected at random, the probability distribution for x, the number of sales that are General Motors vehicles, is given in the following table.

| x | 0 | 1 | 2 | 3 |
|---|---|---|---|---|
| P(x) | .5697 | .3526 | .0727 | .0050 |

Find the expected number of sales among the three sales selected that are General Motors vehicles. (Data from: www.wsj.com.)

**40. Gaming Device** According to a recent report of the Entertainment Software Association, 80% of U.S. households own a gaming device of some sort. If five households are selected at random, the probability distribution for x, the number of households that own a gaming device, is given in the following table.

| x | 0 | 1 | 2 | 3 | 4 | 5 |
|---|---|---|---|---|---|---|
| P(x) | .0003 | .0064 | .0512 | .2048 | .4096 | .3277 |

Find the expected number of households among the 5 selected that own a gaming device. (Data from: venturebeat.com.)

**41. Drug Cost Effectiveness** Two antibiotics are used to treat common infections. Researchers wish to compare the two antibiotics—Drug A and Drug B—for their cost effectiveness. Drug A is inexpensive, safe, and effective. Drug B is also safe. However, it is considerably more expensive and is generally more effective. Use the given tree diagram (in which costs are estimated as the total cost of the medication, a visit to a medical office, and hours of lost work) to complete the following tasks:

**(a)** Find the expected cost of using each antibiotic to treat a middle-ear infection.

**(b)** To minimize the total expected cost, which antibiotic should be chosen?

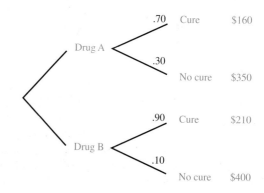

**42. Hurricane Seeding** One of the few methods that can be used in an attempt to cut the severity of a hurricane is to *seed* the storm. In this process, silver iodide crystals are dropped into the storm in order to decrease the wind speed. Unfortunately, silver iodide crystals sometimes cause the storm to *increase* its speed. Wind speeds may also increase or decrease even with no seeding. Use the given tree diagram to complete the following tasks.

**(a)** Find the expected amount of damage under each of the options, "seed" and "do not seed."

**(b)** To minimize total expected damage, which option should be chosen?

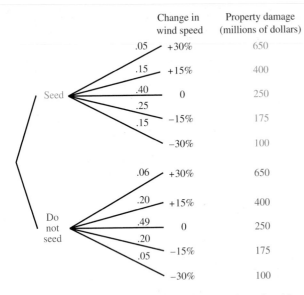

| | Change in wind speed | Property damage (millions of dollars) |
|---|---|---|

43. In the 2016 Gentlemen's Singles Wimbledon Championships, Andy Murray and Milos Raonic played in the finals. The prize money for the winner was £2,000,000 (British pounds sterling), and the prize money for the runner-up was £1,000,000. Find the expected winnings for Andy Murray if we

   (a) assume both players had an equal chance of winning;

   (b) use the players' prior head-to-head match record whereby Murray had a .64 probability of winning.

44. In the 2016 Ladies' Singles Wimbledon Championships, Serena Williams and Angelique Kerber played in the finals. The prize money for the winner was £2,000,000 (British pounds sterling), and the prize money for the runner-up was £1,000,000. Find the expected winnings for Serena Williams if we use the players' prior head-to-head match record whereby Williams had a .71 probability of winning.

 **Checkpoint Answers**

1. (a)

| $x$ | 0 | 1 | 2 | 3 |
|---|---|---|---|---|
| $P(x)$ | 1/8 | 3/8 | 3/8 | 1/8 |

   (b) 1/2

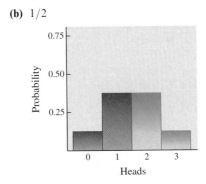

2. 2

3. −$.50. On the average, you lose $.50 per ticket purchased.

4. 0

5. −$4

6. Job $A$ has an expected salary of $52,500, and job $B$ has an expected salary of $48,500. Take job $A$.

# 9.2 The Multiplication Principle, Permutations, and Combinations

We begin with a simple example. If there are three roads from town $A$ to town $B$ and two roads from town $B$ to town $C$, in how many ways can someone travel from $A$ to $C$ by way of $B$? We can solve this simple problem with the help of Figure 9.5, which lists all the possible ways to go from $A$ to $C$.

The possible ways to go from $A$ through $B$ to $C$ are a1, a2, b1, b2, c1, and c2. So there are 6 possible ways. Note that 6 is the product of $3 \cdot 2$, where 3 is the number of ways to go from $A$ to $B$, and 2 is the number of ways to go from $B$ to $C$.

Another way to solve this problem is to use a tree diagram, as shown in Figure 9.6. This diagram shows that, for each of the 3 roads from $A$, there are 2 different routes leading from $B$ to $C$, making $3 \cdot 2 = 6$ different ways.

This example is an illustration of the *multiplication principle*.

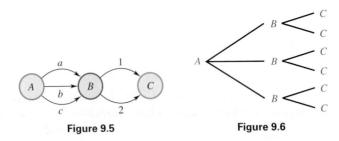

**Figure 9.5**          **Figure 9.6**

## Multiplication Principle

Suppose $n$ choices must be made, with

$$m_1 \text{ ways to make choice 1,}$$
$$m_2 \text{ ways to make choice 2,}$$
$$\vdots$$
$$m_n \text{ ways to make choice } n.$$

Then there are

$$m_1 \cdot m_2 \cdot \ldots \cdot m_n$$

different ways to make the entire sequence of choices.

**EXAMPLE 1**    Suppose Angela has 9 skirts, 8 blouses, and 13 different pairs of shoes. If she is willing to wear any combination, how many different skirt–blouse–shoe choices does she have?

**Solution**  By the multiplication principle, there are 9 skirt choices, 8 blouse choices, and 13 shoe choices, for a total of $9 \cdot 8 \cdot 13 = 936$ skirt–blouse–shoe outfits.

**EXAMPLE 2**    **Kitchen Appliances** In May 2017, there were 111 built-in dishwashers, 224 gas stoves, and 82 side-by-side refrigerators available for purchase at the Home Depot website. How many different ways can a homeowner outfit his or her kitchen with these three appliances?

**Solution**  A tree (or other diagram) would be far too complicated to use here, but the multiplication principle answers the question easily. There are

$$111 \cdot 224 \cdot 82 = 2,038,848$$

ways. ✔ 1

**EXAMPLE 3**    A combination lock can be set to open to any 3-letter sequence.

**(a)**  How many sequences are possible?

**Solution**  Since there are 26 letters of the alphabet, there are 26 choices for each of the 3 letters, and, by the multiplication principle, $26 \cdot 26 \cdot 26 = 17,576$ different sequences.

**(b)**  How many sequences are possible if no letter is repeated?

**Solution**  There are 26 choices for the first letter. It cannot be used again, so there are 25 choices for the second letter and then 24 choices for the third letter. Consequently, the number of such sequences is $26 \cdot 25 \cdot 24 = 15,600$. ✔ 2

## Factorial Notation

The use of the multiplication principle often leads to products such as $5 \cdot 4 \cdot 3 \cdot 2 \cdot 1$, the product of all the natural numbers from 5 down to 1. If $n$ is a natural number, the symbol $n!$ (read "$n$ factorial") denotes the product of all the natural numbers from $n$ down to 1. The factorial is an algebraic shorthand. For example, instead of writing $5 \cdot 4 \cdot 3 \cdot 2 \cdot 1$, we simply write $5!$. If $n = 1$, this formula is understood to give $1! = 1$.

✓ **Checkpoint 1**

On the same day described in Example 2, there were 6497 bathroom vanities with tops, 417 medicine cabinets, and 718 bathroom sink faucets available for purchase at the Home Depot website. How many different ways can a homeowner outfit her or his bathroom with these items?

✓ **Checkpoint 2**

**(a)**  In how many ways can 6 business tycoons line up their golf carts at the country club?

**(b)**  How many ways can 4 pupils be seated in a row with 4 seats?

> ### *n*-Factorial
>
> For any natural number $n$,
>
> $$n! = n(n-1)(n-2) \cdots (3)(2)(1).$$
>
> By definition, $0! = 1$.

Note that $6! = 6 \cdot 5 \cdot 4 \cdot 3 \cdot 2 \cdot 1 = 6 \cdot (5 \cdot 4 \cdot 3 \cdot 2 \cdot 1) = 6 \cdot 5!$. Similarly, the definition of $n!$ shows that

$$n! = n \cdot (n-1)!.$$

One reason that $0!$ is defined to be 1 is to make the preceding formula valid when $n = 1$, for when $n = 1$, we have $1! = 1$ and $1 \cdot (1 - 1)! = 1 \cdot 0! = 1 \cdot 1 = 1$, so $n! = n \cdot (n-1)!$. ✔️ 3

Almost all calculators have an $n!$ key. A calculator with a 10-digit display and scientific-notation capability will usually give the exact value of $n!$ for $n \le 13$ and approximate values of $n!$ for $14 \le n \le 69$. The value of $70!$ is approximately $1.198 \cdot 10^{100}$, which is too large for most calculators. So how would you simplify $\dfrac{100!}{98!}$? Depending on the type of calculator, there may be an overflow problem. The next two examples show how to avoid this problem.

✓ **Checkpoint 3**

Evaluate:

(a) 4!

(b) 6!

(c) 1!

(d) 6!/4!

📇 **TECHNOLOGY TIP**

The factorial key on a graphing calculator is usually located in the PRB or PROB submenu of the MATH or OPTN menu.

---

**EXAMPLE 4**    Evaluate

$$\frac{100!}{98!}.$$

**Solution** We use the fact that $n! = n \cdot (n-1)!$ several times:

$$\frac{100!}{98!} = \frac{100 \cdot 99!}{98!} = \frac{100 \cdot 99 \cdot 98!}{98!} = 100 \cdot 99 = 9900.$$

---

**EXAMPLE 5**    Evaluate

$$\frac{5!}{2! \, 3!}.$$

**Solution** $\dfrac{5!}{2! \, 3!} = \dfrac{5 \cdot 4!}{2! \, 3!} = \dfrac{5 \cdot 4 \cdot 3!}{2! \, 3!} = \dfrac{5 \cdot 4}{2 \cdot 1} = 10.$

---

**EXAMPLE 6**    **Morse Code** Morse code uses a sequence of dots and dashes to represent letters and words. How many sequences are possible with at most 3 symbols?

**Solution** "At most 3" means "1 or 2 or 3." Each symbol may be either a dot or a dash. Thus, the following numbers of sequences are possible in each case:

| Number of Symbols | Number of Sequences |
|:---:|:---:|
| 1 | 2 |
| 2 | $2 \cdot 2 = 4$ |
| 3 | $2 \cdot 2 \cdot 2 = 8$ |

✓ **Checkpoint 4**

How many Morse code sequences are possible with at most 4 symbols?

Altogether, $2 + 4 + 8 = 14$ different sequences of at most 3 symbols are possible. Because there are 26 letters in the alphabet, some letters must be represented by sequences of 4 symbols in Morse code. ✔️ 4

## Permutations

A **permutation** of a set of elements is an ordering of the elements. For instance, there are six permutations (orderings) of the letters $A$, $B$, and $C$, namely,

$$ABC, ACB, BAC, BCA, CAB, \text{ and } CBA,$$

as you can easily verify. As this listing shows, order counts when determining the number of permutations of a set of elements. By saying "order counts," we mean that the event $ABC$ is indeed distinct from $CBA$ or any other ordering of the three letters. We can use the multiplication principle to determine the number of possible permutations of any set.

**EXAMPLE 7**   **Batting Order** How many batting orders are possible for a 9-person baseball team?

**Solution**  There are 9 possible choices for the first batter, 8 possible choices for the second batter, 7 for the third batter, and so on, down to the eighth batter (2 possible choices) and the ninth batter (1 possibility). So the total number of batting orders is

$$9 \cdot 8 \cdot 7 \cdot 6 \cdot 5 \cdot 4 \cdot 3 \cdot 2 \cdot 1 = 362{,}880.$$

In other words, the number of permutations of a 9-person set is 9!.

The argument in Example 7 applies to any set, leading to the conclusion that follows.

> The number of permutations of an $n$ element set is $n!$.

Sometimes we want to order only some of the elements in a set, rather than all of them.

**EXAMPLE 8**   A teacher has 5 books and wants to display 3 of them side by side on her desk. How many arrangements of 3 books are possible?

**Solution**  The teacher has 5 ways to fill the first space, 4 ways to fill the second space, and 3 ways to fill the third space. Because she wants only 3 books on the desk, there are only 3 spaces to fill, giving $5 \cdot 4 \cdot 3 = 60$ possible arrangements. ✓ 5

In Example 8, we say that the possible arrangements are *the permutations of 5 things taken 3 at a time*, and we denote the number of such permutations by $_5P_3$. In other words, $_5P_3 = 60$. More generally, an ordering of $r$ elements from a set of $n$ elements is called a **permutation of $n$ things taken $r$ at a time**, and the number of such permutations is denoted $_nP_r$.[*] To see how to compute this number, look at the answer in Example 8, which can be expressed like this:

$$_5P_3 = 5 \cdot 4 \cdot 3 = 5 \cdot 4 \cdot 3 \cdot \frac{2 \cdot 1}{2 \cdot 1} = \frac{5 \cdot 4 \cdot 3 \cdot 2 \cdot 1}{2 \cdot 1} = \frac{5!}{2!} = \frac{5!}{(5-3)!}.$$

A similar analysis in the general case leads to this useful fact:

## Permutations

If $_nP_r$ (where $r \leq n$) is the number of permutations of $n$ elements taken $r$ at a time, then

$$_nP_r = \frac{n!}{(n-r)!}.$$

✓**Checkpoint 5**

How many ways can a merchant with limited space display 4 fabric samples side by side from her collection of 8?

---

[*] Another notation that is sometimes used is $P(n, r)$.

**TECHNOLOGY TIP** The permutation function on a graphing calculator is in the same menu as the factorial key. For large values of $n$ and $r$, the calculator display for $_nP_r$ may be an approximation.

To find $_nP_r$, we can either use the preceding rule or apply the multiplication principle directly, as the next example shows.

**Figure 9.7**

**EXAMPLE 9**    Early in 2016, 12 candidates sought the Republican nomination for president at the Iowa caucus. In a poll, how many ways could voters rank their first, second, and third choices?

**Solution** This is the same as finding the number of permutations of 12 elements taken 3 at a time. Since there are 3 choices to be made, the multiplication principle gives $_{12}P_3 = 12 \cdot 11 \cdot 10 = 1320$. Alternatively, by the formula for $_nP_r$,

$$_{12}P_3 = \frac{12!}{(12-3)!} = \frac{12 \cdot 11 \cdot 10 \cdot 9 \cdot 8 \cdot 7 \cdot 6 \cdot 5 \cdot 4 \cdot 3 \cdot 2 \cdot 1}{9 \cdot 8 \cdot 7 \cdot 6 \cdot 5 \cdot 4 \cdot 3 \cdot 2 \cdot 1} = 12 \cdot 11 \cdot 10 = 1320.$$

Figure 9.7 shows this result on a TI-84+ graphing calculator.

✓ **Checkpoint 6**

Find the number of permutations of

(a)   5 things taken 2 at a time;

(b)   9 things taken 3 at a time.

Find each of the following:

(c)   $_3P_1$;

(d)   $_7P_3$;

(e)   $_{12}P_2$.

**EXAMPLE 10**    In a college admissions forum, 5 female and 4 male sophomore panelists discuss their college experiences with high school seniors.

**(a)**  In how many ways can the panelists be seated in a row of 9 chairs?

**Solution** Find $_9P_9$, the total number of ways to seat 9 panelists in 9 chairs:

$$_9P_9 = \frac{9!}{(9-9)!} = \frac{9!}{0!} = \frac{9!}{1} = 9 \cdot 8 \cdot 7 \cdot 6 \cdot 5 \cdot 4 \cdot 3 \cdot 2 \cdot 1 = 362{,}880.$$

So, there are 362,880 ways to seat the 9 panelists.

**(b)**  In how many ways can the panelists be seated if the males and females are to be alternated?

**Solution** Use the multiplication principle. In order to alternate males and females, a female must be seated in the first chair (since there are 5 females and only 4 males), any of the males next, and so on. Thus, there are 5 ways to fill the first seat, 4 ways to fill the second seat, 4 ways to fill the third seat (with any of the 4 remaining females), and so on, or

$$5 \cdot 4 \cdot 4 \cdot 3 \cdot 3 \cdot 2 \cdot 2 \cdot 1 \cdot 1 = 2880.$$

So, there are 2880 ways to seat the panelists.

**(c)**  In how many ways can the panelists be seated if the males must sit together and the females sit together?

**Solution** Use the multiplication principle. We first must decide how to arrange the two groups (males and females). There are 2! ways of doing this. Next, there are 5! ways of arranging the females and 4! ways of arranging the men, for a total of

$$2! \, 5! \, 4! = 2 \cdot 120 \cdot 24 = 5760$$

✓ **Checkpoint 7**

A collection of 3 paintings by one artist and 2 by another is to be displayed. In how many ways can the paintings be shown

(a)   in a row?

(b)   if the works of the artists are to be alternated?

(c)   if one painting by each artist is displayed?

ways.

## Combinations

In Example 8, we found that there are 60 ways a teacher can arrange 3 of 5 different books on a desk. That is, there are 60 permutations of 5 things taken 3 at a time. Suppose now that the teacher does not wish to arrange the books on her desk, but rather wishes to choose, at random, any 3 of the 5 books to give to a book sale to raise money for her school. In how many ways can she do this?

At first glance, we might say 60 again, but that is incorrect. The number 60 counts all possible *arrangements* of 3 books chosen from 5. However, the following arrangements, for example, would all lead to the same set of 3 books being given to the book sale:

mystery–biography–textbook          biography–textbook–mystery

mystery–textbook–biography          textbook–biography–mystery

biography–mystery–textbook          textbook–mystery–biography

The foregoing list shows 6 different *arrangements* of 3 books, but only one *subset* of 3 books. A subset of items selected *without regard to order* is called a **combination**. The number of combinations of 5 things taken 3 at a time is written $_5C_3$. Since they are subsets, combinations are *not ordered*.

To evaluate $_5C_3$, start with the $5 \cdot 4 \cdot 3$ *permutations* of 5 things taken 3 at a time. Combinations are unordered; therefore, we find the number of combinations by dividing the number of permutations by the number of ways each group of 3 can be ordered—that is, by 3!:

$$_5C_3 = \frac{5 \cdot 4 \cdot 3}{3!} = \frac{5 \cdot 4 \cdot 3}{3 \cdot 2 \cdot 1} = 10.$$

There are 10 ways that the teacher can choose 3 books at random for the book sale.

Generalizing this discussion gives the formula for the number of combinations of $n$ elements taken $r$ at a time, written $_nC_r$.* In general, a set of $r$ elements can be ordered in $r!$ ways, so we divide $_nP_r$ by $r!$ to get $_nC_r$:

$$_nC_r = \frac{_nP_r}{r!}$$

$$= _nP_r \frac{1}{r!}$$

$$= \frac{n!}{(n-r)!} \cdot \frac{1}{r!} \qquad \text{Definition of } _nP_r$$

$$= \frac{n!}{(n-r)!\, r!}.$$

## ✓ Checkpoint 8

Evaluate $\dfrac{_nP_r}{r!}$ for the given values.

**(a)** $n = 6, r = 2$

**(b)** $n = 8, r = 4$

**(c)** $n = 7, r = 0$

This last form is the most useful for setting up the calculation. ✓ 8

## Combinations

The number of combinations of $n$ elements taken $r$ at a time, where $r \leq n$, is

$$_nC_r = \frac{n!}{(n-r)!r!}.$$

**EXAMPLE 11**    **Committee Selection** From a group of 10 students, a committee is to be chosen to meet with the dean. How many different 3-person committees are possible?

**Solution** A committee is not ordered, so we compute

$$_{10}C_3 = \frac{10!}{(10-3)!\,3!} = \frac{10!}{7!\,3!} = \frac{10 \cdot 9 \cdot 8 \cdot 7!}{7!\,3!} = \frac{10 \cdot 9 \cdot 8}{3 \cdot 2 \cdot 1} = 120.$$

▦ **TECHNOLOGY TIP** The key for obtaining $_nC_r$ on a graphing calculator is located in the same menu as the key for obtaining $_nP_r$.

---

*Another notation that is sometimes used in place of $_nC_r$ is $\binom{n}{r}$.

**Figure 9.8**

**Checkpoint 9**

Use $\dfrac{n!}{(n-r)!\,r!}$ to evaluate $_nC_r$.

**(a)** $_6C_2$

**(b)** $_8C_4$

**(c)** $_7C_0$

Compare your answers with the answers to Checkpoint 8.

**Checkpoint 10**

Five orchids from a collection of 20 are to be selected for a flower show.

**(a)** In how many ways can this be done?

**(b)** In how many different ways can the group of 5 be selected if 2 particular orchids must be included?

**(c)** In how many ways can at least 1 and at most 5 orchids be selected? (*Hint:* Use a calculator.)

**EXAMPLE 12** **Jury Selection** In how many ways can a 12-person jury be chosen from a pool of 40 people?

**Solution** Since the order in which the jurors are chosen does not matter, we use combinations. The number of combinations of 40 things taken 12 at a time is

$$_{40}C_{12} = \frac{40!}{(40-12)!\,12!} = \frac{40!}{28!\,12!}.$$

Using a calculator to compute this number (Figure 9.8), we see that there are 5,586,853,480 possible ways to choose a jury. ✓ 9

**EXAMPLE 13** **Selection of Managers** Three managers are to be selected from a group of 30 to work on a special project.

**(a)** In how many different ways can the managers be selected?

**Solution** Here, we wish to know the number of 3-element combinations that can be formed from a set of 30 elements. (We want combinations, not permutations, since order within the group of 3 does not matter.) So, we calculate

$$_{30}C_3 = \frac{30!}{27!\,3!} = 4060.$$

There are 4060 ways to select the project group.

**(b)** In how many ways can the group of 3 be selected if a certain manager must work on the project?

**Solution** Since 1 manager has already been selected for the project, the problem is reduced to selecting 2 more from the remaining 29 managers:

$$_{29}C_2 = \frac{29!}{27!\,2!} = 406.$$

In this case, the project group can be selected in 406 ways.

**(c)** In how many ways can a nonempty group of at most 3 managers be selected from these 30 managers?

**Solution** The group is to be nonempty; therefore, "at most 3" means "1 or 2 or 3." Find the number of ways for each case:

| Case | Number of Ways |
|---|---|
| 1 | $_{30}C_1 = \dfrac{30!}{29!\,1!} = \dfrac{30\cdot 29!}{29!\,1!} = 30$ |
| 2 | $_{30}C_2 = \dfrac{30!}{28!\,2!} = \dfrac{30\cdot 29\cdot 28!}{28!\cdot 2\cdot 1} = 435$ |
| 3 | $_{30}C_3 = \dfrac{30!}{27!\,3!} = \dfrac{30\cdot 29\cdot 28\cdot 27!}{27!\cdot 3\cdot 2\cdot 1} = 4060$ |

The total number of ways to select at most 3 managers will be the sum

$$30 + 435 + 4060 = 4525. ✓ 10$$

## Choosing a Method

The formulas for permutations and combinations given in this section will be very useful in solving probability problems in later sections. Any difficulty in using these formulas usually comes from being unable to differentiate between them. Both permutations and combinations give the number of ways to choose $r$ objects from a set of $n$ objects. The differences between permutations and combinations are outlined in the following summary.

| Permutations | Combinations |
|---|---|
| Different orderings or arrangements of the $r$ objects are different permutations. | Each choice or subset of $r$ objects gives 1 combination. Order within the $r$ objects does not matter. |
| $$_nP_r = \frac{n!}{(n-r)!}$$ | $$_nC_r = \frac{n!}{(n-r)!\,r!}$$ |
| Clue words: arrangement, schedule, order Order matters! | Clue words: group, committee, set, sample Order does not matter! |

In the examples that follow, concentrate on recognizing which of the formulas should be applied.

**EXAMPLE 14**    For each of the given problems, tell whether permutations or combinations should be used to solve the problem.

**(a)** How many 4-digit numbers are possible if no digits are repeated?

**Solution** Since changing the order of the 4 digits results in a different number, we use permutations.

**(b)** A sample of 3 lightbulbs is randomly selected from a batch of 15 bulbs. How many different samples are possible?

**Solution** The order in which the 3 lightbulbs are selected is not important. The sample is unchanged if the bulbs are rearranged, so combinations should be used.

**(c)** In a basketball conference with 8 teams, how many games must be played so that each team plays every other team exactly once?

**Solution** The selection of 2 teams for a game is an *unordered* subset of 2 from the set of 8 teams. Use combinations again.

**(d)** In how many ways can 4 patients be assigned to 6 hospital rooms so that each patient has a private room?

**Solution** The room assignments are an *ordered* selection of 4 rooms from the 6 rooms. Exchanging the rooms of any 2 patients within a selection of 4 rooms gives a different assignment, so permutations should be used. ✓11

✓**Checkpoint 11**

Solve the problems in Example 14.

**EXAMPLE 15**    A manager must select 4 employees for promotion. Twelve employees are eligible.

**(a)** In how many ways can the 4 employees be chosen?

**Solution** Because there is no reason to consider the order in which the 4 are selected, we use combinations:

$$_{12}C_4 = \frac{12!}{4!\,8!} = 495.$$

**(b)** In how many ways can 4 employees be chosen (from 12) to be placed in 4 different jobs?

**Solution** In this case, once a group of 4 is selected, its members can be assigned in many different ways (or arrangements) to the 4 jobs. Therefore, this problem requires permutations:

$$_{12}P_4 = \frac{12!}{8!} = 11{,}880. \quad \checkmark_{12}$$

✓**Checkpoint 12**

A postal worker has special-delivery mail for 7 customers.

**(a)** In how many ways can he arrange his schedule to deliver to all 7?

**(b)** In how many ways can he schedule deliveries if he can deliver to only 4 of the 7?

(69 ncr 5)*26
            292201338
69!*26
64!5!
            292201338

**Figure 9.9**

✓**Checkpoint 13**

Under earlier Powerball rules, you had to choose five different numbers from 1 to 59 and then choose one powerball number from 1 to 35. Under those rules, how many different selections were possible?

**EXAMPLE 16**   **Powerball** Powerball is a lottery game played in 44 states (plus the District of Columbia, Puerto Rico, and the U.S. Virgin Islands). For a $2 ticket, a player selects five different numbers from 1 to 69 and one powerball number from 1 to 26 (which may be the same as one of the first five chosen). A match of all six numbers wins the jackpot. How many different selections are possible?

**Solution** The order in which the first five numbers are chosen does not matter. So we use combinations to find the number of combinations of 69 things taken 5 at a time—that is, $_{69}C_5$. There are 26 ways to choose one powerball number from 1 to 26. By the multiplication principle, the number of different selections is

$$(_{69}C_5)(26) = \frac{69!}{(69 - 5)!5!}(26) = \frac{69!(26)}{64!5!} = 292{,}201{,}338,$$

as shown in two ways on a graphing calculator in Figure 9.9. $\checkmark_{13}$

**EXAMPLE 17**   A male student going on spring break at Daytona Beach has 8 tank tops and 12 pairs of shorts. He decides he will need 5 tank tops and 6 pairs of shorts for the trip. How many ways can he choose the tank tops and the shorts?

**Solution** We can break this problem into two parts: finding the number of ways to choose the tank tops, and finding the number of ways to choose the shorts. For the tank tops, the order is not important, so we use combinations to obtain

$$_8C_5 = \frac{8!}{3!\,5!} = 56.$$

Likewise, order is not important for the shorts, so we use combinations to obtain

$$_{12}C_6 = \frac{12!}{6!\,6!} = 924.$$

We now know there are 56 ways to choose the tank tops and 924 ways to choose the shorts. The total number of ways to choose the tank tops and shorts can be found using the multiplication principle to obtain $56 \cdot 924 = 51{,}744.$ $\checkmark_{14}$

✓**Checkpoint 14**

Lacy wants to pack 4 of her 10 blouses and 2 of her 4 pairs of jeans for her trip to Europe. How many ways can she choose the blouses and jeans?

As Examples 16 and 17 show, often both combinations and the multiplication principle must be used in the same problem.

**EXAMPLE 18**   To illustrate the differences between permutations and combinations in another way, suppose 2 cans of soup are to be selected from 4 cans on a shelf: noodle ($N$), bean ($B$), mushroom ($M$), and tomato ($T$). As shown in Figure 9.10(a), there are 12 ways to select 2 cans from the 4 cans if the order matters (if noodle first and bean second is considered different from bean and then noodle, for example). However, if order is unimportant, then there are 6 ways to choose 2 cans of soup from the 4, as illustrated in Figure 9.10(b).

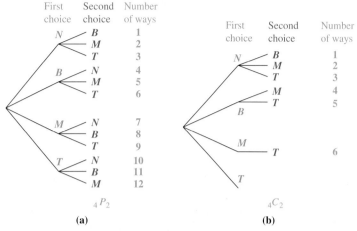

**Figure 9.10**

⚠ **CAUTION** It should be stressed that not all counting problems lend themselves to either permutations or combinations.

# 9.2 Exercises

*Evaluate the given factorials, permutations, and combinations.*

**1.** $_4P_2$            **2.** $3!$            **3.** $_8C_5$

**4.** $7!$            **5.** $_8P_1$            **6.** $_7C_2$

**7.** $4!$            **8.** $_4P_4$            **9.** $_9C_6$

**10.** $_8C_2$            **11.** $_{13}P_3$            **12.** $_9P_5$

*Use a calculator to find values for Exercises 13–20.*

**13.** $_{25}P_5$            **14.** $_{40}P_5$            **15.** $_{14}P_5$

**16.** $_{17}P_8$            **17.** $_{18}C_5$            **18.** $_{32}C_9$

**19.** $_{28}C_{14}$            **20.** $_{35}C_{30}$

**21.** Some students find it puzzling that $0! = 1$, and they think that $0!$ should equal 0. If this were true, what would be the value of $_4P_4$ according to the permutations formula?

**22.** If you already knew the value of $8!$, how could you find the value of $9!$ quickly?

*Use the multiplication principle to solve the given problems. (See Examples 1–6.)*

**23.** **Tile Project** As of May 2017, the Lowe's home improvement store website listed 264 different kinds of glass wall tile, 31 colors of acrylic grout, 12 kinds of mortar trowels, and 17 different kinds of mortar. How many ways are there to buy the products needed for a tile project?

**24.** **Heating and Cooling** How many different heating–cooling units are possible if a home owner has 3 choices for the efficiency rating of the furnace, 3 options for the fan speed, and 6 options for the air condenser?

**25.** **Auto Manufacturing** An auto manufacturer produces 6 models, each available in 8 different colors, with 4 different upholstery fabrics and 3 interior colors. How many varieties of the auto are available?

**26.** **Radio Call Letters** How many different 4-letter radio station call letters can be made

(a) if the first letter must be K or W and no letter may be repeated?

(b) if repeats are allowed (but the first letter still must be K or W)?

(c) How many of the 4-letter call letters (starting with K or W) with no repeats end in R?

**27.** **Social Security Numbers** A Social Security number has 9 digits. How many Social Security numbers are possible? The U.S. population in 2016 was approximately 323 million. Was it possible for every U.S. resident to have a unique Social Security number? (Assume no restrictions.)

**28.** **Zip Codes** The United States Postal Service currently uses 5-digit zip codes in most areas. How many zip codes are possible if there are no restrictions on the digits used? How many would be possible if the first number could not be 0?

**29.** **Zip Codes** The Postal Service is encouraging the use of 9-digit zip codes in some areas, adding 4 digits after the usual 5-digit code. How many such zip codes are possible with no restrictions?

**30.** **License Plates** For many years, the state of California used 3 letters followed by 3 digits on its automobile license plates.

(a) How many different license plates are possible with this arrangement?

(b) When the state ran out of new numbers, the order was reversed to 3 digits followed by 3 letters. How many new license plate numbers were then possible?

(c) Several years ago, the numbers described in part (b) were also used up. The state then issued plates with 1 letter followed by 3 digits and then 3 letters. How many new license plate numbers will this arrangement provide?

**31.** **Hair Products** A recent trip to the drug store revealed 12 different kinds of Pantene® shampoo and 10 different kinds of Pantene® conditioner. How many ways can Sherri buy 1 Pantene® shampoo and 1 Pantene® conditioner?

**32. Sales Route** A pharmaceutical salesperson has 6 doctors' offices to call on.

(a) In how many ways can she arrange her schedule if she calls on all 6 offices?

(b) In how many ways can she arrange her schedule if she decides to call on 4 of the 6 offices?

**Phone Numbers** *The United States is rapidly running out of telephone numbers. In large cities, telephone companies have introduced new area codes as numbers are used up.*

**33.** (a) Until recently, all area codes had a 0 or a 1 as the middle digit and the first digit could not be 0 or 1. How many area codes were possible with this arrangement? How many telephone numbers does the current 7-digit sequence permit per area code? (The 3-digit sequence that follows the area code cannot start with 0 or 1. Assume that there are no other restrictions.)

(b) The actual number of area codes under the previous system was 152. Explain the discrepancy between this number and your answer to part (a).

**34.** The shortage of area codes under the previous system was avoided by removing the restriction on the second digit. How many area codes are available under this new system?

**35.** A problem with the plan in Exercise 34 was that the second digit in the area code had been used to tell phone company equipment that a long-distance call was being made. To avoid changing all equipment, an alternative plan proposed a 4-digit area code and restricted the first and second digits as before. How many area codes would this plan have provided?

**36.** Still another solution to the area-code problem is to increase the local dialing sequence to 8 digits instead of 7. How many additional numbers would this plan create? (Assume the same restrictions.)

**37.** Define permutation in your own words.

**Use permutations to solve each of the given problems. (See Examples 7–10.)**

**38.** A baseball team has 15 players. How many 9-player batting orders are possible?

**39.** Tim is a huge fan of the latest album by country-music singer Kenny Chesney. If Tim has time to listen to only 5 of the 12 songs on the album, how many ways can he listen to the 5 songs?

**40.** From a cooler with 8 cans of different kinds of soda, 3 are selected for 3 people. In how many ways can this be done?

**41.** The Greek alphabet has 24 letters. How many ways can one name a fraternity using 3 Greek letters (with no repeats)?

**42.** A customer speaks to his financial advisor about investment products. The advisor has 9 products available, but knows he will only have time to speak about 4. How many different ways can he give the details on 4 different investment products to the customer?

**43.** The student activity club at the college has 32 members. In how many different ways can the club select a president, a vice president, a treasurer, and a secretary?

**44.** A student can take only one class a semester, and she needs to take 4 more electives in any order. If there are 20 courses from which she can choose, how many ways can she take her 4 electives?

**45.** In a club with 17 members, how many ways can the club elect a president and a treasurer?

**Use combinations to solve each of the given problems. (See Examples 11–13.)**

**46. Manufacturing** Four items are to be randomly selected from the first 25 items on an assembly line in order to determine the defect rate. How many different samples of 4 items can be chosen?

**47. Student Selection** A group of 4 students is to be selected from a group of 10 students to take part in a class in cell biology.

(a) In how many ways can this be done?

(b) In HOW many ways can the group that will *not* take part be chosen?

**48. Study Design** From a group of 15 smokers and 21 nonsmokers, a researcher wants to randomly select 7 smokers and 6 nonsmokers for a study. In how many ways can the study group be selected?

**49. Co-Captain Selection** The college football team has 11 seniors. The team needs to elect a group of 4 senior co-captains. How many different 4-person groups of co-captains are possible?

**50. Drama Role Selection** The drama department holds auditions for a play with a cast of 5 roles. If 33 students audition, how many casts of 5 people are possible?

**51.** Explain the difference between a permutation and a combination.

**52.** Padlocks with digit dials are often referred to as "combination locks." According to the mathematical definition of combination, is this an accurate description? Explain.

**Exercises 53–70 are mixed problems that may require permutations, combinations, or the multiplication principle. (See Examples 14–18.)**

**53.** Use a tree diagram to find the number of ways 2 letters can be chosen from the set $\{P, Q, R\}$ if order is important and

(a) if repetition is allowed;

(b) if no repeats are allowed.

(c) Find the number of combinations of 3 elements taken 2 at a time. Does this answer differ from that in part (a) or (b)?

**54.** Repeat Exercise 53, using the set $\{P, Q, R, S\}$ and 4 in place of 3 in part (c).

**55. Delegation Selection** The U.S. Senate Foreign Relations Committee in 2017 had 10 Democrats and 11 Republicans. A delegation of 5 people is to be selected to visit Iraq.

(a) How many delegations are possible?

(b) How many delegations would have all Republicans?

(c) How many delegations would have 3 Democrats and 2 Republicans?

(d) How many delegations would have at least one Democrat?

**56. Plant Research** In an experiment on plant hardiness, a researcher gathers 6 wheat plants, 5 barley plants, and 3 rye plants. She wishes to select 4 plants at random.

(a) In how many ways can this be done?

(b) In how many ways can this be done if exactly 2 wheat plants must be included?

**57. Ice Cream** According to the Baskin-Robbins® Web site, there are 25 "classic flavors" of ice cream.

(a) How many different double-scoop cones (of 2 different flavors) can be made if order does not matter (for example, putting chocolate on top of vanilla is equivalent to putting vanilla on top of chocolate)?

(b) How many different triple-scoop (of 3 different flavors) cones can be made if order does matter?

**58. Finance** A financial advisor offers 8 mutual funds in the high-risk category, 7 in the moderate-risk category, and 10 in the low-risk category. An investor decides to invest in 3 high-risk funds, 4 moderate-risk funds, and 3 low-risk funds. How many ways can the investor do this?

**59. Lottery Game** A lottery game requires that you pick 6 different numbers from 1 to 99. If you pick all 6 winning numbers, you win $4 million.

(a) How many ways are there to choose 6 numbers if order is not important?

(b) How many ways are there to choose 6 numbers if order matters?

**60. Lottery Game** In Exercise 59, if you pick 5 of the 6 numbers correctly, you win $5,000. In how many ways can you pick exactly 5 of the 6 winning numbers without regard to order?

**61.** The game of Set\* consists of a special deck of cards. Each card has on it either one, two, or three shapes. The shapes on each card are all the same color, either green, purple, or red. The shapes on each card are the same style, either solid, shaded, or outline. There are three possible shapes—squiggle, diamond, and oval—and only one type of shape appears on a card. The deck consists of all possible combinations of shape, color, style, and number. How many cards are in a deck?

**62. Patient Care** Over the course of the previous nursing shift, 16 new patients were admitted onto a hospital ward. If a nurse begins her shift by caring for 6 of the new patients, how many possible ways could the 6 patients be selected from the 16 new arrivals?

**63. Insect Classification** A biologist is attempting to classify 52,000 species of insects by assigning 3 initials to each species. Is it possible to classify all the species in this way? If not, how many initials should be used?

**64. Lottery Game** One play in a state lottery consists of choosing 6 numbers from 1 to 44. If your 6 numbers are drawn (in any order), you win the jackpot.

(a) How many possible ways are there to draw the 6 numbers?

(b) If you get 2 plays for a dollar, how much would it cost to guarantee that one of your choices would be drawn?

(c) Assuming that you work alone and can fill out a betting ticket (for 2 plays) every second, and assuming that the lotto drawing will take place 3 days from now, can you place enough bets to guarantee that 1 of your choices will be drawn?

**65.** A cooler contains 5 cans of Pepsi®, 1 can of Diet Coke®, and 3 cans of 7UP®; you pick 3 cans at random. How many samples are possible in which the soda cans picked are

(a) only Pepsi;

(b) only Diet Coke;

(c) only 7UP;

(d) 2 Pepsi, 1 Diet Coke;

(e) 2 Pepsi, 1 7UP;

(f) 2 7UP, 1 Pepsi;

(g) 2 Diet Coke, 1 7UP?

**66.** A class has 9 male students and 8 female students. How many ways can the class select a committee of four people to petition the teacher not to make the final exam cumulative if the committee has to have 2 males and 2 females?

**67. Study Enrollment** A hospital wants to test the viability of a new medication for attention deficit disorder. It has 35 adults volunteer for the study, but can only enroll 20 in the study. How many ways can it choose the 20 volunteers to enroll in the study?

**68. Pizza Making** Suppose a pizza shop offers 4 choices of cheese and 9 toppings. If the order of the cheeses and toppings does not matter, how many different pizza selections are possible when choosing 2 cheeses and 2 toppings?

*If the n objects in a permutations problem are not all distinguishable—that is, if there are $n_1$ of type 1, $n_2$ of type 2, and so on, for r different types—then the number of distinguishable permutations is*

$$\frac{n!}{n_1!\, n_2! \cdots n_r!}.$$

**Example** *In how many ways can you arrange the letters in the word Mississippi?*

*This word contains 1 m, 4 i's, 4 s's, and 2 p's. To use the formula, let $n = 11$, $n_1 = 1$, $n_2 = 4$, $n_3 = 4$, and $n_4 = 2$ to get*

$$\frac{11!}{1!\,4!\,4!\,2!} = 34,650$$

*arrangements. The letters in a word with 11 different letters can be arranged in $11! = 39{,}916{,}800$ ways.*

**69.** Find the number of distinguishable permutations of the letters in each of the given words.

(a) martini

(b) nunnery

(c) grinding

---

\*SET is a registered trademark of Cannei, LLC.

**70.** A printer has 5 $W$'s, 4 $X$'s, 3 $Y$'s, and 2 $Z$'s. How many different "words" are possible that use all these letters? (A "word" does not have to have any meaning here.)

**71.** Shirley is a shelf stocker at the local grocery store. She has 4 varieties of Stouffer's® frozen dinners, 3 varieties of Lean Cuisine® frozen dinners, and 5 varieties of Weight Watchers® frozen dinners. In how many distinguishable ways can she stock the shelves if

   **(a)** the dinners can be arranged in any order?

   **(b)** dinners from the same company are considered alike and have to be shelved together?

   **(c)** dinners from the same company are considered alike, but do not have to be shelved together?

**72.** A child has a set of different-shaped plastic objects. There are 2 pyramids, 5 cubes, and 6 spheres. In how many ways can she arrange them in a row

   **(a)** if they are all different colors?

   **(b)** if the same shapes must be grouped, but each object is a different color?

   **(c)** In how many distinguishable ways can they be arranged in a row if objects of the same shape are also the same color, but need not be grouped?

✓ **Checkpoint Answers**

**1.** 1,945,240,782

**2.** **(a)** 720     **(b)** 24

**3.** **(a)** 24     **(b)** 720     **(c)** 1     **(d)** 30

**4.** 30

**5.** $8 \cdot 7 \cdot 6 \cdot 5 = 1680$

**6.** **(a)** 20     **(b)** 504     **(c)** 3     **(d)** 210     **(e)** 132

**7.** **(a)** 120     **(b)** 12     **(c)** 6

**8.** **(a)** 15     **(b)** 70     **(c)** 1

**9.** **(a)** 15     **(b)** 70     **(c)** 1

**10.** **(a)** 15,504     **(b)** 816     **(c)** 21,699

**11.** **(a)** 5040     **(b)** 455     **(c)** 28     **(d)** 360

**12.** **(a)** 5040     **(b)** 840

**13.** 175,223,510

**14.** 1260

# 9.3 Applications of Counting

Ⓒ Many of the probability problems involving *dependent* events that were solved with probability trees in Chapter 8 can also be solved by using counting principles—that is, permutations and combinations. Permutations and combinations are especially helpful when the number of choices is large. The use of counting rules to solve probability problems depends on the basic probability principle introduced in Section 8.3 and repeated here.

If event $E$ is a subset of sample space $S$ with equally likely outcomes then the probability that event $E$ occurs, written $P(E)$, is

$$P(E) = \frac{n(E)}{n(S)}.$$

**EXAMPLE 1**   **Jury Selection**   From a potential jury pool with 1 Hispanic, 3 Caucasian, and 2 African-American members, 2 jurors are selected one at a time without replacement. Find the probability that 1 Caucasian and 1 African-American are selected.

**Solution**   In Example 6 of Section 8.5, it was necessary to consider the order in which the jurors were selected. With combinations, it is not necessary: Simply count the number of ways in which 1 Caucasian and 1 African-American juror can be selected. The Caucasian can be selected in $_3C_1$ ways, and the African-American juror can be selected in $_2C_1$ ways. By the multiplication principle, both results can occur in

$$_3C_1 \cdot {_2C_1} = 3 \cdot 2 = 6 \text{ ways,}$$

giving the numerator of the probability fraction. For the denominator, 2 jurors are selected from a total of 6 candidates. This can occur in $_6C_2 = 15$ ways. The required probability is

$$P(1 \text{ Caucasian and 1 African American}) = \frac{_3C_1 \cdot {_2C_1}}{_6C_2} = \frac{3 \cdot 2}{15} = \frac{6}{15} = \frac{2}{5} = .40.$$

This result agrees with the answer found earlier.

**EXAMPLE 2**    **Grievance Committee** From a baseball team of 15 players, 4 are to be selected to present a list of grievances to the coach.

**(a)** In how many ways can this be done?

**Solution** Four players from a group of 15 can be selected in $_{15}C_4$ ways. (Use combinations, since the order in which the group of 4 is selected is unimportant.) So,

$$_{15}C_4 = \frac{15!}{4!\ 11!} = \frac{15(14)(13)(12)}{4(3)(2)(1)} = 1365.$$

There are 1365 ways to choose 4 players from 15.

**(b)** One of the players is Michael Branson. Find the probability that Branson will be among the 4 selected.

**Solution** The probability that Branson will be selected is the number of ways the chosen group includes him, divided by the total number of ways the group of 4 can be chosen. If Branson must be one of the 4 selected, the problem reduces to finding the number of ways the additional 3 players can be chosen. There are 3 chosen from 14 players; this can be done in

$$_{14}C_3 = \frac{14!}{3!\ 11!} = 364$$

ways. The number of ways 4 players can be selected from 15 is

$$n = {}_{15}C_4 = 1365.$$

The probability that Branson will be one of the 4 chosen is

$$P(\text{Branson is chosen}) = \frac{364}{1365} \approx .2667.$$

**(c)** Find the probability that Branson will not be selected.

**Solution** The probability that he will not be chosen is $1 - .2667 = .7333$. ✓₁

✓ **Checkpoint 1**

The ski club has 8 women and 7 men. What is the probability that if the club elects 3 officers at random, all 3 of them will be women?

**EXAMPLE 3**    **Quality Control** A manufacturing company performs a quality-control analysis on the ceramic tile it produces. It produces the tile in batches of 24 pieces. In the quality-control analysis, the company tests 3 pieces of tile per batch. Suppose a batch of 24 tiles has 4 defective tiles.

**(a)** What is the probability that exactly 1 of the 3 tested tiles is defective?

**Solution** Let $P(1\ \text{defective})$ represent the probability of there being exactly 1 defective tile among the 3 tested tiles. To find this probability, we need to know how many ways we can select 3 tiles for testing. Since order does not matter, there are $_{24}C_3$ ways to choose 3 tiles:

$$_{24}C_3 = \frac{24!}{21!\ 3!} = \frac{24 \cdot 23 \cdot 22}{3 \cdot 2 \cdot 1} = 2024.$$

There are $_4C_1$ ways of choosing 1 defective tile from the 4 in the batch. If we choose 1 defective tile, we must then choose 2 good tiles among the 20 good tiles in the batch. We can do this in $_{20}C_2$ ways. By the multiplication principle, there are

$$_4C_1 \cdot {}_{20}C_2 = \frac{4!}{3!1!} \cdot \frac{20!}{18!\ 2!} = 4 \cdot 190 = 760$$

ways to choose exactly 1 defective tile.
   Thus,

$$P(1\ \text{defective}) = \frac{760}{2024} \approx .3755.$$

**(b)** If at least one of the tiles in a batch is defective, the company will not ship the batch. What is the probability that the batch is not shipped?

**Solution** The batch will not be shipped if 1, 2, or 3 of the tiles sampled are defective. We already found the probability of there being exactly 1 defective tile in part (a). We now need to find $P(2\text{ defective})$ and $P(3\text{ defective})$. To find $P(2\text{ defective})$, we need to count the number of ways to choose 2 from the 4 defective tiles in the batch and choose 1 from the 20 good tiles in the batch:

$$_4C_2 \cdot {}_{20}C_1 = \frac{4!}{2!\,2!} \cdot \frac{20!}{19!\,1!} = 6 \cdot 20 = 120.$$

To find $P(3\text{ defective})$, we need to count the number of ways to choose 3 from the 4 defective tiles in the batch and choose 0 from the 20 good tiles in the batch:

$$_4C_3 \cdot {}_{20}C_0 = \frac{4!}{1!\,3!} \cdot \frac{20!}{20!\,0!} = 4 \cdot 1 = 4.$$

We now have

$$P(2\text{ defective}) = \frac{120}{2024} \approx .0593 \text{ and } P(3\text{ defective}) = \frac{4}{2024} \approx .0020.$$

Thus, the probability of rejecting the batch because 1, 2, or 3 tiles are defective is

$$P(1\text{ defective}) + P(2\text{ defective}) + P(3\text{ defective}) \approx .3755 + .0593 + .0020$$
$$= .4368.$$

**(c)** Use the complement rule to find the probability the batch will be rejected.

**Solution** We reject the batch if at least 1 of the sampled tiles is defective. The opposite of at least 1 tile being defective is that none are defective. We can find the probability that none of the 3 sampled tiles is defective by choosing 0 from the 4 defective tiles and choosing 3 from the 20 good tiles:

$$_4C_0 \cdot {}_{20}C_3 = 1 \cdot 1140 = 1140.$$

Therefore, the probability that none of the sampled tiles is defective is

$$P(0\text{ defective}) = \frac{1140}{2024} \approx .5632.$$

Using the complement rule, we have

$$P(\text{at least 1 defective}) \approx 1 - .5632 = .4368,$$

the same answer as in part (b). Using the complement rule can often save time when multiple probabilities need to be calculated for problems involving "at least 1." ✔️2

✓ **Checkpoint 2**

A batch of 15 granite slabs is mined, and 4 have defects. If the manager spot-checks 3 slabs at random, what is the probability that at least 1 slab is defective?

**EXAMPLE 4    Poker** In a common form of 5-card draw poker, a hand of 5 cards is dealt to each player from a deck of 52 cards. (For a review of a standard deck, see Figure 8.20 in Section 8.3.) There is a total of

$$_{52}C_5 = \frac{52!}{5!\,47!} = 2,598,960$$

such hands possible. Find the probability of being dealt each of the given hands.

**(a)** Heart-flush hand (5 hearts)

**Solution** There are 13 hearts in a deck; there are

$$_{13}C_5 = \frac{13!}{5!\,8!} = \frac{13(12)(11)(10)(9)}{5(4)(3)(2)(1)} = 1287$$

different hands containing only hearts. The probability of a heart flush is

$$P(\text{heart flush}) = \frac{1287}{2{,}598{,}960} \approx .000495.$$

**(b)** A flush of any suit (5 cards, all from 1 suit)

**Solution** There are 4 suits to a deck, so

$$P(\text{flush}) = 4 \cdot P(\text{heart flush}) = 4(.000495) \approx .00198.$$

**(c)** A full house of aces and eights (3 aces and 2 eights)

**Solution** There are $_4C_3$ ways to choose 3 aces from among the 4 in the deck and $_4C_2$ ways to choose 2 eights, so

$$P(3 \text{ aces}, 2 \text{ eights}) = \frac{_4C_3 \cdot {}_4C_2}{_{52}C_5} \approx .00000923.$$

**(d)** Any full house (3 cards of one value, 2 of another)

**Solution** There are 13 values in a deck, so there are 13 choices for the first value mentioned, leaving 12 choices for the second value. (Order *is* important here, since a full house of aces and eights, for example, is not the same as a full house of eights and aces.)

$$P(\text{full house}) = \frac{13 \cdot {}_4C_3 \cdot 12 \cdot {}_4C_2}{_{52}C_5} \approx .00144. \quad \boxed{\checkmark}_3$$

✓**Checkpoint 3**

Find the probability of being dealt a poker hand (5 cards) with 4 kings.

---

**EXAMPLE 5** **Soda Selection** A cooler contains 8 different kinds of soda, among which 3 cans are Pepsi®, Coke®, and Sprite®. What is the probability, when picking at random, of selecting the 3 cans in the particular order listed in the previous sentence?

**Solution** Use permutations to find the number of arrangements in the sample, because order matters:

$$n = {}_8P_3 = 8(7)(6) = 336.$$

Since each can is different, there is only 1 way to choose Pepsi, Coke, and Sprite in that order, so the probability is

$$\frac{1}{336} \approx .0030. \quad \boxed{\checkmark}_4$$

✓**Checkpoint 4**

Martha, Leonard, Calvin, and Sheila will be handling the officer duties of president, vice president, treasurer, and secretary.

**(a)** If the offices are assigned randomly, what is the probability that Calvin is the president?

**(b)** If the offices are assigned randomly, what is the probability that Sheila is president, Martha is vice president, Calvin is treasurer, and Leonard is secretary?

---

**EXAMPLE 6** **Birthday Matches** Suppose a group of 5 people is in a room. Find the probability that at least 2 of the people have the same birthday.

**Solution** "Same birthday" refers to the month and the day, not necessarily the same year. Also, ignore leap years, and assume that each day in the year is equally likely as a birthday. First find the probability that *no 2 people* among 5 people have the same birthday. There are 365 different birthdays possible for the first of the 5 people, 364 for the second (so that the people have different birthdays), 363 for the third, and so on. The number of ways the 5 people can have different birthdays is thus the number of permutations of 365 things (days) taken 5 at a time, or

$$_{365}P_5 = 365 \cdot 364 \cdot 363 \cdot 362 \cdot 361.$$

The number of ways that the 5 people can have the same or different birthdays is

$$365 \cdot 365 \cdot 365 \cdot 365 \cdot 365 = (365)^5.$$

Finally, the *probability* that none of the 5 people have the same birthday is

$$\frac{{}_{365}P_5}{(365)^5} = \frac{365 \cdot 364 \cdot 363 \cdot 362 \cdot 361}{365 \cdot 365 \cdot 365 \cdot 365 \cdot 365} \approx .973.$$

The probability that at least 2 of the 5 people *do* have the same birthday is $1 - .973 = .027$.

Example 6 can be extended for more than 5 people. In general, the probability that no 2 people among *n* people have the same birthday is

$$\frac{{}_{365}P_n}{(365)^n}.$$

The probability that at least 2 of the *n* people *do* have the same birthday is

$$1 - \frac{{}_{365}P_n}{(365)^n}. \quad ☑ \boxed{5}$$

The following table shows this probability for various values of *n*:

✓ **Checkpoint 5**

Evaluate $1 - \dfrac{{}_{365}P_n}{(365)^n}$ for

**(a)** $n = 3$;

**(b)** $n = 6$.

| Number of People, *n* | Probability That at Least 2 Have the Same Birthday |
|:---:|:---:|
| 5 | .027 |
| 10 | .117 |
| 15 | .253 |
| 20 | .411 |
| 22 | .476 |
| **23** | **.507** |
| 25 | .569 |
| 30 | .706 |
| 35 | .814 |
| 40 | .891 |
| 50 | .970 |

✓ **Checkpoint 6**

Set up (but do not calculate) the probability that at least 2 of the 9 members of the Supreme Court have the same birthday.

**Figure 9.11**

The probability that 2 people among 23 have the same birthday is .507, a little more than half. Many people are surprised at this result; somehow it seems that a larger number of people should be required. ☑ \boxed{6}

Using a graphing calculator, we can graph the probability formula in the previous example as a function of *n*, but the graphing calculator must be set to evaluate the function at integer points. Figure 9.11 was produced on a TI-84+ by letting $Y_1 = 1 - (365 \, \text{nPr} \, X)/365^X$ on the interval $0 \leq x \leq 47$. (This domain ensures integer values for *x*.) Notice that the graph does not extend past $x = 39$. This is because $P(365, n)$ and $365^n$ are too large for the calculator when $n \geq 40$.

# 9.3 Exercises

**Business** *Suppose in Example 3 that the number of defective tiles in a batch of 24 is 5 rather than 4. If 3 tiles are sampled at random, the management would like to know the probability that*

1. exactly 1 of the sampled tiles is defective;

2. the batch is rejected (that is, at least 1 sampled tile is defective).

**Computer Defects** *A shipment of 8 computers contains 3 with defects. Find the probability that a sample of the given size, drawn from the 8, will not contain a defective computer. (See Example 3.)*

**3.** 1

**4.** 2

**5.** 3

**6.** 5

**Radio Promotion** *A radio station runs a promotion at an auto show with a money box with 10 $100 tickets, 12 $50 tickets, and 20 $25 tickets. The box contains an additional 200 "dummy" tickets with no value. Three tickets are randomly drawn. Find the given probabilities. (See Examples 1 and 2.)*

7. All $100 tickets

8. All $50 tickets

9. Exactly two $25 tickets and no other money winners

10. One ticket of each money amount

11. No tickets with money

12. At least one money ticket

*Two cards are drawn at random from an ordinary deck of 52 cards. (See Example 4.)*

13. How many 2-card hands are possible?

*Find the probability that the 2-card hand in Exercise 13 contains the given cards.*

14. 2 kings

15. No deuces (2's)

16. 2 face cards

17. Different suits

18. At least 1 black card

19. No more than 1 diamond

20. Discuss the relative merits of using probability trees versus combinations to solve probability problems. When would each approach be most appropriate?

21. Several examples in this section used the rule $P(E') = 1 - P(E)$. Explain the advantage (especially in Example 6) of using this rule.

**Geology** *A shipment contains 8 igneous, 7 sedimentary, and 7 metamorphic rocks. If we select 5 rocks at random, find the probability that*

22. all 5 are igneous;

23. exactly 3 are sedimentary;

24. only 1 is metamorphic.

25. **Lottery Game** In Exercise 59 in Section 9.2, we found the number of ways to pick 6 different numbers from 1 to 99 in a state lottery.

    (a) Assuming that order is unimportant, what is the probability of picking all 6 numbers correctly to win the big prize?

    (b) What is the probability if order matters?

26. **Lottery Game** In Exercise 25 (a), what is the probability of picking exactly 5 of the 6 numbers correctly?

27. Example 16 in Section 9.2 shows that the probability of winning the Powerball lottery is 1/292,201,338. If Juanita and Michelle each play Powerball on one particular evening, what is the probability that both will select the winning numbers if they make their selections independently of each other?

28. **Cell Phone Manuracturing** A cell phone manufacturer randomly selects 5 phones from every batch of 50 produced. If at least one of the phones is found to be defective, then each phone in the batch is tested individually. Find the probability that the entire batch will need testing if the batch contains

    (a) 8 defective phones;

    (b) 2 defective phones.

29. In 2017, the House of Representatives Financial Services committee had 60 members, of which 9 were women. If 8 members are selected at random, find the probability that the group of 8 would be composed as follows:

    (a) 7 men and 1 woman;

    (b) 5 men and 3 women;

    (c) at least one woman.

30. **Car Sales** A car dealership has 8 red, 9 silver, and 5 black cars on the lot. Ten cars are randomly chosen to be displayed in front of the dealership. Find the probability that

    (a) 4 are red and the rest are silver;

    (b) 5 are red and 5 are black;

    (c) exactly 8 are red.

31. **Study Volunteers** Twenty subjects volunteer for a study of a new cold medicine. Ten of the volunteers are ages 20–39, 8 are ages 40–59, and 2 are age 60 or older. If we select 7 volunteers at random, find the probability that

    (a) all the volunteers selected are ages 20–39;

    (b) 5 of the volunteers are ages 20–39 and 2 are age 60 or older;

    (c) exactly 3 of the volunteers are ages 40–59.

**Birthday Problem** *For Exercises 32–34, refer to Example 6 in this section.*

32. Set up the probability that at least 2 of the 44 men who have served as president of the United States have had the same birthday.*

33. Set up the probability that at least 2 of the 100 U.S. senators have the same birthday.

34. Set up the probability that at least 2 of the 50 U.S. governors have the same birthday.

**Lottery Game** *One version of the New York State lottery game Quick Draw has players selecting 4 numbers at random from the numbers 1–80.*

---

*In fact, James Polk and Warren Harding were both born on November 2. Although Donald Trump is the 45th president, the 22nd and 24th presidents were the same man: Grover Cleveland.

*The state picks 20 winning numbers. If the player's 4 numbers are selected by the state, the player wins $55. (Data from: www.nylottery.org.)*

**35.** What is the probability of winning?

**36.** If the state picks 3 of the player's numbers, the player wins $5. What is the probability of winning $5?

**37.** What is the probability of having none of your 4 numbers selected by the state?

**38.** What is the probability of having the state pick 2 of the player's numbers?

✓ **Checkpoint Answers**

1. About .1231

2. About .6374

3. .00001847

4. (a) 1/4      (b) 1/24

5. (a) .008      (b) .040

6. $1 - {}_{365}P_9/365^9$

# 9.4 Binomial Probability

In Section 9.1, we learned about probability distributions where we listed each outcome and its associated probability. After learning in Sections 9.2 and 9.3 how to count the number of possible outcomes, we are now ready to understand a special probability distribution known as the *binomial distribution.* This distribution occurs when the same experiment is repeated many times and each repetition is independent of previous ones. One outcome is designated a success and any other outcome is considered a failure. For example, you might want to find the probability of rolling 8 twos in 12 rolls of a die (rolling two is a success; rolling anything else is a failure). The individual trials (rolling the die once) are called **Bernoulli trials**, or **Bernoulli processes**, after the Swiss mathematician Jakob Bernoulli (1654–1705).

If the probability of a success in a single trial is $p$, then the probability of failure is $1 - p$ (by the complement rule). When Bernoulli trials are repeated a fixed number of times, and each trial's outcome is independent of the other trials' outcomes, the resulting distribution of outcomes is called a **binomial distribution**, or a **binomial experiment**. A binomial experiment must satisfy the following conditions.

## Binomial Experiment

**1.** The same experiment is repeated a fixed number of times.

**2.** There are only two possible outcomes: success and failure.

**3.** The probability of success for each trial is constant.

**4.** The repeated trials are independent.

The basic characteristics of binomial experiments are illustrated by a recent look at the New York Stock Exchange (NYSE). On May 23, 2017, the probability that a stock advanced (rather than declined or stayed at the same value) was .57. (Data from: www.wsj.com.) We use $Y$ to denote the event that a stock on the exchange advanced, and $N$ to denote the event that the stock did not advance. If we sample 5 stocks at random and use .57 as the probability for $Y$, we will generate a binomial experiment because all of the requirements are satisfied:

The sampling is repeated a fixed number of times (5);

there are only two outcomes of interest ($Y$ or $N$);

the probability of success is constant ($p = .57$);

"at random" guarantees that the trials are independent.

To calculate the probability that all 5 randomly chosen stocks advance, we use the product rule for independent events (Section 8.5) and $P(Y) = .57$ to obtain

$$P(YYYYY) = P(Y) \cdot P(Y) \cdot P(Y) \cdot P(Y) \cdot P(Y) = (.57)^5 \approx .0602.$$

Determining the probability that 4 out of 5 stocks advanced is slightly more complicated. The stock that did not advance could be the first, second, third, fourth, or fifth stock chosen. So we have the following possible outcomes:

$$
\begin{array}{c}
\text{N Y Y Y Y} \\
\text{Y N Y Y Y} \\
\text{Y Y N Y Y} \\
\text{Y Y Y N Y} \\
\text{Y Y Y Y N}
\end{array}
$$

So the total number of ways in which 4 successes (and 1 failure) can occur is 5, which is the number ${}_5C_4$. The probability of each of these 5 outcomes is

$$P(Y) \cdot P(Y) \cdot P(Y) \cdot P(Y) \cdot P(N) = (.57)^4 (1 - .57)^1 = (.57)^4 (.43).$$

Since the 5 outcomes where there are 4 $Y$s and one $N$ represent disjoint events, we multiply by ${}_5C_4 = 5$:

$$P(4 \ Y\text{s out of 5 trials}) = {}_5C_4 (.57)^4 (.43)^{5-4} = 5(.57)^4 (.43)^1 \approx .2270.$$

The probability of obtaining exactly 3 $Y$s and 2 $N$s can be computed in a similar way. The probability of any one way of achieving 3 $Y$s and 2 $N$s will be

$$(.57)^3 (.43)^2.$$

Again, the desired outcome can occur in more than one way. Using combinations, we find that the number of ways in which 3 $Y$s and 2 $N$s can occur is ${}_5C_3 = 10$. So we have

$$P(3 \ Y\text{s out of 5 trials}) = {}_5C_3 (.57)^3 (.43)^{5-3} = 10(.57)^3 (.43)^2 \approx .3424.$$

✓ **Checkpoint 1**

Find the probability of obtaining

(a) exactly 2 stocks advance;

(b) exactly 1 stock advances;

(c) exactly no stocks advance.

With the probabilities just generated and the answers to Checkpoint 1, we can write the probability distribution for the number of stocks that advance when 5 stocks are selected at random:

| $x$ | 0 | 1 | 2 | 3 | 4 | 5 |
|---|---|---|---|---|---|---|
| $P(x)$ | .0147 | .0974 | .2583 | .3424 | .2270 | .0602 |

When the outcomes and their associated probabilities are written in this form, it is very easy to calculate answers to questions such as, What is the probability that 3 or more stocks advance? We see from the table that

$$P(3 \text{ or more } Y\text{s}) = .3424 + .2270 + .0602 = .6296.$$

Similarly, the probability of one or fewer stocks advancing is

$$P(1 \text{ or fewer } Y\text{s}) = .0974 + .0147 = .1121.$$

The example illustrates the following fact.

**TECHNOLOGY TIP**

On the TI-84+ calculator, use "binompdf($n,p,x$)" in the DISTR menu to compute the probability of exactly $x$ successes in $n$ trials (where $p$ is the probability of success in a single trial). Use "binomcdf($n,p,x$)" to compute the probability of at most $x$ successes in $n$ trials. Figure 9.12 shows the probability of exactly 3 successes in 5 trials and the probability of at most 3 successes in 5 trials, with the probability of success set at .57 for each case.

```
binompdf(5,.57,3)
            .342421857
binomcdf(5,.57,3)
            .7128767728
```

**Figure 9.12**

## Binomial Probability

If $p$ is the probability of success in a single trial of a binomial experiment, the probability of $x$ successes and $n - x$ failures in $n$ independent repeated trials of the experiment is

$$_nC_x p^x (1 - p)^{n-x}.$$

**EXAMPLE 1**   **Business** According to a 2016 study conducted by the Society for Human Resource Management, 38% of U.S. employees reported being very satisfied with their job. Suppose a random sample of 6 employees is chosen. Find the probability of the given scenarios. (Data from: www.shrm.org.)

**(a)** Exactly 4 of the 6 employees were very satisfied.

**Solution** We can think of the 6 employees as independent trials, and a success occurs if a worker is very satisfied with their job. This is a binomial experiment with $p = .38, n = 6$, and $x = 4$. By the binomial probability rule,

$$P(\text{exactly } 4) = {}_6C_4(.38)^4(1 - .38)^{6-4}$$
$$= 15(.38)^4(.62)^2$$
$$\approx .1202.$$

**(b)** All 6 employees are very satisfied with their job.

**Solution** Let $x = 6$. Then we have

$$P(\text{exactly } 6) = {}_6C_6(.38)^6(1 - .38)^{6-6}$$
$$= 1(.38)^6(.62)^0$$
$$\approx .0030.$$

✓ **Checkpoint 2**

According to the study in Example 1, 65% of employees indicated that respectful treatment of all employees was a very important contributor to their job satisfaction. If 4 employees are selected at random, find the probability that exactly the given number feel respectful treatment of all employees contributes to their job satisfaction.

**(a)** 1 of the 4;

**(b)** 3 of the 4.

**EXAMPLE 2**    Suppose a family has 3 children.

**(a)** Find the probability distribution for the number of girls.

**Solution** Let $x = $ the number of girls in three births. According to the binomial probability rule, the probability of exactly one girl being born is

$$P(x = 1) = {}_3C_1\left(\frac{1}{2}\right)^1\left(\frac{1}{2}\right)^2 = 3\left(\frac{1}{2}\right)^3 = \frac{3}{8}.$$

The other probabilities in this distribution are found similarly, as shown in the following table:

| $x$ | 0 | 1 | 2 | 3 |
|---|---|---|---|---|
| $P(x)$ | ${}_3C_0\left(\frac{1}{2}\right)^0\left(\frac{1}{2}\right)^3 = \frac{1}{8}$ | ${}_3C_1\left(\frac{1}{2}\right)^1\left(\frac{1}{2}\right)^2 = \frac{3}{8}$ | ${}_3C_2\left(\frac{1}{2}\right)^2\left(\frac{1}{2}\right)^1 = \frac{3}{8}$ | ${}_3C_3\left(\frac{1}{2}\right)^3\left(\frac{1}{2}\right)^0 = \frac{1}{8}$ |

**(b)** Find the expected number of girls in a 3-child family.

**Solution** For a binomial distribution, we can use the following method (which is presented here with a "plausibility argument," but not a full proof): Because 50% of births are girls, it is reasonable to expect that 50% of a sample of children will be girls. Since 50% of 3 is $3(.50) = 1.5$, we conclude that the expected number of girls is 1.5.

✓ **Checkpoint 3**

Find the probability of getting 2 fours in 8 tosses of a die.

The expected value in Example 2(b) was the product of the number of births and the probability of a single birth being a girl—that is, the product of the number of trials and the probability of success in a single trial. The same conclusion holds in the general case.

## Expected Value for a Binomial Distribution

When an experiment meets the four conditions of a binomial experiment with $n$ fixed trials and constant probability of success $p$, the expected value is

$$E(x) = np.$$

**EXAMPLE 3**   **Student Loan Defaults**  In September 2016, the U.S. Department of Education released the latest data on student loan defaults. The national student loan default rate for the 2013 national cohort of students was 11.3%. If we select 10 students at random, find the probability of the following events.

**(a)**  Exactly 3 of the students defaulted on their student loan.

**Solution**  The experiment is repeated 10 times with defaulting on the loan considered a success. The probability of success is .113. Because the selection is done at random, the trials are considered independent. Thus, we have a binomial experiment, and

$$P(x = 3) = {}_{10}C_3(.113)^3(1 - .113)^7 = 120(.113)^3(.887)^7 \approx .0748.$$

**(b)**  At most, 3 students defaulted on their student loan.

**Solution**  "At most 3" means 0, 1, 2, or 3 successes. We have already calculated the probability of exactly 3 successes in part (a), so we need to calculate the probability of 0, 1, or 2 successes. We can then add those probabilities because they are disjoint events. We have

$$P(x = 0) = {}_{10}C_0(.113)^0(.887)^{10} \approx .3015;$$
$$P(x = 1) = {}_{10}C_1(.113)^1(.887)^9 \approx .3841;$$
$$P(x = 2) = {}_{10}C_2(.113)^2(.887)^8 \approx .2202.$$

Thus, using our answer to part (a) and the above, we have

$$P(\text{at most } 3) \approx .3015 + .3841 + .2202 + .0748 = .9806.$$

**(c)**  The expected number of students who defaulted.

**Solution**  Because this is a binomial experiment, we can use the formula $E(x) = np = 10(.113) = 1.13$. In repeated samples of 10 student borrowers, the average number of those who defaulted is 1.13. ✓ 4

**EXAMPLE 4**   **New Business Failures**  According to the U.S. Bureau of Labor Statistics, about 33% (.33) of new businesses fail within two years.

**(a)**  If 7 new businesses are selected at random, find the probability that at least 3 will fail within two years.

**Solution**  Because the businesses are selected at random, we can treat this problem as a binomial experiment, with $n = 7, p = .33$, and $x$ representing the number of new businesses that fail. "At least 3 of 7" means 3 or 4 or 5, and so on, up to all 7. Thus, we have

$$P(\text{at least } 3) = {}_7C_3(.33)^3(.67)^4 + {}_7C_4(.33)^4(.67)^3 + {}_7C_5(.33)^5(.67)^2 +$$
$$ {}_7C_6(.33)^6(.67)^1 + {}_7C_7(.33)^7(.67)^0$$
$$\approx .2535 + .1248 + .0369 + .0061 + .0004$$
$$= .4217.$$

**(b)**  If 7 new businesses are selected at random, what is the probability that at least one of the new businesses will fail?

**Solution**  As above, "At least 1 of 7" means 1 or 2 or 3, and so on, up to all 7. However, with problems that ask for "at least 1," we can employ the complement rule to save time. It is simpler to find the probability that *none* of the 7 selected start-ups succeed [namely: $P(x = 0)$] and then find the probability that at least one of the new businesses fails by using the complement rule and calculating $1 - P(x = 0)$. The calculations are:

$$P(x = 0) = {}_7C_0(.33)^0(.67)^7 \approx .0606;$$
$$P(x \geq 1) = 1 - P(x = 0) \approx 1 - .0606 = .9394.$$ ✓ 5

**TECHNOLOGY TIP** Some spreadsheets provide binomial probabilities. In Microsoft Excel, for example, the command "=BINOM.DIST (0,7,.33,false)" gives .060607, which is the probability $x = 0$ in Example 4(b). Alternatively, the command "=BINOM.DIST (2,7,.33,true)" gives .578326 as the probability of 2 or fewer businesses failing. Subtract .578326 from 1 to get .421674 as the probability that at least 3 businesses will fail. This agrees with the answer we found for Example 4(a).

# 9.4 Exercises

*In Exercises 1–39, see Examples 1– 4.*

**Student Loan Debt** *According to a report from the American Association of University Women released in 2017, 44% of female undergraduates take on debt. Find the probability that the given number of female undergraduates have taken on debt if 10 female undergraduates were selected at random. (Data from: www.aauw.org.)*

1. Exactly 6
2. Exactly 5
3. None
4. All
5. At least 1
6. At most 4

**Student Loan Debt** *The report cited for Exercises 1–6 reported that 39% of male undergraduates take on debt. Find the probability that the given number of male undergraduates have taken on debt if 8 male undergraduates were selected at random. (Data from: www.aauw.org.)*

7. Exactly 2
8. Exactly 1
9. None
10. All
11. At least 1
12. At most 2

*A coin is tossed 5 times. Find the probability of getting*

13. all heads;
14. exactly 3 heads;
15. no more than 3 heads;
16. at least 3 heads.

17. How do you identify a probability problem that involves a binomial experiment?

18. Why do combinations occur in the binomial probability formula?

**Labor Force** *According to data from the U.S. Bureau of Labor Statistics, 27.2% of the civilian labor force had some college or an associate degree. For Exercises 19–21, find the probability for the described number of workers that had some college or an associate degree if 15 workers are selected at random.*

19. exactly 3;
20. none;
21. at most 2.
22. If 200 workers are selected at random, what would be the expected number of workers with some college or an associate degree?

**Stock Ownership** *In a poll conducted by Gallup in early 2017, 31% of 18- to 29-year-olds had money invested in the stock market (either in an individual stock, a stock market fund, or a self-directed 401(k) or IRA). If 9 adults age 18–29 are selected at random, find the probability that the given number of them have money invested in the stock market. (Data from: www.gallup.com.)*

23. none of them;
24. at most 3;
25. 5 or more;
26. at least 1.
27. If 500 adults aged 18–29 are selected at random, what is the expected number who will have money invested in the stock market?
28. In Exercise 27, what would the expected number be if 1250 adults in that age range are selected at random?

**Twins** *The probability that a birth will result in twins is .027. Assuming independence (perhaps not a valid assumption), what are the probabilities that, out of 100 births in a hospital, there will be the given numbers of sets of twins? (Data from: The World Almanac and Book of Facts, 2001.)*

29. Exactly 2 sets of twins
30. At most 2 sets of twins

**Lefties** *According to the Web site Answers.com, 10–13% of Americans are left handed. Assume that the percentage is 11%. If we select 9 people at random, find the probability that the number who are left handed is*

31. exactly 2;
32. at least 2;
33. none;
34. at most 3.
35. In a class of 35 students, how many left-handed students should the instructor expect?

**Confidence in Banks** *The 2016 General Social Survey indicated that 14% of U.S. residents have a great deal of confidence in banks and financial institutions. If 16 U.S. residents were chosen at random, find the probability that the number given had a great deal of confidence in banks and financial institutions. (Data from: gss.norc.org.)*

36. exactly 2;
37. at most 3;
38. at least 4.
39. If we select 300 U.S. residents at random, what is the expected number who have a great deal of confidence in banks and financial institutions?
40. If we select 800 U.S. residents at random, what is the expected number who have a great deal of confidence in banks and financial institutions?

**Vehicle Sales** *In April 2017, 44% of car and light truck sales were for vehicles made by the "Big 3" American auto makers: General Motors, Ford, and Chrysler. If 100 vehicle sales are selected at random, find the probability of each of the following. (Data from: www.wsj.com.)*

**41.** What is the probability that at most 35 of the selected vehicles were made by one of the Big 3 automakers?

**42.** What is the probability that 50 or more of the selected vehicles were made by one of the Big 3 automakers?

**43.** What is the expected number of vehicles that were made by one of the Big 3 automakers?

**44.** If 2000 vehicles sales are selected at random, what is the expected number made by one of the Big 3 automakers?

✓ **Checkpoint Answers**

**1. (a)** About .2583 **(b)** About .0974 **(c)** About .0147

**2. (a)** About .1115 **(b)** About .3845

**3.** About .2605

**4. (a)** About .2808 **(b)** About .8290 **(c)** 1.48

**5. (a)** About .7304 **(b)** About .9567 **(c)** 2.31

# 9.5 Markov Chains

In Section 8.5, we touched on **stochastic processes**—mathematical models that evolve over time in a probabilistic manner. In the current section, we study a special kind of stochastic process called a **Markov chain**, in which the outcome of an experiment depends only on the outcome of the previous experiment. In other words, the next state of the system depends only on the present state, not on preceding states. Such experiments are common enough in applications to make their study worthwhile. Markov chains are named after the Russian mathematician A. A. Markov (1856–1922), who started the theory of stochastic processes. To see how Markov chains work, we look at an example.

**EXAMPLE 1** **Wireless Customer Churn** The wireless industry uses the term *churn* to describe customers who switch wireless firms. The discount wireless firm BUU hopes to lure customers from another discount wireless company (UMobile). BUU hires a market research firm to determine if an extensive marketing campaign will be effective. After the campaign, the market research firm finds that there is a probability of .8 that a BUU customer will stay with BUU and a .35 chance that a UMobile customer will switch to BUU. Assume that, because there is a .8 probability that a BUU customer will stay with BUU, there is a $1 - .8 = .2$ probability that the customer will switch to UMobile. In a similar way, let us assume there is a $1 - .35 = .65$ chance that a UMobile customer will stay with UMobile. If a customer currently with BUU is said to be in state 1 and a customer currently with UMobile is said to be in state 2, then these probabilities of change from one company to the other are as shown in the following table:

|  | State | Carrier after Campaign | |
| --- | --- | --- | --- |
|  |  | 1 | 2 |
| **Current Carrier** | 1 | .8 | .2 |
|  | 2 | .35 | .65 |

The information from the table can be written in other forms. Figure 9.13 is a **transition diagram** that shows the two states and the probabilities of going from one to another.

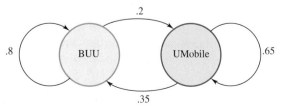

**Figure 9.13**

In a **transition matrix**, the states are indicated at the side and top, as follows:

$$\begin{array}{cc} & \textit{Carrier after Campaign} \\ & \begin{array}{cc} \text{BUU} & \text{UMobile} \end{array} \\ \textit{Current Carrier} \quad \begin{array}{c} \text{BUU} \\ \text{UMobile} \end{array} & \begin{bmatrix} .8 & .2 \\ .35 & .65 \end{bmatrix} \end{array} \cdot ✓_1$$

✓ **Checkpoint 1**

You are given the transition matrix

$$\begin{array}{cc} & \textit{State} \\ & \begin{array}{cc} 1 & 2 \end{array} \\ \textit{State} \quad \begin{array}{c} 1 \\ 2 \end{array} & \begin{bmatrix} .3 & .7 \\ .1 & .9 \end{bmatrix} \end{array}.$$

**(a)** What is the probability of changing from state 1 to state 2?

**(b)** What does the number .1 represent?

**(c)** Draw a transition diagram for this information.

A **transition matrix** has the following features:

1. It is square, since all possible states must be used both as rows and as columns.

2. All entries are between 0 and 1, inclusive, because all entries represent probabilities.

3. The sum of the entries in any row must be 1, because the numbers in the row give the probability of changing from the state at the left to one of the states indicated across the top.

**EXAMPLE 2**    **Wireless Customer Churn**  Suppose that when the new promotional campaign began, BUU had 40% of the market and UMobile had 60%. Use the probability tree in Figure 9.14 to find how these proportions would change after the marketing campaign.

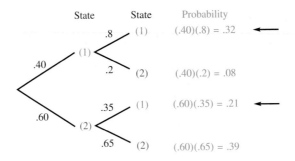

State 1: using carrier BUU
State 2: using carrier UMobile

**Figure 9.14**

**Solution**  Add the numbers indicated with arrows to find the proportion of customers using BUU after the marketing campaign:

$$.32 + .21 = .53$$

Similarly, the proportion using UMobile after the marketing campaign is

$$.08 + .39 = .47.$$

The initial distribution of 40% and 60% becomes 53% and 47%, respectively, after the marketing campaign.

These distributions can be written as the *probability vectors*

$$[.40 \quad .60] \quad \text{and} \quad [.53 \quad .47].$$

A **probability vector** is a one-row matrix, with nonnegative entries, in which the sum of the entries is equal to 1.

The results from the probability tree of Figure 9.14 are exactly the same as the result of multiplying the initial probability vector by the transition matrix (multiplication of matrices was discussed in Section 6.5):

$$[.4 \quad .6]\begin{bmatrix} .8 & .2 \\ .35 & .65 \end{bmatrix} = [.53 \quad .47].$$

If $v$ denotes the original probability vector [.4  .6] and $P$ denotes the transition matrix, then the market share vector after one week is $vP = [.53\ .47]$. To find the market share vector after two weeks, multiply the vector $vP = [.53\ .47]$ by $P$; this amounts to finding $vP^2$. ✓2

Checkpoint 2 shows that after 2 weeks, the market share vector is $vP^2 = [.59\ .41]$. To get the market share vector after three weeks, multiply this vector by $P$; that is, find $vP^3$. Do not use the rounded answer from Checkpoint 2. ✓3

Continuing this process gives each company's share of the market after additional weeks:

| Number of Marketing Campaigns | BUU | UMobile | |
|---|---|---|---|
| 0 | .4 | .6 | $v$ |
| 1 | .53 | .47 | $vP^1$ |
| 2 | .59 | .41 | $vP^2$ |
| 3 | .61 | .39 | $vP^3$ |
| 4 | .63 | .37 | $vP^4$ |
| 12 | .64 | .36 | $vP^{12}$ |
| 13 | .64 | .36 | $vP^{13}$ |

The results seem to approach the probability vector [.64  .36].

What happens if the initial probability vector is different from [.4  .6]? Suppose [.75  .25] is used; then the same powers of the transition matrix as before give the following results:

| Number of Marketing Campaigns | BUU | UMobile | |
|---|---|---|---|
| 0 | .75 | .25 | $v$ |
| 1 | .69 | .31 | $vP^1$ |
| 2 | .66 | .34 | $vP^2$ |
| 3 | .65 | .35 | $vP^3$ |
| 4 | .64 | .36 | $vP^4$ |
| 5 | .64 | .36 | $vP^5$ |
| 6 | .64 | .36 | $vP^6$ |

The results again seem to be approaching the numbers in the probability vector [.64  .36], the same numbers approached with the initial probability vector [.4  .6]. In either case, the long-range trend is for a market share of about 64% for BUU and 36% for UMobile. The example suggests that this long-range trend does not depend on the initial distribution of market shares. This means that if the initial market share for BUU was less than 64%, the advertising campaign has paid off in terms of a greater long-range market share. If the initial share was more than 64%, the campaign did not pay off.

## Regular Transition Matrices

One of the many applications of Markov chains is in finding long-range predictions. It is not possible to make long-range predictions with all transition matrices, but for a large set of transition matrices, long-range predictions *are* possible. Such predictions are always possible with **regular transition matrices**. A transition matrix is **regular** if some power of the matrix contains all positive entries. A Markov chain is a **regular Markov chain** if its transition matrix is regular.

**EXAMPLE 3**   Decide whether the given transition matrices are regular.

(a) $A = \begin{bmatrix} .3 & .1 & .6 \\ 0 & .2 & .8 \\ .3 & .7 & 0 \end{bmatrix}$.

**Solution**  Square $A$:

$$A^2 = \begin{bmatrix} .27 & .47 & .26 \\ .24 & .60 & .16 \\ .09 & .17 & .74 \end{bmatrix}.$$

Since all entries in $A^2$ are positive, matrix $A$ is regular.

(b) $B = \begin{bmatrix} .3 & 0 & .7 \\ 0 & 1 & 0 \\ 0 & 0 & 1 \end{bmatrix}$.

**Solution**  Find various powers of $B$:

$$B^2 = \begin{bmatrix} .09 & 0 & .91 \\ 0 & 1 & 0 \\ 0 & 0 & 1 \end{bmatrix}; \quad B^3 = \begin{bmatrix} .027 & 0 & .973 \\ 0 & 1 & 0 \\ 0 & 0 & 1 \end{bmatrix}; \quad B^4 = \begin{bmatrix} .0081 & 0 & .9919 \\ 0 & 1 & 0 \\ 0 & 0 & 1 \end{bmatrix}.$$

Notice that all of the powers of $B$ shown here have zeros in the same locations. Thus, further powers of $B$ will still give the same zero entries, so that no power of matrix $B$ contains all positive entries. For this reason, $B$ is not regular. ✔️ 4

✔ **Checkpoint 4**

Decide whether the given transition matrices are regular.

(a) $\begin{bmatrix} 0 & 1 \\ 1 & 0 \end{bmatrix}$

(b) $\begin{bmatrix} .45 & .55 \\ 1 & 0 \end{bmatrix}$

📄 **NOTE**  If a transition matrix $P$ has some zero entries, and $P^2$ does as well, you may wonder how far you must compute $P^n$ to be certain that the matrix is not regular. The answer is that if zeros occur in identical places in both $P^n$ and $P^{n+1}$ for any $n$, then they will appear in those places for all higher powers of $P$, so $P$ is not regular.

Suppose that $v$ is any probability vector. It can be shown that, for a regular Markov chain with a transition matrix $P$, there exists a single vector $V$ that does not depend on $v$, such that $v \cdot P^n$ gets closer and closer to $V$ as $n$ gets larger and larger.

## Equilibrium Vector of a Markov Chain

If a Markov chain with transition matrix $P$ is regular, then there is a unique vector $V$ such that, for any probability vector $v$ and for large values of $n$,

$$v \cdot P^n \approx V.$$

Vector $V$ is called the **equilibrium vector**, or the **fixed vector**, of the Markov chain.

In the example with BUU, the equilibrium vector $V$ is approximately [.64   .36]. Vector $V$ can be determined by finding $P^n$ for larger and larger values of $n$ and then looking for a vector that the product $v \cdot P^n$ approaches. Such a strategy can be very tedious, however, and is prone to error. To find a better way, start with the fact that, for a large value of $n$,

$$v \cdot P^n \approx V,$$

as mentioned in the preceding box. We can multiply both sides of this result by $P$, $v \cdot P^n \cdot P \approx V \cdot P$, so that

$$v \cdot P^n \cdot P = v \cdot P^{n+1} \approx VP.$$

Since $v \cdot P^n \approx V$ for large values of $n$, it is also true that $v \cdot P^{n+1} \approx V$ for large values of $n$. (The product $v \cdot P^n$ approaches $V$, so that $v \cdot P^{n+1}$ must also approach $V$.) Thus, $v \cdot P^{n+1} \approx V$ and $v \cdot P^{n+1} \approx VP$, which suggests that

$$VP = V.$$

> If a Markov chain with transition matrix $P$ is regular, then the equilibrium vector $V$ satisfies
>
> $$VP = V.$$

The equilibrium vector $V$ can be found by solving a system of linear equations, as shown in the remaining examples.

**EXAMPLE 4**   Find the long-range trend for the Markov chain in Examples 1 and 2, with transition matrix

$$P = \begin{bmatrix} .8 & .2 \\ .35 & .65 \end{bmatrix}.$$

**Solution**  This matrix is regular, since all entries are positive. Let $P$ represent this transition matrix and let $V$ be the probability vector $[v_1 \quad v_2]$. We want to find $V$ such that

$$VP = V,$$

or

$$[v_1 \quad v_2] \begin{bmatrix} .8 & .2 \\ .35 & .65 \end{bmatrix} = [v_1 \quad v_2].$$

Multiply on the left to get

$$[.8v_1 + .35v_2 \quad .2v_1 + .65v_2] = [v_1 \quad v_2].$$

Set corresponding entries from the two matrices equal to obtain

$$.8v_1 + .35v_2 = v_1; \quad .2v_1 + .65v_2 = v_2.$$

Simplify each of these equations:

$$-.2v_1 + .35v_2 = 0; \quad .2v_1 - .35v_2 = 0.$$

These last two equations are really the same. (The equations in the system obtained from $VP = V$ are always dependent.) To find the values of $v_1$ and $v_2$, recall that $V = [v_1 \quad v_2]$ is a probability vector, so that

$$v_1 + v_2 = 1.$$

Find $v_1$ and $v_2$ by solving the system

$$-.2v_1 + .35v_2 = 0$$
$$v_1 + \quad v_2 = 1.$$

We can rewrite the second equation as $v_1 = 1 - v_2$. Now substitute for $v_1$ in the first equation:

$$-.2(1 - v_2) + .35v_2 = 0.$$

Solving for $v_2$ yields

$$-.2 + .2v_2 + .35v_2 = 0$$
$$-.2 + .55v_2 = 0$$
$$.55v_2 = .2$$
$$v_2 \approx .364.$$

Since $v_2 \approx .364$ and $v_1 = 1 - v_2$, it follows that $v_1 \approx 1 - .364 = .636$, and the equilibrium vector is $[.636 \quad .364] \approx [.64 \quad .36]$.

**EXAMPLE 5** **Manufacturing** The probability that a complex assembly line works correctly depends on whether the line worked correctly the last time it was used. The various probabilities are as given in the following transition matrix:

|  | Works Properly Now | Does Not |
|---|---|---|
| Worked Properly Before | .79 | .21 |
| Did Not | .68 | .32 |

Find the long-range probability that the assembly line will work properly.

**Solution** Begin by finding the equilibrium vector $[v_1 \quad v_2]$, where

$$[v_1 \quad v_2]\begin{bmatrix} .79 & .21 \\ .68 & .32 \end{bmatrix} = [v_1 \quad v_2].$$

Multiplying on the left and setting corresponding entries equal gives the equations

$$.79v_1 + .68v_2 = v_1 \quad \text{and} \quad .21v_1 + .32v_2 = v_2,$$

or

$$-.21v_1 + .68v_2 = 0 \quad \text{and} \quad .21v_1 - .68v_2 = 0.$$

Substitute $v_1 = 1 - v_2$ into the first of these equations to get

$$-.21(1 - v_2) + .68v_2 = 0$$
$$-.21 + .21v_2 + .68v_2 = 0$$
$$-.21 + .89v_2 = 0$$
$$.89v_2 = .21$$
$$v_2 = \frac{.21}{.89} = \frac{21}{89},$$

and $v_1 = 1 - \dfrac{21}{89} = \dfrac{68}{89}$. The equilibrium vector is $[68/89 \quad 21/89]$. In the long run, the company can expect the assembly line to run properly $\dfrac{68}{89} \approx 76\%$ of the time. ✔5

✔ **Checkpoint 5**

In Example 5, suppose the company modifies the line so that the transition matrix becomes

$$\begin{bmatrix} .85 & .15 \\ .75 & .25 \end{bmatrix}.$$

Find the long-range probability that the assembly line will work properly.

**C** **EXAMPLE 6** **Vehicle Sales** Data from the *Wall Street Journal* web site showed that the probability that a vehicle purchased in the United States in April 2017 was from General Motors (*GM*) was .171, from Ford (*F*) was .150, from Chrysler (*C*) was .121, and from other car manufacturers (*O*) was .558. The following transition matrix indicates market share changes from year to year.

|  | GM | F | C | O |
|---|---|---|---|---|
| GM | .85 | .04 | .05 | .06 |
| F | .02 | .91 | .03 | .04 |
| C | .01 | .01 | .95 | .03 |
| O | .03 | .02 | .06 | .89 |

**(a)** Find the probability that a vehicle purchased in the next year was from Ford.

**Solution** To find the market share for each company in the next year we multiply the current market share vector $[.171 \quad .150 \quad .121 \quad .558]$ by the transition matrix. We obtain

$$[.171 \quad .150 \quad .121 \quad .558]\begin{bmatrix} .85 & .04 & .05 & .06 \\ .02 & .91 & .03 & .04 \\ .01 & .01 & .95 & .03 \\ .03 & .02 & .06 & .89 \end{bmatrix}$$

$$= [.16630 \quad .15571 \quad .16148 \quad .51651].$$

The second entry (.15571) indicates that the probability that the vehicle purchased in the next year was from Ford is .15571.

**(b)** If the trend continues, find the long-term probability that when a new vehicle is purchased, it is from Ford.

**Solution** We need to find the equilibrium vector $[v_1 \quad v_2 \quad v_3 \quad v_4]$ where

$$[v_1 \quad v_2 \quad v_3 \quad v_4]\begin{bmatrix} .85 & .04 & .05 & .06 \\ .02 & .91 & .03 & .04 \\ .01 & .01 & .95 & .03 \\ .03 & .02 & .06 & .89 \end{bmatrix} = [v_1 \quad v_2 \quad v_3 \quad v_4].$$

Multiplying on the left side and setting corresponding entries equal gives the equations:

$$.85v_1 + .02v_2 + .01v_3 + .03v_4 = v_1$$
$$.04v_1 + .91v_2 + .01v_3 + .02v_4 = v_2$$
$$.05v_1 + .03v_2 + .95v_3 + .06v_4 = v_3$$
$$.06v_1 + .04v_2 + .03v_3 + .89v_4 = v_4.$$

Simplifying these equations yields:

$$-.15v_1 + .02v_2 + .01v_3 + .03v_4 = 0$$
$$.04v_1 - .09v_2 + .01v_3 + .02v_4 = 0$$
$$.05v_1 + .03v_2 - .05v_3 + .06v_4 = 0$$
$$.06v_1 + .04v_2 + .03v_3 - .11v_4 = 0.$$

Since $V$ is a probability vector, the entries have to add up to 1, so we have

$$v_1 + v_2 + v_3 + v_4 = 1.$$

This gives five equations and four unknown values:

$$v_1 + v_2 + v_3 + v_4 = 1$$
$$-.15v_1 + .02v_2 + .01v_3 + .03v_4 = 0$$
$$.04v_1 - .09v_2 + .01v_3 + .02v_4 = 0$$
$$.05v_1 + .03v_2 - .05v_3 + .06v_4 = 0$$
$$.06v_1 + .04v_2 + .03v_3 - .11v_4 = 0.$$

The system can be solved with the Gauss-Jordan method set forth in Section 6.2. Start with the augmented matrix

$$\begin{bmatrix} 1 & 1 & 1 & 1 & 1 \\ -.15 & .02 & .01 & .03 & 0 \\ .04 & -.09 & .01 & .02 & 0 \\ .05 & .03 & -.05 & .06 & 0 \\ .06 & .04 & .03 & -.11 & 0 \end{bmatrix}.$$

The solution of this system is

$$V \approx [.103 \quad .156 \quad .494 \quad .248].$$

The second entry (.156) tells us the long-term probability that a purchased vehicle is from Ford.

**✓ Checkpoint 6**

Find the equilibrium vector for the transition matrix

$$P = \begin{bmatrix} .3 & .7 \\ .5 & .5 \end{bmatrix}.$$

In Example 4, we found that $[.64 \quad .36]$ was the equilibrium vector for the regular transition matrix

$$P = \begin{bmatrix} .8 & .2 \\ .35 & .65 \end{bmatrix}.$$

Observe what happens when you take powers of the matrix $P$ (the displayed entries have been rounded for easy reading, but the full decimals were used in the calculations):

$$P^2 = \begin{bmatrix} .71 & .29 \\ .51 & .49 \end{bmatrix}; \quad P^3 = \begin{bmatrix} .67 & .33 \\ .58 & .42 \end{bmatrix}; \quad P^4 = \begin{bmatrix} .65 & .35 \\ .61 & .39 \end{bmatrix};$$

$$P^5 = \begin{bmatrix} .64 & .36 \\ .62 & .38 \end{bmatrix}; \quad P^6 = \begin{bmatrix} .64 & .36 \\ .63 & .37 \end{bmatrix}; \quad P^{10} = \begin{bmatrix} .64 & .36 \\ .64 & .36 \end{bmatrix}.$$

As these results suggest, higher and higher powers of the transition matrix $P$ approach a matrix having all identical rows—rows that have as entries the entries of the equilibrium vector $V$.

If you have the technology to compute matrix powers easily (such as a graphing calculator), you can approximate the equilibrium vector by taking higher and higher powers of the transition matrix until all its rows are identical. Figure 9.15 shows part of this process for the transition matrix

$$B = \begin{bmatrix} .79 & .21 \\ .68 & .32 \end{bmatrix}$$

```
[B]²
     [.7669    .2331]
     [.7548    .2452]
[B]³
     [.764359  .235641]
     [.763028  .236972]
```

```
[B]¹⁰
     [.7640...  .2359...]
     [.7640...  .2359...]
```

**Figure 9.15**

Figure 9.15 indicates that the equilibrium vector is approximately [.764  .236], which is what was found algebraically in Example 5.

The results of this section can be summarized as follows.

## Properties of Regular Markov Chains

Suppose a regular Markov chain has a transition matrix $P$.

**1.** As $n$ gets larger and larger, the product $v \cdot P^n$ approaches a unique vector $V$ for any initial probability vector $v$. Vector $V$ is called the *equilibrium vector*, or *fixed vector*.

**2.** Vector $V$ has the property that $VP = V$.

**3.** To find $V$, solve a system of equations obtained from the matrix equation $VP = V$ and from the fact that the sum of the entries of $V$ is 1.

**4.** The powers $P^n$ come closer and closer to a matrix whose rows are made up of the entries of the equilibrium vector $V$.

# 9.5 Exercises

*Decide which of the given vectors could be a probability vector.*

**1.** $\begin{bmatrix} \frac{1}{4} & \frac{3}{4} \end{bmatrix}$

**2.** $\begin{bmatrix} \frac{11}{16} & \frac{5}{16} \end{bmatrix}$

**3.** $\begin{bmatrix} 0 & 1 \end{bmatrix}$

**4.** $\begin{bmatrix} .3 & .3 & .3 \end{bmatrix}$

**5.** $\begin{bmatrix} .3 & -.1 & .6 \end{bmatrix}$

**6.** $\begin{bmatrix} \frac{2}{5} & \frac{3}{10} & .3 \end{bmatrix}$

*Decide which of the given matrices could be a transition matrix. Sketch a transition diagram for any transition matrices.*

**7.** $\begin{bmatrix} .7 & .2 \\ .5 & .5 \end{bmatrix}$

**8.** $\begin{bmatrix} \frac{1}{4} & \frac{3}{4} \\ 0 & 1 \end{bmatrix}$

**9.** $\begin{bmatrix} \frac{4}{9} & \frac{1}{3} \\ \frac{1}{5} & \frac{7}{10} \end{bmatrix}$

**10.** $\begin{bmatrix} 0 & 1 & 0 \\ .3 & .3 & .3 \\ 1 & 0 & 0 \end{bmatrix}$

**11.** $\begin{bmatrix} \frac{1}{2} & \frac{1}{4} & 1 \\ \frac{2}{3} & 0 & \frac{1}{3} \\ \frac{1}{3} & 1 & 0 \end{bmatrix}$

**12.** $\begin{bmatrix} .2 & .3 & .5 \\ 0 & 0 & 1 \\ .1 & .9 & 0 \end{bmatrix}$

*In Exercises 13–15, write any transition diagrams as transition matrices.*

**13.**

**14.**

**15.**

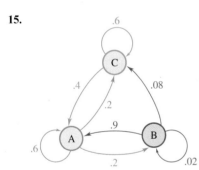

*Decide whether the given transition matrices are regular. (See Example 3.)*

**16.** $\begin{bmatrix} 1 & 0 \\ .25 & .75 \end{bmatrix}$

**17.** $\begin{bmatrix} .2 & .8 \\ .9 & .1 \end{bmatrix}$

**18.** $\begin{bmatrix} .3 & .5 & .2 \\ 1 & 0 & 0 \\ .5 & .1 & .4 \end{bmatrix}$

**19.** $\begin{bmatrix} .25 & .40 & .30 & .05 \\ .18 & .23 & .59 & 0 \\ 0 & .15 & .36 & .49 \\ .28 & .32 & .24 & .16 \end{bmatrix}$

**20.** $\begin{bmatrix} .23 & .41 & 0 & .36 \\ 0 & .27 & .21 & .52 \\ 0 & 0 & 1 & 0 \\ .48 & 0 & .39 & .13 \end{bmatrix}$

*Find the equilibrium vector for each of the given transition matrices. (See Examples 4 and 5.)*

**21.** $\begin{bmatrix} .3 & .7 \\ .4 & .6 \end{bmatrix}$

**22.** $\begin{bmatrix} .55 & .45 \\ .19 & .81 \end{bmatrix}$

**23.** $\begin{bmatrix} \frac{5}{8} & \frac{3}{8} \\ \frac{7}{9} & \frac{2}{9} \end{bmatrix}$

**24.** $\begin{bmatrix} \frac{2}{3} & \frac{1}{3} \\ \frac{1}{8} & \frac{7}{8} \end{bmatrix}$

**25.** $\begin{bmatrix} .25 & .35 & .4 \\ .1 & .3 & .6 \\ .55 & .4 & .05 \end{bmatrix}$

**26.** $\begin{bmatrix} .16 & .28 & .56 \\ .43 & .12 & .45 \\ .86 & .05 & .09 \end{bmatrix}$

**27.** $\begin{bmatrix} .15 & .15 & .70 \\ .42 & .38 & .20 \\ .16 & .28 & .56 \end{bmatrix}$

**28.** $\begin{bmatrix} .44 & .31 & .25 \\ .80 & .11 & .09 \\ .26 & .31 & .43 \end{bmatrix}$

*For each of the given transition matrices, use a graphing calculator or computer to find the first five powers of the matrix. Then find the probability that state 2 changes to state 4 after 5 repetitions of the experiment.*

**29.** $\begin{bmatrix} .3 & .2 & .3 & .1 & .1 \\ .4 & .2 & .1 & .2 & .1 \\ .1 & .3 & .2 & .2 & .2 \\ .2 & .1 & .3 & .2 & .2 \\ .1 & .1 & .4 & .2 & .2 \end{bmatrix}$

**30.** $\begin{bmatrix} .1 & .2 & .2 & .3 & .2 \\ .2 & .1 & .1 & .2 & .4 \\ .2 & .1 & .4 & .2 & .1 \\ .3 & .1 & .1 & .2 & .3 \\ .1 & .3 & .1 & .1 & .4 \end{bmatrix}$

**31. Flu Shots** In a recent year, the percentage of patients at a doctor's office who received a flu shot was 26%. A campaign by the doctors and nurses was designed to increase the percentage of patients who obtain a flu shot. The doctors and nurses believed there was an 85% chance that someone who received a shot in year 1 would obtain a shot in year 2. They also believed that there was a 40% chance that a person who did not receive a shot in year 1 would receive a shot in year 2.

**(a)** Give the transition matrix for this situation.

**(b)** Find the percentages of patients in year 2 who received, and did not receive, a flu shot.

**(c)** Find the long-range trend for the Markov chain representing receipt of flu shots at the doctor's office.

**32. Media Impact** Six months prior to an election, a poll found that only 35% of state voters planned to vote for a casino gambling initiative. After a media blitz emphasizing the new jobs that are created as a result of casinos, a new poll found that among those who did not favor it previously, 30% now favored the initiative. Among those who favored the initiative initially, 90% still favored it.

**(a)** Give the transition matrix for this situation.

**(b)** Find the percentage who favored the gambling initiative after the media blitz.

**(c)** Find the long-term percentage who favor the initiative if the trends and media blitz continue.

**33. Manufacturing** The probability that a complex assembly line works correctly depends on whether the line worked correctly the last time it was used. There is a .91 chance that the line will work correctly if it worked correctly the time before and a .68 chance that it will work correctly if it did *not* work correctly the time before. Set up a transition matrix with this information, and find the long-run probability that the line will work correctly. (See Example 5.)

**34. Manufacturing** Suppose something unplanned occurred to the assembly line of Exercise 33, so that the transition matrix becomes

$$\begin{array}{c} \\ \text{Works} \\ \text{Doesn't Work} \end{array} \begin{array}{c} \text{Works} \quad \text{Doesn't Work} \\ \begin{bmatrix} .81 & .19 \\ .77 & .23 \end{bmatrix} \end{array}.$$

Find the new long-run probability that the line will work properly.

**35. Genetics** In Exercises 59 and 60 of Section 8.4 (pp. 435–436), we discussed the effect on flower color of cross-pollinating pea plants. As shown there, since the gene for red is dominant and the gene for white is recessive, 75% of the pea plants have red flowers and 25% have white flowers, because plants with 1 red and 1 white gene appear red. If a red-flowered plant is crossed with a red-flowered plant known to have 1 red and 1 white gene, then 75% of the offspring will be red and 25% will be white. Crossing a red-flowered plant that has 1 red and 1 white gene with a white-flowered plant produces 50% red-flowered offspring and 50% white-flowered offspring.

(a) Write a transition matrix using this information.

(b) Write a probability vector for the initial distribution of colors.

(c) Find the distribution of colors after 4 generations.

(d) Find the long-range distribution of colors.

**36. Genetics** Snapdragons with 1 red gene and 1 white gene produce pink-flowered offspring. If a red snapdragon is crossed with a pink snapdragon, the probabilities that the offspring will be red, pink, or white are $1/2$, $1/2$, and 0, respectively. If 2 pink snapdragons are crossed, the probabilities of red, pink, or white offspring are $1/4$, $1/2$, and $1/4$, respectively. For a cross between a white and a pink snapdragon, the corresponding probabilities are $0$, $1/2$, and $1/2$. Set up a transition matrix and find the long-range prediction for the fraction of red, pink, and white snapdragons.

**37. Housing** Approximately 58.2% of the residents of San Francisco are homeowners (*O*), 41.0% are renters (*R*), and .8% are homeless. Assume the following transition probabilities for residents of San Francisco per year. (Data from: www.nmhc.org and sfist.com.)

$$\begin{array}{c} \\ O \\ R \\ H \end{array} \begin{array}{c} \quad O \qquad R \qquad H \\ \begin{bmatrix} .90 & .10 & 0 \\ .09 & .909 & .001 \\ 0 & .34 & .66 \end{bmatrix} \end{array}$$

(a) Find the probability that residents own, rent, and are homeless after one year.

(b) Find the long-range probabilities for the three categories.

**38. Earthquakes** Markov chains can be utilized in research into earthquakes. Researchers in Italy give the following example of a transition table in which the rows are magnitudes of an earthquake and the columns are magnitudes of the next earthquake in the sequence.[*]

---

|     | 2.5 | 2.6 | 2.7 | 2.8 |
|-----|-----|-----|-----|-----|
| 2.5 | 3/7 | 1/7 | 2/7 | 1/7 |
| 2.6 | 1/2 | 0   | 1/4 | 1/4 |
| 2.7 | 1/3 | 1/3 | 0   | 1/3 |
| 2.8 | 1/4 | 1/2 | 0   | 1/4 |

Thus, the probability of a 2.5-magnitude earthquake being followed by a 2.8-magnitude earthquake is $1/7$. If these trends were to persist, find the long-range trend for the probabilities of each magnitude for the subsequent earthquake.

**39. Car Insurance** An insurance company classifies its drivers into three groups: $G_0$ (no accidents), $G_1$ (one accident), and $G_2$ (more than one accident). The probability that a driver in $G_0$ will stay in $G_0$ after 1 year is .85, that he will become a $G_1$ is .10, and that he will become a $G_2$ is .05. A driver in $G_1$ cannot move to $G_0$. (This insurance company has a long memory!) There is a .80 probability that a $G_1$ driver will stay in $G_1$ and a .20 probability that he will become a $G_2$. A driver in $G_2$ must stay in $G_2$.

(a) Write a transition matrix using this information.

Suppose that the company accepts 50,000 new policy-holders, all of whom are in group $G_0$. Find the number in each group

(b) after 1 year;    (c) after 2 years;

(d) after 3 years;    (e) after 4 years.

(f) Find the equilibrium vector here.

**40. Car Insurance** The difficulty with the mathematical model of Exercise 39 is that no "grace period" is provided; there should be a certain probability of moving from $G_1$ or $G_2$ back to $G_0$ (say, after 4 years with no accidents). A new system with this feature might produce the following transition matrix:

$$\begin{bmatrix} .85 & .10 & .05 \\ .15 & .75 & .10 \\ .10 & .30 & .60 \end{bmatrix}.$$

Suppose that when this new policy is adopted, the company has 50,000 policyholders in group $G_0$. Find the number in each group

(a) after 1 year;

(b) after 2 years;

(c) after 3 years.

(d) Find the equilibrium vector here.

**41. Wireless Carriers** In the final quarter of 2016, Verizon had 35.1% of the wireless market, while AT&T had 32.4%, and all other carriers had 32.5% of the market. Assume the following transition matrix for customers switching carriers in a year. (Data from: www.statistica.com.)

$$\begin{array}{cc} & \begin{array}{ccc} & \text{Will Switch to} & \\ \text{Verizon} & \text{AT\&T} & \text{Other} \end{array} \\ \text{Now has} \begin{array}{c} \text{Verizon} \\ \text{AT\&T} \\ \text{Other} \end{array} & \begin{bmatrix} .91 & .07 & .02 \\ .03 & .87 & .10 \\ .14 & .04 & .82 \end{bmatrix}. \end{array}$$

---

[*]Michele Lovallo, Vincenzo Lapenna, and Luciano Telesca, "Transition matrix analysis of earthquake magnitude sequences," *Chaos, Solitons, and Fractals* 24 (2005): 33–43.

Find the share of the market held by each company after

**(a)** 1 year; **(b)** 2 years; **(c)** 3 years.

**(d)** What is the long-range prediction?

**42. Changing Majors** At one liberal arts college, students are classified as humanities majors, science majors, or undecided. There is a 23% chance that a humanities major will change to a science major from one year to the next and a 40% chance that a humanities major will change to undecided. A science major will change to humanities with probability .12 and to undecided with probability .38. An undecided student will switch to humanities or science with probabilities of .45 and .28, respectively. Find the long-range prediction for the fraction of students in each of these three majors.

**43. Vehicle Sales** Using data similar to those in Example 6, the probability that a new vehicle purchased was from Toyota ($T$) was .142, American Honda Motor Corporation ($H$) was .097, and all other manufacturers ($O$) was .761. Use the transition matrix below for market share changes from year to year.

$$\begin{array}{c} \\ T \\ H \\ O \end{array} \begin{array}{ccc} T & H & O \\ \left[\begin{array}{ccc} .95 & .02 & .03 \\ .04 & .92 & .04 \\ .09 & .07 & .84 \end{array}\right] \end{array}$$

**(a)** Find the probability that a new vehicle purchased in the next year was from Toyota.

**(b)** Find the long-term probability that a new vehicle purchase was from Honda.

**44. Queuing** In a queuing chain, we assume that people are queuing up to be served by, say, a bank teller. For simplicity, let us assume that once two people are in line, no one else can enter the line. Let us further assume that one person is served every minute, as long as someone is in line. Assume further that, in any minute, there is a probability of .4 that no one enters the line, a probability of .3 that exactly one person enters the line, and a probability of .3 that exactly two people enter the line, assuming that there is room. If there is not enough room for two people, then the probability that one person enters the line is .5. Let the state be given by the number of people in line.

**(a)** Give the transition matrix for the number of people in line:

$$\begin{array}{c} \\ 0 \\ 1 \\ 2 \end{array} \begin{array}{ccc} 0 & 1 & 2 \\ \left[\begin{array}{ccc} ? & ? & ? \\ ? & ? & ? \\ ? & ? & ? \end{array}\right] \end{array}.$$

**(b)** Find the transition matrix for a 2-minute period.

**(c)** Use your result from part (b) to find the probability that a queue with no one in line has two people in line 2 minutes later.

*Use a graphing calculator or computer for Exercises 45 and 46.*

**45. Corporate Training** A company with a new training program classified each employee in one of four states: $s_1$, never in the program; $s_2$, currently in the program; $s_3$, discharged; $s_4$, completed the program. The transition matrix for this company is as follows.

$$\begin{array}{c} \\ s_1 \\ s_2 \\ s_3 \\ s_4 \end{array} \begin{array}{cccc} s_1 & s_2 & s_3 & s_4 \\ \left[\begin{array}{cccc} .4 & .2 & .05 & .35 \\ 0 & .45 & .05 & .5 \\ 0 & 0 & 1 & 0 \\ 0 & 0 & 0 & 1 \end{array}\right] \end{array}.$$

**(a)** What percentage of employees who had never been in the program (state $s_1$) completed the program (state $s_4$) after the program had been offered five times?

**(b)** If the initial percentage of employees in each state was [.5 .5 0 0], find the corresponding percentages after the program had been offered four times.

**46. Corporate Training** Find the long-range prediction for the percentage of employees in each state for the company training program in Exercise 45.

## ✓ Checkpoint Answers

**1. (a)** .7

**(b)** The probability of changing from state 2 to state 1

**(c)**
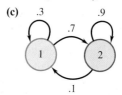

**2.** [.59 .41] (rounded)

**3.** [.61 .39] (rounded)

**4. (a)** No **(b)** Yes

**5.** $5/6 \approx 83\%$

**6.** [5/12 7/12]

# 9.6 Decision Making

John F. Kennedy once remarked that he had assumed that, as president, it would be difficult to choose between distinct, opposite alternatives when a decision needed to be made. Actually, he found that such decisions were easy to make; the hard decisions came when he was faced with choices that were not as clear cut. Most decisions fall into this last category—decisions that must be made under conditions of uncertainty. In Section 9.1, we saw how to use expected values to help make a decision. Those ideas are extended in this section, where we consider decision making in the face of uncertainty. We begin with an example.

**EXAMPLE 1**     **Fruit Production** Freezing temperatures are endangering the orange crop in central California. A farmer can protect his crop by burning smudge pots; the heat from the pots keeps the oranges from freezing. However, burning the pots is expensive, costing $20,000. The farmer knows that if he burns smudge pots, he will be able to sell his crop for a net profit (after the costs of the pots are deducted) of $50,000, provided that the freeze does develop and wipes out other orange crops in California. If he does nothing, he will either lose the $10,000 he has already invested in the crop if it does freeze or make a profit of $46,000 if it does not freeze. (If it does not freeze, there will be a large supply of oranges, and thus his profit will be lower than if there were a small supply.) What should the farmer do?

**Solution**  He should begin by carefully defining the problem. First, he must decide on the **states of nature**—the possible alternatives over which he has no control. Here, there are two: freezing temperatures and no freezing temperatures. Next, the farmer should list the things he can control—his actions or **strategies.** He has two possible strategies: to use smudge pots or not. The consequences of each action under each state of nature, called **payoffs,** are summarized in a **payoff matrix,** as follows, where the payoffs in this case are the profits for each possible combination of events:

<br>

|  |  | *States of Nature* | |
|---|---|---|---|
|  |  | Freeze | No Freeze |
| *Strategies of Farmer* | Use Smudge Pots | $50,000 | $26,000 |
|  | Do Not Use Pots | −$10,000 | $46,000 |

**✓Checkpoint 1**

Explain how each of the given payoffs in the matrix were obtained.

(a)  −$10,000

(b)  $50,000

To get the $26,000 entry in the payoff matrix, use the profit if there is no freeze, namely, $46,000, and subtract the $20,000 cost of using the pots. ✓₁

Once the farmer makes the payoff matrix, what then? The farmer might be an optimist (some might call him a gambler); in this case, he might assume that the best will happen and go for the biggest number of the matrix ($50,000). For that profit, he must adopt the strategy "use smudge pots."

On the other hand, if the farmer is a pessimist, he would want to minimize the worst thing that could happen. If he uses smudge pots, the worst thing that could happen to him would be a profit of $26,000, which will result if there is no freeze. If he does not use smudge pots, he might face a loss of $10,000. To minimize the worst, he once again should adopt the strategy "use smudge pots."

Suppose the farmer decides that he is neither an optimist nor a pessimist, but would like further information before choosing a strategy. For example, he might call the weather forecaster and ask for the probability of a freeze. Suppose the forecaster says that this probability is only .2. What should the farmer do? He should recall the discussion of expected value and work out the expected profit for each of his two possible strategies. If the probability of a freeze is .2, then the probability that there is no freeze is .8. This information leads to the following expected values:

If smudge pots are used:      $50,000(.2) + 26,000(.8) = 30,800;$

If no smudge pots are used:   $-10,000(.2) + 46,000(.8) = 34,800.$

**✓Checkpoint 2**

What should the farmer do if the probability of a freeze is .6? What is his expected profit?

Here, the maximum expected profit, $34,800, is obtained if smudge pots are not used. ✓₂

As the example shows, the farmer's beliefs about the probabilities of a freeze affect his choice of strategies.

**EXAMPLE 2**     **Merchandising** An owner of several greeting-card stores must decide in July about the type of displays to emphasize for Sweetest Day in October. She has three possible choices: emphasize chocolates, emphasize collectible gifts, or emphasize gifts that can be engraved. Her success is dependent on the state of the economy in October. If the economy is strong, she will do well with the collectible gifts, while in a weak economy, the chocolates do very well. In a mixed economy, the gifts that can be engraved will do well. She first prepares a payoff matrix for all three possibilities, where the numbers in the matrix represent her profits in thousands of dollars:

|  |  | *States of Nature* |  |  |
|---|---|---|---|---|
|  |  | Weak Economy | Mixed | Strong Economy |
| *Strategies* | Chocolates | 85 | 30 | 75 |
|  | Collectibles | 45 | 45 | 110 |
|  | Engraved | 60 | 95 | 85 |

**(a)** What would an optimist do?

**Solution** If the owner is an optimist, she should aim for the biggest number on the matrix, 110 (representing $110,000 in profit). Her strategy in this case would be to display collectibles.

**(b)** How would a pessimist react?

**Solution** A pessimist wants to find the best of the worst things that can happen. If she displays collectibles, the worst that can happen is a profit of $45,000. For displaying engravable items, the worst is a profit of $60,000, and for displaying chocolates, the worst is a profit of $30,000. Her strategy here is to use the engravable items.

**(c)** Suppose the owner reads in a business magazine that leading experts believe that there is a 50% chance of a weak economy in October, a 20% chance of a mixed economy, and a 30% chance of a strong economy. How might she use this information?

✓ **Checkpoint 3**

Suppose the owner reads another article, which gives the following predictions: a 35% chance of a weak economy, a 25% chance of an in-between economy, and a 40% chance of a strong economy. What is the best strategy now? What is the expected profit?

**Solution** The owner can now find her expected profit for each possible strategy.

Chocolates $\quad 85(.5) + 30(.20) + 75(.30) \quad = 71;$

Collectibles $\quad 45(.5) + 45(.20) + 110(.30) = 64.5;$

Engraved $\quad 60(.5) + 95(.20) + 85(.30) \quad = 74.5.$

Here, the best strategy is to display gifts that can be engraved; the expected profit is 74.5, or $74,500. ✓₃

# 9.6 Exercises

1. **Real Estate Development** A developer has $100,000 to invest in land. He has a choice of two parcels (at the same price): one on the highway and one on the coast. With both parcels, his ultimate profit depends on whether he faces light opposition from environmental groups or heavy opposition. He estimates that the payoff matrix is as follows (the numbers represent his profit):

|  | *Opposition* | |
|---|---|---|
|  | Light | Heavy |
| Highway | $70,000 | $30,000 |
| Coast | $150,000 | −$40,000 |

What should the developer do if he is

**(a)** an optimist?

**(b)** a pessimist?

**(c)** Suppose the probability of heavy opposition is .8. What is his best strategy? What is the expected profit?

**(d)** What is the best strategy if the probability of heavy opposition is only .4?

2. **Concert Seating** Mount Union College has sold out all tickets for a jazz concert to be held in the stadium. If it rains, the show will have to be moved to the gym, which has a much smaller seating capacity. The dean must decide in advance whether to set up the seats and the stage in the gym, in the stadium, or in both, just in case. The following payoff matrix shows the net profit in each case:

|  |  | *States of Nature* | |
|---|---|---|---|
|  |  | Rain | No Rain |
| *Strategies* | Set up in Stadium | −$1550 | $1500 |
|  | Set up in Gym | $1000 | $1000 |
|  | Set up in Both | $750 | $1400 |

What strategy should the dean choose if she is

**(a)** an optimist?

**(b)** a pessimist?

**(c)** If the weather forecaster predicts rain with a probability of .6, what strategy should she choose to maximize the expected profit? What is the maximum expected profit?

3. **Tire Manufacturing** An analyst must decide what fraction of the automobile tires produced at a particular manufacturing plant are defective. She has already decided that there are three possibilities for the fraction of defective items: .02, .09, and .16. She may recommend two courses of action: upgrade the equipment at the plant or make no upgrades. The following payoff matrix represents the *costs* to the company in each case, in hundreds of dollars:

$$\begin{array}{c} \qquad\qquad\qquad Defectives \\ \qquad\qquad .02 \quad\ .09 \quad\ .16 \\ Strategies \begin{array}{r} \text{Upgrade} \\ \text{No Upgrade} \end{array} \left[ \begin{array}{ccc} 130 & 130 & 130 \\ 28 & 180 & 450 \end{array} \right]. \end{array}$$

What strategy should the analyst recommend if she is

(a) an optimist?

(b) a pessimist?

(c) Suppose the analyst is able to estimate probabilities for the three states of nature as follows:

| Fraction of Defectives | Probability |
|---|---|
| .02 | .70 |
| .09 | .20 |
| .16 | .10 |

Which strategy should she recommend? Find the expected cost to the company if that strategy is chosen.

4. **Marketing** The research department of the Allied Manufacturing Company has developed a new process that it believes will result in an improved product. Management must decide whether to go ahead and market the new product or not. The new product may be better than the old one, or it may not be better. If the new product is better and the company decides to market it, sales should increase by $50,000. If it is not better and the old product is replaced with the new product on the market, the company will lose $25,000 to competitors. If management decides not to market the new product, the company will lose $40,000 if it is better and will lose research costs of $10,000 if it is not.

(a) Prepare a payoff matrix.

(b) If management believes that the probability that the new product is better is .4, find the expected profits under each strategy and determine the best action.

5. **Machine Productivity** A businessman is planning to ship a used machine to his plant in Nigeria. He would like to use it there for the next 4 years. He must decide whether to overhaul the machine before sending it. The cost of overhaul is $2600. If the machine fails when it is in operation in Nigeria, it will cost him $6000 in lost production and repairs. He estimates that the probability that it will fail is .3 if he does not overhaul it and .1 if he does overhaul it. Neglect the possibility that the machine might fail more than once in the 4 years.

(a) Prepare a payoff matrix.

(b) What should the businessman do to minimize his expected costs?

6. **Business** A contractor prepares to bid on a job. If all goes well, his bid should be $25,000, which will cover his costs plus his usual profit margin of $4000. However, if a threatened labor strike actually occurs, his bid should be $35,000 to give him the same profit. If there is a strike and he bids $25,000, he will lose $5500. If his bid is too high, he may lose the job entirely, while if it is too low, he may lose money.

(a) Prepare a payoff matrix.

(b) If the contractor believes that the probability of a strike is .6, how much should he bid?

7. **Shelter Costs** An artist travels to craft fairs all summer long. She must book her booth at a June craft show six months in advance and decide if she wishes to rent a tent for an extra $500 in case it rains on the day of the show. If it does not rain, she believes she will earn $3000 at the show. If it rains, she believes she will earn only $2000, provided she has a tent. If she does not have a tent and it does rain, she will have to pack up and go home and will thus earn $0. Weather records over the last 10 years indicate that there is a .4 probability of rain in June.

(a) Prepare a profit matrix.

(b) What should the artist do to maximize her expected revenue?

8. **Investing** An investor has $50,000 to invest in stocks. She has two possible strategies: buy conservative blue-chip stocks or buy highly speculative stocks. There are two states of nature: the market goes up and the market goes down. The following payoff matrix shows the net amounts she will have under the various circumstances.

$$\begin{array}{c} \qquad\qquad\quad \text{Market Up} \quad \text{Market Down} \\ \begin{array}{r} \text{Buy Blue Chip} \\ \text{Buy Speculative} \end{array} \left[ \begin{array}{cc} \$60,000 & \$46,000 \\ \$80,000 & \$32,000 \end{array} \right]. \end{array}$$

What should the investor do if she is

(a) an optimist?

(b) a pessimist?

(c) Suppose there is a .6 probability of the market going up. What is the best strategy? What is the expected profit?

(d) What is the best strategy if the probability of a market rise is .2?

*Sometimes the numbers (or payoffs) in a payoff matrix do not represent money (profits or costs, for example). Instead, they may represent utility. A utility is a number that measures the satisfaction (or lack of it) that results from a certain action. Utility numbers must be assigned by each individual, depending on how he or she feels about a situation. For example, one person might assign a utility of +20 for a week's vacation in San Francisco, with −6 being assigned if the vacation were moved to Sacramento. Work the problems that follow in the same way as the preceding ones.*

9. **Campaign Strategizing** A politician must plan her reelection strategy. She can emphasize jobs or she can emphasize the environment. The voters can be concerned about jobs or about the environment. Following is a payoff matrix showing the utility of each possible outcome.

$$\begin{array}{c} \qquad\qquad\qquad\qquad Voters \\ \qquad\qquad\qquad \text{Jobs} \quad\ \text{Environment} \\ Candidate \begin{array}{r} \text{Jobs} \\ \text{Environment} \end{array} \left[ \begin{array}{cc} +40 & -10 \\ -12 & +30 \end{array} \right] \end{array}$$

The political analysts feel that there is a .35 chance that the voters will emphasize jobs. What strategy should the candidate adopt? What is its expected utility?

**10. Learning Aides** In an accounting class, the instructor permits the students to bring a calculator or a reference book (but not both) to an examination. The examination itself can emphasize either numbers or definitions. In trying to decide which aid to take to an examination, a student first decides on the utilities shown in the following payoff matrix:

*Exam Emphasizes*

|  |  | Numbers | Definition |
|---|---|---|---|
| *Student Chooses* | Calculator | +50 | 0 |
|  | Book | +15 | +35 |

**(a)** What strategy should the student choose if the probability that the examination will emphasize numbers is .6? What is the expected utility in this case?

**(b)** Suppose the probability that the examination emphasizes numbers is .4. What strategy should the student choose?

## ✓ Checkpoint Answers

**1. (a)** If the crop freezes and smudge pots are not used, the farmer's profit is $-\$10{,}000$ for labor costs.

   **(b)** If the crop freezes and smudge pots are used, the farmer makes a profit of \$50,000.

**2.** Use smudge pots; \$40,400

**3.** Engravable; \$78,750

# CHAPTER 9 Summary and Review

## Key Terms and Symbols

**9.1** random variable
probability distribution
histogram
expected value
fair game

**9.2** $n!$ ($n$ factorial)
multiplication principle

permutations
combinations

**9.4** Bernoulli trials (processes)
binomial experiment
binomial probability

**9.5** stochastic processes
Markov chain

state
transition diagram
transition matrix
probability vector
regular transition
   matrix
regular Markov chain

equilibrium vector
   (fixed vector)

**9.6** states of nature
strategies
payoffs
payoff matrix

## Chapter 9 Key Concepts

**Expected Value**

For a random variable $x$ with values $x_1, x_2, \ldots, x_n$ and probabilities $p_1, p_2, \ldots, p_n$, the expected value is

$$E(x) = x_1p_1 + x_2p_2 + \cdots + x_np_n.$$

**Multiplication Principle**

If there are $m_1$ ways to make a first choice, $m_2$ ways to make a second choice, and so on, then there are $m_1m_2 \cdots m_n$ different ways to make the entire sequence of choices.

**Permutations**

The number of **permutations** of $n$ elements taken $r$ at a time is $_nP_r = \dfrac{n!}{(n-r)!}$.

**Combinations**

The number of **combinations** of $n$ elements taken $r$ at a time is

$$_nC_r = \frac{n!}{(n-r)!r!}.$$

**Binomial Experiments**

Binomial Experiments have the following characteristics: (1) The same experiment is repeated a finite number of times; (2) there are only *two* outcomes, labeled success and failure; (3) the probability of success is the same for each trial; and (4) the trials are independent. If the probability of success in a single trial is $p$, the probability of $x$ successes in $n$ trials is

$$_nC_xp^x(1-p)^{n-x}.$$

**Markov Chains**

A **transition matrix** must be square, with all entries between 0 and 1 inclusive, and the sum of the entries in any row must be 1. A Markov chain is *regular* if some power of its transition matrix $P$ contains all positive entries. The long-range probabilities for a regular Markov chain are given by the **equilibrium,** or **fixed, vector** $V$, where, for any initial probability vector $v$, the products $vP^n$ approach $V$ as $n$ gets larger and $VP = V$. To find $V$, solve the system of equations formed by $VP = V$ and the fact that the sum of the entries of $V$ is 1.

**Decision Making**

A **payoff matrix,** which includes all available strategies and states of nature, is used in decision making to define the problem and the possible solutions. The expected value of each strategy can help to determine the best course of action.

## Chapter 9 Review Exercises

*In Exercises 1–4, (a) sketch the histogram of the given probability distribution, and (b) find the expected value.*

**1.**

| x | 0 | 1 | 2 | 3 |
|------|-----|-----|-----|-----|
| P(x) | .22 | .54 | .16 | .08 |

**2.**

| x | −3 | −2 | −1 | 0 | 1 | 2 | 3 |
|------|-----|-----|-----|-----|-----|-----|-----|
| P(x) | .15 | .20 | .25 | .18 | .12 | .06 | .04 |

**3.**

| x | −10 | 0 | 10 |
|------|-----|-----|-----|
| P(x) | $\frac{1}{3}$ | $\frac{1}{3}$ | $\frac{1}{3}$ |

**4.**

| x | 0 | 2 | 4 | 6 |
|------|-----|-----|-----|-----|
| P(x) | .35 | .15 | .2 | .3 |

**5. Vehicles per Household** The probability distribution for the number of vehicles in a single household is given below. Find the expected number of vehicles in a household. (Data from: www.census.gov.)

| x | 0 | 1 | 2 | 3 | 4 | 5 | 6 |
|------|------|------|------|------|------|------|------|
| P(x) | .079 | .321 | .381 | .151 | .049 | .013 | .006 |

**6. House Size** The probability distribution of the number of bedrooms in a given household is presented below. Find the expected number of bedrooms. (Data from: www.census.gov.)

| x | 0 | 1 | 2 | 3 | 4 | 5 | 6 | 7 | 8 |
|------|-------|-------|-------|-------|-------|-------|-------|-------|-------|
| P(x) | .0216 | .0939 | .2510 | .4136 | .1740 | .0390 | .0042 | .0026 | .0001 |

*In Exercises 7 and 8, (a) give the probability distribution, and (b) find the expected value.*

**7.** A grocery store has 10 bouquets of flowers for sale, 3 of which are rose displays. Two bouquets are selected at random, and the number of rose bouquets is noted.

**8.** In a class of 10 students, 3 did not do their homework. The professor selects 3 members of the class to present solutions to homework problems on the board and records how many of those selected did not do their homework.

### Games of Chance *Solve the given problems.*

**9.** Suppose someone offers to pay you $100 if you draw 3 cards from a standard deck of 52 cards and all the cards are hearts. What should you pay for the chance to win if it is a fair game?

**10.** You pay $2 to play a game of "Over/Under," in which you will roll two dice and note the sum of the results. You can bet that the sum will be less than 7 (under), exactly 7, or greater than 7 (over). If you bet "under" and you win, you get your $2 back, plus $2 more. If you bet 7 and you win, you get your $2 back, plus $4, and if you

bet "over" and win, you get your $2 back, plus $2 more. What are the expected winnings for each type of bet?

**11. Business** In 2016, Apple's estimated share of the personal computer market was 7.1%. If we select 3 new computer purchases at random and define x to be the number of computers that are Apple, the probability distribution is as given below. Find the expected value for x. (Data from: www.businessinsider.com.)

| x | 0 | 1 | 2 | 3 |
|------|-------|-------|-------|-------|
| P(x) | .8018 | .1838 | .0140 | .0004 |

**12. Smartwatches** In the third quarter of 2016, Apple had 41.4% of the worldwide smartwatch market. If we select 5 smartwatch sales at random and define x to be the number that are Apple watches, the probability distribution for x appears below. Find the expected value for x. (Data from: www.icd.com.)

| x | 0 | 1 | 2 | 3 | 4 | 5 |
|------|------|------|------|------|------|------|
| P(x) | .069 | .244 | .345 | .244 | .086 | .012 |

**13.** In how many ways can 8 different taxis line up at the airport?

**14.** How many variations are possible for gold, silver, and bronze medalists in the 50-meter swimming race if there are 8 finalists?

**15.** In how many ways can a sample of 3 computer monitors be taken from a batch of 12 identical monitors?

**16.** If 4 of the 12 monitors in Exercise 15 are broken, in how many ways can the sample of 3 include the following?

   **(a)** 1 broken monitor;

   **(b)** no broken monitors;

   **(c)** at least 1 broken monitor.

**17.** In how many ways can 6 students from a class of 30 be arranged in the first row of seats? (There are 6 seats in the first row.)

**18.** In how many ways can the six students in Exercise 17 be arranged in a row if a certain student must be first?

**19.** In how many ways can the 6 students in Exercise 17 be arranged if half the students are science majors and the other half are business majors and if

   **(a)** like majors must be together?

   **(b)** science and business majors are alternated?

**20.** Explain under what circumstances a permutation should be used in a probability problem and under what circumstances a combination should be used.

**21.** Discuss under what circumstances the binomial probability formula should be used in a probability problem.

*Suppose 2 cards are drawn without replacement from an ordinary deck of 52 cards. Find the probabilities of the given results.*

**22.** Both cards are black.

**23.** Both cards are hearts.

**24.** Exactly 1 is a face card.

**25.** At most 1 is an ace.

*An ice cream stand contains 4 custard flavors, 6 ice cream flavors, and 2 frozen yogurt selections. Three customers come to the window. If each customer's selection is random, find the probability that the selections include*

**26.** all ice cream;

**27.** all custard;

**28.** at least one frozen yogurt;

**29.** one custard, one ice cream, and one frozen yogurt;

**30.** at most one ice cream.

**C** *In Exercises 31 and 32, we study the connection between sets (from Chapter 8) and combinations.*

**31.** Given a set with $n$ elements,

  **(a)** what is the number of subsets of size 0? of size 1? of size 2? of size $n$?

  **(b)** Using your answer from part (a) give an expression for the total number of subsets of a set with $n$ elements.

**32.** Using your answers from Exercise 31 and the results from Chapter 8

  **(a)** explain why the following equation must be true:

$$ {}_nC_0 + {}_nC_1 + {}_nC_2 + \cdots + {}_nC_n = 2^n. $$

  **(b)** verify the equation in part (a) for $n = 4$ and $n = 5$.

**33. Stock Performance** On May 25, 2016, the percentage of stocks on the NASDAQ stock exchange that advanced was 50.7%. (Data from. www.wsj.com.) Suppose 7 stocks from the NASDAQ are selected at random, find

  **(a)** the probability distribution for $x$, the number of advancing stocks among the 7 selected;

  **(b)** the expected value of $x$.

**34. Stock Performance** On May 25, 2016, the percentage of stocks on the NYSE stock exchange that declined was 45.7%. (Data from: www.wsj.com.) Suppose 6 stocks from the NYSE are selected at random, find

  **(a)** the probability distribution for $x$, the number of declining stocks among the 6 selected;

  **(b)** the expected value of $x$.

**35. Internet Access** In 2016, the percentage of the U.S. population with access to the Internet at home was 88.5%. If 4 people are selected at random, find the probability that the given numbers have Internet access at home. (Data from: www.internetlivestats.com.)

  **(a)** None

  **(b)** At least 2 people

  **(c)** All 4 people

**36. Smartphone Ownership** The percentage of the U.S. population who own a smartphone in 2016 was 77%. If 5 people are

selected at random, find the probability that the given numbers own a smartphone. (Data from: www.pewinternet.org.)

  **(a)** Exactly 4

  **(b)** At most 4

  **(c)** At least 1

**37. Bank Earnings** From 2015 to 2016, 65.2% of banks insured by the Federal Deposit Insurance Corporation (FDIC) reported higher earnings. Suppose we select 5 banks insured by the FDIC at random. (Data from: www.fdic.gov.)

  **(a)** Give the probability distribution for $x$, the number of banks that reported higher earnings.

  **(b)** Find the probability that at least 2 banks reported higher earnings.

  **(c)** Give the expected value for the number of banks reporting higher earnings.

  **(d)** Explain what your answer in part (c) means.

**38. Household Earnings** Approximately 22.5% of American households earn $100,000 or more a year. Suppose we select 5 households at random. (Data from: statisticalatlas.com.)

  **(a)** Give the probability distribution for $x$, the number of households earning $100,000 or more.

  **(b)** Find the probability that at most 3 households earned $100,000 or more.

  **(c)** Give the expected value for the number of households earning $100,000 or more.

  **(d)** Explain what your answer in part (c) means.

*Decide whether each matrix is a regular transition matrix.*

**39.** $\begin{bmatrix} 0 & 1 \\ .77 & .23 \end{bmatrix}$  **40.** $\begin{bmatrix} -.2 & .4 \\ .3 & .7 \end{bmatrix}$

**41.** $\begin{bmatrix} .21 & .15 & .64 \\ .50 & .12 & .38 \\ 1 & 0 & 0 \end{bmatrix}$  **42.** $\begin{bmatrix} .22 & 0 & .78 \\ .40 & .33 & .27 \\ 0 & .61 & .39 \end{bmatrix}$

**43. E-Mail Usage** Using e-mail for professional correspondence has become a major component of a worker's day. A study classified e-mail use into 3 categories for an office day: no use, light use (1–60 minutes), and heavy use (more than 60 minutes). Researchers observed a pool of 100 office workers over a month and developed the following transition matrix of probabilities from day to day:

|  | Current Day | | |
|---|---|---|---|
|  | No Use | Light Use | Heavy Use |
| No use | .35 | .15 | .50 |
| Previous Day Light Use | .30 | .35 | .35 |
| Heavy Use | .15 | .30 | .55 |

Suppose the initial distribution for the three categories is [.2, .4, .4]. Find the distribution after

  **(a)** 1 day;

  **(b)** 2 days.

  **(c)** What is the long-range prediction for the distribution of e-mail use?

**44. International Investing** An analyst at a major brokerage firm that invests in Europe, North America, and Asia has examined the investment records for a particular international stock mutual fund over several years. The analyst constructed the following transition matrix for the probability of switching the location of an equity from year to year:

*Current Year*

|  |  | North |  |
|---|---|---|---|
|  | Europe | America | Asia |

|  |  |  |  |  |
|---|---|---|---|---|
|  | Europe | .80 | .14 | .06 |
| *Previous* North America | | .04 | .85 | .11 |
| *Year*  Asia | | .03 | .13 | .84 |

If the initial investment vector is 15% in Europe, 60% in North America, and 25% in Asia,

(a) find the percentages in Europe, North America, and Asia after 1 year;

(b) find the percentages in Europe, North America, and Asia after 3 years;

(c) find the long-range percentages in Europe, North America, and Asia.

45. **Campaign Strategy**  A candidate for city council can come out in favor of a new factory, be opposed to it, or waffle on the issue. The change in votes for the candidate depends on what her opponent does, with payoffs as shown in the following matrix:

*Opponent*

|  | Favors | Waffles | Opposes |
|---|---|---|---|
| Favors | 0 | −1000 | −4000 |
| *Candidate*  Waffles | 1000 | 0 | −500 |
| Opposes | 5000 | 2000 | 0 |

(a) What should the candidate do if she is an optimist?

(b) What should she do if she is a pessimist?

(c) Suppose the candidate's campaign manager feels that there is a 40% chance that the opponent will favor the plant and a

35% chance that he will waffle. What strategy should the candidate adopt? What is the expected change in the number of votes?

(d) The opponent conducts a new poll that shows strong opposition to the new factory. This changes the probability that he will favor the factory to 0 and the probability that he will waffle to .7. What strategy should our candidate adopt? What is the expected change in the number of votes now?

46. **Teaching Strategies**  When teaching, an instructor can adopt a strategy using either active learning or lecturing to help students learn best. A class often reacts very differently to these two strategies. A class can prefer lecturing or active learning. A department chair constructs the following payoff matrix of the average point gain (out of 500 possible points) on the final exam after studying many classes that use active learning and many that use lecturing and polling students as to their preference:

*Students in class prefer*

|  |  | Lecture | Active Learning |
|---|---|---|---|
| *Instructor*  Lecture | | 50 | −80 |
| *uses*  Active Learning | | −30 | 100 |

(a) If the department chair uses the preceding information to decide how to teach her own classes, what should she do if she is an optimist?

(b) What about if she is a pessimist?

(c) If the polling data shows that there is a 75% chance that a class will prefer the lecture format, what strategy should she adopt? What is the expected payoff?

(d) If the chair finds out that her next class has had more experience with active learning, so that there is now a 60% chance that the class will prefer active learning, what strategy should she adopt? What is the expected payoff?

# Case Study 9  Quick Draw® from the New York State Lottery

At bars and restaurants in the state of New York, patrons can play an electronic lottery game called Quick Draw.* A similar game is available in many other states. There are 10 ways for a patron to play this game. Prior to the draw, a person may bet $1 on games called 10-spot, 9-spot, 8-spot, 7-spot, 6-spot, 5-spot, 4-spot, 3-spot, 2-spot, and 1-spot. Depending on the game, the player will choose numbers from 1 to 80. For the 10-spot game, the player chooses 10 numbers; for a 9-spot game, the player chooses 9 numbers; etc. Every four minutes, the State of New York chooses 20 numbers at random from the numbers 1 to 80. For example, if a player chose the 6-spot game, he or she will have picked 6 numbers. If 3, 4, 5, or 6 of the numbers the player picked are also numbers the state picked randomly, then the player will win money. Each game has different ways to win, with differing payoff amounts. Notice with the 10-spot, 9-spot, 8-spot, and 7-spot, a player can win by matching 0 numbers correctly. The accompanying tables show the

payoffs for the different games. Notice that a player does not have to match all the numbers he or she picked in order to win.

| 10-spot Game | | 9-spot Game | |
|---|---|---|---|
| **Numbers Matched** | **Winnings per $1 Played** | **Numbers Matched** | **Winnings per $1 Played** |
| 10 | $100,000 | 9 | $30,000 |
| 9 | $5000 | 8 | $3000 |
| 8 | $300 | 7 | $125 |
| 7 | $45 | 6 | $20 |
| 6 | $10 | 5 | $5 |
| 5 | $2 | 0 | $2 |
| 0 | $5 | | |

*More information on Quick Draw can be found at www.nylottery.org; click on "Daily Games."

| 8-spot Game | |
|---|---|
| Numbers Matched | Winnings per $1 Played |
| 8 | $10,000 |
| 7 | $550 |
| 6 | $75 |
| 5 | $6 |
| 0 | $2 |

| 7-spot Game | |
|---|---|
| Numbers Matched | Winnings per $1 Played |
| 7 | $5000 |
| 6 | $100 |
| 5 | $20 |
| 4 | $2 |
| 0 | $1 |

| 6-spot Game | |
|---|---|
| Numbers Matched | Winnings per $1 Played |
| 6 | $1000 |
| 5 | $55 |
| 4 | $6 |
| 3 | $1 |

| 5-spot Game | |
|---|---|
| Numbers Matched | Winnings per $1 Played |
| 5 | $300 |
| 4 | $20 |
| 3 | $2 |

| 4-spot Game | |
|---|---|
| Numbers Matched | Winnings per $1 Played |
| 4 | $55 |
| 3 | $5 |
| 2 | $1 |

| 3-spot Game | |
|---|---|
| Numbers Matched | Winnings per $1 Played |
| 3 | $23 |
| 2 | $2 |

| 2-spot Game | |
|---|---|
| Numbers Matched | Winnings per $1 Played |
| 2 | $10 |

| 1-spot Game | |
|---|---|
| Numbers Matched | Winnings per $1 Played |
| 1 | $2 |

With our knowledge of counting, it is possible for us to calculate the probability of winning for these different games.

## EXAMPLE 1
Find the probability distribution for the number of matches for the 6-spot game.

**Solution** Let us define $x$ to be the number of matches when playing 6-spot. The outcomes of $x$ are then 0, 1, 2, . . . , 6. To find the probabilities of these matches, we need to do a little thinking. First, we need to know how many ways a player can pick 6 numbers from the selection of 1 to 80. Since the order in which the player picks the numbers does not matter, the number of ways to pick 6 numbers is

$$_{80}C_6 = \frac{80!}{74! \, 6!} = 300,500,200.$$

To find the probability of the outcomes of 0 to 6, we can think of the 80 choices broken into groups: 20 winning numbers the state picked and 60 losing numbers the state did not pick. If $x$, the number of matches, is 0, then the player picked 0 numbers from the 20 winning

numbers and 6 from the 60 losing numbers. Using the multiplication principle, we find that this quantity is

$$_{20}C_0 \cdot {}_{60}C_6 = \left(\frac{20!}{20! \, 0!}\right)\left(\frac{60!}{54! \, 6!}\right) = (1)(50,063,860) = 50,063,860.$$

Therefore,

$$P(x = 0) = \frac{_{20}C_0 \cdot {}_{60}C_6}{_{80}C_6} = \frac{50,063,860}{300,500,200} \approx .16660.$$

Similarly for $x = 1, 2, \ldots, 6$, and completing the probability distribution table, we have

| $x$ | $P(x)$ |
|---|---|
| 0 | $\frac{_{20}C_0 \cdot {}_{60}C_6}{_{80}C_6} \approx .16660$ |
| 1 | $\frac{_{20}C_1 \cdot {}_{60}C_5}{_{80}C_6} \approx .36349$ |
| 2 | $\frac{_{20}C_2 \cdot {}_{60}C_4}{_{80}C_6} \approx .30832$ |
| 3 | $\frac{_{20}C_3 \cdot {}_{60}C_3}{_{80}C_6} \approx .12982$ |
| 4 | $\frac{_{20}C_4 \cdot {}_{60}C_2}{_{80}C_6} \approx .02854$ |
| 5 | $\frac{_{20}C_5 \cdot {}_{60}C_1}{_{80}C_6} \approx .00310$ |
| 6 | $\frac{_{20}C_6 \cdot {}_{60}C_0}{_{80}C_6} \approx .00013$ |

## EXAMPLE 2
Find the expected winnings for a $1 bet on the 6-spot game.

**Solution** To find the expected winnings, we take the winnings for each number of matches and subtract our $1 initial payment fee. Thus, we have the following:

| $x$ | Net Winnings | $P(x)$ |
|---|---|---|
| 0 | −$1 | .16660 |
| 1 | −$1 | .36349 |
| 2 | −$1 | .30832 |
| 3 | $0 | .12982 |
| 4 | $5 | .02854 |
| 5 | $54 | .00310 |
| 6 | $999 | .00013 |

The expected winnings are

$$\begin{aligned}E(\text{winnings}) = & (-1)(.16660) + (-1)(.36349) \\ & + (-1)(.30832) + 0(.12982) + 5(.02854) \\ & + 54(.00310) + 999(.00013) \\ = & -.39844.\end{aligned}$$

Thus, for every $1 bet on the 6-spot game, a player would lose about 40 cents. Put another way, the state gets about 40 cents, on average, from every $1 bet on 6-spot.

## Exercises

1. If New York State initiates a promotion where players earn "double payoffs" for the 6-spot game (that is, if a player matched 3 numbers, she would win $2; if she matched 4 numbers, she would win $12; etc.), find the expected winnings.

2. Would it be in the state's interest to offer such a promotion? Why or why not?

3. Find the probability distribution for the 4-spot game.

4. Find the expected winnings for the 4-spot game.

5. If the state offers double payoffs for the 4-spot game, what are the expected winnings?

# Extended Projects

1. In Ohio, a game similar to Quick Draw is KENO. Investigate the rules, bets, and payoffs of this game. Determine in what ways it is similar to Quick Draw and in what way it differs. Examine the expected value if any of the games were to offer double payoffs. Are any of the expected values with double payoffs positive?

2. Investigate whether your own state or a state near you has a game similar to Quick Draw. Determine if the rules, bets, and payoffs of this game are similar to Quick Draw. Determine in what ways the game in your state is similar to Quick Draw and in what way it differs. Examine the consequences on the expected value if the games were to offer double payoffs.

# Introduction to Statistics

# 10
## CHAPTER

Statistics has applications to almost every aspect of modern life. The digital age is creating a wealth of data that needs to be summarized, visualized, and analyzed, from the payroll of major-league baseball teams to movies' box-office receipts and the revenue of the home improvement industry.

Statistics is the science that deals with the collection and summarization of data. Methods of statistical analysis make it possible to draw conclusions about a population on the basis of data from a sample of the population. Statistical models have become increasingly useful in manufacturing, government, agriculture, medicine, and the social sciences and in all types of research. An Indianapolis race-car team, for example, is using statistics to improve its performance by gathering data on each run around the track. The team samples data 300 times a second and uses computers to process the data. In this chapter, we give a brief introduction to some of the key topics from statistical methodology.

## 10.1 Frequency Distributions

Researchers often wish to learn characteristics or traits of a specific **population** of individuals, objects, or units. The traits of interest are called **variables,** and it is these that we measure or label. Often, however, a population of interest is very large or constantly changing, so measuring each unit is impossible. Thus, researchers are forced to collect data on a subset of the population of interest, called a **sample.**

Sampling is a complex topic, but the universal aim of all sampling methods is to obtain a sample that "represents" the population of interest. One common way of obtaining a representative sample is to perform simple random sampling, in which every unit of the population has an equal chance to be selected to be in the sample. Suppose we wanted to study the height of students enrolled in a class. To obtain a random sample, we could place slips of paper containing the names of everyone in class in a hat, mix the papers, and draw 10 names blindly. We would then record the height (the variable of interest) for each student selected.

A simple random sample can be difficult to obtain in real life. For example, suppose you want to take a random sample of voters in your congressional district to see which candidate they prefer in the next election. If you do a telephone survey, you have a representative sample of people who are at home to answer the telephone, but those who are rarely home to answer the phone, those who have an unlisted number or only a cell phone, those who cannot afford a telephone, and those who refuse to answer telephone surveys are underrepresented. Such people may have an opinion different from those of the people you interview.

A famous example of an inaccurate poll was made by the *Literary Digest* in 1936. Its survey indicated that Alfred Landon would win the presidential election; in fact, Franklin Roosevelt won with 62% of the popular vote. The *Digest*'s major error was mailing its surveys to a sample of those listed in telephone directories. During the Depression, many poor people did not have telephones, and the poor voted overwhelmingly for Roosevelt. Modern pollsters use sophisticated techniques in an attempt to make their sample as representative as possible.

Once a sample has been collected and all data of interest are recorded, the data must be organized so that conclusions may be more easily drawn. With numeric responses, one method of organization is to group the data into intervals, usually of equal size.

---

**EXAMPLE 1**     **Tuition Costs** The following list consists of the 2017–2018 annual tuition (in thousands of dollars) for a random sample of 30 community colleges. (Data from: www.collegetuitioncompare.com.)

| 1.1 | 1.4 | 1.4 | 3.7 | 3.6 | 2.4 | 4.2 | 2.8 | 3.5 | 3.8 |
| 5.9 | 2.8 | 3.5 | 3.7 | 3.7 | 4.4 | 3.8 | 2.7 | 4.7 | 1.2 |
| 2.8 | 4.2 | 4.9 | 5.3 | 4.4 | 2.9 | 3.1 | 1.4 | 3.9 | 5.1 |

Identify the population and the variable, group the data into intervals, and find the frequency of each interval.

**Solution** The population is all community colleges. The variable of interest is the tuition (in thousands of dollars). The lowest value is 1.1 (corresponding to $1100) and the highest value is 5.9 (corresponding to $5900). One convenient way to group the intervals is in intervals of size .5, starting with 1.0–1.4 and ending with 5.0–5.4. This grouping provides an interval for each value in the list and results in 9 equally sized intervals of a convenient size. Too many intervals of smaller size would not simplify the data enough, while too few intervals of larger size would conceal information that the data might provide.

The first step in summarizing the data is to tally the number of schools in each interval. Then total the tallies in each interval, as in the following table:

| Tuition Amount (in thousands of dollars) | Tally | Frequency |
|---|---|---|
| 1.0–1.4 | Ⅳ | 5 |
| 1.5–1.9 | | 0 |
| 2.0–2.4 | Ⅰ | 1 |
| 2.5–2.9 | ⅣⅠ | 5 |
| 3.0–3.4 | Ⅰ | 1 |
| 3.5–3.9 | ⅣⅢ ⅢⅠ | 9 |
| 4.0–4.4 | ⅢⅠ | 4 |
| 4.5–4.9 | ‖ | 2 |
| 5.0–5.4 | ‖ | 2 |
| 5.5–5.9 | Ⅰ | 1 |

✓ **Checkpoint 1**

A restaurant trade group commissioned a study to find how often families eat out. The following data gives the number of times a family ate a meal at a restaurant in the last month for 24 families surveyed at random.

| 8 | 12 | 0 | 6 | 10 | 8 | 0 | 14 |
| 8 | 12 | 14 | 16 | 4 | 14 | 7 | 11 |
| 9 | 12 | 7 | 15 | 11 | 21 | 22 | 19 |

Prepare a grouped frequency distribution for these data. Use intervals 0–4, 5–9, and so on.

*Answers to Checkpoint exercises are found at the end of the section.*

This table is an example of a **grouped frequency distribution.**

The frequency distribution in Example 1 shows information about the data that might not have been noticed before. For example, the interval with the largest number of colleges is 3.5–3.9. However, some information has been lost; for example, we no longer know exactly how many colleges charged 3.9 ($3900) for tuition.

## Picturing Data

The information in a grouped frequency distribution can be displayed graphically with a **histogram,** which is similar to a bar graph. In a histogram, the number of observations in each interval determines the height of each bar, and the size of each interval determines the width of each bar. If equally sized intervals are used, all the bars have the same width.

A **frequency polygon** is another form of graph that illustrates a grouped frequency distribution. The polygon is formed by joining consecutive midpoints of the tops of the histogram bars with straight-line segments. Sometimes the midpoints of the first and last bars are joined to endpoints on the horizontal axis where the next midpoint would appear. (See Figure 10.1.)

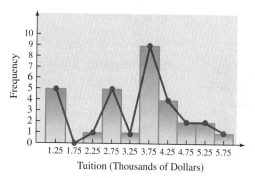

**Figure 10.1**

---

**EXAMPLE 2**   **Tuition Costs**  A grouped frequency distribution of community college tuition was found in Example 1. Draw a histogram and a frequency polygon for this distribution.

**Solution**  First, draw a histogram, shown in blue in Figure 10.1. To get a frequency polygon, connect consecutive midpoints of the tops of the bars. The frequency polygon is shown in red. ✓₂

✓**Checkpoint 2**

Make a histogram and a frequency polygon for the distribution found in Checkpoint 1.

**TECHNOLOGY TIP** As noted in Section 9.1, most graphing calculators can display histograms. Many will also display frequency polygons (which are usually labeled LINE or xyLINE in calculator menus). When dealing with grouped frequency distributions, however, certain adjustments must be made on a calculator:

1.   *The histogram's bar width affects the shape of the graph.* If you use a bar width of .4 in Example 1, the calculator may produce a histogram with gaps in it. To avoid this, use the interval $1.0 \leq x < 1.5$ in place of $1.1 \leq x \leq 1.4$, and similarly for other intervals, and make .5 the bar width.

2.   *A calculator list of outcomes must consist of single numbers, not intervals.* The table in Example 1, for instance, cannot be entered as shown. To convert the first column of the table for calculator use, choose the midpoint number in each interval—1.25 for the first interval, 1.75 for the second interval, 2.25 for the third interval, etc. Then use 1.25, 1.75, 2.25, . . . , 5.75 as the list of outcomes to be entered into the calculator. The frequency list (the last column of the table) remains the same.

**Figure 10.2**

Following this procedure, we obtain the calculator-generated histogram and frequency polygon in Figure 10.2 for the data from Example 1. Note that the width of each histogram bar is .5. Some calculators cannot display both the histogram and the frequency polygon on the same screen, as is done here.

Stem-and-leaf plots allow us to organize the data into a distribution without the disadvantage of losing the original information. In a **stem-and-leaf plot,** we separate the digits in each data point into two parts consisting of the first one or two digits (the stem) and the remaining digit (the leaf). We also provide a key to show the reader the units of the data that were recorded.

---

**EXAMPLE 3**    **Tuition Costs**  Construct a stem-and-leaf plot for the data in Example 1.

**Solution**  Since the data is made up of two-digit numbers, we use the first digit for the stems: 1, 2, 3, 4, and 5 (which in this case represents thousands of dollars). The second digit provides the leaves (which here represents hundreds of dollars). For example, if we look at the seventh row of the stem-and-leaf plot, we have a stem value of 4 and leaf values of 2, 2, 4, and 4. These values correspond to entries of 4.2, 4.2, 4.4, and 4.4, meaning that one college had tuition of $4200, another had tuition of $4200, a third college had tuition of $4400, and a fourth college also had tuition of $4400. In this example, each row corresponds to an interval in the frequency table. The stems and leaves are separated by a vertical line.

| Stem | Leaves |
|------|--------|
| 1 | 12444 |
| 1 | |
| 2 | 4 |
| 2 | 78889 |
| 3 | 1 |
| 3 | 556777889 |
| 4 | 2244 |
| 4 | 79 |
| 5 | 13 |
| 5 | 9 |

*Units:* 5|9 = $5900

If we turn the page on its side, the distribution looks like a histogram, but a stem-and-leaf plot still retains each of the original data values. We used each stem digit twice because, as with a histogram, using too few intervals conceals useful information about the shape of the distribution. ✔3

✓ **Checkpoint 3**

Make a stem-and-leaf plot for the data in Example 1, using one stem each for 1, 2, 3, 4, and 5.

📄 **NOTE**  In this example we have split the leaves into two groups: those with leaf values of 0–4 and 5–9. We leave a row empty if there are no data values. For example, since there were no schools in Example 1 with tuition between $1500 and $1900, that row is left blank. We do not put a zero, however, as that would imply that a school had tuition of $1000. Not all stem-and-leaf plots split the leaves with two rows per stem, but it can help the reader discern the values more easily. Often there is only a single row per stem, but another common approach is to use five rows per stem where leaves of 0–1, 2–3, 4–5, 6–7, and 8–9 are grouped together.

EXAMPLE 4    **Resting Pulse**  List the original data for the following stem-and-leaf plot of resting pulses taken on the first day of class for 36 students:

| Stem | Leaves |
|------|--------|
| 4 | 8 |
| 5 | 278 |
| 6 | 034455688888 |
| 7 | 02222478 |
| 8 | 2269 |
| 9 | 00002289 |

*Units:* $9|0 = 90$ beats per minute

The first stem and its leaf correspond to the data point 48 beats per minute. Similarly, the rest of the data are 52, 57, 58, 60, 63, 64, 64, 65, 65, 66, 68, 68, 68, 68, 68, 70, 72, 72, 72, 72, 74, 77, 78, 82, 82, 86, 89, 90, 90, 90, 90, 92, 92, 98, and 99 beats per minute. ✔️4

✔️ **Checkpoint 4**

List the original data for the following heights (inches) of students:

| Stem | Leaves |
|------|--------|
| 5 | 9 |
| 6 | 00012233334444 |
| 6 | 55567777799 |
| 7 | 0111134 |
| 7 | 558 |

*Units:* $7|5 = 75$ inches

## Assessing the Shape of a Distribution

Histograms and stem-and-leaf plots are very useful in assessing what is called the **shape** of the distribution. One common shape of data is seen in Figure 10.3(a). When all the bars of a histogram are approximately the same height, we say the data has a **uniform** shape. In Figure 10.3(b), we see a histogram that is said to be bell shaped, or **normal.** We use the "normal" label when the frequency peaks in the middle and tapers off equally on each side. When the data do not taper off equally on each side, we say the data are **skewed.** If the data taper off further to the left, we say the data are **left skewed** (Figure 10.3(c)). When the data taper off further to the right, we say the data are **right skewed** (Figure 10.3(d)). (Notice that with skewed data, we say "left skewed" or "right skewed" to refer to the tail, and not the peak of the data.)

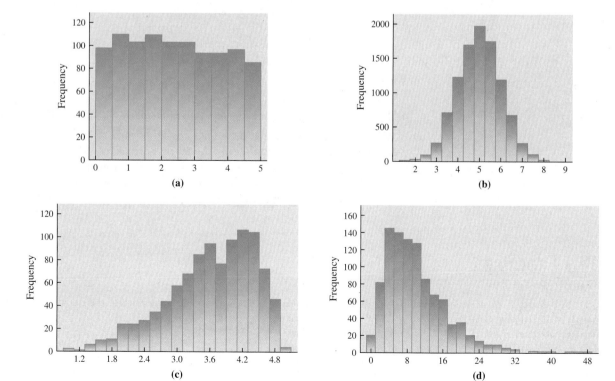

**Figure 10.3**

EXAMPLE 5    **Height and BMI** Characterize the shapes of the given distributions for 1000 adult males. (Data from: www.cdc.gov/nchs/nhanes.htm.)

**(a)** Height (inches); see Figure 10.4.

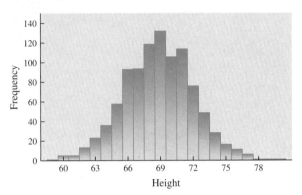

**Figure 10.4**

## Checkpoint 5

Characterize the shape of the distribution from the following stem-and-leaf plot of ages (years):

| Stem | Leaves |
|------|--------|
| 1 | 88888 |
| 2 | 22334 |
| 2 | 5579 |
| 3 | 333344 |
| 3 | 59 |
| 4 | 3344 |
| 4 | 667779 |
| 5 | 34 |
| 5 | 5679 |
| 6 | 1 |
| 6 | |
| 7 | 13 |
| 7 | 678 |

*Units:* 7|6 = 76 years

**Solution** The shape is **normal** because the shape peaks in the middle and tapers equally on each side.

**(b)** Body mass index ($kg/m^2$); see Figure 10.5.

**Figure 10.5**

**Solution** The shape is **right skewed** because the tail is to the right. ✓5

It is important to note that most data are *not* normal, as we will see in the upcoming exercises and the next section. Using the label "normal" is a bit of a misnomer, because many important distributions, such as income, house prices, and infant birth weights, are skewed. It is also important to know that not all distributions have an easy-to-classify shape. This is especially true when samples are small.

# 10.1 Exercises

**S&P 500 Stocks** *The data for Exercises 1–4 consist of a random sample of 20 companies that were part of the S&P 500 and information was current as of March 7, 2017. For each variable, (a) group the data as indicated; (b) prepare a frequency distribution with columns for intervals and frequencies; (c) construct a histogram. (See Examples 1 and 2.) (Data from: data.okfn.org.)*

**1.** The variable is the price of the stock. Use 6 intervals starting with 0–39.99, 40–79.99, and so on.

| | | | | |
|---|---|---|---|---|
| 71.95 | 239.60 | 182.02 | 78.94 | 73.72 |
| 167.00 | 75.97 | 27.20 | 71.36 | 62.55 |
| 82.77 | 58.56 | 67.39 | 143.60 | 37.52 |
| 126.26 | 41.47 | 26.66 | 30.81 | 23.50 |

2. The variable is the dividend yield (percent). Use 10 intervals with 0–.49, .5–.99, and so on.

| | | | | |
|---|---|---|---|---|
| 2.38 | 1.14 | 3.12 | 2.26 | 2.12 |
| 1.06 | 1.50 | 1.62 | 1.37 | 3.23 |
| 1.64 | 3.26 | 0 | 2.14 | 3.98 |
| 2.11 | 2.66 | 0.94 | 4.75 | 2.95 |

3. The variable is the price/earnings ratio (calculated as the market price per share divided by the earnings per share). Use 6 intervals starting with 0–14.99, 15–29.99, and so on.

| | | | | |
|---|---|---|---|---|
| 11.21 | 6.28 | 23.92 | 21.99 | 27.10 |
| 30.81 | 16.55 | 13.90 | 16.74 | 5.36 |
| 28.60 | 19.78 | 12.50 | 25.92 | 6.25 |
| 20.36 | 33.7 | 79.11 | 15.48 | 23.02 |

4. The variable is market capitalization (in billions of dollars).

| | | | | |
|---|---|---|---|---|
| 28.88 | 80.32 | 111.48 | 11.15 | 65.08 |
| 73.30 | 20.48 | 15.92 | 19.71 | 22.96 |
| 30.33 | 18.56 | 40.82 | 9.30 | 56.20 |
| 96.11 | 38.93 | 9.59 | 9.37 | 7.60 |

*The data for Exercises 5–10 consist of random samples of 20 households from the 2015 American Community Survey. Construct a frequency distribution and a histogram for each data set.*

5. Annual household income (in thousands of dollars).

| | | | | | | | | | |
|---|---|---|---|---|---|---|---|---|---|
| 146 | 100 | 15 | 9 | 129 | 49 | 3 | 75 | 62 | 141 |
| 65 | 25 | 117 | 157 | 98 | 45 | 29 | 221 | 180 | 74 |

6. Amount (in dollars) of a monthly mortgage payment for homeowners.

| | | | | | | | | | |
|---|---|---|---|---|---|---|---|---|---|
| 930 | 280 | 600 | 2100 | 670 | 700 | 1300 | 640 | 1600 | 3200 |
| 390 | 1100 | 500 | 1200 | 2600 | 310 | 980 | 900 | 370 | 350 |

7. Amount (in dollars) of monthly electric bill.

| | | | | | | | | | |
|---|---|---|---|---|---|---|---|---|---|
| 180 | 110 | 180 | 350 | 200 | 360 | 60 | 230 | 100 | 100 |
| 60 | 460 | 270 | 30 | 150 | 90 | 170 | 100 | 140 | 140 |

8. Amount (in dollars) of annual water and sewage bill.

| | | | | | | | | | |
|---|---|---|---|---|---|---|---|---|---|
| 1200 | 380 | 940 | 750 | 300 | 1200 | 150 | 600 | 600 | 600 |
| 600 | 1000 | 500 | 960 | 700 | 1100 | 360 | 750 | 2000 | 1100 |

9. Amount (in dollars) of a monthly rent payment for nonhomeowners.

| | | | | | | | | | |
|---|---|---|---|---|---|---|---|---|---|
| 900 | 860 | 780 | 1900 | 1500 | 1400 | 600 | 800 | 600 | 1200 |
| 720 | 1300 | 800 | 1500 | 1500 | 480 | 800 | 300 | 1200 | 910 |

10. Amount (in dollars) of a monthly gas payment.

| | | | | | | | | | |
|---|---|---|---|---|---|---|---|---|---|
| 180 | 40 | 170 | 140 | 130 | 90 | 200 | 200 | 50 | 40 |
| 90 | 50 | 90 | 100 | 20 | 80 | 30 | 90 | 40 | 130 |

*For Exercises 11–20, construct a stem-and-leaf plot for the data in the indicated exercise. (See Examples 3 and 4.)*

11. **Stocks** Stock prices (rounded to the nearest dollar).

| | | | | | | | | | |
|---|---|---|---|---|---|---|---|---|---|
| 25 | 64 | 32 | 99 | 74 | 27 | 66 | 72 | 83 | 54 |
| 59 | 67 | 72 | 66 | 54 | 52 | 33 | 75 | 43 | 39 |

12. **Stocks** Stock prices (rounded to the nearest dollar).

| | | | | | | | | | |
|---|---|---|---|---|---|---|---|---|---|
| 85 | 110 | 76 | 98 | 120 | 105 | 84 | 77 | 94 | 100 |
| 102 | 125 | 84 | 93 | 98 | 132 | 84 | 139 | 112 | 77 |

13. **Height** Heights (in inches) of adult females.

| | | | | | | | | | |
|---|---|---|---|---|---|---|---|---|---|
| 62 | 64 | 64 | 69 | 66 | 62 | 65 | 68 | 61 | 64 |
| 60 | 61 | 64 | 66 | 54 | 61 | 62 | 62 | 63 | 65 |

14. **Height** Heights (in inches) of adult males

| | | | | | | | | | |
|---|---|---|---|---|---|---|---|---|---|
| 69 | 70 | 68 | 72 | 72 | 70 | 73 | 71 | 72 | 69 |
| 68 | 72 | 73 | 69 | 75 | 71 | 71 | 66 | 68 | 73 |

15. **Stocks** The price/earnings ratio for the 20 stocks in Exercise 3 (rounded to the nearest integer).

| | | | | | | | | | |
|---|---|---|---|---|---|---|---|---|---|
| 11 | 6 | 24 | 22 | 27 | 31 | 17 | 14 | 17 | 5 |
| 29 | 20 | 13 | 26 | 6 | 20 | 34 | 79 | 15 | 23 |

16. **Stocks** The dividend yield for the 20 stocks in Exercise 2 (rounded to the nearest tenth).

| | | | | | | | | | |
|---|---|---|---|---|---|---|---|---|---|
| 2.4 | 1.1 | 3.1 | 2.3 | 2.1 | 1.1 | 1.5 | 1.6 | 1.4 | 3.2 |
| 1.6 | 3.3 | 0 | 2.1 | 4.0 | 2.1 | 2.7 | .9 | 4.8 | 3.0 |

17. **High School Graduates** The percentage of residents in the first 25 states alphabetically with a high school diploma. (Data from: National Center for Education Statistics.)

| | | | | |
|---|---|---|---|---|
| 85 | 91 | 91 | 90 | 90 |
| 92 | 89 | 90 | 84 | 90 |
| 86 | 90 | 88 | 84 | 93 |
| 86 | 87 | 89 | 92 | 83 |
| 82 | 86 | 92 | 90 | 89 |

18. **High School Graduates** The percentage of residents in the latter 25 states alphabetically with a high school diploma. (Data from: National Center for Education Statistics.)

| | | | | |
|---|---|---|---|---|
| 92 | 85 | 88 | 91 | 89 |
| 90 | 86 | 90 | 86 | 90 |
| 85 | 86 | 90 | 82 | 86 |
| 93 | 93 | 85 | 91 | 91 |
| 89 | 89 | 86 | 92 | 92 |

19. **Bachelor's Degrees** The percentage of residents in the first 25 states alphabetically with a bachelor's degree or higher. (Data from: National Center for Education Statistics.)

| | | | | |
|---|---|---|---|---|
| 23 | 38 | 32 | 32 | 41 |
| 29 | 38 | 25 | 23 | 27 |
| 27 | 30 | 33 | 23 | 34 |
| 21 | 27 | 25 | 30 | 21 |
| 32 | 29 | 28 | 38 | 28 |

20. **Bachelor's Degrees** The percentage of residents in the latter 25 states alphabetically with a bachelor's degree or higher. (Data from: National Center for Education Statistics.)

| | | | | |
|---|---|---|---|---|
| 28 | 27 | 24 | 27 | 37 |
| 30 | 35 | 31 | 25 | 33 |
| 23 | 29 | 29 | 28 | 19 |
| 35 | 26 | 30 | 31 | 29 |
| 37 | 27 | 26 | 36 | 26 |

*Describe the shape of each of the given histograms. (See Example 5.)*

**21.**

**22.**

**23.**

**24.**

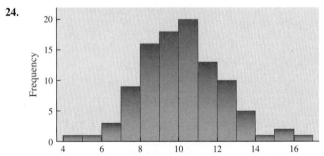

**25. Student Loan Defaults** The following histogram shows the student loan default rates for the 50 states and the District of Columbia reported in 2017. (Data from: www.studentloans.net.)

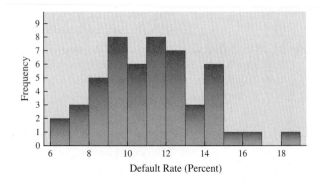

Default Rate (Percent)

**(a)** Describe the shape of the histogram.

**(b)** How many states had a student loan default rate between 6 and 6.99%?

**(c)** How many states had a student loan default rate of 14% or higher?

**26. Stocks** The following histogram shows the stock price of 50 randomly selected stocks from the S&P 500 on March 7, 2017. (Data from: data.okfn.org.)

Stock Price (Dollars)

**(a)** Describe the shape of the histogram.

**(b)** How many stocks had prices between $0 and $99.99?

**(c)** How many stocks had prices below $200?

**27. Cat Ownership** The stem-and-leaf plot below summarizes the percentage of households that own cats for the 50 U.S. states (excluding Alaska). (Data from: American Veterinary Medical Association.)

| Stem | Leaves |
|---|---|
| 2 | 455 |
| 2 | 677777 |
| 2 | 8899999999 |
| 3 | 0001111 |
| 3 | 22223333333 |
| 3 | 4444 |
| 3 | 6 |
| 3 | 899 |
| 4 | 0 |
| 4 | |
| 4 | |
| 4 | 6 |
| 4 | 9 |

*Units:* 4|9 = 49%

**(a)** Describe the shape of the distribution.

**(b)** How many states have the percentage of cat ownership in the thirties?

**(c)** How many states have the percentage of cat ownership above 35%?

**28. Personal Bankruptcies** The stem-and-leaf plot below gives the number of personal bankruptcies per 100,000 residents for the 50 U.S. states and the District of Columbia. (Data from: www.worldatlas.com.)

| Stem | Leaves |
|---|---|
| 0 | 59 |
| 1 | 01234 |
| 1 | 5555666678888 |
| 2 | 2 |
| 2 | 55567899 |
| 3 | 00112444 |
| 3 | 5666779 |
| 4 | 0 |
| 4 | 567 |
| 5 | 12 |
| 5 | |
| 6 | 1 |

*Units:* 6|1 = 610

**(a)** Describe the shape of the distribution.

**(b)** How many states have 400 or more bankruptcies per 100,000 residents?

**(c)** How many states have 150 or less bankruptcies per 100,000 residents?

**29. Test Scores** The grade distribution for scores on a final exam is shown in the following stem-and-leaf plot:

| Stem | Leaves |
|---|---|
| 2 | 7 |
| 3 | |
| 3 | |
| 4 | 01 |
| 4 | 899 |
| 5 | 4 |
| 5 | 5 |
| 6 | 122 |
| 6 | 58 |
| 7 | 00124 |
| 7 | 9 |
| 8 | 0044 |
| 8 | 5679 |
| 9 | 00223334 |
| 9 | 5788 |

*Units:* 9|5 = 95%

**(a)** What is the shape of the grade distribution?

**(b)** How many students earned 90% or better?

**(c)** How many students earned less than 60%?

**30. Chicken Production** For the 20 states that produce a large number of broiler chickens, the following stem-and-leaf plot gives the production in billions of pounds for the year 2015. (Data from: U.S. Department of Agriculture.)

| Stem | Leaves |
|---|---|
| 0 | 223349 |
| 1 | 04447777 |
| 2 | |
| 3 | 7 |
| 4 | 5 |
| 5 | |
| 6 | 114 |
| 7 | 9 |

*Units:* 7|9 = 7.9 billion

**(a)** Describe the shape of the distribution.

**(b)** How many states had production below 1 billion pounds?

**(c)** How many states had production above 2 billion pounds?

## ✓Checkpoint Answers

**1.**

| Interval | Frequency |
|---|---|
| 0–4 | 3 |
| 5–9 | 7 |
| 10–14 | 9 |
| 15–19 | 3 |
| 20–24 | 2 |
| | Total: 24 |

**2.**

**3.**

| Stem | Leaves |
|---|---|
| 1 | 12444 |
| 2 | 478889 |
| 3 | 1556777889 |
| 4 | 224479 |
| 5 | 139 |

*Units:* 5|1 = $5,100

**4.** 59, 60, 60, 60, 61, 62, 62, 63, 63, 63, 63, 64, 64, 64, 64, 65, 65, 65, 66, 67, 67, 67, 67, 67, 69, 69, 70, 71, 71, 71, 71, 73, 74, 75, 75, 78

**5.** Right skewed

# 10.2 Measures of Center

Often, we want to summarize data numerically with a measure that represents a "typical" outcome. There are several ways to do this, and we generally call such a summary a "measure of center." In this section, we learn about the three most common measures of center: the mean, median, and mode.

## Mean

The three most important measures of center are the mean, the median, and the mode. The most used of these is the mean, which is similar to the expected value of a probability distribution. The **arithmetic mean** (or just the "mean" or sometimes called the "average") of a set of numbers is the sum of the numbers, divided by the total number of numbers. We write the sum of $n$ numbers $x_1, x_2, x_3, \ldots, x_n$ in a compact way with **summation notation**, also called **sigma notation.** With the Greek letter $\Sigma$ (sigma), the sum

$$x_1 + x_2 + x_3 + \cdots + x_n$$

is written

$$x_1 + x_2 + x_3 + \cdots + x_n = \sum_{i=1}^{n} x_i.$$

In statistics, $\sum_{i=1}^{n} x_i$ is often abbreviated as just $\sum x_i$. The symbol $\bar{x}$ (read "$x$-bar") is used to represent the mean of a sample.

> ## Mean
>
> The mean of the $n$ numbers $x_1, x_2, x_3, \ldots, x_n$ is
>
> $$\bar{x} = \frac{x_1 + x_2 + \cdots + x_n}{n} = \frac{\sum x_i}{n}.$$

**TECHNOLOGY TIP** Computing the mean is greatly simplified by the statistical capabilities of most scientific and graphing calculators. Calculators vary considerably in how data are entered, so read your instruction manual to learn how to enter lists of data and the corresponding frequencies. On scientific calculators with statistical capabilities, there are keys for finding most of the measures of center discussed in this section. On graphing calculators, most or all of these measures can be obtained with a single keystroke. (Look for a *one-variable statistics* option, which is often labeled 1-VAR, in the STAT menu or its CALC submenu.)

**TECHNOLOGY TIP**
The mean of the five numbers in Example 1 is easily found by using the $\bar{x}$ key on a scientific calculator or the one-variable statistics key on a graphing calculator. A graphing calculator will also display additional information, which will be discussed in the next section.

✓ **Checkpoint 1**

Find the mean dollar amount of the following purchases of eight students selected at random at the campus bookstore during the first week of classes:

| | |
|---|---|
| $250.56 | $567.32 |
| $45.29 | $321.56 |
| $120.22 | $561.04 |
| $321.07 | $226.90 |

**EXAMPLE 1**    **Apple Revenue** The revenue, in billions of dollars, for Apple Inc. for the last five years is given in the following table. (Data from: www.morningstar.com.)

| Year | 2012 | 2013 | 2014 | 2015 | 2016 |
|---|---|---|---|---|---|
| Revenue | 157 | 171 | 183 | 234 | 216 |

Find the mean of revenue for the five years.

**Solution** Let $x_1 = 157, x_2 = 171, x_3 = 183, x_4 = 234,$ and $x_5 = 216$. Here, $n = 5$ because there are 5 numbers in the list. Thus,

$$\bar{x} = \frac{157 + 171 + 183 + 234 + 216}{5} = 192.2.$$

The mean revenue for the 5-year period is $192,200,000,000.

The mean of data that have been arranged into a frequency distribution is found in a similar way. For example, suppose the following quiz score data are collected:

| Value | Frequency |
|-------|-----------|
| 84 | 2 |
| 87 | 4 |
| 88 | 7 |
| 93 | 4 |
| 99 | 3 |
|  | Total: 20 |

The value 84 appears twice, 87 four times, and so on. To find the mean, first add 84 two times, 87 four times, and so on; or get the same result faster by multiplying 84 by 2, 87 by 4, and so on, and then adding the results. Dividing the sum by 20, the total of the frequencies, gives the mean:

$$\bar{x} = \frac{(84 \cdot 2) + (87 \cdot 4) + (88 \cdot 7) + (93 \cdot 4) + (99 \cdot 3)}{20}$$

$$= \frac{168 + 348 + 616 + 372 + 297}{20}$$

$$= \frac{1801}{20}$$

$$\bar{x} = 90.05.$$

Verify that your calculator gives the same result.

### EXAMPLE 2    Students' Ages An instructor of a finite-mathematics class at a small liberal-arts college collects data on the age of her students. The data are recorded in the following frequency distribution:

| Age | Frequency | Age × Frequency |
|-----|-----------|-----------------|
| 18 | 12 | $18 \cdot 12 = 216$ |
| 19 | 9 | $19 \cdot 9 = 171$ |
| 20 | 5 | $20 \cdot 5 = 100$ |
| 21 | 2 | $21 \cdot 2 = 42$ |
| 22 | 2 | $22 \cdot 2 = 44$ |
|  | Total: 30 | Total: 573 |

Find the mean age.

**Solution** The age 18 appears 12 times, 19 nine times, and so on. To find the mean, first multiply 18 by 12, 19 by 9, and so on, to get the column "Age × Frequency," which has been added to the frequency distribution. Adding the products from this column gives a total of 573. The total from the frequency column is 30. The mean age is

$$\bar{x} = \frac{573}{30} = 19.1.$$ ✔2

The mean of grouped data is found in a similar way. For grouped data, intervals are used, rather than single values. To calculate the mean, it is assumed that all of the values in a given interval are located at the midpoint of the interval. We use $x_i$ to represent the midpoints, and $f_i$ represents the frequencies, as shown in the next example.

✓ **Checkpoint 2**

Find $\bar{x}$ for the following frequency distribution for the variable of years of schooling for a sample of construction workers.

| Years | Frequency |
|-------|-----------|
| 7 | 2 |
| 9 | 3 |
| 11 | 6 |
| 13 | 4 |
| 15 | 4 |
| 16 | 1 |

**EXAMPLE 3**    **College Tuition** The grouped frequency distribution for annual tuition (in thousands of dollars) for the 30 community colleges described in Example 1 of Section 10.1 is as follows:

| Tuition Amount (in thousands of dollars) | Frequency, $f_i$ | Midpoint, $x_i$ | Product, $f_i x_i$ |
|---|---|---|---|
| 1.0–1.49 | 5 | 1.25 | 6.25 |
| 1.5–1.99 | 0 | 1.75 | 0 |
| 2.0–2.49 | 1 | 2.25 | 2.25 |
| 2.5–2.99 | 5 | 2.75 | 13.75 |
| 3.0–3.49 | 1 | 3.25 | 3.25 |
| 3.5–3.99 | 9 | 3.75 | 33.75 |
| 4.0–4.49 | 4 | 4.25 | 17 |
| 4.5–4.99 | 2 | 4.75 | 9.5 |
| 5.0–5.49 | 2 | 5.25 | 10.5 |
| 5.5–5.99 | 1 | 5.75 | 5.75 |
| | Total = 30 | | Total = 102 |

Find the mean from the grouped frequency distribution.

**Solution** A column for the midpoint of each interval has been added. The numbers in this column are found by adding the endpoints of each interval and dividing by 2. For the interval 1.0–1.49, the midpoint is $(1.0 + 1.49)/2 = 1.245$ (which we will round to 1.25). The numbers in the product column on the right are found by multiplying each frequency by its corresponding midpoint. Finally, we divide the total of the product column by the total of the frequency column to get

$$\bar{x} = \frac{102}{30} = 3.4.$$

It is important to know that information is always lost when the data are grouped. It is more accurate to use the original data, rather than the grouped frequency, when calculating the mean, but the original data might not be available. Furthermore, the mean based upon the grouped data is typically not too different from the mean based upon the original data, and there may be situations in which the extra accuracy is not worth the effort. ✔️3

**NOTE** If we had used different intervals in Example 3, the mean would have come out to be a slightly different number. This is demonstrated in Checkpoint 4. ✔️4

The formula for the mean of a grouped frequency distribution is as follows.

## Mean of a Grouped Distribution

The mean of a distribution where $x_i$ represents the midpoint of interval $i$, $f_i$ represents the frequency of observations in interval $i$, and $n = \sum f_i x_i$ is

$$\bar{x} = \frac{\sum f_i x_i}{n}.$$

The mean of a random sample is a random variable, and for this reason it is sometimes called the **sample mean**. The sample mean is a random variable because it assigns a number to the experiment of taking a random sample. If a different random sample were taken, the mean would probably have a different value, with some values being more probable than

✔️**Checkpoint 3**

Find the mean of the following grouped frequency distribution for the number of classes completed thus far in the college careers of a random sample of 52 students:

| Classes | Frequency |
|---|---|
| 0–5 | 6 |
| 6–10 | 10 |
| 11–20 | 12 |
| 21–30 | 15 |
| 31–40 | 9 |

✔️**Checkpoint 4**

Find the mean for the college tuition data using the following intervals for the grouped frequency distribution:

| Tuition Amount | Frequency |
|---|---|
| 1.0–1.9 | 5 |
| 2.0–2.9 | 6 |
| 3.0–3.9 | 10 |
| 4.0–4.9 | 6 |
| 5.0–5.9 | 3 |

others. For example, if another set of 30 community colleges were selected in Example 3, the mean tuition amount might not have been 3.4.

We saw in Section 9.1 how to calculate the expected value of a random variable when we know its probability distribution. The expected value is sometimes called the **population mean,** denoted by the Greek letter $\mu$. In other words,

$$E(x) = \mu.$$

Furthermore, it can be shown that the expected value of $\overline{x}$ is also equal to $\mu$; that is,

$$E(\overline{x}) = \mu.$$

For instance, consider again the 30 community colleges in Example 3. We found that $\overline{x} = 3.4$, but the value of $\mu$, the average for all tuition amounts, is unknown. If a good estimate of $\mu$ were needed, the best guess (based on these data) is 3.4.

## Median

Asked by a reporter to give the average height of the players on his team, a Little League coach lined up his 15 players by increasing height. He picked out the player in the middle and pronounced this player to be of average height. This kind of average, called the **median,** is defined as the middle entry in a set of data arranged in either increasing or decreasing order. If the number of entries is even, the median is defined to be the mean of the two middle entries. The following table shows how to find the median for the two sets of data {8, 7, 4, 3, 1} and {2, 3, 4, 7, 9, 12} after each set has been arranged in increasing order.

| Odd Number of Entries | Even Number of Entries |
|---|---|
| 1 | 2 |
| 3 | 3 |
| Median = 4 | 4⎫ |
| 7 | 7⎭ Median = $\dfrac{4 + 7}{2}$ = 5.5 |
| 8 | 9 |
| | 12 |

**NOTE** As shown in the table, when the number of entries is even, the median is not always equal to one of the data entries.

**EXAMPLE 4**   Find the median number of hours worked per week

**(a)** for a sample of 7 male students whose work hours were

$$0, 7, 10, 20, 22, 25, 30.$$

**Solution** The median is the middle number, in this case 20 hours per week. (Note that the numbers are already arranged in numerical order.) In this list, three numbers are smaller than 20 and three are larger.

**(b)** for a sample of 11 female students whose work hours were

$$20, 0, 20, 30, 35, 30, 20, 23, 16, 38, 25.$$

**Solution** First, arrange the numbers in numerical order, from smallest to largest, or vice versa:

$$0, 16, 20, 20, 20, 23, 25, 30, 30, 35, 38.$$

The middle number can now be determined; the median is 23 hours per week.

**(c)** for a sample of 10 students of either gender whose work hours were

$$25, 18, 25, 20, 16, 12, 10, 0, 35, 32.$$

**Solution** Write the numbers in numerical order:

$$0, 10, 12, 16, \mathbf{18}, \mathbf{20}, 25, 25, 32, 35.$$

There are 10 numbers here; the median is the mean of the two middle numbers, or

$$\text{median} = \frac{18 + 20}{2} = 19.$$

The median is 19 hours per week.

> **⚠ CAUTION** Remember, the data must be arranged in numerical order before you locate the median. ✓₅

With grouped data, we find the interval in which the median value would occur. In Example 3, there are 30 values, so the median would occur as the average of the 15th and 16th values. We have both of these values occurring in the interval between 3.5 and 3.99. Thus, the median is the midpoint of that interval—in this case 3.75. If there are an even number of observations, and the two values for which you need to take the mean fall into different intervals, then take the mean of the two midpoints.

Both the mean and the median of a sample are examples of a **statistic,** which is simply a number that gives summary information about a sample. In some situations, the median gives a truer representative or typical element of the data than the mean does. For example, suppose that in an office there are 10 salespersons, 4 secretaries, the sales manager, and Ms. Daly, who owns the business. Their annual salaries are as follows: support staff, $30,000 each; salespersons, $50,000 each; manager, $70,000; and owner, $400,000. The mean salary is

$$\overline{x} = \frac{(30,000)4 + (50,000)10 + 70,000 + 400,000}{16} = \$68,125.$$

However, since 14 people earn less than $68,125 and only 2 earn more, the mean does not seem very representative. The median salary is found by ranking the salaries by size: $30,000, $30,000, $30,000, $30,000, $50,000, $50,000, . . . , $400,000. There are 16 salaries (an even number) in the list, so the mean of the 8th and 9th entries will give the value of the median. The 8th and 9th entries are both $50,000, so the median is $50,000. In this example, the median is more representative of the distribution than the mean is.

When the data include extreme values (such as $400,000 in the preceding example), the mean may not provide an accurate picture of a typical value. So the median is often a better measure of center than the mean for data with extreme values, such as income levels and house prices. *In general, the median is a better measure of center whenever we see right-skewed or left-skewed distributions.*

## Mode

Sue's scores on 10 class quizzes include one 7, two 8's, six 9's, and one 10. She claims that her average grade on quizzes is 9, because most of her scores are 9's. This kind of "average," found by selecting the most frequent entry, is called the **mode.**

> **EXAMPLE 5**   Find the mode for the given data sets.
>
> **(a)**  Ages of retirement: 55, 60, 63, 63, 70, 55, 60, 65, 68, 65, 65, 71, 65, 65

**Solution** The number 65 occurs more often than any other, so it is the mode. It is sometimes convenient, but not necessary, to place the numbers in numerical order when looking for the mode.

**(b)**  Total cholesterol score: 180, 200, 220, 260, 220, 205, 255, 240, 190, 300, 240

**Solution** Both 220 and 240 occur twice. This list has *two* modes, so it is bimodal.

---

✓ **Checkpoint 5**

Find the median for the given heights in inches.

**(a)**  60, 72, 64, 75, 72, 65, 68, 70

**(b)**  73, 58, 77, 66, 69, 69, 66, 68, 67

---

⌨ **TECHNOLOGY TIP**

Many graphing calculators display the median when doing one-variable statistics. You may have to scroll down to a second screen to find it.

✓**Checkpoint 6**

Find the mode for each of the given data sets.

(a) Highway miles per gallon of an automobile: 25, 28, 32, 19, 15, 25, 30, 25

(b) Price paid for last haircut or styling: $11, $35, $35, $10, $0, $12, $0, $35, $38, $42, $0, $25

(c) Class enrollment in six sections of calculus: 30, 35, 26, 28, 29, 19

(c) Prices of new cars: $25,789, $43,231, $33,456, $19,432, $22,971, $29,876

**Solution** No number occurs more than once. This list has no mode.

The mode has the advantages of being easily found and not being influenced by data that are extreme values. It is often used in samples where the data to be "averaged" are not numerical. A major disadvantage of the mode is that there may be more than one, in case of ties, or there may be no mode at all when all entries occur with the same frequency.

The mean is the most commonly used measure of center. Its advantages are that it is easy to compute, it takes all the data into consideration, and it is reliable—that is, repeated samples are likely to give similar means. A disadvantage of the mean is that it is influenced by extreme values, as illustrated in the salary example.

The median can be easy to compute and is influenced very little by extremes. A disadvantage of the median is the need to rank the data in order; this can be tedious when the number of items is large.

---

**EXAMPLE 6** **Business** A sample of 10 working adults was asked "How many hours did you work last week?" Their responses appear below. (Data from: gss.norc.org.)

$$40, 35, 43, 40, 30, 40, 45, 40, 55, 20.$$

Find the mean, median, and mode of the data.

**Solution** The mean number of hours worked is

$$\bar{x} = \frac{40 + 35 + 43 + 40 + 30 + 40 + 45 + 40 + 55 + 20}{10} = 38.8 \text{ hours.}$$

After the numbers are arranged in order from smallest to largest, the middle number, or median, is 40 hours.

The number 40 occurs more often than any other, so it is the mode.

✓**Checkpoint 7**

Following is a list of the number of movies seen at a theater in the last three months by nine students selected at random:

$$1, 0, 2, 5, 2, 0, 0, 1, 4$$

(a) Find the mean.

(b) Find the median.

(c) Find the mode.

## Weighted Averages

Often analysts, researchers, and teachers, among others, want to count some values with more importance than other values when calculating a mean. For example, teachers in colleges and universities often assign more importance to the final examination for a course than they assign to the midterm exams. To do this, we want to calculate what is called a weighted average, or a weighted mean.

For example, if an instructor has three midterm exams that each count 20% toward the final grade and a final exam that counts 40% toward the final grade, then the final grade can be thought of as a weighted average. The weights are the amounts that each exam "counts" toward the final average expressed as a decimal. So the weights would be .2, .2, .2, and .4, respectively. The weights add up to 1.

When we calculate the weighted average, we multiply the weights by the actual exam scores and then sum them to get the final grade. We again use $x_1, x_2, x_3, \ldots, x_n$ to describe the data points and use $w_1, w_2, w_3, \ldots, w_n$ to describe the weights associated with each data point.

---

## Weighted Average (Weighted Mean)

The weighted average of the $n$ numbers $x_1, x_2, x_3, \ldots, x_n$, with corresponding weights $w_1, w_2, w_3, \ldots, w_n$, is $\bar{x} = w_1x_1 + w_2x_2 + w_3x_3 + \cdots + w_nx_n = \sum_{i=1}^{n} w_ix_i.$

**EXAMPLE 7** **Final Grades** A teacher designs a course so that the homework counts 20% toward the final exam, quizzes count 10%, three midterm exams count 15% each, and the final exam counts 25%. Marissa earned 92% on the homework; her quiz average was 81%; her midterm exam scores were 75%, 88%, and 92%; and her final exam was 87%. Find her final average (weighted average).

**Solution** The weights for the weighted average are how much each assessment counts toward the final grade. Thus, the homework weight is .20, the quizzes are weighted as .10, exam 1 is weighted as .15, exam 2 is weighted as .15, exam 3 is weighted as .15, and the final exam is weighted as .25. We can make a table to correspond these weights with the scores earned on each component.

| Component | Weight $w_i$ | Score $x_i$ | Product $w_i x_i$ |
|---|---|---|---|
| Homework | .20 | 92 | 18.4 |
| Quizzes | .10 | 81 | 8.1 |
| Exam 1 | .15 | 75 | 11.25 |
| Exam 2 | .15 | 88 | 13.2 |
| Exam 3 | .15 | 92 | 13.8 |
| Final Exam | .25 | 87 | 21.75 |
| | | | Total = 86.5 |

✓ **Checkpoint 8**

Manuel's professor weighs the homework 30%, quiz average 10%, four midterm exams 10% each, and the final exam 20%. Manuel earned 81% on his homework; his quiz average was 92%; his exam scores were 91, 72, 79, and 84; and his final exam was 78. Find his final average.

Thus, the final average (weighted average) for Marissa is 86.5% ✓8

# 10.2 Exercises

*Find the mean for each data set. Round to the nearest tenth. (See Example 1.)*

1. **Annual Salaries** Secretarial salaries (U.S. dollars):

   $21,900, $22,850, $24,930, $29,710, $28,340, $40,000.

2. **Annual Salaries** Starting teaching salaries (U.S. dollars):

   $38,400, $39,720, $28,458, $29,679, $33,679.

3. **Earthquakes** Earthquakes on the Richter scale:

   3.5, 4.2, 5.8, 6.3, 7.1, 2.8, 3.7, 4.2, 4.2, 5.7.

4. **Body Temperature** Body temperatures of self-classified "healthy" students (degrees Fahrenheit):

   96.8, 94.1, 99.2, 97.4, 98.4, 99.9, 98.7, 98.6.

5. **Foot Length** Lengths of foot (inches) for adult men:

   9.2, 10.4, 13.5, 8.7, 9.7.

*Find the mean for each distribution. Round to the nearest tenth. (See Examples 2 and 3.)*

6. **Quizzes** Scores on a quiz, on a scale from 0 to 10:

   | Value | 7 | 8 | 9 | 10 |
   |---|---|---|---|---|
   | Frequency | 4 | 6 | 7 | 11 |

7. **Students' Ages** Age (years) of students in an introductory accounting class:

   | Value | 19 | 20 | 21 | 22 | 23 | 24 | 28 |
   |---|---|---|---|---|---|---|---|
   | Frequency | 3 | 5 | 25 | 8 | 2 | 1 | 1 |

8. **Commuting Distance** Commuting distance (miles) for students at a university:

   | Value | 0 | 1 | 2 | 5 | 10 | 17 | 20 | 25 |
   |---|---|---|---|---|---|---|---|---|
   | Frequency | 15 | 12 | 8 | 6 | 5 | 2 | 1 | 1 |

**9. Fuel Efficiency** Estimated miles per gallon of automobiles:

| Value | Frequency |
|---|---|
| 9 | 5 |
| 11 | 10 |
| 15 | 12 |
| 17 | 9 |
| 20 | 6 |
| 28 | 1 |

**10–14.** *Find the median of the data in Exercises 1–5. (See Example 4.)*

*Find the mode or modes for each of the given lists of numbers. (See Example 5.)*

**15. Children's Ages** Ages (years) of children in a day-care facility:

1, 2, 2, 1, 2, 2, 1, 1, 2, 2, 3, 4, 2, 3, 4, 2, 3, 2, 3.

**16. Health** Ages (years) in the intensive care unit at a local hospital:

68, 64, 23, 68, 70, 72, 72, 68.

**17. Students' Heights** Heights (inches) of students in a statistics class:

62, 65, 71, 74, 71, 76, 71, 63, 59, 65, 65, 64, 72, 71, 77, 63, 65.

**18. Pain Relief** Minutes of pain relief from acetaminophen after childbirth:

60, 240, 270, 180, 240, 210, 240, 300, 330, 360, 240, 120.

**19. GPA** Grade point averages for 5 students:

3.2, 2.7, 1.9, 3.7, 3.9.

**20.** When is the median the most appropriate measure of center?

**21.** When would the mode be an appropriate measure of center?

**Pet Ownership** *For Exercises 22 and 23, the frequency tables give the frequencies of the percentages of residents that own dogs or cats for all 50 U.S. states except Alaska. For grouped data, the modal class is the interval containing the most data values. Give the mean and modal class for each of the given collections of grouped data. (Round the midpoint to the nearest tenth.) (Data from: American Veterinary Medical Association.)*

**22.** The variable is the percentage of dog owners.

| Interval | Frequency |
|---|---|
| 10–14.99 | 1 |
| 15–19.99 | 0 |
| 20–24.99 | 1 |
| 25–29.99 | 4 |
| 30–34.99 | 13 |
| 35–39.99 | 12 |
| 40–44.99 | 12 |
| 45–49.99 | 6 |

**23.** The variable is the percentage of cat owners.

| Interval | Frequency |
|---|---|
| 10–14.99 | 1 |
| 15–19.99 | 0 |
| 20–24.99 | 1 |
| 25–29.99 | 18 |
| 30–34.99 | 22 |
| 35–39.99 | 4 |
| 40–44.99 | 1 |
| 45–49.99 | 2 |

**24.** To predict the outcome of the next congressional election, you take a survey of your friends. Is this a random sample of the voters in your congressional district? Explain why or why not.

*Work each problem. (See Example 6.)*

**25. MLB Payrolls** The following table gives the payroll (in millions of dollars) of the 10 Major League Baseball teams with the highest payrolls, as estimated by CBS Sports in 2017.

| Team | Payroll |
|---|---|
| Los Angeles Dodgers | 242 |
| New York Yankees | 202 |
| Boston Red Sox | 200 |
| Detroit Tigers | 200 |
| Toronto Blue Jays | 178 |
| Texas Rangers | 176 |
| San Francisco Giants | 172 |
| Chicago Cubs | 172 |
| Washington Nationals | 168 |
| Baltimore Orioles | 164 |

Find the following statistics for these data:

**(a)** mean; **(b)** median.

**26. NFL Team Values** The following table gives the value (in billions of dollars) of the 10 most valued National Football League (NFL) teams, as estimated by *Forbes* in 2016.

| Team | Value |
|---|---|
| Dallas Cowboys | 4.20 |
| New England Patriots | 3.40 |
| New York Giants | 3.10 |
| San Francisco 49ers | 3.00 |
| Washington Redskins | 2.95 |
| Los Angeles Rams | 2.90 |
| New York Jets | 2.75 |
| Chicago Cubs | 2.70 |
| Houston Texans | 2.60 |
| Philadelphia Eagles | 2.50 |

Find the following statistics for these data:

(a) mean;        (b) median.

27. **Business**  The 12 movies that have earned the most revenue (in millions of dollars) from U.S. domestic box-office receipts in 2016 are given in the table. (Data from: www.boxofficemojo.com.)

| Movie | Revenue |
|---|---|
| *Finding Dory* | 486 |
| *Rogue One: A Star Wars Story* | 425 |
| *Captain America: Civil War* | 408 |
| *The Secret Life of Pets* | 368 |
| *The Jungle Book* | 364 |
| *Deadpool* | 363 |
| *Zootopia* | 341 |
| *Batman v Superman: Dawn of Justice* | 330 |
| *Suicide Squad* | 325 |
| *Doctor Strange* | 230 |
| *Fantastic Beasts and Where to Find Them* | 224 |
| *Moana* | 210 |

(a) Find the mean value in dollars for this group of movies.

(b) Find the median value in dollars for this group of movies.

28. **Sirius XM Revenue**  The revenue (in millions of dollars) for Sirius XM Holdings Inc. from 2007 to 2016 is given in the following table. (Data from: www.morningstar.com.)

| Year | Revenue |
|---|---|
| 2007 | 922 |
| 2008 | 1664 |
| 2009 | 2473 |
| 2010 | 2817 |
| 2011 | 3015 |
| 2012 | 3402 |
| 2013 | 3799 |
| 2014 | 4181 |
| 2015 | 4570 |
| 2016 | 5017 |

(a) Find the mean and median for these data.

(b) What year's revenue is closest to the mean?

29. **Dr Pepper and Snapple Revenue**  The revenue (in millions of dollars) for Dr Pepper Snapple Group Inc. from 2007 to 2016 is given in the following table. (Data from: www.morningstar.com.)

| Year | Revenue |
|---|---|
| 2007 | 5748 |
| 2008 | 5710 |
| 2009 | 5531 |
| 2010 | 5636 |
| 2011 | 5903 |
| 2012 | 5995 |
| 2013 | 5997 |
| 2014 | 6121 |
| 2015 | 6282 |
| 2016 | 6440 |

(a) Find the mean and median for these data.

(b) What year's revenue is closest to the mean?

**Monthly Temperatures**  *The table gives the average monthly high and low temperatures, in degrees Fahrenheit, for Raleigh, North Carolina, over the course of a year. (Data from: www.weather.com.)*

| Month | High | Low |
|---|---|---|
| January | 49 | 30 |
| February | 53 | 32 |
| March | 61 | 40 |
| April | 71 | 48 |
| May | 78 | 57 |
| June | 84 | 65 |
| July | 88 | 69 |
| August | 86 | 68 |
| September | 80 | 62 |
| October | 70 | 49 |
| November | 61 | 42 |
| December | 52 | 33 |

*Find the mean and median for each of the given subgroups.*

30. The high temperatures

31. The low temperatures

*For Exercises 32–33 determine the shape of the distribution from the histogram and then decide whether the mean or median is a better measure of center. If the mean is the better measure, calculate the value. If the median is the better measure, give the midpoint of the interval that contains it.*

32. **Market Capitalization**  The variable is market capitalization (in billions of dollars) of 20 randomly selected stocks from the S&P 500 on March 7, 2017. (Round to nearest tenth.) (Data from: data.okfn.org.)

| Interval | Frequency |
|---|---|
| 0–19.99 | 8 |
| 20.00–39.99 | 5 |
| 40.00–59.99 | 2 |
| 60.00–79.99 | 2 |
| 80.00–99.99 | 2 |
| 100.00–119.99 | 1 |

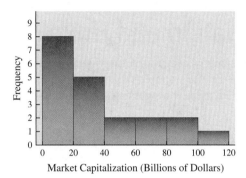

**33. Annual Household Income** The annual household income (in thousands of dollars) for 20 randomly chosen households from the 2015 American Community Survey. (Data from: www.census.gov.)

| Interval | Frequency |
|---|---|
| 0–24 | 3 |
| 25–49 | 4 |
| 50–74 | 3 |
| 75–99 | 2 |
| 100–124 | 2 |
| 125–149 | 3 |
| 150–174 | 1 |
| 175–199 | 1 |
| 200–224 | 1 |

Income (Thousands of Dollars)

**34. Bachelor's Degrees** The stem-and-leaf plot summarizes the percentage of residents in the first 25 states alphabetically with a bachelor's degree or higher. (Data from: National Center for Education Statistics.)

| Stem | Leaves |
|---|---|
| 2 | 11333 |
| 2 | 557778899 |
| 3 | 0022234 |
| 3 | 888 |
| 4 | 1 |

*Units:* 4|1 = 41%

**(a)** Describe the shape of the distribution.

**(b)** Find the median.

**35. Bachelor's Degrees** The stem-and-leaf plot summarizes the percentage of residents in the latter 25 states alphabetically with a bachelor's degree or higher. (Data from: National Center for Education Statistics.)

| Stem | Leaves |
|---|---|
| 1 | 9 |
| 2 | 34 |
| 2 | 566677788999 |
| 3 | 00113 |
| 3 | 55677 |

*Units:* 3|9 = 39%

**(a)** Describe the shape of the distribution.

**(b)** Find the median.

*For Exercises 36–38, calculate the weighted average. (See Example 7.) For Exercises 39 and 40, use the formula for the weighted average to solve for the final examination score.*

**36. Final Average** Gonzalo enrolled in a class where the homework counts as 30% of the final grade, two midterm exams count as 15% each, and a final exam counts as 40% of the final grade. Gonzalo earned 88% on the homework, 85% and 81% on the two midterm exams, and 93% on the final exam. What is Gonzalo's final average?

**37. Final Average** Nan enrolled in a class where the homework counts as 15% of the final grade, quizzes count 15%, three midterms exams count as 12% each, and the final exam counts 34%. Nan earned 73% on her homework; 78% on her quizzes; 91%, 93%, and 83% on her three midterm exams; and 88% on the final exam. What is Nan's final average?

**38. Final Average** Henrietta enrolled in a class where attendance counted as 10% of the final grade, a midterm paper counted as 30% of the final grade, and a final paper counted as 60% of the final grade. She had 100% attendance. She earned a score of 85% on her midterm paper and a 91% on her final paper. What is Henrietta's final average?

**39. Final Exam Score** Trey enrolled in a class where attendance counted as 5% of the final grade, quizzes counted as 25%, two midterm exams counted as 20% each, and the final exam counted as 30%. Trey earned 100% on attendance, 95% on the quizzes, 81% on midterm exam 1, and 89% on midterm exam 2. If 90%

is needed to earn a grade of $A-$, what is the minimum score Trey needs to earn on the final exam in order to earn a final grade of $A-$?

**40. Final Exam Score** Mirabella enrolled in a class where attendance counted as 10% of the final grade, quizzes counted as 15%, one midterm exam counted as 35%, and the final exam counted as 40%. Mirabella earned 80% on attendance, 85% on the quizzes, and 75% on the midterm exam. If 80% is needed to earn a grade of $B-$, what is the minimum score Mirabella needs to earn on the final exam in order to earn a final grade of $B-$?

✓**Checkpoint Answers**

1. $301.75   2. $\bar{x} = 11.75$   3. 18.90   4. About 3.32

5. **(a)** 69 inches   **(b)** 68 inches

6. **(a)** 25 miles per gallon   **(b)** $0 and $35
   **(c)** No mode

7. **(a)** About 1.7   **(b)** 1   **(c)** 0

8. 81.7

# 10.3 Measures of Variation and Boxplots

The mean, median, and mode are measures of center for a list of numbers, but tell nothing about the *spread* of the numbers in the list. For example, look at the following data sets of number of times per week three people ate meals at restaurants over the course of five weeks:

| Jill: | 3 | 5 | 6 | 3 | 3 |
|---|---|---|---|---|---|
| Miguel: | 4 | 4 | 4 | 4 | 4 |
| Sharille: | 10 | 1 | 0 | 0 | 9 |

Each of these three data sets has a mean of 4, but the amount of dispersion or variation within the lists is different. This difference may reflect different dining patterns over time. Thus, in addition to a measure of center, another kind of measure is needed that describes how much the numbers vary.

The largest number of restaurant meals for Jill is 6, while the smallest is 3, a difference of 3. For Miguel, the difference is 0; for Sharille, it is 10. The difference between the largest and smallest number in a sample is called the **range,** one example of a measure of variation. The range is 3 for Jill, 0 for Miguel, and 10 for Sharille. The range has the advantage of being very easy to compute and gives a rough estimate of the variation among the data in the sample. However, it depends only on the two extremes and tells nothing about how the other data are distributed between the extremes.

📷 **TECHNOLOGY TIP**

Many graphing calculators show the largest and smallest numbers in a list when displaying one-variable statistics, usually on the second screen of the display.

**EXAMPLE 1**   **Consumer Expenses** Find the range for each given data set for a small sample of people.

**(a)** Price paid for last haircut (with tip): 10, 0, 15, 30, 20, 18, 50, 120, 75, 95, 0, 5

**Solution** The highest number here is 120; the lowest is 0. The range is the difference of these numbers, or

$$120 - 0 = 120.$$

**(b)** Amount spent for last vehicle servicing: 30, 19, 125, 150, 430, 50, 225

**Solution** Range $= 430 - 19 = 411.$ ✓₁

✓**Checkpoint 1**

Find the range for this sample of the number of miles from students' homes to college: 15, 378, 5, 210, 125.

To find another useful measure of variation, we begin by finding the **deviations from the mean**—the differences found by subtracting the mean from each number in a distribution.

**EXAMPLE 2**   Find the deviations from the mean for the following sample of ages.

$$32, \quad 41, \quad 47, \quad 53, \quad 57.$$

**Solution** Adding these numbers and dividing by 5 gives a mean of 46 years. To find the deviations from the mean, subtract 46 from each number in the sample. For example, the first deviation from the mean is $32 - 46 = -14$; the last is $57 - 46 = 11$ years. All of the deviations are listed in the following table.

| Age | Deviation from Mean |
|-----|---------------------|
| 32 | −14 |
| 41 | −5 |
| 47 | 1 |
| 53 | 7 |
| 57 | 11 |

✓ **Checkpoint 2**

Find the deviations from the mean for the following sample of number of miles traveled by various people to a vacation location:

135, 60, 50, 425, 380.

To check your work, find the sum of the deviations. It should always equal 0. (The answer is always 0 because the positive and negative deviations cancel each other out.) ✓₂

To find a measure of variation, we might be tempted to use the mean of the deviations. However, as just mentioned, this number is always 0, no matter how widely the data are dispersed. To avoid the problem of the positive and negative deviations averaging to 0, we could take absolute values and find $\sum |x_i - \bar{x}|$ and then divide it by $n$ to get the *mean absolute deviation*. However, statisticians generally prefer to square each deviation to get non-negative numbers and then take the square root of the mean of the squared variations in order to preserve the units of the original data (such as inches, pounds). (Using squares instead of absolute values allows us to take advantage of some algebraic properties that make other important statistical methods much easier.) The squared deviations for the data in Example 2 are shown in the following table:

| Number | Deviation from Mean | Square of Deviation |
|--------|---------------------|---------------------|
| 32 | −14 | 196 |
| 41 | −5 | 25 |
| 47 | 1 | 1 |
| 53 | 7 | 49 |
| 57 | 11 | 121 |

In this case, the mean of the squared deviations is

$$\frac{196 + 25 + 1 + 49 + 121}{5} = \frac{392}{5} = 78.4.$$

This number is called the **population variance,** because the sum was divided by $n = 5$, the number of items in the original list.

Since the deviations from the mean must add up to 0, if we know any 4 of the 5 deviations, the 5th can be determined. That is, only $n - 1$ of the deviations are free to vary, so we really have only $n - 1$ independent pieces of information, or *degrees of freedom*. Using $n - 1$ as the divisor in the formula for the mean gives

$$\frac{196 + 25 + 1 + 49 + 121}{5 - 1} = \frac{392}{4} = 98.$$

This number, 98, is called the **sample variance** of the distribution and is denoted $s^2$, because it is found by averaging a list of squares. In this case, the population and sample variances differ by quite a bit. But when $n$ is relatively large, as is the case in real-life applications, the difference between them is rather small.

## Sample Variance

The variance of a sample of $n$ numbers $x_1, x_2, x_3, \ldots, x_n$, with mean $\bar{x}$, is

$$s^2 = \frac{\sum(x_i - \bar{x})^2}{n - 1}.$$

When computing the sample variance by hand, it is often convenient to use the following shortcut formula, which can be derived algebraically from the definition in the preceding box:

$$s^2 = \frac{\sum x_i^2 - n\bar{x}^2}{n - 1}.$$

To find the sample variance, we square the deviations from the mean, so the variance is in squared units. To return to the same units as the data, we use the *square root* of the variance, called the **sample standard deviation,** denoted $s$.

## Sample Standard Deviation

The standard deviation of a sample of $n$ numbers $x_1, x_2, x_3, \ldots, x_n$, with mean $\bar{x}$, is

$$\sqrt{\frac{\sum(x_i - \bar{x})^2}{n - 1}}.$$

**NOTE** The **population standard deviation** is

$$\sigma = \sqrt{\frac{\sum(x_i - \mu)^2}{n}},$$

where $n$ is the population size.

**TECHNOLOGY TIP** When a graphing calculator computes one-variable statistics for a list of data, it usually displays the following information (not necessarily in this order, and sometimes on two screens) and possibly other information as well:

| Information | Notation |
|---|---|
| Number of data entries | $n$ or $N\Sigma$ |
| Mean | $\bar{x}$ or mean $\Sigma$ |
| Sum of all data entries | $\Sigma x$ or TOT $\Sigma$ |
| Sum of the squares of all data entries | $\Sigma x^2$ |
| Sample standard deviation | $Sx$ or $sx$ or $x\sigma_{n-1}$ or SSDEV |
| Population standard deviation | $\sigma x$ or $x\sigma_n$ or PSDEV |
| Largest/smallest data entries | maxX/minX or MAX$\Sigma$/MIN$\Sigma$ |
| Median | Med or MEDIAN |

**NOTE** In the rest of this section, we shall deal exclusively with the sample variance and the sample standard deviation. So whenever standard deviation is mentioned, it means "sample standard deviation," not population standard deviation.

As its name indicates, the standard deviation is the most commonly used measure of variation. The standard deviation is a measure of the variation from the mean. The size of the standard deviation indicates how spread out the data are from the mean.

| Time | Square of the Time |
|------|--------------------|
| 2 | 4 |
| 8 | 64 |
| 3 | 9 |
| 2 | 4 |
| 6 | 36 |
| 11 | 121 |
| 31 | 961 |
| 9 | 81 |
| 72 | 1280 |

✓**Checkpoint 3**

Find the standard deviation for a sample of the number of miles traveled by various people to a vacation location:

135, 60, 50, 425, 380.

**EXAMPLE 3**   **Mobile Phone Conversations** Find the standard deviation for the following sample of the lengths (in minutes) of eight consecutive cell phone conversations by one person:

$$2, \ 8, \ 3, \ 2, \ 6, \ 11, \ 31, \ 9.$$

Work by hand, using the shortcut variance formula on page 532.

**Solution**  Arrange the work in columns, as shown in the table in the margin. Now use the first column to find the mean:

$$\overline{x} = \frac{\sum x_i}{8} = \frac{72}{8} = 9 \text{ minutes.}$$

The total of the second column gives $\sum x_i^2 = 1280$. The variance is

$$s^2 = \frac{\sum x_i^2 - n\overline{x}^2}{n - 1}$$

$$= \frac{1280 - 8(9)^2}{8 - 1}$$

$$\approx 90.3$$

and the standard deviation is

$$s \approx \sqrt{90.3} \approx 9.5 \text{ minutes.} \quad \text{✓}_3$$

**TECHNOLOGY TIP** The screens in Figure 10.6 show two ways to find variance and standard deviation on a TI-84+ calculator: with the LIST menu and with the STAT menu. The data points are first entered in a list—here, $L_5$. See your instruction book for details.

```
variance(L₅)
             90.28571429
stdDev(L₅)
             9.50187951
```

```
1-Var Stats
 x̄=9
 Σ=72
 Σx²=1288
 Sx=9.50187951
 σx=8.88819442
↓n=8
```

**Figure 10.6**

In a spreadsheet, enter the data in cells A1 through A8. Then, in cell A9, type "=VAR(A1:A8)" and press Enter. The standard deviation can be calculated either by taking the square root of cell A9 or by typing "=STDEV (A1:A8)" in cell A10 and pressing Enter.

⚠ **CAUTION** We must be careful to divide by $n - 1$, not $n$, when calculating the standard deviation of a sample. Many calculators are equipped with statistical keys that compute the variance and standard deviation. Some of these calculators use $n - 1$, and others use $n$ for these computations; some may have keys for both. Check your calculator's instruction book before using a statistical calculator for the exercises.

One way to interpret the standard deviation uses the fact that, for many populations, most of the data are within three standard deviations of the mean. (See Section 10.4.) This implies that, in Example 3, most of the population data from which this sample is taken are between

$$\overline{x} - 3s = 9 - 3(9.5) = -19.5$$

and

$$\overline{x} + 3s = 9 + 3(9.5) = 37.5.$$

For Example 3, the preceding calculations imply that most phone conversations are less than 37.5 minutes long. This approach of determining whether sample observations are beyond 3 standard deviations of the mean is often employed in conducting quality control in many industries.

For data in a grouped frequency distribution, a slightly different formula for the standard deviation is used.

---

## Standard Deviation for a Grouped Distribution

The standard deviation for a sample distribution with mean $\bar{x}$, where $x_i$ is an interval midpoint with frequency $f_i$ and $n = \Sigma f_i$, is

$$s = \sqrt{\frac{\Sigma f_i x_i^2 - n\bar{x}^2}{n - 1}}.$$

---

The formula indicates that the product $f_i x_i^2$ is to be found for each interval. Then all the products are summed, $n$ times the square of the mean is subtracted, and the difference is divided by one less than the total frequency—that is, by $n - 1$. The square root of this result is $s$, the standard deviation. The standard deviation found by this formula may (and probably will) differ somewhat from the standard deviation found from the original data.

**!  CAUTION** In calculating the standard deviation for either a grouped or an ungrouped distribution, using a rounded value for the mean or variance may produce an inaccurate value.

**EXAMPLE 4**    **College Tuition** The following frequency distribution gives the 2017–2018 annual tuition (in thousands of dollars) for a random sample of 30 community colleges. (Data from: www.collegetuitioncompare.com.)
Find the sample standard deviation $s$ for these data.

**Solution** We first need to find the mean $\bar{x}$ for these grouped data. We find the midpoint of each interval and label it $x_i$. We multiply the frequency by the midpoint $x_i$ to obtain $x_i f_i$:

| Tuition Amount (in thousands of dollars) | Frequency, $f_i$ | Midpoint, $x_i$ | Product, $x_i f_i$ |
|---|---|---|---|
| 1.0–1.49 | 5 | 1.25 | 6.25 |
| 1.5–1.99 | 0 | 1.75 | 0 |
| 2.0–2.49 | 1 | 2.25 | 2.25 |
| 2.5–2.99 | 5 | 2.75 | 13.75 |
| 3.0–3.49 | 1 | 3.25 | 3.25 |
| 3.5–3.99 | 9 | 3.75 | 33.75 |
| 4.0–4.49 | 4 | 4.25 | 17 |
| 4.5–4.99 | 2 | 4.75 | 9.5 |
| 5.0–5.49 | 2 | 5.25 | 10.5 |
| 5.5–5.99 | 1 | 5.75 | 5.75 |
| | Total = 30 | | Total = 102 |

Therefore,

$$\bar{x} = \frac{102}{30} = 3.4.$$

Now that we have the mean value, we can modify our table to include columns for $x_i^2$ and $f_i x_i^2$. We obtain the following results:

| Tuition Amount (in thousands of dollars) | Frequency, $f_i$ | Midpoint, $x_i$ | Product, $f_i x_i$ | $x_i^2$ | $f_i x_i^2$ |
|---|---|---|---|---|---|
| 1.0–1.49 | 5 | 1.25 | 6.25 | 1.5625 | 7.8125 |
| 1.5–1.99 | 0 | 1.75 | 0 | 3.0625 | 0 |
| 2.0–2.49 | 1 | 2.25 | 2.25 | 5.0625 | 5.0625 |
| 2.5–2.99 | 5 | 2.75 | 13.75 | 7.5625 | 37.8125 |
| 3.0–3.49 | 1 | 3.25 | 3.25 | 10.5625 | 10.5625 |
| 3.5–3.99 | 9 | 3.75 | 33.75 | 14.0625 | 126.5625 |
| 4.0–4.49 | 4 | 4.25 | 17 | 18.0625 | 72.2500 |
| 4.5–4.99 | 2 | 4.75 | 9.5 | 22.5625 | 45.1250 |
| 5.0–5.49 | 2 | 5.25 | 10.5 | 27.5625 | 55.1250 |
| 5.5–5.99 | 1 | 5.75 | 5.75 | 33.0625 | 33.0625 |
| | Total = 30 | | Total = 102 | | Total = 393.3750 |

We now use the formula for the standard deviation with $n = 30$ to find $s$:

$$s = \sqrt{\frac{\sum f_i x_i^2 - n\bar{x}^2}{n-1}}$$

$$= \sqrt{\frac{393.3750 - 30(3.4)^2}{30 - 1}}$$

$$\approx 1.3 \qquad \checkmark_4$$

✓ **Checkpoint 4**

Find the standard deviation for the following grouped frequency distribution of the number of classes completed thus far in the college careers of a random sample of 52 students:

| Classes | Frequency |
|---|---|
| 0–5 | 6 |
| 6–10 | 10 |
| 11–20 | 12 |
| 21–30 | 15 |
| 31–40 | 9 |

Another way to interpret the standard deviation is to think of it as an *approximation* of the average deviation from the mean. In Example 4, the mean tuition was $3400 and the standard deviation was $1300. One way to think of what the value $1300 represents is as the approximate average deviation from $3400 for the 30 community colleges in the sample. Some colleges charge tuition close to $3400 and some charge tuition further from $3400, but the approximate average deviation from $3400 is $1300.

📝 **NOTE** A calculator is almost a necessity for finding a standard deviation. With a non-graphing calculator, a good procedure to follow is first to calculate $\bar{x}$. Then, for each $x$, square that number, and multiply the result by the appropriate frequency. If your calculator has a key that accumulates a sum, use it to accumulate the total in the last column of the table. With a graphing calculator, simply enter the midpoints and the frequencies, and then ask for the one-variable statistics.

## Boxplots

Boxplots are a graphical means of presenting key characteristics of a data set. The idea is to arrange the data in increasing order and choose three numbers $Q_1$, $Q_2$, and $Q_3$ that divide it into four equal parts, as indicated schematically in the following diagram:

The number $Q_1$ is called the **first quartile**, the median $Q_2$ is called the **second quartile**, and $Q_3$ is called the **third quartile.** The minimum, $Q_1$, $Q_2$, $Q_3$, and the maximum are often called the **five-number summary,** and they are used to construct a boxplot, as illustrated in Examples 5 and 6.

**EXAMPLE 5** The following table gives the revenue (in billions of dollars) for Home Depot and Lowe's home improvement corporations for the given years. Construct a boxplot for the Home Depot revenue data. (Data from: www.morningstar.com.)

| Company | 2008 | 2009 | 2010 | 2011 | 2012 | 2013 | 2014 | 2015 | 2016 | 2017 |
|---|---|---|---|---|---|---|---|---|---|---|
| **Home Depot** | 77.3 | 71.3 | 66.2 | 68.0 | 70.4 | 74.8 | 78.8 | 83.2 | 88.5 | 94.6 |
| **Lowe's** | 48.2 | 48.2 | 47.2 | 48.8 | 50.2 | 50.5 | 53.4 | 56.2 | 59.1 | 65.0 |

**Solution** First, we order the Home Depot revenue from low to high.

$$66.2, 68.0, 70.4, 71.3, 74.8, 77.3, 78.8, 83.2, 88.5, 94.6$$

The minimum revenue is 66.2 and the maximum is 94.6. Because there is an even number of data points, the median revenue $Q_2$ is 76.05 (the average of the two center entries). To find $Q_1$, which separates the lower 25% of the data from the rest, we first calculate

25% of $n$ (rounded up to the nearest integer),

where $n$ is the number of data points. Here, $n = 10$, and so $.25(10) = 2.5$, which rounds up to 3. Now count to the *third* revenue (in order) to get $Q_1 = 70.4$.

Similarly, since $Q_3$ separates the lower 75% of the data from the rest, we calculate

75% of $n$ (rounded up to the nearest integer).

When $n = 10$, we have $.75(10) = 7.5$, which rounds up to 8. Count to the *eighth* observation (in order), getting $Q_3 = 83.2$.

The five key numbers for constructing the boxplot are as follows:

Draw a horizontal axis that will include the minimum and maximum values. Draw a box with left boundary at $Q_1$ and a right boundary at $Q_3$. Mark the location of the median $Q_2$ by drawing a vertical line in the box Extend a horizontal line from the vertical line at $Q_1$ to the minimum value and and a similar line from the vertical line at $Q_3$ to the maximum value as shown in Figure 10.7.

**Figure 10.7**

Boxplots are useful in showing the location of the middle 50% of the data. This is the region of the box between $Q_1$ and $Q_3$. Boxplots are also quite useful in comparing two distributions that are measured on the same scale, as we will see in Example 6.

**EXAMPLE 6** **Business** Construct a boxplot for the Lowe's revenue data in Example 5 with the Home Depot revenue boxplot on the same graph.

**Solution** The Lowe's data first need to be sorted low to high:

$$47.2, 48.2, 48.2, 48.8, \mathbf{50.2}, \mathbf{50.5}, 53.4, 56.2, 59.1, 65.0.$$

The minimum is 47.2, the maximum is 65.0, and the median $Q_2 = 50.35$. Again, $n = 10$, and so

$$25\% \text{ of } n = .25(10) = 2.5, \text{ which rounds up to } 3,$$

so that $Q_1 = 48.2$ (the *third* revenue). Similarly,

$$75\% \text{ of } n = .75(10) = 7.5, \text{ which rounds up to } 8,$$

so that $Q_3 = 56.2$ (the *eighth* revenue).

We can now use the minimum, $Q_1$, $Q_2$, and $Q_3$, and the maximum to place the boxplot of Lowe's data above the Home Depot boxplot, as shown in Figure 10.8.

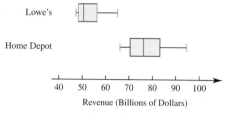

**Figure 10.8**

✓ **Checkpoint 5**

Create a boxplot for the weights (in kilograms) for eight first-year students:

88, 79, 67, 69, 58, 53, 89, 57.

**⊞▯ TECHNOLOGY TIP**
Many spreadsheets and graphing calculators can graph boxpots for single-variable data. With a calculator the procedure is similar to plotting data with the STAT PLOT menu.

When we compare the revenue distributions in this way, we see how much more revenue Home Depot generated in the studied years than Lowe's. Because the box for Home Depot is wider, and the distance from the minimum to the maximum is also wider for Home Depot, we can also say that there is a greater degree of variability for Home Depot. ✓ 5

# 10.3 Exercises

◩ 1. How are the variance and the standard deviation related?

◩ 2. Why can't we use the sum of the deviations from the mean as a measure of dispersion of a distribution?

**State Expenditures** *Use the following table for Exercises 3–10, which lists expenditures (in billions of dollars with the exception of Parks and Recreation, which is in tens of millions of dollars) for a random sample of seven states. Find the range and standard deviation among the seven states for each category of government expenditure. (See Examples 1–3.) (Data from: U.S. Census Bureau.)*

| Type of Expenditure | State | | | | | | |
|---|---|---|---|---|---|---|---|
| | AZ | CA | IL | MA | MI | NJ | PA |
| Education | 9.7 | 79.8 | 17.7 | 13.5 | 23.5 | 17.1 | 22.9 |
| Public Welfare | 9.1 | 83.5 | 21.2 | 16.7 | 15.7 | 15.9 | 24.3 |
| Hospitals | .9 | 9.8 | 1.1 | .5 | 3.2 | 2.3 | 4.0 |
| Health | 2.0 | 8.8 | 2.2 | 1.1 | 1.3 | 1.5 | 3.0 |
| Highways | 1.8 | 12.6 | 5.3 | 2.2 | 2.6 | 3.6 | 6.9 |
| Police Protection | .2 | 1.6 | .5 | .8 | .4 | .9 | .9 |
| Corrections | 1.1 | 8.9 | 1.4 | 1.2 | 1.9 | 1.4 | 2.0 |
| Parks and Recreation | 8 | 55 | 25 | 21 | 14 | 21 | 26 |

3. Education

4. Public Welfare

5. Hospitals

6. Health

7. Highways

8. Police Protection

9. Corrections

10. Parks and Recreation

**Education** *Find the standard deviation for the grouped data in Exercises 11 and 12. (See Example 4.)*

11. Number of credits for a sample of college students:

| College Credits | Frequency |
|---|---|
| 0–24 | 4 |
| 25–49 | 3 |
| 50–74 | 6 |
| 75–99 | 3 |
| 100–124 | 5 |
| 125–149 | 9 |

12. Scores on a calculus exam:

| Scores | 30–39 | 40–49 | 50–59 | 60–69 | 70–79 | 80–89 | 90–99 |
|---|---|---|---|---|---|---|---|
| Frequency | 1 | 6 | 13 | 22 | 17 | 13 | 8 |

**S&P 500 Stocks** *The data for Exercises 13–14 come from a random sample of 10 companies that were part of the S&P 500 and information* *was current as of March 7, 2017. (Hint: Do not round the mean when calculating the standard deviation.) (Data from: data.okfn.org.)*

13. Find the mean and standard deviation of the following stock prices (rounded to the nearest dollar).

| | | | | |
|---|---|---|---|---|
| 25 | 64 | 32 | 99 | 74 |
| 27 | 66 | 72 | 83 | 54 |

14. Find the mean and standard deviation of the following dividend yields (percent).

| | | | | |
|---|---|---|---|---|
| 2.38 | 1.14 | 3.12 | 2.26 | 2.12 |
| 1.06 | 1.50 | 1.62 | 1.37 | 3.23 |

Find the mean and standard deviation of the market capitalization.

*An application of standard deviation is given by Chebyshev's theorem. (P. L. Chebyshev was a Russian mathematician who lived from 1821 to 1894.) This theorem, which applies to any distribution of numerical data, states,*

*For any distribution of numerical data, at least $1 - 1/k^2$ of the numbers lie within k standard deviations of the mean.*

**Example** *For any distribution, at least*

$$1 - \frac{1}{3^2} = 1 - \frac{1}{9} = \frac{8}{9}$$

*of the numbers lie within 3 standard deviations of the mean. Find the fraction of all the numbers of a data set lying within the given numbers of standard deviations from the mean.*

15. 2                16. 4                17. 1.5

*In a certain distribution of numbers, the mean is 50, with a standard deviation of 6. Use Chebyshev's theorem to tell what percent of the numbers are*

18. between 32 and 68;

19. between 26 and 74;

20. less than 38 or more than 62;

21. less than 32 or more than 68;

22. less than 26 or more than 74.

**Movie Studios' Revenue** *For Exercises 23–28, use the following table, which gives the revenue (in millions of dollars) for four major movie studios for the years 2010–2016. (Hint: do not round the mean value when calculating the standard deviation.) (Data from: www.boxofficemojo.com.)*

| Year | Columbia | 20th Century Fox | Warner Bros. | Paramount |
|---|---|---|---|---|
| 2016 | 911 | 1469 | 1902 | 877 |
| 2015 | 966 | 1302 | 1604 | 675 |
| 2014 | 1262 | 1791 | 1563 | 1053 |
| 2013 | 1445 | 1064 | 1864 | 967 |
| 2012 | 1792 | 1025 | 1665 | 914 |
| 2011 | 1274 | 978 | 1826 | 1957 |
| 2010 | 1283 | 1482 | 1924 | 1714 |

*Find the mean and standard deviation for the studio listed.*

23. Columbia                24. 20th Century Fox

25. Warner Bros.                26. Paramount

27. Which studio shows greater variation: Columbia or Warner Bros.?

28. Which studio shows greater variation: 20th Century Fox or Paramount?

*For Exercises 29–32, see Example 4.*

29. **Household Income** A sample of 40 households from the American Housing Survey generates the following frequency table of household income (in thousands). Find the standard deviation.

| Interval | 0–29 | 30–59 | 60–89 | 90–119 | 120–149 | 150–179 | 180–209 |
|---|---|---|---|---|---|---|---|
| Frequency | 12 | 3 | 10 | 2 | 0 | 1 | 2 |

30. **Business** The number of hours worked in a week for 30 workers selected at random from the 2016 General Social Survey appears in the following frequency table. Find the standard deviation. (Data from: gss.norc.org.)

| Interval | Frequency |
|---|---|
| 0–9 | 2 |
| 10–19 | 4 |
| 20–29 | 0 |
| 30–39 | 5 |
| 40–49 | 8 |
| 50–59 | 7 |
| 60–69 | 3 |
| 70–79 | 1 |

**Reading Scores** *Shown in the following table are the reading scores of a second-grade class given individualized instruction and the reading scores of a second-grade class given traditional instruction in the same school:*

| Scores | Individualized Instruction | Traditional Instruction |
|---|---|---|
| 50–59 | 2 | 5 |
| 60–69 | 4 | 8 |
| 70–79 | 7 | 8 |
| 80–89 | 9 | 7 |
| 90–99 | 8 | 6 |

31. Find the mean and standard deviation for the individualized-instruction scores.

32. Find the mean and standard deviation for the traditional-instruction scores.

33. Discuss a possible interpretation of the differences in the means and the standard deviations in Exercises 35 and 36.

34. Discuss what the standard deviation tells us about a distribution.

**IBM and Microsoft Revenue** *The following table gives the annual revenue for IBM Corp. and Microsoft Corp. (in billions of dollars) for a 10-year period. (Data from: www.morningstar.com.) (See Examples 5 and 6.)*

| Year | IBM | Microsoft |
|------|-----|-----------|
| 2007 | 99  | 51 |
| 2008 | 104 | 60 |
| 2009 | 96  | 58 |
| 2010 | 100 | 62 |
| 2011 | 107 | 70 |
| 2012 | 105 | 73 |
| 2013 | 100 | 78 |
| 2014 | 93  | 86 |
| 2015 | 82  | 94 |
| 2016 | 80  | 85 |

**35.** What is the five-number summary for IBM?

**36.** What is the five-number summary for Microsoft?

**37.** Construct a boxplot for IBM.

**38.** Construct a boxplot for Microsoft.

**39.** Graph the boxplots for IBM and Microsoft on the same scale. Are the distributions similar?

**40.** Which company has the higher median of sales over the 10-year period?

**State Expenditures** *Use the following table for Exercises 41–44. The table contains data from a random sample of 12 states and lists expenditures (in billions of dollars) by category spent by states and local communities. (Data from: U.S. Census Bureau.)*

| State | Education | Public Welfare |
|-------|-----------|----------------|
| AL | 10.4 | 6.6 |
| AZ | 9.7 | 9.1 |
| CA | 79.8 | 83.5 |
| FL | 25.1 | 24.4 |
| GA | 17.4 | 11.7 |
| IL | 17.7 | 21.2 |
| MA | 13.5 | 16.7 |
| MD | 12.0 | 10.8 |
| MI | 23.5 | 15.7 |
| NJ | 17.1 | 15.9 |
| PA | 22.9 | 24.3 |
| TX | 50.4 | 32.8 |

**41.** What is the five-number summary for the expenditures on education?

**42.** What is the five-number summary for the expenditures on public welfare?

**43.** Construct a boxplot for the expenditures on education.

**44.** Construct a boxplot for the expenditures on public welfare.

**State Expenditures** *For Exercises 45 and 46, use the boxplots below from data similar to Exercises 41–44 for state and local community expenditures on hospitals and highways.*

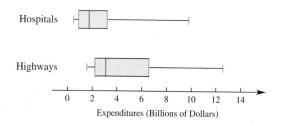

**45.** Estimate $Q_1$, $Q_2$, and $Q_3$ for spending on hospitals and highways.

**46.** Is there greater variability in spending on hospitals or highway? How do you know?

**State Expenditures** *For Exercises 47 and 48, use the boxplots below from data similar to Exercises 41–44 for state and local community expenditures on health and corrections.*

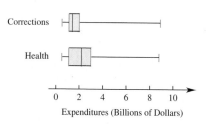

**47.** Is there greater variability in spending on health or corrections? How do you know?

**48.** Estimate $Q_1$, $Q_2$, and $Q_3$ for spending on health and corrections.

✓ **Checkpoint Answers**

**1.** 373

**2.** Mean is 210; deviations are $-75$, $-150$, $-160$, 215, and 170.

**3.** 179.5 miles

**4.** 10.91 classes

**5.**

# 10.4   Normal Distributions

Suppose a bank is interested in improving its services to customers. The manager decides to begin by finding the amount of time tellers spend on each transaction, rounded to the nearest minute. The times for 75 different transactions are recorded, with the results shown in the following table, where the frequencies listed in the second column are divided by 75 to find the empirical probabilities:

| Time | Frequency | Probability |
|------|-----------|-------------|
| 1 | 3 | $3/75 = .04$ |
| 2 | 5 | $5/75 \approx .07$ |
| 3 | 9 | $9/75 = .12$ |
| 4 | 12 | $12/75 = .16$ |
| 5 | 15 | $15/75 = .20$ |
| 6 | 11 | $11/75 \approx .15$ |
| 7 | 10 | $10/75 \approx .13$ |
| 8 | 6 | $6/75 = .08$ |
| 9 | 3 | $3/75 = .04$ |
| 10 | 1 | $1/75 \approx .01$ |

Figure 10.9(a), on the following page, shows a histogram and frequency polygon for the data. The heights of the bars are the empirical probabilities, rather than the frequencies. The transaction times are given to the nearest minute. Theoretically, at least, they could have been timed to the nearest tenth of a minute, or hundredth of a minute, or even more precisely. In each case, a histogram and frequency polygon could be drawn. If the times are measured with smaller and smaller units, there are more bars in the histogram and the frequency polygon begins to look more and more like the curve in Figure 10.9(b) instead of a polygon. Actually, it is possible for the transaction times to take on any real-number value greater than 0. A distribution in which the outcomes can take on any real-number value within some interval is a **continuous distribution.** The graph of a continuous distribution is a curve.

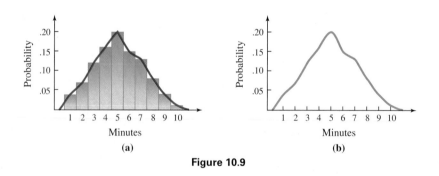

**Figure 10.9**

The distribution of heights (in inches) of adult women is another example of a continuous distribution, since these heights include infinitely many possible measurements, such as 53, 58.5, 66.3, 72.666, and so on. Figure 10.10 shows the continuous distribution of heights of adult women. Here, the most frequent heights occur near the center of the interval displayed.

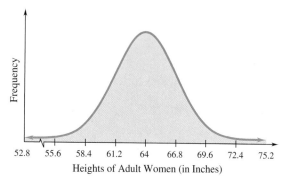

Figure 10.10

## Normal Distributions

As discussed on page 515, we say that data are normal (or normally distributed) when their graph is well approximated by a bell-shaped curve. (See Figure 10.11.) We call the graphs of such distributions **normal curves.** Examples of distributions that are approximately normal are the heights of adult women and cholesterol levels in adults. We use the Greek letters $\mu$ (mu) to denote the mean and $\sigma$ (sigma) to denote the standard deviation of a normal distribution.

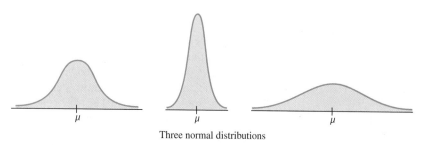

Three normal distributions

Figure 10.11

There are many normal distributions, depending on $\mu$ and $\sigma$. Some of the corresponding normal curves are tall and thin, and others short and wide, as shown in Figure 10.11. But every normal curve has the following properties:

1. Its peak occurs directly above the mean $\mu$.

2. The curve is symmetric about the vertical line through the mean. (That is, if you fold the graph along this line, the left half of the graph will fit exactly on the right half.)

3. The curve never touches the $x$-axis—it extends indefinitely in both directions.

4. The area under the curve (and above the horizontal axis) is 1. (As can be shown with calculus, this is a consequence of the fact that the sum of the probabilities in any distribution is 1.)

A normal distribution is completely determined by its mean $\mu$ and standard deviation $\sigma$.[*] A small standard deviation leads to a tall, narrow curve like the one in the center of Figure 10.11, because most of the data are close to the mean. A large standard deviation means the data are very spread out, producing a flat, wide curve like the one on the right in Figure 10.11.

---

[*]As shown in more advanced courses, its graph is the graph of the function

$$f(x) = \frac{1}{\sigma\sqrt{2\pi}}e^{-(x-\mu)^2/(2\sigma^2)},$$

where $e \approx 2.71828$ is the real number introduced in Section 4.1.

Since the area under a normal curve is 1, parts of this area can be used to determine certain probabilities. For instance, Figure 10.12(a) is the probability distribution of the annual rainfall in a certain region. The probability that the annual rainfall will be between 25 and 60 inches is the area under the curve from 25 to 60. The general case, shown in Figure 10.12(b), can be stated as follows.

> The area of the shaded region under the normal curve from $a$ to $b$ is the probability that an observed data value will be between $a$ and $b$.

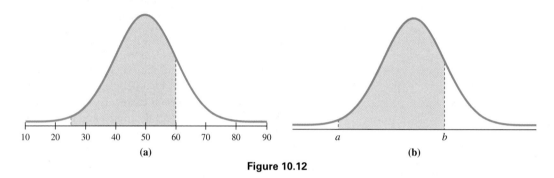

(a)                    (b)

**Figure 10.12**

To use normal curves effectively, we must be able to calculate areas under portions of them. These calculations have already been done for the normal curve with mean $\mu = 0$ and standard deviation $\sigma = 1$ (which is called the **standard normal curve**) and are available in Appendix A at the back of the book. Examples 1 and 2 demonstrate how to use Appendix A to find such areas. Later, we shall see how the standard normal curve may be used to find areas under any normal curve.

The horizontal axis of the standard normal curve is usually labeled $z$. Since the standard deviation of the standard normal curve is 1, the numbers along the horizontal axis (the $z$-values) measure the number of standard deviations above or below the mean $z = 0$.

**Figure 10.13**

▦ **TECHNOLOGY TIP** Some graphing calculators have the ability to graph a normal distribution, given its mean and standard deviation, and to find areas under the curve between two $x$-values. For an area under the curve, some calculators will give the corresponding $z$-value. For details, see your instruction book. (Look for "distribution" or "probability distribution.") A calculator-generated graph of the standard normal curve is shown in Figure 10.13.

**EXAMPLE 1**    Find the given areas under the standard normal curve.

**(a)**  The area between $z = 0$ and $z = 1$, the shaded region in Figure 10.14

**Solution** Find the entry 1 in the $z$-column of Appendix A. The entry next to it in the $A$-column is .3413, which means that the area between $z = 0$ and $z = 1$ is .3413. Since the total area under the curve is 1, the shaded area in Figure 10.14 is 34.13% of the total area under the normal curve.

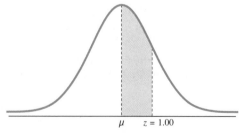

**Figure 10.14**

✓ **Checkpoint 1**

Find the percent of the area between the mean and

**(a)** $z = 1.51$;

**(b)** $z = -2.04$.

**(c)** Find the percent of the area in the shaded region.

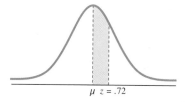

$\mu$   $z = .72$

✓ **Checkpoint 2** 🔲

If your calculator can graph probability distributions and find areas, use it to find the areas requested in Example 1.

**(b)** The area between $z = -2.43$ and $z = 0$

**Solution** Appendix A lists only positive values of $z$. But the normal curve is symmetric around the mean $z = 0$, so the area between $z = 0$ and $z = -2.43$ is the same as the area between $z = 0$ and $z = 2.43$. Find 2.43 in the $z$-column of Appendix A. The entry next to it in the $A$-column shows that the area is .4925. Hence, the shaded area in Figure 10.15 is 49.25% of the total area under the curve. ✓1 ✓2

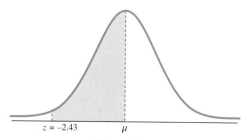

$z = -2.43$   $\mu$

**Figure 10.15**

🔲 **TECHNOLOGY TIP** Because of their convenience and accuracy, graphing calculators and computers have made normal-curve tables less important. Figure 10.16 shows how part (b) of Example 1 can be done on a TI-84+ calculator using a command from the DISTR menu. The second result in the calculator screen gives the area between $-\infty$ and $z = -2.43$; the entry $-1\text{E}99$ represents $-1 \cdot 10^{99}$, which is used to approximate $-\infty$.

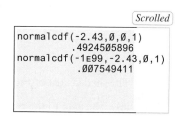

*Scrolled*

```
normalcdf(-2.43,0,0,1)
             .4924505896
normalcdf(-1E99,-2.43,0,1)
             .007549411
```

**Figure 10.16**

Many statistical software packages are widely used today. All of these packages are set up in a way that is similar to a spreadsheet, and they all can be used to generate normal curve values. In addition, most spreadsheets can perform a wide range of statistical calculations.

**EXAMPLE 2**   Use technology or Appendix A to find the percent of the total area for the given areas under the standard normal curve.

**(a)** The area between .88 standard deviations *below* the mean and 2.35 standard deviations *above* the mean (that is, between $z = -.88$ and $z = 2.35$)

**Solution** First, draw a sketch showing the desired area, as in Figure 10.17. From Appendix A, the area between the mean and .88 standard deviations below the mean is .3106. Also, the area from the mean to 2.35 standard deviations above the mean is .4906. As the figure shows, the total desired area can be found by *adding* these numbers:

$$.3106$$
$$+.4906$$
$$.8012.$$

The shaded area in Figure 10.17 represents 80.12% of the total area under the normal curve.

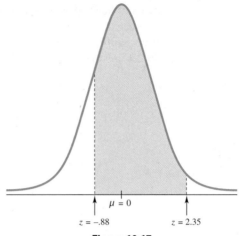

**Figure 10.17**

✓ **Checkpoint 3**

Find the given standard normal-curve areas as percentages of the total area.

**(a)** Between .31 standard deviations below the mean and 1.01 standard deviations above the mean

**(b)** Between .38 standard deviations and 1.98 standard deviations below the mean

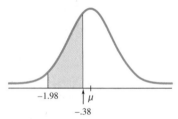

**(c)** To the right of 1.49 standard deviations above the mean

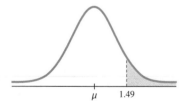

**(d)** What percent of the area is within 1 standard deviation of the mean? within 2 standard deviations of the mean? within 3 standard deviations of the mean? What can you conclude from the last answer?

**(b)** The area between .58 standard deviations above the mean and 1.94 standard deviations above the mean

**Solution** Figure 10.18 shows the desired area. The area between the mean and .58 standard deviations above the mean is .2190. The area between the mean and 1.94 standard deviations above the mean is .4738. As the figure shows, the desired area is found by *subtracting* one area from the other:

$$\begin{array}{r} .4738 \\ -.2190 \\ \hline .2548. \end{array}$$

The shaded area of Figure 10.18 represents 25.48% of the total area under the normal curve.

**(c)** The area to the right of 2.09 standard deviations above the mean

**Solution** The total area under a normal curve is 1. Thus, the total area to the right of the mean is 1/2, or .5000. From Appendix A, the area from the mean to 2.09 standard deviations above the mean is .4817. The area to the right of 2.09 standard deviations is found by subtracting .4817 from .5000:

$$\begin{array}{r} .5000 \\ -.4817 \\ \hline .0183. \end{array}$$

A total of 1.83% of the total area is to the right of 2.09 standard deviations above the mean. Figure 10.19 shows the desired area. ✓ 3

**Figure 10.18**

**Figure 10.19**

The key to finding areas under *any* normal curve is to express each number $x$ on the horizontal axis in terms of standard deviations above or below the mean. The **z-score** for $x$ is the number of standard deviations that $x$ lies from the mean (positive if $x$ is above the mean, negative if $x$ is below the mean).

**EXAMPLE 3**  If a normal distribution has mean 60 and standard deviation 5, find the given $z$-scores.

**(a)** The $z$-score for $x = 65$

**Solution** Since 65 is 5 units above 60 and the standard deviation is 5, 65 is 1 standard deviation above the mean. So its $z$-score is 1.

**(b)** The $z$-score for $x = 52.5$

**Solution** The $z$-score is $-1.5$, because 52.5 is 7.5 units below the mean (since $52.5 - 60 = -7.5$) and 7.5 is 1.5 standard deviations (since $7.5/5 = 1.5$). ✓ 4

In Example 3(b) we found the $z$-score by taking the difference between 52.5 and the mean and dividing this difference by the standard deviation. The same procedure works in the general case.

✓**Checkpoint 4**

Find each $z$-score, using the information in Example 3.

**(a)** $x = 36$

**(b)** $x = 55$

## Z-Score

If a normal distribution has mean $\mu$ and standard deviation $\sigma$, then the $z$-score for the number $x$ is

$$z = \frac{x - \mu}{\sigma}.$$

The importance of $z$-scores is the following fact, whose proof is omitted.

## Area under a Normal Curve

The area under a normal curve between $x = a$ and $x = b$ is the same as the area under the standard normal curve between the $z$-score for $a$ and the $z$-score for $b$.

Therefore, by converting to $z$-scores and using a graphing calculator or Appendix A for the standard normal curve, we can find areas under any normal curve. Since these areas are probabilities (as explained earlier), we can now handle a variety of applications.

Graphing calculators, computer programs, and CAS programs (such as DERIVE) can be used to find areas under the normal curve and, hence, probabilities. The equation of the standard normal curve, with $\mu = 0$ and $\sigma = 1$, is

$$f(x) = (1/\sqrt{2\pi})e^{-x^2/2}.$$

A good approximation of the area under this curve (and above $y = 0$) can be found by using the $x$-interval $[-4, 4]$. However, calculus is needed to find such areas.

**EXAMPLE 4**  **Mileage** Dixie Office Supplies finds that its sales force drives an average of 1200 miles per month per person, with a standard deviation of 150 miles. Assume that the number of miles driven by a salesperson is closely approximated by a normal distribution.

**(a)** Find the probability that a salesperson drives between 1200 and 1600 miles per month.

**Solution** Here, $\mu = 1200$ and $\sigma = 150$, and we must find the area under the normal distribution curve between $x = 1200$ and $x = 1600$. We begin by finding the $z$-score for $x = 1200$:

$$z = \frac{x - \mu}{\sigma} = \frac{1200 - 1200}{150} = \frac{0}{150} = 0.$$

The $z$-score for $x = 1600$ is

$$z = \frac{x - \mu}{\sigma} = \frac{1600 - 1200}{150} = \frac{400}{150} = 2.67.^{*}$$

So the area under the curve from $x = 1200$ to $x = 1600$ is the same as the area under the standard normal curve from $z = 0$ to $z = 2.67$, as indicated in Figure 10.20. A graphing calculator or Appendix A shows that this area is .4962. Therefore, the probability that a salesperson drives between 1200 and 1600 miles per month is .4962.

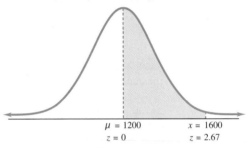

**Figure 10.20**

**(b)** Find the probability that a salesperson drives between 1000 and 1500 miles per month.

**Solution** As shown in Figure 10.21, $z$-scores for both $x = 1000$ and $x = 1500$ are needed.

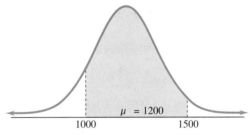

**Figure 10.21**

For $x = 1000$,                    For $x = 1500$,

$$z = \frac{1000 - 1200}{150} \qquad\qquad z = \frac{1500 - 1200}{150}$$

$$= \frac{-200}{150} \qquad\qquad\qquad = \frac{300}{150}$$

$$= -1.33. \qquad\qquad\qquad = 2.00.$$

From Appendix A , $z = 1.33$ leads to an area of .4082, while $z = 2.00$ corresponds to .4773. A total of .4082 + .4773 = .8855, or 88.55%, of all drivers travel between 1000 and 1500 miles per month. From this calculation, the probability that a driver travels between 1000 and 1500 miles per month is .8855.  ✔ 5

**Checkpoint 5**

With the data from Example 4, find the probability that a salesperson drives between 1000 and 1450 miles per month.

📄 **NOTE** Graphing calculators and spreadsheets can calculate normal probabilities with given means and standard deviations. The answers, however, might differ slightly from our answers because the z-scores are not rounded to two decimal places.

---

$^{*}$All $z$-scores here are rounded to two decimal places.

**EXAMPLE 5** **Cholesterol** With data from a recent National Health and Nutritional Examination Survey (NHANES), we can use 186 (mg/dL) as an estimate of the mean total cholesterol level for all Americans and 41 (mg/dL) as an estimate of the standard deviation. Assuming total cholesterol levels to be normally distributed, what is the probability that an American chosen at random has a cholesterol level higher than 250? If 200 Americans are chosen at random, how many can we expect to have total cholesterol higher than 250? (Data from: www.cdc.gov/nchs/nhanes.htm.)

**Solution** Here, $\mu = 186$ and $\sigma = 41$. The probability that a randomly chosen American has cholesterol higher than 250 is the area under the normal curve to the right of $x = 250$. The $z$-score for $x = 250$ is

$$z = \frac{x - \mu}{\sigma} = \frac{250 - 186}{41} = \frac{64}{41} = 1.56.$$

From Appendix A, we see that the area to the right of 1.56 is $.5 - .4406 = .0594$, which is 5.94% of the total area under the curve. Therefore, the probability of a randomly chosen American having cholesterol higher than 250 is .0594.

With 5.94% of Americans having total cholesterol higher than 250, selecting 200 Americans at random yields

$$5.94\% \text{ of } 200 = .0594 \cdot 200 = 11.88.$$

Approximately 12 of these Americans can be expected to have a total cholesterol level higher than 250. ✓ 6

✓**Checkpoint 6**

Using the mean and standard deviation from Example 5, find the probability that an adult selected at random has a cholesterol level below 150.

**NOTE** Notice in Example 5 that $P(z \geq 1.56) = P(z > 1.56)$. The area under the curve is the same whether we include the endpoint or not. Notice also that $P(z = 1.56) = 0$, because no area is included.

⚠️ **CAUTION** When calculating the normal probability, it is wise to draw a normal curve with the mean and the $z$-scores every time. This practice will avoid confusion as to whether you should add or subtract probabilities.

As mentioned earlier, $z$-scores are standard deviations, so $z = 1$ corresponds to 1 standard deviation above the mean, and so on. As found in Checkpoint 3(d) of this section, 68.26% of the area under a normal curve lies within 1 standard deviation of the mean. Also, 95.46% lies within 2 standard deviations of the mean, and 99.74% lies within 3 standard deviations of the mean. These results, summarized in Figure 10.22, can be used to get quick estimates when you work with normal curves.

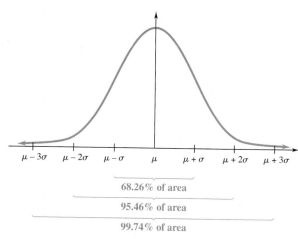

Figure 10.22

As we saw in Section 9.4, many practical experiments have only two possible outcomes. Example 3 from Section 9.4 presented a result from the U.S. Department of Education that 11.3% of student borrowers defaulted on their student loans in the first two years after college. If we use the binomial formula for 100 students chosen at random, we can create the histogram of the probabilities for the number of students who defaulted that appears in Figure 10.23.

Number of Defaulters

**Figure 10.23**

Notice that Figure 10.23 looks very much like the normal curves we have been studying in this section. In fact, we can use the normal distribution to approximate binomial probabilities. To do this, first we need to know the mean and standard deviation of the binomial distribution. Recall from Section 9.4 that, for the binomial distribution $E(x) = np$. In Section 10.2, we referred to $E(x)$ as $\mu$, and that notation will be used here. It is shown in more advanced courses in statistics that the standard deviation of the binomial distribution is given by $\sigma = \sqrt{np(1 - p)}$.

---

## Mean and Standard Deviation for the Binomial Distribution

For the binomial distribution, the mean and standard deviation, respectively, are given by

$$\mu = np \quad \text{and} \quad \sigma = \sqrt{np(1 - p)},$$

where $n$ is the number of trials and $p$ is the probability of success on a single trial.

---

With a repeated binomial experiment, we count the number of successes, which will be an integer. As we see in Figure 10.23, the bars represent the probability of each integer outcome from 0 to 22. When we use a normal distribution to approximate a binomial probability, we subtract .5 from the integer that is our lower bound of interest and we add .5 to the integer of the upper bound interest. Thus, we if we wanted to approximate the probability of between 8 and 13 successes, we would use 7.5 and 13.5 as our bounds for the normal distribution.

**⚠ CAUTION** The normal-curve approximation to a binomial distribution is quite accurate *provided* that n is large and p is not close to 0 or 1. As a rule of thumb, the normal–curve approximation can be used as long as both np and n(1 − p) are at least 5.

**EXAMPLE 6** **Student Loan Defaults** Consider the previously discussed sample of 100 students chosen at random from a population where 11.3% of the students defaulted on their student loans within two years after college.

**(a)** Find the mean and standard deviation for this distribution.

**Solution**  We have $n = 100$ and $p = .113$. Thus,

$$\mu = 100(.113) \qquad \sigma = \sqrt{100(.113)(1 - .113)}$$
$$= 11.3 \qquad\qquad = \sqrt{100(.113)(.887)}$$
$$\approx 3.17.$$

**(b)** Use the normal distribution to approximate the probability that between 8 and 13 (inclusive) students selected at random defaulted on their loans.

**Solution**  We have $np = 100(.113) = 11.3$ and $n(1 - p) = 88.7$. Both numbers are greater than 5, so we can use a normal approximation. As the graph of Figure 10.24 shows, we need to find the area between 7.5 and 13.5. The $z$-scores corresponding to $x = 7.5$ and $x = 13.5$ are

$$z_1 = \frac{7.5 - 11.3}{3.17} = -1.20 \qquad z_2 = \frac{13.5 - 11.3}{3.17} = .69.$$

From Appendix A, a $z$-score of $-1.20$ corresponds to .3849, and a $z$-score of .69 corresponds to .2549. Since $z_1$ is on the left side of 0 and $z_2$ is on the right side of 0, we add these probabilities. We have

$$P(-1.20 \le z \le .69) = .3849 + .2549 = .6398.$$

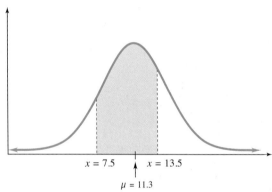

$x = 7.5$    $x = 13.5$

$\mu = 11.3$

**Figure 10.24**

**(c)** Find the probability that 6 or fewer students default on their lawns.

**Solution**  As the graph of Figure 10.25 shows, we need to find the area below 6.5. The $z$-score corresponding to $x = 6.5$ is

$$z_1 = \frac{6.5 - 11.3}{3.17} = -1.51.$$

From Appendix A, we have

$$P(z \le -1.51) = .5 - .4345 = .0655.$$

The probability of having 6 or fewer students default on their loans is approximately .0655.

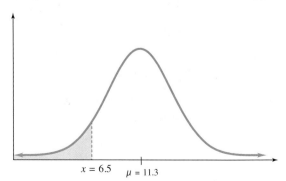

$x = 6.5$    $\mu = 11.3$

**Figure 10.25**

✓ **Checkpoint 7**

On May 23, 2017, the probability that a stock advanced on the New York Stock Exchange (NYSE) was .57. If we select 50 stocks at random, find the approximate probability of the following. (Data from: www.wsj.com.)

**(a)**  30 or more

**(b)**  Between 20 and 30 (inclusive)

## 10.4 Exercises

1. The peak in a normal curve occurs directly above _____.

2. The total area under a normal curve (above the horizontal axis) is _____.

3. How are $z$-scores found for normal distributions with $\mu \neq 0$ or $\sigma \neq 1$?

4. How is the standard normal curve used to find probabilities for normal distributions?

*Find the percentage of the area under a normal curve between the mean and the given number of standard deviations from the mean. (See Example 1.)*

5. 1.75                    6. .26

7. −.43                   8. −2.4

*Find the percentage of the total area under the standard normal curve between the given z-scores. (See Examples 1 and 2.)*

9. $z = 1.41$ and $z = 2.83$

10. $z = .64$ and $z = 2.11$

11. $z = -2.48$ and $z = -.05$

12. $z = -1.63$ and $z = -1.08$

13. $z = -3.05$ and $z = 1.36$

14. $z = -2.91$ and $z = -.51$

*Find a z-score satisfying each of the given conditions. (Hint: Use Appendix A backward or a graphing calculator.)*

15. 5% of the total area is to the right of $z$.

16. 1% of the total area is to the left of $z$.

17. 15% of the total area is to the left of $z$.

18. 25% of the total area is to the right of $z$.

19. For any normal distribution, what is the value of $P(x \leq \mu)$? of $P(x \geq \mu)$?

20. Using Chebyshev's theorem and the normal distribution, compare the probability that a number will lie within 2 standard deviations of the mean of a probability distribution. (See Exercises 15–22 of Section 10.3.) Explain what you observe.

21. Repeat Exercise 20, using 3 standard deviations.

**Heights** *Data from a recent National Health and Nutrition Examination Survey (NHANES) indicates the mean heights of adult females and males, age 18 to 60, is approximately normally distributed. The mean height for women is 63.7 inches, with a standard deviation of 2.8 inches. The mean height of men is 69.1 inches, with a standard deviation of 3.1 inches. Use this information for Exercises 22–26. (See Example 4). (Data from: www.cdc.gov/nchs/nhanes.htm.)*

*Find the probability of each of the following.*

22. That a female is taller than 70 inches

23. That a female is shorter than 60 inches

24. That a male is shorter than 66 inches

25. That a female is between 5 feet and 6 feet (*Hint:* Convert feet to inches.)

26. That a male is between 5 feet and 6 feet (*Hint:* Convert feet to inches.)

*Assume the distributions in Exercises 27–30 are all normal. (See Examples 4 and 5.)*

27. **Cholesterol** Using the data from the same study as in Example 5, we find that the average HDL cholesterol level is 52.6 mg/dL, with a standard deviation of 15.5 mg/dL. Find the probability that an individual will have an HDL cholesterol level greater than 60 mg/dL.

28. **Manufacturing** The production of cars per day at an assembly plant has mean 120.5 and standard deviation 6.2. Find the probability that fewer than 100 cars are produced on a random day.

29. **Starting Salary** Starting salaries for accounting majors have mean $53,300, with standard deviation $3,200. What is the probability an individual will start at a salary above $56,000?

30. **Driving Distance** The driving distance to work for residents of a certain community has mean 21 miles and standard deviation 3.6 miles. What is the probability that an individual drives between 10 and 20 miles to work?

**Business** *Scores on the Graduate Management Association Test (GMAT) are approximately normally distributed. The mean score for 2013–2015 was 552 with a standard deviation of 121. For the following exercises, find the probability that a GMAT test taker selected at random earns a score in the given range, using the normal distribution as a model. (Data from: www.gmac.com.)*

31. Between 540 and 700          32. Between 300 and 540

33. Between 300 and 700          34. Less than 400

35. Greater than 750             36. Between 600 and 700

37. Between 300 and 400

**Traffic Speed** *New studies by Federal Highway Administration traffic engineers suggest that speed limits on many thoroughfares are set arbitrarily and often are artificially low. According to traffic engineers, the ideal limit should be the "85th-percentile speed," the speed at or below which 85% of the traffic moves. Assuming that speeds are normally distributed, find the 85th-percentile speed for roads with the given conditions.*

**38.** The mean speed is 55 mph, with a standard deviation of 10 mph.

**39.** The mean speed is 40 mph, with a standard deviation of 5 mph.

**Education** *One professor uses the following system for assigning letter grades in a course:*

| Grade | Total Points |
|-------|-------------|
| A | Greater than $\mu + \frac{3}{2}\sigma$ |
| B | $\mu + \frac{1}{2}\sigma$ to $\mu + \frac{3}{2}\sigma$ |
| C | $\mu - \frac{1}{2}\sigma$ to $\mu + \frac{1}{2}\sigma$ |
| D | $\mu - \frac{3}{2}\sigma$ to $\mu - \frac{1}{2}\sigma$ |
| F | Below $\mu - \frac{3}{2}\sigma$ |

*What percentage of the students receive the given grades?*

**40.** A          **41.** B          **42.** C

**43.** Do you think the system in Exercises 40–42 would be more likely to be fair in a large first-year class in psychology or in a graduate seminar of five students? Why?

**Nutrition** *In nutrition, the recommended daily allowance of vitamins is a number set by the government as a guide to an individual's daily vitamin intake. Actually, vitamin needs vary drastically from person to person, but the needs are very closely approximated by a normal curve. To calculate the recommended daily allowance, the government first finds the average need for vitamins among people in the population and then the standard deviation. The recommended daily allowance is defined as the mean plus 2.5 times the standard deviation.*

**44.** What percentage of the population will receive adequate amounts of vitamins under this plan?

*Find the recommended daily allowance for the following vitamins.*

**45.** Mean = 550 units, standard deviation = 46 units

**46.** Mean = 1700 units, standard deviation = 120 units

**47.** Mean = 155 units, standard deviation = 14 units

**48.** Mean = 1080 units, standard deviation = 86 units

**Education** *The mean performance score of a large group of fifth-grade students on a math achievement test is 88. The scores are known to be normally distributed. What percentage of the students had scores as follows?*

**49.** More than 1 standard deviation above the mean

**50.** More than 2 standard deviations above the mean

**Job Satisfaction** *According to a 2016 study conducted by the Society for Human Resources Management, 38% of U.S. employees reported being very satisfied with their job. In the same study, 51% indicated they are satisfied but to a lesser degree. If 200 employees are selected at random, calculate the following. (See Example 6.) (Data from: www.shrm.org.)*

**51.** The mean and standard deviation for the number of very satisfied employees (round to one decimal place)

**52.** The mean and standard deviation for the number of employees who are satisfied to a lesser degree (round to one decimal place)

**53.** The probability that between 70 and 80 employees are very satisfied

**54.** The probability that between 90 and 115 employees are satisfied to a lesser degree

**55.** The probability that at least 88 employees are very satisfied

**56.** The probability that 85 or less employees are satisfied to a lesser degree

**Student Loan Debt** *According to a report from the American Association of University Women released in 2017, 44% of female undergraduates and 39% of male undergraduates take on debt. Calculate the following if we select 75 females and 75 males at random. (Data from: www.aauw.org.)*

**57.** The mean and standard deviation for the number of female undergraduates who take on debt (round to the nearest tenth)

**58.** The mean and standard deviation for the number of male undergraduates who take on debt (round to the nearest tenth)

**59.** The probability that between 25 and 35 (inclusive) female undergraduates take on debt

**60.** The probability that between 35 and 40 (inclusive) male undergraduates take on debt

**61.** The probability that 27 or fewer female undergraduates take on debt

**62.** The probability that at least 29 male undergraduates take on debt

✓**Checkpoint Answers**

**1.** (a) 43.45%        (b) 47.93%        (c) 26.42%

**2.** (a) 34.13%        (b) 49.25%

**3.** (a) 46.55%        (b) 32.82%        (c) 6.81%

(d) 68.26%, 95.46%, 99.74%; almost all the data lie within 3 standard deviations of the mean.

**4.** (a) −4.8          (b) −1

**5.** .8607

**6.** .1894

**7.** (a) .3859          (b) .7106

# CHAPTER 10 Summary and Review

## Key Terms and Symbols

**10.1** random sample
grouped frequency
distribution
histogram
frequency polygon
stem-and-leaf-plot
uniform distribution
right-skewed distribution
left-skewed distribution
normal distribution

**10.2** $\Sigma$, summation (sigma)
notation
$\bar{x}$, sample mean
$\mu$, population mean
(arithmetic) mean
median
statistic
mode

**10.3** $s^2$, sample variance
$s$, sample standard deviation

$\sigma$, population standard
deviation
range
deviations from the
mean
variance
standard deviation
boxplot
quartile
five-number summary

**10.4** $\mu$, mean of a normal
distribution
$\sigma$, standard deviation of a
normal distribution
continuous distribution
normal curves
standard normal curve
$z$-score
normal approximation to
the binomial distribution

## Chapter 10 Key Concepts

**Frequency Distributions**

To organize the data from a sample, we use a **grouped frequency distribution**—a set of intervals with their corresponding frequencies. The same information can be displayed with a **histogram**—a type of bar graph with a bar for each interval. The height of each bar is equal to the frequency of the corresponding interval. A **stem-and-leaf plot** presents the individual data in a similar form, so it can be viewed as a bar graph as well. Another way to display this information is with a **frequency polygon**, which is formed by connecting the midpoints of consecutive bars of the histogram with straight-line segments.

**Measures of Center**

The **mean** $\bar{x}$ of a frequency distribution is the expected value.

For $n$ numbers $x_1, x_2, \ldots, x_n$,     For a grouped distribution,

$$\bar{x} = \frac{\Sigma x_i}{n} \qquad\qquad \bar{x} = \frac{\Sigma f_i x_i}{n}$$

The **median** is the middle entry (or mean of the two middle entries) in a set of data arranged in either increasing or decreasing order.

The **mode** is the most frequent entry in a set of numbers.

**Measures of Variability**

The **range** of a distribution is the difference between the largest and smallest numbers in the distribution.

The **sample standard deviation** $s$ is the square root of the sample **variance**.

For $n$ numbers,     For a grouped distribution,

$$s = \sqrt{\frac{\Sigma x_i^2 - n\bar{x}}{n-1}} \qquad\qquad s = \sqrt{\frac{\Sigma f_i x_i^2 - n\bar{x}}{n-1}}$$

**Boxplots**

A **boxplot** organizes a list of data using the minimum and maximum values, the median, and the first and third quartiles to give a visual overview of the distribution.

**The Normal Distribution**

A **normal distribution** is a continuous distribution with the following properties: The highest frequency is at the mean; the graph is symmetric about a vertical line through the mean; the total area under the curve, above the $x$-axis, is 1. If a normal distribution has mean $\mu$ and standard deviation $\sigma$, then the $z$-score for the number $x$ is $z = \dfrac{x - \mu}{\sigma}$. The **area under a normal curve** between $x = a$ and $x = b$ is found by calculating the area under the standard normal curve from $z_a$ to $z_b$.

**Normal Approximation to the Binomial Distribution**

The **binomial distribution** is a distribution with the following properties: For $n$ independent repeated trials, in which the probability of success in a single trial is $p$, the probability of $x$ successes is $_nC_x p^x (1 - p)^{n-x}$. The mean is $\mu = np$, and the standard deviation is

$$\sigma = \sqrt{np(1 - p)}.$$

For a large number of trials, the normal distribution can be used to approximate binomial probabilities.

# Chapter 10 Review Exercises

**NASDAQ 100 Stocks** *For Exercises 1–10, the data come from a random sample of 20 companies of the NASDAQ 100 taken on June 8, 2017. (Data from: www.morningstar.com.)*

*For Exercises 1 and 2, (a) write a frequency distribution; (b) draw a histogram.*

1. The variable is the price of the stock (rounded to the nearest dollar).

| | | | |
|---|---|---|---|
| 123 | 35 | 76 | 38 |
| 45 | 62 | 50 | 111 |
| 180 | 108 | 65 | 46 |
| 66 | 100 | 64 | 144 |
| 57 | 107 | 48 | 110 |

2. The variable is the volume of stock (the number of shares that traded hands on a given day, in thousands).

| | | | |
|---|---|---|---|
| 241 | 7300 | 1700 | 1800 |
| 1400 | 1800 | 5700 | 1000 |
| 2900 | 411 | 4400 | 5800 |
| 2600 | 2900 | 674 | 2300 |
| 2500 | 2300 | 1500 | 4600 |

*For Exercises 3 and 4, draw a stem-and-leaf plot for the data.*

3. The variable is the forward price-to-earnings ratio rounded to the nearest integer. (The forward price-to-earnings ratio is the current stock's price over its predicted earnings per share.)

| | | | |
|---|---|---|---|
| 22 | 16 | 15 | 28 |
| 21 | 18 | 8 | 20 |
| 28 | 21 | 26 | 20 |
| 16 | 25 | 15 | 20 |
| 14 | 15 | 16 | 22 |

4. The variable is the price-per-sales ratio. (The price-per-sales ratio is calculated by dividing the per-share stock price by the per-share revenue.)

| | | | |
|---|---|---|---|
| 2.6 | 4.5 | 0.8 | 3.7 |
| 4.4 | 1.8 | 0.7 | 6.5 |
| 6.1 | 2.7 | 2.1 | 2.8 |
| 2.9 | 3.7 | 1.3 | 2.5 |
| 1.1 | 6.2 | 3.6 | 7.4 |

5. Find the mean and median for the first 10 stock prices of Exercise 1.

123, 45, 180, 66, 57, 35, 62, 108, 100, 107

6. Find the mean and median for the first 10 stock volumes of Exercise 2.

241, 1400, 2900, 2600, 2500, 7300, 1800, 411, 2900, 2300

7. Find the 5-number summary and create a boxplot for the data in Exercise 5.

8. Find the 5-number summary and create a boxplot for the data in Exercise 6.

9. The following table gives the frequency counts for the price-per-book ratio. (This ratio is calculated by dividing the current closing price of the stock by the latest quarter's book value per share.) Find the mean and median for these data. (Round midpoint to the nearest tenth.)

| Interval | Frequency |
|---|---|
| 2–2.99 | 2 |
| 3–3.99 | 5 |
| 4–4.99 | 2 |
| 5–5.99 | 5 |
| 6–6.99 | 1 |
| 7–7.99 | 3 |
| 8–8.99 | 2 |

10. The following table gives the frequency counts for the price/cash flow ratio. (This ratio is calculated by dividing the company's capitalization by the company's operating cash flow in the most recent fiscal year.) Find the mean and median for these data. (Round midpoint to the nearest tenth.)

| Interval | Frequency |
|---|---|
| 0–4.99 | 1 |
| 5–9.99 | 1 |
| 10–14.99 | 5 |
| 15–19.99 | 8 |
| 20–24.99 | 3 |
| 25–29.99 | 2 |

11. Discuss some reasons for organizing data into a grouped frequency distribution.

12. What do the mean, median, and mode of a distribution have in common? How do they differ? Describe each in a sentence or two.

*Find the median and the mode (or modes) for each of the given data sets.*

13. **Hours at Work** Hours worked for 10 adults. (Data from: gss.norc.org.)

30, 45, 16, 20, 70, 40, 48, 40, 50, 50

14. **Education Attainment** The percentage of residents with a bachelor's degree or higher for 10 states. (Data from: U.S. Census Bureau.)

22, 28, 26, 20, 30, 36, 36, 28, 26, 27

*The modal class is the interval containing the most data values. Find the modal class for the distributions of the given data sets.*

15. The data in Exercise 9

16. The data in Exercise 10

*For the given histograms, identify the shape of the distribution.*

**17.**

**18.**

**19.**

**20.**

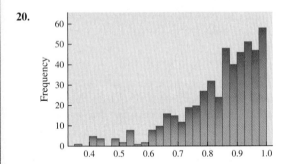

**21.** What is meant by the range of a distribution?

**22.** How are the variance and the standard deviation of a distribution related? What is measured by the standard deviation?

*Find the range and the standard deviation for each of the given distributions.*

**23.** The data in Exercise 13

**24.** The data in Exercise 14

*Find the standard deviation for each of the given distributions.*

**25.** The data summarized in the frequency table of Exercise 9

**26.** The data summarized in the frequency table of Exercise 10

**27. Macy's Revenue** The annual reported revenue (in billions of dollars) for Macy's Inc. is given in the following table. Find the mean and standard deviation for the 5-year period.

| Year | Revenue |
|------|---------|
| 2013 | 27.7 |
| 2014 | 27.9 |
| 2015 | 28.1 |
| 2016 | 27.1 |
| 2017 | 25.8 |

**28. Kohl's Revenue** The annual reported revenue (in billions of dollars) for Kohl's Corp. is given in the following table. Find the mean and standard deviation for the 5-year period.

| Year | Revenue |
|------|---------|
| 2013 | 19.3 |
| 2014 | 19.0 |
| 2015 | 19.0 |
| 2016 | 19.2 |
| 2017 | 18.7 |

**29.** Describe the characteristics of a normal distribution.

**30.** What is meant by a skewed distribution?

*Find the given areas under the standard normal curve.*

**31.** Between $z = 0$ and $z = 1.35$

**32.** To the left of $z = .38$

**33.** Between $z = -1.88$ and $z = 2.41$

**34.** Between $z = 1.53$ and $z = 2.82$

**35.** Find a $z$-score such that 8% of the area under the curve is to the right of $z$.

**36.** Why is the normal distribution not a good approximation of a binomial distribution that has a value of $p$ close to 0 or 1?

**Arm Circumference** *Data from a recent National Health and Nutrition Examination Survey (NHANES) indicated that the distribution of upper arm circumference (in cm) is approximately normally distributed. The mean circumference for adult females is 38.6 cm, with a standard deviation of 2.3 cm. The mean circumference for adult males is 41.5 cm, with a standard deviation of 2.5 cm. Find the following. (Data from: www. cdc.gov/nchs/nhanes.htm.)*

**37.** A randomly selected female has an upper arm circumference of less than 45 cm.

**38.** A randomly selected female has an upper arm circumference between 35 and 45 cm.

**39.** A randomly selected male has an upper arm circumference of more than 38 cm.

**40.** A randomly selected male has an upper arm circumference between 38 and 46 cm.

**Auto Sales** *The table gives the number of vehicles (in thousands) sold within the United States in May 2016 and May 2017 for 12 auto manufacturers. (Data from: Wall Street Journal.)*

| Auto Manufacturer | May 2016 Sales | May 2017 Sales |
|---|---|---|
| General Motors Corp. | 240 | 237 |
| Ford Motor Corp. | 235 | 240 |
| Chrysler LLC | 192 | 189 |
| Toyota Motor Sales USA | 219 | 218 |
| American Honda Motor Co. Inc. | 147 | 148 |
| Nissan North America Inc. | 133 | 137 |
| Hyundai Motor America | 71 | 60 |
| Mazda | 28 | 26 |
| Kia Motors America Inc. | 63 | 59 |
| Subaru of America Inc. | 50 | 56 |
| Mercedes-Benz | 32 | 30 |
| Volkswagen of America Inc. | 29 | 30 |

**41. (a)** Find the mean and standard deviation of sales for the May 2016 data.

**(b)** Which company is closest to the mean sales for the May 2016 data?

**42. (a)** Find the mean and standard deviation of sales for the May 2017 data.

**(b)** Which company is closest to the mean sales for the May 2017 data?

**43.** Find the five-number summary for the May 2016 sales.

**44.** Find the five-number summary for the May 2017 sales.

**45.** Construct a boxplot for the May 2016 sales data.

**46.** Construct a boxplot for the May 2017 sales data.

**College or University Education** *A recent Gallup.com poll found that 51% of Americans who pursued or completed a college or university degree would change at least one aspect of their education experience if they could do it all over again, including their major or field of study, the institution they attended, or the type of degree they obtained. For Exercises 47–50, assume a random sample of 500 adults who pursued a college or university degree was selected. Approximate the probability of the following given events. (Round standard deviation to one decimal place.) (Data from: www.gallup.com.)*

**47.** More than 260 would change at least one aspect of their education experience.

**48.** Between 265 and 280 (inclusive) would change at least one aspect of their education experience.

**49.** Between 240 and 280 (inclusive) would change at least one aspect of their education experience.

**50.** Fewer than 240 would change at least one aspect of their education experience.

**Living at Home** *According to the Pew Research Center, in 2016, 15% of young adults age 25–35 were living in their parents' home. If we select 1000 young adults age 25–35 at random, find the probability of the following given events. (Round standard deviation to one decimal place.) (Data from: pewresearch.org.)*

**51.** More than 130 live in their parents' home.

**52.** Between 120 and 130 (inclusive) live in their parents' home.

**53.** Between 120 and 165 (inclusive) live in their parents' home.

**54.** 165 or fewer live in their parents' home.

# Case Study 10 Standard Deviation as a Measure of Risk

When investing, the idea of risk is difficult to define. Some individuals may deem an investment product to be of high risk if it loses all of its value in a short time, others may define a product as risky if it loses some portion of its value in a short time, and still others may define a product as risky if there is even a remote chance of it losing its value over a long period of time.

One way investors try to quantify risk is to examine the volatility of the return for an investment product. Volatility refers to how the return on the investment fluctuates from the average level of return. One simple measure of volatility is to calculate the standard deviation of the annual rate of return.

For example, if a fund has an annual return of 2% each and every year, then the mean is 2% and the standard deviation would be 0 (because there is no variation from the mean). This fund would have very low volatility. What is also interesting is that a fund that *lost* 2% each and every year would also have a standard deviation of 0 (because it too does not ever vary from its mean of –2%). Nevertheless, the standard deviation can be helpful in assessing the

risk and volatility related to an investment fund.

Let us examine two mutual funds—Fund A and Fund B. The following table lists the annual rate of return for these two funds.

| Fund A | Fund B |
|---|---|
| 3.2 | 9.1 |
| 6.6 | −3.4 |
| 3.9 | 2.1 |
| 5.3 | 11.4 |
| 5.2 | 1.1 |
| 5.8 | 9.7 |

If we calculate the average rate of return for both funds, we obtain 5%. We can notice, however, even by casual inspection, that there is greater variability in Fund B than in Fund A. If we use the notation $s_A$ to denote the standard deviation of fund A and $s_B$ to denote the standard deviation of fund B, we can use the methods of Section 10.3 to obtain the following values for the standard deviation.

$$s_A = \sqrt{\frac{(3.2 - 5)^2 + (6.6 - 5)^2 + (3.9 - 5)^2 + (5.3 - 5)^2 + (5.2 - 5)^2 + (5.8 - 5)^2}{5}} \approx 1.25$$

$$s_B = \sqrt{\frac{(9.1 - 5)^2 + (-3.4 - 5)^2 + (2.1 - 5)^2 + (11.4 - 5)^2 + (1.1 - 5)^2 + (9.7 - 5)^2}{5}} \approx 5.90$$

Our calculations confirm that there is greater variability in Fund B. Thus, Fund B is said to carry more risk.

The way that standard deviation is used in the financial products industry is to examine monthly returns. For example, the website morningstar.com explains that it uses a 36-month window of monthly returns. They then calculate the average monthly return and the standard deviation from those returns. They then report these values (the mean return and the standard deviation) corrected for an annualized return rate. This should be clear in the following example.

To calculate the standard deviation for the Fidelity® Contrafund® Fund (FCNTX), we examine the monthly returns for 36 months from May 2017 to June 2014, going back in time.

| Month | Return | Month | Return | Month | Return |
|---|---|---|---|---|---|
| MAY 2017 | 3.59 | MAY 2016 | 1.67 | MAY 2015 | 2.16 |
| APR 2017 | 2.82 | APR 2016 | 0.25 | APR 2015 | −0.84 |
| MAR 2017 | 1.56 | MAR 2016 | 5.59 | MAR 2015 | −0.49 |
| FEB 2017 | 3.85 | FEB 2016 | −1.18 | FEB 2015 | 5.98 |
| JAN 2017 | 4.38 | JAN 2016 | −5.68 | JAN 2015 | −1.34 |
| DEC 2016 | 0.53 | DEC 2015 | −1.36 | DEC 2014 | −0.53 |
| NOV 2016 | 0.60 | NOV 2015 | 0.64 | NOV 2014 | 2.15 |
| OCT 2016 | −1.67 | OCT 2015 | 7.08 | OCT 2014 | 1.48 |
| SEP 2016 | 0.43 | SEP 2015 | −2.06 | SEP 2014 | −1.13 |
| AUG 2016 | 0.27 | AUG 2015 | −5.96 | AUG 2014 | 4.44 |
| JUL 2016 | 4.48 | JUL 2015 | 3.47 | JUL 2014 | −1.44 |
| JUN 2016 | −1.51 | JUN 2015 | −0.30 | JUN 2014 | 2.41 |

Our first step is to calculate the mean return. We find the average monthly rate of return to be .9539. We then calculate the standard deviation value and obtain 2.9361. To convert these values from monthly return rates to annual return rates, we multiply the mean by 12 and the standard deviation by $\sqrt{12}$. The annual return rate is $12(.9539) = 11.4468$, and the annual standard deviation is $\sqrt{12}(2.9361) = 10.1709$. These values are often reported rounded to two decimal places (11.45 and 10.17, respectively).

Many financial websites report these values for return rates based on monthly calculations. For example, on the morningstar.com website, clicking on a link for "Ratings and Risk" will display the mean and standard deviation values under a heading of "Volatility Measures." With this information, it is easier to make direct comparisons to other funds. For the same period, the S&P 500 had a mean return of 10.14 and a standard deviation of 10.38. Thus, we see that the Fidelity® fund had a slightly higher average return and a slightly lower level of risk.

An additional way that investors interpret the standard deviation is that they assume future returns will follow an approximate normal distribution. We know from Section 10.4 (and, in particular, Figure 10.22) that 68% of data that is normally distributed will fall within one standard deviation of the mean. With the Fidelity® fund example, we then expect that the probability of a future return being within one standard deviation of the mean of 11.45% is 68%. With the standard deviation value of 10.17, our bounds on this interval are

$$(11.45 - 10.17, 11.45 + 10.17) = (1.28, 21.62).$$

Furthermore, we expect the annual return to fall within 2 standard deviation of the mean approximately 95% of the time:

$$(11.45 - 2*10.17, 11.45 + 2*10.17) = (-8.89, 31.79).$$

The standard deviation is not a perfect measure of risk. It is possible to invest in a fund with a low standard deviation and still lose money. It is, however, a useful way to compare funds in similar categories.

## Exercises

*For Exercises 1–10, use the following table, which gives the monthly returns for the Vanguard PRIMECAP Fund Investor Shares (VPMCX), the S&P 500, and the sector returns for Large Growth funds. (Data from: www.morningstar.com.)*

| Month | Vanguard | S&P | Large Growth | Month | Vanguard | S&P | Large Growth |
|---|---|---|---|---|---|---|---|
| MAY 2017 | 2.95 | 1.41 | 2.53 | NOV 2015 | 0.37 | 0.30 | 0.64 |
| APR 2017 | 1.48 | 1.03 | 2.34 | OCT 2015 | 8.84 | 8.44 | 7.87 |
| MAR 2017 | 0.72 | 0.12 | 1.03 | SEP 2015 | −2.05 | −2.47 | −3.19 |
| FEB 2017 | 4.25 | 3.97 | 3.58 | AUG 2015 | −5.81 | −6.03 | −6.38 |
| JAN 2017 | 3.10 | 1.90 | 3.79 | JUL 2015 | 1.51 | 2.10 | 2.95 |
| DEC 2016 | 1.72 | 1.98 | 0.55 | JUN 2015 | −2.77 | −1.94 | −1.17 |
| NOV 2016 | 3.88 | 3.70 | 1.65 | MAY 2015 | 1.59 | 1.29 | 1.73 |
| OCT 2016 | −2.98 | −1.82 | −2.46 | APR 2015 | −0.29 | 0.96 | −0.02 |
| SEP 2016 | 1.76 | 0.02 | 0.64 | MAR 2015 | −1.27 | −1.58 | −0.96 |
| AUG 2016 | 1.27 | 0.14 | −0.09 | FEB 2015 | 5.20 | 5.75 | 6.38 |
| JUL 2016 | 6.54 | 3.69 | 5.00 | JAN 2015 | −1.56 | −3.00 | −1.81 |
| JUN 2016 | −2.35 | 0.26 | −1.47 | DEC 2014 | −0.35 | −0.25 | −0.72 |
| MAY 2016 | 3.10 | 1.80 | 2.24 | NOV 2014 | 3.32 | 2.69 | 2.57 |
| APR 2016 | −0.41 | 0.39 | −0.21 | OCT 2014 | 2.22 | 2.44 | 2.51 |
| MAR 2016 | 6.16 | 6.78 | 6.17 | SEP 2014 | −0.34 | −1.40 | −1.89 |
| FEB 2016 | −1.16 | −0.13 | −1.14 | AUG 2014 | 3.89 | 4.00 | 4.21 |
| JAN 2016 | −6.57 | −4.96 | −7.10 | JUL 2014 | −0.84 | −1.38 | −1.53 |
| DEC 2015 | −.42 | −1.58 | −1.67 | JUN 2014 | 2.44 | 2.07 | 2.46 |

1. Find the mean monthly return for the Vanguard fund. Convert this value to the annual return.

2. Find the mean monthly return for the S&P 500. Convert this value to the annual return.

3. Find the mean monthly return for the Large Growth funds. Convert this value to the annual return.

4. Which fund has the highest annual rate of return?

5. Find the standard deviation of the monthly returns for the Vanguard fund. Convert this value to the annual return.

6. Find the standard deviation of the monthly returns for the S&P 500. Convert this value to the annual return.

7. Find the standard deviation of the monthly returns for the Large Growth funds. Convert this value to the annual return.

8. Which fund has the highest level of risk?

9. Determine the bounds for 68% of future annual returns for the Vanguard Fund.

10. Determine the bounds for 95% of future annual returns for the Vanguard Fund.

# Extended Projects

1. Investigate several mutual funds of your choice. Use a financial data website and obtain the monthly returns for the funds of your choice. Calculate the mean return and the standard deviation of the monthly data and convert the values to the annual values. Determine which funds have the highest return values and which have the lowest standard deviation values.

2. Investigate other measures of risk for investment products. Compare these measures with the standard deviation discussed here.

# Differential Calculus

# 11

How fast are cell phone subscriptions growing? At what rate is the number of Internet users increasing? How are home prices changing? Is childhood mortality decreasing? Are corporate profits increasing? Examples and exercises in this chapter demonstrate that these questions and many others in the fields of business, finance, health, political science, psychology, sociology, and economics can be answered by using calculus.

The algebraic problems considered in earlier chapters dealt with *static* situations:

What is the revenue when $x$ items are sold?

How much interest is earned in 2 years?

What is the equilibrium price?

Calculus, on the other hand, deals with *dynamic* situations:

At what rate is the economy growing?

How fast is revenue changing?

How quickly can production be increased without adversely affecting profits?

The techniques of calculus will allow us to answer many questions like these that deal with rates of change.

The key idea underlying the development of calculus is the concept of limit, so we begin by studying limits.

# 11.1 Limits

We have often dealt with a problem like this: "Find the value of the function $f$ when $x = a$." The underlying idea of "limit," however, is to examine what the function does *near* $x = a$, rather than what it does *at* $x = a$. If you would like to refresh your understanding of functions and functional notation, see Chapter 3.

**EXAMPLE 1**  The function

$$f(x) = \frac{2x^2 - 3x - 2}{x - 2}$$

is not defined when $x = 2$. (Why?) What happens to the values of $f(x)$ when $x$ is *very close* to 2?

**Solution**  Evaluate $f$ at several numbers that are very close to $x = 2$, as in the following table:

| $x$ | 1.9 | 1.99 | 1.999 | 2 | 2.001 | 2.01 | 2.1 |
|-----|-----|------|-------|---|-------|------|-----|
| $f(x)$ | 4.8 | 4.98 | 4.998 | — | 5.002 | 5.02 | 5.2 |

The table suggests that

as $x$ gets closer and closer to 2 from either direction, the corresponding value of $f(x)$ gets closer and closer to 5.

In fact, by experimenting further, you can convince yourself that the values of $f(x)$ can be made *as close as you want* to 5 by taking values of $x$ close enough to 2. This situation is usually described by saying "The *limit* of $f(x)$ as $x$ approaches 2 is the number 5," which is written symbolically as

$$\lim_{x \to 2} f(x) = 5, \quad \text{or equivalently,} \quad \lim_{x \to 2} \frac{2x^2 - 3x - 2}{x - 2} = 5.$$

The graph of $f$ shown in Figure 11.1 also shows that $\lim_{x \to 2} f(x) = 5$.

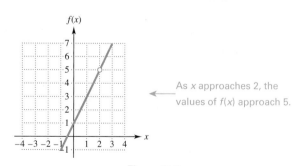

As $x$ approaches 2, the values of $f(x)$ approach 5.

**Figure 11.1**

✓ **Checkpoint 1**

Use a calculator to estimate

$$\lim_{x \to 1} \frac{x^3 + x^2 - 2x}{x - 1}$$

by completing the following table:

| $x$ | $f(x)$ |
|-----|--------|
| .9 | |
| .99 | |
| .999 | |
| 1.001 | |
| 1.01 | |
| 1.1 | |

*Answers to Checkpoint exercises are found at the end of the section.*

The informal definition of "limit" that follows is similar to the situation in Example 1, but now $f$ is any function, and $a$ and $L$ are fixed real numbers (in Example 1, $a = 2$ and $L = 5$).

## Limit of a Function

Let $f$ be a function, and let $a$ and $L$ be real numbers. Assume that $f(x)$ is defined for all $x$ near $x = a$. Suppose that

as $x$ takes values very close (but not equal) to $a$ (on both sides of $a$), the corresponding values of $f(x)$ are very close (and possibly equal) to $L$

and that

the values of $f(x)$ can be made as close as you want to $L$ for all values of $x$ that are close enough to $a$.

Then the number $L$ is the **limit** of the function $f(x)$ as $x$ approaches $a$, which is written

$$\lim_{x \to a} f(x) = L.$$

This definition is *informal* because the expressions "near," "very close," and "as close as you want" have not been precisely defined. In particular, the tables used in Example 1 and the next set of examples provide strong intuitive evidence, but not a proof, of what the limits must be.

---

**EXAMPLE 2**    If $f(x) = x^2 + x + 1$, what is $\lim_{x \to 3} f(x)$?

**Solution**  Make a table showing the values of the function at numbers very close to 3:

| | | *x* approaches 3 from the left | | | | *x* approaches 3 from the right | | |
|---|---|---|---|---|---|---|---|---|
| $x$ | 2.9 | 2.99 | 2.9999 | 3 | 3.0001 | | 3.01 | 3.1 |
| $f(x)$ | 12.31 | 12.9301 | 12.9993 … | | 13.0007 … | | 13.0701 | 13.71 |
| | | | $f(x)$ approaches 13 | | $f(x)$ approaches 13 | | | |

The table suggests that as $x$ approaches 3 from either direction, $f(x)$ gets closer and closer to 13 and, hence, that

$$\lim_{x \to 3} f(x) = 13, \qquad \text{or equivalently,} \qquad \lim_{x \to 3}(x^2 + x + 1) = 13.$$

We did not need to include the value of $f(3)$ in the table above because the limit only concerns values of the function *near* $x = 3$, not *at* $x = 3$. However, note that the function $f(x)$ is defined when $x = 3$ and that $f(3) = 3^2 + 3 + 1 = 13$. So in this case, the limit of $f(x)$ as $x$ approaches 3 is $f(3)$, the value of the function at 3.

### 🖩 TECHNOLOGY TIP

To produce Figure 11.2 on a TI-84+, use Y = to enter the function; press WINDOW to set Xmin = 2.8, Xmax = 3.2, Ymin = −.5, and Ymax = 1.5; then press GRAPH. Use TRACE and the arrow keys to move along the graph.

(a)

(b)

**Figure 11.2**

**EXAMPLE 3**    Use a graphing calculator to find

$$\lim_{x \to 3} \frac{x - 3}{e^{x-3} - 1}.$$

**Solution**  There are two ways to estimate the limit.

**Graphical Method**  Graph $f(x) = \dfrac{x - 3}{e^{x-3} - 1}$ in a very narrow window near $x = 3$. Use the trace feature to move along the graph and observe the $y$-coordinates as $x$ gets very close to 3 from either side. Figure 11.2 suggests that $\lim_{x \to 3} f(x) = 1$.

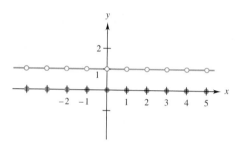

Figure 11.3

**TECHNOLOGY TIP**

To create the table in Figure 11.3 on a TI-84+, press TBLSET (second WINDOW), select "Ask" for the independent variable (X) and "Auto" for the dependent variable (Y), then enter the X-values listed.

**Numerical Method** Use the table feature to make a table of values for $f(x)$ when $x$ is very close to 3. Figure 11.3 shows that when $x$ is very close to 3, $f(x)$ is very close to 1. (The table displays "error" at $x = 3$ because the function is not defined when $x = 3$.) Thus, it appears that

$$\lim_{x \to 3} \frac{x-3}{e^{x-3}-1} = 1.$$

The function has the limit 1 as $x$ approaches 3, even though $f(3)$ is not defined.

**EXAMPLE 4**    Find $\lim_{x \to 4} f(x)$, where $f$ is the function whose rule is

$$f(x) = \begin{cases} 0 & \text{if } x \text{ is an integer} \\ 1 & \text{if } x \text{ is not an integer} \end{cases}$$

and whose graph is shown in Figure 11.4.

Figure 11.4

**Solution** The definition of the limit as $x$ approaches 4 involves only values of $x$ that are close, but not equal, to 4—corresponding to the part of the graph on either side of 4, but not at 4 itself. Now, $f(x) = 1$ for all these numbers (because the numbers very near 4, such as 3.99995 and 4.00002, are not integers). Thus, for all $x$ very close to 4, the corresponding value of $f(x)$ is 1, so $\lim_{x \to 4} f(x) = 1$. However, since 4 is an integer, $f(4) = 0$. Therefore, $\lim_{x \to 4} f(x) \neq f(4)$.

Examples 1–4 illustrate the following facts.

**Limits and Function Values**

If the limit of a function $f(x)$ as $x$ approaches $a$ exists, then there are three possibilities:

1. $f(a)$ is not defined, but $\lim_{x \to a} f(x)$ is defined.    (Examples 1 and 3)

2. $f(a)$ is defined and $\lim_{x \to a} f(x) = f(a)$.    (Example 2)

3. $f(a)$ is defined, but $\lim_{x \to a} f(x) \neq f(a)$.    (Example 4)

## Finding Limits Algebraically

As we have seen, tables and graphs are very useful for estimating limits. However, it is often more efficient and accurate to find limits algebraically. We begin with two simple functions.

Consider the constant function $f(x) = 5$. To compute $\lim_{x \to a} f(x)$, you must ask "When $x$ is very close to $a$, what is the value of $f(x)$?" The answer is easy because no matter what $x$

is, the value of $f(x)$ is *always* the number 5. As $x$ gets closer and closer to $a$, the value of $f(x)$ is always 5. Hence,

$$\lim_{x \to a} f(x) = 5, \qquad \text{which is usually written} \qquad \lim_{x \to a} 5 = 5.$$

The same thing is true for *any* constant function.

## Limit of a Constant Function

If $d$ is a constant, then, for any real number $a$,

$$\lim_{x \to a} d = d.$$

Now consider the *identity function*, whose rule is $f(x) = x$. When $x$ is very close to a number $a$, the corresponding value of $f(x)$ (namely, $x$ itself) is very close to $a$. So we have the following conclusion.

## Limit of the Identity Function

For every real number $a$,

$$\lim_{x \to a} x = a.$$

The facts in the two preceding boxes, together with the properties of limits that follow, will enable us to find a wide variety of limits.

## Properties of Limits

Let $a, r, A,$ and $B$ be real numbers, and let $f$ and $g$ be functions such that

$$\lim_{x \to a} f(x) = A \qquad \text{and} \qquad \lim_{x \to a} g(x) = B.$$

Then the following properties hold:

1. $\displaystyle\lim_{x \to a} [f(x) + g(x)] = A + B = \lim_{x \to a} f(x) + \lim_{x \to a} g(x)$     *Sum Property*
   (The limit of a sum is the sum of the limits.)

2. $\displaystyle\lim_{x \to a} [f(x) - g(x)] = A - B = \lim_{x \to a} f(x) - \lim_{x \to a} g(x)$     *Difference Property*
   (The limit of a difference is the difference of the limits.)

3. $\displaystyle\lim_{x \to a} [f(x) \cdot g(x)] = A \cdot B = \lim_{x \to a} f(x) \cdot \lim_{x \to a} g(x)$     *Product Property*
   (The limit of a product is the product of the limits.)

4. $\displaystyle\lim_{x \to a} \frac{f(x)}{g(x)} = \frac{A}{B} = \frac{\displaystyle\lim_{x \to a} f(x)}{\displaystyle\lim_{x \to a} g(x)} \quad (B \neq 0)$     *Quotient Property*
   (The limit of a quotient is the quotient of the limits, provided that the limit of the denominator is nonzero.)

5. For any real number for which $A^r$ exists,

   $$\lim_{x \to a} [f(x)]^r = A^r = [\lim_{x \to a} f(x)]^r.$$     *Power Property*
   (The limit of a power is the power of the limit.)

Although we won't prove these properties (a rigorous definition of limit is needed for that), you should find most of them plausible. For instance, if the values of $f(x)$ get very close to $A$ and the values of $g(x)$ get very close to $B$ when $x$ approaches $a$, it is reasonable to expect that the corresponding values of $f(x) + g(x)$ will get very close to $A + B$ (Sum Property) and that the corresponding values of $f(x)g(x)$ will get very close to $AB$ (Product Property).

---

**EXAMPLE 5**    Let $f(x) = 4x^2 + 2x - 9$. Find $\lim\limits_{x \to 5} f(x)$.

**Solution**
$$\lim_{x \to 5} f(x) = \lim_{x \to 5} (4x^2 + 2x - 9)$$
$$= \lim_{x \to 5} [4x^2 + (2x - 9)]$$
$$= \lim_{x \to 5} 4x^2 + \lim_{x \to 5} (2x - 9) \qquad \text{Sum property}$$
$$= \lim_{x \to 5} 4x^2 + [\lim_{x \to 5} 2x - \lim_{x \to 5} 9] \qquad \text{Difference property}$$
$$= \lim_{x \to 5} 4 \cdot \lim_{x \to 5} x^2 + \lim_{x \to 5} 2 \cdot \lim_{x \to 5} x - \lim_{x \to 5} 9 \qquad \text{Product property}$$
$$= \lim_{x \to 5} 4 \cdot (\lim_{x \to 5} x)^2 + \lim_{x \to 5} 2 \cdot \lim_{x \to 5} x - \lim_{x \to 5} 9 \qquad \text{Power property}$$
$$= 4 \cdot (5)^2 + 2 \cdot 5 - 9 = 101. \qquad \text{Constant-function limit and identity-function limit}$$

---

Example 5 shows that $\lim\limits_{x \to 5} f(x) = 101$. Note that $f(5) = 4 \cdot 5^2 + 2 \cdot 5 - 9 = 101$. In other words, the limit of the function as $x$ approaches 5 is equal to the value of the function at 5:

$$\lim_{x \to 5} f(x) = f(5).$$

Recall that the function in Example 5 is called a quadratic (degree 2) polynomial and that a polynomial function of degree $n$ is of the form $f(x) = a_n x^n + \cdots + a_2 x^2 + a_1 x + a_0$. The same analysis used in Example 5 works with any polynomial function and leads to the following conclusion.

## Polynomial Limits

If $f$ is a polynomial function and $a$ is a real number, then

$$\lim_{x \to a} f(x) = f(a).$$

In other words, the limit is the value of the function at $x = a$.

---

**EXAMPLE 6**    Find each limit.

**(a)** $\lim\limits_{x \to 2} (x^3 - x + 3)$

**Solution** $\lim\limits_{x \to 2} (x^3 - x + 3)$
$$2^3 - 2 + 3 = 9. \quad \checkmark_2 \qquad \text{Polynomial limit}$$

**(b)** $\lim\limits_{x \to -1} (x^3 + 4x)(2x^2 - 3x)$

**Solution** $\lim\limits_{x \to -1} (x^3 + 4x)(2x^2 - 3x)$
$$= \lim_{x \to -1} (x^3 + 4x) \cdot \lim_{x \to -1} (2x^2 - 3x) \qquad \text{Product property}$$
$$= [(-1)^3 + 4(-1)] \cdot [2(-1)^2 - 3(-1)] \qquad \text{Polynomial limit}$$
$$= (-1 - 4)(2 + 3) = -25.$$

---

✓ **Checkpoint 2**

If $f(x) = 2x^3 - 5x^2 + 8$, find the given limits.

**(a)** $\lim\limits_{x \to 2} f(x)$

**(b)** $\lim\limits_{x \to -3} f(x)$

**(c)** $\lim\limits_{x \to 9} \sqrt{4x - 11}$

**Solution** Recall that the square root function $\sqrt{x}$ is *not* a polynomial. However, we can write $\sqrt{x}$ in exponential form as the power function $x^{1/2}$. Similarly, we can begin evaluating this limit by writing the square root in exponential form.

$$\lim_{x \to 9} \sqrt{4x - 11} = \lim_{x \to 9} [4x - 11]^{1/2}$$
$$= [\lim_{x \to 9} (4x - 11)]^{1/2} \qquad \text{Power property}$$
$$= [4 \cdot 9 - 11]^{1/2} \qquad \text{Polynomial limit}$$
$$= [25]^{1/2} = \sqrt{25} = 5. \quad \boxed{\checkmark}_3$$

✓ **Checkpoint 3**

Use the limit properties to find the given limit.

$$\lim_{x \to 2} \sqrt{3x + 3}$$

Recall that a rational function is the quotient of two polynomials. When a rational function $f(x)$ is defined at $x = a$, it is easy to find the limit of $f(x)$ as $x$ approaches $a$.

**EXAMPLE 7** Find

$$\lim_{x \to 5} \frac{x}{x + 3}.$$

**Solution** $\lim\limits_{x \to 5} \dfrac{x}{x + 3} = \dfrac{\lim\limits_{x \to 5} x}{\lim\limits_{x \to 5} (x + 3)}$     Quotient property

$$= \frac{5}{5 + 3} = \frac{5}{8}.$$     Polynomial limit

Note that $f(5) = \dfrac{5}{5 + 3} = \dfrac{5}{8}$. So the limit of $f(x)$ as $x$ approaches 5 is the value of the function at 5:

$$\lim_{x \to 5} f(x) = f(5). \quad \boxed{\checkmark}_4$$

✓ **Checkpoint 4**

Find the given limits.

**(a)** $\lim\limits_{x \to 2} \dfrac{2x - 1}{3x + 4}$

**(b)** $\lim\limits_{x \to -1} \dfrac{x - 2}{3x - 1}$

The argument used in Example 7 works in the general case, as described in the following result.

**Rational Limits**

If $f$ is a rational function and $a$ is a real number *such that $f(a)$ is defined,* then

$$\lim_{x \to a} f(x) = f(a).$$

In other words, the limit is the value of the function at $x = a$.

If a rational function is not defined at the number $a$, the preceding property cannot be used. Different techniques are needed in such cases to find the limit (if one exists), as we shall see in Examples 8 and 9.

The definition of the limit as $x$ approaches $a$ involves only the values of the function when $x$ is *near* $a$, but not the value of the function *at* $a$. So two functions that agree for all values of $x$, except possibly at $x = a$, will necessarily have the same limit when $x$ approaches $a$. Thus, we have the following fact.

## Limit Theorem

If $f$ and $g$ are functions that have limits as $x$ approaches $a$, and $f(x) = g(x)$ for all $x$ near $a$, then

$$\lim_{x \to a} f(x) = \lim_{x \to a} g(x).$$

**EXAMPLE 8**   Find

$$\lim_{x \to 2} \frac{x^2 + x - 6}{x - 2}.$$

**Solution**   The quotient property cannot be used here, because

$$\lim_{x \to 2} (x - 2) = 0.$$

We can, however, simplify the function by rewriting the fraction as

$$\frac{x^2 + x - 6}{x - 2} = \frac{(x + 3)(x - 2)}{x - 2}.$$

When $x \neq 2$, the quantity $x - 2$ is nonzero and may be cancelled, so that

$$\frac{x^2 + x - 6}{x - 2} = x + 3 \qquad \text{for all } x \neq 2.$$

Now the limit theorem can be used:

$$\lim_{x \to 2} \frac{x^2 + x - 6}{x - 2} = \lim_{x \to 2} (x + 3) = 2 + 3 = 5. \quad ✔_5$$

✔ **Checkpoint 5**

Find $\displaystyle\lim_{x \to 1} \frac{2x^2 + x - 3}{x - 1}$.

**EXAMPLE 9**   Find

$$\lim_{x \to 1} \frac{2x^2 - 5x + 3}{x^2 + 5x - 6}.$$

**Solution**   As $x$ approaches 1, both the numerator and denominator approach 0, giving the meaningless expression 0/0 (called an indeterminate form). Therefore, we cannot use the quotient property as a first step. However, because the limit is concerned only with values of $x$ *near* 1, not *at* $x = 1$, we can simplify using the limit theorem:

$$\lim_{x \to 1} \frac{2x^2 - 5x + 3}{x^2 + 5x - 6}$$

$$= \lim_{x \to 1} \frac{(2x - 3)(x - 1)}{(x + 6)(x - 1)} \qquad \text{Factor numerator and denominator}$$

$$= \lim_{x \to 1} \frac{2x - 3}{x + 6} \qquad \text{Limit theorem } (x - 1 \neq 0 \text{ when } x \neq 1)$$

$$= \frac{2(1) - 3}{1 + 6} = -\frac{1}{7}. \qquad \text{Quotient property and polynomial limit} \quad ✔_6$$

✔ **Checkpoint 6**

Find $\displaystyle\lim_{x \to 2} \frac{x^2 + 3x - 10}{2x^2 - 3x - 2}$.

**EXAMPLE 10**   Find

$$\lim_{x \to 4} \frac{\sqrt{x} - 2}{x - 4}.$$

**Solution** As $x \to 4$, both the numerator and the denominator approach 0, giving the meaningless expression $0/0$. To change the form of the expression, algebra can be used to rationalize the numerator by multiplying both the numerator and the denominator by $\sqrt{x} + 2$. This gives

$$\frac{\sqrt{x} - 2}{x - 4} = \frac{\sqrt{x} - 2}{x - 4} \cdot \frac{\sqrt{x} + 2}{\sqrt{x} + 2} = \frac{\sqrt{x} \cdot \sqrt{x} + 2\sqrt{x} - 2\sqrt{x} - 4}{(x - 4)(\sqrt{x} + 2)}$$

$$= \frac{x - 4}{(x - 4)(\sqrt{x} + 2)} = \frac{1}{\sqrt{x} + 2} \qquad \text{for all } x \neq 4.$$

Now use the limit theorem and the properties of limits:

$$\lim_{x \to 4} \frac{\sqrt{x} - 2}{x - 4} = \lim_{x \to 4} \frac{1}{\sqrt{x} + 2} = \frac{1}{\sqrt{4} + 2} = \frac{1}{2 + 2} = \frac{1}{4}. \quad ✓_7$$

**Checkpoint 7**

Find the given limits.

(a) $\displaystyle\lim_{x \to 1} \frac{\sqrt{x} - 1}{x - 1}$

(b) $\displaystyle\lim_{x \to 9} \frac{\sqrt{x} - 3}{x - 9}$

## Existence of Limits

It is possible that $\lim_{x \to a} f(x)$ may not exist; that is, there may be no number $L$ satisfying the definition of $\lim_{x \to a} f(x) = L$. This can happen in many ways, two of which are illustrated next.

**EXAMPLE 11** Given that

$$g(x) = \frac{x^2 + 4}{x - 2}, \text{ find } \lim_{x \to 2} g(x).$$

**Solution** The quotient property cannot be used, since $\lim_{x \to 2} (x - 2) = 0$, and the limit theorem does not apply because $x^2 + 4$ does not factor. So we use the following table and the graph of $g$ in Figure 11.5:

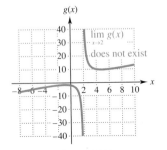
**Figure 11.5**

|  | x approaches 2 from the left | | | | | x approaches 2 from the right | | |
|---|---|---|---|---|---|---|---|---|
| $x$ | 1.8 | 1.9 | 1.99 | 1.999 | 2 | 2.001 | 2.01 | 2.05 |
| $g(x)$ | −36.2 | −76.1 | −796 | −7996 | | 8004 | 804 | 164 |
| | | | $g(x)$ gets smaller and smaller | | | $g(x)$ gets larger and larger | | |

The table and the graph both show that as $x$ approaches 2 from the left, $g(x)$ gets smaller and smaller, but as $x$ approaches 2 from the right, $g(x)$ gets larger and larger. Since $g(x)$ does not get closer and closer to a single real number as $x$ approaches 2 from either side,

$$\lim_{x \to 2} \frac{x^2 + 4}{x - 2} \text{ does not exist.} \quad ✓_8$$

**Checkpoint 8**

Let $f(x) = \dfrac{x^2 + 9}{x - 3}$. Find the given limits.

(a) $\displaystyle\lim_{x \to 3} f(x)$

(b) $\displaystyle\lim_{x \to 0} f(x)$

**EXAMPLE 12** What is

$$\lim_{x \to 0} \frac{|x|}{x}?$$

**Solution** The function $f(x) = \dfrac{|x|}{x}$ is not defined when $x = 0$. Recall the definition of absolute value:

$$|x| = \begin{cases} x & \text{if } x \geq 0 \\ -x & \text{if } x < 0 \end{cases}.$$

Consequently, when $x > 0$,

$$f(x) = \frac{|x|}{x} = \frac{x}{x} = 1,$$

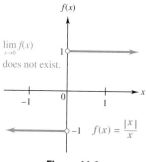

**Figure 11.6**

$$\lim_{x \to 0} f(x)$$
does not exist.

$$f(x) = \frac{|x|}{x}$$

and when $x < 0$,

$$f(x) = \frac{|x|}{x} = \frac{-x}{x} = -1.$$

The graph of $f$ is shown in Figure 11.6. As $x$ approaches 0 from the right, $x$ is always positive, and the corresponding value of $f(x)$ is 1. But as $x$ approaches 0 from the left, $x$ is always negative, and the corresponding value of $f(x)$ is $-1$. Thus, as $x$ approaches 0 from *both* sides, the corresponding values of $f(x)$ do not get closer and closer to a *single* real number. Therefore, the limit does not exist.*

The function $f$ whose graph is shown in Figure 11.7 illustrates various facts about limits that were discussed earlier in this section.

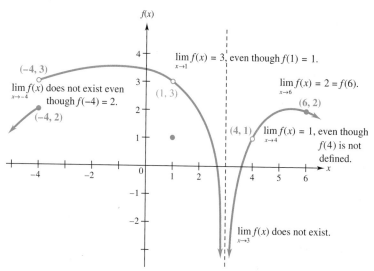

**Figure 11.7**

# 11.1 Exercises

*Use the given graph to determine the value of the indicated limits. (See Examples 4, 11, and 12 and Figure 11.7.)*

**1. (a)** $\lim_{x \to 3} f(x)$  **(b)** $\lim_{x \to -1} f(x)$

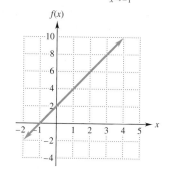

**2. (a)** $\lim_{x \to 3} g(x)$  **(b)** $\lim_{x \to 0} g(x)$

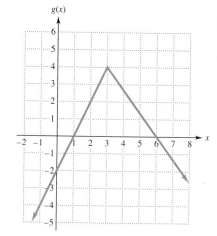

---

*However, the behavior of this function near $x = 0$ can be described by *one-sided limits*, which are discussed in the next section.

**3. (a)** $\lim_{x \to 0} F(x)$      **(b)** $\lim_{x \to 1} F(x)$

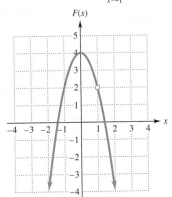

**4. (a)** $\lim_{x \to 1} g(x)$      **(b)** $\lim_{x \to 0} g(x)$

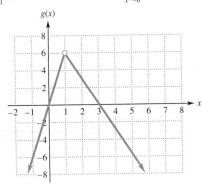

**5. (a)** $\lim_{x \to -3} f(x)$      **(b)** $\lim_{x \to 2} f(x)$

**6. (a)** $\lim_{x \to 4} f(x)$      **(b)** $\lim_{x \to 9} f(x)$

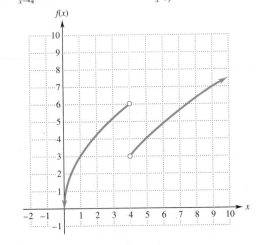

**7. (a)** $\lim_{x \to -2} h(x)$      **(b)** $\lim_{x \to 0} h(x)$

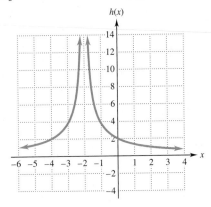

**8. (a)** $\lim_{x \to 3} f(x)$      **(b)** $\lim_{x \to 2} f(x)$

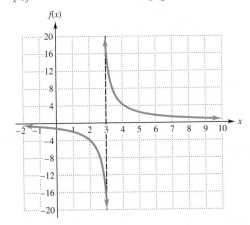

**9.** Explain why $\lim_{x \to 1} F(x)$ in Exercise 3(b) exists, but $\lim_{x \to -3} f(x)$ in Exercise 5(a) does not.

**10.** In Exercise 4(a), why does $\lim_{x \to 1} g(x)$ exist even though $g(1)$ is not defined?

*Use a calculator to estimate the given limits numerically. (See Examples 1–3.)*

**11.** $\displaystyle \lim_{x \to 1} \frac{\ln x}{x - 1}$      **12.** $\displaystyle \lim_{x \to 5} \frac{\ln x - \ln 5}{x - 5}$

**13.** $\displaystyle \lim_{x \to 0} (x \ln|x|)$      **14.** $\displaystyle \lim_{x \to 0} \frac{x}{\ln|x|}$

**15.** $\displaystyle \lim_{x \to 0} \frac{e^{3x} - 1}{x}$      **16.** $\displaystyle \lim_{x \to 0} \frac{x}{e^x - 1}$

**17.** $\displaystyle \lim_{x \to 3} \frac{x^3 - 3x^2 - 4x + 12}{x - 3}$    **18.** $\displaystyle \lim_{x \to 1} \frac{x^3 - x^2 + x - 1}{x - 1}$

**19.** $\displaystyle \lim_{x \to -2} \frac{x^4 + 2x^3 - x^2 + 3x + 1}{x + 2}$

**20.** $\displaystyle \lim_{x \to 4} \frac{.1x^4 - .8x^3 + 1.6x^2 + 2x - 8}{x - 4}$

*Suppose $\lim_{x \to 4} f(x) = 25$ and $\lim_{x \to 4} g(x) = 10$. Use the limit properties to find the given limits. (See Examples 6 and 7.)*

**21.** $\lim_{x \to 4} [f(x) - g(x)]$      **22.** $\lim_{x \to 4} [g(x) \cdot f(x)]$

23. $\lim\limits_{x \to 4} \dfrac{f(x)}{g(x)}$

24. $\lim\limits_{x \to 4} [3 \cdot f(x)]$

25. $\lim\limits_{x \to 4} \sqrt{f(x)}$

26. $\lim\limits_{x \to 4} [g(x)]^3$

27. $\lim\limits_{x \to 4} \dfrac{f(x) + g(x)}{2g(x)}$

28. $\lim\limits_{x \to 4} \dfrac{5g(x) + 2}{1 - f(x)}$

29. **(a)** Graph the function $f$ whose rule is

$$f(x) = \begin{cases} 3 - x & \text{if } x < -2 \\ x + 2 & \text{if } -2 \le x < 2. \\ 1 & \text{if } x \ge 2 \end{cases}$$

Use the graph in part (a) to find the following limits:

**(b)** $\lim\limits_{x \to -2} f(x)$;
**(c)** $\lim\limits_{x \to 1} f(x)$;
**(d)** $\lim\limits_{x \to 2} f(x)$.

30. **(a)** Graph the function $g$ whose rule is

$$g(x) = \begin{cases} x^2 & \text{if } x < -1 \\ x + 2 & \text{if } -1 \le x < 1. \\ 3 - x & \text{if } x \ge 1 \end{cases}$$

Use the graph in part (a) to find the following limits:

**(b)** $\lim\limits_{x \to -1} g(x)$;
**(c)** $\lim\limits_{x \to 0} g(x)$;
**(d)** $\lim\limits_{x \to 1} g(x)$.

*Use algebra and the properties of limits as needed to find the given limits. If the limit does not exist, say so. (See Examples 5–10.)*

31. $\lim\limits_{x \to 2} (3x^3 - 4x^2 - 5x + 2)$

32. $\lim\limits_{x \to -1} (2x^3 - x^2 - 6x + 1)$

33. $\lim\limits_{x \to 3} \dfrac{4x + 7}{10x + 1}$

34. $\lim\limits_{x \to -2} \dfrac{x + 6}{8x - 5}$

35. $\lim\limits_{x \to 5} \dfrac{x^2 - 25}{x - 5}$

36. $\lim\limits_{x \to -4} \dfrac{x^2 - 16}{x + 4}$

37. $\lim\limits_{x \to 4} \dfrac{x^2 - x - 12}{x - 4}$

38. $\lim\limits_{x \to 5} \dfrac{x^2 - 3x - 10}{x - 5}$

39. $\lim\limits_{x \to 2} \dfrac{x^2 - 5x + 6}{x^2 - 6x + 8}$

40. $\lim\limits_{x \to -2} \dfrac{x^2 + 3x + 2}{x^2 - x - 6}$

41. $\lim\limits_{x \to -4} \dfrac{(x + 4)^2(x - 5)}{(x - 4)(x + 4)^2}$

42. $\lim\limits_{x \to -3} \dfrac{(x + 3)(x - 3)(x + 4)}{(x + 8)(x + 3)(x - 4)}$

43. $\lim\limits_{x \to 3} \sqrt{x^2 - 3}$

44. $\lim\limits_{x \to 3} \sqrt{x^2 - 7}$

45. $\lim\limits_{x \to 4} \dfrac{-6}{(x - 4)^2}$

46. $\lim\limits_{x \to -2} \dfrac{3x}{(x + 2)^3}$

47. $\lim\limits_{x \to 0} \dfrac{[1/(x + 3)] - 1/3}{x}$

48. $\lim\limits_{x \to 0} \dfrac{[-1/(x + 2)] + 1/2}{x}$

49. $\lim\limits_{x \to 25} \dfrac{\sqrt{x} - 5}{x - 25}$

50. $\lim\limits_{x \to 36} \dfrac{\sqrt{x} - 6}{x - 36}$

51. $\lim\limits_{x \to 5} \dfrac{\sqrt{x} - \sqrt{5}}{x - 5}$

52. **(a)** Approximate $\lim\limits_{x \to 0} (1 + x)^{1/x}$ to five decimal places.

(Evaluate the function at numbers closer and closer to 0 until successive approximations agree in the first five places.)

**(b)** Find the decimal expansion of the number $e$ to as many places as your calculator can manage.

**(c)** What do parts (a) and (b) suggest about the exact value of $\lim\limits_{x \to 0} (1 + x)^{1/x}$?

*Work these exercises.*

53. **Employee Training** A company training program has determined that a new employee can process an average of $P(s)$ pieces of work per day after $s$ days of on-the-job training, where

$$P(s) = \dfrac{105s}{s + 7}.$$

Find the given quantities.

**(a)** $P(1)$
**(b)** $P(13)$
**(c)** $\lim\limits_{s \to 13} P(s)$

54. **Drug Absorption** The concentration of a drug in a patient's bloodstream $h$ hours after it was injected is given by

$$A(h) = \dfrac{.3h}{h^2 + 3}.$$

Find the given quantities.

**(a)** $A(.5)$
**(b)** $A(1)$
**(c)** $\lim\limits_{h \to 1} A(h)$

55. **Average Cost** The cost of manufacturing a certain mountain bike is $C(x) = 100{,}000 + 200x$, where $x$ is the number of mountain bikes produced. The average cost per mountain bike, denoted by $\overline{C}(x)$, is found by dividing $C(x)$ by $x$:

$$\overline{C}(x) = \dfrac{C(x)}{x}.$$

Find and interpret the given quantities.

**(a)** $\overline{C}(1000)$
**(b)** $\overline{C}(10{,}000)$
**(c)** $\lim\limits_{x \to 100{,}000} \overline{C}(x)$

56. **Average Cost** An independent driver working for a New York City taxicab company pays for the cost of fuel and a maximum of \$115 to lease a vehicle and taxi medallion for a 12-hour shift. Suppose the driver's cost for the shift is approximated by $C(x) = 115 + .20x$, where $x$ is the number of miles driven. The average cost per mile, denoted by $\overline{C}(x)$, is found by dividing $C(x)$ by $x$. Find and interpret the given quantities. (Data from: www.nycgov.com.)

**(a)** $\overline{C}(50)$
**(b)** $\overline{C}(230)$
**(c)** $\lim\limits_{x \to 250} \overline{C}(x)$

57. **Treasury Rates** Treasury maturity rates are shown in the figure on the next page, where $S(t)$ represents the short-term 1-year treasury bill, $M(t)$ represents the medium-term 10-year treasury note, and $L(t)$ represents the long-term 30-year treasury bond. Use the figure to approximate the following quantities. (Data from: Federal Reserve Economic Research.)

**(a)** $\lim\limits_{t \to 2015} [L(t) - M(t)]$

**(b)** $\lim\limits_{t \to 2012} [L(t) - M(t)]$

**(c)** $\lim\limits_{t \to 2009} [M(t) - S(t)]$

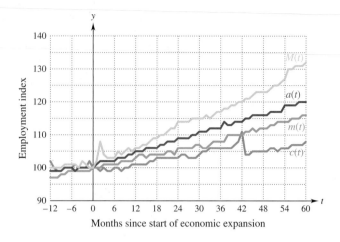

58. **Economic Expansion** A period of U.S. economic expansion began in June 2009, indicated by $t = 0$ in the accompanying figure. Employment figures are indexed to 100 at the start of the expansion. The function $c(t)$ represents the employment index during the current economic expansion; $m(t)$ represents the minimum values from the previous ten expansions; $M(t)$ represents the maximum values from the previous ten expansions; and $a(t)$ represents the average values from the previous ten expansions. Use the figure to approximate the following quantities. (Data from: Federal Reserve Economic Research.)

(a) $\lim_{t \to 42} [c(t) - a(t)]$

(b) $\lim_{t \to 42} [c(t) - m(t)]$

(c) $\lim_{t \to 42} [c(t) - M(t)]$

✓ **Checkpoint Answers**

1. 2.61; 2.9601; 2.996; 3.004; 3.0401; 3.41; the limit appears to be 3.

2. (a) 4    (b) −91    3. 3

4. (a) $\dfrac{3}{10}$    (b) $\dfrac{3}{4}$

5. 5    6. $\dfrac{7}{5}$

7. (a) $\dfrac{1}{2}$    (b) $\dfrac{1}{6}$

8. (a) Does not exist    (b) −3

# 11.2 One-Sided Limits and Limits Involving Infinity

In addition to the limits introduced in Section 11.1, which will be used frequently, there are several other kinds of limits that will appear briefly in Sections 11.9 and 13.4. We begin with one-sided limits.

## One-Sided Limits

Example 12 of Section 11.1 showed that the limit as $x$ approaches 0 of $f(x) = \dfrac{|x|}{x}$ does not exist. However, we can adapt the limit concept to describe the behavior of $f(x)$ near $x = 0$ as follows: First, look at the right half of the graph of $f(x)$ in Figure 11.8.

As x takes values very close to 0 (with $x > 0$), the corresponding values of $f(x)$ are very close (in fact, equal) to 1.

We express this fact in symbols by writing

$$\lim_{x \to 0^+} f(x) = 1,$$

which is read,

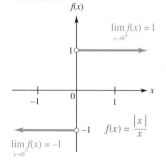

**Figure 11.8**

The limit of $f(x)$ as $x$ approaches 0 from the right is 1.

Now consider the left half of the graph in Figure 11.8.

As $x$ takes values very close to 0 (with $x < 0$), the corresponding values of $f(x)$ are very close (in fact, equal) to $-1$.

We express this fact by writing

$$\lim_{x \to 0^-} f(x) = -1,$$

which is read,

The limit of $f(x)$ as $x$ approaches 0 from the left is $-1$.

The same idea carries over to the general case.

## One-Sided Limits

Let $f$ be a function, and let $a$, $K$, and $M$ be real numbers. Assume that $f(x)$ is defined for all $x$ near $a$, with $x > a$. Suppose that

as $x$ takes values very close (but not equal) to $a$ (with $x > a$), the corresponding values of $f(x)$ are very close (and possibly equal) to $K$,

and that

the values of $f(x)$ can be made as close as you want to $K$ for all values of $x$ (with $x > a$) that are close enough to $a$.

Then the number $K$ is the **limit of $f(x)$ as $x$ approaches $a$ from the right,** which is written

$$\lim_{x \to a^+} f(x) = K.$$

The statement "$M$ is the **limit of $f(x)$ as $x$ approaches $a$ from the left,**" which is written

$$\lim_{x \to a^-} f(x) = M,$$

is defined in a similar fashion. (Just replace $K$ by $M$ and "$x > a$" by "$x < a$" in the previous definition.)

We sometimes refer to a limit of the form $\lim_{x \to a^-} f(x)$ as a **left-hand limit** and a limit of the form $\lim_{x \to a^+} f(x)$ as a **right-hand limit**. The limit $\lim_{x \to a} f(x)$, as defined in Section 11.1, is sometimes called a **two-sided limit**.

### EXAMPLE 1
The graph of a function $f$ is shown in Figure 11.9. Use it to find the following quantities.

**(a)** $\lim_{x \to -3^-} f(x)$  **(b)** $\lim_{x \to -3^+} f(x)$  **(c)** $\lim_{x \to -3} f(x)$  **(d)** $f(-3)$

**(e)** For which value(s) of $x$ is $f(x) = -3$?

**(f)** $\lim_{x \to 2^-} f(x)$  **(g)** $\lim_{x \to 2^+} f(x)$  **(h)** $\lim_{x \to 2} f(x)$  **(i)** $f(2)$

**(j)** For which value(s) of $x$ is $f(x) = 2$?

## Solution

**(a)** The graph shows that as $x$ approaches $-3$ from the left, the corresponding values of $f(x)$ approach 2, therefore $\lim_{x \to -3^-} f(x) = 2$.

**(b)** As $x$ approaches $-3$ from the right, the corresponding values of $f(x)$ approach 1, therefore $\lim_{x \to -3^+} f(x) = 1$.

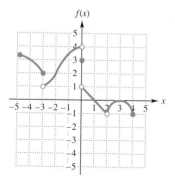

**Figure 11.9**

(c) Since the left- and right-hand limits are not the same, as $x$ approaches $-3$ from *both* sides, the corresponding function values do not get closer and closer to a *single* real number. Therefore, the two-sided limit $\lim_{x \to -3} f(x)$ does not exist.

(d) When $x = -3$, the corresponding function value is $f(-3) = 2$.

(e) There are no $x$-values for which the value of the function is $f(x) = -3$.

In a manner similar to parts (a)–(e), the graph shows the following answers to parts (f)–(j).

(f) $\lim_{x \to 2^-} f(x) = -1$

(g) $\lim_{x \to 2^+} f(x) = -1$

(h) $\lim_{x \to 2} f(x) = -1$

(i) $f(2)$ is undefined.

(j) The $x$-values corresponding to when $f(x) = 2$ are $x = -3$ and $x = -2$. ✔1

In parts (a)–(c) of Example 1, the left- and right-hand limits exist, but are not equal, and the two-sided limit does not exist. In parts (f)–(h), however, all three limits exist and are equal to one another. This case is an example of the following fact.

## ✔ Checkpoint 1

Use Figure 11.9 to find the given quantities.

(a) $\lim_{x \to 0^-} f(x)$

(b) $\lim_{x \to 0^+} f(x)$

(c) $\lim_{x \to 0} f(x)$

(d) $f(0)$

(e) For which value(s) of $x$ is $f(x) = 0$?

## Two-Sided Limits

Let $f$ be a function, and let $a$ and $L$ be real numbers. Then

$$\lim_{x \to a} f(x) = L \quad \text{exactly when} \quad \lim_{x \to a^-} f(x) = L \quad \text{and} \quad \lim_{x \to a^+} f(x) = L.$$

In other words, $f$ has a two-sided limit at $a$ exactly when it has both a left-hand and right-hand limit at $a$ *and* these two limits are the same.

Although its proof is omitted, the following fact will be used frequently:

**All of the properties and theorems about limits in Section 11.1 are valid for left-hand limits and for right-hand limits.**

**EXAMPLE 2**   Find each of the given limits.

(a) $\lim_{x \to 2^+} \sqrt{4 - x^2}$ and $\lim_{x \to 2^-} \sqrt{4 - x^2}$

**Solution** Since $f(x) = \sqrt{4 - x^2}$ is not defined when $x > 2$, the right-hand limit (which requires that $x > 2$) does not exist. For the left-hand limit, write the square root in exponential form and apply the appropriate limit properties (see pages 562 and 563).

$$\lim_{x \to 2^-} \sqrt{4 - x^2} = \lim_{x \to 2^-} (4 - x^2)^{1/2} \qquad \text{Exponential form}$$

$$= [\lim_{x \to 2^-} (4 - x^2)]^{1/2} \qquad \text{Power property}$$

$$= 0^{1/2} = 0. \qquad \text{Polynomial limit}$$

(b) $\lim_{x \to 3^+} [\sqrt{x - 3} + x^2 + 1]$

**Solution**

## ✔ Checkpoint 2

Find the given limits.

(a) $\lim_{x \to 3^+} \sqrt{x^2 - 9}$

(b) $\lim_{x \to -2^-} [\sqrt{x^2 - 4} + x^2 + 4]$

$$\lim_{x \to 3^+} [\sqrt{x - 3} + x^2 + 1]$$

$$= \lim_{x \to 3^+} [(x - 3)^{1/2} + x^2 + 1] \qquad \text{Exponential form}$$

$$= \lim_{x \to 3^+} (x - 3)^{1/2} + \lim_{x \to 3^+} x^2 + \lim_{x \to 3^+} 1 \qquad \text{Sum property}$$

$$= [\lim_{x \to 3^+} (x - 3)]^{1/2} + \lim_{x \to 3^+} x^2 + \lim_{x \to 3^+} 1 \qquad \text{Power property}$$

$$= 0^{1/2} + 9 + 1 = 10. \qquad \text{Polynomial limits} ✔2$$

**EXAMPLE 3**   **First-Class Domestic Mail** The cost $c(x)$ of mailing a first-class letter weighing $x$ ounces on January 1, 2017, was $.47 for the first ounce plus $.21 for each additional ounce or fraction of an ounce, up to a maximum of 3.5 ounces, as shown in the accompanying figure. Find and interpret the following quantities. (Data from: U.S. Postal Service.)

**Figure 11.10**

**(a)** $c(.5)$   **(b)** $c(1)$   **(c)** $c(1.5)$   **(d)** $\lim\limits_{x \to 1^-} c(x)$   **(e)** $\lim\limits_{x \to 1^+} c(x)$   **(f)** $\lim\limits_{x \to 1} c(x)$

**Solution**

**(a)** The graph shows that the cost of mailing a letter weighing .5 ounces is $c(.5) = \$.47$.

**(b)** The closed circle at $(1, .47)$ indicates that this point is part of the graph (whereas the open circles are not part of the graph); hence. $c(1) = \$.47$.

**(c)** According to the graph, $c(1.5) = \$.68$.

**(d)** As $x$ *approaches* 1 *from the left*, the cost is $.47; therefore, $\lim\limits_{x \to 1^-} c(x) = \$.47$.

**(e)** As $x$ *approaches* 1 *from the right*, the cost is $.68; therefore, $\lim\limits_{x \to 1^+} c(x) = \$.68$. Note that the value of the cost function *at* $x = 1$ does not matter as far as the limit is concerned.

**(f)** Because the left- and right-hand limits are not equal, the two-sided limit $\lim\limits_{x \to 1} c(x)$ does not exist. ✔3

✓ **Checkpoint 3**

Use Figure 11.10 to find the given quantities.

**(a)** $c(2)$

**(b)** $\lim\limits_{x \to 2^-} c(x)$

**(c)** $\lim\limits_{x \to 2^+} c(x)$

**(d)** $\lim\limits_{x \to 2} c(x)$

## Infinite Limits

In the next part of the discussion, it is important to remember that

**the symbol $\infty$, which is read "infinity," does *not* represent a real number**.

Nevertheless, the word "infinity" and the symbol $\infty$ are often used as a convenient shorthand to describe the behavior of some functions. Consider the function $f$ whose graph is shown in Figure 11.11 on the following page.

The graph shows that as $x$ approaches 3 (on either side), the corresponding values of $f(x)$ get larger and larger without bound. We express this behavior by writing

$$\lim\limits_{x \to 3} f(x) = \infty,$$

which is read, "The limit of $f(x)$ as $x$ approaches 3 is infinity."

Similarly, as $x$ approaches 1 (on either side), the corresponding values of $f(x)$ get more and more negative (smaller and smaller) without bound. We write

$$\lim\limits_{x \to 1} f(x) = -\infty,$$

which is read, "The limit of $f(x)$ as $x$ approaches 1 is negative infinity."

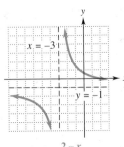

$$\lim_{x \to 3} f(x) = \infty$$

$$\lim_{x \to 1} f(x) = -\infty$$

**Figure 11.11**

The behavior of $f$ near $x = 5$ can be described in a similar fashion with the use of one-sided limits. When $x$ is close to 5, the corresponding values of $f(x)$ get very large on the left side of 5 and very small on the right side of 5, so we write

$$\lim_{x \to 5^-} f(x) = \infty \qquad \text{and} \qquad \lim_{x \to 5^+} f(x) = -\infty$$

and say, "The limit of $f(x)$ as $x$ approaches 5 from the left is infinity" and "The limit of $f(x)$ as $x$ approaches 5 from the right is negative infinity." Because the left- and right-hand limits are not the same, the two-sided limit $\lim_{x \to 5} f(x)$ does not exist.

In all the situations just described, the language of limits and the word "infinity" are useful for describing the behavior of a function that does not have a limit in the sense discussed in Section 11.1. In particular, this language provides an algebraic way to describe precisely the fact that the function in Figure 11.11 has different types of vertical asymptotes at $x = 1$, $x = 3$, and $x = 5$.

✓**Checkpoint 4**

Use "infinite limits" to describe the behavior of the function whose graph is given, near the specified number.

**(a)** $f(x) = \dfrac{3}{5 - x}$ near $x = 5$

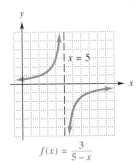

$f(x) = \dfrac{3}{5 - x}$

**(b)** $f(x) = \dfrac{2 - x}{x + 3}$ near $x = -3$

$f(x) = \dfrac{2 - x}{x + 3}$

**EXAMPLE 4** Let $f(x) = \dfrac{3}{5 - x}$. Find the given limits.

**(a)** $\lim_{x \to 5^-} f(x)$

**Solution** As $x$ approaches 5 from the left, the numerator of $f(x)$ remains constant at 3, and the denominator is positive and gets closer and closer to zero. A positive constant divided by a positive number very close zero produces a very large positive quotient, for example, $f(4.99999) = \dfrac{3}{5 - 4.99999} = \dfrac{3}{.00001} = 300,000$. Therefore, the values of $f(x)$ get larger and larger, and hence $\lim_{x \to 5^-} \dfrac{3}{5 - x} = \infty$.

**(b)** $\lim_{x \to 5^+} f(x)$

**Solution** As $x$ approaches 5 from the right, the numerator of $f(x)$ is again a positive constant, but the denominator is negative and gets closer and closer to zero. For example, $f(5.00001) = \dfrac{3}{5 - 5.00001} = \dfrac{3}{-.00001} = -300,000$. Therefore, the values of $f(x)$ get smaller and smaller (more and more negative), and hence $\lim_{x \to 5^+} \dfrac{3}{5 - x} = -\infty$.

**(c)** $\lim_{x \to 5} f(x)$

**Solution** Because the left- and right-hand limits are not the same, the two-sided limit does not exist. For graphical verification of these limits, see Checkpoint 4(a). ✓ 4

## Limits at Infinity

The word "limit" has been used thus far to describe the values of a function $f(x)$ when $x$ is near a particular number $a$. Now we consider the behavior of a function when $x$ is very large or very small.

**EXAMPLE 5**   Figure 11.12 shows the graph of

$$f(x) = \frac{3}{1 + e^{-x}} + 1.$$

**(a)** Describe the behavior of $f$ when $x$ is very large.

**Solution** The right side of the graph appears to coincide with the horizontal line through 4. Actually, as $x$ gets larger and larger, the corresponding values of $f(x)$ are very close (but not equal) to 4, as you can verify with a calculator or by using the trace feature of a graphing calculator (Figure 11.13 below).* We describe this situation by writing

$$\lim_{x \to \infty} f(x) = 4,$$

which is read "The limit of $f(x)$ as $x$ approaches infinity is 4."

**Figure 11.12**

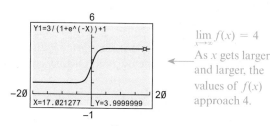

**Figure 11.13**

$\lim_{x \to \infty} f(x) = 4$
As $x$ gets larger and larger, the values of $f(x)$ approach 4.

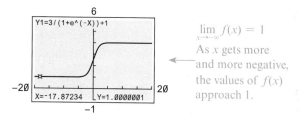

**Figure 11.14**

$\lim_{x \to -\infty} f(x) = 1$
As $x$ gets more and more negative, the values of $f(x)$ approach 1.

**(b)** Describe the behavior of $f$ when $x$ is very small.

**Solution** To say that $x$ gets very small means that it moves to the left on the axis (that is, $x$ gets more and more negative). The graph suggests that when $x$ is very small, the corresponding values of $f(x)$ are very close to 1. You can verify this with a calculator or the trace feature on a graphing calculator (Figure 11.14).† We describe this situation by writing

$$\lim_{x \to -\infty} f(x) = 1,$$

which is read, "The limit of $f(x)$ as $x$ approaches negative infinity is 1." ✓ 5

Once again, the words "infinity" and "negative infinity" and the symbols $\infty$ and $-\infty$ do *not* denote numbers. They are just a convenient shorthand to express the idea that $x$ takes very large or very small values.

A general definition of the kind of limits illustrated in Example 4 is given next. It is an informal one, as was the definition of limit in Section 11.1.

✓ **Checkpoint 5**

The graph of a function $g$ is shown. Use the language of limits, as in Example 4, to describe the behavior of $g$ when

**(a)** $x$ is very small;

**(b)** $x$ is very large.

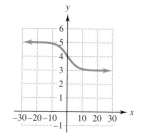

---

*Because of round-off, a TI-84+ will say that $f(x) = 4$ when $x > 28$. Actually $f(x)$ is always less than 4. Similar remarks apply to other calculators.

†Because of round-off, a TI-84+ will say that $f(x) = 1$ when $x < -28$. In fact, $f(x)$ is always more than 1. Similar remarks apply to other calculators.

## Limits at Infinity

Let $f$ be a function that is defined for all large positive values of $x$, and let $L$ be a real number. Suppose that

> as $x$ takes larger and larger positive values, increasing without bound, the corresponding values of $f(x)$ are very close (and possibly equal) to $L$

and that

> the values of $f(x)$ can be made arbitrarily close (as close as you want) to $L$ by taking large enough values of $x$.

Then we say that

**the limit of $f(x)$ as $x$ approaches infinity is $L$,**

which is written

$$\lim_{x \to \infty} f(x) = L.$$

Similarly, if $f$ is defined for all small negative values of $x$ and $M$ is a real number, then the statement

$$\lim_{x \to -\infty} f(x) = M,$$

which is read

**the limit of $f(x)$ as $x$ approaches negative infinity is $M$,**

means that

> as $x$ takes smaller and smaller negative values, decreasing without bound, the corresponding values of $f(x)$ are very close (and possibly equal) to $M$

and

> the values of $f(x)$ can be made arbitrarily close (as close as you want) to $M$ by taking small enough values of $x$.

---

**EXAMPLE 6** Find the given limits.

**(a)** $\displaystyle \lim_{x \to \infty} \frac{1}{x}$.

**Solution** Make a table of values when $x$ is very large (or use the table feature of a graphing calculator):

| $x$ gets larger and larger | | | |
|---|---|---|---|
| $x$ | 50 | 100 | 1000 | 10,000 |
| $1/x$ | .02 | .01 | .001 | .0001 |

$\dfrac{1}{x}$ approaches 0

The table shows that $1/x$ gets very close to 0 as $x$ takes larger and larger values, which suggests that $\displaystyle \lim_{x \to \infty} \frac{1}{x} = 0$. You can reach the same conclusion by examining the right side of the graph of $f(x) = 1/x$ in Figure 11.15.

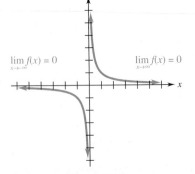

$\lim_{x \to -\infty} f(x) = 0$ $\qquad$ $\lim_{x \to \infty} f(x) = 0$

**Figure 11.15**

**(b)** $\displaystyle\lim_{x \to -\infty} \frac{1}{x}$

**Solution** Make a table of values when $x$ is very small:

| | x gets smaller and smaller → | | | |
|---|---|---|---|---|
| $x$ | $-50$ | $-100$ | $-1000$ | $-10,000$ |
| $1/x$ | $-.02$ | $-.01$ | $-.001$ | $-.0001$ |

$\frac{1}{x}$ approaches 0 →

The table shows that $1/x$ gets very close to 0 as $x$ takes smaller and smaller values, which suggests that $\displaystyle\lim_{x \to -\infty} \frac{1}{x} = 0$. The left side of the graph of $f(x) = 1/x$ in Figure 11.15 shows this result graphically.

If the graph of a function approaches a horizontal line very closely when $x$ is very large or very small, we say that this line is a **horizontal asymptote** of the graph. In Example 5, we see that the horizontal line $y = 0$ (the $x$-axis) is a horizontal asymptote for the graph of $f(x) = \dfrac{1}{x}$.

Limits at infinity have many of the properties of ordinary limits. Consider, for instance, the constant function $f(x) = 6$. As $x$ gets larger and larger (or smaller and smaller), the corresponding value of $f(x)$ is *always* 6. Hence, $\displaystyle\lim_{x \to \infty} f(x) = 6$ and $\displaystyle\lim_{x \to -\infty} f(x) = 6$. Similarly, for any constant $d$,

$$\lim_{x \to \infty} d = d \quad \text{and} \quad \lim_{x \to -\infty} d = d.$$

Finally, we note the following:

**All of the properties of limits in Section 11.1 (page 562) are valid for limits at infinity.**

**EXAMPLE 7**   Find $\displaystyle\lim_{x \to \infty} \frac{4x + 3}{5x - 2}$.

**Solution** To find the limit of a rational function at infinity, first divide both numerator and denominator by the highest power of $x$ appearing in either one (in this example, divide both numerator and denominator by $x$).[*] Then simplify and rewrite the result as follows:

$$\frac{4x + 3}{5x - 2} = \frac{\dfrac{4x + 3}{x}}{\dfrac{5x - 2}{x}} = \frac{\dfrac{4x}{x} + \dfrac{3}{x}}{\dfrac{5x}{x} - \dfrac{2}{x}} = \frac{4 + \dfrac{3}{x}}{5 - \dfrac{2}{x}}.$$

---

[*]In Examples 7 and 8, we are interested only in nonzero values of $x$, and dividing both numerator and denominator by a nonzero quantity will not change the value of the function.

Now use the limit properties and Example 6 to compute the limit.

$$\lim_{x \to \infty} \frac{4x + 3}{5x - 2} = \lim_{x \to \infty} \frac{4 + \dfrac{3}{x}}{5 - \dfrac{2}{x}} \qquad \text{Preceding equation}$$

$$= \frac{\lim\limits_{x \to \infty} \left[4 + \dfrac{3}{x}\right]}{\lim\limits_{x \to \infty} \left[5 - \dfrac{2}{x}\right]} \qquad \text{Quotient property}$$

$$= \frac{\lim\limits_{x \to \infty} 4 + \lim\limits_{x \to \infty} \dfrac{3}{x}}{\lim\limits_{x \to \infty} 5 - \lim\limits_{x \to \infty} \dfrac{2}{x}} \qquad \text{Sum and difference properties}$$

$$= \frac{\lim\limits_{x \to \infty} 4 + \lim\limits_{x \to \infty} 3 \cdot \lim\limits_{x \to \infty} \dfrac{1}{x}}{\lim\limits_{x \to \infty} 5 - \lim\limits_{x \to \infty} 2 \cdot \lim\limits_{x \to \infty} \dfrac{1}{x}} \qquad \text{Product property}$$

$$= \frac{4 + 3 \cdot 0}{5 - 2 \cdot 0} = \frac{4}{5}. \qquad \text{Constant-function limits and Example 6}$$

## EXAMPLE 8    Find

$$\lim_{x \to -\infty} \frac{2x^2}{3 - x^3}.$$

**Solution**  Divide both numerator and denominator by $x^3$, the highest power of $x$ appearing in either one.* Then simplify and rewrite:

$$\frac{2x^2}{3 - x^3} = \frac{\dfrac{2x^2}{x^3}}{\dfrac{3 - x^3}{x^3}} = \frac{\dfrac{2x^2}{x^3}}{\dfrac{3}{x^3} - \dfrac{x^3}{x^3}} = \frac{\dfrac{2}{x}}{\dfrac{3}{x^3} - 1} = \frac{2 \cdot \dfrac{1}{x}}{3 \cdot \dfrac{1}{x} \cdot \dfrac{1}{x} \cdot \dfrac{1}{x} - 1}.$$

Now proceed as in Example 7:

$$\lim_{x \to -\infty} \frac{2x^2}{3 - x^3} = \lim_{x \to -\infty} \frac{2 \cdot \dfrac{1}{x}}{3 \cdot \dfrac{1}{x} \cdot \dfrac{1}{x} \cdot \dfrac{1}{x} - 1} \qquad \text{Preceding equation}$$

$$= \frac{\lim\limits_{x \to -\infty} \left[2 \cdot \dfrac{1}{x}\right]}{\lim\limits_{x \to -\infty} \left[3 \cdot \dfrac{1}{x} \cdot \dfrac{1}{x} \cdot \dfrac{1}{x} - 1\right]} \qquad \text{Quotient property}$$

$$= \frac{\lim\limits_{x \to -\infty} \left[2 \cdot \dfrac{1}{x}\right]}{\lim\limits_{x \to -\infty} \left[3 \cdot \dfrac{1}{x} \cdot \dfrac{1}{x} \cdot \dfrac{1}{x}\right] - \lim\limits_{x \to -\infty} 1} \qquad \text{Difference property}$$

$$= \frac{\lim\limits_{x \to -\infty} 2 \cdot \lim\limits_{x \to -\infty} \dfrac{1}{x}}{\lim\limits_{x \to -\infty} 3 \cdot \lim\limits_{x \to -\infty} \dfrac{1}{x} \cdot \lim\limits_{x \to -\infty} \dfrac{1}{x} \cdot \lim\limits_{x \to -\infty} \dfrac{1}{x} - \lim\limits_{x \to -\infty} 1} \qquad \text{Product property}$$

$$= \frac{2 \cdot 0}{3 \cdot 0 \cdot 0 \cdot 0 - 1} = \frac{0}{-1} = 0. \qquad \text{Constant-function limits and Example 6}$$ ☑ 6

## ✓ Checkpoint 6

Find the given limits.

(a)  $\lim\limits_{x \to -\infty} \dfrac{5x^3}{2x^3 - x + 4}$

(b)  $\lim\limits_{x \to \infty} \dfrac{10x^3 - 9x}{28 + 11x^4}$

The technique in Examples 6 and 7 carries over to any rational function $f(x) = g(x)/h(x)$, **where the degree of $g(x)$ is less than or equal to the degree of $h(x)$.** These limits provide a mathematical justification for the treatment of horizontal asymptotes in Section 3.6. In particular, see the box on page 170.

# 11.2 Exercises

*The graph of the function f is shown. Use it to compute the given limits. (See Examples 1 and 3.)*

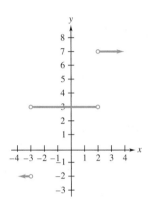

**1.** $\lim\limits_{x \to 2^-} f(x)$       **2.** $\lim\limits_{x \to 2^+} f(x)$

**3.** $\lim\limits_{x \to -3^-} f(x)$      **4.** $\lim\limits_{x \to -3^+} f(x)$

*The graph of the function f is shown. Use it to compute the given limits. (See Examples 1 and 3.)*

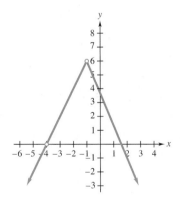

**5.** $\lim\limits_{x \to -1^-} f(x)$      **6.** $\lim\limits_{x \to -1^+} f(x)$

**7.** $\lim\limits_{x \to -4^-} f(x)$      **8.** $\lim\limits_{x \to -4^+} f(x)$

*In Exercises 9–12, use the graph of the function f to determine the given limits. (See Example 1 and Figure 11.11.)*

**(a)** $\lim\limits_{x \to -2} f(x)$       **(b)** $\lim\limits_{x \to 0^+} f(x)$

**(c)** $\lim\limits_{x \to 3^-} f(x)$       **(d)** $\lim\limits_{x \to 3^+} f(x)$

**9.**

**10.**

**11.**

**12.**

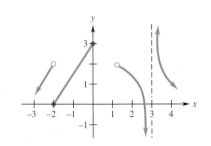

*Find the given limits. (See Example 2.)*

13. $\lim\limits_{x \to 2^+} \sqrt{x^2 - 4}$

14. $\lim\limits_{x \to 3^-} \sqrt{9 - x^2}$

15. $\lim\limits_{x \to 0^+} (\sqrt{x} + x + 1)$

16. $\lim\limits_{x \to 1^+} (\sqrt{x - 1} + 1)$

17. $\lim\limits_{x \to -2^+} (x^3 - x^2 - x + 1)$

18. $\lim\limits_{x \to -2^-} (2x^3 + 3x^2 + 5x + 2)$

19. $\lim\limits_{x \to -5^+} \dfrac{\sqrt{x + 5} + 5}{x^2 - 5}$

20. $\lim\limits_{x \to 1^+} \dfrac{x^3 - 2}{\sqrt{x^3 - 1} + 2}$

21. Graph the function $f$ whose rule is

$$f(x) = \begin{cases} 3 - x & \text{if } x < -2 \\ x + 2 & \text{if } -2 \le x < 2 \\ 1 & \text{if } x = 2 \\ 4 - x & \text{if } x > 2 \end{cases}$$

Use the graph to evaluate the given limits.

(a) $\lim\limits_{x \to -2^-} f(x)$

(b) $\lim\limits_{x \to -2^+} f(x)$

(c) $\lim\limits_{x \to -2} f(x)$

22. Let $f$ be the function in Exercise 21. Evaluate the given limits.

(a) $\lim\limits_{x \to 2^-} f(x)$

(b) $\lim\limits_{x \to 2^+} f(x)$

(c) $\lim\limits_{x \to 2} f(x)$

23. For the function $f$ whose graph is shown, find

(a) $\lim\limits_{x \to 2} f(x)$;

(b) $\lim\limits_{x \to -2^+} f(x)$;

(c) $\lim\limits_{x \to -2^-} f(x)$.

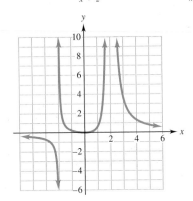

24. For the function $g$ whose graph is shown, find

(a) $\lim\limits_{x \to 3} g(x)$;

(b) $\lim\limits_{x \to -2^+} g(x)$;

(c) $\lim\limits_{x \to -2} g(x)$.

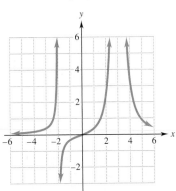

*Find the given limits. (See Example 4.)*

25. $\lim\limits_{x \to 4} \dfrac{-6}{(x - 4)^2}$

26. $\lim\limits_{x \to -2} \dfrac{2x}{(x + 2)^2}$

27. $\lim\limits_{x \to -3^+} \dfrac{2 - x}{x + 3}$

28. $\lim\limits_{x \to -1^+} \dfrac{2}{1 + x}$

*Use the given graph of the function $f$ to find $\lim\limits_{x \to \infty} f(x)$ and $\lim\limits_{x \to -\infty} f(x)$. (See Example 5 and 6.)*

29.

30.

31.

32.

33.

**34.**

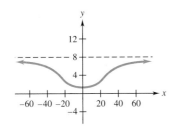

**Use a calculator to estimate the limits given numerically. (See Example 6.)**

**35.** $\lim\limits_{x \to \infty} [\sqrt{x^2 + 1} - (x + 1)]$  **36.** $\lim\limits_{x \to -\infty} [\sqrt{x^2 + x + 1} + x]$

**37.** $\lim\limits_{x \to -\infty} \dfrac{x^{2/3} - x^{4/3}}{x^3}$  **38.** $\lim\limits_{x \to \infty} \dfrac{x^{5/4} + x}{2x - x^{5/4}}$

**39.** $\lim\limits_{x \to -\infty} e^{1/x}$  **40.** $\lim\limits_{x \to \infty} \dfrac{e^{1/x}}{x}$

**41.** $\lim\limits_{x \to \infty} \dfrac{\ln x}{x}$  **42.** $\lim\limits_{x \to \infty} \dfrac{5}{1 + (1.1)^{-x/20}}$

**Use the properties of limits to find the given limits. (See Examples 7 and 8.)**

**43.** $\lim\limits_{x \to \infty} \dfrac{10x^2 - 5x + 8}{3x^2 + 8x - 29}$  **44.** $\lim\limits_{x \to \infty} \dfrac{44x^2 - 12x + 82}{11x^2 - 35x - 37}$

**45.** $\lim\limits_{x \to -\infty} \dfrac{5x + 9x^2 - 17}{21x + 5x^3 + 12x^2}$  **46.** $\lim\limits_{x \to -\infty} \dfrac{11x + 21}{7 + 67x - x^2}$

**47.** $\lim\limits_{x \to -\infty} \dfrac{8x^5 + 7x^4 - 10x^3}{25 - 4x^5}$

**48.** $\lim\limits_{x \to \infty} \dfrac{7x^4 + 3x^3 - 12x^2 + x - 76}{x^4 - 33x^3 - 17x^2 + 6}$

**49.** $\lim\limits_{x \to -\infty} \dfrac{(4x - 1)(x + 2)}{5x^2 - 3x + 1}$  **50.** $\lim\limits_{x \to \infty} \dfrac{(3x - 1)(4x + 3)}{6x^2 - x + 7}$

**51.** $\lim\limits_{x \to \infty} \left( \dfrac{1}{4x^2} \right)$  **52.** $\lim\limits_{x \to -\infty} (7x^2 + 2)^{-2}$

**53.** $\lim\limits_{x \to -\infty} \left( \dfrac{18x}{x - 9} - \dfrac{3x}{x + 4} \right)$  **54.** $\lim\limits_{x \to \infty} \left( \dfrac{2x}{x^2 + 2} + \dfrac{17x}{x^3 - 11} \right)$

**Use the definition of absolute value to find the given limits.**

**55.** $\lim\limits_{x \to \infty} \dfrac{x}{|x|}$  **56.** $\lim\limits_{x \to -\infty} \dfrac{x}{|x|}$

**57.** $\lim\limits_{x \to -\infty} \dfrac{x}{|x| + 1}$  **58.** $\lim\limits_{x \to \infty} \dfrac{x}{|x| + 1}$

**Work these exercises.**

**59. Water Bill** A metropolitan waterworks normally charges a flat monthly rate of $20 for up to 4000 gallons of water consumption and a rate of $5 per thousand gallons of water for consumption in excess of 4000 gallons. To encourage conservation during a severe drought, the waterworks imposed an excessive use surcharge and an additional rate increase for residential customers who consumed more than 10,000 gallons of water in a month. Use the accompanying graph, which shows the monthly

cost, $c(x)$, of consuming $x$ gallons of water to find the given quantities.

**(a)** $\lim\limits_{x \to 4^-} c(x)$  **(b)** $\lim\limits_{x \to 4^+} c(x)$  **(c)** $\lim\limits_{x \to 4} c(x)$

**(d)** $\lim\limits_{x \to 10^-} c(x)$  **(e)** $\lim\limits_{x \to 10^+} c(x)$  **(f)** $\lim\limits_{x \to 10} c(x)$

**(g)** What rate does the waterworks charge for consumption in excess of 10,000 gallons?

**(h)** What is the amount of the surcharge for customers who consume more than 10,000 gallons of water?

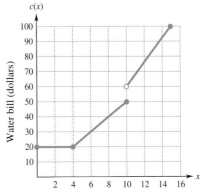

**60. Stock Prices** The opening price of a stock typically differs somewhat from the closing price the previous day. The accompanying graph shows the stock price $f(t)$ for Visa Inc. over two trading days, where $t = 0$ corresponds to the opening of trading on April 12, 2016. Each trading day lasts 6.5 hours, beginning at 9:30 a.m. and concluding at 4:00 p.m., Eastern time. Let $t$ represent the time (in hours) that the market has been open for trading, so $t = 13$ represents the close of trading on April 13, 2016. Approximate the given limits. (Data from: www.morningstar.com.)

**(a)** $\lim\limits_{t \to 6.5^-} f(t)$  **(b)** $\lim\limits_{t \to 6.5^+} f(t)$  **(c)** $\lim\limits_{t \to 6.5} f(t)$

**61. Daycare Providers** According to the Virginia Department of Social Services, an individual daycare provider in Virginia may care for up to eight children (ages two to four). If a daycare provider hires an additional worker, it increases capacity but also incurs additional expense. The accompanying graph approximates the monthly profit, $p(x)$, from providing care for $x$ children. Find the given quantities.

**(a)** $\lim\limits_{x \to 8^-} p(x)$  **(b)** $\lim\limits_{x \to 8^+} p(x)$  **(c)** $\lim\limits_{x \to 8} p(x)$

**(d)** How many children must be cared for to break even (zero profit)?

**(e)** How many children must be cared for to make hiring a second provider beneficial?

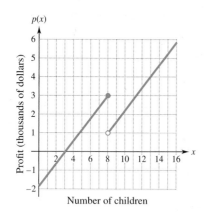

Number of children

62. **First-Class Stamps** The cost of a first-class stamp increased slowly between 1919 and 1968 and has increased more rapidly since 1968. The cost (in cents) can be approximated by

$$c(x) = \begin{cases} .08x + .56 & \text{if } 19 \leq x \leq 68 \\ .89x - 54.52 & \text{if } x > 68 \end{cases},$$

where $x = 0$ corresponds to the year 1900. Find the given limits. (Data from: United States Postal Service.)

**(a)** $\lim_{x \to 68^-} c(x)$    **(b)** $\lim_{x \to 68^+} c(x)$    **(c)** $\lim_{x \to 68} c(x)$

63. **Parking Rates** An aquarium parking garage charges \$3 for up to and including one hour and \$2 for each additional hour (or fraction of an hour), up to a daily maximum of \$15. Let $c(t)$ represent the daily cost of parking, where $t$ is the number of hours parked. Find the given quantities.

**(a)** $c(1.5)$    **(b)** $c(2)$    **(c)** $c(2.5)$

**(d)** $c(3)$    **(e)** $c(8)$    **(f)** $\lim_{t \to 1.5} c(t)$

**(g)** $\lim_{t \to 2^-} c(t)$    **(h)** $\lim_{t \to 2^+} c(t)$    **(i)** $\lim_{t \to 2} c(t)$

**(j)** $\lim_{t \to 8} c(t)$

64. **Travel Time** If you travel from Birmingham, Alabama to Atlanta, Georgia at an average speed of $r$ miles per hour, the time (in hours) it takes to complete the trip is given by $T(r) = \dfrac{150}{r}$. Find and interpret the given limits.

**(a)** $\lim_{r \to 75} T(r)$    **(b)** $\lim_{r \to \infty} T(r)$    **(c)** $\lim_{r \to 0^+} T(r)$

65. **On-Demand Publishing** The cost of printing a certain paperback book using a publish-on-demand service is $c(x) = 6000 + 5x$, where $x$ is the number of books printed. The average cost per book, denoted by $\bar{c}(x)$, is found by dividing $c(x)$ by $x$. Find and interpret the given quantities.

**(a)** $\bar{c}(100)$    **(b)** $\bar{c}(1000)$    **(c)** $\lim_{x \to \infty} \bar{c}(x)$

66. **Drug Concentration** The concentration of a drug in a patient's bloodstream $h$ hours after it was injected is given by

$$A(h) = \frac{.17h}{h^2 + 2}.$$

Find and interpret $\lim_{h \to \infty} A(h)$.

✓**Checkpoint Answers**

1. **(a)** 4                              **(b)** 1
   **(c)** Does not exist                 **(d)** 3
   **(e)** $x = 1$ and $x = 3$

2. **(a)** 0                              **(b)** 8

3. **(a)** \$.68      **(b)** \$.68      **(c)** \$.89
   **(d)** Does not exist

4. **(a)** $\lim_{x \to 5^-} \dfrac{3}{5-x} = \infty$ and $\lim_{x \to 5^+} \dfrac{3}{5-x} = -\infty$

   **(b)** $\lim_{x \to -3^-} \dfrac{2-x}{x+3} = -\infty$ and $\lim_{x \to -3^+} \dfrac{2-x}{x+3} = \infty$

5. **(a)** $\lim_{x \to -\infty} g(x) = 5$    **(b)** $\lim_{x \to \infty} g(x) = 3$

6. **(a)** $\dfrac{5}{2}$                  **(b)** 0

# 11.3  Rates of Change

One of the main applications of calculus is determining how one variable changes in relation to another. A person in business wants to know how profit changes with respect to changes in advertising, while a person in medicine wants to know how a patient's reaction to a drug changes with respect to changes in the dose.

We begin the discussion with a familiar situation. A driver makes the 168-mile trip from Cleveland to Columbus, Ohio, in 3 hours. The following table shows how far the driver has traveled from Cleveland at various times:

| Time (in hours) | 0 | .5 | 1 | 1.5 | 2 | 2.5 | 3 |
|---|---|---|---|---|---|---|---|
| Distance (in miles) | 0 | 22 | 52 | 86 | 118 | 148 | 168 |

If $f$ is the function whose rule is

$$f(x) = \text{distance from Cleveland at time } x,$$

then the table shows, for example, that $f(2) = 118$ and $f(3) = 168$. So the distance traveled from time $x = 2$ to $x = 3$ is $168 - 118$—that is, $f(3) - f(2)$. In a similar fashion, we obtain the other entries in the following chart:

| Time Interval | Distance Traveled |
|---|---|
| $x = 2$ to $x = 3$ | $f(3) - f(2) = 168 - 118 = 50$ miles |
| $x = 1$ to $x = 3$ | $f(3) - f(1) = 168 - 52 = 116$ miles |
| $x = 0$ to $x = 2.5$ | $f(2.5) - f(0) = 148 - 0 = 148$ miles |
| $x = .5$ to $x = 1$ | $f(1) - f(.5) = 52 - 22 = 30$ miles |
| $x = a$ to $x = b$ | $f(b) - f(a)$ miles |

The last line of the chart shows how to find the distance traveled in any time interval $(0 \le a < b \le 3)$.

Since distance = average speed × time,

$$\text{average speed} = \frac{\text{distance traveled}}{\text{time interval}}.$$

In the preceding chart, you can compute the length of each time interval by taking the difference between the two times. For example, from $x = 1$ to $x = 3$ is a time interval of length $3 - 1 = 2$ hours; hence, the average speed over this interval is $116/2 = 58$ mph. Similarly, we have the following information:

| Time Interval | Average Speed $= \dfrac{\text{Distance Traveled}}{\text{Time Interval}}$ |
|---|---|
| $x = 2$ to $x = 3$ | $\dfrac{f(3) - f(2)}{3 - 2} = \dfrac{168 - 118}{3 - 2} = \dfrac{50}{1} = 50$ mph |
| $x = 1$ to $x = 3$ | $\dfrac{f(3) - f(1)}{3 - 1} = \dfrac{168 - 52}{3 - 1} = \dfrac{116}{2} = 58$ mph |
| $x = 0$ to $x = 2.5$ | $\dfrac{f(2.5) - f(0)}{2.5 - 0} = \dfrac{148 - 0}{2.5 - 0} = \dfrac{148}{2.5} = 59.2$ mph |
| $x = .5$ to $x = 1$ | $\dfrac{f(1) - f(.5)}{1 - .5} = \dfrac{52 - 22}{1 - .5} = \dfrac{30}{.5} = 60$ mph |
| $x = a$ to $x = b$ | $\dfrac{f(b) - f(a)}{b - a}$ mph |

✓ **Checkpoint 1**

Find the average speed

(a)  from $t = 1.5$ to $t = 2$;

(b)  from $t = s$ to $t = r$.

The last line of the chart shows how to compute the average speed over any time interval $(0 \le a < b \le 3)$. ✓₁

Now, speed (miles per hour) is simply the *rate of change* of distance with respect to time, and what was done for the distance function $f$ in the preceding discussion can be done with any function.

| Quantity | Meaning for the Distance Function | Meaning for an Arbitrary Function $f$ |
|---|---|---|
| $b - a$ | Time interval = change in time from $x = a$ to $x = b$ | Change in $x$ from $x = a$ to $x = b$ |
| $f(b) - f(a)$ | Distance traveled = corresponding change in distance as time changes from $a$ to $b$ | Corresponding change in $f(x)$ as $x$ changes from $a$ to $b$ |
| $\dfrac{f(b) - f(a)}{b - a}$ | **Average speed** = average rate of change of distance with respect to time as time changes from $a$ to $b$ | **Average rate of change** of $f(x)$ with respect to $x$ as $x$ changes from $a$ to $b$ (where $a < b$) |

**EXAMPLE 1**   If $f(x) = x^2 + 4x + 5$, find the average rate of change of $f(x)$ with respect to $x$ as $x$ changes from $-2$ to 3.

**Solution** This is the situation described in the last line of the preceding chart, with $a = -2$ and $b = 3$. The average rate of change is

$$\frac{f(3) - f(-2)}{3 - (-2)} = \frac{(3)^2 + 4(3) + 5 - [(-2)^2 + 4(-2) + 5]}{3 - (-2)} = \frac{26 - 1}{5}$$

$$= \frac{25}{5} = 5. \ \checkmark_2$$

✓ **Checkpoint 2**

Find the average rate of change of $f(x)$ in Example 1 when $x$ changes from

(a)  0 to 4;

(b)  2 to 7.

**EXAMPLE 2**   **Dow Jones** Suppose $D(t)$ is the value of the Dow Jones Industrial Average (DJIA) at time $t$. Figure 11.16 shows a graph and table of values for $D(t)$ at the start of the given year over a ten-year period. (Data from: www.morningstar.com.)

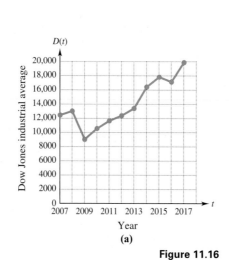

| Year | DJIA |
|------|------|
| 2007 | 12,475 |
| 2008 | 13,044 |
| 2009 | 9,035 |
| 2010 | 10,584 |
| 2011 | 11,671 |
| 2012 | 12,397 |
| 2013 | 13,413 |
| 2014 | 16,441 |
| 2015 | 17,833 |
| 2016 | 17,149 |
| 2017 | 19,882 |

(a)

(b)

**Figure 11.16**

Approximate the average rate of change of the DJIA with respect to time over the given intervals.

**(a)** From the start of 2013 to the start of 2015.

**Solution** The table shows that the DJIA stood at 13,413 at the start of 2013 and 17,833 at the start of 2015; that is $D(2013) = 13,413$ and $D(2015) = 17,833$. The average rate of change over this period is

$$\frac{D(2015) - D(2013)}{2015 - 2013} = \frac{17,833 - 13,413}{2015 - 2013} = \frac{4420}{2} = 2210.$$

Thus the Dow was increasing at an average rate of 2210 index points per year during this period.

✓ **Checkpoint 3**

Use Figure 11.16 to find the average rate of change of the DJIA with respect to time over the given interval.

(a)  From the start of 2007 to the start of 2009

(b)  From the start of 2009 to the start of 2016

**(b)** From the start of 2015 to the start of 2016.

**Solution** The table indicates that $D(2016) = 17,149$, and we saw earlier that $D(2015) = 17,833$. The average rate of change is

$$\frac{D(2016) - D(2015)}{2016 - 2015} = \frac{17,149 - 17,833}{2016 - 2015} = \frac{-684}{1} = -684.$$

The negative number implies that the DJIA was decreasing at an average rate of 684 index points per year. ✓₃

Example 2 also illustrates the geometric interpretation of the average rate of change. In part (a), for instance,

$$\frac{\text{average rate of change}}{\text{from 2013 to 2015}} = \frac{D(2015) - D(2013)}{2015 - 2013} = \frac{17{,}833 - 13{,}413}{2015 - 2013} = 2210 \, \frac{\text{index points}}{\text{year}}.$$

This last number is precisely the *slope* of the line from (2013, 13,413) to (2015, 17,833) on the graph of $D$, as shown in red in Figure 11.17. The same thing is true in general.

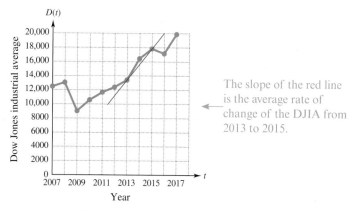

The slope of the red line is the average rate of change of the DJIA from 2013 to 2015.

**Figure 11.17**

## Geometric Meaning of Average Rate of Change

If $f$ is a function, then the average rate of change of $f(x)$ with respect to $x$, as $x$ changes from $a$ to $b$, namely,

$$\frac{f(b) - f(a)}{b - a},$$

is the slope of the line joining the points $(a, f(a))$ and $(b, f(b))$ on the graph of $f$. Examples are shown in Figure 11.18.

**Figure 11.18**

## Instantaneous Rate of Change

Suppose a car is stopped at a traffic light. When the light turns green, the car begins to move along a straight road. Assume that the distance traveled by the car is given by the function

$$s(t) = 2t^2 \qquad (0 \le t \le 30),$$

where time $t$ is measured in seconds and the distance $s(t)$ at time $t$ is measured in feet. We know how to find the *average* speed of the car over any time interval, so we now turn to a different problem: determining the *exact* speed of the car at a particular instant—say, $t = 10$.[*]

The intuitive idea is that the exact speed at $t = 10$ is very close to the average speed over a very short time interval near $t = 10$. If we take shorter and shorter time intervals near $t = 10$, the average speeds over these intervals should get closer and closer to the exact speed at $t = 10$. In other words,

> the exact speed at $t = 10$ is the limit of the average speeds over shorter and shorter time intervals near $t = 10$.

The following chart illustrates this idea:

| Intervals | Average Speed |
|---|---|
| $t = 10$ to $t = 10.1$ | $\dfrac{s(10.1) - s(10)}{10.1 - 10} = \dfrac{204.02 - 200}{.1} = 40.2 \text{ ft/sec}$ |
| $t = 10$ to $t = 10.01$ | $\dfrac{s(10.01) - s(10)}{10.01 - 10} = \dfrac{200.4002 - 200}{.01} = 40.02 \text{ ft/sec}$ |
| $t = 10$ to $t = 10.001$ | $\dfrac{s(10.001) - s(10)}{10.001 - 10} = \dfrac{200.040002 - 200}{.001} = 40.002 \text{ ft/sec}$ |

The chart suggests that the exact speed at $t = 10$ is 40 ft/sec. We can confirm this intuition by computing the average speed between $t = 10$ and $t = 10 + h$, where $h$ is any very small (positive or negative) nonzero number. (The chart does this for $h = .1$, $h = .01$, and $h = .001$.) The average speed between $t = 10$ and $t = 10 + h$ is

$$
\begin{aligned}
\frac{s(10 + h) - s(10)}{(10 + h) - 10} &= \frac{s(10 + h) - s(10)}{h} \\
&= \frac{2(10 + h)^2 - 2 \cdot 10^2}{h} \\
&= \frac{2(100 + 20h + h^2) - 200}{h} \qquad (a + b)^2 = a^2 + 2ab + b^2 \\
&= \frac{200 + 40h + 2h^2 - 200}{h} \\
&= \frac{40h + 2h^2}{h} = \frac{h(40 + 2h)}{h} \\
&= 40 + 2h. \qquad\qquad\qquad (h \neq 0)
\end{aligned}
$$

Saying that the time interval between $t = 10$ and $t = 10 + h$, gets shorter and shorter is equivalent to saying that $h$ gets closer and closer to 0 (from either side). Hence, the exact speed at $t = 10$ is the limit, as $h$ approaches 0, of the average speeds over the intervals between $t = 10$ and $t = 10 + h$; that is,

$$
\lim_{h \to 0} \frac{s(10 + h) - s(10)}{h} = \lim_{h \to 0} (40 + 2h)
$$
$$
= 40 \text{ ft/sec.}
$$

In the preceding example, the car moved in one direction along a straight line. Now suppose that an object is moving back and forth along a number line, with its position at time $t$ given by the function $s(t)$. The **velocity** of the object is the rate of motion in the direction in which it is moving. **Speed** is the absolute value of velocity. For example, a velocity of $-30$ ft/sec indicates a speed of 30 ft/sec in the negative direction, while a velocity of 40 ft/sec indicates a speed of 40 ft/sec in the positive direction.

---

[*]As distance is measured in feet and time in seconds here, speed is measured in feet per second. It may help to know that 15 mph is equivalent to 22 ft/sec and 60 mph to 88 ft/sec.

Hereafter, we deal with **average velocity** and **instantaneous velocity** instead of average speed and exact speed, respectively. When the term "velocity" is used alone, it means instantaneous velocity.

Let $a$ be a fixed number. By replacing 10 by $a$ in the preceding discussion, we see that the average velocity of the object between time $t = a$ and time $t = a + h$; is the quotient

$$\frac{s(a + h) - s(a)}{(a + h) - a} = \frac{s(a + h) - s(a)}{h}.$$

The instantaneous velocity at time $a$ is the limit of this quotient as $h$ approaches 0.

## Velocity

If an object moves along a straight line, with position $s(t)$ at time $t$, then the **velocity** of the object at $t = a$ is

$$\lim_{h \to 0} \frac{s(a + h) - s(a)}{h},$$

provided that this limit exists.

### EXAMPLE 3   Average and Instantaneous Velocity

The position, in feet, of an object moving along a straight line is given by $s(t) = 2t^2 - 5t + 40$, where $t$ is time in seconds.

**(a)** Find the average velocity of the object from 2 seconds to 4 seconds.

**Solution**   The average velocity is

$$\frac{s(4) - s(2)}{4 - 2} = \frac{52 - 38}{2} = \frac{14}{2} = 7$$

feet per second.

**(b)** Find the instantaneous velocity at 4 seconds.

**Solution**   For $t = 4$, the instantaneous velocity is

$$\lim_{h \to 0} \frac{s(4 + h) - s(4)}{h}$$

feet per second. We have

$$s(4 + h) = 2(4 + h)^2 - 5(4 + h) + 40$$
$$= 2(16 + 8h + h^2) - 20 - 5h + 40$$
$$= 32 + 16h + 2h^2 - 20 - 5h + 40$$
$$= 2h^2 + 11h + 52$$

and

$$s(4) = 2(4)^2 - 5(4) + 40 = 52.$$

Thus,

$$s(4 + h) - s(4) = (2h^2 + 11h + 52) - 52 = 2h^2 + 11h,$$

and the instantaneous velocity at $t = 4$ is

$$\lim_{h \to 0} \frac{2h^2 + 11h}{h} = \lim_{h \to 0} \frac{h(2h + 11)}{h}$$
$$= \lim_{h \to 0} (2h + 11) = 11 \text{ ft/sec.}$$

✓ **Checkpoint 4**

In Example 3, if $s(t) = t^2 + 3$, find

**(a)** the average velocity from 1 second to 5 seconds;

**(b)** the instantaneous velocity at 5 seconds.

**EXAMPLE 4** **Blood Flow** The velocity of blood cells is of interest to physicians; a velocity slower than normal might indicate a constriction, for example. Suppose the position of a red blood cell in a capillary is given by

$$s(t) = 1.2t + 5,$$

where $s(t)$ gives the position of the cell in millimeters from some reference point and $t$ is time in seconds. Find the velocity $v(a)$ of this cell at time $t = a$.

**Solution** We have $s(a) = 1.2a + 5$. To find $s(a + h)$, substitute $a + h$ for the variable $t$ in $s(t) = 1.2t + 5$:

$$s(a + h) = 1.2(a + h) + 5.$$

Now use the definition of velocity:

$$v(a) = \lim_{h \to 0} \frac{s(a + h) - s(a)}{h}$$

$$= \lim_{h \to 0} \frac{1.2(a + h) + 5 - (1.2a + 5)}{h}$$

$$= \lim_{h \to 0} \frac{1.2a + 1.2h + 5 - 1.2a - 5}{h} = \lim_{h \to 0} \frac{1.2h}{h} = 1.2 \text{ mm/sec.}$$

The velocity of the blood cell at $t = a$ is 1.2 millimeters per second, regardless of the value of $a$. In other words, the blood velocity is a constant 1.2 millimeters per second at any time. ✓5

✓ **Checkpoint 5**

Repeat Example 4 with $s(t) = .3t - 2$.

The ideas underlying the concept of the velocity of a moving object can be extended to any function $f$. In place of average velocity between time $t = a$ and time $t = a + h$, we have **the average rate of change** of $f(x)$ with respect to $x$ between $x = a$ and $x = a + h$, which is the quotient

$$\text{average rate of change} = \frac{f(a + h) - f(a)}{h}.$$

The **instantaneous rate of change** of $f(x)$ at $x = a$ is the limit of this quotient as $h \to 0$, as summarized in the following box.

---

## Instantaneous Rate of Change

The **instantaneous rate of change** for a function $f$ when $x = a$ is

$$\lim_{h \to 0} \frac{f(a + h) - f(a)}{h},$$

provided that this limit exists.

---

**EXAMPLE 5** **Manufacturing Costs** A company determines that the cost (in hundreds of dollars) of manufacturing $x$ cases of USB drives is

$$C(x) = -.2x^2 + 8x + 40 \qquad (0 \le x \le 20).$$

**(a)** Find the average rate of change of cost for manufacturing between 5 and 10 cases.

**Solution** Use the formula for average rate of change. The cost to manufacture 5 cases is

$$C(5) = -.2(5^2) + 8(5) + 40 = 75,$$

or $7500. The cost to manufacture 10 cases is

$$C(10) = -.2(10^2) + 8(10) + 40 = 100,$$

or $10,000. The average rate of change of cost is

$$\frac{C(10) - C(5)}{10 - 5} = \frac{100 - 75}{5} = 5.$$

Thus, on the average, cost is increasing at the rate of $500 per case when production is increased from 5 to 10 cases.

**(b)** Find the instantaneous rate of change with respect to the number of cases produced when 5 cases are produced.

**Solution** The instantaneous rate of change when $x = 5$ is given by

$$\lim_{h \to 0} \frac{C(5 + h) - C(5)}{h}$$

$$= \lim_{h \to 0} \frac{[-.2(5 + h)^2 + 8(5 + h) + 40] - [-.2(5^2) + 8(5) + 40]}{h}$$

$$= \lim_{h \to 0} \frac{[-5 - 2h - .2h^2 + 40 + 8h + 40] - [75]}{h}$$

$$= \lim_{h \to 0} \frac{6h - .2h^2}{h} \quad \text{Combine terms.}$$

$$= \lim_{h \to 0} (6 - .2h) \quad \text{Divide by } h.$$

$$= 6. \quad \text{Calculate the limit.}$$

When 5 cases are manufactured, the cost is increasing at the rate of $600 per case.

The rate of change of the cost function is called the **marginal cost**.[*] Similarly, **marginal revenue** and **marginal profit** are the rates of change of the revenue and profit functions, respectively. Part (b) of Example 5 shows that the marginal cost when 5 cases are manufactured is $600 per case. Marginal analysis is an important application of calculus to business and will be discussed in detail in Section 11.5.

**EXAMPLE 6** **Health Professions** The total number of bachelor's degrees (in thousands) earned in the health professions in year $t$ can be approximated by

$$f(t) = 52e^{.094t} \qquad (3 \le t \le 15),$$

where $t = 3$ corresponds to the year 2003. (Data from: National Center for Education Statistics.)

**(a)** Find the instantaneous rate of change of the number of health professions bachelor's degrees in 2005.

**Solution** Because $t = 3$ corresponds to the year 2003, we must find the instantaneous rate of change of $f(t)$ at $t = 5$. The algebraic techniques used in the preceding examples will not work with an exponential function, but the rate of change can be approximated by a graphing calculator or spreadsheet program. The average rate of change from 5 to $5 + h$ is:

$$\text{average rate of change} = \frac{f(5 + h) - f(5)}{h} = \frac{52e^{.094(5+h)} - 52e^{.094(5)}}{h}.$$

---
[*]Marginal cost for linear cost functions was discussed in in Section 3.3.

| X | Y1 |
|---|---|
| -.001 | 7.8204 |
| -1E-4 | 7.8207 |
| -1E-5 | 7.8208 |
| 0 | ERROR |
| 1E-5 | 7.8208 |
| 1E-4 | 7.8208 |
| .001 | 7.8211 |

X=.001

**Figure 11.19**

To approximate the instantaneous rate of change, we evaluate this quantity for very small values of $h$, as shown in Figure 11.19 (in which X is used in place of $h$, and Y1 is the average rate of change). The table suggests that, at $t = 5$, the instantaneous rate of change can be approximated by:

$$\text{instantaneous rate of change} = \lim_{h \to 0} \frac{f(5 + h) - f(5)}{h} = \lim_{h \to 0} \frac{52e^{.094(5+h)} - 52e^{.094(5)}}{h}$$

$$\approx 7.820.$$

Therefore, the number of bachelor's degrees in the health professions was increasing at a rate of about 7820 degrees per year in 2005.

**(b)** Find the instantaneous rate of change in 2015.

| X | Y1 |
|---|---|
| -.001 | 20.02 |
| -1E-4 | 20.021 |
| -1E-5 | 20.021 |
| 0 | ERROR |
| 1E-5 | 20.021 |
| 1E-4 | 20.021 |
| .001 | 20.022 |

X=.001

**Figure 11.20**

**Solution** The instantaneous rate of change of $f(t)$ when $t = 15$ is

$$\lim_{h \to 0} \frac{f(15 + h) - f(15)}{h} = \lim_{h \to 0} \frac{52e^{.094(15+h)} - 52e^{.094(15)}}{h} \approx 20.021,$$

as shown in Figure 11.20. In 2015, the number of health professions degrees was increasing at a rate of about 20,021 degrees per year.

## 11.3 Exercises

*Find the average rate of change for the given functions. (See Example 1.)*

1. $f(x) = x^2 + 2x$ between $x = 0$ and $x = 6$

2. $f(x) = -4x^2 - 6$ between $x = 1$ and $x = 7$

3. $f(x) = 2x^3 - 4x^2 + 6$ between $x = -1$ and $x = 2$

4. $f(x) = -3x^3 + 2x^2 - 4x + 2$ between $x = 0$ and $x = 2$

5. $f(x) = \sqrt{x}$ between $x = 1$ and $x = 9$

6. $f(x) = \sqrt{3x - 2}$ between $x = 2$ and $x = 6$

7. $f(x) = \dfrac{1}{x - 1}$ between $x = -2$ and $x = 0$

8. $f(x) = .4525\, e^{1.556\sqrt{x}}$ between $x = 4$ and $x = 4.5$

*Work these exercises. (See Examples 2 and 3.)*

9. **Catalog Sales** The accompanying graph shows the total sales, in thousands of dollars, from the distribution of $x$ thousand catalogs. Find and interpret the average rate of change of sales with respect to the number of catalogs distributed for the given changes in $x$.

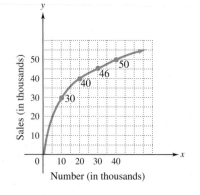

Number (in thousands)

(a) 10 to 20

(b) 20 to 30

(c) 30 to 40

(d) What is happening to the average rate of change of sales as the number of catalogs distributed increases?

(e) Explain why what you have found in part (d) might happen.

10. **Health** The accompanying graph shows the relationship between waist circumference (in inches) and weight (in pounds) for American males (blue) and females (red). Find the average rate of change in weight for changes in waist circumference from

(a) 32 to 44 inches for males;

(b) 28 to 40 inches for females.

(c) Do we need to calculate the rate of change at any other points for males or females? Explain. (Data from: National Health and Nutrition Examination Study (NHANES) 2005–2006.)

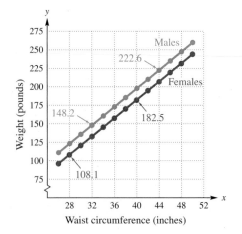

Waist circumference (inches)

11. **S&P 500 Index** The accompanying graph gives the value of the Standard and Poor's 500 Index at the start of each year over a ten-year period. (Data from: www.morningstar.com.)

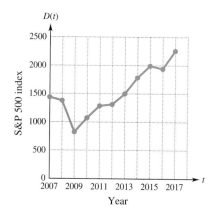

D(t)

S&P 500 index

Year

(a) Which time period has the greater average rate of change: 2007 to 2009, 2011 to 2012, or 2013 to 2016?

(b) Which time period has the greater average rate of change: 2010 to 2013 or 2013 to 2015?

(c) Which time period has the greater average rate of change: 2008 to 2009 or 2015 to 2016?

12. **Facebook Users** The accompanying figure shows the number of Facebook users (in millions) from its launch in 2004 until the start of 2017. (Data from: Facebook.)

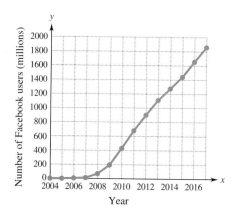

Number of Facebook users (millions)

Year

(a) Approximately when did Facebook reach 200 million users?

(b) Had Facebook exceeded 1 billion users by the start of 2013?

(c) Which time period had the greatest average rate of change: 2004 to 2009, 2009 to 2016, or 2004 to 2016?

13. **Apple Employees and Revenue** The accompanying figure shows the number of full-time employees of Apple, Inc. (in thousands) as a function of the revenue generated by Apple, Inc. (in billions). Find and interpret the average rate of change of employees with respect to revenue for the following changes in revenue. (Data from: www.morningstar.com.)

(a) $5 billion to $108 billion

(b) $108 billion to $234 billion

(c) $5 billion to $234 billion

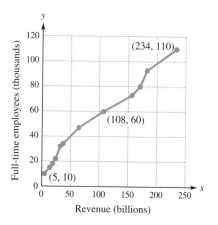

Full-time employees (thousands)

(234, 110)

(108, 60)

(5, 10)

Revenue (billions)

14. **Postsecondary Enrollment** The accompanying graph shows enrollment (in millions) in degree-granting public and private postsecondary institutions (colleges and universities) in the United States, including projections through the year 2022. (Data from: National Center for Educational Statistics.)

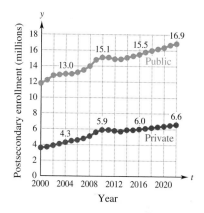

Postsecondary enrollment (millions)

16.9

15.1   15.5

13.0         Public

5.9   6.0   6.6
4.3          Private

Year

(a) Find the projected average rate of change in enrollment at public institutions over the period 2010 to 2022.

(b) Find the projected average rate of change in enrollment at private institutions over the period 2016 to 2022.

15. **Velocity** A car is moving along a straight test track. The position (in feet) of the car, $s(t)$, at various times $t$ is measured, with the following results:

| $t$ (seconds) | 0 | 2 | 4 | 6 | 8 | 10 |
|---|---|---|---|---|---|---|
| $s(t)$ (feet) | 0 | 10 | 14 | 20 | 30 | 36 |

Find and interpret the average velocities for the following changes in $t$:

(a) 0 to 2 seconds;

(b) 2 to 4 seconds;

(c) 4 to 6 seconds;

(d) Estimate the instantaneous velocity at 4 seconds

  (i) by using the formula for estimating the instantaneous rate (with $h = 2$) and

  (ii) by averaging the answers for the average velocity in the 2 seconds before and the 2 seconds afterwards (that is, the answers to parts (b) and (c)).

**16. Velocity** Use the table of data in the previous exercise to find and interpret the average velocity for the following changes in $t$:

(a) 6 to 8 seconds;

(b) 8 to 10 seconds.

(c) Estimate the instantaneous velocity at 8 seconds

(i) by using the formula for estimating the instantaneous rate (with $h = 2$) and

(ii) by averaging the answers for the average velocity in the 2 seconds before and the 2 seconds afterward (that is, the answers to parts (a) and (b)).

**17.** Explain the difference between the average rate of change of $y = f(x)$ as $x$ changes from $a$ to $b$ and the instantaneous rate of change of $y$ at $x = a$.

**18.** If the instantaneous rate of change of $f(x)$ with respect to $x$ is positive when $x = 1$, is $f$ increasing or decreasing there?

*Exercises 19–22 deal with a car moving along a straight road. (See Example 3.) At time $t$ seconds, the position of the car (in feet) is $s(t) = 3t^2$.*

**19.** Find the average velocity of the car over the time interval from

(a) $t = 10$ to $t = 10.1$ seconds;

(b) $t = 10$ to $t = 10.01$ seconds;

(c) $t = 10$ to $t = 10.001$ seconds.

**20.** Find the average velocity of the car over the time interval from

(a) $t = 5$ to $t = 5.1$ seconds;

(b) $t = 5$ to $t = 5.01$ seconds;

(c) $t = 5$ to $t = 5.001$ seconds.

**21.** Find the instantaneous velocity of the car at $t = 10$ seconds. Compare your answer with Exercise 19.

**22.** Find the instantaneous velocity of the car at $t = 5$ seconds. Compare your answer with Exercise 20.

*An object moves along a straight line; its position (in feet) at time $t$ seconds is $s(t) = t^2 + 4t + 3$. Find the instantaneous velocity of the object at the given times. (See Example 3.)*

**23.** $t = 1$

**24.** $t = 6$

*In Exercises 25–30, find (a) $f(a + h)$; (b) $\dfrac{f(a + h) - f(a)}{h}$; (c) the instantaneous rate of change of $f$ when $a = 5$. (See Examples 3–5.)*

**25.** $f(x) = x^2 - 3x$

**26.** $f(x) = x^2 + x$

**27.** $f(x) = x^2 - x - 1$

**28.** $f(x) = x^2 + 2x + 2$

**29.** $f(x) = x^3$

**30.** $f(x) = x^3 - x$

*Solve Exercises 31–34 by algebraic methods. (See Examples 3–5.)*

**31. Revenue** The revenue (in thousands of dollars) from producing $x$ units of an item is

$$R(x) = 10x - .002x^2.$$

(a) Find the average rate of change of revenue when production is increased from 1000 to 1001 units.

(b) Find the marginal revenue (the instantaneous rate of change of the revenue function) when 1000 units are produced.

**32. Demand Function** Suppose customers in a hardware store are willing to buy $N(p)$ boxes of nails at $p$ dollars per box, as given by

$$N(p) = 80 - 5p^2 \quad (1 \le p \le 4).$$

(a) Find the average rate of change of demand for a change in price from \$2 to \$3.

(b) Find the instantaneous rate of change of demand when the price is \$2.

(c) Find the instantaneous rate of change of demand when the price is \$3.

**(d)** As the price is increased from \$2 to \$3, how is demand changing? Is the change to be expected?

**33. Epidemiology** Epidemiologists in College Station, Texas, estimate that $t$ days after the flu begins to spread in town, the percentage of the population infected by the flu is approximated by

$$p(t) = t^2 + t \quad (0 \le t \le 5).$$

(a) Find the average rate of change of $p$ with respect to $t$ over the interval from 1 to 4 days.

(b) Find the instantaneous rate of change of $p$ with respect to $t$ at $t = 3$.

**34. Heart Rate** The maximum recommended heart rate (in beats per minute) is given by $h(x) = 220 - x$ where $x$ represents the age of the individual. (Data from: American Heart Association.)

(a) Find and interpret the average rate of change of $h(x)$ with respect to $x$ for $18 \le x \le 28$ years.

(b) Find and interpret the instantaneous rate of change of $h(x)$ with respect to $x$ at $x = 23$ years.

(c) Explain why your answers in parts (a) and (b) are the same in this case.

*Use technology to work Exercises 35–38. (See Example 6.)*

**35. Unemployment by Age for Women** Unemployment varies by age for both men and women. The unemployment rate (as a percentage) for women as a function of age can be approximated by

$$f(t) = 21e^{-.03t},$$

where $t$ is age in years between 16 and 64 years old. Use a graphing calculator to estimate the instantaneous rate of change of the unemployment rate for women with respect to age at the given age (give your answer to two decimal places). (Data from: U.S. Bureau of Labor Statistics, 2014 Current Population Survey.)

(a) 20 years old

(b) 50 years old

**36. Unemployment by Age for Men** The unemployment rate (as a percentage) for men as a function of age can be approximated by

$$f(t) = 25e^{-.034t},$$

where $t$ is age in years between 16 and 64 years old. Use a graphing calculator to estimate the instantaneous rate of change of the unemployment rate for men with respect to age at the given age (give your answer to two decimal places). (Data from: U.S. Bureau of Labor Statistics, 2014 Current Population Survey.)

(a) 25 years old

(b) 45 years old

**37. Metabolic Rate** The metabolic rate of a person who has just eaten a meal tends to go up and then, after some time has passed, returns to a resting metabolic rate. This phenomenon is known as the *thermic effect of food.* Researchers have indicated that the thermic effect of food (in kJ/hr) for a particular person is

$$F(x) = -10.28 + 175.9xe^{-x/1.3},$$

where $x$ is the number of hours that have elapsed since eating a meal.[*]

(a) Graph the given function in a window with $0 \leq x \leq 6$ and $-20 \leq y \leq 100$.

(b) Find the average rate of change of the thermic effect of food during the first hour after eating.

(c) Use a graphing calculator to find the instantaneous rate of change of the thermic effect of food exactly 1 hour after eating.

(d) Use a graphing calculator to estimate when the function stops increasing and begins to decrease.

**38. Crown Length** The mesiodistal crown length (as shown in the accompanying diagram) of deciduous mandibular first molars in fetuses is related to the postconception age of the tooth as

$$L(x) = -.01x^2 + .788x - 7.048,$$

where $L(x)$ is the crown length, in millimeters, of the molar $x$ weeks after conception.[†]

---

[*]Reed, G. and J. Hill, "Measuring the Thermic Effect of Food," *American Journal of Clinical Nutrition,* Vol. 63, 1996, pp. 164–169.)

[†]Harris, E. F., J. D. Hicks, and B. D. Barcroft, "Tissue Contributions to Sex and Race: Differences in Tooth Crown Size of Deciduous Molars," *American Journal of Physical Anthropology,* Vol. 115, 2001, pp. 223–237.

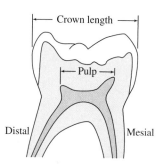

(a) Find the average rate of growth in mesiodistal crown length during weeks 22 through 28.

(b) Find the instantaneous rate of growth in mesiodistal crown length when the tooth is exactly 22 weeks of age.

(c) Graph the given function in a window with $0 \leq x \leq 50$ and $0 \leq y \leq 9$.

(d) Does a function that increases and then begins to decrease make sense for this particular application? What do you suppose is happening during the first 11 weeks? Does this function accurately model crown length during those weeks?

✓**Checkpoint Answers**

1. (a) 64 mph       (b) $\dfrac{f(r) - f(s)}{r - s}$ mph

2. (a) 8       (b) 13

3. (a) Changing at the average rate of $-1720$ index points per year
   (b) Changing at the average rate of about 1159.1 index points per year

4. (a) 6 ft per second       (b) 10 ft per second

5. The velocity is .3 millimeter per second.

# 11.4 Tangent Lines and Derivatives

We now develop a geometric interpretation of the rates of change considered in the previous section. In geometry, a tangent line to a circle at a point $P$ is defined to be the line through $P$ that is perpendicular to the radius $OP$, as in Figure 11.21 (which shows only the top half of the circle). If you think of this circle as a road on which you are driving at night, then the tangent line indicates the direction of the light beam from your headlights as you pass through the point $P$. This analogy suggests a way of extending the idea of a tangent line to any curve: the tangent line to the curve at a point $P$ indicates the "direction" of the curve as it passes through $P$. Using this intuitive idea of direction, we see, for example, that the lines through $P_1$ and $P_3$ in Figure 11.22 appear to be tangent lines, whereas the lines through $P_2$ and $P_4$ do not.

**Figure 11.21**

**Figure 11.22**

We can use these ideas to develop a precise definition of the tangent line to the graph of a function $f$ at the point $R$. As shown in Figure 11.23, choose a second point $S$ on the graph and draw the line through $R$ and $S$; this line is called a **secant line** ($S$ may lie on either side of $R$). You can think of this secant line as a rough approximation of the tangent line.

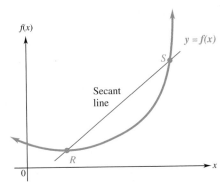

**Figure 11.23**

Now suppose that the point $S$ slides down the curve closer to $R$. Figure 11.24 below shows successive positions $S_2$, $S_3$, and $S_4$ of the point $S$. The closer $S$ gets to $R$, the better the secant line $RS$ approximates our intuitive idea of the tangent line at $R$.

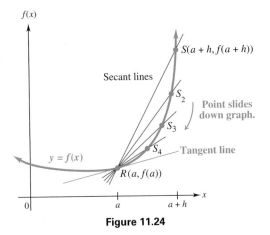

**Figure 11.24**

In particular, the closer $S$ gets to $R$, the closer the slope of the secant line gets to the slope of the tangent line. Informally, we say that

the slope of tangent line at $R$ = the limit of the slope of secant line $RS$ as

$S$ gets closer and closer to $R$.

In order to make this statement more precise, suppose the first coordinate of $R$ is $a$. Then the first coordinate of $S$ can be written as $a + h$ for some (positive or negative) number $h$. (In Figure 11.24, $h$ is the distance on the $x$-axis between the two first coordinates.) Thus, $R$ has coordinates $(a, f(a))$ and $S$ has coordinates $(a + h, f(a + h))$, as shown in Figure 11.24. Consequently, the slope of the secant line $RS$ is

$$\frac{f(a + h) - f(a)}{(a + h) - a} = \frac{f(a + h) - f(a)}{h}.$$

Now, as $S$ moves closer to $R$, their first coordinates move closer to each other, that is, $h$ gets smaller and smaller. Hence,

$$\text{slope of tangent line at } R = \text{limit of slope of secant line } RS$$
$$\text{as } S \text{ gets closer and closer to } R$$
$$= \text{limit of } \frac{f(a+h) - f(a)}{h}$$
$$\text{as } h \text{ gets closer and closer to } 0$$
$$= \lim_{h \to 0} \frac{f(a+h) - f(a)}{h}.$$

This intuitive development suggests the following formal definition.

## Tangent Line

The **tangent line** to the graph of $y = f(x)$ at the point $(a, f(a))$ is the line through this point having slope

$$\lim_{h \to 0} \frac{f(a+h) - f(a)}{h},$$

provided that this limit exists. If this limit does not exist, then there is no tangent line at the point.

The slope of the tangent line at a point is also called the **slope of the curve** at that point. Since the slope of a line indicates its direction (see the table on page 78), the slope of the tangent line at a point indicates the direction of the curve at that point. For a review of slope and equations of lines, see Section 2.2.

**EXAMPLE 1**  Consider the graph of $f(x) = x^2 + 2$.

**(a)** Find the slope of the tangent line to the graph when $x = -1$.

**Solution** Use the preceding definition with $f(x) = x^2 + 2$ and $a = -1$. The slope of the tangent line is calculated as follows:

$$\text{Slope of tangent line} = \lim_{h \to 0} \frac{f(a+h) - f(a)}{h}$$
$$= \lim_{h \to 0} \frac{[(-1+h)^2 + 2] - [(-1)^2 + 2]}{h}$$
$$= \lim_{h \to 0} \frac{[1 - 2h + h^2 + 2] - [1 + 2]}{h}$$
$$= \lim_{h \to 0} \frac{-2h + h^2}{h} = \lim_{h \to 0} \frac{h(-2+h)}{h}$$
$$= \lim_{h \to 0} (-2 + h) = -2. V$$

So the slope of the tangent line at $x = -1$ is $m = -2$.

**(b)** Find the equation of the tangent line when $x = -1$.

**Solution** Recall from Chapter 2 that we can find the equation of a line using point–slope form. The $x$-value of the point on the tangent line is given to be $x = -1$. The corresponding $y$-value is the value of the function when $x = -1$; namely, $y = f(-1) = (-1)^2 + 2 = 3$. So the tangent line passes through the point $(-1, 3)$. In part (a), we already found the slope of the tangent line when $x = -1$ to be $m = -2$. So now we can find the equation of the tangent line using the point-slope form of the equation of a line (see page 81):

$$y - y_1 = m(x - x_1) \qquad \text{Point-slope form}$$
$$y - 3 = -2[x - (-1)] \qquad \text{Substitute.}$$
$$y - 3 = -2(x + 1) \qquad \text{Simplify.}$$
$$y - 3 = -2x - 2 \qquad \text{Distributive property}$$
$$y = -2x + 1. \qquad \text{Solve for } y.$$

Figure 11.25 shows a graph of $f(x) = x^2 + 2$ along with a graph of the tangent line at $x = -1$. ✔1

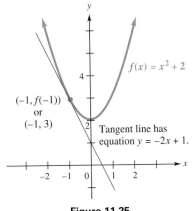

**Figure 11.25**

✔️ **Checkpoint 1**

Let $f(x) = x^2 + 2$.
Find the equation of the tangent line to the graph at the point where $x = 1$.

🖩 **TECHNOLOGY TIP** When finding the equation of a tangent line algebraically, you can confirm your answer with a graphing calculator by graphing both the function and the tangent line on the same screen to see if the tangent line appears to be correct. ✔2

✔️ **Checkpoint 2**

Use a graphing calculator to confirm your answer to Checkpoint 1 by graphing $f(x) = x^2 + 2$ and the tangent line at the point where $x = 1$ on the same screen.

Most graphing calculators can draw the tangent line at any specified point. On the TI-84 Plus, to graph the function and tangent line shown in Example 1, select the DRAW menu (2nd PRGM), then enter TANGENT ($x^2 + 2, -1$). Alternatively, first enter the function in the Y= menu and press GRAPH, then select TANGENT in the DRAW menu and enter $-1$ (the $x$-value). The calculator will display the equation of the tangent line along with the graph. (On some other TI calculators, the DRAW menu is a submenu of the GRAPH menu.)

In Figure 11.25, the tangent line $y = -2x + 1$ appears to coincide with the graph of $f(x) = x^2 + 2$ near the point $(-1, 3)$. This becomes more obvious when both $f(x)$ and the tangent line are graphed in a very small window on a graphing calculator (Figure 11.26). The graph and the tangent line now appear virtually identical near $(-1, 3)$. The table of values in Figure 11.27 (in which $Y_1$ is $f(x)$ and $Y_2$ is the $y$-value of the tangent line) confirms the fact that the tangent line is a good approximation of the function when $x$ is very close to $-1$.

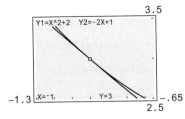

**Figure 11.26**

| X | Y1 | Y2 |
|------|--------|------|
| -1.1 | 3.21 | 3.2 |
| -1.05 | 3.1025 | 3.1 |
| -1.02 | 3.0404 | 3.04 |
| -1 | 3 | 3 |
| -.97 | 2.9409 | 2.94 |
| -.94 | 2.8836 | 2.88 |
| -.9 | 2.81 | 2.8 |

X=-.9

**Figure 11.27**

The same thing is true in the general case.

## Tangent Line

If it exists, the tangent line to the graph of a function $f$ at $x = a$ is a good approximation of the function $f$ near $x = a$.

Suppose the graph of a function $f$ is a straight line. The fact in the preceding box suggests that the tangent line to $f$ at any point should be the graph of $f$ itself (because it is certainly the best possible linear approximation of $f$). The next example shows that this is indeed the case.

**EXAMPLE 2**    Let $a$ be any real number. Find the equation of the tangent line to the graph of $f(x) = 7x + 3$ at the point where $x = a$.

**Solution**  According to the definition, the slope of the tangent line is

$$\lim_{h \to 0} \frac{f(a + h) - f(a)}{h} = \lim_{h \to 0} \frac{[7(a + h) + 3] - [7a + 3]}{h}$$

$$= \lim_{h \to 0} \frac{[7a + 7h + 3] - 7a - 3}{h}$$

$$= \lim_{h \to 0} \frac{7h}{h} = \lim_{h \to 0} 7 = 7.$$

Hence, the equation of the tangent line at the point $(a, f(a))$ is

$$y - y_1 = m(x - x_1) \qquad \text{Point–slope form}$$
$$y - f(a) = 7(x - a)$$
$$y = 7x - 7a + \overbrace{f(a)}$$
$$y = 7x - 7a + \overbrace{7a + 3}$$
$$y = 7x + 3.$$

Thus, the tangent line *is* the graph of the function $f(x) = 7x + 3$.

Secant lines and tangent lines (or, more precisely, their slopes) are the geometric analogues of the average and instantaneous rates of change studied in the previous section, as summarized in the following chart:

| Quantity | Algebraic Interpretation | Geometric Interpretation |
|---|---|---|
| $\dfrac{f(a + h) - f(a)}{h}$ | Average rate of change of $f$ from $x = a$ to $x = a + h$ | Slope of the secant line through $(a, f(a))$ and $(a + h, f(a + h))$ |
| $\lim\limits_{h \to 0} \dfrac{f(a + h) - f(a)}{h}$ | Instantaneous rate of change of $f$ at $x = a$ | Slope of the tangent line to the graph of $f$ at $(a, f(a))$ |

## The Derivative

Finding the instantaneous rate of change of a function has many important applications; therefore, we define the following special notation and terminology. If $y = f(x)$ is a function and $a$ is a number in its domain, then we shall use the symbol $f'(a)$ to denote the special limit

$$\lim_{h \to 0} \frac{f(a + h) - f(a)}{h},$$

provided that it exists. In other words, to each number $a$, we can assign the number $f'(a)$ obtained by calculating the preceding limit. This process defines an important new function.

## Derivative

The **derivative** of the function $f$ is the function denoted $f'$ whose value at the number $x$ is defined to be the number

$$f'(x) = \lim_{h \to 0} \frac{f(x + h) - f(x)}{h},$$

provided that this limit exists.

The derivative function $f'$ has as its domain all the points at which the specified limit exists, and the value of the derivative function at the number $x$ is the number $f'(x)$. Using $x$ instead of $a$ here is similar to the way that $g(x) = 2x$ denotes the function that assigns to each number $a$ the number $2a$.

If $y = f(x)$ is a function, then its derivative is denoted either by $f'$ or by $y'$. If $x$ is a number in the domain of $y = f(x)$ such that $y' = f'(x)$ is defined, then the function $f$ is said to be **differentiable** at $x$. The process that produces the function $f'$ from the function $f$ is called **differentiation.**

The derivative function may be interpreted in many ways, two of which were already discussed:

1. The derivative function $f'$ gives the *instantaneous rate of change* of $y = f(x)$ with respect to $x$. This instantaneous rate of change can be interpreted as marginal cost, marginal revenue, or marginal profit (if the original function represents cost, revenue, or profit, respectively) or as velocity (if the original function represents displacement along a line). From now on, we will use "rate of change" to mean "instantaneous rate of change."

2. The derivative function $f'$ gives the *slope* of the graph of $f$ at any point. If the derivative is evaluated at $x = a$, then $f'(a)$ is the slope of the tangent line to the curve at the point $(a, f(a))$.

---

**EXAMPLE 3**    Use the graph of the function $f(x)$ in Figure 11.28 to answer the given questions.

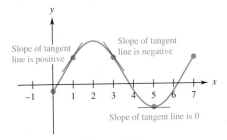

**Figure 11.28**

**(a)** Is $f'(3)$ positive or negative?

**Solution** We know that $f'(3)$ is the slope of the tangent line to the graph at the point where $x = 3$. Figure 11.28 shows that this tangent line slants downward from left to right, meaning that its slope is negative. Hence, $f'(3) < 0$.

**(b)** Which is larger, $f'(1)$ or $f'(5)$?

**Solution** Figure 11.28 shows that the tangent line to the graph at the point where $x = 1$ slants upward from left to right, meaning that its slope, $f'(1)$, is a positive number. The tangent line at the point where $x = 5$ is horizontal, so that it has slope 0. (That is, $f'(5) = 0$). Therefore, $f'(1) > f'(5)$.

**11.4** Tangent Lines and Derivatives **599**

✓**Checkpoint 3**

The graph of a function $g$ is shown. Determine whether the given numbers are positive, negative, or zero.

(a) $g'(0)$

(b) $g'(-1)$

(c) $g'(3)$

(c) For what values of $x$ is $f'(x)$ positive?

**Solution** On the graph, find the points where the tangent line has positive slope (slants upward from left to right). At each such point, $f'(x) > 0$. Figure 11.28 shows that this occurs when $0 < x < 2$ and when $5 < x < 7$. ✓3

**EXAMPLE 4** **Natural Science** A student brings a cold soft drink to a 50-minute math class but is too busy during class to drink it. If $C(t)$ represents the temperature of the soft drink (in degrees Fahrenheit) $t$ minutes after the start of class, interpret the meaning of the following statements, including units.

(a) Interpret $C(0) = 38$.

**Solution** Since $C(t)$ represents the temperature of the drink at time $t$, $C(0) = 38$ means that the temperature of the drink was 38°F at the start of class.

(b) Interpret $C(50) - C(0) = 30$.

**Solution** Since $C(b) - C(a)$ represents the change in temperature from time $t = a$ to time $t = b$ minutes, $C(50) - C(0) = 30$ means that the drink warmed up by 30°F between the start and end of class.

(c) Interpret $\dfrac{C(50) - C(0)}{50 - 0} = .6$.

**Solution** Since $\dfrac{C(b) - C(a)}{b - a}$ represents the average rate of change of $C(t)$ from time $t = a$ to time $t = b$ minutes, $\dfrac{C(50) - C(0)}{50 - 0} = .6$ means that the drink warmed up at an average rate of .6°F per minute during class.

(d) Interpret $C'(0) = 1.5$.

**Solution** Since $C'(t)$ represents the instantaneous rate of change of the temperature of the drink with respect to time, $C'(0) = 1.5$ means that, at the start of the class, the temperature is changing at an instantaneous rate of 1.5°F per minute. Since the average rate of change over a very small time interval is a good *approximation* of the instantaneous rate of change, we can think of $C'(0) = 1.5$ as saying that, at the start of class, the drink will warm up by *approximately* 1.5°F in the next minute. ✓4

✓**Checkpoint 4**

For the function $C(t)$ in Example 4, interpret the following, including units.

(a) $C(50) = 68$

(b) $C(50) - C(25) = 7.6$

(c) $\dfrac{C(50) - C(25)}{50 - 25} = .3$

(d) $C'(50) = .2$

The rule of a derivative function can be found by using the definition of the derivative and the following four-step procedure.

## Finding $f'(x)$ from the Definition of the Derivative

**Step 1**    Find $f(x + h)$.

**Step 2**    Find $f(x + h) - f(x)$.

**Step 3**    Divide by $h$ to get $\dfrac{f(x + h) - f(x)}{h}$.

**Step 4**    Treat $x$ as a constant and let $h \to 0$.

$$f'(x) = \lim_{h \to 0} \frac{f(x + h) - f(x)}{h} \text{ if this limit exists.}$$

**EXAMPLE 5** Let $f(x) = -10x^2 + 3x - 7$.

**(a)** Find the derivative $f'(x)$.

**Solution** By definition,

$$f'(x) = \lim_{h \to 0} \frac{f(x + h) - f(x)}{h}.$$

**Step 1** Find $f(x + h)$.

Replace $x$ with $x + h$ in the rule of $f(x)$:

$$f(x) = -10x^2 + 3x - 7.$$
$$f(x + h) = -10(x + h)^2 + 3(x + h) - 7$$
$$= -10(x^2 + 2xh + h^2) + 3(x + h) - 7 \qquad \text{FOIL}$$
$$= -10x^2 - 20xh - 10h^2 + 3x + 3h - 7. \qquad \text{Distributive property}$$

**Step 2** Find $f(x + h) - f(x)$.

Since $f(x) = -10x^2 + 3x - 7$,

$$f(x + h) - f(x) = (-10x^2 - 20xh - 10h^2 + 3x + 3h - 7) - (-10x^2 + 3x - 7)$$
$$= -10x^2 - 20xh - 10h^2 + 3x + 3h - 7 + 10x^2 - 3x + 7 \quad \text{Distributive property}$$
$$= -20xh - 10h^2 + 3h \qquad \text{Combine like terms.}$$

**Step 3** Form and simplify the quotient $\dfrac{f(x + h) - f(x)}{h}$.

$$\frac{f(x + h) - f(x)}{h} = \frac{-20xh - 10h^2 + 3h}{h}$$
$$= \frac{h(-20x - 10h + 3)}{h} \qquad \text{Factor out } h.$$
$$= -20x - 10h + 3.$$

**Step 4** Find the limit as $h$ approaches 0 of the result in Step 3, treating $x$ as a constant:

$$f'(x) = \lim_{h \to 0} \frac{f(x + h) - f(x)}{h} = \lim_{h \to 0}(-20x - 10h + 3) = -20x + 3.$$

Therefore, the derivative of $f(x) = -10x^2 + 3x - 7$ is $f'(x) = -20x + 3$.

**(b)** Calculate and interpret $f'(1)$.

**Solution** The procedure in part (a) works for *every* $x$ and $f'(x) = -20x + 3$. Hence, when $x = 1$,

$$f'(1) = -20(1) + 3 = -17.$$

The number $-17$ is the slope of the tangent line to the graph of $f(x) = -10x^2 + 3x - 7$ at the point where $x = 1$.

**(c)** Find the equation of the tangent line to the graph of $f(x) = -10x^2 + 3x - 7$ at the point where $x = 1$.

**Solution** When $x = 1$, the $y$-value of the point on the graph of $f(x) = -10x^2 + 3x - 7$ is $f(1) = -10(1)^2 + 3(1) - 7 = -14$; hence the point is $(1, f(1)) = (1, -14)$. By part (b), the slope of the tangent line is $f'(1) = -17$. Therefore, the equation of the tangent line is

$$y - (-14) = -17(x - 1) \qquad \text{Point–slope form}$$
$$y + 14 = -17x + 17$$
$$y = -17x + 3.$$

Both $f(x)$ and the tangent line are shown in Figure 11.29, on the next page.  5

 **Checkpoint 5**

Let $f(x) = -5x^2 + 4$. Find the given expressions.

**(a)** $f(x + h)$

**(b)** $f(x + h) - f(x)$

**(c)** $\dfrac{f(x + h) - f(x)}{h}$

**(d)** $f'(x)$

**(e)** $f'(3)$

**(f)** $f'(0)$

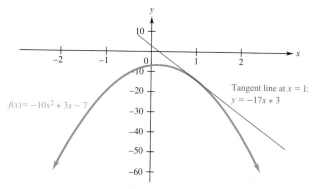

**Figure 11.29**

## ⚠ CAUTION

**1.** In Example 5(a), note that $f(x + h) \neq f(x) + h$ since, by Step 1,

$$f(x + h) = -10x^2 - 20xh - 10h^2 + 3x + 3h - 7,$$

but

$$f(x) + h = (-10x^2 + 3x - 7) + h = -10x^2 + 3x - 7 + h.$$

**2.** In Example 5(c), do not confuse $f(1)$ and $f'(1)$. $f(1)$ is the value of the original function $f(x) = -10x^2 + 3x - 7$ at $x = 1$, namely, $f(1) = -14$, whereas $f'(1)$ is the value of the derivative function $f'(x) = -20x + 3$ at $x = 1$, namely, $f'(1) = -17$.

---

**EXAMPLE 6**    Let $f(x) = 1/x$. Find $f'(x)$.

**Solution**

**Step 1**    $f(x + h) = \dfrac{1}{x + h}$.

**Step 2**    $f(x + h) - f(x) = \dfrac{1}{x + h} - \dfrac{1}{x}$

$$= \frac{1}{x + h} \cdot \frac{x}{x} - \frac{1}{x} \cdot \frac{x + h}{x + h} \qquad \text{Find a common denominator.}$$

$$= \frac{x - (x + h)}{x(x + h)}$$

$$= \frac{x - x - h}{x(x + h)} \qquad \text{Simplify the numerator.}$$

$$= \frac{-h}{x(x + h)}.$$

**Step 3**    $\dfrac{f(x + h) - f(x)}{h} = \dfrac{\dfrac{-h}{x(x + h)}}{h}$

$$= \frac{-h}{x(x + h)} \cdot \frac{1}{h} \qquad \text{Invert and multiply.}$$

$$= \frac{-1}{x(x + h)}.$$

**Step 4**    $f'(x) = \displaystyle\lim_{h \to 0} \dfrac{f(x + h) - f(x)}{h} = \lim_{h \to 0} \dfrac{-1}{x(x + h)}$

$$= \frac{-1}{x(x + 0)} = \frac{-1}{x(x)} = \frac{-1}{x^2}. \quad \boxed{\checkmark}_6$$

✓ **Checkpoint 6**

Let $f(x) = -5/x$. Find the given expressions.

**(a)** $f(x + h)$

**(b)** $f(x + h) - f(x)$

**(c)** $\dfrac{f(x + h) - f(x)}{h}$

**(d)** $f'(x)$

**(e)** $f'(-1)$

**EXAMPLE 7** Let $g(x) = \sqrt{x}$. Find $g'(x)$.

**Solution**

**Step 1** $g(x + h) = \sqrt{x + h}$.

**Step 2** $g(x + h) - g(x) = \sqrt{x + h} - \sqrt{x}$.

**Step 3** $\dfrac{g(x + h) - g(x)}{h} = \dfrac{\sqrt{x + h} - \sqrt{x}}{h}$.

At this point, in order to be able to divide by $h$, multiply both numerator and denominator by $\sqrt{x + h} + \sqrt{x}$, that is, rationalize the *numerator*:

$$\frac{g(x + h) - g(x)}{h} = \frac{\sqrt{x + h} - \sqrt{x}}{h} \cdot \frac{\sqrt{x + h} + \sqrt{x}}{\sqrt{x + h} + \sqrt{x}}$$

$$= \frac{(\sqrt{x + h})^2 - (\sqrt{x})^2}{h(\sqrt{x + h} + \sqrt{x})} \qquad (a - b)(a + b) = a^2 - b^2$$

$$= \frac{x + h - x}{h(\sqrt{x + h} + \sqrt{x})} = \frac{1}{\sqrt{x + h} + \sqrt{x}}.$$

**Step 4** $g'(x) = \lim\limits_{h \to 0} \dfrac{1}{\sqrt{x + h} + \sqrt{x}} = \dfrac{1}{\sqrt{x} + \sqrt{x}} = \dfrac{1}{2\sqrt{x}}$.

## Existence of the Derivative

The definition of the derivative includes the phrase "provided that this limit exists." If the limit used to define $f'(x)$ does not exist, then, of course, the derivative does not exist at that $x$. For example, a derivative cannot exist at a point where the function itself is not defined. If there is no function value for a particular value of $x$, there can be no tangent line for that value. This was the case in Example 6: There was no tangent line (and no derivative) when $x = 0$.

Derivatives also do not exist at "corners" or "sharp points" on a graph. For example, the function graphed in Figure 11.30 is the *absolute-value function*, defined by

$$f(x) = \begin{cases} x & \text{if } x \geq 0 \\ -x & \text{if } x < 0 \end{cases}$$

and written $f(x) = |x|$. The graph has a "corner point" when $x = 0$.

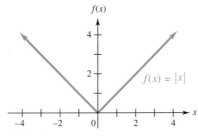

**Figure 11.30**

By the definition of derivative, the derivative at any value of $x$ is given by

$$\lim\limits_{h \to 0} \frac{f(x + h) - f(x)}{h} = \lim\limits_{h \to 0} \frac{|x + h| - |x|}{h}$$

provided that this limit exists. To find the derivative at 0 for $f(x) = |x|$, replace $x$ with 0 to get

$$\lim\limits_{h \to 0} \frac{|0 + h| - |0|}{h} = \lim\limits_{h \to 0} \frac{|h|}{h}.$$

In Example 11 of Section 11.1 (with $x$ in place of $h$), we showed that

$$\lim_{h \to 0} \frac{|h|}{h} \text{ does not exist.}$$

Therefore, there is no derivative at 0. However, the derivative of $f(x) = |x|$ *does* exist for all values of $x$ other than 0.

Since a vertical line has an undefined slope, the derivative cannot exist at any point where the tangent line is vertical, as at $x_5$ in Figure 11.31. This figure summarizes various ways that a derivative can fail to exist.

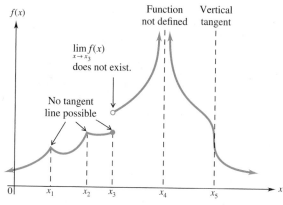

**Figure 11.31**

## Derivatives and Technology

Many computer programs (such as Mathematica and Maple) and a few graphing calculators (such as the TI-89) can find symbolic formulas for the derivatives of most functions. Although other graphing calculators cannot find the rule of a derivative function, most of them can approximate the numerical value of the derivative function at any number where it is defined by using the **numerical derivative** feature. (See Exercise 48 at the end of this section for an explanation of computational technique used by the numerical derivative feature.)

### EXAMPLE 8    Memorization Experiment  A psychology experiment found that the number of facts $N(t)$ remembered $t$ days after the facts were memorized was given by

$$N(t) = \frac{10.003}{1 + .0003e^{.8t}}.$$

Use a graphing calculator to find an approximate value of $N'(10)$ and interpret the answer.

**Solution**  The numerical derivative feature is labeled nDeriv, or d/dx, or nDer and is usually in the MATH or CALC menu or one of its submenus. Most calculators require that we use $x$ as the independent variable instead of $t$, so we will write the function as $N(x) = \frac{10.003}{1 + .0003e^{.8x}}$. Check your instruction manual for the correct syntax, but on many graphing calculators, entering either

$$\text{nDeriv}(10.003/(1 + .0003e^{.8x}), x, 10) \quad \text{or}$$
$$\text{d/dx}(10.003/(1 + .0003e^{.8x}), x, 10)$$

produces an (approximate) value of $N'(10)$. Answers will vary slightly depending on the calculator, but you are likely to get an answer close to $-1.9943694066$.

The negative sign implies that the number of facts remembered is decreasing over time. One interpretation of $N'(10) \approx -2$ is that 10 days after the facts were memorized, they are being forgotten at a rate of about 2 facts per day. In other words, 10 days after the facts were memorized, approximately 2 facts will be forgotten during the next day. ✔️ 7

✔️ **Checkpoint 7**

If $f(x) = x^{2/3} - 8x^3 + 4^x$, use a graphing calculator to find the (approximate) value of

(a)  $f'(2)$;

(b)  $f'(3.2)$.

**TECHNOLOGY TIP** On many graphing calculators, if you have the function $f(x)$ stored in the function memory as, say, $Y_1$, you can use $Y_1$ with the nDeriv key instead of typing in the rule of $f(x)$.

 **EXAMPLE 9**    **Finance** If \$15,000 is deposited in a money market account that pays 1.2% per year compounded monthly, then the amount in the account after $n$ months is

$$A(n) = 15{,}000\left(1 + \frac{.012}{12}\right)^n = 15{,}000(1.001)^n.$$

Use a graphing calculator to find an approximate value of $A'(120)$ and interpret the answer.

**Solution** Using the numerical derivative feature on a typical graphing calculator gives an approximate value of $A'(120) = 16.90$. This means that at time $t = 120$ months, the instantaneous rate of change of the account balance with respect to time is 16.90 dollars per month. We can also describe this by saying that 10 years (120 months) after making the deposit, the balance on the account will increase by *approximately* \$16.90 in the next month. Note that the *actual* change in the balance over the next one month is given by:

$$A(121) - A(120) = 15{,}000(1.001)^{121} - 15{,}000(1.001)^{120} = \$16.91.$$ ✔ 8

✓ **Checkpoint 8**

For the account in Example 9, suppose that one year has passed since the initial deposit. Use a derivative to approximate how much the balance will change in the next month.

🛑 **CAUTION** Because of the approximation methods used, the nDeriv key may display an answer at numbers where the derivative is not defined. For instance, we saw earlier that the derivative of $f(x) = |x|$ is not defined when $x = 0$, but the nDeriv key on most calculators produces 1 or 0 or $-1$ as $f'(0)$.

# 11.4 Exercises

*Find $f'(x)$ for each function. Then find $f'(2)$, $f'(0)$, and $f'(-3)$. (See Examples 5–7.)*

**1.** $f(x) = 8x + 6$

**2.** $f(x) = 9 - 2x$

**3.** $f(x) = -4x^2 + 11x$

**4.** $f(x) = x^2 - 3x + 7$

**5.** $f(x) = \dfrac{-2}{x}$

**6.** $f(x) = \dfrac{4}{x-1}$

**7.** $f(x) = 2\sqrt{x}$

**8.** $f(x) = 4\sqrt{x}$

*The derivatives of each of the given functions were found in Examples 5–7. Use them to find the equation of the tangent line to the graph of the function at the given point. (See Examples 1 and 2.)*

**9.** $f(x) = -10x^2 + 3x - 7$ at $x = 0$

**10.** $f(x) = -10x^2 + 3x - 7$ at $x = -1$

**11.** $f(x) = \dfrac{1}{x}$ at $x = -2$

**12.** $f(x) = \dfrac{1}{x}$ at $x = -5$

**13.** $g(x) = \sqrt{x}$ at $x = 9$

**14.** $g(x) = \sqrt{x}$ at $x = 36$

*For each of the given functions, (a) find the slope of the tangent line to the graph at the given point; (b) find the equation of the tangent line. (See Examples 1 and 2.)*

**15.** $f(x) = x^2 + 3$ at $x = 2$

**16.** $g(x) = 1 - 2x^2$ at $x = -1$

**17.** $h(x) = \dfrac{7}{x}$ at $x = 3$

**18.** $f(x) = \dfrac{-3}{x}$ at $x = -4$

**19.** $g(x) = 5\sqrt{x}$ at $x = 9$

**20.** $g(x) = \sqrt{x+1}$ at $x = 15$ (*Hint:* In Step 3, multiply numerator and denominator by $\sqrt{16+h} + \sqrt{4}$.)

*Use the fact that $f'(c)$ is the slope of the tangent line to the graph of $f(x)$ at $x = c$ to work these exercises. (See Example 3.)*

**21.** In the graph of the function $f$, at which of the labeled $x$-values is

(a) $f(x)$ the largest?

(b) $f(x)$ the smallest?

(c) $f'(x)$ the smallest?

(d) $f'(x)$ the closest to 0?

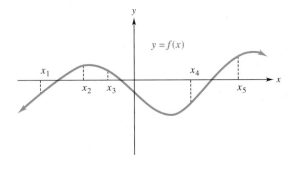

**22.** Sketch the graph of the *derivative* of the function g whose graph is shown. (*Hint:* Consider the slope of the tangent line at each point along the graph of g. Are there any points where there is no tangent line?)

**(b)**

**23.** Sketch the graph of a function g with the property that $g'(x) > 0$ for $x < 0$ and $g'(x) < 0$ for $x > 0$. Many correct answers are possible.

**24.** Sketch the graph of a function g with the property that $g'(x) < 0$ for $x < 0$ and $g'(x) > 0$ for $x > 0$. Many correct answers are possible.

**28. (a)**

**25. Self-Cleaning Oven** The accompanying graph shows the temperature in an oven during a self-cleaning cycle.* (The open circles on the graph are not points of discontinuity, but merely the times when the thermal door lock turns on and off.) The oven temperature is 100° when the cycle begins and 600° after half an hour. Let $T(x)$ be the temperature (in degrees Fahrenheit) after $x$ hours.

**(b)**

**(a)** Find and interpret $T'(.5)$.

**(b)** Find and interpret $T'(2)$.

**(c)** Find and interpret $T'(3.5)$.

**26.** At what $x$-values does the derivative of the function graphed in the previous exercise fail to exist?

*In Exercises 27 and 28, tell which graph, (a) or (b), represents velocity and which represents distance from a starting point. (Hint: Consider where the derivative is zero, positive, or negative.)*

**27. (a)**

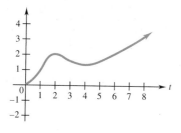

*Find all points where the functions whose graphs are shown do not have derivatives. (See Figure 11.31 and the preceding discussion.)*

**29.**

**30.**

**31.**

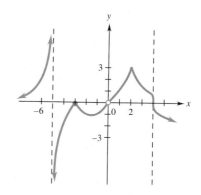

---

**32. (a)** Sketch the graph of $g(x) = \sqrt[3]{x}$ for $-1 \leq x \leq 1$.

**(b)** Explain why the derivative of $g(x)$ is not defined at $x = 0$. (*Hint:* What is the slope of the tangent line at $x = 0$?)

*Work these exercises. (See Example 4.)*

**33. Free Fall** A tourist drops a stone off the Cliffs of Moher, Ireland and it falls into the Atlantic Ocean. Let $s(t)$ represent the position of the stone (in feet above sea level) $t$ seconds after it was dropped. Explain what the equations $s(7) = 0$ and $s'(7) = -224$ mean in practical terms.

**34. Kayaking** Kayakers like to paddle Slippery Rock Creek, Pennsylvania when the river is at a safe level between 1 and 4 feet high (the river level is measured at a fixed location). Let $H(t)$ represent yesterday's river level (in feet) $t$ hours after midnight. Due to a flash flood, it happened that $H(13) = 3.5$ and $H'(13) = 3$. Explain what this means in practical terms for a kayaker.

**35. Political Polling** A polling firm determines that the percent of likely voters who support a certain mayoral candidate can be represented by a function $P(t)$ where $t$ is measured in weeks since the candidate entered the race. Interpret the given equations, including units.

**(a)** $P(6) = 25$    **(b)** $P'(6) = 3$

**36. Hiking** On a hiking trail up Mount LeConte, Tennessee, the trailhead has an elevation of 3830 feet and the summit has an elevation of 6953 feet (above sea level). Let $T(x)$ represent the record high temperature on the trail (in degrees Fahrenheit) at an elevation $x$ feet above sea level.

**(a)** What are the units for $T'(x)$?

**(b)** Would you expect the sign of $T'(x)$ to be positive, negative, or zero? Why?

**37. Home Mortgage** You plan to take out a 30-year fixed-rate mortgage for $300,000. Let $P(r)$ be your monthly payment if the interest rate is $r\%$ per year, compounded monthly. Interpret the given equations, including units.

**(a)** $P(5) = 1,610.46$    **(b)** $P'(5) = 183.37$

**38. Home Mortgage** You plan to take out a 30-year fixed-rate mortgage at an interest rate of 5.25% per year, compounded monthly. Let $P(A)$ be your monthly payment if $A$ is the initial amount of the loan. Would you expect the sign of $P'(A)$ to be positive, negative, or zero? Why?

**39. Home Mortgage** You plan to take out a 30-year fixed-rate mortgage and you want your monthly payment to be $1200. The amount of the loan you can afford depends on the interest rate you will have to pay. Let $A(r)$ be the amount of the money you can borrow if the interest rate is $r\%$ per year, compounded monthly. Interpret the given equations, including units.

**(a)** $A(5) = 223,537.94$

**(b)** Would you expect the sign of $A'(r)$ to be positive, negative, or zero? Why?

**(c)** What are the units for $A'(r)$?

**(d)** $A'(5) = -25,449.21$

**40. Credit Card Payments** A credit card statement includes information on how long it will take to pay off the card if you make only the minimum payments and how long it will take if you make larger payments. Let $M(p)$ represent the number of months it will take to pay off your credit card if you make a monthly payment of $\$p$ (assuming no additional charges are made).

**(a)** Explain what $M(25) = 50$ means.

**(b)** Would you expect $M'(p)$ to be positive, negative, or zero? Why?

**(c)** What are the units for $M'(p)$?

**(d)** Explain what $M'(25) = -3$ means.

**41. Population Growth** In one research study, the population of a certain shellfish in an area at time $t$ was closely approximated by the accompanying graph. The tangent line is shown at $t = 2$, $t = 10$, and $t = 13$. Estimate and interpret the derivative (the slope of the tangent line) at each of these points.

**42. Food Intake** The eating behavior of a typical human during a meal can be described by

$$I(t) = 27 + 72t - 1.5t^2,$$

where $t$ is the number of minutes since the meal began and $I(t)$ represents the amount (in grams) that the person has eaten at time $t$.*

**(a)** Find the instantaneous rate of change of the intake of food for a person 5 minutes into a meal, and interpret it.

**(b)** Verify that the rate at which food is consumed is zero 24 minutes after the meal starts.

**(c)** Comment on the assumptions and usefulness of this function after 24 minutes. On the basis of your answer, determine a logical range for the function.

⊞ *Use numerical derivatives to work these exercises. (See Examples 8 and 9.)*

**43. Research and Development in China** The amount of money (in billions of U.S. dollars) that China spends on research and development can be approximated by the function

$$f(t) = 42e^{.15t},$$

where $t = 0$ corresponds to the year 2000. (Data from: Organization for Economic Cooperation and Development.)

**(a)** Find the value of the derivative in the years 2005, 2010, and 2015.

**(b)** What does your answer in part (a) suggest about the rate at which research and development spending is changing in China?

---

*Kissileff, H. R. and J. L. Guss, "Microstructure of Eating Behaviour in Humans," *Appetite*, Vol. 36, No. 1, Feb. 2001, pp. 70–78.

**44. Bicycle Lanes** The number of protected bicycle lanes in the United States can be approximated by

$$f(t) = 1.51e^{.347t},$$

where $t = 9$ corresponds to the year 2009. Round your answers to the nearest integer. (Data from: www.peopleforbikes.org.)

**(a)** Find and interpret $f(15)$, including units.

**(b)** Find and interpret $f'(15)$, including units.

**45. Licensed Drivers** The number of licensed drivers in the United States (in millions) can be approximated by the function

$$g(x) = 183x^{.06},$$

where $x = 5$ corresponds to the year 2005. Estimate and interpret the value of the derivative in 2018. (Data from: www.fhwa.dot.gov.)

**46. Patient Body Temperature** The temperature of a patient (in degrees Fahrenheit) during a 12-hour hospital shift is modeled by $P(t) = 98.6 + 2.3te^{-.25t}$, where $t$ is measured in hours since the start of the shift.

**(a)** Use a calculator to sketch $y = P(t)$, and give a general description of the graph in terms of what it means for the patient.

**(b)** Find and interpret $P(1)$ and $P'(1)$.

**(c)** Find and interpret $P(8)$ and $P'(8)$.

*Use technology for Exercises 47–48.*

**47. (a)** Graph the (numerical) derivative of

$$f(x) = .5x^5 - 2x^3 + x^2 - 3x + 2 \text{ for } -3 \le x \le 3.$$

**(b)** Graph $g(x) = 2.5x^4 - 6x^2 + 2x - 3$ on the same screen.

**(c)** How do the graphs of $f'(x)$ and $g(x)$ compare? What does this suggest that the derivative of $f(x)$ is?

**48.** If $f$ is a function such that $f'(x)$ is defined, then it can be proved that

$$f'(x) = \lim_{h \to 0} \frac{f(x + h) - f(x - h)}{2h}.$$

Consequently, when $h$ is very small, say, $h = .001$,

$$f'(x) \approx \frac{f(x + h) - f(x - h)}{2h}$$

$$= \frac{f(x + .001) - f(x - .001)}{.002}.$$

**(a)** In Example 7, we saw that the derivative of $f(x) = \sqrt{x}$ is the function $f'(x) = \dfrac{1}{2\sqrt{x}}$. Make a table in which the first

column lists $x = 1, 6, 11, 16,$ and $21$; the second column lists the corresponding values of $f'(x)$; and the third column lists the corresponding values of

$$\frac{f(x + .001) - f(x - .001)}{.002}.$$

**(b)** How do the second and third columns of the table compare? (If you use the table feature on a graphing calculator, the entries in the table will be rounded off, so move the cursor over each entry to see it fully displayed at the bottom of the screen.) Your answer to this question may explain why most graphing calculators use the method in the third column to compute numerical derivatives.

✓ **Checkpoint Answers**

**1.** $y = 2x + 1$

**2.**

**3.** **(a)** Positive     **(b)** Zero     **(c)** Negative

**4.** **(a)** At the end of class, the drink was 68°F.

**(b)** The drink warmed up by 7.6°F between $t = 25$ and $t = 50$ minutes.

**(c)** The average rate of change of the temperature of the drink between $t = 25$ and $t = 50$ minutes was .3°F/min.

**(d)** At the end of class, the drink will warm up by approximately .2°F in the next minute.

**5.** **(a)** $-5x^2 - 10xh - 5h^2 + 4$

**(b)** $-10xh - 5h^2$     **(c)** $-10x - 5h$     **(d)** $-10x$

**(e)** $-30$     **(f)** $0$

**6.** **(a)** $\dfrac{-5}{x + h}$     **(b)** $\dfrac{5h}{x(x + h)}$     **(c)** $\dfrac{5}{x(x + h)}$

**(d)** $\dfrac{5}{x^2}$     **(e)** $5$

**7.** **(a)** $-73.29$     **(b)** $-128.2$

**8.** $15.17

# 11.5 Techniques for Finding Derivatives

In the previous section, the derivative of a function was defined as a special limit. The mathematical process of finding this limit, called *differentiation*, resulted in a new function that was interpreted in several different ways. Using the definition to calculate the derivative of a function is a very involved process, even for simple functions. In this section, we develop rules that make the calculation of derivatives much easier. Keep in mind that even though the process of finding a derivative will be greatly simplified with these rules, *the interpretation of the derivative will not change.*

In addition to $y'$ and $f'(x)$, there are several other commonly used notations for the derivative.

## Notations for the Derivative

The derivative of the function $y = f(x)$ may be denoted in any of the following ways:

$$f'(x), \quad y', \quad \frac{dy}{dx}, \quad \text{or} \quad \frac{d}{dx}[f(x)].$$

The $dy/dx$ notation for the derivative is sometimes referred to as *Leibniz notation*, named after one of the coinventors of calculus, Gottfried Wilhelm Leibniz (1646–1716). (The other was Sir Isaac Newton (1642–1727).)

For example, the derivative of $f(x) = -10x^2 + 3x - 7$, which we found in Example 5 of the last section to be $f'(x) = -20x + 3$, can also be written in the following ways:

$$y' = -20x + 3,$$

$$\frac{dy}{dx} = -20x + 3, \text{ or}$$

$$\frac{d}{dx}(-10x^2 + 3x - 7) = -20x + 3.$$

A variable other than $x$ may be used as the independent variable. For example, if $y = f(t)$ gives population as a function of time, then the derivative of $y$ with respect to $t$ could be written as

$$y', f'(t), \frac{dy}{dt}, \text{ or } \frac{d}{dt}[f(t)].$$

In this section, the definition of the derivative,

$$f'(x) = \lim_{h \to 0} \frac{f(x + h) - f(x)}{h},$$

is used to develop some rules for finding derivatives more easily than by the four-step process of the previous section. The first rule tells how to find the derivative of a constant function, such as $f(x) = 5$. Since $f(x + h)$ is also 5, $f'(x)$ is by definition

$$f'(x) = \lim_{h \to 0} \frac{f(x + h) - f(x)}{h}$$

$$= \lim_{h \to 0} \frac{5 - 5}{h} = \lim_{h \to 0} \frac{0}{h} = \lim_{h \to 0} 0 = 0.$$

The same argument works for $f(x) = k$, where $k$ is a constant real number, establishing the following rule.

## Constant Rule

If $f(x) = k$, where $k$ is any real number, then

$$f'(x) = 0.$$

(The derivative of a constant function is 0.)

**Figure 11.32**

Figure 11.32 illustrates the constant rule; it shows a graph of the horizontal line $y = k$. At any point $P$ on this line, the tangent line at $P$ is the line itself. Since a horizontal line has a slope of 0, the slope of the tangent line is 0. This finding agrees with the constant rule: The derivative of a constant is 0.

EXAMPLE 1    Find the derivative of the given function.

**(a)** $f(x) = 25$

**Solution** Since 25 is a constant, we can write $f'(x) = 0$ or $\dfrac{d}{dx}[f(x)] = 0$.

**(b)** $y = \pi$

**Solution** Since $\pi$ is a constant with value equal to $3.14159265\ldots$, we have $y' = 0$ or $dy/dx = 0$.

**(c)** $y = 4^3$

**✓ Checkpoint 1**

Find the derivative of the given function.

**(a)** $y = -4$

**(b)** $f(x) = \pi^3$

**(c)** $y = 0$

**Solution** Since $4^3$ is the constant 64, we have $y' = 0$ or $dy/dx = 0$. ✓₁

Functions of the form $y = x^n$, where $n$ is a fixed real number, are common in applications. We now find the derivative functions when $n = 2$ and $n = 3$:

$$f(x) = x^2 \qquad\qquad\qquad f(x) = x^3$$

$$f'(x) = \lim_{h\to 0}\frac{f(x+h)-f(x)}{h} \qquad f'(x) = \lim_{h\to 0}\frac{f(x+h)-f(x)}{h}$$

$$= \lim_{h\to 0}\frac{(x+h)^2 - x^2}{h} \qquad\qquad = \lim_{h\to 0}\frac{(x+h)^3 - x^3}{h}$$

$$= \lim_{h\to 0}\frac{(x^2+2xh+h^2)-x^2}{h} \qquad = \lim_{h\to 0}\frac{(x^3+3x^2h+3xh^2+h^3)-x^3}{h}$$

$$= \lim_{h\to 0}\frac{2xh+h^2}{h} \qquad\qquad = \lim_{h\to 0}\frac{3x^2h+3xh^2+h^3}{h}$$

$$f'(x) = \lim_{h\to 0}(2x+h) = 2x \qquad f'(x) = \lim_{h\to 0}(3x^2+3xh+h^2) = 3x^2.$$

Similar calculations show that

$$\frac{d}{dx}(x^4) = 4x^3 \qquad \text{and} \qquad \frac{d}{dx}(x^5) = 5x^4.$$

The pattern here (the derivative is the product of the exponent on the original function and a power of $x$ that is one less) suggests that the derivative of $y = x^n$ is $y' = nx^{n-1}$, which is indeed the case. (See Exercise 77 at the end of this section for an outline of the proof.)

Furthermore, the pattern holds even when $n$ is not a positive integer. For instance, in Example 6 of the last section we showed that

$$\text{if } f(x) = \frac{1}{x}, \qquad \text{then } f'(x) = -\frac{1}{x^2}.$$

This is the same result that the preceding pattern produces:

$$\text{If } f(x) = \frac{1}{x} = x^{-1}, \qquad \text{then } f'(x) = (-1)x^{-1-1} = -x^{-2} = -\frac{1}{x^2}.$$

Similarly, Example 7 in the last section showed that the derivative of $g(x) = \sqrt{x}$ is $g'(x) = \dfrac{1}{2\sqrt{x}}$, and the preceding pattern gives the same answer:

$$\text{If } g(x) = \sqrt{x} = x^{1/2}, \qquad \text{then } g'(x) = \frac{1}{2}x^{\frac{1}{2}-1} = \frac{1}{2}x^{-1/2} = \frac{1}{2}\cdot\frac{1}{x^{1/2}} = \frac{1}{2\sqrt{x}}.$$

Consequently, the following statement should be plausible.

## Power Rule

If $f(x) = x^n$ for any number $n$, then

$$f'(x) = nx^{n-1}.$$

(The derivative of $f(x) = x^n$ is found by multiplying the exponent $n$ on the original function by a power of $x$ that is one less.)

**EXAMPLE 2**   For each of the given functions, find the derivative.

**(a)** $y = x^8$

**Solution** The exponent for $x$ is $n = 8$. Using the power rule, we multiply by 8 and decrease the power on $x$ by 1 to obtain $y' = 8x^7$.

**(b)** $y = x$

**Solution** The exponent for $x$ is $n = 1$. Using the power rule, we multiply by 1 and decrease the power on $x$ by 1 to obtain $y' = 1x^0$. Recall that $x^0 = 1$ if $x \neq 0$. So we have $y' = 1$.

**(c)** $y = t^{3/2}$

**Solution** The exponent for $t$ is $n = \frac{3}{2}$. Using the power rule, we multiply by $\frac{3}{2}$ and decrease the power on $t$ by 1 to obtain $\frac{dy}{dt} = \frac{3}{2}t^{3/2-1} = \frac{3}{2}t^{1/2}$.

**(d)** $y = \sqrt[3]{x}$

**Solution** First we rewrite $y = \sqrt[3]{x}$ as $y = x^{\frac{1}{3}}$. Using the power rule, we multiply by $\frac{1}{3}$ and decrease the power on $x$ by 1 to obtain

$$\frac{dy}{dx} = \frac{1}{3}x^{1/3-1} = \frac{1}{3}x^{-2/3} = \frac{1}{3x^{2/3}} = \frac{1}{3\sqrt[3]{x^2}}.$$

**(e)** $y = \dfrac{1}{x^3}$

**Solution** Rewrite $y = \dfrac{1}{x^3}$ as $y = x^{-3}$. Using the power rule, we multiply by $-3$ and decrease the power on $x$ by 1 to obtain $\dfrac{dy}{dx} = -3x^{-4} = -\dfrac{3}{x^4}$. ✔️ 2

✔️ **Checkpoint 2**

**(a)** If $y = x^4$, find $y'$.

**(b)** If $y = x^{17}$, find $y'$.

**(c)** If $y = x^{-2}$, find $dy/dx$.

**(d)** If $y = t^{-5}$, find $dy/dt$.

**(e)** If $y = t^{5/4}$, find $y'$.

The next rule shows how to find the derivative of the product of a constant and a function. We omit the straightforward proof.

## Constant Times a Function Rule

Let $k$ be a real number. If $g'(x)$ exists, then the derivative of $f(x) = k \cdot g(x)$ is

$$f'(x) = k \cdot g'(x).$$

(The derivative of a constant times a function is the constant times the derivative of the function.)

## EXAMPLE 3

**(a)** If $y = 8x^4$, find $y'$.

**Solution** Since the derivative of $g(x) = x^4$ is $g'(x) = 4x^3$ and $y = 8x^4 = 8g(x)$,

$$y' = 8g'(x) = 8(4x^3) = 32x^3.$$

**(b)** If $y = -\dfrac{3}{4}t^{12}$, find $dy/dt$.

**Solution** $\dfrac{dy}{dt} = \dfrac{d}{dt}\left[-\dfrac{3}{4}t^{12}\right] = -\dfrac{3}{4}\dfrac{d}{dt}[t^{12}] = -\dfrac{3}{4}(12t^{11}) = -9t^{11}.$

**(c)** Find $\dfrac{d}{dx}[15x]$.

**Solution** $\dfrac{d}{dx}[15x] = 15 \cdot \dfrac{d}{dx}[x] = 15 \cdot 1 = 15.$

**(d)** If $y = 6/x^2$, find $y'$.

**Solution** Rewrite $y = \dfrac{6}{x^2}$ as $y = 6 \cdot \dfrac{1}{x^2} = 6x^{-2}$. Then

$$y' = 6(-2x^{-3}) = -12x^{-3} = -\dfrac{12}{x^3}.$$

**(e)** If $y = \dfrac{1}{6x^2}$, find $y'$.

**Solution** Be careful here. Rewrite $y = \dfrac{1}{6x^2}$ as $y = \dfrac{1}{6} \cdot \dfrac{1}{x^2} = \dfrac{1}{6}x^{-2}$. Then

$$y' = \dfrac{1}{6}(-2x^{-3}) = -\dfrac{2}{6}x^{-3} = -\dfrac{1}{3} \cdot \dfrac{1}{x^3} = -\dfrac{1}{3x^3}.$$

| X | Y1 | Y2 |
|---|------|------|
| 5 | 33.541 | 33.541 |
| 6 | 36.742 | 36.742 |
| 7 | 39.686 | 39.686 |
| 8 | 42.426 | 42.426 |
| 9 | 45 | 45 |
| 10 | 47.434 | 47.434 |
| 11 | 49.749 | 49.749 |

X=11

(a)

80

0
0                                    25

(b)

**Figure 11.33**

**(f)** If $y = \dfrac{6x^4}{\sqrt{x}}$, find $\dfrac{dy}{dx}$.

**Solution** First use algebra to simplify $y = \dfrac{6x^4}{\sqrt{x}} = \dfrac{6x^4}{x^{1/2}} = 6x^{7/2}$. Then use calculus to find

$$\dfrac{dy}{dx} = 6\left(\dfrac{7}{2}x^{5/2}\right) = 21\sqrt{x^5}.$$

**(g)** Find $\dfrac{d}{dx}(10x^{3/2})$, and use a graphing calculator to confirm your answer numerically and graphically.

**Solution** $\dfrac{d}{dx}(10x^{3/2}) = 10\left(\dfrac{3}{2}x^{1/2}\right) = 15x^{1/2}.$

To confirm this result numerically, make a table of values for $y_1 = 15x^{1/2}$ and $y_2 =$ the numerical derivative of $10x^{3/2}$; check your instruction manual for the correct syntax for $y_2$, which is probably one of the following:

$$\text{nDeriv}(10x^{3/2}, x), \quad \text{Nderiv}(10x^{3/2}, x, x),$$
$$d/dx(10x^{3/2}), \quad \text{or} \quad d/dx(10x^{3/2}, x).$$

Figure 11.33(a) indicates that the corresponding values are identical to three decimal places. To confirm this result graphically, graph $y_1$ and $y_2$ on the same screen and verify that the graphs appear to be the same, as shown in Figure 11.33(b). ✓ 3

## ✓ Checkpoint 3

Find the derivative of the given function.

**(a)** $y = 12x^3$

**(b)** $f(t) = 30t^7$

**(c)** $y = -35t$

**(d)** $y = 5\sqrt{x}$

**(e)** $y = -10/t$

Confirming your calculations numerically or graphically, as in part (e) of Example 3, is a good way to detect algebraic errors. If you compute the rule of the derivative $f'(x)$, but its graph differs from the graph of the numerical derivative of $f(x)$, then you have made a mistake. If the two graphs appear to be identical, then you are *probably* correct. (The fact that two graphs appear identical on a calculator screen does not *prove* that they really are identical.)

The final rule in this section is for the derivative of a function that is a sum or difference of functions.

## Sum and Difference Rules

If $f(x) = u(x) + v(x)$, and if $u'(x)$ and $v'(x)$ exist, then

$$f'(x) = u'(x) + v'(x).$$

If $f(x) = u(x) - v(x)$, and if $u'(x)$ and $v'(x)$ exist, then

$$f'(x) = u'(x) - v'(x).$$

(The derivative of a sum or difference of two functions is the sum or difference of the derivatives of the functions.)

For a proof of this rule, see Exercise 78 at the end of this section. This rule also works for sums and differences with more than two terms.

---

**EXAMPLE 4**     Find the derivatives of the given functions.

**(a)** $y = 6x^3 + 15x^2$

**Solution** Let $u(x) = 6x^3$ and $v(x) = 15x^2$. Then

$$y = u(x) + v(x) \qquad \text{and}$$
$$y' = u'(x) + v'(x)$$
$$= 6(3x^2) + 15(2x) = 18x^2 + 30x.$$

**(b)** $p(t) = 8t^4 - 6\sqrt{t} + \dfrac{5}{t}$

**Solution** Rewrite $p(t)$ as $p(t) = 8t^4 - 6t^{1/2} + 5t^{-1}$; then $p'(t) = 32t^3 - 3t^{-1/2} - 5t^{-2}$, which also may be written as $p'(t) = 32t^3 - \dfrac{3}{\sqrt{t}} - \dfrac{5}{t^2}$.

**(c)** $f(x) = 5\sqrt[3]{x^2} + \dfrac{4}{x^2} + 7$

**Solution** Rewrite $f(x)$ as $f(x) = 5x^{2/3} + 4x^{-2} + 7$. Then

$$f'(x) = \frac{10}{3}x^{-1/3} - 8x^{-3} + 0$$

$$= \frac{10}{3\sqrt[3]{x}} - \frac{8}{x^3}.$$

**(d)** $f(x) = \dfrac{1 + 3x^4 + 10x^8}{x^4}$

**Solution** First rewrite $f(x)$ as the sum of three terms and simplify to get

$$f(x) = \frac{1 + 3x^4 + 10x^8}{x^4} = \frac{1}{x^4} + \frac{3x^4}{x^4} + \frac{10x^8}{x^4} = \frac{1}{x^4} + 3 + 10x^4 = x^{-4} + 3 + 10x^4.$$

Then we are able to find

$$f'(x) = -4x^{-5} + 0 + 40x^3$$

$$= -\frac{4}{x^5} + 40x^3. \quad ✓_4$$

✓ **Checkpoint 4**

Find the derivatives of the given functions.

(a)  $y = -10x^3 - 6x + 12$

(b)  $y = 5t^{8/7} + 5t^{1/2}$

(c)  $f(t) = -2\sqrt{t} + 4/t$

The rules developed in this section make it possible to find the derivative of a function more directly than with the definition of the derivative, so that applications can be dealt with more effectively. As we saw in the previous section, for any function $f(x)$, the derivative $f'(x)$ is the instantaneous rate of change of $f(x)$ with respect to $x$. In particular, velocity is the instantaneous rate of change in position with respect to time.

**EXAMPLE 5**   **Traveling Salesperson** A sales representative for a textbook publishing company frequently makes a 4-hour drive from her home in a large city to a university in another city. Let

$$s(t) = -5t^3 + 30t^2$$

represent her position $t$ hours into the trip (where position is measured in miles on the highway, with position 0 corresponding to her home).

**(a)** What is her position after 1 hour? After 2 hours?

**Solution** Her position after 1 hour is

$$s(1) = -5(1)^3 + 30(1)^2 = 25 \text{ miles.}$$

After 2 hours, it is

$$s(2) = -5(2)^3 + 30(2)^2 = 80 \text{ miles.}$$

**(b)** Find the velocity function $v(t)$.

**Solution** Velocity is the instantaneous rate of change in position with respect to time, so we need to find the derivative of the position function. Since

$$s(t) = -5t^3 + 30t^2,$$

the velocity function is

$$v(t) = s'(t) = -15t^2 + 60t.$$

**(c)** What is her velocity 1 hour into the trip? 2 hours into the trip? At the end of the trip?

**Solution** At $t = 1$, the velocity is

$$v(1) = s'(1) = -15(1)^2 + 60(1) = 45 \text{ miles per hour.}$$

At $t = 2$, the velocity is

$$v(2) = s'(2) = -15(2)^2 + 60(2) = 60 \text{ miles per hour.}$$

Since the trip takes 4 hours, the velocity at the end of the trip is

$$v(4) = s'(4) = -15(4)^2 + 60(4) = 0 \text{ miles per hour.} \quad ✓_5$$

✓ **Checkpoint 5**

In Example 5, if $s(t) = 3t^4 - 12t^2$, find the velocity after 2 hours.

The examples that follow illustrate some important applications of derivatives to business.

## Marginal Analysis

In business and economics, the rates of change of such variables as cost, revenue, and profit are important considerations. Economists use the word *marginal* to refer to rates of change: for example, *marginal cost* refers to the rate of change of cost with respect to the number of items produced. Since the derivative of a function gives the rate of change of the function, a marginal cost (or revenue, or profit) function is found by taking the derivative of the cost (or revenue, or profit) function. Roughly speaking, the marginal cost at some level of production $x$ is the cost of producing the $(x + 1)$st item, as we now show. (Similar statements could be made for revenue or profit.)

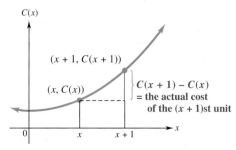

**Figure 11.34**

Look at Figure 11.34, where $C(x)$ represents the cost of producing $x$ units of some item. Then the cost of producing $x + 1$ units is $C(x + 1)$. The cost of the $(x + 1)$st unit is, therefore, $C(x + 1) - C(x)$. This quantity is shown on the graph in Figure 11.34.

Now, if $C$ is the cost function, then the marginal cost $C'$ represents the slope of the tangent line at any point $(x, C(x))$. The graph in Figure 11.35 shows the cost function $C$ and the tangent line at point $P = (x, C(x))$. We know that the slope of the tangent line at $P$ is $C'(x)$ and that the slope can be computed using the triangle $PQR$ in Figure 11.35:

$$C'(x) = \text{slope} = \frac{QR}{PR} = \frac{QR}{1} = QR.$$

So the length of the line segment $QR$ is the number $C'(x)$.

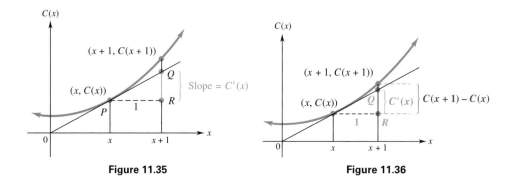

**Figure 11.35**                    **Figure 11.36**

Superimposing the graphs from Figures 11.34 and 11.35, as in Figure 11.36, shows that $C'(x)$ is indeed very close to $C(x + 1) - C(x)$. Therefore, we have the conclusion in the following box.

## Marginal Cost

If $C(x)$ is the cost function, then the **marginal cost** (rate of change of cost) is given by the derivative $C'(x)$:

$$C'(x) \approx \text{cost of making one more item after } x \text{ items have been made.}$$

The marginal revenue $R'(x)$ and marginal profit $P'(x)$ are interpreted similarly.

EXAMPLE 6     **Marginal Cost** A land developer has purchased 115 acres in the Blue Ridge Mountains to subdivide and sell as one-acre lots for mountain homes. Construction of gravel roads to access the lots is more expensive at higher elevations on the property. The cost of access roads (in thousands of dollars) for $x$ lots can be approximated by

$$C(x) = 150 + 9x + .5x^{1.6} \qquad (0 \le x \le 115).$$

Find and interpret the marginal cost for the given values of $x$.

**(a)**  $x = 25$

**Solution** To find the marginal cost, first find the derivative of the cost function:

$$C'(x) = 9 + .5(1.6x^{.6}) = 9 + .8x^{.6}.$$

When $x = 25$,

$$C'(25) = 9 + .8(25)^{.6} = 14.519.$$

After 25 access roads have been constructed, the cost of constructing one more access road will be *approximately* $14,519.

Note that the *actual* cost of constructing one more access road is:

$$C(26) - C(25) = (150 + 9(26) + .5(26)^{1.6}) - (150 + 9(25) + .5(25)^{1.6})$$
$$\approx 14.585.$$

So the actual cost to construct one additional road is $14,585.

**(b)**  $x = 50$

**Solution** After 50 access roads have been constructed, the cost of constructing an additional access road will be *approximately*

$$C'(50) = 9 + .8(50)^{.6} = 17.365,$$

or $17,365. Compare this result with that for part (a). The cost of accessing one additional lot is almost $3000 more after 50 lots have road access than the cost of accessing one additional lot when only 25 lots have road access. Management must be careful to keep track of marginal costs. If the marginal cost of producing an extra unit exceeds the revenue received from selling it, the company will lose money on that unit. ✓ 6

✓**Checkpoint 6**

A cost function is given by $C(x) = 2790 + 30x + 12\sqrt{x}$. Find the marginal cost at the given production levels.

**(a)**  $x = 1$

**(b)**  $x = 4$

## Demand Functions

The **demand function**, defined by $p = f(x)$, relates the number of units, $x$, of an item that consumers are willing to purchase at the price $p$. (Demand functions were also discussed in Section 3.3.) The total revenue $R(x)$ is related to the price per unit and the amount demanded (or sold) by the equation

Revenue = (number of items) · (price per item) or
$$R(x) = xp = x \cdot f(x).$$

EXAMPLE 7     **Marginal Revenue** The demand function for a certain product is given by

$$p = \frac{50,000 - x}{25,000}.$$

Find the marginal revenue when $x = 10,000$ units and $p$ is in dollars.

**Solution** From the function $p$, the revenue function is given by

$$R(x) = xp$$

$$= x\left(\frac{50,000 - x}{25,000}\right)$$

$$= \frac{50,000x - x^2}{25,000} = 2x - \frac{1}{25,000}x^2.$$

The marginal revenue is

$$R'(x) = 2 - \frac{2}{25,000}x.$$

When $x = 10,000$, the marginal revenue is

$$R'(10,000) = 2 - \frac{2}{25,000}(10,000) = 1.2,$$

or $1.20 per unit. Thus, the next unit sold (at sales of 10,000) will produce additional revenue of about $1.20. ✔ 7

✓ **Checkpoint 7**

Suppose the demand function for $x$ units of an item is

$$p = 5 - \frac{x}{1000},$$

where $p$ is the price in dollars. Find

**(a)** the marginal revenue;

**(b)** the marginal revenue at $x = 500$;

**(c)** the marginal revenue at $x = 1000$.

In economics, the demand function is written in the form $p = f(x)$, as in Example 7. From the perspective of a consumer, it is probably more reasonable to think of the quantity demanded as a function of price. Mathematically, these two viewpoints are equivalent. In Example 6, the demand function could have been written from the consumer's viewpoint as

$$x = 50,000 - 25,000p.$$

The next example considers marginal profit, the rate of change of the profit function. Recall from Section 3.3 that

$$\text{Profit} = \text{revenue} - \text{cost or}$$
$$P(x) = R(x) - C(x).$$

**EXAMPLE 8** **Marginal Profit** Suppose that the cost function for the product in Example 6 is given by

$$C(x) = 2100 + .25x \qquad (0 \le x \le 30,000).$$

Find the marginal profit from the production of the given numbers of units.

**(a)** 15,000

**Solution** From Example 7, the revenue from the sale of $x$ units is

$$R(x) = 2x - \frac{1}{25,000}x^2.$$

Since profit $P$ is given by $P = R - C$,

$$P(x) = R(x) - C(x)$$

$$= \left(2x - \frac{1}{25,000}x^2\right) - (2100 + .25x)$$

$$= 2x - \frac{1}{25,000}x^2 - 2100 - .25x$$

$$= 1.75x - \frac{1}{25,000}x^2 - 2100.$$

The marginal profit from the sale of $x$ units is

$$P'(x) = 1.75 - \frac{2}{25,000}x = 1.75 - \frac{1}{12,500}x.$$

At $x = 15,000$, the marginal profit is

$$P'(15,000) = 1.75 - \frac{1}{12,500}(15,000) = .55,$$

or $.55 per unit. This means that the next unit sold (at sales of 15,000) will produce additional profit of about 55¢.

**(b)** 21,875

**Solution** When $x = 21,875$, the marginal profit is

$$P'(21,875) = 1.75 - \frac{1}{12,500}(21,875) = 0.$$

**(c)** 25,000

**Solution** When $x = 25,000$, the marginal profit is

$$P'(25,000) = 1.75 - \frac{1}{12,500}(25,000) = -.25,$$

or $-$.25 per unit.

As shown by parts (b) and (c), if more than 21,875 units are sold, the marginal profit is negative. This indicates that increasing production beyond that level will *reduce* profit. ✔8

The final example shows a medical application of the derivative as the rate of change of a function.

**✔ Checkpoint 8**

For a certain product, the cost is $C(x) = 1300 + .80x$ and the revenue is $R(x) = 6x - \frac{2x^2}{11,000}$ for $x$ units.

**(a)** Find the profit $P(x)$.

**(b)** Find $P'(12,000)$.

**(c)** Find $P'(25,000)$.

**(d)** Interpret the results of parts (b) and (c).

---

**EXAMPLE 9**    **Tumor Growth**  A tumor has the approximate shape of a cone. (See Figure 11.37.) The radius of the tumor is fixed by the bone structure at 2 centimeters, but the tumor is growing along the height of the cone. The formula for the volume of a cone is $V = \frac{1}{3}\pi r^2 h$, where $r$ is the radius of the base and $h$ is the height of the cone. Find the rate of change of the volume of the tumor with respect to the height.

**Solution**  To emphasize that the rate of change of the volume is found with respect to the height, we use the symbol $dV/dh$ for the derivative. For this tumor, $r$ is fixed at 2 cm. By substituting 2 for $r$,

$$V = \frac{1}{3}\pi r^2 h \quad \text{becomes} \quad V = \frac{1}{3}\pi \cdot 2^2 \cdot h, \quad \text{or} \quad V = \frac{4}{3}\pi h.$$

Since $4\pi/3$ is constant,

$$\frac{dV}{dh} = \frac{4\pi}{3} \approx 4.2 \text{ cu cm per cm}.$$

For each additional centimeter that the tumor grows in height, its volume will increase by approximately 4.2 cubic centimeters. ✔9

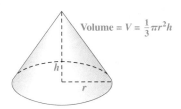

Volume $= V = \frac{1}{3}\pi r^2 h$

**Figure 11.37**

**✔ Checkpoint 9**

A balloon is spherical. The formula for the volume of a sphere is $V = (4/3)\pi r^3$, where $r$ is the radius of the sphere. Find the given quantities.

**(a)** $dV/dr$

**(b)** The rate of change of the volume when $r = 3$ inches

---

# 11.5 Exercises

*Find the derivatives of the given functions. (See Examples 1–4.)*

**1.** $f(x) = 4x^2 - 3x + 5$

**2.** $g(x) = 8x^2 + 2x - 12$

**3.** $y = 2x^3 + 3x^2 - 6x + 2$

**4.** $y = 4x^3 + 24x + 4$

**5.** $g(x) = x^4 + 3x^3 - 8x - 7$

6. $f(x) = 6x^6 - 3x^4 + x^3 - 3x + 9$

7. $f(x) = 6x^{1.5} - 4x^{.5}$      8. $f(x) = -2x^{2.5} + 8x^{.5}$

9. $y = -15x^{3/2} + 2x^{1.9} + x$      10. $y = 5x^{5/2} - 4x^{3.1} + 9x$

11. $y = 24t^{3/2} + 4t^{1/2}$      12. $y = -24t^{5/2} - 6t^{1/2}$

13. $y = 8\sqrt{x} + 6x^{3/4}$      14. $y = -100\sqrt{x} - 11x^{2/3}$

15. $g(x) = 6x^{-5} - 2x^{-1}$      16. $y = 8x^{-2} + 6x^{-1} + 1$

17. $y = 10x^{-2} + 6x^{-4} + 3x$

18. $y = 3x^{-7} + 4x^{-5} - 9x^{-4} + 21x^{-1}$

19. $f(t) = \dfrac{-4}{t} + \dfrac{8}{t^2}$      20. $f(t) = \dfrac{8}{t} - \dfrac{5}{t^2}$

21. $f(t) = -\dfrac{1}{4t} + \dfrac{1}{8t^2}$      22. $f(t) = \dfrac{1}{8t} - \dfrac{1}{5t^2}$

23. $g(x) = 5x^{-1/2} - 7x^{1/2} + 101x$

24. $f(x) = -8x^{-1/2} - 8x^{1/2} + 12x$

25. $y = 7x^{-3/2} + 8x^{-1/2} + x^3 - 9$

26. $y = 2x^{-3/2} + 9x^{-1/2} + x^{-4} - x^2$

27. $y = \dfrac{19}{\sqrt[4]{x}}$      28. $y = \dfrac{-5}{\sqrt[3]{x}}$

29. $y = \dfrac{-8t}{\sqrt[3]{t^2}}$      30. $g(t) = \dfrac{10t}{\sqrt[3]{t^8}}$

*Find each of the given derivatives. (See Examples 1–4.)*

31. $\dfrac{dy}{dx}$ if $y = 8x^{-5} - 9x^{-4} + 9x^4$

32. $\dfrac{dy}{dx}$ if $y = -3x^{-2} - 4x^{-5} + 5x^{-7}$

33. $\dfrac{d}{dx}\left(9x^{-1/2} + \dfrac{2}{x^{3/2}}\right)$      34. $\dfrac{d}{dx}\left(8x^{-1/4} - \dfrac{3}{\sqrt{x^3}}\right)$

35. $f'(-2)$ if $f(x) = 6x^2 - 2x$      36. $f'(3)$ if $f(x) = 9x^3 - 6x^2$

37. $f'(4)$ if $f(t) = 2\sqrt{t} - \dfrac{3}{\sqrt{t}}$      38. $f'(8)$ if $f(t) = -5\sqrt[3]{t} + \dfrac{6}{\sqrt[3]{t}}$

39. $y'$ if $y = \dfrac{12 - 7x + 6x^3}{x^4}$      40. $y'$ if $y = \dfrac{102 - 6x + 17x^4}{x^6}$

C *Find the slope and the equation of the tangent line to the graph of each function at the given value of x.*

41. $f(x) = x^4 - 2x^2 + 1; \ x = 1$

42. $g(x) = -x^5 + 4x^2 - 2x + 2; \ x = 2$

43. $y = 4x^{1/2} + 2x^{3/2} + 1; \ x = 4$

44. $y = -x^{-3} + 5x^{-1} + x; \ x = 2$

*Work these exercises. (See Examples 6–9.)*

45. **Marginal Cost** The total cost to produce $x$ Bluetooth speakers is

$$C(x) = 100 + 12x + .1x^2 + .001x^3.$$

Find the marginal cost for the given values of $x$.

(a) $x = 0$      (b) $x = 10$

(c) $x = 30$      (d) $x = 50$

46. **Marginal Cost** The total cost in dollars to produce $x$ hundred T-shirts imprinted with a college logo can be approximated by

$$C(x) = 1500 + 1000x^{.9} \qquad (0 \le x \le 50).$$

Find the marginal cost for the given production levels.

(a) 1500 shirts      (b) 2000 shirts

47. **Marginal Profit** The profit in dollars from the sale of $x$ smart-watches is

$$P(x) = .03x^2 - 4x + 3x^{.8} - 5000.$$

Find the marginal profit for the given values of $x$.

(a) $x = 100$      (b) $x = 1000$

(c) $x = 5000$      (d) $x = 10,000$

48. **Marginal Profit** The profit in dollars from the sale of $x$ drones can be approximated by

$$P(x) = 400x - 490\sqrt{x} - 600 \qquad (5 \le x \le 25).$$

Find the marginal profit for the given values of $x$.

(a) $x = 10$      (b) $x = 15$

49. **Marginal Revenue** The revenue equation (in hundreds of millions of dollars) for barley production in the United States is approximated by

$$R(x) = .0608x^2 + 1.4284x + 2.3418,$$

where $x$ is in hundreds of millions of bushels. (Data from: www.usda.gov.)

(a) Find the marginal-revenue function.

(b) Find the marginal revenue for the production of 500,000,000 bushels.

(c) Find the marginal revenue for the production of 700,000,000 bushels.

50. **Marginal Revenue** The revenue equation (in billions of dollars) for corn production in the United States is approximated by

$$R(x) = .016x^2 + 1.034x + 11.338,$$

where $x$ is in billions of bushels. (Data from: www.usda.gov.)

(a) Find the marginal-revenue function.

(b) Find the marginal revenue for the production of 10 billion bushels.

(c) Find the marginal revenue for the production of 20 billion bushels.

51. **Trending Products** Often, sales of a new trending product grow rapidly at first and then slow down with time. This is the case with the sales represented by the function

$$S(t) = 10,000 - 10,000t^{-.2} + 100t^{.1},$$

where $t$ represents time in years. Find the rate of change of sales for the given values of $t$.

(a) $t = 1$      (b) $t = 10$

**52. Beer Production** The profit (in dollars) from producing $x$ hundred kegs of beer is given by $P(x) = 1500 + 30x - 2x^2$. Find the marginal profit at the given production levels. In each case, decide whether the firm should increase production of beer.

**(a)** 700          **(b)** 650

**(c)** 1200        **(d)** 1800

**53. Matcha Tea** The revenue generated from the sale of $x$ containers of matcha tea is given by

$$R(x) = 20x - \frac{x^2}{500}.$$

**(a)** Find the marginal revenue when $x = 1000$ units.

**(b)** Determine the actual revenue from the sale of the 1001st unit.

**(c)** Compare the answers to parts (a) and (b). How are they related?

**54. Charge Sticks** The demand for a certain portable USB battery charger is given by $D(p) = -p^2 + 5p + 1$, where $p$ represents the price in dollars.

**(a)** Find the rate of change of demand with respect to price.

**(b)** Find and interpret the rate of change of demand when the price is $12.

**Demand Functions** *Work these exercises. (See Example 7.)*

**55.** The demand equation for handcrafted violins by a certain violin maker can be approximated by $p = 24 - x$ where $p$ is the price in thousands of dollars and $x$ is the quantity of violins demanded. Find and interpret the marginal revenue for each of the given production levels.

**(a)** $x = 5$          **(b)** $x = 10$          **(c)** $x = 12$

**56.** The cost (in thousands of dollars) of making $x$ violins described in the previous exercise can be approximated by $C(x) = 15 + 1.2x$. Find the marginal profit for each of the given production levels.

**(a)** $x = 5$          **(b)** $x = 10$          **(c)** $x = 12$

**57.** Assume that a demand equation is given by $x = 5000 - 100p$. Find the marginal revenue for the given production levels (values of $x$). (*Hint:* Solve the demand equation for $p$ and use $R(x) = xp$.)

**(a)** 1000 units    **(b)** 2500 units    **(c)** 3000 units

**58.** Suppose that, for the situation in Exercise 57, the cost of producing $x$ units is given by $C(x) = 3000 - 20x + .03x^2$. Find the marginal profit for each of the given production levels.

**(a)** 500 units      **(b)** 815 units      **(c)** 1000 units

*Work these exercises.*

**59. Insurance Rates for Women** An insurance company typically offers discounts for purchasing larger amounts of life insurance coverage. For a female under 30 years of age who qualifies for the lowest premiums, the annual cost (in dollars) of a life insurance policy can be approximated by

$$C(x) = 3\sqrt[4]{x^3},$$

where $x$ is the amount of life insurance coverage (in thousands of dollars). Find and interpret the given quantities. (Data from: www.abendowment.org/insurance/life20_rate.asp.)

**(a)** $C(100)$ and $C'(100)$          **(b)** $C(1000)$ and $C'(1000)$

**60. Insurance Rates for Men** For a male 50 years of age who qualifies for the lowest premiums, the annual cost (in dollars) of a life insurance policy can be approximated by

$$C(x) = 5.60x^{.88},$$

where $x$ is the amount of life insurance coverage (in thousands of dollars). Find and interpret the given quantities. (Data from: www.abendowment.org/insurance/life20_rate.asp.)

**(a)** $C(100)$ and $C'(100)$          **(b)** $C(1000)$ and $C'(1000)$

**61. Marketing to Youth** A marketing firm is interested in the percentage of the total United States population that is age 0–17 years. This percentage can be estimated by

$$P(x) = 26x^{-.03},$$

where $x = 0$ corresponds to the year 2000 and the estimate is valid through the year 2025. Find and interpret the given quantities. (Data from: www.childstats.gov.)

**(a)** $P(14)$ and $P'(14)$          **(b)** $P(24)$ and $P'(24)$

**62. Marketing to Seniors** A marketing firm is interested in the percentage of the total United States population that is age 65 years or older. This percentage can be estimated by

$$P(x) = .01x^2 - .01x + 12,$$

where $x = 0$ corresponds to the year 2000 and the estimate is valid through the year 2025. Find and interpret the given quantities. (Data from: www.childstats.gov.)

**(a)** $P(14)$ and $P'(14)$          **(b)** $P(24)$ and $P'(24)$

**63. Case-Shiller National Index** The Case-Shiller U.S. National Home Price Index can be approximated by

$$C(x) = 2x^2 - 47x + 415,$$

where $x = 8$ corresponds to the year 2008. (Data from: Standard and Poor's.)

**(a)** Find and interpret $C(10)$ and $C'(10)$.

**(b)** Find and interpret $C(16)$ and $C'(16)$.

**64. Case-Shiller City Index** The Case-Shiller Seattle Home Price Index can be approximated by

$$C(x) = 3x^2 - 71x + 555,$$

where $x = 8$ corresponds to the year 2008. (Data from: Standard and Poor's.)

**(a)** Find and interpret $C(10)$ and $C'(10)$.

**(b)** Find and interpret $C(16)$ and $C'(16)$.

**65. Centenarians** According to the U.S. Census Bureau, the number of Americans (in thousands) who are expected to be over 100 years old in year $x$ is approximated by the function $f(x) = .27x^2 + 3.52x + 51.78$, where $x = 0$ corresponds to the year 2000 and the formula is valid through 2045.

**(a)** Find a formula giving the rate of change in the number of Americans over 100 years old.

**(b)** What is the rate of change in the number of Americans expected to be over 100 years old in the year 2020?

**(c)** Is the number of Americans expected to be over 100 years old increasing or decreasing in 2020?

66. **100-Meter Dash** The women's 100-meter dash was introduced to the Olympic Games in 1928. Between 1928 and 2012, the winning time (in seconds) can be approximated by

$$W(x) = .000178x^2 - .0404x + 13.068,$$

where $x = 0$ corresponds to the year 1900. (Data from: www.Olympic.org.)

**(a)** Would you expect the derivative of $W(x)$ to be positive, negative, or zero? Why?

**(b)** Find and interpret the rate of change of $W(x)$ in 1928.

**(c)** Find and interpret the rate of change of $W(x)$ in 2012.

67. **Blood Vessels** A short length of blood vessel has a cylindrical shape. The volume of a cylinder is given by $V = \pi r^2 h$. Suppose an experimental device is set up to measure the volume of blood in a blood vessel of fixed length 80 mm as the radius changes. Then the volume of this particular blood vessel is

$$V = \pi r^2(80) = 80\pi r^2.$$

**(a)** Find $dV/dr$.

Suppose a drug is administered that causes the blood vessel to expand. Evaluate $dV/dr$ for the following values of $r$, and interpret your answers:

**(b)** 4 mm;          **(c)** 6 mm;          **(d)** 8 mm.

68. **Birth Weight** The birth weight (in pounds) of an infant can be approximated by the function

$$g(x) = .0011x^{2.394},$$

where $x$ represents the number of completed weeks of gestation.

**(a)** Find the function $g'(x)$.

**(b)** Evaluate $g'(32)$.

**(c)** Evaluate $g'(40)$.

**(d)** Interpret the results of parts (b) and (c).

69. **Body Mass Index** The body mass index (BMI) is a number that can be calculated for any individual using the equation BMI $= 703w/h^2$, where $w$ is weight in pounds and $h$ is height in inches. For 100-pound individuals, this formula becomes

$$f(h) = \frac{703(100)}{h^2} = \frac{70,300}{h^2}.$$

**(a)** Find $f'(h)$, the rate of change of BMI with respect to height for 100-pound individuals.

**(b)** Would you expect $f'(h)$ to be positive or negative? Why?

**(c)** Find $f'(60)$, and round the answer to two decimal places.

70. **Body Mass Index** For 200-pound individuals, the BMI function is

$$f(h) = \frac{140,600}{h^2},$$

where $h$ is height in inches. Find and interpret the following.

**(a)** Find $f'(h)$, the rate of change of BMI with respect to height for 200-pound individuals.

**(b)** Would you expect $f'(h)$ to be positive or negative? Why?

**(c)** Find $f'(72)$, and round the answer to two decimal places.

**Velocity** *For each of the given position functions $s(t)$, find (a) the velocity function $v(t) = s'(t)$; and (b) the velocity when $t = 0, t = 5$, and $t = 10$. (See Example 5.)*

71. $s(t) = 8t^2 + 3t + 1$          72. $s(t) = 10t^2 - 5t + 6$

73. $s(t) = 2t^3 + 6t^2$          74. $s(t) = -t^3 + 3t^2 + t - 1$

75. **Velocity** If a rock is dropped from a 144-foot-high building, its position (in feet above the ground) is given by $s(t) = -16t^2 + 144$, where $t$ is the time in seconds since it was dropped.

**(a)** What is its velocity 1 second after being dropped? 2 seconds after being dropped?

**(b)** When will it hit the ground? (Hint: Solve $s(t) = 0$.)

**(c)** What is its velocity upon impact (at the instant when it hits the ground)?

76. **Velocity** A ball is thrown vertically upward from the ground at a velocity of 64 feet per second. Its distance from the ground at $t$ seconds is given by $s(t) = -16t^2 + 64t$.

**(a)** How fast is the ball moving 2 seconds after being thrown? 3 seconds after being thrown?

**(b)** When will it hit the ground?

**(c)** What is its velocity upon impact?

77. Perform each step and give reasons for your results in the following proof that the derivative of $y = x^n$ is $y' = n \cdot x^{n-1}$ (we prove this result only for positive integer values of $n$, but it is valid for all values of $n$):

**(a)** Recall the binomial theorem from algebra:

$$(p + q)^n = p^n + n \cdot p^{n-1}q + \frac{n(n-1)}{2}p^{n-2}q^2 + \cdots + q^n.$$

Evaluate $(x + h)^n$.

**(b)** Find the quotient $\dfrac{(x + h)^n - x^n}{h}$.

**(c)** Use the definition of the derivative to find $y'$.

78. Perform each step and give reasons for your results in the following proof that the derivative of $y = f(x) + g(x)$ is

$$y' = f'(x) + g'(x).$$

**(a)** Let $s(x) = f(x) + g(x)$. Show that

$$s'(x) = \lim_{h \to 0} \frac{[f(x + h) + g(x + h)] - [f(x) + g(x)]}{h}.$$

**(b)** Show that

$$s'(x) = \lim_{h \to 0} \left[ \frac{f(x + h) - f(x)}{h} + \frac{g(x + h) - g(x)}{h} \right].$$

**(c)** Finally, show that $s'(x) = f'(x) + g'(x)$.

*Use a graphing calculator or computer to graph each function and its derivative on the same screen. Determine the values of $x$ where the derivative is (a) positive, (b) zero, and (c) negative. (d) What is true of the graph of the function in each case?*

79. $g(x) = 6 - 4x + 3x^2 - x^3$          80. $k(x) = 2x^4 - 3x^3 + x$

## Checkpoint Answers

1. (a) 0    (b) 0    (c) 0

2. (a) $y' = 4x^3$    (b) $y' = 17x^{16}$    (c) $dy/dx = -2/x^3$

   (d) $dy/dt = -5/t^6$    (e) $y' = \dfrac{5}{4}t^{1/4}$

3. (a) $36x^2$    (b) $210t^6$    (c) $-35$

   (d) $(5/2)x^{-1/2}$, or $5/(2\sqrt{x})$    (e) $10t^{-2}$, or $10/t^2$

4. (a) $y' = -30x^2 - 6$

   (b) $y' = \dfrac{40}{7}t^{1/7} + \dfrac{5}{2}t^{-1/2}$, or $y' = \dfrac{40}{7}t^{1/7} + \dfrac{5}{2t^{1/2}}$

   (c) $f'(t) = \dfrac{-1}{\sqrt{t}} - \dfrac{4}{t^2}$

5. 48 miles per hour

6. (a) \$36 per unit    (b) \$33 per unit

7. (a) $R'(x) = 5 - \dfrac{x}{500}$

   (b) \$4    (c) \$3

8. (a) $P(x) = 5.2x - \dfrac{2x^2}{11{,}000} - 1300$

   (b) .84    (c) $-3.89$

   (d) Profit is increasing by \$.84 per unit at 12,000 units in part (b) and decreasing by \$3.89 per unit at 25,000 units in part (c).

9. (a) $4\pi r^2$

   (b) $36\pi$ cubic inches per inch

# 11.6 Derivatives of Products and Quotients

In the last section, we saw that the derivative of the sum of two functions can be obtained by taking the sum of the derivatives. What about products? Is the derivative of a product of two functions equal to the product of their derivatives? For example, if

$$u(x) = 2x + 4 \quad \text{and} \quad v(x) = 3x^2,$$

then the product of $u$ and $v$ is

$$f(x) = u(x) \cdot v(x) = (2x + 4)(3x^2) = 6x^3 + 12x^2.$$

Using the rules of the last section, we have

$$u'(x) = 2, \quad v'(x) = 6x, \quad \text{and} \quad f'(x) = 18x^2 + 24x,$$

so that

$$u'(x) \cdot v'(x) = 12x \quad \text{and} \quad f'(x) = [u(x) \cdot v(x)]' = 18x^2 + 24x.$$

Obviously, these two functions are *not* the same, which shows that $[u(x) \cdot v(x)]' \neq u'(x) \cdot v'(x)$, as cautioned below. ✔1

## ✔ Checkpoint 1

Let $u(x) = x^2$ and $v(x) = x^3$.

(a) Find $f(x) = u(x) \cdot v(x)$.

(b) Find $f'(x) = [u(x) \cdot v(x)]'$.

(c) Find $u'(x)$.

(d) Find $v'(x)$.

(e) Find $u'(x) \cdot v'(x)$.

(f) Check that $[u(x) \cdot v(x)]' \neq u'(x) \cdot v'(x)$.

⚠ **CAUTION** The derivative of a product is *not* equal to the product of the derivatives.

The correct rule for finding the derivative of a product is as follows.

### Product Rule

If $f(x) = u(x) \cdot v(x)$, and if both $u'(x)$ and $v'(x)$ exist, then

$$f'(x) = u(x) \cdot v'(x) + v(x) \cdot u'(x).$$

(The derivative of a product of two functions is the first function times the derivative of the second, plus the second function times the derivative of the first.)

To sketch the method used to prove the product rule, let

$$f(x) = u(x) \cdot v(x).$$

Then $f(x + h) = u(x + h) \cdot v(x + h)$, and, by definition,

$$f'(x) = \lim_{h \to 0} \frac{f(x + h) - f(x)}{h}$$

$$= \lim_{h \to 0} \frac{u(x + h) \cdot v(x + h) - u(x) \cdot v(x)}{h}.$$

Now subtract and add $u(x + h) \cdot v(x)$ in the numerator, giving

$$f'(x) = \lim_{h \to 0} \frac{u(x + h) \cdot v(x + h) - u(x + h) \cdot v(x) + u(x + h) \cdot v(x) - u(x) \cdot v(x)}{h}.$$

Factor $u(x + h)$ from the first two terms of the numerator, and factor $v(x)$ from the last two terms. Then use the properties of limits (Section 11.1) as follows:

$$f'(x) = \lim_{h \to 0} \frac{u(x + h)[v(x + h) - v(x)] + v(x)[u(x + h) - u(x)]}{h}$$

$$= \lim_{h \to 0} u(x + h)\left[\frac{v(x + h) - v(x)}{h}\right] + \lim_{h \to 0} v(x)\left[\frac{u(x + h) - u(x)}{h}\right]$$

$$= \lim_{h \to 0} u(x + h) \cdot \lim_{h \to 0} \frac{v(x + h) - v(x)}{h} + \lim_{h \to 0} v(x) \cdot \lim_{h \to 0} \frac{u(x + h) - u(x)}{h}. \quad (*)$$

If $u'$ and $v'$ both exist, then

$$\lim_{h \to 0} \frac{u(x + h) - u(x)}{h} = u'(x) \quad \text{and} \quad \lim_{h \to 0} \frac{v(x + h) - v(x)}{h} = v'(x).$$

The fact that $u'$ exists can be used to prove that

$$\lim_{h \to 0} u(x + h) = u(x),$$

and since no $h$ is involved in $v(x)$,

$$\lim_{h \to 0} v(x) = v(x).$$

Substituting these results into equation (*) gives

$$f'(x) = u(x) \cdot v'(x) + v(x) \cdot u'(x),$$

the desired result.

**EXAMPLE 1**   Let $f(x) = (2x + 4)(3x^2)$. Use the product rule to find $f'(x)$.

**Solution**  Here, $f$ is given as the product of $u(x) = 2x + 4$ and $v(x) = 3x^2$. By the product rule and the fact that $u'(x) = 2$ and $v'(x) = 6x$,

$$f'(x) = u(x) \cdot v'(x) + v(x) \cdot u'(x)$$

$$= (2x + 4)(6x) + (3x^2)(2)$$

$$= 12x^2 + 24x + 6x^2 = 18x^2 + 24x.$$

This result is the same as that found at the beginning of the section. ✓ 2

✓ **Checkpoint 2**

Use the product rule to find the derivatives of the given functions.

**(a)** $f(x) = (5x^2 + 6)(3x)$

**(b)** $g(x) = (8x)(4x^2 + 5x)$

✓ **Checkpoint 3**

Find the derivatives of the given functions.

**(a)** $f(x) = (x^2 - 3)(\sqrt{x} + 5)$

**(b)** $g(x) = (\sqrt{x} + 4)(5x^2 + x)$

**EXAMPLE 2**   Find the derivative of $y = (\sqrt{x} + 3)(x^2 - 5x)$.

**Solution**  Let $u(x) = \sqrt{x} + 3 = x^{1/2} + 3$ and $v(x) = x^2 - 5x$. Then

$$y' = u(x) \cdot v'(x) + v(x) \cdot u'(x)$$

$$= (x^{1/2} + 3)(2x - 5) + (x^2 - 5x)\left(\frac{1}{2}x^{-1/2}\right)$$

$$= 2x^{3/2} - 5x^{1/2} + 6x - 15 + \frac{1}{2}x^{3/2} - \frac{5}{2}x^{1/2}$$

$$= \frac{5}{2}x^{3/2} + 6x - \frac{15}{2}x^{1/2} - 15. \quad ✓ 3$$

We could have found the derivatives in Examples 1 and 2 by multiplying out the original functions. The product rule then would not have been needed. In the next section, however, we shall see products of functions where the product rule is essential.

What about *quotients* of functions? To find the derivative of the quotient of two functions, use the next rule. The proof of the quotient rule is similar to that of the product rule and is omitted here.

> ## Quotient Rule
>
> If $f(x) = \dfrac{u(x)}{v(x)}$, if all indicated derivatives exist, and if $v(x) \neq 0$, then
>
> $$f'(x) = \frac{v(x) \cdot u'(x) - u(x) \cdot v'(x)}{[v(x)]^2}.$$
>
> (The derivative of a quotient is the denominator times the derivative of the numerator, minus the numerator times the derivative of the denominator, all divided by the square of the denominator.)

⬡ **CAUTION** Just as the derivative of a product is *not* the product of the derivatives, the derivative of a quotient is *not* the quotient of the derivatives. If you are asked to take the derivative of a product or a quotient, it is essential that you recognize that the function contains a product or quotient and then use the appropriate rule.

**EXAMPLE 3**    Find $f'(x)$ if

$$f(x) = \frac{2x - 1}{4x + 3}.$$

**Solution**  Let $u(x) = 2x - 1$, with $u'(x) = 2$. Also, let $v(x) = 4x + 3$, with $v'(x) = 4$. Then, by the quotient rule,

$$f'(x) = \frac{v(x) \cdot u'(x) - u(x) \cdot v'(x)}{[v(x)]^2}$$

$$= \frac{(4x + 3)(2) - (2x - 1)(4)}{(4x + 3)^2}$$

$$= \frac{8x + 6 - 8x + 4}{(4x + 3)^2}$$

$$f'(x) = \frac{10}{(4x + 3)^2}. \quad ✔_4$$

✓ **Checkpoint 4**

Find the derivatives of the given functions.

(a)  $f(x) = \dfrac{3x + 7}{5x + 8}$

(b)  $g(x) = \dfrac{2x + 11}{5x - 1}$

⬡ **CAUTION** In the second step of Example 3, we had the expression

$$\frac{(4x + 3)(2) - (2x - 1)(4)}{(4x + 3)^2}.$$

Students often incorrectly "cancel" the $4x + 3$ in the numerator with one factor of the denominator. Because the numerator is a *difference* of two products, however, you must multiply and combine terms *before* looking for common factors in the numerator and denominator.

**EXAMPLE 4**    Find the derivative

$$\frac{d}{dx}\left(\frac{x - 2x^2}{4x^2 + 1}\right).$$

**Solution** Using the quotient rule,

$$\frac{d}{dx}\left(\frac{x - 2x^2}{4x^2 + 1}\right) = \frac{(4x^2 + 1)\dfrac{d}{dx}(x - 2x^2) - (x - 2x^2)\dfrac{d}{dx}(4x^2 + 1)}{(4x^2 + 1)^2}$$

$$= \frac{(4x^2 + 1)(1 - 4x) - (x - 2x^2)(8x)}{(4x^2 + 1)^2}$$

$$= \frac{4x^2 - 16x^3 + 1 - 4x - 8x^2 + 16x^3}{(4x^2 + 1)^2}$$

$$= \frac{-4x^2 - 4x + 1}{(4x^2 + 1)^2}. \quad ✓\,5$$

✓ **Checkpoint 5**

Find the derivative of each function. Write your answer with positive exponents.

(a) $f(x) = \dfrac{x^{-2} - 1}{x^{-1} + 2}$

(b) $f(x) = \dfrac{2 + x^{-1}}{x^3 + 1}$

**EXAMPLE 5**  Find $f'(x)$, if

$$f(x) = \frac{(3 - 4x)(5x + 1)}{7x - 9}.$$

**Solution** This function has a product within a quotient. Instead of multiplying the factors in the numerator first (which is an option), we can use the quotient rule together with the product rule. Use the quotient rule first to get

$$f'(x) = \left[\frac{(3 - 4x)(5x + 1)}{7x - 9}\right]'$$

$$= \frac{(7x - 9)[(3 - 4x)(5x + 1)]' - [(3 - 4x)(5x + 1)](7x - 9)'}{(7x - 9)^2}.$$

Now use the product rule to find $[(3 - 4x)(5x + 1)]'$ in the numerator.

$$= \frac{(7x - 9)[(3 - 4x)5 + (5x + 1)(-4)] - (3 + 11x - 20x^2)(7)}{(7x - 9)^2}$$

$$= \frac{(7x - 9)(15 - 20x - 20x - 4) - (21 + 77x - 140x^2)}{(7x - 9)^2}$$

$$= \frac{(7x - 9)(11 - 40x) - 21 - 77x + 140x^2}{(7x - 9)^2}$$

$$= \frac{-280x^2 + 437x - 99 - 21 - 77x + 140x^2}{(7x - 9)^2}$$

$$= \frac{-140x^2 + 360x - 120}{(7x - 9)^2}. \quad ✓\,6$$

✓ **Checkpoint 6**

Find the derivative of each function.

(a) $f(x) = \dfrac{(3x - 1)(4x + 2)}{2x}$

(b) $f(x) = \dfrac{5x^2}{(2x + 1)(x - 1)}$

## Average Cost

Suppose $y = C(x)$ gives the total cost of manufacturing $x$ items. As mentioned earlier, the average cost per item is found by dividing the total cost by the number of items. The rate of change of average cost, called the *marginal average cost*, is the derivative of the average cost.

### Average Cost and Marginal Average Cost

If the total cost of manufacturing $x$ items is given by $C(x)$, then the **average cost per item** is

$$\overline{C}(x) = \frac{C(x)}{x}.$$

The **marginal average cost** is the derivative of the average-cost function $\overline{C}'(x)$.

**Solution**   One way is to let $f(x) = 4x^2 + 5$ and $g(x) = \sqrt{x}$, so that

$$g[f(x)] = g[4x^2 + 5] = \sqrt{4x^2 + 5} = h(x).$$

Another way is to let $k(x) = 4x^2$ and $t(x) = \sqrt{x + 5}$; then

$$t[k(x)] = t[4x^2] = \sqrt{4x^2 + 5} = h(x). \quad \checkmark_3$$

✓ **Checkpoint 3**

Express the given function as a composite of two other functions.

**(a)**   $h(x) = (7x^2 + 5)^4$

**(b)**   $h(x) = \sqrt{15x^2 + 1}$

## The Chain Rule

The product and quotient rules tell how to find the derivative of $fg$ and $f/g$ from the derivatives of $f$ and $g$. To develop a way to find the derivative of the composite function $f[g(x)]$ from the derivatives of $f$ and $g$, we consider an example.

Metal expands or contracts as the temperature changes, and the temperature changes over a period of time. Think of the length $y$ of a metal bar as a function of the temperature $d$, say, $y = f(d)$, and the temperature $d$ as a function of time $x$, say, $d = g(x)$. Then

$$y = f(d) = f(g(x)).$$

So the composite function $f(g(x))$ gives the length $y$ as a function of time $x$.

Suppose that the length of the bar is increasing at the rate of 2 mm for every degree increase in temperature and that the temperature is increasing at the rate of 4° per hour. In functional notation, this means that $f'(d) = 2$ and $g'(x) = 4$. During the course of an hour, the length of the bar will increase by $2 \cdot 4 = 8$ mm. So the rate of change of the length $y$ with respect to time $x$ is 8 mm/hr; that is, $y' = 8$. Consequently,

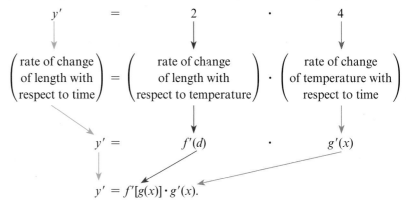

This example shows how the derivative of the composite function $y$ is related to the derivatives of $f$ and $g$. Although we omit the proof, the same result holds for any composite function.

> ## Chain Rule
>
> If $f$ and $g$ are functions and $y = f[g(x)]$, then
>
> $$y' = f'[g(x)] \cdot g'(x),$$
>
> provided that $f'[g(x)]$ and $g'(x)$ exist.
>
> (To find the derivative of $f[g(x)]$, find the derivative of $f(x)$, replace each $x$ with $g(x)$, and multiply the result by the derivative of $g(x)$.)

**EXAMPLE 4**   Use the chain rule to find the derivative of $y = \sqrt{15x^2 + 1}$.

**Solution**   First write $y = \sqrt{15x^2 + 1}$ as $y = (15x^2 + 1)^{1/2}$. Next, to write $y$ as the composition of two functions in a way that will be useful in applying the chain rule, we let $f(x) = x^{1/2}$ and $g(x) = 15x^2 + 1$. Then $\sqrt{15x^2 + 1} = f[g(x)]$ and

$$y' = f'[g(x)] \cdot g'(x).$$

Here, $f'(x) = \frac{1}{2}x^{-1/2}$, with $f'[g(x)] = \frac{1}{2}[g(x)]^{-1/2}$, and $g'(x) = 30x$ so that

$$
\begin{aligned}
y' = f'[g(x)] \cdot g'(x) &= \frac{1}{2}[g(x)]^{-1/2} \cdot g'(x) \\
&= \frac{1}{2}(15x^2 + 1)^{-1/2} \cdot (30x) \\
&= \frac{15x}{(15x^2 + 1)^{1/2}}. \quad \text{✓}_4
\end{aligned}
$$

✓ **Checkpoint 4**

Let $y = \sqrt{2x^4 + 3}$.

Find $y'$.

The chain rule can also be stated in the Leibniz notation for derivatives. If $y$ is a function of $u$, say, $y = f(u)$, and $u$ is a function of $x$, say, $u = g(x)$, then

$$
f'(u) = \frac{dy}{du} \quad \text{and} \quad g'(x) = \frac{du}{dx}.
$$

Now $y$ can be considered a function of $x$, namely, $y = f(u) = f[g(x)]$. According to the chain rule, the derivative of $y$ is

$$
\frac{dy}{dx} = f'[g(x)] \cdot g'(x) = f'(u) \cdot g'(x) = \frac{dy}{du} \cdot \frac{du}{dx}.
$$

Thus, we have the following alternative version of the chain rule.

## The Chain Rule (Alternative Form)

If $y$ is a function of $u$, say, $y = f(u)$, and if $u$ is a function of $x$, say, $u = g(x)$, then $y = f(u) = f[g(x)]$ and

$$
\frac{dy}{dx} = \frac{dy}{du} \cdot \frac{du}{dx},
$$

provided that $dy/du$ and $du/dx$ exist.

One way to remember the chain rule is to *pretend* that $dy/du$ and $du/dx$ are fractions, with $du$ "canceling out." We caution that this is merely a memory aid.

**EXAMPLE 5**    Find $dy/dx$ if $y = (3x^2 - 5x)^7$.

**Solution** Let $y = u^7$, where $u = 3x^2 - 5x$. Then $\dfrac{dy}{du} = 7u^6$ and $\dfrac{du}{dx} = 6x - 5$.

Hence,

$$
\begin{aligned}
\frac{dy}{dx} &= \frac{dy}{du} \cdot \frac{du}{dx} \\
&= 7u^6 \cdot (6x - 5).
\end{aligned}
$$

✓ **Checkpoint 5**

Use the chain rule to find $dy/dx$ if $y = 10(2x^2 + 1)^4$.

Replacing $u$ with $3x^2 - 5x$ gives

$$
\frac{dy}{dx} = 7(3x^2 - 5x)^6(6x - 5). \quad \text{✓}_5
$$

Each of the functions in Examples 4 and 5 has the form $y = u^n$, with $u$ a function of $x$:

$$
y = (15x^2 + 1)^{1/2} \quad \text{and} \quad y = (3x^2 - 5x)^7.
$$
$$
y = u^{1/2} \qquad\qquad\qquad\qquad y = u^7.
$$

Their derivatives are

$$y' = \frac{1}{2}(15x^2 + 1)^{-1/2}(30x) \quad \text{and} \quad y' = 7(3x^2 - 5x)^6(6x - 5)$$

$$y' = \frac{1}{2}u^{-1/2}\,u' \qquad\qquad\qquad\qquad y' = 7u^6\,u'.$$

Thus, these functions are examples of the following special case of the chain rule.

## Generalized Power Rule

Let $u$ be a function of $x$, and let $y = u^n$, for any real number $n$. Then

$$y' = n \cdot u^{n-1} \cdot u',$$

provided that $u'$ exists.

(The derivative of $y = u^n$ is found by decreasing the exponent on $u$ by 1 and multiplying the result by the exponent $n$ and by the derivative of $u$ with respect to $x$.)

### EXAMPLE 6

**(a)** Use the generalized power rule to find the derivative of $y = (3 + 5x)^2$.

**Solution** Let $u = 3 + 5x$ and $n = 2$. Then $u' = 5$. By the generalized power rule,

$$y' = n \cdot u^{n-1} \cdot u'$$

$$\begin{array}{cccc} n & u & n-1 & u' \\ \downarrow & \downarrow & \downarrow & \downarrow \end{array}$$

$$= 2 \cdot (3 + 5x)^{2-1} \cdot \frac{d}{dx}(3 + 5x)$$

$$= 2(3 + 5x)^{2-1} \cdot 5$$

$$= 10(3 + 5x)$$

$$= 30 + 50x.$$

**(b)** Find $y'$ if $y = (3 + 5x)^{-3/4}$.

**Solution** Use the generalized power rule with $n = -\frac{3}{4}$, $u = 3 + 5x$, and $u' = 5$:

$$y' = -\frac{3}{4}(3 + 5x)^{(-3/4)-1}(5)$$

$$= -\frac{15}{4}(3 + 5x)^{-7/4}.\ \boxed{\checkmark}_6$$

## ✓ Checkpoint 6

Find $dy/dx$ for the given functions.

**(a)** $y = (2x + 5)^6$

**(b)** $y = (4x^2 - 7)^3$

**(c)** $f(x) = \sqrt{3x^2 - x}$

**(d)** $g(x) = (2 - x^4)^{-3}$

### EXAMPLE 7    Find the derivative of the given function.

**(a)** $y = 2(7x^2 + 5)^4$

**Solution** Let $u = 7x^2 + 5$. Then $u' = 14x$, and

$$\begin{array}{cccc} n & u & n-1 & u' \\ \downarrow & \downarrow & \downarrow & \downarrow \end{array}$$

$$y' = 2 \cdot 4(7x^2 + 5)^{4-1} \cdot \frac{d}{dx}(7x^2 + 5)$$

$$= 2 \cdot 4(7x^2 + 5)^3(14x)$$

$$= 112x(7x^2 + 5)^3.$$

**(b)** $y = \sqrt{9x + 2}$

**Solution** Write $y = \sqrt{9x + 2}$ as $y = (9x + 2)^{1/2}$. Then

$$y' = \frac{1}{2}(9x + 2)^{-1/2}(9) = \frac{9}{2}(9x + 2)^{-1/2}.$$

The derivative also can be written as

$$y' = \frac{9}{2(9x + 2)^{1/2}} \quad \text{or} \quad y' = \frac{9}{2\sqrt{9x + 2}}. \enspace ✓_7$$

✓ **Checkpoint 7**

Find $dy/dx$ for the given functions.

**(a)** $y = 12(x^2 + 6)^5$

**(b)** $y = 8(4x^2 + 2)^{3/2}$

**⚠ CAUTION**

**(a)** A common error is to forget to multiply by $g'(x)$ when using the generalized power rule. Remember, the generalized power rule is an example of the chain rule, so the derivative must involve a "chain," or product, of derivatives.

**(b)** Another common mistake is to write the derivative as $n[g'(x)]^{n-1}$. Remember to leave $g(x)$ unchanged and then to multiply by $g'(x)$.

Sometimes both the generalized power rule and either the product or quotient rule are needed to find a derivative, as the next two examples show.

**EXAMPLE 8** Find the derivative of $y = 4x(3x + 5)^5$.

**Solution** Write $4x(3x + 5)^5$ as the product

$$y = 4x \cdot (3x + 5)^5.$$

According to the product rule,

$$y' = 4x \cdot \frac{d}{dx}(3x + 5)^5 + (3x + 5)^5 \cdot \frac{d}{dx}(4x).$$

Use the generalized power rule to find the derivative of $(3x + 5)^5$;

$$\begin{array}{c} \overbrace{\text{Derivative of } (3x + 5)^5} \qquad \overbrace{\text{Derivative of } 4x} \\ y' = 4x[5(3x + 5)^4 \cdot 3] + (3x + 5)^5(4) \end{array}$$

$$= 60x(3x + 5)^4 + 4(3x + 5)^5$$

$$= 4(3x + 5)^4[15x + (3x + 5)^1] \qquad \text{Factor out the greatest common}$$

$$= 4(3x + 5)^4(18x + 5). \enspace ✓_8 \qquad \text{factor, } 4(3x + 5)^4.$$

✓ **Checkpoint 8**

Find the derivatives of the given functions.

**(a)** $y = 9x(x + 7)^2$

**(b)** $y = -3x(4x^2 + 7)^3$

**EXAMPLE 9** Find the derivative of

$$y = \frac{(3x + 2)^7}{x - 1}.$$

**Solution** Use the quotient rule and the generalized power rule:

$$\frac{dy}{dx} = \frac{(x - 1) \cdot \dfrac{d}{dx}(3x + 2)^7 - (3x + 2)^7 \dfrac{d}{dx}(x - 1)}{(x - 1)^2} \qquad \text{Quotient rule}$$

$$= \frac{(x - 1)[7(3x + 2)^6 \cdot 3] - (3x + 2)^7(1)}{(x - 1)^2} \qquad \text{Generalized power rule}$$

$$= \frac{21(x - 1)(3x + 2)^6 - (3x + 2)^7}{(x - 1)^2}$$

$$= \frac{(3x + 2)^6[21(x - 1) - (3x + 2)]}{(x - 1)^2} \qquad \text{Factor out the greatest} \\ \text{common factor, } (3x + 2)^6.$$

$$= \frac{(3x + 2)^6[21x - 21 - 3x - 2]}{(x - 1)^2} \qquad \text{Simplify inside brackets.}$$

$$= \frac{(3x + 2)^6(18x - 23)}{(x - 1)^2}. \enspace ✓_9$$

✓ **Checkpoint 9**

Find the derivatives of the given functions.

**(a)** $y = \dfrac{(5x + 2)^3}{11x}$

**(b)** $y = \dfrac{(x - 12)^7}{6x + 1}$

**EXAMPLE 10**   Consider that we may rewrite the following rational function as a product in the form

$$y = \frac{3x - 7}{5x + 1} = (3x - 7) \cdot (5x + 1)^{-1}$$

To find the derivative, we can use either the quotient rule or the product rule combined with the chain rule (in the form of the generalized power rule).

**(a)**  Use the quotient rule to find the derivative of

$$y = \frac{3x - 7}{5x + 1}.$$

**Solution**  According to the quotient rule,

$$y' = \frac{(5x + 1)\dfrac{d}{dx}(3x - 7) - (3x - 7)\dfrac{d}{dx}(5x + 1)}{(5x + 1)^2} \qquad \text{Quotient rule}$$

$$= \frac{(5x + 1)(3) - (3x - 7)(5)}{(5x + 1)^2} \qquad \text{Find derivatives in numerator.}$$

$$= \frac{15x + 3 - 15x + 35}{(5x + 1)^2} \qquad \text{Distributive property}$$

$$= \frac{38}{(5x + 1)^2}. \qquad \text{Simplify numerator}$$

**(b)**  Use the product rule and the generalized power rule to find the derivative of

$$y = (3x - 7) \cdot (5x + 1)^{-1}.$$

**Solution**  First use the product rule, then find the indicated derivatives and simplify to get

$$y' = (3x - 7) \cdot \frac{d}{dx}(5x + 1)^{-1} + (5x + 1)^{-1} \cdot \frac{d}{dx}(3x - 7) \qquad \text{Product rule}$$

$$= (3x - 7) \cdot (-1)(5x + 1)^{-2}(5) + (5x + 1)^{-1} \cdot (3) \qquad \text{Generalized power rule}$$

$$= (3x - 7) \cdot \frac{-1}{(5x + 1)^2}(5) + \frac{1}{5x + 1} \cdot (3)$$

$$= (3x - 7) \cdot \frac{-5}{(5x + 1)^2} + \frac{3}{5x + 1} \cdot \frac{5x + 1}{5x + 1} \qquad \text{Find a common denominator.}$$

$$= \frac{-15x + 35}{(5x + 1)^2} + \frac{15x + 3}{(5x + 1)^2}$$

$$= \frac{38}{(5x + 1)^2}.$$

Note that the derivative is the same as was found in part (a). ✓10

✓ **Checkpoint 10**

Use any method to find the derivative of

$$y = \frac{x}{5x + 3} = x(5x + 3)^{-1}.$$

## Applications

Some applications requiring the use of the chain rule or the generalized power rule are illustrated in the next examples.

**EXAMPLE 11**   **Finance**  A generous aunt deposits $20,000 in an account to be used by her newly born niece to attend college. The account earns interest at the rate of $r$ percent per year, compounded monthly. At the end of 18 years, the balance in the account is given by

$$A = 20{,}000\left(1 + \frac{r}{1200}\right)^{216}.$$

Find the rate of change of $A$ with respect to $r$ if $r = 1.5, 2.5,$ or $3$.

**Solution** First find $dA/dr$, using the generalized power rule:

$$\frac{dA}{dr} = (20{,}000)(216)\left(1 + \frac{r}{1200}\right)^{215}\left(\frac{1}{1200}\right) = 3600\left(1 + \frac{r}{1200}\right)^{215}.$$

If $r = 1.5$, we obtain

$$\frac{dA}{dr} = 3600\left(1 + \frac{1.5}{1200}\right)^{215} = 4709.19,$$

or \$4709.19 per percentage point. If $r = 2.5$, we obtain

$$\frac{dA}{dr} = 3600\left(1 + \frac{2.5}{1200}\right)^{215} = 5631.55,$$

or \$5631.55 per percentage point. If $r = 3$, we obtain

$$\frac{dA}{dr} = 3600\left(1 + \frac{3}{1200}\right)^{215} = 6158.07,$$

or \$6158.07 per percentage point. This means that when the interest rate is 3%, an increase of 1% in the interest rate will produce an increase in the balance of approximately \$6158.07. ✓ 11

✓ **Checkpoint 11**

If the initial donation in Example 11 is \$25,000, find the rate of change of $A$ with respect to $r$ if $r = 2$.

The chain rule can be used to develop the formula for the **marginal-revenue product,** an economic concept that approximates the change in revenue when a manufacturer hires an additional employee. Start with $R = px$, where $R$ is total revenue from the daily production of $x$ units and $p$ is the price per unit. The demand function is $p = f(x)$, as before. Also, $x$ can be considered a function of the number of employees, $n$. Since $R = px$, and $x$—and therefore, $p$—depends on $n$, $R$ can also be considered a function of $n$. To find an expression for $dR/dn$, use the product rule for derivatives on the function $R = px$ to get

$$\frac{dR}{dn} = p \cdot \frac{dx}{dn} + x \cdot \frac{dp}{dn}. \qquad (*)$$

By the chain rule,

$$\frac{dp}{dn} = \frac{dp}{dx} \cdot \frac{dx}{dn}.$$

Substituting for $dp/dn$ in equation (*) yields

$$\frac{dR}{dn} = p \cdot \frac{dx}{dn} + x\left(\frac{dp}{dx} \cdot \frac{dx}{dn}\right).$$

Factoring out $\frac{dx}{dn}$ from the equation above shows that the marginal-revenue product is

$$\frac{dR}{dn} = \left(p + x \cdot \frac{dp}{dx}\right)\frac{dx}{dn}.$$

**EXAMPLE 12** **Marginal-Revenue Product** Find and interpret the marginal-revenue product $dR/dn$ (in dollars per employee) when $n = 20$ if the demand function is $p = 600/\sqrt{x}$ and $x = 5n$.

**Solution** As shown above,

$$\frac{dR}{dn} = \left(p + x \cdot \frac{dp}{dx}\right)\frac{dx}{dn}.$$

Find $dp/dx$ and $dx/dn$. From

$$p = \frac{600}{\sqrt{x}} = 600x^{-1/2},$$

we have the derivative

$$\frac{dp}{dx} = -300x^{-3/2}.$$

Also, from $x = 5n$, we have

$$\frac{dx}{dn} = 5.$$

Then, by substitution,

$$\frac{dR}{dn} = \left[\frac{600}{\sqrt{x}} + x(-300x^{-3/2})\right]5 = \left[\frac{600}{\sqrt{x}} - \frac{300}{\sqrt{x}}\right]5 = \frac{1500}{\sqrt{x}}.$$

If $n = 20$, then $x = 100$, and

$$\frac{dR}{dn} = \frac{1500}{\sqrt{100}}$$

$$= 150.$$

✓**Checkpoint 12**

Find the marginal-revenue product at $n = 10$ if the demand function is $p = 1000/x^2$ and $x = 8n$. Interpret your answer.

This means that hiring an additional employee when production is at a level of 100 items will produce an approximate increase in revenue of $150. ✓₁₂

# 11.7 Exercises

*Let $f(x) = 2x^2 + 3x$ and $g(x) = 4x - 1$. Find each of the given values. (See Example 1.)*

**1.** $f[g(5)]$

**2.** $f[g(-4)]$

**3.** $g[f(5)]$

**4.** $g[f(-4)]$

*Find $f[g(x)]$ and $g[f(x)]$ for the given functions. (See Example 2.)*

**5.** $f(x) = 8x + 12; g(x) = 2x + 3$

**6.** $f(x) = -6x + 9; g(x) = 5x + 7$

**7.** $f(x) = 3x^2 - 8x + 1; g(x) = x - 4$

**8.** $f(x) = x^2 + 5x - 9; g(x) = 2x + 7$

**9.** $f(x) = \dfrac{1}{x}; g(x) = x^2$

**10.** $f(x) = \dfrac{2}{x^4}; g(x) = 2 - x$

**11.** $f(x) = \sqrt{2x + 5}; g(x) = 3x - 8$

**12.** $f(x) = \sqrt{6 - x}; g(x) = 7x - 9$

*Write each function as a composition of two functions. (There may be more than one way to do this.) (See Example 3.)*

**13.** $y = (4x + 3)^5$

**14.** $y = (x^2 + 2)^{1/5}$

**15.** $y = \sqrt{6 + 3x^2}$

**16.** $y = \sqrt[3]{8x - 3}$

**17.** $y = \dfrac{1}{(4x - 1)^2}$

**18.** $y = \dfrac{1}{(2x + 7)^5}$

**19.** $y = (x^{1/2} - 3)^3 + (x^{1/2} - 3) + 5$

**20.** $y = (x^2 + 5x)^{1/3} - 2(x^2 + 5x)^{2/3} + 7$

*Find the derivative of each of the given functions. (See Examples 4–7.)*

**21.** $y = (7x - 12)^5$

**22.** $y = (13x - 3)^4$

**23.** $y = 7(4x + 3)^4$

**24.** $y = -4(3x - 2)^7$

**25.** $y = -4(10x^2 + 8)^5$

**26.** $y = -3(x^4 + 7x^2)^3$

**27.** $y = (4x - 9)^{3/2}$

**28.** $y = (8x + 7)^{5/2}$

**29.** $y = -7(4x^2 + 9x)^{3/2}$

**30.** $y = 11(5x^2 + 6x)^{3/2}$

**31.** $y = 8\sqrt{4x + 7}$

**32.** $y = -3\sqrt{7x - 1}$

**33.** $y = -2\sqrt{x^2 + 4x}$

**34.** $y = 4\sqrt{2x^2 + 3}$

*Use the product or quotient rule or the generalized power rule to find the derivative of each of the given functions. (See Examples 8–10.)*

**35.** $y = (x + 1)(x - 3)^2$

**36.** $y = (2x + 1)^3(x - 5)$

**37.** $y = 5(x + 3)^2(2x - 1)^5$

**38.** $y = -9(x + 4)^2(2x - 3)^3$

**39.** $y = (3x + 1)^3\sqrt{x}$

**40.** $y = (3x + 5)^2\sqrt{x + 1}$

**41.** $y = \dfrac{1}{(x - 2)^3}$

**42.** $y = \dfrac{-3}{(4x + 1)^2}$

**43.** $y = \dfrac{(2x - 7)^2}{4x - 1}$

**44.** $y = \dfrac{(x - 4)^2}{5x + 2}$

**45.** $y = \dfrac{x^2 + 5x}{(3x + 1)^3}$

**46.** $y = \dfrac{4x^3 - x}{(x - 3)^2}$

**47.** $y = (x^{1/2} + 2)(2x + 3)$

**48.** $y = (4 - x^{2/3})(x + 1)^{1/2}$

**G** *Find the equation of the tangent line to the graph of $f(x)$ at the given point.*

**49.** $f(x) = (5x - 1)^3$ at $(0, -1)$

**50.** $f(x) = (3x + 2)^4$ at $(0, 16)$

**51.** $f(x) = \sqrt{20x + 5}$ at $(1, 5)$

**52.** $f(x) = \sqrt{6x + 3}$ at $(1, 3)$

*Find the indicated composition of functions. (See Example 2.)*

**53. Home Robotics** Suppose the demand for a certain robotic vacuum cleaner is given by

$$D(p) = \frac{-p^2}{100} + 500,$$

where $p$ is the price in dollars. If the price, in terms of the cost $c$, is expressed as

$$p(c) = 2c - 10,$$

find $D[p(c)]$, the demand in terms of the cost.

**54. Oil Spill** An oil well off the Gulf Coast is leaking, spreading oil in a circle over the surface. At any time $t$, in minutes, after the beginning of the leak, the radius of the circular oil slick on the surface is $r(t) = t^2$ feet. Let $A(r) = \pi r^2$ represent the area of a circle of radius $r$. Find and interpret $A[r(t)]$.

*Work these exercises. (see Example 11.)*

**55. Mortgage Debt** The average mortgage debt in dollars can be approximated by

$$M(x) = -(10x - 470)^2 + 55,000,$$

where $x$ is the age of the borrower in years. (Data from: Federal Reserve Bank of New York.)

**(a)** Find the average mortgage debt for 57-year-old borrowers.

**(b)** Find $M'(x)$, the rate of change of mortgage debt with respect to age.

**(c)** Find $M'(57)$ and explain whether mortgage debt increases or decreases with respect to age for 57-year-old borrowers.

**56. Electronic Cigarettes** The percentage of U.S. high school students using electronic cigarettes (e-cigarettes) between the years 2011 and 2015 can be approximated by

$$f(x) = (.8x - 8)^2,$$

where $x = 11$ corresponds to the year 2011. (Data from: Centers for Disease Control and Prevention.)

**(a)** Find and interpret $f(15)$.

**(b)** Find and interpret $f'(15)$.

**57. Marginal Cost** The cost of producing $x$ bags of dog food is given by

$$C(x) = 600 + \sqrt{200 + 25x^2 - x} \qquad (0 \le x \le 5000).$$

Find the marginal-cost function.

**58. Marginal Cost** The cost (in hundreds of dollars) of producing $x$ smart phones is given by

$$C(x) = 900 + \sqrt{5x^2 + 4x + 1200}.$$

**(a)** Find the marginal-cost function.

**(b)** Find and interpret $C'(200)$.

**59. Marginal Revenue** The revenue function from the sale of $x$ earbuds is given by

$$R(x) = 15\sqrt{500x - 2x^2} \qquad (0 \le x \le 200).$$

**(a)** Find the marginal-revenue function.

**(b)** Evaluate the marginal-revenue function at $x = 40, 80, 120,$ and $160$.

**(c)** As the number sold increases, does the marginal revenue increase or decrease?

**60. Marginal Profit** The profit (in thousands of dollars) of producing $x$ thousand pairs of running shoes is given by

$$P(x) = 45\sqrt[3]{x^3 + 246x + 500} - 500 \qquad (0 \le x \le 100).$$

**(a)** Is the company profitable or losing money when producing 2000 pairs of running shoes? 6000 pairs?

**(b)** Find the marginal-profit function.

**61. Tablet Depreciation** A tablet computer is purchased for $500. Its value in dollars after $t$ years can be approximated by

$$V(t) = \frac{500}{(2t + 1)^2} \qquad (0 \le t \le 4).$$

**(a)** Find the value of the tablet computer 2 years after purchase.

**(b)** At time $t = 2$ years, use a derivative to estimate how much the value will change within the next year.

**62. Patient Recovery Rate** The percent of patients who report being symptom-free $t$ days after the onset of a certain illness can be approximated by

$$F(t) = 100 - \frac{200}{\sqrt{t^2 + 4}} \qquad (0 \le t \le 7).$$

Find the rate at which patients are recovering after the given number of days.

**(a)** 2 days

**(b)** 5 days

**63. Theater Tickets** The demand function for tickets to the student theatrical production is given by

$$p = \sqrt{900 - 2x} \qquad (0 \le x \le 450),$$

where $x$ is the number of tickets demanded and $p$ is the price in dollars.

**(a)** Find the revenue as a function of $x$.

**(b)** Find the marginal-revenue function.

**(c)** Find and interpret the marginal revenue at $x = 250$.

**64. Smart Phone Apps** The demand (in thousands) for a certain smart phone app as a function of the price $\$p$ is given by

$$x = 30(8 - .5p^2)^{1/2} \qquad (0 \le p \le 4).$$

**(a)** Find the demand when the price is $2.

**(b)** Find the rate of change of demand with respect to price when the price is $2.

**65. Finance** An amount of $12,000 is deposited in an account with an interest rate of $r$ percent per year, compounded monthly. At the end of 8 years, the balance in the account is given by

$$A = 12{,}000\left(1 + \frac{r}{1200}\right)^{96}.$$

**(a)** Find $\dfrac{dA}{dr}$, the rate of change of $A$ with respect to $r$.

**(b)** Find $\dfrac{dA}{dr}$ when $r = 4$, and interpret your answer.

**66. Finance** An amount of $8000 is invested in a 5-year certificate of deposit with an interest rate of $r$ percent per year, compounded quarterly. On the maturity date, the value of the certificate of deposit is given by

$$A = 8000\left(1 + \frac{r}{400}\right)^{20}.$$

**(a)** Would you expect the rate of change of $A$ with respect to $r$ to be positive, negative, or zero? Why?

**(b)** Find $\dfrac{dA}{dr}$.

**(c)** Find and interpret $\dfrac{dA}{dr}$ at $r = 2$.

**67. Finance** An amount of $24,000 is invested in a money market account with an interest rate of $r$ percent per year, compounded monthly. At the end of 2 years, the balance on the account is given by

$$A = 24{,}000\left(1 + \frac{r}{1200}\right)^{24}.$$

**(a)** Find $\dfrac{dA}{dr}$.

**(b)** Find $\dfrac{dA}{dr}$ when $r = 2$, and interpret your answer.

**68. Finance** The price (present value) of a zero coupon bond with a face value of $5000, a maturity date of 10 years, and an interest rate of $r$ percent compounded annually is given by

$$P = \frac{5000}{\left(1 + \dfrac{r}{100}\right)^{10}}.$$

**(a)** Find the price of this bond, assuming an interest rate of $r = 4$.

**(b)** Would you expect the rate of change of $P$ with respect to $r$ to be positive, negative, or zero? Why?

**(c)** Find $\dfrac{dP}{dr}$.

**(d)** Find and interpret $\dfrac{dP}{dr}$ at an interest rate of $r = 4$.

*Work these exercises. (See Example 12.)*

**69. Marginal-Revenue Product** Find the marginal-revenue product for a manufacturer with 10 workers if the demand function is $p = \dfrac{200 - x}{4}$ and if $x = 6n$.

**70. Marginal-Revenue Product** Find the marginal-revenue product for a manufacturer with 5 workers if the demand function is $p = \dfrac{210}{\sqrt{x}}$ and if $x = 45n$.

*Work the following exercises.*

**71. Average Cost** A cost function is given by $C(x) = \sqrt{6x + 5}$, where $x$ is the number of items produced.

**(a)** Find the average-cost function $\overline{C}(x) = \dfrac{C(x)}{x}$.

**(b)** Find the marginal average-cost function $\overline{C}'(x)$.

**72. Media Coverage** The percent of the adult population of a city who have heard about a scandal in the city government $t$ days after it occurs can be approximated by

$$F(t) = \frac{85t}{\sqrt{t^2 + 2}}.$$

**(a)** What percent of the population is aware of the scandal after 1 day?

**(b)** Find $F'(t)$.

**(c)** Find and interpret $F'(1)$.

**73. Calcium Uptake** To test an individual's use of calcium, a researcher injects a small amount of radioactive calcium into the person's bloodstream. The calcium remaining in the bloodstream is measured each day for several days. Suppose the amount of the calcium remaining in the bloodstream, in milligrams per cubic centimeter, $t$ days after the initial injection is approximated by

$$C(t) = \frac{1}{2}(2t + 1)^{-1/2}.$$

Find the rate of change of $C$ with respect to time for each of the given number of days.

**(a)** $t = 0$          **(b)** $t = 4$

**(c)** $t = 6$          **(d)** $t = 7.5$

**74. Drug Reaction** The strength of a person's reaction to a certain drug is given by

$$R(Q) = Q\left(C - \frac{Q}{3}\right)^{1/2},$$

where $Q$ represents the quantity of the drug given to the patient and $C$ is a constant. The derivative $R(Q)$ is called the *sensitivity* to the drug.

**(a)** Find $R'(Q)$.

**(b)** Find $R'(Q)$ if $Q = 87$ and $C = 59$.

**75. On-the-Job Training** Studies show that after $t$ hours on the job, the number of items a supermarket cashier can scan per minute is given by

$$F(t) = 60 - \frac{150}{\sqrt{8 + t^2}}.$$

**(a)** Find $F'(t)$, the rate at which the cashier's speed is increasing.

**(b)** Find $F'(5)$.

**(c)** Find $F'(40)$.

**(d)** Is the cashier's speed increasing more rapidly after 5 hours or after 40 hours?

**76. Natural Science** The volume and surface area of a jaw-breaker candy of any radius $r$ are given by the formulas

$$V(r) = \frac{4}{3}\pi r^3 \quad \text{and} \quad S(r) = 4\pi r^2,$$

respectively. It is estimated that the radius of a jawbreaker while in a person's mouth is

$$r(t) = 6 - \frac{3}{17}t,$$

where $r(t)$ is in mm and $t$ is in min.[*]

**(a)** What is the life expectancy of a jawbreaker?

**(b)** Find $\dfrac{dV}{dt}$ and $\dfrac{dS}{dt}$ when $t = 17$, and interpret your answer.

✓**Checkpoint Answers**

**1. (a)** 243      **(b)** 1

**2.** $f[g(x)] = \sqrt{x^2 + 5x + 5};$

$\qquad g[f(x)] = (\sqrt{x+4})^2 + 5\sqrt{x+4} + 1$

$\qquad\qquad = x + 5 + 5\sqrt{x+4}$

---

*Roger Guffey, "The Life Expectancy of a Jawbreaker: An Application of the Composition of Functions," *Mathematics Teacher* 92, no. 2 (February 1999): 125–127.

**3.** There are several correct answers, including

**(a)** $h(x) = f[g(x)]$, where $f(x) = x^4$ and $g(x) = 7x^2 + 5;$

**(b)** $h(x) = f[g(x)]$, where $f(x) = \sqrt{x}$ and $g(x) = 15x^2 + 1.$

**4.** $\dfrac{dy}{dx} = \dfrac{4x^3}{\sqrt{2x^4 + 3}}$

**5.** $160x(2x^2 + 1)^3$

**6. (a)** $12(2x + 5)^5$      **(b)** $24x(4x^2 - 7)^2$

    **(c)** $\dfrac{6x - 1}{2\sqrt{3x^2 - x}}$      **(d)** $\dfrac{12x^3}{(2 - x^4)^4}$

**7. (a)** $120x(x^2 + 6)^4$      **(b)** $96x(4x^2 + 2)^{1/2}$

**8. (a)** $9(x + 7)(3x + 7)$      **(b)** $-21(4x^2 + 7)^2(4x^2 + 1)$

**9. (a)** $\dfrac{11(5x + 2)^2(10x - 2)}{121x^2}$      **(b)** $\dfrac{(x - 12)^6(36x + 79)}{(6x + 1)^2}$

**10.** $y' = \dfrac{3}{(5x + 3)^2}$

**11.** \$6437.32 per percentage point

**12.** $-$\$1.25; hiring an additional employee will produce a decrease in revenue of \$1.25.

# 11.8 Derivatives of Exponential and Logarithmic Functions

$$y_1 = \frac{e^h - 1}{h}$$

**Figure 11.39**

The exponential function $f(x) = e^x$ and the logarithmic function to the base $e$, $g(x) = \ln x$, were studied in Chapter 4. (Recall that $e \approx 2.71828$.) We are interested in exponential and logarithmic functions because they arise naturally in a variety of applications in business, economics, and the social and life sciences. In this section, we shall find the derivatives of these functions. In order to do that, we must first find a limit that will be needed in our calculations.

We claim that

$$\lim_{h \to 0} \frac{e^h - 1}{h} = 1.$$

Although a rigorous proof of this fact is beyond the scope of this book, a graphing calculator provides both graphical and numerical support, as shown in Figure 11.39.[†]

To find the derivative of $f(x) = e^x$, we use the definition of the derivative function,

$$f'(x) = \lim_{h \to 0} \frac{f(x + h) - f(x)}{h},$$

provided that this limit exists. (Remember that $h$ is the variable here and $x$ is treated as a constant.)

---

†As usual, the variable X is used in place of $h$ on a calculator.

For $f(x) = e^x$, we see that

$$f'(x) = \lim_{h \to 0} \frac{e^{x+h} - e^x}{h}$$

$$= \lim_{h \to 0} \frac{e^x e^h - e^x}{h} \qquad \text{Product property of exponents}$$

$$= \lim_{h \to 0} \frac{e^x(e^h - 1)}{h} \qquad \text{Factor out } e^x.$$

$$= \lim_{h \to 0} e^x \cdot \lim_{h \to 0} \frac{e^h - 1}{h}, \qquad \text{Product property of limits}$$

provided that the last two limits exist. But $h$ is the variable here and $x$ is constant; therefore,

$$\lim_{h \to 0} e^x = e^x.$$

Combining this fact with our previous work, we see that

$$f'(x) = \lim_{h \to 0} e^x \cdot \lim_{h \to 0} \frac{e^h - 1}{h} = e^x \cdot 1 = e^x.$$

In other words, *the exponential function $f(x) = e^x$ is its own derivative.*

---

**EXAMPLE 1**    Find each derivative.

**(a)** $y = x^3 e^x$

**Solution** The product rule and the fact that $f(x) = e^x$ is its own derivative show that

$$y' = x^3 \frac{d}{dx}(e^x) + e^x \frac{d}{dx}(x^3)$$

$$= x^3 \cdot e^x + e^x \cdot 3x^2 = e^x(x^3 + 3x^2).$$

**(b)** $y = (2e^x + x)^5$

**Solution** By the generalized power, sum, and constant times a function rules,

$$y' = 5(2e^x + x)^4 \frac{d}{dx}(2e^x + x)$$

$$= 5(2e^x + x)^4 \left[ \frac{d}{dx}(2e^x) + \frac{d}{dx}(x) \right]$$

$$= 5(2e^x + x)^4 \left[ 2\frac{d}{dx}(e^x) + 1 \right]$$

$$= 5(2e^x + x)^4(2e^x + 1). \quad ✓_1$$

✓ **Checkpoint 1**

Differentiate the given expressions.

**(a)** $(2x^2 - 1)e^x$

**(b)** $(1 - e^x)^{1/2}$

---

**EXAMPLE 2**    Find the derivative of $y = e^{x^2 - 3x}$.

**Solution** Let $f(x) = e^x$ and $g(x) = x^2 - 3x$. Then

$$y = e^{x^2 - 3x} = e^{g(x)} = f[g(x)],$$

and $f'(x) = e^x$ and $g'(x) = 2x - 3$. By the chain rule,

$$y' = f'[g(x)] \cdot g'(x)$$

$$= e^{g(x)} \cdot (2x - 3)$$

$$= e^{x^2 - 3x} \cdot (2x - 3) = (2x - 3)e^{x^2 - 3x}.$$

The argument used in Example 2 can be used to find the derivative of $y = e^{g(x)}$ for any differentiable function $g$. Let $f(x) = e^x$; then $y = e^{g(x)} = f[g(x)]$. By the chain rule,

$$y' = f'[g(x)] \cdot g'(x) = e^{g(x)} \cdot g'(x) = g'(x)e^{g(x)}.$$

We summarize these results as follows.

---

## Derivatives of $e^x$ and $e^{g(x)}$

If $y = e^x$, then $y' = e^x$.
If $y = e^{g(x)}$, then $y' = g'(x) \cdot e^{g(x)}$.

---

**! CAUTION**    Notice the difference between the derivative of a variable to a constant power, such as $\frac{d}{dx}(x^3) = 3x^2$, and the derivative of a constant to a variable power, like $\frac{d}{dx}(e^x) = e^x$.

Remember, $\frac{d}{dx}(e^x) \neq xe^{x-1}$.

---

**EXAMPLE 3**    Find the derivatives of each of the given functions.

**(a)** $f(x) = e^x + x^{10} + x^e + e^{10}$

**Solution**  The first term of $f$ is an *exponential function* (constant to a variable power), but the second and third terms are *power functions* (variable to a constant power), and the final term is just a *constant* (constant to a constant power). So the derivative is

$$f'(x) = e^x + 10x^9 + ex^{e-1}.$$

**(b)** $y = e^{7x}$

**Solution**  Let $g(x) = 7x$, with $g'(x) = 7$. Then

$$y' = g'(x)e^{g(x)} = 7e^{7x}.$$

**(c)** $y = 5e^{-3x}$

**Solution**  Here, let $g(x) = -3x$ and $g'(x) = -3$, so that

$$y' = 5(-3e^{-3x}) = -15e^{-3x}.$$

**(d)** $y = 8e^{4x^3}$

**Solution**  Let $g(x) = 4x^3$, with $g'(x) = 12x^2$. Then

$$y' = 8(12x^2e^{4x^3}) = 96x^2e^{4x^3}. \ \blacksquare_2$$

**✓ Checkpoint 2**

Find the derivative of the given function.

**(a)** $y = 4e^{11x}$

**(b)** $y = -5e^{(-9x+2)}$

**(c)** $y = e^{-x^3}$

---

**EXAMPLE 4**    Find $y'$ given that

$$y = \frac{250{,}000}{2 + 15e^{-4x}}.$$

**Solution**  Use the quotient rule:

$$y' = \frac{(2 + 15e^{-4x})(0) - 250{,}000(-60e^{-4x})}{(2 + 15e^{-4x})^2}$$

$$= \frac{15{,}000{,}000e^{-4x}}{(2 + 15e^{-4x})^2}. \ \blacksquare_3$$

**✓ Checkpoint 3**

Find the derivative of the given function.

**(a)** $y = \dfrac{e^x}{12 + x}$

**(b)** $y = \dfrac{9000}{15 + 3e^x}$

**EXAMPLE 5**    **Radioactive Decay** A 100-gram sample of a radioactive substance decays exponentially. The amount left after $t$ years is given by $A(t) = 100e^{-.12t}$. Find the rate of change of the amount after 3 years.

**Solution** The rate of change is given by the derivative

$$A'(t) = 100(-.12e^{-.12t}) = -12e^{-.12t}.$$

After 3 years ($t = 3$), the rate of change is

$$A'(3) = -12e^{-.12(3)} = -12e^{-.36} \approx -8.4$$

grams per year.

**EXAMPLE 6**    If $y = 10^x$, find $y'$.

**Solution** By one of the basic properties of logarithms (page 200), we have $10 = e^{\ln 10}$. Therefore,

$$y = 10^x = (e^{\ln 10})^x = e^{(\ln 10)x},$$

and

$$y' = \frac{d}{dx}[(\ln 10)x] \cdot e^{(\ln 10)x} \qquad \text{By the preceding box, with } g(x) = (\ln 10)x$$

$$= (\ln 10)e^{(\ln 10)x} \qquad \text{Remember that ln 10 is a constant.}$$

$$= (\ln 10)(e^{\ln 10})^x$$

$$= (\ln 10)10^x. \qquad e^{\ln 10} = 10 \quad \boxed{4}$$

✓ **Checkpoint 4**

Adapt the solution of Example 6 to find the derivative of $y = 25^x$.

## Derivatives of Logarithmic Functions

To find the derivative of $g(x) = \ln x$, we use the definitions and properties of natural logarithms that were developed in Section 4.3:

$$g(x) = \ln x \qquad \text{means} \qquad e^{g(x)} = x,$$

and for all $x > 0$, $y > 0$, and every real number $r$,

$$\ln xy = \ln x + \ln y, \qquad \ln \frac{x}{y} = \ln x - \ln y, \qquad \text{and} \qquad \ln x^r = r \ln x.$$

Note that we must have $x > 0$ and $y > 0$ because logarithms of zero and negative numbers are not defined.

Differentiating with respect to $x$ on each side of $e^{g(x)} = x$ shows that

$$\frac{d}{dx}(e^{g(x)}) = \frac{d}{dx}(x)$$

$$g'(x) \cdot e^{g(x)} = 1.$$

Because $e^{g(x)} = x$, this last equation becomes

$$g'(x) \cdot x = 1$$

$$g'(x) = \frac{1}{x}$$

$$\frac{d}{dx}(\ln x) = \frac{1}{x} \qquad \text{for all } x > 0.$$

**EXAMPLE 7**   Assume that $x > 0$, and use the chain rule to find the derivative of $y = \ln 6x$.

**Solution** Let $f(x) = \ln x$ and $g(x) = 6x$, so that $y = \ln 6x = \ln g(x) = f(g(x))$. Then, by the chain rule,

$$y' = f'[g(x)] \cdot g'(x) = \frac{1}{g(x)} \cdot \frac{d}{dx}(6x) = \frac{1}{6x} \cdot 6 = \frac{1}{x}.$$

The argument used in Example 6 applies equally well in the general case. The derivative of $y = \ln g(x)$, where $g(x)$ is a function and $g(x) > 0$, can be found by letting $f(x) = \ln x$, so that $y = f[g(x)]$, and applying the chain rule:

$$y' = f'[g(x)] \cdot g'(x) = \frac{1}{g(x)} \cdot g'(x) = \frac{g'(x)}{g(x)}.$$

We summarize these results as follows.

## Derivatives of ln $x$ and ln $g(x)$

If $y = \ln x$, then $y' = \dfrac{1}{x}$   $(x > 0)$.

If $y = \ln g(x)$, then $y' = \dfrac{g'(x)}{g(x)}$   $(g(x) > 0)$.

**EXAMPLE 8**   Find the derivatives of the given functions.

**(a)** $y = \ln(3x^2 - 4x)$

**Solution** Let $g(x) = 3x^2 - 4x$, so that $g'(x) = 6x - 4$. From the second formula in the preceding box,

$$y' = \frac{g'(x)}{g(x)} = \frac{6x - 4}{3x^2 - 4x}.$$

**(b)** $y = 3x \ln x^2$

**Solution** Since $3x \ln x^2$ is the product of $3x$ and $\ln x^2$, use the product rule:

$$y' = (3x)\left(\frac{d}{dx} \ln x^2\right) + (\ln x^2)\left(\frac{d}{dx}(3x)\right)$$

$$= 3x\left(\frac{2x}{x^2}\right) + (\ln x^2)(3) \qquad \text{Take derivatives.}$$

$$= 6 + 3 \ln x^2.$$

**(c)** $y = \ln[(x^2 + x + 1)(4x - 3)^5]$

**Solution** Here, we use the properties of logarithms before taking the derivative (the same thing could have been done in part (b) by writing $\ln x^2$ as $2 \ln x$):

$$y = \ln[(x^2 + x + 1)(4x - 3)^5]$$

$$= \ln(x^2 + x + 1) + \ln(4x - 3)^5 \qquad \text{Property of logarithms}$$

$$= \ln(x^2 + x + 1) + 5 \ln (4x - 3); \qquad \text{Property of logarithms}$$

$$y' = \frac{2x + 1}{x^2 + x + 1} + 5 \cdot \frac{4}{4x - 3} \qquad \text{Take derivatives.}$$

$$= \frac{2x + 1}{x^2 + x + 1} + \frac{20}{4x - 3}. \quad \checkmark\ 5$$

✓ **Checkpoint 5**

Find $y'$ for the given functions.

**(a)** $y = \ln(7 + x)$

**(b)** $y = \ln(4x^2)$

**(c)** $y = \ln(8x^3 - 3x)$

**(d)** $y = x^2 \ln x$

The function $y = \ln(-x)$ is defined for all $x < 0$ (since $-x > 0$ when $x < 0$). Its derivative can be found by applying the derivative rule for $\ln g(x)$ with $g(x) = -x$:

$$y' = \frac{g'(x)}{g(x)}$$

$$= \frac{-1}{-x}$$

$$= \frac{1}{x}.$$

This is the same as the derivative of $y = \ln x$, with $x > 0$. Since

$$|x| = \begin{cases} x & \text{if } x \geq 0 \\ -x & \text{if } x < 0, \end{cases}$$

we can combine two results into one in the manner that follows.

---

### Derivative of $\ln|x|$

If $y = \ln|x|$, then $y' = \dfrac{1}{x}$   $(x \neq 0)$.

---

**EXAMPLE 9**   Let $y = e^x \ln|x|$. Find $y'$.

**Solution**  Use the product rule:

$$y' = e^x \cdot \frac{1}{x} + \ln|x| \cdot e^x = e^x \left( \frac{1}{x} + \ln|x| \right). \ \ \checkmark_6$$

✓ **Checkpoint 6**

Find the derivative of each function.

(a)  $y = e^{x^2} \ln|x|$

(b)  $y = x^2/\ln|x|$

**EXAMPLE 10**   If $f(x) = \log x$, find $f'(x)$.

**Solution**  By the change-of-base theorem (page 202),

$$f(x) = \log x = \frac{\ln x}{\ln 10} = \frac{1}{\ln 10} \cdot \ln x.$$

Since $1/\ln 10$ is a constant,

$$f'(x) = \frac{1}{\ln 10} \cdot \frac{d}{dx}(\ln x) = \frac{1}{\ln 10} \cdot \frac{1}{x} = \frac{1}{(\ln 10)x}.$$

**EXAMPLE 11**   **Finance**  If a person borrows $120,000 on a 30-year fixed-rate mortgage at 5.7% interest per year, compounded monthly, then the required monthly payment is $696. However, the borrower would like to pay off the loan early by making larger monthly payments. The actual number of months it will take to pay off the mortgage if a payment of $p$ dollars is made each month is approximated by

$$f(p) = 211 \ln\left( \frac{p}{p - 570} \right) \quad (p \geq 696).$$

Find $f(710)$ and $f'(710)$ and explain what your answers mean for the borrower.

**Solution**  Evaluating $f(p)$ at $p = 710$ yields:

$$f(710) = 211 \ln\left( \frac{710}{710 - 570} \right) = 211 \ln\left( \frac{710}{140} \right) \approx 342.6.$$

Making monthly payments of $710 means that the borrower will pay off the loan in 343 months instead of 360 months (the payment in the final month will be somewhat smaller). In other words, increasing the monthly payment by just $14 (from $696 to $710) means that the borrower will pay off the loan 17 months faster: 343 months instead of 360 months, or 30 years!

Prior to taking the derivative, we may choose to use properties of logarithms to rewrite $f(p)$ as follows:

$$f(p) = 211 \ln\left(\frac{p}{p-570}\right) = 211[\ln p - \ln(p-570)] = 211 \ln(p) - 211 \ln(p-570).$$

Differentiating $f(p)$ with respect to $p$ yields

$$f'(p) = 211\left(\frac{1}{p}\right) - 211\left(\frac{1}{p-570}\right) = \frac{211}{p} - \frac{211}{p-570}.$$

By evaluating the derivative at $p = 710$ we find that

$$f'(710) = \frac{211}{710} - \frac{211}{710-570} = \frac{211}{710} - \frac{211}{140} \approx -1.2.$$

This means that, when $p = \$710$, increasing the monthly payment by 1 dollar will shorten the length of time needed to pay off the loan by approximately 1 month. ✔7

✓**Checkpoint 7**

If $f(x) = 6\ln\left(\dfrac{x}{7-3x}\right)$,

find $f'(2)$.

Often, a population or the sales of a certain product will start growing slowly, then grow more rapidly, and then gradually level off. Such growth can frequently be approximated by a *logistic function* of the form

$$f(x) = \frac{c}{1 + ae^{kx}}$$

for appropriate constants $a$, $c$, and $k$. (Logistic functions were introduced in Section 4.2.)

---

**EXAMPLE 12** **Cell Phone Subscriptions** The number of cell phone subscriptions worldwide (in millions) in year $x$ can be approximated by

$$f(t) = \frac{8300}{1 + 12e^{-.3t}},$$

where $t = 0$ corresponds to the year 2000. Find the rate of change of cell phone subscriptions in 2000, 2008, and 2014. (Data from: www.worldbank.org.)

**Solution** Use the quotient rule to find the derivative of $f(t)$:

$$f'(t) = \frac{(1 + 12e^{-.3t})(0) - 8300[12(-.3e^{-.3t})]}{(1 + 12e^{-.3t})^2}$$

$$= \frac{29{,}880e^{-.3t}}{(1 + 12e^{-.3t})^2}.$$

The rate of change in 2000 ($t = 0$) is

$$f'(0) = \frac{29{,}880e^{-.3(0)}}{(1 + 12e^{-.3(0)})^2} \approx 177.$$

✓**Checkpoint 8**

Suppose that a fish population is given by

$$f(x) = \frac{4000}{1 + e^{.5x}},$$

where $x$ is time in years. Find the rate of change of the population when

**(a)** $x = 0$

**(b)** $x = 3$

**(c)** Is the population increasing or decreasing?

Similarly, we compute that $f'(8) \approx 621$ and $f'(14) \approx 322$. This means that, in 2000, cell phone subscriptions were increasing at a rate of about 177 million per year. By 2008, the rate of change had jumped dramatically, to about 621 million per year. By 2014, the rate of increase had slowed to about 322 million per year. ✔8

# 11.8 Exercises

*Find the derivatives of the given functions. (See Examples 1–4 and 7–9.)*

**1.** $y = e^{5x}$

**2.** $y = e^{-4x}$

**3.** $f(x) = 5e^{2x}$

**4.** $f(x) = 4e^{-2x}$

**5.** $g(x) = -4e^{-7x}$

**6.** $g(x) = 6e^{x/2}$

**7.** $y = e^{x^2}$

**8.** $y = e^{-x^2}$

**9.** $f(x) = e^{x^3/3}$

**10.** $y = e^{4-x^3}$

**11.** $y = -3e^{3x^2+5}$

**12.** $y = 4e^{2x^2-4}$

**13.** $y = \ln(-10x^2 + 7x)$

**14.** $y = \ln(11 - 2x)$

**15.** $y = \ln\sqrt{5x + 1}$

**16.** $y = \ln\sqrt{x - 8}$

**17.** $f(x) = \ln[(5x + 9)(x^2 + 1)]$

**18.** $y = \ln[(4x + 1)(5x - 2)]$

**19.** $y = \ln\left(\dfrac{4 - x}{3x + 8}\right)$

**20.** $f(x) = \ln\left(\dfrac{7x - 1}{12x + 5}\right)$

**21.** $y = x^4 e^{-3x}$

**22.** $y = xe^{3x}$

**23.** $y = (6x^2 - 5x)e^{-3x}$

**24.** $y = (x - 5)^2 e^{3x}$

**25.** $y = \ln(8x^3 - 3x)^{1/2}$

**26.** $y = \ln(x^5 + 8x^2)^{3/2}$

**27.** $y = x \ln(9 - x^4)$

**28.** $y = -4x \ln(x + 12)$

**29.** $y = x^4 \ln|x|$

**30.** $y = (2x^3 - 1) \ln|x|$

**31.** $y = \dfrac{-5 \ln|x|}{5 - 2x}$

**32.** $y = \dfrac{3 \ln|x|}{3x + 4}$

**33.** $y = \sqrt{\ln(x - 3)}$

**34.** $y = [\ln(x + 1)]^4$

**35.** $y = \dfrac{e^x - 1}{\ln|x|}$

**36.** $y = \dfrac{e^x}{\ln|x|}$

**37.** $y = \dfrac{e^x - e^{-x}}{x}$

**38.** $y = \dfrac{e^x + e^{-x}}{x}$

**39.** $f(x) = e^{3x+2} \ln(4x - 5)$

**40.** $y = e^{x^3} \ln|x|$

**41.** $y = \dfrac{700}{7 - 10e^{.4x}}$

**42.** $f(x) = \dfrac{2400}{3 + 8e^{.2x}}$

**43.** $y = \dfrac{500}{12 + 5e^{-.5x}}$

**44.** $y = \dfrac{10,000}{9 + 4e^{-.2x}}$

**45.** $y = \ln(\ln|x|)$

**46.** $y = 5e^{e^x}$

*Find the derivatives of the given functions. (See Examples 6 and 10.) The chain rule is needed in Exercises 48–52.*

**47.** $y = 8^x$

**48.** $y = 8^{x^2}$

**49.** $y = 15^{2x}$

**50.** $f(x) = 2^{-x}$

**51.** $g(x) = \log 6x$

**52.** $y = \log(x^2 + 1)$

**C** *For Exercises 53–56, find the equation of the tangent line to the graph of f at the given point.*

**53.** $f(x) = \dfrac{e^x}{2x + 1}$ at $(0, 1)$

**54.** $f(x) = \dfrac{x^2}{e^x}$ when $x = 1$

**55.** $f(x) = \ln(1 + x^2)$ at $(0, 0)$

**56.** $f(x) = \ln(2x^2 - 7x)$ when $x = -1$

**57.** If $f(x) = e^{2x}$, find $f'(\ln 5)$.

**58.** If $g(x) = 3e \ln[\ln x]$, find $g'(e)$.

*Work these exercises. (See Example 5.)*

**59. Gross National Income of China** The gross national income (GNI) of China (in billions of U.S. dollars) can be approximated by

$$f(t) = 1000e^{.165t},$$

where $t = 0$ corresponds to the year 2000. (Data from: www.worldbank.org.)

(a) Find $f'(t)$.

(b) At what rate was the GNI changing in 2000? Round to the nearest integer.

(c) Repeat part (b) for 2016.

**60. Internet Users** The number of Internet users worldwide (in billions) is projected to be approximately

$$f(x) = 1.2e^{.062x},$$

where $x = 15$ corresponds to the year 2015. (Data from: Cisco Visual Networking Index.)

(a) How many Internet users are projected for the year 2020? (Give your answer to one decimal place.)

(b) According to this model, what will the rate of growth of Internet users be in 2020? (Give your answer to two decimal places.)

**61. Endangered Species** The black-footed ferret is one of the most endangered mammals in the United States. According to the U.S. Fish and Wildlife Service Black-Footed Ferret Recovery Plan, the goal is to have 1500 breeding adults in the wild by 2023. If progress toward this goal continues, the population of breeding adults in the wild in year $x$ can be approximated by

$$P(x) = 76e^{.13x},$$

where $x = 12$ corresponds to the year 2012. Use this model to find the following.

(a) $P(12)$      (b) $P'(x)$ and $P'(12)$

(c) Interpret your answers in parts (a) and (b).

**62. Life Insurance Costs** A business provides a \$50,000 life insurance policy for each of its employees. The annual cost in dollars of a policy (which depends on the age of the employee) can be approximated by $C(x) = 4.55e^{.075x}$, where $x$ is the age of the employee in years ($18 \le x \le 65$). Find and interpret the given quantities. (Data from: Internal Revenue Code section 79, Table 01.)

(a) $C(20)$ and $C'(20)$      (b) $C(60)$ and $C'(60)$

**63. Childhood Mortality** The preschool mortality rate (number of deaths under age 5 per 1000 people) in Russia can be approximated by

$$M(t) = 23e^{-.062t},$$

where $t = 0$ corresponds to the year 2000. Assuming this model remains accurate, find the following. Give your answers to one decimal place. (Data from: www.childmortality.org.)

**(a)** The preschool mortality rate in 2022.

**(b)** The rate at which preschool mortality is projected to be changing in 2022.

**(c)** Is preschool mortality projected to be increasing or decreasing in 2022?

**64. Nonperforming Bank Loans** The percentage of loans that are nonperforming in year $x$ can be approximated by

$$N(x) = 30e^{-.19x},$$

where $x = 9$ corresponds to the year 2009. Use this model to find the following. Give your answers to two decimal places. (Data from: www.worldbank.org.)

**(a)** The percentage of loans that were nonperforming in 2015.

**(b)** The rate at which nonperforming loans were changing in 2015.

**(c)** Were nonperforming loans increasing or decreasing in 2015?

**65. Response to Infection** The temperature of a certain patient (in degrees Fahrenheit) $t$ hours after the onset of an infection can be approximated by

$$P(t) = 98.6 + te^{-.2t}.$$

Find how the patient's temperature is changing at the given times.

**(a)** $t = 1$            **(b)** $t = 12$

**66. Website Traffic** The number of hourly hits (in thousands) on a company's website $x$ hours after a news story breaks can be approximated by

$$H(x) = 1 + 30xe^{-.5x}.$$

Find the rate of change of hits to this website for the given number of hours after the story breaks.

**(a)** 1 hour            **(b)** 12 hours

*Work these exercises. (See Examples 11 and 12.)*

**67. Cell Phone Subscriptions** The number of cell phone subscriptions per 100 people in middle income countries can be approximated by

$$f(t) = \frac{100}{1 + 25e^{-.4t}},$$

where $t = 0$ corresponds to the year 2000. Find the following (round to one decimal place). (Data from: www.worldbank.org.)

**(a)** $f'(t)$            **(b)** $f'(17)$

**(c)** Interpret your answer in part (b).

**68. Primary School Completion** The percentage of children who complete primary school in the world's poorest countries can be approximated by

$$f(t) = \frac{85}{1 + .67e^{-.1t}},$$

where $t = 0$ corresponds to the year 2000. Find the following (round to one decimal place). (Data from: International Development Association.)

**(a)** $f'(t)$            **(b)** $f'(14)$

**(c)** Interpret your answer in part (b).

**69. Logistic Population Growth** The population of a bed of clams in the Great South Bay off Long Island is approximated by the logistic function

$$G(t) = \frac{52,000}{1 + 12e^{-.52t}},$$

where $t$ is measured in years. Find the clam population and its rate of growth after

**(a)** 1 year     **(b)** 4 years     **(c)** 10 years.

**(d)** What happens to the rate of growth over time?

**70. Tumor Growth** The function

$$V(t) = \frac{1100}{(1 + 1023e^{-.02415t})^4},$$

where $t$ is in months and $V(t)$ is in cubic centimeters, models the relationship between the volume of a breast tumor and the amount of time it has been growing.[*]

**(a)** Find the tumor volume at 240 months.

**(b)** Assuming that the shape of a tumor is spherical, find the radius of the tumor from part (a). (*Hint:* The volume of a sphere is given by the formula $v = \frac{4}{3}\pi r^3$.)

**(c)** If a tumor of volume .5 cm$^3$ is detected, according to the formula, how long has it been growing? What does this imply?

**(d)** Calculate the rate of change of tumor volume at 240 months. Interpret this rate.

**71. United States Population** The projected population of the United States (in millions) can be approximated by

$$f(x) = 66 \ln (x) + 135,$$

where $x = 18$ corresponds to the year 2018. Use this model to find the following. (Data from: U.S. Census Bureau.)

**(a)** The projected population in 2033, to the nearest million.

**(b)** The projected population *growth rate* in year $x$.

**(c)** The projected population *growth rate* in 2033.

**72. Renewable Energy** Renewable energy consumption in the United States (as a percentage of total energy consumption) can be approximated by

$$f(x) = 1.5 \ln (x) + 4,$$

where $x = 0$ corresponds to the year 2000. Find the following (round to two decimal places). (Data from: www.worldbank.org.)

**(a)** $f(10)$            **(b)** $f'(10)$

**(c)** Interpret your answers to parts (a) and (b).

*Work the following exercises.*

**73. Marginal Revenue** Suppose the demand function for $x$ thousand of a certain item is

$$p = 100 + \frac{50}{\ln x} \quad (x > 1),$$

---

[*]Data from John Spratt et al., "Decelerating Growth and Human Breast Cancer," *Cancer* 71, no. 6 (1993): 2013–2019.

where $p$ is in dollars.

(a) Find the marginal revenue.

(b) Find the revenue from the next thousand items at a demand of 8000 ($x = 8$).

**74. Finance** A person owes $4000 on a credit card that charges an annual interest rate of 15.9%, compounded monthly. If the card holder makes no additional charges and makes a payment of $p$ dollars each month, the number of months it will take to pay off the debt is given by

$$f(p) = 76 \ln\left(\frac{p}{p - 53}\right) \qquad (p > 53).$$

(a) Find and interpret $f(78)$.

(b) Would you expect $f'(p)$ to be positive or negative?

(c) Find and interpret $f'(78)$.

**75. Finance** An individual deposits $250 in an account that earns 2.8% interest per year, compounded monthly. The number of months needed for the balance in the account to reach an amount of $A$ dollars is given by

$$f(A) = 429 \ln\left(\frac{A}{250}\right).$$

(a) Find and interpret $f(300)$.

(b) Would you expect $f'(A)$ to be positive or negative?

(c) Find and interpret $f'(300)$.

**76. Richter Scale** The Richter scale provides a measure of the magnitude of an earthquake. One of the largest Richter numbers $M$ ever recorded for an earthquake was 8.9, from the 1933 earthquake in Japan. The following formula shows a relationship between the amount of energy released, $E$, and the Richter number:

$$M = \frac{2 \ln E - 2 \ln .007}{3 \ln 10}.$$

Here, $E$ is measured in kilowatt-hours (kWh).*

---
*Christopher Bradley, "Media Clips," *Mathematics Teacher* 93, no. 4 (April 2000): 300–303.

(a) For the 1933 earthquake in Japan, what value of $E$ gives a Richter number $M = 8.9$?

(b) If the average household uses 247 kilowatt-hours per month, for how many months would the energy released by an earthquake of this magnitude power 10 million households?

(c) Find the rate of change of the Richter number $M$ with respect to energy when $E = 70,000$ kWh.

(d) What happens to $\dfrac{dM}{dE}$ as $E$ increases?

✓**Checkpoint Answers**

1. (a) $(2x^2 - 1)e^x + 4xe^x$  (b) $\dfrac{-e^x}{2(1 - e^x)^{1/2}}$

2. (a) $y' = 44e^{11x}$  (b) $y' = 45e^{(-9x+2)}$
   (c) $y' = -3x^2 e^{-x^3}$

3. (a) $y' = \dfrac{e^x(11 + x)}{(12 + x)^2}$  (b) $y' = \dfrac{-27,000e^x}{(15 + 3e^x)^2}$

4. $y' = (\ln 25)25^x$

5. (a) $y' = \dfrac{1}{7 + x}$  (b) $y' = \dfrac{2}{x}$
   (c) $y' = \dfrac{24x^2 - 3}{8x^3 - 3x}$  (d) $y' = x(1 + 2\ln x)$

6. (a) $y' = e^{x^2}\left(\dfrac{1}{x} + 2x \ln|x|\right)$  (b) $y' = \dfrac{2x \ln|x| - x}{(\ln|x|)^2}$

7. 21

8. (a) $-500$ fish per year  (b) About $-298$ fish per year
   (c) Decreasing

# 11.9 Continuity and Differentiability

Intuitively speaking, a function is **continuous** at a point if you can draw the graph of the function near that point without lifting your pencil from the paper. Conversely, a function is **discontinuous** at a point if the pencil *must* be lifted from the paper in order to draw the graph on both sides of the point.

Looking at some graphs that have points of discontinuity will clarify the idea of continuity at a point. Each of the three graphs in Figure 11.40, on the next page, is continuous *except* at $x = 2$. As usual, an open circle indicates that the point is not part of the graph. You cannot draw these graphs without lifting your pencil at $x = 2$, at least for an instant.

However, there are different reasons why each function is discontinuous at $x = 2$:

1. In graph (a), $f(2)$ is not defined (because there is a hole in the graph).

2. In graph (b), $\lim_{x \to 2} f(x)$ does not exist (because the left-hand limit does not equal the right-hand limit—see Section 11.2).

3. In graph (c), $f(2)$ is defined and $\lim_{x \to 2} f(x)$ does exist, but $\lim_{x \to 2} f(x) \neq f(2)$ (because $\lim_{x \to 2} f(x) = 4$ whereas $f(2) = 6$).

By phrasing each of these considerations positively, we obtain the following definition.

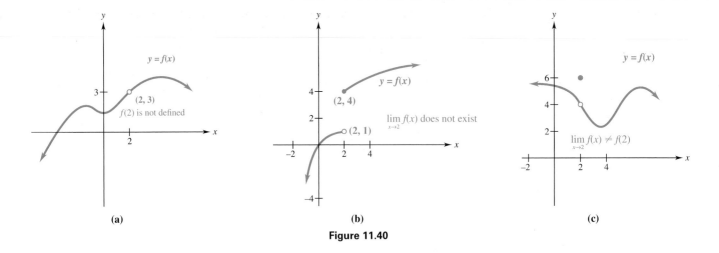

**Figure 11.40**

## Continuity at a Point

A function $f$ is **continuous** at $x = c$ if

**(a)** $f(c)$ is defined;

**(b)** $\lim_{x \to c} f(x)$ exists;

**(c)** $\lim_{x \to c} f(x) = f(c)$.

If $f$ is not continuous at $x = c$, it is **discontinuous** there.

The idea behind condition (c) is this: To draw the graph near $x = c$ without lifting your pencil, it must be the case that when $x$ is very close to $c$, $f(x)$ must be very close to $f(c)$—otherwise, you would have to lift the pencil at $x = c$, as happens in Figure 11.40 at $x = 2$. ✔₁

**Checkpoint 1**

Find any points of discontinuity for the given functions.

**(a)**

**(b)**

**(c)**

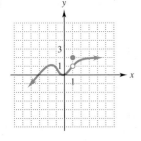

**EXAMPLE 1** Tell why the given functions are discontinuous at the indicated point.

**(a)** $f(x)$ in Figure 11.41 at $x = 3$

**Solution** The open circle on the graph in Figure 11.41 at the point where $x = 3$ means that $f(3)$ is not defined. Because of this, part (a) of the definition fails.

**Figure 11.41**

**Figure 11.42**

**(b)** $h(x)$ in Figure 11.42 at $x = 0$

**Solution** The graph in Figure 11.42 shows that $h(0) = -1$. Also, as $x$ approaches 0 from the left, $h(x)$ is $-1$. However, as $x$ approaches 0 from the right, $h(x)$ is 1. As discussed in Section 11.1, for a limit to exist at a particular value of $x$, the values of $h(x)$ must approach a single number. Since no single number is approached by the values of $h(x)$ as $x$ approaches 0, $\lim_{x \to 0} h(x)$ does not exist, and part (b) of the definition fails.

**(c)** $g(x)$ at $x = 4$ in Figure 11.43

**Solution** In Figure 11.43, the heavy dot above $x = 4$ shows that $g(4)$ is defined. In fact, $g(4) = 1$. However, the graph also shows that

$$\lim_{x \to 4} g(x) = -2,$$

so $\lim_{x \to 4} g(x) \neq g(4)$, and part (c) of the definition fails.

**Figure 11.43**

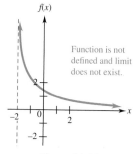

**Figure 11.44**

**(d)** $f(x)$ in Figure 11.44 at $x = -2$

**Solution** The function $f$ graphed in Figure 11.44 is not defined at $-2$, and $\lim_{x \to -2} f(x)$ does not exist. Either of these reasons is sufficient to show that $f$ is not continuous at $-2$. (Function $f$ is continuous at any value of $x$ greater than $-2$, however.)

**Checkpoint 2**

Tell why the given functions are discontinuous at the indicated point.

**(a)**

**(b)**

## Continuity at Endpoints

One-sided limits (see Section 11.2) can be used to define continuity at an endpoint of a graph.* Consider the endpoint $(1, 0)$ of the graph of the half-circle function $f(x) = \sqrt{1 - x^2}$ in Figure 11.45. Note that $f$ has a left-hand limit at $x = 1$. Furthermore,

$$f(1) = 0 \quad \text{and} \quad \lim_{x \to 1^-} f(x) = 0, \quad \text{so that} \quad \lim_{x \to 1^-} f(x) = f(1).$$

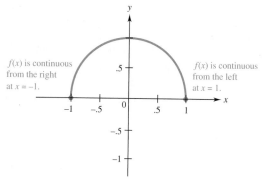

**Figure 11.45**

We say that $f(x)$ is *continuous from the left* at $x = 1$. A similar situation occurs with the other endpoint $(-1, 0)$: There is a right-hand limit at $x = -1$, and we have $\lim_{x \to -1^+} f(x) = 0 = f(-1)$. We say that $f(x)$ is *continuous from the right* at $x = -1$.

---

*The preceding definition of continuity does not apply to an endpoint because the two-sided limit $\lim_{x \to c} f(x)$ does not exist when $(c, f(c))$ is an endpoint.

This situation is an example of the following definition.

## Continuity from the Left and from the Right

A function $f$ is **continuous from the left** at $x = c$ if

**(a)** $f(c)$ is defined;

**(b)** $\lim\limits_{x \to c^-} f(x)$ exists;

**(c)** $\lim\limits_{x \to c^-} f(x) = f(c)$.

Similarly, $f$ is **continuous from the right** at $x = c$ if

**(a)** $f(c)$ is defined;

**(b)** $\lim\limits_{x \to c^+} f(x)$ exists;

**(c)** $\lim\limits_{x \to c^+} f(x) = f(c)$.

✓ **Checkpoint 3**

State whether the given function is continuous from the left, continuous from the right, or neither at the given value of $x$.

**(a)** $f(x)$ in Figure 11.40(b) at $x = 2$

**(b)** $h(x)$ in Figure 11.42 at $x = 0$

**(c)** $g(x)$ in Figure 11.43 at $x = 4$

Note that a function which is continuous at $x = c$ is automatically continuous from the left and from the right at $x = c$. ✓₃

## Continuity on an Interval

When discussing the continuity of a function, it is often helpful to use interval notation, which was introduced in Chapter 1. The following chart should help you recall how it is used.

| Interval | Name | Description | Interval Notation |
|---|---|---|---|
| | Open interval | $-2 < x < 3$ | $(-2, 3)$ |
| | Closed interval | $-2 \le x \le 3$ | $[-2, 3]$ |
| | Open interval | $x < 3$ | $(-\infty, 3)$ |
| | Open interval | $x > -5$ | $(-5, \infty)$ |

✓ **Checkpoint 4**

Write each of the following in interval notation.

**(a)**

**(b)**

**(c)**

Remember, the symbol $\infty$ does not represent a number; $\infty$ is used for convenience in interval notation to indicate that the interval extends without bound in the positive direction. Also, $-\infty$ indicates no bound in the negative direction. ✓₄

Continuity at a point was defined previously. Continuity on an interval is defined as follows.

## Continuity on Open and Closed Intervals

A function $f$ is **continuous on the open interval $(a, b)$** if $f$ is continuous at every $x$-value in the interval $(a, b)$.

A function $f$ is **continuous on the closed interval $[a, b]$** if

**1.** $f$ is continuous on the open interval $(a, b)$;

**2.** $f$ is continuous from the right at $x = a$;

**3.** $f$ is continuous from the left at $x = b$.

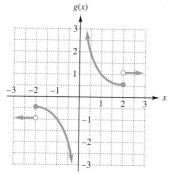

**Figure 11.46**

Intuitively, a function is continuous on an interval if you can draw the entire graph of $f$ over that interval without lifting your pencil from the paper.

**EXAMPLE 2**  Is the function of Figure 11.46 continuous on the given $x$-intervals?

**(a)** $(0, 2)$

**Solution**  The function $g$ is discontinuous only at $x = -2$, 0, and 2. Hence, $g$ is continuous at every point of the open interval $(0, 2)$, which does not include 0 or 2.

**(b)** $[0, 2]$

**Solution**  The function $g$ is not defined at $x = 0$, so it is not continuous from the right there. Therefore, it is not continuous on the closed interval $[0, 2]$.

**(c)** $(1, 3)$

**Solution**  This interval contains a point of discontinuity at $x = 2$. So $g$ is not continuous on the open interval $(1, 3)$.

**(d)** $[-2, -1]$

**Solution**  The function $g$ is continuous on the open interval $(-2, -1)$, continuous from the right at $x = -2$, and continuous from the left at $x = -1$. Hence, $g$ is continuous on the closed interval $[-2, -1]$. ✅5

✓**Checkpoint 5**

Are the functions whose graphs are shown continuous on the indicated intervals?

**(a)**  $[-4, -1]; [2, 5]$

**(b)**  $(1, 3); [4, 6]$

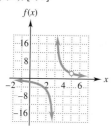

**EXAMPLE 3**  **Cell Phone Data Plan**  The monthly cost $C(x)$ (in dollars) of a popular smartphone data plan for the use of $x$ gigabytes (GB) of data is $45 for the first 3 GB of data plus $15 for each additional GB of data (or fraction thereof). (Data from: Verizon Wireless, 2016 prices.)

**(a)**  Sketch the graph of $y = C(x)$ on the interval $0 \le x \le 6$.

**Solution**  For any amount of data usage up to and including 3 GB, the charge is $45. The charge for using more than 3 GB but less than 4 GB of data is $60. Similar results lead to the graph in Figure 11.47.

**Figure 11.47**

**(b)**  Find any points of discontinuity for $C$ on the interval $(0, 6)$.

**Solution**  As the graph suggests, $C$ is discontinuous at $x = 3$, 4, 5 gigabytes. However, $C$ is continuous from the left at each of these $x$-values. ✅6

✓**Checkpoint 6**

Suppose a smartphone data plan for heavy users charges $60 for the first 6 GB of data and then $20 for each additional 2 GB of data (or fraction thereof). If $C(x)$ is the cost (in dollars) of using $x$ GB of data, find any points of discontinuity for $C$ on the interval $(0, 12)$.

**Figure 11.48**

**Figure 11.49**

## Continuity and Technology

Because of the way a graphing calculator or computer graphs functions, the graph of a continuous function may look like disconnected, closely spaced line segments (Figure 11.48). Furthermore, calculators often have trouble accurately portraying graphs at points of discontinuity. For instance, "jump discontinuities" (such as those at $x = 3, 4$, and 5 in Figure 11.47) may appear as "stair steps," with vertical line segments connecting separate pieces of the graph (Figure 11.49). Similarly, a hole in the graph (such as the one at $x = 4$ in Figure 11.43) may not be visible on a calculator screen, depending on what viewing window is used. Nor can a calculator indicate whether an endpoint is included in the graph.

One reason these inaccuracies appear is due to the manner in which a calculator graphs: It plots points and connects them with line segments. In effect, the calculator assumes that the function is continuous unless it actually computes a function value and determines that it is undefined. In that case, it will skip a pixel and not connect points on either side of it.

The moral of this story is that a calculator or computer is only a tool. In order to use it correctly and effectively, you must understand the mathematics involved. When you do, it is usually easy to interpret screen images correctly.

## Continuity and Differentiability

As shown earlier in this chapter, a function fails to have a derivative at a point where the function is not defined, where the graph of the function has a "sharp point," or where the graph has a vertical tangent line. (See Figure 11.50.)

The function graphed in Figure 11.50 is continuous on the interval $(x_1, x_2)$ and has a derivative at each point on this interval. On the other hand, the function is also continuous on the interval $(0, x_2)$, but does *not* have a derivative at a $x_1$.

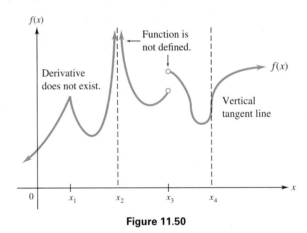

**Figure 11.50**

A similar situation holds in the general case.

> If the derivative of a function exists at a point, then the function is continuous at that point. However, a function may be continuous at a point and not have a derivative there.

**EXAMPLE 4**    **Overtime Pay** The Fair Labor Standards Act requires overtime pay for covered, nonexempt employees at a rate of not less than one and one-half times an employee's regular rate of pay after 40 hours of work in a workweek. (Data from: U.S. Department of Labor.) A fast-food restaurant pays an entry-level employee $10 per hour for regular pay and time-and-a-half for overtime pay. Let $P(x)$ represent the

✓ **Checkpoint 7**

State whether the function whose graph is shown is continuous and/or differentiable at the given values of $x$.

**(a)** $x = 2$

**(b)** $x = 4$

**(c)** $x = 6$

employee's gross pay prior to taxes and other withholdings (in dollars) for working $x$ hours in a week.

Then

$$P(x) = \begin{cases} 10x & \text{if } 0 \leq x \leq 40 \\ 15x - 200 & \text{if } x > 40. \end{cases}$$

Sketch the graph of $y = P(x)$ for $0 \leq x \leq 60$. Find any points where $P$ is discontinuous. Also find any points where $P$ is not differentiable.

**Solution** Since $P$ is a piecewise-linear function, we can sketch the graph by plotting two points for each line segment. Using the first line of the function, we find that when $x = 0$, the gross pay is $P(0) = 0$, and when $x = 40$, the gross pay is $P(40) = 400$. Using the second line of the function, we find that $P(50) = 550$ and $P(60) = 700$. Sketching the first line segment for $0 \leq x \leq 40$ and the second line segment for $40 < x \leq 60$ leads to the graph in Figure 11.51. The graph indicates that $P$ is a continuous function because it can be drawn without lifting pencil from paper. However, $P$ is not differentiable at $x = 40$ hours, where the graph has a sharp corner. ✓ 7

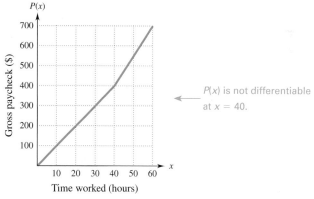

P(x) is not differentiable at x = 40.

**Figure 11.51**

⊞ **TECHNOLOGY TIP** In some viewing windows, a calculator graph may appear to have a sharp corner when, in fact, the graph is differentiable at that point. When in doubt, try a different window to see if the corner disappears.

# 11.9 Exercises

*Find all points of discontinuity for the functions whose graphs are shown. (See Example 1.)*

**1.**

**2.**

**3.**

**4.**

**5.**

**6.**

**7.**

**8.**

*Are the given functions continuous at the given values of x?*

**9.** $f(x) = \dfrac{4}{x - 2}; x = 0, x = 2$

**10.** $g(x) = \dfrac{5}{x + 5}; x = -5, x = 5$

**11.** $h(x) = \dfrac{1}{x(x - 3)}; x = 0, x = 3, x = 5$

**12.** $f(x) = \dfrac{-1}{x(x - 1)}; x = -1, x = 0, x = 1$

**13.** $g(x) = \dfrac{x + 2}{x^2 - x - 2}; x = 1, x = 2, x = -2$

**14.** $h(x) = \dfrac{3x}{6x^2 + 15x + 6}; x = 0, x = -\dfrac{1}{2}, x = 3$

**15.** $g(x) = \dfrac{x^2 - 4}{x - 2}; x = 0, x = 2, x = -2$

**16.** $h(x) = \dfrac{x^2 - 25}{x + 5}; x = 0, x = 5, x = -5$

**17.** $f(x) = \begin{cases} x - 2 & \text{if } x \le 3 \\ 2 - x & \text{if } x > 3 \end{cases}; x = 2, x = 3$

**18.** $f(x) = \begin{cases} 2x + 6 & \text{if } x < 0 \\ 4 - x & \text{if } x \ge 0 \end{cases}; x = 0, x = 4$

*Find all x-values where the function is discontinuous.*

**19.** $g(x) = x^2 - 5x + 12$

**20.** $f(x) = -x^2 + 6x - 3$

**21.** $h(x) = \dfrac{|x + 4|}{x + 4}$

**22.** $r(x) = \dfrac{x + 5}{|x + 5|}$

**23.** $k(x) = e^{\sqrt{x - 2}}$

**24.** $s(x) = \ln(x - 7)$

*In Exercises 25–28, (a) graph the given function, and (b) find all values of x where the function is discontinuous. (See Example 4.)*

**25.** $f(x) = \begin{cases} 2 & \text{if } x < 1 \\ x + 3 & \text{if } 1 \le x < 4 \\ 8 & \text{if } x \ge 4 \end{cases}$

**26.** $g(x) = \begin{cases} 5 & \text{if } x < -1 \\ 4 - x & \text{if } -1 \le x \le 1 \\ 1 & \text{if } x > 1 \end{cases}$

**27.** $f(x) = \begin{cases} 4 & \text{if } x \le -2 \\ 5 - x^2 & \text{if } -2 < x < 2 \\ 1 & \text{if } x \ge 2 \end{cases}$

**28.** $g(x) = \begin{cases} -10 & \text{if } x < -2 \\ x^2 + 4x - 6 & \text{if } -2 \le x < 2 \\ 10 & \text{if } x \ge 2 \end{cases}$

*Work these exercises. (See Examples 3 and 4.)*

**29. State Sales Tax** Officials in California have raised and lowered the state sales tax in various years, as shown in the accompanying figure. Let $T(x)$ represent the sales tax rate (as a percentage) in year $x$. Find the given quantities. (Data from: www.boe.ca.gov/sutax/taxrateshist.htm.)

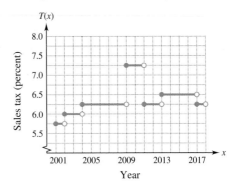

**(a)** $\lim\limits_{x \to 2010^-} T(x)$ **(b)** $\lim\limits_{x \to 2010^+} T(x)$ **(c)** $\lim\limits_{x \to 2010} T(x)$

**(d)** $\lim\limits_{x \to 2017^-} T(x)$ **(e)** $\lim\limits_{x \to 2017^+} T(x)$ **(f)** $\lim\limits_{x \to 2017} T(x)$

**(g)** Find all values on the interval (2001, 2018) where $T$ is discontinuous.

**(h)** Find all values on the interval (2001, 2018) where $T$ is not differentiable.

**30. Natural Science** Suppose a gram of ice is at a temperature of $-100°C$. The accompanying graph shows the temperature of the ice as an increasing number of calories of heat are applied. Where is this function discontinuous? Where is it differentiable?

**31. Learning Rate** With certain skills (such as music), learning is rapid at first and then levels off. However, sudden insights may cause learning to speed up sharply. A typical graph of such learning is shown in the accompanying figure. Where is the function discontinuous? Where is it differentiable?

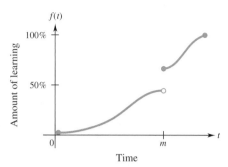

Time

**32. Produce Pricing** A pick-your-own strawberry farm charges $10 per gallon of strawberries. The cost of partial gallons is pro-rated except that there is a minimum charge of $10. Customers who pick more than 4 gallons receive a $10 discount off their total order. The accompanying figure shows the price (in dollars), $P(x)$, of picking $x$ gallons of strawberries. Find the given quantities.

**(a)** $P(1)$  **(b)** $\lim_{x \to 1} P(x)$

**(c)** $P(4)$  **(d)** $\lim_{x \to 4} P(x)$

**(e)** Find all values on the interval $(0, 8)$ where $P$ is discontinuous.

**(f)** Find all values on the interval $(0, 8)$ where $P$ is not differentiable.

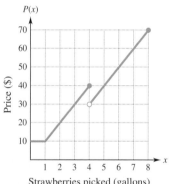

Strawberries picked (gallons)

*Work the following exercises.*

**33. First-Class Postage** The accompanying table shows the cost $C(x)$ (in dollars) of mailing a first-class letter weighing $x$ ounces domestically on January 1, 2017. (Data from: www.usps.gov.)

| Weight not over (ounces) | Price ($) |
|---|---|
| 1 | 0.47 |
| 2 | 0.68 |
| 3 | 0.89 |
| 3.5 | 1.10 |

**(a)** Sketch a graph of $y = C(x)$ on the interval $[0, 3.5]$.

**(b)** $\lim_{x \to 1^-} C(x)$  **(c)** $\lim_{x \to 1^+} C(x)$  **(d)** $\lim_{x \to 1} C(x)$

**(e)** $\lim_{x \to 2.5^-} C(x)$  **(f)** $\lim_{x \to 2.5^+} C(x)$  **(g)** $\lim_{x \to 2.5} C(x)$

**(h)** Find all values on the interval $(0, 3.5)$ where $C$ is discontinuous.

**34. Bulk Pricing** A nut supplier offers discounts for buying in bulk. The accompanying table shows the price per pound for jumbo raw Georgia pecans (with the cost of partial pounds being prorated). Let $C(x)$ represent the cost (in dollars) of purchasing $x$ pounds of nuts. Find the cost of buying the given amounts.

**(a)** ½ pound  **(b)** 10 pounds  **(c)** 11 pounds

**(d)** Find all values where $C$ is discontinuous.

| Pounds Purchased | Price per Pound |
|---|---|
| $0 < x \le 1$ | $16.99 |
| $1 < x \le 10$ | $15.99 |
| $10 < x \le 30$ | $14.50 |

**35. Towing Charges** The state of Connecticut has established maximum charges, $C(x)$, (in dollars) for nonconsensual towing of vehicles (up to 10,000 lbs) a distance of $x$ miles. The base charge of $88 includes mileage to the scene, preparing the vehicle, and the first 2 miles of the tow (loaded miles). In addition there is a mileage charge of $4.75 per mile (or fraction of a mile) in excess of 2 loaded miles. Find the given quantities. (Data from: www.ct.gov.)

**(a)** $C(1)$  **(b)** $\lim_{x \to 1} C(x)$

**(c)** $C(4)$  **(d)** $\lim_{x \to 4} C(x)$

**(e)** Find all values where $C$ is discontinuous on the interval $(0, 5)$.

**36. Airport Parking** The cost of hourly parking at the domestic terminal of Hartsfield-Jackson Atlanta International Airport is $2 per hour (or fraction of an hour) for the first two hours, $3 per hour (or fraction of an hour) for the next 4 hours, and $32 for between 6 and 24 hours. Let $C(x)$ be the cost (in dollars) of parking for $x$ hours. Find the left-hand, right-hand, and two-sided limits and determine whether $C(x)$ is continuous at the given $x$-values. (Data from: www.atlanta-airport.com, 2017 prices.)

**(a)** $x = 1$  **(b)** $x = 5$  **(c)** $x = 10$

*Write each of the given ranges in interval notation.*

**37.** ←——(————|————)——→
        $-5$       $7$

**38.** ←——[————|————]——→
        $-3$       $16$

**39.** ←———————————|]——→
                    $-3$

**40.** ←——(————————————→
        $12$

**41.** On which of the following intervals is the function in Exercise 6 continuous: $(-6, 0)$; $[0, 3]$; $(4, 8)$?

**42.** On which of the following intervals is the function in Exercise 5 continuous: $(-3, 0)$; $[0, 3]$; $[1, 3]$?

**43. Weight during Pregnancy** During pregnancy, a woman's weight naturally increases during the course of the event. When she delivers, her weight immediately decreases by the approximate weight of the child. Suppose that a 120-lb woman gains

27 lb during pregnancy, delivers a 7-lb baby, and then, loses the remaining weight during the next 20 weeks.

(a) Graph the weight gain and loss during the pregnancy and the 20 weeks following the birth of the baby. Assume that the pregnancy lasts 40 weeks, that delivery occurs immediately after this time interval, and that the weight gain and loss before and after birth, respectively, are linear.

(b) Is this function continuous? If not, then find the value(s) of $t$ where the function is discontinuous.

**44. Gas Company Revenue** The Lacledge Gas Company in St. Louis, Missouri, charges customers $19.50 per month plus $.31290 per therm for the first 30 therms of gas used and $.15297 for each therm above 30, as summarized by the function

$$R(x) = \begin{cases} 19.50 + .31290x & \text{if } x \leq 30 \\ 24.2979 + .15297x & \text{if } x > 30, \end{cases}$$

where $R(x)$ is the revenue generate by one customer who uses $x$ therms of gas. Find the following quantities. (Data from: http://www.lacledegas.com, 2017 summer rates.)

(a) $\lim\limits_{x \to 30^-} R(x)$       (b) $\lim\limits_{x \to 30^+} R(x)$

(c) $\lim\limits_{x \to 30} R(x)$       (d) $R(30)$

(e) Is $R(x)$ continuous at $x = 30$?

(f) Is $R(x)$ differentiable at $x = 30$?

✓**Checkpoint Answers**

1. (a) $x = -1, 1$       (b) $x = -2$       (c) $x = 1$

2. (a) $f(a)$ does not exist.       (b) $\lim\limits_{x \to b^-} f(x)$ does not exist.

3. (a) Continuous from the right
   (b) Continuous from the left
   (c) Neither

4. (a) $(-5, 3)$       (b) $[4, 7]$       (c) $(-\infty, -1]$

5. (a) Yes; no       (b) Yes; no

6. $x = 6, 8, 10$ GB

7. (a) Continuous, differentiable
   (b) Continuous, not differentiable
   (c) Not continuous, not differentiable

# CHAPTER 11 Summary and Review

## Key Terms and Symbols

**11.1** $\lim\limits_{x \to a} f(x)$ limit of $f(x)$ as $x$ approaches $a$
   limit of a constant function
   limit of the identity function
   properties of limits
   polynomial limits
   limit theorem

**11.2** $\lim\limits_{x \to a^+} f(x)$ limit of $f(x)$ as $x$ approaches $a$ from the right
   $\lim\limits_{x \to a^-} f(x)$ limit of $f(x)$ as $x$ approaches $a$ from the left
   $\lim\limits_{x \to \infty} f(x)$ limit of $f(x)$ as $x$ approaches infinity

$\lim\limits_{x \to -\infty} f(x)$ limit of $f(x)$ as $x$ approaches negative infinity
   infinite limits

**11.3** average rate of change
   geometric meaning of average rate of change
   velocity
   instantaneous rate of change
   marginal cost, revenue, profit

**11.4** $y'$, derivative of $y$
   $f'(x)$, derivative of $f(x)$
   secant line
   tangent line

derivative
   differentiable

**11.5** $\dfrac{dy}{dx}$, derivative of $y$
   $\dfrac{d}{dx}[f(x)]$,
      derivative of $f(x)$
   derivative rules
   demand function

**11.6** $\overline{C}(x)$, average cost
   product rule
   quotient rule
   marginal average cost

**11.7** $g[f(x)]$, composite function
   chain rule

generalized power rule
   marginal-revenue product

**11.8** derivatives of exponential functions
   derivatives of logarithmic functions

**11.9** continuous at a point
   discontinuous
   continuous from the left and from the right
   interval notation
   continuous on an open or closed interval

## Chapter 11 Key Concepts

**Limit of a Function**

Let $f$ be a function, and let $a$ and $L$ be real numbers. Suppose that as $x$ takes values very close (but not equal) to $a$ (on both sides of $a$), the corresponding values of $f(x)$ are very close (and possibly equal) to $L$ and that the values of $f(x)$ can be made arbitrarily close to $L$ for all values of $x$ that are close enough to $a$. Then $L$ is the **limit** of $f$ as $x$ approaches $a$, written $\lim\limits_{x \to a} f(x) = L$.

**One-Sided Limits**

The **limit of $f$ as $x$ approaches $a$ from the right**, written $\lim\limits_{x \to a^+} f(x)$, is defined by replacing the phrase "on both sides of $a$" by "with $x > a$" in the definition of the limit of a function.

The **limit of $f$ as $x$ approaches $a$ from the left**, written $\lim\limits_{x \to a^-} f(x)$, is defined by replacing the phrase "on both sides of $a$" by "with $x < a$" in the definition of the limit of a function.

**Properties of Limits**

Let $a$, $k$, $A$, and $B$ be real numbers, and let $f$ and $g$ be functions such that

$$\lim_{x \to a} f(x) = A \quad \text{and} \quad \lim_{x \to a} g(x) = B.$$

Then

1. $\lim_{x \to a} k = k$ (for any constant k);

2. $\lim_{x \to a} x = a$ (for any real number $a$);

3. $\lim_{x \to a} [f(x) \pm g(x)] = A \pm B = \lim_{x \to a} f(x) \pm \lim_{x \to a} g(x)$;

4. $\lim_{x \to a} [f(x) \cdot g(x)] = A \cdot B = \lim_{x \to a} f(x) \cdot \lim_{x \to a} g(x)$;

5. $\lim_{x \to a} \dfrac{f(x)}{g(x)} = \dfrac{A}{B} = \dfrac{\lim_{x \to a} f(x)}{\lim_{x \to a} g(x)} \quad (B \neq 0)$;

6. for any real number $r$ for which $A^r$ exists,

$$\lim_{x \to a} [f(x)]^r = A^r = [\lim_{x \to a} f(x)]^r.$$

**Polynomial Limits**

If $f$ is a polynomial function, then $\lim_{x \to a} f(x) = f(a)$.

**Limit Theorem**

If $f$ and $g$ are functions that have limits as $x$ approaches $a$, and $f(x) = g(x)$ for all $x \neq a$, then $\lim_{x \to a} f(x) = \lim_{x \to a} g(x)$.

**Limits at Infinity**

Let $f$ be a function that is defined for all large values of $x$, and let $L$ be a real number.

Suppose that as $x$ takes larger and larger values, increasing without bound, the corresponding values of $f(x)$ are very close (and possibly equal) to $L$, and suppose that the values of $f(x)$ can be made arbitrarily close to $L$ by taking large enough values of $x$. Then $L$ is the **limit of $f$ as $x$ approaches infinity**, written $\lim_{x \to \infty} f(x) = L$.

Let $f$ be a function that is defined for all small negative values of $x$, and let $M$ be a real number. Suppose that as $x$ takes smaller and smaller negative values, decreasing without bound, the corresponding values of $f(x)$ are very close (and possibly equal) to $M$, and suppose that the values of $f(x)$ can be made arbitrarily close to $M$ by taking small enough values of $x$. Then $M$ is the **limit of $f$ as $x$ approaches negative infinity**, written $\lim_{x \to -\infty} f(x) = M$.

**Derivatives**

The **instantaneous rate of change** of a function $f$ when $x = a$ is

$$\lim_{h \to 0} \frac{f(a + h) - f(a)}{h},$$

provided that this limit exists.

The **tangent line** to the graph of $y = f(x)$ at the point $(a, f(a))$ is the line through this point and having slope $\lim_{h \to 0} \dfrac{f(a + h) - f(a)}{h}$, provided that this limit exists.

The **derivative** of the function $f$ is the function denoted $f'$ whose value at the number $x$ is $f'(x) = \lim_{h \to 0} \dfrac{f(x + h) - f(x)}{h}$, provided that this limit exists.

**Rules for Derivatives**

(Assume that all indicated derivatives exist.)

**Constant Function** If $f(x) = k$, where $k$ is any real number, then $f'(x) = 0$.

**Power Rule** If $f(x) = x^n$, for any real number $n$, then $f'(x) = n \cdot x^{n-1}$.

**Constant Times a Function** Let $k$ be a real number. Then the derivative of $y = k \cdot f(x)$ is $y' = k \cdot f'(x)$.

**Sum and Difference Rules** If $f(x) = u(x) \pm v(x)$, then $f'(x) = u'(x) \pm v'(x)$.

**Product Rule** If $f(x) = u(x) \cdot v(x)$, then $f'(x) = u(x) \cdot v'(x) + v(x) \cdot u'(x)$.

**Quotient Rule** If $f(x) = \dfrac{u(x)}{v(x)}$, and $v(x) \neq 0$, then $f'(x) = \dfrac{v(x) \cdot u'(x) - u(x) \cdot v'(x)}{[v(x)]^2}$.

**Chain Rule** Let $y = f[g(x)]$. Then $y' = f'[g(x)] \cdot g'(x)$.

**Chain Rule (alternative form)** If $y$ is a function of $u$, say, $y = f(u)$, and if $u$ is a function of $x$, say, $u = g(x)$, then $y = f(u) = f[g(x)]$, and

$$\frac{dy}{dx} = \frac{dy}{du} \cdot \frac{du}{dx}.$$

**Generalized Power Rule**  Let $u$ be a function of $x$, and let $y = u^n$ for any real number $n$. Then

$$y' = n \cdot u^{n-1} \cdot u'.$$

**Exponential Function**  If $y = e^{g(x)}$, then $y' = g'(x) \cdot e^{g(x)}$.

**Natural-Logarithm Function**  If $y = \ln|x|$, then $y' = \dfrac{1}{x}$.

If $y = \ln[g(x)]$, then $y' = \dfrac{g'(x)}{g(x)}$.

**Continuity**  A function $f$ is **continuous** at $x = c$ if $f(c)$ is defined, $\lim_{x \to c} f(x)$ exists, and $\lim_{x \to c} f(x) = f(c)$.

# Chapter 11 Review Exercises

*In Exercises 1–4, determine graphically or numerically whether the limit exists. If the limit exists, find its (approximate) value.*

**1.** $\lim_{x \to -3} f(x)$; $\lim_{x \to -3^-} f(x)$; $\lim_{x \to -3^+} f(x)$

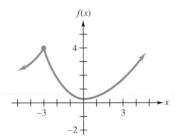

**2.** $\lim_{x \to -1} g(x)$; $\lim_{x \to -1^-} g(x)$; $\lim_{x \to -1^+} g(x)$

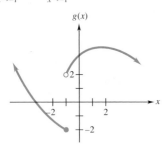

**3.** $\lim_{x \to 0} \dfrac{\sqrt{2-x} - \sqrt{2}}{x}$

**4.** $\lim_{x \to -1} \dfrac{10^x - .1}{x + 1}$

*Find the given limit if it exists.*

**5.** $\lim_{x \to 5} \dfrac{5x - 4}{x - 2}$

**6.** $\lim_{x \to -1} \dfrac{7 - 3x}{x - 4}$

**7.** $\lim_{x \to 3} \dfrac{x^2 - 9}{x + 3}$

**8.** $\lim_{x \to 7} \dfrac{49 - x^2}{x + 7}$

**9.** $\lim_{x \to 4} \dfrac{x^2 - 16}{x - 4}$

**10.** $\lim_{x \to -2} \dfrac{x^2 - x - 6}{x + 2}$

**11.** $\lim_{x \to 9} \dfrac{\sqrt{x} - 3}{x - 9}$

**12.** $\lim_{x \to 16} \dfrac{\sqrt{x} - 4}{x - 16}$

*Find the given limit if it exists.*

**13.** $\lim_{x \to 4^+} \sqrt{x^2 - x - 12}$

**14.** $\lim_{x \to 3^-} (5x + \sqrt{9 - x^2})$

**15.** $\lim_{x \to 7} \dfrac{|x - 7|}{x - 7}$

**16.** $\lim_{x \to -9} \dfrac{|x + 9|}{x + 9}$

**17.** $\lim_{x \to 6^-} \dfrac{x^2 + 1}{x - 6}$

**18.** $\lim_{x \to -5} \dfrac{8x}{(x + 5)^2}$

**19.** $\lim_{x \to \infty} \dfrac{9x + 1}{3x - 4x^2}$

**20.** $\lim_{x \to \infty} \dfrac{5x^2 - 8x + 3}{5 - 7x^3}$

**21.** $\lim_{x \to \infty} \dfrac{6x^2 - x + 50}{3 - 2x^2}$

**22.** $\lim_{x \to \infty} \dfrac{12x^3 + 3x^2 - 7x + 1}{6 - 3x - 4x^3}$

*Use the accompanying graph to find the average rate of change of $f$ on the given intervals.*

**23.** $x = 0$ to $x = 4$

**24.** $x = 2$ to $x = 8$

*Find the average rate of change for each of the given functions.*

**25.** $f(x) = 3x^2 - 5$, from $x = 1$ to $x = 5$

**26.** $g(x) = -x^3 + 2x^2 + 1$, from $x = -2$ to $x = 3$

**27.** $h(x) = \dfrac{6 - x}{2x + 3}$, from $x = 0$ to $x = 6$

**28.** $f(x) = e^{2x} + 5 \ln x$, from $x = 1$ to $x = 7$

*Use the definition of the derivative to find the derivative of each of the given functions.*

**29.** $y = 2x + 3$

**30.** $y = 4 - 3x$

**31.** $y = 2x^2 - x - 1$

**32.** $y = x^2 + 2x$

*Find the slope of the tangent line to the given curve at the given value of $x$. Find the equation of each tangent line.*

**33.** $y = x^2 - 6x$; at $x = 2$

**34.** $y = 8 - x^2$; at $x = 1$

**35.** $y = \dfrac{-2}{x + 5}$; at $x = -2$

**36.** $y = \sqrt{6x - 2}$; at $x = 3$

*Find the derivative of each of the given functions.*

**37.** $f(x) = 7x^8 - 8x + 3$     **38.** $f(x) = 2 + 9x - 7x^6$

**39.** $f(x) = e + ex + e^x + x^e$

**40.** $f(x) = 1 + \pi - \pi^2 + \sqrt{\pi} + \dfrac{1}{\pi}$

**41.** $f(x) = 8x^{1/2} + 7x^{-3}$     **42.** $f(x) = 9x^{1/3} - 5x^{-7}$

**43.** $f(x) = 6\sqrt{x} - \dfrac{3}{x}$     **44.** $f(x) = 4\sqrt{x} - \dfrac{3}{x^2}$

**45.** $y = (4t^2 + 9)(t^3 - 5)$     **46.** $y = (-5t + 3)(t^3 - 4t)$

**47.** $y = 8x^{3/4}(5x + 1)$     **48.** $y = 40x^{-1/4}(x^2 + 1)$

**49.** $f(x) = \dfrac{4x}{x^2 - 8}$     **50.** $g(x) = \dfrac{-5x^2}{3x + 1}$

**51.** $y = \dfrac{\sqrt{x} - 6}{3x + 7}$     **52.** $y = \dfrac{\sqrt{x} + 9}{x - 4}$

**53.** $y = \dfrac{x^2 - x + 1}{x - 1}$     **54.** $y = \dfrac{2x^3 - 5x^2}{x + 2}$

**55.** $f(x) = (4x - 2)^4$     **56.** $k(x) = (5x - 1)^6$

**57.** $y = \sqrt{2t - 5}$     **58.** $y = -3\sqrt{8t - 1}$

**59.** $y = 2x(3x - 4)^3$     **60.** $y = 5x^2(2x + 3)^5$

**61.** $f(u) = \dfrac{3u^2 - 4u}{(2u + 3)^3}$     **62.** $g(t) = \dfrac{t^3 + t - 2}{(2t - 1)^5}$

**63.** $y = -6e^{2x}$     **64.** $y = 8e^{.5x}$

**65.** $y = e^{-2x^3}$     **66.** $y = -5e^{x^2}$

**67.** $y = 5x \cdot e^{2x}$     **68.** $y = -7x^2 \cdot e^{-3x}$

**69.** $y = \ln(x^2 + 4x - 1)$     **70.** $y = \ln(4x^3 + 2x)$

**71.** $y = \dfrac{\ln 6x}{x^2 - 1}$     **72.** $y = \dfrac{\ln(3x + 5)}{x^2 + 5x}$

*Find all points of discontinuity for the given graphs.*

**73.**

**74.**

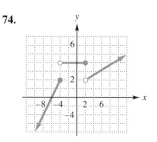

*Are the given functions continuous at the given points?*

**75.** $f(x) = \dfrac{x - 6}{x + 5}; x = 6, x = -5, x = 0$

**76.** $f(x) = \dfrac{x^2 - 9}{x + 3}; x = 3, x = -3, x = 0$

**77.** $h(x) = \dfrac{2 - 3x}{2 - x - x^2}; x = -2, x = \dfrac{2}{3}, x = 1$

**78.** $f(x) = \dfrac{x^2 - 4}{x^2 - x - 6}; x = 2, x = 3, x = 4$

*Work these problems.*

**79. Business** Shipping charges at an online pet supply store are 10% of the total purchase price except that there is a minimum charge of $5 and orders over $100 receive free shipping. The accompanying figure shows the charge, $S(x)$, for shipping an order in the amount of $x$ dollars. Find the given limits.

(a) $\lim\limits_{x \to 50} S(x)$     (b) $\lim\limits_{x \to 100} S(x)$

**80. Tax Revenue** The Iowa Department of Revenue held a back-to-school sales tax holiday (during which clothing and footwear were exempt from state sales tax) on August 3–4, 2018. Suppose that the accompanying figure shows the sales tax receipts $T(x)$ (in dollars) from a certain apparel store at time $x$, where $x = 1$ corresponds to August 1. Find the given limits.

(a) $\lim\limits_{x \to 2} T(x)$     (b) $\lim\limits_{x \to 3} T(x)$

(c) $\lim\limits_{x \to 4} T(x)$     (d) $\lim\limits_{x \to 5} T(x)$

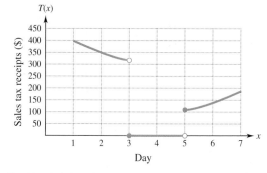

**81. Car Rental** An airport car rental agency charges $65 per day (or fraction of a day) to rent a full-size car, plus a one-time airport surcharge of $5. Let $C(x)$ be the cost of renting a full-size car from this agency for $x$ days. Find the given quantities.

(a) $C(2)$     (b) $\lim\limits_{x \to 2^-} C(x)$

(c) $\lim\limits_{x \to 2^+} C(x)$     (d) $\lim\limits_{x \to 2} C(x)$

**82. Printing Costs** The average cost (in dollars per hundred brochures) of printing $x$ hundred glossy brochures (one page, two-sided, tri-fold) can be approximated by $\overline{C}(x) = \dfrac{450 + 115x}{x}$. Find $\lim\limits_{x \to \infty} \overline{C}(x)$.

**83. Target Stock** The accompanying figure shows the stock price for Target Corp. at the start of each year. Find the average rate of change of the price of Target stock between the years 2014 and 2016. (Data from: www.morningstar.com.)

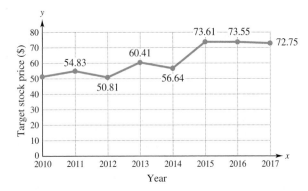

**84.** Use the figure in the preceding exercise to find the average rate of change of Target stock price between the years 2012 and 2015.

**85. Marginal Cost** If $C(x)$ is the cost of producing $x$ paddleboards, explain what each of the given equations means in practical terms, including units.
 (a) $C(100) = 80,000$     (b) $C'(100) = 300$

**86. Natural Science** Suppose that $D(x)$ represents the depth of a certain lake (in feet) at a distance $x$ feet from the bank (in the direction toward the center of the lake). Explain what each of the given equations means in practical terms for a person standing in the lake.
 (a) $D(20) = 5$     (b) $D'(20) = .3$

**87. McDonald's Revenue** The revenue (in billions of dollars) generated by McDonald's Corp. for the years 2008 to 2015 can be approximated by

$$R(t) = -.1t^3 + 3.25t^2 - 3.5t + 134,$$

where $t = 8$ corresponds to the year 2008. Find the rate of change of revenue with respect to time. (Data from: www.morningstar.com.)

**88. McDonald's Profits** The profit (in billions of dollars) generated by McDonald's Corp. for the years 2008 to 2015 can be approximated by

$$P(t) = -.024t^3 + .7t^2 - 6.2t + 26,$$

where $t = 8$ corresponds to the year 2008. Find the rate of change of profit with respect to time. (Data from: www.morningstar.com.)

**89. Cost** A popular food truck that drives to various locations to sell lunch items has determined that the cost (in hundreds of dollars) of making $x$ hundred sandwiches can be approximated by

$$C(x) = \frac{250x + 45}{2x + 40}.$$

 (a) Find the marginal-cost function.
 (b) If the food truck operators are currently making 1000 sandwiches per week, find the approximate cost of making an additional 100 sandwiches.

**90. Average Cost** Consider the cost function given in the previous exercise.
 (a) Find the average-cost function.
 (b) Find the marginal average-cost function.
 (c) If the food truck operators are currently making 1000 sandwiches per week, approximately how will the average cost change if they make an additional 100 sandwiches?

**91. Demand** The daily demand function for coffee at the campus coffee cart can be approximated by $p = \sqrt{25 - .1x}$, where $p$ is the price (in dollars) and $x$ is the number of cups of coffee demanded.
 (a) Find the revenue function.
 (b) Find the marginal revenue function.
 (c) If the demand is currently 90 cups of coffee, find the revenue.
 (d) If the demand is currently 90 cups of coffee, find the approximate revenue from selling one additional cup of coffee.

**92. Profit** The cost function for the coffee cart described in the previous exercise can be approximated by $C(x) = 120 + .75x$, where $C(x)$ is the cost (in dollars) of producing $x$ cups of coffee.
 (a) Find the profit function.
 (b) Find the marginal profit function.
 (c) If the demand is currently 120 cups of coffee, find the approximate profit from selling one additional cup of coffee.

**93. Total Life Expectancy** Total life expectancy as a function of birth year can be approximated by

$$T(x) = 15 \ln(x + 100) + 7.7,$$

where $x$ is the year of birth, with $x = 0$ corresponding to the year 2000. Give your answers to one decimal place. (Data from: U.S. Census Bureau.)
 (a) Would you expect the derivative of $T$ to be positive or negative? Why?
 (b) Find the total life expectancy for a person born in 2018.
 (c) Find $T'(18)$, and use your answer to explain approximately how the total life expectancy would be different for an individual born one year later.

**94. Remaining Life Expectancy** Remaining life expectancy as a function of current age can be approximated by

$$R(x) = -37 \ln(x) + 172,$$

where $x$ is the current age of the individual in years. Give your answers to one decimal place. (Data from: Social Security Administration.)
 (a) Would you expect the derivative of $R$ to be positive or negative? Why?
 (b) Find the remaining life expectancy for a person who is 69 years old.
 (c) Find $R'(69)$, and use your answer to explain approximately how the remaining life expectancy would be different for an individual who is one year older.

95. **Business** Use the graph of $y = S(x)$ shown in Exercise 79 to answer the given questions.

   (a) At what $x$-values is the function $S$ not continuous?

   (b) At what $x$-values is the function $S$ not differentiable?

96. **Business** The monthly cost of an international cell phone plan is $50 for the first 450 minutes of airtime plus $.50 per minute (or fraction of a minute) for usage over 450 minutes. Let $C(x)$ be the cost of using $x$ minutes of airtime. Find the given quantities.

   (a) $C(400)$

   (b) $\lim\limits_{x \to 400} C(x)$

   (c) Is $C(x)$ continuous at $x = 400$?

   (d) $C(460)$

   (e) $\lim\limits_{x \to 460} C(x)$

   (f) Is $C(x)$ continuous at $x = 460$?

# Case Study 11  Price Elasticity of Demand

Any retailer who sells a product or a service is concerned with how a change in price affects demand for the article. The sensitivity of demand to price changes varies with different items. For smaller items, such as soft drinks, food staples, and lightbulbs, small percentage changes in price will not affect the demand for the item much. However, sometimes a small percentage change in price on big-ticket items, such as cars, homes, and furniture, can have significant effects on demand.

One way to measure the sensitivity of changes in price to demand is by the ratio of percent change in demand to percent change in price. If $q$ represents the quantity demanded and $p$ the price of the item, this ratio can be written as

$$\frac{\Delta q / q}{\Delta p / p},$$

where $\Delta q$ represents the change in $q$ and $\Delta p$ represents the change in $p$. The ratio is usually negative, because $q$ and $p$ are positive, while $\Delta q$ and $\Delta p$ typically have opposite signs. (An *increase* in price causes a *decrease* in demand.) If the absolute value of this quantity is large, it shows that a small increase in price can cause a relatively large decrease in demand.

Applying some algebra, we can rewrite this ratio as

$$\frac{\Delta q / q}{\Delta p / p} = \frac{\Delta q}{q} \cdot \frac{p}{\Delta p} = \frac{p}{q} \cdot \frac{\Delta q}{\Delta p}.$$

Suppose $q = f(p)$. (Note that this is the inverse of the way our demand functions have been expressed so far; previously, we had $p = D(q)$.) Then $\Delta q = f(p + \Delta p) - f(p)$. It follows that

$$\frac{\Delta q}{\Delta p} = \frac{f(p + \Delta p) - f(p)}{\Delta p}.$$

As $\Delta p \to 0$, this quotient becomes

$$\lim_{\Delta p \to 0} \frac{\Delta q}{\Delta p} = \lim_{\Delta p \to 0} \frac{f(p + \Delta p) - f(p)}{\Delta p} = \frac{dq}{dp},$$

and

$$\lim_{\Delta p \to 0} \frac{p}{q} \cdot \frac{\Delta q}{\Delta p} = \frac{p}{q} \cdot \frac{dq}{dp}.$$

The quantity

$$E = -\frac{p}{q} \cdot \frac{dq}{dp}$$

is positive because $dq/dp$ is negative. $E$ is called **elasticity of demand** and measures the instantaneous responsiveness of demand to price. For example, $E$ may be .6 for physician services (expenses that have considerable price increases each year, but still have a high demand), but may be 2.3 for restaurant meals (highly enjoyable, but high-cost items and not necessities). These numbers indicate that the demand for physician services is much less responsive to price changes than the demand for restaurant meals. Another factor that impacts elasticity of demand is the availability of substitute products. For example, if one airline increases prices on a particular route but other airlines serving the route do not, then rather than paying the higher prices, consumers could fly with a different airline (the substitute product).

If $E < 1$, the relative change in demand is less than the relative change in price, and the demand is called **inelastic.** If $E > 1$, the relative change in demand is greater than the relative change in price, and the demand is called **elastic.** When $E = 1$, the percentage changes in price and demand are relatively equal, and the demand is said to have **unit elasticity.**

Sometimes elasticity is counterintuitive. The addiction to illicit drugs is an excellent example. The quantity of the drug demanded by addicts, if anything, increases, no matter what the cost. Thus, illegal drugs are an inelastic commodity.

**EXAMPLE 1**  Suppose that the demand for flat screen televisions is expressed by the equation

$$q = -.025p + 20.45,$$

where $q$ is the annual demand (in millions of televisions) and $p$ is the price of the product (in dollars).

(a) Calculate and interpret the elasticity of demand when $p = \$200$ and when $p = \$500$.

**Solution** Since $q = -.025p + 20.45$, we have $dq/dp = -.025$, so that

$$E = -\frac{p}{q} \cdot \frac{dq}{dp}$$

$$= -\frac{p}{-.025p + 20.45} \cdot (-.025)$$

$$= \frac{.025p}{-.025p + 20.45}.$$

Let $p = 200$ to get

$$E = \frac{.025(200)}{-.025(200) + 20.45} \approx .324.$$

Since $.324 < 1$, the demand was inelastic, and a percentage change in price resulted in a smaller percentage change in demand. For example, a 10% increase in price will cause a 3.24% decrease in demand.

If $p = 500$, then

$$E = \frac{.025(500)}{-.025(500) + 20.45} \approx 1.57.$$

Since $1.57 > 1$, the price is elastic. At this point, a percentage increase in price resulted in a greater percentage decrease in demand. A 10% increase in price resulted in a 15.7% decrease in demand.

**(b)** Determine the price at which demand had unit elasticity ($E = 1$). What is the significance of this price?

**Solution** Demand had unit elasticity at the price $p$ that made $E = 1$, so we must solve the equation

$$E = \frac{.025p}{-.025p + 20.45} = 1$$

$$.025p = -.025p + 20.45$$

$$.05p = 20.45$$

$$p = 409.$$

Demand had unit elasticity at a price of $409 per flat screen television. Unit elasticity indicates that the changes in price and demand are about the same.

The definitions from the preceding discussion can be expressed in the manner that follows.

---

## Elasticity of Demand

Let $q = f(p)$, where $q$ is the demand at a price $p$. The **elasticity of demand** is defined to be:

$$E = -\frac{p}{q} \frac{dq}{dp}.$$

---

We say that:

Demand is **inelastic** if $E < 1$.

Demand is **elastic** if $E > 1$.

Demand has **unit elasticity** if $E = 1$.

## Exercises

1. The monthly demand for beef in a given region can be expressed by the equation $q = -3.003p + 675.23$, where $q$ is the monthly demand in tons and $p$ is the price in dollars per 100 pounds. Determine the elasticity of demand when the price is $70.[*]

2. Acme Stationery sells designer-brand pens. The demand equation for annual sales of these pens is $q = -1000p + 70,000$, where $p$ is the price per pen. Normally, the pens sell for $30 each. They are very popular, and Acme has been thinking of raising the price by one third.[†]

   (a) Find the elasticity of demand if $p = $30.

   (b) Find the elasticity of demand if the price is raised by one third. Is raising the price a good idea?

3. The monthly demand for lodging in a certain city is given by $q = -2481.52p + 472,191.2$, where $p$ is the nightly rate.[§]

   (a) Find $E$ when $p = $100 and when $p = $75.

   (b) At what price is there unit elasticity?

4. The following multiple-choice question is a released sample question from the Major Field Test in Economics.[**] The demand for a good is given by $q = 100 - 4p^2$, where $q = $ quantity demanded per unit of time and $p = $ the price per unit. At a price of $4, the absolute value of the price elasticity of demand is approximately equal to which of the following?

   (a) $\infty$      (b) 36.0      (c) 3.6

   (d) 1.0      (e) 0.28

5. What must be true about demand if $E = 0$ everywhere?

---

[*]Adapted from *How Demand and Supply Determine Price*, Agricultural Marketing Manual, Alberta, Canada, February 1999.

[†]Taken from R. Horn, *Economics 331: Warm-up Problems: Supply, Demand, Elasticity.*

[§]Extracted from Bjorn Hanson, *Price Elasticity of Lodging Demand* (Pricewaterhouse-Coopers, 2000).

[**]The Major Field Tests are developed by ETS (http://www.ets.org/Media/Tests/MFT/pdf/2008/MFT_Economics_Sample_Items_08.pdf).

# Extended Project

Sensitivity to higher prices was evident in the tobacco market following increases in taxes that included large disparities in tax rates for cigarettes and pipe tobacco versus roll-your-own tobacco. Data for prices and quantities demanded are summarized in the table on the following page for the year prior to and following the implementation of the tax increases.[*] In this project, the price data includes only the cost of the tax rather than the total cost of the product plus tax since the increases in taxes were the significant aspect of the changes in price.

---

[*]Tobacco Taxes: Large Disparities in Rates for Smoking Products Trigger Significant Market Shifts to Avoid Higher Taxes, United States Government Accountability Office Report to Congressional Committees, April 2012.

| Year | Cigarettes | | Roll-Your-Own Tobacco | | Pipe Tobacco | |
|---|---|---|---|---|---|---|
| | Price (dollars per thousand sticks) | Quantity (billions of sticks) | Price (dollars per pound) | Quantity (thousands of pounds) | Price (dollars per pound) | Quantity (thousands of pounds) |
| 2008 | 19.50 | 337.64 | 1.10 | 19,673 | 1.10 | 3150 |
| 2010 | 50.33 | 296.23 | 24.78 | 6152 | 2.83 | 20,817 |

**(a)** The data show that the percent increase in the tax on cigarettes was $\dfrac{50.33 - 19.50}{19.50} = 158\%$. Find the percent increase for pipe tobacco and roll-your-own tobacco. Were the increases similar for all products or were there large disparities?

**(b)** For cigarettes, use a linear model (the equation of a line) to express the quantity demanded as a function of price.

**(c)** Repeat part (b) for roll-your-own tobacco.

**(d)** For cigarettes, determine the elasticity of demand $E = -\dfrac{p}{q}\dfrac{dq}{dp}$ at the lower price prior to the tax increases and also at the higher price following the tax increases. Was the elasticity of demand for cigarettes inelastic ($E < 1$) or elastic ($E > 1$) at the lower price? At the higher price?

**(e)** Repeat part (d) for roll-your-own tobacco.

**(f)** Factors that impact elasticity of demand include price, the necessity of the product, and availability of substitute products. What is a possible explanation for the large difference in elasticity of demand for cigarettes and roll-your-own tobacco at the higher price?

**(g)** You work for a tobacco store that sells a wide variety of products including those listed in the table above. Your employer has asked you to write a report that summarizes the elasticity of demand information you have calculated, provides a possible explanation for the large differences in elasticity of demand for cigarettes and roll-your-own tobacco after the price increases, and comments on the implications for the tobacco store.

# Applications of the Derivative

## 12

**CHAPTER**

Functions and their derivatives can be used to solve a variety of optimization problems. For example, business owners want to maximize revenue and minimize costs. Automotive engineers want to design engines with maximum fuel efficiency. Manufacturers want packaging that requires the smallest amount of cardboard (and hence is cheapest). A hospital or a business may want to determine the optimum intervals at which supplies should be ordered to minimize the total cost of ordering, delivery, and storage. Examples and exercises in this chapter illustrate these and many other applications of the derivative.

In the last chapter, we defined and interpreted the derivative and developed formulas for the derivatives of many functions. In this chapter, we investigate the connection between the derivative of a function and the graph of the function. In particular, we shall see how to use the derivative to determine algebraically the maximum and minimum values of a function, as well as the intervals on which the function is increasing or decreasing.

## 12.1  Local Extrema

Informally, we say that a function is **increasing** on an interval if its graph is *rising* from left to right over the interval and that a function is **decreasing** on an interval if its graph is *falling* from left to right over the interval.

**EXAMPLE 1**    For which *x*-intervals is the function graphed in Figure 12.1, on the following page, increasing? For which intervals is it decreasing?

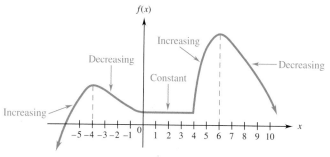

**Figure 12.1**

**Solution** Moving from left to right, the function is increasing up to −4, then decreasing from −4 to 0, constant (neither increasing nor decreasing) from 0 to 4, increasing from 4 to 6, and, finally, decreasing from 6 on. In interval notation, the function is increasing on $(-\infty, -4)$ and $(4, 6)$, decreasing on $(-4, 0)$ and $(6, \infty)$, and constant on $(0, 4)$. ✓₁

In order to examine the connection between the graph of a function $f$ and the derivative of $f$, it is sometimes helpful to think of the graph of $f$ as a roller-coaster track, with a roller-coaster car moving from left to right along the graph, as shown in Figure 12.2. At any point along the graph, the floor of the car (a straight-line segment) represents the tangent line to the graph at that point.

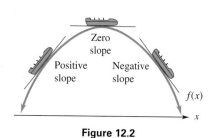

**Figure 12.2**

The slope of the tangent line is positive when the car travels uphill (the function is *increasing*), and the slope of the tangent line is negative when the car travels downhill (the function is *decreasing*). Since the slope of the tangent line at the point $(x, f(x))$ is given by the derivative $f'(x)$, we have the following useful facts.

## Increasing and Decreasing Functions

Suppose a function $f$ has a derivative at each point in an open interval. Then

**1.** if $f'(x) > 0$ for each $x$ in the interval, $f$ is *increasing* on the interval;
**2.** if $f'(x) < 0$ for each $x$ in the interval, $f$ is *decreasing* on the interval;
**3.** if $f'(x) = 0$ for each $x$ in the interval, $f$ is *constant* on the interval.

**EXAMPLE 2** Find the intervals on which the function

$$f(x) = x^3 + 3x^2 - 9x + 4,$$

whose graph is shown in Figure 12.3, is increasing, decreasing, and constant.

**Solution** The graph in Figure 12.3 does not clearly show the intervals on which $f$ is increasing or decreasing. To get more precise information, we first find the derivative:

$$f'(x) = 3x^2 + 6x - 9.$$

---

✓ **Checkpoint 1**

For what values of $x$ is the function whose graph is shown increasing? decreasing?

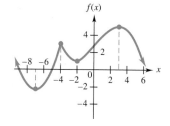

*Answers to Checkpoint exercises are found at the end of the section.*

---

$f(x) = x^3 + 3x^2 - 9x + 4$

**Figure 12.3**

According to the box preceding the example, the function $f$ is increasing on the intervals where $f'(x) > 0$ and decreasing on the intervals where $f'(x) < 0$. So we must solve the following inequalities:

$$f'(x) > 0 \quad\text{and}\quad f'(x) < 0$$
$$3x^2 + 6x - 9 > 0 \qquad\qquad 3x^2 + 6x - 9 < 0.$$

As shown in Section 2.5, the first step is to solve the equation $f'(x) = 0$:

$$3x^2 + 6x - 9 = 0$$
$$3(x^2 + 2x - 3) = 0$$
$$3(x + 3)(x - 1) = 0$$
$$x = -3 \quad\text{or}\quad x = 1.$$

Now use either graphical or algebraic methods to solve the inequalities.

**Graphical**  The graph of the derivative $f'(x) = 3x^2 + 6x - 9$ in Figure 12.4 lies above the $x$-axis on the intervals $(-\infty, -3)$ and $(1, \infty)$, meaning that $f'(x) > 0$ on these intervals. Similarly, $f'(x) < 0$ on $(-3, 1)$ because the graph of $f'(x)$ is below the $x$-axis on $(-3, 1)$.

**Algebraic**  The numbers $-3$ and $1$ divide the $x$-axis into three intervals: $(-\infty, -3)$, $(-3, 1)$, and $(1, \infty)$. Determine the sign of $f'(x)$ on each interval by testing a number in that interval, as summarized next.

$f'(x) = 3x^2 + 6x - 9$

**Figure 12.4**

| Interval | $(-\infty, -3)$ | $(-3, 1)$ | $(1, \infty)$ |
|---|---|---|---|
| **Test Value in Interval** | $-4$ | $0$ | $2$ |
| **Value of $f'(x) = 3x^2 + 6x - 9$** | $15$ | $-9$ | $15$ |
| **Graph of $f'$** | Above $x$-axis | Below $x$-axis | Above $x$-axis |
| **Sign of $f'(x)$** | $f'(x) > 0$ | $f'(x) < 0$ | $f'(x) > 0$ |
| **Conclusion about $f$** | $f$ increasing | $f$ decreasing | $f$ increasing |

Regardless of method, the conclusion is the same: The original function $f$ is increasing on $(-\infty, -3)$ and $(1, \infty)$, which are the intervals where $f'(x) > 0$, and decreasing on $(-3, 1)$, where $f'(x) < 0$. ☑2

✓ **Checkpoint 2**

Find all intervals on which
$f(x) = 4x^3 + 3x^2 - 18x + 1$
is increasing or decreasing.

## Critical Numbers

In Example 2, the numbers for which $f'(x) = 0$ were essential for determining exactly where the function was increasing and decreasing. The situation is a bit different with the absolute-value function $f(x) = |x|$, whose graph is shown in Figure 12.5. Clearly, $f$ is decreasing on the left of $x = 0$ and increasing on the right of $x = 0$. But as we saw on pages 602–603, the derivative of $f(x) = |x|$ does not exist at $x = 0$. These examples suggest that the points where the derivative is $0$ or undefined play an important role.

In view of the preceding discussion, we have the following definition.

**Figure 12.5**

### Critical Numbers

If $f$ is a function, then a number $c$ for which $f(c)$ is defined and

$$\text{either } f'(c) = 0 \quad\text{or}\quad f'(c) \text{ does not exist}$$

is called a **critical number** of $f$. The corresponding point $(c, f(c))$ on the graph of $f$ is called a **critical point**.

The procedure used in Example 2, which applies to all functions treated in this text, can now be summarized in the manner that follows.

## Increasing/Decreasing Test

To find the intervals on which a function $f$ is increasing or decreasing, do the following:

**Step 1**  Compute the derivative $f'$.

**Step 2**  Find all numbers for which $f'(x) = 0$ or $f'(x)$ does not exist.

**Step 3**  Solve the inequalities $f'(x) > 0$ and $f'(x) < 0$ graphically or algebraically (by testing a number in each of the intervals determined by the numbers found in Step 2).

The solutions of $f'(x) > 0$ are intervals on which $f$ is increasing, and the solutions of $f'(x) < 0$ are intervals on which $f$ is decreasing.

## Maxima and Minima

We have seen that the graph of a typical function may have "peaks" or "valleys," as illustrated in Figure 12.6.

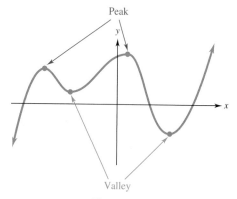

**Figure 12.6**

Peaks and valleys are sometimes called **turning points** of the graph, because as you go from left to right, the graph changes from increasing to decreasing at a peak and from decreasing to increasing at a valley.

A peak is the highest point in its neighborhood, but not necessarily the highest point on the graph. Similarly, a valley is the lowest point in its neighborhood, but not necessarily the lowest point on the graph. Consequently, a peak is called a *local maximum* (or *relative maximum*) and a valley a *local minimum* (or *relative minimum*). More precisely, we have the following definitions.

## Local Extrema

Let $c$ be a number in the domain of a function $f$. Then

**1.** $f$ has a **local maximum** at $c$ if $f(x) \le f(c)$ for all $x$ near $c$;

**2.** $f$ has a **local minimum** at $c$ if $f(x) \ge f(c)$ for all $x$ near $c$.

The function $f$ is said to have a **local extremum** at $c$ if it has a local maximum or minimum there. A local extremum is also called a **relative extremum**.

**NOTE**  The plurals of *maximum*, *minimum*, and *extremum* are, respectively, *maxima*, *minima*, and *extrema*.

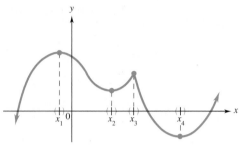

**EXAMPLE 3** Identify the locations of the local extrema of the function whose graph is shown in Figure 12.7.

**Figure 12.7**

**✓Checkpoint 3**

Identify the *x*-values of all points where the given graphs have local maxima and local minima.

**(a)**

**(b)**

**(c)**

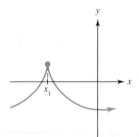

**Solution** The function has local maxima at $x_1$ and $x_3$ and local minima at $x_2$ and $x_4$. ✓ 3

The *exact* location of a local extremum (rather than a calculator's approximation) can normally be found by using derivatives. To see why this is so, let *f* be a function and, once again, think of the graph of *f* as a roller-coaster track, as shown in Figure 12.8.

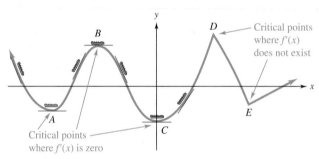

**Figure 12.8**

As the car passes through the local extrema at *A*, *B*, and *C*, the tangent line is horizontal and has slope 0. At *D* and *E*, however, a real roller-coaster car would have trouble. It would fly off the track at *D* and be unable to make the 90° change of direction at *E*. Notice that the graph does not have tangent lines at points *D* and *E*. (See the discussion of the existence of derivatives in Section 11.4.) Thus, a point $(c, f(c))$ where a local extremum occurs has the following property: The tangent line has slope 0, *or* there is no tangent line—that is, $f'(c) = 0$ or $f'(c)$ does not exist. In other words, *c* is a critical number.

## Local Extrema and Critical Numbers

If *f* has a local extremum at $x = c$, then *c* is a critical number of *f*.

**⚠ CAUTION** This result says that every local extremum occurs at a critical number, *not* that every critical number produces a local extremum. Thus, the critical numbers provide a list of *possibilities*: If there is a local extremum, it must occur at a number on the list, but the list may include numbers at which there is no local extremum.

**Figure 12.9**

**Figure 12.10**

**Figure 12.11**

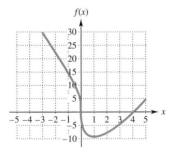

**Figure 12.12**

## ✔ Checkpoint 4

Find the critical numbers for each of the given functions.

**(a)** $\dfrac{1}{3}x^3 - x^2 - 15x + 6$

**(b)** $6x^{2/3} - 4x$

**EXAMPLE 4**  Find the critical numbers of the given functions.

**(a)** $f(x) = 2x^3 - 3x^2 - 72x + 15$

**Solution** Critical numbers are numbers in the domain of $f$ where $f'(x) = 0$ or $f'(x)$ does not exist. We have $f'(x) = 6x^2 - 6x - 72$, so $f'(x)$ exists for every $x$. Setting $f'(x) = 0$ shows that

$$6x^2 - 6x - 72 = 0$$
$$6(x^2 - x - 12) = 0$$
$$x^2 - x - 12 = 0$$
$$(x + 3)(x - 4) = 0$$
$$x + 3 = 0 \quad \text{or} \quad x - 4 = 0$$
$$x = -3 \quad \text{or} \quad x = 4.$$

Therefore, $-3$ and $4$ are the critical numbers of $f$; these are the only *possible* locations where local extrema could occur. The graph of $f$ in Figure 12.9 shows that there is a local maximum at $x = -3$ and a local minimum at $x = 4$.

**(b)** $f(x) = x^3$

**Solution** The derivative $f'(x) = 3x^2$ exists for every $x$. Setting $f'(x) = 0$ shows that $3x^2 = 0$ exactly when $x = 0$. Therefore, $x = 0$ is the only critical number of $f$ and the only possible location for a local maximum or minimum. In this case, however, we know what the graph of $f$ looks like. (See Figure 12.10.) The graph shows that there is no local maximum or minimum at $x = 0$. We emphasize that $x = 0$ is a critical number of $f$, so it is a *possible* location for a local extremum; however, a local extremum does *not* exist.

**(c)** $f(x) = \sqrt[3]{x^2}$

**Solution** First, do a bit of algebra to rewrite $f(x) = \sqrt[3]{x^2} = x^{2/3}$. Then it is easy to find that the derivative is

$$f'(x) = \frac{2}{3}x^{-\frac{1}{3}} = \frac{2}{3} \cdot \frac{1}{x^{1/3}} = \frac{2}{3} \cdot \frac{1}{\sqrt[3]{x}} = \frac{2}{3\sqrt[3]{x}}.$$

The derivative is never zero because the numerator is never zero. The derivative does not exist when the denominator is zero: $3\sqrt[3]{x} = 0$ when $x = 0$. Therefore, $x = 0$ is the only critical number and the only *possible* location of a local extremum. The graph of $f$ in Figure 12.11 suggests that a local minimum occurs at $x = 0$.

**(d)** $f(x) = 3x^{4/3} - 12x^{1/3}$

**Solution** We first compute the derivative:

$$f'(x) = 3 \cdot \frac{4}{3}x^{1/3} - 12 \cdot \frac{1}{3}x^{-2/3}$$
$$= 4x^{1/3} - \frac{4}{x^{2/3}}$$
$$= \frac{4x^{1/3}x^{2/3}}{x^{2/3}} - \frac{4}{x^{2/3}}$$
$$= \frac{4x - 4}{x^{2/3}}. \quad \longleftarrow \quad f'(x) = 0 \text{ when } 4x - 4 = 0.$$
$$\longleftarrow \quad f'(x) \text{ does not exist when } x^{2/3} = 0.$$

The derivative fails to exist when $x = 0$, since the denominator is zero there. Since the original function $f$ is defined when $x = 0$, it follows that $0$ is a critical number of $f$. If $x \neq 0$, then $f'(x)$ is $0$ only when the numerator $4x - 4 = 0$, that is, when $x = 1$. Therefore, the critical numbers of $f$ are $0$ and $1$. These numbers are the *possible* locations for local extrema. However, the graph of $f$ in Figure 12.12 suggests that there is a local minimum at $x = 1$, but no local extremum at $x = 0$. ✔₄

**TECHNOLOGY TIP** Most graphing calculators have a maximum/minimum finder that can approximate local extrema to a high degree of accuracy. On a TI-84, for the function in Example 4(a), enter Y1 = 2X³ − 3X² − 72X + 15. To graph the function, go to the ZOOM menu, and select ZStandard followed by ZoomFit (or else manually select an appropriate viewing window). Next look in the CALC menu (2nd TRACE) and choose the MINIMUM option. When the calculator asks for a Left Bound, use the graph to pick an appropriate entry, such as −10. Similarly, enter a suitable Right Bound, such as 10. Finally, the calculator will ask for a Guess. Again use the graph to enter a reasonable guess, such as 0. The calculator finds that the minimum occurs at X = 4.0 and the minimum function value is Y = −58. Similarly, in the CALC menu, the MAXIMUM option finds that the maximum occurs at X = −3.0 and the maximum value is Y = 285.

## The First-Derivative Test

When all the critical numbers of a function $f$ have been found, you must then determine which ones lead to local extrema. Sometimes this can be done graphically, as in Example 4. It can also be done algebraically by using the following observation:

> At a local maximum, $f$ changes from increasing to decreasing, and at a local minimum, $f$ changes from decreasing to increasing, as illustrated in Figure 12.13.

When $f$ changes from increasing to decreasing, its derivative $f'$ changes from positive to negative. Similarly, when $f$ changes from decreasing to increasing, $f'$ changes from negative to positive. These facts lead to the following test for local extrema; a formal proof of the test is omitted.

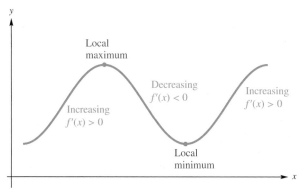

**Figure 12.13**

---

### First-Derivative Test

Assume that $a < c < b$ and that $c$ is the only critical number for a function $f$ in the interval $[a, b]$. Assume that $f$ is differentiable for all $x$ in $[a, b]$, except possibly at $x = c$.

1. If $f'(a) > 0$ and $f'(b) < 0$, then there is a local maximum at $c$.
2. If $f'(a) < 0$ and $f'(b) > 0$, then there is a local minimum at $c$.
3. If $f'(a)$ and $f'(b)$ are both positive or both negative, then there is no local extremum at $c$.

---

The sketches in the following table show how the first-derivative test works (assume the same conditions on $a$, $b$, and $c$ as those stated in the box):

| $f(x)$ has ... | Sign of $f'(a)$ | Sign of $f'(b)$ | Sketches | |
|---|---|---|---|---|
| Local Maximum | + | − | | |
| Local Minimum | − | + | | |
| No Local Extrema | + | + | | |
| No Local Extrema | − | − | | |

The local extrema of the functions in Example 4(a) and 4(d) were determined graphically (once the critical numbers had been found algebraically). Examples 5 and 6 show how the first-derivative test can be used in place of graphing.

**EXAMPLE 5** Use the first-derivative test to find the local extrema of

$$f(x) = 2x^3 - 3x^2 - 72x + 15.$$

**Solution** Example 4(a) shows that the critical numbers of $f$ are $-3$ and $4$. To use the first-derivative test on $-3$, we must choose an interval $[a, b]$ containing $-3$, but not containing the other critical number. We will pick $a = -4$ and $b = 0$. Many other choices of $a$ and $b$ are possible, but we try to select ones that will make the computations easy. Since

$$f'(x) = 6x^2 - 6x - 72 = 6(x^2 - x - 12) = 6(x + 3)(x - 4),$$

it follows that

$$f'(-4) = 6(-4 + 3)(-4 - 4) = 6(-1)(-8) > 0$$

and

$$f'(0) = 6(0 + 3)(0 - 4) = 6(3)(-4) < 0.$$

(Note that it is not necessary to finish calculating the exact value of $f'(x)$ in order to determine its sign.) Thus, the value of the derivative is positive to the left of $-3$ and negative to the right of $-3$, as shown in Figure 12.14. By part 1 of the first-derivative test, there is a local maximum at $x = -3$, and it is $f(-3) = 150$.

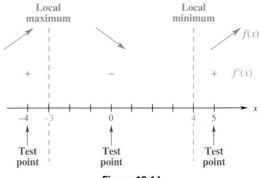

**Figure 12.14**

Similarly, we can use $a = 0$ and $b = 5$ to test the critical number $c = 4$. We just saw that $f'(0) < 0$; also,

$$f'(5) = 6(5 + 3)(5 - 4) = 6(8)(1) > 0.$$

Hence, by part 2 of the first-derivative test, there is a local minimum at $x = 4$, given by $f(4) = -193.$ ✓₅

✓ **Checkpoint 5**

Find the location of all local extrema of the given functions.

(a) $f(x) = 2x^2 + 12x - 15$

(b) $g(x) = \frac{1}{3}x^3 - x^2 - 8x + 57$

**EXAMPLE 6**  Use the first-derivative test to find the local extrema of

$$f(x) = 3x^{4/3} - 12x^{1/3}.$$

**Solution**  Example 4(d) shows that the critical numbers of $f$ are 0 and 1. We can use the intervals $[-1, 1/2]$ and $[1/2, 2]$ to test these critical numbers:

$$f'(x) = \frac{4x - 4}{x^{2/3}} = \frac{4x - 4}{\sqrt[3]{x^2}}$$

$$f'(-1) = \frac{4(-1) - 4}{\sqrt[3]{(-1)^2}} = \frac{-4 - 4}{1} < 0$$

$$f'\left(\frac{1}{2}\right) = \frac{4(1/2) - 4}{\sqrt[3]{(1/2)^2}} = \frac{-2}{\sqrt[3]{1/4}} < 0$$

$$f'(2) = \frac{4(2) - 4}{\sqrt[3]{2^2}} = \frac{4}{\sqrt[3]{4}} > 0.$$

Since $f'(x)$ is negative at both $-1$ and $\frac{1}{2}$, part 3 of the first-derivative test shows that there is no local extremum at $x = 0$. Since $f'(\frac{1}{2}) < 0$ and $f'(2) > 0$, there is a local minimum at $x = 1$, by part 2 of the first-derivative test. These results are shown schematically in Figure 12.15. ✓₆

✓ **Checkpoint 6**

Find any local extrema for $f(x) = x^3 + 4x^2 - 3x + 5$.

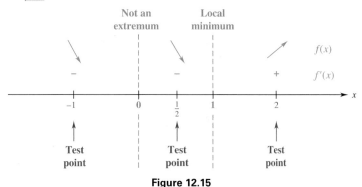

**Figure 12.15**

Although graphing technology sometimes eliminates the need to use the first-derivative test, this is not always the case, as the next example illustrates.

**EXAMPLE 7**  Use the first-derivative test to find the local extrema of

$$f(x) = \frac{x}{x^2 + 1}.$$

**Solution**  Using the quotient rule, we find that the derivative of $f$ is

$$f'(x) = \frac{(x^2 + 1)(1) - (x)(2x)}{(x^2 + 1)^2} \qquad \left[\frac{u(x)}{v(x)}\right]' = \frac{v(x)u'(x) - u(x)v'(x)}{[v(x)]^2}$$

$$= \frac{x^2 + 1 - 2x^2}{(x^2 + 1)^2}$$

$$= \frac{1 - x^2}{(x^2 + 1)^2}.$$

The critical numbers are numbers in the domain of $f$ where the derivative $f'$ is zero or does not exist. The derivative is zero when the numerator is zero:

$$1 - x^2 = 0$$
$$(1 + x)(1 - x) = 0$$
$$1 + x = 0 \qquad 1 - x = 0$$
$$x = -1 \qquad x = 1.$$

The derivative does not exist if the denominator is zero. However, the denominator is never zero because $x^2 + 1$ is always positive. Therefore, the only critical numbers are $x = -1$ and $x = 1$.

To use the first-derivative test on the critical number $x = -1$, we can choose the interval $[-2, 0]$, and to test $x = 1$, we can choose the interval $[0, 2]$:

$$f'(-2) = \frac{1 - (-2)^2}{[(-2)^2 + 1]^2} = \frac{-3}{[(-2)^2 + 1]^2} < 0$$

$$f'(0) = \frac{1 - (0)^2}{[(0)^2 + 1]^2} > 0$$

$$f'(2) = \frac{1 - (2)^2}{[(2)^2 + 1]^2} = \frac{-3}{[(2)^2 + 1]^2} < 0.$$

Because $f'(x)$ is negative to the left of the critical number $x = -1$ and positive to the right of $x = -1$, part 2 of the first-derivative test shows that there is a local minimum at $x = -1$. Similarly, because $f'(x)$ is positive to the left of and negative to the right of the critical number of $x = 1$, there is a local maximum at $x = 1$ by part 1 of the first-derivative test. These results are shown schematically in Figure 12.16.

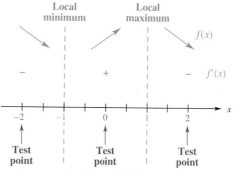

**Figure 12.16**

The values of the local minimum and maximum are, respectively,

$$f(-1) = \frac{-1}{(-1)^2 + 1} = -\frac{1}{2} \quad \text{and} \quad f(1) = \frac{1}{(1)^2 + 1} = \frac{1}{2}.$$

These extrema can be seen easily in the graph of $f$ in Figure 12.17. ✓ 7

✓**Checkpoint 7**

Find the locations of the local extrema of $f(x) = \dfrac{x}{x^2 + 4}$.

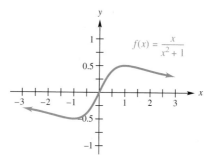

**Figure 12.17**

**EXAMPLE 8**    Find all local extrema of $f(x) = x^2 e^{5x}$.

**Solution**  Although the graph of $f$ in Figure 12.18 suggests that there are no local extrema, algebraic analysis shows otherwise. Using the product rule for derivatives, we have

$$f'(x) = x^2(5e^{5x}) + e^{5x}(2x) = 5x^2 e^{5x} + 2xe^{5x},$$

which exists for every $x$.

So the critical numbers are the solutions of

$$5x^2 e^{5x} + 2xe^{5x} = 0$$
$$e^{5x}(5x^2 + 2x) = 0$$
$$e^{5x}(5x + 2)x = 0.$$

Since $e^{5x}$ is never 0, this derivative can equal 0 only when

$$5x + 2 = 0 \qquad \text{or} \qquad x = 0$$
$$x = -2/5 = -.4 \qquad \text{or} \qquad x = 0.$$

We choose the interval $[-1, -.2]$ to test $-.4$ and the interval $[-.2, 1]$ to test 0. Using a calculator to compute values of $y_1 = f'(x)$, we obtain Figure 12.19. By the first-derivative test, there is a local maximum at $x = -.4$ and there is a local minimum at $x = 0$, as indicated schematically in Figure 12.20.

Figure 12.18

Figure 12.19

**Figure 12.20**

**Figure 12.21**

✓ **Checkpoint 8**

Find all local extrema of
$g(x) = x^3 e^x$.

The local maximum value is $f(-.4) \approx .022$, and the local minimum value is $f(0) = 0$. Using this information, we can select a viewing window on a graphing calculator that clearly shows these local extrema (see Figure 12.21). ✓8

**EXAMPLE 9**    **Priceline Costs**  The cost function (in millions of dollars) for Priceline Group Inc. can be approximated by

$$C(x) = -\frac{7}{3}x^3 + 53x^2 - 295x + 1202 \qquad (0 \le x \le 15),$$

where $x = 0$ corresponds to the year 2000. Determine when costs are increasing and when costs are decreasing. Also find all local extrema of the cost function. (Data from: www.morningstar.com.)

**Solution** To determine the critical numbers, we find the derivative of the cost function

$$C'(x) = -7x^2 + 106x - 295.$$

Because the derivative exists for every $x$, the only critical numbers occur where the derivative is zero. Solve $C'(x) = 0$ using the quadratic formula:

$$-7x^2 + 106x - 295 = 0$$

$$x = \frac{-b \pm \sqrt{b^2 - 4ac}}{2a}$$

$$x = \frac{-(106) \pm \sqrt{(106)^2 - 4(-7)(-295)}}{2(-7)} \approx \begin{cases} 3.7 \\ 11.5. \end{cases} \longleftarrow \text{The critical numbers are approximately 3.7 and 11.5.}$$

Because the domain of the cost function is [0, 15], both critical numbers lie within the domain. We choose the interval [0, 10] to test the critical number 3.7 and the interval [10, 15] to test the critical number 11.5:

$$C'(0) = -295 < 0$$
$$C'(10) = 65 > 0$$
$$C'(15) = -280 < 0.$$

Therefore, costs are decreasing on the intervals (0, 3.7) and (11.5, 15) and increasing on the interval (3.7, 11.5). By the first-derivative test, there is a local minimum at $x = 3.7$ and a local maximum at $x = 11.5$. In other words, costs reach a local minimum toward the end of 2003 and a local maximum in the middle of 2011. The local minimum value is $C(3.7) \approx 718$ (minimum costs of approximately \$718 million in 2003) and the local maximum value is $C(11.5) \approx 1270$ (maximum costs of approximately \$1270 million in 2011), as can be seen in Figure 12.22. ✓ 9

 **Checkpoint 9**

If a sales function is given by $S(x) = x^3 - 21x^2 + 120x$, determine when sales are decreasing.

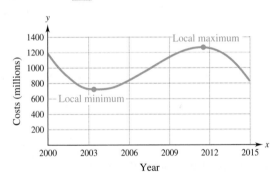

**Figure 12.22**

# 12.1 Exercises

*For each function, list the intervals on which the function is increasing, the intervals on which it is decreasing, and the location of all local extrema. (See Examples 1 and 3.)*

**1.**

**2.**

**3.**

**4.**

**5.**

**6.**

**7.**

**8.**

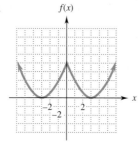

**Find the intervals on which each function is increasing or decreasing. (See Example 2.)**

**9.** $f(x) = \frac{1}{3}x^3 - 25x + 7$

**10.** $f(x) = \frac{1}{3}x^3 - 4x^2 + 11$

**11.** $f(x) = 2x^3 + 3x^2 - 12x + 5$

**12.** $f(x) = 4x^3 - 9x^2 - 12x + 7$

**13.** $f(x) = 2x^3 - 5x^2 - 4x + 2$

**14.** $f(x) = 4x^3 - 15x^2 - 72x + 5$

**15.** $f(x) = \frac{x+1}{x+3}$   **16.** $f(x) = \frac{x^2+4}{x}$

**17.** $f(x) = \sqrt{x^2+9}$   **18.** $f(x) = \sqrt{6-x}$

**Determine the location of each local extremum of the function. (See Examples 4–8.)**

**19.** $f(x) = x^3 + 3x^2 - 3$

**20.** $f(x) = x^3 - x^2 - 5x + 1$

**21.** $f(x) = x^3 + 6x^2 + 9x + 2$

**22.** $f(x) = x^3 - 3x^2 - 24x + 5$

**23.** $f(x) = \frac{4}{3}x^3 - \frac{21}{2}x^2 + 5x + 3$

**24.** $f(x) = -\frac{2}{3}x^3 - \frac{1}{2}x^2 - 3x - 4$

**25.** $f(x) = \frac{2}{3}x^3 - x^2 - 12x + 2$

**26.** $f(x) = \frac{4}{3}x^3 + 10x^2 - 24x - 3$

**27.** $f(x) = \frac{3}{2}x^4 + 4x^3 - 24x^2 + 20$

**28.** $f(x) = -3x^4 - 4x^3 + 12x^2 - 1$

**29.** $f(x) = 5 - x^{2/3}$   **30.** $f(x) = x^{2/5} + 3$

**31.** $f(x) = \frac{x}{x^2+49}$   **32.** $f(x) = \frac{x^2+4}{x}$

**Use the first-derivative test to determine the location of each local extremum and the value of the function at this extremum. (See Examples 5–8.)**

**33.** $f(x) = \frac{x^3}{x^3+1}$

**34.** $f(x) = \frac{x^2-2x+1}{x-3}$

**35.** $f(x) = xe^{-x}$   **36.** $f(x) = -xe^x$

**37.** $f(x) = x \cdot \ln|x|$   **38.** $f(x) = x - \ln|x|$

**39.** $f(x) = xe^{3x} - 2$   **40.** $f(x) = x^3e^{4x} + 1$

⊞ **Use the maximum/minimum finder on a graphing calculator to determine the approximate location of all local extrema of these functions.**

**41.** $f(x) = .2x^4 - x^3 - 12x^2 + 99x - 5$

**42.** $f(x) = x^5 - 12x^4 - x^3 + 232x^2 + 260x - 600$

**43.** $f(x) = .01x^5 + .2x^4 - x^3 - 6x^2 + 5x + 40$

**44.** $f(x) = .1x^5 + 3x^4 - 4x^3 - 11x^2 + 3x + 2$

**Work the given exercises. (See Examples 1 and 3.)**

**45. Sports Car Power and Torque** The following graph shows the power (measured in horsepower) and torque (in foot-pounds) for a certain sports car as a function of the engine speed (in thousands of revolutions per minute [RPM]).

(a) On what intervals is power increasing?

(b) On what intervals is power decreasing?

(c) On what intervals is torque increasing?

(d) On what intervals is torque decreasing?

**46. 10-Year Treasury Notes** The figure on the next page approximates the interest rate on 10-year treasury notes for the years 2006 through 2016. (Data from: www.federalreserve.gov.)

(a) On what intervals is the graph increasing?

(b) On what intervals is the graph decreasing?

(c) In what year does a local minimum occur?

(d) In what year does a local maximum occur?

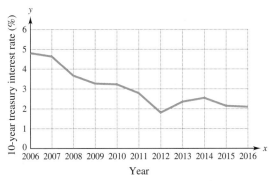

**47. 911 Call Volume** The graph shows the average number of 911 calls for police service for the Cleveland Police Department during the 24-hour day, with $x = 0$ corresponding to the hour 12:00 to 12:59 A.M.[*]

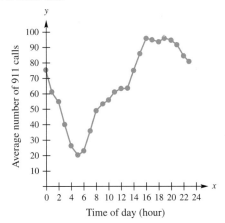

(a) List the intervals on which the average number of 911 calls is increasing.

(b) List the intervals on which the average number of 911 calls is decreasing.

(c) List the times of day at which this function has a local maximum or minimum.

**48. Pfizer Profits** Profits (in billions of dollars) generated by Pfizer Inc. for the years 2007 through 2015 are shown in the following figure. (Data from: www.morningstar.com.)

(a) On what intervals is the graph increasing?

(b) On what intervals is the graph decreasing?

(c) In what year does a local minimum occur?

(d) In what year does a local maximum occur?

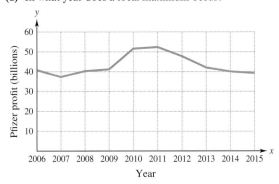

*Work these exercises. (See Examples 5 and 9.)*

**49. Monster Beverage Costs** The cost function (in millions) for the energy drink company Monster Beverage Corp. for the years 2010 through 2015 can be approximated by

$$P(x) = -29x^2 + 812x - 4559,$$

where $x = 10$ corresponds to the year 2010. (Data from: www.morningstar.com.)

(a) During what year did a local maximum profit occur?

(b) What was the maximum profit?

**50. ExxonMobil Revenue** Total revenue (in billions) for Exxon-Mobil Corp. can be approximated for the years 2009 through 2015 by

$$R(x) = -21.5x^2 + 516x - 2614,$$

where $x = 9$ corresponds to the year 2009. (Data from: www.morningstar.com.)

(a) During what year did a local maximum revenue occur?

(b) What was the maximum revenue?

**51. Pandora Stock Price** The stock price for Pandora Media Inc. can be approximated for the years 2011 through 2016 by

$$f(x) = 6.5x^2 - 158x + 975,$$

where $x = 11$ corresponds to the year 2011. During what year did a local minimum stock price occur? (Data from: www.morningstar.com.)

**52. Procter & Gamble Revenue** Total revenue (in billions) for Procter & Gamble Co. can be approximated for the years 2010 through 2015 by

$$R(x) = -1.1x^2 + 27x - 81,$$

where $x = 10$ corresponds to the year 2010. During what year did revenue have a local maximum? (Data from: www.morningstar.com.)

**53. Baseball Toss** You stand on a hotel balcony 48 feet above the ground and lob a baseball directly upward. You miss your catch as the ball comes down, so it falls to the vacant parking lot below. The height of the ball (in feet above the ground) after $t$ seconds is given by

$$h(t) = -16t^2 + 32t + 48.$$

(a) At what *time* does the ball reach its maximum height?

(b) What is the maximum *height* of the ball?

(c) At what *time* does the ball hit the ground? [*Hint:* Solve $h(t) = 0$.]

(d) What is the *velocity* of the ball when it hits the ground (this is called the impact velocity)?

**54. The Green Monster** A Boston Red Sox pitcher stands on top of the 37-foot-high left-field wall (the "Green Monster") in Fenway Park and fires a fastball straight up. The position function that gives the height of the ball (in feet) at time $t$ seconds is given by $s(t) = -16t^2 + 140t + 37$.[†] Find

(a) the maximum height of the ball;

[*]Holcomb, J., and Radke Sharpe, N. "Forecasting Police Calls during Peak Times for the City of Cleveland," *Case Studies in Business, Industry, and Government Statistics*, 1(1), 2007, pp. 47–53.

[†]Exercise provided by Frederick Russell of Charles County Community College.

**(b)** the time when the ball hits the ground;

**(c)** the velocity of the ball when it hits the ground.

**55. Microbe Concentration** The microbe concentration $B(x)$, in appropriate units, of Lake Tom depends approximately on the oxygen concentration $x$, again in appropriate units, according to the function

$$B(x) = x^3 - 7x^2 - 160x + 1800 \qquad (0 \le x \le 20).$$

**(a)** Find the oxygen concentration that will lead to the minimum microbe concentration.

**(b)** What is the minimum microbe concentration?

**56. Seasonal Sales** A company found that sales (in thousands of dollars) at time $t$ (in months) of fall seasonal items were approximated by

$$S(t) = -2t^3 + 33t^2 - 108t + 100,$$

where $t = 1$ represents January, $t = 2$ represents February, and so on.

**(a)** Find the value of $t$ when the local minimum occurs. What month is this?

**(b)** Find the value of $t$ when the local maximum occurs. What month is this?

**57. Building Products** Sales (in millions of dollars) for Huttig Building Products Inc. for the years 2002 through 2015 can be approximated by

$$C(x) = \frac{7}{3}x^3 - 57x^2 + 360x + 377,$$

where $x = 2$ corresponds to the year 2002. Find the locations of all local extrema and the value of the function at each extremum. (Data from: www.morningstar.com.)

**58. IBM Revenue** Revenue (in billions of dollars) for International Business Machines Corp. (IBM) for the years 2008 through 2015 can be approximated by

$$R(x) = -\frac{1}{3}x^3 + 10.5x^2 - 107x + 455,$$

where $x = 8$ corresponds to the year 2008. Find the locations of all local extrema and the value of the function at each extremum. (Data from: www.morningstar.com.)

*Work these exercises. You may need to use the quadratic formula to find some critical numbers. (See Example 9.)*

**59. Salmon Spawning** The number of salmon swimming upstream to spawn is approximated by

$$S(x) = -x^3 + 3x^2 + 400x + 5000 \quad (6 \le x \le 20),$$

where $x$ represents the temperature of the water in degrees Celsius. Find the water temperature that produces the maximum number of salmon swimming upstream.

**60. Southwest Airlines Costs** Costs for Southwest Airlines (in billions of dollars) for the years 2007 through 2015 can be approximated by

$$C(x) = -.108x^3 + 3.5x^2 - 36x + 122.5 \quad (7 \le x \le 15),$$

where $x = 7$ corresponds to the year 2007. Find the locations of all local extrema and the value of the function at each extremum. (Data from: www.morningstar.com.)

**61. Blood Alcohol Concentration** The function

$$A(x) = .004x^3 - .05x^2 + .16x + .05$$

approximates the blood alcohol concentration in an average person's bloodstream $x$ hours after drinking 8 ounces of 100-proof whiskey. The function applies only to the interval [0, 6].

**(a)** On what time intervals is the blood alcohol concentration increasing?

**(b)** On what intervals is it decreasing?

**62. Animal Activity Level** In the summer, the activity level of a certain type of lizard varies according to the time of day. A biologist has determined that the activity level is given by the function

$$a(t) = .008t^3 - .27t^2 + 2.02t + 7 \qquad (0 \le t \le 24),$$

where $t$ is the number of hours after noon. When is the activity level highest? When is it lowest?

*Work the following exercises.*

**63. Bell-Shaped Curve** The standard normal probability function is used to describe many different populations. Its graph is the well-known normal curve. This function is defined by

$$f(x) = \frac{1}{\sqrt{2\pi}}e^{-x^2/2}.$$

Give the intervals where the function is increasing and decreasing.

**64. Website Traffic** The number of hourly hits (in thousands) on a company's website $x$ hours after a certain news story breaks can be approximated by

$$H(x) = 1 + 30xe^{-.5x}.$$

Find the locations of all local extrema and the value of the function at each extremum.

**65. Training Videos** A group of researchers found that people prefer training films of moderate length; shorter films contain too little information, while longer films are boring. For a training film on the care of exotic birds, the researchers determined that the ratings people gave for the film could be approximated by

$$R(t) = \frac{20t}{t^2 + 100},$$

where $t$ is the length of the film (in minutes). Find the film length that received the highest rating.

**66. Decontamination** A marshy region used for agricultural drainage has become contaminated with selenium. It has been determined that flushing the area with clean water will reduce the selenium for a while, but it will then begin to build up again. A biologist has found that the percentage of selenium in the soil $x$ months after the flushing begins is given by

$$f(x) = \frac{x^2 + 36}{2x} \qquad (1 \le x \le 12).$$

When will the selenium be reduced to a minimum? What is the minimum percentage?

**67. Drug Concentration** The percent of concentration of a drug in the bloodstream $x$ hours after the drug is administered is given by

$$K(x) = \frac{4x}{3x^2 + 27}.$$

(a) On what time intervals is the concentration of the drug increasing?

(b) On what intervals is it decreasing?

**68. Disease Epidemic** The number of people $P(t)$ (in hundreds) infected $t$ days after an epidemic begins is approximated by

$$P(t) = \frac{10 \ln(.19t + 1)}{.19t + 1}.$$

When will the number of people infected start to decline?

**69. JP Morgan Chase Income** Interest income generated by JP Morgan Chase & Company (in billions) can be approximated for the years 2000 through 2015 by

$$R(x) = .0245x^4 - .8x^3 + 8x^2 - 22x + 45 \qquad (0 \le x \le 15),$$

where $x = 0$ corresponds to the year 2000. Use technology to approximate the local minima. (Data from: www.morningstar.com.)

**70. JP Morgan Chase Income** For the function given in the previous exercise, use technology to approximate the local maxima. (Data from: www.morningstar.com.)

✓ **Checkpoint Answers**

1. Increasing on $(-7, -4)$ and $(-2, 3)$; decreasing on $(-\infty, -7)$, $(-4, -2)$, and $(3, \infty)$

2. Increasing on $(-\infty, -3/2)$ and $(1, \infty)$; decreasing on $(-3/2, 1)$

3. (a) Local maximum at $x_2$; local minima at $x_1$ and $x_3$

   (b) No local maximum; local minimum at $x_1$

   (c) Local maximum at $x_1$; no local minimum

4. (a) $-3, 5$       (b) $0, 1$

5. (a) Local minimum at $-3$

   (b) Local maximum at $-2$; local minimum at 4

6. Local maximum at $-3$: $f(-3) = 23$; local minimum at $1/3$:

$$f\left(\frac{1}{3}\right) = \frac{121}{27}$$

7. Local minimum at $x = -2$; local maximum at $x = 2$

8. Local minimum at $x = -3$; no local extremum at $x = 0$

9. Decreasing on $(4, 10)$

# 12.2 The Second Derivative

The first derivative of a function indicates when the function is increasing or decreasing. In many applications, however, this information isn't enough. For example, Figure 12.23 shows the prices of two different stocks over a period of months. Both stocks are worth $5 per share at the beginning and are continually increasing in price.

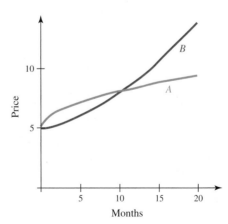

**Figure 12.23**

It is easy to see from the graph that stock $B$ is the better long-term investment. In real life, however, you see only *part* of the graph (the past performance of a stock). How can that help you to predict future performance? To answer this question, look at the graphs again. Both are always increasing, but after the first few months, stock $B$ increases at a faster rate than stock $A$. So the *rate* of increase plays a crucial role here.

In mathematical terms, if the price of the stock in month $x$ is $f(x)$, then the derivative $f'(x)$ tells you when the price is increasing (or decreasing). The *rate* of increase (or decrease) of the derivative function $f'$ is given by *its* derivative. In other words, the rate at which the stock price is increasing (or decreasing) is given by the derivative of the derivative of the price function. In order to deal with such situations, we need some additional terminology and notation.

## Higher Derivatives

If a function $f$ has a derivative $f'$, then the derivative of $f'$, if it exists, is the **second derivative** of $f$, written $f''$. The derivative of $f''$, if it exists, is called the **third derivative** of $f$, and so on. By continuing this process, we can find **fourth derivatives** and other higher derivatives. For example, if $f(x) = x^4 + 2x^3 + 3x^2 - 5x + 7$, then

$$f'(x) = 4x^3 + 6x^2 + 6x - 5, \quad \text{First derivative of } f$$
$$f''(x) = 12x^2 + 12x + 6, \quad \text{Second derivative of } f$$
$$f'''(x) = 24x + 12, \quad \text{Third derivative of } f$$
$$f^{(4)}(x) = 24. \quad \text{Fourth derivative of } f$$
$$f^{(5)}(x) = 0 \quad \text{Fifth derivative of } f$$

✓ **Checkpoint 1**

Let $f(x) = x^3 - 5x^2 - 7x + 99$.
Find

**(a)** $f'(x)$;

**(b)** $f''(x)$;

**(c)** $f'''(x)$;

**(d)** $f^{(4)}(x)$.

The second derivative of $y = f(x)$ can be written in any of the following notations:

$$f''(x), \quad y'', \quad \text{or} \quad \frac{d^2y}{dx^2}.$$

The third derivative can be written in a similar way. For $n \geq 4$, the $n$th derivative is written $f^{(n)}(x)$.

**EXAMPLE 1**  Find the second derivative of the given functions.

**(a)** $f(x) = 8x^3 - 9x^2 + 6x + 4$

**Solution** Here, $f'(x) = 24x^2 - 18x + 6$. The second derivative is the derivative of $f'(x)$, or

$$f''(x) = 48x - 18.$$

**(b)** $y = \dfrac{4x + 2}{3x - 1}$

**Solution** Use the quotient rule to find $y'$:

$$y' = \frac{(3x - 1)(4) - (4x + 2)(3)}{(3x - 1)^2} = \frac{12x - 4 - 12x - 6}{(3x - 1)^2} = \frac{-10}{(3x - 1)^2}.$$

Use the quotient rule again to find $y''$:

$$y'' = \frac{(3x - 1)^2(0) - (-10)(2)(3x - 1)(3)}{[(3x - 1)^2]^2}$$
$$= \frac{60(3x - 1)}{(3x - 1)^4} = \frac{60}{(3x - 1)^3}.$$

**(c)** $y = xe^x$

**Solution** Using the product rule gives

$$\frac{dy}{dx} = x \cdot e^x + e^x \cdot 1 = xe^x + e^x.$$

Differentiate this result to get $\dfrac{d^2y}{dx^2}$:

$$\frac{d^2y}{dx^2} = (xe^x + e^x) + e^x = xe^x + 2e^x = (x + 2)e^x. \quad ☑_2$$

### ✓ Checkpoint 2

Find the second derivatives of the given functions.

**(a)**  $y = -9x^3 + 11x$

**(b)**  $y = -2x^4 + 6x^2$

**(c)**  $y = \dfrac{x + 2}{5x - 1}$

**(d)**  $y = e^x + \ln x$

## Applications

In the previous chapter, we saw that the first derivative of a function represents the rate of change of the function. The second derivative, then, represents the rate of change of the first derivative. This fact has a variety of applications. For one such application, we take another look at stocks $A$ and $B$ from the beginning of this section.

**EXAMPLE 2**   **Finance** Suppose that, over a 10-month period, the price of stock $A$ is given by $f(x) = x^{1/2} + 5$ and the price of stock $B$ is given by $g(x) = .1x^{3/2} + 5$.

**(a)** When are stocks $A$ and $B$ increasing in price?

**Solution** The first derivatives of the price functions are, respectively,

$$f'(x) = \frac{1}{2}x^{-1/2} = \frac{1}{2\sqrt{x}} \quad \text{and} \quad g'(x) = .1\left(\frac{3}{2}x^{1/2}\right) = \frac{.3}{2}\sqrt{x}.$$

Both derivatives are always positive (because $\sqrt{x}$ is positive for $x > 0$). So both price functions are increasing for all $x > 0$.

**(b)** At what rate are these stock prices increasing in the 10th month? What does this suggest about their future performance?

**Solution** The rate at which $f'(x) = \dfrac{1}{2}x^{-1/2}$ and $g'(x) = \dfrac{.3}{2}x^{1/2}$ are changing is given by *their* derivatives (the second derivatives of $f$ and $g$):

$$f''(x) = \frac{1}{2} \cdot \frac{-1}{2}x^{-3/2} = \frac{-1}{4\sqrt{x^3}} \quad \text{and} \quad g''(x) = \frac{.3}{2} \cdot \frac{1}{2}x^{-1/2} = \frac{3}{40\sqrt{x}}.$$

When $x = 10$,

$$f''(10) = \frac{-1}{4\sqrt{10^3}} \approx -.0079 \quad \text{and} \quad g''(10) = \frac{3}{40\sqrt{10}} \approx .0237.$$

The rate of increase for stock $A$ is negative, meaning that its price is increasing at a decreasing rate. The rate of increase for stock $B$ is positive, meaning that its price is increasing at an increasing rate. In other words, the price of stock $A$ is increasing more and more slowly, while the price of stock $B$ is increasing faster and faster. This result suggests that stock $B$ is probably a better investment for the future.

The preceding discussion assumes that present trends continue (which is not guaranteed in the stock market). If they do continue for another 10 months, then Figure 12.23 shows what will happen to the stocks during this period.

Although Example 2 is much simpler than real life, second derivatives are actually used by some investors. According to an article in the *Wall Street Journal* (August 1, 2001), "many investors hope it will help them get into the market ahead of any big rallies. In essence, the second-derivative approach involves looking at such things as profit warnings, analysts' earning estimates and equipment orders (including cancellations) for signs that the downward momentum is slowing."

The second derivative also plays a role in the physics of a moving particle. If a function describes the position of a moving object (along a straight line) at time $t$, then the first derivative gives the velocity of the object. That is, if $y = s(t)$ describes the position (along a straight line) of the object at time $t$, then $v(t) = s'(t)$ gives the velocity at time $t$.

The rate of change of velocity is called **acceleration**. Since the second derivative gives the rate of change of the first derivative, the acceleration is the derivative of the velocity. Thus, if $a(t)$ represents the acceleration at time $t$, then

$$a(t) = \frac{d}{dt}v(t) = s''(t).$$

**EXAMPLE 3**   **Velocity and Acceleration**  Suppose that an object is moving along a straight line, with its position, in feet, at time $t$, in seconds, given by

$$s(t) = t^3 - 2t^2 - 7t + 9.$$

**(a)**  Find the velocity at any time $t$.

**Solution**  The velocity is given by

$$v(t) = s'(t) = 3t^2 - 4t - 7.$$

**(b)**  Find the acceleration at any time $t$.

**Solution**  Acceleration is given by

$$a(t) = v'(t) = s''(t) = 6t - 4.$$

**(c)**  When the object stops, its velocity is zero. For $t \geq 0$, when does that occur?

**Solution**  Set $v(t) = 0$:

$$3t^2 - 4t - 7 = 0$$
$$(3t - 7)(t + 1) = 0$$
$$3t - 7 = 0 \quad \text{or} \quad t + 1 = 0$$
$$t = \frac{7}{3} \quad\quad\quad t = -1.$$

✓ **Checkpoint 3**

Rework Example 3 if
$s(t) = t^4 - t^3 + 10.$

Since we want $t \geq 0$, only $t = \frac{7}{3}$ is acceptable here. The object will stop in $\frac{7}{3}$ seconds.  ✓₃

## Concavity

We shall now see how the second derivative provides information about how the graph of a function "bends." A graph is **concave upward** on an interval if it bends upward over the interval and **concave downward** if it bends downward, as shown in Figure 12.24. The graph is concave downward on the interval $(a, b)$ and concave upward on the interval $(b, c)$.[*] A point on the graph where the concavity changes (such as the point where $x = b$ in Figure 12.24) is called a **point of inflection**.

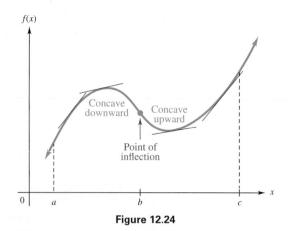

**Figure 12.24**

---

[*]Figure 12.24 also illustrates the formal definition of concavity: A function is *concave downward* on an interval if its graph lies below the tangent line at each point in the interval and *concave upward* if its graph is above the tangent line at each point in the interval.

A function that is increasing on an interval may have either kind of concavity; the same is true for a decreasing function. Some of the possibilities are illustrated in Figure 12.25.

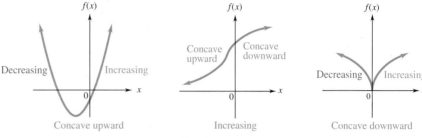

**Figure 12.25**

Next, we examine the relationship between the second derivative of a function $f$ and the concavity of the graph of $f$. We have seen that when the derivative of any function is positive, that function is increasing. Consequently, if the *second* derivative of $f$ (the derivative of the first derivative) is positive, then the *first* derivative of $f$ is increasing. Since the first derivative gives the slope of the tangent line to the graph of $f$ at each point, the fact that the first derivative is increasing means that the tangent-line slopes are increasing as you move from left to right along the graph of $f$, as illustrated in Figure 12.26.

**Figure 12.26**

In Figure 12.26(a), the slopes of the tangent lines increase from negative at the left, to 0 in the center, to positive at the right. In Figure 12.26(b), the slopes are all positive, but are increasing as the tangent lines get steeper. Note that both graphs in Figure 12.26 are *concave upward*.

Similarly, when the second derivative is negative, the first derivative (slope of the tangent line) is decreasing, as illustrated in Figure 12.27. In Figure 12.27(a), the tangent-line slopes decrease from positive, to 0, to negative. In Figure 12.27(b), the slopes get more and more negative as the tangent lines drop downward more steeply. Note that both graphs are *concave downward*.

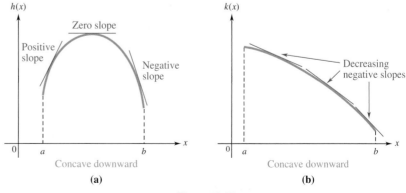

**Figure 12.27**

The preceding discussion suggests the following result.

## Concavity Test

Let $f$ be a function whose first and second derivatives exist at all points in the interval $(a, b)$.

1. If $f''(x) > 0$ for all $x$ in $(a, b)$, then $f$ is concave upward on $(a, b)$.
2. If $f''(x) < 0$ for all $x$ in $(a, b)$, then $f$ is concave downward on $(a, b)$.

**Figure 12.28**

**Figure 12.29**

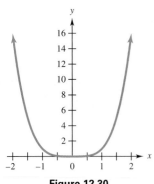

**Figure 12.30**

**EXAMPLE 4**    Find the intervals over which the given function is concave upward or downward, and find any points of inflection.

**(a)** $f(x) = .5x^3 - 2x^2 + 5x + 1$

**Solution** The graph of $f$ (Figure 12.28) suggests that $f$ is concave upward to the right of the $y$-axis, but the concavity near the $y$-axis is not clear. Even in a window where the concavity is visible, the location of the point of inflection is not clear. (Try it!) So we use the test in the preceding box. The first derivative is $f'(x) = 1.5x^2 - 4x + 5$, and the second derivative is $f''(x) = 3x - 4$. The function is concave upward when $f''(x) > 0$—that is, when

$$3x - 4 > 0$$
$$3x > 4$$
$$x > \frac{4}{3}.$$

Similarly, $f$ is concave downward when $3x - 4 < 0$—that is, when $x < \frac{4}{3}$. Therefore, $f$ is concave downward on $(-\infty, \frac{4}{3})$ and concave upward on $(\frac{4}{3}, \infty)$, with a point of inflection when $x = \frac{4}{3}$.

**(b)** $g(x) = x^{1/3}$

**Solution** The graph of $g$ in Figure 12.29 suggests that $g$ is concave upward when $x < 0$ and concave downward when $x > 0$. We can confirm this conclusion algebraically by noting that $g'(x) = \frac{1}{3}x^{-2/3}$ and

$$g''(x) = -\frac{2}{9}x^{-5/3} = \frac{-2}{9\sqrt[3]{x^5}}.$$

When $x < 0$, the denominator of $g''(x)$ is negative (why?), so $g''(x)$ is positive and $g$ is concave upward on $(-\infty, 0)$. When $x > 0$, the denominator of $g''(x)$ is positive, so $g''(x)$ is negative and $g$ is concave downward on $(0, \infty)$. The graph changes concavity at $(0, 0)$, which is a point of inflection.

**(c)** $h(x) = x^4$

**Solution** The graph of $h$ (Figure 12.30) suggests that $h$ has no inflection points. We check this algebraically by finding $h'(x) = 4x^3$ and $h''(x) = 12x^2$. We see that $h''(x) = 0$ when $x = 0$, but that $h''(x)$ is positive both when $x < 0$ and when $x > 0$. Therefore $h$ is concave up on the intervals $(-\infty, 0)$ and $(0, \infty)$, and there is no point of inflection.

In Example 4(a), the point of inflection on the graph of $f$ occurs at $x = \frac{4}{3}$, the number at which the second derivative $f''(x) = 3x - 4$ is 0. In Example 4(b), the point of inflection occurs when $x = 0$, the number at which the second derivative $g''(x) = \frac{-2}{9\sqrt[3]{x^5}}$ does not exist. These facts suggest the following result.

If a function $f$ has a point of inflection at $x = c$, then $f''(c) = 0$ or $f''(c)$ does not exist.

Find the intervals where the given functions are concave upward. Identify any inflection points.

(a) $f(x) = 6x^3 - 24x^2 + 9x$

(b) $f(x) = 2x^2 - 4x + 8$

**⬢ CAUTION** The converse of the preceding statement is not always true: The second derivative may be 0 at a point that is not a point of inflection. In Example 4(c), $h''(x) = 0$ when $x = 0$. However, $h$ has no inflection point at $x = 0$ (or anywhere else). ✓ 4

**EXAMPLE 5** **Sirius Revenue** Total revenue (in millions) for Sirius XM satellite radio can be approximated for the period 2004–2015 by

$$R(x) = -4x^3 + 108x^2 - 440x + 300 \qquad (4 \le x \le 15),$$

where $x = 4$ corresponds to the year 2004. It can be shown using the methods of Section 12.1 that revenue was increasing throughout this period. Find the intervals over which the revenue function is concave upward or downward and any points of inflection. Interpret your answers in practical terms. (Data from: www.morningstar.com.)

**Solution** The *possible* locations of the points of inflection are where $R''(x) = 0$ or $R''(x)$ does not exist. The first and second derivatives of the revenue function are, respectively,

$$R'(x) = -12x^2 + 216x - 440 \text{ and}$$
$$R''(x) = -24x + 216.$$

Because $R''(x)$ never fails to exist, we need only solve the equation $R''(x) = 0$:

$$-24x + 216 = 0$$
$$x = \frac{216}{24} = 9.$$

So the only possible location where $R(x)$ can change concavity is at $x = 9$. We test the second derivative at an $x$-value in the domain ($4 \le x \le 15$) to the left of $x = 9$, say,

$$R''(8) = -24(8) + 216 = 24 > 0.$$

This shows that the original function $R(x)$ is concave up on the interval $4 < x < 9$. Next we test an $x$-value to the right of $x = 9$, say,

$$R''(10) = -24(10) + 216 = -24 < 0.$$

Therefore, $R(x)$ is concave down on the interval $9 < x < 15$. Because the concavity changes at $x = 9$, it is the location of a point of inflection. To find the corresponding $y$-value, we evaluate

$$R(9) = -4(9)^3 + 108(9)^2 - 440(9) + 300 = 2172.$$

So the point of inflection is $(9, 2172)$. This corresponds to the year 2009, when the revenue was approximately \$2,172 million.

In practical terms, we know that revenue was increasing throughout this period, so this means that revenue was increasing at an increasing rate up until the year 2009 (increasing and concave up), and increasing at a decreasing rate thereafter (increasing and concave down), as shown in Figure 12.31.

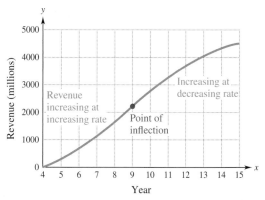

**Figure 12.31**

**EXAMPLE 6**    **Unemployment** The unemployment rate rose rapidly after the 2008 recession before starting to decline, falling from 9.8% in 2010 to 4.9% in 2016. For these years, the unemployment rate can be approximated by

$$f(x) = .045x^3 - 1.755x^2 + 21.6x - 75.7 \qquad (10 \leq x \leq 16),$$

where $x = 10$ corresponds to the year 2010. It can be shown that $f(x)$ is decreasing throughout this period. Find and interpret the point of inflection. (Data from: U.S. Bureau of Labor Statistics.)

**Solution** We find the first and second derivatives to be, respectively, $f'(x) = .135x^2 - 3.51x + 21.6$ and $f''(x) = .27x - 3.51$. We see that the second derivative always exists and that $f''(x) = 0$ when $.27x + 3.51 = 0$ or $x = 13$. By testing the second derivative at numbers to the left and right of $x = 13$, we find that the graph is concave down on the left and concave up on the right, so a point of inflection occurs at $x = 13$. In practical terms, this means that unemployment was declining throughout this period, but the rate of decline started to slow down in 2013, as can be seen in Figure 12.32.

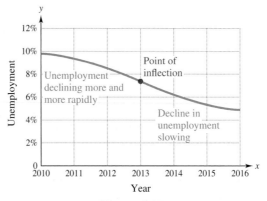

**Figure 12.32**

The **law of diminishing returns** in economics is related to the idea of concavity. The graph of the function $f$ in Figure 12.33 shows the output $y$ from a given input $x$. For instance, the input might be advertising costs and the output the corresponding revenue from sales.

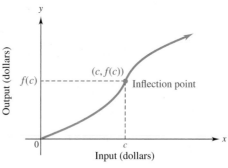

**Figure 12.33**

The graph in Figure 12.33 shows an inflection point at $(c, f(c))$. For $x < c$, the graph is concave upward, so the rate of change of the slope is increasing. This indicates that the output $y$ is increasing at a faster rate with each additional dollar spent. When $x > c$, however, the graph is concave downward, the rate of change of the slope is decreasing, and the increase in $y$ is smaller with each additional dollar spent. Thus, further input beyond $c$ dollars produces diminishing returns. The point of inflection at $(c, f(c))$ is called the **point of diminishing returns**. The *rate of increase* reaches its maximum at the point of diminishing returns, so any investment beyond the value $c$ is not considered a good use of capital.

**EXAMPLE 7**   **Point of Diminishing Returns** The revenue $R(x)$ generated from airline passengers is related to the number of available seat miles $x$ by

$$R(x) = -.928x^3 + 31.492x^2 - 326.80x + 1143.88 \qquad (8 \le x \le 12),$$

where $x$ is measured in hundreds of billions of miles and $R(x)$ is in billions of dollars. Is there a point of diminishing returns for this function? If so, what is it? (Data from: Air Transport Association of America.)

**Solution** Since a point of diminishing returns occurs at an inflection point, we must find $R''(x)$. We have

$$R'(x) = -.928(3x^2) + 31.492(2x) - 326.80$$
$$= -2.784x^2 + 62.984x - 326.80$$

and

$$R''(x) = -2.784(2x) + 62.984$$
$$= -5.568x + 62.984,$$

which exists for every $x$.

When we set $R''(x) = 0$ and solve for $x$, we obtain

$$-5.568x + 62.984 = 0$$
$$-5.568x = -62.984$$
$$x \approx 11.31.$$

Test a number in the interval $(8, 11.31)$ to see that $R''(x)$ is positive there. Then test a number in the interval $(11.31, 12)$ to see that $R''(x)$ is negative in that interval. Since the sign of $R''(x)$ changes from positive to negative at $x = 11.31$, the graph changes from concave upward to concave downward at that point, and there is a point of diminishing returns at the inflection point $(11.31, 133.54)$. Increasing capacity beyond 1.131 trillion seat miles would not pay off. ✔ 5

## ✓ Checkpoint 5

The revenue $R(x)$ generated from sales of a certain product is related to the amount $x$ spent on advertising by

$$R(x) = \frac{1}{75,000}(125x^3 - x^4)$$
$$(0 < x \le 100),$$

where $x$ and $R(x)$ are in thousands of dollars. What is the point of diminishing returns?

## Maxima and Minima

If a function $f$ has a local maximum at $c$ and $f'(c)$ is defined, then the graph of $f$ is necessarily concave downward near $x = c$. Similarly, if $f$ has a local minimum at $d$ and $f'(d)$ exists, then the graph is concave upward near $x = d$, as shown in Figure 12.34.

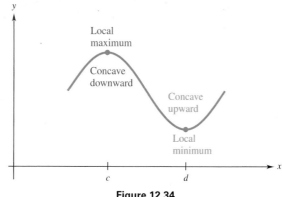

**Figure 12.34**

These facts should make the following result plausible.

## The Second-Derivative Test

Let $c$ be a critical number of the function $f$ such that $f'(c) = 0$ and $f''(x)$ exists for all $x$ in some open interval containing $c$.

**1.** If $f''(c) > 0$, then $f$ has a local minimum at $c$.

**2.** If $f''(c) < 0$, then $f$ has a local maximum at $c$.

**3.** If $f''(c) = 0$, then this test gives no information; use the first-derivative test.

**EXAMPLE 8**    **Consumption and Investment** Government consumption expenditures and gross investment (in billions of dollars) can be approximated for the years 2006 through 2016 by

$$g(x) = \frac{8}{3}x^3 - 92x^2 + 1040x - 704 \qquad (6 \leq x \leq 16),$$

where $x = 6$ corresponds to the year 2006. Find all local extrema of this function, and interpret your answers. (Data from: U.S. Bureau of Economic Analysis.)

**Solution** First find the critical numbers. The first derivative is

$$g'(x) = 8x^2 - 184x + 1040.$$

Because $g'(x)$ always exists, critical numbers can occur only where $g'(x) = 0$. We can solve this equation either by factoring or by using the quadratic formula:

$$8x^2 - 184x + 1040 = 0$$

$$x = \frac{-(-184) \pm \sqrt{(-184)^2 - 4(8)(1040)}}{2(8)} \qquad x = \frac{-b \pm \sqrt{b^2 - 4ac}}{2a}$$

$$= \frac{184 \pm \sqrt{576}}{16} = \begin{cases} 10 \\ 13 \end{cases}. \qquad \text{The critical numbers are 10 and 13.}$$

Now use the second-derivative test on each critical number. The second derivative is

$$g''(x) = 16x - 184.$$

For the critical number $x = 10$, we have

$$g''(10) = 16(10) - 184 = -24 < 0,$$

which means that there is a local maximum at $x = 10$. The local maximum value is

$$g(10) = \frac{8}{3}(10)^3 - 92(10)^2 + 1040(10) - 704 \approx 3162.67.$$

For the critical number $x = 13$, we have

$$g''(13) = 16(13) - 184 = 24 > 0;$$

hence there is a local minimum at $x = 13$. The local minimum value is

$$g(13) = \frac{8}{3}(13)^3 - 92(13)^2 + 1040(13) - 704 \approx 3126.67.$$

In practical terms, this means that government consumption and investment were at a local maximum of approximately \$3163 billion in the year 2010 and a local minimum of approximately \$3127 billion in the year 2013. ✓⁶

✓ **Checkpoint 6**

Use the second-derivative test to find all local maxima and local minima for $f(x) = x^3 - 3x^2 - 6x$.

📄 **NOTE** The second-derivative test works only for those critical numbers $c$ which make $f'(c) = 0$. This test does not work for those critical numbers $c$ for which $f'(c)$ does not exist (since $f''(c)$ would not exist there either). Also, the second-derivative test does not work for critical numbers $c$ that make $f''(c) = 0$. In both of these cases, use the first-derivative test.

# 12.2 Exercises

*For each of these functions, find $f''(x), f''(0), f''(2),$ and $f''(-3)$. (See Examples 1 and 2.)*

**1.** $f(x) = x^3 - 6x^2 + 1$

**2.** $f(x) = -2x^3 + 5x^2 - 10x + 13$

**3.** $f(x) = 2e^x$

**4.** $f(x) = 13 - 5e^x$

**5.** $f(x) = 6e^{2x}$

**6.** $f(x) = 2 + e^{-x}$

**7.** $f(x) = (x + 3)^4$

**8.** $f(x) = (2x + 1)^4$

**9.** $f(x) = \sqrt{4x + 1}$

**10.** $f(x) = \sqrt{2x + 9}$

**11.** $f(x) = -x^{2/3}$

**12.** $f(x) = 5x^{4/5}$

**13.** $f(x) = \ln(2x + 1)$

**14.** $f(x) = \ln(3x - 1)$

**15.** $f(x) = xe^{-x}$

**16.** $f(x) = xe^{2x}$

**17.** $f(x) = \dfrac{x^2}{1 + x}$

**18.** $f(x) = \dfrac{-x}{1 - x^2}$

*In Exercises 19 and 20, P(t) is the price of a certain stock at time t during a particular day. (See Example 2.)*

**19. Finance** If the price of the stock is falling faster and faster, are $P'(t)$ and $P''(t)$ positive or negative? Explain your answer.

**20. Finance** When the stock reaches its highest price during the day, are $P'(t)$ and $P''(t)$ positive or negative? Explain your answer.

**Velocity and Acceleration** *Each of the functions in Exercises 21–24 gives the position at time t of a particle moving along a line. Find the velocity and acceleration functions. Then find the velocity and acceleration at $t = 0$ and $t = 4$. Assume that time is measured in seconds and distance is measured in centimeters. Velocity will be in centimeters per second (cm/sec) and acceleration in centimeters per second per second (cm/sec$^2$). (See Example 3.)*

**21.** $s(t) = 6t^2 + 5t$

**22.** $s(t) = 7t^2 + 2t + 10$

**23.** $s(t) = 3t^3 - 4t^2 + 8t - 9$

**24.** $s(t) = 3t^3 - 6t^2 + 8t - 4$

*Find the largest open intervals on which each function is concave upward or concave downward, and find the location of any points of inflection. (See Example 4.)*

**25.** $f(x) = x^3 + 3x - 5$

**26.** $f(x) = -x^3 + 8x - 7$

**27.** $f(x) = x^3 + 4x^2 - 6x + 3$

**28.** $f(x) = 5x^3 + 12x^2 - 32x - 14$

**29.** $f(x) = \dfrac{2}{x - 4}$

**30.** $f(x) = \dfrac{-2}{x + 1}$

**31.** $f(x) = x^4 + 8x^3 - 30x^2 + 24x - 3$

**32.** $f(x) = x^4 + 8x^3 + 18x^2 + 12x - 84$

**Business** *In Exercises 33–36, R(x) represents revenue, in thousands of dollars, and x represents the amount spent on advertising, in thousands of dollars. Find (a) $R''(x)$ and (b) the point of diminishing returns. (See Example 7.)*

**33.** $R(x) = -x^3 + 30x^2 + 1000; 0 \le x \le 15$

**34.** $R(x) = -\dfrac{1}{3}x^3 + 3x^2 + 1; 0 \le x \le 5$

**35.** $R(x) = 10{,}000 - x^3 + 42x^2 + 800x; 0 \le x \le 20$

**36.** $R(x) = 1000 + 36x^2 - x^3$

*Find all critical numbers of the functions in Exercises 37–48. Then use the second-derivative test on each critical number to determine whether it leads to a local maximum or minimum. (See Example 8.)*

**37.** $f(x) = -2x^3 - 3x^2 + 72x + 1$

**38.** $f(x) = \dfrac{2}{3}x^3 + \dfrac{1}{2}x^2 - x - \dfrac{1}{4}$

**39.** $f(x) = (x - 3)^5$

**40.** $f(x) = (2x - 1)^5$

**41.** $f(x) = x^4 - 8x^2$

**42.** $f(x) = x^4 - 32x^2 + 7$

**43.** $f(x) = x + \dfrac{4}{x}$

**44.** $f(x) = x + \dfrac{4}{x^2}$

**45.** $f(x) = \dfrac{x^2 + 9}{2x}$

**46.** $f(x) = \dfrac{x^2 + 16}{2x}$

**47.** $f(x) = \dfrac{2 - x}{2 + x}$

**48.** $f(x) = \dfrac{x + 2}{x - 1}$

*Use the given graphs to work these exercises.*

**49.** In each part, list the points ($A$–$F$) on the graph of $f$ whose $x$-coordinates satisfy the given conditions.

**(a)** $f'(x) > 0$ and $f''(x) > 0$

**(b)** $f'(x) < 0$ and $f''(x) > 0$

**(c)** $f'(x) = 0$ and $f''(x) < 0$

**(d)** $f'(x) = 0$ and $f''(x) > 0$

**(e)** $f'(x) < 0$ and $f''(x) = 0$

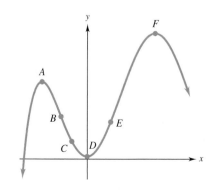

**50.** In each part, list the point (*A–E*) on the graph of *f* whose *x*-coordinate satisfies the given conditions.

(a) $f'(x) > 0$ and $f''(x) > 0$

(b) $f'(x) < 0$ and $f''(x) > 0$

(c) $f'(x) = 0$ and $f''(x) < 0$

(d) $f'(x) = 0$ and $f''(x) > 0$

(e) $f'(x) < 0$ and $f''(x) = 0$

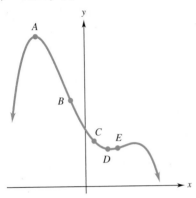

*Work these problems. (See Example 2.)*

**51. Semiconductors** The stock price for the semiconductor company Silicon Motion Technology Corporation over the 12-week period at the beginning of 2016 can be approximated by the function

$$f(x) = 28e^{.02x},$$

where $x = 1$ designates the beginning of week 1. (Data from: www.morningstar.com.)

(a) When is the stock increasing in price?

(b) What is the rate of change of the increase in stock price at the beginning of the 5th week?

(c) What is the rate of change of the increase in stock price at the beginning of the 10th week?

**52. 3D Printing** The stock price for the 3D printing company 3D Systems Corporation over the 12-week period at the beginning of 2016 can be approximated by the function

$$f(x) = 6.4e^{.07x},$$

where $x = 1$ designates the beginning of week 1. (Data from: www.morningstar.com.)

(a) When is the stock increasing in price?

(b) What is the rate of change of the increase in stock price at the beginning of the 3rd week?

(c) What is the rate of change of the increase in stock price at the beginning of the 10th week?

*Work these problems. (See Example 3.)*

**53. Velocity and Acceleration** When an object is dropped straight down, its position, in feet, after *t* seconds is given by

$$s(t) = -16t^2.$$

Note that negative velocity indicates downward motion. Find the velocity at each of the following times.

(a) After 3 seconds

(b) After 5 seconds

(c) After 8 seconds

(d) Find the acceleration. (The answer here is a constant, the acceleration due to the influence of gravity alone.)

**54.** When an object is dropped straight down, its position, in meters, after *t* seconds is given by

$$s(t) = -4.9t^2 + s_0,$$

where $s_0$ represents the initial height from which the object is dropped. Find the velocity at each of the following times. Note that negative velocity indicates downward motion.

(a) After 5 seconds

(b) After 3 seconds

(c) After 10 seconds

(d) Find the acceleration. (The answer here is a constant, the acceleration due to the influence of gravity alone.)

**55.** If an object is thrown directly upward with a velocity of 256 ft/sec, its height, in feet, above the ground after *t* seconds is given by $s(t) = 256t - 16t^2$. Find the velocity and the acceleration after *t* seconds. When does the object reach its maximum height? What is the maximum height the object reaches? When does it hit the ground?

**56.** If a slingshot loaded with a rubber ball is aimed directly upward and shot with an initial velocity of 128 feet per second, the height of the ball, in feet, above the ground after *t* seconds is given by

$$s(t) = 128t - 16t^2.$$

Find the velocity and the acceleration after *t* seconds. When does the rubber ball reach its maximum height? What is its maximum height? When does it hit the ground?

*Work these exercises. (See Example 7.)*

**57. Point of Diminishing Returns** A local automobile dealer has found that advertising produces sales but that too much advertising tends to turn consumers off, resulting in reduced sales. On the basis of past experience, the dealer expects that the number $V(x)$ of vehicles sold during a month is related to the amount spent on advertising by the function

$$V(x) = -.1x^3 + 2.4x^2 + 1.2x + 15 \qquad (0 \le x \le 15),$$

where *x* is the amount spent on advertising (in tens of thousands of dollars). What is the point of diminishing returns?

**58. Point of Diminishing Returns** The revenue $R(x)$ (in millions of dollars) generated from cattle farming can be approximated by the function

$$R(x) = -3.21x^3 + 62.14x^2 + 172.92x + 48.65$$
$$(0 \le x \le 14),$$

where *x* represents the number of farms (in millions). What is the point of diminishing returns? (Data from: U.S. Department of Agriculture.)

*Work these problems. (See Examples 5 and 6.)*

**59. Exelixis Stock** The stock price for the cancer-fighting company Exelixis, Inc. for the second quarter of 2016 (weeks 14 through 26) can be approximated by

$$f(x) = -.002x^3 + .12x^2 - 2x + 14 \qquad (14 \le x \le 26),$$

where $x = 14$ corresponds to the beginning of week 14. (Data from: www.morningstar.com.)

(a) Find $f''(x)$.

(b) Find the *location* of the point of inflection.

(c) On what interval was $f(x)$ concave up?

(d) On what interval was $f(x)$ concave down?

(e) Given that $f(x)$ was increasing during this period, interpret your answers.

60. **Pharmaceuticals Stock** The stock price for Valeant Pharmaceuticals International Inc. for the first 20 weeks of 2016 can be approximated by

$$f(x) = .02x^3 - .6x^2 + 99 \qquad (1 \le x \le 20),$$

where $x = 1$ corresponds to the beginning of week 1. (Data from: www.morningstar.com.)

(a) Find $f''(x)$.

(b) Find the *location* of the point of inflection.

(c) On what interval was $f(x)$ concave up?

(d) On what interval was $f(x)$ concave down?

(e) Given that $f(x)$ was decreasing during this period, interpret your answers.

*Work these exercises.*

61. **Starbucks Stock** The price of Starbucks Corporation stock for the years 2003 through 2015 can be approximated by

$$f(x) = \frac{1}{6}x^3 - \frac{7}{2}x^2 + \frac{45}{2}x - 12 \qquad (3 \le x \le 15),$$

where $x = 3$ corresponds to the year 2003. (Data from: www.morningstar.com.)

(a) Find $f'(x)$.

(b) Find the critical numbers of $f(x)$. (*Hint:* Multiply both sides of $f'(x) = 0$ by 2.)

(c) Find $f''(x)$.

(d) Use the second-derivative test to determine in what years the price was at a local maximum or minimum.

62. **Computing Stock** The stock price for the visual computing company NVIDIA Corporation for the years 2010 through 2016 can be approximated by

$$f(x) = .55x^3 - 20.3x^2 + 249x - 1000 \qquad (10 \le x \le 16),$$

where $x = 10$ corresponds to the year 2010. (Data from: www.morningstar.com.)

(a) Find $f'(x)$.

(b) Use the quadratic formula to find the critical numbers of $f(x)$.

(c) Find $f''(x)$.

(d) Use the second-derivative test to determine in what years the price was at a local maximum or minimum.

63. **Personal Income** Per capita personal income (in dollars) can be approximated for the years 2002 through 2015 by

$$f(x) = 7877 \ln(x) + 24,215 \qquad (2 \le t \le 15),$$

where $x = 2$ corresponds to the year 2002. (Data from: U.S. Bureau of Economic Analysis.)

(a) Find $f'(x)$.

(b) Was personal income increasing or decreasing during this period?

(c) Find $f''(x)$.

(d) Determine whether the graph of $f$ is concave up or concave down, and explain what that means in practical terms.

64. **Social Science** The population of Wyoming (in thousands) is projected to be approximately

$$f(x) = 18 \ln(x) + 478 \qquad (5 \le x \le 20),$$

where $x = 5$ corresponds to the year 2005. (Data from: www.census.gov.)

(a) Is the population of Wyoming projected to increase or decrease through 2020?

(b) Determine whether the graph of this population function is concave up or concave down, and explain what that means for Wyoming.

65. **Business** One store's revenue from selling $x$ cell phones is given by

$$R(x) = (200x)e^{-.01x}.$$

(a) How many phones should be sold to maximize revenue?

(b) What is the maximum revenue?

(c) When revenue is at a maximum, what is the average price per phone?

66. **Influenza (Flu) Outbreak** The number of people infected by the flu during a particular outbreak is approximated by

$$p(t) = 4t^3 - \frac{t^4}{10} \qquad (0 \le t \le 40),$$

where $t$ is the number of days after the flu was first observed.

(a) After how many days was the number of infected people at a maximum?

(b) What was the maximum number of people with the flu?

## ✓ Checkpoint Answers

1. (a) $f'(x) = 3x^2 - 10x - 7$  (b) $f''(x) = 6x - 10$
   (c) $f'''(x) = 6$  (d) $f^{(4)}(x) = 0$

2. (a) $y'' = -54x$  (b) $y'' = -24x^2 + 12$
   (c) $y'' = \dfrac{110}{(5x - 1)^3}$  (d) $y'' = e^x - \dfrac{1}{x^2}$

3. (a) $v(t) = 4t^3 - 3t^2$
   (b) $a(t) = 12t^2 - 6t$
   (c) At 0 and $\frac{3}{4}$ second

4. (a) Concave upward on $(\frac{4}{3}, \infty)$; point of inflection is $(\frac{4}{3}, -\frac{175}{9})$
   (b) $f''(x) = 4$, which is always positive; function is always concave upward, no inflection point

5. $(62.5, 203.45)$

6. Local maximum of about 2.39 at $x \approx -.73$; local minimum of about $-18.39$ at $x \approx 2.73$

# 12.3 Optimization Applications

In most applications, the domains of the functions involved are restricted to numbers in a particular interval. For example, a factory that can produce a maximum of 40 units (because of market conditions, availability of labor, etc.) might have this cost function:

$$C(x) = -3x^3 + 135x^2 + 3600x + 12,000 \qquad (0 \le x \le 40).$$

Even though the rule of $C$ is defined for all numbers $x$, only the numbers in the interval $[0, 40]$ are relevant, because the factory cannot produce a negative number of units or more than 40 units. In such applications, we often want to find a smallest or largest quantity—for instance, the minimum cost or the maximum profit—when $x$ is restricted to the relevant interval. So we begin with the mathematical description of such a situation.

Let $f$ be a function that is defined for all $x$ in the closed interval $[a, b]$. Let $c$ be a number in the interval. We say that $f$ has an **absolute maximum on the interval** at $c$ if

$$f(x) \le f(c) \quad \text{for all } x \text{ with } a \le x \le b,$$

that is, if $(c, f(c))$ is the highest point on the graph of $f$ over the interval $[a, b]$. Similarly, $f$ has an **absolute minimum on the interval** at $c$ if

$$f(x) \ge f(c) \quad \text{for all } x \text{ with } a \le x \le b,$$

that is, if $(c, f(c))$ is the lowest point on the graph of $f$ over the interval $[a, b]$.

> **EXAMPLE 1**  Figure 12.35 shows the graph of a function $f$. Consider the function $f$ on the interval $[-2, 6]$. Since we are interested only in the interval $[-2, 6]$, the values of the function outside this interval are irrelevant. On the interval $[-2, 6]$, $f$ has an absolute minimum at $x = 3$ (which is also a local minimum) and an absolute maximum at $x = 6$ (which is not a local maximum of the entire function). ✓1

✓ **Checkpoint 1**

Find the location of the absolute maximum and absolute minimum of the function $f$ in Figure 12.35 on the interval $[-2, \frac{1}{2}]$.

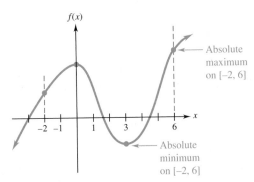

**Figure 12.35**

The absolute maximum in Example 1 occurred at $x = 6$, which is an endpoint of the interval, and the absolute minimum occurred at $x = 3$, which is a critical number of $f$ (because $f$ has a local minimum there). Similarly, in Checkpoint 1, the absolute maximum occurred at a critical number and the absolute minimum at an endpoint. These examples illustrate the following result, whose proof is omitted.

## Extreme-Value Theorem

If a function $f$ is continuous on a closed interval $[a, b]$, then $f$ has both an absolute maximum and an absolute minimum on the interval. Each of these values occurs either at an endpoint of the interval or at a critical number of $f$.

⚠ **CAUTION** The extreme-value theorem may not hold on intervals that are not closed (that is, intervals that exclude one or both endpoints). For example, $f(x) = 1/x$ does not have an absolute maximum on the interval $(0, 1)$; the values of $f(x)$ get larger and larger as $x$ approaches 0, as you can easily verify with a calculator.

**EXAMPLE 2** Use the extreme-value theorem to find the absolute extrema of $f(x) = 4x + \dfrac{36}{x}$ on the interval $[1, 6]$.

**Solution** According to the extreme-value theorem, we need to consider only the critical numbers of $f$ and the endpoints, 1 and 6, of the interval. Begin by finding the derivative and determining the critical numbers:

$$f'(x) = 4 - \frac{36}{x^2} = 0$$

$$\frac{4x^2 - 36}{x^2} = 0.$$

Since $f'(x)$ is not defined only at $x = 0$, $f'(x)$ exists for every $x$ in the interval $[1, 6]$. The derivative is zero when the numerator is zero:

$$4x^2 - 36 = 0$$
$$4x^2 = 36$$
$$x^2 = 9$$
$$x = -3 \quad \text{or} \quad x = 3.$$

Since $-3$ is not in the interval $[1, 6]$, disregard it; 3 is the only critical number of interest. By the extreme-value theorem, the absolute maximum and minimum must occur at critical numbers or endpoints—that is, at 1, 3, or 6. Evaluate $f$ at these three numbers to see which ones give the largest and smallest values:

| $x$-Value | Value of Function | |
|---|---|---|
| 1 | 40 | ← Absolute maximum |
| 3 | 24 | ← Absolute minimum |
| 6 | 30 | |

✓ **Checkpoint 2**

Find the absolute maximum and absolute minimum values for $f(x) = -x^2 + 4x - 8$ on $[-4, 4]$.

A graphing calculator or other graphing technology sometimes makes it easy to find absolute extrema without taking the derivative. Technology, however, has its limits, so you should confirm your results algebraically whenever feasible.

**EXAMPLE 3** Find the absolute extrema of $g(x) = -.02x^3 + 600x - 20{,}000$ on $[60, 135]$.

**Solution** The graph of $g$ on $[60, 135]$ in Figure 12.36 shows that the absolute maximum occurs at the local maximum, which is close to $x = 100$. The absolute minimum occurs at one of the endpoints, but you cannot determine which one from the graph.

**Figure 12.36**

We first determine the critical number exactly. Since $g'(x) = -.06x^2 + 600$ exists for every $x$, the critical numbers are the solutions of

$$-.06x^2 + 600 = 0.$$

We solve as follows:

$$-.06x^2 = -600$$
$$x^2 = 10{,}000$$
$$x = -100 \quad \text{or} \quad x = 100.$$

We disregard $x = -100$ because it is not in [60, 135]. According to the extreme-value theorem, we only need to evaluate $g$ at the endpoints and the critical number 100, as shown in the next table. The absolute maximum of 20,000 occurs at $x = 100$ (as we suspected from the graph), and the absolute minimum of 11,680 occurs at $x = 60$.

| $x$-Value | Value of Function | |
|---|---|---|
| 60 | 11,680 | ← Absolute minimum |
| 100 | 20,000 | ← Absolute maximum |
| 135 | 11,792.5 | |

✓ **Checkpoint 3**

Find the absolute extrema of $g(x) = x^3 - 15x^2 + 48x + 50$ on [1, 6].

## Absolute Extrema on Other Intervals

The extreme-value theorem is very useful on closed intervals, but may be *false* on an open interval. For example, the function

$$f(x) = \frac{x}{(x + 2)(x - 3)}$$

has vertical asymptotes at $x = -2$ and $x = 3$. On the open interval $(-2, 3)$ between the asymptotes, the function has neither an absolute maximum nor an absolute minimum, as shown in Figure 12.37. The graph rises higher and higher near $x = -2$ and falls lower and lower near $x = 3$.

It is sometimes possible to identify an absolute maximum or minimum on an interval that is not closed by using the following theorem, whose proof is omitted.

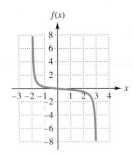

**Figure 12.37**

### Critical-Point Theorem

Suppose that a function $f$ is continuous on an interval $I$ and that $f$ has exactly one critical number in the interval $I$, say, $x = c$.*

If $f$ has a local maximum at $x = c$, then this local maximum is the absolute maximum of $f$ on the interval $I$.

If $f$ has a local minimum at $x = c$, then this local minimum is the absolute minimum of $f$ on the interval $I$.

**EXAMPLE 4**    Consider the function $f(x) = x^3 - 3x + 1$.

**(a)** Without graphing, show that $f$ has an absolute minimum on the interval (0, 2).

**Solution**  The derivative is $f'(x) = 3x^2 - 3$, which is defined everywhere, so the critical numbers are the solutions of $f'(x) = 0$:

$$3x^2 - 3 = 0$$
$$3x^2 = 3$$
$$x^2 = 1$$
$$x = -1 \quad \text{or} \quad x = 1.$$

The only critical number in the interval (0, 2) is $x = 1$. Use the second-derivative test to determine whether there is a local extremum at $x = 1$:

$$f'(x) = 3x^2 - 3$$
$$f''(x) = 6x$$
$$f''(1) = 6(1) = 6 > 0.$$

Hence, $f$ has a local minimum at $x = 1$. Therefore, by the critical-point theorem, the absolute minimum of $f$ on the interval (0, 2) occurs at $x = 1$.

_____

* $I$ may be open, closed, or neither.

**Figure 12.38**

 **(b)** Confirm the result of part (a) by graphing.

**Solution** The graph of $f$ in Figure 12.38 shows that $(1, -1)$ is the lowest point on the graph in the interval $(0, 2)$.

## Applications

When solving applied problems that involve maximum and minimum values, follow these guidelines:

### Solving Applied Problems

**Step 1**  Read the problem carefully. Make sure you understand what is given and what is asked for.

**Step 2**  If possible, sketch a diagram and label the various parts.

**Step 3**  Decide which variable is to be maximized or minimized. Express that variable as a function of *one* other variable. Be sure to determine the domain of this function.

**Step 4**  Find the critical numbers for the function in Step 3.

**Step 5**  If the domain is a closed interval, apply the extreme-value theorem: evaluate the function at the endpoints and at each critical number to see which yields the absolute maximum or minimum. If the domain is a (not necessarily closed) interval in which there is exactly one critical number, $x = c$, apply the critical-point theorem: if $f$ has a local extremum at $x = c$, then this local extremum is the absolute extremum.

**⊘ CAUTION** Do not skip Step 5 in the preceding box. If you are looking for a maximum and you find a single critical number in Step 4, do not automatically assume that the maximum occurs there. It may occur at an endpoint or may not exist at all.

**EXAMPLE 5**  **Average Cost** The annual cost (in thousands of dollars) of manufacturing $x$ thousand sets of wireless earbuds is given by

$$C(x) = .001x^3 + 3x + 100,$$

and no more than 60,000 earbuds can be produced in a year.

**(a)** Find the average cost function.

**Solution** As we saw in Section 11.6, the average cost function $\overline{C}$ is given by

$$\overline{C}(x) = \frac{C(x)}{x} = \frac{.001x^3 + 3x + 100}{x}$$

$$= .001x^2 + 3 + \frac{100}{x}.$$

**(b)** How many sets of earbuds should be made in order to minimize the average cost per set? What is the minimum average cost?

**Solution** The average cost function $\overline{C}(x) = .001x^2 + 3 + 100x^{-1}$ is defined for $0 < x \le 60$ (since the function is undefined at $x = 0$ and no more than 60,000 earbuds can be produced in a year). The derivative of the average cost function is

$$\overline{C}'(x) = .001(2x) + 100(-x^{-2})$$

$$= .002x - \frac{100}{x^2}$$

$$= \frac{.002x^3 - 100}{x^2}.$$

The critical numbers occur at the numbers in the domain of the average cost function where the value of the derivative $\overline{C}'(x)$ is zero or does not exist. Since $\overline{C}'(x)$ always exists for $0 < x \le 60$, the only critical numbers occur where $\overline{C}'(x)$ is zero:

$$\overline{C}'(x) = 0$$

$$\frac{.002x^3 - 100}{x^2} = 0 \quad\longleftarrow$$

A rational expression can be zero only when the numerator is zero.

$$.002x^3 - 100 = 0 \quad\longleftarrow$$

$$.002x^3 = 100$$

$$x^3 = \frac{100}{.002} = 50{,}000$$

$$x = \sqrt[3]{50{,}000} \approx 36.840. \quad\longleftarrow$$

The critical number is $x \approx 36.840$ or 36,840 earbuds.

Since $\overline{C}(x)$ is defined on $0 < x \le 60$, which is *not* a closed interval, we *cannot* use the extreme-value theorem. However, since there is exactly one critical number in the interval $0 < x \le 60$, we can apply the critical-point theorem. Use the second-derivative test to determine whether a local minimum occurs at this critical number:

$$\overline{C}'(x) = .002x - 100x^{-2}$$

$$\overline{C}''(x) = .002 - 100(-2x^{-3}) = .002 + \frac{200}{x^3}.$$

The value of the second derivative $\overline{C}''(x)$ is positive for all $x > 0$, so in particular, $\overline{C}''(36.840) > 0$, hence $\overline{C}$ has a local minimum at $x \approx 36.840$. By the critical-point theorem, an absolute minimum occurs at $x \approx 36.840$. Therefore, 36,840 sets of earbuds should be made in order to minimize the average cost per set. The minimum average cost is $\overline{C}(36.840) = 7.072$. In other words, the minimum average cost is $7072 per 1000 earbuds or $7.07 per set of earbuds. ✔ 4

✓ **Checkpoint 4**

For the cost function given by $C(x) = .02x^3 + 7x + 5000$, find the production level that leads to the minimum average cost.

**EXAMPLE 6** **Business** An open box is to be made by cutting a square from each corner of a 12-inch-by-12-inch piece of metal and then folding up the sides. The finished box must be at least 1.5 inches deep, but not deeper than 3 inches. What size square should be cut from each corner in order to produce a box of maximum volume?

**Solution** Let $x$ represent the length of a side of the square that is cut from each corner, as shown in Figure 12.39(a). The width of the box is $12 - 2x$, as is the length. As shown in Figure 12.39(b), the depth of the box is $x$ inches.

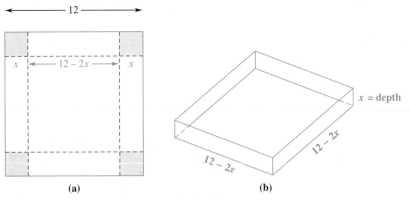

**(a)**          **(b)**

**Figure 12.39**

We must maximize the volume $V(x)$ of the box, which is given by

Volume = length · width · height

$$V(x) = (12 - 2x) \cdot (12 - 2x) \cdot x = (144 - 48x + 4x^2) \cdot x = 144x - 48x^2 + 4x^3.$$

Since the height $x$ must be between 1.5 and 3 inches, the domain of this volume function is the closed interval $[1.5, 3]$. To find the critical numbers, first take the derivative of $V$:

$$V'(x) = 12x^2 - 96x + 144.$$

The derivative always exists, so the only critical numbers occur when the derivative is equal to 0. Therefore, we set the derivative equal to 0 and solve:

$$12x^2 - 96x + 144 = 0$$
$$12(x^2 - 8x + 12) = 0$$
$$12(x - 2)(x - 6) = 0$$
$$x - 2 = 0 \quad \text{or} \quad x - 6 = 0$$
$$x = 2 \quad \text{or} \quad x = 6.$$

Since 6 is not in the domain, the only critical number of interest here is $x = 2$. By the extreme-value theorem, the maximum volume must occur at $x = 2$ or at the endpoints $x = 1.5$ or $x = 3$.

| $x$ | $V(x)$ | |
|-----|--------|---|
| 1.5 | 121.5 | |
| 2 | 128 | ←—Absolute maximum |
| 3 | 108 | |

The preceding table shows that the box has maximum volume when $x = 2$ inches and that this maximum volume is 128 cubic inches. ✔ 5

### ✓ Checkpoint 5

An open box is to be made by cutting squares from each corner of a 20-cm-by-32-cm piece of metal and folding up the sides. Let $x$ represent the length of the side of the square to be cut out. Find

(a) an expression for the volume $V(x)$ of the box;

(b) $V'(x)$;

(c) the value of $x$ that leads to maximum volume (*Hint*: The solutions of the equation $V'(x) = 0$ are 4 and $40/3$);

(d) the maximum volume.

**Figure 12.40**

**EXAMPLE 7** **Shipping** The U.S. Postal Service requires that boxes sent by Priority Mail have a length plus girth of no more than 108 inches, as shown in Figure 12.40. Find the dimensions of the box with largest volume that can be sent by Priority Mail, assuming that its width and height are equal.

**Solution** Let $x$ be the width and $y$ the length of the box. Since the height is the same as the width, the volume of the box is

$$V = \text{length} \cdot \text{width} \cdot \text{height} = y \cdot x \cdot x = yx^2.$$

Now express $V$ in terms of just *one* variable. Use the facts that the girth is $x + x + x + x = 4x$ and that the length plus girth is 108, so that

$$y + 4x = 108, \quad \text{or equivalently,} \quad y = 108 - 4x.$$

Substitute for $y$ in the expression for $V$ to get

$$V = yx^2 = (108 - 4x)x^2 = 108x^2 - 4x^3.$$

Since $x$ and $y$ are dimensions, we must have $x > 0$ and $y > 0$. Now, $y = 108 - 4x > 0$ implies that

$$4x < 108, \quad \text{or equivalently,} \quad x < 27.$$

Therefore, the domain of the volume function $V$ (the set of values of $x$ that make sense in the situation) is the open interval $(0, 27)$. Since its derivative always exists, we find the critical numbers for $V$ by setting its derivative equal to 0 and solving the equation:

$$V' = 216x - 12x^2 = 0$$
$$x(216 - 12x) = 0$$
$$x = 0 \quad \text{or} \quad 12x = 216$$
$$x = 18.$$

Because $x > 0$, the only critical number in the domain of $V$ is $x = 18$. Use the second-derivative test to check this critical number. Since $V'(x) = 216x - 12x^2$,

$$V''(x) = 216 - 24x.$$

Hence, $V''(18) = 216 - 24 \cdot 18 = -216$, and $V$ has a local maximum when $x = 18$. By the critical-point theorem, the absolute maximum of $V$ on $(0, 27)$ is at $x = 18$. In this case, $y = 108 - 4 \cdot 18 = 36$. Therefore, a box with dimensions 18 inches by 18 inches by 36 inches satisfies postal regulations and yields the maximum volume of $18^2 \cdot 36 = 11{,}664$ cubic inches.

---

**EXAMPLE 8**   **Landscaping**  A landscape gardener wants to build an 8000-square-foot rectangular display garden along the side of a river. The garden will be fenced on three sides. (No fence is necessary along the river.) She plans to use ornamental steel fencing at $12 per foot on the side opposite the river and chain-link fencing at $3 per foot on the other two sides. What dimensions for the garden will minimize her costs?

**Solution**  First draw a sketch of the situation and label the sides of the garden, as in Figure 12.41.

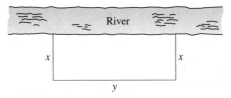

**Figure 12.41**

We see that $2x$ feet of fencing (at $3 per foot) are needed for the parallel sides and $y$ feet (at $12 per foot) for the side opposite the river. So the cost is

$$C = 2x(3) + y(12) = 6x + 12y.$$

We must find the values of $x$ and $y$ that make $C$ as small as possible. First, however, we must express $C$ in terms of a single variable. Since the garden is to have an area of 8000 square feet and the area of a rectangle is its length times its width, we have

$$\text{Area} = \text{length} \cdot \text{width} = xy = 8000$$

$$y = \frac{8000}{x}.$$

Substituting this expression for $y$ in the cost function, we obtain

$$C = 6x + 12y = 6x + 12\left(\frac{8000}{x}\right) = 6x + \frac{96{,}000}{x}.$$

Since $x$ is a length, it is positive. Hence, the domain of the cost function is the open interval $(0, \infty)$, and we need look only at the critical numbers. The derivative is

$$C'(x) = 6 - 96{,}000x^{-2} = 6 - \frac{96{,}000}{x^2}.$$

Since $C'(x)$ exists for all $x$ in the domain, the critical numbers are the solutions of

$$6 - \frac{96{,}000}{x^2} = 0$$

$$6x^2 - 96{,}000 = 0 \qquad \text{\textcolor{gray}{Multiply both sides by } } x^2.$$

$$6x^2 = 96{,}000$$

$$x^2 = 16{,}000$$

$$x = \pm\sqrt{16{,}000} \approx \pm 126.5.$$

The only critical number in the domain $(0, \infty)$ is $x = \sqrt{16{,}000}$.

Next we apply the second-derivative test. Since

$$C''(x) = -96{,}000(-2x^{-3}) = 192{,}000/x^3$$

is positive for all $x > 0$, a local minimum must occur at the critical number $x = \sqrt{16{,}000}$. Since this is the only critical number in the domain, the critical-point theorem ensures that an absolute minimum occurs there. Therefore, the garden with the least expensive fence has dimensions

$$x = \sqrt{16{,}000} \approx 126.5 \text{ ft and } y = \frac{8000}{x} = \frac{8000}{\sqrt{16{,}000}} \approx 63.2 \text{ ft.}$$

The preceding examples illustrate some of the factors that may affect applications in the real world. First, you must be able to find a function that models the situation. The rule of this function may be defined for values of $x$ that do not make sense in the context of the application, so the domain must be restricted to the relevant values of $x$.

The techniques of calculus apply to functions that are defined and continuous at every real number in some interval, so the maximum or minimum for the mathematical model (function) may not be feasible in the setting of the problem. For instance, if $C(x)$ has a minimum at $x = 80\sqrt{3}$ ($\approx 138.564$), where $C(x)$ is the cost of hiring $x$ employees, then the real-life minimum occurs at either 138 or 139, whichever one leads to lower cost.

## Economic Lot Size

Suppose that a company manufactures a constant number of units of a product per year and that the product can be manufactured in several batches of equal size during the year. On the one hand, if the company were to manufacture the item only once per year, it would minimize setup costs, but incur high warehouse costs. On the other hand, making many small batches would increase setup costs. Calculus can be used to find the number of batches per year that should be manufactured in order to minimize total cost. This number is called the **economic lot size** or **economic order quantity**.

Figure 12.42 shows several of the possibilities for a product having an annual demand of 12,000 units. The top graph shows the results if only one batch of the product is made annually; in this case, an average of 6000 items will be held in a warehouse. If four batches (of 3000 each)

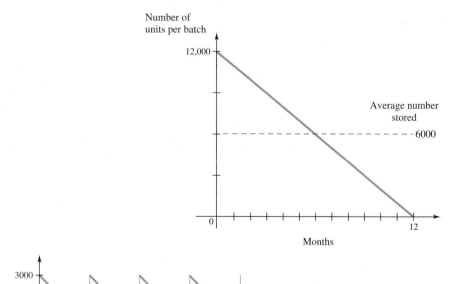

**Figure 12.42**

are made at equal time intervals during a year, the average number of units in the warehouse falls to only 1500. If 12 batches are made, an average of 500 items will be in the warehouse.

The following variables will be used in our discussion of economic lot size:

$x$ = number of batches to be manufactured annually;

$k$ = cost of storing one unit of the product for one year;

$a$ = fixed setup cost to manufacture the product;

$b$ = variable cost of manufacturing a single unit of the product;

$M$ = total number of units produced annually.

The company has two types of costs associated with the production of its product: a cost associated with manufacturing the item and a cost associated with storing the finished product.

During a year, the company will produce $x$ batches of the product, with $M/x$ units of the product produced per batch. Each batch has a fixed cost $a$ and a variable cost $b$ per unit, so the manufacturing cost per batch is

$$a + b\left(\frac{M}{x}\right).$$

There are $x$ batches per year, so the total annual manufacturing cost is

$$\left[a + b\left(\frac{M}{x}\right)\right]x. \tag{1}$$

Each batch consists of $M/x$ units, and demand is constant; therefore, it is common to assume an average inventory of

$$\frac{1}{2}\left(\frac{M}{x}\right) = \frac{M}{2x}$$

units per year. The cost to store one unit of the product for a year is $k$, so the total storage cost is

$$k\left(\frac{M}{2x}\right) = \frac{kM}{2x}. \tag{2}$$

The total production cost is the sum of the manufacturing and storage costs, or the sum of expressions (1) and (2). If $T(x)$ is the total cost of producing $x$ batches,

$$T(x) = \left[a + b\left(\frac{M}{x}\right)\right]x + \frac{kM}{2x} = ax + bM + \left(\frac{kM}{2}\right)x^{-1}.$$

Now find the value of $x$ that will minimize $T(x)$. (Remember that $a$, $b$, $k$, and $M$ are constants.) To do so, first find $T'(x)$:

$$T'(x) = a - \frac{kM}{2}x^{-2}.$$

Set this derivative equal to 0 and solve for $x$ (remember that $x > 0$):

$$a - \frac{kM}{2}x^{-2} = 0$$

$$a = \frac{kM}{2x^2}$$

$$2ax^2 = kM$$

$$x^2 = \frac{kM}{2a}$$

$$x = \sqrt{\frac{kM}{2a}}. \tag{3}$$

The second-derivative test can be used to show that $\sqrt{kM/(2a)}$ is the annual number of batches that gives the minimum total production cost.

**EXAMPLE 9** **Economic Lot Size** A garden supplier has steady annual demand for 50,000 cubic feet of potting soil. The cost accountant for the company says that it costs $2.50 to store 1 cubic foot of potting soil for 1 year and $850 to set up the production facility to mix the ingredients for the potting soil. Find the number of batches of potting soil that should be produced for the minimum total production cost.

**Solution** Use equation (3), given previously:

$$x = \sqrt{\frac{kM}{2a}}$$

$$x = \sqrt{\frac{(2.50)(50,000)}{2(850)}} \qquad \text{Let } k = 2.50, M = 50,000, \text{ and } a = 850.$$

$$x \approx 9.$$

Nine batches of potting soil per year will lead to minimum production costs. ✓ 6

✓ **Checkpoint 6**

In Example 9, suppose annual demand is 95,000 cubic feet, the annual cost of storage is $2.75 per cubic foot, and setup costs are $1100. Find the number of batches that should be made annually to minimize total costs.

# 12.3 Exercises

*Find the location of the absolute maximum and absolute minimum of the function on the given interval. (See Example 1.)*

**1.** $[0, 4]$

**2.** $[2, 5]$

**3.** $[-1, 2]$

**4.** $[-4, 2]$

**5.** $[-8, 0]$

**6.** $[-4, 4]$

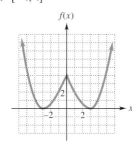

*Find the locations of the absolute extrema of each function on the given interval. (See Examples 2 and 3.)*

**7.** $f(x) = x^3 - x^2 - x + 10; [0, 2]$

**8.** $f(x) = 2x^3 - 3x^2 - 12x + 4; [-2, 0]$

**9.** $f(x) = x^4 - 18x^2 + 1; [-4, 4]$

**10.** $f(x) = x^4 - 32x^2 - 7; [-5, 6]$

**11.** $f(x) = \dfrac{8 + x}{8 - x}; [4, 6]$

**12.** $f(x) = \dfrac{1 - x}{3 + x}; [0, 3]$

**13.** $f(x) = \dfrac{x}{x^2 + 4}; [0, 4]$

**14.** $f(x) = \dfrac{-2x}{x^2 + 1}; [0, 3]$

**15.** $f(x) = (x^2 + 18)^{2/3}; [-3, 2]$

**16.** $f(x) = (x^2 + 4)^{1/3}; [-1, 2]$

**17.** $f(x) = \dfrac{1}{\sqrt{x^2 + 1}}; [-1, 1]$

**18.** $f(x) = \dfrac{3}{\sqrt{x^2 + 4}}; [-2, 2]$

*If possible, find the location of an absolute extremum of each function on the given interval. (See Example 4.)*

**19.** $g(x) = 2x^3 - 3x^2 - 12x + 1; (0, 4)$

**20.** $g(x) = x^3 - 2x^2 - 4x + 1; (0, \infty)$

**21.** $g(x) = 3x - 4x^{3/4}; (0, \infty)$

22. $g(x) = 6x^{2/3} - 4x; (0, \infty)$

23. $g(x) = 2xe^{-x/3}; (-\infty, \infty)$

24. $g(x) = 2xe^{3-x}; (-\infty, \infty)$

25. $g(x) = \dfrac{x}{e^x}; (0, 3)$

26. $g(x) = \dfrac{x^2 + 1}{x}; (0, \infty)$

*Work these problems. (See Example 5.)*

27. **Business** The daily cost of producing $x$ high-performance wheels for racing bicycles is given by

$$C(x) = .01x^3 - 1.5x^2 + 175x,$$

and no more than 100 wheels can be produced each day. What production level will give the lowest *average* cost per wheel? What is the minimum *average* cost?

28. **Business** The monthly cost of producing $x$ thousand ultimate flying discs is given by

$$C(x) = 9750e^{.04x},$$

and at most 30,000 discs can be produced in a month. What production level will give the lowest *average* cost per disc? What is the minimum *average* cost?

*Work these exercises.*

29. **Corporate Profits** Total corporate profits (before taxes) in billions of dollars can be approximated for the years 2006 through 2015 by

$$P(x) = 17x^2 - 306x + 2324,$$

where $x = 6$ corresponds to the year 2006. Find the absolute extrema of $P(x)$ on the closed interval $[6, 15]$. (Data from: Pro-Quest Statistical Abstract of the U.S.: 2016.)

30. **Corporate Taxes** For the years 2006 through 2015, corporate taxes in billions of dollars can be approximated by

$$T(x) = 5.8x^2 - 116x + 795,$$

where $x = 6$ corresponds to the year 2006. Find the absolute extrema of $T(x)$ on the closed interval $[6, 15]$. (Data from: Pro-Quest Statistical Abstract of the U.S.: 2016.)

31. **Business** A manufacturer produces gas grills that sell for $400 each. The total cost of producing $x$ grills is approximated by the function $C(x) = 525{,}000 - 30x + .012x^2$.

 (a) Write the revenue function in this situation.

 (b) Write the profit function in this situation.

 (c) What number of grills should be made to guarantee maximum profit? What will that profit be?

32. **Business** Saltwater taffy can be sold wholesale for $45 per thousand individual candies. The cost of producing $x$ thousand candies is $C(x) = .001x^3 + .045x^2 - 1.75x$.

 (a) What is the revenue function in this situation?

 (b) What is the profit function in this situation?

 (c) What number of candies will produce the largest possible profit?

33. **Entertainment Expenditures** Entertainment expenditures (on average, per consumer) can be approximated for the years 2003 through 2015 by

$$E(x) = 2x^3 - 63x^2 + 624x + 816,$$

where $x = 3$ corresponds to the year 2003. Find and interpret the absolute extrema of $E(x)$ for $3 \le x \le 15$. (Data from: ProQuest Statistical Abstract of the U.S.: 2016.)

34. **Consumer Spending** Average consumer expenditures on all items (in dollars) during the years 2004 through 2014 can be approximated by

$$S(x) = 38x^3 - 1064x^2 + 9728x + 20{,}680,$$

where $x = 4$ corresponds to the year 2004. Find and interpret the absolute extrema of $S(x)$ for $4 \le x \le 14$. (Data from: ProQuest Statistical Abstract of the U.S.: 2016.)

35. **Natural Science** A lake polluted by bacteria is treated with an antibacterial chemical. After $t$ days, the number $N$ of bacteria per ml of water is approximated by

$$N(t) = 20\left(\frac{t}{12} - \ln\left(\frac{t}{12}\right)\right) + 30 \qquad (1 \le t \le 15).$$

 (a) When during this period will the number of bacteria be a minimum?

 (b) What is this minimum number of bacteria?

 (c) When during the period will the number of bacteria be a maximum?

 (d) What is this maximum number of bacteria?

36. **Organ Transplants** The number of people waiting for organ transplants can be approximated for the years 2006 through 2016 by

$$f(t) = 35\ln(t) - \frac{t}{4} + 29 \qquad (6 \le t \le 16),$$

where $t = 6$ corresponds to 2006. (Data from: Organ Procurement and Transplant Network.)

 (a) During this period, when was the number of people waiting at a minimum?

 (b) What was the minimum number of people waiting? Round to the nearest thousand.

 (c) During this period, when was the number of people waiting at a maximum?

 (d) What was the maximum number of people waiting? Round to the nearest thousand.

37. **Health** A disease has hit College Station, Texas. The percentage of the population infected $t$ days after the disease has arrived is approximated by $p(t) = 10te^{-t/8}$ for $0 \le t \le 40$.

 (a) After how many days is the percentage of infected people a maximum?

 (b) What is this maximum percentage of the population infected?

38. **Unemployment for People Age 20 to 24** The unemployment rate (as a percentage) for individuals age 20 to 24 years spiked after the 2008 recession, then started to decline. For the years 2008 through 2016, it can be approximated by

$$f(t) = 13te^{-\frac{t}{2}} + 6.3,$$

where $t = 0$ corresponds to the year 2008. (Data from: Bureau of Labor Statistics.)

(a) How many years after 2008 was the unemployment rate for people age 20 to 24 at a maximum?

(b) What was the maximum unemployment rate for this age group? Round to one decimal place.

*Work these problems. (See Examples 6 and 7.)*

**39. Business** An open box is to be made by cutting a square from each corner of a 3-foot-by-8-foot piece of cardboard and then folding up the sides. What size square should be cut from each corner in order to produce a box of maximum volume?

**40. Business** An artist makes a closed box with a square base. The box is to have a volume of 16,000 cubic centimeters. The material for the top and bottom of the box costs 3 cents per square centimeter, while the material for the sides costs 1.5 cents per square centimeter. Find the dimensions of the box that will lead to the minimum total cost. What is the minimum total cost?

**41. Business** A watch manufacturing firm needs to design an open-topped box with a square base. The box must hold 32 cubic inches. Find the dimensions of the box that can be built with the minimum amount of materials.

**42. Business** A company wishes to manufacture a rectangular box with a volume of 36 cubic feet that is open on top and that is twice as long as it is wide. Find the dimensions of the box produced from the minimum amount of material.

**43. Business** A cylindrical box will be tied up with ribbon as shown in the accompanying figure. The longest piece of ribbon available is 130 cm long, and 10 cm of that is required for the bow. Find the radius and height of the box with the largest possible volume.

**44. Business** A cylindrical can of volume 58 cubic inches (approximately 1 quart) is to be designed. For convenient handling, it must be at least 1 inch high and 2 inches in diameter. What dimensions (radius of top, and height of can) will use the least amount of material?

*Work these problems. (See Example 8.)*

**45. Business** A farmer has 1200 m of fencing and wants to enclose a rectangular field bordering a river, with no fencing needed along the river. Let $x$ represent the width of the field.

(a) Write an expression for the length of the field.

(b) Find the area of the field.

(c) Find the value of $x$ leading to the maximum area.

(d) Find the maximum area.

**46. Business** A rectangular field is to be enclosed with a fence. One side of the field is against an existing fence, so that no fence is needed on that side. If material for the fence costs $2 per foot for the two ends and $4 per foot for the side parallel to the existing fence, find the dimensions of the field of largest area that can be enclosed for $1000.

**47. Business** A fence must be built to enclose a rectangular area of 20,000 ft². Fencing material costs $3 per foot for the two sides facing north and south and $6 per foot for the other two sides. Find the cost of the least expensive fence.

**48. Business** A rectangular field is to be enclosed on all four sides with a fence. Fencing material costs $3 per foot for two opposite sides and $6 per foot for the other two sides. Find the maximum area that can be enclosed for $2400.

**49. Business** A mathematics book is to contain 36 square inches of printed matter per page, with margins of 1 inch along the sides and $1\frac{1}{2}$ inches along the top and bottom. Find the dimensions of the page that will lead to the minimum amount of paper being used for a page.

**50. Business** A fence must be built in a large field to enclose a rectangular area of 15,625 m². One side of the area is bounded by an existing fence; no fence is needed there. Material for the fence costs $2 per meter for the two ends and $4 per meter for the side opposite the existing fence. Find the cost of the least expensive fence.

*Work these problems.*

**51. Business** If the price charged for a candy bar is $p(x)$ cents, where

$$p(x) = 100 - \frac{x}{10},$$

then $x$ thousand candy bars will be sold in a certain city.

(a) Find an expression for the total revenue from the sale of $x$ thousand candy bars. (*Hint*: Find the product of $p(x)$, $x$, and 1000.)

(b) Find the value of $x$ that leads to the maximum revenue.

(c) Find the maximum revenue.

**52. Business** A company makes beach balls that sell wholesale for \$75 per hundred. The total cost (in dollars) of making $x$ hundred beach balls is $C(x) = 5x^2 - 20x + 12$. Assume that the company can sell all the beach balls it makes. How many beach balls should be made to maximize profit?

**53. Business** We can use the function

$$f(x) = 55.44(54.3x - 143.9)e^{-.15x}$$

to model the revenue (in millions of U.S. dollars) from the cell phone industry in India, where $x = 2$ corresponds to the year 2002. According to this model, in what year does revenue reach its maximum? (Data from: www.researchinchina.com.)

**54. Business** A band travels between gigs in a large bus. While traveling $x$ miles per hour, this bus burns fuel at the rate of $G(x)$ gallons per mile, where

$$G(x) = \frac{1}{50}\left(\frac{200}{x} + \frac{x}{15}\right).$$

(a) If fuel costs \$3 per gallon, find the speed that will produce the minimum total cost for a 250-mile trip.

(b) Find the minimum total cost.

**55. Natural Science** Homing pigeons avoid flying over large bodies of water, preferring to fly around them instead. (One possible explanation is the fact that extra energy is required to fly over water because air pressure drops over water in the daytime.) Assume that a pigeon released from a boat 1 mile from the shore of a lake (point $B$ in the accompanying figure) flies first to point $P$ on the shore and then along the straight edge of the lake to reach its home at $L$. Assume that $L$ is 2 miles from point $A$, the point on the shore closest to the boat, and that a pigeon needs $\frac{4}{3}$ as much energy to fly over water as over land. Find the location of point $P$ if the pigeon uses the least possible amount of energy.

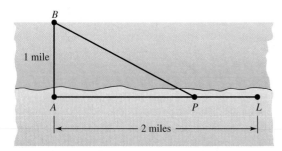

**56. Business** A company wishes to run a utility cable from point $A$ on the shore to an installation at point $B$ on the island. The island is 6 miles from the shore (at point $C$), and point $A$ is 9 miles from point $C$. It costs \$400 per mile to run the cable on land and \$500 per mile underwater. Assume that the cable starts at $A$, runs along the shoreline, and then angles and runs underwater to the island. Find the point at which the line should begin to angle in order to yield the minimum total cost. (*Hint*: The length of the line underwater is $\sqrt{x^2 + 36}$.) See the accompanying figure.

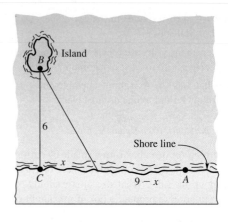

*The remaining exercises refer to economic lot size. (See Example 9.)*

**57. Business** A manufacturer of porcelain sinks has an annual demand of 12,300 sinks. It costs \$2.25 to store a sink for a year, and it costs \$350 to set up a factory to produce each batch. Find the number of batches of sinks that should be produced for the minimum total production cost.

**58. Business** How many sinks per batch will be produced in the previous exercise?

**59. Business** A regional market has a steady annual demand for 15,900 cases of beer. It costs \$3.25 to store a case for a year. The market pays \$5.50 for each order that is placed. Find the number of orders for cases of beer that should be placed each year. (*Hint*: Use the formula for economic lot size, with ordering cost in place of setup cost.)

**60. Business** Find the number of cases per order in the previous exercise.

**61. Business** A large hospital has an annual demand for 50,000 booklets on healthy eating. It costs \$.75 to store one booklet for a year, and it costs \$80 to place an order for a new batch of booklets. Find the optimum number of copies per order.

**62. Business** A restaurant has an annual demand for 900 bottles of a California wine. It costs \$1 to store one bottle for one year, and it costs \$5 to place a reorder. Find the number of orders that should be placed annually.

✓ **Checkpoint Answers**

1. Absolute maximum at $x = 0$; absolute minimum at $x = -2$

2. Absolute maximum of $-4$ at $x = 2$; absolute minimum of $-40$ at $x = -4$

3. Absolute maximum of 94 at $x = 2$; absolute minimum of 14 at $x = 6$

4. $x = 50$

5. (a) $V(x) = 640x - 104x^2 + 4x^3$
   (b) $V'(x) = 640 - 208x + 12x^2$
   (c) $x = 4$
   (d) $V(4) = 1152$ cubic centimeters

6. 11 batches

# 12.4   Implicit Differentiation

So far we have studied functions of the form $y = f(x)$ that are *explicitly* defined. This means that $y$ is isolated on one side of the equation. In other words, $y$ appears only once, and it appears by itself (not in combination with other functions). For example, the following equations define $y$ explicitly as a function of $x$:

*y appears only once by itself.*

$$y = 5x - 1;$$
$$y = 2x^2 - 3x + 100;$$
$$y = 7e^x.$$

In the above examples, $y$ appears alone on the left side of the equation. If we substitute a given value of $x$, each equation above yields the corresponding value of $y$ (the function value).

It is sometimes necessary or convenient to consider **implicitly defined functions**. For example, the following equations define $y$ implicitly as a function of $x$:

*y appears in combination with other functions.*

$$x + y = 7;$$
$$e^y = 2x + 5;$$
$$y = y^3 + x. \quad \longleftarrow \text{ y appears more than once.}$$

If we substitute a given value of $x$, we would need to do some algebra to solve each equation above in order to find the corresponding value of $y$.

---

**EXAMPLE 1**   The graph of the equation $x^2 + y^2 = 1$ for $y \geq 0$ is a semicircle (the top half of the unit circle). This equation defines $y$ implicitly as a function of $x$.

**(a)**  Rewrite the equation so that it defines $y$ explicitly as a function of $x$.

**Solution**  To write $y$ explicitly as a function of $x$, we must get $y$ by itself on one side of the equation:

$$x^2 + y^2 = 1$$
$$y^2 = 1 - x^2 \qquad \text{Subtract } x^2 \text{ from both sides.}$$
$$y = \pm\sqrt{1 - x^2} \qquad \text{Take the square root of both sides.}$$
$$y = \sqrt{1 - x^2} \qquad \text{We are given that } y \geq 0.$$

The last equation defines $y$ explicitly as a function of $x$.

**(b)**  Find the derivative of $y$ with respect to $x$.

**Solution**  To differentiate the last equation above, we first use rules of exponents to rewrite the square root as a power:

$$y = (1 - x^2)^{\frac{1}{2}}.$$

We are now able to find the derivative:

$$\frac{dy}{dx} = \frac{1}{2}(1 - x^2)^{-\frac{1}{2}} \frac{d}{dx}(1 - x^2) \qquad \text{Chain rule: } \frac{d}{dx}(u^{\frac{1}{2}}) = \frac{1}{2}u^{-\frac{1}{2}} \cdot \frac{du}{dx}$$

$$= \frac{1}{2}(1 - x^2)^{-\frac{1}{2}}(-2x)$$

$$= \frac{1}{2}\frac{1}{\sqrt{1 - x^2}}(-2x) \qquad \text{Rules of exponents}$$

$$= \frac{-x}{\sqrt{1 - x^2}}. \qquad \text{Simplify.}$$

Rather than solving the equation in the last example for $y$ explicitly in terms of $x$, it would be more convenient if we could differentiate the original function that was implicitly defined. We now introduce the method for doing so, called **implicit differentiation**.

## Implicit Differentiation

To find $\dfrac{dy}{dx}$ by implicit differentiation:

**Step 1**    Differentiate both sides of the equation with respect to $x$. Remember to use the chain rule since $y$ is a function of $x$.

**Step 2**    Solve the resulting equation for $\dfrac{dy}{dx}$.

**EXAMPLE 2**    Given that $x^2 + y^2 = 1$ and $y \geq 0$, find $\dfrac{dy}{dx}$ by implicit differentiation.

**Solution**  When using implicit differentiation, there is no need to solve the equation explicitly for $y$.

**Step 1**    The first step is to differentiate both sides of the equation with respect to $x$. Recall that $\dfrac{d}{dx}[f(x)]$ is another notation for the derivative of $f(x)$ with respect to $x$.

$$\frac{d}{dx}(x^2 + y^2) = \frac{d}{dx}(1)$$

$$\frac{d}{dx}(x^2) + \frac{d}{dx}(y^2) = \frac{d}{dx}(1)$$

$$2x + 2y\frac{dy}{dx} = 0. \qquad \text{Chain rule: } \frac{d}{dx}(u^2) = 2u\frac{du}{dx}$$

On the line above, it is easy to see that the derivative of $x^2$ *with respect to* $x$ is $2x$, that is, $\dfrac{d}{dx}(x^2) = 2x$. However, when differentiating $y^2$ *with respect to* $x$, we must remember that $y$ is an implicitly defined function of $x$, say $y = u(x)$. Thus $y^2$ is actually a composition of functions, $y^2 = [u(x)]^2$. Therefore, to find the derivative, we must apply the chain rule:

$$\frac{d}{dx}(y^2) = \frac{d}{dx}([u(x)]^2) = 2[u(x)]\frac{d}{dx}[u(x)] = 2y\frac{dy}{dx}.$$

**Step 2**    Since our goal is to find the derivative of $y$ with respect to $x$, we now solve for $\dfrac{dy}{dx}$:

$$2x + 2y\frac{dy}{dx} = 0$$

$$2y\frac{dy}{dx} = -2x \qquad \text{Move terms without } \frac{dy}{dx} \text{ to the right side.}$$

$$\frac{dy}{dx} = \frac{-2x}{2y} = \frac{-x}{y} \qquad \text{Divide both sides by } 2y \text{ and simplify.}$$

 **Checkpoint 1**

Find $\dfrac{dy}{dx}$ for $x^4 + y^5 = 6$.

In the line above, we can divide by $2y$ only if $y$ is nonzero; therefore, the domain of $\dfrac{dy}{dx}$ is restricted to the values of $x$ for which $y$ is nonzero. ✓ 1

Note that in Example 1 we found that $y = \sqrt{1 - x^2}$ and $\dfrac{dy}{dx} = \dfrac{-x}{\sqrt{1 - x^2}}$. In Example 2 we found that $\dfrac{dy}{dx} = -\dfrac{x}{y}$. These answers are equivalent since $\dfrac{dy}{dx} = -\dfrac{x}{y} = \dfrac{-x}{\sqrt{1 - x^2}}$. It is often much easier to find a derivative by implicit differentiation (as in Example 2) rather than solving an equation explicitly for the desired variable and then taking the derivative (as in Example 1).

In Examples 1 and 2, suppose we did not have the restriction that $y \geq 0$. Then the equation $x^2 + y^2 = 1$ defines $y$ as two functions of $x$, namely, $y = \sqrt{1 - x^2}$ and $y = -\sqrt{1 - x^2}$. For the first of these functions, we have already seen that $\dfrac{dy}{dx} = \dfrac{-x}{\sqrt{1 - x^2}} = \dfrac{-x}{y}$. For the second function, $y = -\sqrt{1 - x^2}$, so $\dfrac{dy}{dx} = -\left( \dfrac{-x}{\sqrt{1 - x^2}} \right) = \dfrac{-x}{-\sqrt{1 - x^2}} = \dfrac{-x}{y}$ (notice that the derivatives are the same). For the remainder of this section, we will assume that appropriate restrictions are implied so that the given equation defines $y$ implicitly as a function of $x$, and we can use implicit differentiation.

**EXAMPLE 3**   Find $\dfrac{dy}{dx}$ given that $e^{1-5x} + e^y = 7x - 8y$.

**Solution**

**Step 1**   Take the derivative of both sides of the equation with respect to $x$:

$$\frac{d}{dx}(e^{1-5x}) + \frac{d}{dx}(e^y) = \frac{d}{dx}(7x) - \frac{d}{dx}(8y)$$

$$e^{1-5x}(-5) + e^y\frac{dy}{dx} = 7 - 8\frac{dy}{dx} \qquad \text{Chain rule: } \frac{d}{dx}(e^u) = e^u\frac{du}{dx}$$

$$-5e^{1-5x} + e^y\frac{dy}{dx} = 7 - 8\frac{dy}{dx}. \qquad \text{Simplify.}$$

**Step 2**   Solve the last equation for $\dfrac{dy}{dx}$:

$$-5e^{1-5x} + e^y\frac{dy}{dx} = 7 - 8\frac{dy}{dx}$$

$$e^y\frac{dy}{dx} = 7 - 8\frac{dy}{dx} + 5e^{1-5x} \qquad \text{Move terms without } \frac{dy}{dx} \text{ to the right side.}$$

$$e^y\frac{dy}{dx} + 8\frac{dy}{dx} = 7 + 5e^{1-5x} \qquad \text{Move terms involving } \frac{dy}{dx} \text{ to the left side.}$$

$$\frac{dy}{dx}(e^y + 8) = 7 + 5e^{1-5x} \qquad \text{Factor out } \frac{dy}{dx}.$$

$$\frac{dy}{dx} = \frac{7 + 5e^{1-5x}}{e^y + 8}. \qquad \text{Divide both sides by } e^y + 8. \;\; \boxed{\checkmark}_2$$

✓ **Checkpoint 2**

Find $\dfrac{dy}{dx}$ for $e^{9x^2} - e^y = 6y^{10}$.

Note that when finding $\dfrac{dy}{dx}$ by implicit differentiation, the answer will typically involve both $x$ and $y$. This is to be expected; it is acceptable to leave the answer in this form.

**EXAMPLE 4**   Find $\dfrac{dy}{dx}$ for $10x^5y^8 = 3y + \pi$.

**Solution**

**Step 1**   First we take the derivative of both sides of the equation with respect to $x$. Because $y$ is implicitly a function of $x$, the left side of the equation is a product of two functions of $x$. Therefore we will need the product rule to differentiate the left side of the equation:

$$\frac{d}{dx}[10x^5y^8] = \frac{d}{dx}[3y + \pi]$$

$$[10x^5]\frac{d}{dx}[y^8] + [y^8]\frac{d}{dx}[10x^5] = \frac{d}{dx}[3y + \pi] \qquad \text{Product rule}$$

$$[10x^5]\left[8y^7\frac{dy}{dx}\right] + [y^8][50x^4] = 3\frac{dy}{dx} \qquad \text{Power and chain rules}$$

$$80x^5y^7\frac{dy}{dx} + 50x^4y^8 = 3\frac{dy}{dx}. \qquad \text{Simplify.}$$

**Step 2**    Solve the last equation for $\dfrac{dy}{dx}$:

$$80x^5y^7\frac{dy}{dx} + 50x^4y^8 = 3\frac{dy}{dx}$$

$$80x^5y^7\frac{dy}{dx} = 3\frac{dy}{dx} - 50x^4y^8 \qquad \text{Move terms without } \frac{dy}{dx} \text{ to the right side.}$$

$$80x^5y^7\frac{dy}{dx} - 3\frac{dy}{dx} = -50x^4y^8 \qquad \text{Move terms involving } \frac{dy}{dx} \text{ to the left side.}$$

$$\frac{dy}{dx}(80x^5y^7 - 3) = -50x^4y^8 \qquad \text{Factor out } \frac{dy}{dx}.$$

$$\frac{dy}{dx} = -\frac{50x^4y^8}{80x^5y^7 - 3}. \qquad \text{Divide both sides by } 80x^5y^7 - 3.$$  ✓3

✓ **Checkpoint 3**

Find $\dfrac{dy}{dx}$ for $3x^6y^7 = 5$.

---

**EXAMPLE 5**    Find $\dfrac{dy}{dx}$ at the point (0, 4) given that $\sqrt{y} - \ln(3x + 1) = 2$.

**Solution**

**Step 1**    We want to take the derivative of both sides of the equation with respect to $x$. First we must rewrite $\sqrt{y}$ as $y^{\frac{1}{2}}$, so the given equation becomes

$$y^{\frac{1}{2}} - \ln(3x + 1) = 2.$$

Now differentiating both sides with respect to $x$ yields

$$\frac{1}{2}y^{-\frac{1}{2}}\left(\frac{dy}{dx}\right) - \frac{1}{3x + 1}(3) = 0 \qquad \text{Chain rule}$$

$$\frac{1}{2\sqrt{y}}\frac{dy}{dx} - \frac{3}{3x + 1} = 0. \qquad \text{Rules of exponents}$$

**Step 2**    Since we want to find $\dfrac{dy}{dx}$ at the given point (0, 4), it is easier to substitute these values in the equation above prior to solving it. Substituting $x = 0$ and $y = 4$ in the equation above yields

$$\frac{1}{2\sqrt{4}}\frac{dy}{dx} - \frac{3}{3(0) + 1} = 0 \qquad \text{Substitute.}$$

$$\frac{1}{2 \cdot 2}\frac{dy}{dx} - \frac{3}{1} = 0 \qquad \text{Simplify.}$$

$$\frac{1}{4}\frac{dy}{dx} - 3 = 0. \qquad \text{Simplify.}$$

It is now easy to solve the last equation for $\dfrac{dy}{dx}$. Thus at the point $(0, 4)$ we have

$$\frac{1}{4}\frac{dy}{dx} = 3$$

$$\frac{dy}{dx} = 12. \qquad \text{Solve for } \tfrac{dy}{dx}. \quad ✓_4$$

✓ **Checkpoint 4**

Find $\dfrac{dy}{dx}$ at the point $(0, 1)$ for
$x^4 + 7x + \ln y = 0$.

**EXAMPLE 6**   Find an equation of the tangent line to the curve $xe^y = 1$ at the point $(1, 0)$.

**Solution**  To find the slope of the tangent line, we must find $\dfrac{dy}{dx}$. Using implicit differentiation, we first differentiate both sides of the equation with respect to $x$:

$$\frac{d}{dx}(xe^y) = \frac{d}{dx}(1)$$

$$x\frac{d}{dx}(e^y) + e^y\frac{d}{dx}(x) = 0 \qquad \text{Product rule}$$

$$x\left(e^y\frac{dy}{dx}\right) + e^y(1) = 0 \qquad \text{Chain rule}$$

$$xe^y\frac{dy}{dx} + e^y = 0. \qquad \text{Simplify.}$$

Since we want to find $\dfrac{dy}{dx}$ at the point $(1, 0)$, it is easier to substitute $x = 1$ and $y = 0$ into the equation above before solving the equation:

$$1 \cdot e^0 \cdot \frac{dy}{dx} + e^0 = 0 \qquad \text{Substitute.}$$

$$\frac{dy}{dx} + 1 = 0 \qquad \text{Simplify.}$$

$$\frac{dy}{dx} = -1. \qquad \text{Solve for } \tfrac{dy}{dx}.$$

Thus the slope of the tangent line at $(1, 0)$ is $m = -1$.

To find the equation of the tangent line at the point $(x_1, y_1) = (1, 0)$, we use point-slope form:

$$y - y_1 = m(x - x_1) \qquad \text{Point–slope form}$$
$$y - 0 = -1(x - 1) \qquad \text{Substitute.}$$
$$y = -x + 1. \qquad \text{Simplify.}$$

✓ **Checkpoint 5**

Find the equation of the tangent line to $x^2 + y^2 = 1$ at the point $(0, 1)$.

The graph of the curve $xe^y = 1$ and its tangent line $y = -x + 1$ at the point $(1, 0)$ are shown in Figure 12.43.  ✓_5

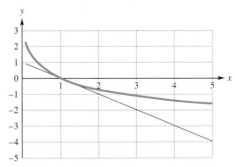

**Figure 12.43**

**EXAMPLE 7**    **Paraphernalia Demand**    The demand function for bumper stickers for an elementary school is given by

$$p = 9 - q^2,$$

where $p$ is the price (in dollars) and $q$ is the quantity demanded (in hundreds).

**(a)**  Find the derivative of $q$ with respect to $p$ by implicit differentiation.

**Solution**

**Step 1**    Differentiate both sides of the equation with respect to $p$:

$$\frac{d}{dp}(p) = \frac{d}{dp}(9) - \frac{d}{dp}(q^2)$$

$$1 = 0 - 2q\frac{dq}{dp} \qquad \text{Chain rule: } \frac{d}{dp}(q^2) = 2q\frac{dq}{dp}$$

$$1 = -2q\frac{dq}{dp}. \qquad \text{Simplify.}$$

**Step 2**    Solve for $\dfrac{dq}{dp}$:

$$1 = -2q\frac{dq}{dp}$$

$$-\frac{1}{2q} = \frac{dq}{dp} \qquad \text{Divide both sides by } -2q.$$

$$\frac{dq}{dp} = -\frac{1}{2q}. \qquad \text{Reverse the equation.}$$

**(b)**  Find the derivative of $q$ with respect to $p$ when the demand is 200 bumper stickers.

**Solution**  Substituting $q = 2$ (since $q$ is measured in hundreds) in $\dfrac{dq}{dp} = \dfrac{-1}{2q}$ shows that

$$\frac{dq}{dp} = -\frac{1}{2 \cdot 2} = -\frac{1}{4} = -.25$$

hundred bumper stickers per dollar.

This means that when the demand is 200 bumper stickers, increasing the price by \$1 will reduce the demand by approximately 25 bumper stickers (one quarter of a hundred). ✓ 6

✓ **Checkpoint 6**

Find $\dfrac{dq}{dp}$ by implicit differentiation for $p = 15 - 10q^3$.

**EXAMPLE 8**    **Pop-Up Ads**    A baseball team finds that the relationship between the amount $a$ (in thousands of dollars) spent on pop-up ads on a popular search engine and the amount of online sales $s$ (in thousands of dollars) of team paraphernalia can be approximated by

$$s^2 = 125a^3 + 100.$$

**(a)**  Find the sales when the team is spending \$1000 on advertising.

**Solution**  Given that \$1000 is being spent on advertising ($a = 1$), we need to find the corresponding value of $s$. Substituting $a = 1$ into $s^2 = 125a^3 + 100$ yields

$$s^2 = 125(1)^3 + 100$$

$$s^2 = 225$$

$$s = \pm\sqrt{225}$$

$$s = \sqrt{225} \qquad \text{Sales cannot be negative.}$$

$$s = 15.$$

This means that spending \$1000 on advertising generates \$15,000 in online sales.

**(b)** Find the rate of change of sales with respect to advertising when the team is spending $1000 on advertising.

**Solution**

**Step 1**  Differentiate both sides of the given equation with respect to $a$:

$$\frac{d}{da}(s^2) = \frac{d}{da}(125a^3 + 100)$$

$$2s\left(\frac{ds}{da}\right) = 375a^2. \qquad \text{Chain rule}$$

**Step 2**  Substituting $a = 1$ and $s = 15$ into the last equation and solving for $\dfrac{ds}{da}$ yields

$$2(15)\left(\frac{ds}{da}\right) = 375(1)^2 \qquad \text{Substitute.}$$

$$30\frac{ds}{da} = 375 \qquad \text{Simplify.}$$

$$\frac{ds}{da} = 12.5. \qquad \text{Solve for } \frac{ds}{da}.$$

✓ **Checkpoint 7**

Suppose the sales function in Example 8 was given by $s^4 = 16a^3$. Find the derivative of sales with respect to advertising when $a = 1$.

Therefore, when the team is spending $1000 on advertising, spending an additional $1000 on advertising would yield approximately an additional $12,500 in online sales. ✓7

# 12.4 Exercises

*Find* $\dfrac{dy}{dx}$ *by implicit differentiation. (See Examples 2–4.)*

**1.** $x + y = 100$

**2.** $-15x + y = 77$

**3.** $x^4 + y^4 = 13$

**4.** $x^3 + y^5 = 97$

**5.** $10x^3 - 2y^5 = -6$

**6.** $5x^4 - 3y^2 = 1$

**7.** $x^2y^5 = 19$

**8.** $x^3y^4 = 33$

**9.** $5\sqrt{x} = 2\sqrt{y} + 1$

**10.** $3\sqrt{y} = \sqrt{x} - 49$

**11.** $5x^2 = 1 - e^y$

**12.** $5y^2 = 1 - e^x$

**13.** $2xe^y = 7$

**14.** $x^3e^y = 13$

**15.** $3 \ln x = 4 - \ln y$

**16.** $2 \ln y = 7x^6 + 5$

**17.** $8y \ln x = 15$

**18.** $7y^2 \ln x = 2$

*Find* $\dfrac{dy}{dx}$ *at the given point. (See Example 5.)*

**19.** $2x^2 - 3y^3 = 5x - 3y$ at $(0, 1)$

**20.** $3x^3 - 5y^2 = 10x - 5y$ at $(0, 1)$

**21.** $\sqrt{x^3} = \sqrt[3]{y^2}$ at $(1, 1)$

**22.** $\sqrt{2x - 1} = \sqrt[3]{2y - 1}$ at $(1, 1)$

**23.** $7x^2y^5 = 1 - x^2 - y$ at $(0, 1)$

**24.** $5x^6y^2 = 6 - 2x - 6y^2$ at $(0, 1)$

**25.** $2xe^{5y} = 2$ at $(1, 0)$

**26.** $x^2e^{1-y} = 1$ at $(1, 1)$

**27.** $x \ln y = 1$ at $(1, e)$

**28.** $y \ln x = 1$ at $(e, 1)$

*Find the equation of the tangent line to the curve at the given point. (See Example 6.)*

**29.** $x^2 + y^2 = 25$ at $(0, 5)$

**30.** $x^2 + y^2 = 9$ at $(0, -3)$

**31.** $y^3 = 8x^4$ at $(1, 2)$

**32.** $y^5 = 32x^4$ at $(1, 2)$

**33.** $5x - xy + y^2 = 5$ at $(1, 1)$

**34.** $x^3 + 3xy^2 + 3x^2y + y^3 = 8$ at $(1, 1)$

*Work these exercises. (See Example 1.)*

**35.** Consider $xe^y = 1$.

(a) Find $\dfrac{dy}{dx}$ by first solving the given equation explicitly for $y$ in terms of $x$ and then differentiating $y$.

(b) Find $\dfrac{dy}{dx}$ by implicit differentiation.

(c) Compare your answers from parts (a) and (b).

**36.** Consider $e^y = x$.

(a) Find $\dfrac{dy}{dx}$ by first solving the given equation explicitly for $y$ in terms of $x$ and then differentiating $y$.

**(b)** Find $\dfrac{dy}{dx}$ by implicit differentiation.

**(c)** Compare your answers from parts (a) and (b).

*Work these exercises. (See Example 7.)*

37. **Golfing** The demand for tee times at a public golf course can be approximated by

$$p = 200 - \frac{q}{2},$$

where $p$ is the price of the greens fee (in dollars) and $q$ is the quantity demanded. Use implicit differentiation to find $\dfrac{dq}{dp}$ when $q = 200$ and explain what your answer means in practical terms.

38. **Rock Climbing** The demand for entry to a popular rock climbing gym can be approximated by

$$p = 25 - \frac{q}{10},$$

where $p$ is the admission price (in dollars) and $q$ is the quantity demanded. Use implicit differentiation to find $\dfrac{dq}{dp}$ when $q = 150$ and explain what your answer means in practical terms.

39. **Hotel Bookings** A casino hotel manager has approximated demand for two-room suites to be

$$p = 5000 - \frac{q^2}{20},$$

where $p$ is the price (in dollars) and $q$ is the quantity demanded. Use implicit differentiation to find and interpret $\dfrac{dq}{dp}$ when $q = 5$.

40. **Food Truck Sales** A food truck owner has approximated demand for burritos by

$$10p + 3q^2 = 75,$$

where $p$ is the price (in dollars) and $q$ is the quantity demanded (in hundreds). Use implicit differentiation to find and interpret $\dfrac{dq}{dp}$ when $q = 3$.

41. **Drink Specials** A night club has approximated the demand for their Thursday night drink special by

$$p^2 + 100q^2 = 100,$$

where $p$ is the price (in dollars) and $q$ is the quantity demanded (in thousands). Find and interpret $\dfrac{dq}{dp}$ when the price is $6 and the demand is 800 drinks.

42. **Downloading Tunes** The demand to download a hit single by a certain band can be approximated by

$$100p^2 + 9q^2 = 900,$$

where $p$ is the price (in dollars) and $q$ is the quantity demanded (in thousands). Find and interpret $\dfrac{dq}{dp}$ when the price is $1.80 and the demand is 8000 downloads.

*Work these exercises.*

43. **Bank of America** For Bank of America Corp., the relationship between revenue and the number of full-time employees can be approximated by

$$2000R^2 = n^3,$$

where $R$ is revenue in billions of dollars and $n$ is the number of employees in thousands. (Data from: www.mergentonline.com.)

**(a)** Would you expect $\dfrac{dn}{dR}$ to be positive, negative, or zero? Why?

**(b)** Find $\dfrac{dn}{dR}$. What are the units?

**(c)** Find $\dfrac{dn}{dR}$ when $R = 128$ and $n = 320$, and interpret your answer.

44. For the equation given in the previous exercise, find $\dfrac{dR}{dn}$ when $R = 128$ and $n = 320$, and interpret your answer.

45. **Walt Disney Company** The revenue generated by the Walt Disney Company is approximately related to size of its workforce according to the equation

$$3n^3 - 7000R^2 = 0,$$

where $R$ is revenue in billions and $n$ is the number of employees in thousands. (Data from: www.mergentonline.com.)

**(a)** Find $\dfrac{dR}{dn}$, including units.

**(b)** Find $\dfrac{dR}{dn}$ when $R = 63$ and $n = 210$, and interpret your answer.

46. For the equation given in the previous exercise, find $\dfrac{dn}{dR}$ when $R = 63$ and $n = 210$, and interpret your answer.

47. **Drug Absorption** The amount $A$ of a certain drug (in mg) that remains in the blood stream $t$ hours after taking a 100 mg tablet can be approximated by

$$Ae^{2t} = 99.$$

Use implicit differentiation to find and interpret $\dfrac{dA}{dt}$ when $t = 12$ and $A = 9$.

48. **Printing Services** At a certain online printing service, the cost $y$ (in dollars) of purchasing $x$ thousand business cards can be approximated by

$$xy + y = 90x.$$

Use implicit differentiation to find and interpret $\dfrac{dy}{dx}$ when $x = 1$ and $y = 45$.

## ✓Checkpoint Answers

1. $\dfrac{dy}{dx} = -\dfrac{4x^3}{5y^4}$    2. $\dfrac{dy}{dx} = \dfrac{18xe^{9x^2}}{60y^9 + e^y}$

3. $\dfrac{dy}{dx} = -\dfrac{6y}{7x}$    4. $\dfrac{dy}{dx} = -7$

5. $y = 1$    6. $\dfrac{dq}{dp} = -\dfrac{1}{30q^2}$

7. $\dfrac{ds}{da} = 1.5$

# 12.5    Related Rates

In many applications, two quantities that are related to each other are also changing over time. For example, the revenue $R$ generated by a business is related to the number of employees $n$ the business has. Both revenue and number of employees also change over time. Therefore it makes sense to consider their derivatives with respect to time, $\dfrac{dR}{dt}$ and $\dfrac{dn}{dt}$. Since $R$ and $n$ are related, their rates of change $\dfrac{dR}{dt}$ and $\dfrac{dn}{dt}$ are also related. These are called **related rates.**

In general, whenever two quantities $A$ and $B$ vary over time and are related to each other, their derivatives $\dfrac{dA}{dt}$ and $\dfrac{dB}{dt}$ are called related rates. If we have information about $A$ and $B$ and one of these rates of change, we can determine how the other quantity is changing. The following box summarizes the procedure for solving related rates problems.

## Related Rates Applications

**Step 1**    Determine an equation that relates the two variables of interest. This equation may be given or may need to be developed from information in the problem. If the equation includes an extraneous variable (other than the two variables of interest), eliminate it using given information that holds at all times.

**Step 2**    Differentiate both sides of the equation found in Step 1 with respect to time. Remember to use the chain rule since each variable is a function of time.

**Step 3**    Substitute given values that hold at the particular time of interest and solve for the desired rate of change.

---

**EXAMPLE 1**    **Revenue and Employees**  For the Walt Disney Company, the relationship between revenue $R$ (in billions) and number of employees $n$ (in thousands) can be approximated by

$$7000R^2 = 3n^3.$$

In 2015, Disney generated \$52 billion in revenue (rounded to the nearest billion), had 185,000 employees, and was adding employees at the rate of 5000 employees per year. Find the rate of change of revenue with respect to time in 2015. (Data from: www.mergentonline.com.)

**Solution**  Since revenue $R$ and number of employees $n$ are related, the derivatives $\dfrac{dR}{dt}$ and $\dfrac{dn}{dt}$ are related rates.

**Step 1**    We are given the relationship between $R$ and $n$:

$$7000R^2 = 3n^3.$$

**Step 2**    Our goal is to find $\dfrac{dR}{dt}$, the rate of change of revenue with respect to time. Differentiate both sides of this equation with respect to time. Remember to use the chain rule because both $R$ and $n$ are functions of time:

$$\frac{d}{dt}(7000R^2) = \frac{d}{dt}(3n^3) \qquad \text{Differentiate both sides with respect to } t.$$

$$7000(2R)\frac{dR}{dt} = 3(3n^2)\frac{dn}{dt} \qquad \text{Chain rule}$$

$$14{,}000R\frac{dR}{dt} = 9n^2\frac{dn}{dt} \qquad \text{Simplify.}$$

**Step 3**    Next we substitute the given values for the year 2015. The revenue is $52 billion and the number of employees is 185,000, so we know that $R = 52$ (in billions) and $n = 185$ (in thousands). Since the company is *adding* 5000 employees per year, the *rate of change* of the number of employees with respect to time is $\dfrac{dn}{dt} = 5$ (in thousands per year). Substituting these values into the equation above and solving for $\dfrac{dR}{dt}$ yields

$$14{,}000(52)\frac{dR}{dt} = 9(185)^2(5) \qquad \text{Substitute.}$$

$$\frac{dR}{dt} = \frac{9(185)^2(5)}{14{,}000(52)} \approx 2.116 \qquad \text{Solve for } \frac{dR}{dt}.$$

In 2015, the revenue for the Walt Disney Company was increasing at a rate of over $2 billion per year. ☑ 1

✓ **Checkpoint 1**

Given that $R^3 = 25n^4$, find $\dfrac{dR}{dt}$ when $n = 5$, $R = 25$, and $\dfrac{dn}{dt} = 2$.

**❗ CAUTION**    In Example 1, it was important that we first differentiated both sides of the equation and then substituted in the given values for the specified time (the year 2015). It would not be accurate to first substitute the values for 2015 since these values do not hold at all times. It is easy to catch this common mistake since, if we substituted the given values first, both sides of the equation would involve only constants (no variables), so the derivative of each side would be zero.

**EXAMPLE 2**    Suppose $x$ and $y$ both vary over time and are related by the equation $3x^3 + 12y = 10xy$. Find $\dfrac{dy}{dt}$ given that $x = 2$, $y = 3$, and $\dfrac{dx}{dt} = -4$.

**Solution**

**Step 1**    We are given the equation relating $x$ and $y$:

$$3x^3 + 12y = 10xy.$$

**Step 2**    Since both $x$ and $y$ vary over time, we differentiate both sides of the given equation with respect to $t$. Remember to use the product rule on the right side of the equation and also to use the chain rule since both $x$ and $y$ are functions of time. This yields

$$\overbrace{\phantom{9x^2\frac{dx}{dt} + 12\frac{dy}{dt} = (10x)\left(\frac{dy}{dt}\right) + (y)\left(10\frac{dx}{dt}\right)}}^{\text{Product Rule}}$$

$$9x^2\frac{dx}{dt} + 12\frac{dy}{dt} = (10x)\left(\frac{dy}{dt}\right) + (y)\left(10\frac{dx}{dt}\right).$$

  Chain rule

**Step 3**    Substitute the given values of $x = 2$, $y = 3$, and $\dfrac{dx}{dt} = -4$ into the equation above and then solve for the desired derivative $\dfrac{dy}{dt}$:

$$9(2)^2(-4) + 12\frac{dy}{dt} = (10(2))\left(\frac{dy}{dt}\right) + (3)(10(-4)) \qquad \text{Substitute.}$$

$$-144 + 12\frac{dy}{dt} = 20\frac{dy}{dt} - 120 \qquad \text{Simplify.}$$

$$12\frac{dy}{dt} - 20\frac{dy}{dt} = -120 + 144 \qquad \text{Isolate terms with } \frac{dy}{dt} \text{ on the left.}$$

$$-8\frac{dy}{dt} = 24 \qquad \text{Simplify.}$$

$$\frac{dy}{dt} = -\frac{24}{8} = -3. \qquad \text{Solve for } \frac{dy}{dt}. \; ☑ 2$$

✓ **Checkpoint 2**

Given the equation in Example 2 and that $x = 2$, $y = 3$, and $\dfrac{dy}{dt} = -3$, find $\dfrac{dx}{dt}$.

**EXAMPLE 3**   **Tumor Growth Rate** A cancerous tumor in the shape of a sphere is undergoing radiation treatment. The radius of the tumor is decreasing at the rate of 2 millimeters per week. How fast is the volume of the tumor changing when the radius of the tumor is 10 millimeters? The volume (in cubic millimeters) of a sphere of radius $r$ (in millimeters) is given by $V = \frac{4}{3}\pi r^3$.

**Solution**

**Step 1**   The relationship between the volume and radius of the tumor was given:

$$V = \frac{4}{3}\pi r^3.$$

**Step 2**   To find $\frac{dV}{dt}$, differentiate both sides of the given equation with respect to time (remember to use the chain rule since $r$ is a function of time):

$$\frac{dV}{dt} = \frac{4}{3}\pi(3r^2)\frac{dr}{dt} \qquad \text{Chain rule}$$

$$\frac{dV}{dt} = 4\pi r^2\frac{dr}{dt}. \qquad \text{Simplify.}$$

**Step 3**   Substitute the given values of $r = 10$ and $\frac{dr}{dt} = -2$ (since the radius is decreasing by 2 mm per week) into the equation above to find that

$$\frac{dV}{dt} = 4\pi(10)^2(-2) = -800\pi \approx -2513. \qquad \text{Substitute and simplify.}$$

The volume of the tumor is decreasing at a rate of about 2513 cubic millimeters per week.

✓ **Checkpoint 3**

The volume (in cubic centimeters) of a cube of side length $x$ centimeters is given by $V = x^3$. Find $\frac{dV}{dt}$ given that $x = 5$ cm and $\frac{dx}{dt} = 4$ cm per minute.

**EXAMPLE 4**   **Population Growth** A metropolitan area is approximately in the shape of a circle and the population density is roughly constant throughout. Because the metropolitan area is already densely populated, it is growing outwards in all directions to accommodate population growth of 13,000 people per year. When the radius of the metro-politan area is $r$ miles, its population is approximated by $P = 5000\pi r^2$. Find the rate of change of the radius of the metropolitan area with respect to time when the radius is 4 miles.

**Solution**

**Step 1**   We are given the relationship $P = 5000\pi r^2$ between the population and the ra-dius of the metropolitan area.

**Step 2**   To find $\frac{dr}{dt}$, we differentiate both sides of the given equation with respect to time:

$$\frac{dP}{dt} = 5000\pi(2r)\frac{dr}{dt} \qquad \text{Chain rule}$$

$$\frac{dP}{dt} = 10,000\,\pi r\frac{dr}{dt}. \qquad \text{Simplify.}$$

**Step 3**   Substitute the given values of $\frac{dP}{dt} = 13,000$ (since the population is increasing by 13,000 people per year) and $r = 4$ into the equation above and solve for $\frac{dr}{dt}$:

$$13,000 = 10,000\pi(4)\frac{dr}{dt} \qquad \text{Substitute.}$$

$$\frac{dr}{dt} = \frac{13,000}{10,000\pi(4)} \approx .10 \qquad \text{Solve for } \frac{dr}{dt}.$$

✓ **Checkpoint 4**

The area of a circle is $A = \pi r^2$ square inches where $r$ is the radius of the circle in inches. Find $\frac{dr}{dt}$ given that $r = 3$ inches and $\frac{dA}{dt} = 10$ square inches per minute.

The radius of the metropolitan area is growing at the rate of about .10 mile per year.

In Examples 1–4, we were given the relationship between two quantities of interest (such as revenue and number of employees). In some application problems, it is necessary to determine the relationship between two quantities using information given in the problem. The next example illustrates this.

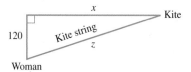    **EXAMPLE 5**    **Physical Science**  A woman is flying a kite at a constant altitude of 120 feet. How fast is the kite traveling at the instant when 130 feet of string have already been let out and the woman is letting out the string at a rate of 10 feet per second?

**Solution**

**Step 1**    We need to determine a relationship between the variables of interest. Drawing a diagram such as the one in Figure 12.44 is often helpful.

**Figure 12.44**

Using the Pythagorean Theorem, we see that $x^2 + 120^2 = z^2$.

**Step 2**    We want to find $\dfrac{dx}{dt}$, the speed of the kite (which is flying horizontally at a constant altitude). Differentiating both sides of the equation in Step 1 with respect to $t$ yields

$$2x\frac{dx}{dt} + 0 = 2z\frac{dz}{dt} \qquad \text{Chain rule}$$

$$x\frac{dx}{dt} = z\frac{dz}{dt}. \qquad \text{Simplify.}$$

**Step 3**    We were given that the speed at which the string is being let out is $\dfrac{dz}{dt} = 10$ feet per second. We were also given that the length of string that has been let out is $z = 130$ feet, but we need to find the corresponding value of $x$. Substituting $z = 130$ into the equation $x^2 + 120^2 = z^2$ yields

$$x^2 + 120^2 = 130^2$$
$$x^2 + 14{,}400 = 16{,}900$$
$$x^2 = 16{,}900 - 14{,}400$$
$$x^2 = 2500$$
$$x = \pm\sqrt{2500}$$
$$x = 50. \qquad \text{Since } x \text{ cannot be negative.}$$

We now substitute $x = 50$, $z = 130$, and $\dfrac{dz}{dt} = 10$ into $x\dfrac{dx}{dt} = z\dfrac{dz}{dt}$, which yields

$$50\frac{dx}{dt} = 130(10) \qquad \text{Substitute.}$$

$$\frac{dx}{dt} = \frac{130(10)}{50} = 26. \qquad \text{Solve for } \frac{dx}{dt}.$$

Therefore, the kite is flying at 26 feet per second.  ✓5

✓ **Checkpoint 5**

Suppose that in Example 5 the kite was flying at a constant altitude of 40 feet and the string was being let out at a rate of 6 feet per second. Find the speed of the kite when 50 feet of string have been let out.

**EXAMPLE 6**   **Physical Science**   In a certain automatic coffee maker, hot water passes through coffee grounds in a conical filter and then pours out the bottom of the cone into a coffee pot. However, the cone has become blocked so that no coffee is flowing into the coffee pot and water is rising in the filter. The shape of the filter is such that the radius of the surface of the water in the cone is always half of the height of the water in the cone. If water is entering the cone at a rate of 6 cubic centimeters per second, how fast is the height of the water rising when the height of the water is 2 cm?

**Solution**

**Step 1**   The volume $V$ of water in the cone (in cubic centimeters) when the radius of the surface of the water is $r$ centimeters and the height of the water is $h$ centimeters is $V = \frac{1}{3}\pi r^2 h$ (see Figure 12.45). However, we need a relationship between $V$ and $h$ that does not involve $r$. We say that $r$ is an extraneous variable and we need to eliminate $r$ from the equation using information that holds at all times. Since we are given that the radius of the surface of the water in the cone is always half of the height of the water, we may substitute $r = \frac{1}{2}h = \frac{h}{2}$ into the equation above:

$$V = \frac{1}{3}\pi\left(\frac{h}{2}\right)^2 h = \frac{1}{3}\pi\left(\frac{h^2}{4}\right)h = \frac{1}{12}\pi h^3.$$

**Figure 12.45**

**Step 2**   Now that we have a relation between $V$ and $h$, we differentiate both sides of the equation with respect to time:

$$\frac{dV}{dt} = \frac{1}{12}\pi(3h^2)\frac{dh}{dt} \qquad \text{Chain rule}$$

$$\frac{dV}{dt} = \frac{1}{4}\pi h^2\frac{dh}{dt}. \qquad \text{Simplify.}$$

**Step 3**   Finally, we substitute the given values of $\frac{dV}{dt} = 6$ (since the volume of water entering the cone is 6 cm³ per second) and $h = 2$ into the equation above and solve for $\frac{dh}{dt}$:

$$6 = \frac{1}{4}\pi(2)^2\frac{dh}{dt} \qquad \text{Substitute.}$$

$$\frac{dh}{dt} = \frac{6}{\pi} \approx 1.9. \qquad \text{Solve for } \frac{dh}{dt}.$$

When the height of water in the cone is 2 cm, water is rising in the cone at a rate of about 1.9 cm per second.  ✓ 6

✓ **Checkpoint 6**

Water is draining from a cone whose height is three times the radius. Find $\frac{dV}{dr}$ at the instant when $r = 5$ cm and $\frac{dr}{dt} = -2$ cm per minute.

**EXAMPLE 7**   **Demand and Profit**   At a local clothing store, the daily demand for a certain brand of blue jeans is given by $p = 200 - 2x$, where $p$ is the price in dollars and $x$ is the quantity demanded. The cost of purchasing these jeans at wholesale is $25 per pair (ignore any fixed costs). If daily demand for this brand of jeans is currently 35 pairs of jeans and declining by 2 pairs per day, how is the profit changing?

**Solution**

**Step 1**   We must determine an equation that relates profit and the quantity demanded. Since profit is the difference between revenue and cost, we must find the revenue and cost functions. The revenue generated by selling $x$ pairs of jeans is the product of the price per item and the number of items demanded:

$$R = xp = x(200 - 2x) = 200x - 2x^2.$$

The cost of purchasing $x$ pairs of jeans at the wholesale price of \$25 per pair is

$$C = 25x.$$

Therefore, the profit when $x$ pairs of jeans are demanded is

$$P = R - C = (200x - 2x^2) - (25x) = 175x - 2x^2.$$

**Step 2**    Differentiate both sides of the equation in Step 1 with respect to time.

$$\frac{dP}{dt} = 175\frac{dx}{dt} - 4x\frac{dx}{dt}. \qquad \text{Chain rule}$$

**Step 3**    Substitute the given values of $x = 35$ and $\frac{dx}{dt} = -2$ (since demand is decreasing by 2 pairs of jeans per day):

$$\frac{dP}{dt} = 175(-2) - 4(35)(-2) = -70. \qquad \text{Substitute and simplify.}$$

This means that when the daily demand is 35 pairs of jeans and dropping at 2 pairs per day, the profit is decreasing by \$70 per day. ✔️ 7

✓ **Checkpoint 7**

Suppose that in Example 7
$p = 50 - x$ and $C = 10x$. Find
$\frac{dP}{dt}$ when $x = 15$ and $\frac{dx}{dt} = -1$.

# 12.5 Exercises

*Given that x and y are functions of time, find the indicated rate of change. (See Example 2.)*

1.  Find $\frac{dy}{dt}$ when $x = 1, y = 2,$ and $\frac{dx}{dt} = 3,$ given that $y^3 = 2x^4 + 6.$

2.  Find $\frac{dx}{dt}$ when $x = 1, y = 2,$ and $\frac{dy}{dt} = 2,$ given that $y^3 = 2x^4 + 6.$

3.  Find $\frac{dx}{dt}$ when $x = 0, y = 1,$ and $\frac{dy}{dt} = 5,$ given that $2x^5 + x = y^2 - 1.$

4.  Find $\frac{dy}{dt}$ when $x = 0, y = 1,$ and $\frac{dx}{dt} = 10,$ given that $2x^5 + x = y^2 - 1.$

5.  Find $\frac{dx}{dt}$ when $x = 2, y = 2,$ and $\frac{dy}{dt} = -1,$ given that $xy = x + y.$

6.  Find $\frac{dy}{dt}$ when $x = 2, y = 2,$ and $\frac{dx}{dt} = 1,$ given that $xy = x + y.$

7.  Find $\frac{dy}{dt}$ when $x = 1, y = 2,$ and $\frac{dx}{dt} = -4,$ given that $x^3 + y^3 = 9.$

8.  Find $\frac{dx}{dt}$ when $x = 1, y = 2,$ and $\frac{dy}{dt} = 1,$ given that $x^3 + y^3 = 9.$

9.  Find $\frac{dx}{dt}$ when $x = 0, y = 1,$ and $\frac{dy}{dt} = 1,$ given that $x^3 = 9y^5 - 1.$

10. Find $\frac{dy}{dt}$ when $x = 0, y = 1,$ and $\frac{dx}{dt} = \frac{15}{4},$ given that $x^3 = 9y^5 - 1.$

*Work these exercises. (See Examples 1, 3, and 4.)*

11. **Tesla Motors** The relationship between revenue generated and the number of full-time employees for Tesla Motors Inc. can be approximated by

$$10n + 31R = 3500,$$

where $R$ is revenue in millions of dollars and $n$ is the number of employees. In 2015, the rate of change of revenue with respect to time was \$848 million per year; find the rate of change of the number of employees with respect to time. (Data from: www.mergentonline.com.)

12. Using the equation in the previous exercise, find the rate of change of revenue with respect to time. Assume that the company was adding employees at a rate of 2790 employees per year.

13. **Bank of America** For Bank of America Corp., the relationship between revenue generated and the number of full-time employees can be approximated by

$$2000R^2 = n^3,$$

where $R$ is revenue in billions of dollars and $n$ is the number of employees in thousands. At one time, the company had 180,000 employees, revenue of \$54 billion, and revenue was increasing at a rate of \$7 billion per year. How many employees would expect to be hired in the following year? (Data from: www.mergentonline.com.)

14. Using the equation in the previous exercise, find the rate of change of revenue with respect to time. Assume that the company generated \$128 billion in revenue, had 320,000 employees, and was adding employees at a rate of 10,000 employees per year.

15. **National Football League** During the 2015 National Football League regular season, the relationship between a team's winning percentage $W$ (wins divided by total games, expressed as a decimal) and turnover differential $T$ (turnovers taken away minus turnovers given away) can be approximated by the equation $187W - 5T = 95.$ (Data from: www.nfl.com.) Suppose that a team's turnover differential dropped by 2 turnovers one week.

(a) Would you expect the team's winning percentage to increase or decrease that week?

(b) Find the rate of change of winning percentage with respect to time.

16. **Weight Loss** A man who is 5 foot 10 inches tall is losing weight at the rate of 4 pounds per month. For a person of this height, the relationship between body mass index $B$ and weight $w$ (in pounds) is given by $4900B = 703w$. Find the rate of change of body mass index with respect to time.

*Work these exercises.*

17. **Bookstore Sales** The campus bookstore has estimated that its profit (in dollars) from selling $x$ hundred basketball conference championship T-shirts is given by

$$P = -48x^2 + 576x - 528.$$

The demand is currently 400 T-shirts, but euphoria over the championship is subsiding so the demand is dropping by 100 T-shirts per day. How is the profit changing with respect to time?

18. **Downloadable Ring Tones** Following a tweet by a celebrity, hourly demand for downloading a certain ring tone is currently 9000 and increasing by 2000 downloads per hour. Determine how the profit is changing assuming that the profit (in dollars) from selling $x$ thousand downloads is given by

$$P = -x^2 + 2000x - 50{,}000.$$

19. **Political Paraphernalia** During a local political race, the demand $x$ (in thousands) for campaign buttons is related to the price $p$ (in dollars) by the equation

$$25p^2 + 49x^2 = 1225.$$

If the quantity demanded is currently 3000 at a price of \$5.60 and demand is increasing by 400 buttons per month, find the rate of change of price with respect to time.

20. **Production Costs** The cost (in hundreds of dollars) of producing $x$ electronic readers is given by

$$C(x) = 1000 + \sqrt{2x^2 + 3x + 25}.$$

If the quantity demanded is currently 50,000 and increasing by 1000 per month, find the rate of change of cost with respect to time.

21. **Electricity from Coal and Renewable Sources** The relationship between the percentage of electricity generated by coal $C$ and the percentage of electricity generated by renewable sources $R$ (excluding hydroelectric) can be approximated by

$$102 \ln(R) + 10C = 586.$$

If the percentage of electricity generated by renewable sources is 7% and increasing at a rate of 1% per year, find the rate of change of the percentage of electricity generated by coal with respect to time. (Data from: International Energy Agency.)

22. **Savings and Unemployment** The personal savings rate (savings as a percentage of disposable personal income) $P$ is approximately related to the civilian unemployment rate $U$ according to the equation

$$13 \ln(U) - 4P = 4.$$

At the start of 2015, the civilian unemployment rate was 5.7% and decreasing at .8% per year. Find the rate of change of the personal savings rate with respect to time. (Data from: Federal Reserve Economic Data.)

23. **Dripping Fluid** A parked car dripping transmission fluid is creating a circular puddle whose area (in square inches) is $A = \pi r^2$ where $r$ is the radius of the circle in inches. Find the rate at which the area of the puddle is increasing at the instant when the radius is 3 inches and increasing at a rate of 2 inches per minute.

24. **Oil Slick** An oil production company is concerned about an oil slick that is forming on the surface of the water in the Gulf of Mexico. The slick is roughly in the shape of a circle with area $A = \pi r^2$ square kilometers where $r$ is the radius in kilometers. Find the rate at which the area of the oil slick is changing at the instant when the radius is 2 kilometers and increasing at .3 kilometer per day.

25. **Tumor Growth** The radius of an inoperable spherical tumor is increasing by .5 millimeter per month. How fast is the volume of the tumor changing when the radius is 30 millimeters? The volume of a sphere of radius $r$ is given by $V = \dfrac{4}{3}\pi r^3$.

26. **Angioplasty** An angioplasty balloon is roughly in the shape of a sphere and being inflated at 25 cubic millimeters per second. How fast is the radius of the balloon changing when the radius is 1 millimeter? The volume of a sphere of radius $r$ is given by $V = \dfrac{4}{3}\pi r^3$.

*Work these exercises. (See Example 5.)*

27. **Quarterback Throw** At a certain point during a football game, the quarterback is standing motionless while the intended receiver is running directly downfield toward the goal line. Let $z$ represent the distance between quarterback and receiver as measured by a straight line. The lateral distance between quarterback and receiver is constant at 6 yards. Let $x$ represent the downfield distance between quarterback and receiver (see the accompanying figure). Find $\dfrac{dz}{dt}$ at the instant when $x = 8$ yards and $\dfrac{dx}{dt} = 3$ yards per second.

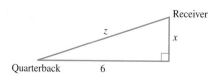

28. **Aircraft Flight** An aircraft flying at a constant altitude of 5 miles and a constant speed of 325 mph is heading away from John F. Kennedy airport in New York. Its path is being tracked by an air traffic controller on the ground. (Ignore the height of air traffic control tower.) How fast is the distance between aircraft and air traffic controller changing at the instant when this distance is 13 miles?

29. **Architectural Model** An architectural firm must decide on the size of a scale model to present to a client. The area of the base $A$ (in ft$^2$) is related to the weight of the model $W$ (in lbs) according to the equation $W^2 = 9A^3$. If the area of the base is currently

4 ft$^2$ and is increased by 2 ft$^2$ (while keeping the model to scale), determine the corresponding change in the weight of the model.

**30. Batting Average** During a six-game hitless slump early in the season, the number of hits for a certain softball player remained constant at 5. The relationship between her batting average $B$ and number of at bats $A$ is given by $B = \dfrac{5}{A}$. Determine how her batting average was changing when she had 20 at bats and went to the plate one more time and struck out.

*Work these exercises. (See Example 6.)*

**31. Physical Science** Sand falling through an hourglass is forming a conical pile such that the height is always twice the radius. Find $\dfrac{dV}{dt}$ at the instant when $r = 1$ cm and $\dfrac{dr}{dt} = 2$ cm per minute.

**32. Physical Science** Water from a drinking fountain is filling a paper cup. The cup is in the shape of a cone whose radius is half of its height. Find $\dfrac{dV}{dt}$ at the instant when $h = 2$ cm and $\dfrac{dh}{dt} = 4$ cm per minute.

✓**Checkpoint Answers**

1. About 13.33
2. $\dfrac{dx}{dt} = -4$
3. 300 cm$^3$/min
4. About .53 inch per minute
5. 10 feet per second
6. About $-471.2$ cm$^3$/min
7. $-\$10$ per day

# 12.6 Curve Sketching

In earlier sections, we saw that the first and second derivatives of a function provide a variety of information about the graph of the function, such as the location of its local extrema, the concavity of the graph, and the intervals on which it is increasing and decreasing. This information can be very helpful for interpreting misleading screen images on a graphing calculator or computer. It also enables us to make reasonably accurate graphs of many functions by hand if graphing technology is not available.

When graphing functions by hand, you should use the following guidelines. It may not always be feasible to carry out all the steps, but you should do as many as necessary, in any convenient order, to obtain a reasonable graph.

> ## To sketch the graph of a function $y = f(x)$, execute the following steps:
>
> 1. Find the $y$-intercept (if it exists) by letting $x = 0$ and computing $y = f(0)$.
> 2. Find the $x$-intercepts (if any) by letting $y = 0$ and solving the equation $f(x) = 0$ if doing so is not too difficult.
> 3. If $f$ is a rational function, find any vertical asymptotes by finding the numbers for which the denominator is 0, but the numerator is nonzero. Find any horizontal asymptotes by using the techniques of Section 3.6, as summarized in the box on page 172.
> 4. Find $f'(x)$ and $f''(x)$.
> 5. Locate any critical numbers by solving the equation $f'(x) = 0$ and determining where $f'(x)$ does not exist, but $f(x)$ does. Find the local extrema by using the first- or second-derivative test. Find the intervals where $f$ is increasing or decreasing by solving the inequalities $f'(x) > 0$ and $f'(x) < 0$.
> 6. Locate potential points of inflection by solving the equation $f''(x) = 0$ and determining where $f''(x)$ does not exist, but $f(x)$ does. Find the intervals where $f$ is concave upward or downward by solving the inequalities $f''(x) > 0$ and $f''(x) < 0$. Use this information to determine the points of inflection.
> 7. Use the preceding results and any other information that may be available to determine the general shape of the graph.

8. Plot the intercepts, critical points, points of inflection, and other points as needed. Connect the points with a smooth curve, using correct concavity and being careful not to draw a connected graph through points where the function is not defined.

9. If feasible, verify your results with a graphing calculator or computer. If there are significant differences between your graph and the calculator's, check for mistakes in the hand-drawn graph or adjust the calculator viewing window appropriately.

**EXAMPLE 1**   Graph $f(x) = 2x^3 - 3x^2 - 12x + 1$.

**Solution**

**Step 1**   The $y$-intercept is $f(0) = 2 \cdot 0^3 - 3 \cdot 0^2 - 12 \cdot 0 + 1 = 1$.

**Step 2**   To find the $x$-intercepts, we must solve the equation

$$2x^3 - 3x^2 - 12x + 1 = 0.$$

There is no easy way to do this by hand, so skip this step.

**Step 3**   Since $f$ is a polynomial function, the graph has no asymptotes.

**Step 4**   The first derivative is $f'(x) = 6x^2 - 6x - 12$, and the second derivative is $f''(x) = 12x - 6$.

**Step 5**   The first derivative is defined for all $x$, so the only critical numbers are the solutions of $f'(x) = 0$:

$$6x^2 - 6x - 12 = 0$$
$$x^2 - x - 2 = 0 \qquad \text{Divide both sides by 6.}$$
$$(x + 1)(x - 2) = 0 \qquad \text{Factor.}$$
$$x = -1 \quad \text{or} \quad x = 2.$$

Using the second-derivative test on the critical number $x = -1$, we have

$$f''(-1) = 12(-1) - 6 = -18 < 0.$$

Hence, there is a local maximum when $x = -1$, that is, at the point $(-1, f(-1)) = (-1, 8)$. Similarly,

$$f''(2) = 12(2) - 6 = 18 > 0,$$

so there is a local minimum when $x = 2$ (at the point $(2, f(2)) = (2, -19)$).

Next, we determine the intervals on which $f$ is increasing or decreasing by solving the inequalities

$$f'(x) > 0 \qquad \text{and} \qquad f'(x) < 0$$
$$6x^2 - 6x - 12 > 0 \qquad\qquad 6x^2 - 6x - 12 < 0.$$

The critical numbers divide the $x$-axis into three regions. Testing a number from each region, as indicated in Figure 12.46, we conclude that $f$ is increasing on the intervals $(-\infty, -1)$ and $(2, \infty)$ and decreasing on $(-1, 2)$.

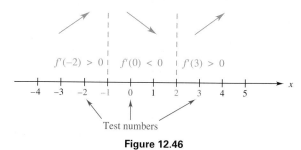

**Figure 12.46**

**Step 6** The second derivative is defined for all $x$, so the possible points of inflection are determined by the solutions of $f''(x) = 0$:

$$12x - 6 = 0$$

$$x = \frac{1}{2}.$$

Determine the concavity of the graph by solving

$$f''(x) > 0 \quad \text{and} \quad f''(x) < 0$$
$$12x - 6 > 0 \quad\quad\quad 12x - 6 < 0$$
$$x > \frac{1}{2} \quad\quad\quad\quad x < \frac{1}{2}.$$

Therefore, $f$ is concave upward on the interval $(\frac{1}{2}, \infty)$ and concave downward on $(-\infty, \frac{1}{2})$. Consequently, the only point of inflection is $(\frac{1}{2}, f(\frac{1}{2})) = (\frac{1}{2}, -5.5)$.

**Step 7** Since $f$ is a third-degree polynomial function, we know from Section 3.6 that when $x$ is very large in absolute value, its graph must resemble the graph of its highest-degree term, $2x^3$; that is, the graph must rise sharply on the right side and fall sharply on the left. Combining this fact with the information obtained in the preceding steps, we see that the graph of $f$ must have the general shape shown in Figure 12.47.

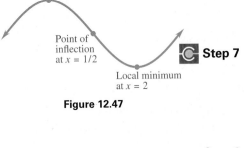

Local maximum at $x = -1$

Point of inflection at $x = 1/2$

Local minimum at $x = 2$

**Figure 12.47**

**Step 8** Now we plot the points determined in Steps 1, 5, and 6, together with a few additional points, to obtain the graph in Figure 12.48. ✔1 ✔2

✔ **Checkpoint 1** 🔲

**Step 9** Use technology to verify that the graph in Figure 12.48 is correct.

✔ **Checkpoint 2**

Sketch the graph of

$$f(x) = x^3 - 3x^2.$$

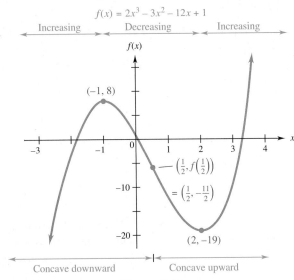

$f(x) = 2x^3 - 3x^2 - 12x + 1$

Increasing    Decreasing    Increasing

$f(x)$

$(-1, 8)$

$-10$

$\left(\frac{1}{2}, f\left(\frac{1}{2}\right)\right)$

$= \left(\frac{1}{2}, -\frac{11}{2}\right)$

$-20$

$(2, -19)$

Concave downward    Concave upward

**Figure 12.48**

**EXAMPLE 2** Graph

$$f(x) = \frac{3x^2}{x^2 + 5}.$$

**Solution**

**Step 1** The $y$-intercept is $f(0) = 0/5 = 0$.

**Step 2** To find the $x$-intercepts, note that $f(x)$ is always defined, because $x^2 + 5$ is always positive. Hence, $f(x) = 0$ when the numerator $3x^2 = 0$; this occurs when $x = 0$. So the point $(0, f(0)) = (0, 0)$ is both the $x$- and $y$-intercept.

**Step 3**  This is a rational function, but its denominator is always nonzero (why?), so there are no vertical asymptotes. Using the techniques presented in Section 3.6, we see that

$$f(x) = \frac{3x^2}{x^2 + 5} = \frac{\dfrac{3x^2}{x^2}}{\dfrac{x^2}{x^2} + \dfrac{5}{x^2}} = \frac{3}{1 + \dfrac{5}{x^2}}.$$

When $x$ is very large in absolute value, so is $x^2$; thus, $5/x^2$ is very close to 0, and hence $f(x)$ is very close to $3/(1 + 0) = 3$. Consequently, the horizontal line $y = 3$ is a horizontal asymptote.

**Step 4**  The first derivative is

$$f'(x) = \frac{(x^2 + 5)(6x) - (3x^2)(2x)}{(x^2 + 5)^2} = \frac{30x}{(x^2 + 5)^2}.$$

The second derivative is

$$f''(x) = \frac{(x^2 + 5)^2\, 30 - (30x)(2)(x^2 + 5)(2x)}{(x^2 + 5)^4}.$$

Factor $30(x^2 + 5)$ out of the numerator:

$$f''(x) = \frac{30(x^2 + 5)[(x^2 + 5) - (x)(2)(2x)]}{(x^2 + 5)^4}.$$

Divide a factor of $(x^2 + 5)$, out of the numerator and denominator and simplify the numerator:

$$f''(x) = \frac{30[(x^2 + 5) - (x)(2)(2x)]}{(x^2 + 5)^3}$$

$$= \frac{30[(x^2 + 5) - 4x^2]}{(x^2 + 5)^3}$$

$$= \frac{30(5 - 3x^2)}{(x^2 + 5)^3}.$$

**Step 5**  Since $f'(x) = \dfrac{30x}{(x^2 + 5)^2}$ and $x^2 + 5 \neq 0$ for all $x$, $f'(x)$ is always defined. $f'(x) = 0$ when its numerator, $30x$, is 0; this occurs when $x = 0$. The critical number 0 divides the $x$-axis into two regions (Figure 12.49). Testing a number in each region shows that $f$ is decreasing on $(-\infty, 0)$ and increasing on $(0, \infty)$. By the first-derivative test, $f$ has a local minimum at $x = 0$.

Local minimum

$f'(-1) < 0$  $f'(1) > 0$

$-1$  $0$  $1$  $x$

Test numbers

**Figure 12.49**

**Step 6**  Since $f''(x) = \dfrac{30(5 - 3x^2)}{(x^2 + 5)^3}$ and $x^2 + 5 \neq 0$ for all $x$, $f''(x)$ is always defined. $f''(x) = 0$ when

$$30(5 - 3x^2) = 0$$
$$3x^2 = 5$$
$$x = \pm\sqrt{5/3} \approx \pm 1.29.$$

Testing a point in each of the three intervals defined by these points shows that $f$ is concave downward on $(-\infty, -1.29)$ and $(1.29, \infty)$ and concave upward on $(-1.29, 1.29)$. The graph has inflection points at $(\pm\sqrt{5/3}, f(\pm\sqrt{5/3})) \approx (\pm 1.29, .75)$.

**Step 7** The information about the shape of the graph obtained in Steps 5 and 6 is summarized in the following chart:

| Interval | $(-\infty, -1.29)$ | $(-1.29, 0)$ | $(0, 1.29)$ | $(1.29, \infty)$ |
|---|---|---|---|---|
| **Sign of** $f'$ | – | – | + | + |
| **Sign of** $f''$ | – | + | + | – |
| $f$ **Increasing or Decreasing** | Decreasing | Decreasing | Increasing | Increasing |
| **Concavity of** $f$ | Downward | Upward | Upward | Downward |
| **Shape of Graph** | ⌢ | ⌣ | ⌣ | ⌢ |

**Step 8** Plot some points (several are needed near the origin), including the intercept at the origin, and use the fact that $y = 3$ is a horizontal asymptote to obtain the graph in Figure 12.50.

**Figure 12.50**

✓ **Checkpoint 3**

**Step 9** Use technology to verify that the graph in Figure 12.50 is correct.

**EXAMPLE 3** Graph $f(x) = x + 1/x$.

**Solution**

**Step 1** Since $x = 0$ is not in the domain of the function (why?), there is no $y$-intercept.

**Step 2** To find the $x$-intercepts, solve $f(x) = 0$:

$$x + \frac{1}{x} = 0$$

$$x = -\frac{1}{x}$$

$$x^2 = -1.$$

Since $x^2$ is never negative, there is also no $x$-intercept.

**Step 3** Note that the rule of $f$ can be written as

$$f(x) = x + \frac{1}{x} = \frac{x^2 + 1}{x}.$$

When $x = 0$, the denominator is 0, but the numerator is nonzero, so there is a vertical asymptote at $x = 0$. Since the numerator of $f(x)$ is of a higher degree than the denominator, there is no horizontal asymptote.

**Step 4** Because $f(x) = x + 1/x = x + x^{-1}$, we have

$$f'(x) = 1 - x^{-2} = 1 - \frac{1}{x^2},$$

so

$$f''(x) = 2x^{-3} = \frac{2}{x^3}.$$

**Step 5**  $f'(x) = 0$ when

$$\frac{1}{x^2} = 1$$

$$x^2 = 1$$

$$x = 1 \quad \text{or} \quad x = -1.$$

Hence, $x = -1$ and $x = 1$ are critical numbers. The derivative does not exist when $x = 0$, but the function is not defined there either, so $x = 0$ is not a critical number. Evaluating $f'(x)$ in each of the regions determined by the critical numbers and the asymptote shows that $f$ is increasing on $(-\infty, -1)$ and $(1, \infty)$ and decreasing on $(-1, 0)$ and $(0, 1)$, as summarized in the chart in Step 7. By the first-derivative test, $f$ has a relative maximum of $y = f(-1) = -2$ when $x = -1$ and a relative minimum of $y = f(1) = 2$ when $x = 1$.

**Step 6**  The second derivative $f''(x) = 2/x^3$ is never equal to 0 and does not exist when $x = 0$. (The function itself also does not exist at 0.) Because of this, there may be a change in concavity, but not an inflection point, when $x = 0$. The second derivative is negative when $x$ is negative, making $f$ concave downward on $(-\infty, 0)$. Also, $f''(x) > 0$ when $x > 0$, making $f$ concave upward on $(0, \infty)$, as indicated in the chart in Step 7.

**Step 7**  The preceding information is summarized in the following chart:

| Interval | $(-\infty, -1)$ | $(-1, 0)$ | $(0, 1)$ | $(1, \infty)$ |
|---|---|---|---|---|
| Sign of $f'$ | $+$ | $-$ | $-$ | $+$ |
| Sign of $f''$ | $-$ | $-$ | $+$ | $+$ |
| $f$ Increasing or Decreasing | Increasing | Decreasing | Decreasing | Increasing |
| Concavity of $f$ | Downward | Downward | Upward | Upward |
| Shape of Graph | ⌒ | ⌐ | ⌣ | ⌡ |

We can determine the shape of the graph when $x$ is very large in absolute value by noting that as $x$ gets very large, the second term of its rule, $1/x$, gets very small, so that $f(x) = x + 1/x \approx x$. Hence, the graph gets closer and closer to the straight line $y = x$ as $x$ becomes larger and larger. The line $y = x$ is called an **oblique**, or **slant asymptote**.

**Step 8**  Plot several points and use the preceding information to obtain the graph of $f(x)$ in Figure 12.51.

✓ **Checkpoint 4** ▦

**Step 9** Use technology to verify that the graph in Figure 12.51 is correct.

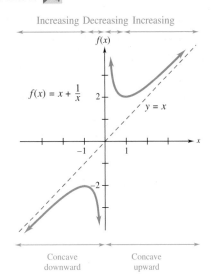

**Figure 12.51**

# 12.6 Exercises

*Sketch the graph of the function. Identify any local extrema and points of inflection. (See Examples 1–3.)*

**1.** $f(x) = -x^2 - 10x - 25$      **2.** $f(x) = x^2 - 12x + 36$

**3.** $f(x) = 3x^3 - 3x^2 + 1$      **4.** $f(x) = 2x^3 - 4x^2 + 2$

**5.** $f(x) = -2x^3 - 9x^2 + 108x - 10$

**6.** $f(x) = -2x^3 - 9x^2 + 60x - 8$

**7.** $f(x) = 2x^3 + \dfrac{7}{2}x^2 - 5x + 3$

**8.** $f(x) = x^3 - \dfrac{15}{2}x^2 - 18x - 1$

**9.** $f(x) = (x + 3)^4$      **10.** $f(x) = (x - 2)^4$

**11.** $f(x) = x^4 - 18x^2 + 5$      **12.** $f(x) = x^4 - 8x^2$

**13.** $f(x) = x - \dfrac{1}{x}$      **14.** $f(x) = 2x + \dfrac{8}{x}$

**15.** $f(x) = \dfrac{x^2 + 25}{x}$      **16.** $f(x) = \dfrac{x^2 + 4}{x}$

**17.** $f(x) = \dfrac{x - 1}{x + 1}$      **18.** $f(x) = \dfrac{x}{1 + x}$

*Sketch the graph of the function. Identify any local extrema and inflection points.*

**19.** $y = x - \ln|x|$      **20.** $y = \dfrac{\ln x}{x}$

**21.** $y = xe^{-x}$      **22.** $f(x) = xe^{-\frac{x}{2}}$

**C** *In Exercises 23–28, sketch the graph of a function f that has all of the properties listed. There are many correct answers, and your graph need not be given by an algebraic formula.*

**23.** **(a)** The domain of $f$ is $[0, 10]$.

   **(b)** $f'(x) > 0$ and $f''(x) > 0$ for all $x$ in the domain of $f$.

**24.** **(a)** The domain of $f$ is $[0, 10]$.

   **(b)** $f'(x) > 0$ and $f''(x) < 0$ for all $x$ in the domain of $f$.

**25.** **(a)** Continuous and differentiable for all real numbers

   **(b)** Increasing on $(-\infty, -3)$ and $(1, 4)$

   **(c)** Decreasing on $(-3, 1)$ and $(4, \infty)$

   **(d)** Concave downward on $(-\infty, -1)$ and $(2, \infty)$

   **(e)** Concave upward on $(-1, 2)$

   **(f)** $f'(-3) = f'(4) = 0$

   **(g)** Inflection points at $(-1, 3)$ and $(2, 4)$

**26.** **(a)** Continuous for all real numbers

   **(b)** Increasing on $(-\infty, -2)$ and $(0, 3)$

   **(c)** Decreasing on $(-2, 0)$ and $(3, \infty)$

   **(d)** Concave downward on $(-\infty, 0)$ and $(0, 5)$

   **(e)** Concave upward on $(5, \infty)$

   **(f)** $f'(-2) = f'(3) = 0$

**27.** **(a)** Continuous for all real numbers

   **(b)** Decreasing on $(-\infty, -6)$ and $(1, 3)$

   **(c)** Increasing on $(-6, 1)$ and $(3, \infty)$

   **(d)** Concave upward on $(-\infty, -6)$ and $(3, \infty)$

   **(e)** Concave downward on $(-6, 3)$

   **(f)** $y$-intercept at $(0, 2)$

**28.** **(a)** Continuous and differentiable everywhere except at $x = -1$, where it has a vertical asymptote

   **(b)** Decreasing everywhere it is defined

   **(c)** Concave downward on $(-\infty, -1)$ and $(1, 2)$

   **(d)** Concave upward on $(-1, 1)$ and $(2, \infty)$

*Work these exercises.*

**29.** **Business** The accompanying figure shows the *product life cycle* graph, with typical products marked on it. It illustrates the fact that a new product is often purchased at a faster and faster rate as people become familiar with it. In time, saturation is reached and the purchase rate stays constant until the product is made obsolete by newer products, after which it is purchased less and less often. (Data from: www.tutor2u.net/business/marketing/products_lifecycle.asp.)

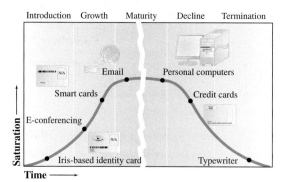

Which product on the left side of the graph is closest to the left-hand point of inflection? What does the point of inflection mean here?

**30.** Refer to the graph in Exercise 29. Which product on the right side of the graph is closest to the right-hand point of inflection? What does the point of inflection mean here?

**31.** **Chevron Profits** The profit function (in billions) for Chevron Corp. for the years 2009 through 2016 can be approximated by

$$P(x) = -2.3x^2 + 55.2x - 252.6,$$

where $x = 9$ corresponds to the year 2009.

   **(a)** During what year did the local maximum profit occur?

   **(b)** What was the maximum profit?

   **(c)** Does the function have any inflection points?

**32.** **Average Temperature** During this century, the average monthly temperature $T(x)$ in degrees Fahrenheit in month $x$ in the United States can be approximated by

$$T(x) = -1.4x^2 + 19.6x - 6.1,$$

where $x = 1$ corresponds to January, $x = 2$ to February, and so on. (Data from: www.worldbank.org.) During which month does the maximum average monthly temperature occur? What is the maximum average monthly temperature?

📷 *Use a graphing calculator or computer to find the approximate location of all local extrema and points of inflection of the given functions.*

**33. Physical Science** The pressure of the oil in a reservoir tends to drop with time. By taking sample pressure readings for a particular oil reservoir, petroleum engineers have found that the change in pressure is given by

$$P(t) = t^3 - 18t^2 + 81t \qquad (0 \le t \le 15),$$

where $t$ is time in years from the date of the first reading.

**34. Health** The cost–benefit curve for pollution control is given by

$$y = \frac{9.2x}{106 - x} \qquad (0 < x \le 100),$$

where $y$ is the cost, in thousands of dollars, or removing $x$ percent of a specified industrial pollutant.

📷 *Use a graphing calculator or computer to find the approximate location of all local extrema of the given functions.*

**35. Dupont Sales** Sales (in millions of dollars) for Dupont Inc. for the years 2007 through 2015 can be approximated by

$$S(x) = -.072x^3 + 2.45x^2 - 26.5x + 106,$$

where $x = 7$ corresponds to the year 2007. (Data from: www.morningstar.com.)

**36. Lifepoint Revenue** Revenue (in billions of dollars) for Lifepoint Health Inc. for the years 2010 through 2017 can be approximated by

$$R(x) = -.028x^3 + 1.16x^2 - 15.3x + 68.2,$$

where $x = 10$ corresponds to the year 2010. (Data from: www.morningstar.com.)

✓**Checkpoint Answers**

**1.**

**2.**

**3.**

**4.**

---

# CHAPTER 12 Summary and Review

## Key Terms and Symbols

**12.1** increasing function on an interval
decreasing function on an interval
critical number
critical point
local maximum (maxima)
local minimum (minima)
local extremum (extrema)
first-derivative test

**12.2** $f''(x)$ or $y''$ or $\dfrac{d^2y}{dx^2}$, second derivative of $y = f(x)$
$f'''(x)$, third derivative of $f$
$f^{(n)}(x)$, $n$th derivative of $f$
acceleration
concave upward

concave downward
point of inflection
point of diminishing returns
second-derivative test

**12.3** absolute maximum on an interval
absolute minimum on an interval
extreme-value theorem

critical-point theorem
economic lot size

**12.4** implicitly defined function
implicit differentiation

**12.5** related rates

**12.6** curve sketching
oblique asymptote

## Chapter 12 Key Concepts

| | |
|---|---|
| **Increasing and Decreasing Functions** | If $f'(x) > 0$ for each $x$ in an interval, then $f$ is **increasing** on the interval; if $f'(x) < 0$ for each $x$ in the interval, then $f$ is **decreasing** on the interval; if $f'(x) = 0$ for each $x$ in the interval, then $f$ is **constant** on the interval. |
| **Critical Numbers** | If $f$ is a function, then a number $c$ for which $f(c)$ is defined and either $f'(c) = 0$ or $f'(c)$ does not exist is called a crucial number of $f$. The corresponding point $(c, f(c))$ on the graph of $f$ is called a critical point. |
| **Local Extrema** | Let $c$ be a number in the domain of a function $f$. Then $f$ has a **local maximum** at $c$ if $f(x) \leq f(c)$ for all $x$ near $c$, and $f$ has a **local minimum** at $c$ if $f(x) \geq f(c)$ for all $x$ near $c$. If $f$ has a local extremum at $c$, then $c$ is a critical number of $f$. |
| **First-Derivative Test** | Let $f$ be a differentiable function for all $x$ in $[a, b]$, except possibly at $x = c$. Assume that $a < c < b$ and that $c$ is the only critical number for $f$ in $[a, b]$. If $f'(a) > 0$ and $f'(b) < 0$, then there is a local maximum at $c$. If $f'(a) < 0$ and $f'(b) > 0$, then there is a local minimum at $c$. |
| **Concavity** | Let $f$ have derivatives $f'$ and $f''$ for all $x$ in $(a, b)$. Then $f$ is **concave upward** on $(a, b)$ if $f''(x) > 0$ for all $x$ in $(a, b)$, $f$ is **concave downward** on $(a, b)$ if $f''(x) < 0$ for all $x$ in $(a, b)$, and $f$ has a **point of inflection** at $x = c$ if $f''(x)$ changes sign at $x = c$. |
| **Second-Derivative Test** | Let $c$ be a critical number of $f$ such that $f'(c) = 0$ and $f''(x)$ exists for all $x$ in some open interval containing $c$. If $f''(c) > 0$, then there is a local minimum at $c$. If $f''(c) < 0$, then there is a local maximum at $c$. If $f''(c) = 0$, then the test gives no information. |
| **Absolute Extrema** | Let $c$ be in an interval $[a, b]$ where $f$ is defined. Then $f$ has an **absolute maximum** on the interval at $c$ if $f(x) \leq f(c)$ for all $x$ in $[a, b]$, and $f$ has an **absolute minimum** on the interval at $c$ if $f(x) \geq f(c)$ for all $x$ in $[a, b]$. |
| **Extreme-Value Theorem** | If a function $f$ is continuous on a closed interval $[a, b]$, then $f$ has both an absolute maximum and an absolute minimum on the interval. Each of these values occurs either at an endpoint of the interval or at a critical number of $f$. |
| **Critical-Point Theorem** | Suppose that a function $f$ is continuous on an interval $I$ (which may be open, closed, or neither) and that $f$ has exactly one critical number in the interval $I$, say, $x = c$. If $f$ has a local maximum at $x = c$, then this local maximum is the absolute maximum of $f$ on the interval $I$. If $f$ has a local minimum at $x = c$, then this local minimum is the absolute minimum of $f$ on the interval $I$. |
| **Implicit Differentiation** | Let $y$ be implicitly defined as a function of $x$ by a given equation. To find $\dfrac{dy}{dx}$ by implicit differentiation, first differentiate both sides of the equation with respect to $x$, then solve for $\dfrac{dy}{dx}$. |
| **Related Rates** | To solve a related rates problem, first determine an equation that relates only the two variables of interest. Next, differentiate both sides of the equation with respect to time. Finally, substitute the given values, and solve for the desired rate of change. |

## Chapter 12 Review Exercises

1. When the rule of a function is given, how can you determine where it is increasing and where it is decreasing?

2. When the rule of a function is given, how can you determine where the local extrema are located? State two algebraic ways to test whether a local extremum is a maximum or a minimum.

3. What is the difference between a local extremum and an absolute extremum? Can a local extremum be an absolute extremum? Is a local extremum necessarily an absolute extremum?

4. What information about a graph can be found from the first derivative? from the second derivative?

*Find the largest open intervals on which the given functions are increasing or decreasing.*

5. $f(x) = x^2 + 9x - 9$

6. $f(x) = -3x^2 - 3x + 11$

7. $g(x) = 2x^3 - x^2 - 4x + 7$

8. $g(x) = -4x^3 - 5x^2 + 8x + 1$

9. $f(x) = \dfrac{4}{x - 4}$

10. $f(x) = \dfrac{-6}{3x - 5}$

*Find the locations and the values of all local maxima and minima for the given functions.*

11. $f(x) = 2x^3 - 3x^2 - 36x + 10$

12. $f(x) = 2x^3 - 3x^2 - 12x + 2$

13. $f(x) = x^4 - \dfrac{8}{3}x^3 - 6x^2 + 2$

14. $f(x) = x \cdot e^x$

15. $f(x) = 3x \cdot e^{-x}$

16. $f(x) = \dfrac{e^x}{x - 1}$

*Find the second derivatives of the given functions; then find $f''(1)$ and $f''(-2)$.*

17. $f(x) = 2x^5 - 5x^3 + 3x - 1$

18. $f(x) = \dfrac{3 - 2x}{x + 3}$

19. $f(x) = -5e^{2x}$          20. $f(x) = \ln|5x + 2|$

*Sketch the graph of each of the given functions. List the location of each local extremum and point of inflection, the intervals on which the function is increasing and decreasing, and the intervals on which it is concave upward and concave downward.*

21. $f(x) = -2x^3 - \dfrac{1}{2}x^2 - x - 3$

22. $f(x) = -\dfrac{4}{3}x^3 + x^2 + 30x - 7$

23. $f(x) = x^4 - \dfrac{4}{3}x^3 - 4x^2 + 1$

24. $f(x) = -\dfrac{2}{3}x^3 + \dfrac{9}{2}x^2 + 5x + 1$

25. $f(x) = \dfrac{x - 1}{2x + 1}$          26. $f(x) = \dfrac{2x - 5}{x + 3}$

27. $f(x) = -4x^3 - x^2 + 4x + 5$

28. $f(x) = x^3 + \dfrac{5}{2}x^2 - 2x - 3$

29. $f(x) = x^4 + 2x^2$          30. $f(x) = 6x^3 - x^4$

31. $f(x) = \dfrac{x^2 + 4}{x}$          32. $f(x) = x + \dfrac{8}{x}$

*Find the locations and values of all absolute maxima and absolute minima for the given functions on the given intervals.*

33. $f(x) = -x^2 + 6x + 1; [2, 4]$

34. $f(x) = 4x^2 - 4x - 7; [-1, 3]$

35. $f(x) = x^3 + 2x^2 - 15x + 3; [-.5, 3.3]$

36. $f(x) = -2x^3 - x^2 + 4x - 1; [-3, 1]$

*If possible, find an absolute extremum of each function on the given interval.*

37. $g(x) = x^3 + 6x^2 - 15x + 1; (0, 9)$

38. $g(x) = x^3 + 6x^2 - 15x + 1; (-9, 0)$

39. $g(x) = xe^{-x/3}; (-\infty, \infty)$          40. $g(x) = -5xe^{-2x}; (-\infty, \infty)$

*Find $\dfrac{dy}{dx}$ by implicit differentiation.*

41. $10xy = 2x + 5y$          42. $12xy = 4y + 3x$

43. $ye^x - 10x = y$          44. $5ye^x + x^2 = -10y$

*Find the equation of the tangent line to the curve at the given point.*

45. $5 - y^3 = 6 - 2x^6$ at $(1, 1)$

46. $5 - y^2 = 6 - 2x^4$ at $(1, 1)$

*Work these exercises.*

47. **Chicago Cubs** The accompanying figure shows average attendance (in thousands per game) at Chicago Cubs baseball games held at Wrigley Field.

 (a) In 2016, the Cubs won the World Series for the first time in 108 years. What was the average attendance per game that year? Give your answer to the nearest thousand.

 (b) During this time period, when was attendance decreasing?

 (c) What was the absolute minimum attendance? The absolute maximum?

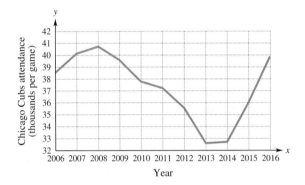

48. **Academy Awards** According to Google Trends, search interest in the Academy Awards can be calculated on a scale of 0 to 100 for the months January 2015 through January 2017, as shown in the accompanying figure.

 (a) In what month does there appear to be a local maximum? Why might this be the case?

 (b) Is search interest in the Academy Awards increasing or decreasing in March?

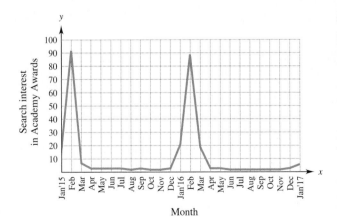

49. **Population Growth** The population of Arizona (in millions) is projected to be approximately

$$f(x) = 5.18e^{.024x} \qquad (0 \le x \le 30),$$

where $x = 0$ corresponds to the year 2000. (Data from: www.census.gov.)

**(a)** Is the population of Arizona projected to increase or decrease through 2030?

**(b)** Determine whether the graph of this population function is concave up or concave down and explain what that means for Arizona.

**50. Olympic High Jump** The gold medal height (in centimeters) in the women's Olympic high jump can be approximated for the years 1948 through 2016 by

$$f(x) = 42 \ln x + 6 \quad (48 \le x \le 116),$$

where $x = 48$ corresponds to the year 1948. (Data from: www.olympic.org.)

**(a)** According to this model, is the winning height increasing or decreasing during this period?

**(b)** Determine whether the graph of the winning height is concave up or concave down, and explain what that means about the high jump.

**51. Gross Income** Average adjusted gross income as a function of age (between the ages of 25 and 75) can be approximated by

$$I(x) = -68x^2 + 7480x - 118,000 \quad (25 \le x \le 75),$$

where $x = 25$ corresponds to 25 years of age. (Data from: Internal Revenue Service.)

**(a)** For what ages is average adjusted gross income increasing?

**(b)** For what ages is average adjusted gross income decreasing?

**(c)** Find and interpret all local extrema.

**52. Fitbit IPO** The initial public offering (IPO) for Fitbit Inc. was held on the New York Stock Exchange on June 18, 2015. For the first 60 trading days after the IPO, the closing stock price can be approximated by

$$I(x) = -.02x^2 + 1.32x + 29.86 \quad (0 \le x \le 60),$$

where $x = 0$ corresponds to the date of the IPO. Find and interpret all local extrema. (Data from: www.morningstar.com.)

**53. Population of Japan** Using historical data and future projections, the population of Japan (in millions) for the years 2000 through 2020 can be approximated by

$$P(x) = -.02x^2 + .32x + 127 \quad (0 \le x \le 20),$$

where $x = 0$ corresponds to the year 2000. (Data from: www.worldometers.info.)

**(a)** To the nearest million, what was the population of Japan in 2000? In 2020?

**(b)** Find and interpret the absolute extrema during this period.

**54. Women in the U.S. Senate** The number of women in the U.S. Senate for the years 1959 through 2017 can be approximated by

$$W(x) = .011x^2 - 1.606x + 58.7 \quad (59 \le x \le 117),$$

where $x = 59$ corresponds to the year 1959. Round your answers to the nearest integer. (Data from: womenincongress.house.gov.)

**(a)** How many female senators were there in 1959? In 2017?

**(b)** Find and interpret the absolute extrema during this period.

*Work these exercises. You may need the quadratic formula to find some of the critical numbers.*

**55. IBM Profits** The profits (in billions of dollars) for the International Business Machines (IBM) Corp. between the years 2008 and 2015 can be approximated by

$$P(t) = -.2t^3 + 1.5t^2 - 2.4t + 48 \quad (0 \le t \le 7),$$

where $t = 0$ corresponds to 2008, $t = 1$ corresponds to 2009, and so on. (Data from: www.morningstar.com.)

**(a)** Find the critical numbers.

**(b)** When did a local minimum occur?

**(c)** When did a local maximum occur?

**56. Walt Disney Stock** The price of stock for the Walt Disney Company during the 12 months of 2016 can be approximated by

$$P(t) = .1t^3 - 1.8t^2 + 8.1t + 93 \quad (1 \le t \le 12),$$

where $t = 1$ corresponds to January, $t = 2$ corresponds to February, and so on. (Data from: www.morningstar.com.)

**(a)** Find the critical numbers.

**(b)** When did a local minimum occur?

**(c)** When did a local maximum occur?

**57. Unemployment** Long-term unemployment (expressed as a percentage of the total unemployed) can be approximated by

$$f(x) = -.02x^4 + .55x^3 - 4.8x^2 + 16x - 7 \quad (1 \le x \le 14),$$

where $x = 1$ corresponds to the year 2001. What was the absolute maximum long-term unemployment as a percentage of total unemployment during this period? Give your answer to the nearest whole percent. During what year did this occur? (Data from: www.worldbank.org.)

**58. Cell Phones** The number of cell phone subscriptions per 100 people in Greece can be approximated by

$$f(x) = -.31x^4 + 13.2x^3 - 206x^2 + 1394x - 3332 \quad (7 \le x \le 14),$$

where $x = 7$ corresponds to the year 2007. What was the absolute minimum number of cell phones per 100 people during this period? Give your answer to the nearest integer. What was the absolute maximum? (Data from: www.worldbank.org.)

*Work these exercises.*

**59. Landscaping** A landscaper needs to design an enclosed rectangular garden with one side against an existing garage, with no fence needed there. Find the dimensions of the rectangular space of maximum area that can be enclosed with 500 meters of fence.

**60. Landscaping** If the garden described in the previous exercise needs fencing on all four sides, find the dimensions of the maximum rectangular area that can be made with 500 meters of fence.

**61. Packaging** The packaging department of a corporation is designing a box with a square base and top. The volume is to be 27 cubic meters. To reduce the cost, the surface area of the box is to be minimized. What dimensions (height, length, and width) should the box have?

62. **Packaging** A company plans to package its product in a cylinder that is open at one end. The cylinder is to have a volume of $27\pi$ cubic inches. What radius should the circular bottom of the cylinder have in order to minimize the cost of the material? (*Hint*: The volume of a circular cylinder is $\pi r^2 h$, where $r$ is the radius of the circular base and $h$ is the height; the surface area of an open cylinder is $2\pi rh + \pi r^2$.)

63. **Cell Phones** A cell phone manufacturer produces 145,000 phones a year. It costs \$3.25 to store each phone for one year and \$230 to produce each batch. Find the number of batches that should be produced annually.

64. **Cell Phones** How many phones need to be produced for each batch in the previous exercise?

65. **Candy Production** A chocolate maker has annual demand of 250,000 pounds per year. It costs \$1.25 to store a pound of chocolate for a year and \$350 to produce a batch. Find the number of batches to be produced annually.

66. **Candy Production** How many pounds should be produced for each batch in the previous exercise?

*Find the indicated rate of change.*

67. Find $\dfrac{dx}{dt}$ when $x = 0$, $y = 1$, and $\dfrac{dy}{dt} = 5$, given that $2x^5 + x = y^2 - 1$.

68. Find $\dfrac{dy}{dt}$ when $x = 0$, $y = 1$, and $\dfrac{dx}{dt} = 10$, given that $2x^5 + x = y^2 - 1$.

*Work these exercises.*

69. **Natural Science** The relationship between temperature $F$ measured in degrees Fahrenheit and temperature $C$ measured in degrees Celsius is given by $F = \dfrac{9}{5}C + 32$.

    (a) If the temperature in Fahrenheit is dropping by 4 degrees per hour $\left(\dfrac{dF}{dt} = -4\right)$, find $\dfrac{dC}{dt}$, the rate of change of temperature in degrees Celsius per hour.

    (b) If the temperature in Celsius is rising by 3 degrees per hour $\left(\dfrac{dC}{dt} = 3\right)$, find $\dfrac{dF}{dt}$, the rate of change of temperature in degrees Fahrenheit per hour.

70. **Health** A personal trainer instructs clients interested in high-intensity cardiac training to workout at 80% of their maximum heart rate. This desired heart rate $H$ (in beats per minute) is related to age $A$ (in years) according to the formula $4A + 5H = 880$. As a client becomes one year older, how much should her desired heart rate change?

71. **Coffee Shop Profits** Suppose that the daily profit (in dollars) from selling $x$ large cups of latte at the Crestwood Coffee Company can be approximated by $P = -\dfrac{x^2}{360} + 6x - 300$. At the start of final exams, the demand is $x = 450$ cups and increasing at a rate of $\dfrac{dx}{dt} = 90$ cups per day. Find $\dfrac{dP}{dt}$ at the start of final exams.

72. **Bottled Water Demand** The demand $x$ for cold bottles of water from a street vendor outside a sports stadium can be approximated by $3x + 160p^2 = 1440$, where $p$ is the price in dollars. If the street vendor increases the price from \$2.00 per bottle to \$2.50 per bottle, how would she expect the demand to change?

73. **Rowing** A woman rowing a single skull is racing at 221 meters per minute in a straight line. A spectator is watching from shore, 10 meters in a perpendicular direction from the path of the competitor. How fast is the distance between the spectator and competitor changing at the instant when this distance is 26 meters?

74. **Baseball** A baseball player hits the ball and runs toward first base. Let $x$ represent the distance in feet between the player and first base and $z$ represent the distance in feet between the player and second base (measured by a straight line, not by the path the player must run). Find $\dfrac{dz}{dt}$ at the instant when $x = 45$, $z = 45\sqrt{5}$, and $\dfrac{dx}{dt} = -9$ ft/sec. Note that a baseball diamond is a square that is 90 feet on each side.

# Case Study 12  A Total Cost Model for a Training Program*

In this application, we set up a mathematical model for determining the total cost required to set up a training program. Then we use calculus to find the time interval between training programs that produces the minimum total cost. The model assumes that the demand for trainees is constant and that the fixed cost of training a batch of trainees is known. Also, it is assumed that people who are trained, but for whom no job is readily available, will be paid a fixed amount per month while waiting for a job to open up.

The model uses the following variables:

$D$ = demand for trainees per month;

$N$ = number of trainees per batch;

$C_1$ = fixed cost of training a batch of trainees;

$C_2$ = variable cost of training per trainee per month;

$C_3$ = salary paid monthly to a trainee who has not yet been given a job after training;

$m$ = time interval in months between successive batches of trainees;

---
*Based on "A Total Cost Model for a Training Program," by P. L. Goyal and S. K. Goyal, Faculty of Commerce and Administration, Concordia University. Used with permission.

$t$ = length of training program in months;

$Z(m)$ = total monthly cost of program.

The total cost of training a batch of trainees is given by $C_1 + NtC_2$. However, $N = mD$, so that the total cost per batch is $C_1 + mDtC_2$.

After training, personnel are given jobs at the rate of $D$ per month. Thus, $N - D$ of the trainees will not get a job the first month, $N - 2D$ will not get a job the second month, and so on. The $N - D$ trainees who do not get a job the first month produce total costs of $(N - D)C_3$, those not getting jobs during the second month produce costs of $(N - 2D)C_3$, and so on. Since $N = mD$, the costs during the first month can be written as

$$(N - D)C_3 = (mD - D)C_3 = (m - 1)DC_3,$$

while the costs during the second month are $(m - 2)DC_3$, and so on. The total cost for keeping the trainees without a job is thus

$$(m - 1)DC_3 + (m - 2)DC_3$$
$$+ (m - 3)DC_3 + \cdots + 2DC_3 + DC_3,$$

which can be factored to give

$$DC_3[(m - 1) + (m - 2) + (m - 3) + \cdots + 2 + 1].$$

The expression in brackets is the sum of the terms of an arithmetic sequence discussed in most algebra texts. Using formulas for arithmetic sequences, we can show that the expression in brackets is equal to $m(m - 1)/2$, so we have

$$DC_3\left[\frac{m(m - 1)}{2}\right] \tag{1}$$

as the total cost for keeping jobless trainees.

The total cost per batch is the sum of the training cost per batch, $C_1 + mDtC_2$, and the cost of keeping trainees without a proper job, given by equation (1). Since we assume that a batch of trainees is trained every $m$ months, the total cost per month, $Z(m)$, is given by

$$Z(m) = \frac{C_1 + mDtC_2}{m} + \frac{DC_3\left[\dfrac{m(m - 1)}{2}\right]}{m}$$
$$= \frac{C_1}{m} + DtC_2 + DC_3\left(\frac{m - 1}{2}\right).$$

## Exercises

1. Find $Z'(m)$.

2. Solve the equation $Z'(m) = 0$. As a practical matter, it is usually required that $m$ be a positive integer. If $m$ does not come out to be an integer, then $m^+$ and $m^-$, the two integers closest to $m$, must be chosen. Calculate both $Z(m^+)$ and $Z(m^-)$; the smaller of the two provides the optimum value of $Z$.

3. Suppose a company finds that its demand for trainees is three per month, that a training program requires 12 months, that the fixed cost of training a batch of trainees is $15,000, that the marginal cost per trainee per month is $100, and that trainees are paid $900 per month after training, but before going to work. Use your result from Exercise 2 to find $m$.

4. Since $m$ is not an integer, find $m^+$ and $m^-$.

5. Calculate $Z(m^+)$ *and* $Z(m^-)$.

6. What is the optimum time interval between successive batches of trainees? How many trainees should be in a batch?

# Extended Project

A pharmaceutical company is planning to gradually expand its service area over the next year and needs to hire and train new sales representatives. The company is considering two options for training programs. The first option is to bring new employees from various sales regions to the company headquarters for an intensive two-week training program with top company experts. This option is expensive due to travel and lodging costs. The second option involves online training through the use of webinars. Although this option avoids travel expenses, the same level of intensity is not possible and the training requires four weeks to be effective. The details for the two options are below.

|       | On-site Training | Online Training |
|-------|------------------|-----------------|
| $D$   | 4                | 4               |
| $C_1$ | 6000             | 6000            |
| $C_2$ | 3500             | 2000            |
| $C_3$ | 3000             | 3000            |
| $t$   | .5               | 1               |

(a) For the on-site training option, what is the optimum time interval between successive batches of trainees? How many trainees should be in a batch? What is the total monthly cost of this training program?

(b) Repeat part (a) for the online training option.

(c) Your employer has asked you to write a report summarizing your findings about the cost of the two training options over the next year, and to make and justify a recommendation about which program to select.

# Integral Calculus

# 13

## CHAPTER

In business and finance, integrals are used to find total revenue from marginal revenue functions. Integral calculus is also used to determine the point where profitability fails for a company. Social scientists use integrals to estimate the total population size from a rate-of-growth function. Integral calculus is also needed to model and solve problems involving rates of changes, such as the rate of growth of funds in a financial investment or the bacteria count in a dairy product. Exercises and examples in this chapter show how integral calculus allows us to tackle all these problems and more.

The derivative and its applications, which were studied in Chapters 11 and 12, are part of what is called *differential calculus*. This chapter is devoted to the other main branch of calculus, *integral calculus*. Integrals are used to find areas (which could represent total revenue or total population), to solve complicated probability problems, to calculate the location of an object (such as the height of a newly launched GPS satellite) from its velocity, and in a variety of other ways. Such topics may seem quite different from those studied earlier, but the fundamental theorem of calculus in Section 13.5 will reveal a surprisingly close connection between differential and integral calculus.

## 13.1 Antiderivatives

Functions used in applications in previous chapters have provided information about the *total amount* of a quantity, such as cost, revenue, profit, data usage, gallons of oil, or distance. Derivatives of these functions provided information about the rate of change of these quantities and allowed us to answer important questions about the extrema of the functions. It is not always possible to find ready-made functions that provide information about the total amount of a quantity, but it is often possible to collect enough data to come up with a function that gives the *rate of change* of a quantity. We know that derivatives give the rate of change when the total amount is known. In this section, we shall see that this process can be reversed: When the rate of change is known, a function that gives the total amount can be obtained by a process called *antidifferentiation*. Here is the basic definition.

## Antiderivatives

If $F$ and $f$ are functions such that $F' = f$, then $F$ is said to be an **antiderivative** of $f$.

## EXAMPLE 1

**(a)** If $F(x) = 10x$, then $F'(x) = 10$, so $F(x) = 10x$ is an antiderivative of $f(x) = 10$.

**(b)** For $F(x) = x^5$, $F'(x) = 5x^4$, which means that $F(x) = x^5$ is an antiderivative of $f(x) = 5x^4$.

## EXAMPLE 2 Find an antiderivative of $f(x) = 2x$.

**Solution** By remembering formulas for derivatives, it is easy to see that $F(x) = x^2$ is an antiderivative of $f(x)$ because $F'(x) = 2x = f(x)$. Note that $G(x) = x^2 + 2$ and $H(x) = x^2 - 4$ are also antiderivatives of $f(x)$, because

$$G'(x) = 2x + 0 = f(x) \quad \text{and} \quad H'(x) = 2x - 0 = f(x). \; ✓_1$$

Any two of the antiderivatives of $f(x) = 2x$ that were found in Example 2 differ by a constant. For instance, $G(x) - F(x) = 2$ and $H(x) - G(x) = -6$. The same thing is true in general.

> If $F$ and $G$ are both antiderivatives of $f$, then there is a constant $C$ such that
> $$F(x) - G(x) = C.$$
> (Two antiderivatives of a function can differ only by a constant.)

The statement in the box reflects a geometric fact about derivatives, namely, that the derivative of a function gives the slope of the tangent line at any number $x$. For example, if you graph the three antiderivatives of $f(x) = 2x$ found in Example 2, you will see that the graphs all have the same shape because, at any $x$-value, their tangent lines all have the same slope, as shown in Figure 13.1.

The family of all antiderivatives of the function $f$ is indicated by

$$\int f(x) \, dx.$$

The symbol $\int$ is the **integral sign**, $f(x)$ is the **integrand**, and $\int f(x) \, dx$ is called an **indefinite integral**. Since any two antiderivatives of $f$ differ by a constant (which means that one is a constant plus the other), we can describe the indefinite integral as follows.

✓ **Checkpoint 1**

Find an antiderivative for each of the given expressions.

**(a)** $3x^2$

**(b)** $5x$

**(c)** $8x^7$

*Answers to Checkpoint exercises are found at the end of the section.*

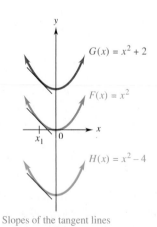

Slopes of the tangent lines at $x = x_1$ are the same.

**Figure 13.1**

## Indefinite Integral

If $F'(x) = f(x)$, then

$$\int f(x)\, dx = F(x) + C,$$

Integral sign

Integrand

Constant

Antiderivative

for any real number $C$.

For example, with this notation,

$$\int 2x\, dx = x^2 + C.$$

**NOTE** The $dx$ in the indefinite integral $\int f(x)\, dx$ indicates that $x$ is the variable of the function whose antiderivative is to be found, in the same way that $dy/dx$ denotes the derivative when $y$ is a function of the variable $x$. For example, in the indefinite integral $\int 2ax\, dx$, the variable of the function is $x$ (that is, the integrand is $f(x) = 2ax$), whereas in the indefinite integral $\int 2ax\, da$, the variable is $a$ (that is, the integrand is $f(a) = 2ax$).

The symbol $\int f(x)\, dx$ was created by G. W. Leibniz (1646–1716) in the latter part of the 17th century. The $\int$ is an elongated S from *summa*, the Latin word for *sum*. The word *integral* as a term in calculus was coined by Jakob Bernoulli (1654–1705), a Swiss mathematician who corresponded frequently with Leibniz. The relationship between sums and integrals will be clarified in Section 13.4.

Because finding an antiderivative is the inverse of finding a derivative, each formula for derivatives leads to a rule for antiderivatives. For instance, the power rule for derivatives tells us that

$$\text{if } F(x) = x^4, \qquad \text{then } F'(x) = 4x^3.$$

Consequently,

$$\text{if } F(x) = \frac{1}{4}x^4, \qquad \text{then } F'(x) = \frac{1}{4}(4x^3) = x^3.$$

In other words, an antiderivative of $f(x) = x^3$ is $F(x) = \frac{1}{4}x^4$. Similarly,

$$\text{if } G(x) = \frac{1}{8}x^8, \qquad \text{then } G'(x) = \frac{1}{8}(8x^7) = x^7,$$

so $G(x) = \frac{1}{8}x^8$ is an antiderivative of $g(x) = x^7$. The same pattern holds in the general case: To find the antiderivative of $x^n$, increase the exponent by 1 and divide by that same number.

## Power Rule for Antiderivatives

For any real number $n \neq -1$,

$$\int x^n\, dx = \frac{1}{n+1}x^{n+1} + C.$$

This result can be verified by differentiating the expression on the right in the box:

$$\frac{d}{dx}\left(\frac{1}{n+1}x^{n+1} + C\right) = \frac{n+1}{n+1}x^{(n+1)-1} + 0 = x^n.$$

(If $n = -1$, the expression in the denominator is 0, and the power rule cannot be used. We will see later in this section how to find an antiderivative in this case.)

---

**EXAMPLE 3**   Find each indefinite integral.

**(a)** $\displaystyle\int x^5\, dx$

**Solution**   Use the power rule with $n = 5$:

$$\int x^5 dx = \frac{1}{5+1} x^{5+1} + C = \frac{1}{6} x^6 + C.$$

Notice that, in the first step (in blue) on the line above, the integral sign and the $dx$ are no longer written because we have found the antiderivative. It is also important to remember to add the constant.

**(b)** $\displaystyle\int \frac{1}{t^2}\, dt$

**Solution**   To be able to use the power rule, we must first write $1/t^2$ as $t^{-2}$. Then

$$\int \frac{1}{t^2}\, dt = \int t^{-2}\, dt = \frac{1}{-2+1} t^{-2+1} = \frac{t^{-1}}{-1} + C = \frac{-1}{t} + C.$$

**(c)** $\displaystyle\int \sqrt{u}\, du$

**Solution**   Since $\sqrt{u} = u^{1/2}$,

$$\int \sqrt{u}\, du = \int u^{1/2} du = \frac{1}{1/2+1} u^{1/2+1} + C = \frac{1}{3/2} u^{3/2} + C = \frac{2}{3} u^{3/2} + C.$$

To check this result, differentiate $(2/3)u^{3/2} + C$; the derivative is $u^{1/2}$, the original function.

**(d)** $\displaystyle\int dx$

**Solution**   Writing $dx$ as $1 \cdot dx$, and using the fact that $x^0 = 1$ for any nonzero number $x$, we obtain

$$\int dx = \int 1\, dx = \int x^0\, dx = \frac{1}{1} x^1 + C = x + C.$$

✓ **Checkpoint 2**

Find each of the following.

**(a)** $\displaystyle\int x^6\, dx$

**(b)** $\displaystyle\int \sqrt[4]{x}\, dx$

**(c)** $\displaystyle\int 8\, dx$

As shown in Chapter 11, the derivative of the product of a constant and a function is the product of the constant and the derivative of the function. A similar rule applies to antiderivatives. Also, since derivatives of sums and differences are found term by term, antiderivatives can also be found term by term.

---

**Properties of Antiderivatives**

Let $f$ and $g$ be functions that have antiderivatives and let $k$ be any real number.

**Constant-multiple rule:**  $\displaystyle\int k \cdot f(x)\, dx = k \int f(x)\, dx.$

**Sum-or-difference rule:**  $\displaystyle\int [f(x) \pm g(x)]\, dx = \int f(x)\, dx \pm \int g(x)\, dx.$

🛑 **CAUTION** The constant-multiple rule requires that $k$ be a *number*. The rule does not apply to a *variable*. For example,

$$\int x\sqrt{x-1}\, dx \neq x\int \sqrt{x-1}\, dx.$$

**EXAMPLE 4**    Find each of the given indefinite integrals.

**(a)** $\displaystyle\int 4x^7\, dx$

**Solution** By the constant-multiple rule and the power rule,

$$\int 4x^7 = 4\int x^7\, dx = 4\left(\frac{1}{8}x^8\right) + C = \frac{1}{2}x^8 + C.$$

Since $C$ represents any real number, it is not necessary to multiply it by 4 in the next-to-last step. In other words, if we add a constant (let's call it $C_1$) inside the parentheses above, and then distribute the 4, we get

$$4\int x^7\, dx = 4\left(\frac{1}{8}x^8 + C_1\right) = \frac{1}{2}x^8 + 4C_1 = \frac{1}{2}x^8 + C,$$

where we simply rename the constant $4C_1$ as $C$. In practice, $C$ represents an arbitrary constant, so we omit the intermediate step and label the final constant as $C$.

**(b)** $\displaystyle\int \frac{10}{z^6}\, dz$

**Solution** Convert to a negative exponent, and use the constant-multiple rule and the power rule:

$$\int \frac{10}{z^6}\, dz = \int 10\, z^{-6}\, dz$$

$$= 10\int z^{-6}\, dz \qquad \text{Constant-multiple rule}$$

$$= 10\left(\frac{z^{-5}}{-5}\right) + C \qquad \text{Power rule}$$

$$= \frac{-2}{z^5} + C.$$

**(c)** $\displaystyle\int (5y^2 - 6y + 2)\, dy$

By extending the sum-or-difference rule to more than two terms, we get

$$\int (5y^2 - 6y + 2)\, dy = \int 5y^2\, dy - \int 6y\, dy + \int 2\, dy$$

$$= 5\int y^2\, dy - 6\int y\, dy + 2\int dy$$

$$= 5\left(\frac{y^3}{3}\right) - 6\left(\frac{y^2}{2}\right) + 2y + C$$

$$= \frac{5}{3}y^3 - 3y^2 + 2y + C.$$

Only one constant $C$ is needed in the answer, because the three constants from the term-by-term antiderivatives are combined. ✔3

✔ **Checkpoint 3**

Find each of the following.

**(a)** $\displaystyle\int (-6x^8)\, dx$

**(b)** $\displaystyle\int 7x^{2/3}\, dx$

**(c)** $\displaystyle\int \frac{5}{x^3}\, dx$

**(d)** $\displaystyle\int (5x^4 - 3x^2 + 8)\, dx$

**(e)** $\displaystyle\int \left(4\sqrt[3]{x} + \frac{2}{x^2}\right) dx$

The nice thing about working with antiderivatives is that you can always check your work by taking the derivative of the result. For instance, in Example 4(c), check that $\frac{5}{3}y^3 - 3y^2 + 2y + C$ is the required antiderivative by taking the derivative:

$$\frac{d}{dy}\left(\frac{5}{3}y^3 - 3y^2 + 2y + C\right) = 5y^2 - 6y + 2.$$

The result is the original function to be integrated, so the work checks out.

**EXAMPLE 5** Find each of the given indefinite integrals.

**(a)** $\displaystyle\int \frac{x^2 + 1}{\sqrt{x}}\, dx$

**Solution** It is *not* correct to find the antiderivative of the numerator and denominator separately. We must do some algebra to rewrite the integrand as follows:

$$\int \frac{x^2 + 1}{\sqrt{x}}\, dx = \int \left(\frac{x^2}{\sqrt{x}} + \frac{1}{\sqrt{x}}\right) dx$$

$$= \int \left(\frac{x^2}{x^{1/2}} + \frac{1}{x^{1/2}}\right) dx$$

$$= \int (x^{3/2} + x^{-1/2})\, dx. \qquad \text{Quotient rule for exponents}$$

Now find the antiderivative:

$$\int (x^{3/2} + x^{-1/2})\, dx = \int x^{3/2}\, dx + \int x^{-1/2}\, dx$$

$$= \frac{x^{5/2}}{5/2} + \frac{x^{1/2}}{1/2} + C$$

$$= \frac{2}{5}x^{5/2} + 2x^{1/2} + C.$$

**(b)** $\displaystyle\int (x^2 - 1)^2\, dx$

**Solution** First, use algebra to rewrite the integrand, and then find the antiderivative.

$$\int (x^2 - 1)^2\, dx = \int (x^4 - 2x^2 + 1)\, dx$$

$$= \int x^4\, dx - \int 2x^2\, dx + \int 1\, dx$$

$$= \frac{x^5}{5} - \frac{2x^3}{3} + x + C \quad ✔ 4$$

✓ **Checkpoint 4**

Find each of the following.

**(a)** $\displaystyle\int \frac{\sqrt{x} + 1}{x^2}\, dx$

**(b)** $\displaystyle\int (\sqrt{x} + 2)^2\, dx$

As shown in Chapter 11, the derivative of $f(x) = e^x$ is $f'(x) = e^x$. Also, the derivative of $f(x) = e^{kx}$ is $f'(x) = k \cdot e^{kx}$. These results lead to the following formulas for antiderivatives of exponential functions.

## Antiderivatives of Exponential Functions

If $k$ is a real number, $k \neq 0$, then

$$\int e^x\, dx = e^x + C;$$

$$\int e^{kx}\, dx = \frac{1}{k} \cdot e^{kx} + C.$$

**EXAMPLE 6** Here are some indefinite integrals involving exponential functions:

**(a)** $\int 8e^x \, dx = 8 \int e^x \, dx = 8e^x + C;$

**(b)** $\int e^{8t} \, dt = \frac{1}{8}e^{8t} + C;$

**(c)** $\int 3e^{(5/6)u} \, du = 3\left(\frac{1}{5/6}e^{(5/6)u}\right) + C = 3\left(\frac{6}{5}\right)e^{(5/6)u} + C$

$\qquad\qquad = \frac{18}{5}e^{(5/6)u} + C.$ ✓₅

✓ **Checkpoint 5**

Find each of the following.

**(a)** $\int (-7e^x) \, dx$

**(b)** $\int e^{5x} \, dx$

**(c)** $\int (e^{4x} - 4e^x) \, dx$

**(d)** $\int (-11e^{-x}) \, dx$

The antiderivative formula $\int x^n \, dx = \dfrac{x^{n+1}}{n+1} + C$ does not hold when $n = -1$, because the denominator is 0 in that case. Nevertheless, the function $g(x) = x^{-1}$ does have an antiderivative. Recall that the derivative of $f(x) = \ln|x|$ is $f'(x) = 1/x = x^{-1}$.

**Antiderivative of $x^{-1}$**

$$\int x^{-1} \, dx = \int \frac{1}{x} \, dx = \ln|x| + C, \qquad \text{where } x \neq 0.$$

⚠ **CAUTION** The domain of the logarithmic function is the set of positive real numbers. However, $y = x^{-1} = 1/x$ has the set of all nonzero real numbers as its domain, so the absolute value of $x$ *must* be used in the antiderivative.

**EXAMPLE 7** Here are some indefinite integrals involving logarithmic functions:

**(a)** $\int \frac{4}{x} \, dx = 4 \int \frac{1}{x} \, dx = 4 \cdot \ln|x| + C;$

**(b)** $\int \left(-\frac{5}{x} + e^{-2x}\right) dx = -5 \cdot \ln|x| - \frac{1}{2}e^{-2x} + C.$ ✓₆

✓ **Checkpoint 6**

Find each of the following.

**(a)** $\int (-9/x) \, dx$

**(b)** $\int (8e^{4x} - 3x^{-1}) \, dx$

## Applications

In the preceding examples, a family of antiderivative functions was found. In many applications, however, the given information allows us to determine the value of the constant $C$. The next three examples illustrate this idea.

**EXAMPLE 8** **Cell Phones** The *rate of change* of the number of cell phone subscriptions worldwide (in millions per year) can be approximated for the years 2000 through 2016 by

$$\frac{dS}{dx} = -10x^2 + 162x - 34.$$

In this model, $x = 0$ corresponds to the year 2000, when there were 738 million cell phone subscriptions worldwide. Find the function $S(x)$ that gives the number of subscribers (in millions) in year $x$. (Data from: www.worldbank.org.)

**Solution**  Because $\dfrac{dS}{dx}$ gives the *rate of change* of cell phone subscriptions, we find the number of subscribers by taking the antiderivative:

$$S(x) = \int \frac{dS}{dx} dx$$

$$= \int (-10x^2 + 162x - 34)dx$$

$$= -\frac{10}{3}x^3 + \frac{162}{2}x^2 - 34x + C$$

$$= -\frac{10}{3}x^3 + 81x^2 - 34x + C.$$

To find the value of $C$, use the fact that $S(0) = 738$ because there were 738 million subscribers in the year 2000:

$$S(0) = 738$$

$$-\frac{10}{3}(0)^3 + 81(0)^2 - 34(0) + C = 738$$

$$C = 738.$$

Thus, the number of subscribers in year $x$ is

$$S(x) = -\frac{10}{3}x^3 + 81x^2 - 34x + 738.\ \boxed{7}$$

✓ **Checkpoint 7**

Find $S(x)$ given that
$S'(x) = 7x + 3$ and $S(4) = 76$.

---

**EXAMPLE 9**    **Marginal Analysis**  Suppose the marginal revenue from selling $x$ units of a product is given by $R'(x) = 40/e^{.05x} + 10$.

**(a)**  Find the revenue function.

**Solution**  The marginal revenue is the derivative of the revenue function, so

$$R'(x) = \frac{40}{e^{.05x}} + 10$$

$$R(x) = \int \left( \frac{40}{e^{.05x}} + 10 \right) dx = \int (40e^{-.05x} + 10)\, dx$$

$$= 40\left( \frac{-1}{.05} \right)e^{-.05x} + 10x + C = -800e^{-.05x} + 10x + C,$$

where $C$ is a constant. If $x = 0$, then $R(0) = 0$ (no items sold means no revenue), so

$$-800e^{0} + 10(0) + C = 0$$

$$-800 + C = 0$$

$$C = 800.$$

Thus,

$$R(x) = -800e^{-.05x} + 10x + 800$$

is the revenue function.

**(b)**  Find the cost function given that the marginal cost is $C'(x) = \dfrac{9}{\sqrt[4]{x}}$, and the cost of 16 units is $200.

**Solution** Because marginal cost is the derivative of the cost function, we have

$$C(x) = \int C'(x)\, dx$$

$$= \int \frac{9}{\sqrt[4]{x}}\, dx = 9 \int \frac{1}{x^{1/4}}\, dx = 9 \int x^{-1/4}\, dx$$

$$= 9\left(\frac{4}{3} x^{3/4}\right) + D = 12x^{3/4} + D,$$

where $D$ is a constant (we use $D$ for the name of the constant because we had already used $C$ as the traditional name for the cost function). We find $D$ by using the fact that 16 units cost \$200, hence $C(16) = 200$:

$$C(16) = 12(16)^{3/4} + D = 12(8) + D = 96 + D = 200$$

$$D = 200 - 96 = 104.$$

Therefore the cost function is $C(x) = 12x^{3/4} + 104.$ ◢8

**Checkpoint 8**

The marginal cost at a level of production of $x$ items is

$$C'(x) = 2x^3 + 6x - 5.$$

The fixed cost is \$800. Find the cost function $C(x)$.

---

**EXAMPLE 10** **Population Growth** The population of the southern region of the United States was 101 million at the start of the millennium. For the years 2000 through 2017, its *growth rate* (in millions per year) can be approximated by the function

$$f(t) = 1.212e^{.012t},$$

where $t = 0$ corresponds to the year 2000. (Data from: www.census.gov.)

**(a)** Find the function $F(t)$ that gives the population (in millions) in year $t$.

**Solution** The derivative of the population function $F(t)$ is the rate at which the population is growing, namely, $F'(t) = f(t) = 1.212e^{.012t}$. Therefore,

$$F(t) = \int F'(t)\, dt = \int 1.212e^{.012t}\, dt = 1.212\left(\frac{1}{.012}\right)e^{.012t} + C = 101e^{.012t} + C.$$

Because the population was 101 million in 2000 (that is, $101 = F(0)$) we have

$$101 = F(0) = 101e^{.012(0)} + C = 101e^0 + C = 101 + C,$$

so that $C = 0$. Therefore, the population function is $F(t) = 101e^{.012t}$.

**(b)** According to this model, what is the population of the South projected to be (to the nearest million) in the year 2023?

**Solution** Because 2023 corresponds to $t = 23$, the population is projected to be

$$F(23) = 101e^{.012(23)} \approx 133 \text{ million}.$$ ◢9

✓**Checkpoint 9**

Suppose a population growth rate (in millions per year) is given by $f(t) = 5e^{.04t}$ and that the population is 125 million in the year $t = 0$.

**(a)** Find the population function $F(t)$ in millions.

**(b)** Find the population in year $t = 20$.

# 13.1 Exercises

**1.** What must be true of $F(x)$ and $G(x)$ if both are antiderivatives of $f(x)$?

◣ **2.** How is the antiderivative of a function related to the function?

◣ **3.** In your own words, describe what is meant by an integrand.

◣ **4.** Explain why the restriction $n \neq -1$ is necessary in the rule

$$\int x^n\, dx = \frac{1}{n+1} x^{n+1} + C.$$

*Find each of the given indefinite integrals. (See Examples 3–7.)*

**5.** $\int 12x\, dx$

**6.** $\int 25r^2\, dr$

**7.** $\int 8p^3\, dp$

**8.** $\int 5t^5\, dt$

**9.** $\int 105\, dx$

**10.** $\int 35\, dt$

**11.** $\int (5z - 1)\,dz$

**12.** $\int (2m^2 + 3)\,dm$

**13.** $\int (z^2 - 4z + 6)\,dz$

**14.** $\int (2y^2 + 10y + 7)\,dy$

**15.** $\int (x^3 - 14x^2 + 22x + 8)\,dx$

**16.** $\int (5x^3 + 5x^2 - 10x - 6)\,dx$

**17.** $\int 6\sqrt{y}\,dy$

**18.** $\int (7 + 12\sqrt{t})\,dt$

**19.** $\int (6t\sqrt{t} + 3\sqrt[4]{t})\,dt$

**20.** $\int (12\sqrt{x} - x\sqrt{x})\,dx$

**21.** $\int (56t^{1/2} + 18t^{7/2})\,dt$

**22.** $\int (10u^{3/2} - 8u^{5/2})\,du$

**23.** $\int \dfrac{24}{x^3}\,dx$

**24.** $\int \dfrac{-20}{x^2}\,dx$

**25.** $\int \left(\dfrac{1}{y^3} - \dfrac{2}{\sqrt{y}}\right) dy$

**26.** $\int \dfrac{1}{\sqrt{x}} - \dfrac{6}{x^4}\,dx$

**27.** $\int (6x^{-3} + 2x^{-1})\,dx$

**28.** $\int (3x^{-1} - 10x^{-2})\,dx$

**29.** $\int 4e^{3u}\,du$

**30.** $\int -e^{-3x}\,dx$

**31.** $\int 3e^{-.8x}\,dx$

**32.** $\int -4e^{.2v}\,dv$

**33.** $\int \left(\dfrac{6}{x} + 4e^{-.5x}\right) dx$

**34.** $\int \left(\dfrac{9}{x} - 3e^{-.4x}\right) dx$

**35.** $\int \dfrac{1 + 2t^4}{t}\,dt$

**36.** $\int \dfrac{2y^{1/2} - 3y^2}{y^3}\,dy$

**37.** $\int \left(e^{2u} + \dfrac{4}{u}\right) du$

**38.** $\int \left(\dfrac{2}{v} - e^{3v}\right) dv$

**39.** $\int (5x + 1)^2\,dx$

**40.** $\int (2y - 1)^2\,dy$

**41.** $\int \dfrac{\sqrt{x} + 1}{\sqrt[3]{x}}\,dx$

**42.** $\int \dfrac{1 - 2\sqrt[3]{z}}{\sqrt[3]{z}}\,dz$

**43.** Find the equation of the curve whose tangent line has a slope of

$$f'(x) = x^{2/3}$$

if the point $(1, \frac{3}{5})$ is on the curve.

**44.** The slope of the tangent line to a curve is given by

$$f'(x) = 6x^2 - 4x - 3.$$

If the point $(0, 1)$ is on the curve, find the equation of the curve.

*Work the given problems. (See Examples 8 and 10.)*

**45.** **Homeownership** Homeownership (as a percentage) in the United States at the beginning of the millennium was 67.1%. The *rate of change* of homeownership over time can be approximated by

$$f'(x) = -.1x + .59,$$

where $x = 0$ corresponds to the year 2000. (Data from: www.census.gov.)

(a) Find the function $f(x)$ that gives the homeownership in year $x$.

(b) What was homeownership in 2016 (round to one decimal place)?

**46.** **Crowd Size** The *rate of change* of the size of the crowd (in people per hour) at a certain theme park is given by

$$P'(t) = 2800 - 400t,$$

where $t = 0$ corresponds to the opening of the park at 8:00 a.m.

(a) Find the function $P(t)$ that gives the size of the crowd $t$ hours after opening.

(b) When is the size of the crowd at a maximum?

(c) What is the size of the maximum crowd?

**47.** **NVIDIA Stock** The semiconductor corporation NVIDIA had a banner year in 2016. The *growth* in its stock price per month can be approximated by

$$p'(x) = 3e^{.12x},$$

where $x$ is in months and $x = 0$ corresponds to the start of 2016, when the stock price was about \$29. (Data from: www.morningstar.com.)

(a) Find the stock price $p(x)$.

(b) To the nearest dollar, find $p(12)$, the stock price just 12 months later.

**48.** **Natural Science** If the rate of excretion of a biochemical compound is given by

$$f'(t) = .01e^{-.01t},$$

the total amount excreted by time $t$ (in minutes) is $f(t)$.

(a) Find an expression for $f(t)$.

(b) If 0 units are excreted at time $t = 0$, how many units are excreted in 10 minutes?

*Work the given problems. (See Example 9.)*

**49.** **Marginal Profit** The marginal profit at a sales level of $x$ hundred items of a product is $P'(x) = 4 - 6x + 3x^2$, and the profit when no items are sold is $-\$40$ (recall that a negative profit indicates a loss). Find the profit function.

**50.** **Marginal Revenue** The marginal revenue at a sales level of $x$ thousand items of a product is $R'(x) = 50 - 4x - 3x^2$. Find the revenue function.

**Marginal Cost** *Find the cost function for each of the given marginal cost functions.*

**51.** $C'(x) = x^{2/3} + 5$; 8 units cost \$58.

**52.** $C'(x) = x^{1/2}$; 16 units cost \$70.

**53.** $C'(x) = x^2 - 2x + 3$; 3 units cost \$25.

**54.** $C'(x) = .2x^2 + .4x + .8$; 6 units cost \$32.50.

**55.** $C'(x) = -\dfrac{40}{e^{.05x}} + 100$; 5 units cost \$1400.

**56.** $C'(x) = 1.2e^{.02x}$; 2 units cost \$105.

*Work the given problems.*

**57. Population Growth** The population of the United States was about 326 million in the year 2017. The projected growth rate of the population (in millions per year) can be approximated by

$$f(t) = \frac{66}{t} \quad (17 \le t \le 47),$$

where $t = 17$ corresponds to the year 2017. (Data from: www.census.gov.)

(a) Find the function $F(t)$ that gives the projected population in year $t$. (Round the constant $C$ to the nearest integer.)

(b) Find and interpret $F(47)$.

**58. Driving Accidents** Recently, there were approximately 200 million licensed drivers in the United States. Sixteen-year-olds had 20 accidents per 100 drivers. The rate at which the number of accidents per 100 drivers changed as the driver's age increased was approximated by

$$g(x) = -\frac{10.2}{x} \quad (16 \le x \le 80),$$

where $x$ is the age of the driver (in years). (Data from: National Safety Council.)

(a) Find a function that gives the number of accidents per 100 drivers of age $x$ years. (Round the constant $C$ to two decimal places.)

(b) Approximately how many accidents per 100 drivers did drivers of age 18 have? of age 20? of age 50? of age 70?

(c) What reasons might account for the answers in part (b)?

## ✓ Checkpoint Answers

1. Only one possible antiderivative is given for each.

   (a) $x^3$       (b) $\frac{5}{2}x^2$       (c) $x^8$

2. (a) $\frac{1}{7}x^7 + C$       (b) $\frac{4}{5}x^{5/4} + C$       (c) $8x + C$

3. (a) $-\frac{2}{3}x^9 + C$       (b) $\frac{21}{5}x^{5/3} + C$

   (c) $-\frac{5}{2}x^{-2} + C$ or $-\frac{5}{2x^2} + C$       (d) $x^5 - x^3 + 8x + C$

   (e) $3x^{4/3} - \frac{2}{x} + C$

4. (a) $-\frac{2}{\sqrt{x}} - \frac{1}{x} + C$       (b) $\frac{x^2}{2} + \frac{8}{3}x^{3/2} + 4x + C$

5. (a) $-7e^x + C$       (b) $\frac{1}{5}e^{5x} + C$

   (c) $\frac{1}{4}e^{4x} - 4e^x + C$       (d) $11e^{-x} + C$

6. (a) $-9 \cdot \ln|x| + C$       (b) $2e^{4x} - 3 \cdot \ln|x| + C$

7. $S(x) = \frac{7}{2}x^2 + 3x + 8$

8. $C(x) = \frac{1}{2}x^4 + 3x^2 - 5x + 800$

9. (a) $F(t) = 125e^{.04t}$       (b) $F(20) \approx 278.2$ million

# 13.2 Integration by Substitution

In Section 13.1, we saw how to find the antiderivatives of a few simple functions. The antiderivatives of more complicated functions can sometimes be found by a process called *integration by substitution*. The technique depends on the following concept: If $u = f(x)$, then the **differential** of $u$, written $du$, is defined as

$$du = f'(x)\,dx.$$

## ✓ Checkpoint 1

Find $du$ for the given functions.

(a) $u = 9x$

(b) $u = 5x^3 + 2x^2$

(c) $u = e^{-2x}$

For example, if $u = 6x^4$, then $du = 24x^3dx$. ✓₁

Differentials have many useful interpretations that are studied in more advanced courses. We shall use them only as a convenient notational device when finding an antiderivative such as

$$\int (3x^2 + 4)^6 6x\,dx.$$

The function $(3x^2 + 4)^4\, 6x$ is reminiscent of the chain rule, so we shall try to use differentials and the chain rule in *reverse* to find the antiderivative. Let $u = 3x^2 + 4$; then $du = 6x\,dx$. Now substitute $u$ for $3x^2 + 4$ and $du$ for $6x\,dx$ in the indefinite integral:

$$\int (3x^2 + 4)^4\, 6x\,dx = \int \overset{u}{\overbrace{(3x^2 + 4)^4}}\,\overset{du}{\overbrace{(6x\,dx)}}$$

$$= \int u^4\,du.$$

This last integral can now be found by the power rule:

$$\int u^4 \, du = \frac{u^5}{5} + C.$$

Finally, substitute $3x^2 + 4$ for $u$ back into the antiderivative above to get:

$$\frac{u^5}{5} + C = \frac{(3x^2 + 4)^5}{5} + C.$$

We can check the accuracy of this result by using the chain rule to take the derivative. We get

$$\frac{d}{dx}\left[\frac{(3x^2 + 4)^5}{5} + C\right] = \frac{1}{5} \cdot 5(3x^2 + 4)^4 (6x) + 0$$

$$= (3x^2 + 4)^4 \, 6x,$$

which is the original function.

This method of integration is called **integration by substitution.** As just shown, it is simply the chain rule for derivatives in reverse. The results can always be verified by differentiation.

---

**EXAMPLE 1**    Find

$$\int (4x + 5)^9 \, dx.$$

**Solution**  We choose $4x + 5$ as $u$. Then $du = 4 \, dx$. We are missing the constant 4. Using the fact that $4(1/4) = 1$, we can rewrite the integral:

$$\int (4x + 5)^9 \, dx = \frac{1}{4} \cdot 4 \int (4x + 5)^9 \, dx$$

$$= \frac{1}{4} \int \overset{u}{\overbrace{(4x + 5)^9}}\overset{du}{\overbrace{(4 \, dx)}} \qquad k \int f(x) \, dx = \int kf(x) \, dx$$

$$= \frac{1}{4} \int u^9 \, du \qquad\qquad \text{Substitute.}$$

$$= \frac{1}{4} \cdot \frac{u^{10}}{10} + C = \frac{u^{10}}{40} + C \qquad \text{Find the antiderivative.}$$

$$= \frac{(4x + 5)^{10}}{40} + C. \qquad\qquad \text{Substitute.} \;✓_2$$

---

✓ **Checkpoint 2**

Find the given antiderivatives.

(a) $\displaystyle \int 8x(4x^2 - 1)^5 \, dx$

(b) $\displaystyle \int (3x - 8)^4 \, dx$

(c) $\displaystyle \int 18x^2(x^3 - 7)^{3/2} \, dx$

---

⊘ **CAUTION**  When changing the $x$-problem to the $u$-problem, make sure that the change is complete—that is, that no $x$'s are left in the $u$-problem.

---

**EXAMPLE 2**    Find

$$\int x^2 \sqrt{x^3 + 5} \, dx.$$

**Solution**  Rewrite the function as $x^2(x^3 + 5)^{1/2}$. An expression raised to a power is usually a good choice for $u$, so we let $u = x^3 + 5$; then $du = 3x^2 \, dx$. The integrand does not contain the constant 3, which is part of $du$. One way to take care of this is to solve the differential $du = 3x^2 \, dx$ for $x^2 \, dx$:

$$du = 3x^2 \, dx$$

$$\frac{1}{3} du = x^2 \, dx.$$

Substitute $u$ for $x^3 + 5$ and $\frac{1}{3}\,du$ for $x^2\,dx$:

$$\int x^2 \sqrt{x^3 + 5}\,dx = \int \sqrt{\underbrace{x^3 + 5}_{u}}\underbrace{(x^2\,dx)}_{\frac{1}{3}\,du} = \int \sqrt{u} \cdot \frac{1}{3}\,du.$$

Now use the constant-multiple rule to bring the $\frac{1}{3}$ outside the integral sign and then find the antiderivative:

$$\int x^2 \sqrt{x^3 + 5}\,dx = \int \sqrt{u} \cdot \frac{1}{3}\,du = \frac{1}{3}\int u^{1/2}\,du$$

$$= \frac{1}{3} \cdot \frac{u^{3/2}}{3/2} + C \qquad \text{Find the antiderivative.}$$

$$= \frac{2}{9}u^{3/2} + C \qquad \text{Simplify.}$$

$$= \frac{2}{9}(x^3 + 5)^{3/2} + C \qquad \text{Substitute.} \; \boxed{✓_3}$$

✓ **Checkpoint 3**

Find the given antiderivatives.

(a) $\displaystyle\int 4x(6x^2 + 3)^2\,dx$

(b) $\displaystyle\int x\sqrt{x^2 + 16}\,dx$

🛑 **CAUTION** The substitution method given in the preceding examples *will not always work*. For example, we might try to find

$$\int x^3 \sqrt{x^5 + 1}\,dx$$

by substituting $u = x^5 + 1$, so that $du = 5x^4\,dx$. However, there is no *constant* that can be inserted inside the integral sign to give $5x^4$. This integral, and a great many others, cannot be evaluated by substitution.

With practice, choosing $u$ will become easy if you keep two principles in mind. First, $u$ should equal some expression in the integral that, when replaced with $u$, tends to make the integral simpler. Second, and more importantly, $u$ must be an expression whose derivative is also present in the integral. The substitution should include as much of the integral as possible, so long as its derivative is still present. In Example 2, we could have chosen $u = x^3$, but $u = x^3 + 1$ is better, because it has the same derivative as $x^3$ and captures more of the original integral. If we carry this reasoning further, we might try $u = \sqrt{x^3 + 1} = (x^3 + 1)^{1/2}$, but this is a poor choice because $du = (1/2)(x^3 + 1)^{-1/2}(3x^2)\,dx$, an expression not present in the original integral.

**EXAMPLE 3**    Find

$$\int \frac{x^2 + 5}{(x^3 + 15x)^2}\,dx.$$

**Solution**   Let $u = x^3 + 15x$, so that $du = (3x^2 + 15)\,dx = 3(x^2 + 5)\,dx$. The integrand is missing the 3, so multiply by $\frac{3}{3}$, putting 3 inside the integral sign and $\frac{1}{3}$ outside:

$$\int \frac{x^2 + 5}{(x^3 + 15x)^2}\,dx = \frac{1}{3}\int \frac{3(x^2 + 5)}{(x^3 + 15x)^2}\,dx$$

$$= \frac{1}{3}\int \frac{du}{u^2} \qquad \text{Substitute.}$$

$$= \frac{1}{3}\int u^{-2}\,du \qquad \text{Rules of exponents}$$

$$= \frac{1}{3} \cdot \frac{u^{-1}}{-1} + C \qquad \text{Find the antiderivative.}$$

$$= \frac{-1}{3u} + C \qquad \text{Simplify.}$$

$$= \frac{-1}{3(x^3 + 15x)} + C \qquad \text{Substitute.} \; \boxed{✓_4}$$

✓ **Checkpoint 4**

Find the given antiderivatives.

(a) $\displaystyle\int z(z^2 + 1)^2\,dz$

(b) $\displaystyle\int \frac{x^2 + 3}{\sqrt{x^3 + 9x}}\,dx$

Recall that if $f(x)$ is a function, then by the chain rule, the derivative of the exponential function $y = e^{f(x)}$ is

$$\frac{dy}{dx} = e^{f(x)} \cdot f'(x).$$

Both the function $f(x)$ and its derivative appear on the right. This suggests that the anti-derivative of a function of the form $e^{f(x)} \cdot f'(x)$ can be found by letting $u$ be the exponent.

**EXAMPLE 4** Find

$$\int 2e^{x^2}x \, dx.$$

Let $u = x^2$, so that $du = 2x \, dx$. Then

$$\int 2e^{x^2}x \, dx = \int e^{x^2} \, 2x \, dx$$

$$= \int e^u \, du$$

$$= e^u + C = e^{x^2} + C.$$

Check this answer by using the chain rule to take the derivative:

$$\frac{d}{dx}(e^{x^2} + C) = e^{x^2}(2x) + 0 = 2e^{x^2} x.$$

**EXAMPLE 5** Find the given indefinite integrals.

**(a)** $\displaystyle\int e^{-12x+5} \, dx$

**Solution** Choose $u = -12x + 5$, so $du = -12 \, dx$. Multiply the integral by $(-\frac{1}{12})(-12)$, and use the rule for $\int e^u \, du$:

$$\int e^{-12x+5} \, dx = -\frac{1}{12} \cdot -12 \int e^{-12x+5} \, dx$$

$$= -\frac{1}{12} \int e^{-12x+5}(-12 \, dx)$$

$$= -\frac{1}{12} \int e^u \, du \qquad \text{Substitute.}$$

$$= -\frac{1}{12} e^u + C \qquad \text{Find the antiderivative.}$$

$$= -\frac{1}{12} e^{-12x+5} + C. \qquad \text{Substitute.}$$

**(b)** $\displaystyle\int 6x^2 \cdot e^{x^3} \, dx$

**Solution** Let $u = x^3$, the exponent on $e$. Then $du = 3x^2 \, dx$, and $(\frac{1}{3}) \, du = x^2 \, dx$.

$$\int 6x^2 \cdot e^{x^3} \, dx = \int 6e^{x^3}(x^2 \, dx)$$

$$= \int 6e^u \left(\frac{1}{3} \, du\right) \qquad \text{Substitute.}$$

$$= 2 \int e^u \, du \qquad \text{Constant-multiple rule}$$

$$= 2e^u + C \qquad \text{Find the antiderivative.}$$

$$= 2e^{x^3} + C. \qquad \text{Substitute.} \quad \boxed{\checkmark}_5$$

✓ **Checkpoint 5**

Find the given antiderivatives.

**(a)** $\displaystyle\int e^{5x} \, dx$

**(b)** $\displaystyle\int 10xe^{3x^2} \, dx$

**(c)** $\displaystyle\int 2x^3 e^{x^4-1} \, dx$

Recall that an antiderivative of $f(x) = 1/x$ is $\ln|x|$. The next example uses $\int x^{-1}\, dx = \ln|x| + C$ and the method of substitution.

**EXAMPLE 6**  Find the following.

**(a)** $\displaystyle\int \frac{dx}{9x + 6}$

**Solution**  Choose $u = 9x + 6$, so $du = 9\, dx$. Multiply by $(\frac{1}{9})(9)$:

$$\int \frac{dx}{9x + 6} = \frac{1}{9} \cdot 9 \int \frac{dx}{9x + 6} = \frac{1}{9} \int \frac{1}{9x + 6}(9\, dx)$$

$$= \frac{1}{9} \int \frac{1}{u}\, du = \frac{1}{9}\ln|u| + C = \frac{1}{9}\ln|9x + 6| + C.$$

**(b)** $\displaystyle\int \frac{(2x - 3)\, dx}{x^2 - 3x}$

**Solution**  Let $u = x^2 - 3x$, so that $du = (2x - 3)\, dx$. Then

$$\int \frac{(2x - 3)\, dx}{x^2 - 3x} = \int \frac{du}{u} = \ln|u| + C = \ln|x^2 - 3x| + C. \quad \checkmark_6$$

✓ **Checkpoint 6**

Find the given antiderivatives.

**(a)** $\displaystyle\int \frac{4\, dx}{x - 6}$

**(b)** $\displaystyle\int \frac{(3x^2 + 5)\, dx}{x^3 + 5x + 4}$

The techniques used in the preceding examples can be summarized as follows.

## Substitution Method

Let $u$ be some function of $x$.

| Form of the Integral | Form of the Antiderivative |
|---|---|
| **1.** $\displaystyle\int [u(x)]^n \cdot u'(x)\, dx$, where $n \neq -1$ | $\dfrac{[u(x)]^{n+1}}{n + 1} + C$ |
| **2.** $\displaystyle\int e^{u(x)} \cdot u'(x)\, dx$ | $e^{u(x)} + C$ |
| **3.** $\displaystyle\int \frac{u'(x)\, dx}{u(x)}$ | $\ln|u(x)| + C$ |

A slightly more complicated substitution technique is needed in some cases.

**EXAMPLE 7**  Find

$$\int x\sqrt[3]{(1 - x)}\, dx.$$

**Solution**  Let $u = 1 - x$. Then solve for $x$ to get $x = 1 - u$ and $dx = -du$. Now substitute:

$$\int x\sqrt[3]{(1 - x)}\, dx = \int (1 - u)\sqrt[3]{u}(-du) = \int (u - 1)u^{1/3}\, du$$

$$= \int (u^{4/3} - u^{1/3})\, du = \frac{3u^{7/3}}{7} - \frac{3u^{4/3}}{4} + C$$

$$= \frac{3}{7}(1 - x)^{7/3} - \frac{3}{4}(1 - x)^{4/3} + C. \quad \checkmark_7$$

✓ **Checkpoint 7**

Find $\displaystyle\int x(x + 1)^{2/3}\, dx$.

## Applications

Integration by substitution enables you to handle a wider variety of real-life applications.

---

**EXAMPLE 8** **Retail Sales** The rate of change of retail spending per capita in health and personal care stores (in hundreds of dollars per year) can be approximated by

$$S'(x) = \frac{2.1}{\sqrt{4.2x + 30.4}},$$

where $x = 0$ corresponds to the year 2000. Retail spending per capita in such stores in 2000 was \$551. (Data from: www.census.gov.)

**(a)** Find the function giving retail spending per capita in health and personal care stores (in hundreds of dollars). Round the constant $C$ to two decimal places.

**Solution** The retail spending per capita function is

$$S(x) = \int S'(x)\, dx = \int \frac{2.1}{\sqrt{4.2x + 30.4}}\, dx.$$

Let $u = 4.2x + 30.4$. Then $du = 4.2\, dx$ and so $\frac{1}{4.2}\, du = dx$. Thus we have

$$S(x) = \int \frac{2.1}{\sqrt{4.2x + 30.4}}\, dx$$

$$= \int \frac{2.1}{\sqrt{u}} \left( \frac{1}{4.2}\, du \right) \qquad \text{Substitute.}$$

$$= \frac{2.1}{4.2} \int \frac{1}{\sqrt{u}}\, du \qquad \text{Constant-multiple rule}$$

$$= \frac{1}{2} \int u^{-\frac{1}{2}}\, du \qquad \text{Property of exponents}$$

$$= \frac{1}{2}\, \frac{u^{-\frac{1}{2}+1}}{-\frac{1}{2} + 1} + C \qquad \text{Find the antiderivative.}$$

$$= \frac{1}{2} \cdot \frac{2}{1} u^{\frac{1}{2}} + C \qquad \text{Simplify.}$$

$$= \sqrt{u} + C$$

$$= \sqrt{4.2x + 30.4} + C. \qquad \text{Substitute.}$$

Find the value of $C$ by using the fact that retail spending per capita was \$551 in 2000; that is, $S(0) = 5.51$ (recall that $S(x)$ is in hundreds of dollars). Thus,

$$S(x) = \sqrt{4.2x + 30.4} + C$$

$$S(0) = \sqrt{4.2(0) + 30.4} + C$$

$$5.51 = 5.5136 + C$$

$$C \approx 0.$$

Therefore retail spending per capita in health and personal care stores is given by

$$S(x) = \sqrt{4.2x + 30.4}.$$

**(b)** If this trend continues, what will retail spending per capita in health and personal care stores be in the year 2022?

**Solution** In 2022 ($x = 22$) we have

$$S(22) = \sqrt{4.2(22) + 30.4} \approx 11.08 \text{ (in hundreds of dollars)}.$$

Thus, retail spending per capita in health and personal care stores is projected to be approximately \$1108 by the year 2022.

✓ **Checkpoint 8**

Find $S(x)$, given that

$$S'(x) = \frac{15}{\sqrt{5x + 1}} \text{ and } S(0) = 6.$$

# 13.2 Exercises

◤ **1.** Integration by substitution is related to what differentiation method? What type of integrand suggests using integration by substitution?

**2.** For each of the given integrals, decide what expression should be $u$. Then find $du$.

   **(a)** $\displaystyle\int (3x^2 - 5)^4 2x \, dx$      **(b)** $\displaystyle\int \sqrt{1 - x} \, dx$

   **(c)** $\displaystyle\int \frac{x^2}{2x^3 + 1} \, dx$      **(d)** $\displaystyle\int (8x - 8)(4x^2 - 8x) \, dx$

*Use substitution to find the given indefinite integrals. (See Examples 1–6 for Exercises 3–36 and Example 7 for Exercises 37–40.)*

**3.** $\displaystyle\int 3(12x - 1)^2 \, dx$       **4.** $\displaystyle\int 5(4 + 9t)^3 \, dt$

**5.** $\displaystyle\int \frac{5}{(3t - 6)^2} \, dt$       **6.** $\displaystyle\int \frac{8}{(7 - 2x)^5} \, dx$

**7.** $\displaystyle\int \sqrt{5 - 2x} \, dx$       **8.** $\displaystyle\int \sqrt{5x - 8} \, dx$

**9.** $\displaystyle\int r^2 \sqrt{r^3 + 3} \, dr$       **10.** $\displaystyle\int y^3 \sqrt{y^4 - 8} \, dy$

**11.** $\displaystyle\int (-4e^{5k}) \, dk$       **12.** $\displaystyle\int (-2e^{-3z} dz)$

**13.** $\displaystyle\int 4w^3 e^{2w^4} \, dw$       **14.** $\displaystyle\int 10ze^{-z^2} \, dz$

**15.** $\displaystyle\int (2 - t)e^{4t - t^2} \, dt$       **16.** $\displaystyle\int (3 - x^2)e^{9x - x^3} \, dx$

**17.** $\displaystyle\int \frac{e^{\sqrt{y}}}{\sqrt{y}} \, dy$       **18.** $\displaystyle\int \frac{e^{1/z^2}}{z^3} \, dz$

**19.** $\displaystyle\int \frac{-5}{12 + 6x} \, dx$       **20.** $\displaystyle\int \frac{9}{3 - 4x} \, dx$

**21.** $\displaystyle\int \frac{e^{2t}}{e^{2t} + 1} \, dt$       **22.** $\displaystyle\int \frac{e^{5w+1}}{2 - e^{5w+1}} \, dw$

**23.** $\displaystyle\int \frac{x + 2}{(2x^2 + 8x)^3} \, dx$       **24.** $\displaystyle\int \frac{4y - 2}{(y^2 - y)^4} \, dy$

**25.** $\displaystyle\int \frac{1}{\sqrt{2x + 9}} \, dx$       **26.** $\displaystyle\int \frac{1}{\sqrt{9 - 7x}} \, dx$

**27.** $\displaystyle\int \frac{x^2 + 1}{(x^3 + 3x)^{2/3}} \, dx$       **28.** $\displaystyle\int \frac{B^3 - 1}{(2B^4 - 8B)^{3/2}} \, dB$

**29.** $\displaystyle\int \frac{6x + 7}{3x^2 + 7x + 8} \, dx$       **30.** $\displaystyle\int \frac{x^2}{x^3 + 5} \, dx$

**31.** $\displaystyle\int 2x(x^2 + 5)^3 \, dx$       **32.** $\displaystyle\int 2y^2(y^3 - 8)^3 \, dy$

**33.** $\displaystyle\int (\sqrt{x^2 + 12x})(x + 6) \, dx$    **34.** $\displaystyle\int (\sqrt{x^2 - 6x})(x - 3) \, dx$

**35.** $\displaystyle\int \frac{1}{x(\ln x)} \, dx$       **36.** $\displaystyle\int \frac{1}{x(\ln x)^2} \, dx$

**37.** $\displaystyle\int \frac{5u}{\sqrt{u - 1}} \, du$       **38.** $\displaystyle\int \frac{2x}{(x + 5)^6} \, dx$

**39.** $\displaystyle\int t\sqrt{5t - 1} \, dt$       **40.** $\displaystyle\int 3r\sqrt{6 - r} \, dr$

*Work these problems. Round the constant C to two decimal places.*

**41. Production Costs** A company has found that the rate of change in the cost of a new production line (in thousands of dollars per year) is

$$C'(x) = \frac{60x}{5x^2 + 1},$$

where $x$ is the number of years the line is in use.

   **(a)** If the fixed costs are \$10,000, find the total cost function for the production line.

   **(b)** The company will add a new production line if the total cost of this one for the first five years stays below \$40,000. Should the company add the new line?

**42. Maintenance Costs** The rate of expenditure (in dollars per year) for maintenance of a particular machine is given by

$$M'(x) = \sqrt{x^2 + 8x}(2x + 8),$$

where $x$ is time measured in years. Total maintenance costs through the fourth year are \$540.

   **(a)** Find the total maintenance function.

   **(b)** How many years must pass before the total maintenance costs reach \$2000?

**43. Bicycle Shops** The rate of change of the number of retail bicycle stores can be approximated by

$$f(x) = -185.85e^{-.03x},$$

where $x = 0$ corresponds to the year 2000. The total number of stores was 6192 in the year 2000. (Data from: National Bicycle Dealers Association.)

   **(a)** Find the function giving the total number of stores in year $x$.

   **(b)** According to this model, how many stores will there be in 2020?

**44. Manufacturing** The rate of growth of profits (in millions of dollars per year) in year $x$ from a new manufacturing process is approximated by

$$f(x) = xe^{-x^2}.$$

The total profit in the third year of the new process is $10,000.

(a) Find the total profit function. (Remember that profits are given in millions of dollars.)

(b) As time goes on, what will happen to the total profit?

**45. Marginal Revenue** The marginal revenue (in dollars per unit) when producing $x$ units of a certain energy bar is given by

$$R'(x) = \frac{50x}{(x^2 + 10,000)^{3/4}}.$$

(a) Find the revenue function $R(x)$. (*Hint:* There is no revenue if no units are sold.)

(b) What is the revenue from selling 1000 units?

(c) How many units must be sold to produce revenue of at least $2000?

**46. Marginal Profit** The marginal profit (in thousands of dollars per unit) from the sale of a certain video game console is given by

$$P'(x) = 1.8x(x^2 + 27,000)^{-2/3}.$$

The profit from 150 units is $32,000.

(a) Find the profit function.

(b) What is the profit from selling 250 units?

(c) How many units must be sold to produce a profit of at least $100,000?

**47. Human Resources** The number of jobs in the mining industry is changing at a rate (in thousands of jobs per year) approximated by

$$f(x) = \frac{53}{x + 1},$$

where $x = 0$ corresponds to the year 2000. There were 520,000 mining industry jobs in 2000. (Data from: U.S. Bureau of Labor Statistics.)

(a) Find the function giving the number of mining industry jobs in year $x$.

(b) Find the projected number of mining industry jobs in the year 2020.

**48. Human Resources** For Nike Inc., the rate of change of revenue with respect to the number of employees (in billions of dollars of revenue per million employees) can be approximated by

$$f(x) = \frac{44}{2x + 1},$$

where $x$ is the number of employees in millions. When the number of employees was 34.4 million, the revenue was $19 billion. (Data from: www.mergentonline.com.)

(a) Find the revenue (in billions) as a function of $x$.

(b) Find the revenue (to the nearest billion) when there were 44 million employees.

✓**Checkpoint Answers**

1. (a) $du = 9\,dx$  (b) $du = (15x^2 + 4x)\,dx$
   (c) $du = -2e^{-2x}\,dx$

2. (a) $\dfrac{(4x^2 - 1)^6}{6} + C$  (b) $\dfrac{(3x - 8)^5}{15} + C$
   (c) $\dfrac{12(x^3 - 7)^{5/2}}{5} + C$

3. (a) $\dfrac{1}{9}(6x^2 + 3)^3 + C$  (b) $\dfrac{1}{3}(x^2 + 16)^{3/2} + C$

4. (a) $\dfrac{(z^2 + 1)^3}{6} + C$  (b) $\dfrac{2}{3}\sqrt{x^3 + 9x} + C$

5. (a) $\dfrac{1}{5}e^{5x} + C$  (b) $\dfrac{5}{3}e^{3x^2} + C$
   (c) $\dfrac{1}{2}e^{x^4-1} + C$

6. (a) $4\ln|x - 6| + C$  (b) $\ln|x^3 + 5x + 4| + C$

7. $\dfrac{3}{8}(x + 1)^{8/3} - \dfrac{3}{5}(x + 1)^{5/3} + C$

8. $S(x) = 6\sqrt{5x + 1}$

# 13.3 Integration by Parts

In Section 13.2, we saw that the method of integration by substitution allows us to use the chain rule in reverse to find certain antiderivatives. In this section, we will learn the method of **integration by parts**, which sometimes allows us to use the product rule in reverse to find an antiderivative.

To illustrate this idea, suppose that we want to find an antiderivative that involves a product such as the following:

$$\int xe^x\,dx.$$

We do not currently have a strategy for tackling this antiderivative. However, using the product rule, we see that

$$(xe^x)' = x(e^x)' + e^x(x)'$$
$$(xe^x)' = xe^x + e^x.$$

Now, taking the antiderivative of each side of this equation yields

$$\int (xe^x)' \, dx = \int (xe^x + e^x) \, dx.$$

On the left side, the *antiderivative of the derivative* of $xe^x$ is just $xe^x$ (ignoring the constant for the moment). On the right side, we use the sum rule for antiderivatives to get

$$xe^x = \int xe^x \, dx + \int e^x \, dx.$$

Notice that the antiderivative that we are trying so hard to find is part of the equation above (in blue). So we can just solve this equation for the antiderivative we want:

$$\int xe^x \, dx = xe^x - \int e^x \, dx.$$

Note that the remaining antiderivative on the right side is easy to find. So now we have the answer we desire:

$$\int xe^x \, dx = xe^x - \int e^x \, dx = xe^x - e^x + C.$$

There was no harm in ignoring the constant of integration along the way because there is a constant of integration in the final step. The point to this method is to *reduce a difficult antiderivative* (such as the one in blue above that involves a product) *to an easier antiderivative* that we can find. In general, for differentiable functions $u$ and $v$, this method is called integration by parts and can be summarized as follows.

## Integration by Parts

$$\int u \, dv = uv - \int v \, du.$$

To prove the method of integration by parts, let $u$ and $v$ be differentiable functions of $x$, and use the product rule to get

$$(uv)' = uv' + vu'.$$

Taking the antiderivative of each side with respect to $x$ yields

$$\int (uv)' \, dx = \int (uv' + vu') \, dx.$$

On the left side, the antiderivative of the derivative of $uv$ is just $uv$ (ignoring the constant). On the right side, we use the sum rule for antiderivatives to get

$$uv = \int uv' \, dx + \int vu' \, dx.$$

Note that, in the alternative notation for the derivative, this would be written

$$uv = \int u \frac{dv}{dx} \, dx + \int v \frac{du}{dx} \, dx.$$

Recalling that the differential of $v$ is $dv = v' \, dx$ and, similarly, $du = u' \, dx$ yields

$$uv = \int u \, dv + \int v \, du.$$

Rearranging the terms gives us exactly the method of integration by parts:

$$\int u \, dv = uv - \int v \, du.$$

---

**EXAMPLE 1**    Use integration by parts to find

$$\int 10xe^{2x} \, dx.$$

**Solution**  When we use integration by parts, the indefinite integral

$$\int u \, dv$$

represents what we are trying to find. Therefore, we must choose $u$ and $dv$. This takes a little practice, but in general, $u$ should be something that is easy to differentiate and $dv$ should be something that is easy to antidifferentiate. In this problem, we choose $u = 10x$. Then we no longer have a choice because $dv$ must be everything that remains: $dv = e^{2x}dx$.

$$\int \overset{u}{\overbrace{10x}} \, \overset{dv}{\overbrace{e^{2x}dx}}.$$

Now find $du$ by taking a derivative:

$$u = 10x$$

$$\frac{du}{dx} = 10$$

$$du = 10 \, dx.$$

Typically, we omit the middle equation above, going straight from $u = 10x$ to $du = 10 \, dx$.
Next find $v$ by taking an antiderivative (ignoring the constant for the moment):

$$dv = e^{2x}dx$$

$$\int dv = \int e^{2x}dx \qquad \text{Take the antiderivative of each side.}$$

$$\int 1 \, dv = \int e^{2x}dx \qquad \text{There is an implied 1 because } 1 \cdot dv = dv.$$

$$v = \frac{1}{2}e^{2x}. \qquad \int e^{kx}dx = \frac{1}{k}e^{kx} + C.$$

We usually omit the steps with the integration signs above, going straight from the first equation to the last.
Now we are ready to use the integration by parts formula:

$$\int u \, dv = uv - \int v \, du$$

$$\int \overset{u}{\overbrace{10x}} \, \overset{dv}{\overbrace{e^{2x}dx}} = \qquad uv \qquad - \int \quad v \quad du \qquad \text{Integration by parts}$$

$$= (10x)\left(\frac{1}{2}e^{2x}\right) - \int \frac{1}{2}e^{2x} \cdot 10 \, dx \qquad \text{Replace } u, v, \text{ and } du.$$

$$= (10x)\left(\frac{1}{2}e^{2x}\right) - 5\int e^{2x} \, dx \qquad \text{Constant-multiple rule}$$

$$= (10x)\left(\frac{1}{2}e^{2x}\right) - 5\left(\frac{1}{2}e^{2x}\right) + C \qquad \text{Find the antiderivative.}$$

$$= 5xe^{2x} - \frac{5}{2}e^{2x} + C. \qquad \text{Simplify.}$$

Notice in the second-to-last step above that we did still need to find an antiderivative, but it was a much easier antiderivative to find than the one in the original problem. Integration by parts is useful (when it can be applied) because it reduces finding a difficult antiderivative to finding an easier one. You can always check your work by taking the derivative of your answer. ✓₁

✓ **Checkpoint 1**

Find the antiderivative $\int xe^{7x}dx$.

Integration by parts is a useful technique, but it can be used to find only certain antiderivatives. Determining whether a particular antiderivative requires integration by substitution, integration by parts, some other rule, or simply can't be found is a matter of judgment that will improve as you gain experience.

**EXAMPLE 2**   Find the given indefinite integrals.

**(a)** $\int 10xe^{x^2}dx$

**Solution**   Although this looks similar to Example 1, integration by substitution is what is needed here. If we let $u = x^2$, then $du = 2x\, dx$. Because $x\, dx$ is already in the integral, we solve the last equation to get $x\, dx = du/2$. Then we can find the indefinite integral:

$$\int 10xe^{x^2}\, dx = \int 10e^{x^2}(x\, dx)$$

$$= \int 10e^{u}\left(\frac{du}{2}\right) \qquad \text{Integration by substitution}$$

$$= \int 5e^{u}du \qquad \text{Simplify.}$$

$$= 5e^{u} + C \qquad \text{Find the antiderivative.}$$

$$= 5e^{x^2} + C. \qquad \text{Substitute.}$$

**(b)** $\int e^{4-3x}\, dx$

**Solution**   An easy substitution works here with $u = 4 - 3x$. Then $du = -3\, dx$, so solving for $dx$ gives $dx = du/(-3)$. Then

$$\int e^{4-3x}\, dx = \int e^{u}\left(\frac{du}{-3}\right) \qquad \text{Integration by substitution}$$

$$= -\frac{1}{3}\int e^{u}\, du \qquad \text{Constant-multiple rule}$$

$$= -\frac{1}{3}e^{u} + C \qquad \text{Find the antiderivative.}$$

$$= -\frac{1}{3}e^{4-3x} + C. \qquad \text{Substitute.}$$

**(c)** $\int xe^{4-3x}\, dx$

**Solution**   You can easily check that integration by substitution does not work here. We will try integration by parts by choosing $u = x$ and $dv = e^{4-3x}\, dx$. Now we find the differential of $u$ (below on the left) and the antiderivative of $dv$ (below on the right—recall that we found this antiderivative in part (b) above):

$$u = x \qquad\qquad dv = e^{4-3x}\, dx$$

$$du = dx \qquad\qquad v = -\frac{1}{3}e^{4-3x}.$$

Now use the integration by parts formula:

$$\int u\, dv = uv - \int v\, du$$

$$\int \overset{u}{\overset{|}{x}} \overset{dv}{\overset{|}{e^{4-3x}dx}} = \qquad uv \qquad - \int \quad v \quad du \qquad \text{Integration by parts}$$

$$= (x)\left(-\frac{1}{3}e^{4-3x}\right) - \int -\frac{1}{3}e^{4-3x}\, dx$$

$$= (x)\left(-\frac{1}{3}e^{4-3x}\right) + \frac{1}{3}\int e^{4-3x}\, dx \qquad \text{Constant-multiple rule}$$

$$= (x)\left(-\frac{1}{3}e^{4-3x}\right) + \frac{1}{3}\left(-\frac{1}{3}e^{4-3x}\right) \qquad \text{Find the antiderivative.}$$

$$= -\frac{1}{3}xe^{4-3x} - \frac{1}{9}e^{4-3x} + C. \qquad \text{Simplify.}$$

**(d)** $\displaystyle\int \frac{5x-7}{e^x}\, dx$

**Solution** First, we can rewrite the integrand as a product:

$$\int \frac{5x-7}{e^x}\, dx = \int (5x-7)e^{-x}\, dx.$$

Then we can use integration by parts with $u = 5x - 7$ and $dv = e^{-x}\, dx$ to get

$$u = 5x - 7 \qquad dv = e^{-x}\, dx$$
$$du = 5\, dx \qquad v = -e^{-x}.$$

Then the integration by parts formula yields:

$$\int \overset{u}{\overset{|}{(5x-7)}} \overset{dv}{\overset{|}{e^{-x}dx}} = \qquad uv \qquad - \int \quad v \quad du \qquad \text{Integration by parts}$$

$$= (5x - 7)(-e^{-x}) - \int -e^{-x}(5\, dx)$$

$$= (5x - 7)(-e^{-x}) + 5\int e^{-x}\, dx \qquad \text{Constant-multiple rule}$$

$$= (5x - 7)(-e^{-x}) + 5(-e^{-x}) + C \qquad \text{Find the antiderivative.}$$

$$= -5xe^{-x} + 7e^{-x} - 5e^{-x} + C \qquad \text{Distributive property}$$

$$= -5xe^{-x} + 2e^{-x} + C. \ \boxed{✓}_2 \qquad \text{Collect like terms.}$$

✓**Checkpoint 2**

Find $\displaystyle\int \frac{x}{e^{x+1}}\, dx$.

---

**EXAMPLE 3**    Find $\int 8x^2 e^{4x-9}\, dx$.

**Solution** Integration by substitution does not work here. For example, if we try $u = 4x - 9$, then $du = 4\, dx$, but substitution does not eliminate the pesky $x^2$ in the integrand. Try integration by parts with $u = 8x^2$ and $dv = e^{4x-9}\, dx$. Then

$$u = 8x^2 \qquad dv = e^{4x-9}\, dx$$

$$du = 16x\, dx \qquad v = \frac{1}{4}e^{4x-9}.$$

The integration by parts formula yields

$$\int u \, dv = uv - \int v \, du$$

$$\int \overset{u}{8x^2} \overset{dv}{e^{4x-9}dx} = \quad uv \quad - \int v \quad du \qquad \text{Integration by parts}$$

$$= (8x^2)\left(\frac{1}{4}e^{4x-9}\right) - \int \frac{1}{4}e^{4x-9} \cdot 16x \, dx$$

$$= 2x^2 e^{4x-9} \quad - \int 4xe^{4x-9} \, dx. \qquad (*)$$

The point to integration by parts is to reduce a difficult antiderivative to an easier one, and certainly the antiderivative in $(*)$ on the last line above looks *easier* than the one in the original problem. However, we still need to use integration by parts again in order to find it. For the moment, we will focus on just this indefinite integral. Using *integration by parts a second time*, let $U = 4x$ and $dV = e^{4x-9} \, dx$ (we use uppercase letters here for $U$ and $V$ to distinguish them from $u$ and $v$ above, but in practice, lowercase letters are often used again without causing confusion). Then

$$U = 4x \qquad dV = e^{4x-9} \, dx$$

$$dU = 4 \, dx \qquad V = \frac{1}{4}e^{4x-9}.$$

The integration by parts formula yields

$$\int U \, dV = UV - \int V \, dU$$

$$\int \overset{U}{4x} \overset{dV}{e^{4x-9}dx} = \quad UV \quad - \int V \quad dU \qquad \text{Integration by parts}$$

$$= (4x)\left(\frac{1}{4}e^{4x-9}\right) - \int \frac{1}{4}e^{4x-9} \cdot 4 \, dx$$

$$= xe^{4x-9} \quad - \int e^{4x-9} \, dx$$

$$= xe^{4x-9} \quad -\frac{1}{4}e^{4x-9} + C. \qquad (**)$$

Returning to the line above labeled $(*)$, and putting this all together, we have

$$\int 8x^2 e^{4x-9} \, dx = 2x^2 e^{4x-9} - \int 4xe^{4x-9} \, dx \qquad \text{Using } (*)$$

$$= 2x^2 e^{4x-9} - \left[ xe^{4x-9} - \frac{1}{4}e^{4x-9} + C \right] \qquad \text{Using } (**)$$

$$= 2x^2 e^{4x-9} - xe^{4x-9} + \frac{1}{4}e^{4x-9} + C. \qquad \text{Distributive property}$$

Don't forget to distribute the negative sign in the last step above because it applied to the entire indefinite integral and hence to every term inside the square brackets. ✓ 3

✓ **Checkpoint 3**

Use integration by parts twice to find $\int 4x^2 e^{2x} \, dx$.

**EXAMPLE 4** Find each indefinite integral.

**(a)** $\int \ln x \, dx$

**Solution** Although there is no product in this integrand, we now see that integration by parts also works for many integrals involving logarithms. If we were to choose $dv = \ln x \, dx$, then we would not be able to take the antiderivative to find $v$. Therefore, we choose $u = \ln x$ and $dv = dx$. Then we find that

$$u = \ln x \qquad dv = dx$$
$$du = \frac{1}{x} \, dx \qquad v = x.$$

Using integration by parts we get

$$\int u \, dv = uv - \int v \, du$$

$$\int \overset{u}{\overbrace{\ln x}} \overset{dv}{\overbrace{dx}} = uv \; - \int v \quad du \qquad \text{Integration by parts}$$

$$= (\ln x)(x) - \int x \cdot \frac{1}{x} \, dx$$

$$= x \ln x \; - \int 1 \, dx \qquad \text{Simplify.}$$

$$= x \ln x \; - \; x + C. \qquad \text{Find the antiderivative.}$$

As with any indefinite integral, we can check our work by differentiating the answer:

$$(x \ln x - x + C)' = [(x)(\ln x)' + (\ln x)(x)'] - 1 + 0 \qquad \text{Product rule in brackets}$$

$$= \left[ x\left(\frac{1}{x}\right) + (\ln x)(1) \right] - 1 \qquad \text{Find the derivatives.}$$

$$= [1 + \ln x] - 1 = \ln x. \qquad \text{Simplify.}$$

**(b)** $\displaystyle\int (3x^2 - 5x + 1) \ln x \, dx$

**Solution** We will again try integration by parts with $u = \ln x$. Then we must have $dv = (3x^2 - 5x + 1) \, dx$. This gives

$$u = \ln x \qquad dv = (3x^2 - 5x + 1) \, dx$$
$$du = \frac{1}{x} \, dx \qquad v = x^3 - \frac{5}{2}x^2 + x.$$

Using integration by parts we get

$$\int \overset{u}{\overbrace{\ln x}} \overset{dv}{\overbrace{(3x^2 - 5x + 1) \, dx}} = uv \; - \int v \quad du \qquad \text{Integration by parts}$$

$$= (\ln x)\left( x^3 - \frac{5}{2}x^2 + x \right) - \int \left( x^3 - \frac{5}{2}x^2 + x \right) \cdot \frac{1}{x} \, dx$$

$$= (\ln x)\left( x^3 - \frac{5}{2}x^2 + x \right) - \int \left( x^2 - \frac{5}{2}x + 1 \right) dx \qquad \text{Simplify.}$$

$$= (\ln x)\left( x^3 - \frac{5}{2}x^2 + x \right) - \left( \frac{1}{3}x^3 - \frac{5}{4}x^2 + x \right) + C. \qquad \text{Find the antiderivative.} \quad \text{✓}_{4}$$

✓ **Checkpoint 4**

Find $\displaystyle\int 2x \ln(x) \, dx$.

The result in Example 4 (a) is highlighted below.

## Antiderivative of the Logarithmic Function

$$\int \ln x \, dx = x \ln x - x + C.$$

**EXAMPLE 5**   **Velocity** The driver of a car stopped on the shoulder of a highway speeds up rapidly at first, then continues to speed up, but at a slower rate. After a minute, the velocity of the car is about the same as other cars on the highway. The velocity of the car (in feet per second) after $t$ seconds can be approximated by

$$v(t) = 25 \ln(t + 1) \qquad (0 \le t \le 60).$$

**(a)** Find and interpret $v(60)$.

**Solution** The velocity of the car after 60 seconds is $v(60) = 25 \ln(60 + 1) \approx 103$ feet per second. Because there are 5280 feet in a mile, 60 seconds in a minute, and 60 minutes in an hour, we can convert units to get

$$v(60) \approx 103 \frac{\text{ft}}{\text{sec}} \cdot \frac{1 \text{ mile}}{5280 \text{ ft}} \cdot \frac{60 \text{ sec}}{1 \text{ min}} \cdot \frac{60 \text{ min}}{1 \text{ hr}} \approx 70 \frac{\text{mile}}{\text{hr}}.$$

This means that, after a minute, the car is traveling with the other cars at about 70 miles per hour.

**(b)** Find a formula for the function $s(t)$ that gives the position of the car in relation to the starting point (in feet) after $t$ seconds.

**Solution** Velocity is the rate of change of position, hence $v(t) = s'(t)$. To find the position function, we must find the antiderivative of the velocity function. We know the antiderivative of the logarithmic function, but here we have $\ln(t + 1)$, so we need to use integration by substitution first. Let $u = t + 1$, then $du = dt$, and the position function is

$$s(t) = \int s'(t)\, dt = \int v(t)\, dt$$

$$= \int 25 \ln(t + 1)\, dt$$

$$= \int 25 \ln(u)\, du \qquad \text{Substitute.}$$

$$= 25[u \ln(u) - u] + C \qquad \text{Antiderivative of } \ln(u)$$

$$= 25[(t + 1) \ln(t + 1) - (t + 1)] + C \qquad \text{Substitute.}$$

We can find the value of $C$ by using the fact that $s(0) = 0$, hence

$$25[(0 + 1) \ln(0 + 1) - (0 + 1)] + C = 0$$

$$25[-1] + C = 0 \qquad \text{Recall that } \ln(1) = 0.$$

$$C = 25.$$

So the position function is

$$s(t) = 25[(t + 1) \ln(t + 1) - (t + 1)] + 25 = 25(t + 1) \ln(t + 1) - 25t.$$

**(c)** Find a formula for the function $a(t)$ that gives the acceleration of the car (in feet per second per second) after $t$ seconds.

**Solution** The acceleration function is the rate of change of the velocity function, so

$$a(t) = v'(t) = [25 \ln(t + 1)]' = 25 \cdot \frac{1}{t + 1} = \frac{25}{t + 1}.$$ ✓ 5

✓ **Checkpoint 5**

Find $s(t)$ given that
$s'(t) = 3 \ln(3t + 1)$.

**EXAMPLE 6**   **Restaurant Crowds** The *rate* at which customers are entering a popular lunch spot (in people per minute) can be approximated by

$$f(t) = \frac{1}{5} t e^{\frac{-t}{50}},$$

where $t$ is in minutes, with $t = 0$ corresponding to 11:00 a.m., the time that the restaurant opens.

**(a)** Find a formula for the function $F(t)$ that gives the total number of customers who have entered by time $t$.

**Solution** The function giving the total number of customers who have walked in the door is the antiderivative of the rate of change given; therefore

$$F(t) = \int f(t)\, dt = \int \frac{1}{5} t e^{\frac{-t}{50}}\, dt.$$

Using integration by parts with $u = t/5$ and $dv = e^{-t/50} dt$ gives us $du = dt/5$ and $v = -50e^{-t/50}$. Thus

$$\int \overset{u}{\overbrace{\frac{1}{5}t}}\, \overset{dv}{\overbrace{e^{\frac{-t}{50}}}}\,dt = \left(\frac{1}{5}t\right)(-50e^{\frac{-t}{50}}) - \int -50e^{\frac{-t}{50}} \cdot \frac{1}{5}\, dt \qquad \int u\, dv = uv - \int v\, du$$

$$= -10te^{\frac{-t}{50}} + 10(-50e^{\frac{-t}{50}}) + C \qquad \text{Find the antiderivative.}$$

$$= -10te^{\frac{-t}{50}} - 500e^{\frac{-t}{50}} + C. \qquad \text{Simplify.}$$

Because no customers have entered the restaurant at the instant the doors open, we have

$$0 = F(0) = -10(0)e^{\frac{-0}{50}} - 500e^{\frac{-0}{50}} + C, \text{ hence } C = 500.$$

Therefore,

$$F(t) = -10te^{\frac{-t}{50}} - 500e^{\frac{-t}{50}} + 500.$$

**(b)** How many customers have entered by 2:00 p.m.?

✓ **Checkpoint 6**

In Example 6, how many customers would have entered if we were given $f(t) = te^{\frac{-t}{45}}$.

**Solution** The total number of customers who came in the door in the 180 minutes between 11:00 a.m. and 2:00 p.m. is

$$F(180) = -10(180)e^{\frac{-180}{50}} - 500e^{\frac{-180}{50}} + 500 \approx 437 \text{ customers.} \quad \boxed{\checkmark\ 6}$$

# 13.3 Exercises

*Find the given indefinite integrals. State whether integration by substitution or integration by parts was used. (See Examples 1–3.)*

**1.** $\int 2e^{-35x}\, dx$

**2.** $\int 5e^{-14x}\, dx$

**3.** $\int 10xe^{5x}\, dx$

**4.** $\int 45xe^{9x}\, dx$

**5.** $\int xe^{x/4}\, dx$

**6.** $\int xe^{x/8}\, dx$

**7.** $\int \frac{5x+1}{e^x}\, dx$

**8.** $\int \frac{8x-4}{e^{2x}}\, dx$

**9.** $\int 7xe^{1-3x}\, dx$

**10.** $\int 3xe^{7-9x}\, dx$

**11.** $\int (8x-3)e^{10x-17}\, dx$

**12.** $\int (7x-2)e^{9x+5}\, dx$

**13.** $\int 4xe^{2x^2}\, dx$

**14.** $\int 10xe^{5x^2}\, dx$

**15.** $\int 4x^2 e^{2x}\, dx$

**16.** $\int 10x^2 e^{5x}\, dx$

**17.** $\int (x^2+1)e^x\, dx$

**18.** $\int (x^2-1)e^{3x}\, dx$

**19.** $\int xe^{x^2+1}\, dx$

**20.** $\int 3xe^{x^2-1}\, dx$

*Find each indefinite integral. (See Example 4.)*

**21.** $\int (3\ln x - 7)\, dx$

**22.** $\int (1 - 9\ln x)\, dx$

**23.** $\int \ln(8x-5)\, dx$

**24.** $\int \ln(4-9x)\, dx$

**25.** $\int 26x \ln x\, dx$

**26.** $\int 12x^2 \ln x\, dx$

**27.** $\int x^3 \ln x\, dx$

**28.** $\int x^4 \ln x\, dx$

**29.** $\int 9\sqrt{x} \ln x\, dx$

**30.** $\int \frac{\ln x}{\sqrt{x}}\, dx$

**31.** $\int (9-2x)\ln x\, dx$

**32.** $\int (14x+5)\ln x\, dx$

**33.** $\int (18x^2+12x+4)\ln x\, dx$ **34.** $\int (45x^2-16x+3)\ln x\, dx$

*Use integration by parts with the given u to find the indefinite integrals.*

**35.** $\int 15(x+1)\sqrt{x+2}\, dx; \quad u = 15(x+1)$

**36.** $\int \dfrac{6x - 3}{\sqrt{x - 7}} \, dx; \qquad u = 6x - 3$

**Velocity** *Work these exercises. (See Example 5.)*

**37.** A car was driving on a two-lane highway at 80 feet per second (about 55 mph) when it abruptly speeds up to pass a slow-moving truck. The velocity of the car (in ft/sec) after $t$ seconds (see accompanying figure) is given by

$$v(t) = 80 + 10te^{-t/4} \qquad (0 \le t \le 30).$$

Time (seconds)

**(a)** Find and interpret $v(0)$, $v(4)$, and $v(30)$.

**(b)** Find a formula for the position function $s(t)$.

**38.** A student takes off running to class, but slows down when she realizes that she will make it on time. The velocity of the student after $t$ seconds is

$$v(t) = 7te^{-t/5} \qquad (0 \le t \le 15).$$

**(a)** Find a formula for the position function $s(t)$.

**(b)** If it took 15 seconds to get to class, find and interpret $s(15)$ to the nearest integer.

**39.** The velocity of the a car (in feet per second) $t$ seconds after starting is given by

$$v(t) = 18 \ln(3t + 1) \qquad (0 \le t \le 60).$$

**(a)** Find a formula for the position function $s(t)$.

**(b)** Find the position of the car (in feet from the starting position) after 30 seconds.

**(c)** Find the velocity of the car (in feet per second) after 30 seconds, then convert to mph (there are 5280 feet in a mile and 3600 seconds in an hour).

**40.** A bicycle has velocity (in feet per second) after $t$ seconds given by

$$v(t) = 6 \ln(2t + 1) \qquad (0 \le t \le 60).$$

**(a)** Find a formula for the position function $s(t)$.

**(b)** Find the position of the car (in feet from the starting position) after 30 seconds.

**(c)** Find the velocity of the car (in feet per second) after 30 seconds, then convert to mph (there are 5280 feet in a mile and 3600 seconds in an hour).

**41.** Find the position function if a velocity function is given by

$$v(t) = t^2 e^{-2t}.$$

**42.** Find the position of an object moving with velocity

$$v(t) = 135t^2 e^{-3t}.$$

*Work these exercises. (See Example 6.)*

**43. Disease Outbreak** The *rate* at which people are becoming infected (in people per day) during the outbreak of a disease was approximated by

$$f(x) = 2xe^{-x/7},$$

where $x$ is in days and $x = 0$ corresponds to the start of the outbreak.

**(a)** Find the function $F(x)$ that gives total number of people who have become infected by day $x$ given that 12 people were infected when $x = 0$.

**(b)** Find the total number of people who have become infected by the end of 4 weeks.

**44. Total Sales** The *rate* of sales at a fireworks outlet (in thousands of dollars per day) can be approximated by

$$f(x) = 5xe^{-x/3},$$

where $x$ is in days and $x = 0$ corresponds to the start of July.

**(a)** Find the function $F(x)$ that gives total sales since the start of July through day $x$. (*Hint:* $F(0) = 0$.)

**(b)** Find the total sales since the start of July through $x = 4$.

**45. Total Cost** Find the total cost function $C(x)$ (in thousands of dollars) if the marginal cost (in thousands of dollars per unit) at a production level of $x$ units is

$$C'(x) = \dfrac{5x + 1}{e^x} \qquad (0 \le x \le 4),$$

and the fixed costs are \$10,000 ($C(0) = 10$).

**46. Total Revenue** Find the total revenue function $R(x)$ (in thousands of dollars) if the marginal revenue (in thousands of dollars per unit) at a production level of $x$ units is

$$R'(x) = \dfrac{16x + 3}{e^x} \qquad (0 \le x \le 4),$$

and $R(0) = 0$.

**47. Coal Consumption** According to Energy Information Administration projections for the years 2020 through 2040, the *rate of consumption* of coal (in quadrillion British thermal units [BTUs] per year) can be approximated by

$$q(t) = 15e^{-.02t} \qquad (0 \le t \le 20),$$

where $t = 0$ corresponds to the year 2020. Projections for the *price* of coal (in millions of dollars per quadrillion BTUs) can be approximated by

$$p(t) = 75t + 2700 \qquad (0 \le t \le 20).$$

**(a)** Find the function $f(t) = p(t)q(t)$ that gives the *rate of spending* on coal (in millions of dollars per year).

**(b)** Find the function $F(t)$ that gives the total spending on coal (in millions of dollars) through year $t$.

**48. Propane Consumption** The *rate of consumption* of residential propane (in trillions of BTUs per year) is projected for the years 2020 through 2040 to be approximately

$$q(t) = 450e^{-.008t} \qquad (0 \le t \le 20),$$

and the *price* (in millions of dollars per 1 trillion BTUs) is projected to be approximately

$$p(t) = .16t + 17.4 \qquad (0 \le t \le 20),$$

where $t = 0$ corresponds to the year 2020. (Data from: U.S. Energy Information Administration.)

(a) Find the function $f(t) = p(t)q(t)$ that gives the *rate of spending* (in millions of dollars per year) on residential propane.

(b) Find the function $F(t)$ that gives the total spending on residential propane (in millions of dollars) through year $t$.

49. **Pharmacy Sales**  The *rate* of sales (in billions of dollars per year) in pharmacies and drugstores can be approximated by

$$f(x) = 42 \ln(x) + 124 \qquad (x \ge 1),$$

where $x = 1$ corresponds to the year 2001. Find the function $F(x)$ that gives *total sales* between the year 2001 and year $x$. (*Hint:* $F(1) = 0$ because we are counting only total sales since 2001.) (Data from: Annual Retail Trade Survey.)

50. **Specialty Foods**  In specialty food stores, the *rate* of sales (in billions of dollars per year) can is approximately

$$f(x) = 2.25 \ln(x) + 14 \qquad (x \ge 1),$$

where $x = 1$ corresponds to the year 2001. Find the function $F(x)$ that gives *total sales* between the year 2001 and year $x$. (Data from: Annual Retail Trade Survey.)

✓**Checkpoint Answers**

1. $\dfrac{1}{7}xe^{7x} - \dfrac{1}{49}e^{7x} + C$

2. $-xe^{-(x+1)} - e^{-(x+1)} + C$

3. $2x^2e^{2x} - 2xe^{2x} + e^{2x} + C$

4. $x^2 \ln(x) - \dfrac{1}{2}x^2 + C$

5. $s(t) = (3t + 1) \ln(3t + 1) - 3t$

6. About 1840 customers

# 13.4  Area and the Definite Integral

Suppose a car travels along a straight road at a constant speed of 50 mph. The speed of the car at time $t$ is given by the constant function $v(t) = 50$, whose graph is a horizontal straight line, as shown in Figure 13.2(a).

(a)                                    (b)

**Figure 13.2**

How far does the car travel from time $t = 2$ to $t = 6$? Since this is a 4-hour period, the answer, of course, is $4 \cdot 50 = 200$ miles. Note that 200 is precisely the *area* under the graph of the speed function $v$ and above the $t$-axis from $t = 2$ to $t = 6$, as shown in Figure 13.2(b).

As we saw in Chapter 12, the speed function $v$ is the rate of change of distance with respect to time—that is, the rate of change of the distance function $s$ (which gives the distance traveled by the car at time $t$). Now, the distance traveled from time $t = 2$ to $t = 6$ is the amount that the distance has changed from $t = 2$ to $t = 6$. In other words, the *total change* in distance from $t = 2$ to $t = 6$ is the area under the graph of the speed (rate of change) function from $t = 2$ to $t = 6$.

A more complicated argument (omitted here) shows that a similar situation holds in general for other nonnegative rate of change functions.

## Total Change in $F(x)$

Let $f$ be a function such that $f$ is continuous on the interval $[a, b]$ and $f(x) \ge 0$ for all $x$ in $[a, b]$. If $f$ is the rate of change of a function $F$, then the **total change in** $F(x)$ as $x$ goes from $a$ to $b$ is the area between the graph of $f(x)$ and the $x$-axis from $x = a$ to $x = b$.

**EXAMPLE 1** **Maintenance Charges** Figure 13.3 shows the graph of the function that gives the rate of change of the annual maintenance charges for a certain machine. The rate function is increasing because maintenance tends to cost more as the machine gets older. Estimate the total maintenance charges over the 10-year life of the machine.

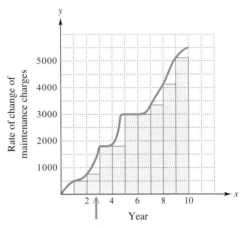

**Figure 13.3**

**Solution** This is the situation described in the preceding box, with $F$ being the maintenance cost function and $f$, whose graph is given, the rate-of-change function. The total maintenance charges are the total change in $F(x)$ from $x = 0$ to $x = 10$—that is, the area between the graph of the rate function and the $x$-axis from $x = 0$ to $x = 10$. We can approximate this area by using the shaded rectangles in Figure 13.3. For instance, the rectangle marked with an arrow has base 1 (from year 2 to year 3) and height 750 (the rate of change at $x = 2$), so its area is $1 \times 750 = 750$. Similarly, each of the other rectangles has base 1 and height determined by the rate of change at the beginning of the year. Consequently, we estimate the area to be the sum

$$1 \cdot 0 + 1 \cdot 500 + 1 \cdot 750 + 1 \cdot 1800 + 1 \cdot 1800 + 1 \cdot 3000 + 1 \cdot 3000$$
$$+ 1 \cdot 3400 + 1 \cdot 4200 + 1 \cdot 5200 = 23{,}650.$$

Hence, the total maintenance charges over the 10 years are at least \$23,650. (The unshaded areas under the rate graph have not been accounted for in this estimate). ✓ 1

✓ **Checkpoint 1**

Use Figure 13.3 to estimate the maintenance charge during

(a) the first 6 years of the machine's life;

(b) the first 8 years of the machine's life.

## Area

The preceding examples show that the area between a graph and the $x$-axis has useful interpretations. In this section and the next, we develop a means of measuring such areas precisely when the function is given by an algebraic formula. The underlying idea is the same as in Example 1: Use rectangles to approximate the area under the graph.

**EXAMPLE 2** Find the area under the graph of $f(x) = \sqrt{4 - x^2}$ from $x = 0$ to $x = 2$, shown in Figure 13.4.

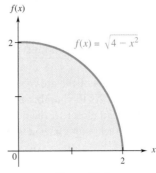

**Figure 13.4**

**Solution** A very rough approximation of the area of this region can be found by using two rectangles, as in Figure 13.5. The height of the rectangle on the left is $f(0) = 2$, and the height of the rectangle on the right is $f(1) = \sqrt{3}$. The width of each rectangle is 1, making the total area of the two rectangles

$$1 \cdot f(0) + 1 \cdot f(1) = 2 + \sqrt{3} \approx 3.7321 \text{ square units.}$$

 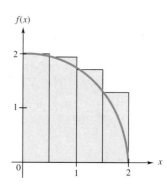

| Figure 13.5 | Figure 13.6 |

As Figure 13.5 suggests, this approximation is greater than the actual area. To improve the accuracy of the approximation, we could divide the interval from $x = 0$ to $x = 2$ into four equal parts, each of width $\frac{1}{2}$, as shown in Figure 13.6. As before, the height of each rectangle is given by the value of $f$ at the left-hand side of the rectangle, and its area is the width, $\frac{1}{2}$, multiplied by the height. The total area of the four rectangles is

$$\frac{1}{2} \cdot f(0) + \frac{1}{2} \cdot f\left(\frac{1}{2}\right) + \frac{1}{2} \cdot f(1) + \frac{1}{2} \cdot f\left(\frac{3}{2}\right)$$

$$= \frac{1}{2}(2) + \frac{1}{2}\left(\frac{\sqrt{15}}{2}\right) + \frac{1}{2}(\sqrt{3}) + \frac{1}{2}\left(\frac{\sqrt{7}}{2}\right)$$

$$= 1 + \frac{\sqrt{15}}{4} + \frac{\sqrt{3}}{2} + \frac{\sqrt{7}}{4} \approx 3.4957 \text{ square units.}$$

This approximation looks better, but it is still greater than the actual area desired. To improve the approximation, divide the interval from $x = 0$ to $x = 2$ into eight parts with equal widths of $\frac{1}{4}$. (See Figure 13.7.) The total area of all these rectangles is

$$\frac{1}{4} \cdot f(0) + \frac{1}{4} \cdot f\left(\frac{1}{4}\right) + \frac{1}{4} \cdot f\left(\frac{1}{2}\right) + \frac{1}{4} \cdot f\left(\frac{3}{4}\right) + \frac{1}{4} \cdot f(1) + \frac{1}{4} \cdot f\left(\frac{5}{4}\right)$$

$$+ \frac{1}{4} \cdot f\left(\frac{3}{2}\right) + \frac{1}{4} \cdot f\left(\frac{7}{4}\right) \approx 3.33982 \text{ square units.}$$

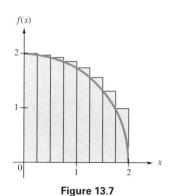

**Figure 13.7**

This process of approximating the area under the curve by using more and more rectangles to get a better and better approximation can be extended. To do this, divide the interval from $x = 0$ to $x = 2$ into $n$ equal parts. Each of these $n$ intervals has width

$$\frac{2 - 0}{n} = \frac{2}{n},$$

so each rectangle has width $2/n$ and height determined by the value of the function $f(x) = \sqrt{4 - x^2}$ at the left side of the rectangle. A computer was used to find approximations of the area for several values of $n$ given in the table at the side.

As the number $n$ of rectangles gets larger and larger, the sum of their areas gets closer and closer to the actual area of the region. In other words, the actual area is the *limit* of these sums as $n$ gets larger and larger without bound, which can be written

$$\text{Area} = \lim_{n \to \infty} (\text{sum of areas of } n \text{ rectangles}).$$

| $n$ | Area |
|---|---|
| 125 | 3.15675 |
| 2000 | 3.14257 |
| 8000 | 3.14184 |
| 32,000 | 3.14165 |
| 128,000 | 3.14160 |
| 512,000 | 3.14159 |

The table suggests that this limit is a number whose decimal expansion begins $3.14159\ldots$, which is the same as the beginning of the decimal approximation of $\pi$. Therefore, it seems plausible that

$$\text{Area} = \lim_{n \to \infty}(\text{sum of areas of } n \text{ rectangles}) = \pi.$$

It can be shown that the region whose area was found in Example 2 is one-fourth of the interior of a circle of radius 2 with center at the origin (see Figure 13.4). Hence, its area is

$$\frac{1}{4}(\pi r^2) = \frac{1}{4}(\pi \cdot 2^2) = \pi,$$

which agrees with our answer in Example 2.

The method used in Example 2 can be generalized to find the area under the curve $y = f(x)$, above the $x$-axis, and between the vertical lines $x = a$ and $x = b$, as shown in Figure 13.8. To approximate this area, we could divide the region under the curve first into 10 rectangles (Figure 13.8(a)) and then into 20 rectangles (Figure 13.8(b)). In each case, the sum of the areas of the rectangles gives an approximation of the area under the curve.

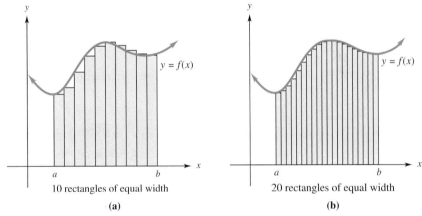

10 rectangles of equal width    20 rectangles of equal width
**(a)**                          **(b)**

**Figure 13.8**

We can get better and better approximations by increasing the number $n$ of rectangles. For simplicity, we choose that all rectangles have the same width. Here is a description of the general procedure: Let $n$ be a positive integer. Divide the interval from $a$ to $b$ into $n$ pieces of equal length. The symbol $\Delta x$ is traditionally used to denote the length of each piece. Since the length of the entire interval is $b - a$, each of the $n$ pieces has length

$$\Delta x = \frac{b - a}{n}.$$

Use each of these pieces as the base of a rectangle, as shown in Figure 13.9, on the following page, where the endpoints of the $n$ intervals are labeled $x_1, x_2, x_3, \ldots, x_{n+1}$. A typical rectangle, the one whose lower left corner is at $x_i$, has red shading. The base of this rectangle is $\Delta x$, and its height is the height of the graph over $x_i$, namely, $f(x_i)$, so

$$\text{Area of } i\text{th rectangle} = f(x_i) \cdot \Delta x.$$

The total area under the curve is approximated by the sum of the areas of all $n$ of the rectangles, namely,

$$f(x_1) \cdot \Delta x + f(x_2) \cdot \Delta x + f(x_3) \cdot \Delta x + \cdots + f(x_n) \cdot \Delta x.$$

The exact area is defined to be the limit of this sum (if it exists) as the number of rectangles gets larger and larger, without bound. Hence,

$$\text{Exact area} = \lim_{n \to \infty}(f(x_1) \cdot \Delta x + f(x_2) \cdot \Delta x + f(x_3) \cdot \Delta x + \cdots + f(x_n) \cdot \Delta x)$$

$$= \lim_{n \to \infty}([f(x_1) + f(x_2) + f(x_3) + \cdots + f(x_n)] \cdot \Delta x).$$

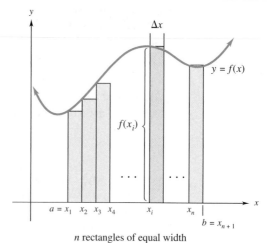

**Figure 13.9**

This limit is called the *definite integral* of $f$ from $a$ to $b$ and is denoted by the symbol

$$\int_a^b f(x)\, dx.$$

The preceding discussion can be summarized as follows.

---

## The Definite Integral

If $f$ is a continuous function on the interval $[a, b]$, then the **definite integral** of $f$ from $a$ to $b$ is the number

$$\int_a^b f(x)\, dx = \lim_{n \to \infty}[(f(x_1) + f(x_2) + f(x_3) + \cdots + f(x_n))\,\Delta x],$$

where $\Delta x = (b - a)/n$ and $x_i$ is the left-hand endpoint of the $i$th interval.

---

For instance, the area of the region in Example 2 could be written as the definite integral

$$\int_0^2 \sqrt{4 - x^2}\, dx = \pi.$$

Although the definition in the box is obviously motivated by the problem of finding areas, it is applicable to many other situations, as we shall see in later sections. In particular, *the definition of the definite integral is valid even when $f(x)$ takes negative values* (that is, when the graph goes below the $x$-axis). In that case, however, the resulting number is not the area between the graph and the $x$-axis. In this section, we shall deal only with the area interpretation of the definite integral.

The elongated "S" in the notation for the definite integral stands for the word "Sum," which plays a crucial role in the definition. This notation looks very similar to that used for antiderivatives (also called *indefinite* integrals) in earlier sections. The connection between the definite integral and antiderivatives, which is the reason for the similar terminology and notation, will be explained in the next section.

⚠️ **CAUTION** The symbols $\int f(x)\, dx$ and $\int_a^b f(x)\, dx$ have totally different meanings—the $a$ and $b$ make all the difference. The indefinite integral $\int f(x)\, dx$ denotes a set of *functions* (the antiderivatives of $f(x)$), whereas the definite integral $\int_a^b f(x)\, dx$ represents a *number* (which can be interpreted as the area under the graph when $f(x) \geq 0$).

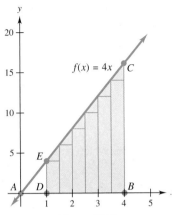

**Figure 13.10**

**EXAMPLE 3**   Approximate $\int_1^4 4x\,dx$, which is the area under the graph of $f(x) = 4x$, above the $x$-axis, and between $x = 1$ and $x = 4$, by using six rectangles of equal width whose heights are the values of the function at the left endpoint of each rectangle.

**Solution**   We want to find the area of the shaded region in Figure 13.10.
    The heights of the six rectangles given by $f(x_i)$ are as follows:

| $i$ | $x_i$ | $f(x_i)$ |
|---|---|---|
| 1 | $x_1 = 1$ | $f(1) = 4$ |
| 2 | $x_2 = 1.5$ | $f(1.5) = 6$ |
| 3 | $x_3 = 2$ | $f(2) = 8$ |
| 4 | $x_4 = 2.5$ | $f(2.5) = 10$ |
| 5 | $x_5 = 3$ | $f(3) = 12$ |
| 6 | $x_6 = 3.5$ | $f(3.5) = 14$ |

The width of each rectangle is $\Delta x = \dfrac{4-1}{6} = \dfrac{1}{2} = .5$. The sum of the areas of the six rectangles is

$$f(x_1)\Delta x + f(x_2)\Delta x + f(x_3)\Delta x + f(x_4)\Delta x + f(x_5)\Delta x + f(x_6)\Delta x$$
$$= f(1)\Delta x + f(1.5)\Delta x + f(2)\Delta x + f(2.5)\Delta x + f(3)\Delta x + f(3.5)\Delta x$$
$$= (4)(.5) + (6)(.5) + (8)(.5) + (10)(.5) + (12)(.5) + (14)(.5)$$
$$= 27.$$

We can check the accuracy of this approximation by noting that the area of the shaded region in Figure 13.10 is the difference of the areas of triangle $ABC$ and triangle $ADE$. Triangle $ABC$ has base 4 and height 16, and triangle $ADE$ has base 1 and height 4. Using the formula for the area of a triangle, $A = \frac{1}{2}bh$, we see that the area of the shaded region is

$$\text{Area } ABC - \text{Area } ADE = (1/2)(4)(16) - (1/2)(1)(4) = 32 - 2 = 30.$$

So our approximation is a bit low. Using more rectangles will produce a better approximation of the area. ✔️ 2

### ✓ Checkpoint 2

Divide the region of Figure 13.10 into 12 rectangles of equal width whose heights are the values of the function at the left endpoint of each rectangle.

**(a)**   Complete this table.

| $i$ | $x_i$ | $f(x_i)$ |
|---|---|---|
| 1 | 1 | 4 |
| 2 | 1.25 | |
| 3 | 1.5 | |
| 4 | 1.75 | |
| 5 | 2 | |
| 6 | 2.25 | |
| 7 | | |
| 8 | | |
| 9 | | |
| 10 | | |
| 11 | | |
| 12 | | |

**(b)**   Use the results from the table to approximate $\int_1^4 4x\,dx$.

### 🖩 Technology and Definite Integrals

The preceding examples illustrate the difficulty of evaluating definite integrals by direct computation. In the next section, an algebraic evaluation technique that works in many cases will be developed. In all cases, technology can be used to evaluate definite integrals. Computer programs and calculators that have symbolic capabilities, such as Maple, Mathematica, the TI-89, and the TI-Nspire will generally produce exact results. Other graphing calculators and spreadsheets usually provide highly accurate approximations.

**TECHNOLOGY TIP**   There are two ways to approximate a definite integral with a graphing calculator, each of which is illustrated here for $\int_{-2}^{1}(x^3 - 5x + 6)\,dx$.

**1.**   Approximate the limit in the definition by computing the sum for a large value of $n$, say, $n = 100$. Then $\Delta x = (b-a)/n = (1-(-2))/100 = 3/100 = .03$, and the sum is

$$[f(x_1) + f(x_2) + f(x_3) + \cdots + f(x_n)] \cdot \Delta x$$
$$= [f(x_1) + f(x_2) + f(x_3) + \cdots + f(x_{100})](.03).$$

To compute this sum automatically on most TI calculators, look in the submenus of the LIST menu and enter

$$\text{sum(seq}(x^3 - 5x + 6, x, -2, 1 - .03, .03)) \times (.03).$$

Left endpoint    Right endpoint $-\Delta x$    $\Delta x$

The calculator produces the approximation 21.8393.

2.  Use **numerical integration.** This feature approximates the definite integral by using more complicated summing techniques than used previously (which typically involve trapezoids or regions with parabolic boundaries). On the new edition of the TI-84+, go to the MATH menu and enter fnInt. The screen will then show a definite integral with blanks to fill in. Alternatively, enter the function as Y1, and graph the function. Then go to the CALC menu, enter $\int f(x)\,dx$, and follow the on-screen instructions. On older TI calculators, look in the MATH or CALC menu and enter

$$\text{fnInt}(x^3 - 5x + 6, x, -2, 1).$$

In each case, the calculator approximates the integral as 21.75, which actually is its exact value, as we shall see in the next section.

**EXAMPLE 4**    Figure 13.11(a) shows the graph of

$$f(x) = x^3 - 8x^2 + 18x + 5.$$

(a)                              (b)

**Figure 13.11**

The area under the graph from $x = 1$ to $x = 4.5$, which is shown in Figure 13.11(b), is given by

$$\int_{1}^{4.5} (x^3 - 8x^2 + 18x + 5)\,dx.$$

**(a)**  Use a calculator and the definition presented earlier, together with 100 rectangles, to approximate this integral.

**Solution**  For $n = 100$ rectangles, $\Delta x = (b - a)/n = (4.5 - 1)/100 = .035$. The left-hand endpoints of the bases of the rectangles are $1, 1.035, 1.07, 1.105, 1.14, \ldots, 4.43$, and $4.465$. (The right-hand endpoint of the base of the last rectangle is 4.5.) We must compute the following sum:

$$[f(1) + f(1.035) + f(1.07) + \cdots + f(4.43) + f(4.465)](.035).$$

The TI-84+ (see part 1 of the previous Technology Tip) produces the answer 52.69778281 square units.

**(b)**  Approximate the integral by using numerical integration on a calculator.

**Solution**  The numerical integration program in the MATH menu of the TI-84+ (see part 2 of the previous Technology Tip) produces the answer in Figure 13.12. This answer is more accurate than the estimate in part (a) and is correct to seven decimal places, as the techniques in the next section will show. Alternatively, sketching the graph and using the CALC menu produces the result in Figure 13.11(b).

$$\int_{1}^{4.5} (X^3 - 8X^2 + 18X + 5)\,dx$$
$$52.68229167$$

**Figure 13.12**

At the beginning of this section, we saw that if $f$ (with $f$ continuous and $f(x) \geq 0$) is the rate of change of a function $F$, then the total change in $F(x)$ as $x$ goes from $a$ to $b$ is the area between the graph of $f(x)$ and the $x$-axis from $x = a$ to $x = b$. In other words,

$$\text{Total change in } F(x) = \int_{a}^{b} f(x)\,dx.$$

$$\int_{350}^{400} \frac{(X-600)^2}{450} \, dx$$
$$5648.148148$$

**Figure 13.13**

 **EXAMPLE 5** **Marginal Cost** At a production level of $x$ snowboards, the marginal cost (in dollars per snowboard) is given by

$$C'(x) = \frac{(x - 600)^2}{450}.$$

Find the amount added to the total cost when production goes from 350 to 400 snowboards.

**Solution** The amount added is the total change in cost from $x = 350$ to $x = 400$. Since the marginal cost function is the derivative of the cost function,

$$\text{Total change in } C(x) = \int_{350}^{400} C'(x) \, dx$$

$$= \int_{350}^{400} \frac{(x - 600)^2}{450} \, dx.$$

Numerical integration (Figure 13.13) shows that the total change in cost from $x = 350$ to $x = 400$ is \$5648.15. ☑3

✓ **Checkpoint 3** 🖩

If the marginal revenue from selling $x$ of the snowboards in Example 5 is given by

$$R'(x) = .001x^2 + .35x,$$

find the total change in revenue when production goes from 350 to 400 snowboards.

# 13.4 Exercises

*In Exercises 1–4, estimate the required areas by using rectangles whose height is given by the value of the function at the left side of each rectangle. (See Example 1.)*

1. **Energy Consumption** The accompanying graph shows the rate of use of electrical energy (in millions of kilowatts per hour) in a certain city on a hot day. Estimate the total usage of electricity on that day. Let the width of each rectangle be 2 hours.

2. **Energy Consumption** According to the U.S. Energy Information Administration, the rate of residential energy consumption (in quadrillion BTUs per year) is projected to be as shown in the accompanying graph. Estimate the total residential energy consumption that is projected for the period from 2015 to 2040 by using rectangles of width 5 years.

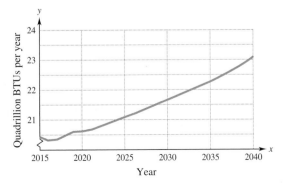

3. **Enrollment in Postsecondary Institutions** The accompanying graph shows data and projections for the rate of enrollment (in millions of students per year) for first-time degree-seeking students in postsecondary institutions (including 2-year colleges, 4-year institutions, and universities). Estimate the total number of students who have started (or will start) a degree program between the years 2010 and 2022. Let the width of each rectangle be 1 year. (Data from: National Center for Education Statistics.)

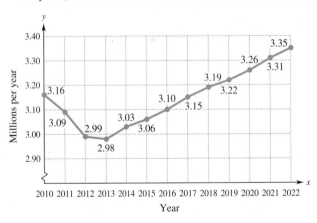

4. **Natural Science** The accompanying graph shows the rate of inhalation of oxygen by a person riding a bicycle very rapidly for 10 minutes. Estimate the total volume of oxygen inhaled in the first 20 minutes after the beginning of the ride. Use rectangles of width 1 minute.

**5.** Explain the difference between an indefinite integral and a definite integral.

**6.** Complete the following statement:

$$\int_0^3 (x^2 + 2)\, dx = \lim_{n \to \infty} \underline{\hspace{1cm}}, \text{ where } \Delta x = \underline{\hspace{1cm}}.$$

*Approximate the area under each curve and above the x-axis on the given interval, using rectangles whose height is the value of the function at the left side of the rectangle.*

**(a)** *Use two rectangles. (See Examples 2 and 3.)*

**(b)** *Use four rectangles.*

**(c)** *Use a graphing calculator (or other technology) and 40 rectangles. (See Example 4.)*

**7.** $f(x) = 3x + 8;\ [0, 4]$    **8.** $f(x) = 9 - 2x;\ [0, 4]$

**9.** $f(x) = 4 - x^2;\ [-2, 2]$    **10.** $f(x) = x^2 + 5;\ [-2, 2]$

**11.** $f(x) = e^{2x} - .5;\ [0, 2]$    **12.** $f(x) = e^{-3x};\ [0, 2]$

**13.** $f(x) = \dfrac{60}{x + 5};\ [-2, 2]$    **14.** $f(x) = \dfrac{2}{x + 1};\ [1, 9]$

*Work the given exercises. (See Example 3.)*

**15.** Find $\int_0^5 (5 - x)\, dx$ by using the formula for the area of a triangle.

**16.** Consider the region below $f(x) = x/2$, above the x-axis, between $x = 0$ and $x = 4$. Let $x_i$ be the left endpoint of the $i$th subinterval.

**(a)** Use four rectangles to approximate the area of the region.

**(b)** Use eight rectangles to approximate the area of the region.

**(c)** Find $\int_0^4 f(x)\, dx$ by using the formula for the area of a triangle.

*Use the numerical integration feature on a graphing calculator to approximate the value of the definite integral. (See Example 4.)*

**17.** $\displaystyle\int_{-5}^{0} (x^3 + 6x^2 - 10x + 2)\, dx$

**18.** $\displaystyle\int_{-1}^{2} (-x^4 + 3x^3 - 4x^2 + 20)\, dx$

**19.** $\displaystyle\int_{2}^{7} 5\ln(2x^2 + 1)\, dx$    **20.** $\displaystyle\int_{1}^{5} 3x \ln x\, dx$

**21.** $\displaystyle\int_{0}^{3} 4x^2 e^{-3x}\, dx$    **22.** $\displaystyle\int_{0}^{1} 3x^2 e^{x^3}\, dx$

**Business** *A marginal revenue function $R'(x)$ (in dollars per unit) is given. Use numerical integration on a graphing calculator or computer to find the total revenue over the given range. (See Example 5.)*

**23.** $R'(x) = .04x^3 - .5x^2 + 2x \quad (0 \le x \le 99)$

**24.** $R'(x) = .04x^4 - .5x^3 + 2x^2 - x \quad (0 \le x \le 60)$

**25.** $R'(x) = \dfrac{175}{1 + 3e^{-.7x}} - 20 \quad (0 \le x \le 310)$

**26.** $R'(x) = \dfrac{50}{1 + e^{-.1x}} \quad (0 \le x \le 200)$

**27. Distance Traveled** An insurance company collected data on the speed of two cars during a particular three-hour window.

The accompanying graph shows that the red car was driving at a constant 25 mph throughout this period, whereas the speed of the blue car was increasing.

**(a)** At time $t = 1$, was the red car traveling faster than, slower than, or the same speed as the blue car? At time $t = 1$, was the total distance traveled by the red car (starting at $t = 0$ when data collection began) greater than, less than, or the same as that of the blue car?

**(b)** Repeat part (a) for time $t = 2$.

**(c)** Repeat part (a) for time $t = 3$.

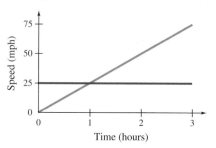

**28. Distance Traveled** Two cars start from rest at a traffic light and accelerate for several minutes. The accompanying graph shows the speed of each car (in feet per second) as a function of time (in seconds). Car A is the one that initially has the greater speed.[*]

**(a)** How far has car A traveled after 2 seconds? (*Hint:* Use formulas from geometry.)

**(b)** When is car A farthest ahead of car B?

**(c)** Estimate the farthest that car A gets ahead of car B. For car A, use formulas from geometry. For car B, use $n = 4$ and the value of the function at the midpoint of each interval for the height of the rectangle.

**(d)** Give a rough estimate of when car B catches up with car A.

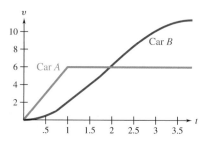

**Distance Traveled** *The graphs for Exercises 29 and 30 show the speed, in miles per hour, at time t, in seconds, when a car accelerates from a dead stop. To find the total distance traveled by the car in reaching 100 miles per hour, we must estimate the definite integral*

$$\int_0^T v(t)\, dt,$$

*where T represents the number of seconds it takes for the car to reach 100 mph.*

*Use the graphs to estimate this distance by adding the areas of rectangles with width 5 seconds and whose height is the value of the function at the left side of the rectangle. To adjust your answer to miles per hour, divide by 3600 (the number of seconds in an hour). You then have the number of miles that the car traveled in reaching 100 mph. Finally, multiply by 5280 feet per mile to convert the answers to feet.*

---

[*] Based on an example given by Steve Monk of the University of Washington.

**29.** Estimate the distance traveled by the car in the following graph:

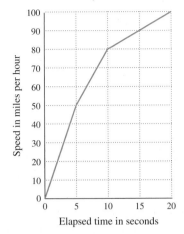

**30.** Estimate the distance traveled by the car in the following graph:

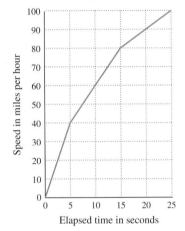

✓ **Checkpoint Answers**

1. (a) $7850

   (b) $14,250

2. (a)

| $i$ | $x_i$ | $f(x_i)$ |
| --- | --- | --- |
| 1 | 1 | 4 |
| 2 | 1.25 | 5 |
| 3 | 1.5 | 6 |
| 4 | 1.75 | 7 |
| 5 | 2 | 8 |
| 6 | 2.25 | 9 |
| 7 | 2.5 | 10 |
| 8 | 2.75 | 11 |
| 9 | 3 | 12 |
| 10 | 3.25 | 13 |
| 11 | 3.5 | 14 |
| 12 | 3.75 | 15 |

   (b) 28.5

3. $13,604.17

# 13.5 The Fundamental Theorem of Calculus

We now develop the connection between definite integrals and antiderivatives, which will explain the similar notation for these two concepts. More importantly, it provides a way to calculate definite integrals exactly.

In the last section, we saw that when $f$ is the rate of change of the function $F$ and $f$ is continuous and nonnegative on $[a, b]$, the definite integral has this interpretation:

$$\int_a^b f(x)\, dx = \text{total change in } F(x) \text{ as } x \text{ changes from } a \text{ to } b. \qquad (*)$$

The total change in $F(x)$ as $x$ changes from $a$ to $b$ is the difference between the value of $F$ at the end and the value of $F$ at the beginning—that is, $F(b) - F(a)$. So equation (*) can be stated as

$$\int_a^b f(x)\, dx = F(b) - F(a).$$

Now, $f$ is the rate of change of $F$, which means that $f$ is the derivative of $F$, or equivalently, $F$ is an antiderivative of $f$. This relationship is an example of the following result.

> ### Fundamental Theorem of Calculus
>
> Suppose $f$ is continuous on the interval $[a, b]$ and $F$ is *any* antiderivative of $f$. Then
>
> $$\int_a^b f(x)\,dx = F(b) - F(a).$$

The proof of the fundamental theorem is discussed at the end of this section. For now, we shall concentrate on applying it. Note these facts:

1. The fundamental theorem applies to every continuous function $f$. It does not require that $f(x) \geq 0$.

2. If the antiderivative $F(x)$ is replaced by $F(x) + C$ for any constant $C$, the conclusion of the fundamental theorem is the same because $C$ is eliminated in the final answer:

$$\int_a^b f(x)\,dx = (F(b) + C) - (F(a) + C)$$

$$= F(b) - F(a).$$

3. The variable used in the integrand does not matter. Each of the following integrals represents the number $F(b) - F(a)$:

$$\int_a^b f(x)\,dx = \int_a^b f(t)\,dt = \int_a^b f(u)\,du.$$

4. In the definition of $\int_a^b f(x)\,dx$, we call $a$ and $b$ the limits of integration. Typically, $a < b$ (in other words, the lower limit of integration is a smaller number than the upper limit of integration). When $a > b$, we make this definition:

$$\int_a^b f(x)\,dx = -\int_b^a f(x)\,dx.$$

For example, $\int_3^1 x^2\,dx = -\int_1^3 x^2\,dx$. Similarly, when the limits of integration are the same, we define $\int_a^a f(x)\,dx$ to be 0. The fundamental theorem is valid for such integrals.

The fundamental theorem of calculus certainly deserves its name, since it is the key connection between differential calculus and integral calculus, which were originally developed separately without knowledge of this connection between them. Most importantly for our purposes, it provides an algebraic means to evaluate definite integrals exactly.

### Applying the Fundamental Theorem

When evaluating definite integrals, the number $F(b) - F(a)$ is sometimes denoted by the symbol

$$F(x)\Big|_a^b$$

For example, if $F(x) = x^3$, then

$$F(x)\Big|_1^2 \quad \text{means } F(2) - F(1) = 2^3 - 1^3 = 7.$$

✓ **Checkpoint 1**

Let $C(x) = x^3 + 4x^2 - x + 3$.
Find the following.

(a)  $C(x)\Big|_1^5$

(b)  $C(x)\Big|_3^4$

**EXAMPLE 1**    Evaluate each of the given integrals.

**(a)** $\displaystyle\int_1^2 4t^3\,dt.$

**Solution**  We know that $F(t) = t^4$ is an antiderivative of $4t^3$. By the fundamental theorem,

$$\int_1^2 4t^3\,dt = t^4\Big|_1^2 = 2^4 - 1^4 = 15.$$

**(b)** $\displaystyle\int_0^5 e^{2x}\,dx$

**Solution**  By the antiderivative rules in Section 13.1, we know that an antiderivative of $e^{2x}$ is $\frac{1}{2}e^{2x}$. Therefore,

$$\int_0^5 e^{2x}\,dx = \frac{1}{2}e^{2x}\Big|_0^5 = \frac{1}{2}e^{10} - \frac{1}{2}e^0 = \frac{1}{2}e^{10} - \frac{1}{2}. \ \checkmark_2$$

To evaluate more complicated integrals, we need the following properties.

## Properties of Definite Integrals

For any real numbers $a$ and $b$ for which the definite integrals exist,

1. $\displaystyle\int_a^b k \cdot f(x)\,dx = k \cdot \int_a^b f(x)\,dx,$ for any real constant $k$
   (constant multiple of a function);

2. $\displaystyle\int_a^b [f(x) \pm g(x)]\,dx = \int_a^b f(x)\,dx \pm \int_a^b g(x)\,dx$
   (sum or difference of functions);

3. $\displaystyle\int_a^b f(x)\,dx = \int_a^c f(x)\,dx + \int_c^b f(x)\,dx,$ for any real number $c$.

Properties 1 and 2 follow directly from the similar properties of antiderivatives. Property 3 won't be proved here, but Figure 13.14 illustrates it when $f(x) > 0$ and $a < c < b$. In this case,

$$\int_a^b f(x)\,dx = \text{area of the entire shaded region}$$

$$= \text{area of blue region} + \text{area of red region}$$

$$= \int_a^c f(x)\,dx \qquad + \int_c^b f(x)\,dx.$$

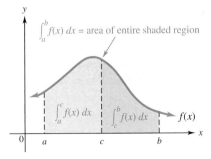

**Figure 13.14**

✓ **Checkpoint 2**

Evaluate each of the given integrals.

**(a)** $\displaystyle\int_4^6 5z\,dz$

**(b)** $\displaystyle\int_0^1 e^{3t}\,dt$

**(c)** $\displaystyle\int_1^9 \sqrt{z}\,dz$

**EXAMPLE 2**   Evaluate

$$\int_1^7 (4x^2 - 6x + 7)\, dx.$$

**Solution** To evaluate this integral, use property 2 of definite integrals to obtain

$$\int_1^7 (4x^2 - 6x + 7)\, dx = \int_1^7 4x^2\, dx - \int_1^7 6x\, dx + \int_1^7 7\, dx.$$

Next, find an antiderivative for each of the three functions on the right-hand side:

$$4x^2 \text{ has antiderivative } 4\frac{x^3}{3} = \frac{4}{3}x^3;$$

$$6x \text{ has antiderivative } 6\frac{x^2}{2} = 3x^2;$$

$$7 \text{ has antiderivative } 7x.$$

Using the fundamental theorem on the three integrals on the right-hand side above, we see that

$$\int_1^7 (4x^2 - 6x + 7)\, dx = \frac{4}{3}x^3 \Big|_1^7 - 3x^2 \Big|_1^7 + 7x \Big|_1^7$$

$$= \frac{4}{3}(7^3 - 1^3) - 3(7^2 - 1^2) + 7(7 - 1)$$

$$= \frac{4}{3}(343 - 1) - 3(49 - 1) + 7(6)$$

$$= 456 - 144 + 42 = 354. \quad \boxed{3}$$

✓ **Checkpoint 3**

Evaluate each definite integral.

(a) $\displaystyle\int_1^3 (4x + 3x^2)\, dx$

(b) $\displaystyle\int_{-2}^3 (6k^2 - 2k + 1)\, dk$

**EXAMPLE 3**   Find

$$\int_1^2 \frac{dy}{y}.$$

**Solution** In Section 13.1, we saw that an antiderivative of $\dfrac{1}{y}$ is $\ln|y|$. Hence,

$$\int_1^2 \frac{dy}{y} = \ln|y| \Big|_1^2 = \ln|2| - \ln|1| \approx .6931 - 0 = .6931. \quad \boxed{4}$$

✓ **Checkpoint 4**

Evaluate the given integrals.

(a) $\displaystyle\int_3^5 \frac{dx}{x}$

(b) $\displaystyle\int_2^7 \frac{4}{x}\, dx$

**EXAMPLE 4**   Evaluate

$$\int_0^5 x\sqrt{25 - x^2}\, dx.$$

**Solution** Use substitution to find the antiderivative. Let $u = 25 - x^2$, so that $du = -2x\,dx$. The integrand is missing the $-2$, so multiply by $\dfrac{-2}{-2}$, putting $-2$ inside the integral and $-\dfrac{1}{2}$ outside:

$$\int x\sqrt{25 - x^2}\,dx = -\frac{1}{2}\int \sqrt{25 - x^2}(-2x\,dx)$$

$$= -\frac{1}{2}\int \sqrt{u}\,du \qquad \text{Substitute.}$$

$$= -\frac{1}{2}\int u^{1/2}\,du \qquad \text{Fractional exponent}$$

$$= -\frac{1}{2}\cdot\frac{u^{3/2}}{3/2} \qquad \text{Use antiderivative formula.}$$

$$= -\frac{1}{2}\cdot\frac{2}{3}u^{3/2} = -\frac{1}{3}u^{3/2} \qquad \text{Simplify.}$$

$$= -\frac{1}{3}(25 - x^2)^{3/2}. \qquad \text{Express in terms of } x.$$

Now evaluate the definite integral using the fundamental theorem:

$$\int_0^5 x\sqrt{25 - x^2}\,dx = -\frac{1}{3}(25 - x^2)^{3/2}\Big|_0^5$$

$$= -\frac{1}{3}(25 - 5^2)^{3/2} - \left[-\frac{1}{3}(25 - 0^2)^{3/2}\right]$$

$$= 0 + \frac{1}{3}\cdot 25^{3/2} = \frac{125}{3}.$$

**✓ Checkpoint 5**

Find $\displaystyle\int_0^2 \frac{x}{x^2 + 1}\,dx$.

**⬢ CAUTION** When using substitution, be sure to express the antiderivative in terms of $x$ before evaluating it. In Example 4, for instance, if you evaluate the antiderivative $-\dfrac{1}{3}u^{3/2}$ at 5 and 0, you get a wrong answer.

**📄 NOTE** An *alternate approach* to evaluating a definite integral when using integration by substitution is to change the limits of integration to correspond to $u$ instead of $x$. In the definite integral in Example 4, the lower limit of integration was $x = 0$. We used the substitution $u = 25 - x^2$, which corresponds to a $u$-value of $u = 25 - (0)^2 = 25$. The upper limit of integration was $x = 5$, which corresponds to $u = 25 - (25)^2 = 0$. If we change the limits of integration, the definite integral becomes

$$\int_{x=0}^{x=5} x\sqrt{25 - x^2}\,dx = -\frac{1}{2}\int_{u=25}^{u=0} \sqrt{u}\,du \qquad \text{Same substitution as in Example 4.}$$

$$= -\frac{1}{3}u^{3/2}\Big|_{u=25}^{u=0} \qquad \text{Notice the limits of integration match the variable.}$$

$$= -\left[\frac{1}{3}(0)^{3/2} - \frac{1}{3}(25)^{3/2}\right] \qquad \text{Substitute } u \text{ limits of integration for variable } u.$$

$$= -\left[0 - \frac{1}{3}(125)\right] = \frac{125}{3}. \qquad \text{The result is the same as in Example 4.}$$

## Area

When the graph of $f(x)$ lies above the $x$-axis between $a$ and $b$, the definite integral $\int_a^b f(x)\,dx$ gives the area between the graph of $f(x)$ and the $x$-axis from $a$ to $b$. Now consider a function $f$ whose graph lies below the $x$-axis from $a$ to $b$. The shaded area in Figure 13.15, can be approximated by a sum of areas of rectangles. In this case, however, because $f$ is negative, a sum of the form

$$f(x_1) \cdot \Delta x + f(x_2) \cdot \Delta x + f(x_3) \cdot \Delta x + \cdots + f(x_n) \cdot \Delta x,$$

represents the negative of the sum of the areas of the rectangles. Consequently, the definite integral $\int_a^b f(x)\,dx$, which is the limit of such sums as $n$ gets larger and larger, is the *negative* of the shaded area in Figure 13.15.

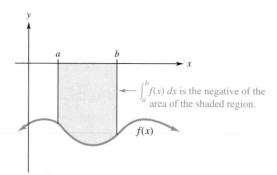

**Figure 13.15**

---

✓ **Checkpoint 6**

Find each shaded area.

(a)

(b)

---

**EXAMPLE 5**     Find the area between the $x$-axis and the graph

$$f(x) = \frac{1}{2}x^2 - 4x \qquad \text{from} \qquad x = 2 \text{ to } x = 8.$$

**Solution** That region, which is shaded in Figure 13.16, lies below the $x$-axis, and the definite integral gives the negative of its area:

$$\int_2^8 \left(\frac{1}{2}x^2 - 4x\right) dx = \left(\frac{x^3}{6} - 2x^2\right)\Big|_2^8 = \left(\frac{512}{6} - 128\right) - \left(\frac{8}{6} - 8\right) = -36.$$

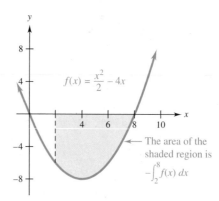

**Figure 13.16**

Therefore, the area of the region is 36. ☑ 6

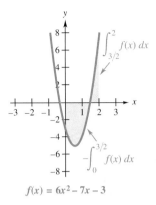

$f(x) = 6x^2 - 7x - 3$

**Figure 13.17**

**EXAMPLE 6**    Find the area between the graph of $f(x) = 6x^2 - 7x - 3$ and the $x$-axis from $x = 0$ to $x = 2$, which is the shaded region in Figure 13.17.

**Solution**  To find the total area, compute the area under the $x$-axis and the area over the $x$-axis separately. Start by finding the $x$-intercepts of the graph by solving

$$6x^2 - 7x - 3 = 0$$
$$(2x - 3)(3x + 1) = 0$$
$$x = 3/2 \quad \text{or} \quad x = -1/3.$$

Since we are concerned just with the graph between 0 and 2, the only relevant $x$-intercept is $\frac{3}{2}$. The area below the $x$-axis is the negative of

$$\int_0^{3/2} (6x^2 - 7x - 3)\, dx = \left( 2x^3 - \frac{7}{2}x^2 - 3x \right) \Big|_0^{3/2}$$

$$= \left( 2\left(\frac{3}{2}\right)^3 - \frac{7}{2}\left(\frac{3}{2}\right)^2 - 3\left(\frac{3}{2}\right) \right) - (0 - 0 - 0) = -\frac{45}{8}.$$

So the area of the lower region is $\frac{45}{8}$. The area above the $x$-axis is

$$\int_{3/2}^2 (6x^2 - 7x - 3)\, dx = \left( 2x^3 - \frac{7}{2}x^2 - 3x \right) \Big|_{3/2}^2$$

$$= \left( 2(2)^3 - \frac{7}{2}(2)^2 - 3(2) \right) - \left( 2\left(\frac{3}{2}\right)^3 - \frac{7}{2}\left(\frac{3}{2}\right)^2 - 3\left(\frac{3}{2}\right) \right)$$

$$= \frac{13}{8}.$$

The area of the upper region is $\frac{13}{8}$, and the total area between the graph and the $x$-axis is

$$\frac{45}{8} + \frac{13}{8} = \frac{58}{8} = 7.25 \text{ square units.}$$

⚠ **CAUTION**  In Example 6, it is essential to compute the upper and lower areas separately, using the two integrals over $\left[0, \frac{3}{2}\right]$ and $\left[\frac{3}{2}, 2\right]$. If you attempt to find the area by using a single integral over the entire interval [0, 2], you will get a wrong answer because

$$\int_0^2 (6x^2 - 7x - 3)\, dx = \left( 2x^3 - \frac{7}{2}x^2 - 3x \right) \Big|_0^2 = -4 - 0 = -4.$$

Although it may be helpful, it is not necessary to graph a function in order to find the entire area between its graph and the $x$-axis. As Example 6 shows, the procedure depends only on finding the $x$-intercepts of the graph that are the limits of integration on the various integrals. Here is a summary of the procedure used there.

## Finding Area

To find the area bounded by $y = f(x)$, the vertical lines $x = a$ and $x = b$, and the $x$-axis, use the following steps:

**Step 1**    Sketch a graph if convenient.

**Step 2**    Find any $x$-intercepts in $[a, b]$. These divide the total region into subregions.

**Step 3**    The definite integral will be *positive* for subregions above the $x$-axis and *negative* for subregions below the $x$-axis. Use separate integrals to find the areas of the subregions.

**Step 4**    The total area is the sum of the areas of all of the subregions.

We have seen a variety of interpretations of certain definite integrals in terms of area. We now present an interpretation that applies to all definite integrals. Consider, for example, $\int_0^2 (6x^2 - 7x - 3)\, dx$. By property 3 of definite integrals,

$$\int_0^2 (6x^2 - 7x - 3)\, dx = \int_0^{3/2} (6x^2 - 7x - 3)\, dx + \int_{3/2}^2 (6x^2 - 7x - 3)\, dx.$$

Each of the integrals on the right-hand side was evaluated in Example 6. Using those results, we see that

$$\int_0^2 (6x^2 - 7x - 3)\, dx = -\frac{45}{8} + \frac{13}{8} = \frac{13}{8} - \frac{45}{8}.$$

As Example 6 shows, this last difference is precisely

(area above the $x$-axis) − (area below the $x$-axis).

The same result holds in general.

---

If $f$ is a continuous function on $[a, b]$, then

$$\int_a^b f(x)\, dx = \left( \begin{array}{c} \text{area between the} \\ \text{graph and the } x\text{-axis} \\ \textit{above} \text{ the axis} \end{array} \right) - \left( \begin{array}{c} \text{area between the} \\ \text{graph and the } x\text{-axis} \\ \textit{below} \text{ the axis} \end{array} \right)$$

---

## Applications

In the last section, we saw that the area under the graph of a rate-of-change function $f'(x)$ from $x = a$ to $x = b$ gives the total change in $f(x)$ from $a$ to $b$. We can now use the fundamental theorem to compute this change.

**EXAMPLE 7**    **Renewable Energy** Data and projections from the U.S. Energy Information Administration were used to generate the following model for the rate of consumption of marketed renewable energy (in trillion BTUs per year), where $x = 10$ corresponds to the beginning of the year 2010:

$$C'(x) = -1.86x^2 + 386x + 4400 \qquad (10 \le x \le 35).$$

At this consumption rate, what is the amount of energy that will be used between the beginning of 2010 and the beginning of 2030?

**Solution** The amount of marketed renewable energy used is the total change in consumption from year 10 to year 30, so it is given by the definite integral

$$\int_{10}^{30} C'(x) = \int_{10}^{30} (-1.86x^2 + 386x + 4400)\, dx$$

$$= \left( -\frac{1.86}{3} x^3 + \frac{386}{2} x^2 + 4400x \right) \Bigg|_{10}^{30} \qquad \text{Find the antiderivative.}$$

$$= (-.62x^3 + 193x^2 + 4400x) \Bigg|_{10}^{30} \qquad \text{Simplify.}$$

$$= (-.62(30)^3 + 193(30)^2 + 4400(30))$$
$$\qquad -(-.62(10)^3 + 193(10)^2 + 4400(10)) = 226{,}280.$$

Therefore, 226,280 trillion BTUs of marketed renewable energy are projected to be used between 2010 and 2030. ✔ 7

✔ **Checkpoint 7**

Use the function in Example 7 to find the amount of marketable renewable energy projected to be used between the beginning of 2010 and the beginning of 2020.

In Chapter 10, you may have studied the normal probability distribution. There are many other distributions that can be used to model probability. One of these is the exponential distribution, which has the general form

$$f(x) = ae^{-ax} \qquad (0 \le x < \infty),$$

where $a$ is a positive constant that changes with the particular model. With this distribution—and, indeed, all probability distributions—the function is greater than or equal to zero, and the area between the curve and the $x$-axis equals 1. Figure 13.18 shows the graph of $f(x)$ for several values of $a$.

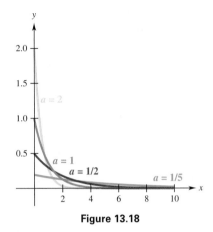

**Figure 13.18**

The next example shows how to use the exponential distribution and the fundamental theorem of calculus to estimate probabilities.

**EXAMPLE 8**    **Probability Models** Biology students were interested in analyzing the amount of time hummingbirds spend at feeders before flying away. They found that an exponential model $f(x) = ae^{-ax}$ with $a = \frac{1}{4}$ approximated the probability of feeding time very well. Use the model to determine the probability that a hummingbird will feed for less than 5 seconds.

**Solution** With $a = \frac{1}{4}$, we can calculate the probability of feeding for less than 5 seconds by determining the area beneath the curve

$$f(x) = \frac{1}{4}e^{-x/4}$$

from $x = 0$ to $x = 5$ seconds. Figure 13.19 shows this area shaded in blue.

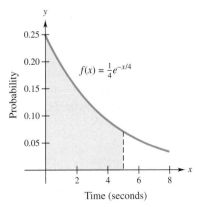

**Figure 13.19**

Thus, we have

$$\int_0^5 \frac{1}{4} e^{-x/4}\, dx = -e^{-x/4} \Big|_0^5 = -e^{-5/4} - (-e^0) \approx -.2865 + 1 = .7135.$$

Therefore, the probability that a hummingbird feeds for less than 5 seconds at the feeder is .7135. ☑ 8

✓ **Checkpoint 8**

The feeding time for a different species of hummingbird can be modeled using the function from Example 8 with $a = .303$. With this model, find the probability that a hummingbird feeds for between 1 and 4 seconds.

## Proof of the Fundamental Theorem

Although we cannot give a rigorous proof of the fundamental theorem of calculus here, we can indicate why it is true in an important case. Suppose $f(x) > 0$ when $a \le x \le b$, so that the definite integral represents the area under the curve and above the $x$-axis.

Define a new function $A(x)$ by this rule:

$$A(x) = \text{area between the graph of } f(x) \text{ and the } x\text{-axis from } a \text{ to } x,$$

as shown in Figure 13.20. For instance, if the area under the graph from $a$ to 4 is 35, then $A(4) = 35$.

We now show that $A(x)$ is an antiderivative of $f(x)$. We must show that

$$A'(x) = \lim_{h \to 0} \frac{A(x + h) - A(x)}{h}$$

is actually $f(x)$. Look at the situation geometrically. When $h$ is a small positive number, $A(x + h)$ is the area under the graph of $f$ from $a$ to $x + h$, and $A(x)$ is the area from $a$ to $x$. Consequently, $A(x + h) - A(x)$ is the area of the shaded strip in Figure 13.21. This area can be approximated with a rectangle having base $h$ and height $f(x)$. The area of the rectangle is $h \cdot f(x)$, and

$$A(x + h) - A(x) \approx h \cdot f(x).$$

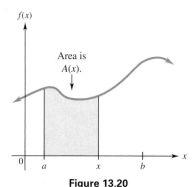

**Figure 13.20**  **Figure 13.21**

Dividing both sides by $h$ gives

$$\frac{A(x + h) - A(x)}{h} \approx f(x).$$

This approximation improves as $h$ gets smaller and smaller. Take the limit on the left as $h$ approaches 0:

$$\lim_{h \to 0} \frac{A(x + h) - A(x)}{h} = f(x).$$

The limit is simply $A'(x)$, so

$$A'(x) = f(x).$$

This result means that $A$ is an antiderivative of $f$, as we set out to show.

Since $A(x)$ is the area under the graph of $f$ from $a$ to $x$, it follows that $A(a) = 0$ and $A(b)$ is the area under the graph from $a$ to $b$. But this last area is the definite integral $\int_a^b f(x)\,dx$. Putting these facts together, we have

$$\int_a^b f(x)\,dx = A(b) = A(b) - 0 = A(b) - A(a).$$

This argument suggests that the fundamental theorem is true when $f(x) > 0$ and the area function $A(x)$ is used as the antiderivative. Other arguments, which are omitted here, handle the case when $f(x)$ may not be positive and any antiderivative $F(x)$ is used.

# 13.5 Exercises

*Evaluate each of the given definite integrals. (See Examples 1–4.)*

**1.** $\int_{-1}^{3} (6x^2 - 7x + 8)\,dx$

**2.** $\int_{0}^{2} (-3x^2 + 6x + 5)\,dx$

**3.** $\int_{0}^{2} 3\sqrt{4u + 1}\,du$

**4.** $\int_{4}^{14} \sqrt{2r + 8}\,dr$

**5.** $\int_{0}^{1} 3(t^{1/2} + 4t)\,dt$

**6.** $\int_{0}^{4} -(3x^{3/2} - x^{1/2})\,dx$

**7.** $\int_{1}^{4} (10y\sqrt{y} + 3\sqrt{y})\,dy$

**8.** $\int_{4}^{9} (4\sqrt{r} - .5r\sqrt{r})\,dr$

**9.** $\int_{4}^{7} \frac{11}{(x - 3)^2}\,dx$

**10.** $\int_{1}^{4} \frac{-3}{(2p + 1)^2}\,dp$

**11.** $\int_{1}^{5} (6n^{-1} + n^{-4})\,dn$

**12.** $\int_{2}^{3} (3x^{-1} - 5x^{-4})\,dx$

**13.** $\int_{0}^{5} 6e^{2x}\,dx$

**14.** $\int_{1}^{2} 8e^{5x}\,dx$

**15.** $\int_{2}^{3} \left(.1e^{-.1A} + \frac{3}{A}\right)\,dA$

**16.** $\int_{1}^{2} \left(\frac{-1}{B} + 3e^{.2B}\right)\,dB$

**17.** $\int_{-1}^{0} y(2y^2 - 3)^5\,dy$

**18.** $\int_{0}^{2} m^2(5m^3 - 2)^3\,dm$

**19.** $\int_{1}^{64} \frac{\sqrt{z} - 2}{\sqrt[3]{z}}\,dz$

**20.** $\int_{1}^{8} \frac{9 - y^{1/3}}{y^{2/3}}\,dy$

**21.** $\int_{1}^{3} \frac{\ln x}{4x}\,dx$

**22.** $\int_{1}^{9} \frac{\sqrt{\ln x}}{x}\,dx$

**23.** $\int_{0}^{8} x^{1/3}\sqrt{x^{4/3} + 9}\,dx$

**24.** $\int_{0}^{4} 9x^{1/2}\sqrt{x^{3/2} + 1}\,dx$

**25.** $\int_{0}^{1} \frac{4e^t}{(3 + e^t)^2}\,dt$

**26.** $\int_{0}^{1} \frac{4e^t}{(e^t + 5)^3}\,dt$

**27.** $\int_{1}^{49} \frac{(1 + \sqrt{x})^{4/3}}{\sqrt{x}}\,dx$

**28.** $\int_{2}^{8} \frac{(1 + x^{1/3})^6}{x^{2/3}}\,dx$

**29.** $\int_{2}^{3} \frac{4x^3 + 6}{x^4 + 6x + 9}\,dx$

**30.** $\int_{0}^{2} \frac{18x^2 - 24x + 6}{x^3 - 2x^2 + x + 12}\,dx$

**31.** Suppose the function in Example 6, $f(x) = 6x^2 - 7x - 3$ from $x = 0$ to $x = 2$, represented the annual rate of change of profit of a company over a two-year period. What might the negative integral for the first year and a half indicate? What integral would represent the overall profit for the two-year period?

**32.** In your own words, describe how the fundamental theorem of calculus relates definite and indefinite integrals.

*Find the area of each shaded region. (See Examples 5 and 6.)*

**33.**

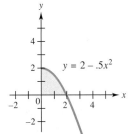

$y = 2 - .5x^2$

**34.**

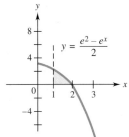

$y = \dfrac{e^2 - e^x}{2}$

**35.**

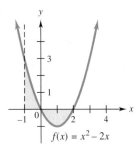

$f(x) = x^2 - 2x$

**36.**

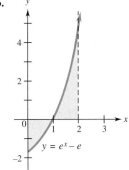

$y = e^x - e$

*Use the definite integral to find the area between the x-axis and f(x)*
*over the indicated interval. Check first to see if the graph crosses the*
*x-axis in the given interval. (See Examples 5 and 6.)*

**37.** $f(x) = 9 - x^2$; $[0, 4]$      **38.** $f(x) = x^2 - 3x - 4$; $[0, 5]$

**39.** $f(x) = x^3 - 1$; $[-1, 2]$      **40.** $f(x) = x^3 - 4x$; $[-2, 4]$

**41.** $f(x) = e^{2x} - 1$; $[-2, 1]$      **42.** $f(x) = 1 - e^{-3x}$; $[-1, 2]$

**43.** $f(x) = x^2 e^{-x^3/2}$; $[0, 3]$      **44.** $f(x) = 5xe^{x^2}$; $[0, 1]$

**45.** $f(x) = \dfrac{1}{x}$; $[1, e]$      **46.** $f(x) = \dfrac{2}{x}$; $[e, e^2]$

**47.** $f(x) = \dfrac{12(\ln x)^3}{x}$; $[1, 4]$

**48.** $f(x) = \dfrac{-(6x - 4)}{(3x^2 - 4x + 4)}$; $[-1, .5]$

**49.** $f(x) = -3\sqrt{x - 5}$; $[5, 9]$    **50.** $f(x) = -\sqrt{3x + 1}$; $[0, 5]$

*Work these problems. (See Example 7.)*

**51. Expenditure Rate**  De Win Enterprises has found that its rate
of change of expenditures (in hundreds of dollars per day) on a
certain type of job is given by

$$E'(x) = 4x + 2,$$

where $x$ is the number of days since the start of the job.

**(a)** Find the total expenditure if the job takes 10 days.

**(b)** How much will be spent on the job from the 10th to the 25th
day?

**52. Learning Rate**  A worker new to a job will improve efficiency
with time so that it takes fewer hours to produce an item with
each day on the job, up to a certain point. Suppose the rate of
change of the number of hours it takes a worker in a certain factory to produce the $x$th item is given by

$$H'(x) = 20 - 2x.$$

**(a)** What is the total number of hours required to produce the
first 5 items?

**(b)** What is the total number of hours required to produce the
first 10 items?

**53. Electronic Shopping**  The rate of spending (in billions of
dollars per year) through electronic shopping and mail-order
houses can be approximated by

$$f(x) = 114e^{.086x}      (0 \le x \le 15),$$

where $x = 0$ corresponds to the start of the year 2000. Find
total sales (to the nearest billion) between the start of 2000 and
the start of 2015. (Data from: Annual Retail Trade Survey.)

**54. Hospital Care**  The expenditure rate on hospital care (in billions
of dollars per year) through the year 2024 is projected to be
approximately

$$f(x) = 426e^{.059x}      (15 \le x \le 24),$$

where $x = 15$ corresponds to the start of the year 2015. Find
the total expenditures (to the nearest billion) between the start
of 2015 and the start of 2024. (Data from: Centers for Medicare
and Medicaid Services.)

**55. Millennials**  The 2010 U.S. Census gives an age distribution
approximated by

$$f(x) = -.81x^2 + 38x + 3904,$$

where $x$ is age in years and $f(x)$ gives the number of people in
thousands who are of age $x$. The population of an age group
can be found by integrating this function over the interval for
that age group.[*] (Data from: www.census.gov.)

**(a)** Find $\displaystyle\int_{20}^{30} f(x)\, dx$. What does this integral represent?

**(b)** Find $\displaystyle\int_{0}^{90} f(x)\, dx$. What does this integral represent?

**(c)** The millennial generation (also known as Generation Y)
includes people who were between 10 and 30 years old at the
time of the census. Find the number of millennials.

**56. Natural Gas**  The rate at which natural gas was consumed (in
billions of cubic feet per year) by the industrial sector for lease
and plant fuel can be approximated by

$$f(x) = 3x^2 - 14x + 1131      (0 \le x \le 15),$$

where $x = 0$ corresponds to the start of the year 2000. Find the
total consumption (to the nearest billion cubic feet) between the
start of 2000 and the start of 2015. (Data from: U.S. Energy
Information Administration.)

**57. Depreciation Rate**  The rate of depreciation (in dollars per
year) for a certain truck is

$$f(t) = \dfrac{6000(.3 + .28t)}{(1 + .3t + .14t^2)^2},$$

where $t$ is in years and $t = 0$ is the year of purchase. Find the
total depreciation at the end of 3 years.

**58. Growth Rate**  A colony of *H. pylori* bacteria grows at the rate
(in milligrams per hour)

$$w' = \sqrt[3]{2t + 5},$$

where $w$ is the weight (in milligrams) after $t$ hours. Find the
change in weight of the colony from $t = 0$ to $t = 4$ hours.

**59. Oil Spill**  An oil tanker is leaking oil at a rate given in barrels
per hour by

$$L'(t) = \dfrac{70 \ln(t + 1)}{t + 1},$$

where $t$ is the time in hours after the tanker hits a hidden rock
(when $t = 0$).

**(a)** Find the total number of barrels that the ship will leak on
the first day.

**(b)** Find the total number of barrels that the ship will leak on
the second day.

**60. Food Service**  The rate of sales (in billions of dollars per
year) in drinking places that serve alcoholic beverages can be
approximated by

$$f(x) = 2.3 \ln(x) + 15      (1 \le x \le 15),$$

[*] Exercise suggested by Ralph DeMarr, University of New Mexico.

where $x = 1$ corresponds to the start of the year 2001. Find total sales (to the nearest billion) between the start of 2001 and the start of 2015. (Data from: Annual Retail Trade Survey.)

**61. Total Profit** A marginal profit function (in dollars per unit) at a production level of $x$ units is given by

$$P'(x) = xe^{-.04x} \qquad (0 \le x \le 100).$$

Find and interpret $P(100)$.

**62. Coal Projections** The rate of spending on coal (in millions of dollars per year) is projected for the years 2020 through 2030 to be approximately

$$f(t) = (1125t + 40{,}500)e^{-.02t} \qquad (0 \le t \le 10),$$

where $t = 0$ corresponds to the year 2020. Find the projected total spending (to the nearest million) on coal between 2020 and 2030. (Data from: U.S. Energy Information Administration.)

**Business** *A model for the amount of time it takes for the first customer to arrive at a coffee shop in the morning is modeled by the exponential distribution and the function*

$$f(x) = \frac{1}{3}e^{-x/3}.$$

*(See Example 8.)*

**63.** Find the probability that the first customer arrives in the first 5 minutes.

**64.** Find the probability that the first customer arrives between 1 and 3 minutes after the shop opens.

✓ **Checkpoint Answers**

1. (a) 216      (b) 64

2. (a) 50      (b) $\dfrac{1}{3}e^3 - \dfrac{1}{3}$      (c) $52/3$

3. (a) 42      (b) 70

4. (a) .51083      (b) 5.01105

5. $\dfrac{1}{2}\ln 5$

6. (a) $8/3$      (b) $7/4$

7. 97,560 trillion BTUs      8. .4410

# 13.6 Applications of Integrals

If a function gives the rate of change of a quantity, then the area between the graph of the rate function and the $x$-axis is the total amount of the quantity. This fact has numerous applications, some of which are explored here.

**EXAMPLE 1** **Extended Warranty** When you buy a used car, you may be offered the opportunity to buy a warranty that covers repair costs for a certain period. A company that provides such warranties uses current data and past experience to determine that the annual rate of repair costs for a particular model is given in year $x$ by

$$r(x) = 120\,e^{.4x}.$$

What will the total repair costs be for one year and for three years?

**Solution** The total repair costs are represented by the area under the rate curve over the appropriate interval. For the first three years, the area is shown in Figure 13.22 and is given by the integral

$$\int_0^3 120\,e^{.4x}\,dx = \frac{120}{.4}e^{.4x}\bigg|_0^3 = 300\,e^{.4x}\bigg|_0^3$$

$$= 300\,e^{.4(3)} - 300\,e^{.4(0)}$$

$$= 300\,e^{1.2} - 300 \approx \$696.04.$$

Similarly, the total repair costs for one year are

$$\int_0^1 120\,e^{.4x}\,dx = 300e^{.4x}\bigg|_0^1 = 300\,e^{.4} - 300 \approx \$147.55.$$

In order to make a profit, the company must charge more than $147.55 for a one-year warranty and more than $696.04 for a three-year warranty. ✓₁

$r(x) = 120e^{.4x}$

**Figure 13.22**

✓ **Checkpoint 1**

In Example 1, find the total repair costs for

(a) two years;

(b) four years.

**EXAMPLE 2** **Walmart Profit** The rate at which profit was generated over time (in billions of dollars per year) by Walmart Stores Inc. can be approximated by

$$f(t) = -.35t^2 + 12t + 17 \qquad (7 \le t \le 17),$$

where $t = 7$ corresponds to the start of the year 2007. What was the total profit generated (to the nearest billion) over the 10-year period from the start of the year 2007 to the end of 2016? (Data from: www.morningstar.com.)

**Solution** The end of 2016 coincides with the start of 2017, so the total profit generated is the area under the graph of $f(t)$ and above the horizontal axis from $t = 7$ to $t = 17$, as shown in Figure 13.23. This area is given by

$$\text{Total profit} = \int_{7}^{17} (-.35t^2 + 12t + 17)\, dt$$

$$= \left( -\frac{.35}{3}t^3 + 6t^2 + 17t \right)\Bigg|_{7}^{17} \qquad \text{Find the antiderivative.}$$

$$= \left[ -\frac{.35}{3}(17)^3 + 6(17)^2 + 17(17) \right]$$

$$- \left[ -\frac{.35}{3}(7)^3 + 6(7)^2 + 17(7) \right]$$

$$\approx 1076.8.$$

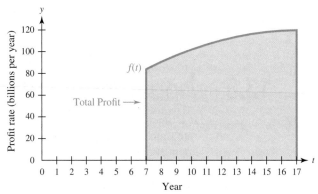

**Figure 13.23**

So the total profit generated over these 10 years was approximately \$1076.8 billion, or just over \$1 trillion. ✔️₂

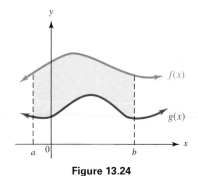

**Figure 13.24**

## Area between Two Curves

In some applications, it is necessary to find the area between two curves. For example, the area between the graphs of $f(x)$ and $g(x)$ from $x = a$ to $x = b$ in Figure 13.24 is shaded. This area is the area under the graph of $f(x)$ *minus* the area under the graph of $g(x)$, that is,

$$\int_{a}^{b} f(x)\, dx - \int_{a}^{b} g(x)\, dx,$$

which can be written as

$$\int_{a}^{b} [f(x) - g(x)]\, dx.$$

Similar arguments in other cases (even when the graphs are not above the $x$-axis) produce the following important result.

### Area between Two Curves

If $f$ and $g$ are continuous functions and $f(x) \geq g(x)$ on the interval $[a, b]$, then the area between the graphs of $f(x)$ and $g(x)$ from $x = a$ to $x = b$ is given by

$$\int_{a}^{b} [f(x) - g(x)]\, dx.$$

**EXAMPLE 3**    **Amazon Profit** The rate at which revenue was generated over time (in billions of dollars per year) by Amazon.com Inc. can be approximated by

$$f(t) = 1.06t^2 - 11.3t + 39.2 \qquad (7 \le t \le 17),$$

where $t = 7$ corresponds to the start of the year 2007. The rate at which costs were incurred (in billions of dollars per year) during that time was approximately

$$g(t) = .43t^2 - 1.5t \qquad (7 \le t \le 17).$$

Find the total profit (to the nearest billion) earned over the 10-year period between the start of the year 2007 and the end of 2016. (Data from: www.morningstar.com.)

**Solution** The rate of revenue and rate of cost functions are graphed in Figure 13.25. Total revenue during this period is the area under the rate of revenue curve, and total cost is the area under the rate of cost curve, so total profit is the area between the two curves. Because the end of 2016 coincides with the start of 2017, we want the area between the curves from $t = 7$ to $t = 17$, as shaded in Figure 13.25.

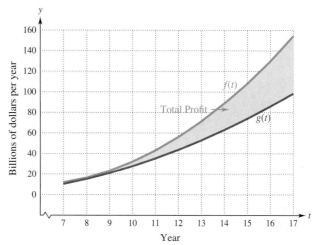

**Figure 13.25**

The area of the shaded region is

$$\text{Total profit} = \int_7^{17} [f(t) - g(t)]\, dt$$

$$= \int_7^{17} [(1.06t^2 - 11.3t + 39.2) - (.43t^2 - 1.5t)]\, dt$$

**✓ Checkpoint 3**

Find the total profit in Example 3 if we were given $f(t) = 10e^{.1t}$ and $g(t) = 3 + 2t$.

$$= \int_7^{17} (.63t^2 - 9.8t + 39.2)\, dt$$

$$= \left( \frac{.63}{3}t^3 - \frac{9.8}{2}t^2 + 39.2t \right)\Bigg|_7^{17}$$

$$= (.21t^3 - 4.9t^2 + 39.2t)\Bigg|_7^{17}$$

$$= [.21(17)^3 - 4.9(17)^2 + 39.2(17)] - [.21(7)^3 - 4.9(7)^2 + 39.2(7)]$$

$$= 175.7.$$

So the total profit for Amazon was approximately $176 billion over this period.

If revenue, cost, and profit at a production level of $x$ units are denoted by $R(x)$, $C(x)$, and $P(x)$, respectively, then

$$P(x) = R(x) - C(x),$$

so that

$$P'(x) = R'(x) - C'(x).$$

As we saw in Chapter 12, maximal profit occurs at a value of $x$ for which $P'(x) = 0$. The previous equation shows that $P'(x) = 0$ exactly when $R'(x) = C'(x)$. In other words, maximal profit occurs at a production level for which marginal revenue and marginal cost are equal.

**EXAMPLE 4**    **Maximizing Profit** A manufacturer produces a fabric similar to nylon. The product has been selling well, with a marginal revenue (in hundreds of dollars per hundred bolts) given by

$$R'(x) = -.3x^2 + 9x + 11,$$

where $x$ is measured in hundreds of bolts of fabric. The marginal cost (in hundreds of dollars per hundred bolts) is given by

$$C'(x) = 2x + 6.$$

**(a)** To maximize profit, how many bolts should the company produce?

**Solution** The manufacturer should continue production until the marginal costs equal the marginal revenue. Find this point by solving the equation $R'(x) = C'(x)$ as follows:

$$-.3x^2 + 9x + 11 = 2x + 6$$
$$-3x^2 + 90x + 110 = 20x + 60 \qquad \text{Multiply by 10.}$$
$$-3x^2 + 70x + 50 = 0.$$

Figure 13.26

Find the positive solution by using a graphing calculator (Figure 13.26) or the quadratic formula:

$$x = \frac{-70 - \sqrt{(70)^2 - 4(-3)(50)}}{2(-3)} \approx 24.0.$$

To maximize profit, the company should produce 2400 bolts of the fabric.

**(b)** What will the profit be on 2400 bolts?

**Solution** To find the profit on 2400 bolts, find the area between the graphs of the marginal revenue and marginal cost functions, shown in Figure 13.27.

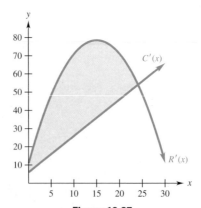

**Figure 13.27**

The area (profit) is

$$\text{Profit} = \int_0^{24} [R'(x) - C'(x)]\, dx$$

$$= \int_0^{24} [(-.3x^2 + 9x + 11) - (2x + 6)]\, dx$$

$$= \int_0^{24} (-.3x^2 + 7x + 5)\, dx \qquad \text{Combine terms.}$$

$$= \left( \frac{-.3x^3}{3} + \frac{7x^2}{2} + 5x \right) \Big|_0^{24} \qquad \text{Find the antiderivative.}$$

$$= 753.6.$$

The profit on 2400 bolts will be $75,360.  ☑ 4

## ✓ Checkpoint 4

In Example 4, suppose we were given $R'(x) = 20 - x$ and $C'(x) = 3x + 8$.

**(a)** To maximize profit, how many bolts should the company produce?

**(b)** What will the maximum profit be?

**EXAMPLE 5**  **Maximizing Savings** A company is considering a new manufacturing process in one of its plants. The new process provides substantial initial savings, with the savings declining with time $x$ according to the rate-of-savings function

$$S'(x) = 100 - x^2,$$

where $x$ is measured in years and $S'(x)$ in thousands of dollars per year. At the same time, the cost of operating the new process increases with time $x$ according to the rate-of-cost function (in thousands of dollars per year)

$$C'(x) = x^2 + \frac{14}{3}x.$$

**(a)** To maximize its savings, for how many years should the company use the new process?

**Solution** Figure 13.28 shows the graphs of the rate-of-savings and the rate-of-cost functions. As was the case in Example 4, maximum savings will occur when the rate of savings equals the rate of cost, that is, the time at which these graphs intersect.

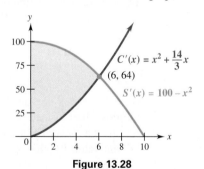

**Figure 13.28**

The graphs intersect when

$$S'(x) = C'(x),$$

or

$$100 - x^2 = x^2 + \frac{14}{3}x.$$

Solve this equation as follows:

$$0 = 2x^2 + \frac{14}{3}x - 100$$

$$0 = 3x^2 + 7x - 150 \qquad \text{Multiply by } \tfrac{3}{2}.$$

$$0 = (x - 6)(3x + 25) \qquad \text{Factor.}$$

Set each factor equal to 0 and solve both equations to get

$$x = 6 \quad \text{or} \quad x = -\frac{25}{3}.$$

Only 6 is a meaningful solution here. Alternatively, the intersection point can be found by using the intersection finder on a calculator, as in Figure 13.29. The company should use the new process for 6 years.

**(b)** Taking the additional costs into account, how much will the company actually save during the 6-year period?

**Solution** The actual savings (or net savings) are the difference between the savings and the additional costs. Total savings are given by the area under the rate-of-savings curve, and total additional costs are given by the area under the rate-of-cost curve, from $x = 0$ to $x = 6$. So total net savings are given by the difference between these two areas—that is, by the shaded area in Figure 13.28. This area is given by the integral

$$\text{Net savings} = \int_0^6 [S'(x) - C'(x)]\, dx$$

$$= \int_0^6 \left[ (100 - x^2) - \left( x^2 + \frac{14}{3}x \right) \right] dx$$

$$= \int_0^6 \left( 100 - \frac{14}{3}x - 2x^2 \right) dx \qquad \text{Combine terms.}$$

$$= \left( 100x - \frac{7}{3}x^2 - \frac{2}{3}x^3 \right)\Big|_0^6 \qquad \text{Find the antiderivative.}$$

$$= 100(6) - \frac{7}{3}(36) - \frac{2}{3}(216) = 372.$$

The company will save a total of $372,000 over the 6-year period. ✔5

110

0 ──────────── 10
Intersection
X=6          Y=64
-30

**Figure 13.29**

✔ **Checkpoint 5**

If a company's rate-of-savings and rate-of-cost functions are, respectively,

$$S'(x) = 150 - 2x^2$$

and

$$C'(x) = x^2 + 15x,$$

where $S'(x)$ and $C'(x)$ give amounts in thousands of dollars per year, find the given quantities.

**(a)** Number of years to maximize savings

**(b)** Net savings during the period in part (a)

We saw in Chapter 5 that a sum of money that can be deposited today to yield some larger amount in the future is called the **present value** of that future amount. The present value $PV$ of a future value of $FV$ dollars at an interest rate of $r$ per year compounded continuously for $t$ years is

$$PV = \frac{FV}{e^{rt}} = FV \cdot e^{-rt}.$$

Suppose that, instead of a single future payment of $FV$ dollars, we want a future income stream at a *rate of $FV(t)$ dollars per year* over the next $T$ years. In that case, to find the total present value $PV$ that we must deposit today, we need to find the definite integral of the income stream rate over the given time period:

$$PV = \int_0^T FV(t) \cdot e^{-rt}\, dt.$$

**EXAMPLE 6**    **Present Value** Assuming an interest rate of 2% ($r = .02$) per year compounded continuously, find the present value in each of the following situations.

**(a)** Find the present value of a single future payment of $FV = \$10,000$ 5 years from today.

**Solution** This is a single future payment, so we do not need to integrate over a period of time but rather just use the present value formula:

$$PV = FV \cdot e^{-rt} = 10,000 \cdot e^{-.02(5)} = \$9048.37.$$

This means that a deposit of $9048.37 is required today in order to have one future payment of $10,000 in 5 years.

**(b)** Find the present value of a future income stream of $FV(t) = \$10,000$ per year over the next 5 years.

**Solution** Because we have a stream of future payments, we must find a definite integral over the period of time starting at $t = 0$ and ending at $t = 5$:

$$PV = \int_0^T FV(t) \cdot e^{-rt}\, dt = \int_0^5 10{,}000 e^{-.02t}\, dt = 47{,}581.29.$$

This means that a deposit of $47,581.29 is required today in order to support a series of future payments at a rate of $10,000 per year for the next 5 years. ✔️ 6

## Consumers' and Producers' Surplus

The market determines the price at which a product is sold. As indicated earlier, the point of intersection of the demand curve and the supply curve for a product gives the equilibrium price. At the equilibrium price, consumers will purchase the same amount of the product that the manufacturers want to sell. Some consumers, however, will be willing to spend more for an item than the equilibrium price. The total of the differences between the equilibrium price of the item and the higher prices all those individuals would be willing to pay is called the **consumers' surplus.**

In Figure 13.30, the total shaded area (red and blue) under the demand curve is the total amount consumers are willing to spend for $q_0$ items. The red area under the line $y = p_0$ shows the total amount consumers actually will spend at the equilibrium price of $p_0$. The blue area represents the consumers' surplus. As the figure suggests, the consumers' surplus is given by the area between the two curves $p = D(q)$ and $p = p_0$, so its value can be found with a definite integral as follows.

### ✓ Checkpoint 6

Find the present value of an income stream of $50,000 per year over the next 20 years if the interest rate is 6% per year compounded continuously.

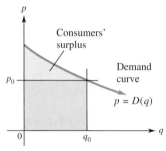

**Figure 13.30**

---

### Consumers' Surplus

If $D(q)$ is a demand function with equilibrium price $p_0$ and equilibrium demand $q_0$, then

$$\text{Consumers' surplus} = \int_0^{q_0} [D(q) - p_0]\, dq.$$

---

Similarly, if some manufacturers would be willing to supply a product at a price *lower* than the equilibrium price $p_0$, the total of the differences between the equilibrium price and the lower prices at which the manufacturers would sell the product is called the **producers' surplus.** Figure 13.31 shows the (red) area under the supply curve from $q = 0$ to $q = q_0$, which is the minimum total amount the manufacturers are willing to realize from the sale of $q_0$ items. The total shaded area (red and blue) under the line $p = p_0$ is the amount actually realized. The difference between these two areas, the producers' surplus (blue), is also given by a definite integral.

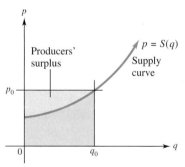

**Figure 13.31**

---

### Producers' Surplus

If $S(q)$ is a supply function with equilibrium price $p_0$ and equilibrium supply $q_0$, then

$$\text{Producers' surplus} = \int_0^{q_0} [p_0 - S(q)]\, dq.$$

**EXAMPLE 7** **Supply and Demand** Suppose the price (in dollars per ton) for steel-cut oatmeal is

$$D(q) = 900 - 20q - q^2$$

when the demand for the product is $q$ tons. Also, suppose the function

$$S(q) = q^2 + 10q$$

gives the price (in dollars per ton) when the supply is $q$ tons. Find the consumers' surplus and the producers' surplus.

**Solution** We begin by finding the equilibrium quantity by setting the two prices equal:

$$D(q) = S(q)$$
$$900 - 20q - q^2 = q^2 + 10q$$
$$0 = 2q^2 + 30q - 900$$
$$0 = q^2 + 15q - 450$$
$$0 = (q - 15)(q + 30)$$
$$q = 15 \quad \text{or} \quad q = -30.$$

Either factor (as shown above) or use the quadratic formula to see that the only positive solution of this equation is $q = 15$. At the equilibrium point where the supply and demand are both $q_0 = 15$ tons, the price is

$$\text{Equilibrium price} = p_0 = S(q_0) = S(15) = 15^2 + 10(15) = 375,$$

or \$375. Verify that the same answer is found by computing $D(15)$. The consumers' surplus, represented by the blue area shown in Figure 13.32, is

$$\text{Consumers' surplus} = \int_0^{q_0} [D(q) - p_0]\, dq = \int_0^{15} [(900 - 20q - q^2) - 375]\, dq$$

$$= \int_0^{15} (525 - 20q - q^2)\, dq.$$

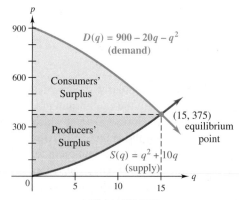

**Figure 13.32**

Evaluating this definite integral gives

$$\left(525q - 10q^2 - \frac{1}{3}q^3\right)\Big|_0^{15} = \left[525(15) - 10(15)^2 - \frac{1}{3}(15)^3\right] - 0$$

$$= 4500.$$

Here, the consumers' surplus is \$4500. The producers' surplus (the red area in Figure 13.32) is given by

$$\text{Producers' surplus} = \int_0^{q_0} [p_0 - S(q)]\, dq = \int_0^{15} [375 - (q^2 + 10q)]\, dq$$

$$= \int_0^{15} (375 - q^2 - 10q)\, dq$$

$$= \left(375q - \frac{1}{3}q^3 - 5q^2\right)\Big|_0^{15}$$

$$= \left[375(15) - \frac{1}{3}(15)^3 - 5(15)^2\right] - 0$$

$$= 3375.$$

The producers' surplus is \$3375.

## ✓ Checkpoint 7

Given the demand function $D(q) = 12 - .07q$ and the supply function $S(q) = .05q$, where $D(q)$ and $S(q)$ are in dollars, find

(a) the equilibrium point;

(b) the consumers' surplus;

(c) the producers' surplus.

# 13.6 Exercises

*Work the given exercises. (See Examples 1 and 2.)*

1. **Maintenance Costs** A car-leasing firm must decide how much to charge for maintenance on the cars it leases. After careful study, the firm determines that the rate of maintenance, $M(x)$, on a new car will approximate $M(x) = 60(1 + x^2)$ dollars per year, where $x$ is the number of years the car has been in use. What total maintenance cost can the company expect for a 2-year lease? What minimum amount should be added to the monthly lease payments to pay for maintenance on a 2-year lease?

2. **Maintenance Costs** Using the function of Exercise 1, find the total maintenance cost during the first three years. What minimum monthly charge should be added to cover a 3-year lease?

3. **Total Savings** A company is considering a new manufacturing process. It knows that the rate of savings (in dollars per year) from the process will be about $S(t) = 1000(t + 2)$, where $t$ is the number of years the process has been in use. Find the total savings during the first year. Find the total savings during the first six years.

4. **Production Capacity** A company is introducing a new product. Production is expected to grow slowly because of difficulties in the start-up process. It is expected that the rate of production will be approximated by $P(x) = 1000e^{.2x}$, where $x$ is the number of years since the introduction of the product. Will the company be able to supply 20,000 units during the first four years?

5. **Procter & Gamble Revenue** The rate at which revenue was generated (in billions of dollars per year) for Procter & Gamble Co. can be approximated by

$$f(t) = -1.38t^2 + 34t - 123 \qquad (10 \le t \le 17),$$

where $t = 10$ corresponds to the year 2010. What was the total revenue generated (to the nearest billion) between the start of 2010 and the start of 2017? (Data from: www.morningstar.com.)

6. **Procter & Gamble Costs** The rate at which costs were incurred (in billions of dollars per year) for Procter & Gamble Co. can be approximated by

$$g(t) = -.83t^2 + 21t - 89 \qquad (10 \le t \le 17),$$

where $t = 10$ corresponds to the start of 2010. What was the total cost incurred (to the nearest billion) between the start of 2010 and the start of 2017? (Data from: www.morningstar.com.)

7. **Priceline Revenue** The rate at which revenue was generated (in millions of dollars per year) for The Priceline Group Inc. for the years 2010 through 2016 can be approximated by

$$f(t) = 340e^{.22t} \qquad (10 \le t \le 16),$$

where $t = 10$ corresponds to the start of the year 2010. (Data from: www.morningstar.com.) Find and interpret

$$\int_{10}^{16} f(t)\, dt.$$

8. **Autodesk Costs** The rate at which costs were incurred (in millions of dollars per year) for the NASDAQ technology company Autodesk Inc. can be approximated by

$$f(t) = 53e^{.116t} \qquad (10 \le t \le 17),$$

where $t = 10$ corresponds to the start of the year 2010. (Data from: www.morningstar.com.) Find and interpret

$$\int_{10}^{17} f(t)\, dt.$$

*Work these exercises. (See Example 3.)*

9. Consider the graphs of $f(x) = x^2$ and $g(x) = 8 - 2x$ shown in the accompanying figure.

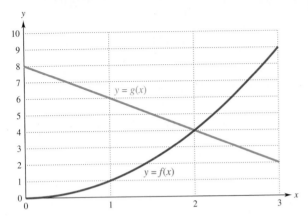

(a) Use algebra to find where the two curves intersect, and confirm that your answer agrees with the graph.

(b) Find the area between the curves on the interval [0, 2].

(c) Find the area between the curves on the interval [2, 3].

(d) Find the area between the curves on the interval [0, 3].

10. Use the figure in the previous exercise to find the area between the curves on the interval [1, 3].

11. Consider the graphs of $f(x) = \frac{x}{2}$ and $g(x) = \sqrt{x}$ shown in the accompanying figure.

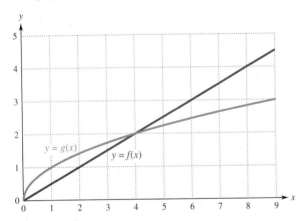

(a) Use algebra to find where the two curves intersect, and confirm that your answer agrees with the graph.

(b) Find the area between the curves on the interval [0, 4].

(c) Find the area between the curves on the interval [4, 9].

(d) Find the area between the curves on the interval [0, 9].

12. Use the figure in the previous exercise to find the area between the curves on the interval [1, 9].

*Find the area between the two curves. (See Example 3.)*

13. $y = 3x$ and $y = x^2 - 4$ from $x = -1$ to $x = 3$

14. $y = x^2 - 30$ and $y = 10 - 3x$ from $x = 1$ to $x = 5$

15. $y = x^2$ and $y = x^3$ from $x = 0$ to $x = \frac{1}{2}$

16. $y = 6x^2$ and $y = 8x^3$ from $x = 0$ to $x = 2$

17. $y = e^x$ and $y = x$ from $x = 0$ to $x = 1$

18. $y = e^x$ and $y = 3 - e^x$ from $x = -3$ to $x = 0$

🖩 *Use a graphing calculator to approximate the area between the graphs of each pair of functions on the given interval.*

19. $y = \ln x$ and $y = 2xe^x$; [1, 4]

20. $y = \ln x$ and $y = 4 - x^2$; [2, 5]

21. $y = \sqrt{9 - x^2}$ and $y = \sqrt{x + 1}$; [−1, 2]

22. $y = \sqrt{4 - 4x^2}$ and $y = \sqrt{\dfrac{9 - x^2}{7}}$; [−.5, .5]

*Work these problems. (See Examples 3–5.)*

23. **Maximizing Savings** Suppose a company wants to introduce a new machine that will produce annual savings in dollars per year at the rate given by

$$S'(x) = 150 - x^2,$$

where $x$ is the number of years of operation of the machine, while producing annual costs in dollars per year at the rate of

$$C'(x) = x^2 + \frac{11}{4}x.$$

(a) To maximize its net savings, for how many years should the company use this new machine?

(b) What are the net savings during the first year of use of the machine?

(c) What are the net savings over the period determined in part (a)?

24. **Smog Control** A new smog-control device will reduce the output of sulfur oxides from automobile exhausts. It is estimated that the rate of savings (in millions of dollars per year) to the community from the use of this device in year $x$ will be approximated by

$$S'(x) = -x^2 + 4x + 8.$$

The new device cuts down on the production of sulfur oxides, but it causes an increase in the production of nitrous oxides. The rate of costs (in millions of dollars per year) to the community in year $x$ is approximated by

$$C'(x) = \frac{3}{25}x^2.$$

(a) To maximize net savings, for how many years should the new device be used?

(b) What will be the net savings over the period found in part (a)?

25. **Maximizing Profit** A factory at Levinson Industries has installed a new process that will produce an increased rate of revenue (in thousands of dollars per year) of

$$r(t) = 104 - .4e^{t/2},$$

where $t$ is time measured in years. The new process produces additional costs (in thousands of dollars per year) at the rate of

$$c(t) = .3e^{t/2}.$$

**(a)** To maximize this additional profit, how long should the new process be used?

**(b)** Find the additional profit during this period.

26. **Construction Expenditures** Richardson Construction Company has an expenditure rate of $E(x) = e^{.15x}$ dollars per day on a particular paving job and an income rate of $I(x) = 120.3 - e^{.15x}$ dollars per day on the same job, where $x$ is the number of days from the start of the job. The company's profit on that job will equal total income less total expenditures. Profit will be maximized if the job ends at the optimum time, which is the point where the two curves meet. Find the following:

**(a)** the optimum number of days for the job to last;

**(b)** the total income for the optimum number of days;

**(c)** the total expenditures for the optimum number of days;

**(d)** the maximum profit for the job.

27. **Oil Production** After $t$ years, an oil refinery is producing at the rate of

$$P(t) = \frac{20}{1.2t + 1.6}$$

trillions of gallons per year. At the same time, the oil is consumed at a rate of $C(t) = t + .8$ trillions of gallons per year.

**(a)** In how many years will the rate of consumption equal the rate of production?

**(b)** What is the total excess production before consumption and production are equal?

28. **Maximizing Savings** The rate of expenditure (in dollars per year) for maintenance of a certain machine is given by

$$m(x) = x^2 + 6x,$$

where $x$ is time measured in years. The machine produces a rate of savings (in dollars per year) given by

$$s(x) = 360 - 2x^2.$$

**(a)** In how many years will the maintenance rate equal the savings rate?

**(b)** What will be the net savings over this period?

29. **Water Pollution** Polluted fluid from a factory begins to enter a lake at the rate

$$f(t) = 8(1 - e^{-.25t}),$$

where $t$ is time (in hours) and $f(t)$ is in gallons per hour. At the same time, a pollution filter begins to remove the pollution at the rate of

$$g(t) = .2t,$$

where $t$ is in hours and $g(t)$ is in gallons per hour.

**(a)** How much pollution is in the lake after 20 hours?

**(b)** At what time will the rate at which pollution enters the lake equal the rate at which pollution is being removed?

**(c)** How much pollution is in the lake at the time found in part (b)?

**(d)** At what time will all the pollution be removed from the lake?

30. **Blood Alcohol Content** Suppose that, over a 4-hour period on a certain evening, an individual is consuming alcohol at the rate

$$f(x) = 1 - e^{-x},$$

where $x$ is in hours and $f(x)$ is in ounces per hour. At the same time, this individual is metabolizing alcohol at the rate

$$m(x) = .6$$

ounces per hour. (Note that one 12-ounce beer that is 5% alcohol by volume contains .6 ounce of pure alcohol.)

**(a)** What was the total amount of alcohol consumed during the 4-hour period?

**(b)** What was the total amount of alcohol metabolized during the 4-hour period?

**(c)** How much alcohol remains in the body at $x = 4$ hours?

**Present Value** *Work these exercises. (See Example 6.)*

31. Find the present value of a single future payment of $FV = \$40,000$ to be made 8 years from today, assuming an interest rate of 1.5% per year compounded continuously.

32. What amount must be deposited today at an interest rate of .9% per year compounded continuously to have a future amount worth $7000 in 15 years?

33. Find the present value of a future income stream of $FV(t) = \$25,000$ per year over the next 4 years, assuming an interest rate of 3% per year compounded continuously.

34. Assuming an interest rate of 2.5% per year compounded continuously, what initial deposit is required to fund a future income stream of $10,000 per year for the next 20 years?

35. It is possible (and often desirable) to have a future income stream for which the size of the payments fluctuates over time. Suppose you want a future income stream of $FV(t) = 5000 + 500t$ over the next 10 years. Find the present value of this income stream, assuming an interest rate of 2% per year compounded continuously.

36. Suppose you want a future income stream of $FV(t) = 30,000 + 1000t$ for 20 years starting today. Find the present value $PV$ of this income stream, assuming an interest rate of 4% per year compounded continuously.

**Consumers' and Producers' Surplus** *Work the given supply-and-demand exercises, where the price is given in dollars. (See Example 7.)*

37. Suppose the supply function of a certain item is given by

$$S(q) = \frac{7}{5}q$$

and the demand function is given by

$$D(q) = -\frac{3}{5}q + 10.$$

**(a)** Graph the supply and demand curves.

**(b)** Find the point at which supply and demand are in equilibrium.

**(c)** Find the consumers' surplus.

**(d)** Find the producers' surplus.

**38.** Find the consumers' surplus and the producers' surplus for an item having supply function

$$S(q) = 3q^2$$

and demand function

$$D(q) = 120 - \frac{q^2}{6}.$$

**39.** Suppose the supply function for concrete is given by

$$S(q) = 100 + 3q^{3/2} + q^{5/2}$$

and that supply and demand are in equilibrium at $q = 9$. Find the producers' surplus.

**40.** Find the producers' surplus if the supply function for digital cameras sold by an office supply chain is given by

$$S(q) = q^{5/2} + 2.1q^{3/2} + 48.$$

Assume that supply and demand are in equilibrium at $q = 16$.

**41.** Find the consumers' surplus if the demand function for a particular wine is given by

$$D(q) = \frac{15{,}500}{(3.2q + 7)^3}$$

and if supply and demand are in equilibrium at $q = 5$.

**42.** Find the consumers' surplus if the demand function for grass seed is given by

$$D(q) = \frac{80}{(3q + 1)^2},$$

assuming that supply and demand are in equilibrium at $q = 3$.

**43.** Suppose the supply function of a certain item is given by

$$S(q) = e^{q/2} - 1$$

and the demand function is given by

$$D(q) = 400 - e^{q/2}.$$

**(a)** Graph the supply and demand curves.

**(b)** Find the point at which supply and demand are in equilibrium.

**(c)** Find the consumers' surplus.

**(d)** Find the producers' surplus.

**44.** Repeat the four parts in the previous exercise for the supply function

$$S(q) = q^2 + \frac{11}{4}q$$

and the demand function

$$D(q) = 150 - q^2.$$

✓**Checkpoint Answers**

**1. (a)** $367.66    **(b)** $1185.91

**2.** About $1486 billion

**3.** About $100 billion

**4. (a)** 300 bolts    **(b)** $1800

**5. (a)** 5    **(b)** $437,500

**6.** $582,338.16

**7. (a)** $q = 100$    **(b)** $350    **(c)** $250

# 13.7  Differential Equations

A **differential equation** is an equation that involves an unknown function $y = f(x)$ and a finite number of its derivatives. Solving the differential equation for $y$ would give the unknown function. Differential equations have been important in the physical sciences and engineering for several centuries. More recently, they have been used in the social sciences, life sciences, and economics to solve a variety of problems involving population growth, biological balance, interest rates, and more.

Up to now, we have considered only equations whose solutions are *numbers*. The solutions of differential equations, however, are *functions*. For example, the solutions of the differential equation

$$\frac{dy}{dx} = 3x^2 - 2x \qquad (1)$$

consist of all functions $y$ that satisfy the equation. Since the left side of the equation is the derivative of $y$ with respect to $x$, the equation says that the derivative of $y$ is $3x^2 - 2x$, which means that $y$ is an *antiderivative* of $3x^2 - 2x$. In other words, the solutions are all functions $y$ such that

$$y = \int (3x^2 - 2x)\, dx = x^3 - x^2 + C. \qquad (2)$$

Each different value of $C$ in equation (2) leads to a different solution of equation (1), showing that a differential equation can have an infinite number of solutions. Equation (2) is the **general solution** of the differential equation **(1)**. Some of the solutions of equation **(1)** are graphed in Figure 13.33.

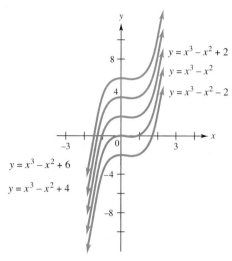

$$y = x^3 - x^2 + 2$$
$$y = x^3 - x^2$$
$$y = x^3 - x^2 - 2$$

$$y = x^3 - x^2 + 6$$

$$y = x^3 - x^2 + 4$$

**Figure 13.33**

Equation **(1)** has the form

$$\frac{dy}{dx} = f(x),$$

so its solution suggests the following generalization.

## General Solution of *dy/dx* = *f(x)*

The general solution of the differential equation $dy/dx = f(x)$ is

$$y = \int f(x)\, dx.$$

**EXAMPLE 1** **Population Growth** The population $P$ of a flock of birds is growing exponentially so that

$$\frac{dP}{dx} = 20e^{.05x},$$

where $x$ is time in years. Find $P$ in terms of $x$ if there were 20 birds in the flock initially.

**Solution** Solve the differential equation:

$$P = \int 20e^{.05x}\, dx$$

$$= \frac{20}{.05} e^{.05x} + C$$

$$= 400e^{.05x} + C.$$

Since $P$ is 20 when $x$ is 0,

$$20 = 400e^{0} + C$$

$$-380 = C,$$

and

$$P = 400e^{.05x} - 380.$$

In Example 1, the given information was used to produce a solution with a specific value of $C$. Such a solution is called a **particular solution** of the given differential equation. The given information, $P = 20$ when $x = 0$, is called an **initial condition.** A differential equation together with an initial condition is called an initial value problem.  ☑1

Sometimes a differential equation must be rewritten in the form

$$\frac{dy}{dx} = f(x)$$

before it can be solved.

✓ **Checkpoint 1**

Find the particular solution in Example 1 if there were 100 birds in the flock after 2 years.

**EXAMPLE 2**    Find the particular solution of the differential equation

$$\frac{dy}{dx} - 3x = 7,$$

given that $y = 2$ when $x = -2$.

**Solution**  Add $3x$ to both sides of the equation to get

$$\frac{dy}{dx} = 3x + 7.$$

The general solution is

$$y = \int (3x + 7)\, dx$$

$$y = \frac{3x^2}{2} + 7x + C.$$

Substituting 2 for $y$ and $-2$ for $x$ gives

$$2 = \frac{3(-2)^2}{2} + 7(-2) + C$$

$$C = 10.$$

✓ **Checkpoint 2**

Find the particular solution of

$$2\frac{dy}{dx} - 4 = 9x^2,$$

given that $y = 4$ when $x = 1$.

The particular solution is $y = \dfrac{3x^2}{2} + 7x + 10$.  ☑2

**Marginal productivity** is the rate at which production changes (increases or decreases) for a unit change in investment. Thus, marginal productivity can be expressed as the first derivative of the function that gives production in terms of investment.

**EXAMPLE 3**    **Marginal Productivity**  Suppose the marginal productivity of a manufacturing process is given by

$$P'(x) = 3x^2 - 10, \tag{3}$$

where $x$ is the amount of the investment in hundreds of thousands of dollars. If the process produces 100 units with the present investment of \$300,000 (that is, $x = 3$), how would production increase if the investment were increased to \$500,000?

**Solution**  To obtain an equation for production, we solve the differential equation $P'(x) = 3x^2 - 10$. The general solution is:

$$P(x) = \int (3x^2 - 10)\, dx$$

$$= x^3 - 10x + C.$$

To find $C$, use the given initial values—that is, $P(3) = 100$:

$$100 = 3^3 - 10(3) + C$$
$$C = 103.$$

Production is thus represented by

$$P(x) = x^3 - 10x + 103,$$

and if investment is increased to \$500,000, production becomes

$$P(5) = 5^3 - 10(5) + 103 = 178.$$

An increase to \$500,000 in investment will increase production from 100 units to 178 units. ✔ 3

## ✔ Checkpoint 3

In Example 3, if marginal productivity is changed to

$$P'(x) = 3x^2 + 8x,$$

with the same initial conditions,

(a) find an equation for production;

(b) find the increase in production if investment increases to \$500,000.

## Separation of Variables

The solution method used so far in this section is essentially the same as that in Sections 13.1 through 13.3, where we studied antiderivatives. But not all differential equations can be solved so easily. Suppose, for example, that $y$ is a function of $x$ such that

$$\frac{dy}{dx} = \frac{-6x}{y}.$$

We cannot simply find the antiderivative since the right side involves both variables. Nevertheless, there is a solution method that we now develop.

Consider a differential equation of the form

$$\frac{dy}{dx} = \frac{f(x)}{g(y)}. \tag{4}$$

(The example in the preceding paragraph is the case when $f(x) = -6x$ and $g(y) = y$.) Up to this point, we have used the symbol $dy/dx$ to denote the derivative function of the function $y$. In advanced courses, it is shown that $dy/dx$ can also be interpreted as a quotient of two differentials, $dy$ divided by $dx$. (You may recall that differentials were used in integration by substitution.) Then we can multiply both sides of equation **(4)** by $g(y)\, dx$ to obtain

$$g(y)\, dy = f(x)\, dx.$$

In this form, all terms involving $y$ (including $dy$) are on one side of the equation, and all terms involving $x$ (and $dx$) are on the other side. Such a differential equation is said to be **separable,** since the variables $x$ and $y$ can be separated. A separable differential equation may be solved by integrating each side after separating the variables.

**EXAMPLE 4**    Find the general solution of the differential equation

$$\frac{dy}{dx} = \frac{14x}{6y^2 + 1}.$$

**Solution**  Separate the variables by multiplying both sides by $(6y^2 + 1)\, dx$:

$$(6y^2 + 1)\, dy = 14x\, dx.$$

To solve this equation, take antiderivatives on each side:

$$\int (6y^2 + 1)\, dy = \int 14x\, dx$$

$$\frac{6y^3}{3} + y + C_1 = \frac{14}{2}x^2 + C_2$$

$$2y^3 + y = 7x^2 + C_2 - C_1.$$

We can replace the constant $C_2 - C_1$ by the single constant $C$ to obtain

$$\frac{y^3}{3} + y = 7x^2 + C.$$

From now on, we will add just one constant, with the understanding that it represents the difference between the two constants obtained from the two antiderivatives. Because several powers of $y$ are involved in the final equation above, it is better to leave the solution in this form rather than trying to solve for $y$. ✔️4

✓**Checkpoint 4**

Find the general solution of
$\dfrac{dy}{dx} = \dfrac{27x^2}{10y^4}$.

**EXAMPLE 5**    **Sales** Sales at a small store were initially \$20,000 and have been continuously increasing at an annual rate of 5%.

**(a)** Find a function $y = f(t)$ that gives the amount of sales $y$ after $t$ years.

**Solution** To say that sales are continuously increasing at an annual rate of 5% means that the rate of increase at any time $t$ is

$$\text{rate of increase} = 5\% \text{ of } y.$$

Since the rate of increase is $dy/dt$, we have

$$\frac{dy}{dt} = .05y.$$

To solve this differential equation for $y$, begin by separating the variables in order to have all terms involving $y$ (including $dy$) on the left and all terms involving $t$ (including $dt$) on the right. Multiply both sides of the preceding equation by $\frac{dt}{y}$ (note that $y \neq 0$) to obtain

$$\frac{dy}{y} = .05 \, dt.$$

Now take antiderivatives on both sides and use properties of logarithms and exponents to simplify the result:

$$\int \frac{dy}{y} = \int .05 \, dt$$

$$\ln|y| = .05t + C$$

$$\ln y = .05t + C \qquad \text{\small $|y| = y$ since $y$ is positive}$$

$$e^{\ln y} = e^{.05t + C}$$

$$y = e^{.05t + C} \qquad \text{\small Property of logarithms}$$

$$y = e^{.05t} e^{C}. \qquad \text{\small Property of exponents}$$

If we denote the constant $e^C$ by $M$, the sales function has the form

$$y = Me^{.05t},$$

the same as the exponential growth functions considered in Section 4.2. To find the value of $M$, use the fact that $y = 20{,}000$ when $t = 0$, that is,

$$20{,}000 = Me^{.05(0)} = Me^0 = M.$$

Therefore, the sales function is $y = 20{,}000e^{.05t}$.

**(b)** What was the amount of sales after 4.5 years?

**Solution** Evaluate $y$ when $t = 4.5$:

$$y = 20{,}000e^{.05(4.5)} \approx \$25{,}046.45.$$

**(c)** After how many years did sales reach $30,000?

**Solution** To find when the sales will be $30,000, we solve the following equation for $t$:

$$y = 30,000 \qquad \text{Sales} = \$30,000.$$

$$20,000e^{.05t} = 30,000$$

$$e^{.05t} = \frac{30,000}{20,000} = 1.5 \qquad \text{Divide both sides by 20,000.}$$

$$\ln(e^{.05t}) = \ln(1.5) \qquad \text{Take the natural log of both sides.}$$

$$.05t = \ln(1.5) \qquad \text{Properties of logarithms}$$

$$t = \frac{\ln(1.5)}{.05} \approx 8.1. \qquad \text{Solve for } t.$$

Thus sales reach $30,000 after about 8.1 years.

**(d)** If the trend continues, when will sales be twice the initial amount?

**Solution** Since the sales were initially $M = 20,000$, to find when the sales will be twice the initial level, we solve the following equation for $t$:

$$y = 2M \qquad \text{Sales} = \text{twice initial sales.}$$

$$Me^{.05t} = 2M$$

$$e^{.05t} = 2 \qquad \text{Divide both sides by } M.$$

$$\ln(e^{.05t}) = \ln(2) \qquad \text{Take the natural log of both sides.}$$

$$.05t = \ln(2) \qquad \text{Properties of logarithms}$$

$$t = \frac{\ln(2)}{.05} \approx 13.9. \qquad \text{Solve for } t.$$

Thus sales will double in about 13.9 years, as shown in Figure 13.34. Note that the time it takes for sales to double does not depend on the initial level of sales $M$. We can see this in our work above since $M$ was eliminated from the equation in the third step of the solution. ✔5

✔ **Checkpoint 5**

Suppose that sales are currently $5000 and continuously increasing at an annual rate of 10%.

**(a)** What will the amount of sales be in 4 years?

**(b)** When will sales be double the current level?

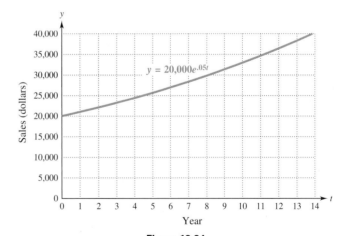

**Figure 13.34**

We can now explain how the exponential growth and decay functions presented in Section 4.2 were obtained. In the absence of inhibiting conditions, a population $y$ (which might be human, animal, bacterial, etc.) grows in such a way that the rate of change of the population is proportional to the population at time $x$; that is, there is a constant $k$ such that

$$\frac{dy}{dx} = ky.$$

The constant $k$ is called the **growth-rate constant.** Example 5 is the case when $k = .05$. The same argument used there (with $k$ in place of .05) shows that the population $y$ at time $x$ is given by

$$y = Me^{kx},$$

where $M$ is the population at time $x = 0$. A positive value of $k$ indicates growth while a negative value indicates decay.

---

**EXAMPLE 6**    **Mortality Rate**  The mortality rate for children under 5 in China (deaths of children under 5 years of age per 1000 live births) has been dropping since the start of the millennium at a continuous rate of approximately 8.3% per year. In the year 2000, the mortality rate was 37 deaths per 1000 live births. (Data from: United Nations Inter-Agency Group for Child Mortality Estimation.)

**(a)**  Find the function representing the child mortality rate in China.

**Solution**  If $y$ represents the child mortality rate in China in year $t$ and $k$ is the growth-rate constant, then $y$ satisfies the differential equation

$$\frac{dy}{dt} = ky.$$

As in Example 5, the solution to this differential equation is

$$y = Me^{kt},$$

where $M$ is the initial value of the function. The year 2000 corresponds to $t = 0$, so we have $M = 37$. We were also told that the child mortality is *dropping* at 8.3% per year, so the growth-rate constant is $k = -.083$. Putting this together we have

$$y = 37e^{-.083t}.$$

**(b)**  How many years after the start of the millennium did it take for the child mortality rate to be cut in half?

**Solution**  Because we want to find when the mortality rate was half the initial value, we solve the equation

$$y = \frac{1}{2}M \qquad\qquad \text{Mortality = half initial mortality}$$

$$Me^{-.083t} = \frac{1}{2}M \qquad\qquad \text{Substitute } y = Me^{-.083t}.$$

$$e^{-.083t} = \frac{1}{2} \qquad\qquad \text{Divide both sides by } M.$$

$$\ln(e^{-.083t}) = \ln\left(\frac{1}{2}\right) \qquad\qquad \text{Take logarithms of both sides.}$$

$$-.083t = \ln\left(\frac{1}{2}\right) \qquad\qquad \text{Properties of logarithms}$$

$$t = \frac{1}{-.083}\ln\left(\frac{1}{2}\right) \approx 8.4 \text{ years.} \qquad \text{Solve for } t.$$

**(c)**  If the model remains accurate, what will the mortality rate be in 2020?

**Solution**  Evaluate $y$ when $t = 20$ to get

$$y = 37e^{-.083(20)} \approx 7.0 \text{ deaths per 1000 live births.}$$

✓ **Checkpoint 6**

Given that $\dfrac{dy}{dx} = .2y$ and

$y(0) = 50$, find $y(10)$.

Figure 13.35 shows a graph of child mortality over time. The result we found algebraically above is confirmed graphically in the figure. ✓ 6

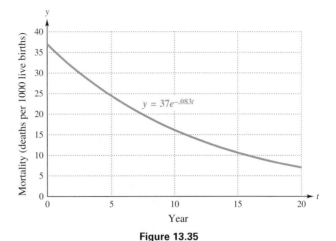

**Figure 13.35**

---

**EXAMPLE 7** **Growth Rate** Suppose that an investment grew at a continuous rate from an initial deposit of $1000 to the amount of $3000 in 20 years. Find the growth-rate constant.

**Solution** If $M$ represents the initial investment (the present value), $y$ represents the amount the investment is worth after a period of $t$ years (the future value), and $k$ represents the continuous growth-rate constant (the interest rate), then $y$ satisfies the differential equation

$$\frac{dy}{dt} = ky.$$

We know from Example 5 that the solution to this differential equation is

$$y = Me^{kt}.$$

If we use the standard notation of $PV$ for present value, $FV$ for future value, and $r$ for the interest rate, this equation becomes the future value formula found in Chapter 5:

$$FV = PVe^{rt}.$$

We were given that $PV = 1000$. To find the value of $r$, use the fact that $FV = 3000$ when $t = 20$, and solve for $r$:

| | |
|---|---|
| $3000 = 1000e^{r(20)}$ | Substitute 3000 for *FV*, 1000 for *PV*, and 20 for *t*. |
| $\dfrac{3000}{1000} = e^{r(20)}$ | Divide both sides by 1000. |
| $3 = e^{20r}$ | Simplify. |
| $\ln 3 = \ln e^{20r}$ | Take logarithms of both sides. |
| $\ln 3 = 20r$ | Properties of logarithms |
| $r = \dfrac{\ln 3}{20} \approx .0549.$ | Solve for *r*. |

So the continuous growth-rate constant is .0549, which represents an interest rate of 5.49% per year compounded continuously. ✓ 7

✓ **Checkpoint 7**

Find the growth-rate constant if an investment in Example 7 doubled in 20 years.

## Limited Growth

As a model of population growth, the equation $y = Me^{kx}$ is not realistic over the long run. As shown by graphs of functions of the form $y = Me^{kx}$, with both $M$ and $k$ positive, growth would be unbounded. Additional factors, such as space restrictions or a limited

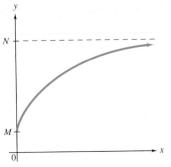

**Figure 13.36**

amount of food, tend to inhibit the growth of populations as time goes on. In an alternative model that assumes a maximum population of size $N$, the rate of growth of a population is proportional to how close the population is to that maximum. These assumptions lead to the differential equation

$$\frac{dy}{dx} = k(N - y)$$

and the limited-growth function mentioned in Chapter 4. Graphs of limited-growth functions look like the graph in Figure 13.36, where $M$ is the initial population. In the case at the end of this chapter, we will investigate another population growth model called logistic growth.

---

**EXAMPLE 8**     **Limited Population Growth**   Researchers believe that a certain lake can support no more than 3000 fish. There are approximately 500 fish in the lake at present, with a continuous annual growth constant of .25.

**(a)** Write and solve a differential equation for the rate of growth of this population using the limited growth model.

**Solution**  Let $N = 3000$ and $k = .25$. The rate of growth of the population $y$ is given by

$$\frac{dy}{dx} = k(N - y)$$

$$\frac{dy}{dx} = .25(3000 - y).$$

To solve for $y$, first separate the variables. (Assume $y < 3000$, so that $3000 - y \neq 0$.)

$$\frac{dy}{3000 - y} = .25 \, dx \qquad \text{Multiply both sides by } \frac{dx}{3000 - y}.$$

$$\int \frac{dy}{3000 - y} = \int .25 \, dx \qquad \text{Take antiderivatives on both sides.}$$

$$-\ln(3000 - y) = .25x + C$$

$$\ln(3000 - y) = -.25x - C$$

$$3000 - y = e^{-.25x - C} = (e^{-.25x})(e^{-C}).$$

The absolute-value bars are not needed for $\ln(3000 - y)$ because $y$ must be less than 3000 for this population, so that $3000 - y$ is always positive. If we denote the constant $e^{-C}$ by $B$, then

$$3000 - y = Be^{-.25x}$$

$$y = 3000 - Be^{-.25x}.$$

Find $B$ by using the fact that $y = 500$ when $x = 0$:

$$500 = 3000 - Be^{0}$$

$$500 = 3000 - B$$

$$B = 2500.$$

Notice that the value of $B$ is the difference between the maximum population and the initial population. Substituting 2500 for $B$ in the equation for $y$ gives

$$y = 3000 - 2500e^{-.25x}.$$

**(b)** What will the fish population be in four years?

**Solution** In four years, the population will be

$$y = 3000 - 2500e^{-.25(4)} = 3000 - 2500e^{-1}$$
$$\approx 2080.3,$$

or about 2080 fish.

**(c)** After how many years will the population of fish be 2000?

**Solution** To find when the population will be 2000, we solve the following equation for $x$:

| | |
|---|---|
| $y = 2000$ | Population = 2000. |
| $3000 - 2500e^{-.25x} = 2000$ | |
| $-2500e^{-.25x} = -1000$ | Subtract 3000 from both sides. |
| $e^{-.25x} = \dfrac{-1000}{-2500} = .4$ | Divide both sides by $-2500$. |
| $\ln(e^{-.25x}) = \ln(.4)$ | Take the natural log of both sides. |
| $-.25x = \ln(.4)$ | Properties of logarithms |
| $x = \dfrac{\ln(.4)}{-.25} \approx 3.7.$ | Solve for $x$. |

So the population will reach 2000 fish after about 3.7 years. ✓ 8

✓ **Checkpoint 8**

An animal population is growing at a continuous rate of 4%. The habitat will support no more than 10,000 animals. There are 3000 animals present now.

**(a)** Write an equation giving the population $y$ in $x$ years.

**(b)** Estimate the animal population in five years.

**(c)** When will the population be 6500?

# 13.7 Exercises

*Find general solutions for the given differential equations. (See Examples 1 and 4.)*

**1.** $\dfrac{dy}{dx} = -3x^2 + 7x$

**2.** $\dfrac{dy}{dx} = 5 - 9x^2$

**3.** $\dfrac{dy}{dx} = 6e^{-2x}$

**4.** $\dfrac{dy}{dx} = 3e^{-4x}$

**5.** $3x^3 - 2\dfrac{dy}{dx} = 0$

**6.** $3x^2 - 3\dfrac{dy}{dx} = 8$

**7.** $5y^4\dfrac{dy}{dx} = 6x$

**8.** $5y^4\dfrac{dy}{dx} = 7 - 10x$

**9.** $\dfrac{dy}{dx} = \dfrac{1}{21y^2 + 8}$

**10.** $\dfrac{dy}{dx} = \dfrac{1}{5 - 12y^2}$

**11.** $\dfrac{dy}{dx} = \dfrac{10x}{e^y}$

**12.** $\dfrac{dy}{dx} = \dfrac{26x}{e^y}$

**13.** $\dfrac{dy}{dx} = \dfrac{1}{3y^2}$

**14.** $\dfrac{x}{3y^2} = \dfrac{dy}{dx}$

**15.** $\dfrac{dy}{dx} = -3y$

**16.** $\dfrac{dy}{dx} = 7y$

**17.** $\dfrac{dy}{dx} = y - 7$

**18.** $\dfrac{dy}{dx} = y - 4$

**19.** $\dfrac{dy}{dx} = 1 - y$

**20.** $\dfrac{dy}{dx} = 9 - y$

**21.** $\dfrac{dy}{dx} = 4xy$

**22.** $\dfrac{dy}{dx} = x^2y$

*Find particular solutions for the given equations. (See Examples 2 and 5.)*

**23.** $\dfrac{dy}{dx} + 2x = 3x^2; y = 4$ when $x = 0$

**24.** $\dfrac{dy}{dx} = 8x^3 - 3x^2 + 2x; y = -1$ when $x = 1$

**25.** $\dfrac{dy}{dx} = 3.7y; y = 91$ when $x = 0$

**26.** $\dfrac{dy}{dx} = -.09y; y = 100$ when $x = 0$

**27.** $\dfrac{dy}{dx} = \dfrac{x^2}{y}; y = 4$ when $x = 0$

**28.** $\dfrac{dy}{dx} = \dfrac{5x^4}{2y}; y = 1$ when $x = 0$

**29.** $(5x + 3)y = \dfrac{dy}{dx}; y = 1$ when $x = 0$

**30.** $x\dfrac{dy}{dx} - y\sqrt{x} = 0; y = 1$ when $x = 0$

**31.** $\dfrac{dy}{dx} = \dfrac{7x + 1}{y - 3}; y = 4$ when $x = 0$

**32.** $\dfrac{dy}{dx} = \dfrac{x^2 + 5}{2y - 1}; y = 11$ when $x = 0$

**33.** What is the difference between a general solution and a particular solution of a differential equation?

**34.** What is meant by a separable differential equation?

*Work the given problems. (See Example 3.)*

35. **Marginal Productivity** The marginal productivity of a manufacturing process is given by

$$\frac{dy}{dx} = \frac{3x^2 - 4x}{25},$$

where $x$ is the amount of investment (in thousands of dollars) and $y$ is the number of units produced per month (in hundreds). Production is 2300 units when investment is $5000. Find the production level if investment is increased to

(a) $8000;  (b) $11,000.

36. **Marginal Productivity** The marginal productivity of a process is given by

$$\frac{dy}{dx} = -\frac{40}{32 - 4x},$$

where $x$ represents the investment (in thousands of dollars) and $y$ is the number of units produced per month. Find the productivity for each of the given investments if productivity is 100 units when the investment is $1000.

(a) $3000  (b) $5000

(c) Can investments ever reach $8000 according to this model? Why?

*Work these problems. (See Examples 4–6.)*

37. **Internet Users** The percentage $y$ of the population who are Internet users in low-income countries around the world approximately satisfies the differential equation

$$\frac{dy}{dt} = .23y,$$

where $t = 0$ corresponds to the year 2000. (Data from: www.worldbank.org.)

(a) Solve the differential equation for $y$, given the initial condition that $y = .3$ when $t = 0$.

(b) What percentage of the population in low-income countries were Internet users in 2015?

(c) How many years did it take for $y$ to reach 10%?

38. **Mortality Rate** The mortality rate $y$ (deaths of children under 5 years of age per 1000 live births) in low-income countries around the world approximately satisfies the differential equation

$$\frac{dy}{dt} = -.045y,$$

where $t = 0$ corresponds to the year 2000. (Data from: United Nations.)

(a) Solve the differential equation for $y$, given the initial condition that $y = 150$ when $t = 0$.

(b) What was this mortality rate in 2015?

(c) How many years did it take for the mortality rate to be cut in half?

39. **Food Safety** The sell-by dating of dairy products depends on the solution of a differential equation. The rate of growth of bacteria in such products increases with time. If $y$ is the number of bacteria (in thousands) present at a time $t$ (in days), then the rate of growth of bacteria for a particular dairy product can be expressed as

$$\frac{dy}{dt} = 8t.$$

(a) Suppose that the initial amount of bacteria present is $y = 50$ (in thousands) when $t = 0$. Solve the differential equation for $y$.

(b) Suppose the maximum allowable value for $y$ is 450 (thousand). How many days will it take for the number of bacteria to reach this level (this is called the sell-by date)?

40. **Food Safety** Find the sell-by date (see the previous exercise) of a dairy product if the rate of growth of bacteria is

$$\frac{dy}{dt} = 3t,$$

where $y$ is the number of bacteria (in thousands) present at a time $t$ (in days). Assume that initially there were 100 (in thousands) bacteria present and the maximum allowable is 700 (thousand).

41. **Sales** Sales of a particular product have been declining continuously at the rate of 15% per year.

(a) Express the rate of sales decline as a differential equation. (Here, $k$ will be negative.)

(b) Find the general solution for the equation in part (a).

(c) When will sales decrease to 25% of their original level?

42. **Natural Science** The amount of a tracer dye injected into the bloodstream decreases exponentially, with a decay constant of 3% per minute. If 6 cc are present initially, how many cc are present after 8 minutes? (Here, $k$ will be negative.)

43. **Inflation** Suppose that the inflation rate is 2% per year compounded continuously. If an item costs $1 today, how long will it take until the item costs twice as much?

44. **Deflation** Suppose that deflation is occurring at a rate of 1% per year compounded continuously. If an item costs $1 today, how long will it take until the item costs only 90¢?

*Work these exercises. (See Example 7.)*

45. **Social Science** Suppose the rate at which a rumor spreads over time is proportional to the number of people who have heard it. If $y$ is the number of people who have heard the rumor, then

$$\frac{dy}{dt} = ky,$$

where $t$ is the time in days and $k$ is a constant.

(a) If $y$ is 1 when $t = 0$, and $y$ is 5 when $t = 2$, find $k$.

Using the value of $k$ from part (a), find $y$ for each of the following times:

(b) $t = 3$;  (c) $t = 5$;  (d) $t = 10$.

46. **Depreciation** Suppose that an asset is depreciating in value at a continuous rate $r$ per year (expressed as a decimal).

(a) If the initial value of the asset was $10,000, and it was worth $8000 after 5 years, find $r$ to three decimal places.

(b) Using the value of $k$ found in part (a), find the value of the asset after 10 years.

47. **Finance** A life insurance company invests $5000 to fund a death benefit of $20,000. Growth in the investment over time can be modeled by the differential equation

$$\frac{dA}{dt} = Ai,$$

where $i$ is the interest rate and $A(t)$ is the amount invested at time $t$ (in years). Calculate the interest rate that the investment must earn in order for the company to fund the death benefit in 24 years. (Choose one of the following.)*

(a) $\dfrac{-\ln 2}{12}$   (b) $\dfrac{-\ln 2}{24}$   (c) $\dfrac{\ln 2}{24}$

(d) $\dfrac{\ln 2}{12}$   (e) $\dfrac{\ln 2}{6}$

48. **Finance** A company invests $1,000,000 today to fund a $4,000,000 obligation in 24 years. The growth of the investment over time can be modeled by the differential equation

$$\frac{dFV}{dt} = rFV,$$

where $t$ is time in years, $r$ is the continuous interest rate, and $FV$ is the future value of the investment at time $t$. Find $r$ to three decimal places.

*Work these problems. (See Example 8.)*

49. **Limited Growth** A bacteria population (in thousands) initially has 1000 cells and is growing according to the limited growth model

$$\frac{dy}{dx} = 30 - y.$$

(a) Solve the differential equation for $y$.

(b) How many bacteria cells will be present at time $x = 10$?

50. **Limited Growth** An isolated fish population is limited to 4000 by the amount of food available. If there are now 320 fish and the population is growing continuously at the rate of 2% per year, find the population at the end of 10 years, assuming the limited growth model.

✓ **Checkpoint Answers**

1. $P = 400e^{.05x} - 342$

2. $y = \dfrac{3}{2}x^3 + 2x + \dfrac{1}{2}$

3. (a) $P(x) = x^3 + 4x^2 + 37$
   (b) Production goes from 100 units to 262 units, an increase of 162 units.

4. $2y^5 = 9x^3 + C.$

5. (a) $7459.12   (b) About 6.9 years

6. $y(10) \approx 369.5$

7. 3.47%

8. (a) $y = 10,000 - 7000e^{-.04x}$
   (b) About 4269 animals
   (c) About 17.3 years

---

*Problem 47 is from the May 2003 Course 1 Examination of the *Education and Examination Committee of the Society of Actuaries*. Reprinted by permission of the Society of Actuaries.

# CHAPTER 13 Summary and Review

## Key Terms and Symbols

**13.1** $\int f(x)\,dx$, indefinite integral of $f$
antiderivative
integral sign
integrand
power rule
constant-multiple rule

sum-or-difference rule
**13.2** differential
integration by substitution
**13.3** integration by parts
**13.4** $\int_a^b f(x)\,dx$, definite integral of $f$
total change in $F(x)$

**13.5** Fundamental Theorem of Calculus
$$F(x)\Big|_a^b = F(b) - F(a)$$
**13.6** consumers' surplus
producers' surplus
present value
**13.7** differential equation

general solution
particular solution
initial condition
marginal productivity
separable differential equation
growth-rate constant

## Chapter 13 Key Concepts

**Antiderivative**  $F$ is an antiderivative of $f$ if $F' = f$.

**Indefinite Integral**  If $F'(x) = f(x)$, then $\int f(x)\,dx = F(x) + C$, for any real number $C$.

**Properties of Integrals**  $\int k \cdot f(x)\,dx = k \cdot \int f(x)\,dx$, for any real number $k$.

$\int [f(x) \pm g(x)]\,dx = \int f(x)\,dx \pm \int g(x)\,dx.$

**Rules for Integrals**

$$\int x^n \, dx = \frac{x^{n+1}}{n+1} + C \quad \text{for } n \neq -1$$

$$\int x^{-1} \, dx = \int \frac{1}{x} \, dx = \ln|x| + C \quad \text{for } n \neq -1$$

$$\int \ln x \, dx = x \ln x - x + C$$

$$\int e^x \, dx = e^x + C$$

$$\int e^{kx} \, dx = \frac{1}{k} e^x + C$$

$$\int u \, dv = uv - \int v \, du$$

**Definite Integral**

If $f$ is continuous on $[a, b]$, the definite integral of $f$ from $a$ to $b$ is

$$\int_a^b f(x) \, dx = \lim_{n \to \infty} ([f(x_1) + f(x_2) + f(x_3) + \cdots + f(x_n)] \cdot \Delta x),$$

provided that this limit exists, where $\Delta x = \dfrac{b-a}{n}$ and $x_i$ is the left endpoint of the $i$th interval.

**Total Change in $F(x)$**

Let $f$ be continuous on $[a, b]$ and $f(x) \geq 0$ for all $x$ in $[a, b]$. If $f$ is the rate of change of $F$, then the total change in $F(x)$ as $x$ goes from $a$ to $b$ is given by

$$\int_a^b f(x) \, dx.$$

**Fundamental Theorem of Calculus**

Let $f$ be continuous on $[a, b]$ and let $F$ be any antiderivative of $f$. Then

$$\int_a^b f(x) \, dx = F(x) \Big|_a^b = F(b) - F(a).$$

**General Solution of $\dfrac{dy}{dx} = f(x)$**

The general solution of the differential equation $dy/dx = f(x)$ is

$$y = \int f(x) \, dx.$$

**General Solution of $\dfrac{dy}{dx} = ky$**

The general solution of the differential equation $dy/dx = ky$ is

$$y = Me^{kx}.$$

# Chapter 13 Review Exercises

*Find each indefinite integral.*

1. $\displaystyle\int (x^2 - 8x - 7) \, dx$

2. $\displaystyle\int (6 - x^2) \, dx$

3. $\displaystyle\int 7\sqrt{x} \, dx$

4. $\displaystyle\int \frac{\sqrt{x}}{8} \, dx$

5. $\displaystyle\int (x^{1/2} + 3x^{-2/3}) \, dx$

6. $\displaystyle\int (8x^{4/3} + x^{-1/2}) \, dx$

7. $\displaystyle\int \frac{-6}{x^3} \, dx$

8. $\displaystyle\int \frac{9}{x^4} \, dx$

9. $\displaystyle\int -3e^{2x} \, dx$

10. $\displaystyle\int 8e^{-x} \, dx$

11. $\displaystyle\int \frac{10}{x-3} \, dx$

12. $\displaystyle\int \frac{-14}{x+2} \, dx$

13. $\displaystyle\int \frac{1}{3-x} \, dx$

14. $\displaystyle\int \frac{5}{2-x} \, dx$

15. $\displaystyle\int \frac{3x}{x^2-1} \, dx$

16. $\displaystyle\int \frac{-6x}{4-x^2} \, dx$

17. $\displaystyle\int \frac{12x^2 \, dx}{(x^3+5)^4}$

18. $\displaystyle\int (x^2 - 5x)^4(2x-5) \, dx$

19. $\displaystyle\int \frac{4x-5}{2x^2-5x} \, dx$

20. $\displaystyle\int \frac{8(2x+9)}{x^2+9x+1} \, dx$

21. $\displaystyle\int 5xe^{3x^2} dx$

22. $\displaystyle\int 4xe^{-3x^2+7} \, dx$

23. $\displaystyle\int \frac{x^2}{e^{3x^3}} \, dx$

24. $\displaystyle\int \frac{-2x}{e^{x^2}} \, dx$

25. $\displaystyle\int \frac{3(\ln x)^5}{x} \, dx$

26. $\displaystyle\int \frac{10(\ln(5x+3))^2}{5x+3} \, dx$

27. $\displaystyle\int \sqrt{2x+9}\,dx$

28. $\displaystyle\int \sqrt{1+5x}\,dx$

29. $\displaystyle\int 7xe^{4-x}\,dx$

30. $\displaystyle\int xe^{3x-5}\,dx$

31. $\displaystyle\int 28x^2e^{2x}\,dx$

32. $\displaystyle\int 9x^2e^{-x}\,dx$

33. $\displaystyle\int 6\ln(6x+1)\,dx$

34. $\displaystyle\int \ln(8-x)\,dx$

*Use rectangles to approximate the given area.*

35. The curve shown in the accompanying figure gives the rate at which an investment accumulates income (in dollars per year). Use rectangles of width 2 units and height determined by the value of the function at the left endpoint to approximate the total income accumulated over 10 years.

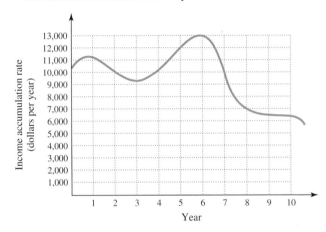

36. Use four rectangles to approximate the area under the graph of $f(x) = 2x + 3$ and above the $x$-axis from $x = 0$ to $x = 4$. Let the height of each rectangle be the value of the function on the left side.

37. (a) Use five rectangles to approximate the area between the graph of $f(x) = 16x^2 - x^4 + 2$ and the $x$-axis from $x = -2$ to $x = 3$.

    (b) Use numerical integration to approximate this area.

38. Repeat the previous exercise using six rectangles for the function $g(x) = -x^4 + 12x^2 + x + 5$ from $x = -3$ to $x = 3$.

*Find each definite integral.*

39. $\displaystyle\int_0^4 (x^3 - x^2)\,dx$

40. $\displaystyle\int_0^1 (5 - 10x + 15x^2)\,dx$

41. $\displaystyle\int_2^5 (6x^{-2} + x^{-3})\,dx$

42. $\displaystyle\int_2^3 (5x^{-2} + 7x^{-4})\,dx$

43. $\displaystyle\int_1^5 15x^{-1}\,dx$

44. $\displaystyle\int_1^6 \frac{8x^{-1}}{3}\,dx$

45. $\displaystyle\int_0^4 2e^{-5x}\,dx$

46. $\displaystyle\int_1^2 \frac{7}{2}e^{4x}\,dx$

47. $\displaystyle\int_{\sqrt{5}}^5 2x\sqrt{x^2 - 3}\,dx$

48. $\displaystyle\int_0^1 x\sqrt{5x^2 + 4}\,dx$

49. $\displaystyle\int_1^6 \frac{8x+12}{x^2+3x+9}\,dx$

50. $\displaystyle\int_0^3 \frac{6x}{x^2+1}\,dx$

51. $\displaystyle\int_1^e \frac{9(\ln x)^2}{x}\,dx$

52. $\displaystyle\int_1^4 \frac{3(\ln x)^5}{x}\,dx$

53. $\displaystyle\int_0^1 (3x+7)e^x\,dx$

54. $\displaystyle\int_0^5 (9-2x)e^x\,dx$

55. $\displaystyle\int_0^3 4x^2e^x\,dx$

56. $\displaystyle\int_0^1 5x^2e^x\,dx$

57. $\displaystyle\int_{1/8}^{e/8} \ln(8x)\,dx$

58. $\displaystyle\int_1^e 8\ln x\,dx$

*Find the area between the $x$-axis and $f(x)$ over each of the given intervals.*

59. $f(x) = e^x$; $[0, 2]$

60. $f(x) = 1 + e^{-x}$; $[0, 4]$

**Business** *Find the cost function for each of the given marginal cost functions.*

61. $C'(x) = 10 - 5x$; fixed cost is \$4.

62. $C'(x) = 2x + 3x^2$; 2 units cost \$12.

63. $C'(x) = 3\sqrt{2x-1}$; 13 units cost \$270.

64. $C'(x) = \dfrac{6}{x+1}$; fixed cost is \$18.

*Work the given exercises.*

65. **PayPal Revenue** The rate at which revenue was generated (in millions of dollars per year) by PayPal Holdings Inc. can be approximated by
$$f(t) = 1288t - 10{,}576 \qquad (12 \le t \le 17),$$
where $t = 12$ corresponds to the start of the year 2012. What was the total profit generated over the 5-year period from the start of the year 2012 to the start of 2017? (Data from: www.morningstar.com.)

66. **Research and Development** The rate of spending (in billions of dollars per year) on research and development (R&D) in the United States for the years 2016 through 2026 is projected to be approximately
$$f(t) = 17.2t + 238.8 \qquad (16 \le t \le 26),$$
where $t = 16$ corresponds to the start of the year 2016. What is the total projected spending on R&D from the start of 2016 to the start of 2026? (Data from: R&D Magazine.)

67. **Hormel Profit** The rate at which profit was generated (in millions of dollars per year) by Hormel Foods Corp. can be approximated by
$$f(t) = 32t^2 - 715t + 5240 \qquad (10 \le t \le 17),$$
where $t = 10$ corresponds to the start of the year 2007. Find and interpret $\displaystyle\int_{10}^{17} f(t)\,dt$. (Data from: www.morningstar.com.)

68. **Petroleum Consumption** The rate of petroleum consumption (in trillions of BTUs per year) for the years 2015 through 2035 is projected to be approximately
$$f(t) = 2.27t^3 - 82t^2 + 700t + 36{,}200 \qquad (0 \le t \le 20),$$

where $t = 0$ corresponds to the start of the year 2015. Find and interpret $\int_0^{20} f(t)\, dt$. (Data from: U.S. Energy Information Administration.)

**69. Sales** The rate of change of sales of a new brand of tomato soup, in thousands per month, is given by

$$S(x) = \sqrt{x} + 3,$$

where $x$ is the time in months that the new product has been on the market. Find the total sales after 9 months.

**70. Population Growth** The rate of change of the population of a rare species of Australian spider is given by

$$f'(t) = 150 - \sqrt{3.2t + 4},$$

where $f(t)$ is the number of spiders present at time $t$, measured in months. Find the total number of additional spiders in the first 10 months.

**71. Maximizing Savings** A company has installed new machinery that will produce a savings rate (in thousands of dollars per year) of

$$S'(x) = 225 - x^2,$$

where $x$ is the number of years the machinery is to be used. The rate of additional costs (in thousands of dollars per year) to the company because of the new machinery is expected to be

$$C'(x) = x^2 + 25x + 150.$$

To maximize net savings, for how many years should the company use the new machinery? Find the net savings (in thousands of dollars) over this period.

**72. Maximizing Savings** A firm has decided to use more energy-efficient equipment that is expected to produce a savings rate (in thousands of dollars per year) of

$$S'(x) = 60 - 4x,$$

where $x$ is the number of years the equipment has been in place. There will be new costs associated with the equipment that are projected to be at a rate (in thousands of dollars per year) of

$$C'(x) = x^2 + x + 10.$$

To maximize savings, how long should this firm use the new equipment? What are the net savings (in thousands of dollars) over this period?

**73. Producers' Surplus** Suppose that the supply function of some commodity is

$$S(q) = q^2 + 5q + 100$$

and the demand function for the commodity is

$$D(q) = 350 - q^2.$$

Find the producers' surplus.

**74. Consumers' Surplus** For the supply and demand functions given in the previous exercise, find the consumers' surplus.

**75. Disease Spread** The rate of infection of a disease (in people per month) is given by the function

$$I'(t) = \frac{100t}{t^2 + 2},$$

where $t$ is the time in months since the disease broke out. Find the total number of infected people over the first four months of the disease.

**76. Drug Absorption** For a certain drug, the rate of release of the active ingredient into the bloodstream (in milligrams per hour) is given by

$$R(t) = \frac{4}{3t} + \frac{2}{t^2},$$

where $t$ is measured in hours after the drug is administered. Find the total amount released from

**(a)** $t = 1$ to 10 hours;      **(b)** $t = 10$ to 20 hours.

**77. Waiting Time** Suppose waiting times (in minutes) in a teller line at a bank can be estimated by an exponential distribution and the function

$$f(x) = \frac{1}{2} e^{-x/2}.$$

**(a)** Find the probability that a customer in line waits less than 2 minutes. (*Hint:* Find the area under $f(x)$ on $[0, 2]$.)

**(b)** Find the probability that a customer in line waits more than 3 minutes.

**78. Natural Science** Suppose the amount of time between blinks of an eye for a given individual can be modeled by an exponential distribution and the function

$$f(t) = .44e^{-.44t},$$

where $t$ is measured in seconds.

**(a)** Find the probability that the person will take between 1 and 2 seconds between blinks.

**(b)** Find the probability that the person will take more than 4 seconds between blinks.

*Find general solutions for the given differential equations.*

**79.** $\dfrac{dy}{dx} = -.07y$          **80.** $\dfrac{dy}{dx} = \dfrac{y}{13}$

**81.** $\dfrac{dy}{dx} = \dfrac{1 - 20x^4}{3y^2}$          **82.** $\dfrac{dy}{dx} = \dfrac{14e^{2x}}{y^2 + 5}$

*Find particular solutions for the given differential equations.*

**83.** $\dfrac{dy}{dx} = 5(e^{-x} - 1)$; $y = 17$ when $x = 0$

**84.** $\dfrac{dy}{dx} = \dfrac{x}{x^2 - 3} + 7$; $y = 52$ when $x = 2$

**85.** $(5 - 2x)y = \dfrac{dy}{dx}$; $y = 2$ when $x = 0$

**86.** $\sqrt{x}\,\dfrac{dy}{dx} = xy$; $y = 4$ when $x = 1$

**87. Antique Guitars** One of the guitars that John Lennon played while he was with the Beatles was a 1958 Rickenbacker 325. In 2017, you could purchase this model of guitar for $13,000 on eBay. Assume this instrument appreciates in value at a continuous rate of 2% per year.

**(a)** Write and solve a differential equation to find the value $y$ of the guitar in year $t$, where $t = 0$ corresponds to 2017.

**(b)** According to this model, what will be the value of the instrument in 2020?

**(c)** When is it projected to be worth $15,000?

**88. Library Wi-Fi** The number of Wi-Fi sessions provided at Chicago Public Library locations was about 22,000 in the month of January, 2011. This number grew for at a continuous rate of approximately 5.8% per month until reaching 300,000. (Data from: City of Chicago Data Portal.)

**(a)** Write and solve a differential equation to find the number $y$ of Wi-Fi sessions (in thousands) in month $t$, where $t = 0$ corresponds to January, 2011.

**(b)** How many monthly Wi-Fi sessions (to the nearest thousand) were there at time $t = 7$?

**(c)** How many months did it take for the number of monthly Wi-Fi sessions to grow to 300,000?

**89. Social Science** The percentage $y$ of 3- to 5-year-old children enrolled in preprimary programs in year $x$ approximately satisfies the differential equation

$$\frac{dy}{dx} = .05(70 - y).$$

Let $x = 0$ correspond to the year 1965 when enrollment was 27%. (Data from: National Center for Education Statistics.)

**(a)** Solve the differential equation for $y$.

**(b)** If this trend continues, what percent of 3- to 5-year-old children will be enrolled in preprimary programs in 2015 ($x = 50$)?

**90. Physical Science** A roast at a temperature of 40° is put in a 300° oven. After 1 hour, the roast has reached a temperature of 150°. Newton's law of cooling states that

$$\frac{dT}{dt} = k(T - T_F),$$

where $T$ is the temperature of an object at time $t$, the surrounding medium has temperature $T_F$, and $k$ is a constant.

**(a)** Use Newton's law of cooling to find the temperature of the roast after 2.5 hours.

**(b)** How long does it take for the roast to reach a temperature of 250°?

# Case Study 13  Bounded Population Growth

The population growth model $\frac{dP}{dt} = kP$, whose solution is the exponential growth function $P = P_0 e^{kt}$, can work well in a situation where a population is changing rapidly over a short period of time. But the model may not be useful in the long term. Exponential functions grow very large very quickly and will eventually exceed any values that the population under consideration may actually attain. For this reason, the exponential function is said to be *unbounded*. A more realistic growth model would take into account any natural bounds on the population size.

For example, consider a population of cattle or sheep that lives and grazes on a piece of farmland. Sheep, in particular, obtain the majority of their food intake from foraging in their environment. The sheep happily graze on the grass, plants, and weeds without exhausting the supply. This allows time for the plant life to grow back, and the eating–growing cycle continues. If the farmer populates the farmland with too many sheep, however, the plants may be eaten too quickly, failing to grow enough before some of the sheep develop health problems and possibly die. The maximum number of animals the farmland can comfortably sustain over a long period of time is called the *carrying capacity* of the land.

One model used for bounded population growth is the *logistic differential equation*

$$\frac{dP}{dt} = kP(C - P),$$

where $k$ is a positive proportionality constant and $C$ is the carrying capacity of the environment under consideration. When the population satisfies

$$0 < P < C,$$

the expression $kP(C - P)$ will be positive, and a positive derivative indicates that the population is increasing. If the population $P$ exceeds the carrying capacity $C$, the factor $C - P$ becomes negative and

$$\frac{dP}{dt} < 0,$$

indicating that the population is decreasing, which must be the case when the carrying capacity is exceeded. The model thus agrees with the expected behavior of a bounded population.

For example, given an initial population of 10 sheep on land with carrying capacity $C = 200$ and a proportionality factor $k = .005$, the logistic differential equation becomes

$$\frac{dP}{dt} = .005P(200 - P).$$

Separating variables and integrating gives

$$\int \frac{1}{P(200 - P)} \, dP = \int .005 \, dt. \tag{1}$$

The right side of equation **(1)** is easily integrated:

$$\int .005 \, dt = .005t + K. \tag{2}$$

The left side of equation **(1)** does not match the standard integration formulas, but can be integrated by using the following identity, which you are asked to verify in Exercise 1:

$$\frac{1}{P(C - P)} = \frac{1/C}{P} + \frac{1/C}{C - P}. \tag{3}$$

Applying this identity with $C = 200$ allows the left-hand side of equation **(1)** to be easily integrated:

$$\int \frac{1}{P(200 - P)} \, dP = \int \frac{1/200}{P} \, dP + \int \frac{1/200}{200 - P} \, dP$$

$$= \frac{1}{200} \ln(P) - \frac{1}{200} \ln(200 - P). \qquad \textbf{(4)}$$

Substituting equations **(2)** and **(4)** into equation **(1)** shows that

$$\frac{1}{200} \ln(P) - \frac{1}{200} \ln(200 - P) = .005t + K. \qquad \textbf{(5)}$$

In Exercise 2, you are asked to solve equation **(5)** for $P(t)$ and use the fact that $P(0) = 10$ to obtain the population function

$$P(t) = \frac{200e^t}{e^t + 19}. \qquad \textbf{(6)}$$

After one year, the population would consist of $P(1) = 25$ sheep.

The graph of the population function in the accompanying figure illustrates the typical S-shaped curve for a population modeled by the logistic differential equation. The population increases rapidly at first, and then the rate of increase decreases as the population approaches the carrying capacity of its environment.

Before the logistic model can be applied, an estimate of the carrying capacity is needed. In the case of cattle and sheep, veterinary and agricultural science, along with hundreds of years of recorded experience, tell a farmer how much livestock can survive comfortably on a given plot of land. A much more difficult problem is to determine the carrying capacity of a population when many factors contribute to the carrying capacity.

## Exercises

1. Verify equation **(3)**.

2. Use the techniques of Section 13.7 and the fact that $P(0) = 10$ to solve equation **(5)**.

3. The carrying capacity of Atlantic harp seals has been estimated to be $C = 10$ million seals. Let $t = 0$ correspond to the year 1980 when this seal population was estimated to be about 2 million. (Data from: Fisheries and Oceans Canada.)

   (a) Use a logistic growth model $\frac{dP}{dt} = kP(C - P)$ with $k = .007$ to find a population function similar to the one given in equation **(6)**.

   (b) According to this model, what is the seal population in the year 2018?

   (c) If the model remains accurate, when will the seal population reach 8 million?

4. The logistic growth model can be used in other contexts where growth is bounded. For example, the percentage of women participating in the Olympic Games was about 20% in 1976 and grew rapidly for a period of time, but the rate of growth slowed as the percentage of women participants drew closer to 50%. (Data from: www.olympic.org.)

   (a) Let $P$ represent the percentage of women in the Olympic Games in year $t$, with $t = 0$ corresponding to the year 1976. Use a logistic growth model $\frac{dP}{dt} = kP(C - P)$, with $k = .065$ and a carrying capacity of $C = 50$, to find a function for $P$ similar to the one given in equation **(6)**.

   (b) According to this model, what percentage of Olympians were women (rounded to the nearest percent) in the year 2000 (when $t = 24$)? What percentage were women in 2016? What is the projected percentage of women athletes for the 2020 games?

# Extended Project

The logistic growth model $\frac{dP}{dt} = kP(C - P)$ can be applied to the population of the world. You will need to do some research to find the current world population and an estimate of the carrying capacity of the Earth. (Internet sites for the U.S. Census Bureau and the United Nations may be helpful.) Let $t = 0$ correspond to the current year. Use $k = .0011$ and the values you found for the current world population and carrying capacity to find a function for the population of the world. If the model remains accurate, what will the world population be in 2025? When will the world population reach 9 billion? According to the model, will the world population ever reach 20 billion? Write a report explaining your findings. Include the logistic growth model, the values you found for the current population and carrying capacity (including your sources), the estimates you found for future world population, and an explanation that the model is an estimate and has limitations.

# Multivariate Calculus

**14**

**CHAPTER**

Real-world phenomena typically depend on more than one factor. A baseball player's batting average depends on both the number of hits and the number of at bats. Body mass index depends on height and weight. Anyone outside during cold weather may experience windchill, which is determined by air temperature and wind speed. The price of a bond depends on the face value, interest rate, and time to maturity. The profit a business makes depends on a variety of factors, such as the cost of materials, wages, competitive price pressures, and interest rates. Examples and exercises in this chapter illustrate that accurate models of these and other real-life situations require functions of two or more variables.

Many of the ideas developed for functions of one variable also apply to functions of more than one variable. In particular, the fundamental idea of the derivative generalizes in a very natural way to functions of more than one variable.

# 14.1 Functions of Several Variables

Recall from Section 3.1 that a **function** consists of a set of inputs called the **domain,** a set of outputs called the **range,** and a **rule** by which each input determines exactly one output. For all of the functions studied previously in this text, both the inputs and outputs were real numbers. We now extend the concept of function to the case when the inputs are *ordered pairs* of real numbers.

Suppose, for example, that a gourmet popsicle company makes fruity and creamy popsicles. It costs $10 per dozen to produce fruit-based popsicles and $15 per dozen to make cream-based popsicles. So the cost of producing $x$ dozen fruity popsicles and $y$ dozen creamy popsicles is $10x + 15y$. We can think of this cost function as a function of *two* variables whose rule is

$$C(x, y) = 10x + 15y.$$

For instance, the cost of producing 250 dozen fruity popsicles and 100 dozen creamy popsicles is denoted $C(250, 100)$ and is found by letting $x = 250$ and $y = 100$ in the rule of the function:

$$C(250, 100) = 10 \cdot 250 + 15 \cdot 100 = 2500 + 1500 = \$4000.$$

Here is the general definition.

> **A function of two variables** consists of
>
> a set of inputs that are *ordered pairs* of real numbers (the **domain**);
> a set of outputs that are real numbers (the **range**); and
> a rule by which each input pair produces exactly one output number.

Essentially the same definition works for functions of three, four, or more variables—just change the inputs to ordered *triples* of real numbers for functions of three variables, to ordered *quadruples* of real numbers for functions of four variables, and so on. Functions of more than one variable are called **multivariate functions.**

Many familiar formulas can be considered as the rules of multivariate functions. For example,

- the volume $V$ of a cone is given by a function of two variables,

$$V(r, h) = \frac{1}{3}\pi r^2 h,$$

  where $r$ is the radius of the base and $h$ is the height of the cone;

- the future value $A$ of an amount deposited at compound interest is given by a function of three variables,

$$A(P, i, n) = P(1 + i)^n,$$

  where $P$ is the amount deposited, $i$ is the interest rate per period, and $n$ is the number of periods.

Unless otherwise stated, the domain of a multivariate function defined by a formula or equation is assumed to be the set of all inputs for which the rule of the function produces a real number as output.

---

**EXAMPLE 1**    Let $f(x, y) = 4x^2 + 2xy + 9/y$, and find each of the given quantities.

**(a)** $f(-1, 3)$

**Solution** Replace $x$ with $-1$ and $y$ with 3:

$$f(-1, 3) = 4(-1)^2 + 2(-1)(3) + \frac{9}{3} = 4 - 6 + 3 = 1.$$

**(b)** The domain of $f$

**Solution** Because of the quotient $9/y$, it is not possible to replace $y$ with 0, so the domain of the function $f$ consists of all ordered pairs $(x, y)$ such that $y \neq 0$. ✓₁

✓ **Checkpoint 1**

Let $f(x, y) = x^3 - 5x^2 + xy$. Find

**(a)** $f(2, 4)$;

**(b)** $f(-2, 3)$.

*Answers to Checkpoint exercises are found at the end of the section.*

**EXAMPLE 2**    **Earned Run Average** In major league baseball, a pitcher's earned run average is the mean number of earned runs given up per nine innings pitched. Unearned runs, which result from defensive errors, are excluded from this statistic.

**(a)** Write the rule of the function $f(r, p)$ that gives the earned run average (ERA) for a pitcher who allowed $r$ earned runs in $p$ innings pitched.

**Solution** Since the pitcher has given up $r$ earned runs in $p$ innings, this corresponds to a mean of $\dfrac{r}{p}$ earned runs per inning. Since the ERA is defined to be the mean number of earned runs allowed in nine innings, the rule for the function is $f(r, p) = \dfrac{9r}{p}$.

**(b)** The pitcher with the lowest ERA in the 2016 preseason was Aaron Sanchez of the Toronto Blue Jays. He pitched 20 innings and allowed only 3 runs, all of which were earned. Find his ERA. (Data from: www.mlb.com.)

**Solution** Replace $r$ with 3 and $p$ with 20 to get

$$f(3, 20) = \frac{9(3)}{20} = 1.35 \text{ earned runs per nine innings pitched.} \quad \boxed{✓}_2$$

✓ **Checkpoint 2**

Find the ERA for Tanner Roark of the Washington Nationals, who pitched 18 innings during spring training in 2016 and gave up 4 earned runs.

## Graphing Equations in Three Variables

Equations in two variables, such as $3x + 2y = 5$, are graphed by using an $x$-axis and a $y$-axis to locate points in a plane. The plane determined by the $x$- and $y$-axes is called the **xy-plane.** In order to graph an equation in three variables, such as $2x + y + z = 6$, a third axis is needed—the $z$-axis, which goes through the origin in the $xy$-plane and is perpendicular to both the $x$-axis and the $y$-axis.

Figure 14.1 shows one possible way to draw the three axes. In Figure 14.1, the $yz$-plane is in the plane of the page, with the $x$-axis perpendicular to the plane of the page.

✓ **Checkpoint 3**

Locate $(3, 2, 4)$ on the coordinate system shown here.

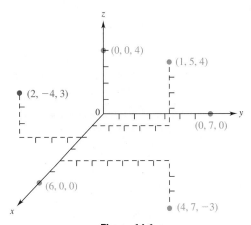

**Figure 14.1**

Just as we graphed ordered pairs earlier, we can now graph **ordered triples** of the form $(x, y, z)$. For example, to locate the point corresponding to the ordered triple $(2, -4, 3)$, start at the origin and go 2 units along the positive $x$-axis. Then go 4 units in a negative direction (to the left), parallel to the $y$-axis. Finally, go up 3 units, parallel to the $z$-axis. The point representing $(2, -4, 3)$ is shown in Figure 14.1, along with several other points. The region of three-dimensional space where all coordinates are positive is called the **first octant.** $\boxed{✓}_3$

Some simple equations in three variables can be graphed by hand. In Chapter 2, we saw that the graph of $ax + by = c$ (where $a$, $b$, and $c$ are constants and not both $a$ and $b$ are zero) is a straight line. This result generalizes to three dimensions.

## Planes

If $a$, $b$, $c$, and $d$ are real numbers, with $a$, $b$, and $c$ not all zero, then the graph of

$$ax + by + cz = d$$

is a plane.

**EXAMPLE 3**    Graph $2x + y + z = 6$.

**Solution** By the result shown in the preceding box, the graph of this equation is a plane. Earlier, we graphed straight lines by finding $x$- and $y$-intercepts. A similar idea helps graph a plane. To find the $x$-intercept—the point where the graph crosses the $x$-axis—let $y = 0$ and $z = 0$:

$$2x + 0 + 0 = 6$$
$$x = 3.$$

The point $(3, 0, 0)$ is on the graph. To find the $y$-intercept, let $x = 0$ and $z = 0$, which gives the point $(0, 6, 0)$. Letting $x = 0$ and $y = 0$ leads to the point $(0, 0, 6)$, which is the $z$-intercept. The plane through these three points includes the triangular surface shown in Figure 14.2. This region is the first-octant part of the plane that is the graph of $2x + y + z = 6$.

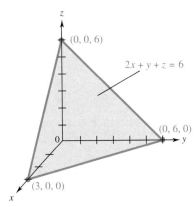

**Figure 14.2**

Throughout this discussion, we assume that all equations involve three variables. Consequently, an equation such as $x + z = 6$ is understood to have a $y$-term with zero coefficient: $x + 0y + z = 6$.

**EXAMPLE 4**    Graph $x + z = 6$.

**Solution** To find the $x$-intercept, let $z = 0$, giving $(6, 0, 0)$. If $x = 0$, we get the $z$-intercept $(0, 0, 6)$. Because there is no $y$ in the equation $x + z = 6$, there can be no $y$-intercept. A plane that has no $y$-intercept is parallel to the $y$-axis. The first-octant portion of the graph of $x + z = 6$ is shown in Figure 14.3.

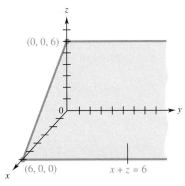

**Figure 14.3**

EXAMPLE 5   Graph the three planes whose equations are respectively

$$x = 0, \quad y = 0, \quad \text{and} \quad z = 0.$$

**Solution**  The graph of $x = 0$ consists of all points with first coordinate 0. These are the points in the **yz-plane** shown in Figure 14.4. Similarly, the graph of $y = 0$ consists of all points with second coordinate 0, that is, all points in the **xz-plane** in Figure 14.4. Finally, the graph of $z = 0$ consists of all points with third coordinate 0—the **xy-plane** in Figure 14.4.

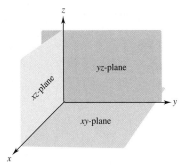

**Figure 14.4**

EXAMPLE 6   Graph each of the given equations.

**(a)**  $x = 3$

**Solution**  This graph, which goes through (3, 0, 0), can have no $y$-intercept and no $z$-intercept. It is, therefore, a plane parallel to the $y$-axis and the $z$-axis and, therefore, to the $yz$-plane. The first-octant portion of the graph is shown in Figure 14.5.

**Figure 14.5**

**(b)** $y = 4$

**Solution** This graph goes through $(0, 4, 0)$ and is parallel to the $xz$-plane. The first-octant portion of the graph is shown in Figure 14.6.

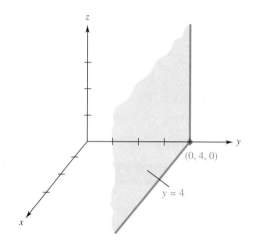

**Figure 14.6**

**(c)** $z = 1$.

**Solution** The graph is a plane parallel to the $xy$-plane and passing through $(0, 0, 1)$. Its first-octant portion is shown in Figure 14.7. ✔ 4

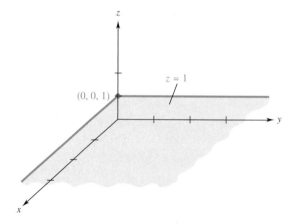

**Figure 14.7**

✔**Checkpoint 4**

Describe each graph and give any intercepts.

**(a)** $2x + 3y - z = 4$

**(b)** $x + y = 3$

**(c)** $z = 2$

## Graphing Functions of Two Variables

The graph of a function of one variable, $y = f(x)$, is a curve in the plane. If $x_0$ is in the domain of $f$, the point $(x_0, f(x_0))$ on the graph lies directly above or directly below (or possibly on) the number $x_0$ on the $x$-axis, as shown in Figure 14.8.

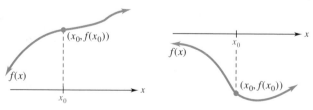

**Figure 14.8**

The graph of a function of two variables, $z = f(x, y)$, is a **surface** in three-dimensional space. If $(x_0, y_0)$ is in the domain of $f$, the point $(x_0, y_0, f(x_0, y_0))$ lies directly above or directly below (or possibly on) the point $(x_0, y_0)$ in the $xy$-plane, as shown in Figure 14.9.

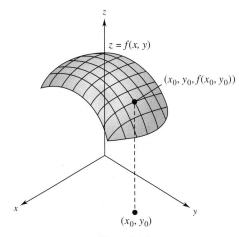

**Figure 14.9**

Computer software is available for drawing the graphs of functions of two independent variables, but you can sometimes get a good picture of a graph without it by finding various **traces**—the curves that result when a surface is cut by a plane. The **$xy$-trace** is the intersection of the surface with the $xy$-plane. The **$yz$-trace** and **$xz$-trace** are defined similarly. You can also determine the intersection of the surface with planes parallel to the $xy$-plane. Such planes are of the form $z = k$, where $k$ is a constant, and the curves that result when they cut the surface are called **level curves.**

---

**EXAMPLE 7**    Graph $z = x^2 + y^2$.

**Solution** The $yz$-plane is the plane in which every point has first coordinate 0, so its equation is $x = 0$. When $x = 0$, the equation becomes $z = y^2$, which is the equation of a parabola in the $yz$-plane, as shown in Figure 14.10(a). Similarly, to find the intersection of the surface with the $xz$-plane (whose equation is $y = 0$), let $y = 0$ in the equation. It then becomes $z = x^2$, which is the equation of a parabola in the $xz$-plane (shown in Figure 14.10(a)). The $xy$-trace (the intersection of the surface with the plane $z = 0$) is the single point $(0, 0, 0)$, because $x^2 + y^2$ is never negative and is equal to 0 only when $x = 0$ and $y = 0$.

(a)

(b)

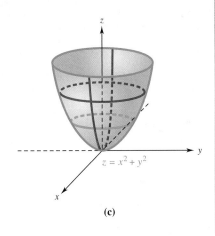

(c)

**Figure 14.10**

Next, we find the level curves by intersecting the surface with the planes $z = 1$, $z = 2$, $z = 3$, etc. (all of which are parallel to the $xy$-plane). In each case, the result is a circle,

$$x^2 + y^2 = 1, \qquad x^2 + y^2 = 2, \qquad x^2 + y^2 = 3,$$

and so on, as shown in Figure 14.10(b). Drawing the traces and level curves on the same set of axes suggests that the graph of $z = x^2 + y^2$ is the bowl-shaped figure, called a **paraboloid,** that is shown in Figure 14.10(c).

Figure 14.11 shows the level curves from Example 7 plotted in the $xy$-plane. The picture can be thought of as a topographical map that describes the surface generated by $z = x^2 + y^2$, just as the topographical map in Figure 14.12 describes the surface of the land in a part of New York State.

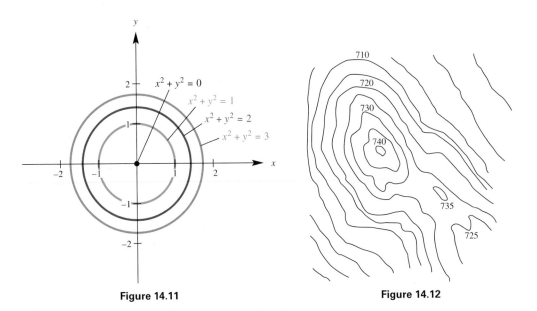

**Figure 14.11**                    **Figure 14.12**

## Applications

One application of level curves in economics occurs with production functions. A **production function** $z = f(x, y)$ is a function that gives the quantity $z$ of an item produced as a function of $x$ and $y$, where $x$ is the amount of labor and $y$ is the amount of capital (in appropriate units) needed to produce $z$ units. If the production function has the special form $z = P(x, y) = Ax^a y^{1-a}$, where $A$ and $a$ are constants and $0 < a < 1$, the function is called a **Cobb–Douglas production function.**

**EXAMPLE 8**    Find the level curve at a production of 100 items for the Cobb–Douglas production function $z = x^{2/3}y^{1/3}$.

**Solution** Let $z = 100$ and solve for $y$ to get

$$100 = x^{2/3}y^{1/3}$$

$$\frac{100}{x^{2/3}} = y^{1/3}.$$

Now reverse the equation and cube both sides to express $y$ as a function of $x$:

$$(y^{1/3})^3 = \frac{100^3}{(x^{2/3})^3}$$

$$y = \frac{1,000,000}{x^2}.$$

✓ **Checkpoint 5**

Find the equation of the level curve for the production of 100 items if the production function is $z = 5x^{1/4}y^{3/4}$.

The level curve of height 100 is graphed in three dimensions in Figure 14.13(a) and on the familiar $xy$-plane in Figure 14.13(b). The points on the graph correspond to those values of $x$ and $y$ that lead to the production of 100 items. ✓ 5

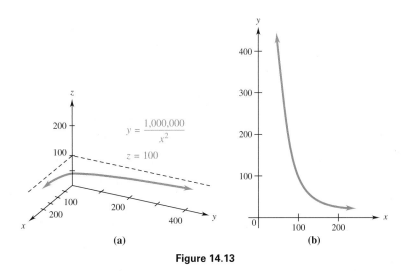

**Figure 14.13**

The curve in Figure 14.13 is called an *isoquant*, for *iso* (equal) and *quant* (amount). In Example 8, the "amounts" all "equal" 100.

## Other Three-Dimensional Graphs

**Figure 14.14**

The graphing techniques discussed up to now may not always be effective. So technology is normally used to graph equations in three variables. Figure 14.14 shows the graph of $z = x^2 + y^2$, which was discussed in Example 7, on a TI-89 graphing calculator. (Most graphing calculators cannot do three-dimensional graphing.) Figure 14.15 shows the same graph drawn by the computer program Maple™. The computer-generated graph has more detail than a graphing calculator can produce.

**Paraboloid,**
$z = x^2 + y^2$

---

$xy$-trace: point
$yz$-trace: parabola
$xz$-trace: parabola

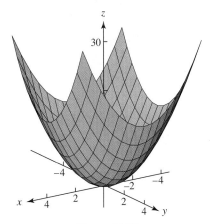

**Figure 14.15**

Figure 14.16 shows some other computer-generated graphs.

**Ellipsoid,**

$$\frac{x^2}{a^2} + \frac{y^2}{b^2} + \frac{z^2}{c^2} = 1$$

*xy*-trace: ellipse
*yz*-trace: ellipse
*xz*-trace: ellipse

**Hyperboloid of Two Sheets,**

$$-x^2 - y^2 + z^2 = 1$$

*xy*-trace: none
*yz*-trace: hyperbola
*xz*-trace: hyperbola

**Hyperbolic Paraboloid,**

$$x^2 - y^2 = z$$

(sometimes called a **saddle**)

*xy*-trace: two intersecting lines
*yz*-trace: parabola
*xz*-trace: parabola

**Figure 14.16**

# 14.1 Exercises

*For each of the given functions, find $f(2, -1)$, $f(-4, 1)$, $f(-2, -3)$, and $f(0, 8)$. (See Examples 1 and 2.)*

1. $f(x, y) = 5x + 2y - 4$

2. $f(x, y) = 7x - 3y + 1$

3. $f(x, y) = 2x^2 - xy + y^2$

4. $f(x, y) = 10x^2 + xy - y^2$

5. $f(x, y) = \sqrt{y^2 + 2x^2}$

6. $f(x, y) = \sqrt{3 + xy}$

*Find the domain of the given function. (See Example 1.)*

7. $f(x, y) = e^x - 7y^3$

8. $f(x, y) = 5x^2y + e^y$

9. $f(x, y) = \sqrt{x} - \sqrt{y}$

10. $f(x, y) = \sqrt{x - 1} + \sqrt{y + 1}$

11. $h(u, v, w) = \left| \dfrac{u + v}{v - w} \right|$

12. $f(x, y) = \dfrac{3x + 7}{2y - 6}$

13. $f(x, y) = \ln(x) + y - 5$

14. $f(x, y) = x + \ln(y - 5)$

*Write the rule of the function. (See Example 2.)*

15. Write the rule of a function that gives the perimeter of a rectangle of length *l* and width *w*.

16. Write the rule of a function that gives the area of a triangle of base *b* and height *h*.

17. **Landscaping** Find a formula $C(x, y)$ that gives the cost of materials for a rectangular fence, assuming there is decorative fencing costing \$20 per foot along the front and back (measuring *x* feet), and privacy fencing costing \$7 per foot along the sides (measuring *y* feet).

18. **Aquarium** Find the cost of materials $C(x, y, z)$ for a rectangular aquarium of width *x*, length *y*, and height *z* if the base is to be made of slate costing \$25 per square foot, the sides are made of glass costing \$5 per square foot, and there is no top.

19. If labor (*x*) costs \$200 per unit, materials (*y*) cost \$100 per unit, and capital (*z*) costs \$50 per unit, write a function for total cost.

20. Write the rule of the function that gives the batting average *a* of a baseball player who has had *h* hits in *b* at bats.

21. What are the *xy*-, *xz*-, and *yz*-traces of a graph?

22. What is a level curve?

*Graph the first-octant portion of each of the given planes. (See Examples 3, 4, and 6.)*

**23.** $3x + 2y + z = 12$   **24.** $2x + 3y + 3z = 18$

**25.** $x + y = 4$   **26.** $y + z = 3$

**27.** $z = 4$   **28.** $y = 3$

**Body Mass Index**  *A person's body mass index (BMI) is given by*

$$\text{BMI} = \frac{703w}{h^2},$$

*where w is weight in pounds and h is height in inches. An individual is considered overweight if he or she has a BMI over 25. Find the BMI for an individual with the given weight and height, and determine whether the individual is overweight. Give your answers to one decimal place.*

**29.** (a) 5 feet tall weighing 100 pounds
   (b) 5 feet tall weighing 150 pounds

**30.** (a) 5 foot 10 inches tall weighing 200 pounds
   (b) 6 foot 4 inches tall weighing 200 pounds

**Temperature–Humidity Index**  *The temperature–humidity index (THI), sometimes called the discomfort index, is given by*

$$f(d, w) = 15 + .4(d + w),$$

*where d is the "dry-bulb temperature" (essentially the ambient-air temperature) and w is the "wet-bulb temperature" (which reflects the relative humidity as well as the temperature). When the THI is 75 or more, most people are uncomfortable. Find the THI and determine whether most people feel comfortable or uncomfortable under the given conditions.*

**31.** (a) $d = 72°$ and $w = 70°$   (b) $d = 72°$ and $w = 78°$

**32.** (a) $d = 85°$ and $w = 60°$   (b) $d = 85°$ and $w = 70°$

**Zero-Coupon Discount Bonds**  *The price p (in dollars) of a zero-coupon discount bond is calculated according to the formula*

$$p = \frac{v}{(1 + r)^m}$$

*where v is the face value of the bond at maturity (in dollars), r is the annual interest rate (expressed as a decimal), and m is the time to maturity (in years). Find the price of the given zero-coupon bonds.*

**33.** (a) $v = \$1000$, $m = 10$ years, $r = .05$
   (b) $v = \$1000$, $m = 10$ years, $r = .06$

**34.** (a) $v = \$10,000$, $m = 10$ years, $r = .05$
   (b) $v = \$10,000$, $m = 20$ years, $r = .05$

**Tax Free Savings**  *The multiplier function*

$$M(n, i, t) = \frac{(1 + i)^n(1 - t) + t}{[1 + (1 - t)i]^n}$$

*compares the growth of an Individual Retirement Account (IRA) with the growth of the same amount (at the same interest rate for the same period of time) in an ordinary savings account. The savings account is subject to federal income tax, but the IRA is tax free (until money is withdrawn). The three variables are the number of years n; the annual interest rate i; and the marginal income tax rate t of the saver. When* $M > 1$, *the IRA grows faster than the savings account. In each of the next two exercises, find the multiplier and determine which account grows faster.*

**35.** Money can be invested for 25 years at 5%, and the marginal income tax rate is 33%.

**36.** Money is invested for 40 years at 6%, and the marginal income tax rate is 28%.

**Melanoma**  *The number of deaths from malignant melanoma of the skin in each state of the United States per 10 million residents can be approximated by the function*

$$f(l, o) = 349.1 - 5.17l + 36.6o - .45lo,$$

*where l is the degrees latitude and o is 1 if the state borders an ocean and 0 if not.*

**37.** Find the number of deaths for Florida (latitude is 28 degrees and it borders an ocean).

**38.** Find the number of deaths for Iowa (latitude 42.2 degrees and it does not border an ocean).

*Work these exercises.*

**39.** **Checkout Lines**  According to the article "How to Pick the Fastest Line at the Supermarket," the amount of time (in seconds) you will be waiting in a checkout line has been approximated by Dan Meyer to be 41 seconds for each person in line ahead of you plus 3 seconds for each item ahead of you. (Data from: *The New York Times*, September 7, 2016.)

   (a) Find a formula $f(p, n)$ for your total waiting time if there are $p$ people and $n$ items ahead of you.

   (b) Find your waiting time in an express checkout line that has 5 people ahead of you, and each person has 10 items.

   (c) Find your waiting time in a regular checkout line that has only 2 people ahead of you, but each person has 40 items.

   (d) Is the express line in part (b) faster than the regular checkout line in part (c)?

**40.** **Oversized Packages**  Extra postage is charged for parcels sent by U.S. mail that are more than 108 inches in length and girth combined. (Girth is the distance around the parcel perpendicular to its length. See the accompanying figure.) Express the combined length and girth as a function of $L$, $W$, and $H$.

*By considering traces, match each equation in Exercises 41–46 with its graph in (a)–(f), on the following page. (See Example 7.)*

**41.** $z = x^2 + y^2$   **42.** $z^2 - y^2 - x^2 = 1$

**43.** $x^2 - y^2 = z$   **44.** $z = y^2 - x^2$

**45.** $\dfrac{x^2}{16} + \dfrac{y^2}{25} + \dfrac{z^2}{4} = 1$   **46.** $z = 5(x^2 + y^2)^{-1/2}$

**(a)**

**(b)**

**(c)**

**(d)**

**(e)**

**(f)**

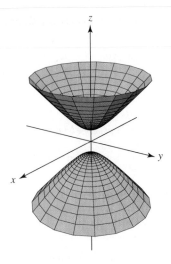

*Find the level curve at a production of z = 500 for each of the production functions in Exercises 47 and 48. Graph each function on the xy-plane. (See Example 8.)*

47. The production function $z$ for the United States was once estimated as $z = x^{.7}y^{.3}$, where $x$ stands for the amount of labor and $y$ stands for the amount of capital.

48. A study published in the *Journal of Political Economy* considered a production function of the form $z = x^{.6}y^{.4}$, where $x$ represents the amount of labor and $y$ the amount of capital.[*]

✓**Checkpoint Answers**

1. **(a)** $-4$      **(b)** $-34$

2. ERA of 2

3.

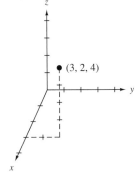

4. **(a)** A plane; $(2, 0, 0)$, $(0, 4/3, 0)$, $(0, 0, -4)$

  **(b)** A plane parallel to the $z$-axis; $(3, 0, 0)$, $(0, 3, 0)$

  **(c)** A plane parallel to the $x$-axis and the $y$-axis; $(0, 0, 2)$

5. $y = \dfrac{20^{4/3}}{x^{1/3}}$

[*]Storesletten, Kjetil, "Sustaining Fiscal Policy Through Immigration," *Journal of Political Economy*, Vol. 108, No. 2, April 2000, pp. 300–323.

# 14.2 Partial Derivatives

A start-up cell phone accessory company makes only two products: colorful phone chargers and decorative phone covers that are almost indestructible. The profits of the firm are given by

$$P(x, y) = 40x^2 - 10xy + 5y^2 - 80,$$

where $x$ is the number of units of chargers sold and $y$ is the number of units of covers sold. How will a change in $x$ or $y$ affect $P$?

Suppose that sales of chargers have been steady at 10 units; only the sales of covers vary. The management would like to find the marginal profit with respect to $y$, the number of covers sold. Recall that marginal profit is given by the derivative of the profit function. Here, $x$ is fixed at 10. Using this information, we begin by defining a new function $f$ such that $f(y) = P(10, y)$. Let $x = 10$ to get

$$f(y) = P(10, y) = 40(10)^2 - 10(10)y + 5y^2 - 80$$
$$= 3920 - 100y + 5y^2.$$

The function $f$ shows the profit from the sale of $y$ covers, assuming that $x$ is fixed at 10 units. Find the derivative $df/dy$ to get the marginal profit with respect to $y$:

$$\frac{df}{dy} = -100 + 10y.$$

In this example, the derivative of the function $f$ was taken with respect to $y$ only; we assumed that $x$ was fixed. To generalize, let $z = f(x, y)$. An intuitive definition of the *partial derivatives* of $f$ with respect to $x$ and $y$ follows.

## Partial Derivatives (Informal Definition)[*]

The **partial derivative of $f$ with respect to $x$** is the derivative of $f$ obtained by treating $x$ as a variable and $y$ as a constant.

The **partial derivative of $f$ with respect to $y$** is the derivative of $f$ obtained by treating $y$ as a variable and $x$ as a constant.

The symbols $f_x(x, y)$ (no prime used), $\partial z/\partial x$, and $\partial f/\partial x$ are used to represent the partial derivative of $z = f(x, y)$ with respect to $x$, with similar symbols used for the partial derivative with respect to $y$. The symbol $f_x(x, y)$ is often abbreviated as just $f_x$, with $f_y(x, y)$ abbreviated $f_y$.

---

**EXAMPLE 1**  Let $f(x, y) = 4x^2 - 9xy + 6y^3$. Find $f_x$ and $f_y$.

**Solution**  To find $f_x$, treat $x$ as a variable and $y$ as a constant. Then find the derivative term by term. The derivative of $4x^2$ is $8x$. The term $-9xy$ can be written as $(-9y)x$. Since $y$ is being treated as a constant, this term is the constant $-9y$ times $x$, and its derivative with respect to $x$ is $-9y$. The last term, $6y^3$, is a constant, so its derivative is 0. In sum,

$$f(x, y) = 4x^2 - 9xy + 6y^3$$
$$f_x = 8x - 9y + 0$$
$$f_x = 8x - 9y.$$

---

[*]A formal definition is given at the end of the section.

To find $f_y$, treat $y$ as a variable and $x$ as a constant. Proceeding term by term to find the derivative, we have

$$f(x, y) = 4x^2 - 9xy + 6y^3$$
$$f_y = 0 - 9x + 18y^2$$
$$f_y = -9x + 18y^2. \quad ✓_1$$

✓ **Checkpoint 1**

Find $f_x$ and $f_y$.

**(a)** $f(x, y) = -x^2y + 3xy + 2xy^2$

**(b)** $f(x, y) = x^3 + 2x^2y + xy$

**EXAMPLE 2**    Find the indicated partial derivatives.

**(a)** Find $f_x$ and $f_y$ for $f(x, y) = e^x + e^{y^3} + e^{5x} + e^{xy}$.

**Solution** Recall that if $y = e^x$, then $y' = e^x$. By the chain rule, if $y = e^{g(x)}$, then $y' = g'(x)e^{g(x)}$. Treating $x$ as a variable and $y$ as a constant and differentiating $f$ term by term we obtain

$$f(x, y) = e^x + e^{y^3} + e^{5x} + e^{xy}$$
$$f_x = e^x + 0 + 5e^{5x} + ye^{xy}.$$

Similarly, treating $y$ as the variable and $x$ as a constant we see that

$$f(x, y) = e^x + e^{y^3} + e^{5x} + e^{xy}$$
$$f_y = 0 + 3y^2e^{y^3} + 0 + xe^{xy}.$$

**(b)** Find $f_x$ and $f_y$ for $f(x, y) = 8ye^{2x+3y}$.

**Solution** To find $f_x$, we treat $8y$ as a constant, so we have

$$f_x = 8y(2e^{2x+3y}) = 16ye^{2x+3y}.$$

To find $f_y$, we must use the product rule since both factors of $f$ involve the variable $y$. Therefore,

$$f_y = (8y)\frac{\partial}{\partial y}(e^{2x+3y}) + (e^{2x+3y})\frac{\partial}{\partial y}(8y)$$
$$f_y = (8y)(3e^{2x+3y}) + (e^{2x+3y})(8)$$
$$f_y = 24ye^{2x+3y} + 8e^{2x+3y}. \quad ✓_2$$

✓ **Checkpoint 2**

Find $f_x$ and $f_y$, given that $f(x, y) = 5x^3e^{10x-2y^4}$.

**EXAMPLE 3**    Let $f(x, y) = \ln(x^2 + y)$. Find $f_x$ and $f_y$.

**Solution** Recall the formula for the derivative of a natural logarithmic function. If $y = \ln[g(x)]$ then $y' = g'(x)/g(x)$. Using this formula and treating $y$ as a constant, we obtain

$$f_x = \frac{\frac{\partial}{\partial x}(x^2 + y)}{x^2 + y} = \frac{2x}{x^2 + y}.$$

Similarly, treating $x$ as a constant leads to the following result:

$$f_y = \frac{\frac{\partial}{\partial y}(x^2 + y)}{x^2 + y} = \frac{1}{x^2 + y}. \quad ✓_3$$

✓ **Checkpoint 3**

Find $f_x$ and $f_y$, given that $f(x, y) = \ln(2x + 3y)$.

The notation

$$f_x(a, b) \quad \text{or} \quad \frac{\partial f}{\partial x}(a, b)$$

represents the value of a partial derivative when $x = a$ and $y = b$, as shown in the next example.

**EXAMPLE 4** Let $f(x, y) = 2x^2 + 3xy^3 + 2y + 5$. Find the following.

**(a)** $f_x(-1, 2)$

**Solution** First, find $f_x$ by holding $y$ constant in $f(x, y)$:

$$f(x, y) = 2x^2 + 3xy^3 + 2y + 5$$
$$f_x = 4x + 3y^3.$$

Now let $x = -1$ and $y = 2$:

$$f_x(-1, 2) = 4(-1) + 3(2)^3 = -4 + 24 = 20.$$

**(b)** $\dfrac{\partial f}{\partial y}(-4, -3)$

**Solution** Hold $x$ constant in $f(x, y)$ to find $\dfrac{\partial f}{\partial y}$:

$$f(x, y) = 2x^2 + 3xy^3 + 2y + 5$$
$$\frac{\partial f}{\partial y} = 9xy^2 + 2.$$

Then

$$\frac{\partial f}{\partial y}(-4, -3) = 9(-4)(-3)^2 + 2 = 9(-36) + 2 = -322.$$

✓ **Checkpoint 4**

Let $f(x, y) = x^2 + xy^2 + 5y - 10$.
Find the given partial derivatives.

**(a)** $f_x(2, 1)$

**(b)** $\dfrac{\partial f}{\partial y}(-1, 0)$

 The derivative of a function of one variable can be interpreted as the slope of the tangent line to the graph at that point. With some modification, the same is true of partial derivatives of functions of two variables. At a point on the graph of a function of two variables, $z = f(x, y)$, there may be many tangent lines, all of which lie in the same tangent plane, as shown in Figure 14.17.

In any particular direction, however, there will be only one tangent line. We now demonstrate how to use partial derivatives to find the slope of the tangent lines in the $x$- and $y$-directions.

Figure 14.18 shows a surface $z = f(x, y)$ and a plane that is parallel to the $xz$-plane. The equation of the plane is $y = b$. (This equation corresponds to holding $y$ fixed.)

**Figure 14.17**

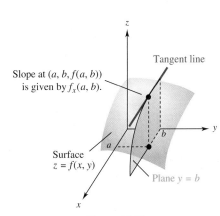

**Figure 14.18**

Every point $(x, y, z)$ on the plane has $y = b$. Hence, every point on the curve that is the intersection of the plane and the surface must have the form $(x, b, f(x, b))$. Thus, the curve is the graph of $z = f(x, b)$. Since $b$ is a constant, $z = f(x, b)$ is a function of the single variable $x$, whose derivative with respect to $x$ is $f_x(x, b)$. When this derivative is evaluated at $x = a$, it gives the slope of the tangent line to the curve at the point $(a, b, f(a, b))$—that is, the tangent line to the surface in the $x$-direction at the point $(a, b, f(a, b))$. In the same way, the partial derivative with respect to $y$ will give the slope of the tangent line to the surface in the $y$-direction at the point $(a, b, f(a, b))$.

## Rate of Change

 The derivative of $y = f(x)$ gives the rate of change of $y$ with respect to $x$. In the same way, if $z = f(x, y)$, then $f_x$ gives the rate of change of $z$ with respect to $x$ if $y$ is held constant and $f_y$ gives the rate of change of $z$ with respect to $y$ if $x$ is held constant.

**EXAMPLE 5**    **Marginal Cost** A souvenir vendor in the college football stadium pays a fee to use the souvenir stall. The vendor buys T-shirts and sweatshirts by the hundred from a wholesaler to resell at retail prices at the game. The cost function is given by

$$C(x, y) = 500 + 700x + 1800y - 2x^2 - 3y^2 - xy,$$

where $x$ is the number of T-shirts stocked (in hundreds) and $y$ is the number of sweatshirts stocked (in hundreds).

**(a)** In $C(x, y)$, what does 500 represent?

**Solution** In the cost function, the 500 represents the fixed costs: $C(0, 0) = \$500$ is the fee for renting the souvenir stall even if no T-shirts or sweatshirts are stocked.

**(b)** Find and interpret $C_x(20, 15)$.

**Solution** First, find the partial derivative $C_x(x, y)$:

$$C_x(x, y) = 700 - 4x - y.$$

This partial derivative gives the rate of change of $C(x, y)$ with respect to $x$. Replacing $x$ with 20 and $y$ with 15 gives

$$C_x(20, 15) = 700 - 4(20) - 15 = 605.$$

This is the marginal cost of stocking one more unit of T-shirts. In other words, $605 is the approximate change in cost if the number of units of T-shirts increases by 1 unit from $x = 20$ (2000 T-shirts) to $x = 21$ (2100 T-shirts) while the number of units of sweatshirts remains constant at $y = 15$ (1500 sweatshirts).

**(c)** Find and interpret $C_y(20, 15)$.

**Solution** The partial derivative $C_y(x, y)$ is

$$C_y(x, y) = 1800 - 6y - x.$$

This partial derivative gives the rate of change of $C(x, y)$ with respect to $y$. Letting $x = 20$ and $y = 15$ gives

$$C_y(20, 15) = 1800 - 6(15) - 20 = 1690.$$

This shows that $1690 is the marginal cost of one additional unit of sweatshirts (increasing from $y = 15$ to $y = 16$) while keeping the number of units of T-shirts constant at $x = 20$ (2000 T-shirts). ✓5

✓**Checkpoint 5**

Use the function in Example 5 to find and interpret the given partial derivatives.

**(a)** $C_x(25, 10)$.

**(b)** $C_y(25, 10)$.

As mentioned in the previous section, if $P(x, y)$ gives the output $P$ produced by $x$ units of labor and $y$ units of capital, then $P(x, y)$ is a production function. The partial derivatives of this production function have practical implications. For example, $\partial P/\partial x$ gives the **marginal productivity** of labor. This represents the rate at which the output is changing with respect to changes in labor for a fixed capital investment. That is, if capital investment is held constant and labor is increased by 1 unit, $\partial P/\partial x$ will yield the approximate change in the production level. Likewise, $\partial P/\partial y$ gives the marginal productivity of capital, which represents the rate at which the output is changing with respect to changes in capital for a fixed labor value. So if the labor force is held constant and the capital investment is increased by 1 unit, $\partial P/\partial y$ will approximate the corresponding change in the production level.

---

**EXAMPLE 6**   **Marginal Productivity** A company that manufactures drones has determined that its production function is given by

$$P(x, y) = 500x + 800y + 3x^2 y - \frac{11x^3}{15} - \frac{y^4}{4},$$

where $x$ is the size of the labor force (in thousands of work hours per week) and $y$ is the amount of capital (in units of $1000) invested. Find the marginal productivity of labor and the marginal productivity of capital when $x = 50$ and $y = 20$, and interpret the results.

**Solution** The marginal productivity of labor is found by taking the derivative of $P$ with respect to $x$:

$$\frac{\partial P}{\partial x} = 500 + 6xy - \frac{33x^2}{15}$$

$$\frac{\partial P}{\partial x}(50, 20) = 500 + 6(50)(20) - \frac{33(50^2)}{15}$$

$$= 1000.$$

Thus, if capital investment is held constant at $20,000 and labor is increased from 50 thousand to 51 thousand work hours per week, production will increase by about 1000 units. In the same way, the marginal productivity of capital is $\partial P/\partial y$:

$$\frac{\partial P}{\partial y} = 800 + 3x^2 - y^3$$

$$\frac{\partial P}{\partial y}(50, 20) = 800 + 3(50)^2 - 20^3$$

$$= 300.$$

If work hours are held constant at 50 thousand hours per week and capital investment is increased from $20,000 to $21,000, production will increase by about 300 units. ✓ 6

✓ **Checkpoint 6**

Suppose a production function is given by $P(x, y) = 10x^2 y + 100x + 400y - 5xy^2$, where $x$ and $y$ are defined as in Example 6. Find the marginal productivity of labor and capital when $x = 20$ and $y = 10$.

## Second-Order Partial Derivatives

The second derivative of a function of one variable is very useful in determining local maxima and minima. **Second-order partial derivatives** (partial derivatives of a partial derivative) are used in a similar way for functions of two or more variables. The situation is somewhat more complicated, however, with more independent variables. For example, $f(x, y) = 4x + x^2 y + 2y$ has two first-order partial derivatives:

$$f_x = 4 + 2xy \quad \text{and} \quad f_y = x^2 + 2.$$

Because each of these partial derivatives has two partial derivatives, one with respect to $x$ and one with respect to $y$, there are *four* second-order partial derivatives of the function $f$.

The same is true for any function $z = f(x, y)$. The following schematic diagram shows the partial derivatives and the second-order partial derivatives of $f$ and the various notations used for each:

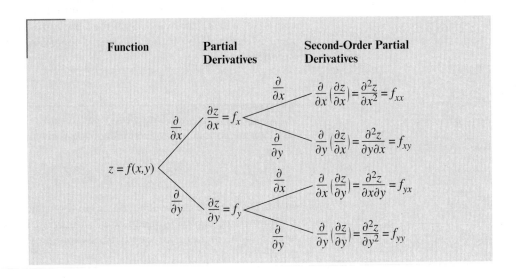

As shown in the diagram, $f_{xx}$ is used as an abbreviation for $f_{xx}(x, y)$, with $f_{yy}, f_{xy}$, and $f_{yx}$ used in a similar way. The symbol $f_{xx}$ is read, "the partial derivative of $f_x$ with respect to $x$," and $f_{xy}$ is read, "the partial derivative of $f_x$ with respect to $y$." Also, the symbol $\partial^2 z / \partial y^2$ is read, "the partial derivative of $\partial z / \partial y$ with respect to $y$."

**NOTE** For most functions found in applications, and for all the functions in this text, the second-order partial derivatives $f_{xy}$ and $f_{yx}$ are equal. Therefore, we need not be particular about the order in which these derivatives are found.

**EXAMPLE 7** Find all second-order partial derivatives of
$$f(x, y) = -4x^3 - 3x^2y^3 + 2y^2.$$

**Solution** First find $f_x$ and $f_y$:
$$f_x = -12x^2 - 6xy^3 \quad \text{and} \quad f_y = -9x^2y^2 + 4y.$$

To find $f_{xx}$, take the partial derivative of $f_x$ with respect to $x$:
$$f_{xx} = -24x - 6y^3.$$

Now find $f_{yy}$ by taking the partial derivative of $f_y$ with respect to $y$:
$$f_{yy} = -18x^2y + 4.$$

Next, find $f_{xy}$ by taking the partial derivative of $f_x$ with respect to $y$:
$$f_{xy} = -18xy^2.$$

Finally, find $f_{yx}$ by taking the partial derivative of $f_y$ with respect to $x$:
$$f_{yx} = -18xy^2.$$

✓ **Checkpoint 7**

Let $f(x, y) = 4x^2y^2 - 9xy + 8x^2 - 3y^4$. Find all second-order partial derivatives.

Observe that $f_{xy} = f_{yx}$.

**EXAMPLE 8** Let

$$f(x, y) = \sqrt{x} + \frac{1}{y} - 2x^3y^5.$$

Find all second-order partial derivatives.

**Solution** Using laws of exponents, rewrite the function as

$$f(x, y) = x^{\frac{1}{2}} + y^{-1} - 2x^3y^5.$$

Then the first-order partial derivatives are

$$f_x = \frac{1}{2}x^{-\frac{1}{2}} - 6x^2y^5 \quad \text{and} \quad f_y = -y^{-2} - 10x^3y^4.$$

Now find the second-order partial derivatives:

$$f_{xx} = -\frac{1}{4}x^{-\frac{3}{2}} - 12xy^5; \quad f_{xy} = -30x^2y^4;$$

$$f_{yy} = 2y^{-3} - 40x^3y^3; \quad f_{yx} = -30x^2y^4. \quad \checkmark_8$$

✓ **Checkpoint 8**

Find the first- and second-order partial derivatives of $f(x, y) = 4x^6\sqrt{y}$.

Partial derivatives of multivariate functions with more than two variables are found in a way similar to that for functions with two variables. For example, to find $f_x$ for $w = f(x, y, z)$, treat $y$ and $z$ as constants and differentiate with respect to $x$.

**EXAMPLE 9** Let $f(x, y, z) = xy^2z + 2x^2y - 4xz^2$. Find $f_x, f_y, f_z, f_{xy}$, and $f_{yz}$.

**Solution** We find the first three partial derivatives as follows:

$$f_x = y^2z + 4xy - 4z^2$$
$$f_y = 2xyz + 2x^2$$
$$f_z = xy^2 - 8xz.$$

To find $f_{xy}$, differentiate $f_x$ with respect to $y$:

$$f_{xy} = 2yz + 4x.$$

In the same way, differentiate $f_y$ with respect to $z$ to get

$$f_{yz} = 2xy. \quad \checkmark_9$$

✓ **Checkpoint 9**

Let $f(x, y, z) = xyz + x^2yz + xy^2z^3$. Find $f_x, f_y, f_z$, and $f_{xz}$.

**EXAMPLE 10** **Duration of a Bond** A zero-coupon discount bond is a bond that does not pay periodic coupon payments but instead pays one lump sum at maturity. The amount paid at maturity is called the face value. The price of the bond (the initial cost in dollars to buy the bond) is calculated according to the formula

$$p = \frac{v}{(1 + r)^m} = v(1 + r)^{-m},$$

where $v$ is the face value of the bond (in dollars), $r$ is the annual interest rate (expressed as a decimal), and $m$ is the time to maturity (in years).

**(a)** Would you expect $p_r$ to be positive, negative, or zero? Why?

**Solution** The partial derivative $p_r$ is the rate of change of price with respect to interest rate. If the interest rate is higher, a smaller initial investment (price of the bond) is needed to generate the face value of the bond by the maturity date. Therefore, $p_r$ is negative.

**(b)** Find $p_r$.

**Solution** To find $p_r$, we treat $r$ as the variable and $v$ and $m$ as constants; therefore,

$$p_r = v(-m)(1 + r)^{-m-1} = \frac{-vm}{(1 + r)^{m+1}}.$$

**(c)** The *modified duration* of a bond is defined to be the partial derivative of the price with respect to the interest rate divided by the negative of the price. In other words,

$$\text{Modified duration} = \frac{p_r}{-p}.$$

Modified duration can be interpreted as the approximate percentage by which the price of the bond will decrease with a 1% increase in interest rate. Find the modified duration.

**Solution** Using the formula for $p$ given in the problem and the formula for $p_r$ found in part (b), we see that

$$\text{Modified duration} = \frac{p_r}{-p} = \frac{\dfrac{-vm}{(1 + r)^{m+1}}}{\dfrac{-v}{(1 + r)^m}} = \frac{-vm}{(1 + r)^{m+1}} \cdot \frac{(1 + r)^m}{-v} = \frac{m}{(1 + r)}.$$

Notice that the modified duration depends only on $m$ and $r$. The modified duration gives the *percentage change* in the price of the bond as the interest rate changes, so it does not depend on the face value of the bond.

**(d)** Assume that the interest rate is $r = 5\% = .05$. Calculate the modified duration for a 1-year bond and a 10-year bond. Interpret your answer.

**Solution** For a 1-year bond at 5% interest, we have

$$\text{Modified duration} = \frac{m}{(1 + r)} = \frac{1}{(1 + .05)} \approx .952.$$

For a 10-year bond at 5% interest, we have

$$\text{Modified duration} = \frac{m}{(1 + r)} = \frac{10}{(1 + .05)} \approx 9.52.$$

For a 1-year bond, a 1% increase in the interest rate decreases the price of the bond by slightly less than 1% (by .952%), whereas for a 10-year bond, a 1% increase in interest rate decreases the price of the bond by almost 10% (by 9.52%). Bonds with longer-dated maturities are more sensitive to interest rate shifts.

For those who may be interested, we close this section with the formal definition of partial derivatives. As you can see, it is very similar to the definition of the derivative of a function of one variable.

## Partial Derivatives (Formal Definition)

Let $z = f(x, y)$ be a function of two variables. Then the **partial derivative of $f$ with respect to $x$** is

$$f_x(x, y) = \lim_{h \to 0} \frac{f(x + h, y) - f(x, y)}{h},$$

and the **partial derivative of $f$ with respect to $y$** is

$$f_y(x, y) = \lim_{h \to 0} \frac{f(x, y + h) - f(x, y)}{h},$$

provided that these limits exist.

Similar definitions could be given for functions of more than two independent variables.

# 14.2 Exercises

*For each of the given functions, find the following. (See Examples 1 and 4.)*

(a) $\dfrac{\partial z}{\partial x}$    (b) $\dfrac{\partial z}{\partial y}$    (c) $\dfrac{\partial z}{\partial x}(2, 3)$    (d) $\dfrac{\partial z}{\partial y}(1, -2)$

1. $z = f(x, y) = 8x^3 - 4x^2y + 9y^2$

2. $z = f(x, y) = -3x^2 - 2xy^2 + 5y^3$

*In Exercises 3–8, find $f_x$ and $f_y$. Then find $f_x(2, -1)$ and $f_y(-4, 3)$. Leave the answers in terms of e in Exercises 5–6. (See Examples 1–4.)*

3. $f(x, y) = 6y^3 - 4xy + 5$    4. $f(x, y) = -3x^4y^3 + 10$

5. $f(x, y) = e^{2x+y}$    6. $f(x, y) = -4e^{x-y}$

7. $f(x, y) = \ln(x - 2y)$    8. $f(x, y) = \ln(y - 8x)$

*In Exercises 9–14, find $f_x$ and $f_y$.*

9. $f(x, y) = \sqrt{8x^2y + 4}$    10. $f(x, y) = \sqrt{2xy^5 + 3}$

11. $f(x, y) = x^2e^{2xy}$    12. $f(x, y) = ye^{5x+2y}$

13. $f(x, y) = \dfrac{3x}{1 + 6y}$    14. $f(x, y) = \dfrac{5y}{1 - 2x}$

*Find all second-order partial derivatives of the given functions. (See Examples 7 and 8.)*

15. $f(x, y) = 4x^2y^2 - 16x^2 + 4y$

16. $R(x, y) = 4x^2 - 5xy^3 + 12x^2y^2$

17. $h(x, y) = -3y^2 - 4x^2y^2 + 7xy^2$

18. $P(x, y) = -16x^3 + 3xy^2 - 12x^4y^2$

19. $f(x, y) = x^3y^{-1}$    20. $f(x, y) = x^{-2}y^7$

21. $f(x, y) = \dfrac{1}{x} - \dfrac{5}{y^2}$    22. $f(x, y) = \dfrac{1}{x^5} + \dfrac{7}{y}$

23. $f(x, y) = e^{4y-3x}$    24. $f(x, y) = e^{2x-7y}$

25. $z = 4xe^y$    26. $z = -3ye^x$

27. $f(x, y) = \ln(3xy)$    28. $f(x, y) = \ln(8xy)$

29. $f(x, y) = \dfrac{3x}{y - 1}$    30. $f(x, y) = \dfrac{10y}{1 - x}$

31. $z = x\ln(xy)$    32. $z = (y + 1)\ln(x^3y)$

*In Exercises 33 and 34, evaluate $f_{xy}(1, 2)$ and $f_{yy}(3, 1)$.*

33. $f(x, y) = x\ln(xy)$    34. $f(x, y) = (y + 1)\ln(x^3y)$

*Find $f_x, f_y, f_z,$ and $f_{yz}$ for the given functions. In Exercises 35 and 36, also find $f_y(3, 4, -2)$ and $f_{yz}(1, -1, 0)$. (See Example 9.)*

35. $f(x, y, z) = x^2 + yz + z^4$    36. $f(x, y, z) = 3x^5 - x^2 + y^5$

37. $f(x, y, z) = 9ye^{3xz^4}$    38. $f(x, y, z) = 2ye^{7xz}$

39. How many partial derivatives does a function in three variables have? How many second-order partial derivatives does it have? Explain why.

40. Suppose $z = f(x, y)$ describes the cost to build a certain structure, where $x$ represents the labor costs and $y$ represents the cost of materials. Describe what $f_x$ and $f_y$ represent.

*Work the following applied problems. (See Example 5.)*

41. **Bicycle Manufacturing** A bicycle manufacturer produces two kinds of frames, aluminum and carbon fiber. The cost function for these products is $C(x, y) = 500x + 1100y + 20,000$, where $x$ is the number of aluminum frames and $y$ is the number of carbon fiber frames produced.

    (a) Find $C_x$, the marginal cost of producing one additional aluminum frame when the production of carbon fiber models is held constant.

    (b) Find $C_y$, the marginal cost of producing one additional carbon fiber frame when the production of aluminum models is held constant.

42. **Radio-Controlled Planes** A radio-controlled plane manufacturer produces two models, the Basic and the Ultra. Its cost function for these products is

    $$C(x, y) = 125x + 400y + 10,000,$$

    where $x$ is the number of Basic models and $y$ is the number of Ultra models produced.

    (a) Find $C_x$, the marginal cost of producing one additional Basic model when the production of Ultras is held constant.

    (b) Find $C_y$, the marginal cost of producing one additional Ultra model when the production of Basics is held constant.

43. **Cell Phone Accessories** Suppose that the cost function for the company discussed in the first paragraph of this section is given by:

    $$C(x, y) = 600x + 120y + 4000,$$

    where $x$ is the number of units of phone chargers and $y$ is the number of units of phone covers produced.

    (a) In $C(x, y)$, what does 4000 represent?

    (b) Find $\partial C/\partial x$ and $\partial C/\partial y$. What do these quantities represent?

44. **Souvenir Sales** The profit function for the souvenir vendor in Example 5 is given by

    $$P(x, y) = 2x^2 + 3y^2 + xy + 1300x + 2200y - 500,$$

    where $x$ is the number of T-shirts (in hundreds) and $y$ is the number of sweatshirts (in hundreds).

    (a) Find and interpret $P_x(30, 5)$.

    (b) Find and interpret $P_y(30, 5)$.

45. **Pharmaceutical Profits** The profit (in dollars) from the sale of $x$ units of a tranquilizer and $y$ units of an antibiotic is given by

    $$P(x, y) = 3x^2 + 4y^2 - 4xy.$$

    Suppose 9 units of tranquilizer and 5 units of antibiotic are sold.

    (a) Find $P_x(9, 5)$, the *approximate* effect on profit if the amount of tranquilizer sold is increased from 9 to 10 units, while antibiotic sales remain constant at 5 units.

**(b)** Find $P_y(9, 5)$, the *approximate* effect on profit if the amount of antibiotic sold is increased from 5 to 6 units, while tranquilizer sales remain constant at 9 units.

**46. Pharmaceutical Revenue** The revenue from the sale of $x$ units of a tranquilizer and $y$ units of an antibiotic is given by

$$R(x, y) = 5x^2 + 9y^2 - 4xy.$$

Suppose 9 units of tranquilizer and 5 units of antibiotic are sold. Use partial derivatives to answer these questions.

**(a)** What is the *approximate* effect on revenue if the amount of tranquilizer sold is increased to from 9 to 10 units, while antibiotic sales remain constant?

**(b)** What is the *approximate* effect on revenue if the amount of antibiotic sold is increased to 6 units, while tranquilizer sales remain constant?

**47. Automobile Sales** A car dealership estimates that the total weekly sales of its most popular model are a function of the car's list price $p$ (in dollars) and the interest rate $i$ (as a percent) offered by the manufacturer. The weekly sales are given by

$$f(p, i) = 99p - .5pi - .0025p^2.$$

**(a)** Find the weekly sales if the average list price is $p = \$19,400$ and the manufacturer is offering an 8% interest rate ($i = 8$).

**(b)** Find and interpret $f_p(p, i)$ and $f_i(p, i)$.

**(c)** What would be the approximate effect on weekly sales if the price is $19,400 and interest rates rise from 8% to 9%?

**48. Diet and Exercise** A weight-loss counselor has prepared a program of diet and exercise for a client. If the client sticks to the program, the weight loss that can be expected (in pounds per week) is given by

$$\text{Weight loss} = f(n, c) = \frac{1}{8}n^2 - \frac{1}{5}c + \frac{1937}{8},$$

where $c$ is the average daily calorie intake for the week and $n$ is the number of 40-minute aerobic workouts per week.

**(a)** How many pounds can the client expect to lose by eating an average of 1200 calories per day and participating in four 40-minute workouts in a week?

**(b)** Find and interpret $\partial f / \partial n$.

**(c)** The client currently averages 1100 calories per day and does three 40-minute workouts each week. What would be the *approximate* impact on weekly weight loss of adding a fourth workout per week?

**49. Future Value for Compound Interest** If $P$ dollars are invested at interest rate $i$ per period, then the compound amount (future value) $A$ after $n$ compounding periods is

$$A = P(1 + i)^n.$$

**(a)** Would you expect $A_P$ to be positive or negative? Find $A_P$.

**(b)** Would you expect $A_i$ to be positive or negative? Find $A_i$.

**50. Present Value for Compound Interest** The present value $P$ of $A$ dollars compounded at an interest rate $i$ per period for $n$ periods is

$$P = \frac{A}{(1 + i)^n} = A(1 + i)^{-n}.$$

**(a)** Would you expect $P_A$ to be positive or negative? Find $P_A$.

**(b)** Would you expect $P_i$ to be positive or negative? Find $P_i$.

**51. Future Value for Continuously Compounded Interest** The compound amount (future value) $A$ for a deposit of $P$ dollars at an interest rate $r$ per year compounded continuously for $t$ years is given by

$$A = Pe^{rt}.$$

**(a)** Would you expect $A_P$ to be positive or negative? Find $A_P$.

**(b)** Would you expect $A_r$ to be positive or negative? Find $A_r$.

**(c)** Would you expect $A_t$ to be positive or negative? Find $A_t$.

**52. Present Value for Continuously Compounded Interest** The present value $P$ of $A$ dollars at an interest rate $r$ per year compounded continuously for $t$ years is

$$P = \frac{A}{e^{rt}} = Ae^{-rt}.$$

**(a)** Would you expect $P_A$ to be positive or negative? Find $P_A$.

**(b)** Would you expect $P_r$ to be positive or negative? Find $P_r$.

**(c)** Would you expect $P_t$ to be positive or negative? Find $P_t$.

*Work the following marginal productivity problems. (See Example 6.)*

**53.** A manufacturer of automobile batteries estimates that total production in thousands of units is given by

$$f(x, y) = 3x^{1/3}y^{2/3},$$

where $x$ is the number of units of labor and $y$ is the number of units of capital utilized.

**(a)** Find and interpret the marginal productivity of labor $f_x(64, 125)$ and the marginal productivity of capital $f_y(64, 125)$ if the current level of production uses 64 units of labor and 125 units of capital.

**(b)** What would be the *approximate* effect on production of increasing labor to 65 units while holding capital constant?

**(c)** Suppose that sales have been good and management wants to increase either capital or labor by 1 unit. Which option would result in a larger increase in production?

**54.** A manufacturer estimates that production (in hundreds of units) is a function of the respective amounts $x$ and $y$ of labor and capital used, as follows:

$$f(x, y) = \left[ \frac{1}{3}x^{-1/3} + \frac{2}{3}y^{-1/3} \right]^{-3}.$$

**(a)** Find the number of units produced when 27 units of labor and 64 units of capital are utilized.

**(b)** Find and interpret $f_x(27, 64)$ and $f_y(27, 64)$.

**(c)** What would be the *approximate* effect on production of increasing labor by 1 unit?

**(d)** What would be the *approximate* effect on production of increasing capital by 1 unit?

**55.** The production function $z$ for the United States was once estimated as

$$z = x^{.7}y^{.3},$$

where $x$ stands for the amount of labor and $y$ stands for the amount of capital. Find the marginal productivity of labor (that is, find $\partial z / \partial x$) and of capital (that is, find $\partial z / \partial y$.)

**56.** A similar production function for Canada is

$$z = x^{.4}y^{.6},$$

with $x$, $y$, and $z$ as in Exercise 55. Find the marginal productivity of labor and of capital.

*Work these exercises.*

**57. Body Mass Index** The body mass index (BMI) is a number that can be calculated for any individual as follows: Multiply a person's weight by 703 and divide by the person's height squared. That is,

$$B = \frac{703w}{h^2},$$

where $w$ is in pounds and $h$ is in inches. The National Heart, Lung and Blood Institute uses the BMI to determine whether a person is "overweight" ($25 \le B < 30$) or "obese" ($B \ge 30$). (Data from: National Institutes of Health.)

(a) Calculate the BMI for a person who weighs 220 pounds and is 74 inches tall.

(b) Calculate $\dfrac{\partial B}{\partial w}$ and $\dfrac{\partial B}{\partial h}$.

**58. Gravity of the Earth** The gravitational attraction $F$ on a body a distance $r$ from the center of the earth, where $r$ is greater than the radius of the earth, is a function of the mass $m$ of the body and the distance $r$. Specifically,

$$F = \frac{mgR^2}{r^2},$$

where $R$ is the radius of the earth and $g$ is the force of gravity—about 32 feet per second per second (ft/sec$^2$).

(a) Find and interpret $F_m$ and $F_r$.

(b) Show that $F_m > 0$ and $F_r < 0$. Why is this reasonable?

**59. Human Surface Area** The surface area of a human (in square meters) is approximated by

$$A(M, H) = .202M^{.425}H^{.725},$$

where $M$ is the mass of the person (in kilograms) and $H$ is the height (in meters). Find the *approximate* change in surface area under the given conditions.

(a) The mass changes from 72 kg to 73 kg, while the height remains 1.8 m.

(b) The mass remains stable at 70 kg, while the height changes from 1.6 m to 1.7 m.

**60. Wind Chill Index** In 1941, explorers Paul Siple and Charles Passel discovered that the amount of heat lost when an object is exposed to cold air depends on both the temperature of the air and the velocity of the wind. They developed the windchill index as a way to measure the danger of frostbite while performing outdoor activities. On the basis of new findings and additional data, the windchill index was recalculated in 2001. The new index is given by

$$W(V, T) = 35.74 + .6215T - 35.75V^{.16} + (.4275T)V^{.16},$$

where $V$ is the wind speed in miles per hour and $T$ is the Fahrenheit temperature for wind speeds between 4 and 45 mph.[*]

(a) Find the windchill for a wind speed of 20 mph and 10°F.

(b) A weather report indicates that the windchill is −22°F and the actual outdoor temperature is 5°F. Use a graphing calculator to find the corresponding wind speed to the nearest mile per hour.

(c) Find and interpret $W_V(20, 10)$ and $W_T(20, 10)$.

(d) Using the table command on a graphing calculator or a spreadsheet, develop a windchill chart for wind speeds of 5, 10, 15, and 20 mph and temperatures of 0°, 5°, 15°, and 25°.

**61. Scuba Diving** In 1908, J. Haldane constructed diving tables that give a relationship between the water pressure on body tissues for various water depths and dive times. The tables were used by divers and virtually eliminated the occurrence of decompression sickness. The pressure in atmospheres for a no-stop dive is given by the formula

$$p(l, t) = 1 + \frac{l}{33}(1 - e^{-.1386t}),$$

where $t$ is in minutes, $l$ is in feet, and $p$ is in atmospheres (atm).[†]

(a) Find the pressure at 33 feet for a 10-minute dive.

(b) Find $p_l(33, 10)$ and $p_t(33, 10)$.

(c) Haldane estimated that decompression sickness for no-stop dives could be avoided if the diver's tissue pressure did not exceed 2.15 atmospheres. Find the maximum amount of time (including the time going down and coming back up) that a diver could stay down if he or she wants to dive to a depth of 66 feet.

**62. Drug Reaction** The reaction to $x$ units of a drug $t$ hours after it was administered is given by

$$R(x, t) = x^2(a - x)t^2e^{-t},$$

for $0 \le x \le a$ (where $a$ is a constant). Find the given partial derivatives.

(a) $\dfrac{\partial R}{\partial x}$

(b) $\dfrac{\partial R}{\partial t}$

(c) Interpret your answers to parts (a) and (b).

✓**Checkpoint Answers**

**1.** (a) $f_x = -2xy + 3y + 2y^2$; $f_y = -x^2 + 3x + 4xy$

(b) $f_x = 3x^2 + 4xy + y$; $f_y = 2x^2 + x$

**2.** $f_x = 15x^2e^{10x - 2y^4} + 50x^3e^{10x - 2y^4}$; $f_y = -40x^3y^3e^{10x - 2y^4}$

[*]Joint Action Group of the National Weather Service and the Meteorological Services of Canada, press release, August 2001.

[†]The estimates provided by the formula are conservative. Consult modern diving tables before making a dive. See David Westbrook, "The Mathematics of Scuba Diving," *UMAP Journal* 18, no. 2 (1997): 2–19.

**3.** $f_x = \dfrac{2}{2x + 3y}; f_y = \dfrac{3}{2x + 3y}$

**4. (a)** 5      **(b)** 5

**5. (a)** $C_x(25, 10) = 590$ is the approximate increase in cost if the number of T-shirts stocked is increased from 2500 to 2600 while the number of sweatshirts remains constant at 1000.

**(b)** $C_y(25, 10) = 1715$ is the approximate increase in cost if the number of sweatshirts stocked is increased from 1000 to 1100 while the number of T-shirts remains constant at 2500.

**6.** The marginal productivity of labor is 3600. The marginal productivity of capital is 2400.

**7.** $f_{xx} = 8y^2 + 16$
$f_{yy} = 8x^2 - 36y^2$
$f_{xy} = 16xy - 9$
$f_{yx} = 16xy - 9$

**8.** $f_x = 24x^5 y^{\frac{1}{2}}; f_y = 2x^6 y^{-\frac{1}{2}}; f_{xx} = 120x^4 y^{\frac{1}{2}}; f_{yy} = -x^6 y^{-\frac{3}{2}};$
$f_{xy} = f_{yx} = 12x^5 y^{-\frac{1}{2}}$

**9.** $f_x = yz + 2xyz + y^2 z^3$
$f_y = xz + x^2 z + 2xyz^3$
$f_z = xy + x^2 y + 3xy^2 z^2$
$f_{xz} = y + 2xy + 3y^2 z^2$

# 14.3 Extrema of Functions of Several Variables

 One of the most important applications of calculus is finding maxima and minima of functions. Earlier, we studied this idea extensively for functions of a single variable; now we shall see that extrema can be found for functions of two variables. In particular, an extension of the second-derivative test can be derived and used to identify maxima or minima. We begin with the definitions of local maxima and minima.

---

## Local Maxima and Minima

Let $(a, b)$ be in the domain of a function $f$.

**1.** $f$ has a **local maximum** at $(a, b)$ if there is a circular region in the $xy$-plane with $(a, b)$ in its interior such that

$$f(a, b) \geq f(x, y)$$

for all points $(x, y)$ in the circular region.

**2.** $f$ has a **local minimum** at $(a, b)$ if there is a circular region in the $xy$-plane with $(a, b)$ in its interior such that

$$f(a, b) \leq f(x, y)$$

for all points $(x, y)$ in the circular region.

---

As before, the term *local extremum* (also called a *relative extremum*) is used for either a local maximum or a local minimum. Examples of a local maximum and a local minimum are given in Figures 14.19 and 14.20, respectively.

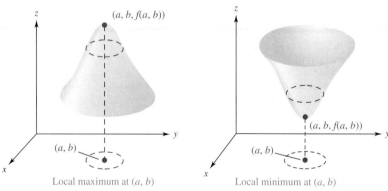

Local maximum at $(a, b)$

**Figure 14.19**

Local minimum at $(a, b)$

**Figure 14.20**

**NOTE** With functions of a single variable, we made a distinction between local extrema and absolute extrema. The methods for finding absolute extrema are quite involved for functions of two variables, so we will discuss only local extrema.

Tangent is horizontal.

Tangent is horizontal.

$(a, b)$

**Figure 14.21**

As suggested by Figure 14.21, at a local maximum the tangent line parallel to the $x$-axis (if it exists) has a slope of 0, as does the tangent line parallel to the $y$-axis (if it exists). (Notice the similarity to functions of one variable.) That is, if the function $z = f(x, y)$ has a local extremum at $(a, b)$, then $f_x(a, b) = 0$ and $f_y(a, b) = 0$, as stated in the following theorem.

## Location of Extrema

If a function $f$ has a local maximum or local minimum at the point $(a, b)$, and $f_x(a, b)$ and $f_y(a, b)$ both exist, then

$$f_x(a, b) = 0 \quad \text{and} \quad f_y(a, b) = 0.$$

Just as with functions of one variable, the fact that the slopes of the tangent lines are 0 is no guarantee that a local extremum has been located. For example, Figure 14.22 shows the graph of $z = f(x, y) = x^2 - y^2$. Both $f_x(0, 0) = 0$ and $f_y(0, 0) = 0$, and yet $(0, 0)$ leads to neither a local maximum nor a local minimum for the function. The point $(0, 0, 0)$ on the graph of this function is called a **saddle point;** it is a minimum when approached from one direction (in this case, the $x$-direction), but a maximum when approached from another direction (in this case, the $y$-direction). A saddle point is neither a maximum nor a minimum.

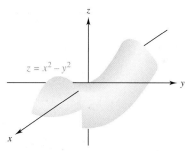

$z = x^2 - y^2$

**Figure 14.22**

## Critical Points

For a function $f$, the points $(a, b)$ such that $f_x(a, b) = 0$ and $f_y(a, b) = 0$ (or such that $f_x(a, b)$ or $f_y(a, b)$ does not exist) are called **critical points.**

**EXAMPLE 1** Find all critical points for

$$f(x, y) = 6x^2 + 6y^2 + 6xy + 36x - 54y - 5.$$

**Solution** Since the partial derivatives always exist, we must find all points $(a, b)$ such that $f_x(a, b) = 0$ and $f_y(a, b) = 0$. Here,

$$f_x = 12x + 6y + 36 \quad \text{and} \quad f_y = 12y + 6x - 54.$$

Set each of these two partial derivatives equal to 0:

$$12x + 6y + 36 = 0 \quad \text{and} \quad 12y + 6x - 54 = 0.$$

These two equations form a system of linear equations that we can rewrite as

$$12x + 6y = -36$$
$$6x + 12y = 54.$$

To solve this system by elimination, multiply the first equation by $-2$ and then add the equations:

$$-24x - 12y = 72$$
$$\underline{6x + 12y = 54}$$
$$-18x = 126$$
$$x = -7.$$

Substituting $x = -7$ in the first equation of the system, we have

$$12(-7) + 6y = -36$$
$$6y = 48$$
$$y = 8.$$

Therefore, $(-7, 8)$ is the solution of the system. Since this is the only solution, $(-7, 8)$ is the only critical point for the given function. By the previous theorem, if the function has a local extremum, it must occur at $(-7, 8)$. ✓₇

The results of the next theorem can be used to decide whether $(-7, 8)$ in Example 1 leads to a local maximum, a local minimum, or neither. The proof of this theorem is beyond the scope of this course.

✓ **Checkpoint 1**

Find all critical points for the given functions.

**(a)** $f(x, y) = 4x^2 + 3xy + 2y^2 + 7x - 6y - 6$

**(b)** $f(x, y) = e^{-2x} + xy$

---

## Test for Local Extrema

Suppose $f$ is a function such that $f_{xx}$, $f_{yy}$, and $f_{xy}$ all exist. Let $(a, b)$ be a critical point for which

$$f_x(a, b) = 0 \quad \text{and} \quad f_y(a, b) = 0.$$

Let $M$ be the number defined by

$$M = f_{xx}(a, b) \cdot f_{yy}(a, b) - [f_{xy}(a, b)]^2.$$

1. If $M > 0$ and $f_{xx}(a, b) < 0$, then $f$ has a **local maximum** at $(a, b)$.

2. If $M > 0$ and $f_{xx}(a, b) > 0$, then $f$ has a **local minimum** at $(a, b)$.

3. If $M < 0$, then $f$ has a **saddle point** at $(a, b)$.

4. If $M = 0$, the test gives **no information.**

---

The following chart summarizes the conclusions of the theorem:

|  | $f_{xx}(a, b) < 0$ | $f_{xx}(a, b) > 0$ |
|---|---|---|
| $M > 0$ | Local maximum | Local minimum |
| $M = 0$ | No information | |
| $M < 0$ | Saddle point | |

---

**EXAMPLE 2**    Example 1 showed that the only critical point of the function

$$f(x, y) = 6x^2 + 6y^2 + 6xy + 36x - 54y - 5$$

is $(-7, 8)$. Does $(-7, 8)$ lead to a local maximum, a local minimum, or neither?

**Solution** We can find out by using the test for local extrema. From Example 1,

$$f_x(-7, 8) = 0 \quad \text{and} \quad f_y(-7, 8) = 0.$$

Now find the various second-order partial derivatives used in finding $M$. From $f_x = 12x + 6y + 36$ and $f_y = 12y + 6x - 54$, we have

$$f_{xx} = 12, \quad f_{yy} = 12, \quad \text{and} \quad f_{xy} = 6.$$

(If these second-order partial derivatives had not all been constants, we would have had to evaluate them at the point $(-7, 8)$.) Now,

$$M = f_{xx}(-7, 8) \cdot f_{yy}(-7, 8) - [f_{xy}(-7, 8)]^2 = 12 \cdot 12 - 6^2 = 108.$$

Since $M > 0$ and $f_{xx}(-7, 8) = 12 > 0$, Part 2 of the theorem applies, showing that $f(x, y) = 6x^2 + 6y^2 + 6xy + 36x - 54y - 5$ has a local minimum at $(-7, 8)$. The value of this local minimum is $f(-7, 8) = -347$. ✔️ ②

✔️ **Checkpoint 2**

Find any local maxima or minima for the functions defined in Checkpoint 1.

---

**EXAMPLE 3**  Find all points where the function

$$f(x, y) = 9xy - x^3 - y^3 - 6$$

has any local maxima or local minima.

**Solution** First find any critical points. Here,

$$f_x = 9y - 3x^2 \quad \text{and} \quad f_y = 9x - 3y^2.$$

Set each of these partial derivatives equal to 0:

$$
\begin{array}{cc}
f_x = 0 & f_y = 0 \\
9y - 3x^2 = 0 & 9x - 3y^2 = 0 \\
9y = 3x^2 & 9x = 3y^2 \\
3y = x^2. & 3x = y^2.
\end{array}
$$

The last two equations provide some useful information. Since $3y = x^2$ and $x^2 \geq 0$, we must have $y \geq 0$. Similarly, since $3x = y^2$ and $y^2 \geq 0$, we must have $x \geq 0$. Consequently,

any critical point $(x, y)$ of $f$ must satisfy $x \geq 0$ and $y \geq 0$.

The substitution method can be used to solve the system of equations

$$3y = x^2$$
$$3x = y^2.$$

The first equation, $3y = x^2$, can be rewritten as $y = x^2/3$. Substitute this expression into the second equation to get

$$3x = y^2 = \left(\frac{x^2}{3}\right)^2$$

$$3x = \frac{x^4}{9}.$$

Solve this equation as follows:

$$
\begin{array}{ll}
27x = x^4 & \text{Multiply both sides by 9.} \\
x^4 - 27x = 0 & \\
x(x^3 - 27) = 0. & \text{Factor} \\
x = 0 \quad \text{or} \quad x^3 - 27 = 0 & \text{Set each factor equal to 0.} \\
\qquad\qquad\quad x^3 = 27 & \\
\qquad\qquad\quad x = 3 & \text{Take the cube root on each side.}
\end{array}
$$

Use these values of $x$, along with the equation $3x = y^2$, to find $y$:

| If $x = 0$, | If $x = 3$, |
|---|---|
| $3x = y^2$ | $3x = y^2$ |
| $3(0) = y^2$ | $3(3) = y^2$ |
| $0 = y^2$ | $9 = y^2$ |
| $0 = y.$ | $3 = y$ or $-3 = y.$ |

So the solutions for the system of equations are $(0, 0)$, $(3, 3)$ and $(3, -3)$. However, as we saw earlier, a critical point here must have *both* coordinates nonnegative. Hence, the only critical points of $f$ are $(0, 0)$ and $(3, 3)$.

To see if either critical point produces an extremum, use the test for local extrema. Here,

$$f_x = 9y - 3x^2 \quad \text{and} \quad f_y = 9x - 3y^2.$$

Hence,

$$f_{xx} = -6x, \quad f_{yy} = -6y, \quad \text{and} \quad f_{xy} = 9.$$

Test each of the critical points.

For $(0, 0)$,

$$f_{xx}(0, 0) = -6(0) = 0, \quad f_{yy}(0, 0) = -6(0) = 0, \quad \text{and} \quad f_{xy}(0, 0) = 9,$$

so that $M = 0 \cdot 0 - 9^2 = -81$. Since $M < 0$, there is a saddle point at $(0, 0)$.

For $(3, 3)$,

$$f_{xx}(3, 3) = -6(3) = -18, \quad f_{yy}(3, 3) = -6(3) = -18, \quad \text{and} \quad f_{xy}(3, 3) = 9,$$

so that $M = (-18)(-18) - 9^2 = 243$. Since $M > 0$ and $f_{xx}(3, 3) = -18 < 0$, there is a local maximum at $(3, 3)$. If you examine Figure 14.23 carefully, you will see both the local maximum at $(3, 3)$ and the saddle point at $(0, 0)$. ✔ 3

 **Checkpoint 3**

Find any local extrema for

$$f(x, y) = \frac{2\sqrt{2}}{3}x^3 - xy$$
$$+ \frac{1}{3}y^3 - 10.$$

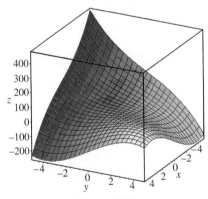

**Figure 14.23**

---

**EXAMPLE 4**  **Business** A company is developing a new soft drink. The cost in dollars to produce a batch of the drink is approximated by

$$C(x, y) = 2200 + 27x^3 - 72xy + 8y^2,$$

where $x$ is the number of kilograms of sugar per batch and $y$ is the number of grams of flavoring per batch.

**(a)** Find the amounts of sugar and flavoring that result in a local minimum cost for a batch.

**Solution** Start with the following partial derivatives:

$$C_x = 81x^2 - 72y \quad \text{and} \quad C_y = -72x + 16y.$$

Set each of these expressions equal to 0 and solve for $y$:

$$81x^2 - 72y = 0 \qquad\qquad -72x + 16y = 0$$
$$-72y = -81x^2 \qquad\qquad 16y = 72x$$
$$y = \frac{9}{8}x^2. \qquad\qquad y = \frac{9}{2}x.$$

From the equation on the left, $y \geq 0$. Since $(9/8)x^2$ and $(9/2)x$ both are equal to $y$, they are equal to each other. Set $(9/8)x^2$ and $(9/2)x$ equal and solve the resulting equation for $x$:

$$\frac{9}{8}x^2 = \frac{9}{2}x$$
$$9x^2 = 36x$$
$$9x^2 - 36x = 0$$
$$9x(x - 4) = 0$$
$$9x = 0 \quad \text{or} \quad x - 4 = 0$$
$$x = 0 \quad \text{or} \quad x = 4.$$

Substitute $x = 0$ and $x = 4$ into $y = (9/2)x$ to find the corresponding $y$ values:

$$y = \left(\frac{9}{2}\right)(0) = 0 \quad \text{and} \quad y = \left(\frac{9}{2}\right)(4) = 18.$$

✓ **Checkpoint 4**

Compute $M$ for the critical point $(0, 0)$ for the function in Example 4. What do you conclude?

So the critical points are $(0, 0)$ and $(4, 18)$. Verify that $(0, 0)$ does not produce a minimum. Now check to see if the critical point $(4, 18)$ leads to a local minimum. Note that

$$C_{xx} = 162x, \qquad C_{yy} = 16, \qquad \text{and} \qquad C_{xy} = -72.$$

At $(4, 18)$, $C_{xx}(4, 18) = 162(4) = 648$, $C_{yy}(4, 18) = 16$, and $C_{xy}(4, 18) = -72$. Therefore,

$$M = (648)(16) - (-72)^2 = 5184.$$

Since $M > 0$ and $C_{xx}(4, 18) > 0$, the cost at $(4, 18)$ is a minimum.

**(b)** What is the minimum cost?

**Solution**  To find the minimum cost, go back to the cost function and evaluate $C(4, 18)$:

$$C(x, y) = 2200 + 27x^3 - 72xy + 8y^2$$
$$C(4, 18) = 2200 + 27(4)^3 - 72(4)(18) + 8(18)^2 = 1336.$$

The minimum cost for a batch is $1336.

⊞ **TECHNOLOGY TIP** Spreadsheet programs have built-in solvers that are able to optimize many functions of two or more variables. The Excel Solver is located on the Data tab in the Analysis group. (If the Solver command or the Analysis group is not available, you need to load the Solver Add-in program—see Excel Help for details.) The Solver requires that cells be identified ahead of time for each variable in the problem. It also requires that another cell be identified where the function to be optimized, in terms of the variable cells, is placed. In Example 4, for instance, we could identify cells A1 and B1 to represent the variables $x$ and $y$, respectively. The Solver requires that we place a guess for the answer in these cells. Thus, for our initial value or guess, we place the number 5 in each of these cells. The function must be placed in a cell in terms of cells A1 and B1. If we choose cell A3 to represent the function, then we would type "=2200 + 27*A1^3 − 72*A1*B1 + 8*B1^2" in cell A3.

We now click on the Data tab and choose Solver from the Analysis group to find a solution that either maximizes or minimizes the value of cell A3, depending on which option we choose. Figure 14.24 illustrates the Solver box in Excel 2016 (Microsoft Corporation Excel © 2016) and the items placed in it.

**Figure 14.24**

To obtain a solution, select the minimize option, and then click on Solve. When Excel shows a dialog box with "Keep Solver Solution" as the default, click OK. The rounded solution $x = 4$ and $y = 18$ is located in cells A1 and B1, respectively. The minimum cost $C(4, 18) = 1336$ is located in cell A3.

*Note:* One must be careful when using the Excel Solver, because it will not find a "maximum or minimum" of a function if the initial guess is the exact place in which a saddle point occurs. For example, in the preceding problem, if our initial guess was (0, 0), the Solver would have returned the value of (0, 0) as the place where a minimum occurs. But (0, 0) is a saddle point. Thus, it is always a good idea to run the Solver for two different initial values and compare the solutions.

# 14.3 Exercises

*Find all critical points of the given functions. (See Example 1.)*

**1.** $f(x, y) = 6x^2 + 6y^2 + 6xy + 36x - 5$

**2.** $f(x, y) = 50 + 8x - 5y + x^2 + y^2 + xy$

**3.** $f(x, y) = 9xy - x^3 - y^3 - 6$

**4.** $f(x, y) = 2200 + 27x^3 + 72xy + 8y^2$

*Find all points where the given functions have any local extrema. Give the values of any local extrema. Identify any saddle points. (See Examples 1–3.)*

**5.** $f(x, y) = 2x^2 + 4xy + 6y^2 - 8x - 10$

**6.** $f(x, y) = x^2 + xy + y^2 - 6x - 3$

**7.** $f(x, y) = x^2 - xy + y^2 + 2x + 2y + 6$

**8.** $f(x, y) = 2x^2 + 3xy + 2y^2 - 5x - 5y$

**9.** $f(x, y) = 3xy + 6y - 5x$

**10.** $f(x, y) = 5xy - 7x^2 - y^2 + 3x - 6y - 4$

**11.** $f(x, y) = 4xy - 10x^2 - 4y^2 + 8x + 8y + 9$

**12.** $f(x, y) = 4y^2 + 2xy + 6x + 4y - 8$

**13.** $f(x, y) = x^2 + xy - 2x - 2y + 2$

**14.** $f(x, y) = x^2 + xy + y^2 - 3x - 5$

**15.** $f(x, y) = x^2 - y^2 - 2x + 4y - 7$

**16.** $f(x, y) = 4x + 2y - x^2 + xy - y^2 + 3$

**17.** $f(x, y) = 2x^3 + 2y^2 - 12xy + 15$

**18.** $f(x, y) = 2x^2 + 4y^3 - 24xy + 18$

**19.** $f(x, y) = x^2 + 4y^3 - 6xy - 1$

**20.** $f(x, y) = 2y^3 + 5x^2 + 60xy + 25$

**21.** $f(x, y) = e^{xy}$

**22.** $f(x, y) = e^{x^2 - y^2}$

*Figures (a)–(f) show the graphs of the functions defined in Exercises 23–28, respectively. Find all local extrema for each function, and then match the equation with its graph. (See Example 3.)*

**23.** $z = -3xy + x^3 - y^3 + \dfrac{1}{8}$

**24.** $z = \dfrac{3}{2}y - \dfrac{1}{2}y^3 - x^2y + \dfrac{1}{16}$

**25.** $z = y^4 - 2y^2 + x^2 - \dfrac{17}{16}$

**26.** $z = -2x^3 - 3y^4 + 6xy^2 + \dfrac{1}{16}$

**27.** $z = -x^4 + y^4 + 2x^2 - 2y^2 + \dfrac{1}{16}$

**28.** $z = -y^4 + 4xy - 2x^2 + \dfrac{1}{16}$

**(a)**

**(b)**

**(c)**

**(d)**

**(e)**

**(f)**
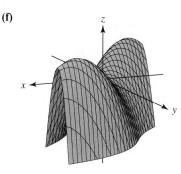

**Business** *In the following exercises, find the local maximum or local minimum, as requested. (See Example 4.)*

**29.** The labor cost in dollars for manufacturing a digital camera can be approximated by

$$L(x, y) = \dfrac{3}{2}x^2 + y^2 - 2x - 2y - 2xy + 68,$$

where $x$ is the number of hours required by a skilled craftsperson and $y$ is the number of hours required by a semiskilled person. Find values of $x$ and $y$ that minimize the labor charge. Find the minimum labor charge.

**30.** Suppose that the profit of a certain firm is approximated by

$$P(x, y) = 1000 + 24x - x^2 + 80y - y^2,$$

where $x$ is the cost of a unit of labor and $y$ is the cost of a unit of goods. Find values of $x$ and $y$ that maximize the profit. Find the maximum profit.

**31.** The most significant expenses for a particular house cleaning service are transportation and labor costs. Daily profit (in dollars) for the business is approximated by

$$P(x, y) = 600 - 2x^3 + 6xy - y^2,$$

where $x$ is the cost of a gallon of gasoline and $y$ is the hourly wage paid to housekeeping employees. Find the maximum profit and the costs of gasoline and labor that produce the maximum profit.

**32.** The total profit from one acre of a certain crop depends on the amounts spent on fertilizer, $x$, and hybrid seed, $y$, according to the model

$$P(x, y) = -x^2 + 3xy + 160x - 5y^2 + 200y + 2{,}600{,}000.$$

Find values of $x$ and $y$ that lead to maximum profit. Find the maximum profit.

**33.** The total cost required to produce $x$ units of electrical tape and $y$ units of packing tape is given by

$$C(x, y) = 2x^2 + 3y^2 - 2xy + 2x - 126y + 3800.$$

Find the number of units of each kind of tape that should be produced so that the total cost is a minimum. Find the minimum total cost.

**34.** The total revenue, in thousands of dollars, from the sale of $x$ spas and $y$ solar heaters is approximated by

$$R(x, y) = 12 + 74x + 85y - 3x^2 - 5y^2 - 5xy.$$

Find the number of each that should be sold to produce maximum revenue. Find the maximum revenue.

**35.** A rectangular closed box is to be built at minimum cost to hold 27 cubic meters. Since the cost will depend on the surface area, find the dimensions that will minimize the surface area of the box.

**36.** Find the dimensions that will minimize the surface area (and hence the cost) of a rectangular fish aquarium, open on top, with a volume of 32 cubic feet.

**37.** The U.S. Postal Service requires that any box sent by Priority Mail have a length plus girth (distance around) totaling no more than 108 inches. (See Example 7 in Section 12.3 for a special case.) Find the dimensions of the box with maximum volume that can be sent by Priority Mail.

**38.** A box that is sent through the U.S. mail by parcel post must have a length plus girth of no more than 130 inches. Find the dimensions of the box with maximum volume that can be sent by parcel post. (See Exercise 37.)

**39.** Find the dimensions of the circular tube with maximum volume that can be sent by U.S. Priority Mail. (See Exercise 37.)

**40.** A piece of carry-on luggage on most airlines (such as American, Delta, and Continental) must satisfy the condition that the sum of its length, width, and height is no more than 45 inches. If no other restrictions apply, find the dimensions of the piece of carry-on luggage with maximum volume that can be brought on these airlines.*

**41.** The monthly revenue, in hundreds of dollars, from the production of $x$ thousand tons of grade A iron ore and $y$ thousand tons of grade B iron ore is given by

$$R(x, y) = 2xy + 2y + 12,$$

and the corresponding cost, in hundreds of dollars, is given by

$$C(x, y) = 2x^2 + y^2.$$

Find the amount of each grade of ore that will produce maximum profit.

**42.** The revenue and cost (in thousands of dollars) from the manufacture of $x$ thousand specialty cupcakes and $y$ thousand pies are, respectively,

$$R(x, y) = 12xy + 5 - 2x^2 \quad \text{and} \quad C(x, y) = x^2 + 2y^3.$$

How many units of each product will produce the maximum profit? What is the maximum profit?

**43.** The profit (in thousands of dollars) from producing $x$ tons of steel and $y$ tons of aluminum can be approximated by

$$P(x, y) = 36xy - x^3 - 8y^3.$$

Find the amounts of steel and aluminum that maximize the profit, and find the value of the maximum profit.

**44.** **Health** Research has shown that the total change in color, $E$ (measured in the form of energy as kJ/mol), when blanched potato strips are fried can be estimated by the function

$$E(t, T) = 436.16 - 10.57t - 5.46T - .02t^2 + .02T^2 + .08Tt,$$

where $T$ is the temperature (in degrees Celsius) and $t$ is the frying time (in minutes).[†]

**(a)** Find the value of $E$ prior to cooking. Assume that $T = 0$.

**(b)** Estimate the total change in color of a potato strip that has been cooked for 10 minutes at 180°C.

**(c)** Determine the critical point of this function and whether a maximum, minimum, or saddle point occurs at that point.

✓**Checkpoint Answers**

**1. (a)** $(-2, 3)$   **(b)** $(0, 2)$

**2. (a)** Local minimum at $(-2, 3)$

   **(b)** Neither a local minimum nor a local maximum at $(0, 2)$

**3.** Local minimum at $\left(\dfrac{1}{2}, \dfrac{\sqrt{2}}{2}\right)$; saddle point at $(0, 0)$

**4.** $M = -5184$, so there is a saddle point at $(0, 0)$.

---

*Depending on the plane and the airline, additional restrictions may apply (for instance, that the luggage fit under the seat or in the overhead compartment).

[†] Hindra, F., and Oon-Doo Baik, "Kinetics of Quality Changes During Food Frying," *Critical Reviews in Food Science and Nutrition*, Vol. 46, 2006, pp. 239–358.

# 14.4 Lagrange Multipliers

It is often necessary to solve optimization problems (such as finding the maximum profit or the minimum cost). We have seen two techniques for doing this with functions of two variables:

1. Change the given problem into an equivalent one-variable problem by making a suitable substitution (as in Example 7 of Section 12.3).

2. Use the test for local extrema (as in Example 4 of Section 14.3).

The first technique works only when algebra can be used to express one variable in terms of the other. The second technique does not work when constraints are present (as is often the case in the real world). Finally, neither method applies to functions of three or more variables.

In this section, we present Lagrange's method for optimizing a function of two or more variables, subject to a constraint.[*] The method is used for problems of the form

Find the optimum value of the function $f(x, y)$,

subject to the constraint $g(x, y) = 0$;

that is, find the point $(x, y)$ satisfying the constraint $g(x, y) = 0$, at which $f(x, y)$ has its optimum value. The optimum value may be either a maximum value or a minimum value, depending on the problem.

Here is a typical example:

Find the minimum value of $f(x, y) = x^2 + y^2$,

subject to the constraint $x + y - 4 = 0$.

In this case, $g(x, y) = x + y - 4$ and the optimum value is the minimum value. Graphically, the point $(x, y, z)$ satisfying $x + y - 4 = 0$ at which $z = f(x, y)$ is smallest is the lowest point on the intersection of the graph of $z = x^2 + y^2$ and the plane $x + y - 4 = 0$, as shown in Figure 14.25.[†] Note that this point is *not* the lowest point on the graph of $f(x, y) = x^2 + y^2$. The minimum value of $f(x, y)$ (with no constraint) occurs at $(0, 0, 0)$, which does not lie on the plane $x + y - 4 = 0$ given by the constraint.

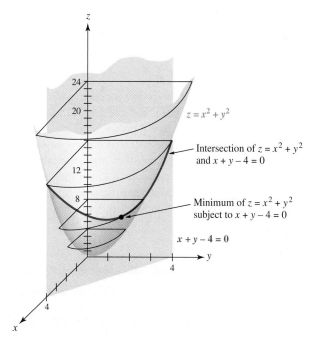

**Figure 14.25**

Lagrange's method provides an algebraic way to solve problems such as this one. The key idea is to introduce an additional variable called the **Lagrange multiplier** and denoted by the lowercase Greek letter $\lambda$ (lambda) and to use the *Lagrange function*

$$F(x, y, \lambda) = f(x, y) - \lambda \cdot g(x, y).$$

In our example,

$$F(x, y, \lambda) = (x^2 + y^2) - \lambda(x + y - 4)$$
$$= x^2 + y^2 - \lambda x - \lambda y + 4\lambda.$$

---

[*]Named for the French mathematician Joseph Louis Lagrange (1736–1813). The method is stated here for functions of two variables, but with obvious modifications, it is valid for any number of variables.

[†]These graphs were found in Example 7 and Exercise 25 of Section 14.1.

Lagrange's method (whose proof is beyond the scope of this text) uses the partial derivatives of the function $F$ to find the point satisfying the constraint $g(x, y) = 0$ at which $f(x, y)$ has its optimum value (if such a point exists).

## Lagrange's Method

To optimize the function $f(x, y)$ subject to the constraint $g(x, y) = 0$,

**Step 1** Form the Lagrange function
$$F(x, y, \lambda) = f(x, y) - \lambda \cdot g(x, y).$$

**Step 2** Find the partial derivatives of $F$: $F_x(x, y, \lambda)$, $F_y(x, y, \lambda)$, and $F_\lambda(x, y, \lambda)$.

**Step 3** Solve the following system of equations:
$$F_x(x, y, \lambda) = 0$$
$$F_y(x, y, \lambda) = 0$$
$$F_\lambda(x, y, \lambda) = 0.$$

If the original problem has a solution, it will occur at a point $(x, y)$ for which there exists a value of $\lambda$ such that $(x, y, \lambda)$ is a solution of the system in Step 3.

The points found in Step 3 of Lagrange's method provide a list of *possibilities.* If the optimum value exists, it must occur at a point on the list, but the list may contain points that do not produce an optimum value.* Thus, to apply Lagrange's method, we must know that the required maximum or minimum actually exists. In many applied problems, the given conditions make it clear that an optimum value must exist. In other cases, graphical information can sometimes be used to make this determination, as Example 1 illustrates. Unless stated otherwise, you may assume that the required optimum exists in each example and exercise of this section.

**EXAMPLE 1** Use Lagrange's method to find the minimum value of
$$f(x, y) = x^2 + y^2, \text{ subject to the constraint } x + y = 4.$$

**Solution** First, rewrite the constraint in the form $g(x, y) = 0$:
$$x + y - 4 = 0.$$

Then follow the steps in the preceding box.

**Step 1** As we saw previously, the Lagrange function is
$$F(x, y, \lambda) = f(x, y) - \lambda \cdot g(x, y).$$
$$= x^2 + y^2 - \lambda x - \lambda y + 4\lambda.$$

**Step 2** Find the partial derivatives of $F$:
$$F_x(x, y, \lambda) = 2x - \lambda$$
$$F_y(x, y, \lambda) = 2y - \lambda$$
$$F_\lambda(x, y, \lambda) = -x - y + 4.$$

---

*This is similar to the situation with functions of one variable. Every local extremum must occur at a critical number (where the derivative is 0 or does not exist), but not every critical number produces a local extremum.

**Step 3** Set each partial derivative equal to 0 and solve the resulting system:

$$2x - \lambda = 0$$
$$2y - \lambda = 0$$
$$-x - y + 4 = 0. \tag{1}$$

Since this is a system of *linear* equations in $x$, $y$, and $\lambda$, it could be solved by the matrix techniques of Section 6.2. However, we shall use a different technique, one that can be used even when the equations of the system are not all linear. Begin by solving the first two equations for $\lambda$:

$$2x - \lambda = 0 \qquad 2y - \lambda = 0$$
$$\lambda = 2x \qquad\qquad \lambda = 2y.$$

Set the two expressions for $\lambda$ equal to obtain

$$2x = 2y$$
$$x = y.$$

Now make the substitution $y = x$ in the third equation of the system **(1)**:

$$-x - y + 4 = 0$$
$$-x - x + 4 = 0$$
$$-2x = -4$$
$$x = 2.$$

Since $y = x$ and $\lambda = 2x$, we see that the only solution of system **(1)** is

$$x = 2, \qquad y = 2, \qquad \lambda = 2 \cdot 2 = 4.$$

Graphical considerations show that the original problem has a solution (see Figure 14.25 and the discussion preceding it), so we conclude that the minimum value of $f(x, y) = x^2 + y^2$, subject to the constraint $x + y = 4$, occurs when $x = 2$ and $y = 2$. (We do not need the value of $\lambda$.) The minimum value is

$$f(2, 2) = 2^2 + 2^2 = 8.$$

**EXAMPLE 2** A builder plans to construct a three-story building with a rectangular floor plan. The cost of the building, which depends on the dimensions of the floor plan, is given by

$$\text{Cost of building} = xy + 30x + 20y + 474{,}000,$$

where $x$ and $y$ are, respectively, the length and width of the rectangular floors. What length and width should be used if the building is to cost $500,000 and have maximum area on each floor?

**Solution** The area of each floor is given by $A(x, y) = xy$. Since the building is to cost $500,000, we have the following constraint:

$$\text{Cost of building} = xy + 30x + 20y + 474{,}000 = 500{,}000$$
$$\underbrace{xy + 30x + 20y - 26{,}000}_{g(x,\,y)} = 0.$$

So we must find the maximum value of $A(x, y)$, subject to $g(x, y) = 0$.

**Step 1** The Lagrange function is

$$F(x, y, \lambda) = A(x, y) - \lambda \cdot g(x, y).$$
$$= xy - \lambda(xy + 30x + 20y - 26{,}000)$$
$$= xy - \lambda xy - 30\lambda x - 20\lambda y + 26{,}000\lambda.$$

**Step 2**    The partial derivatives of $F$ are

$$F_x(x, y, \lambda) = y - \lambda y - 30\lambda$$
$$F_y(x, y, \lambda) = x - \lambda x - 20\lambda$$
$$F_\lambda(x, y, \lambda) = -xy - 30x - 20y + 26{,}000.$$

**Step 3**    Set each partial derivative equal to 0 and solve the system of equations

$$y - \lambda y - 30\lambda = 0$$
$$x - \lambda x - 20\lambda = 0$$
$$-xy - 30x - 20y + 26{,}000 = 0 \qquad \textbf{(2)}$$

as follows. Solve the first two equations for $\lambda$:

$$
\begin{aligned}
y - \lambda y - 30\lambda &= 0 & \qquad x - \lambda x - 20\lambda &= 0 \\
y &= \lambda y + 30\lambda & x &= \lambda x + 20\lambda \\
y &= (y + 30)\lambda & x &= (x + 20)\lambda \\
\lambda &= \frac{y}{y + 30} & \lambda &= \frac{x}{x + 20}.
\end{aligned}
$$

In the equation on the left in the line above, it was permissible to divide both sides by $y + 30$ because $y$ is a positive length and hence $y + 30$ is nonzero. Similarly, $x + 20$ is nonzero. Next, set the two expressions for $\lambda$ equal and solve the resulting equation:

$$\frac{y}{y + 30} = \frac{x}{x + 20}$$

$$y(x + 20) = x(y + 30) \qquad \text{Multiply both sides by } (x + 20)(y + 30).$$
$$xy + 20y = xy + 30x$$
$$20y = 30x \qquad\qquad \text{Subtract } xy \text{ from both sides.}$$
$$y = 1.5x. \qquad\qquad \text{Divide both sides by 20.}$$

Make the substitution $y = 1.5x$ in the third equation of the system **(2)**:

$$-xy - 30x - 20y + 26{,}000 = 0$$
$$-x(1.5x) - 30x - 20(1.5x) + 26{,}000 = 0$$
$$-1.5x^2 - 60x + 26{,}000 = 0.$$

By the quadratic formula,

$$x = \frac{-(-60) \pm \sqrt{(-60)^2 - 4(-1.5)(26{,}000)}}{2(-1.5)} \approx \begin{cases} 113.17 \\ -153.17 \end{cases}.$$

Since $x$ is a length, the negative value does not apply. Since $y = 1.5x$,

$$y \approx 1.5(113.17) \approx 169.75.$$

We now have sufficient information. (There is no need to solve for $\lambda$.) Floors that measure 113.17 ft $\times$ 169.75 ft will result in the largest area per floor, which is

$$A(113.17, 169.75) = (113.17)(169.75) \approx 19{,}211 \text{ sq ft.}$$

📄 **NOTE** In Examples 1 and 2, we solved the system by first solving each equation with a $\lambda$ in it for $\lambda$. Then we set the expressions for $\lambda$ equal and solved for one of the original variables. This is often a good approach, since we are not usually interested in the value of $\lambda$.

With a constrained function of two variables (as in the preceding examples), the Lagrange function has three variables, and a system of three equations in three variables must be solved. Next we consider constrained functions of three variables, in which case the Lagrange function has four variables and a system of four equations in four variables must be solved.

**EXAMPLE 3**   Find three positive numbers $x$, $y$, and $z$ whose sum is 50 and such that $xyz^2$ is as large as possible.

**Solution**   We must find the maximum value of $f(x, y, z) = xyz^2$, subject to the constraint $x + y + z = 50$. Begin by writing the constraint as

$$\underbrace{x + y + z - 50}_{g(x, y, z)} = 0.$$

**Step 1**   The Lagrange function is

$$\begin{aligned} F(x, y, z, \lambda) &= f(x, y, z) - \lambda \cdot g(x, y, z) \\ &= xyz^2 - \lambda(x + y + z - 50) \\ &= xyz^2 - \lambda x - \lambda y - \lambda z + 50\lambda. \end{aligned}$$

**Step 2**   The partial derivatives of $F$ are

$$\begin{aligned} F_x(x, y, z, \lambda) &= yz^2 - \lambda \\ F_y(x, y, z, \lambda) &= xz^2 - \lambda \\ F_z(x, y, z, \lambda) &= 2xyz - \lambda \\ F_\lambda(x, y, z, \lambda) &= -x - y - z + 50. \end{aligned}$$

**Step 3**   Set each partial derivative equal to 0 and solve the system of equations

$$\begin{aligned} yz^2 - \lambda &= 0 \\ xz^2 - \lambda &= 0 \\ 2xyz - \lambda &= 0 \\ -x - y - z + 50 &= 0. \end{aligned} \tag{3}$$

The first three equations are easily solved for $\lambda$:

$$\lambda = yz^2, \qquad \lambda = xz^2, \qquad \lambda = 2xyz.$$

Setting the expressions for $\lambda$ equal produces three equations:

$$yz^2 = xz^2, \qquad yz^2 = 2xyz, \qquad xz^2 = 2xyz.$$

Next, we solve the first of these equations. In doing so, we can divide both sides of the equation by $z^2$ because $z$ is nonzero ($x$, $y$, and $z$ were given to be positive).

$$\begin{aligned} yz^2 &= xz^2 && \text{First equation} \\ \frac{yz^2}{z^2} &= \frac{xz^2}{z^2} && \text{Divide both sides by } z^2. \\ y &= x \end{aligned}$$

Now solve the second equation by dividing both sides by the nonzero quantity $yz$.

$$\begin{aligned} yz^2 &= 2xyz && \text{Second equation} \\ \frac{yz^2}{yz} &= \frac{2xyz}{yz} && \text{Divide both sides by } yz. \\ z &= 2x. \end{aligned}$$

(This information is sufficient, so we need not solve the third equation.) Substitute these values into the last equation of the system **(3)**:

$$\begin{aligned} -x - y - z + 50 &= 0 \\ -x - x - 2x + 50 &= 0 && \text{Substitute } y = x \text{ and } z = 2x. \\ -4x &= -50 \\ x &= 12.5. && \text{Divide both sides by } -4. \end{aligned}$$

Since $y = x$ and $z = 2x$, the three numbers in the original problem are

$$x = 12.5, \qquad y = 12.5, \qquad \text{and} \qquad z = 2(12.5) = 25,$$

and the maximum value of $f(x, y, z) = xyz^2$ is $(12.5)(12.5)(25^2) = 97{,}656.25.$

**Checkpoint 1**

Find three positive numbers $x$, $y$, and $z$ whose sum is 15 and whose product $xyz$ is as large as possible.

**EXAMPLE 4** Find the dimensions of the closed rectangular box of maximum volume that can be constructed from 6 square feet of material.

**Solution** Let the dimensions of the box be $x$, $y$, and $z$, as shown in Figure 14.26. The volume of the box is given by

$$f(x, y, z) = xyz.$$

**Figure 14.26**

By considering the surface area of each side of the box in Figure 14.26, we obtain a constraint:

$$\text{Right end} + \text{left end} + \text{front} + \text{back} + \text{top} + \text{bottom} = 6$$
$$xy \quad + \quad xy \quad + \quad xz + \quad xz + yz + \quad yz \quad = 6$$
$$2xy + 2xz + 2yz = 6$$
$$xy + xz + yz = 3. \qquad \text{\small Divide both sides by 2.}$$

So we must find the maximum value of $f(x, y, z) = xyz$, subject to the constraint $xy + xz + yz - 3 = 0$.

**Step 1** The Lagrange function is

$$\begin{aligned} F(x, y, z, \lambda) &= f(x, y, z) - \lambda \cdot g(x, y, z). \\ &= xyz - \lambda(xy + xz + yz - 3) \\ &= xyz - \lambda xy - \lambda xz - \lambda yz + 3\lambda. \end{aligned}$$

**Step 2** The partial derivatives of $F$ are

$$\begin{aligned} F_x(x, y, z, \lambda) &= yz - \lambda y - \lambda z \\ F_y(x, y, z, \lambda) &= xz - \lambda x - \lambda z \\ F_z(x, y, z, \lambda) &= xy - \lambda x - \lambda y \\ F_\lambda(x, y, z, \lambda) &= -xy - xz - yz + 3. \end{aligned}$$

**Step 3** Set each partial derivative equal to 0 and solve the system of equations

$$\begin{aligned} yz - \lambda y - \lambda z &= 0 \\ xz - \lambda x - \lambda z &= 0 \\ xy - \lambda x - \lambda y &= 0 \\ -xy - xz - yz + 3 &= 0. \end{aligned} \qquad (4)$$

Begin by solving each of the first three equations for $\lambda$ (notice in the last step that, because the dimensions of the box are positive, the quantities $y + x$, $x + z$, and $x + y$ are nonzero):

| | | | |
|---|---|---|---|
| $yz - \lambda y - \lambda z = 0$ | $xz - \lambda x - \lambda z = 0$ | $xy - \lambda x - \lambda y = 0$ | First three equations |
| $-\lambda y - \lambda z = -yz$ | $-\lambda x - \lambda z = -xz$ | $-\lambda x - \lambda y = -xy$ | Isolate terms with $\lambda$ on the left. |
| $\lambda(y + z) = yz$ | $\lambda(x + z) = xz$ | $\lambda(x + y) = xy$ | Multiply both sides by $-1$ and factor out $\lambda$. |
| $\lambda = \dfrac{yz}{y + z}$ | $\lambda = \dfrac{xz}{x + z}$ | $\lambda = \dfrac{xy}{x + y}$. | Divide by nonzero quantities. |

Set these expressions for $\lambda$ equal, and simplify as follows (notice in the second and last steps that we divide by nonzero quantities):

$$\frac{yz}{y+z} = \frac{xz}{x+z} \qquad \text{and} \qquad \frac{xz}{x+z} = \frac{xy}{x+y}$$

$$\frac{y}{y+z} = \frac{x}{x+z} \qquad\qquad \frac{z}{x+z} = \frac{y}{x+y}$$

$$y(x+z) = x(y+z) \qquad\qquad z(x+y) = y(x+z)$$

$$xy + yz = xy + xz \qquad\qquad zx + zy = yx + yz$$

$$yz = xz \qquad\qquad\qquad zx = yx$$

$$y = x \qquad\qquad\qquad z = y.$$

(Setting the first and third expressions equal gives no additional information.) Thus, $x = y = z$. From the last equation in system **(4)**, with $x = y$ and $z = y$,

$$-xy - xz - yz + 3 = 0$$
$$-y^2 - y^2 - y^2 + 3 = 0$$
$$-3y^2 = -3$$
$$y^2 = 1$$
$$y = \pm 1.$$

The negative solution is not applicable, so the solution of the system of equations is $x = 1, y = 1, z = 1$. In other words, the box with maximum volume under the constraint is a cube that measures 1 ft on each side. ✓₂

✓ **Checkpoint 2**

Find the dimensions of the open rectangular box (which does not have a top) of maximum volume that can be constructed from 12 square feet of material.

**TECHNOLOGY TIP** Finding extrema of a constrained function of one or more variables can be done with a spreadsheet. In addition to the requirements stated in the Technology Tip at the end of Section 14.3, the constraint must be input into the Excel Solver. To do this, we need to input the variable part of the constraint into a designated cell. In Example 4, for instance, suppose that A5 is the designated cell. Then, in cell A5, we would type "=A1*B1 + A1*C1 + B1*C1".

We now click on the Data tab and choose Solver from the Analysis group to find a solution that either maximizes or minimizes the value of cell A3, depending on which option we choose. Figure 14.27 illustrates the Solver box in Excel 2016 (Microsoft Corporation Excel © 2016) and the items placed in it.

**Figure 14.27**

To obtain a solution, click on Solve. When Excel shows a dialog box with "Keep Solver Solution" as the default, click OK. The solution $x = 1$, $y = 1$, $z = 1$ is located in cells A1, B1, and C1, respectively. The maximum volume $f(1, 1, 1) = 1$ is located in cell A3.

*Note:* One must be careful when using the Solver because the solution may depend on the initial value. Thus, it is always a good idea to run the Solver for two different initial values and compare the solutions.

# 14.4 Exercises

*Find the maximum or minimum in Exercises 1–12. Assume all variables are positive. (See Examples 1 and 3.)*

1. Maximum of $f(x, y) = 2xy$, subject to $x + y = 20$

2. Maximum of $f(x, y) = 4xy + 8$, subject to $x + y = 16$

3. Maximum of $f(x, y) = xy^2$, subject to $x + 4y = 15$

4. Maximum of $f(x, y) = 2xy^2$, subject to $x + y = 6$

5. Minimum of $f(x, y) = x^2 + 2y^2 - xy$, subject to $x + y = 8$

6. Minimum of $f(x, y) = 3x^2 + 4y^2 - xy - 2$, subject to $2x + y = 21$

7. Maximum of $f(x, y) = x^2 - 10y^2$, subject to $x - y = 18$

8. Maximum of $f(x, y) = 12xy - x^2 - 3y^2$, subject to $x + y = 16$

9. Maximum of $f(x, y) = e^{-x^2 - y^2}$, subject to $x + y = 2$

10. Minimum of $f(x, y) = e^{-xy}$, subject to $x + y = 10$

11. Maximum of $f(x, y, z) = xyz^2$, subject to $x + y + z = 6$

12. Minimum of $f(x, y, z) = xy + 2xz + 2yz$, subject to $xyz = 32$

13. Find positive numbers $x$ and $y$ such that $x + y = 21$ and $2xy^2$ is maximized.

14. Find positive numbers $x$ and $y$ such that $x + y = 24$ and $x^2y + 4$ is maximized.

15. Find three positive numbers whose sum is 102 and whose product is a maximum.

16. Find three positive numbers such that their sum is 75 and the sum of their squares is a maximum.

17. Explain Lagrange's method in your own words.

18. Explain the difference between the two methods we used in Sections 14.3 and 14.4 to solve extrema problems.

19. Why is it unnecessary to find the value of $\lambda$ when using the method explained in this section?

20. Show that the three equations in Step 3 of the box on page 844 are equivalent to the three equations

$$f_x(x, y) = \lambda g_x(x, y),$$
$$f_y(x, y) = \lambda g_y(x, y),$$
$$g(x, y) = 0.$$

**Business** *Solve the given problems. (See Examples 2 and 4.)*

21. The profit from the sale of $x$ units of radiators for automobiles and $y$ units of radiators for generators is given by

$$P(x, y) = -x^2 - y^2 + 4x + 8y.$$

Find values of $x$ and $y$ that lead to a maximum profit if the firm must produce a total of 6 units of radiators.

22. The total cost to produce $x$ small first aid kits and $y$ large ones is given by

$$C(x, y) = 2x^2 + 6y^2 + 4xy + 10.$$

If a total of 10 kits must be made, how should production be allocated so that total cost is minimized?

23. A manufacturing firm estimates that its total production of automobile batteries, in thousands of units, is

$$f(x, y) = 3x^{1/3}y^{2/3},$$

where $x$ is the number of units of labor and $y$ is the number of units of capital utilized. Labor costs are \$80 per unit, and capital costs are \$150 per unit. How many units each of labor and capital will maximize production if the firm can spend \$40,000 for these costs?

24. An outdoor outfitter estimates that its total production of 60-meter rock climbing ropes is

$$f(x, y) = 4x^{3/4}y^{1/4},$$

where $x$ is the number of units of labor and $y$ is the number of units of capital utilized. If \$40,000 is available for labor and capital, and if the firm's costs are \$100 per unit of labor and \$200 per unit of capital, how many units of each will give the maximum production?

25. Because of terrain difficulties, two sides of a fence can be built for \$6 per ft, while the other two sides cost \$4 per ft. (See the accompanying sketch.) Find the field of maximum area that can be enclosed for \$1200.

**26.** To enclose a yard, a fence is built against a large building, so that fencing material is used only on three sides. Material for the ends costs $8 per ft; material for the side opposite the building costs $6 per ft. Find the dimensions of the yard of maximum area that can be enclosed for $1200.

**27.** A farmer has 200 m of fencing. Find the dimensions of the rectangular field of maximum area that can be enclosed by this amount of fencing.

**28.** Find the area of the largest rectangular field that can be enclosed with 600 m of fencing. Assume that no fencing is needed along one side of the field that runs along a straight stretch of river.

**Business** *Work these exercises.*

**29.** A cylindrical can is to be made that will hold $250\pi$ cubic inches of candy. Find the dimensions of the can with minimum surface area.

**30.** An ordinary 12-oz soda pop can holds about 25 cubic inches. Find the dimensions of a can with minimum surface area. Measure a can and see how close its dimensions are to the results you found.

**31.** A rectangular box with no top is to be built from 500 square meters of material. Find the dimensions of such a box that will enclose the maximum volume.

**32.** A 1-lb cracker box has a volume of 185 cubic inches. The end of the box is square. Find the dimensions of such a box that has minimum surface area.

**33.** Find the dimensions that will minimize the surface area (and hence the cost) of a rectangular fish aquarium, open on top, with a volume of 32 cubic feet.

**34.** A rectangular closed box is to be built at minimum cost to hold 27 cubic meters. Since the cost will depend on the surface area, find the dimensions that will minimize the surface area of the box.

**35.** A company needs to construct a box with an open top that will be used to transport 400 cubic yards of material, in several trips, from one place to another. Two of the sides and the bottom of the box can be made of a free, lightweight material, but only 4 square yards of the material is available. Because of the nature of the material to be transported, the two ends of the box must be made from a heavyweight material that costs $20 per square yard. Each trip costs 10 cents.

**(a)** Let $x$, $y$, and $z$ respectively denote the length, width, and height of the box. If we want to use all of the free material, show that the total cost in dollars is given by the function

$$f(x, y, z) = \frac{40}{xyz} + 40yz,$$

subject to the constraint $2xz + xy = 4$.

**(b)** Use the Solver feature on a spreadsheet to find the dimensions of the box that minimize the transportation cost, subject to the given constraint.

**36.** **Social Science** The probability that the majority of a three-person jury will convict a guilty person is given by the formula

$$P(r, s, t) = rs(1 - t) + (1 - r)st + r(1 - s)t + rst$$

subject to the constraint

$$r + s + t = \alpha,$$

where $r$, $s$, and $t$ represent each of the three jury members' probability of reaching a guilty verdict and $\alpha$ is some fixed constant that is generally less than or equal to the number of jurors.

**(a)** Form the Lagrange function.

**(b)** Find the values of $r$, $s$, and $t$ that maximize the probability of convicting a guilty person when $\alpha = .75$.

**(c)** Find the values of $r$, $s$, and $t$ that maximize the probability of convicting a guilty person when $\alpha = 3$.

✓**Checkpoint Answers**

**1.** $x = y = z = 5$

**2.** 2 ft $\times$ 2 ft $\times$ 1 ft

# CHAPTER 14 Summary and Review

## Key Terms and Symbols

**14.1** $z = f(x, y)$, function of two variables
multivariate function
$xy$-plane
ordered triple
first octant
surface
trace
level curves
paraboloid

production function
Cobb–Douglas production function

**14.2** $f_x$ or $\dfrac{\partial f}{\partial x}$, partial derivative of $f$ with respect to $x$
$\dfrac{\partial}{\partial x}\left(\dfrac{\partial z}{\partial x}\right)$ or $\dfrac{\partial^2 z}{\partial x^2}$ or $f_{xx}$,
second-order partial derivative of $f$, partial

derivative of $\partial z/\partial x$ (or $f_x$) with respect to $x$
$\dfrac{\partial}{\partial y}\left(\dfrac{\partial z}{\partial x}\right)$ or $\dfrac{\partial^2 z}{\partial y \partial x}$ or $f_{xy}$,
second-order partial derivative of $f$, partial derivative of $\partial z/\partial x$ (or $f_x$) with respect to $y$

**14.3** saddle point
local maximum

local minimum
critical point

**14.4** constraint
Lagrange multipliers
Lagrange function
Lagrange's method

# Chapter 14 Key Concepts

**Graphs of Multivariate Functions**

The graph of $ax + by + cz = d$ is a **plane.**

The graph of $z = f(x, y)$ is a **surface** in three-dimensional space.

The graph of $z = ax^2 + by^2$ is a **paraboloid.**

**Partial Derivatives**

The **partial derivative of $f(x, y)$ with respect to $x$** is the derivative of $f$ found by treating $x$ as a variable and $y$ as a constant.

The **partial derivative of $f(x, y)$ with respect to $y$** is the derivative of $f$ found by treating $y$ as a variable and $x$ as a constant.

**Second-Order Partial Derivatives**

For a function $z = f(x, y)$, if all indicated partial derivatives exist, then

$$\frac{\partial}{\partial x}\left(\frac{\partial z}{\partial x}\right) = \frac{\partial^2 z}{\partial x^2} = f_{xx} \qquad \frac{\partial}{\partial y}\left(\frac{\partial z}{\partial y}\right) = \frac{\partial^2 z}{\partial y^2} = f_{yy}$$

$$\frac{\partial}{\partial y}\left(\frac{\partial z}{\partial x}\right) = \frac{\partial^2 z}{\partial y \partial x} = f_{xy} \qquad \frac{\partial}{\partial x}\left(\frac{\partial z}{\partial y}\right) = \frac{\partial^2 z}{\partial x \partial y} = f_{yx}.$$

**Local Extrema**

Let $(a, b)$ be in the domain of a function $f$.

1.  $f$ has a **local maximum** at $(a, b)$ if there is a circular region in the $xy$-plane with $(a, b)$ in its interior such that

$$f(a, b) \geq f(x, y)$$

for all points $(x, y)$ in the circular region.

2.  $f$ has a **local minimum** at $(a, b)$ if there is a circular region in the $xy$-plane with $(a, b)$ in its interior such that

$$f(a, b) \leq f(x, y)$$

for all points $(x, y)$ in the circular region.

If $f(a, b)$ is a local extremum, and $f_x(a, b)$ and $f_y(a, b)$ both exist, then $f_x(a, b) = 0$ and $f_y(a, b) = 0$.

**Test for Local Extrema**

Suppose $f(x, y)$ is a function such that $f_{xx}, f_{yy},$ and $f_{xy}$ all exist. Let $(a, b)$ be a critical point for which $f_x(a, b) = 0$ and $f_y(a, b) = 0$. Let $M = f_{xx}(a, b) \cdot f_{yy}(a, b) - [f_{xy}(a, b)]^2$.

If $M > 0$ and $f_{xx}(a, b) < 0$, then $f$ has a **local maximum** at $(a, b)$.

If $M > 0$ and $f_{xx}(a, b) > 0$, then $f$ has a **local minimum** at $(a, b)$.

If $M < 0$, then $f$ has a **saddle point** at $(a, b)$.

If $M = 0$, the test gives **no information.**

**Lagrange's Method**

To optimize the function $f(x, y)$ subject to the constraint $g(x, y) = 0$, form the Lagrange function $F(x, y, \lambda) = f(x, y) - \lambda \cdot g(x, y)$ and find its partial derivatives $F_x(x, y, \lambda)$, $F_y(x, y, \lambda)$, and $F_\lambda(x, y, \lambda)$. Then solve the system of equations

$$F_x(x, y, \lambda) = 0, \qquad F_y(x, y, \lambda) = 0, \qquad F_\lambda(x, y, \lambda) = 0.$$

If the original problem has a solution, it will occur at a point $(x, y)$ for which there exists a value of $\lambda$ such that $(x, y, \lambda)$ is a solution of the preceding system.

# Chapter 14 Review Exercises

*Find $f(-3, 1)$ and $f(5, -2)$ for each of the given functions.*

1.  $f(x, y) = 6y^2 - 5xy + 2x$

2.  $f(x, y) = -3x + 2x^2y^2 + 5y$

3.  $f(x, y) = \dfrac{2x - 4}{x + 3y}$

4.  $f(x, y) = \dfrac{12y - 6x}{x + 3y}$

5.  $f(x, y) = 7x\sqrt{y}$

6.  $f(x, y) = x\sqrt{x^2 + y^2}$

7.  Describe the graph of $2x + y + 4z = 12$.

8.  Describe the graph of $y = 2$ on a three-dimensional grid.

*Graph the first-octant portion of each plane.*

9.  $x + 2y + 4z = 4$

10.  $3x + 2y = 6$

11.  $4x + 5y = 20$

12.  $x = 6$

*Find the indicated partial derivatives.*

**13.** Let $z = f(x, y) = -2x^2 + 5xy + y^2$. Find the following:

(a) $\dfrac{\partial z}{\partial x}$; (b) $\dfrac{\partial z}{\partial y}(-1, 4)$; (c) $f_{xy}(2, -1)$.

**14.** Let $z = f(x, y) = \dfrac{2y + x^2}{3y - x}$. Find the following:

(a) $\dfrac{\partial z}{\partial y}$; (b) $\dfrac{\partial z}{\partial x}(0, 2)$; (c) $f_{yy}(-1, 0)$.

**15.** Explain the difference between $\dfrac{\partial z}{\partial x}$ and $\dfrac{\partial z}{\partial y}$.

**16.** Explain the difference between $f_{xx}$ and $f_{xy}$.

*Find $f_x$ and $f_y$.*

**17.** $f(x, y) = 3y - 7x^2y^3$ **18.** $f(x, y) = 4x^3y + 12xy^3$

**19.** $f(x, y) = \sqrt{3x^2 + 2y^2}$ **20.** $f(x, y) = \sqrt{1 - 4xy^3}$

**21.** $f(x, y) = \dfrac{-x^2y}{5x + 8y^3}$ **22.** $f(x, y) = \dfrac{3x^2 - 2y^2}{x^2 + 4y^2}$

**23.** $f(x, y) = x^3e^{3y}$ **24.** $f(x, y) = (y + 1)^2e^{2x+y}$

**25.** $f(x, y) = \ln|x^2 - 4y^3|$ **26.** $f(x, y) = \ln|1 + x^3y^2|$

*Find $f_{xx}$ and $f_{xy}$.*

**27.** $f(x, y) = 4x^3y^2 - 8xy$ **28.** $f(x, y) = -6xy^3 + 2x^2y$

**29.** $f(x, y) = \dfrac{2x + y}{x - 2y}$ **30.** $f(x, y) = \dfrac{3x + y}{x - 1}$

**31.** $f(x, y) = x^2e^y$ **32.** $f(x, y) = ye^{x^2}$

**33.** $f(x, y) = \ln(2 - x^2y)$ **34.** $f(x, y) = \ln(1 + 3xy^2)$

**35. Business** The cost (in dollars) of repainting a car is given by

$$C(x, y) = 4x^2 + 5y^2 - 4xy + 50,$$

where $x$ is the number of hours of labor and $y$ is the number of gallons of paint used. Find the cost in the given cases.

(a) 10 hours of labor and 8 gallons of paint are needed.

(b) 12 hours of labor and 10 gallons of paint are needed.

(c) 14 hours of labor and 14 gallons of paint are needed.

**36. Business** Write a function in terms of $L$, $W$, and $H$ that gives the total material required to build the closed shipping box shown in the accompanying figure.

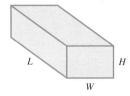

**37. Business** The manufacturing cost (in dollars) for a USB jump drive is given by

$$C(x, y) = \frac{x}{4} + y + \frac{xy}{4} + 1,$$

where $x$ is its memory capacity (in gigabytes) and $y$ is the number of hours of labor required. Find each of the following.

(a) $\dfrac{\partial C}{\partial x}(32, 1)$ (b) $\dfrac{\partial C}{\partial y}(32, 1)$

**38. Business** The production function for a certain company is given by

$$z = x^{.7}y^{.3},$$

where $x$ is the amount of labor (in appropriate units) and $y$ is the amount of capital (in thousands of dollars). Find the marginal productivity of

(a) labor; (b) capital.

*Find all points where these functions have any local extrema. Find any saddle points.*

**39.** $z = x^2 + 2y^2 - 4y$

**40.** $z = x^2 + y^2 + 9x - 8y + 1$

**41.** $f(x, y) = x^2 + 5xy - 10x + 3y^2 - 12y$

**42.** $z = x^3 - 8y^2 + 6xy + 4$

**43.** $z = x^3 + y^2 + 2xy - 4x - 3y - 2$

**44.** $f(x, y) = 7x^2 + y^2 - 3x + 6y - 5xy$

**45.** $f(x, y) = e^{x^2+y^2}$

**46.** $f(x, y) = e^{xy}$

**47. Business** The total cost in dollars to manufacture $x$ solar cells and $y$ solar collectors is

$$C(x, y) = x^2 + 5y^2 + 4xy - 70x - 164y + 1800.$$

(a) Find values of $x$ and $y$ that produce the minimum total cost.

(b) Find the minimum total cost.

**48. Business** The total profit from 1 acre of a certain crop depends on the amount spent on fertilizer, $x$, and on hybrid seed, $y$, according to the model

$$P(x, y) = .01(-x^2 + 3xy + 160x - 5y^2 + 200y + 2600).$$

The budget for fertilizer and seed is limited to $280.

(a) Use the budget constraint to express one variable in terms of the other. Then substitute into the profit function to get a function with one variable. Use the method shown in Chapter 12 to find the amounts spent on fertilizer and seed that will maximize profit. What is the maximum profit per acre? (*Hint:* Throughout this problem, you may ignore the coefficient of .01 until you need to find the maximum profit.)

**(b)** Ignore the budget constraint and find the amounts spent on fertilizer and seed that will maximize profit, using the method shown in Section 14.3.

**(c)** Discuss any relationships between these methods.

**49.** Minimize $f(x, y) = x^2 + y^2$, subject to the constraint $x = y + 2$.

**50.** Find positive numbers $x$ and $y$ that maximize the value of $f(x, y) = x^2y$, subject to the constraint $x + y = 4$.

**51.** Find positive numbers $x$ and $y$ whose sum is 80 such that $x^2y$ is maximized.

**52.** Find positive numbers $x$ and $y$ whose sum is 50 such that $xy^2$ is maximized.

**53.** Notice in the previous two exercises that we specified that $x$ and $y$ must be positive numbers. Does a maximum exist without this requirement? Explain why or why not.

**54.** Use Lagrange's method to find the amounts spent on fertilizer and seeds that will maximize profit in Exercise 48.

# Case Study 14  Global Warming and the Method of Least Squares

The least-squares regression line for modeling data that is approximately linear was introduced in Section 2.3. The discussion there explained why least squares are used to measure how well a line fits the data. In this case study, we show how the theory of extrema of functions of several variables is used to construct the least-squares regression line.

Global warming is a term used to describe the phenomenon in recent years that the atmospheric temperature of Earth is slowly increasing over time. The National Oceanic and Atmospheric Administration (NOAA) has compiled a great deal of historical global temperature data. Data from recent years is compiled by satellite, but a variety of methods, including the analysis of temperature-sensitive phenomena such as tree-ring density, allows for temperature estimates from centuries past. The data is often presented in terms of temperature anomalies—that is, as deviations from some norm.

For example, NOAA reports the temperature anomalies in the table below. The baseline temperature against which the anomalies were determined is the twentieth-century average global temperature over land, an average computed to be 8.5°C. Adding 8.5 to each anomaly gives the actual temperature.

**Global Land Surface Average Temperature**

| Year | Temperature Anomaly (°C) | Temperature (°C) |
|------|--------------------------|------------------|
| 1900 | −0.14 | 8.36 |
| 1960 | −0.03 | 8.47 |
| 2016 | 1.46 | 9.96 |

Figure 1 shows a plot of the temperature data given in the table.

**Figure 1**   Global Land Surface Average Temperature

If the three points lie on a straight line, the rate at which the temperature is changing can be determined by the slope of that line. The points clearly are not on the same line, but can be approximated by a straight line, as in the figure. The purpose of the least-squares method is to find the best such approximation.

Figure 2 shows the vertical distance from each point to the approximating line.

**Figure 2**   Deviations from Approximating Line

If the first point is $(x_1, y_1)$ and the equation of the line is $y = mx + b$, then the first vertical distance is $y_1 - (mx_1 + b) = y_1 - mx_1 - b$. In the least-squares method, the values of $m$ and $b$ are determined so that the sum of the squares of these distances is minimized. For three points $(x_1, y_1)$, $(x_2, y_2)$, and $(x_3, y_3)$, the following function of two variables must be minimized,

$$
\begin{aligned}
f(m, b) = {}& (y_1 - mx_1 - b)^2 \\
& + (y_2 - mx_2 - b)^2 \qquad \text{(1)} \\
& + (y_3 - mx_3 - b)^2.
\end{aligned}
$$

Using the points (1900, 8.36), (1960, 8.47) and (2016, 9.96), we find that the equation above becomes

$$f(m, b) = 3b^2 + 11{,}515{,}856m^2 + 11{,}752bm - 53.58b$$
$$- 105{,}129m + 240.832.$$

The partial derivatives of $f$ with respect to $b$ and $m$ are

$$\frac{\partial f}{\partial b} = 6b + 117{,}52m - 53.58 \quad \text{and}$$

$$\frac{\partial f}{\partial m} = 11{,}752b + 23{,}031{,}712m - 105{,}129.$$

The resulting critical point is approximately $m = .01$ and $b = -17.81$. The line of best fit for these data is $y = .01x - 17.81$, which is the line drawn in Figure 1. According to these data, the global land temperature increases at a rate of approximately $.01°C$ per year.

It is not fair to say that this calculation is evidence of global warming, as only three pieces of data were used. The data set was kept small to illustrate the least-squares method. This method can be generalized to use any amount of data and to fit more complicated functions, including exponential functions.

## Exercises

1. Verify that the critical point $m = .01$ and $b = -17.81$ corresponds to a minimum of the function $f(m, b)$.

2. By differentiating equation (1) with respect to $m$ and $b$ without substituting the specific values for points $(x_1, y_1)$, $(x_2, y_2)$, and $(x_3, y_3)$, the formulas

$$m = \frac{3(x_1y_1 + x_2y_2 + x_3y_3) - (x_1 + x_2 + x_3)(y_1 + y_2 + y_3)}{3(x_1^2 + x_2^2 + x_3^2) - (x_1 + x_2 + x_3)^2}$$

and

$$b = \frac{y_1 + y_2 + y_3}{3} - m\left(\frac{x_1 + x_2 + x_3}{3}\right)$$

can be derived. Show that these formulas give the same values for $m$ and $b$ given earlier.

3. The formulas in Exercise 2 generalize to any number of data points. For example, for four data points, the formulas for $m$ and $b$ respectively become

$$m = \frac{4(x_1y_1 + x_2y_2 + x_3y_3 + x_4y_4)}{4(x_1^2 + x_2^2 + x_3^2 + x_4^2) - (x_1 + x_2 + x_3 + x_4)^2}$$
$$- \frac{(x_1 + x_2 + x_3 + x_4)(y_1 + y_2 + y_3 + y_4)}{4(x_1^2 + x_2^2 + x_3^2 + x_4^2) - (x_1 + x_2 + x_3 + x_4)^2}$$

and

$$b = \frac{y_1 + y_2 + y_3 + y_4}{4} - m\left(\frac{x_1 + x_2 + x_3 + x_4}{4}\right).$$

Add the temperature data point (2010, 9.60) to the preceding three and find the least-squares straight-line fit. Plot the points and the line, and discuss the value of $m$.

# Extended Project

Investigate the most recent data available for global land surface average temperature. Add these data to those given in the table above and then use the method of least squares to find the straight line that best fits these data. According to this linear model, what is the predicted global land surface average temperature for next year? Write a report summarizing the method of least squares, the given data and the most recent data you found, the resulting line of best fit, the interpretation of the slope of the line, and your prediction for the future.

# Areas under the Normal Curve

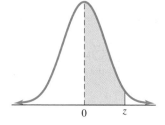

The column under $A$ gives the proportion of the area under the entire curve that is between $z = 0$ and a positive value of $z$.

| z | A | z | A | z | A | z | A |
|---|---|---|---|---|---|---|---|
| .00 | .0000 | .48 | .1844 | .96 | .3315 | 1.44 | .4251 |
| .01 | .0040 | .49 | .1879 | .97 | .3340 | 1.45 | .4265 |
| .02 | .0080 | .50 | .1915 | .98 | .3365 | 1.46 | .4279 |
| .03 | .0120 | .51 | .1950 | .99 | .3389 | 1.47 | .4292 |
| .04 | .0160 | .52 | .1985 | 1.00 | .3413 | 1.48 | .4306 |
| .05 | .0199 | .53 | .2019 | 1.01 | .3438 | 1.49 | .4319 |
| .06 | .0239 | .54 | .2054 | 1.02 | .3461 | 1.50 | .4332 |
| .07 | .0279 | .55 | .2088 | 1.03 | .3485 | 1.51 | .4345 |
| .08 | .0319 | .56 | .2123 | 1.04 | .3508 | 1.52 | .4357 |
| .09 | .0359 | .57 | .2157 | 1.05 | .3531 | 1.53 | .4370 |
| .10 | .0398 | .58 | .2190 | 1.06 | .3554 | 1.54 | .4382 |
| .11 | .0438 | .59 | .2224 | 1.07 | .3577 | 1.55 | .4394 |
| .12 | .0478 | .60 | .2258 | 1.08 | .3599 | 1.56 | .4406 |
| .13 | .0517 | .61 | .2291 | 1.09 | .3621 | 1.57 | .4418 |
| .14 | .0557 | .62 | .2324 | 1.10 | .3643 | 1.58 | .4430 |
| .15 | .0596 | .63 | .2357 | 1.11 | .3665 | 1.59 | .4441 |
| .16 | .0636 | .64 | .2389 | 1.12 | .3686 | 1.60 | .4452 |
| .17 | .0675 | .65 | .2422 | 1.13 | .3708 | 1.61 | .4463 |
| .18 | .0714 | .66 | .2454 | 1.14 | .3729 | 1.62 | .4474 |
| .19 | .0754 | .67 | .2486 | 1.15 | .3749 | 1.63 | .4485 |
| .20 | .0793 | .68 | .2518 | 1.16 | .3770 | 1.64 | .4495 |
| .21 | .0832 | .69 | .2549 | 1.17 | .3790 | 1.65 | .4505 |
| .22 | .0871 | .70 | .2580 | 1.18 | .3810 | 1.66 | .4515 |
| .23 | .0910 | .71 | .2612 | 1.19 | .3830 | 1.67 | .4525 |
| .24 | .0948 | .72 | .2642 | 1.20 | .3849 | 1.68 | .4535 |
| .25 | .0987 | .73 | .2673 | 1.21 | .3869 | 1.69 | .4545 |
| .26 | .1026 | .74 | .2704 | 1.22 | .3888 | 1.70 | .4554 |
| .27 | .1064 | .75 | .2734 | 1.23 | .3907 | 1.71 | .4564 |
| .28 | .1103 | .76 | .2764 | 1.24 | .3925 | 1.72 | .4573 |
| .29 | .1141 | .77 | .2794 | 1.25 | .3944 | 1.73 | .4582 |
| .30 | .1179 | .78 | .2823 | 1.26 | .3962 | 1.74 | .4591 |
| .31 | .1217 | .79 | .2852 | 1.27 | .3980 | 1.75 | .4599 |
| .32 | .1255 | .80 | .2881 | 1.28 | .3997 | 1.76 | .4608 |
| .33 | .1293 | .81 | .2910 | 1.29 | .4015 | 1.77 | .4616 |
| .34 | .1331 | .82 | .2939 | 1.30 | .4032 | 1.78 | .4625 |
| .35 | .1368 | .83 | .2967 | 1.31 | .4049 | 1.79 | .4633 |
| .36 | .1406 | .84 | .2996 | 1.32 | .4066 | 1.80 | .4641 |
| .37 | .1443 | .85 | .3023 | 1.33 | .4082 | 1.81 | .4649 |
| .38 | .1480 | .86 | .3051 | 1.34 | .4099 | 1.82 | .4656 |
| .39 | .1517 | .87 | .3079 | 1.35 | .4115 | 1.83 | .4664 |
| .40 | .1554 | .88 | .3106 | 1.36 | .4131 | 1.84 | .4671 |
| .41 | .1591 | .89 | .3133 | 1.37 | .4147 | 1.85 | .4678 |
| .42 | .1628 | .90 | .3159 | 1.38 | .4162 | 1.86 | .4686 |
| .43 | .1664 | .91 | .3186 | 1.39 | .4177 | 1.87 | .4693 |
| .44 | .1700 | .92 | .3212 | 1.40 | .4192 | 1.88 | .4700 |
| .45 | .1736 | .93 | .3238 | 1.41 | .4207 | 1.89 | .4706 |
| .46 | .1772 | .94 | .3264 | 1.42 | .4222 | 1.90 | .4713 |
| .47 | .1808 | .95 | .3289 | 1.43 | .4236 | 1.91 | .4719 |

(*continued*)

| z | A | z | A | z | A | z | A |
|------|-------|------|-------|------|-------|------|-------|
| 1.92 | .4726 | 2.42 | .4922 | 2.92 | .4983 | 3.42 | .4997 |
| 1.93 | .4732 | 2.43 | .4925 | 2.93 | .4983 | 3.43 | .4997 |
| 1.94 | .4738 | 2.44 | .4927 | 2.94 | .4984 | 3.44 | .4997 |
| 1.95 | .4744 | 2.45 | .4929 | 2.95 | .4984 | 3.45 | .4997 |
| 1.96 | .4750 | 2.46 | .4931 | 2.96 | .4985 | 3.46 | .4997 |
| 1.97 | .4756 | 2.47 | .4932 | 2.97 | .4985 | 3.47 | .4997 |
| 1.98 | .4762 | 2.48 | .4934 | 2.98 | .4986 | 3.48 | .4998 |
| 1.99 | .4767 | 2.49 | .4936 | 2.99 | .4986 | 3.49 | .4998 |
| 2.00 | .4773 | 2.50 | .4938 | 3.00 | .4987 | 3.50 | .4998 |
| 2.01 | .4778 | 2.51 | .4940 | 3.01 | .4987 | 3.51 | .4998 |
| 2.02 | .4783 | 2.52 | .4941 | 3.02 | .4987 | 3.52 | .4998 |
| 2.03 | .4788 | 2.53 | .4943 | 3.03 | .4988 | 3.53 | .4998 |
| 2.04 | .4793 | 2.54 | .4945 | 3.04 | .4988 | 3.54 | .4998 |
| 2.05 | .4798 | 2.55 | .4946 | 3.05 | .4989 | 3.55 | .4998 |
| 2.06 | .4803 | 2.56 | .4948 | 3.06 | .4989 | 3.56 | .4998 |
| 2.07 | .4808 | 2.57 | .4949 | 3.07 | .4989 | 3.57 | .4998 |
| 2.08 | .4812 | 2.58 | .4951 | 3.08 | .4990 | 3.58 | .4998 |
| 2.09 | .4817 | 2.59 | .4952 | 3.09 | .4990 | 3.59 | .4998 |
| 2.10 | .4821 | 2.60 | .4953 | 3.10 | .4990 | 3.60 | .4998 |
| 2.11 | .4826 | 2.61 | .4955 | 3.11 | .4991 | 3.61 | .4999 |
| 2.12 | .4830 | 2.62 | .4956 | 3.12 | .4991 | 3.62 | .4999 |
| 2.13 | .4834 | 2.63 | .4957 | 3.13 | .4991 | 3.63 | .4999 |
| 2.14 | .4838 | 2.64 | .4959 | 3.14 | .4992 | 3.64 | .4999 |
| 2.15 | .4842 | 2.65 | .4960 | 3.15 | .4992 | 3.65 | .4999 |
| 2.16 | .4846 | 2.66 | .4961 | 3.16 | .4992 | 3.66 | .4999 |
| 2.17 | .4850 | 2.67 | .4962 | 3.17 | .4992 | 3.67 | .4999 |
| 2.18 | .4854 | 2.68 | .4963 | 3.18 | .4993 | 3.68 | .4999 |
| 2.19 | .4857 | 2.69 | .4964 | 3.19 | .4993 | 3.69 | .4999 |
| 2.20 | .4861 | 2.70 | .4965 | 3.20 | .4993 | 3.70 | .4999 |
| 2.21 | .4865 | 2.71 | .4966 | 3.21 | .4993 | 3.71 | .4999 |
| 2.22 | .4868 | 2.72 | .4967 | 3.22 | .4994 | 3.72 | .4999 |
| 2.23 | .4871 | 2.73 | .4968 | 3.23 | .4994 | 3.73 | .4999 |
| 2.24 | .4875 | 2.74 | .4969 | 3.24 | .4994 | 3.74 | .4999 |
| 2.25 | .4878 | 2.75 | .4970 | 3.25 | .4994 | 3.75 | .4999 |
| 2.26 | .4881 | 2.76 | .4971 | 3.26 | .4994 | 3.76 | .4999 |
| 2.27 | .4884 | 2.77 | .4972 | 3.27 | .4995 | 3.77 | .4999 |
| 2.28 | .4887 | 2.78 | .4973 | 3.28 | .4995 | 3.78 | .4999 |
| 2.29 | .4890 | 2.79 | .4974 | 3.29 | .4995 | 3.79 | .4999 |
| 2.30 | .4893 | 2.80 | .4974 | 3.30 | .4995 | 3.80 | .4999 |
| 2.31 | .4896 | 2.81 | .4975 | 3.31 | .4995 | 3.81 | .4999 |
| 2.32 | .4898 | 2.82 | .4976 | 3.32 | .4996 | 3.82 | .4999 |
| 2.33 | .4901 | 2.83 | .4977 | 3.33 | .4996 | 3.83 | .4999 |
| 2.34 | .4904 | 2.84 | .4977 | 3.34 | .4996 | 3.84 | .4999 |
| 2.35 | .4906 | 2.85 | .4978 | 3.35 | .4996 | 3.85 | .4999 |
| 2.36 | .4909 | 2.86 | .4979 | 3.36 | .4996 | 3.86 | .4999 |
| 2.37 | .4911 | 2.87 | .4980 | 3.37 | .4996 | 3.87 | .5000 |
| 2.38 | .4913 | 2.88 | .4980 | 3.38 | .4996 | 3.88 | .5000 |
| 2.39 | .4916 | 2.89 | .4981 | 3.39 | .4997 | 3.89 | .5000 |
| 2.40 | .4918 | 2.90 | .4981 | 3.40 | .4997 | | |
| 2.41 | .4920 | 2.91 | .4982 | 3.41 | .4997 | | |

# Solutions to Prerequisite Skills Test and Calculus Readiness Test

## Prerequisite Skills Test

1. We must find a common denominator before subtracting. Since 6 can be written as a fraction as $\dfrac{6}{1}$, then 2 can be the common denominator.

$$\frac{5}{2} - 6 = \frac{5}{2} - \frac{6}{1}$$

$$= \frac{5}{2} - \frac{6(2)}{2} = \frac{5}{2} - \frac{12}{2}$$

$$= \frac{5 - 12}{2} = \frac{-7}{2}.$$

2. Division of fraction implies that we multiply by the reciprocal. In this case, we multiply $\dfrac{1}{2}$ by the reciprocal of $\dfrac{2}{5}$, which is $\dfrac{5}{2}$. We obtain

$$\frac{1}{2} \div \frac{2}{5} = \frac{1}{2} \cdot \frac{5}{2} = \frac{1 \cdot 5}{2 \cdot 2} = \frac{5}{4}.$$

3. We rewrite 3 as a fraction as $\dfrac{3}{1}$ and then multiply by the reciprocal.

$$\frac{1}{3} \div \frac{3}{1} = \frac{1}{3} \cdot \frac{1}{3} = \frac{1 \cdot 1}{3 \cdot 3} = \frac{1}{9}.$$

4. We perform the operation inside the parenthesis first. We have

$$2 \div 6 = \frac{2}{6} = \frac{1}{3}.$$

Thus,

$$7 + 2 - 3(2 \div 6) = 7 + 2 - 3\left(\frac{1}{3}\right).$$

We then perform the multiplication and see that

$$3\left(\frac{1}{3}\right) = \frac{3}{1} \cdot \frac{1}{3} = \frac{3(1)}{1(3)} = \frac{3}{3} = 1.$$

So we now have

$$7 + 2 - 3\left(\frac{1}{3}\right) = 7 + 2 - 1.$$

Performing addition and subtraction from left to right, we obtain $9 - 1 = 8$.

5. We begin by performing the multiplication on the numerator. We obtain

$$\frac{2 \times 3 + 12}{1 + 5} - 1 = \frac{6 + 12}{1 + 5} - 1.$$

Next, we perform addition in the numerator and addition in the denominator.

$$\frac{6+12}{1+5} - 1 = \frac{18}{6} - 1.$$

Next, we perform the division and realize that $\frac{18}{6} = 3$, to obtain $3 - 1 = 2$.

6. False. We add the numerator values together to obtain

$$\frac{4+3}{3} = \frac{7}{3} \ne 5.$$

7. False. We obtain a common denominator of 35 to obtain

$$\frac{5}{7} + \frac{7}{5} = \frac{5(5)}{35} + \frac{7(7)}{35} = \frac{25}{35} + \frac{49}{35} = \frac{25+49}{35} = \frac{74}{35} \ne 1.$$

8. False. We write 1 as a fraction with the same denominator as $\frac{3}{5}$, namely $1 = \frac{5}{5}$. Thus, we have

$$\frac{3}{5} + 1 = \frac{3}{5} + \frac{5}{5} = \frac{8}{5} \ne \frac{6}{5}.$$

9. Since $n$ represents the number of shoes that Alicia has, we can add 2 to $n$ to obtain the number of shoes for Manuel. Thus we are looking for the expression $n + 2$ for the number of shoes for Manuel.

10. We can say $x$ represents David's age. Since Selina's age is 6 more than David, Selina's age is $x + 6$. The two ages together then are

$$x + (x + 6) = 42.$$

If we combine like terms, we have

$$2x + 6 = 42.$$

We can solve for $x$ by adding $-6$ to each side to obtain

$$2x + 6 - 6 = 42 - 6$$
$$2x = 36.$$

We then divide each side by 2 to obtain

$$x = 18.$$

Thus, David's age is 18 and Selina's age is $18 + 6 = 24$.

11. If the sweater is reduced by 20%, then its sale price is $100 - 20 = 80\%$ of the original price. To obtain the new price we first convert 80% to a decimal to obtain .80 and then multiply by the price of the sweater to obtain

$$.80(\$72) = \$57.60$$

12. When we plot points, the first coordinate is the value for the $x$-axis and the second value is for the $y$-axis. For points A, B, and C, we obtain

**13.** Similarly, we plot points D, E, and F to obtain

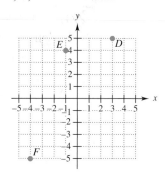

**14. (a)** 4.27659 has a value of 2 in the tenths place, and the value after that is 7, so we round up to obtain 4.3 to the nearest tenth.

**(b)** 245.984 has a 5 in the units place and a 9 in the tenths place, so we round up the units place to obtain 246 to the nearest unit.

**15. (a)** 16.38572 has an 8 in the hundredths place and a 5 in the thousandths place, so we round up to obtain 16.39 to the nearest hundredth.

**(b)** 1,763,304.42 has a 3 in the thousands place and the following digit is also a 3, so we do not round up and obtain 1,763,000 rounded to the nearest thousand.

**16. (a)** Writing 34 million dollars as a numerical value is $34 \cdot 1{,}000{,}000 = \$34{,}000{,}000$.

**(b)** Writing 2.2 thousand dollars as a numerical value is $2.2 \cdot 1000 = \$2{,}200$.

**17. (a)** 17 hundred thousand as a numerical value is $17 \cdot 100{,}000 = 1{,}700{,}000$.

**(b)** Three and a quarter billion as a numerical value is $3.25 \cdot 1{,}000{,}000{,}000 = 3{,}250{,}000{,}000$.

**18.** There is no solution to $\dfrac{5}{0}$ because a 0 in the denominator indicates the fraction is undefined.

**19.** Yes, it is. If a car is traveling 60 miles in one hour, it is traveling 60 miles in 60 minutes, which equates to 1 mile per minute.

**20.** $-9$ is greater than $-900$.

# Calculus Readiness Test

**1.** $(7x^4 - 8x^3 + 2x^2 - 3) - (9x^4 - 2x^2 + 5x + 6)$

$= (7x^4 - 9x^4) + (-8x^3 - 0) + (2x^2 - (-2x^2)) + (0 - 5x) + (-3 - 6)$

$= -2x^4 - 8x^3 + 4x^2 - 5x - 9$

(See Section 1.2, Example 7.)

**2.** $\dfrac{-6}{3x + 4} + \dfrac{2}{x} = \dfrac{-6}{3x + 4} \cdot \dfrac{x}{x} + \dfrac{2}{x} \cdot \dfrac{3x + 4}{3x + 4} = \dfrac{-6x}{x(3x + 4)} + \dfrac{2(3x + 4)}{x(3x + 4)}$

$= \dfrac{-6x}{3x^2 + 4x} + \dfrac{6x + 8}{3x^2 + 4x} = \dfrac{8}{3x^2 + 4x}$

(See Section 1.4, Example 5.)

**3.** $(3x + 7)(5x - 8) = (3x)(5x) + (3x)(-8) + (7)(5x) + (7)(-8)$

$= 15x^2 - 24x + 35x - 56 = 15x^2 + 11x - 56$

(See Section 1.2, Example 10.)

**4.** $(7x + 3)(6x^2 + x - 8) = 7x(6x^2 + x - 8) + 3(6x^2 + x - 8)$

$= (42x^3 + 7x^2 - 56x) + (18x^2 + 3x - 24) = 42x^3 + 25x^2 - 53x - 24$

(See Section 1.2, Example 8.)

**5.** $\dfrac{(x^5 y^{-5})^3}{(x^2\sqrt{y})^6} = \dfrac{(x^5)^3(y^{-5})^3}{(x^2)^6(y^{1/2})^6} = \dfrac{x^{15}y^{-15}}{x^{12}y^3} = \dfrac{x^{15}}{x^{12}y^3 y^{15}} = \dfrac{x^{15-12}}{y^{3+15}} = \dfrac{x^3}{y^{18}}$

(See Section 1.5, Example 8.)

**6.** $(8x - 9)^2 = (8x - 9)(8x - 9) = 64x^2 - 72x - 72x + 81 = 64x^2 - 144x + 81$

(See Section 1.2, Example 10.)

**7.** $9x^2 - 49 = (3x)^2 - 7^2 = (3x + 7)(3x - 7)$

(See Section 1.3, Example 7.)

**8.** $9x - 4 = 8 + 7(x - 5)$

$\Rightarrow 9x - 4 = 8 + 7x - 35$

$\Rightarrow 9x - 7x = 8 - 35 - (-4)$

$\Rightarrow 2x = -23$

$\Rightarrow x = -\dfrac{23}{2}$

(See Section 1.6, Example 2.)

**9.** $3x^2 + 2x + 8 = 20x - 7$

$\Rightarrow 3x^2 + 2x - 20x + 8 + 7 = 0$

$\Rightarrow 3x^2 - 18x + 15 = 0$

$\Rightarrow 3(x^2 - 6x + 5) = 0$

$\Rightarrow 3(x - 5)(x - 1) = 0$

$\Rightarrow x - 5 = 0 \quad \text{or} \quad x - 1 = 0$

$\Rightarrow x = 5 \quad \text{or} \quad x = 1$

(See Section 1.7, Example 2.)

**10.** $x^2 - x - 6 < 0$

First solve $x^2 - x - 6 = 0 \Rightarrow (x + 2)(x - 3) = 0 \Rightarrow x = -2$ or $x = 3$. The numbers $-2$ and $3$ divide the number line into three regions. Test one number in each region to determine that $(x + 2)(x - 3) < 0$ when $-2 < x < 3$.

(See Section 2.5, Example 3.)

**11.** $5e^{x-3} - 1 = 9$

$\Rightarrow 5e^{x-3} = 10$

$\Rightarrow e^{x-3} = 2$

$\Rightarrow \ln(e^{x-3}) = \ln(2)$

$\Rightarrow x - 3 = \ln(2)$

$\Rightarrow x = \ln(2) + 3$

(See Section 4.4, Example 6.)

**12.** $\ln(5x + 1) = 2$

$\Rightarrow e^{\ln(5x+1)} = e^2$

$\Rightarrow 5x + 1 = e^2$

$\Rightarrow 5x = e^2 - 1$

$\Rightarrow x = \dfrac{e^2 - 1}{5}$

(See Section 4.4, Example 2.)

**13.** $f(-2) = 5(-2)^3 - 3(-2)^2 + 9(-2) + 60$

$= 5(-8) - 3(4) + 9(-2) + 60$

$= -40 - 12 - 18 + 60$

$= -10$

(See Section 3.1, Example 8.)

**14.** $f(x) = 0$

$\Rightarrow 6x^2 - 7x - 20 = 0$

$\Rightarrow (3x + 4)(2x - 5) = 0$

$\Rightarrow 3x + 4 = 0 \quad \text{or} \quad 2x - 5 = 0$

$\Rightarrow x = -4/3 \quad \text{or} \quad x = 5/2$

(See Section 1.7, Example 2.)

**15.** $y - y_1 = m(x - x_1)$

$\Rightarrow y - (-3) = -5(x - 8)$

$\Rightarrow y + 3 = -5x + 40$

$\Rightarrow y = -5x + 37$

(See Section 2.2, Example 9.)

# Table of Integrals

**C is an arbitrary constant.**

1. $\displaystyle\int x^n \, dx = \frac{1}{n+1} x^{n+1} + C \qquad (n \neq -1)$

2. $\displaystyle\int e^{kx} \, dx = \frac{1}{k} e^{kx} + C$

3. $\displaystyle\int \frac{a}{x} \, dx = a \ln|x| + C$

4. $\displaystyle\int \ln|ax| \, dx = x(\ln|ax| - 1) + C$

5. $\displaystyle\int \frac{1}{\sqrt{x^2 + a^2}} \, dx = \ln\left|x + \sqrt{x^2 + a^2}\right| + C$

6. $\displaystyle\int \frac{1}{\sqrt{x^2 - a^2}} \, dx = \ln\left|x + \sqrt{x^2 - a^2}\right| + C$

7. $\displaystyle\int \frac{1}{a^2 - x^2} \, dx = \frac{1}{2a} \cdot \ln\left|\frac{a+x}{a-x}\right| + C \qquad (a \neq 0)$

8. $\displaystyle\int \frac{1}{x^2 - a^2} \, dx = \frac{1}{2a} \cdot \ln\left|\frac{x-a}{x+a}\right| + C \qquad (a \neq 0)$

9. $\displaystyle\int \frac{1}{x\sqrt{a^2 - x^2}} \, dx = -\frac{1}{a} \cdot \ln\left|\frac{a + \sqrt{a^2 - x^2}}{x}\right| + C \qquad (a \neq 0)$

10. $\displaystyle\int \frac{1}{x\sqrt{a^2 + x^2}} \, dx = -\frac{1}{a} \cdot \ln\left|\frac{a + \sqrt{a^2 + x^2}}{x}\right| + C \qquad (a \neq 0)$

11. $\displaystyle\int \frac{x}{ax + b} \, dx = \frac{x}{a} - \frac{b}{a^2} \cdot \ln|ax + b| + C \qquad (a \neq 0)$

12. $\displaystyle\int \frac{x}{(ax + b)^2} \, dx = \frac{b}{a^2(ax + b)} + \frac{1}{a^2} \cdot \ln|ax + b| + C \qquad (a \neq 0)$

13. $\displaystyle\int \frac{1}{x(ax + b)} \, dx = \frac{1}{b} \cdot \ln\left|\frac{x}{ax + b}\right| + C \qquad (b \neq 0)$

14. $\displaystyle\int \frac{1}{x(ax + b)^2} \, dx = \frac{1}{b(ax + b)} + \frac{1}{b^2} \cdot \ln\left|\frac{x}{ax + b}\right| \qquad (b \neq 0)$

15. $\displaystyle\int \sqrt{x^2 + a^2} \, dx = \frac{x}{2} \sqrt{x^2 + a^2} + \frac{a^2}{2} \cdot \ln\left|x + \sqrt{x^2 + a^2}\right| + C$

16. $\displaystyle\int x^n \cdot \ln|x| \, dx = x^{n+1}\left[\frac{\ln|x|}{n+1} - \frac{1}{(n+1)^2}\right] + C \qquad (n \neq -1)$

17. $\displaystyle\int x^n e^{ax} \, dx = \frac{x^n e^{ax}}{a} - \frac{n}{a} \cdot \int x^{n-1} e^{ax} \, dx + C \qquad (a \neq 0)$

# Answers to Selected Exercises

## Chapter 1

### Section 1.1 (Page 8)

1. True   3. Answers vary, but 2,508,429,787/798,458,000 is the best.
5. Distributive property   7. Identity property of addition
9. Associative property of addition   11. Answers vary.
13. $-39$   15. $-2$   17. 45.6   19. about .9167   21. $-12$
23. 0   25. 4   27. $-1$   29. $\dfrac{2040}{523}, \dfrac{189}{37}, \sqrt{27}, \dfrac{4587}{691}, 6.735, \sqrt{47}$
31. $12 < 18.5$   33. $x \geq 5.7$   35. $z \leq 7.5$   37. $<$   39. $=$
41. $a$ lies to the right of $b$ or is equal to $b$.   43. $c < a < b$
45.   47.
49.   51. $-3$   53. $-19$   55. $=$
57. $=$   59. $=$   61. $>$   63. $7 - a$   65. Answers vary.
67. Answers vary.   69. 1   71. 9   73. 4   75. 4.3
77. 64.1   79. 2.3   81. 2010, 2015, 2016   83. 2010, 2011, 2012, 2013, 2014, 2015

### Section 1.2 (Page 16)

1. 1,973,822.685   3. 289.0991339   5. Answers vary.   7. $4^5$
9. $(-6)^7$   11. $(5u)^{28}$   13. Degree 4; coefficients: 6.2, $-5$, 4, $-3$, 3.7; constant term 3.7   15. 3   17. $-x^3 + x^2 - 13x$
19. $-6y^2 + 3y + 6$   21. $-6x^2 + 4x - 4$   23. $-18m^3 - 54m^2 + 9m$
25. $12z^3 + 14z^2 - 7z + 5$   27. $12k^2 + 16k - 3$   29. $6y^2 + 13y + 5$
31. $18k^2 - 7kq - q^2$   33. $4.34m^2 + 5.68m - 4.42$   35. $-k + 3$
37. $R = 5000x$; $C = 200,000 + 1800x$; $P = 3200x - 200,000$
39. $R = 9750x$; $C = -3x^2 + 3480x + 259,675$;
$P = 3x^2 + 6270x - 259,675$   41. (a) $673 million
(b) About $505 million   43. (a) $1384 million (b) About $1022 million
45. About $4036 million   47. About $6745 million
49. False   51. False   53. About $20,663 million
55. About $34,896 million   57. 2013   59. $-100,375$; $96,220;
Answers vary.   61. $342,500

### Section 1.3 (Page 23)

1. $12x(x - 2)$   3. $r(r^2 - 5r + 1)$   5. $6z(z^2 - 2z + 3)$
7. $(2y - 1)^2(14y - 4) = 2(2y - 1)^2(7y - 2)$
9. $(x + 5)^4(x^2 + 10x + 28)$   11. $(x + 1)(x + 4)$   13. $(x + 3)(x + 4)$
15. $(x + 3)(x - 2)$   17. $(x - 1)(x + 3)$   19. $(x - 4)(x + 1)$
21. $(z - 7)(z - 2)$   23. $(z + 4)(z + 6)$   25. $(2x - 1)(x - 4)$
27. $(3p - 4)(5p - 1)$   29. $(2z - 5)(2z - 3)$   31. $(2x + 1)(3x - 4)$
33. $(5y - 2)(2y + 5)$   35. $(2x - 1)(3x + 4)$   37. $(3a + 5)(a - 1)$
39. $(x + 9)(x - 9)$   41. $(3p - 2)^2$   43. $(r - 2t)(r + 5t)$
45. $(m - 4n)^2$   47. $(2u + 3)^2$   49. Cannot be factored.
51. $(2r + 3v)(2r - 3v)$   53. $(x + 2y)^2$   55. $(3a + 5)(a - 6)$
57. $(7m + 2n)(3m + n)$   59. $(y - 7z)(y + 3z)$   61. $(11x + 8)(11x - 8)$
63. $(a - 4)(a^2 + 4a + 16)$   65. $(2r - 3s)(4r^2 + 6rs + 9s^2)$
67. $(4m + 5)(16m^2 - 20m + 25)$   69. $(10y - z)(100y^2 + 10yz + z^2)$
71. $(x^2 + 3)(x^2 + 2)$   73. $b^2(b + 1)(b - 1)$
75. $(x + 2)(x - 2)(x^2 + 3)$   77. $(4a^2 + 9b^2)(2a + 3b)(2a - 3b)$
79. $x^2(x^2 + 2)(x^4 - 2x^2 + 4)$   81. Answers vary.
83. Answers vary.

### Section 1.4 (Page 29)

1. $\dfrac{x}{7}$   3. $\dfrac{5}{7p}$   5. $\dfrac{5}{4}$   7. $\dfrac{4}{w + 6}$   9. $\dfrac{y - 4}{3y^2}$   11. $\dfrac{m - 2}{m + 3}$
13. $\dfrac{x + 3}{x + 1}$   15. $\dfrac{3}{16a}$   17. $\dfrac{3y}{x^2}$   19. $\dfrac{5}{4c}$   21. $\dfrac{3}{4}$   23. $\dfrac{3}{10}$
25. $\dfrac{2(a + 4)}{a - 3}$   27. $\dfrac{k + 2}{k + 3}$   29. Answers vary.   31. $\dfrac{3}{35z}$
33. $\dfrac{4}{3}$   35. $\dfrac{20 + x}{5x}$   37. $\dfrac{3m - 2}{m(m - 1)}$   39. $\dfrac{37}{5(b + 2)}$
41. $\dfrac{33}{20(k - 2)}$   43. $\dfrac{7x - 1}{(x - 3)(x - 1)(x + 2)}$   45. $\dfrac{y^2}{(y + 4)(y + 3)(y + 2)}$
47. $\dfrac{x + 1}{x - 1}$   49. $\dfrac{-1}{x(x + h)}$   51. (a) $\dfrac{\pi x^2}{4x^2}$   (b) $\dfrac{\pi}{4}$   53. (a) $\dfrac{x^2}{25x^2}$   (b) $\dfrac{1}{25}$
55. $\dfrac{-7.2x^2 + 6995x + 230,000}{1000x}$   57. About $3.8 million   59. No
61. $2.55   63. $6531

### Section 1.5 (Page 41)

1. 49   3. $16c^2$   5. $\dfrac{32}{x^5}$   7. $108u^{12}$   9. $\dfrac{1}{7}$   11. $-\dfrac{1}{7776}$
13. $-\dfrac{1}{y^3}$   15. $\dfrac{9}{16}$   17. $\dfrac{b^3}{a}$   19. 7   21. About 1.55   23. $-16$
25. $\dfrac{81}{16}$   27. $\dfrac{4^2}{5^3}$   29. $4^3$   31. $4^8$   33. $z^3$   35. $\dfrac{p}{9}$   37. $\dfrac{q^5}{r^3}$
39. $\dfrac{8}{25p^7}$   41. $2^{5/6}p^{3/2}$   43. $2p + 5p^{5/3}$   45. $\dfrac{1}{3y^{2/3}}$   47. $\dfrac{a^{1/2}}{49b^{5/2}}$
49. $x^{7/6} - x^{11/6}$   51. $x - y$   53. (f)   55. (h)   57. (g)
59. (c)   61. 5   63. 5   65. 21   67. $\sqrt{77}$   69. $5\sqrt{3}$
71. $-\sqrt{2}$   73. $15\sqrt{5}$   75. 3   77. $-3 - 3\sqrt{2}$   79. $4 + \sqrt{3}$
81. $\dfrac{7}{11 + 6\sqrt{2}}$   83. (a) 14 (b) 85 (c) 58.0   85. About $11.1 billion
87. About $11.7 billion   89. About $4.2 billion   91. About $5.8 billion
93. About 6.1 million   95. About 6.5 million

### Section 1.6 (Page 50)

1. 4   3. 7   5. $-\dfrac{10}{9}$   7. 4   9. $\dfrac{40}{7}$   11. $\dfrac{26}{3}$   13. $-\dfrac{12}{5}$
15. $-\dfrac{59}{6}$   17. $-\dfrac{9}{4}$   19. $x = .72$   21. $r \approx -13.26$   23. $\dfrac{b - 5a}{2}$
25. $x = \dfrac{3b}{a + 5}$   27. $V = \dfrac{k}{p}$   29. $g = \dfrac{V - V_0}{t}$   31. $B = \dfrac{2A}{h} - b$
or $B = \dfrac{2A - bh}{h}$   33. $-2, 3$   35. $-8, 2$   37. $\dfrac{5}{2}, \dfrac{7}{2}$   39. $23°$
41. $71.6°$   43. 13.12   45. 2018   47. 2021   49. 2011
51. 2018   53. 2013   55. 2020   57. $205.41   59. $21,000
61. $70,000 for the first plot; $50,000 for the second
63. About 8.6 million   65. 301 million   67. $\dfrac{400}{3}L$

### Section 1.7 (Page 58)

1. $-4, 14$   3. $0, -6$   5. $0, 2$   7. $-7, -8$   9. $\dfrac{1}{2}, 3$

**11.** $-\dfrac{1}{2}, \dfrac{1}{3}$    **13.** $\dfrac{5}{2}, 4$    **15.** $-5, -2$    **17.** $\dfrac{4}{3}, -\dfrac{4}{3}$    **19.** $0, 1$

**21.** $2 \pm \sqrt{7}$    **23.** $\dfrac{1 \pm 2\sqrt{5}}{4}$    **25.** $\dfrac{-7 \pm \sqrt{41}}{4}$; $-.1492, -3.3508$

**27.** $\dfrac{-1 \pm \sqrt{5}}{4}$; $.3090, -.8090$    **29.** $\dfrac{-5 \pm \sqrt{65}}{10}$; $.3062, -1.3062$

**31.** No real-number solutions    **33.** $-\dfrac{5}{2}, 1$    **35.** No real-number solutions

**37.** $-5, \dfrac{3}{2}$    **39.** 1    **41.** 2    **43.** $x \approx .4701$ or $1.8240$

**45.** $x \approx -1.0376$ or $.6720$    **47.** \$43.6 thousand    **49.** 2006 ($x \approx 5.46$)
**51.** 2015 ($x \approx 14.42$)    **53.** \$43.19    **55.** First quarter of 2017 ($x \approx 8.14$)
**57.** About 1.046 ft.    **59. (a)** $x + 20$   **(b)** Northbound: $5x$; eastbound:
$5(x + 20)$ or $5x + 100$   **(c)** $(5x)^2 + (5x + 100)^2 = 300^2$
**(d)** About 31.23 mph and 51.23 mph    **61. (a)** $150 - x$
**(b)** $x(150 - x) = 5000$   **(c)** Length 100 m; width 50 m
**63.** 9 ft. by 12 ft.    **65.** 6.25 sec.    **67. (a)** About 3.54 sec.   **(b)** 2.5 sec.

**(c)** 144 ft.    **69. (a)** 2 sec.   **(b)** $\dfrac{3}{4}$ sec. or $\dfrac{13}{4}$ sec.   **(c)** It reaches the given
height twice: once on the way up and once on the way down.

**71.** $t = \dfrac{\sqrt{2Sg}}{g}$    **73.** $h = \dfrac{d^2\sqrt{kL}}{L}$    **75.** $R = \dfrac{-2Pr + E^2 \pm E\sqrt{E^2 - 4Pr}}{2P}$

## Chapter 1 Review Exercises (Page 61)

| Refer to Section | 1.1 | 1.2 | 1.3 | 1.4 | 1.5 | 1.6 | 1.7 |
|---|---|---|---|---|---|---|---|
| For Exercises | 1–18, 81–84 | 19–24, 85–92 | 25–32 | 33–38, 93–96 | 39–60, 97–98 | 61–68, 99–100 | 69–80, 101–106 |

**1.** $0, 6$    **3.** $-12, -6, -\dfrac{9}{10}, -\sqrt{4}, 0, \dfrac{1}{8}, 6$    **5.** Commutative property

of multiplication    **7.** Distributive property    **9.** $x \geq 9$
**11.** $-|3 - (-2)|, -|-2|, |6 - 4|, |8 + 1|$    **13.** $-1$

**15.**     **17.** $-\dfrac{7}{9}$    **19.** $4x^4 - 4x^2 + 11x$

**21.** $25k^2 - 4h^2$    **23.** $9x^2 + 24xy + 16y^2$    **25.** $k(2h^2 - 4h + 5)$
**27.** $a^2(5a + 2)(a + 2)$    **29.** $(12p + 13q)(12p - 13q)$

**31.** $(3y - 1)(9y^2 + 3y + 1)$    **33.** $\dfrac{9x^2}{4}$    **35.** 4    **37.** $\dfrac{(m - 1)^2}{3(m + 1)}$

**39.** $\dfrac{1}{5^3}$ or $\dfrac{1}{125}$    **41.** $-1$    **43.** $4^3$    **45.** $\dfrac{1}{8}$    **47.** $\dfrac{7}{10}$    **49.** $\dfrac{1}{5^{\frac{2}{3}}}$

**51.** $3^{\frac{7}{2}}a^{\frac{5}{2}}$    **53.** 3    **55.** $3pq\sqrt[3]{2q^2}$    **57.** $-21\sqrt{3}$    **59.** $\sqrt{6} - \sqrt{3}$

**61.** $-\dfrac{1}{3}$    **63.** No solution    **65.** $x = \dfrac{3}{8a - 2}$    **67.** $-38, 42$

**69.** $-7 \pm \sqrt{5}$    **71.** $\dfrac{1}{2}, -2$    **73.** $-\dfrac{3}{2}, 7$    **75.** $\pm\dfrac{\sqrt{3}}{3}$

**77.** $r = \dfrac{-Rp \pm E\sqrt{Rp}}{p}$    **79.** $s = \dfrac{a \pm \sqrt{a^2 + 4K}}{2}$    **81.** 82%

**83.** \$1536.25    **85.** 2013    **87.** 2014    **89.** 17.05 million
**91.** March ($x \approx 2.33$)    **93.** About 9.16 million    **95.** 2013 ($x \approx 12.67$)
**97.** About \$30.91 billion    **99.** 11%    **101.** About 69.9%
**103.** 3.2 feet    **105.** About 7.77 seconds

## Case Study 1 Exercises (Page 64)

**1.** $218 + 508x$    **3.** Electric by \$1880    **5.** $1529.10 + 50x$
**7.** LG by \$29.10

# Chapter 2

## Section 2.1 (Page 72)
**1.** IV, II, I, III    **3.** Yes    **5.** No

**7.**

**9.**

**11.**

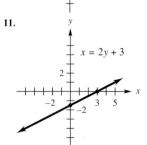

**13.** $x$-intercepts $-2.5, 3$; $y$-intercept 3    **15.** $x$-intercepts $-1, 2$;
$y$-intercept $-2$    **17.** $x$-intercepts 4; $y$-intercept 3    **19.** $x$-intercept 12;
$y$-intercept $-8$    **21.** $x$-intercepts 3, $-3$; $y$-intercept $-9$
**23.** $x$-intercepts $-5, 4$; $y$-intercept $-20$    **25.** No $x$-intercept; $y$-intercept 7

**27.**     **29.**

**31.**     **33.**

**35.**

**37.**

**39.**

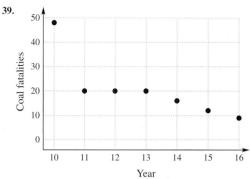

**41.** 1999 **43.** 2011 **45. (a)** About $1,250,000 **(b)** About $1,750,000 **(c)** About $4,250,000 **47. (a)** About $500,000 **(b)** About $1,000,000 **(b)** About $1,500,000 **49.** Beef: about 57 pounds; chicken: about 81 pounds; pork: about 45 pounds; **51.** 2009 **53.** About $115 **55.** Days 6, 12, 14 to 21 **57.** Day 19; about $117.50 **59.** Day 2; about $117.00 **61.** About $8.50

**63.**

**65.**

**67.**

**69.** No;

**71.** $x \approx -1.1038$ **73.** $x \approx 2.1017$ **75.** $x \approx -1.7521$

## Section 2.2 (Page 85)

**1.** $-\dfrac{3}{2}$ **3.** $-2$ **5.** $-\dfrac{5}{2}$ **7.** Not defined **9.** $y = 4x + 5$

**11.** $y = -2.3x + 1.5$ **13.** $y = -\dfrac{3}{4}x + 4$ **15.** $m = 2; b = -9$

**17.** $m = 3; b = -2$ **19.** $m = \dfrac{2}{3}; b = -\dfrac{16}{9}$ **21.** $m = \dfrac{2}{3}; b = 0$

**23.** $m = 1; b = 5$ **25. (a)** C **(b)** B **(c)** B **(d)** D

**27.**

**29.**

**31.**

**33.** Perpendicular **35.** Parallel **37.** Neither **39. (a)** $\dfrac{2}{5}, \dfrac{9}{8}, -\dfrac{5}{2}$

**(b)** Yes **41.** $y = -\dfrac{2}{3}x$ **43.** $y = 3x - 3$ **45.** $y = 1$

**47.** $x = -2$ **49.** $y = 2x + 3$ **51.** $2y = 7x - 3$ **53.** $y = 5x$
**55.** $x = 6$ **57.** $y = 2x - 2$ **59.** $y = x - 6$ **61.** $y = -x + 2$
**63.** $1330.42 **65.** $6715.23 **67.** $-11.7\%$ **69.** 2016 ($x \approx 15.21$)
**71.** 91.2 million **73.** 2018 ($x \approx 17.5$) **75.** 13.104 minutes;
.044 minutes **77. (a)** $y = 4.5x - 44.5$ **(b)** 14% **(c)** 2017 ($x \approx 16.56$)
**79. (a)** $y = 300x + 500$ **(b)** $4700 **(c)** 2019 ($x \approx 18.33$)

## Section 2.3 (Page 93)

**1. (a)** $y = \dfrac{5}{9}(x - 32)$ **(b)** 10°C and 23.89°C **3.** 463.89°C

**5.** 227.92; 236.98 **7.** 13.08 million **9.** 4 ft.
**11. (a)** Equation 1: $-.55, -.17, 0, .35, 0$; sum $= -.37$; Equation
2: $-.12, 0, -.09, 0, -.61$; sum $= -.82$ **(b)** Equation 1: .4539;
Equation 2: .3946 **(c)** Equation 2
**13.** Yes. $r \approx -.9999$

**15. (a)** $y = -170.8x + 6320$ **(b)** About 3246
**17. (a)** $y = .8x + 14.2$ **(b)** About $23 billion
**19. (a)** $y = -1.829x + 98.4$ **(b)** About $76.5 billion
**21. (a)** $y = .295x + 48$ **(b)** About $50.95 billion **(c)** Second quarter
of 2018 ($x \approx 13.56$) **(d)** $r \approx .795$

## Section 2.4 (Page 100)
**1.** Answers vary.
**3.** $[-4, \infty)$

**5.** $(-\infty, 0)$

**7.** $\left(-\infty, \dfrac{10}{3}\right]$

**9.** $(-\infty, -8]$

**11.** $(-\infty, 3)$

**13.** $(-1, \infty)$

**15.** $(-\infty, 1]$

**17.** $\left(\dfrac{1}{5}, \infty\right)$

**19.** $(-5, 7)$

**21.** $\left[\dfrac{7}{3}, 5\right]$

**23.** $\left[-\dfrac{11}{2}, \dfrac{7}{2}\right]$

**25.** $\left[-\dfrac{17}{7}, \infty\right)$

**27.** $x \geq 2$ **29.** $-3 < x \leq 5$ **31.** $x \geq 400$ **33.** $x \geq 50$
**35.** Impossible to break even
**37.** $(-\infty, -7)$ or $(7, \infty)$

**39.** $[-5, 5]$

**41.** All real numbers

**43.** $\left(-\dfrac{3}{2}, \dfrac{5}{2}\right)$

**45.** $\left(-\infty, \dfrac{3}{2}\right]$ or $\left[\dfrac{1}{4}, \infty\right)$

**47.** $76 \le T \le 90$    **49.** $40 \le T \le 82$    **51.** $32.5 \le P \le 38.5$
**53.** $35 \le B \le 43$    **55.** $0 < x \le 9325$;   $9325 < x \le 37{,}950$;
$37{,}950 < x \le 91{,}900$;   $91{,}900 < x \le 191{,}650$;
$191{,}650 < x \le 416{,}700$;   $416{,}700 < x \le 418{,}400$; $x > 418{,}400$

## Section 2.5 (Page 108)

**1.** $\left[-4, \dfrac{3}{2}\right]$

**3.** $(-\infty, -3)$ or $(-1, \infty)$

**5.** $\left[-2, \dfrac{1}{4}\right]$

**7.** $(-\infty, -1)$ or $\left(\dfrac{1}{4}, \infty\right)$

**9.** $[-6, 6]$

**11.** $(-\infty, 0)$ or $(16, \infty)$

**13.** $[-3, 0]$ or $[3, \infty)$    **15.** $[-7, -2]$ or $[2, \infty)$    **17.** $(-\infty, -5)$ or $(-1, 3)$

**19.** $\left(-\infty, -\dfrac{1}{2}\right)$ or $\left(0, \dfrac{4}{3}\right)$    **21.** No    **23.** $(-.1565, 2.5565)$

**25.** $[-2.2635, .7556]$ or $[3.5079, \infty)$    **27.** $(.5, .8393)$

**29.** $(-\infty, 1)$ or $[4, \infty)$    **31.** $\left(\dfrac{7}{2}, 5\right)$    **33.** $(-\infty, 2)$ or $(5, \infty)$

**35.** $(-\infty, -1)$    **37.** $(-\infty, -2)$ or $(0, 3)$    **39.** $[-1, .5]$
**41.** $(8, \infty)$    **43.** $[52, 200]$

## Chapter 2 Review Exercises (Page 110)

| Refer to Section | 2.1 | 2.2 | 2.3 | 2.4 | 2.5 |
|---|---|---|---|---|---|
| For Exercises | 1–10 | 11–34 | 35–38 | 39–54 | 55–68 |

**1.** $(-2, 3), (0, -5), (3, -2), (4, 3)$

**3.**

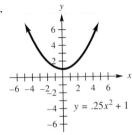

$5x - 3y = 15$

**5.**

$y + 3 = 0$

**7.**

$y = .25x^2 + 1$

**9. (a)** About 11:30 a.m. to about 7:30 p.m.    **(b)** From midnight until about 5 a.m. and after about 10:30 p.m.    **11.** Answers vary.

**13.** $-3$    **15.** $-\dfrac{1}{4}$    **17.** $3$    **19.** $0$    **21.** $-3$

**23.**

$(-3, 4)$
$(-4, 1)$

**25.** $3y = 2x - 13$    **27.** $4y = -5x + 17$    **29.** $x = -1$
**31.** $3y = 5x + 15$    **33. (a)** $y = .08x$   **(b)** Positive; answers vary.
**(c)** \$1.36 trillion    **35. (a)** $y = 590x + 43{,}376$
**(b)** $y = 964.6x + 40{,}833$   **(c)** \$52,226; \$55,302; least squares
predicts better   **(d)** \$58,196    **37. (a)** $y = 93.1x + 2034.8$
**(c)** Yes   **(d)** $r \approx .9993$

**(b)**

**39.** $\left(\dfrac{3}{8}, \infty\right)$    **41.** $\left(-\infty, \dfrac{1}{4}\right]$    **43.** $\left[-\dfrac{1}{2}, 2\right]$    **45.** $[-8, 8]$

**47.** $(-\infty, 2]$ or $[5, \infty)$    **49.** $\left[-\dfrac{9}{5}, 1\right]$    **51. (d)** $|x - 6| > 4$

**53.** Less than or equal to 2 gigabytes    **55.** $(-3, 2)$

**57.** $(-\infty, -5]$ or $\left[\dfrac{3}{2}, \infty\right)$    **59.** $(-\infty, -5]$ or $[-2, 3]$

**61.** $[-2, 0)$    **63.** $\left(-1, \dfrac{3}{2}\right)$    **65.** $[-19, -5)$ or $(2, \infty)$

## Case Study 2 Exercises (Page 113)

**1.**

```
LinReg
 y=ax+b
 a=-.765952381
 b=82.32928571
 r²=.0399034604
 r=-.1997585053
```

**3.** $10.24, -0.73, 4.83, -13.86, -10.49, -1.14, -1.10, 12.63$

**5.**

```
LinReg
 y=ax+b
 a=.3911151515
 b=-24.59715152
 r²=.9918612299
 r=.9959223011
```

**7.** The model predicts a negative wage.

**9.**

```
LinReg
 y=ax+b
 a=1.6242424E-5
 b=.0026575758
 r²=2.101177E-7
 r=4.5838595E-4
```

A flat line with 0 slope is a good fit for the residuals.

# Chapter 3

## Section 3.1 (Page 120)
**1.** Function    **3.** Not a function    **5.** Function
**7.** Not a function    **9.** Function    **11.** $(-\infty, \infty)$    **13.** $(-\infty, \infty)$
**15.** $(-\infty, 5]$    **17.** All real numbers except 2
**19.** All real numbers except 2 and $-2$    **21.** All real numbers
such that $x > -4$ and $x \neq 3$    **23. (a)** 8    **(b)** 8    **(c)** 8    **(d)** 8
**25. (a)** 48    **(b)** 6    **(c)** 25.38    **(d)** 28.42    **27. (a)** 1    **(b)** $\sqrt{8}$
**(c)** $\sqrt{2.3}$    **(d)** $\sqrt{9.9}$    **29. (a)** 12    **(b)** 23    **(c)** 12.91
**(d)** 49.41    **31. (a)** $\dfrac{\sqrt{3}}{15}$    **(b)** Not defined    **(c)** $\dfrac{\sqrt{1.7}}{6.29}$
**(d)** Not defined    **33. (a)** 5    **(b)** 10    **(c)** 3    **(d)** 13.8
**35. (a)** $6 - p$    **(b)** $6 + r$    **(c)** $3 - m$    **37. (a)** $\sqrt{4 - p}\ (p \leq 4)$
**(b)** $\sqrt{4 + r}\ (r \geq -4)$    **(c)** $\sqrt{1 - m}\ (m \leq 1)$    **39. (a)** $p^2 - 1$
**(b)** $r^2 - 1$    **(c)** $m^2 + 6m + 8$    **41. (a)** $\dfrac{3}{p - 1}\ (p \neq 1)$
**(b)** $\dfrac{3}{-r - 1}\ (r \neq -1)$    **(c)** $\dfrac{3}{m + 2}\ (m \neq -2)$    **43.** 2
**45.** $2x + h$    **47. (a)** About \$16,515 billion    **(b)** About \$17,693
billion    **(c)** About \$18,282 billion    **49. (a)** At the start of 2012, the
closing price of Microsoft was \$30.20.    **(b)** At the start of 2014, the closing
price of Microsoft was \$37.80.    **(c)** At the start of 2016, the closing price
of Microsoft was \$51.80.    **51. (a)** A person with a taxable income
of \$100 pays \$2 in state income tax.    **(b)** A person with a taxable
income of \$2500 pays \$90 in state income tax.    **(c)** A person with
a taxable income of \$103,000 pays \$5110 in state income tax.
**53. (a)** About 23.9 years    **(b)** About 24.2 years    **(c)** About 24.4 years
**55.** $f(t) = 2050 - 500t$    **57. (a)** $c(x) = 1800 + .5x$    **(b)** $r(x) = 1.2x$
**(c)** $p(x) = .7x - 1800$

**59.**

## Section 3.2 (Page 130)
**1.**

**3.**
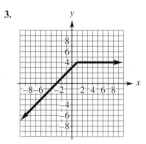
$$f(x) = \begin{cases} x + 3 & \text{if } x \ 1 \\ 4 & \text{if } x > 1 \end{cases}$$

**5.**
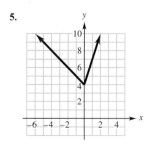
$$y = \begin{cases} 4 - x & \text{if } x \leq 0 \\ 3x + 4 & \text{if } x > 0 \end{cases}$$

**7.**
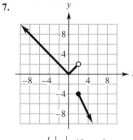
$$f(x) = \begin{cases} |x| & \text{if } x < 2 \\ -2x & \text{if } x \geq 2 \end{cases}$$

**9.**
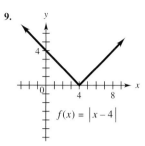
$f(x) = |x - 4|$

**11.**
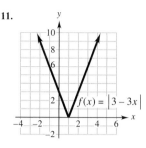
$f(x) = |3 - 3x|$

**13.**

$y = -|x - 1|$

**15.**
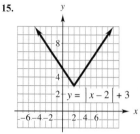
$y = |x - 2| + 3$

**17.**
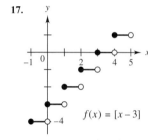
$f(x) = [x - 3]$

**19.**

$g(x) = [-x]$

**21.**

**23.**

$g(x) = \sqrt{-x}$

**25.**
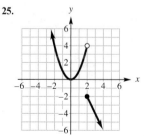

**27.** Function    **29.** Not a function    **31.** Function
**33.**

**35.**

(1, −1) is on the graph; (1, 3) is not on the graph

**37. (a)** $0, $2900, $7600

**(b)**

**39.**

**41. (a)** 15%   **(b)** $f(x) = -.6x + 24$   **(c)** 24%; 13.8%   **(d)** Yes
**43. (a)** 2011   **(b)** 2012   **(c)** 2009 and 2015   **45. (a)** About $120 billion
**(b)** [2011, 2015]   **(c)** [40, 170]   **47. (a)** 44¢; 49¢   **(b)** The figure has
vertical line segments, which can't be part of the graph of a function.
(Why?) To make the figure into the graph of $f$, delete the vertical line
segments; then, for each horizontal segment of the graph, put a closed dot
on the left end and an open circle on the right end (as in Figure 3.8).
**49. (a)** $34.99   **(b)** $24.99   **(c)** $64.99   **(d)** $74.99
**(e)**

**51.** There are many correct answers, including the following:

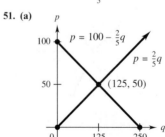

## Section 3.3 (Page 142)

**1.** Let $C(x)$ be the cost of renting a saw for $x$ hours; $C(x) = 25 + 5x$.
**3.** Let $C(x)$ be the cost (in dollars) for $x$ hours; $C(x) = 8 + 2.5x$.
**5.** $C(x) = 36x + 200$   **7.** $C(x) = 120x + 3800$
**9.** $48, $15.60, $13.80   **11.** $55.50, $11.40, $8.46
**13. (a)** $f(x) = -2730x + 25,035$   **(b)** $11,385   **(c)** $2730 per year
**15. (a)** $f(x) = -11,875x + 120,000$   **(b)** [0, 8]   **(c)** $48,750
**17. (a)** $80,000   **(b)** $42.50   **(c)** $122,500; $1,440,000   **(d)** $122.50; $45
**19. (a)** $C(x) = .42x + 37$   **(b)** $37   **(c)** $457   **(d)** $457.42
**(e)** $.42   **(f)** $.42   **21.** $R(x) = 1.87x + 2,816,000$
**23. (a)** $C(x) = 10x + 750$   **(b)** $R(x) = 35x$   **(c)** $P(x) = 25x - 750$
**(d)** $1750   **25. (a)** $C(x) = 18x + 300$   **(b)** $R(x) = 28x$
**(c)** $P(x) = 10x - 300$   **(d)** $700   **27. (a)** Decreased, Increased
**29.** $(3, -1)$   **31.** $\left(-\dfrac{11}{4}, -\dfrac{61}{4}\right)$   **33. (a)** 200,000 policies ($x = 200$)
**(b)**

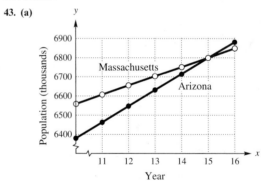

**(c)** Revenue: $12,500; cost: $15,000   **35. (a)** $C(x) = .126x + 1.5$
**(b)** $2.382 million   **(c)** About 17.857 units   **37.** Break-even point is
about 467 units; do not produce the item.   **39.** Break-even point is about
1037 units; produce the item.   **41.** The percentage is approximately
51% in 2009.

**43. (a)**

**(b)** Yes   **(c)** 2015   **45.** $140   **47.** 10 items
**49. (a)** $16   **(b)** $11   **(c)** $6   **(d)** 8 units   **(e)** 4 units   **(f)** 0 units
**(g)**   **(h)** 0 units   **(i)** $\dfrac{40}{3}$ units   **(j)** $\dfrac{80}{3}$ units
**(k)** See part (g).   **(l)** 8 units   **(m)** $6

$p = 16 - \dfrac{5}{4}q$

$(8, 6)$   $p = \dfrac{3}{4}q$

**51. (a)**   **(b)** 125 units   **(c)** 50¢
**(d)** Shortage of supply

$p = 100 - \dfrac{2}{5}q$

$p = \dfrac{2}{5}q$

$(125, 50)$

**53.** Total cost increases when more items are made (because it includes the cost of all previously made items), so the graph cannot move downward. No; the average cost can decrease as more items are made, so its graph can move downward.

## Section 3.4 (Page 154)
**1.** Upward    **3.** Downward    **5.** (5, 7); upward
**7.** $(-1, -9)$; downward    **9.** i    **11.** k    **13.** j    **15.** f
**17.** $f(x) = \frac{1}{4}(x - 1)^2 + 2$    **19.** $f(x) = (x + 1)^2 - 2$
**21.** $(-3, -12)$    **23.** $(2, -7)$    **25.** $x$-intercepts 1, 3; $y$-intercept 9
**27.** $x$-intercepts $5 \pm \sqrt{5}$; $y$-intercept 20
**29.** $(-2, 0)$, $x = -2$              **31.** $(2, 2)$, $x = 2$

$f(x) = (x + 2)^2$

$f(x) = x^2 - 4x + 6$
$f(x) = (x - 2)^2 + 2$

**33.** (a) (13, 27.5)  (b) 2013  (c) About $28 billion
**35.** (a) 10 milliseconds  (b) 40 responses per millisecond
**37.** 54    **39.** (a) About 12 books  (b) 10 books  (c) About 7 books
(d) 0 books  (e) 5 books  (f) About 7 books  (g) 10 books
(h) About 12 books
(i)

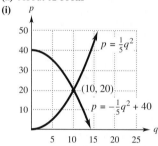

**41.** (a) $640  (b) $515  (c) $140
(d)

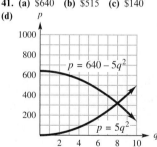

(e) 800 units  (f) $320    **43.** 80; $3600    **45.** 30; $1500
**47.** 20    **49.** 10    **51.** (a) $R(x) = 84x - 2x^2$  (b) 21  (c) $882
**53.** (a) $R(x) = 300x - 10x^2$  (b) 15 nights  (c) $2250
**55.** (a) $R(x) = (100 - x)(200 + 4x) = 20{,}000 + 200x - 4x^2$
(b)

(c) 25 seats  (d) $22,500

**57.** 13 weeks; $96.10/hog    **59.** (a) $f(x) = 68x^2 + 5674$
(b) About $10,026 billion; yes, the estimate is reasonable.
**61.** (a) $f(x) = -37.5(x - 13)^2 + 545$  (b) $395; yes, the estimate is reasonable.
**63.** $f(x) = 29.92x^2 + 305.90x + 5580.91$; about $9943 billion

**65.** $f(x) = -39.71x^2 + 1032.37x - 6151.69$; about $400

**67.** (a) 11.3 and 88.7  (b) 50  (c) $3000  (d) $x < 11.3$ or $x > 88.7$
(e) $11.3 < x < 88.7$

## Section 3.5 (Page 165)
**1.**

$f(x) = x^4$

**3.**

$h(x) = -.2x^5$

**5.** (a) Yes  (b) No  (c) No  (d) Yes    **7.** (a) Yes  (b) No  (c) Yes
(d) No    **9.** d    **11.** b    **13.** a
**15.**

**17.**

**19.**

**21.** (a) In 2004, Home Depot revenue was about $65 billion.
(b) About $78 billion; about $68 billion; about $78 billion

(c)

Year

(d) Revenue fluctuates.

$R(x) = 0.19x^3 - 5.2x^2 + 44x - 40$

**23. (a)** In 2004, Home Depot costs were about $41 billion.
**(b)** About $50 billion; about $44 billion; about $51 billion
**(c)**                                    **(d)** Costs fluctuate

$C(x) = 0.13x^3 - 3.6x^2 + 31x - 34$

**25.** $P(x) = .06x^3 - 1.6x^2 + 13x - 6$; about $27 billion
**27. (a)** $933.33 billion   **(b)** $1200 billion   **(c)** $1145.8 billion
**(d)** $787.5 billion
**29. (a)**

**(b)** $f(x) = x^3 - 33x^2 + 465x + 47274$
**(c)** Yes

Y1=X3-33X2+465X+47274

X=23        Y=52679

**(d)** About 52,679,000
**31. (a)**    130

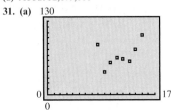

0                              17
0

**(b)** $f(x) = 0.5634x^4 - 26.123x^3 + 449.82x^2 - 3403.8x + 9594$
**(c)** Yes

Y5=.5634X^4-26.123X3+44...

X=15            Y=103.5

**(d)** 103.5; yes    **33.** $-3 \le x \le 5$ and $-20 \le y \le 5$    **35.** $-3 \le x \le 4$
and $-35 \le y \le 20$

## Section 3.6 (Page 174)
**1.** $x = -5,\ y = 0$                **3.** $x = -\dfrac{5}{2},\ y = 0$

$f(x) = \dfrac{1}{x+5}$

$f(x) = \dfrac{-3}{2x+5}$

**5.** $x = 1,\ y = 3$              **7.** $x = -5,\ y = 1$

$f(x) = \dfrac{3x}{x-1}$

$f(x) = \dfrac{x-3}{x+5}$

**9.** $x = 3;\ y = -1$             **11.** $x = -\dfrac{1}{8},\ y = \dfrac{1}{2}$

$f(x) = \dfrac{2-x}{x-3}$

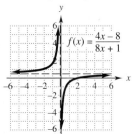

$f(x) = \dfrac{4x-8}{8x+1}$

**13.** $x = -4,\ x = 2,\ y = 0$    **15.** $x = -2,\ x = 2,\ y = 1$

$h(x) = \dfrac{x+1}{x^2 + 2x - 8}$

$f(x) = \dfrac{x^2 + 4}{x^2 - 4}$

**17.** $x = 1, x = -1$    **19.** $x = -1, x = 5$    **21. (a)** $C(x) = 100 + 5x$
**(b)** $\overline{C}(x) = \dfrac{100 + 5x}{x}$    **(c)** $25; When producing 5 pairs, the average cost
is $25 per pair. **(d)** $7; When producing 50 pairs, the average cost is
$7 per pair.  **(e)** $y = 5$; The average cost approaches $5 as more pairs
are produced.    **23. (a)** 0%; 2.5%; 4%; 4.875% In the presence of no food,
there will be no growth; with 2 units of food, the growth rate will be 2.5%;
with 8 units of food, the growth rate will be 4%; with 78 units of food, the
growth rate will be 4.875%.    **(b)** $y = 5$    **(c)** No
**(d)**

$f(x) = \dfrac{5x}{2+x}$

**(e)** The maximum possible growth rate
**25. (a)** $4300   **(b)** $10,033.33   **(c)** $17,200   **(d)** $38,700   **(e)** $81,700
**(f)** $210,700   **(g)** $425,700   **(h)** No
**(i)**

$f(x) = \dfrac{4.3x}{100 - x}$

**27.**

$$y = \frac{125{,}000 - 25x}{125 + 2x}$$

100,000 gal of oil;
500,000 gal of gasoline

**29. (a)** $11.2 billion   **(b)** $12.5 billion   **(c)** $12 billion   **(d)** $9.9 billion
**(e)**   15

**(f)** 28.99%; $12.6 billion

**31. (a)**

**(b)** No; yes

**(c)** Yes

## Chapter 3 Review Exercises (Page 177)

| Refer to Section | 3.1 | 3.2 | 3.3 | 3.4 | 3.5 | 3.6 |
|---|---|---|---|---|---|---|
| For Exercises | 1–12 | 13–24 | 25–36 | 37–60 | 61–70 | 71–80 |

**1.** Not a function    **3.** Function    **5.** Not a function    **7. (a)** 23
**(b)** −9    **(c)** $4p − 1$    **(d)** $4r + 3$    **9. (a)** −28    **(b)** −12
**(c)** $−p^2 + 2p − 4$    **(d)** $−r^2 − 3$    **11. (a)** −13    **(b)** 3    **(c)** $−k^2 − 4k$
**(d)** $−9m^2 + 12m$    **(e)** $−k^2 + 14k − 45$    **(f)** $12 − 5p$

**13.**

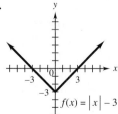

$f(x) = |x| − 3$

**15.**

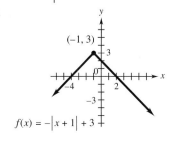

$f(x) = −|x + 1| + 3$

**17.**

$f(x) = [x − 3]$

**19.**

$f(x) = \begin{cases} −4x + 2 & \text{if } x \le 1 \\ 3x − 5 & \text{if } x > 1 \end{cases}$

**21.**

**23.**

$h(x) = \sqrt{x} + 2$

**25. (a)**

**(b)** Domain: $(0, \infty)$ range: {65, 85, 105, 125, ...}    **(c)** 2 days

**27. (a)** Decreasing    **(b)** 45%    **(c)** $f(x) = −x + 49$    **(d)** 32%
**(e)** 2016    **29. (a)** $C(x) = 30x + 60$    **(b)** $30    **(c)** $30.60
**31. (a)** $C(x) = 30x + 85$    **(b)** $30    **(c)** $30.85    **33. (a)** $18,000
**(b)** $R(x) = 28x$    **(c)** 4500 cartridges    **(d)** $126,000    **35.** Equilibrium
quantity is 36 million subscribers at a price of $12.95 per month.
**37.** Upward; (2, 6)    **39.** Downward; (−1, 8)

**41.**

Vertex (0, −9)

**43.**

Vertex (−1, −7)

**45.** Vertex (−3, 14)

**47.**

Vertex (3, −8)

**49.** Minimum value; $-11$   **51.** Maximum value; 7   **53. (a)** 71,741,000
**(b)** (9,74090); maximum   **(c)** 2009   **(d)** 74,090,000   **55.** 2012; 3.4%
**57. (a)** $f(x) = 72.875(x - 6)^2 + 997$   **(b)** \$8284.5 million
**59. (a)** $R(x) = 65x^2 - 724x + 3047$
**(b)** Yes  8000                          **(c)** \$8103 million

**61.**

$g(x) = x^3 - 4x$

**63.**

$f(x) = x(x - 4)(x + 1)$

**65.**

$f(x) = x^4 - 5x^2 - 6$

**67. (a)** $R(x) = -.194x^3 + 6.576x^2 - 69.942x + 256.288$
**(b)** Yes  40                          **(c)** \$28.696 million; yes

**69.** $P(x) = -.041x^3 + 1.439x^2 - 16.040x + 61.120$; \$4.928 million
**71.** $x = 3$, $y = 0$                **73.** $x = 2$, $y = 0$

$f(x) = \dfrac{1}{x - 3}$

$f(x) = \dfrac{-3}{2x - 4}$

**75.** $x = -\dfrac{1}{2}$, $x = \dfrac{3}{2}$, $y = 0$

$g(x) = \dfrac{5x - 2}{4x^2 - 4x - 3}$

**77. (a)** About \$10.83   **(b)** About \$4.64   **(c)** About \$3.61   **(d)** About \$2.71
**(e)**

$C(x) = \dfrac{650}{2x + 40}$

**79.** (10, 50)

Supply
$p = \dfrac{q^2}{4} + 25$

Demand
$p = \dfrac{500}{q}$

## Case Study 3 Exercises (Page 181)

**1.** 550 washing machine loads leads to a maximum profit of \$992.50.
**3.** \$3.25 per load

## Extended Project (Page 181)

**(a)** 500 dryer loads   **(b)** maximum profit of \$105   **(c)** \$1.50 per dryer load

# Chapter 4

## Section 4.1 (Page 188)

**1.** Exponential   **3.** Quadratic   **5.** Exponential   **7. (a)** The graph
is entirely above the $x$-axis and falls from left to right, crossing the $y$-axis at
1 and then getting very close to the $x$-axis.   **(b)** (0, 1), (1, .6)
**9. (a)** The graph is entirely above the $x$-axis and rises from left to right, less
steeply than the graph of $f(x) = 2^x$.   **(b)** (0, 1), (1, $2^{.5}$) = (1, $\sqrt{2}$)
**11. (a)** The graph is entirely above the $x$-axis and falls from left to right,
crossing the $y$-axis at 1 and then getting very close to the $x$-axis.   **(b)** (0, 1),
(1, $e^{-1}$) $\approx$ (1, .367879)

**13.**

$f(x) = 3^x$

**15.**

$f(x) = 2^{x/2}$

**17.**

$f(x) = \left(\dfrac{1}{5}\right)^x$

**19.** (a)–(c)                    (d) Answers vary.

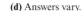

**21.** 2.3    **23.** .75    **25.** .31    **27.** (a) $a > 1$    (b) Domain: $(-\infty, \infty)$, range: $(0, \infty)$

(c)                    (d) Domain: $(-\infty, \infty)$, range: $(-\infty, 0)$

(e)                    (f) Domain: $(-\infty, \infty)$, range: $(0, \infty)$

**29.** (a) 3    (b) $\dfrac{1}{3}$    (c) 9    (d) 1

**31.**                    **33.**

$f(x) = 2^{-x^2 + 2}$

**35.** (a)

| $t$ | 0 | 1 | 2 | 3 | 4 | 5 | 6 | 7 | 8 | 9 | 10 |
|---|---|---|---|---|---|---|---|---|---|---|---|
| $y$ | 1 | 1.06 | 1.12 | 1.19 | 1.26 | 1.34 | 1.42 | 1.50 | 1.59 | 1.69 | 1.79 |

(b)

**37.** (a) About $141,892    (b) About $64.10    **39.** (a) About $616.8 billion    (b) About $729.7 billion    (c) About $863.1 billion    **41.** (a) About .97 kg    (b) About .75 kg    (c) About .65 kg    (d) About 24,360 years    **43.** $34,706.99    **45.** (a) About 6.85 billion    (b) About 7.20 billion    (c) About 8.78 billion    (d) Answers vary.    **47.** About $7.8 trillion; about $20.6 trillion    **49.** 2027 ($x \approx 126.99$)    **51.** (a) About $2.74 trillion; about $5.50 trillion    (b) 2017 $x \approx (16.96)$    **53.** (a) About $4.46 billion    (b) 2018 ($x \approx 17.59$)

## Section 4.2 (Page 195)

**1.** (a) $752.27    (b) $707.39    (c) $432.45    (d) $298.98    (e) Answer vary.    **3.** (a) About 27.17 trillion cubic feet    (b) About 36.24 quadtrillion Btus    **5.** (a) $f(t) = 30.48(1.074)^t$    (b) $71.79    (c) 2010    **7.** (a) $f(t) = 9.6(1.102)^t$    (b) About $98.8 billion    (c) About $176.9 billion    **9.** (a) Two-point: $f(t) = 17(1.178)^t$; regression: $g(t) = 17.4(1.181)^t$    (b) Two-point: 324.4 trillion yuan; 530.0 trillion yuan; regression: 347.6 trillion yuan; 572.5 trillion yuan    (c) Two-point: 2020 ($x \approx 19.998$); regression: 2020 ($x \approx 19.55$)    **11.** (a) Two-point: $f(t) = 257.6(.97)^t$; regression: $g(t) = 252.6(.97)^t$    (b) Two-point: about 144.4 about 127.8; regression: about 141.6; about 125.4    (c) Two-point: 2032 ($x \approx 31.07$); regression: 2031 ($x \approx 30.42$)    **13.** (a) About 6 items    (b) About 23 items    (c) 25 items    **15.** 2.6°C    **17.** (a) About 12.0%; about 47.6%    (b) No    **19.** (a) About $27.4 billion; about $109.1 billion    (b) 2013 ($x \approx 12.15$)    **21.** (a) About 66.1%; about 61.6%

(b)

## Section 4.3 (Page 204)

**1.** $a^y$    **3.** It is missing the value that equals $b^y$. If that value is $x$, the expression should read $y = \log_b x$.    **5.** $10^5 = 100,000$

**7.** $9^2 = 81$    **9.** $\log 96 = 1.9823$    **11.** $\log_3\left(\dfrac{1}{9}\right) = -2$

**13.** 3    **15.** 2    **17.** 3    **19.** $-2$    **21.** $\dfrac{1}{2}$    **23.** 8.77

**25.** 1.724    **27.** $-4.991$    **29.** Because $a^0 = 1$ for every valid base $a$.

**31.** $\log 24$    **33.** $\ln 5$    **35.** $\log\left(\dfrac{u^2 w^3}{v^6}\right)$    **37.** $\ln\left(\dfrac{(x+2)^2}{x+3}\right)$

**39.** $\dfrac{1}{2}\ln 6 + 2\ln m + \ln n$    **41.** $\dfrac{1}{2}\log x - \dfrac{5}{2}\log z$    **43.** $2u + 5v$

**45.** $3u - 2v$    **47.** 3.32112    **49.** 2.429777    **51.** There are many correct answers, including $b = 1, c = 2$.

**53.**                    **55.**

**57.** Answers vary.    **59.** $\ln 2.75 = 1.0116009$; $e^{1.0116009} = 2.75$    **61.** (a) 17.67 yr    (b) 9.01 yr    (c) 4.19 yr    (d) 2.25 yr    (e) The pattern is that it takes about $72/k$ years to double at $k\%$ interest.    **63.** (a) $2089.84    (b) 3000    (c) Gradually increasing

**65.** (a) $399.83; $417.81    (b) 2019 ($x \approx 18.89$)    **67.** (a) About $2.1 billion; about $3.5 billion    (b) 2024 ($x \approx 23.46$)    **69.** (a) About 205.5 million    (b) 2017 ($x \approx 16.77$)

## Section 4.4 (Page 212)

**1.** 8    **3.** 9    **5.** 11    **7.** $\dfrac{11}{6}$    **9.** $\dfrac{4}{9}$    **11.** 10    **13.** 5.2378    **15.** 10

**17.** $\dfrac{4+b}{4}$    **19.** $\dfrac{10^{2-b}-5}{6}$    **21.** Answers vary.    **23.** 4    **25.** $-\dfrac{5}{6}$

**27.** −4   **29.** −2   **31.** 2.3219   **33.** 2.710   **35.** −1.825

**37.** .597253   **39.** −.123   **41.** $\dfrac{\log d + 3}{4}$   **43.** $\dfrac{\ln b + 1}{2}$

**45.** 4   **47.** No solution   **49.** −4, 4   **51.** 9   **53.** 1
**55.** 4, −4   **57.** ±2.0789   **59.** 1.386   **61.** Answers vary.
**63. (a)** 2013 ($x \approx 12.03$)   **(b)** 2016 ($x \approx 15.28$)   **65. (a)** About
$138.0 billion   **(b)** 2011 ($x \approx 20.06$)   **67. (a)** 2021 ($x \approx 20.74$)
**(b)** 2027 ($x \approx 26.87$)   **69. (a)** $2.85   **(b)** 2012 ($x \approx 11.2$)
**71. (a)** About $109.5 billion   **(b)** 2014 ($x \approx 13.8$)   **73.** The seventh
quarter is the third quarter of 2016 ($x \approx 6.44$).   **75. (a)** 25 g
**(b)** About 4.95 yr   **77.** About 3689 yr old   **79. (a)** Approximately
79,432,823 $i_0$   **(b)** Approximately 251,189 $i_0$   **(c)** About 316.23 times
stronger   **81. (a)** 21   **(b)** 100   **(c)** 105   **(d)** 120   **(e)** 140

## Chapter 4 Review Exercises (Page 216)

| Refer to Section | 4.1. | 4.2 | 4.3 | 4.4 |
|---|---|---|---|---|
| For Exercises | 1–10 | 13–14, 55–60, 63–64 | 11–12, 15–40, 65–68 | 41–54, 61–62 |

**1.** (c)   **3.** (d)   **5.** $0 < a < 1$   **7.** All positive real numbers
**9.**   **11.**

 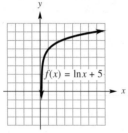

$f(x) = 4^x$

$f(x) = \ln x + 5$

**13. (a)** About $294 billion   **(b)** 2008 ($x \approx 17.95$)
**15.** $\log 340 = 2.53148$   **17.** $\ln 45 = 3.8067$   **19.** $10^4 = 10{,}000$

**21.** $e^{4.3957} = 81.1$   **23.** 5   **25.** 8.9   **27.** $\dfrac{1}{3}$   **29.** $\log 20x^6$

**31.** $\log\left(\dfrac{b^3}{c^2}\right)$   **33.** 4   **35.** 97   **37.** 5   **39.** 5   **41.** −2

**43.** −2   **45.** 1.416   **47.** −2.807   **49.** −3.305   **51.** .747
**53.** 28.463   **55. (a)** C   **(b)** A   **(c)** D   **(d)** B   **57. (a)** About 400
**(b)** 2001 ($x \approx 10.11$)   **59. (a)** $70.69   **(b)** The third quarter of
2015 ($x \approx 6.82$)   **61. (a)** 63,095,734.45   **(b)** 1,000,000$i_0$
**(c)** About 63.1 times stronger   **63.** 81.25°C
**65. (a)** $f(x) = 25(1.057)^x$   **(b)** $g(x) = 25.08(1.052)^x$
**(c)** Two point: about $34.9 billion; regression: about $34.0 billion
**(d)** Two point: 2020 ($x \approx 8.48$); regression: 2021 ($x \approx 9.21$)
**67. (a)** $f(x) = 4.9 + .684 \ln x$   **(b)** $g(x) = 5.1 + .697 \ln x$
**(c)** Two point: about 6.6 million; regression: about 6.9 million
**(d)** Two point: March 2016 ($x \approx 21.5$); regression: Oct 2015 ($x \approx 16.2$)

## Case Study 4 Exercises (Page 221)

**1.**

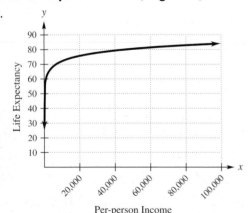

**3.** About 67.3 years   **5.** About $993
**7.**

**9.** About $3564; about $12,251
**11.**

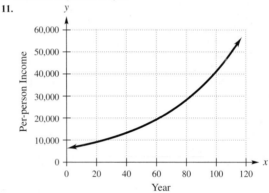

**13.** About $19,365; about $49,629

# Chapter 5

## Section 5.1 (Page 228)
**1.** Time and interest rate   **3.** $133   **5.** $217.48   **7.** $86.26
**9.** $158.82   **11.** $155.50; $9330   **13.** $256.25; $5125
**15.** $39; $390   **17.** $12,105   **19.** $6727.50   **21.** Answers vary.
**23.** $14,354.07   **25.** $15,089.46   **27.** $19,986.50; about .27018 %
**29.** $15,472.10; about .36065%   **31.** $19,746.50; about 5.1351%
**33.** $15,104.75; about 5.2335%   **35.** $652,778.08   **37.** $3090
**39.** About 5.0%   **41.** 4 months   **43.** $5825.24   **45.** $83,533.51
**47.** 3.5%   **49.** 4.25%   **51.** About 11.36%   **53.** About 102.91%
**55.** $13,725; yes   **57. (a)** $y_1 = 8t + 100$; $y_2 = 6t + 200$   **(b)** The graph
of $y_1$ is a line with slope 8 and $y$-intercept 100. The graph of $y_2$ is a line with
slope 6 and $y$-intercept 200.
**(c)**

**(d)** The $y$-intercept of each graph indicates the amount invested. The slope
of each graph is the annual amount of interest paid.

## Section 5.2 (Page 238)
**1.** Answers vary.   **3.** Interest rate and total number of compounding periods
**5.** Answers vary.   **7.** $1223.59; $753.59   **9.** $9297.93; $2797.93

**11.** $10,111.62; $111.62   **13.** $15,456.62; $456.62
**15.** $77,746.04; $2746.04   **17.** About 3.75%   **19.** About 5.25%
**21.** $23,824.92; $3824.92   **23.** $31,664.54; $1664.54   **25.** $10,000
**27.** $20,000   **29.** 4.04%   **31.** 5.095%   **33.** $8954.58
**35.** $11,572.58   **37.** $4203.64   **39.** $9963.10   **41.** $4688.71
**43.** $465.85   **45.** $1000 now   **47.** Treasury note
**49.** The 4.8% bond (by 11 cents).   **51.** About $2,259,696
**53.** About $21,043   **55.** 23.4 years   **57.** 14.2 years
**59.** About 11.9 years   **61. (a)** $16,659.95   **(b)** $21,472.67   **63.** $16,500

## Section 5.3 (Page 247)

**1.** 21.5786   **3.** $73,167.09   **5.** $7679.69   **7.** $105,033.01
**9.** About $205,490   **11.** About $310,831   **13.** $797.36
**15.** $4566.33   **17.** $152.53   **19.** 5.19%   **21.** 5.223%
**23.** Answers vary.   **25.** $5069.27   **27.** $2,133,205.87
**29.** $18,128.80   **31.** $5661.31   **33.** $620.46   **35.** $265.71
**37.** $163,959.61   **39. (a)** $137,895.79   **(b)** $132,318.77   **(c)** $5577.02
**41. (a)** $69,421.34   **(b)** $99,185.58   **43.** $32,426.46   **45.** $284,527.35
**47. (a)** $256.08   **(b)** $247.81   **49.** $863.68   **51. (a)** $1200
**(b)** $3511.58   **53.** 6.5%   **55. (a)** Answers vary.   **(b)** $24,000
**(c)** $603,229   **(d)** $84,000   **(e)** $460,884

## Section 5.4 (Page 258)

**1.** Answers vary.   **3.** $8693.71   **5.** $1,566,346.66   **7.** $11,468.10
**9.** $38,108.61   **11.** $557.68   **13.** $6272.14   **15.** $446.31
**17.** $11,331.18   **19.** $589.31   **21.** $1033.09   **23.** $942.10
**25.** $416.11; $18,524.33   **27.** $958.30; $151,811.44   **29.** $6.80
**31.** $42.04   **33.** $30,669,881   **35.** $23,016,361   **37. (a)** $1465.42
**(b)** $214.58   **39.** $2320.83   **41.** $414.18; $14,701.60   **43. (a)** $2717.36
**(b)** 2   **45. (a)** $969.75   **(b)** $185,658.15   **(c)** $143,510
**47.** About $8143.79   **49.** $406.53   **51.** $663.22   **53. (a)** $566.93
**(b)** $5,818.96   **(c)** 60   **(d)** $4794.04

**55.**

| Payment Number | Amount of Payment | Interest for Period | Portion to Principal | Principal at End of Period |
|---|---|---|---|---|
| 0 | — | — | — | $4000.00 |
| 1 | $1207.68 | $320.00 | $887.68 | 3112.32 |
| 2 | 1207.68 | 248.99 | 958.69 | 2153.63 |
| 3 | 1207.68 | 172.29 | 1035.39 | 1118.24 |
| 4 | 1207.70 | 89.46 | 1118.24 | 0 |

**57.**

| Payment Number | Amount of Payment | Interest for Period | Portion to Principal | Principal at End of Period |
|---|---|---|---|---|
| 0 | — | — | — | $7184.00 |
| 1 | $189.18 | $71.84 | 117.34 | 7066.66 |
| 2 | 189.18 | 70.67 | 118.51 | 6948.15 |
| 3 | 189.18 | 69.48 | 119.70 | 6828.45 |
| 4 | 189.18 | 68.28 | 120.90 | 6707.55 |

## Chapter 5 Review Exercises (Page 262)

| Refer to Section | 5.1 | 5.2 | 5.3 | 5.4 |
|---|---|---|---|---|
| For Exercises | 1–14 | 15–28, 65–66, 71 | 32–41, 67–70, 75–76 | 29–31, 42–64, 72–74, 77–78 |

**1.** $292.08   **3.** $62.05   **5.** $285; $3420   **7.** $7925.67
**9.** $3084.05   **11.** $78,742.54   **13.** About 4.082%
**15.** $5634.15; $2834.15   **17.** $20,402.98; $7499.53   **19.** $8532.58
**21.** $15,000   **23.** 5.0625%   **25.** $18,207.65   **27.** $1088.54
**29.** $9706.74   **31.** Answers vary.   **33.** $162,753.15
**35.** $22,643.29   **37.** Answers vary.   **39.** $2619.29   **41.** $916.12
**43.** $31,921.91   **45.** $14,222.42   **47.** $136,340.32   **49.** $1194.02
**51.** $38,298.04   **53.** $3581.11   **55.** $435.66   **57.** $85,459.20
**59.** $45.04   **61.** $277.13   **63. (a)** $14,320,000   **(b)** $270,088,972.50
**65.** $2298.58   **67.** $24,818.76; $2418.76   **69.** $12,538.59
**71.** $5596.62   **73.** $3560.61   **75.** $165,249.03   **77.** $711.25

## Case Study 5 Exercises (Page 266)

**1.** $150,677.26   **3. (a)** 67   **(b)** $301,354.52   **5.** 3 times; $80,000

# Chapter 6

## Section 6.1 (Page 273)

**1.** Yes   **3.** $(-1, -4)$   **5.** $\left(\dfrac{2}{7}, -\dfrac{11}{7}\right)$   **7.** $\left(\dfrac{11}{5}, -\dfrac{7}{5}\right)$
**9.** $(28, 22)$   **11.** $(2, -1)$   **13.** No solution   **15.** $(4y + 1, y)$ for any real number $y$   **17.** $(12, 6)$   **19.** $(15, 75)$; In 2015, the number of millennials and boomers were each 75 million.   **21.** About 2025
**23.** About 2014   **25.** No   **27.** 145 adults; 55 children
**29.** Plane: 550 mph; wind: 50 mph   **31. (a)** $y = x + 1$
**(b)** $y = 3x + 4$   **(c)** $\left(-\dfrac{3}{2}, -\dfrac{1}{2}\right)$

## Section 6.2 (Page 283)

**1.**
$$\begin{aligned} x \quad\quad - 3z &= 2 \\ 2x - 4y + 5z &= 1 \\ 5x - 8y + 7z &= 6 \\ 3x - 4y + 2z &= 3 \end{aligned}$$
**3.**
$$\begin{aligned} x + 2y + 4z &= 3 \\ x \quad\quad + 2z &= 0 \\ x + 2y + \tfrac{1}{2}z &= \tfrac{3}{2} \end{aligned}$$
**5.**
$$\begin{aligned} x + y + 2z + 3w &= 1 \\ - y - z - 2w &= -1 \\ 3x + y + 4z + 5w &= 2 \end{aligned}$$
**7.**
$$\begin{aligned} x + 12y - 3z + 4w &= 10 \\ 2y + 3z + w &= 4 \\ - z &= -7 \\ 6y - 2z - 3w &= 0 \end{aligned}$$
**9.** $(-68, 13, -6, 3)$   **11.** $\left(20, -9, \dfrac{15}{2}, 3\right)$
**13.** $\begin{bmatrix} 2 & 1 & 1 & | & 3 \\ 3 & -4 & 2 & | & -5 \\ 1 & 1 & 1 & | & 2 \end{bmatrix}$
**15.**
$$\begin{aligned} 2x + 3y + 8z &= 20 \\ x + 4y + 6z &= 12 \\ 3y + 5z &= 10 \end{aligned}$$
**17.** $\begin{bmatrix} 1 & 7 & -3 & | & 5 \\ 2 & 2 & 9 & | & 3 \\ 5 & -1 & 4 & | & 8 \end{bmatrix}$
**19.** $\begin{bmatrix} 2 & 5 & 1 & | & -1 \\ -4 & 0 & 4 & | & 6 \\ \frac{3}{2} & 0 & 2 & | & -1 \end{bmatrix}$
**21.** $\begin{bmatrix} -4 & -3 & 1 & -1 & | & 2 \\ 0 & -4 & 7 & -2 & | & 10 \\ 0 & -2 & 9 & 4 & | & 5 \end{bmatrix}$   **23.** $\left(\dfrac{3}{2}, 17, -5, 0\right)$
**25.** No solution   **27.** $(12 - w, -3 - 2w, -5, w)$ for any real number $w$
**29.** Dependent   **31.** Inconsistent   **33.** Independent
**35.** $\left(\dfrac{3}{2}, \dfrac{3}{2}, -\dfrac{3}{2}\right)$   **37.** $(-.8, -1.8, .8)$   **39.** $(100, 50, 50)$
**41.** $(0, 3.5, 1.5)$   **43.** $(-3, 4, 0)$   **45.** $(1, 0, -1)$   **47.** $(-93, -31, 10)$
**49.** No solution   **51.** $(22 + 7z, 15 + 10z, z)$ for any real number $z$
**53.** $(-7, 5)$   **55.** No solution   **57.** $(-3, z - 17, z)$ for any real number $z$   **59.** $(-11.5, 13.75, -2.25, -5)$   **61.** No solution
**63.** $(z - 3, -z + 5, z)$ for any real number $z$   **65.** $(12, 5)$; In 2012, all three counties had a population of 5000.   **67.** 220 adults; 150 teenagers; 200 children
**69.**

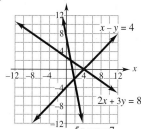

There is no point that is on all three lines.

**71.** $y = .75x^2 + .25x - .5$

## Section 6.3 (Page 292)

**1.** 15 minutes; 5 miles   **3.** Purchase 200 pounds and 600 euros
**5.** 200 of Coca-Cola; 500 of Intel

**7. (a)** $x + 2y + 3z = 31$
$y + z = 12$
$x = 4$

**(b)** The center made 4 free throws, 9 regular baskets, and 3 three-point baskets. **9.** 100 vans, 50 small trucks, 25 large trucks **11.** 10 units of corn; 25 units of soybeans; 40 units of cottonseed **13.** Shirt $1.99; slacks $4.99; sports coat $6.49 **15.** 220 adults; 150 teenagers; 200 preteens **17.** $12,000 at 6%, $7000 at 7%, and $6000 at 10% **19.** $13,750 in the mutual fund, $27,500 in bonds, and $28,750 in the franchise **21.** 40 lb pretzels, 20 lb dried fruit, 80 lb nuts **23.** Four possible solutions: (1) no cases of A and D; 12 cases of B; 8 cases of C; (2) 1 of A; 8 of B; 9 of C; 1 of D; (3) 2 of A; 4 of B; 10 of C; 2 of D; (4) 3 of A; none of B; 11 of C; 3 of D **25. (a)** $f(x) = -.008x^2 + 1.6x + 140.8$ **(b)** 169.6 million **27. (a)** $y = .01x^2 - .3x + 4.24$ **(b)** 15 platters; $1.99 **29.** 4 cup rotini, 5 cups healthy, 6 cups chunky, serving size 1.5 cups

## Section 6.4 (Page 300)

**1.** $2 \times 3$; $\begin{bmatrix} -7 & 8 & -4 \\ 0 & -13 & -9 \end{bmatrix}$ **3.** $3 \times 3$; square matrix; $\begin{bmatrix} 3 & 0 & -11 \\ -1 & -\frac{1}{4} & 7 \\ -5 & 3 & -9 \end{bmatrix}$

**5.** $2 \times 1$; column matrix; $\begin{bmatrix} -7 \\ -11 \end{bmatrix}$ **7.** B is a $5 \times 3$ zero matrix

**9.** $\begin{bmatrix} -7 & 14 & 2 & 4 \\ 6 & -3 & 2 & -4 \end{bmatrix}$ **11.** $\begin{bmatrix} 3 & -1 & 2 \\ 3 & 1 & 5 \end{bmatrix}$ **13.** $\begin{bmatrix} -8 & -8 \\ -8 & -8 \\ -8 & -8 \end{bmatrix}$

**15.** Not defined **17.** $\begin{bmatrix} -4 & 0 \\ 10 & 6 \end{bmatrix}$ **19.** $\begin{bmatrix} 0 & -8 \\ -16 & 24 \end{bmatrix}$

**21.** $\begin{bmatrix} 8 & 10 \\ 0 & -42 \end{bmatrix}$ **23.** $\begin{bmatrix} 4 & -\frac{7}{2} \\ 4 & \frac{21}{2} \end{bmatrix}$

**25.** $X + T = \begin{bmatrix} x & y \\ z & w \end{bmatrix} + \begin{bmatrix} r & s \\ t & u \end{bmatrix} = \begin{bmatrix} x+r & y+s \\ z+t & w+u \end{bmatrix}$; a $2 \times 2$ matrix

**27.** $X + (T + P) = \begin{bmatrix} x + (r+m) & y + (s+n) \\ z + (t+p) & w + (u+q) \end{bmatrix}$
$= \begin{bmatrix} (x+r)+m & (y+s)+n \\ (z+t)+p & (w+u)+q \end{bmatrix} = (X + T) + P$

**29.** $P + O = \begin{bmatrix} m+0 & n+0 \\ p+0 & q+0 \end{bmatrix} = \begin{bmatrix} m & n \\ p & q \end{bmatrix} = P$

**31.** Several possible correct answers, including

| | Basketball | Hockey | Football | Baseball |
|---|---|---|---|---|
| Percent of no shows | 16 | 16 | 20 | 18 |
| Lost revenue per fan ($) | 18.20 | 18.25 | 19 | 15.40 |
| Lost annual revenue (millions $) | 22.7 | 35.8 | 51.9 | 96.3 |

**33.**
| | 2010 | 2016 |
|---|---|---|
| Heart | 2874 | 4154 |
| Lung | 1759 | 1470 |
| Liver | 15,394 | 14,794 |
| Kidney | 83,919 | 100,467 |

**35. (a)**
| | Bread | Milk | Butter | Peanut Cuts | Cold |
|---|---|---|---|---|---|
| I | 88 | 48 | 16 | 112 | |
| II | 105 | 72 | 21 | 147 | |
| III | 60 | 40 | 0 | 50 | |

**(b)** $\begin{bmatrix} 110 & 60 & 20 & 140 \\ 140 & 96 & 28 & 196 \\ 66 & 44 & 0 & 55 \end{bmatrix}$ **(c)** $\begin{bmatrix} 198 & 108 & 36 & 252 \\ 245 & 168 & 49 & 343 \\ 126 & 84 & 0 & 105 \end{bmatrix}$

## Section 6.5 (Page 310)

**1.** $2 \times 2$; $2 \times 2$ **3.** $3 \times 3$; $5 \times 5$ **5.** AB does not exist; $3 \times 2$

**7.** columns; rows **9.** $\begin{bmatrix} 5 \\ 9 \end{bmatrix}$ **11.** $\begin{bmatrix} -2 & 4 \\ 0 & -8 \end{bmatrix}$

**13.** $\begin{bmatrix} -4 & 1 \\ 2 & -3 \end{bmatrix}$ **15.** $\begin{bmatrix} 3 & -5 & 7 \\ -2 & 1 & 6 \\ 0 & -3 & 4 \end{bmatrix}$ **17.** $\begin{bmatrix} 16 & 10 \\ 2 & 28 \\ 58 & 46 \end{bmatrix}$

**19.** $AB = \begin{bmatrix} -30 & -45 \\ 20 & 30 \end{bmatrix}$, but $BA = \begin{bmatrix} 0 & 0 \\ 0 & 0 \end{bmatrix}$.

**21.** $(A + B)(A - B) = \begin{bmatrix} -7 & -24 \\ -28 & -33 \end{bmatrix}$, but $A^2 - B^2 = \begin{bmatrix} -37 & -69 \\ -8 & -3 \end{bmatrix}$.

**23.** $(PX)T = \begin{bmatrix} (mx+nz)r + (my+nw)t & (mx+nz)s + (my+nw)u \\ (px+qz)r + (py+qw)t & (px+qz)s + (py+qw)u \end{bmatrix}$.
$P(XT)$ is the same, so $(PX)T = P(XT)$.

**25.** $k(X + T) = k\begin{bmatrix} x+r & y+s \\ z+t & w+u \end{bmatrix}$
$= \begin{bmatrix} k(x+r) & k(y+s) \\ k(z+t) & k(w+u) \end{bmatrix}$
$= \begin{bmatrix} kx+kr & ky+ks \\ kz+kt & kw+ku \end{bmatrix}$
$= \begin{bmatrix} kx & ky \\ kz & kw \end{bmatrix} + \begin{bmatrix} kr & ks \\ kt & ku \end{bmatrix} = kX + kT$

**27.** No **29.** Yes **31.** Yes **33.** No inverse

**35.** $\begin{bmatrix} 1 & 2 \\ 1 & 1 \end{bmatrix}$ **37.** $\begin{bmatrix} 2 & 3 \\ 1 & 2 \end{bmatrix}$ **39.** $\begin{bmatrix} -4 & -2 & 3 \\ -5 & -2 & 3 \\ 2 & 1 & -1 \end{bmatrix}$

**41.** $\begin{bmatrix} -13 & 8 & 7 \\ -6 & 4 & 3 \\ 2 & -1 & -1 \end{bmatrix}$ **43.** No inverse **45.** $\begin{bmatrix} 6 & -5 & 8 \\ -1 & 1 & -1 \\ -1 & 1 & -2 \end{bmatrix}$

**47.** $\begin{bmatrix} \frac{1}{2} & -1 & -\frac{1}{2} & \frac{1}{2} \\ \frac{1}{2} & 4 & \frac{5}{2} & -\frac{1}{2} \\ -\frac{1}{4} & -\frac{1}{2} & -\frac{1}{4} & \frac{1}{4} \\ \frac{1}{2} & -2 & -\frac{3}{2} & \frac{1}{2} \end{bmatrix}$ **49. (a)** $\begin{bmatrix} \$56 & \$52 \\ \$20 & \$22 \end{bmatrix}$

**(b)** Row 1 represents you, and row 2 represents your roommate.
**(c)** Column 1 represents Timesaver, and column 2 represents Wheels.
**(d)** $76; $74 **51. (a)** $AB = \begin{bmatrix} 705 & 80 \\ 121 & 14 \end{bmatrix}$ **(b)** Row 1 represents the restaurant, and row 2 represents the food truck. Column 1 represents total salary, and column 2 represents total benefits (in thousands) for all employees at those locations. **(c)** $705,000; $121,000; $80,000; $14,000

**53. (a)** No **(b)** $\begin{bmatrix} 4800 \\ 3200 \\ 5600 \end{bmatrix}$ represents the revenue in each state. **(c)** $13,600

**55. (a)** $A = \begin{bmatrix} 195 & 143 & 1225 & 1341 \\ 210 & 141 & 1387 & 1388 \end{bmatrix}$ **(b)** $B = \begin{bmatrix} .016 & .006 \\ .011 & .014 \\ .023 & .008 \\ .013 & .007 \end{bmatrix}$

**(c)** $AB = \begin{bmatrix} 50 & 22 \\ 55 & 24 \end{bmatrix}$ **(d)** The rows represent the years 2010 and 2020. Column 1 gives the total births (in millions) in those years, and column 2 gives the total deaths (in millions). **(e)** About 22 million; about 55 million

## Section 6.6 (Page 321)

**1.** $\begin{bmatrix} 2 \\ -2 \end{bmatrix}$ **3.** $\begin{bmatrix} \frac{1}{2} & 1 \\ \frac{3}{2} & 1 \end{bmatrix}$ **5.** $\begin{bmatrix} -7 \\ 15 \\ -3 \end{bmatrix}$ **7.** $(-63, 50, -9)$

**9.** $(-31, -131, 181)$ **11.** $(-48, 8, 12)$ **13.** $(1, 0, 2, 1)$ **15.** $9000 and 100 transactions **17.** 60 buffets; 600 chairs; 100 tables **19.** Jeans $34.50; jacket $72; sweater $44; shirt $21.75 **21.** 2340 of the first species, 10,128 of the second species, 224 of the third species

**23.** $\begin{bmatrix} 6.43 \\ 26.12 \end{bmatrix}$ **25.** About 1073 metric tons of wheat, about 1431 metric

tons of oil **27. (a)** $\frac{7}{4}$ bushels of yams, $\frac{15}{8} \approx 2$ pigs **(b)** 167.5 bushels of yams, $153.75 \approx 154$ pigs **29.** Gas $98 million; electric $123 million **31. (a)** .40 unit of agriculture, .12 unit of manufacturing, and 3.60 units of households **(b)** 848 units of agriculture, 516 units of manufacturing, and 2970 units of households **(c)** About 813 units **33. (a)** .017 unit of manufacturing and .216 unit of energy **(b)** 123,725,000 pounds of agriculture, 14,792,000 pounds of manufacturing, 1,488,000 pounds of energy

**(c)** 195,492,000 pounds of agriculture, 25,933,000 pounds of manufacturing, 13,580,000 pounds of energy   **35.** $532 million of natural resources, $481 million of manufacturing, $805 million of trade and services, $1185 million of personal consumption   **37.** $\begin{bmatrix} 43 \\ 100 \end{bmatrix}, \begin{bmatrix} 29 \\ 63 \end{bmatrix}, \begin{bmatrix} 54 \\ 117 \end{bmatrix}, \begin{bmatrix} 100 \\ 227 \end{bmatrix}, \begin{bmatrix} 53 \\ 121 \end{bmatrix}, \begin{bmatrix} 28 \\ 61 \end{bmatrix}$

**39.** Love   **41. (a)** 3  **(b)** 3  **(c)** 5  **(d)** 3

**43. (a)**
$C = \begin{array}{c} \\ \text{Dogs} \\ \text{Rats} \\ \text{Cats} \\ \text{Mice} \end{array}\begin{array}{cccc} \text{Dogs} & \text{Rats} & \text{Cats} & \text{Mice} \\ 0 & 1 & 1 & 1 \\ 0 & 0 & 0 & 1 \\ 0 & 1 & 0 & 1 \\ 0 & 0 & 0 & 0 \end{array}$

**(b)** $\begin{bmatrix} 0 & 1 & 0 & 2 \\ 0 & 0 & 0 & 0 \\ 0 & 0 & 0 & 1 \\ 0 & 0 & 0 & 0 \end{bmatrix}$ ; $C^2$ gives the number of food sources once removed from the feeder.

## Chapter 6 Review Exercises (Page 326)

| Refer to Section | 6.1 | 6.2 | 6.3 | 6.4 | 6.5 | 6.6 |
|---|---|---|---|---|---|---|
| For Exercises | 1–10 | 11–16 | 17–20, 79–86 | 21–38 | 39–66 | 67–78, 87–96 |

**1.** $(9, -14)$   **3.** $(2, -2)$   **5.** 100 shares of first stock; 300 shares of second stock   **7.** Inconsistent, no solution   **9.** $(-18, 69, 12)$
**11.** $(-14, 17, 11)$   **13.** No solution   **15.** $(4, 3, 2)$   **17.** 19 ones; 7 fives; 9 tens   **19.** 5 blankets; 3 rugs; 8 skirts   **21.** $2 \times 2$; square
**23.** $1 \times 4$; row   **25.** $2 \times 3$   **27.** $\begin{bmatrix} 162.45 & 164.44 & 163.64 \\ 31.88 & 31.94 & 31.86 \\ 51.66 & 52.19 & 51.95 \end{bmatrix}$
**29.** $\begin{bmatrix} -1 & -2 & 3 \\ -2 & -3 & 0 \\ 0 & -1 & -4 \end{bmatrix}$   **31.** $\begin{bmatrix} 14 & 6 \\ -4 & 4 \\ 13 & 23 \end{bmatrix}$   **33.** Not defined
**35.** $\begin{bmatrix} 2 & 18 \\ -4 & -12 \\ 7 & 13 \end{bmatrix}$   **37.** $\begin{bmatrix} -.08 & 25.1 \\ -1.50 & 12.6 \\ .65 & 3.1 \end{bmatrix}, \begin{bmatrix} .10 & 27.2 \\ .63 & 12.4 \\ .47 & 2.9 \end{bmatrix}, \begin{bmatrix} .02 & 52.3 \\ -.87 & 25.0 \\ 1.12 & 6.0 \end{bmatrix}$
**39.** $\begin{bmatrix} 14 & 56 \\ -6 & -22 \\ 19 & 79 \end{bmatrix}$   **41.** Not defined   **43.** $\begin{bmatrix} 322 & 420 \\ -126 & -166 \\ 455 & 591 \end{bmatrix}$
**45.** About 20 head and face injuries; about 12 concussions; about 3 neck injuries; about 55 other injuries
**47. (a)** $\begin{bmatrix} 60.94 & 1.00 \\ 113.22 & 3.22 \\ 119.11 & 4.24 \end{bmatrix}$   **(b)** $[500 \quad 300 \quad 200]$   **(c)** $[88{,}258 \quad 2314]$
**49.** Many correct answers, including $\begin{bmatrix} 1 & 2 \\ 3 & 4 \end{bmatrix}$   **51.** $\begin{bmatrix} -.5 & .2 \\ 0 & .2 \end{bmatrix}$
**53.** No inverse   **55.** $\begin{bmatrix} \frac{1}{4} & \frac{1}{2} & \frac{1}{2} \\ \frac{1}{4} & -\frac{1}{2} & \frac{1}{2} \\ \frac{1}{12} & -\frac{1}{6} & -\frac{1}{6} \end{bmatrix}$   **57.** No inverse
**59.** $\begin{bmatrix} -\frac{2}{3} & -\frac{17}{3} & -\frac{14}{3} & -3 \\ \frac{1}{3} & \frac{1}{3} & \frac{1}{3} & 0 \\ -\frac{1}{3} & -\frac{10}{3} & -\frac{7}{3} & -2 \\ 0 & 2 & 1 & 1 \end{bmatrix}$   **61.** $\begin{bmatrix} -\frac{7}{19} & \frac{2}{19} \\ \frac{6}{19} & \frac{1}{19} \end{bmatrix}$   **63.** $\begin{bmatrix} -\frac{1}{12} & -\frac{1}{4} \\ \frac{5}{12} & \frac{1}{4} \end{bmatrix}$
**65.** $\begin{bmatrix} -1.2 & 1.1 & -.9 \\ .8 & -.4 & .6 \\ -.2 & .1 & .1 \end{bmatrix}$   **67.** $\begin{bmatrix} -5 \\ -3 \end{bmatrix}$   **69.** $\begin{bmatrix} -22 \\ -18 \\ 15 \end{bmatrix}$   **71.** $(-4, 2)$
**73.** $(4, 2)$   **75.** $(-1, 0, 2)$   **77.** No inverse; no solution for the system   **79.** 16 liters of the 9%; 24 liters of the 14%   **81.** 30 liters of 40% solution; 10 liters of 60% solution   **83.** 80 bowls; 120 plates   **85.** $12,750 at 8%; $27,250 at 8.5%; $10,000 at 11%
**87. (a)** $\begin{bmatrix} 1 & -\frac{1}{4} \\ -\frac{1}{2} & 1 \end{bmatrix}$   **(b)** $\begin{bmatrix} \frac{8}{7} & \frac{2}{7} \\ \frac{4}{7} & \frac{8}{7} \end{bmatrix}$   **(c)** $\begin{bmatrix} 2800 \\ 2800 \end{bmatrix}$   **89.** Agriculture $140,909; manufacturing $95,455   **91. (a)** .4 unit agriculture; .09 unit construction; .4 unit energy; .1 unit manufacturing; .9 unit transportation   **(b)** 2000 units of agriculture; 600 units of construction; 1700 units of energy; 3700 units of manufacturing; 2500 units of transportation

**93. (a)** $\begin{bmatrix} 2 & 1 & 1 & 2 \\ 1 & 3 & 2 & 1 \\ 1 & 2 & 3 & 1 \\ 2 & 1 & 1 & 2 \end{bmatrix}$   **(b)** 2   **(c)** $\begin{bmatrix} 2 & 5 & 5 & 2 \\ 5 & 4 & 5 & 5 \\ 5 & 5 & 4 & 5 \\ 2 & 5 & 5 & 2 \end{bmatrix}$   **(d)** 2
**95.** $\begin{bmatrix} 54 \\ 32 \end{bmatrix}, \begin{bmatrix} 134 \\ 89 \end{bmatrix}, \begin{bmatrix} 172 \\ 113 \end{bmatrix}, \begin{bmatrix} 118 \\ 74 \end{bmatrix}, \begin{bmatrix} 208 \\ 131 \end{bmatrix}$

## Case Study 6 Exercises (Page 331)

**1.** All Stampede Air cities except Lubbock may be reached by a two-flight sequence from San Antonio; all Stampede Air cities may be reached by a three-flight sequence.   **3.** The connection between Lubbock and Corpus Christi and the connection between Lubbock and San Antonio take three flights.
**5.** $B^2 = \begin{bmatrix} 2 & 1 & 1 & 1 & 1 \\ 1 & 1 & 1 & 1 & 0 \\ 1 & 1 & 1 & 1 & 0 \\ 1 & 1 & 1 & 2 & 1 \\ 1 & 0 & 0 & 1 & 4 \end{bmatrix}$ ; an entry is nonzero whenever there is a two-step sequence between the cities.

# Chapter 7

## Section 7.1 (Page 339)

**1.** F   **3.** A   **5.** E

**7.**

**9.**

**11.**

**13.**

**15.**

**17.**

**19.**

**21.**

**23.**

**25.**

**27.** Answers vary.

**29.**

**31.**

**33.**

**35.**

**37.**

**39.**

**41.**

**43.**

**45.**

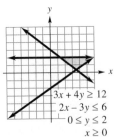

**47.**  $x \geq 0$
$0 \leq y \leq 4$
$4x + 3y \leq 24$

**49.**  $2 < x < 7$
$-1 < y < 3$

**51.**

**53.**

**55.**

## Section 7.2 (Page 347)

**1.** Maximum of 40 at (5, 10); minimum of 7 at (1, 1)   **3.** Maximum of 6 at (0, 12) minimum of 0 at (0, 0)   **5. (a)** No maximum; minimum of 12 at (12, 0)  **(b)** No maximum; minimum of 18 at (3, 4)   **(c)** No maximum; minimum of 21 at $\left(\frac{13}{2}, 2\right)$  **(d)** No maximum; minimum of 8 at (0, 8)
**7.** Maximum of 8.4 at (1.2, 1.2)   **9.** Minimum of 13 at (5, 3)
**11.** Maximum of 68.75 at $\left(\frac{105}{8}, \frac{25}{8}\right)$   **13.** No maximum, minimum of 9
**15.** Maximum of 22; no minimum   **17. (a)** (18, 2)  **(b)** $\left(\frac{12}{5}, \frac{39}{5}\right)$  **(c)** No maximum   **19.** Answers vary.

## Section 7.3 (Page 352)

**1.** $8x + 5y \leq 110$, $x \geq 0$, and $y \geq 0$, where $x$ is the number of canoes, and $y$ is the number of rowboats   **3.** $250x + 750y \leq 9500$, $x \geq 0$, and $y \geq 0$, where $x$ is the number of radio spots, and $y$ is the number of TV ads
**5.** 12 chain saws; no chippers   **7.** 10 lb deluxe; 40 lb regular
**9.** 12 shocks; 6 brakes   **11.** 8 radio spots; 10 TV ads   **13.** 3 Brand X and 2 Brand Z, for a minimum cost $1.05   **15.** 800 type 1 and 1600 type 2,

for a maximum revenue of \$272     **17.** From warehouse I, ship 60 boxes to San Jose and 250 boxes to Memphis; from warehouse II, ship 290 boxes to San Jose and none to Memphis, for a minimum cost of \$136.70.     **19.** \$20 million in bonds and \$24 million in mutual funds, for maximum interest of \$2.24 million
**21.** About 218 shares of General Electric and 66 shares of Procter & Gamble     **23.** 125 shares of Visa stock     **25.** About 109 shares of the Fidelity fund

## Section 7.4 (Page 365)

**1. (a)** 3   **(b)** $s_1, s_2, s_3$   **(c)**
$$4x_1 + 2x_2 + s_1 \qquad\qquad s_3 = 20$$
$$5x_1 + x_2 \qquad + s_2 \qquad\quad = 50$$
$$2x_1 + 3x_2 \qquad\qquad s_2 + s_3 = 25$$

**3. (a)** 3   **(b)** $s_1, s_2, s_3$   **(c)**
$$3x_1 - x_2 + 4x_3 + s_1 \qquad\qquad s_2 = 95$$
$$7x_1 + 6x_2 + 8x_3 \qquad + s_2 \qquad = 118$$
$$4x_1 + 5x_2 + 10x_3 \qquad\qquad s_2 + s_3 = 220$$

**5.**

| $x_1$ | $x_2$ | $s_1$ | $s_2$ | $s_3$ | $z$ | |
|---|---|---|---|---|---|---|
| 2 | 5 | 1 | 0 | 0 | 0 | 6 |
| 4 | 1 | 0 | 1 | 0 | 0 | 6 |
| 5 | 3 | 0 | 0 | 1 | 0 | 15 |
| −5 | −1 | 0 | 0 | 0 | 1 | 0 |

**7.**

| $x_1$ | $x_2$ | $x_3$ | $s_1$ | $s_2$ | $s_3$ | $z$ | |
|---|---|---|---|---|---|---|---|
| 1 | 2 | 3 | 1 | 0 | 0 | 0 | 10 |
| 2 | 1 | 1 | 0 | 1 | 0 | 0 | 8 |
| 3 | 0 | 4 | 0 | 0 | 1 | 0 | 6 |
| −1 | −5 | −10 | 0 | 0 | 0 | 1 | 0 |

**9.** 4 in row 2, column 2     **11.** 6 in row 3, column 1

**13.**

| $x_1$ | $x_2$ | $x_3$ | $s_1$ | $s_2$ | $z$ | |
|---|---|---|---|---|---|---|
| −1 | 0 | 3 | 1 | −1 | 0 | 16 |
| 1 | 1 | $\frac{1}{2}$ | 0 | $\frac{1}{2}$ | 0 | 20 |
| 2 | 0 | $-\frac{1}{2}$ | 0 | $\frac{3}{2}$ | 1 | 60 |

**15.**

| $x_1$ | $x_2$ | $x_3$ | $s_1$ | $s_2$ | $s_3$ | $z$ | |
|---|---|---|---|---|---|---|---|
| $-\frac{1}{2}$ | $\frac{1}{2}$ | 0 | 1 | $-\frac{1}{2}$ | 0 | 0 | 10 |
| $\frac{3}{2}$ | $\frac{1}{2}$ | 1 | 0 | $\frac{1}{2}$ | 0 | 0 | 50 |
| $-\frac{7}{2}$ | $\frac{1}{2}$ | 0 | 0 | $-\frac{3}{2}$ | 1 | 0 | 50 |
| 2 | 0 | 0 | 0 | 1 | 0 | 1 | 100 |

**17. (a)** Basic: $x_3, s_2, z$; nonbasic: $x_1, x_2, s_1$   **(b)** $x_1 = 0, x_2 = 0, x_3 = 16,$ $s_1 = 0, s_2 = 29, z = 11$   **(c)** Not a maximum     **19. (a)** Basic: $x_1, x_2, s_2, z$; nonbasic: $x_3, s_1, s_3$   **(b)** $x_1 = 6, x_2 = 13, x_3 = 0, s_1 = 0, s_2 = 21, s_3 = 0,$ $z = 18$   **(c)** Maximum     **21.** Maximum is 30 when $x_1 = 0, x_2 = 10,$ $s_1 = 0, s_2 = 0,$ and $s_3 = 16$   **23.** Maximum is 8 when $x_1 = 4, x_2 = 0,$ $s_1 = 8, s_2 = 2,$ and $s_3 = 0$   **25.** No maximum     **27.** No maximum
**29.** Maximum is 34 when $x_1 = 17, x_2 = 0, x_3 = 0, s_1 = 0,$ and $s_2 = 14$ or when $x_1 = 0, x_2 = 17, x_3 = 0, s_1 = 0,$ and $s_2 = 14$.     **31.** Maximum is 26,000 when $x_1 = 60, x_2 = 40, x_3 = 0, s_1 = 0, s_2 = 80,$ and $s_3 = 0$.
**33.** Maximum is 64 when $x_1 = 28, x_2 = 16, x_3 = 0, s_1 = 0, s_2 = 28,$ and, $s_3 = 0$   **35.** Maximum is 250 when $x_1 = 0, x_2 = 0, x_3 = 0, x_4 = 50,$ $s_1 = 0,$ and $s_2 = 50$   **37. (a)** Maximum is 24 when $x_1 = 12, x_2 = 0,$ $x_3 = 0, s_1 = 0,$ and $s_2 = 6$   **(b)** Maximum is 24 when $x_1 = 0, x_2 = 12,$ $x_3 = 0, s_1 = 0,$ and $s_2 = 18$   **(c)** The unique maximum value of $z$ is 24, but this occurs at two different basic feasible solutions.

## Section 7.5 (Page 371)

**1.**

| $x_1$ | $x_2$ | $s_1$ | $s_2$ | $s_3$ | |
|---|---|---|---|---|---|
| 2 | 1 | 1 | 0 | 0 | 90 |
| 1 | 2 | 0 | 1 | 0 | 80 |
| 1 | 1 | 0 | 0 | 1 | 50 |
| −12 | −10 | 0 | 0 | 0 | 0 |

where $x_1$ is the number of units of WildSplash and $x_2$ is the number of units of TastyTreat.

**3.**

| $x_1$ | $x_2$ | $x_3$ | $x_4$ | $s_1$ | $s_2$ | $s_3$ | |
|---|---|---|---|---|---|---|---|
| 0 | 0 | .375 | .625 | 1 | 0 | 0 | 500 |
| 0 | .75 | .5 | .375 | 0 | 1 | 0 | 600 |
| 1 | .25 | .125 | 0 | 0 | 0 | 1 | 300 |
| −90 | −70 | −60 | −50 | 0 | 0 | 0 | 0 |

where $x_1$ is the number of kilograms of $P$, $x_2$ is the number of kilograms of $Q$, $x_3$ is the number of kilograms of $R$, and $x_4$ is the number of kilograms of $S$.

**5. (a)** Make no 1-speed or 3-speed bicycles; make 2295 10-speed bicycles; maximum profit is \$550,800.   **(b)** 5280 units of aluminum are unused; all the steel is used.     **7. (a)** 12 minutes to the sports segment, 9 minutes to the news segment, and 6 minutes to the weather segment for a maximum of 1.32 million viewers   **(b)** $s_1 = 0$ means that all of the 27 available minutes were used; $s_2 = 0$ means that sports had exactly twice as much as the weather; $s_3 = 0$ means that the sports and the weather had a total time exactly twice the time of the news.     **9. (a)** 300 Japanese maple trees and 300 tricolor beech trees are sold, for a maximum profit of \$255,000
**(b)** There are 200 unused hours of delivering to the client.     **11.** 4 radio ads, 6 TV ads, and no newspaper ads, for a maximum exposure of 64,800 people
**13. (a)** 22 fund-raising parties, no mailings, and 3 dinner parties, for a maximum of \$6,200,000   **(b)** Answers vary.     **15.** 3 hours running, 4 hours biking, and 8 hours walking, for a maximum calorie expenditure of 6313 calories.
**17.** The juice blender should make 40 units of WildSplash and 10 units of TastyTreat, for a maximum gross income of \$580.     **19.** 163.6 kilograms of food $P$, none of $Q$, 1090.9 kilograms of $R$, 145.5 kilograms of $S$; maximum is 87,454.5

## Section 7.6 (Page 381)

**1.**
$$\begin{bmatrix} 3 & 1 & 0 \\ -4 & 10 & 3 \\ 5 & 7 & 6 \end{bmatrix}$$
**3.**
$$\begin{bmatrix} 3 & 4 \\ 0 & 17 \\ 14 & 8 \\ -5 & -6 \\ 3 & 1 \end{bmatrix}$$

**5.** Maximize $\quad z = 4x_1 + 6x_2$
subject to $\quad 3x_1 - x_2 \le 3$
$\qquad\qquad\quad x_1 + 2x_2 \le 5$
$\qquad\qquad\quad x_1 \ge 0, x_2 \ge 0.$

**7.** Maximize $\quad z = 18x_1 + 15x_2 + 20x_3$
subject to $\quad x_1 + 4x_2 + 5x_3 \le 2$
$\qquad\qquad\quad 7x_1 + x_2 + 3x_3 \le 8$
$\qquad\qquad\quad x_1 \ge 0, x_2 \ge 0, x_3 \ge 0.$

**9.** Maximize $\quad z = 18x_1 + 20x_2$
subject to $\quad 7x_1 + 4x_2 \le 5$
$\qquad\qquad\quad 6x_1 + 5x_2 \le 1$
$\qquad\qquad\quad 8x_1 + 10x_2 \le 3$
$\qquad\qquad\quad x_1 \ge 0, x_2 \ge 0.$

**11.** Maximize $\quad z = 5x_1 + 4x_2 + 15x_3$
subject to $\quad x_1 + x_2 + 2x_3 \le 8$
$\qquad\qquad\quad x_1 + x_2 + x_3 \le 9$
$\qquad\qquad\quad x_1 + 3x_3 \le 3$
$\qquad\qquad\quad x_1 \ge 0, x_3 \ge 0, x_3 \ge 0.$

**13.** $y_1 = 0, y_2 = 4, y_3 = 0$; minimum is 4.   **15.** $y_1 = 4, y_2 = 0, y_3 = \frac{7}{3}$; minimum is 39.   **17.** $y_1 = 0, y_2 = 100, y_3 = 0$, minimum is 100.
**19.** $y_1 = 0, y_2 = 12, y_3 = 0$, minimum is 12.     **21.** $y_1 = 0, y_2 = 0, y_3 = 2$; minimum is 4.     **23.** $y_1 = 10, y_2 = 10, y_3 = 0$; minimum is 320.
**25.** $y_1 = 0, y_2 = 0, y_3 = 4$; minimum is 4.     **27.** 4 servings of $A$ and 2 servings of $B$, for a minimum cost of \$1.76     **29.** 28 units of regular beer and 14 units of light beer, for a minimum cost of \$1,680,000
**31. (a)** 1 bag of feed 1, 2 bags of feed 2   **(b)** 1.4 bags of feed 1 and 1.2 bags of feed 2, for a daily cost of \$6.60
**33. (a)** Minimize $\quad w = 200y_1 + 600y_2 + 90y_3$
subject to $\quad y_1 + 4y_2 \ge 1$
$\qquad\qquad\quad 2y_1 + 3y_2 + y_3 \ge 1.5$
$\qquad\qquad\quad y_1 \ge 0, y_2 \ge 0, y_3 \ge 0.$
**(b)** $y_1 = .6, y_2 = .1, y_3 = 0, w = 180$   **(c)** \$186

## Section 7.7 (Page 391)

1. (a) Maximize   $z = -5x_1 + 4x_2 - 2x_3$
   subject to    $-2x_2 + 5x_3 - s_1 \quad\quad = 8$
   $\quad\quad\quad\quad\quad\quad\quad\quad\quad s_2 = 8$
   $\quad\quad\quad 4x_1 - x_2 + 3x_3 \quad\quad + s_2 = 12$
   $\quad\quad\quad x_1 \geq 0, x_2 \geq 0, x_3 \geq 0, s_1 \geq 0, s_2 \geq 0.$

(b)

| $x_1$ | $x_2$ | $x_3$ | $s_1$ | $s_2$ | |
|---|---|---|---|---|---|
| 0 | -2 | 5 | -1 | 0 | 8 |
| 4 | -1 | 3 | 0 | 1 | 12 |
| 5 | -4 | 2 | 0 | 0 | 0 |

3. (a) Maximize   $z = 2x_1 - 3x_2 + 4x_3$
   subject to   $x_1 + x_2 + x_3 + s_1 \quad\quad\quad = 100$
   $\quad\quad\quad x_1 + x_2 + x_3 \quad - s_2 \quad\quad = 75$
   $\quad\quad\quad x_1 + x_2 \quad\quad\quad\quad - s_3 = 27$
   $\quad\quad\quad x_1 \geq 0, x_2 \geq 0, x_3 \geq 0, s_1 \geq 0, s_2 \geq 0, s_3 \geq 0.$

(b)

| $x_1$ | $x_2$ | $x_3$ | $s_1$ | $s_2$ | $s_3$ | |
|---|---|---|---|---|---|---|
| 1 | 1 | 1 | 1 | 0 | 0 | 100 |
| 1 | 1 | 1 | 0 | -1 | 0 | 75 |
| 1 | 1 | 0 | 0 | 0 | -1 | 27 |
| -2 | 3 | -4 | 0 | 0 | 0 | 0 |

5. Maximize   $z = -2y_1 - 5y_2 + 3y_3$
   subject to   $y_1 + 2y_2 + 3y_3 \geq 115$
   $\quad\quad\quad 2y_1 + y_2 + y_3 \leq 200$
   $\quad\quad\quad y_1 \quad\quad + y_3 \geq 50$
   $\quad\quad\quad y_1 \geq 0, y_2 \geq 0, y_3 \geq 0.$

| $y_1$ | $y_2$ | $y_3$ | $s_1$ | $s_2$ | $s_3$ | |
|---|---|---|---|---|---|---|
| 1 | 2 | 3 | -1 | 0 | 0 | 115 |
| 2 | 1 | 1 | 0 | 1 | 0 | 200 |
| 1 | 0 | 1 | 0 | 0 | -1 | 50 |
| 2 | 5 | -3 | 0 | 0 | 0 | 0 |

7. Maximize $z = -10y_1 - 8y_2 - 15y_3$
   subject to   $y_1 + y_2 + y_3 \geq 12$
   $\quad\quad\quad 5y_1 + 4y_2 + 9y_3 \geq 48$
   $\quad\quad\quad y_1 \geq 0, y_2 \geq 0, y_3 \geq 0.$

| $y_1$ | $y_2$ | $y_3$ | $s_1$ | $s_2$ | |
|---|---|---|---|---|---|
| 1 | 1 | 1 | -1 | 0 | 12 |
| 5 | 4 | 9 | 0 | -1 | 48 |
| 10 | 8 | 15 | 0 | 0 | 0 |

9. Maximum is 480 when $x_1 = 40$ and $x_2 = 0$.
11. Maximum is 114 when $x_1 = 38$, $x_2 = 0$, and $x_3 = 0$.
13. Maximum is 90 when $x_1 = 12$ and $x_2 = 3$ or when $x_1 = 0$ and $x_2 = 9$.
15. Minimum is 40 when $y_1 = 0$ and $y_2 = 20$.
17. Maximum is $133\frac{1}{3}$ when $x_1 = 33\frac{1}{3}$ and $x_2 = 16\frac{2}{3}$.
19. Minimum is 512 when $y_1 = 6$ and $y_2 = 8$.
21. Maximum is 112 when $x_1 = 0$, $x_2 = 36$, and $x_3 = 16$.
23. Maximum is 346 when $x_1 = 27$, $x_2 = 0$, and $x_3 = 73$.
25. Minimum is $-600$ when $y_1 = 0$, $y_2 = 0$, and $y_3 = 200$.
27. Minimum is 96 when $y_1 = 0$, $y_2 = 12$, and $y_3 = 0$.
29.

| $y_1$ | $y_2$ | $y_3$ | $s_1$ | $s_2$ | $s_3$ | $s_4$ | |
|---|---|---|---|---|---|---|---|
| 1 | 1 | 1 | -1 | 0 | 0 | 0 | 10 |
| 1 | 1 | 1 | 0 | 1 | 0 | 0 | 15 |
| 1 | $-\frac{1}{4}$ | 0 | 0 | 0 | -1 | 0 | 0 |
| -1 | 0 | 1 | 0 | 0 | 0 | -1 | 0 |
| .30 | .09 | .27 | 0 | 0 | 0 | 0 | 0 |

31.

| $y_1$ | $y_2$ | $y_3$ | $y_4$ | $s_1$ | $s_2$ | $s_3$ | $s_4$ | $s_5$ | $s_6$ | |
|---|---|---|---|---|---|---|---|---|---|---|
| 1 | 1 | 0 | 0 | -1 | 0 | 0 | 0 | 0 | 0 | 32 |
| 1 | 1 | 0 | 0 | 0 | 1 | 0 | 0 | 0 | 0 | 32 |
| 0 | 0 | 1 | 1 | 0 | 0 | -1 | 0 | 0 | 0 | 20 |
| 0 | 0 | 1 | 1 | 0 | 0 | 0 | 1 | 0 | 0 | 20 |
| 1 | 0 | 1 | 0 | 0 | 0 | 0 | 0 | 1 | 0 | 25 |
| 0 | 1 | 0 | 1 | 0 | 0 | 0 | 0 | 0 | 1 | 30 |
| 14 | 12 | 22 | 10 | 0 | 0 | 0 | 0 | 0 | 0 | 0 |

33. Ship 200 barrels of oil from supplier $S_1$ to distributor $D_1$; ship 2800 barrels of oil from supplier $S_2$ to distributor $D_1$; ship 2800 barrels of oil from supplier $S_1$ to distributor $D_2$; ship 2200 barrels of oil from supplier $S_2$ to distributor $D_2$. Minimum cost is $180,400.
35. Use 1000 lb of bluegrass, 2400 lb of rye, and 1600 lb of Bermuda, for a minimum cost of $560.
37. Allot $3,000,000 in commercial loans and $22,000,000 in home loans, for a maximum return of $2,940,000.
39. Make 32 units of regular beer and 10 units of light beer, for a minimum cost of $1,632,000.
41. $1\frac{2}{3}$ ounces of ingredient I, $6\frac{2}{3}$ ounces of ingredient II, and $1\frac{2}{3}$ ounces of ingredient III produce a minimum cost of $1.55 per barrel.
43. 22 from $W_1$ to $D_1$, 10 from $W_2$ to $D_1$, none from $W_1$ to $D_2$, and 20 from $W_2$ to $D_2$, for a minimum cost of $628.

## Chapter 7 Review Exercises (Page 395)

| Refer to Section | 7.1 | 7.2 | 7.3 | 7.4 | 7.5 | 7.6 | 7.7 |
|---|---|---|---|---|---|---|---|
| For Exercises | 1–10 | 11–16 | 17–18 | 19–32, 37 | 33–36, 57–59 | 38–46, 51–52, 61–62 | 47–50, 53–56, 63–64 |

1.

$y \leq 3x + 2$

3.
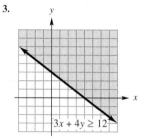
$3x + 4y \geq 12$

5.
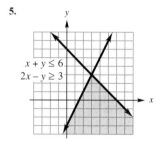
$x + y \leq 6$
$2x - y \geq 3$

7.
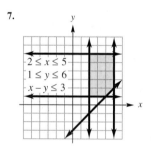
$-2 \leq x \leq 5$
$1 \leq y \leq 6$
$x - y \leq 3$

9. Let $x$ represent the number of summary reports, and let $y$ represent the number of inference reports. Then
$$2x + 4y \leq 15$$
$$1.5x + 2y \leq 9$$
$$x \geq 0, y \geq 0.$$

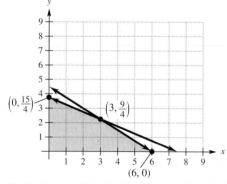

11. Maximum of 46 at (6, 7); minimum of 10 at (2, 1).
13. Maximum of 30 when $x = 5$ and $y = 0$.
15. Minimum of 40 when $x = 0$ and $y = 20$.

**17.** She should complete 3 summary projects and 2.25 inference projects for a maximum profit of $3187.50.    **19.** Donald should buy 367 (rounded) shares from the BlackRock fund for a minimum risk of about 348.

**21. (a)**
$$\begin{aligned}
x_1 + x_2 + x_3 + s_1 &= 100 \\
2x_1 + 3x_2 \quad + s_2 &= 500 \\
x_1 \quad + 2x_3 \quad + s_3 &= 350
\end{aligned}$$

**(b)**

| $x_1$ | $x_2$ | $x_3$ | $s_1$ | $s_2$ | $s_3$ | |
|---|---|---|---|---|---|---|
| 1 | 1 | 1 | 1 | 0 | 0 | 100 |
| 2 | 3 | 0 | 0 | 1 | 0 | 500 |
| 1 | 0 | 2 | 0 | 0 | 1 | 350 |
| −5 | −6 | −3 | 0 | 0 | 0 | 0 |

**23. (a)**
$$\begin{aligned}
x_1 + x_2 + x_3 + s_1 &= 90 \\
2x_1 + 5x_2 + x_3 \quad + s_2 &= 120 \\
x_1 + 3x_2 \quad + s_3 &= 80
\end{aligned}$$

**(b)**

| $x_1$ | $x_2$ | $x_3$ | $s_1$ | $s_2$ | $s_3$ | |
|---|---|---|---|---|---|---|
| 1 | 1 | 1 | 1 | 0 | 0 | 90 |
| 2 | 5 | 1 | 0 | 1 | 0 | 120 |
| 1 | 3 | 0 | 0 | 0 | 1 | 80 |
| −1 | −8 | −2 | 0 | 0 | 0 | 0 |

**25.** Maximum is 80 when $x_1 = 16$, $x_2 = 0$, $x_3 = 0$, $s_1 = 12$, and $s_2 = 0$.
**27.** Maximum is 35 when $x_1 = 5$, $x_2 = 0$, $x_3 = 5$, $s_1 = 35$, $s_2 = 0$, and $s_3 = 0$.    **29.** Maximum of 600 when $x_1 = 0$, $x_2 = 100$, and $x_3 = 0$
**31.** Maximum of 225 when $x_1 = 0$, $x_2 = \frac{15}{2}$, and $x_3 = \frac{165}{2}$
**33. (a)** Let $x_1 = $ the number of item $A$, $x_2 = $ the number of item $B$, and $x_3 = $ the number of item $C$.
**(b)** $z = 4x_1 + 3x_2 + 3x_3$
**(c)** $2x_1 + 3x_2 + 6x_3 \le 1200$
$\quad x_1 + 2x_2 + 2x_3 \le 800$
$\quad 2x_1 + 2x_2 + 4x_3 \le 500$
$\quad x_1 \ge 0, x_2 \ge 0, x_3 \ge 0$
**35. (a)** Let $x_1 = $ the number of gallons of Fruity wine and $x_2 = $ the number of gallons of Crystal wine.    **(b)** $z = 12x_1 + 15x_2$
**(c)** $2x_1 + x_2 \le 110$
$\quad 2x_1 + 3x_2 \le 125$
$\quad 2x_1 + x_2 \le 90$
$\quad x_1 \ge 0, x_2 \ge 0$
**37.** When there are more than 2 variables    **39.** Any standard minimization problem    **41.** Minimum of 172 at (5, 7, 3, 0, 0, 0).
**43.** Minimum of 640 at (7, 2, 0, 0).    **45.** Minimum of 170 when $y_1 = 0$ and $y_2 = 17$

**47.**
| 5 | 10 | 1 | 0 | 120 |
|---|---|---|---|---|
| 10 | 15 | 0 | −1 | 200 |
| −20 | −30 | 0 | 0 | 0 |

**49.**
| 1 | 1 | 2 | −1 | 0 | 0 | 0 | 48 |
|---|---|---|---|---|---|---|---|
| 1 | 1 | 0 | 0 | 1 | 0 | 0 | 12 |
| 0 | 0 | 1 | 0 | 0 | −1 | 0 | 10 |
| 3 | 0 | 1 | 0 | 0 | 0 | −1 | 30 |
| 12 | 20 | −8 | 0 | 0 | 0 | 0 | 0 |

**51.** Minimum of 427 at (8, 17, 25, 0, 0, 0)    **53.** Maximum is 480 when $x_1 = 24$ and $x_2 = 0$.    **55.** Minimum is 40 when $y_1 = 10$ and $y_2 = 0$.
**57.** Get 250 of $A$ and none of $B$ or $C$ for a maximum profit of $1000.
**59.** Make 17.5 gal of Crystal and 36.25 gal of Fruity, for a maximum profit of $697.50.    **61.** Produce 660 cases of corn, no beans, and 340 cases of carrots, for a minimum cost of $15,100.    **63.** Use 1,060,000 kilograms for whole tomatoes and 80,000 kilograms for sauce, for a minimum cost of $4,500,000.

## Case Study 7 Exercises (Page 399)

**1.** The answer in 100-gram units is 0.243037 unit of feta cheese, 2.35749 units of lettuce, 0.3125 unit of salad dressing, and 1.08698 units of tomato. Converting into kitchen units gives approximately $\frac{1}{6}$ cup feta cheese, $4\frac{1}{4}$ cups lettuce, $\frac{1}{8}$ cup salad dressing, and $\frac{7}{8}$ of a tomato.

# Chapter 8

## Section 8.1 (Page 407)

**1.** False    **3.** True    **5.** True    **7.** True    **9.** False
**11.** Answers vary.    **13.** $\subset$    **15.** $\not\subseteq$    **17.** $\subset$    **19.** $\subset$    **21.** 8
**23.** 128    **25.** $\{x \mid x$ is an integer $\le 0$ or $\ge 8\}$    **27.** Answers vary.
**29.** $\cap$    **31.** $\cap$    **33.** $\cup$    **35.** $\cup$    **37.** $\{b, 1, 3\}$
**39.** $\{d, e, f, 4, 5, 6\}$    **41.** $\{e, 4, 6\}$    **43.** $\{a, b, c, d, 1, 2, 3, 5\}$
**45.** All students not taking this course.    **47.** All students taking both accounting and philosophy.    **49.** $C$ and $D$, $A$ and $E$, $C$ and $E$, $D$ and $E$
**51.** $\{\text{Ford}\}$    **53.** $\{\text{Ford}\}$    **55.** $M \cap E$ is the set of all male employed clients.    **57.** $M' \cup S'$ is the set of all female or married clients.
**59.** $\{T\}$    **61.** $\{C\}$    **63.** $\{\text{Direct TV, Comcast, Charter, Dish}\}$
**65.** $\{\text{Charter, Dish, Verizon FIOS, Altice}\}$
**67.** $\{\text{Direct TV, Comcast, Charter, Dish}\}$
**69.** $\{\text{Dish, Verizon FIOS, Altice}\}$    **71.** $\varnothing$    **73.** $\{\text{Corn, Rice}\}$
**75.** $\varnothing$    **77.** $\{\text{Wheat, Corn, Rice}\}$

## Section 8.2 (Page 414)

**1.**     **3.**

**5.** 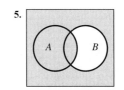    **7.** $\phi$
**9.** 8

**11.**     **13.**

**15.** 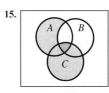    **17.** 13
**19. (a)** 105    **(b)** 39    **(c)** 15    **(d)** 3
**21. (a)** 227    **(b)** 182    **(c)** 27
**23. (a)** 54    **(b)** 17    **(c)** 10    **(d)** 7
**(e)** 15    **(f)** 3    **(g)** 12    **(h)** 1

**25. (a)** 290,793    **(b)** 149,167    **(c)** 81,878    **(d)** 317,794    **27. (a)** 945
**(b)** 55    **(c)** 1340    **(d)** 928    **29. (a)** 507    **(b)** 245    **(c)** The number of females who are completely satisfied or very satisfied with their job.
**31. (a)** 800    **(b)** 411    **(c)** The number of males who are completely, very, or fairly satisfied with their job.    **33.** Answers vary.    **35.** 9    **37.** 27

**39.**     **41.**

**43.**     **45.**

**47.**

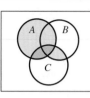

**49.** The complement of $A$ intersect $B$ equals the union of the complement of $A$ and the complement of $B$.
**51.** $A$ union ($B$ intersect $C$) equals ($A$ union $B$) intersect ($A$ union $C$).

## Section 8.3 (Page 425)

**1.** Answers vary.    **3.** {January, February, March,. . . , December}
**5.** {0, 1, 2,. . . , 80}    **7.** {Go ahead, cancel}    **9.** {$Q_1, Q_2, Q_3, Q_4$}
**11.** Answers vary.    **13.** No    **15.** Yes    **17.** No
**19.** $S$ = {C&K, C&L, C&J, C&T, C&N, K&L, K&J, K&T, K&N, L&J, L&T, L&N, J&T, J&N, T&N}    **(a)** {C&T, K&T, L&T, J&T, T&N}
**(b)** {C&J, C&N, J&N}    **21.** $S$ = {forest sage and rag painting, forest sage and colorwash, evergreen and rag painting, evergreen and colorwash, opaque and rag painting, opaque and colorwash}    **(a)** $\frac{1}{2}$    **(b)** $\frac{2}{3}$
**23.** $S$ = {10′ and beige, 10′ and forest green, 10′ and rust, 12′ and beige, 12′ and forest green, 12′ and rust}    **(a)** $\frac{1}{6}$    **(b)** $\frac{1}{2}$    **(c)** $\frac{1}{3}$    **25.** $\frac{1}{2}$    **27.** $\frac{1}{3}$
**29.** S = {male beagle, male boxer, male collie, male Labrador, female beagle, female boxer, female collie, female Labrador}    **31.** $\frac{1}{2}$    **33.** $\frac{1}{8}$
**35.** $\frac{3}{4}$    **37.** $\frac{800}{4836} \approx .1654$    **39. (a)** $\frac{312}{2867} \approx .1088$    **(b)** $\frac{250}{2867} \approx .0872$
**(c)** $\frac{817}{2867} \approx .2850$    **41.** $\frac{35}{505} \approx .0693$    **43.** $\frac{58}{505} \approx .1149$
**45.** $\frac{159}{505} \approx .3149$    **47.** Possible    **49.** Not possible    **51.** Not possible

## Section 8.4 (Page 433)

**1.** $\frac{20}{38} = \frac{10}{19}$    **3.** $\frac{28}{38} = \frac{14}{19}$    **5.** $\frac{5}{38}$    **7.** $\frac{12}{38} = \frac{6}{19}$    **9. (a)** $\frac{5}{36}$
**(b)** $\frac{1}{9}$    **(c)** $\frac{1}{12}$    **(d)** 0    **11. (a)** $\frac{5}{18}$    **(b)** $\frac{5}{12}$    **(c)** $\frac{1}{3}$    **13.** $\frac{1}{3}$
**15. (a)** $\frac{1}{2}$    **(b)** $\frac{2}{5}$    **(c)** $\frac{3}{10}$    **17. (a)** .62    **(b)** .32    **(c)** .11    **(d)** .81
**19. (a)** .3962    **(b)** .7952    **(c)** .2048    **(d)** .9464    **21. (a)** .2600
**(b)** .4580    **(c)** .8373    **23.** 1:5    **25.** 2:1    **27. (a)** 1:4    **(b)** 8:7
**(c)** 4:11    **29.** 2:7    **31.** 19:81    **33.** 1:19    **35.** $\frac{31}{100}$    **37.** $\frac{23}{100}$
**39.** About .8950    **41.** About .0594    **43.** About .9307
**45.** About .6416    **47.** About .8053    **49.** About .2535
**51.** About .0411    **53.** About .8464    **55. (a)** Answers vary but should be close to .2778.    **(b)** Answers vary, but should be close to .4167.
**57. (a)** Answers vary but should be close to .0463.    **(b)** Answers vary but should be close to .2963.    **59.** $\frac{3}{4}$

## Section 8.5 (Page 446)

**1.** $\frac{1}{3}$    **3.** 1    **5.** $\frac{1}{6}$    **7.** $\frac{4}{17}$    **9.** $\frac{32}{2652} \approx .0121$    **11.** Answers vary.
**13.** Answers vary.    **15.** No    **17.** Yes    **19.** No, yes
**21.** About .0787    **23.** About .1266    **25.** About .9401
**27.** About .0204    **29.** About .5065    **31.** About .0737
**33.** About .1648    **35.** .527    **37.** .042    **39.** .857    **41.** Dependent
**43.** $P(C) = .0800, P(D) = .0050, P(C \cap D) = .0004, P(C|D) = .08$; yes; independent    **45.** .05    **47.** About .8019    **49.** About .7961
**51.** About .1497    **53.** About .8232    **55.** Delta    **57.** .4955
**59.** .999985    **61. (a)** 3 backups **(b)** Answers vary.    **63.** Dependent or "No"

## Section 8.6 (Page 452)

**1.** $\frac{1}{3}$    **3.** .0488    **5.** .5122    **7.** .4706    **9.** About .0931
**11.** About .1826    **13.** About .3880    **15.** About .1604
**17.** About .7367    **19.** .519    **21.** About .4856    **23.** About .3328
**25.** About .5543    **27.** About .3696    **29.** About .4457
**31.** About .5009    **33.** .5335    **35.** .4947

## Chapter 8 Review Exercises (Page 454)

| Refer to Section | 8.1 | 8.2 | 8.3 | 8.4 | 8.5 | 8.6 |
|---|---|---|---|---|---|---|
| For Exercises | 1–23 | 24–30 | 31–48 | 49–54, 65–74, 79–82 | 55–64, 75–78, 89–96 | 83–88 |

**1.** False    **3.** False    **5.** True    **7.** True    **9.** {New Year's Day, Martin Luther King Jr.'s Birthday, Presidents' Day, Memorial Day, Independence Day, Labor Day, Columbus Day, Veterans' Day, Thanksgiving, Christmas}    **11.** {1, 2, 3, 4}    **13.** {$B_1, B_2, B_3, B_6, B_{12}$}
**15.** {$A, C, E$}    **17.** {$A, B_3, B_6, B_{12}, C, D, E$}    **19.** Female students older than 22    **21.** Females or students with a GPA $> 3.5$    **23.** Non–finance majors who are 22 or younger
**25.**

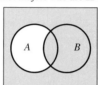

**27.** 4    **29.** 43    **31.** {1, 2, 3, 4, 5, 6}    **33.** {(Rock, Rock), (Rock, Pop), (Rock, Alternative), (Pop, Rock), (Pop, Pop), (Pop, Alternative), (Alternative, Rock), (Alternative, Pop), (Alternative, Alternative)}
**35.** {(Dell, Epson), (Dell, HP), (Apple, Epson), (Apple, HP), (HP, Epson), (HP, HP)}    **37.** $E' \cap F'$ or $(E \cup F)'$    **39.** A probability cannot be $> 1$.    **41.** Answers vary.    **43.** About .2157    **45.** About .8384
**47.** About .7843    **49.** About .4255    **51.** About .1030
**53.** About .8066    **55.** About .1694    **57.** About .1894
**59.** 12 years; 13 – 16 years
**61.**

|  | $N_2$ | $T_2$ |
|---|---|---|
| $N_1$ | $N_1 N_2$ | $N_1 T_2$ |
| $T_1$ | $T_1 N_2$ | $T_1 T_2$ |

**63.** $\frac{1}{2}$    **65.** $\frac{5}{36} \approx .1389$    **67.** $\frac{5}{18} \approx .2778$    **69.** $\frac{1}{3} \approx .3333$
**71.** .79    **73.** .66    **75.** .7    **77.** Answers vary.    **79.** About .5427
**81.** 1363:1137    **83.** $\frac{19}{22} \approx .8636$    **85.** $\frac{1}{3} \approx .3333$    **87.** .0646
**89.** .271    **91.** .625    **93.** .690    **95.** No. Answers vary.

## Case Study 8 Exercises (Page 458)

**1.** .001    **3.** About .331    **5.** About .2094

# Chapter 9

## Section 9.1 (Page 465)

**1.**

| Number of Boys | 0 | 1 | 2 | 3 | 4 |
|---|---|---|---|---|---|
| $P(x)$ | .063 | .25 | .375 | .25 | .063 |

**3.**

| Number of Queens | 0 | 1 | 2 | 3 |
|---|---|---|---|---|
| $P(x)$ | .7826 | .2042 | .0130 | .0002 |

**5.**

**7.**

**9.** 4　　**11.** 5.4　　**13.** 2.7　　**15.** 2.5　　**17.** $-\frac{1}{19} \approx -.0526$

**19.** $-\$.50$　　**21.** $-\$.97$　　**23.** $-\$1.67$　　**25.** 1.1319

**27.** Not valid, cannot have a probability $< 0$　　**29.** Valid　　**31.** .15

**33.** .25　　**35.** Many correct answers, including .15, .15　　**37.** \$4550

**39.** .513　　**41. (a)** Drug A \$217; Drug B \$229　　**(b)** Drug A

**43. (a)** £1,500,000　　**(b)** £1,640,000

## Section 9.2 (Page 477)

**1.** 12　　**3.** 56　　**5.** 8　　**7.** 24　　**9.** 84　　**11.** 1716

**13.** 6,375,600　　**15.** 240,240　　**17.** 8568　　**19.** 40,116,600

**21.** $\frac{24}{0}$ is undefined.　　**23.** 1,669,536　　**25.** 576　　**27.** 1 billion in

theory; however, some numbers will never be used (such as those beginning
with 000); yes　　**29.** 1 billion　　**31.** 120　　**33. (a)** 160; 8,000,000
**(b)** Some, such as 800, 900, and so on, are reserved.　　**35.** 1600
**37.** Answers vary.　　**39.** 95,040　　**41.** 12,144　　**43.** 863,040
**45.** 272　　**47. (a)** 210　　**(b)** 210　　**49.** 330　　**51.** Answers vary.
**53. (a)** 9　　**(b)** 6　　**(c)** 3; yes　　**55. (a)** 20,349　　**(b)** 462　　**(c)** 6600
**(d)** 19,887　　**57. (a)** 300　　**(b)** 13,800　　**59. (a)** 1,120,529,256
**(b)** 806,781,064,320　　**61.** 81　　**63.** Not possible; 4 initials
**65. (a)** 10　　**(b)** 0　　**(c)** 1　　**(d)** 10　　**(e)** 30　　**(f)** 15　　**(g)** 0
**67.** 3,247,943,160　　**69. (a)** 2520　　**(b)** 840　　**(c)** 5040
**71. (a)** 479,001,600　　**(b)** 6　　**(c)** 27,720

## Section 9.3 (Page 484)

**1.** About .4224　　**3.** $\frac{5}{8}$　　**5.** $\frac{5}{28}$　　**7.** About .00005　　**9.** About .0163

**11.** About .5630　　**13.** 1326　　**15.** About .8507　　**17.** About .7647
**19.** About .9412　　**21.** Answers vary.　　**23.** About .1396
**25. (a)** $8.9 \times 10^{-10}$　　**(b)** $1.2 \times 10^{-12}$　　**27.** $1.17 \times 10^{-17}$
**29. (a)** About .4072　　**(b)** About .0771　　**(c)** About .7511
**31. (a)** About .0015　　**(b)** About .0033　　**(c)** About .3576

**33.** $1 - \frac{_{365}P_{100}}{(365)^{100}}$　　**35.** About .0031　　**37.** About .3083

## Section 9.4 (Page 490)

**1.** About .1499　　**3.** About .0030　　**5.** About .9970　　**7.** About .2194

**9.** About .0192　　**11.** About .9808　　**13.** $\frac{1}{32}$　　**15.** $\frac{13}{16}$

**17.** Answers vary.　　**19.** About .2029　　**21.** About .1818
**23.** About .0355　　**25.** About .1115　　**27.** 155　　**29.** .2468
**31.** .1927　　**33.** .3504　　**35.** 3.85　　**37.** About .8237　　**39.** 42
**41.** About .0424　　**43.** 44

## Section 9.5 (Page 498)

**1.** Yes　　**3.** Yes　　**5.** No　　**7.** No　　**9.** No　　**11.** No

**13.** Not a transition diagram　　**15.** $\begin{bmatrix} .6 & .20 & .20 \\ .9 & .02 & .08 \\ .4 & 0 & .6 \end{bmatrix}$

**17.** Yes　　**19.** Yes　　**21.** $\left[\frac{4}{11}, \frac{7}{11}\right]$　　**23.** [.675, .325]

**25.** [.302, .350, .348]　　**27.** [.230, .278, .492]

**29.** $A^2 = \begin{bmatrix} .23 & .21 & .24 & .17 & .15 \\ .26 & .18 & .26 & .16 & .14 \\ .23 & .18 & .24 & .19 & .16 \\ .19 & .19 & .27 & .18 & .17 \\ .17 & .2 & .26 & .19 & .18 \end{bmatrix}$

$A^3 = \begin{bmatrix} .226 & .192 & .249 & .177 & .156 \\ .222 & .196 & .252 & .174 & .156 \\ .219 & .189 & .256 & .177 & .159 \\ .213 & .192 & .252 & .181 & .162 \\ .213 & .189 & .252 & .183 & .163 \end{bmatrix}$

$A^4 = \begin{bmatrix} .2205 & .1916 & .2523 & .1774 & .1582 \\ .2206 & .1922 & .2512 & .1778 & .1582 \\ .2182 & .1920 & .2525 & .1781 & .1592 \\ .2183 & .1909 & .2526 & .1787 & .1595 \\ .2176 & .1906 & .2533 & .1787 & .1598 \end{bmatrix}$

$A^5 = \begin{bmatrix} .21932 & .19167 & .25227 & .17795 & .15879 \\ .21956 & .19152 & .25226 & .17794 & .15872 \\ .21905 & .19152 & .25227 & .17818 & .15898 \\ .21880 & .19144 & .25251 & .17817 & .15908 \\ .21857 & .19148 & .25253 & .17824 & .15918 \end{bmatrix}$; .17794

**31. (a)** $\begin{bmatrix} .85 & .15 \\ .40 & .60 \end{bmatrix}$　　**(b)** [.517, .483]　　**(c)** $\left[\frac{8}{11}, \frac{3}{11}\right]$　　**33.** [.883, .117]

**35. (a)** $\begin{array}{cc} & R \quad\ W \\ \begin{matrix} R \\ W \end{matrix} & \begin{bmatrix} .75 & .25 \\ .50 & .50 \end{bmatrix} \end{array}$　　**(b)** [.75, .25]　　**(c)** [.667, .333]　　**(d)** $\left[\frac{2}{3}, \frac{1}{3}\right]$

**37. (a)** [.5607, .43361, .00569]　　**(b)** [.473, .526, .002]

**39. (a)** $\begin{bmatrix} .85 & .10 & .05 \\ 0 & .8 & .2 \\ 0 & 0 & 1 \end{bmatrix}$　　**(b)** 42,500; 5000; 2500　　**(c)** 36,125; 8250; 5625

**(d)** 30,706; 10,213; 9081　　**(e)** 26,100; 11,241; 12,659　　**(f)** [0 0 1]
**41. (a)** 37.5% Verizon; 31.9% AT&T; 30.6% other　　**(b)** 39.3% Verizon;
31.6% AT&T; 29.0% other　　**(c)** 40.8% Verizon; 31.4% AT&T;
27.8% other　　**(d)** 45.75% Verizon, 31.6% AT&T, 22.64% other
**43. (a)** About .2073 **(b)** About .2865　　**45. (a)** About 85%
**(b)** 1.28%, 5.13%, 9.62%, 83.97%

## Section 9.6 (Page 503)

**1. (a)** Coast　　**(b)** Highway　　**(c)** Highway; \$38,000　　**(d)** Coast
**3. (a)** Do not upgrade.　　**(b)** Upgrade.　　**(c)** Do not upgrade; \$10,600.

**5. (a)** $\begin{array}{c} \\ \text{Overhaul} \\ \text{Don't Overhaul} \end{array} \begin{array}{cc} \text{Fails} & \text{Doesn't Fail} \\ \begin{bmatrix} -\$8600 & -\$2600 \\ -\$6000 & \$0 \end{bmatrix} \end{array}$　　**(b)** Don't overhaul

the machine.　　**7. (a)** $\begin{array}{c} \\ \text{Rent tent} \\ \text{Do not rent tent} \end{array} \begin{array}{cc} \text{No Rain} & \text{Rain} \\ \begin{bmatrix} \$2500 & \$1500 \\ \$3000 & \$0 \end{bmatrix} \end{array}$

**(b)** Rent tent because the expected value is \$2100.　　**9.** Environment, 15.3

## Chapter 9 Review Exercises (Page 506)

| Refer to Section | 9.1 | 9.2 | 9.3 | 9.4 | 9.5 | 9.6 |
|---|---|---|---|---|---|---|
| For Exercises | 1–12 | 13–21, 31–33 | 22–30 | 33–38 | 39–44 | 45–46 |

**1. (a)**　　　　　　　　　　　　　**(b)** 1.1

**3. (a)**                                                  **(b)** 0

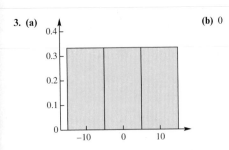

**5.** 1.833    **7. (a)**

| x    | 0              | 1              | 2              |
|------|----------------|----------------|----------------|
| P(x) | $\frac{7}{15}$ | $\frac{7}{15}$ | $\frac{1}{15}$ |

**(b)** .6    **9.** $1.29

**11.** .213    **13.** 40,320    **15.** 220    **17.** 427,518,000
**19. (a)** 14,905,800 **(b)** 14,905,800    **21.** Answers vary.
**23.** .0588    **25.** .9955    **27.** .0182    **29.** .2182
**31. (a)** $1, n, \dfrac{n(n-1)}{2}, 1$    **(b)** $_nC_0 + {_nC_1} + {_nC_2} + \cdots + {_nC_n}$

**33.**

| x    | 0      | 1      | 2      | 3      | 4      | 5      | 6      | 7      |
|------|--------|--------|--------|--------|--------|--------|--------|--------|
| P(x) | 0.0071 | 0.0510 | 0.1572 | 0.2695 | 0.2771 | 0.1710 | 0.0586 | 0.0086 |

$E(x) = 3.549$
**35. (a)** About .0002    **(b)** About .9944    **(c)** About .6134
**37. (a)**

| x    | 0      | 1      | 2      | 3      | 4      | 5      |
|------|--------|--------|--------|--------|--------|--------|
| P(x) | 0.0051 | 0.0478 | 0.1792 | 0.3357 | 0.3144 | 0.1178 |

**(b)** About .9471    **(c)** $E(x) = 3.26$    **(d)** Answers vary.    **39.** Yes
**41.** Yes    **43. (a)** [.25, .29, .46]    **(b)** [.2345, .277, .4795]
**(c)** [.240, .278, .482]    **45. (a)** Oppose **(b)** Oppose **(c)** Oppose; 2700
**(d)** Oppose; 1400

## Case Study 9 Exercises (Page 510)

**1.** +.2031    **3.**

| x    | 0      | 1      | 2      | 3      | 4      |
|------|--------|--------|--------|--------|--------|
| P(x) | .30832 | .43273 | .21264 | .04325 | .00306 |

**5.** $.19438

# Chapter 10

## Section 10.1 (Page 516)

**1.**

| Interval       | Frequency |
|----------------|-----------|
| 0–39.99        | 5         |
| 40.00–79.99    | 9         |
| 80.00–119.99   | 1         |
| 120.00–159.99  | 2         |
| 160.00–199.99  | 2         |
| 200.00–239.99  | 1         |

Price (dollars)

**3.**

| Interval      | Frequency |
|---------------|-----------|
| 0–14.99       | 6         |
| 15.00–29.99   | 11        |
| 30.00–44.99   | 2         |
| 45.00–59.99   | 0         |
| 60.00–74.99   | 0         |
| 75.00–89.99   | 1         |

Price/Earnings ratio

**5.** Answers vary, but one way to create the intervals appears here.

| Interval | Frequency |
|----------|-----------|
| 0–24     | 3         |
| 25–49    | 4         |
| 50–74    | 3         |
| 75–99    | 2         |
| 100–124  | 2         |
| 125–149  | 3         |
| 150–174  | 1         |
| 175–199  | 1         |
| 200–224  | 1         |

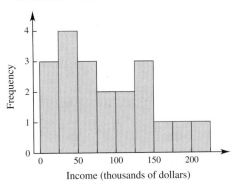
Income (thousands of dollars)

**7.** Answers vary, but one way to create the intervals appears here.

| Interval | Frequency |
|----------|-----------|
| 0–49     | 1         |
| 50–99    | 3         |
| 100–149  | 6         |
| 150–199  | 4         |
| 200–249  | 2         |
| 250–299  | 1         |
| 300–349  | 0         |
| 350–399  | 2         |
| 400–449  | 0         |
| 450–500  | 1         |

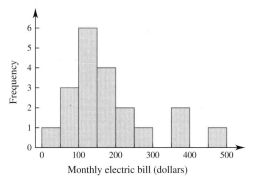

Monthly electric bill (dollars)

**9.** Answers vary, but one way to create the intervals appears here.

| Interval | Frequency |
|---|---|
| 200–399 | 1 |
| 400–599 | 1 |
| 600–799 | 4 |
| 800–999 | 6 |
| 1000–1199 | 0 |
| 1200–1399 | 3 |
| 1400–1599 | 4 |
| 1600–1799 | 0 |
| 1800–1999 | 1 |

Monthly rent payment (dollars)

**11.**

| Stem | Leaves |
|---|---|
| 2 | 57 |
| 3 | 239 |
| 4 | 3 |
| 5 | 2449 |
| 6 | 4667 |
| 7 | 2245 |
| 8 | 3 |
| 9 | 9 |

*Units:* 9|9 = $99

**13.**

| Stem | Leaves |
|---|---|
| 5 | 4 |
| 5 | |
| 5 | |
| 6 | 0111 |
| 6 | 22223 |
| 6 | 444455 |
| 6 | 66 |
| 6 | 89 |

*Units:* 6|8 = 68 inches

**15.**

| Stem | Leaves |
|---|---|
| 0 | 566 |
| 1 | 134577 |
| 2 | 00234679 |
| 3 | 14 |
| 4 | |
| 5 | |
| 6 | |
| 7 | 9 |

*Units:* 7|9 = 79

**17.**

| Stem | Leaves |
|---|---|
| 8 | 23 |
| 8 | 445 |
| 8 | 6667 |
| 8 | 8999 |
| 9 | 00000011 |
| 9 | 2223 |

*Units:* 9|2 = 92%

**19.**

| Stem | Leaves |
|---|---|
| 2 | 11333 |
| 2 | 557778899 |
| 3 | 0022234 |
| 3 | 888 |
| 4 | 1 |

*Units:* 4|1 = 41%

**21.** Uniform    **23.** Left skewed    **25. (a)** Normal  **(b)** 2  **(c)** 9
**27. (a)** Right skewed  **(b)** 26  **(c)** 7    **29. (a)** Left skewed  **(b)** 12  **(c)** 8

## Section 10.2 (Page 526)

**1.** $27,955    **3.** 4.8    **5.** 10.3    **7.** 21.2    **9.** 14.8    **11.** $33,679
**13.** 98.5    **15.** 2    **17.** 65, 71    **19.** No mode    **21.** Answers vary.
**23.** $\bar{x} \approx 31.28$; $30 - 34.99$    **25. (a)** $187.4 million  **(b)** $177 million
**27. (a)** $339.5 million  **(b)** $352 million    **29. (a)** Mean $5936.3 million;
median $5949 million  **(b)** 2011    **31.** $\bar{x} = 49.6$; median $= 48.5$
**33.** Right skewed; median $= $74.5$ thousand    **35. (a)** Normal
**(b)** Median $= 29\%$    **37.** 84.61    **39.** 90.84

## Section 10.3 (Page 537)

**1.** Answers vary.    **3.** 70.1; about 24.1    **5.** 9.3; about 3.2
**7.** 10.8; about 3.8    **9.** 7.8; about 2.8    **11.** 45.2
**13.** $\bar{x} = 59.6$; $s \approx 24.9$    **15.** $\frac{3}{4}$    **17.** $\frac{5}{9}$    **19.** 93.75%
**21.** 11.1%    **23.** $\bar{x} \approx 1276.14286$; $s \approx 295.6$
**25.** $\bar{x} = 1764$; $s \approx 149.6$    **27.** Columbia    **29.** $\bar{x} = 60.5$; $s = 52.1$
**31.** $\bar{x} = 80.17$; $s = 12.2$    **33.** Answers vary.
**35.** Min $= 80$; $Q_1 = 93$; $Q_2 = 99.5$; $Q_3 = 104$; Max $= 107$
**37.**    IBM

75  80  85  90  95  100  105  110
Revenue (billions of dollars)

**39.**  Microsoft

IBM

40  50  60  70  80  90  100  110
Revenue (billions of dollars)

The distributions are not similar. IBM generally has higher revenue and
less variability.
**41.** Min $= 9.7$; $Q_1 = 12$; $Q_2 = 17.55$; $Q_3 = 23.5$; Max $= 79.8$
**43.**  Education

0  10  20  30  40  50  60  70  80  90
Expenditures (billions of dollars)

**45.** Answers vary, but they should be close to the following: Hospitals
$Q_1 = .9$; $Q_2 = 1.7$; $Q_3 = 3.2$; Highways $Q_1 = 2.2$; $Q_2 = 3.1$; $Q_3 = 6.6$.
**47.** Health, because the box is wider.

## Section 10.4 (Page 550)

**1.** The mean    **3.** Answers vary.    **5.** 45.99%    **7.** 16.64%    **9.** 7.7%
**11.** 47.35%    **13.** 91.20%    **15.** 1.64 or 1.65    **17.** −1.04    **19.** .5; .5
**21.** .889; .997    **23.** .0934    **25.** .9051    **27.** .3156    **29.** .2004
**31.** .4286    **33.** .8700    **35.** .0505    **37.** .0850    **39.** 45.2 mph
**41.** 24.17%    **43.** Answers vary.    **45.** 665 units    **47.** 190 units
**49.** 15.87%    **51.** $\mu = 76$; $\sigma \approx 6.9$    **53.** .5686    **55.** .0475
**57.** $\mu = 33$; $\sigma \approx 4.3$    **59.** .6952    **61.** .1003

## Chapter 10 Review Exercises (Page 553)

| Refer to Section | 10.1 | 10.2 | 10.3 | 10.4 |
|---|---|---|---|---|
| For Exercises | 1–4, 11, 17–20, 30 | 5–6, 9–10, 12–16 | 7–8, 21–28, 41–46 | 29, 31–40, 47–54 |

**1.** Answers vary, but one way to create the intervals appears here.

| Interval | Frequency |
|---|---|
| 20–39.99 | 2 |
| 40–59.99 | 5 |
| 60–79.99 | 5 |
| 80–99.99 | 0 |
| 100–119.99 | 5 |
| 120–139.99 | 1 |
| 140–159.99 | 1 |
| 160–179.99 | 0 |
| 180–199.99 | 1 |

**3.**

| Stem | Leaves |
|---|---|
| 0 | 8 |
| 1 | |
| 1 | |
| 1 | 4555 |
| 1 | 666 |
| 1 | 8 |
| 2 | 00011 |
| 2 | 22 |
| 2 | 5 |
| 2 | 6 |
| 2 | 88 |

*Units:* $2|8 = 28$

**5.** Mean \$88.3; median \$83
**7.** Min = 35, $Q_1 = 57$, $Q_2 = 83$, $Q_3 = 108$, Max = 180

Price (dollars)

**9.** Mean 5.25; median 5.5   **11.** Answers vary.   **13.** 42.5 hours;
40, 50 hours   **15.** 3 − 3.99 and 5 − 5.99   **17.** Right skewed
**19.** Normal   **21.** Answers vary.   **23.** Range = 54; $s \approx 15.8$
**25.** $s \approx 1.9$   **27.** $\bar{x} = 27.32$, $s \approx .93$   **29.** Answers vary.
**31.** .4115   **33.** .9620   **35.** 1.41   **37.** .9973   **39.** .9192
**41.** (a) Mean is about 119.9167, $s \approx 84.5$ (b) Nissan
**43.** Min = 28, $Q_1 = 32$, $Q_2 = 102$, $Q_3 = 192$, max = 240

**45.** May 2016

Number of vehicles (thousands)

**47.** .3121   **49.** .9049   **51.** .9582   **53.** .9112

## Case Study 10 Exercises (Page 556)
**1.** 1.032; 12.38   **3.** .8064; 9.68   **5.** 3.2555; 11.28   **7.** 3.2307; 11.19
**9.** (1.10, 23.66)

# Chapter 11
## Section 11.1 (Page 567)
**1.** (a) 8 (b) 0   **3.** (a) 4 (b) 2 **5.** (a) Does not exist (b) 2
**7.** (a) Does not exist (b) 2   **9.** Answers vary.   **11.** 1
**13.** 0   **15.** 3   **17.** 5   **19.** Does not exist   **21.** 15
**23.** $\dfrac{5}{2}$   **25.** 5   **27.** $\dfrac{7}{4}$
**29.** (a)

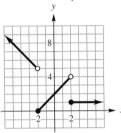

(b) Does not exist (c) 3 (d) Does not exist   **31.** 0   **33.** $\dfrac{19}{31}$

**35.** 10   **37.** 7   **39.** $\dfrac{1}{2}$   **41.** $\dfrac{9}{8}$   **43.** $\sqrt{6}$   **45.** Does not exist

**47.** $-\dfrac{1}{9}$   **49.** $\dfrac{1}{10}$   **51.** $\dfrac{1}{2\sqrt{5}}$ or $\dfrac{\sqrt{5}}{10}$   **53.** (a) 13.125 (b) 68.25

(c) 68.25   **55.** (a) \$300; when manufacturing 1000 bikes, the average
cost is \$300 per bike.   (b) \$210; when manufacturing 10,000 bikes, the
average cost is \$210 per bike.   (c) \$201; the average cost approaches \$201
per bike when the number of bikes manufactured approaches 100,000.
**57.** (a) About .5% (b) About 1% (c) About 2%

## Section 11.2 (Page 579)
**1.** 3   **3.** −2   **5.** 6   **7.** 0   **9.** (a) 0 (b) 1 (c) −1 (d) −1
**11.** (a) 1 (b) 0 (c) ∞ (d) −∞   **13.** 0   **15.** 1   **17.** −9   **19.** $\dfrac{1}{4}$

**21.**

(a) 5 (b) 0 (c) Does not exist   **23.** (a) ∞ (b) ∞ (c) −∞   **25.** −∞
**27.** ∞   **29.** ∞ and 0   **31.** −10 and −10   **33.** ∞ and −∞

**35.** −1   **37.** 0   **39.** 1   **41.** 0   **43.** $\dfrac{10}{3}$   **45.** 0   **47.** −2

**49.** $\dfrac{4}{5}$   **51.** 0   **53.** 15   **55.** 1   **57.** −1   **59.** (a) \$20 (b) \$20

(c) \$20 (d) \$50 (e) \$60 (f) Does not exist (g) \$8 per thousand
(h) \$10   **61.** (a) \$3000 (b) \$1000 (c) Does not exist (d) 3 (e) 12
**63.** (a) \$5 (b) \$5 (c) \$7 (d) \$7 (e) \$15 (f) \$5 (g) \$5 (h) \$7
(i) Does not exist (j) \$15   **65.** (a) \$65 (b) \$11 (c) \$5

## Section 11.3 (Page 590)

**1.** 8   **3.** 2   **5.** $\dfrac{1}{4}$   **7.** $-\dfrac{1}{3}$   **9. (a)** 1; from catalog distributions of 10,000 to 20,000, sales will have an average increase of $1000 for each additional 1000 catalogs distributed.   **(b)** $\dfrac{3}{5}$; from catalog distributions of 20,000 to 30,000, sales will have an average increase of $600 for each additional 1000 catalogs distributed.   **(c)** $\dfrac{2}{5}$; from catalog distributions of 30,000 to 40,000, sales will have an average increase of $400 for each additional 1000 catalogs distributed.   **(d)** As more catalogs are distributed, sales increase at a smaller and smaller rate.   **(e)** Answers vary.
**11. (a)** 2013 to 2016   **(b)** 2013 to 2015   **(c)** 2015 to 2016
**13. (a)** About 485 employees per $1 billion   **(b)** About 397 employees per $1 billion   **(c)** About 437 employees per $1 billion   **15. (a)** 5 ft/sec
**(b)** 2 ft/sec   **(c)** 3 ft/sec   **(d) (i)** 3 ft/sec   **(ii)** 2.5 ft/sec
**17.** Answers vary.   **19. (a)** 60.3 ft/sec   **(b)** 60.03 ft/sec
**(c)** 60.003 ft/sec   **21.** 60 ft/sec   **23.** 6 ft/sec
**25. (a)** $a^2 + 2ah + h^2 - 3a - 3h$   **(b)** $2a + h - 3$   **(c)** 7
**27. (a)** $a^2 + 2ah + h^2 - a - h - 1$   **(b)** $2a + h - 1$   **(c)** 9
**29. (a)** $a^3 + 3a^2h + 3ah^2 + h^3$   **(b)** $3a^2 + 3ah + h^2$   **(c)** 75
**31. (a)** $5998 per unit   **(b)** $6000 per unit   **33. (a)** 6% per day
**(b)** 7% per day   **35. (a)** About $-.35\%$ per year   **(b)** About $-.14\%$ per year
**37. (a)**

**(b)** About 81.51 kilojoules per hour   **(c)** About 18.81 kilojoules per hour
**(d)** At approximately t = 1.3 hours

## Section 11.4 (Page 604)

**1.** $f'(x) = 8; 8; 8; 8$   **3.** $f'(x) = -8x + 11; -5; 11; 35$
**5.** $f'(x) = \dfrac{2}{x^2}; \dfrac{1}{2}$; undefined; $\dfrac{2}{9}$   **7.** $f'(x) = \dfrac{1}{\sqrt{x}}; \dfrac{1}{\sqrt{2}}$; undefined; undefined
**9.** $y = 3x - 7$   **11.** $y = -\dfrac{1}{4}x - 1$   **13.** $y = \dfrac{1}{6}x + \dfrac{3}{2}$   **15. (a)** 4
**(b)** $y = 4x - 1$   **17. (a)** $-\dfrac{7}{9}$   **(b)** $y = -\dfrac{7}{9}x + \dfrac{14}{3}$   **19. (a)** $\dfrac{5}{6}$
**(b)** $y = \dfrac{5}{6}x + \dfrac{15}{2}$.   **21. (a)** $x_5$   **(b)** $x_4$   **(c)** $x_3$   **(d)** $x_2$
**23.** Here is one of many possible graphs:

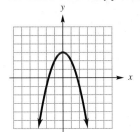

**25. (a)** $T'(5) = 1000$, so the oven temperature is increasing at 1000° per hour.
**(b)** $T'(2) = 0$, so the oven temperature is not changing.
**(c)** $T'(3.5) = -1000$, so the oven temperature is decreasing at 1000° per hour.
**27. (a)** is distance   **(b)** is velocity   **29.** $-6; 6$   **31.** $-5; -3; 0; 2; 4$
**33.** Seven seconds after it is dropped, the stone hits the ocean, and its velocity at that instant is $-224$ ft/s.   **35. (a)** Six weeks after entering the race, 25% of likely voters support the candidate.   **(b)** Six weeks after entering the race, support for the candidate increased by approximately 3 percentage points in the next week.   **37. (a)** If the interest rate on your mortgage is 5%, your monthly payment will be $1610.46.

**(b)** If the interest rate you can get on your mortgage increases from 5% to 6%, your monthly payment will increase by approximately $183.37.
**39. (a)** If the interest rate on your mortgage is 5%, the amount you can afford to borrow is $223,537.94.   **(b)** Negative. If you have to pay a higher interest rate, the amount you can afford to borrow is smaller.
**(c)** The units for $A'(r)$ are dollars/percent.   **(d)** If the interest rate you can get on your mortgage increases from 5% to 6%, the amount you can afford to borrow decreases by approximately $25,449.21.   **41.** 1000; the population is increasing at a rate of 1000 shellfish per time unit. 570; the population is increasing more slowly at 570 shellfish per time unit. 250; the population is increasing at a much slower rate of 250 shellfish per time unit.
**43. (a)** About $13.34 billion per year; about $28.23 billion per year; about $59.77 billion per year   **(b)** The rate of spending on research and development is increasing.   **45.** In 2018, the number of licensed drivers will increase by approximately 726,000 in the next year.
**47. (a)**   **(b)**

| 15 | 15 |
|---|---|

**(c)** The graphs appear identical, which suggests that $f'(x) = g(x)$.

## Section 11.5 (Page 617)

**1.** $f'(x) = 8x - 3$   **3.** $y' = 6x^2 + 6x - 6$   **5.** $g'(x) = 4x^3 + 9x^2 - 8$
**7.** $f'(x) = 9x^{.5} - 2x^{-.5}$   **9.** $y' = -22.5x^{1/2} + 3.8x^{.9} + 1$
**11.** $y' = 36t^{1/2} + 2t^{-1/2}$   **13.** $y' = 4x^{-1/2} + \dfrac{9}{2}x^{-1/4}$
**15.** $g'(x) = -30x^{-6} + 2x^{-2}$   **17.** $y' = -20x^{-3} - 24x^{-5} + 3$
**19.** $f'(t) = \dfrac{4}{t^2} - \dfrac{16}{t^3}$   **21.** $f'(t) = \dfrac{1}{4t^2} - \dfrac{1}{4t^3}$
**23.** $g'(x) = \dfrac{-5}{2}x^{-3/2} - \dfrac{7}{2}x^{-1/2} + 101$   **25.** $y' = \dfrac{-21}{2}x^{-5/2} - 4x^{-3/2} + 3x^2$
**27.** $y' = \dfrac{-19}{4}x^{-5/4}$   **29.** $y' = \dfrac{-8}{3}t^{-2/3}$   **31.** $\dfrac{dy}{dx} = -40x^{-6} + 36x^{-5} + 36x^3$
**33.** $-\dfrac{9}{2x^{3/2}} - \dfrac{3}{x^{5/2}}$   **35.** $-26$   **37.** $\dfrac{11}{16}$
**39.** $y' = -48x^{-5} + 21x^{-4} - 6x^{-2}$   **41.** $0; y = 0$   **43.** $7; y = 7x - 3$
**45. (a)** $12   **(b)** $14.30   **(c)** $20.70   **(d)** $29.50   **47. (a)** $2.96
**(b)** $56.60   **(c)** $296.44   **(d)** $596.38   **49. (a)** $R'(x) = .1216x + 1.4284$
**(b)** $2.0364 hundred million   **(c)** $2.2796 hundred million
**51. (a)** 2010   **(b)** 127.45   **53. (a)** $16 per table   **(b)** $15.998 or $16
**(c)** The marginal revenue found in part (a) approximates the actual revenue from the sale of the 1001st item found in part (b).   **55. (a)** $14,000/violin
**(b)** $4000/violin   **(c)** 0   **57. (a)** 30   **(b)** 0   **(c)** $-10$
**59. (a)** For such an individual, the cost of $100,000 in life insurance is $94.87, and adding an additional $1000 in coverage would increase the cost by approximately $0.71.   **(b)** For such an individual, the cost of $1,000,000 in life insurance is $533.48, and adding an additional $1000 in coverage would increase the cost by approximately $0.40.   **61. (a)** In 2014, the percentage of the population that is age 0–17 years is estimated to be 24.02% and to decrease by approximately .05% in the next year.
**(b)** In 2024, the percentage of the population that is age 0–17 years is estimated to be 23.64% and to decrease by approximately .03% in the next year.
**63. (a)** In 2010, the National Home Price Index was 145 and was decreasing at approximately 7 index points per year.   **(b)** In 2016, the National Home Price Index was 175 and was increasing at approximately 17 index points per year.   **65. (a)** $f'(x) = .54x + 3.52$   **(b)** About 14 persons/year
**(c)** Increasing   **67. (a)** $\dfrac{dV}{dr} = 160\pi r$   **(b)** $640\pi$; the volume of blood is increasing $640\pi$ cu mm per mm of change in the radius.   **(c)** $960\pi$; the volume of blood is increasing $960\pi$ cu mm per mm of change in the radius. **(d)** $1280\pi$; the volume of blood is increasing $1280\pi$ cu mm per mm of change in the radius.   **69. (a)** $f'(h) = -\dfrac{140,600}{h^3}$

**(b)** Negative because taller individuals of the same weight would have lower BMI.  **(c)** $-.65$  **71. (a)** $v(t) = 16t + 3$  **(b)** 3; 83; 163
**73. (a)** $v(t) = 6t^2 + 12t$  **(b)** 0; 210; 720  **75. (a)** $-32$ ft per sec; $-64$ ft per sec  **(b)** In 3 sec  **(c)** $-96$ ft per sec
**77. (a)** $(x + h)^n = x^n + n \cdot x^{n-1}h + \dfrac{n(n-1)}{2}x^{n-2}h^2 + \cdots + h^n$

**(b)** $\dfrac{(x + h)^n - x^n}{h} = n \cdot x^{n-1} + \dfrac{n(n-1)}{2}x^{n-2}h + \cdots + h^{n-1}$

**(c)** $\lim\limits_{h \to 0} \dfrac{(x + h)^n - x^n}{h} = n \cdot x^{n-1} = y'$  **79. (a)–(c)** The derivative is negative on $(-\infty, \infty)$.  **(d)** The graph is always falling.

## Section 11.6 (Page 627)

**1.** $y' = 9x^2 + 4x - 6$  **3.** $y' = 120x^3 - 54x^2 + 10$
**5.** $y' = -27x^8 + 6x^5 - 30x^4 + 48x^3 - 4$  **7.** $y = 18x - 24$
**9.** $y' = 96x^5 + 240x^4 + 144x^3$  **11.** $y' = \dfrac{-4}{(x-3)^2}$
**13.** $y' = \dfrac{-7x^2 + 6x - 56}{(x^2 - 8)^2}$  **15.** $f'(t) = \dfrac{t^2 + 6t - 12}{(t + 3)^2}$
**17.** $g'(x) = \dfrac{-6x^4 - 4x^3 - 12x - 2}{(2x^3 - 2)^2}$  **19.** $y' = \dfrac{x^2 + 6x - 14}{(x + 3)^2}$
**21.** $y' = \dfrac{\dfrac{9\sqrt{x}}{2} + \dfrac{4}{\sqrt{x}}}{x}$ or $\dfrac{9x + 8}{2x\sqrt{x}}$  **23.** $y' = \dfrac{-236}{(4x + 11)^2}$
**25.** $f'(p) = \dfrac{264p^2 + 396p - 277}{(4p + 3)^2}$
**27.** $g'(x) = \dfrac{6x^4 + 30x^3 - 27x^2 + 96x + 120}{(3x + 9)^2(2x - 1)^2}$  **29.** In the first step, the numerator should be $(x^2 - 1)(2) - (2x + 5)(2x)$.  **31.** $y = 13x - 11$
**33.** $y = -2x + 9$  **35. (a)** $\overline{C}(x) = \dfrac{200 + 75x}{x}$  **(b)** $\overline{C}(10) = 95$ means that, at a sales level of 1000 hot dogs, the average cost is $95 per 100 hot dogs (or $.95/hot dog). Similarly, $\overline{C}(50) = 79$.  **(c)** $\overline{C}'(x) = \dfrac{-200}{x^2}$
**(d)** $\overline{C}'(10) = -2$ means that, at a sales level of 1000 hot dogs, if an additional 100 hot dogs are sold, the average cost decreases by approximately $2 per hundred hot dogs. Similarly, $\overline{C}'(50) = -.08$.
**37. (a)** $\dfrac{999}{(x + 100)^2}$; $6.94/basket; $5.10/basket  **(b)** $\dfrac{10x + 1}{x^2 + 100x}$; $8.38/basket; $7.16/basket  **(c)** $-\dfrac{10x^2 + 2x + 100}{(x^2 + 100x)^2}$; $-$.07/basket per basket; $-$.05/basket per basket  **39. (a)** $T'(x) = \dfrac{-10x^2 + 50}{(x^2 + 5)^2}$
**(b)** 2; temperature increasing at 2 degrees per hour  **(c)** About 1.1111; temperature increasing at about 1.1111 degrees per hour  **(d)** About $-.2041$; temperature decreasing at about .2041 degrees per hour  **41. (a)** 8.57 min
**(b)** 16.36 min  **(c)** 6.12 min/(kcal/min) and 2.48 min/(kcal/min)
**43. (a)** .1173  **(b)** 2.625  **45. (a)** $M'(x) = \dfrac{12,000x}{(x^2 + 75)^2}$
**(b)** About 2 codes/shift; 6 codes/shift; 1 code/shift  **(c)** Answers vary.

## Section 11.7 (Page 637)

**1.** 779  **3.** 259  **5.** $16x + 36$; $16x + 27$
**7.** $3x^2 - 32x + 81$; $3x^2 - 8x - 3$  **9.** $\dfrac{1}{x^2}$; $\dfrac{1}{x^2}$
**11.** $\sqrt{6x - 11}$; $3\sqrt{2x + 5} - 8$  **13.** If $f(x) = x^5$ and $g(x) = 4x + 3$, then $y = f[g(x)]$.  **15.** If $(x) = \sqrt{x}$ and $g(x) = 6 + 3x^2$, then $y = f[g(x)]$.  **17.** If $f(x) = x^{-2}$ and $g(x) = 4x - 1$, then $y = f[g(x)]$.
**19.** If $f(x) = x^3 + x + 5$ and $g(x) = x^{1/2} - 3$, then $y = f[g(x)]$
**21.** $y' = 35(7x - 12)^4$  **23.** $y' = 112(4x + 3)^3$
**25.** $y' = -400x(10x^2 + 8)^4$  **27.** $y' = 6(4x - 9)^{1/2}$

**29.** $y' = -\dfrac{21}{2}(8x + 9)(4x^2 + 9x)^{1/2}$  **31.** $y' = \dfrac{16}{\sqrt{4x + 7}}$
**33.** $y' = \dfrac{-2x - 4}{\sqrt{x^2 + 4x}}$  **35.** $y' = 2(x + 1)(x - 3) + (x - 3)^2$
**37.** $y' = 70(x + 3)(2x - 1)^4(x + 2)$  **39.** $y' = \dfrac{(3x + 1)^2(21x + 1)}{2\sqrt{x}}$
**41.** $y' = \dfrac{-3}{(x - 2)^4}$  **43.** $y' = \dfrac{(2x - 7)(8x + 24)}{(4x - 1)^2}$
**45.** $y' = \dfrac{-3x^2 - 28x + 5}{(3x + 1)^4}$  **47.** $y' = \dfrac{6x + 8\sqrt{x} + 3}{2x^{1/2}}$
**49.** $y = 15x - 1$  **51.** $y = 2x + 3$
**53.** $D(c) = \dfrac{-c^2 + 10c - 25}{25} + 500$  **55. (a)** $45,000
**(b)** $M'(x) = -200x + 9400$  **(c)** $-$2000 per year; decreasing
**57.** $C'(x) = \dfrac{(50x - 1)}{2(200 + 25x^2 - x)^{1/2}}$  **59. (a)** $R'(x) = \dfrac{15(250 - 2x)}{(500x - 2x^2)^{1/2}}$
**(b)** $19.67; $8.19; $.85; $-6.19  **(c)** Answers vary
**61. (a)** $20  **(b)** $-$16  **63. (a)** $R(x) = x\sqrt{900 - 2x}$
**(b)** $R'(x) = \dfrac{900 - 3x}{\sqrt{900 - 2x}}$  **(c)** $R'(250) = 7.50$ per ticket
**65. (a)** $\dfrac{dA}{dr} = 960\left(1 + \dfrac{r}{1200}\right)^{95}$  **(b)** $1316.95 per percentage point; when the interest rate is 4%, increasing the interest rate by 1 percentage point would increase the balance by approximately $1316.95.
**67. (a)** $\dfrac{dA}{dr} = 480\left(1 + \dfrac{r}{1200}\right)^{22}$  **(b)** $498.74 per percentage point; when the interest rate is 2%, increasing the interest rate by 1 percentage point would increase the balance by approximately $498.74.
**69.** $120 per worker  **71. (a)** $\overline{C}(x) = \dfrac{\sqrt{6x + 5}}{x}$
**(b)** $\overline{C}'(x) = \dfrac{-3x - 5}{x^2\sqrt{6x + 5}}$  **73. (a)** $-.5$  **(b)** $-\dfrac{1}{54}$
**(c)** About $-.010667$  **(d)** $-\dfrac{1}{128}$  **75. (a)** $F'(t) = \dfrac{150t}{(8 + t^2)^{3/2}}$
**(b)** About 3.96 items scanned per minute per hour on the job  **(c)** About .09 items scanned per minute per hour on the job  **(d)** After 5 hours

## Section 11.8 (Page 647)

**1.** $y' = 5e^{5x}$  **3.** $f'(x) = 10e^{2x}$  **5.** $g'(x) = 28e^{-7x}$
**7.** $y' = 2xe^{x^2}$  **9.** $f'(x) = x^2e^{x^3/3}$  **11.** $y' = -18xe^{3x^2+5}$
**13.** $y' = \dfrac{-20x + 7}{-10x^2 + 7x}$  **15.** $y' = \dfrac{5}{2(5x + 1)}$
**17.** $f'(x) = \dfrac{15x^2 + 18x + 5}{(5x + 9)(x^2 + 1)}$  **19.** $y' = \dfrac{-20}{(4 - x)(3x + 8)}$
**21.** $y' = e^{-3x}(-3x^4 + 4x^3)$  **23.** $y' = e^{-3x}[-18x^2 + 27x - 5]$
**25.** $y' = \dfrac{3(8x^2 - 1)}{2x(8x^2 - 3)}$  **27.** $y' = \dfrac{-4x^4}{9 - x^4} + \ln(9 - x^4)$
**29.** $x^3(1 + 4\ln|x|)$  **31.** $y' = \dfrac{\dfrac{-25}{x} + 10 - 10\ln|x|}{(5 - 2x)^2}$
**33.** $y' = \dfrac{1}{2(x - 3)\sqrt{\ln(x - 3)}}$  **35.** $y' = \dfrac{xe^x \ln|x| - e^x + 1}{x(\ln|x|)^2}$
**37.** $y' = \dfrac{xe^x + xe^{-x} - e^x + e^{-x}}{x^2}$ or $\dfrac{e^x(x - 1) + e^{-x}(x + 1)}{x^2}$
**39.** $f'(x) = \dfrac{4e^{3x+2}}{4x - 5} + 3e^{3x+2}\ln(4x - 5)$  **41.** $y' = \dfrac{2800e^{4x}}{(7 - 10e^{4x})^2}$
**43.** $y' = \dfrac{1250e^{-.5x}}{(12 + 5e^{-.5x})^2}$  **45.** $y' = \dfrac{1}{x \ln|x|}$  **47.** $y' = (\ln 8)8^x$
**49.** $y' = 2(\ln 15) \cdot 15^{2x}$  **51.** $g'(x) = \dfrac{1}{(\ln 10) x}$  **53.** $y = -x + 1$

**55.** $y = 0$    **57.** 50    **59. (a)** $f(t) = 165e^{.165t}$  **(b)** About $165 billion
per year  **(c)** About $2312 billion per year    **61. (a)** About 362 ferrets
**(b)** $P'(x) = 9.88e^{.13}$; about 47 ferrets per year  **(c)** In 2012, there were
about 362 ferrets, and the population was increasing at a rate of about 47
ferrets per year.    **63. (a)** About 5.9 deaths per 1000 people
**(b)** About $-.4$ deaths per 1000 people per year  **(c)** Decreasing
**65. (a)** About .7°F per hour  **(b)** About $-.1$°F per hour

**67. (a)** $f'(t) = \dfrac{1000e^{-.4t}}{(1 + 25e^{-.4t})^2}$  **(b)** About 1.1 subscriptions per 100 people
per year **(c)** In 2017, cell phone subscriptions per 100 people in middle-income
countries were increasing at about 1.1 subscriptions per 100 people per year.
**69. (a)** 6393; increasing about 2916 per year  **(b)** 20,807; increasing about
6490 per year  **(c)** 48,771; increasing about 1575 per year  **(d)** The growth
rate increases rapidly at first and then gradually decreases to almost 0.

**71. (a)** About 366 million  **(b)** $f'(x) = \dfrac{66}{x}$  **(c)** 2 million per year

**73. (a)** $\dfrac{dR}{dx} = 100 + \dfrac{50(\ln x - 1)}{(\ln x)^2}$  **(b)** $112.48    **75. (a)** It takes about
78.2 months to reach a balance of $300.  **(b)** Positive  **(c)** When $A = $300$;
if you want the balance to increase another $1, it will take approximately
1.4 more months.

## Section 11.9 (Page 655)

**1.** $x = 0, 2$    **3.** $x = 1$    **5.** $x = -3, 0, 3$    **7.** $x = 0, 6$
**9.** Yes, no    **11.** No, no, yes    **13.** Yes, no, yes
**15.** Yes, no, yes    **17.** Yes, no    **19.** Continuous everywhere
**21.** $x = -4$    **23.** $x < 2$

**25. (a)**

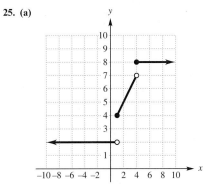

**(b)** $x = 1, 4$

**27. (a)**

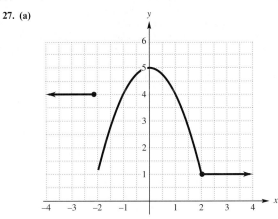

**(b)** $x = -2$    **29. (a)** 7.25%  **(b)** 7.25%  **(c)** 7.25%
**(d)** 6.5%  **(e)** 6.25%  **(f)** Does not exist  **(g)** 2002, 2004, 2009,
2011, 2013, 2017  **(h)** 2002, 2004, 2009, 2011, 2013, 2017
**31.** Discontinuous at $t = m$; differentiable everywhere except $t = m$ and
the endpoints

**33.**

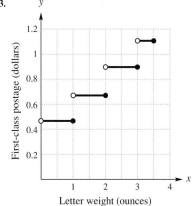

**(a)** $.47  **(b)** $.68  **(c)** Does not exist  **(d)** $.89  **(e)** $.89  **(f)** $.89
**(g)** $x = 1, 2, 3$ ounces    **35. (a)** $88  **(b)** $88  **(c)** $97.50
**(d)** Does not exist  **(e)** $x = 2, 3, 4$ miles    **37.** $(-5, 7)$    **39.** $(-\infty, -3]$
**41.** $(-6, 0), (4, 8)$

**43. (a)**

**(b)** No, at $t = 40$

## Chapter 11 Review Exercises (Page 660)

| Refer to Section | 11.1 | 11.2 | 11.3 | 11.4 | 11.5 | 11.6 | 11.7 | 11.8 | 11.9 |
|---|---|---|---|---|---|---|---|---|---|
| For Exercises | 1–12, 79–80 | 13–22, 81–82 | 23–28, 83–84 | 29–36, 85–86 | 37–44, 87–88 | 45–54, 89–90 | 55–62, 91–92 | 63–72, 93–94 | 73–78, 95–96 |

**1.** 4; 4; 4    **3.** About $-.35$    **5.** 7    **7.** 0    **9.** 8    **11.** $\dfrac{1}{6}$

**13.** 0    **15.** Does not exist    **17.** $-\infty$    **19.** 0    **21.** $-3$    **23.** $\dfrac{1}{4}$

**25.** 18    **27.** $-\dfrac{1}{3}$    **29.** $y' = 2$    **31.** $y' = 4x - 1$    **33.** $-2; y = -2x - 4$

**35.** $\dfrac{2}{9}; y = \dfrac{2}{9}x - \dfrac{2}{9}$    **37.** $f'(x) = 56x^7 - 8$    **39.** $f'(x) = e + e^x + exe^{e-1}$

**41.** $f'(x) = 4x^{-1/2} - 21x^{-4}$    **43.** $f'(x) = \dfrac{3}{\sqrt{x}} + \dfrac{3}{x^2}$

**45.** $y' = 20t^4 + 27t^2 - 40t$    **47.** $y' = 40x^{3/4} + 6(5x + 1)x^{-1/4}$

**49.** $f'(x) = \dfrac{-4(x^2 + 8)}{(x^2 - 8)^2}$    **51.** $y' = \dfrac{-3x + 7 + 36\sqrt{x}}{2\sqrt{x}(3x + 7)^2}$

**53.** $y' = \dfrac{x^2 - 2x}{(x - 1)^2}$    **55.** $f'(x) = 16(4x - 2)^3$    **57.** $y' = \dfrac{1}{\sqrt{2t - 5}}$

**59.** $y' = 18x(3x - 4)^2 + 2(3x - 4)^3$

**61.** $f'(u) = \dfrac{(2u + 3)(6u - 4) - 6(3u^2 - 4u)}{(2u + 3)^4}$

**63.** $y' = -12e^{2x}$    **65.** $y' = -6x^2e^{-2x^3}$    **67.** $y' = 10xe^{2x} + 5e^{2x}$

**69.** $y' = \dfrac{2x + 4}{x^2 + 4x - 1}$    **71.** $y' = \dfrac{x - \dfrac{1}{x} - 2x(\ln 6x)}{(x^2 - 1)^2}$    **73.** None

**75.** Yes, no, yes    **77.** No, yes, no    **79. (a)** $5  **(b)** Does not exist
**81. (a)** $135  **(b)** $135  **(c)** $200  **(d)** Does not exist    **83.** $8.46 per year
**85. (a)** The cost of producing 100 paddleboards is $80,000.

**(b)** When producing 100 paddleboards, the cost of producing one additional paddleboard is approximately \$300.    **87.** $R'(t) = -.3t^2 + 6.5t - 3.5$

**89. (a)** $\overline{C}(x) = \dfrac{9910}{(2x + 40)^2}$   **(b)** \$275.28    **91. (a)** $R(x) = x\sqrt{25 - .1x}$

**(b)** $R'(x) = \dfrac{25 - .15x}{\sqrt{25 - .1x}}$   **(c)** \$360   **(d)** \$2.88    **93. (a)** Positive because

life expectancy has improved over time.   **(b)** 79.3 years   **(c)** .1 years; life expectancy would be approximately .1 years longer for a person born in 2019.    **95. (a)** $x = \$100$   **(b)** $x = \$50, \$100$

## Case Study 11 Exercises (Page 664)

**1.** About .452    **3. (a)** About 1.1; about .65   **(b)** \$95.14    **5.** Demand is constant, no matter what the price.

## Extended Project (Page 664)

**(a)** Roll-your-own tobacco, 2153%; pipe tobacco, 157%. Pipe tobacco had a similar percentage increase compared to cigarettes, but roll-your-own tobacco had a dramatically larger increase.   **(b)** $q = -1.34p + 364$
**(c)** $q = -570.99p + 20{,}301$   **(d)** .08 inelastic; .23 inelastic   **(e)** .03 inelastic; 2.30 elastic   **(f)** At the higher price, demand for cigarettes is inelastic, but demand for roll-your-own tobacco is elastic, maybe because pipe tobacco is a less expensive substitute product for roll-your-own tobacco.
**(g)** Answers vary.

# Chapter 12

## Section 12.1 (Page 677)

**1.** Increasing on $(1, \infty)$; decreasing on $(-\infty, 1)$; local minimum of $-4$ at $x = 1$    **3.** Increasing on $(-\infty, -2)$; decreasing on $(-2, \infty)$; local maximum of 3 at $x = -2$    **5.** Increasing on $(-\infty, -4)$ and $(-2, \infty)$; decreasing on $(-4, -2)$; local maximum of 3 at $x = -4$; local minimum of 1 at $x = -2$
**7.** Increasing on $(-7, -4)$ and $(-2, \infty)$; decreasing on $(-\infty, -7)$ and $(-4, -2)$; local maximum of 3 at $x = -4$; local minima of $-2$ at $x = -7$ and $x = -2$    **9.** Increasing on $(-\infty, -5)$ and $(5, \infty)$; Decreasing on $(-5, 5)$    **11.** Increasing on $(-\infty, -2)$ and $(1, \infty)$; decreasing on $(-2, 1)$
**13.** Increasing on $\left(-\infty, -\dfrac{1}{3}\right)$ and $(2, \infty)$; decreasing on $\left(-\dfrac{1}{3}, 2\right)$
**15.** Increasing on $(-\infty, -3)$ and $(-3, \infty)$    **17.** Increasing on $(0, \infty)$; decreasing on $(-\infty, 0)$    **19.** Local maximum at $-2$; local minimum at 0
**21.** Local maximum at $-3$; local minimum at $-1$    **23.** Local maximum at $\dfrac{1}{4}$; local minimum at 5    **25.** Local maximum at $-2$; local minimum at 3
**27.** Local maximum at $x = 0$; local minima at $x = -4, 2$    **29.** Local maximum at $x = 0$; no local minimum    **31.** Local minimum at $x = -7$; local maximum at $x = 7$    **33.** No local extrema    **35.** Local maximum at 1; $f(1) = \dfrac{1}{e}$    **37.** Local maximum at $-\dfrac{1}{e}$; $f\left(-\dfrac{1}{e}\right) = \dfrac{1}{e}$; local minimum at $\dfrac{1}{e}$; $f\left(\dfrac{1}{e}\right) = -\dfrac{1}{e}$    **39.** Local minimum at $-\dfrac{1}{3}$; $f\left(-\dfrac{1}{3}\right) \approx -2.1226$
**41.** Approximate local minimum at $(-5.5861, 563.4222)$
**43.** Approximate local maxima: $(-18.5239, 5982.7502)$ and $(.3837, 40.9831)$; approximate local minima: $(-2.8304, 11.4750)$ and $(4.9706, -53.7683)$
**45. (a)** $(2.5, 5.0)$   **(b)** $(5.0, 6.0)$   **(c)** $(3.5, 5.0)$   **(d)** $(2.5, 3.5), (5.0, 6.0)$
**47. (a)** $(5, 16), (18, 19)$   **(b)** $(0, 5), (16, 18), (19, 23)$   **(c)** Local minima at 5:00 a.m. and 6:00 p.m.; local maxima at 4:00 p.m. and 7:00 p.m.
**49. (a)** 2014   **(b)** \$1125 million    **51.** 2012    **53. (a)** $t = 1$ second
**(b)** $h(1) = 64$ feet   **(c)** $t = 3$ seconds   **(d)** $h'(3) = -64$ feet per second
**55. (a)** 10   **(b)** 500    **57.** Local minimum at $(12, 521)$; local maximum at about $(4.3, 1056.6)$    **59.** 12.6°C    **61. (a)** $(0, 2.16)$   **(b)** $(2.16, 6)$
**63.** Increasing on $(-\infty, 0)$; decreasing $(0, \infty)$    **65.** 10 min    **67. (a)** $(0, 3)$
**(b)** $(3, \infty)$    **69.** Local minima at about $(1.8, 26.9)$ and $(13.9, 51.0)$

## Section 12.2 (Page 691)

**1.** $f''(x) = 6x - 12$; $-12$; 0; $-30$    **3.** $f''(x) = 2e^x$; 2; $2e^2$; $\dfrac{2}{e^3}$
**5.** $f''(x) = 24e^{2x}$; 24; $24e^4$; $\dfrac{24}{e^6}$    **7.** $f''(x) = 12(x + 3)^2$; 108; 300; 0

**9. (a)** $f''(x) = -\dfrac{4}{(4x + 1)^{\frac{3}{2}}}$   **(b)** $-4$   **(c)** $-4/27$   **(d)** Undefined

**11.** $f''(x) = \dfrac{2}{9x^{4/3}}$; not defined; $\dfrac{1}{9 \cdot 2^{1/3}}$; $\dfrac{2}{9(-3)^{4/3}}$

**13.** $f''(x) = \dfrac{-4}{(2x + 1)^2}$; $-4$; $\dfrac{-4}{25}$; not defined    **15. (a)** $f''(x) = e^{-x}(x - 2)$

**(b)** $f''(0) = -2$   **(c)** $f''(2) = 0$   **(d)** $f''(-3) = -5e^3$

**17.** $f''(x) = \dfrac{2}{(1 + x)^3}$; 2; $\dfrac{2}{27}$; $-\dfrac{1}{4}$    **19.** Answers vary.

**21.** $v(t) = 12t + 5$; $a(t) = 12$; $v(0) = 5$ cm/sec; $v(4) = 53$ cm/sec; $a(0) = a(4) = 12$ cm/sec$^2$    **23.** $v(t) = 9t^2 - 8t + 8$; $a(t) = 18t - 8$; $v(0) = 8$ cm/sec; $v(4) = 120$ cm/sec; $a(0) = -8$ cm/sec$^2$; $a(4) = 64$ cm/sec$^2$
**25.** Concave upward on $(0, \infty)$; concave downward on $(-\infty, 0)$; point of inflection at $(0, -5)$    **27.** Concave upward on $\left(-\dfrac{4}{3}, \infty\right)$; concave downward on $\left(-\infty, -\dfrac{4}{3}\right)$; point of inflection at $\left(-\dfrac{4}{3}, \dfrac{425}{27}\right)$
**29.** Concave upward on $(4, \infty)$; concave downward on $(-\infty, 4)$; no points of inflection    **31.** Concave upward on $(-\infty, -5)$ and $(1, \infty)$; concave downward on $(-5, 1)$; points of inflection at $(-5, -1248)$ and $(1, 0)$
**33. (a)** $R''(x) = 60 - 6x$   **(b)** $(10, 3000)$    **35. (a)** $R''(x) = -6x + 84$
**(b)** $(14, 26{,}688)$    **37.** Local minimum at $x = -4$; $f(-4) = -207$; local maximum at $x = 3$; $f(3) = 136$;    **39.** Critical number 3; no local extrema    **41.** Local maximum at $x = 0$; local minima at $x = 2$ and $x = -2$    **43.** Local maximum at $x = -2$; local minimum at $x = 2$
**45.** Local maximum at $x = -3$; local minimum at $x = 3$    **47.** No critical numbers; no local extrema    **49. (a)** E   **(b)** C   **(c)** A, F   **(d)** D   **(e)** B
**51. (a)** $[1, 12]$   **(b)** \$.0124/week$^2$   **(c)** \$.0137/week$^2$    **53. (a)** $-96$ ft/sec
**(b)** $-160$ ft/sec   **(c)** $-256$ ft/sec   **(d)** $-32$ ft/sec$^2$
**55.** $v(t) = 256 - 32t$ ft/sec; $a(t) = -32$ ft/sec$^2$; 1024 ft; at 16 sec
**57.** $(8, 127)$    **59. (a)** $f''(x) = -.012x + .24$   **(b)** $x = 20$   **(c)** $(14, 20)$
**(d)** $(20, 26)$   **(e)** The stock was increasing at an increasing rate until week 20 and increasing at a decreasing rate thereafter.

**61. (a)** $f'(x) = \dfrac{1}{2}x^2 - 7x + \dfrac{45}{2}$   **(b)** $x = 5, 9$   **(c)** $f''(x) = x - 7$
**(d)** Local maximum in 2005; local minimum in 2009
**63. (a)** $f'(x) = 7877/x$   **(b)** Increasing   **(c)** $f''(x) = -7877/x^2$
**(d)** Concave down; it is increasing at a decreasing rate
**65. (a)** 100   **(b)** About \$7357.59   **(c)** About \$73.58

## Section 12.3 (Page 703)

**1.** Absolute maximum at $x = 4$; absolute minimum at $x = 1$
**3.** Absolute maximum at $x = -1$; absolute minimum at $x = 1$
**5.** Absolute maximum at $x = -4$; absolute minima at $x = -7$ and $x = -2$
**7.** Absolute maximum at $x = 2$; absolute minimum at $x = 1$
**9.** Absolute maximum at $x = 0$; absolute minima at $x = -3$ and $x = 3$
**11.** Absolute maximum at $x = 6$; absolute minimum at $x = 4$
**13.** Absolute maximum at $x = 2$; absolute minimum at $x = 0$
**15.** Absolute maximum at $x = -3$; absolute minimum at $x = 0$
**17.** Absolute maximum at $x = 0$; absolute minima at $x = -1$ and $x = 1$
**19.** Absolute minimum at $x = 2$    **21.** Absolute minimum at $x = 1$; no absolute maximum    **23.** Absolute maximum at $x = 3$; no absolute minimum    **25.** Absolute maximum at $x = 1$    **27.** Absolute minimum average cost of \$118.75 per wheel when $x = 75$ wheels are produced
**29.** Absolute minimum $(9, 947)$ Absolute maximum $(15, 1559)$
**31. (a)** $R(x) = 400x$   **(b)** $P(x) = 430x - 525{,}000 - .012x^2$   **(c)** About 17,917 grills; \$3,327,083 profit    **33.** The minimum average consumer spending on entertainment during the years 2003 to 2015 was \$2175 in 2003, and the maximum \$2800 in 2008.    **35. (a)** After 12 days   **(b)** 50 per ml
**(c)** After one day   **(d)** 81.365 per ml    **37. (a)** 8   **(b)** 29.43%
**39.** 8 in. by 8 in.    **41.** 4 in. by 4 in. by 2 in.    **43.** $r = 10$ cm; $h = 10$ cm
**45. (a)** $1200 - 2x$   **(b)** $A(x) = 1200x - 2x^2$   **(c)** 300 m   **(d)** 180,000 m$^2$
**47.** \$2400    **49.** About 6.90 in. × 10.35 in.
**51. (a)** $R(x) = 100{,}000x - 100x^2$   **(b)** 500   **(c)** 25,000,000¢ or \$250,000
**53.** 2009    **55.** $P$ is about 1.134 miles from point $A$.    **57.** 6    **59.** 69
**61.** 3333 booklets

## Section 12.4 (Page 713)

**1.** $-1$  **3.** $-\dfrac{x^3}{y^3}$  **5.** $\dfrac{3x^2}{y^4}$  **7.** $-\dfrac{2y}{5x}$  **9.** $\dfrac{5\sqrt{y}}{2\sqrt{x}}$  **11.** $\dfrac{-10x}{e^y}$

**13.** $-\dfrac{1}{x}$  **15.** $-\dfrac{3y}{x}$  **17.** $-\dfrac{y}{x\ln x}$  **19.** $-\dfrac{5}{6}$  **21.** $\dfrac{9}{4}$  **23.** 0

**25.** $-\dfrac{1}{5}$  **27.** $-e$  **29.** $y = 5$  **31.** $y = \dfrac{8}{3}x - \dfrac{2}{3}$  **33.** $y = -4x + 5$

**35. (a)** $-\dfrac{1}{x}$  **(b)** $-\dfrac{1}{x}$  **(c)** Answers agree.  **37.** $\dfrac{dq}{dp} = -2$ means that if
the price is increased by \$1, the quantity demanded decreases by approximately
2 tee times.  **39.** $\dfrac{dq}{dp} = -2$ means that if the price is increased by \$1, the
quantity demanded decreases by approximately 2 suites.  **41.** $\dfrac{dq}{dp} = -.075$
means that if the price is increased by \$1, the quantity demanded decreases
by approximately 75 drinks.  **43. (a)** Positive; if revenue grows, we
would expect more employees would be hired.  **(b)** $\dfrac{dn}{dR} = \dfrac{4000R}{3n^2}$ thousand
employees per billion dollars  **(c)** About 1.667; if revenue grows by 1 billion,
we expect a corresponding increase of approximately 1667 employees
(1.667 thousand employees).  **45. (a)** $\dfrac{dR}{dn} = \dfrac{9n^2}{14{,}000R}$ billion dollars per
thousand employees  **(b)** .45; if the number of employees grows by 1000, we
expect a corresponding increase of approximately \$450 million in revenue.

**47.** $\dfrac{dA}{dt} = -1.8$ means that if one more hour passes, approximately 1.8 mg
of drug will be eliminated from the body.

## Section 12.5 (Page 720)

**1.** 2  **3.** 10  **5.** 1  **7.** 1  **9.** $\dfrac{15}{4}$  **11.** About 2629 employees
per year  **13.** About 15,556 employees per year  **15. (a)** Decrease
**(b)** About $-.05$ per week  **17.** $-\$192/$day  **19.** $-\$.42/$month
**21.** About $-1.46\%$ per year  **23.** About 37.7 in$^2$/min  **25.** About
5655 mm$^3$/month  **27.** 2.4 yd/sec  **29.** 18 lbs  **31.** About
12.6 cm$^3$/min

## Section 12.6 (Page 728)

**1.** No points of inflection

$f(x) = -x^2 - 10x - 25$

**3.** Point of inflection at $x = \dfrac{1}{3}$

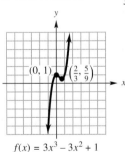

$f(x) = 3x^3 - 3x^2 + 1$

**5.** Point of inflection at $x = -\dfrac{3}{2}$

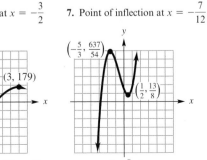

$f(x) = -2x^3 - 9x^2 + 108x - 10$

**7.** Point of inflection at $x = -\dfrac{7}{12}$

$f(x) = 2x^3 + \dfrac{7}{2}x^2 - 5x + 3$

**9.** No points of inflection

$f(x) = (x + 3)^4$

**11.** Points of inflection at $x = -\sqrt{3}$
and $x = \sqrt{3}$

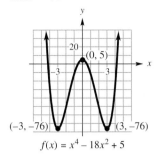

$f(x) = x^4 - 18x^2 + 5$

**13.** No points of inflection

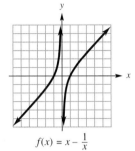

$f(x) = x - \dfrac{1}{x}$

**15.** No points of inflection

$f(x) = \dfrac{x^2 + 25}{x}$

**17.** No points of inflection

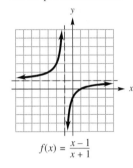

$f(x) = \dfrac{x - 1}{x + 1}$

**19.** No points of inflection

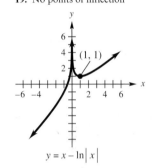

$y = x - \ln|x|$

**21.** Point of inflection at $x = 2$

$y = xe^{-x}$

*For Exercises 23–28, many correct answers are possible, including the following:*

**23.**

**25.**

**27.**

**29.** Smart cards; rate of growth of sales begins to decline at the point of inflection.    **31. (a)** 2012    **(b)** $78.6 billion    **(c)** No
**33.** Local minimum at (9, 0); local maximum at (3, 108); inflection point at (6, 54)    **35.** Local maximum at (13.8, 17.7); local minimum at (8.9, 13.5)

## Chapter 12 Review Exercises (Page 730)

| Refer to Section | 12.1 | 12.2 | 12.3 | 12.4 | 12.5 | 12.6 |
|---|---|---|---|---|---|---|
| For Exercises | 1, 5–16, 48, 51–52, 55–56 | 2, 17–20, 49–50 | 3, 33–40, 47, 53–54, 57–66 | 41–46, 67–68 | 69–74 | 4, 21–32 |

**1.** Answers vary.    **3.** Answers vary.    **5.** Increasing on $\left(-\frac{9}{2}, \infty\right)$; decreasing on $\left(-\infty, -\frac{9}{2}\right)$    **7.** Increasing on $\left(-\infty, -\frac{2}{3}\right)$ and $(1, \infty)$; decreasing on $\left(-\frac{2}{3}, 1\right)$    **9.** Decreasing on $(-\infty, 4)$ and $(4, \infty)$
**11.** Local maximum of 54 at $x = -2$; local minimum of $-71$ at $x = 3$
**13.** Local maximum of 2 at $x = 0$; local minima of $-\frac{1}{3}$ at $x = -1$ and $-43$ at $x = 3$    **15.** Local maximum of $\frac{3}{e}$ at $x = 1$    **17.** $f''(x) = 40x^3 - 30x$; 10; $-260$    **19.** $f''(x) = -20e^{2x}; -20e^2; \frac{-20}{e^4}$    **21.** No local extrema; point of inflection at $x = -\frac{1}{12}$; decreasing on $(-\infty, \infty)$; concave upward on $\left(-\infty, -\frac{1}{12}\right)$; concave downward on $\left(-\frac{1}{12}, \infty\right)$

**23.** Local maximum at $x = 0$; local minima at $x = -1$ and $x = 2$; points of inflection at $x = \frac{1 - \sqrt{7}}{3}$ and $x = \frac{1 + \sqrt{7}}{3}$; increasing on $(-1, 0)$ and $(2, \infty)$; decreasing on $(-\infty, -1)$ and $(0, 2)$; concave upward on $\left(-\infty, \frac{1 - \sqrt{7}}{3}\right)$ and $\left(\frac{1 + \sqrt{7}}{3}, \infty\right)$; concave downward on $\left(\frac{1 - \sqrt{7}}{3}, \frac{1 + \sqrt{7}}{3}\right)$

**25.** No local extrema or points of inflection; increasing on $\left(-\infty, -\frac{1}{2}\right)$; and $\left(-\frac{1}{2}, \infty\right)$; concave upward on $\left(-\infty, -\frac{1}{2}\right)$; concave downward on $\left(-\frac{1}{2}, \infty\right)$

**27.** Local maximum at $x = \frac{1}{2}$; local minimum at $x = -\frac{2}{3}$; point of inflection at $x = -\frac{1}{12}$; increasing on $\left(-\frac{2}{3}, \frac{1}{2}\right)$; decreasing on $\left(-\infty, -\frac{2}{3}\right)$ and $\left(\frac{1}{2}, \infty\right)$; concave upward on $\left(-\infty, -\frac{1}{12}\right)$; concave downward on $\left(-\frac{1}{12}, \infty\right)$

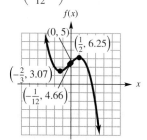

**29.** Local minimum at $x = 0$; increasing on $(0, \infty)$; decreasing on $(-\infty, 0)$; concave upward on $(-\infty, \infty)$. In the graph, the vertical lines are 1 unit apart, and the horizontal lines are 10 units apart.

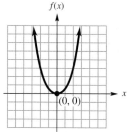

**31.** Local maximum at $x = -2$; local minimum at $x = 2$; no point of inflection; increasing on $(-\infty, -2)$ and $(2, \infty)$; decreasing on $(-2, 0)$ and $(0, 2)$; concave upward on $(0, \infty)$; concave downward on $(-\infty, 0)$. In the graph, the horizontal lines are 2 units apart.

**33.** Absolute maximum of 10 at $x = 3$; absolute minima of 9 at $x = 2$ and $x = 4$    **35.** Absolute maximum of 11.217 at $x = 3.3$; absolute minimum of $-\frac{319}{27}$ at $x = \frac{5}{3}$    **37.** Absolute minimum at $x = 1$    **39.** Absolute maximum at $x = 3$    **41.** $\frac{2 - 10y}{10x - 5}$    **43.** $\frac{10 - ye^x}{e^x - 1}$    **45.** $y = 4x - 3$
**47. (a)** 40,000    **(b)** 2008 through 2013    **(c)** 33,000; 41,000

**49.** **(a)** Increase    **(b)** The graph is increasing and concave up, which means the population is increasing at a faster and faster rate.    **51.** **(a)** 25 to 55 years    **(b)** 55 to 75 years    **(c)** A local maximum income of $87,000 occurs at 55 years of age.    **53.** **(a)** 127 million; 125 million    **(b)** During this period, Japan had an absolute minimum population of 125 million in 2020 and an absolute maximum of 128 million in 2008.    **55.** **(a)** $t = 1$ and $t = 4$    **(b)** $t = 1$ (2009)    **(c)** $t = 4$ (2012)    **57.** 29% in 2012    **59.** 125 m $\times$ 250 m    **61.** 3 m $\times$ 3 m $\times$ 3 m    **63.** 32 batches    **65.** 21 batches    **67.** 10    **69.** **(a)** $-2.2°C/hr$    **(b)** $5.4°F/hr$    **71.** $315/day    **73.** 204 m/min

## Case Study 12 Exercises (Page 734)

**1.** $Z'(m) = -\dfrac{C_1}{m^2} + \dfrac{DC_3}{2}$

**3.** $D = 3, t = 12, C_1 = 15,000, C_3 = 900, m = \dfrac{10}{3}$

**5.** $Z(m^+) = Z(4) = \$11,400; Z(m^-) = Z(3) = \$11,300$

## Extended Project (Page 734)

**(a)** 1 month; 4 employees per batch; $13,000 per month    **(b)** 1 month; 4 employees per batch; $14,000 per month    **(c)** Answers vary.

# Chapter 13

## Section 13.1 (Page 743)

**1.** $F(x)$ and $G(x)$ must differ by a constant    **3.** Answers vary.

**5.** $6x^2 + C$    **7.** $2p^4 + C$    **9.** $105x + C$    **11.** $\dfrac{5z^2}{2} - z + C$

**13.** $\dfrac{z^3}{3} - 2z^2 + 6z + C$    **15.** $\dfrac{x^4}{4} - \dfrac{14x^3}{3} + 11x^2 + 8x + C$

**17.** $4y^{3/2} + C$    **19.** $\dfrac{12t^{5/2}}{5} + \dfrac{12t^{5/4}}{5} + C$    **21.** $\dfrac{112t^{3/2}}{3} + 4t^{9/2} + C$

**23.** $-\dfrac{12}{x^2} + C$    **25.** $-\dfrac{1}{2y^2} - 4y^{1/2} + C$    **27.** $-3x^{-2} + 2\ln|x| + C$

**29.** $\dfrac{4e^{3u}}{3} + C$    **31.** $-\dfrac{15}{4}e^{-.8x} + C$    **33.** $6\ln|x| - 8e^{-.5x} + C$

**35.** $\ln|t| + \dfrac{t^4}{2} + C$    **37.** $\dfrac{e^{2u}}{2} + 4\ln|u| + C$    **39.** $\dfrac{25x^3}{3} + 5x^2 + x + C$

**41.** $\dfrac{6x^{7/6}}{7} + \dfrac{3x^{2/3}}{2} + C$    **43.** $f(x) = \dfrac{3x^{5/3}}{5}$

**45.** **(a)** $f(x) = -.05x^2 + .59x + 67.1$    **(b)** 63.7%

**47.** **(a)** $p(x) = 25e^{.12x} + 4$,    **(b)** $110    **49.** $x^3 - 3x^2 + 4x - 40$

**51.** $C(x) = \dfrac{3x^{5/3}}{5} + 5x - 12$    **53.** $C(x) = \dfrac{x^3}{3} - x^2 + 3x + 16$

**55.** $C(x) = \dfrac{800}{e^{.05x}} + 100x + 276.96$    **57.** **(a)** $F(t) = 66\ln t + 139$

**(b)** The projected population of the United States in 2047 is about 393 million.

## Section 13.2 (Page 751)

**1.** Answers vary.    **3.** $\dfrac{(12x - 1)^3}{12} + C$    **5.** $\dfrac{-5}{3(3t - 6)} + C$

**7.** $-\dfrac{1}{3}(5 - 2x)^{3/2} + C$    **9.** $\dfrac{2(r^3 + 3)^{3/2}}{9} + C$    **11.** $-\dfrac{4}{5}e^{5k} + C$

**13.** $\dfrac{e^{2w^4}}{2} + C$    **15.** $\dfrac{e^{4t - t^2}}{2} + C$    **17.** $2e^{\sqrt{y}} + C$    **19.** $-\dfrac{5\ln|12 + 6x|}{6} + C$

**21.** $\dfrac{1}{2}\ln|e^{2t} + 1| + C$    **23.** $-\dfrac{1}{8(2x^2 + 8x)^2} + C$    **25.** $\sqrt{2x + 9} + C$

**27.** $(x^3 + 3x)^{1/3} + C$    **29.** $\ln|3x^2 + 7x + 8| + C$    **31.** $\dfrac{(x^2 + 5)^4}{4} + C$

**33.** $\dfrac{(x^2 + 12x)^{3/2}}{3} + C$    **35.** $\ln|\ln(x)| + C$

**37.** $\dfrac{10}{3}(u - 1)^{3/2} + 10(u - 1)^{1/2} + C$

**39.** $\dfrac{2(5t - 1)^{5/2}}{125} + \dfrac{2(5t - 1)^{3/2}}{75} + C$    **41.** **(a)** $6\ln(5x^2 + 1) + 10$

**(b)** Yes    **43.** **(a)** $F(x) = 6195e^{-.03x} - 3$    **(b)** About 3397 stores

**45.** **(a)** $R(x) = 100\sqrt[4]{x^2 + 10,000} - 1000$    **(b)** $2170.15    **(c)** 895 units

**47.** **(a)** $F(x) = 53\ln(x + 1) + 520$    **(b)** About 681,000 jobs

## Section 13.3 (Page 760)

**1.** $-\dfrac{2}{35}e^{-35x} + C$; substitution    **3.** $2xe^{5x} - \dfrac{2}{5}e^{5x} + C$; parts

**5.** $4xe^{x/4} - 16e^{x/4} + C$; parts    **7.** $-5xe^{-x} - 6e^{-x} + C$; parts

**9.** $-\dfrac{7}{3}xe^{1-3x} - \dfrac{7}{9}e^{1-3x} + C$; parts    **11.** $\dfrac{4}{5}xe^{10x-17} - \dfrac{19}{50}e^{10x-17} + C$; parts

**13.** $e^{2x^2} + C$; substitution    **15.** $2x^2e^{2x} - 2xe^{2x} + e^{2x} + C$; parts twice

**17.** $x^2e^x - 2xe^x + 3e^x + C$; parts twice    **19.** $\dfrac{1}{2}e^{x^2+1} + C$; substitution

**21.** $3x(\ln(x)) - 10x + C$    **23.** $\dfrac{1}{8}[(8x - 5)\ln(8x - 5) - (8x - 5)] + C$

**25.** $13x^2\ln x - \dfrac{13}{2}x^2 + C$    **27.** $\dfrac{1}{4}x^4\ln x - \dfrac{1}{16}x^4 + C$

**29.** $6x^{3/2}\ln x - 4x^{3/2} + C$    **31.** $(9x - x^2)\ln x - \left(9x - \dfrac{1}{2}x^2\right) + C$

**33.** $(6x^3 + 6x^2 + 4x)\ln x - (2x^3 + 3x^2 + 4x) + C$

**35.** $10(x + 1)(x + 2)^{3/2} - 4(x + 2)^{5/2} + C$

**37.** **(a)** The car started out driving 80 ft/sec, sped up to about 95 ft/sec after 4 seconds, and slowed back down to about 80 ft/sec after 30 seconds.

**(b)** $80t - 40te^{-t/4} - 160e^{-t/4} + 160$

**39.** **(a)** $s(t) = (18t + 6)\ln(3t + 1) - 18t$    **(b)** About 1923 feet

**(c)** About 81 ft/s or 55 mph    **41.** $s(t) = -\dfrac{1}{2}t^2e^{-2t} - \dfrac{1}{2}te^{-2t} - \dfrac{1}{4}e^{-2t} + \dfrac{1}{4}$

**43.** **(a)** $F(x) = -14xe^{-x/7} - 98e^{-x/7} + 110$    **(b)** About 101 people

**45.** $F(x) = -5xe^{-x} - 6e^{-x} + 16$    **47.** **(a)** $f(t) = (1125t + 40,500)e^{-.02t}$

**(b)** $F(t) = -56,250te^{-.02t} - 4,837,500e^{-.02t} + 4,837,500$

**49.** $F(x) = 42x\ln(x) + 82x - 82$

## Section 13.4 (Page 769)

**1.** About 149 kWh    **3.** 37.54 million    **5.** Answers vary.    **7.** **(a)** 44

**(b)** 50    **(c)** 55.4    **9.** **(a)** 8    **(b)** 10    **(c)** 10.66    **11.** **(a)** 7.39

**(b)** 14.60    **(c)** 24.48    **13.** **(a)** 64    **(b)** 57    **(c)** About 51.41    **15.** 12.5

**17.** 228.75    **19.** 90.55    **21.** .294    **23.** $808,680.51    **25.** $47,703.43

**27.** **(a)** At $t = 1$, the red car was traveling at the same speed as the blue car, but it had traveled a longer distance.    **(b)** At $t = 2$, the red car was traveling slower, but it had traveled the same distance.    **(c)** At $t = 3$, the red car was traveling slower and had traveled a shorter distance.

**29.** About 1613 feet

## Section 13.5 (Page 781)

**1.** 60    **3.** 13    **5.** 8    **7.** 138    **9.** $8\dfrac{1}{4}$    **11.** 9.9873    **13.** $3e^{10} - 3$

**15.** 1.294    **17.** $\dfrac{91}{3}$    **19.** 63.857    **21.** .1509    **23.** 49    **25.** .3005

**27.** 105.3946    **29.** $\ln\left(\dfrac{108}{37}\right) \approx 1.0712$

**31.** A loss, $\displaystyle\int_0^2 (6x^2 - 7x - 3)\, dx$    **33.** $\dfrac{8}{3}$    **35.** 2.67    **37.** $\dfrac{64}{3}$

**39.** 4.75    **41.** 3.7037    **43.** .6667    **45.** 1    **47.** 11.08

**49.** 16 square units    **51.** **(a)** $22,000    **(b)** $108,000    **53.** About $3490 billion    **55.** **(a)** 43,410,000; the number of people between 20 and 30    **(b)** 308,430,000; the number of people between 0 and 90    **(c)** 86,260,000 millennials    **57.** $4101.27    **59.** **(a)** 363    **(b)** 167    **61.** The total profit from the sale of 100 units is $576.76.    **63.** .811

## Section 13.6 (Page 791)

**1.** $280; $11.67    **3.** $2500; $30,000    **5.** About $552 billion

**7.** The total revenue generated between the start of 2010 and the start of 2016 was about $38,625 million ($38.265 billion).    **9.** **(a)** $x = 2$    **(b)** $\dfrac{28}{3}$

**(c)** $\dfrac{10}{3}$    **(d)** $\dfrac{38}{3}$    **11.** **(a)** $x = 4$    **(b)** $\dfrac{4}{3}$    **(c)** $\dfrac{43}{12}$    **(d)** $\dfrac{59}{12}$    **13.** $\dfrac{56}{3}$

**15.** $\dfrac{5}{192}$   **17.** About 1.22   **19.** 325.04   **21.** 4.999

**23.** (a) 8 years  (b) $147.96  (c) About $771   **25.** (a) About 10 years
(b) About $833,622   **27.** (a) 3.02 years  (b) About 12.74 trillion gallons
**29.** (a) About 88.22 gallons  (b) About 40 hours  (c) About 128 gallons
(d) About 75.28 hours   **31.** $35,476.82   **33.** $94,223.00   **35.** $67,221.20
**37.** (a)

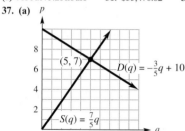

(b) (5, 7)  (c) $7.50  (d) $17.50   **39.** $1999.54   **41.** $38.50
**43.** (a)

(b) (10.60, 199.50)  (c) $1726.63  (d) $1726.63

## Section 13.7 (Page 803)

**1.** $y = -x^3 + \dfrac{7x^2}{2} + C$   **3.** $y = -3e^{-2x} + C$   **5.** $y = \dfrac{3x^4}{8} + C$

**7.** $y = \sqrt[5]{3x^2 + C}$   **9.** $7y^3 + 8y = x + C$   **11.** $y = \ln(5x^2 + C)$
**13.** $y = \sqrt[3]{x + C}$   **15.** $y = Me^{-3x}$   **17.** $y = 7 + Me^x$
**19.** $y = Me^{-x} + 1$   **21.** $y = Me^{2x^2}$   **23.** $y = x^3 - x^2 + 4$

**25.** $y = 91e^{3.7x}$   **27.** $y^2 = \dfrac{2x^3}{3} + 16$   **29.** $y = e^{2.5x^2 + 3x}$

**31.** $y^2 - 6y = 7x^2 + 2x - 8$   **33.** Answers vary.   **35.** (a) 3536 units
(b) 6356 units   **37.** (a) $y = .3e^{.23t}$  (b) About 9.5%  (c) About 15.2 years

**39.** (a) $y = 4t^2 + 50$  (b) 10 days   **41.** (a) $\dfrac{dy}{dt} = -.15y$

(b) $y = Me^{-.15t}$  (c) About 9.2 years   **43.** 34.7 years   **45.** (a) $k = .8$

(b) 11 people  (c) 55 people  (d) 2981 people   **47.** (d) $\dfrac{\ln 2}{12}$

**49.** (a) $y = 30 - 29e^{-x}$  (b) 29,999 cells

## Chapter 13 Review Exercises (Page 806)

| Refer to Section | 13.1 | 13.2 | 13.3 | 13.4 | 13.5 | 13.6 | 13.7 |
|---|---|---|---|---|---|---|---|
| For Exercises | 1–8 | 9–28 | 29–34 | 35–38 | 39–58 | 59–78 | 79–90 |

**1.** $\dfrac{x^3}{3} - 4x^2 - 7x + C$   **3.** $\dfrac{14x^{3/2}}{3} + C$   **5.** $\dfrac{2x^{3/2}}{3} + 9x^{1/3} + C$

**7.** $3x^{-2} + C$   **9.** $-\dfrac{3e^{2x}}{2} + C$   **11.** $10\ln|x - 3| + C$

**13.** $-\ln|3 - x| + C$   **15.** $\dfrac{3\ln|x^2 - 1|}{2} + C$   **17.** $-\dfrac{4}{3}(x^3 + 5)^{-3} + C$

**19.** $\ln|2x^2 - 5x| + C$   **21.** $\dfrac{5e^{3x^2}}{6} + C$   **23.** $-\dfrac{e^{-3x^3}}{9} + C$

**25.** $\dfrac{(\ln x)^6}{2} + C$   **27.** $\dfrac{1}{3}(2x + 9)^{\frac{3}{2}} + C$   **29.** $-7xe^{4-x} - 7e^{4-x} + C$

**31.** $14x^2e^{2x} - 14xe^{2x} + 7e^{2x} + C$   **33.** $(6x + 1)\ln(6x + 1) - (6x + 1) + C$

**35.** About $101,400   **37.** (a) 136  (b) 141.667   **39.** $\dfrac{128}{3}$   **41.** $\dfrac{381}{200}$

**43.** $15\ln 5 \approx 24.1416$   **45.** .40   **47.** 66.907   **49.** 6.313   **51.** 3

**53.** $7e - 4$   **55.** $20e^3 - 8$   **57.** $\dfrac{1}{8}$   **59.** 6.3891

**61.** $C(x) = 10x - \dfrac{5x^2}{2} + 4$   **63.** $C(x) = (2x - 1)^{3/2} + 145$

**65.** $40,500 million ($40.5 billion)   **67.** Total profit generated from the
start of 2007 to the start of 2017 was $10,851 million ($10.851 billion).
**69.** 45,000 units   **71.** 2.5 years, $98,958.33   **73.** 916.67
**75.** About 110 people   **77.** (a) .6321  (b) .2231   **79.** $y = Me^{-.07x}$
**81.** $y = \sqrt[3]{x - 4x^5 + C}$   **83.** $y = -5e^{-x} - 5x + 22$   **85.** $y = 2e^{5x-x^2}$
**87.** (a) $y = 13,000e^{.02t}$  (b) $13,804  (c) About 7.2 years
**89.** (a) $y = 70 - 43e^{-.05x}$  (b) About 66%

## Case Study 13 Exercises (Page 810)

**1.** Answers vary.   **3.** (a) $P(t) = \dfrac{10e^{.07t}}{e^{.07t} + 4}$   (b) About 7.8 million

(c) About 2020

# Chapter 14

## Section 14.1 (Page 820)

**1.** $4; -22; -20; 12$   **3.** $11; 37; 11; 64$   **5.** $3; \sqrt{33}; \sqrt{17}; 8$
**7.** All ordered pairs of real numbers   **9.** All ordered pairs $(x, y)$, where $x$
and $y$ are nonnegative real numbers   **11.** All ordered triples $(u, v, w)$, where
$u$, $v$, and $w$ are real numbers such that $v \neq w$   **13.** All ordered pairs $(x, y)$ such that $x > 0$   **15.** $P(l, w) = 2l + 2w$   **17.** $C(x, y) = 40x + 14y$
**19.** $C(x, y, z) = 200x + 100y + 50z$   **21.** Answers vary.
**23.**                                      **25.**

**27.**

**29.** (a) 19.5; not overweight  (b) 29.3; overweight
**31.** (a) 71.8°; comfortable  (b) 75°; uncomfortable   **33.** (a) $613.91
(b) 558.39   **35.** 1.14; the IRA grows faster.   **37.** About 228
**39.** (a) $f(p, n) = 41p + 3n$  (b) 355 seconds  (c) 322 seconds  (d) No.
**41.** (c)   **43.** (e)   **45.** (b)
**47.**

## Section 14.2 (Page 831)

**1.** (a) $24x^2 - 8xy$  (b) $-4x^2 + 18y$  (c) 48  (d) $-40$
**3.** $f_x(x, y) = -4y; f_y(x, y) = -4x + 18y^2; 4; 178$
**5.** $f_x = 2e^{2x+y}; f_y = e^{2x+y}; 2e^3; e^{-5}$

**7.** $f_x(x, y) = \dfrac{1}{x - 2y}; f_y(x, y) = \dfrac{-2}{x - 2y}; f_x(2, -1) = \dfrac{1}{4}; f_y(-4, 3) = \dfrac{1}{5}$

**9.** $f_x = \dfrac{8xy}{\sqrt{8x^2y + 4}}; f_y = \dfrac{4x^2}{\sqrt{8x^2y + 4}}$

**11.** $f_x = 2x^2ye^{2xy} + 2xe^{2xy}$; $f_y = 2x^3e^{2xy}$; $-4e^{-4}$; $-128e^{-24}$

**13.** $f_x = \dfrac{3}{1 + 6y}$; $f_y = \dfrac{-18x}{(1 + 6y)^2}$

**15.** $f_{xx} = 8y^2 - 32$; $f_{yy} = 8x^2$; $f_{xy} = f_{yx} = 16xy$

**17.** $h_{xx} = -8y^2$; $h_{xy} = h_{yx} = -16xy + 14y$; $h_{yy} = -6 - 8x^2 + 14x$

**19.** $f_{xx} = 6xy^{-1}$; $f_{yy} = 2x^3y^{-3}$; $f_{xy} = f_{yx} = -3x^2y^{-2}$

**21.** $f_{xx} = \dfrac{2}{x^3}$; $f_{yy} = \dfrac{-30}{y^4}$; $f_{xy} = f_{yx} = 0$

**23.** $f_{xx} = 9e^{4y-3x}$; $f_{yy} = 16e^{4y-3x}$; $f_{xy} = f_{yx} = -12e^{4y-3x}$

**25.** $z_{xx} = 0$; $z_{yy} = 4xe^y$; $z_{xy} = z_{yx} = 4e^y$

**27.** $f_{xx} = \dfrac{-1}{x^2}$; $f_{yy} = \dfrac{-1}{y^2}$; $f_{xy} = f_{yx} = 0$

**29.** $f_{xx} = 0$; $f_{yy} = \dfrac{6x}{(y - 1)^3}$; $f_{xy} = f_{yx} = \dfrac{-3}{(y - 1)^2}$

**31.** $z_{xx} = \dfrac{1}{x}$; $z_{yy} = -\dfrac{x}{y^2}$; $z_{xy} = z_{yx} = \dfrac{1}{y}$

**33.** $\dfrac{1}{2}$; $-3$   **35.** $f_x = 2x$; $f_y = z$; $f_z = y + 4z^3$; $f_{yz} = 1$; $-2$; 1

**37.** $f_x = 27yz^4e^{3xz^4}$; $f_y = 9e^{3xz^4}$; $f_z = 108xyz^3e^{3xz^4}$; $f_{yz} = 108xz^3e^{3xz^4}$

**39.** 3; 9; answers vary.   **41. (a)** \$500 **(b)** \$1100   **43. (a)** Fixed costs, that is, the cost when no items have been produced **(b)** $\dfrac{\partial C}{\partial x} = \$600$ is the marginal cost of producing 1 unit of chargers when the production of covers remains constant. $\dfrac{\partial C}{\partial y} = \$120$ is the marginal cost of producing 1 unit of covers when the production of chargers remains constant.

**45. (a)** Profit increases by about \$34. **(b)** Profit increases by about \$4.
**47. (a)** \$902,100 **(b)** $f_p(p, i) = 99 - .5i - .005p$; $f_i(p, i) = -.5p$; $f_p(p, i)$ gives the rate at which weekly sales are changing per unit change in price when the interest rate remains constant; $f_i(p, i)$ gives the rate at which weekly sales are changing per unit change in interest rate when the price remains constant. **(c)** A weekly sales decrease of \$9700
**49. (a)** Positive; $A_p = (1 + i)^n$ **(b)** Positive; $A_i = nP(1 + i)^{n-1}$
**51. (a)** Positive; $A_p = e^{rt}$ **(b)** Positive; $A_r = tPe^{rt}$ **(c)** Positive; $A_t = rPe^{rt}$
**53. (a)** $f_x(64, 125) = 1.5625$, the approximate change in production (in thousands of units) for a 1-unit change in labor; $f_y(64, 125) = 1.6$, the approximate change in production (in thousands of units) for a 1-unit change in capital. **(b)** An increase of approximately 1563 batteries **(c)** Increasing capital by 1 unit

**55.** $.7x^{-.3}y^{.3}$, or $\dfrac{.7y^{.3}}{x^{.3}}$; $.3x^{.7}y^{-.7}$, or $\dfrac{.3x^{.7}}{y^{.7}}$ **57. (a)** 28.24
**(b)** $\dfrac{\partial B}{\partial w} = \dfrac{703}{h^2}$; $\dfrac{\partial B}{\partial h} = \dfrac{-1406w}{h^3}$ **59. (a)** About .0112 **(b)** About .0783
**61. (a)** About 1.75 atm **(b)** About .0227; about .0347 **(c)** About 6.17 min

## Section 14.3 (Page 840)

**1.** $(-4, 2)$ **3.** $(0, 0)$ and $(3, 3)$ **5.** Local minimum of $-22$ at $(3, -1)$
**7.** Local minimum of 2 at $(-2, -2)$ **9.** Saddle point at $\left(-2, \dfrac{5}{3}\right)$
**11.** Local maximum of 17 at $\left(\dfrac{2}{3}, \dfrac{4}{3}\right)$ **13.** Saddle point at $(2, -2)$
**15.** Saddle point at $(1, 2)$ **17.** Saddle point at $(0, 0)$; local minimum of $-201$ at $(6, 18)$ **19.** Saddle point at $(0, 0)$; local minimum of $-7.75$ at $(4.5, 1.5)$ **21.** Saddle point at $(0, 0)$ **23.** Local maximum of $\dfrac{9}{8}$ at $(-1, 1)$; saddle point at $(0, 0)$; (a) **25.** Local minima of $-\dfrac{33}{16}$ at $(0, 1)$ and $(0, -1)$; saddle point at $(0, 0)$; (b) **27.** Local maxima of $\dfrac{17}{16}$ at $(1, 0)$ and $(-1, 0)$; local minima of $-\dfrac{15}{16}$ at $(0, 1)$ and $(0, -1)$; saddle points at $(0, 0), (-1, 1), (1, -1), (1, 1),$ and $(-1, -1)$; (e) **29.** $x = 4, y = 5$; \$59
**31.** $P(3, 9) = \$627$ **33.** $C(12, 25) = 2237$ **35.** 3 m × 3 m × 3 m

**37.** 36 in (length) × 18 in × 18 in **39.** Length 36 in; radius $36/\pi \approx 11.5$ in
**41.** 1000 tons of grade A ore and 2000 tons of grade B ore **43.** 6 tons of steel, 3 tons of aluminum, \$216,000

## Section 14.4 (Page 850)

**1.** 200 when $x = 10$ and $y = 10$ **3.** $\dfrac{125}{4}$ when $x = 5$ and $y = \dfrac{5}{2}$
**5.** 28 when $x = 5$ and $y = 3$ **7.** 360 when $x = 20$ and $y = 2$
**9.** $e^{-2}$ when $x = y = 1$ **11.** $\dfrac{81}{4}$ when $x = \dfrac{3}{2}, y = \dfrac{3}{2}$ and $z = 3$
**13.** $x = 7, y = 14$ **15.** $x = 34, y = 34, z = 34$ **17.** Answers vary.
**19.** Answers vary. **21.** $x = 2, y = 4$ **23.** About 166.67 units of labor and 177.78 units of capital **25.** 60 ft × 60 ft **27.** 50 m × 50 m
**29.** Radius 5 in; height 10 in **31.** About 12.91 m × 12.91 m × 6.45 m
**33.** 4 ft × 4 ft × 2 ft (height) **35. (a)** Let the sides with area $xz$ and the bottom be made from the free material; this gives the constraint. It takes $\dfrac{400}{xyz}$ trips to transport the material at a cost of $\left(\dfrac{400}{xyz}\right)(.10)$. The total cost also includes the material for the ends of the box.
**(b)** $x = 2$ yd; $y = 1$ yd; $z = .5$ yd

## Chapter 14 Review Exercises (Page 852)

| Refer to Section | 14.1 | 14.2 | 14.3 | 14.4 |
|---|---|---|---|---|
| For Exercises | 1–12, 35–36 | 13–34, 37–38 | 39–48 | 49–54 |

**1.** 15; 84 **3.** Not defined; $-6$ **5.** $-21$; not defined **7.** Answers vary.
**9.**  **11.**

**13. (a)** $-4x + 5y$ **(b)** 3 **(c)** 5 **15.** Answers vary.
**17.** $f_x = -14xy^3$; $f_y = 3 - 21x^2y^2$
**19.** $f_x = \dfrac{3x}{\sqrt{3x^2 + 2y^2}}$; $f_y = \dfrac{2y}{\sqrt{3x^2 + 2y^2}}$
**21.** $f_x = \dfrac{-5x^2y - 16xy^4}{(5x + 8y^3)^2}$; $f_y = \dfrac{-5x^3 + 16x^2y^3}{(5x + 8y^3)^2}$
**23.** $f_x = 3x^2e^{3y}$; $f_y = 3x^3e^{3y}$ **25.** $f_x = \dfrac{2x}{x^2 - 4y^3}$; $f_y = \dfrac{-12y^2}{x^2 - 4y^3}$
**27.** $f_{xx} = 24xy^2$; $f_{xy} = 24x^2y - 8$
**29.** $f_{xx} = \dfrac{10y}{(x - 2y)^3}$; $f_{xy} = \dfrac{-5x - 10y}{(x - 2y)^3}$ **31.** $f_{xx} = 2e^y$; $f_{xy} = 2xe^y$
**33.** $f_{xx} = \dfrac{-2x^2y^2 - 4y}{(2 - x^2y)^2}$; $f_{xy} = -\dfrac{4x}{(2 - x^2y)^2}$ **35. (a)** \$450 **(b)** \$646
**(c)** \$1030 **37. (a)** $\dfrac{1}{2}$ **(b)** 9 **39.** Local minimum at $(0, 1)$
**41.** Saddle point at $(0, 2)$ **43.** Local minimum at $\left(1, \dfrac{1}{2}\right)$; saddle point at $\left(-\dfrac{1}{3}, \dfrac{11}{6}\right)$ **45.** Local minimum at $(0, 0)$ **47. (a)** $x = 11, y = 12$
**(b)** \$431 **49.** Minimum of 2 when $x = 1$ and $y = -1$ **51.** $\dfrac{160}{3}$ and $\dfrac{80}{3}$
**53.** Answers vary.

## Case Study 14 Exercises (Page 855)

**1.** The second-derivative test shows the critical point is a local minimum.
**3.** $y = .01x - 17.55$

# Photo Credits

# Index of Companies, Products, and Agencies

# Index of Applications

# Economics

# Education

# Finance

# Subject Index

# Geometric Formulas

## Triangle

- Area: $A = \frac{1}{2}bh$

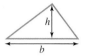

## Rectangle

- Area: $A = lw$
- Perimeter: $P = 2l + 2w$

## Parallelogram

- Area: $A = bh$
- Perimeter: $P = 2a + 2b$

## Circle

- Area: $A = \pi r^2$
- Circumference: $C = 2\pi r$

## Sphere

- Area: $V = \frac{4}{3}\pi r^3$
- Surface area: $A = 4\pi r^2$

## Rectangular Box

- Volume: $V = lwh$
- Surface area: $A = 2lh + 2wh + 2lw$

## Circular Cylinder

- Volume: $V = \pi r^2 h$
- Surface area: $A = 2\pi r^2 + 2\pi rh$

## Triangular Cylinder

- Volume: $V = \frac{1}{2}bhl$

## Cone

- Volume: $V = \frac{1}{3}\pi r^2 h$

# Finite Math Review

## Compound Amount

If $PV$ dollars are invested at interest rate $i$ per period, the **compound amount** (future value, $FV$) after $n$ compounding periods is $FV = PV(1 + i)^n$.

## Present and Future Value

An **ordinary annuity** of $n$ payments of $PMT$ dollars each at the end of consecutive interest periods with interest compounded at a rate of $i$ per period has **future value**

$$FV = PMT\left[\frac{(1 + i)^n - 1}{i}\right] \text{ and } \textbf{present value } PV = PMT\left[\frac{1 - (1 + i)^{-n}}{i}\right].$$

## Matrix Operations

The **matrix** version of the elimination method uses the following **matrix row operations** to obtain the augmented matrix of an equivalent system. They correspond to using elementary row operations on a system of equations.

1. Interchange any two rows.

2. Multiply each element of a row by a nonzero constant.

3. Replace a row by the sum of itself and a constant multiple of another row in the matrix.

## Union Rule

For any events $E$ and $F$ from a sample space $S$: $P(E \cup F) = P(E) + P(F) - P(E \cap F)$.

## Properties of Probability

Let $S$ be a sample space consisting of $n$ distinct outcomes $s_1, s_2, \ldots, s_n$. An acceptable probability assignment consists of assigning to each outcome $s_i$ a number $p_i$ (the probability of $s_i$) according to the following rules.

1. The probability of each outcome is a number between 0 and 1:

$$0 \le p_1 \le 1, 0 \le p_2 \le 1, \ldots, 0 \le p_n \le 1$$

2. The sum of the probabilities of all possible outcomes is 1:

$$p_1 + p_2 + p_3 + \cdots + p_n = 1$$

## Bayes' Formula

For any events $E$ and $F_1, F_2, \ldots F_n$, from a sample space $S$, where $F_1 \cup F_2 \cup \cdots \cup F_n = S$,

$$P(F_i|E) = \frac{P(F_i) \cdot P(E|F_i)}{P(F_1) \cdot P(E|F_1) + \cdots + P(F_n) \cdot P(E|F_n)}.$$